W9-ATH-871

UCSMP The University of Chicago School Mathematics Project

Functions, Statistics, and Trigonometry

Authors

John W. McConnell
Susan A. Brown
Paul J. Karafiol
Sara Brouwer
Mary Ives
Marshall Lassak
Rosa McCullagh
Natalie Jakucyn
Zalman Usiskin

UChicagoSolutions

Authors

3rd EDITION AUTHORS

John W. McConnell Lecturer in Statistics
North Park University, Chicago, IL

Susan A. Brown Mathematics/Science Division Chair
York High School, Elmhurst, IL

Paul J. Karafiol Mathematics Teacher
Walter Payton College Prep High School, Chicago, IL

Sara Brouwer Head of Mathematics
The American School in London, London, England

Mary Ives Mathematics Teacher
University High School, Irvine, CA

Marshall Lassak Associate Professor of Mathematics
and Computer Science
Eastern Illinois University, Charleston, IL

Rosa McCullagh Mathematics Teacher
University of Chicago Laboratory Schools

Natalie Jakucyn Mathematics Teacher
Glenbrook South High School, Glenview, IL

Zalman Usiskin Professor of Education
The University of Chicago

AUTHORS OF EARLIER EDITIONS

Sharon L. Senk Professor of Mathematics
Michigan State University, East Lansing, MI

Steven S. Viktora Mathematics Department Chair
New Trier Township High School, Winnetka, IL

Nils Ahbel Mathematics Teacher
Kent School, Kent, CT

Virginia Highstone Mathematics Teacher
York High School, Elmhurst, IL

David Witonsky
UCSMP

Rheta N. Rubenstein Mathematics Department Head
Renaissance High School, Detroit, MI

James E. Schultz Associate Professor of Mathematics
Ohio State University, Columbus, OH

Margaret Hackworth Mathematics Supervisor
Pinellas County Schools, Largo, FL

Dora Aksoy
UCSMP

James Flanders
UCSMP

Barry Kissane Associate Professor of Education
Murdoch University, Perth, Western Australia

Zalman Usiskin Professor of Education
The University of Chicago

http://ucsmp.uchicago.edu/secondary/overview

UChicagoSolutions

Printed in the United States of America.

Send all inquiries to:
UChicagoSolutions
1427 E 60th St, First Floor
Chicago, IL 60637

ISBN 978-1-943237-00-5
ISBN 1-943237-00-X

2 3 4 5 6 7 8 9 WEBC 22 21 20 19 18

UCSMP EVALUATION, EDITORIAL, DESIGN, AND PRODUCTION

Director of Writing
Natalie Jakucyn

Director of Evaluation
Denisse R. Thompson
Professor of Mathematics Education
University of South Florida, Tampa, FL

Evaluation Consultant
Sharon L. Senk, *Professor of Mathematics*
Michigan State University, East Lansing, MI

Evaluation Assistants
Allison Burlock, Julian Owens

Executive Managing Editor
Clare Froemel

Editorial Staff
Gary Spencer, Carlos Encalada, Kathryn Rich, Catherine Ballway, Carla Agard-Strickland, Isaac Greenspan, Emily Mokros, Currence Thomas Monson, Scott Neff

Technology Assistant
Brian Cordonnier

Design Staff
Susan Davis, Sarah Brechbill, Allison Rothmeier, Susan Zhou

Production Coordinator
Benjamin R. Balskus

Production Assistants
Timothy Arehart, Nathan Bartlett, Gwendolyn Marks, David McQuown, Robyn Mericle, Pamela Olson, Tatiana Colodeeva, Samantha Fukushima, Rachel Huddlestone, Nurit Kirshenbaum, Eric Li, Gretchen Neihardt, Sarah Schieffer, DuanDuan Yang, Don Reneau, Elizabeth Olin, Loren Santow

Coordinator of School Relations
Carol Siegel

The following teachers and schools participated in evaluations of the third edition.

Jason Bridges
Greenwood High School
Greenwood, Arkansas

Beth Hawkins
Greenwood High School
Greenwood, Arkansas

Editha Banaban
University High School
Irvine, California

Mary Ives
University High School
Irvine, California

Brian Ray
Lindblom Math & Science Academy
Chicago, Illinois

Natalie Jakucyn
Glenbrook South High School
Glenview, Illinois

Kenneth Kerr
Glenbrook South High School
Glenview, Illinois

Chris Hayward
Cape Elizabeth High School
Cape Elizabeth, Maine

Jo Guido
Clark Montessori High School
Cincinnati, Ohio

Robert Seitz
John Adams High School
Cleveland, Ohio

James Norman
Tomahawk High School
Tomahawk, Wisconsin

The following schools participated in field studies of the first edition or second edition.

Brentwood School
Los Angeles, California

Thornton Fractional High School North
Calumet City, Illinois

Kenwood Academy
Chicago, Illinois

Thornton Fractional High School South
Lansing, Illinois

Niles Township High School North
Skokie, Illinois

The Culver Academies
Culver, Indiana

M. L. King High School
Detroit, Michigan

Renaissance High School
Detroit, Michigan

Southwestern High School
Detroit, Michigan

Woodward High School
Cincinnati, Ohio

Newark High School
Newark, Ohio

We wish to acknowledge the generous support of the **Amoco (now BP) Foundation** and the **Carnegie Corporation of New York** in helping to make it possible for the first edition of these materials to be developed, tested, and distributed, and the additional support of the **Amoco (now BP) Foundation** for the second edition.

UCSMP The University of Chicago School Mathematics Project

The University of Chicago School Mathematics Project (UCSMP) is a long-term project designed to improve school mathematics in Grades K-12. UCSMP began in 1983 with a 6-year grant from the Amoco Foundation. Additional funding has come from the National Science Foundation, the Ford Motor Company, the Carnegie Corporation of New York, the Stuart Foundation, the General Electric Foundation, GTE, Citicorp/Citibank, the Exxon Educational Foundation, the Illinois Board of Higher Education, the Chicago Public Schools, from royalties, and from publishers of UCSMP materials.

From 1983 to 1987, the director of UCSMP was Paul Sally, Professor of Mathematics. Since 1987, the director has been Zalman Usiskin, Professor of Education.

UCSMP *Functions, Statistics, and Trigonometry*

The text *Functions, Statistics, and Trigonometry* has been developed by the Secondary Component of the project, and constitutes the core of the sixth year in a seven-year middle and high school mathematics curriculum. The names of the seven texts around which these years are built are:

- *Pre-Transition Mathematics*
- *Transition Mathematics*
- *Algebra*
- *Geometry*
- *Advanced Algebra*
- *Functions, Statistics, and Trigonometry*
- *Precalculus and Discrete Mathematics*

Why a Third Edition?

Since the second edition, there has been a substantial increase in the percent of students taking four or more years of high school mathematics and in the general view of the importance of statistics and quantitative literacy for all students. More than double the percent of students now take mathematics beyond two years of algebra and a year of geometry than were taking this mathematics a generation ago. These increased expectations for the performance of all students in high schools require a broad-based, reality-oriented, easy-to-comprehend approach to mathematics. These materials are designed to take into account the wider range of backgrounds and knowledge students bring to the classroom.

The third edition of UCSMP is also motivated by recent advances in technology both inside and outside the classroom, and the widespread availability of computers in schools and homes.

There has also been a substantial increase in the number of students taking a full course in algebra before ninth grade. With the UCSMP curriculum, these students will have four years of mathematics beyond algebra before calculus and other college-level mathematics. Thousands of schools have used the first and second editions and have noted success in student achievement and in teaching practices. Research from these schools shows that the UCSMP materials really work. Many of these schools have made suggestions for additional improvements in future editions of the UCSMP materials. We have attempted to utilize these ideas in the development of the third edition.

UCSMP *Functions, Statistics, and Trigonometry* – Third Edition

All lessons have been reviewed and examined afresh for this edition but the overall content and mathematical prerequisites are the same as in previous editions. Those familiar with earlier editions will notice particularly that the work with statistics has been rewritten with more emphasis on decision-making.

The previous editions of UCSMP courses introduced many features that are retained and sometimes enhanced in this third edition. As the title of the course indicates, this course has a **wider scope** than most courses at this level. A **real-world orientation** has guided both the selection of content and the approaches allowed the student in working out exercises and problems because being able to do mathematics is of little use to an individual unless he or she can apply that content. We ask students to **read mathematics** because students must read to understand mathematics in later courses and must learn to read technical matter in the world at large. The use of **new and powerful technology** is integrated throughout, with computer algebra systems and statistics utilities assumed throughout the materials.

Four dimensions of understanding are emphasized: skill in carrying out various algorithms; developing and using mathematical properties, relationships, and proofs; applying mathematics in realistic situations; and representing concepts with graphs or other diagrams. We call this the SPUR approach: **S**kills, **P**roperties, **U**ses, **R**epresentations.

The **lessons** include prose designed to show why the content is important and to explain how ideas are related to each other. They also include fully-developed examples that often show multiple worked-out solutions and checks.

Each lesson has a question set that begins with **Covering the Ideas** questions that demonstrate the student's knowledge of the overall concepts of the lesson. The **Applying the Mathematics** questions go beyond lesson examples, emphasising real-world problem solving. **Review** questions relate either to previous lessons in the course or to content from earlier courses. **Exploration** questions ask students to explore ideas related to the lesson and frequently have many possible answers.

The **book organization** is designed to maximize the acquisition of both skills and concepts. The daily review feature allows students several nights to learn and practice important concepts and skills. Then, at the end of each chapter, a carefully focused Self-Test and a Chapter Review are keyed to the SPUR objectives. They are used to solidify performance of skills and concepts from the chapter, building confidence for later use. Finally, to increase retention, important ideas are reviewed in later chapters.

New instructional features for this edition include: a **Big Idea** highlighting the key concept of each lesson; **Mental Math** questions at the beginning of lessons to sharpen "in your head" skills; **Activities** in virtually every lesson to develop concepts and skills; **Guided Examples** that provide partially completed solutions to encourage active learning; and **Quiz Yourself (QY)** stopping points to periodically check understanding.

Comments about these materials are welcomed.

Please address them to:

The University of Chicago School Mathematics Project
http://ucsmp.uchicago.edu
ucsmp@uchicago.edu
773-702-1130

Contents

Chapter 3 150
Transformations of Graphs and Data

Chapter 4 220
Trigonometric Functions

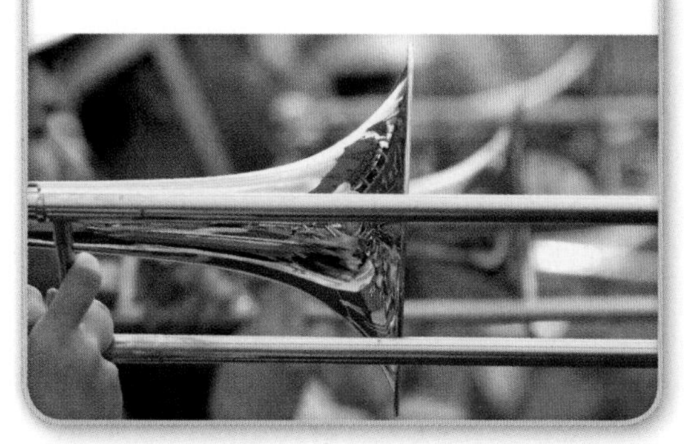

Chapter 5 292
Trigonometry

Chapter 6 358
Counting, Probability, and Inference

Chapter 7 436
Polynomial Functions

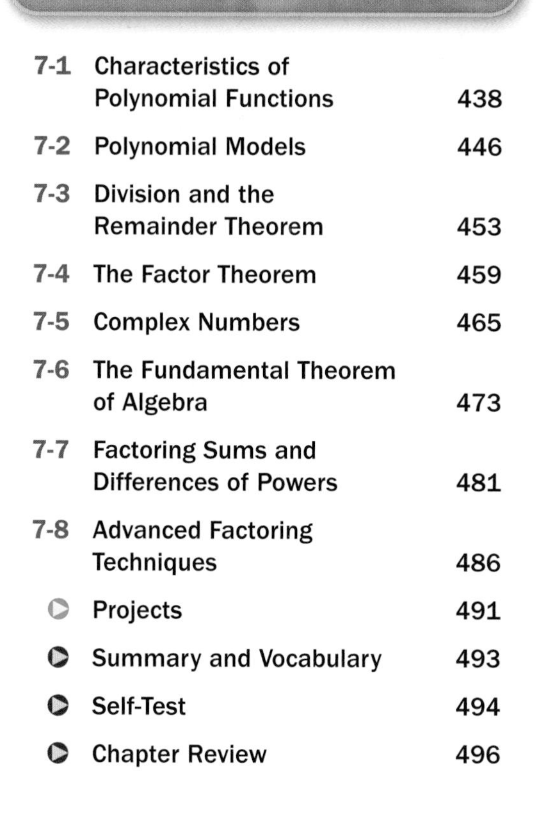

Chapter 8 500
Sequences and Series

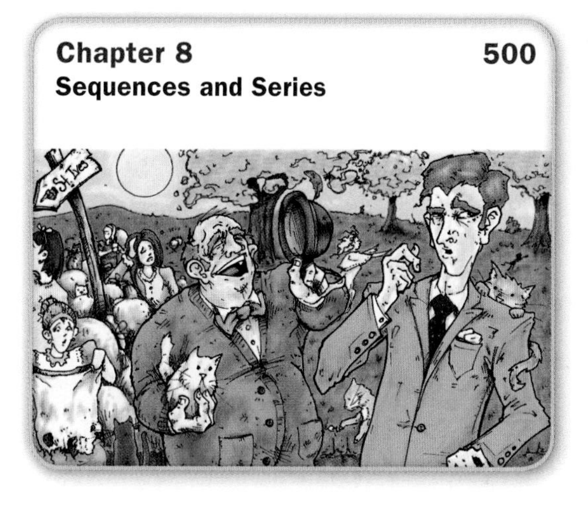

Chapter 9 558
Roots, Powers, and Logarithms

Chapter 10 616
Binomial Distributions

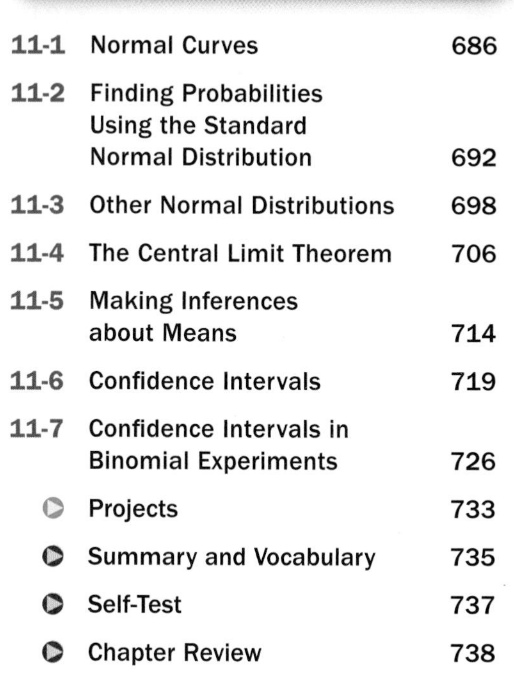

To the Student: Getting Started

Welcome to UCSMP *Functions, Statistics, and Trigonometry*! We hope you enjoy this book—it was written for you.

The Themes of This Book

Functions are correspondences or mappings that relate variables. For many people, functions are the most important content of high school mathematics. In your earlier work, you have studied linear, quadratic, exponential, and logarithm functions, and perhaps also polynomial and trigonometric functions. In this book, you will review and extend ideas about these functions. Because technology has changed the ways in which people use functions, we expect you to have technology that has graphical, statistical, geometric, and algebraic capabilities. This will broaden your ability to solve problems, to make connections among mathematical ideas, and to develop generalizations.

Statistics are used by people who work in government or journalism, who have to make decisions in business, who need to interpret the results of medical or psychological studies, who are responsible for monitoring quality of environment and health, or who wish simply to understand the world. The capability of computers and networks to store, share, and analyze data has made statistics an essential subject for citizenship and careers. Statistics provides ways of extracting information from data and using that information to build knowledge that will affect your life.

Trigonometry receives special emphasis in this book because it is fundamental to higher mathematics and to many applications. Physicists, cartographers, pilots, geologists, and engineers use trigonometry to precisely measure and predict lengths, angles, and locations. Trigonometric functions link algebra with geometry. They are used as models for situations that have nothing to do with geometry or triangles: the heights of tides, the pressure of a sound wave, and the swinging of a pendulum. The fundamental ideas of trigonometry may not be new to you; in this course, you will review those ideas and extend them to new contexts and applications.

Tools Needed for This Book

In addition to the lined and unlined notebook paper, pencils, graph paper, and erasers you typically use when doing mathematics, you need a calculator or computer software that can:

- Perform numerical and algebraic computations;
- Graph functions automatically;
- Create lists and graph them;
- Display certain relations that are not functions with parametric equations;
- Graph in rectangular and polar coordinates;
- Perform operations with matrices;
- Solve equations and transform algebraic expressions;
- Generate random numbers;

- Calculate statistics such as means, standard deviations, and regressions;
- Construct and manipulate geometric figures.

Some calculators do all of these tasks. These calculators have features such as a graphing utility, a computer algebra system (CAS), a statistics package, and a dynamic geometry system (DGS). You will need all of this technology to complete your assignments. In addition, the Internet is an excellent source for data, information, and Web-based software.

Studying Mathematics

A goal of this book is to help you learn mathematics so that you will be able to deal with the mathematics you see in newspapers, in magazines, on television, on any job, and in school. The authors offer the following advice.

1. **Mathematics is not a spectator sport.** You cannot learn much mathematics by watching other people do it. You must participate. You must think through and work problems, and learn from your successful and unsuccessful efforts.

2. **You are expected to read each lesson.** Sometimes you may do this as a class or in a small group; other times you will do the reading on your own. No matter how you do the reading, it is vital for you to understand what you have read. Here are some ways to improve your reading comprehension.

   Give each lesson a quick read to get a sense of what the lesson is about. Some students also scan the first questions at the end of the lesson. This establishes a mind-set that will make studying the lesson easier. Then go back and *reread slowly and thoughtfully*, paying attention to each word, graph and symbol.

   Read actively. The lesson has short **QY** questions ("Quiz Yourself"), examples, guided examples, and activities that build understanding of the lesson topic. Do the examples. Answer the QY questions and check your answers with what is shown at the end of the lesson. Complete the steps in the guided examples and check your answers with what is in the back of the book. Make sure you know how your technology carries out procedures illustrated in the lesson. **Activities** give you a chance to explore mathematical and statistical ideas. Your teacher may do some of the activities in class. You should do the others as you read the lesson.

3. **Writing mathematics is an important skill.** Writing is a tool for communicating your solutions and thoughts to others, and it can help you understand mathematics. In this book you may be asked to justify your solution to a problem or to write a description of data. Writing good explanations takes practice. You should use solutions to the examples in each lesson to guide your writing. What you might write for homework is shown in a special font like this. Some lessons will give you guidance on how to use statistics when you write about patterns in data.

4. **Be persistent.** If you cannot answer a question immediately, don't give up! Read the lesson again. Read the question again. Look for examples. If you can, go away from the problem and come back to it a little later. If the problem still stumps you, write a sentence or two telling how you tried to answer the question and move on to the next problems. Write something for every assigned question.

5. **Develop good study habits.**
Ask questions in class and talk to others when you do not understand something. Get the phone numbers or email addresses of students in your class with whom you can share thoughts about what you are learning. Ask your teacher for help with important ideas that are still not clear after you have worked on them.

Getting Off to a Good Start

The questions that follow are designed to introduce some features of *Functions, Statistics, and Trigonometry.*

We hope you join the hundreds of thousands of students who have enjoyed this book. We wish you much success.

Questions

COVERING THE IDEAS

1. What kinds of functions are studied in this book?
2. What are some fields that use statistics?
3. State four features of technology you need for this course.
4. How can the statement "Mathematics is not a spectator sport" be applied to the study of mathematics?
5. Name the way to improve reading comprehension given in this introduction that you think will be most useful to you.
6. Why is writing important for mathematics?
7. What should you do if you cannot answer an assigned question?

KNOWING YOUR TEXTBOOK

In 8–11, answer the question by looking at the Table of Contents, lessons and chapters in the textbook, or material at the end of the book.

8. Find the first QY question in Lesson 1-1.
   a. What is the purpose of a QY question?
   b. Where are the answers to QY questions?
9. What are the four categories of questions at the end of each lesson?
10. Suppose you are working on Lesson 2-5.
    a. Where can you find the answers to the Guided Examples in the lesson?
    b. Where can you find answers to check your work on the Questions?
11. Find the Self-Test at the end of Chapter 1. What should you do after taking the test?
12. Find the Chapter Review at the end of Chapter 2. What does SPUR stand for?

Chapter

1 Exploring Data

Contents

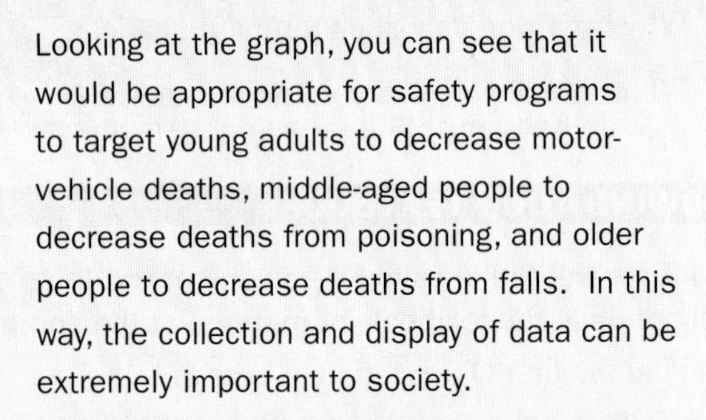

Accidents are the fifth leading cause of death in the United States. In 2004, a reported 112,012 Americans died in accidents. Each of these deaths is particularly tragic because we think of accidental deaths as being preventable and because such deaths occur without warning. The graph on the next page displays the number of deaths from the six leading causes of accidents in the United States by age.

Looking at the graph, you can see that it would be appropriate for safety programs to target young adults to decrease motor-vehicle deaths, middle-aged people to decrease deaths from poisoning, and older people to decrease deaths from falls. In this way, the collection and display of data can be extremely important to society.

Statistics is the science of the collection, organization, analysis, and interpretation of data. For a situation like the causes of accidental deaths, statisticians decide what will constitute an accident, how the accidents will be counted, whether the data suggest particular action be taken, and how to display the data to the public. In this chapter you are asked to act like a statistician and use statistics and graphs to uncover patterns in data, compare groups, and draw conclusions.

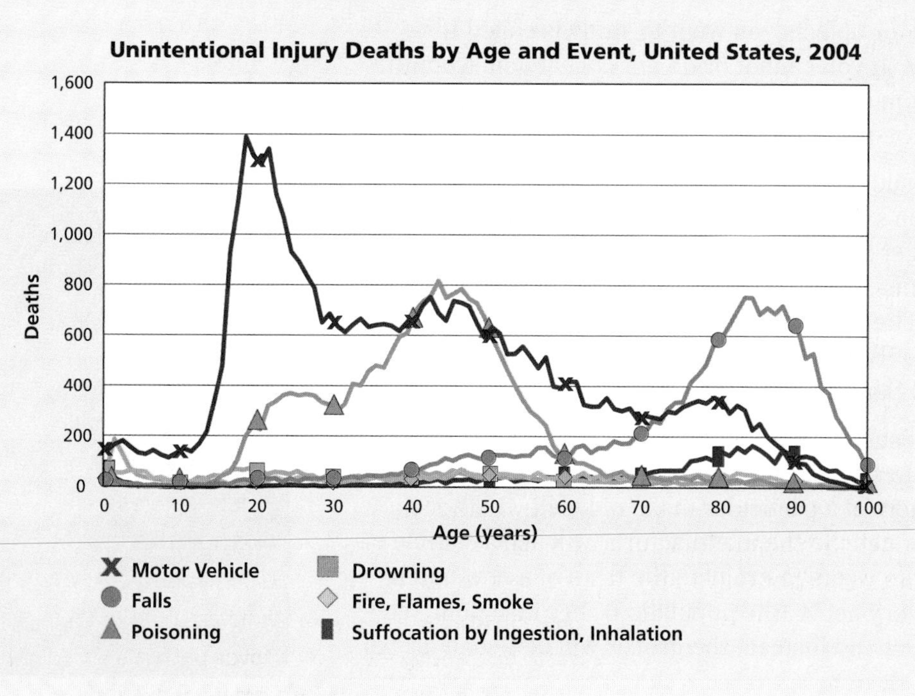

Source: National Safety Council, Injury Facts, 2008 Edition

Lesson 1-1 Variables, Tables, and Graphs

Vocabulary

statistic(s)
data, datum
variable
population
sample
survey
census
representative sample
categorical variable
numerical variable

▶ **BIG IDEA** Data can be presented in tables and graphs; statistics is about understanding and interpreting data to extract information and patterns.

The word *statistics* has two meanings. **Statistics** is the science of the collection, organization, analysis, and interpretation of *data*. (**Data** is the plural of **datum**, a piece of information.) Each of these activities depends on mathematics you will study in this course. A **statistic** also refers to a number used to describe a set of numbers. For example, the mean of a data set is a statistic describing a center of the data set.

Mental Math

In a class of 137 girls and 126 boys, g girls and b boys signed up for the bus to go to a football game.

How many more girls than boys signed up for the bus?

Variables, Populations, and Samples

In statistics, a **variable** is a characteristic of a person or thing that can be classified, counted, ordered, or measured. For instance, some variables that describe a person are gender, religion, number of siblings, height, and family income. Some variables describing a country are population, area, major political parties, number of tons of steel produced, and birth rate.

The set of *all* individuals or objects you want to study is called the **population** for that study. If you cannot or do not collect data from the entire population, but from only a part of it, that part is called a *sample*. A **sample** is a subset of the population.

Sometimes, for reasons such as fairness or legal requirements, the entire population must be studied. For instance, to be fair, the president of a club might want to get opinions from every member of the club. Gathering facts or opinions through an interview or questionnaire is called taking a **survey**. The U.S. Constitution requires that every ten years a **census** be taken. The U.S. census is a survey of the entire population of the United States.

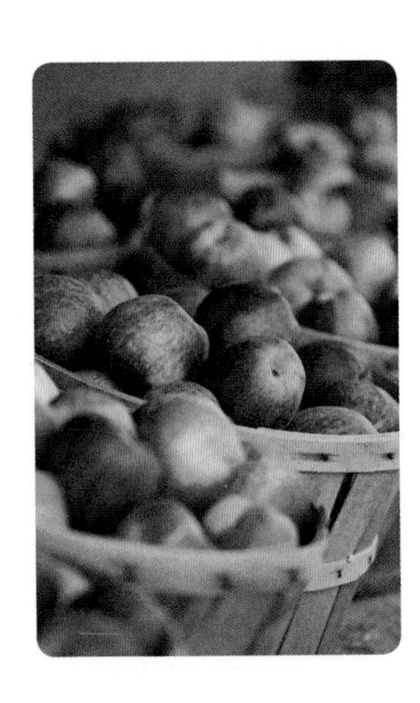

Other times, for reasons such as cost, safety, or preservation of a product, it is preferable to study a sample. For instance, it might be too expensive to ask all owners of a particular make of truck whether they are pleased with the product. So the manufacturer will ask a sample of the owners. A grocer who wants to evaluate the taste of a new batch of apples cannot taste every apple (the population), because doing so would destroy the product. So instead, the grocer will taste one or two of the apples (the sample).

Quiz Yourself (QY) questions help you follow the reading. You should answer each QY question before reading on. The answers to QY questions are at the end of the lesson.

See Quiz Yourself 1 at the right.

The grocer and the medical technician use samples to draw conclusions about populations. Their chances of drawing a valid conclusion are better when their samples are **representative samples**, that is, when the samples have the same characteristics as the population. A blood sample is a representative sample of a patient's blood because the circulatory system thoroughly mixes cells, chemicals, viruses, and wastes that are present in a patient's body. If the grocer samples two or three apples from the top of the box, he may miss rotten apples in the third and fourth layers down. The grocer can make his sample more representative by taking an apple from the top layer, an apple from the bottom layer, and an apple from a middle layer.

QUIZ YOURSELF 1

A medical laboratory technician draws 10 mL of blood in order to determine the number of white blood cells in a patient's bloodstream. Identify the variable, population, and sample.

Reading a Table

Data can come from experiments or surveys. To make sense of data, it helps to organize the data in a table or graph. In Guided Example 1 below, examine the table carefully. Look at the header information and the labels of the table's rows and columns. Try to determine the meaning of every number in the table.

A "Guided Example" is an example in which some, but not all, of the solution is shown. You should try to complete the example before reading on. Answers to Guided Examples are in the Selected Answers section at the back of the book.

GUIDED

Example 1

In Boston, 2145 selected inline skaters were chosen for a survey and asked to classify their ability as "Beginner," "Average," or "Advanced." They were also asked whether they wore protective gear like helmets, knee pads, shin pads, and elbow pads when skating. The data were organized in a table like that shown below.

	Amount of Protective Gear Worn			
Skill Level	**No gear**	**1 piece**	**2-4 pieces**	**Total**
Beginner	188	294	227	709
Average	314	245	123	682
Advanced	372	242	140	754
Total	874	781	490	2145

(continued on next page)

a. Are the data based on a sample or on the entire population?

b. What variables are presented?

c. What percent of skaters wore no gear?

d. What percent of beginners wore no gear?

e. How many skaters said they were advanced?

f. How many advanced skaters wore protective gear?

Solution

a. Only selected skaters were surveyed. The data are based on a sample.

b. The variables are often named as labels of the rows and columns. In this table there are only two variables: skill level and __?__. Both variables are reported in categories, described by the labels of the rows and columns.

c. 874 out of the 2145 skaters wore no gear. $\frac{874}{2145} \approx 0.407$. About __?__% of the skaters wore no gear.

d. __?__ out of the __?__ beginners wore no gear. About __?__% of the beginners wore no gear.

e. You can find the total number of advanced skaters by adding the values in the three categories in the row labeled "Advanced":
$372 + \_\_?\_\_ + 140 = \_\_?\_\_$. You can also find this total in the rightmost column of the table. __?__ skaters said they were advanced.

f. __?__ advanced skaters wore one piece of gear and __?__ wore 2 to 4 pieces of gear for a total of 382. So __?__ advanced skaters wore some protective gear.

The purpose of the study in Example 1 was to get information about a population of skaters. If the sample is representative of inline skaters in Boston, then about 41% of Boston skaters do not wear protective gear.

In the table of Example 1, the skill level is a **categorical variable** because the different values do not correspond to specific numbers. The number of pieces of protective gear is also reported as a categorical variable because different values are grouped into categories together. If the study had not combined 2, 3, and 4 pieces of gear into a single category, then that variable would have been a **numerical variable**. With a numerical variable, you can calculate numerical statistics such as the mean. With a categorical variable, you cannot calculate numerical statistics, though you may be able to estimate them.

Drawing Conclusions from Tables

When reading a table, you should ask three questions:

1. What variables are presented, and how?

You can usually find the variables as headings of rows and columns, but sometimes a single table will present the same variable in more than one way: by giving a value and a corresponding percentage, for example.

2. Are the data trustworthy?

Consider the data source, the accuracy of the data, and the time when the data were collected. Ideally, the data source should be given, allowing you to verify the data if you want, and it should be reputable. The age of the data may matter; even when addressing historical issues, newer data may be more accurate, reliable, or precise than older data. Knowing when and how the data were collected helps you decide what conclusions to draw.

3. What conclusions can you draw from the table?

Some conclusions can be read directly from data in a table. Others can be drawn only after performing calculations using the data in a table, or after looking for patterns or trends in the data.

Below is a table summarizing some data collected by the U.S. Census Bureau. Before going on, read the table carefully and try to make sense of every number in it.

Income of Households by Highest Education Level of Householder in 2005

Highest Level of Education Completed by Householder	Number of Households (thousands)	Percent Distribution by Income Level								Median Income (dollars)
		Under $10,000	$10,000-$14,999	$15,000-$24,999	$25,000-$34,999	$35,000-$49,999	$50,000-$74,999	$75,000-$99,999	$100,000 and over	
Less than 9th grade	6,088	21.7	15.6	21.6	14.7	12.0	9.1	3.3	2.2	20,224
Some high school, but no diploma	9,130	17.3	12.8	20.5	15.1	14.4	11.9	4.7	3.4	24,675
High school graduate	32,345	8.8	7.8	15.8	13.3	16.7	18.9	9.4	9.3	38,191
Some college	28,874	5.7	4.8	10.6	11.7	16.8	21.1	13.2	16.1	50,412
Bachelor's degree or higher	31,153	3.0	2.2	5.0	6.5	11.6	19.9	15.6	36.2	77,179

Source: U.S. Census Bureau, Current Population Survey, 2006 Annual Social and Economic Supplement.

See Quiz Yourself 2 at the right.

QUIZ YOURSELF 2

Are the data in the table trustworthy?

Example 2

Refer to the table of income data above.

a. What variable is represented by the row headings and what kind of variable is it? What is the other variable in the table?

b. What kinds of information are presented in the columns? What columns correspond to each kind?

c. Interpret the table entry 31,153 in a sentence describing its context.

d. Write a statement relating median income to education of householder.

e. How many times as likely was a family to have an income of at least $100,000 if the head of household had graduated college rather than not started high school?

(continued on next page)

Solution

a. The row headings present the highest level of education completed by the householder, which is a categorical variable. The other variable is the annual household income. All other numbers in the table were calculated from the income data.

b. The leftmost data column gives the total number of households (in thousands) at each education level. Data columns 2-9 give the percentage of households within each education level at each income level. The rightmost column gives the median income (in dollars) for all households at each education level.

c. 31,153 occurs in the last row of the first data column, so it represents the total number of households in which the householder had a bachelor's degree or higher. Notice that values in this column are given in thousands, so we have to multiply 31,153 by 1000 to get the actual value. In about 31,153,000 households, the householder had a bachelor's degree or higher.

d. Examine the rightmost column. The numbers increase going down the column. Households headed by people with higher education levels have higher median incomes.

e. The percents of households with an income of at least $100,000 are given in the second column from the right. Families headed by someone who did not start high school are represented by the first row of level of education. About 2.2% of households headed by someone who was not did not start high school had incomes of at least $100,000, compared with 36.2% of households headed by college graduates. Thus, a college graduate-headed household was more than 16 times as likely to have an income of at least $100,000 as a household where the head had not started high school.

See Quiz Yourself 3 at the right.

QUIZ YOURSELF 3

In 2005, how many households with income under $10,000 were headed by a householder with some high school but no diploma?

Bar Graphs

Conclusions can often be determined or supported by graphs. For instance, the *bar graph* at the right pictures the data from the rightmost column of the table on the previous page. It supports the conclusion in Example 2, Part e.

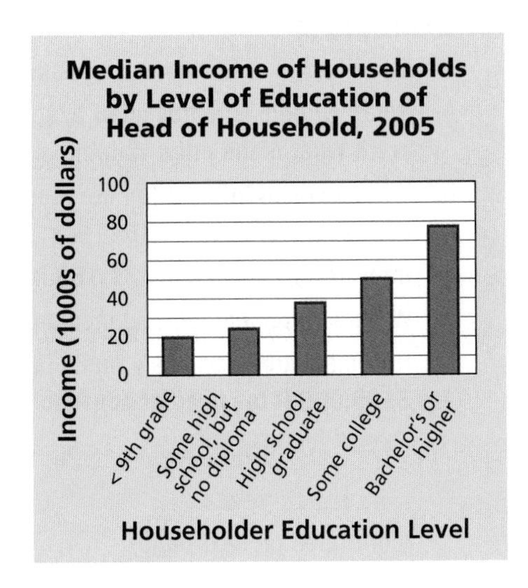

Bar graphs are appropriate when one variable is a categorical variable, and the other variable is numerical. One axis lists the categories, in this case householder education level. The other axis is a numerical scale, typically with counts, percents, or measurements such as income. A well-made bar graph has a descriptive title. In addition, the graph identifies the variables being described, labels the scale in equal intervals, and uses bars of equal widths for each category.

More complex types of bar graphs can be used to show features of data with two categorical variables. The *stacked bar graph* at the right was created from the table in Example 1 in four steps:

- First, each frequency was converted to a percentage of skaters in that experience category. For example, 188 of the 709 beginner skaters wore no gear, giving about 27%.
- At each skill level, a bar was drawn for the percentage of skaters wearing no gear.
- Bars were stacked on top of other bars by adding values. For example, because 27% of beginning skaters wore no gear and 41% wore 1 piece, the stacked bar representing 1 piece starts at 27% on the vertical axis and goes to 27% + 41% = 68% on the vertical axis.
- The parts of each stack were colored differently to distinguish them, and a legend describing each color was placed by the graph.

The stacked bar graph shows that the Average and Advanced skaters are similar in use of protective gear and that Beginners are more likely than other skaters to wear protective gear.

Questions

COVERING THE IDEAS

These questions cover the content of the reading. If you cannot answer a Covering the Ideas question, you should go back to the reading for help in obtaining an answer.

1. Give three examples of variables that can be used to describe a person but were not mentioned in the lesson.
2. Give three characteristics of a situation that might cause a person to study a sample rather than an entire population.

In 3 and 4, a situation is described. Determine the population, whether the data are based on a sample, and the variable of interest.

3. A pastry inspector counts the number of raisins per cookie in 10 oatmeal-raisin cookies in a batch fresh out of the oven.
4. A statistician computes the free-throw percentage for every player in the WNBA.

In 5 and 6, refer to the National Safety Council graph on page 5.

5. What are the three most common types of accidental deaths among 80-year-olds?
6. Estimate the total number of accidental deaths in the U.S. in 2004
   a. among 10-year-olds.
   b. among 20-year-olds.

7. What is a *representative sample*?

In 8–10, refer to Guided Example 1.

8. a. What is the sample? b. What is the population?

9. a. What percent of skaters who wore no protective gear were beginners?

   b. Why is your answer to Part a different from the answer to Guided Example 1 Part d?

10. If you were to notice a skater in Boston without any protective gear, what is the chance the skater is advanced?

11. Tell whether the variable is numerical or categorical.

   a. gender
   b. method of transportation to school
   c. height
   d. distance from school

In 12–15, refer to the table before Example 2.

12. What percent of households in 2005 were headed by people who had some college, but not a bachelor's degree?

13. What percent of households headed by high school graduates (with no college) earned at least $25,000 annually?

14. a. How many households headed by someone who was not a high school graduate earned less than $15,000 in 2005?

   b. What percent of the total number of households is this?

15. Write a sentence or two describing how the incomes of households headed by someone with a bachelor's degree or higher compares to the incomes of households headed by someone with a high school diploma but no college.

APPLYING THE MATHEMATICS

These questions extend the content of the lesson. You should study the examples and explanations if you cannot answer the question. For some questions, you can check your answers with the ones in the Selected Answers section at the back of this book.

16. The *2005–2006 High School Athletics Participation Survey* of the National Federation of State High School Associations listed participation of high school students in 47 sports. The table at the right shows participation for the six sports that had the most participants. The graph on the next page displays the information in the table.

   a. What are the categorical variables in the table?

   b. Compute the missing percents on the graph.

   c. The graph shows that 100% of the participants in baseball are boys. Yet the table shows that there are girls who play baseball. Why is there a difference?

Numbers of Participants in High School Sports 2005-2006

Sport	Boys	Girls	Total
Baseball	470,671	1,382	472,053
Basketball	546,335	452,929	999,264
Football	1,071,775	1,173	1,072,948
Soccer	538,935	321,555	860,490
Track and Field	533,985	439,200	973,185
Volleyball	42,878	390,034	432,912
Total	3,204,579	1,606,273	4,810,852

d. Write a few sentences about the differences between boys' and girls' participation in these sports.

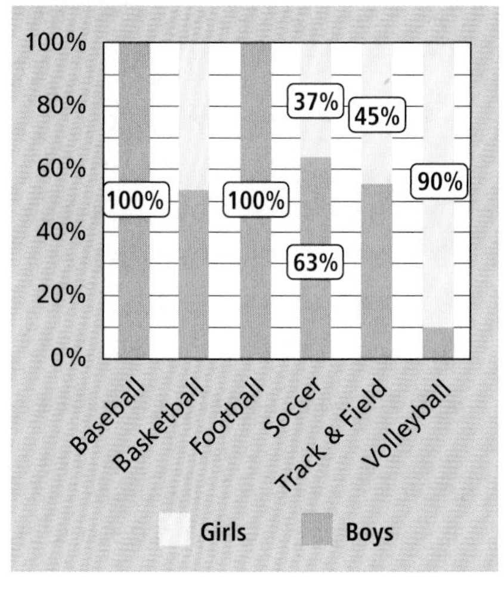

17. A biologist captures 96 trout from different parts of a lake to determine the age distribution. She finds that 53 are less than six months old and 29 are between six months and one year old.
   a. What variable or variables are being studied?
   b. Is the study based on a sample or on a population? If a sample, what is the population, and what steps were taken to make the sample representative?
   c. What percent of the trout are more than one year old?

18. A school district surveys a sample of families to determine how far students must travel to school. They find that of 284 children in elementary school, 72 travel 2 blocks or fewer, 132 travel between 2 and 6 blocks, and 80 travel six blocks or more. Among 224 students in middle school, 34 travel 2 blocks or fewer and 107 travel between 2 and 6 blocks. Among 209 high school students, 12 travel 2 blocks or fewer and 108 travel 6 blocks or more.
   a. Create a table representing these data.
   b. Create a stacked bar graph showing the number (not percentage) of students in each grade band and, within each band, the number at each distance.
   c. Is the distance traveled a numerical or a categorical variable?

REVIEW

Every lesson contains review questions to practice ideas studied earlier.

19. ***Skill Sequence*** Solve each equation. **(Previous Course)**
   a. $13x = 260$ b. $0.13x = 260$
   c. $260x = 13$ d. $260x = 0.13$

20. Identify the slope and y-intercept of the graph of the equation $y = mx + b$. **(Previous Course)**

21. A line has slope -3 and contains the point $(-5, 2)$. **(Previous Course)**
   a. Name two other points on the line.
   b. Find an equation for the line.

22. Find the *mean* (average) of $x + 3$, $x + 7$, and $x + 29$. **(Previous Course)**

EXPLORATION

These questions ask you to explore topics related to the lesson. Sometimes you will need to use references found in a library or on the Internet.

23. Find an example of a bar graph in a newspaper, magazine, or other publication. Identify the variables. What conclusion(s) can you draw from the graph?

QUIZ YOURSELF ANSWERS

1. The variable is the number of white blood cells. The population is all the patient's blood. The sample is the 10 mL of the patient's blood that is drawn.

2. Yes; the data are from the U.S. Census Bureau, a reputable source. The source is shown.

3. about $0.173 \cdot 9{,}130{,}000 \approx 1{,}579{,}490$ households

Lesson

1-2 Centers of Data and Weighted Averages

Vocabulary

mean
median
measures of center, measures of central tendency
mode
subscripted variables
sigma, Σ
index, *i*
summation notation, sigma notation, Σ-notation
weighted average
relative frequency

▸ **BIG IDEA** The mean and median are both measures of center for data.

The word "average" has more than one meaning. When people say that someone is of "average height," there is no calculation. They are saying that the person's height is somewhere in the middle range of all heights. But when people say, "The average height of the students was 162 cm," they are usually speaking of a calculated statistic, the *mean*.

Mental Math

A store has a sale selling 5 baseball caps for a total of *x*. At regular price they each cost *y*.

How much will each of 5 individuals save if they pool their money together and buy the hats on sale?

Measures of Center

The **mean** of a data set is calculated by finding the sum of the numbers in the set and dividing by the number of elements. The **median** is the middle number when the set of numbers is put in increasing or decreasing order; it is the mean of the two middle numbers when the set has an even number of elements. Both the mean and median are **measures of center** or **measures of central tendency**, and at times people will use either the mean or median as the "average" of a data set.

GUIDED

Example 1

The Wacky Widget Company has 15 employees. The jobs and annual salary for each job are given in the table at the right. A newspaper reported, "Average Wacky Widget worker earns $78,000 a year."

a. Which statistic was it reporting, the mean or the median?

b. Why might most employees be upset by the newspaper article?

No.	Job	Salary ($1000)
1	CEO	380
2	CEO Asst	35
3	VP	150
4	VP Asst	30
5	Parts Mgr	90
6	Parts Worker	30
7	Parts Worker	30
8	Custodian	27
9	Custodian	24
10	Custodian	24
11	Sales Mgr	90
12	Sales Rep	65
13	Sales Rep	65
14	Sales Rep	65
15	Sales Rep	65

Solution

a. Calculate the median salary. Since all the numbers are in thousands, ignore the thousand and consider $90,000 as 90. Put the salaries in order, smallest to largest. 24, 24, 27, 30, 30, 30, 35, 65, 65, 65, 65, 90, 90, 150, 380. The value of the middle number is __?__. So the median salary is __?__. The newspaper did not report the median. The mean is the sum of the salaries divided by __?__. The sum is __?__. So the mean is $78,000. The newspaper reported the mean.

b. The newspaper was reporting an average of $78,000, but __?__ of the 15 employees make less than this.

In Example 1, even though the newspaper accurately reported the mean salary, the president's salary alone contributes $\frac{380}{15}$ thousand dollars, or over \$25,000, toward the mean salary of employees. Reporting the mean salary in this situation is misleading. When dealing with data sets that have widely varying values, the median is often more representative of the data, and the **mode**, the most frequent value, is also often reported.

See Quiz Yourself 1 at the right.

> **QUIZ YOURSELF 1**
>
> What is the mode salary of Wacky Widget Company employees?

Summation Notation

Calculating a mean involves finding a sum. Sums are so basic to mathematics that a shorthand notation for representing sums is commonly used. There are three parts to *summation notation*: **subscripted variables** $x_1, x_2, x_3, \ldots, x_n$ to identify each of the n data values; the Greek letter $\sum$ (**sigma**) to indicate there is a sum; and the **index *i*** to indicate which of the subscripted variables are being added.

To indicate the sum of the 15 salaries of the Wacky Widget Company, let x_i be the salary of the ith employee. That is, $x_1 = \$380{,}000$, $x_2 = \$35{,}000$, and so on, until $x_{15} = \$65{,}000$. Notice that the index subscript simply indicates the position in the list. The sum

$$x_1 + x_2 + x_3 + \ldots + x_{15} \text{ is written } \sum_{i=1}^{15} x_i$$

and is read "The sum of x-sub-i as i goes from 1 to 15." This notation is called **summation notation, sigma notation**, or **Σ-notation**. If you want to indicate the sum of the salaries of the three Parts workers at the Wacky Widget Company, you could write $\sum_{i=5}^{7} x_i$, which equals $x_5 + x_6 + x_7$, or \$150,000. This notation may need to be adapted when using technology. If you were to represent these 3 salaries as entries A5 to A7 in a spreadsheet, their sum would be `SUM(A5:A7)`.

See Quiz Yourself 2 at the right.

> **QUIZ YOURSELF 2**
>
> Use Σ-notation to indicate the sum of the salaries of the five people in sales at Wacky Widget Company.

The mean of a data set can be expressed using sigma notation. If $\{x_1, x_2, x_3, \ldots x_n\}$ is a data set of n numbers, the mean is

$$\frac{\sum_{i=1}^{n} x_i}{n} = \frac{1}{n}\sum_{i=1}^{n} x_i.$$

Example 2

A full parking lot has 46 cars. Let p_i = the number of people who rode in the ith car parked in the lot.

a. What does $\sum_{i=1}^{46} p_i$ represent?

b. Use Σ-notation to express the mean number of people per car.

(continued on next page)

Solution

a. $\sum_{i=1}^{46} p_i = p_1 + p_2 + p_3 + \ldots + p_{46}$. This represents the total number of people who rode in the cars parked in the lot.

b. To find the mean, divide the total number of people by the number of parking spots. $\frac{\sum_{i=1}^{46} p_i}{46}$ or $\frac{1}{46}\sum_{i=1}^{46} p_i$.

Most statistics utilities based on spreadsheets have special syntax for measures of center. For example, cell A5 in spreadsheets I and II shows the commands for the median and mean of the entries in cells A1, A2, A3, and A4 (the first four annual salaries at the Wacky Widget Company).

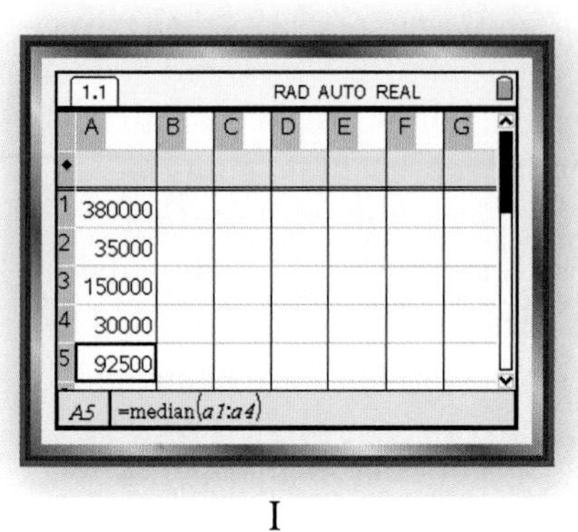

I

1.1
RAD AUTO REAL
A B C D E F G
1 380000
2 35000
3 150000
4 30000
5 148750
A5 =mean(a1:a4)

II

Weighted Averages

In some high schools and colleges, a student's grade point average (GPA) is calculated based on the weights or credits assigned to a letter grade in a specific class. In Susan's high school, AP courses are weighted as 5 credits, Honors courses are 4 credits, regular level courses are 3 credits and elective courses (like keyboarding and home economics) are 1 credit. The grading scale below is used in her school.

Letter	F	D-	D	D+	C-	C	C+	B-	B	B+	A-	A	A+
Grade Points	0.00	0.67	1.00	1.33	1.67	2.00	2.33	2.67	3.0	3.33	3.67	4.00	4.00

Susan's grades are shown on her report card in Example 3.

Example 3

Use Susan's report card at the right to find her GPA.

Solution The weights are the number of credits per course. For example, the B- Susan received in AP Calculus has a weight of 5 credits towards her GPA. To calculate Susan's GPA we use the credits per course as the weight and the grade points as the value being weighted.

Susan's GPA is $\frac{1(4.00) + 5(2.67) + 4(3.33) + 3(0.67) + 3(2.00)}{1 + 5 + 4 + 3 + 3}$

≈ 2.42

Susan's Report Card

Course	Credit	Letter	Grade Points
Keyboarding	1	A+	4.00
AP Calculus	5	B-	2.67
Honors English	4	B+	3.33
Regular Level Art History	3	D-	0.67
Regular Level Statistics	3	C	2.00

Susan's GPA is an example of a *weighted average.* In general,

if x_i = a value in a data set and

w_i = the weight of the value,

then the **weighted average** of the values x_i is

$$\frac{\sum_{i}^{n} w_i x_i}{\sum_{i}^{n} w_i}, \text{which means } \frac{w_1 x_1 + w_2 x_2 + \ldots + w_n x_n}{w_1 + w_2 + \ldots + w_n}.$$

In Example 3, if each of Susan's courses were weighted as 1 credit, then the weighted average would be equivalent to the mean since all of the weights would be equal. In general, every mean is a weighted average in which each weight $w_i = 1$.

CAUTION: When calculating a GPA, it is common to mistake the point value for the weight. Since when calculating the weighted average, the denominator is the sum of the weights, you can see in Example 3 that the credits are the weights.

Thinking of Frequencies as Weights

You can think of frequencies as weights and calculate a mean as a weighted average as in the following situation.

Alexis wanted to buy a ticket to a concert. She saw that ticket prices were \$20, \$30, and \$100. She wrote a letter to the concert promoters. She said, "An average price of \$50 is too much for a concert for teenagers." Would you agree with Alexis that the average price of a ticket is \$50?

The concert promoters responded that most of the tickets were \$20. They reported that there were 1,000 \$20 seats, 500 \$30 seats, and only 20 \$100 seats. These data are arranged in a frequency table and a bar graph below.

Ticket Price	Availability
\$20	1,000
\$30	500
\$100	20

The calculations shown in Example 3 for the weighted average can be used to quickly calculate the mean ticket price. Computing the *mean* involves adding all the \$20, \$30 and \$100 tickets and dividing by the sum of the frequencies:

$$\frac{20(1000) + 30(500) + 100(20)}{1000 + 500 + 20} = \frac{37000}{1520} \approx \$24.34.$$

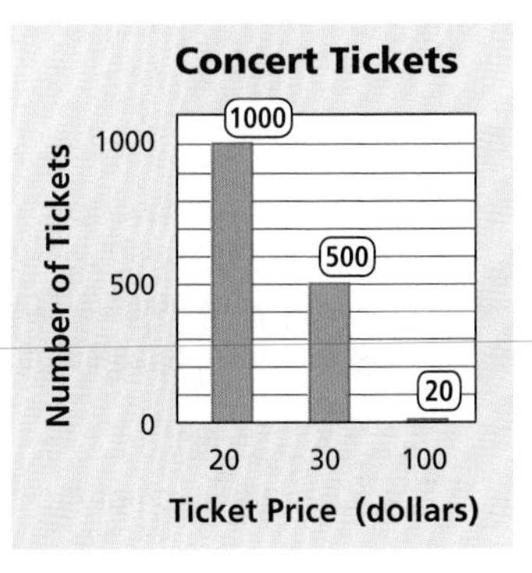

The result is not \$50, but \$24.34 because there are more \$20 and \$30 tickets available. This means that because the \$20 seats have a higher frequency of occurrence, they contribute more in computing the mean. They have more "weight" than the \$100 tickets.

Using Relative Frequencies to Calculate Weighted Averages

The **relative frequency** is the ratio of the number of times a number or event occurs to the total number of numbers or events. The weighted average of a data set is the same whether the weights are frequencies or relative frequencies.

Example 4

a. **Convert the frequency table and graph for the concert tickets to a relative frequency table and graph.**

b. **Compute the weighted average using the relative frequency values and compare the result with that obtained from using the frequencies.**

Solution

a. Convert each frequency to a relative frequency by dividing the frequency by the total number of tickets sold. In this case, relative frequency has been rounded to two decimal places. An advantage to converting the data to a relative frequency is that it is now easy to see that about 66% of the tickets sold were \$20 tickets, about 33% were \$30 tickets, and about 1% were \$100 tickets.

Ticket Price	Relative Frequency
\$20	$\frac{1000}{1520} \approx 0.66$
\$30	$\frac{500}{1520} \approx 0.33$
\$100	$\frac{20}{1520} \approx 0.01$

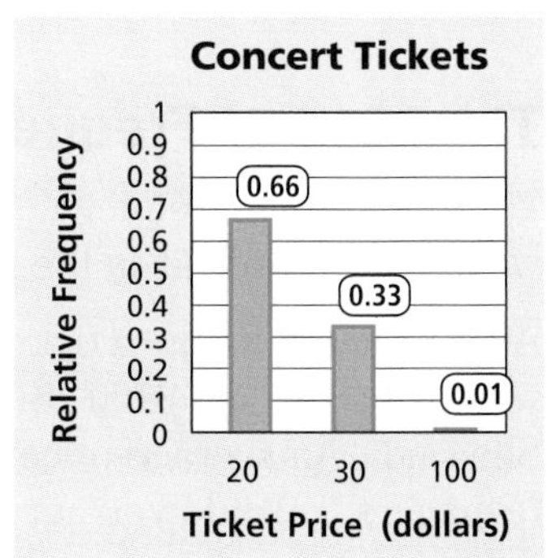

b. The weighted average is computed by multiplying each price by its relative frequency.

$$20 \cdot \frac{100}{1520} + 30 \cdot \frac{500}{1520} + 100 \cdot \frac{20}{1520} = \frac{37000}{1520} \approx \$24.34$$

The result is identical to the weighted average computed by multiplying prices by their frequencies.

Another advantage of using relative frequencies is that the sum of the relative frequencies always equals 1 since each relative frequency is a percent of the number of tickets available.

A spreadsheet or other software can calculate the weighted average using relative frequencies.

Activity

Use software or a calculator to compute the mean of the Wacky Widget salaries as you would a weighted average. A spreadsheet and calculation split-screen are shown for one calculator. With other technology, your work may look different.

Step 1 Enter each of the eight salary levels in column A.

Step 2 Enter the corresponding frequencies in column B.

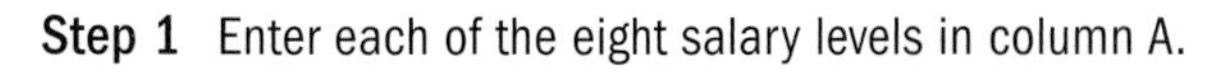

Step 3 In column C, compute the product of each salary level and its corresponding frequency.

Step 4 Compute the sums of the data in columns B and C. The SUM function can be used in the spreadsheet. In cell B9, type =SUM(B1:B8). Enter a corresponding function in C9 or use summation notation on the calculation screen as shown at the right.

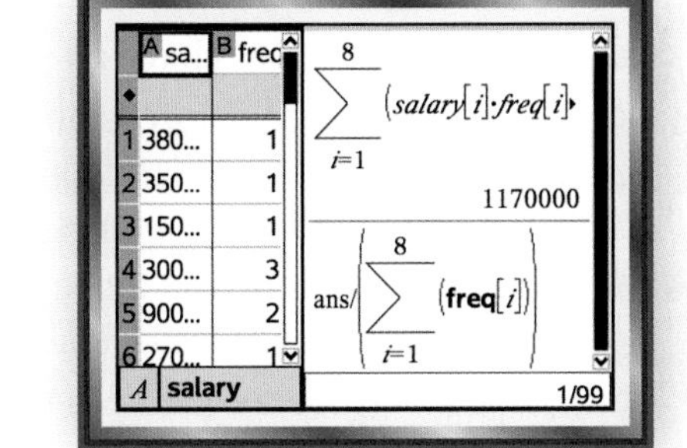

Step 5 Compute the mean using cells C9 and B9. Put the formula in cell C10.

Step 6 Compute the relative frequency of each salary. Confirm the mean using relative frequencies.

Step 7 If everyone at Wacky Widget gets a $5,000 raise, what is the new mean salary? Justify your answer using the spreadsheet or algebra.

Questions

COVERING THE IDEAS

1. What two statistics are considered measures of central tendency?
2. Express the mean of the set {3, 5, 7, 13, 17, 19} using sigma notation.
3. What is a weighted average?
4. Explain the difference between frequency and relative frequency.

In 5–7, refer to Guided Example 1 and the Activity.

5. In one computation of the mean, why can 90 be multiplied by 2 and 65 be multiplied by 4?
6. Consider the salaries of all employees except the CEO.
   a. Find the mean and median of the salaries.
   b. Compare your answers to those in the solution on page 14. By how much has each changed?
   c. In general, which is more affected by extreme values, the mean or the median?
7. A new parts worker is hired.
   a. What is the new relative frequency of the parts worker salary?
   b. What is the new mean salary?
   c. What is the new median salary?
8. What does the number 27 indicate in the expression $\sum_{i=1}^{27} x_i$?
9. Write an expanded expression for $\sum_{i=4}^{8} x_i$.
10. One teacher computes semester grades as follows: 1 grading period (GP) is equivalent to 2 semester tests (ST). One semester contains 3 grading periods and 1 semester test. Elliott's grades are as follows: GP 1: 65; GP 2: 80; GP 3: 90; ST 1: 87. What is Elliott's semester grade?

In 11 and 12, use the dotplot at the right, which shows the distribution of the number of siblings of the students in Mr. Maestro's homeroom.

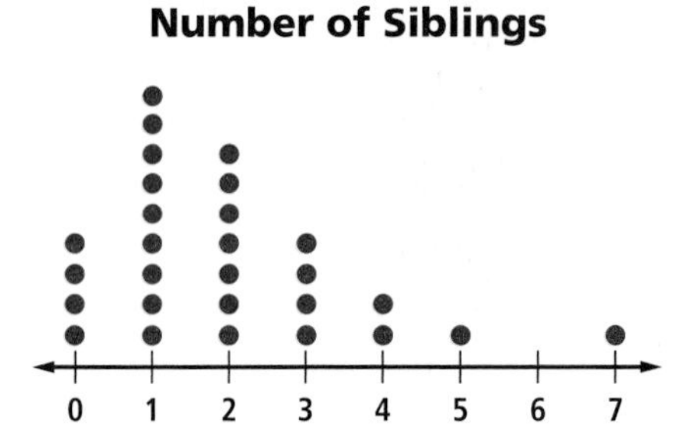

11. a. How many students are in Mr. Maestro's homeroom?
    b. How many students have no siblings?
    c. Write a numerical expression for the mean number of siblings per student. Use your expression to calculate the mean number of siblings per student.
12. a. Give the relative frequencies of 0, 1, 2, 3, 4, 5, 6, and 7 siblings in Mr. Maestro's homeroom.
    b. Use the relative frequencies to calculate the mean number of siblings per student.

APPLYING THE MATHEMATICS

13. A science fair judge rates projects using the weighting scale shown at the right. Brendan finds out that each category can receive a rating of between 0 and 10. If he received a 7 in visual, an 8 in content and a weighted score of about 7.92, what was his score in sources to the nearest digit?

Category	Weight
Visual	4
Content	5
Sources	3

14. Betty bought shares of Statco stock from time to time at various prices as shown below. That is, she bought 300 shares at \$14.23, 200 shares at \$11.55, and so on.

Number of Shares	300	200	500	100	400	200	100	300
Price (\$)	14.23	11.55	12.48	13.92	16.03	15.45	14.85	13.26

    a. What is Betty's average cost per share of Statco stock?
    b. If Betty had purchased 400 shares at \$10.28 instead of \$16.03, how much lower would her average cost per share have been?
15. Mrs. Dalloway weights the unit grades of the students in her math classes as shown in the circle graph on the right. Gyan's scores were 78 on the test, 85 on homework, and 92 in participation.
    a. What is Gyan's unit grade in Mrs. Dalloway's class?
    b. What would Gyan's unit grade be if the three categories were equally weighted?

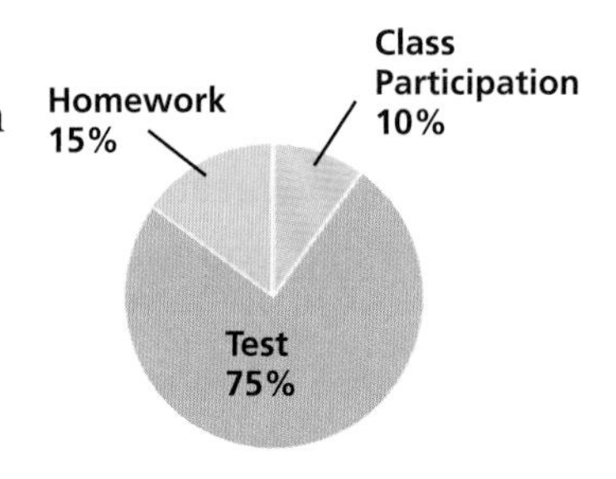

16. The stemplot at the right displays the scores of 24 students on a chemistry quiz. The first row stands for scores of 34 and 39. No student had a score in the 40s or 50s. The highest score obtained by any student was 98.
    a. What does the 7 in the row with a stem of 8 stand for?
    b. Let x_i be the score of the ith student. If the stemplot order is followed, $x_1 = 34$, $x_2 = 39$, …, and $x_{24} = 98$. Find $\sum_{i=1}^{24} x_i$.
    c. Find $\frac{1}{24}\sum_{i=1}^{24} x_i$ and identify what you have calculated.
    d. What is the median quiz score?
    e. What is the mode quiz score?

stems	leaves
3	4 9
4	
5	
6	2 8
7	0 5 5 5 6 8 8
8	0 2 4 4 5 5 7
9	0 3 4 5 8 8

17. Consider $\sum_{i=1}^{100}(x_i + 2)$ and $\sum_{i=1}^{100}x_i + 2$, where x_i is a positive number. Which expression has the greater value? By how much?

18. A bowler has averaged 164 for the first 21 games of a 90-game season. What would the bowler have to average for the rest of the season to bring his 90-game average score up to 170?

19. The mean of a set of n numbers is 15. If the numbers 12 and 24 are added to the set, the mean is 16. What is n?

20. The cost to produce and deliver gasoline to consumers is displayed in the stacked bar graph at the right.
   a. What is the current price per gallon of gas at a gasoline station near where you live?
   b. If a person fills a tank with 12 gallons of gas, how much will be paid for marketing and distributing?
   c. If refinery costs increase by 50%, by what percent should the total cost increase?
   d. Which should affect gas price more, a 10% tax increase or a 5% increase in the price of crude oil? Justify your answer.

What We Pay For in a Gallon of Regular Gasoline (April 2008)

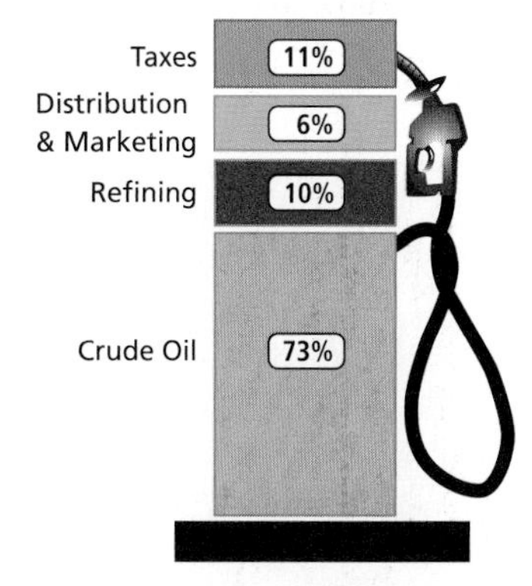

Source: Energy Information Administration

REVIEW

In 21–23, use the table at the right to answer the following questions for 2001. (Lesson 1-1)

21. About how many Black, non-Hispanic children used computers at home?

22. About how many American Indian, non-Hispanic children used computers at school?

23. What percentage of Asian, non-Hispanic children did not use computers at school?

Computer and Internet Use by Children and Adolescents: 2001

Race/Ethnicity	Number of Children (1,000)	Percent using computers at school	Percent using computers at home
White, non-Hispanic	33,433	83.5	76.9
Black, non-Hispanic	8,275	79.8	41.0
Hispanic	8,400	71.8	40.6
Asian, non-Hispanic	2,268	76.1	75.7
American Indian, non-Hispanic	637	83.0	54.1

Source: Statistical Abstract of the United States, 2004-2005

24. ***Skill Sequence*** Rewrite each expression, assuming denominators do not equal zero. **(Previous Course)**
   a. $\frac{16x - x}{3x}$
   b. $\frac{27y + 18y}{3y + 6y}$
   c. $\frac{27z + 18}{3z + 6}$

EXPLORATION

25. Find out how weighted averages are referred to in one of the following areas. A suggestion is indicated in parentheses for each area. If you can find an example of the computation, check to see whether it matches the use of weighted average in this lesson.

   investing (average interest rate)
   chemistry (relative atomic mass)
   government (poverty index)
   sports (slugging percentage in baseball)

QUIZ YOURSELF ANSWERS

1. $65,000
2. $\sum_{i=11}^{15} x_i$

Lesson

1-3 Creating and Using Histograms

Vocabulary

distribution
histogram
bins
frequency histograms
relative frequency histograms
skewed
symmetric
population pyramid

▶ **BIG IDEA** Histograms display data and show how frequently values occur.

The number of representatives that each state has in the U.S. Congress depends on the population of the state. Below are the numbers of representatives from each of the 50 states in the 110th Congress (2007–2009).

7	7	2	4	2	1	3	5	1	11	1	5	5	3	53	13	9
6	15	3	29	5	9	9	4	1	2	7	8	3	8	4	13	18
13	19	32	8	8	25	19	10	9	2	1	2	3	6	1	1	

Mental Math

a. How many thousands are in 1 billion?

b. How many millions are in 1 trillion?

This data set can be turned into a *distribution* by finding the frequency of each data value. A **distribution** is a function whose values are the frequencies, relative frequencies, or probabilities of mutually exclusive (non-overlapping) events. By graphing a distribution, you may see features of the data that are hard to see in a table. One important type of graph is a **histogram**, which is a special type of bar graph. A histogram breaks the range of a numerical variable into non-overlapping intervals of equal width, which are called **bins**. **Frequency histograms** display the number of values that fall into each interval. **Relative frequency histograms** display the percent of values that fall into each interval.

Two histograms showing the congressional data are on the next page. The histogram on the left shows frequency; the one on the right shows relative frequency.

We chose to use bins of width 5 and to make each interval include its left endpoint but not its right. For instance, on each graph the leftmost two intervals are $0 \le x < 5$ and $5 \le x < 10$, so a state with exactly 5 representatives is recorded in the second bar from the left.

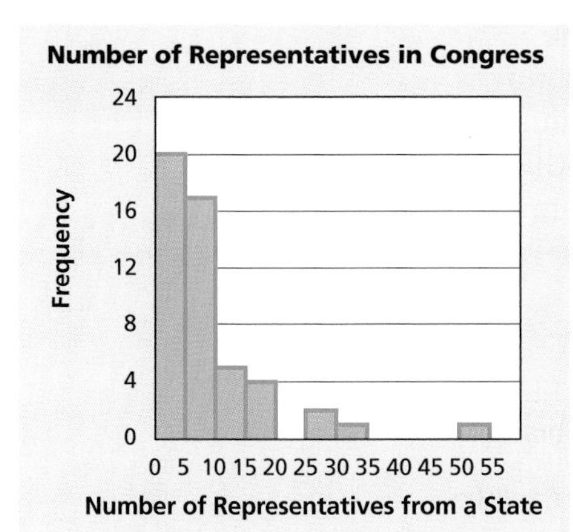

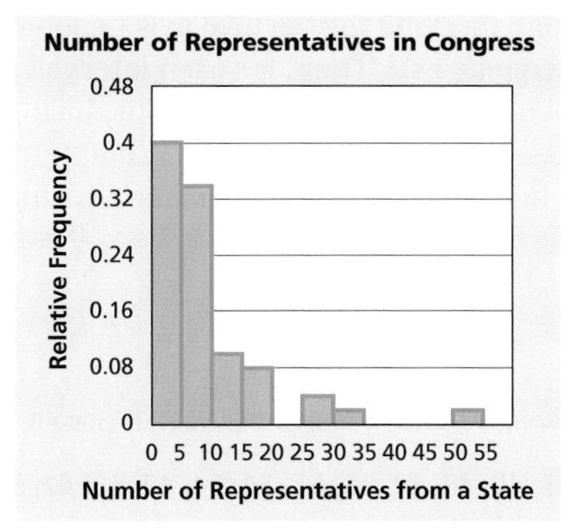

Example 1

Refer to the histograms on congressional data.

a. About how many states have 15 to 19 representatives?

b. What percent of states have 20 or more representatives?

c. In what bin does the median number of representatives fall among the 50 states?

Solution

a. Use the frequency histogram above on the left. Read the height of the bar spanning $15 \le x < 20$. Four states have from 15 to 19 representatives.

b. To find the *percent* of the states rather than the *number* of states, use the relative frequency graph. Read the bars to right of $x = 20$.
$0.04 + 0.02 + 0.02 = 0.08$. So 8% of the states have 20 or more representatives in Congress.

c. When the 50 values are rank-ordered, the median is between the 25th and 26th values. From the frequency histogram, we conclude that there are 20 states with fewer than 5 representatives and 17 with between 5 and 10 representatives. So the 25th and 26th values are between 5 and 10, and the median must be in the bin between 5 and 10.

Drawing a Histogram

To make a histogram, first organize the data into non-overlapping intervals of equal width. Choosing the bin width is a matter of judgment. There is usually not a single best size. Generally, choosing 5 to 10 intervals is about right for a histogram. Too few bins will lump all the data together; too many will result in only a few numbers in each bin.

Second, count the number of observations per bin and record the results in a table that gives the frequency or relative frequency for each of the bins created.

Finally, draw the histogram by first marking the horizontal axis to show the endpoints of the bins and marking the vertical axis to show the frequencies (or relative frequencies). Then, for each interval, draw a bar to represent the frequency. Unlike other bar graphs, histograms are drawn with no gaps between bars unless an interval is empty, in which case its bar has height 0. In the Activity you will make a histogram using technology.

Activity

Below are the daily high temperatures in March one year in Lincoln, Nebraska.

69, 60, 34, 41, 36, 44, 27, 45, 43, 49, 71, 67, 64, 54, 43, 40, 42, 58, 61, 68, 56, 45, 45, 64, 61, 60, 49, 51, 58, 53, 42

Step 1 Enter the temperatures into a statistical package or spreadsheet.

Step 2 Use the technology to make a histogram, using 5 as the bin width. Label your axes as shown.

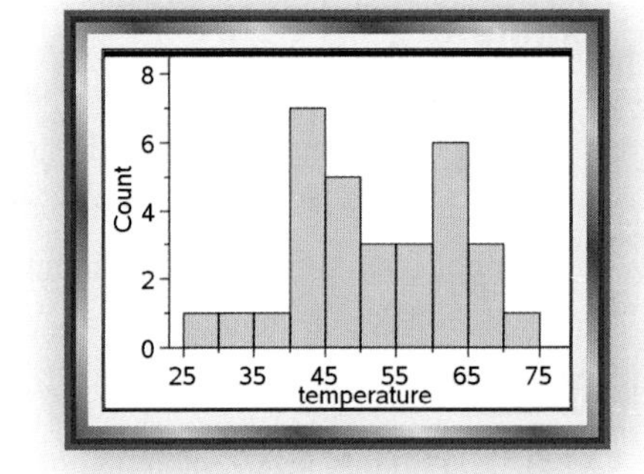

Step 3 Change the bin width from 5 to 2. How does this affect the graph?

Step 4 Change the bin width from 2 to 10. How does this affect the graph? Which of the three bin widths (2, 5, or 10) gives the best description of the data?

Step 5 Return the bin width back to 5. Adjust the settings on your technology so that the vertical axis displays relative frequency rather than frequency. In what ways is this graph different from the one in Step 2?

Analyzing Histograms

Histograms help you to see features of a data set that are hard to capture from a table. The graphs shown here illustrate three different shapes that are common. Histogram (A) at the right shows a **skewed** distribution. It has a *cluster* of high temperatures on the right side of the graph. Histogram (B) on the next page also shows a skewed distribution, but it tapers off toward the right end to form a *tail*. Histogram (C) is close to being **symmetric**, with two sides that are approximately the same shape. So, to describe the shape of a distribution, consider the following questions:

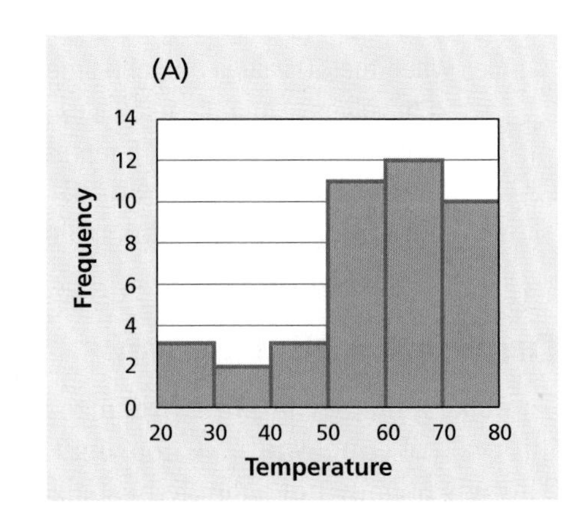

- Is the distribution skewed?
- Does the distribution have a tail? If so, at what end?
- Is the distribution symmetric?

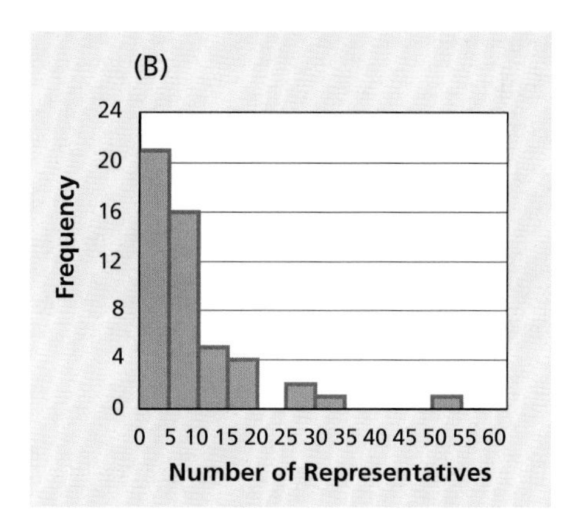

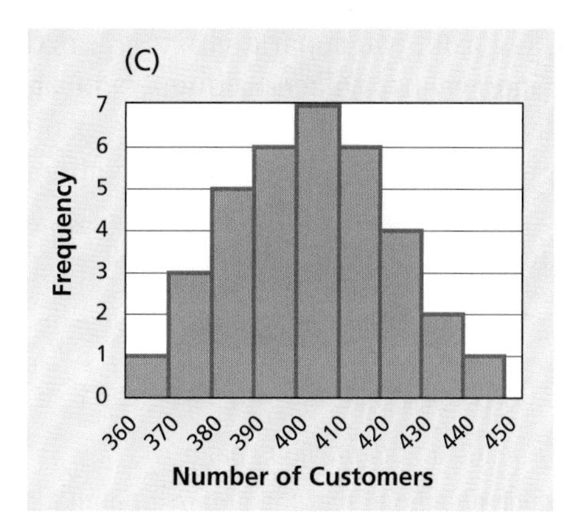

Population Pyramids

Demographers use histograms to analyze populations. One display is a double histogram called a *population pyramid*. Two examples of population pyramids are shown below.

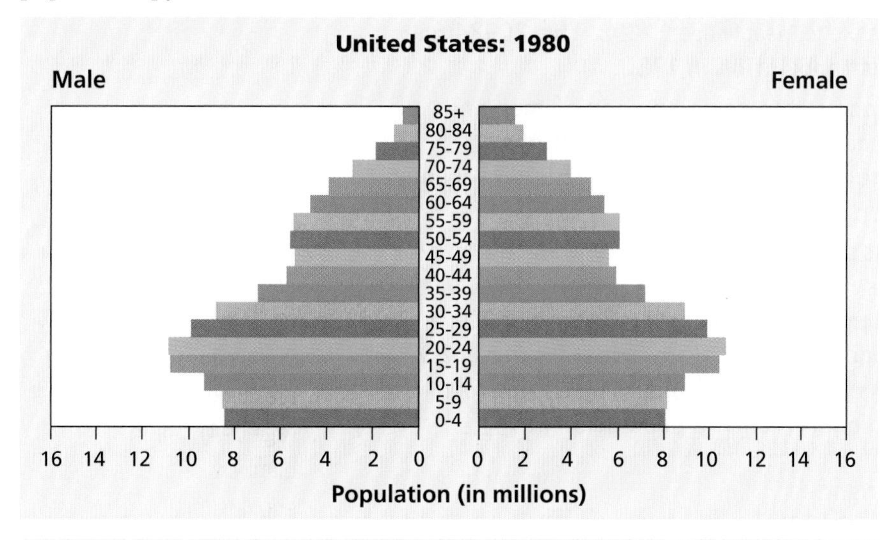

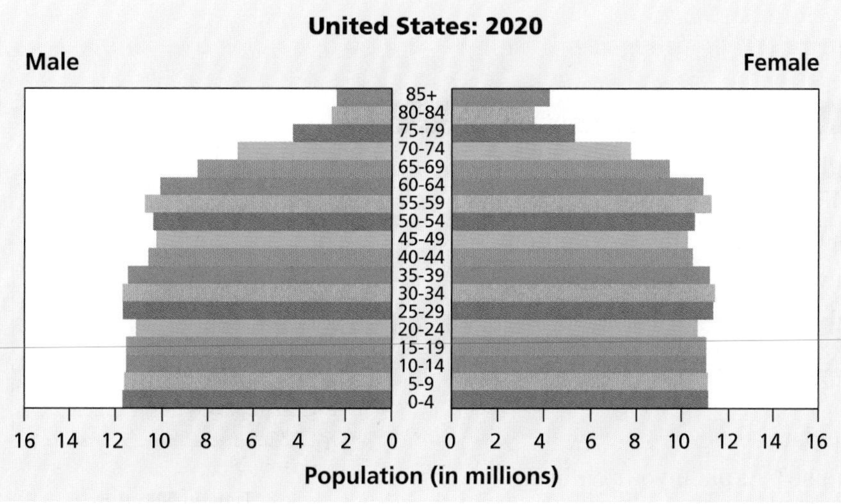

Source: U.S. Census Bureau, International Data Base

A **population pyramid** is made from two separate histograms that have been rotated 90°. The bin intervals are placed along a central vertical axis and the frequencies are on the horizontal axis. The examples on the previous page show age distributions for men and women. The first is from actual U.S. Census Bureau data. The second is a *projection*, based on what demographers expect to happen.

GUIDED

Example 2

Refer to the population pyramids on the previous page.

a. For 1980, estimate which age group had the greatest population.

b. For 2020, compare the distribution for men with the one for women. In what age groups are they most different?

c. Compare the 1980 and 2020 population pyramids. Describe three significant differences.

Solution

a. For 1980, the longest bars are for people age 20–24. The 20–24 age group had the greatest population in 1980.

b. There appears to be about the same number of men and women in most age brackets. Starting at about age _?_, there are clearly more females than males. This difference is very pronounced for the 85+ age group.

c. First, the 1980 population pyramid has a bulge in the 15 to _?_ age range that is not as pronounced in the 2020 pyramid. Second, more people are projected to live longer in 2020. In 1980 there are _?_ million people 80 or older, while in 2020 there are _?_ million over 80 years old. Third, the total population in _?_ is bigger than the population in _?_.

See Quiz Yourself at the right.

QUIZ YOURSELF

According to the 1980 population pyramid on the previous page, for preschool children (age 0–4), are the numbers of boys and girls equal?

Questions

COVERING THE IDEAS

In 1 and 2, the graph at the right shows the lengths of songs that Jessie has stored on an MP3 player.

1. Use two to three sentences to describe the distribution.
2. For this histogram, the bin width is 0.5. About how high would the leftmost bar be if the bin width were 1?

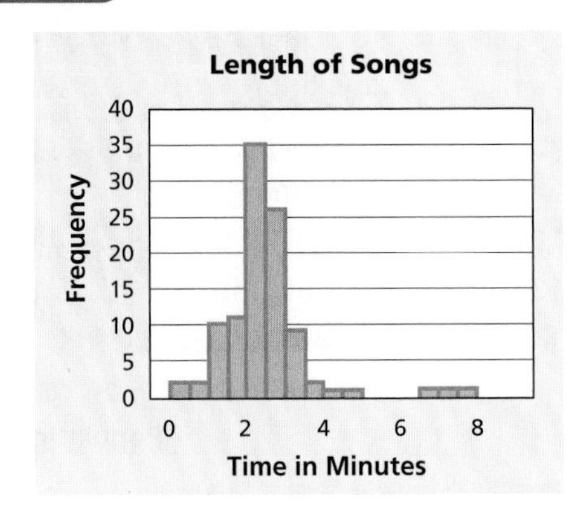

3. Here is a list of heights (in inches) of 15 students: 73, 66, $63\frac{3}{4}$, $68\frac{3}{4}$, 70, 69, 65, 71, 68, $80\frac{1}{2}$, 64, 67, 65, 64, $64\frac{1}{2}$.
   a. Draw a histogram of the data.
   b. What bin width did you use? Explain your choice.

In 4–8, use the histogram at the right of the scores of students taking the SAT Literature Subject Test of the College Entrance Examination Board in 2007.

4. About how many students scored from 500 to 549?
5. What is the bin width?
6. Approximately how many students took the Literature exam in 2007?
7. About what percent of the students scored 700 or better?
8. In what interval is the median score?

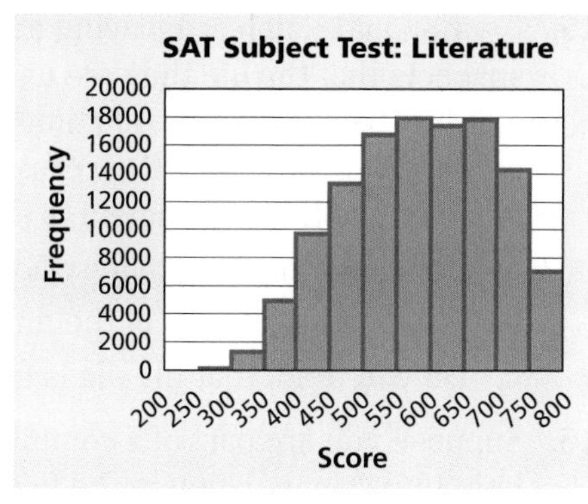

Source: The College Board

In 9 and 10, refer to the graphs of heights of African American males.

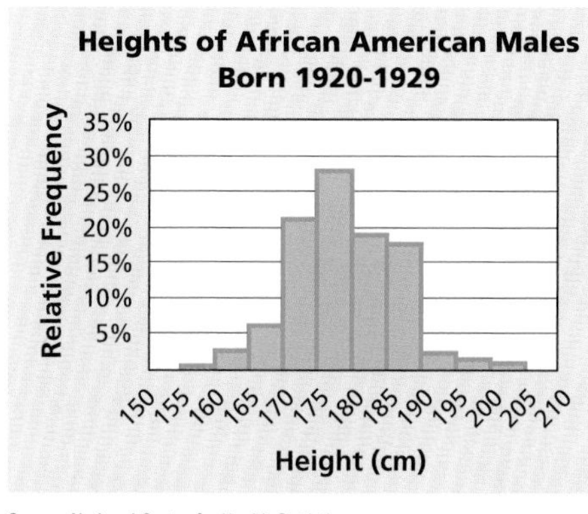

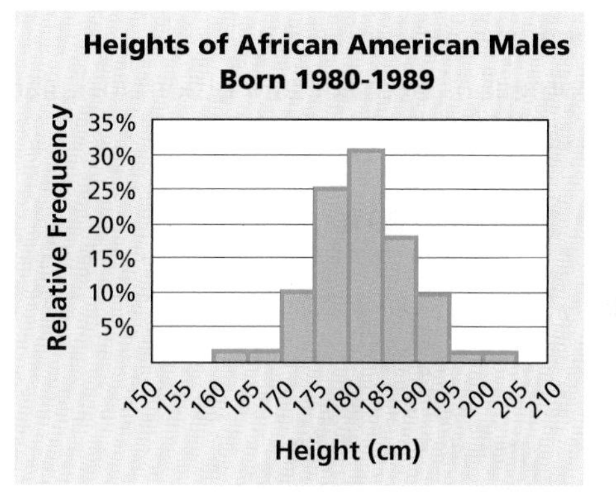

Source: National Center for Health Statistics

9. Use the histograms to estimate which bin contains the median height for each group of African American males. Which group has the larger median height?
10. a. About what percent of African American males born in 1920–1929 were at least 190 cm (about 6'3") tall?
    b. About what percent of African American males born in 1980–1989 were that tall?

APPLYING THE MATHEMATICS

11. At the right are some data regarding accidental deaths in the U.S. in 2004.
    a. Create a histogram of the data on drowning.
    b. At what ages is there the highest risk of drowning?
    c. What is likely to make the histogram in Part a misleading?

Age	Drownings
under 1 yr.	62
1–4	430
5–14	269
15–24	574
25–34	385
35–44	435

Source: National Safety Council

In 12–14, describe a possible shape for the distribution of the variable. Explain your reasoning.

12. the mint dates of U.S pennies that are in a store's cash register
13. the salaries of employees in a large company
14. the scores on an easy test

15. Cynthia and Ralph are playing a board game. Cynthia suspects that the die they are using is biased. She decides to test the die by rolling it 100 times and counting the number of times 1, 2, 3, 4, 5, or 6 occurs. A distribution of the outcomes of this experiment is shown at the right.

a. Use two sentences to describe the distribution.

b. About what percent of the time did a 1 show up?

c. Do you think that the die is fair? Justify your answer.

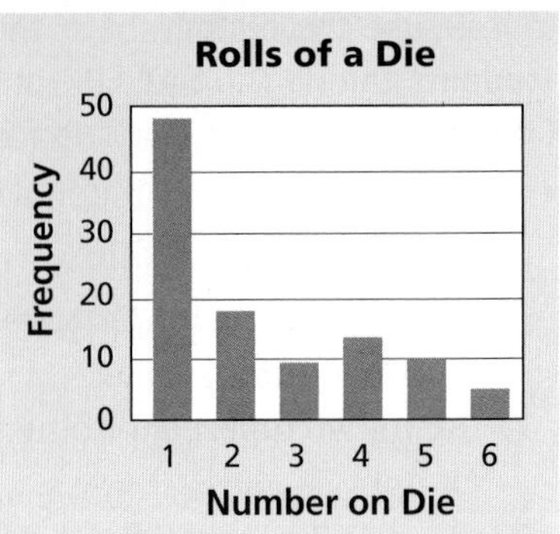

16. Suppose you are part of a committee to plan projects that will help India's population in the future. Using the population projections below, order these three developments in terms of importance, and explain why you have picked the order you made.

(1) Build more elementary and high schools.

(2) Stimulate development of factories and other workplaces.

(3) Build more trains and roads.

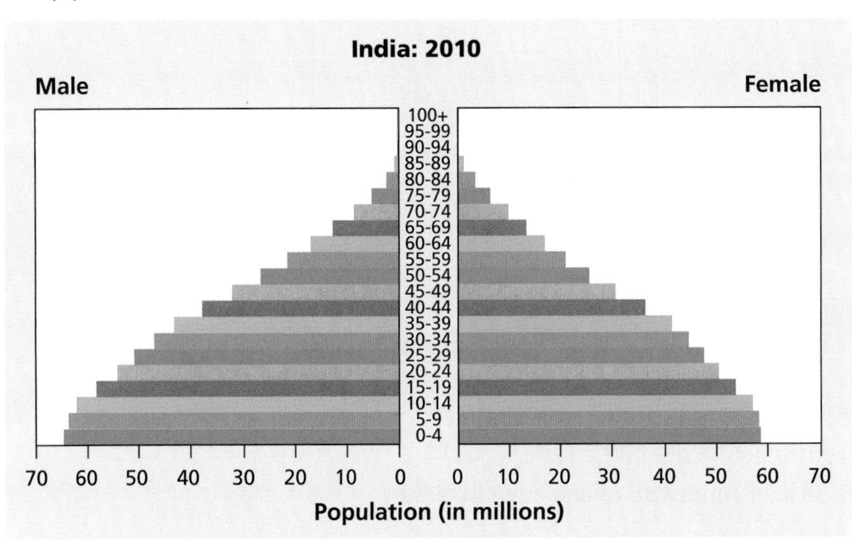

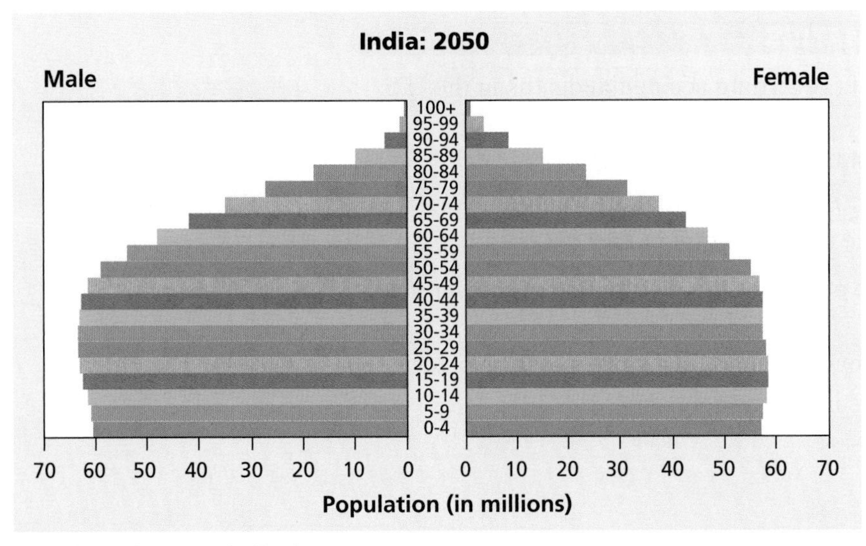

Source: U.S. Census Bureau, International Data Base

REVIEW

17. Monday through Friday a store averaged \$2,100 a day in sales. On Saturday and Sunday the store took in \$7,200 total in sales. What is the store's mean sales per day? **(Lesson 1-2)**

18. In his Biology class, Mr. Boynton covered six chapters in the second semester. He gave two quizzes and a chapter test for each chapter, and a final at the end of the semester. Connie averaged 91 on the quizzes, 73 on the chapter tests, and 85 on the final.

a. Mr. Boynton weights chapter tests twice as much as quizzes, and the final five times as much as each chapter test. Compute Connie's weighted average score in Biology.

b. What would Connie have had to score on the final in order to average at least 85 for the semester? **(Lesson 1-2)**

19. If $\sum_{i=1}^{6} x_i = 15$, $\sum_{i=1}^{5} x_i = 20$, and $\frac{\sum_{i=1}^{7} x_i}{7} = 8$, find $x_6 \cdot x_7$. **(Lesson 1-2)**

20. Describe a situation in which the students in your school would be

a. a sample.

b. a population. **(Lesson 1-1)**

21. Match the graphs to their equations. Do not use a calculator. **(Previous Course)**

a.

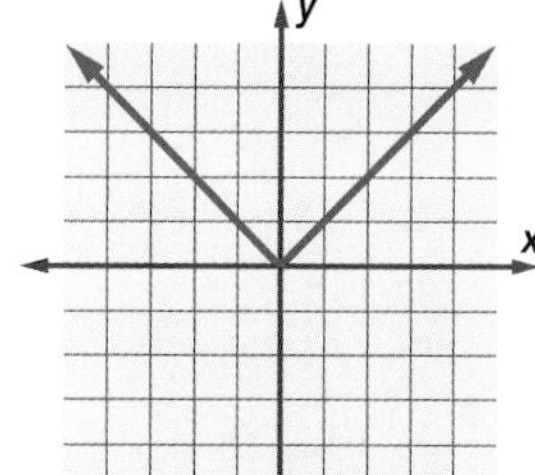

b.

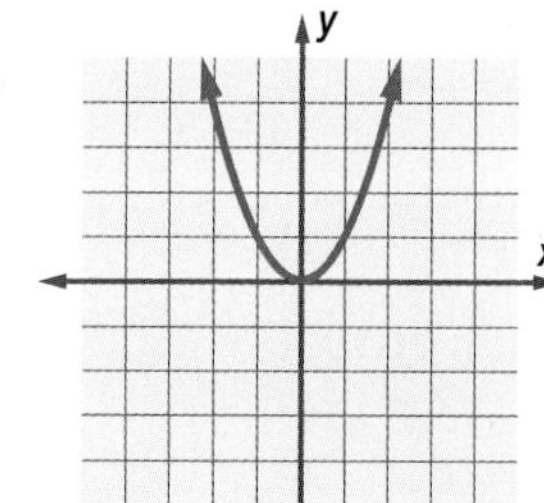

c.

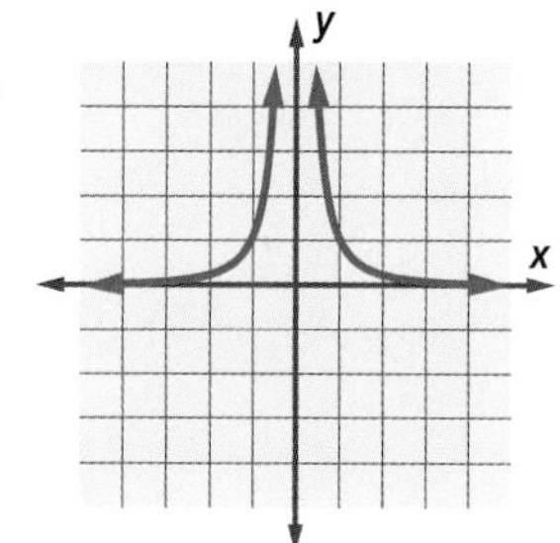

d.

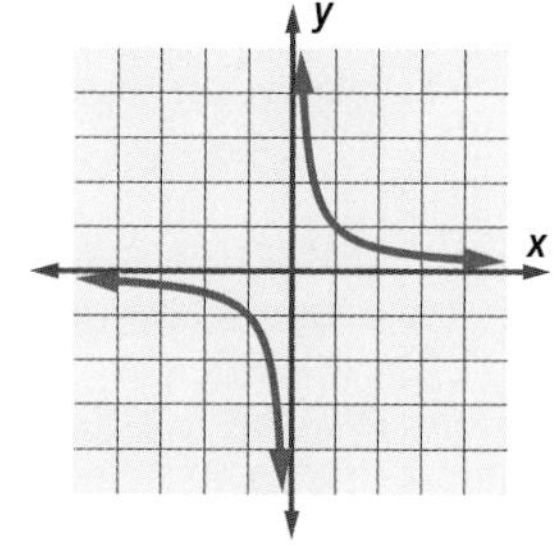

i. $y = \frac{1}{x}$ ii. $y = x^2$ iii. $y = \sqrt{x}$ iv. $y = \frac{1}{x^2}$ v. $y = \sqrt{x^2}$

EXPLORATION

22. Since 1978, the People's Republic of China has had a policy that strongly encourages families to limit themselves to one child, or at most two children. In contrast, Russia is concerned about population decline and is encouraging larger families. The population pyramids below show U.S. Census Department projections for these countries in 2010. For each country make a hypothetical population pyramid for the year 2030 assuming that the government policies are successful. What long-term concerns are raised?

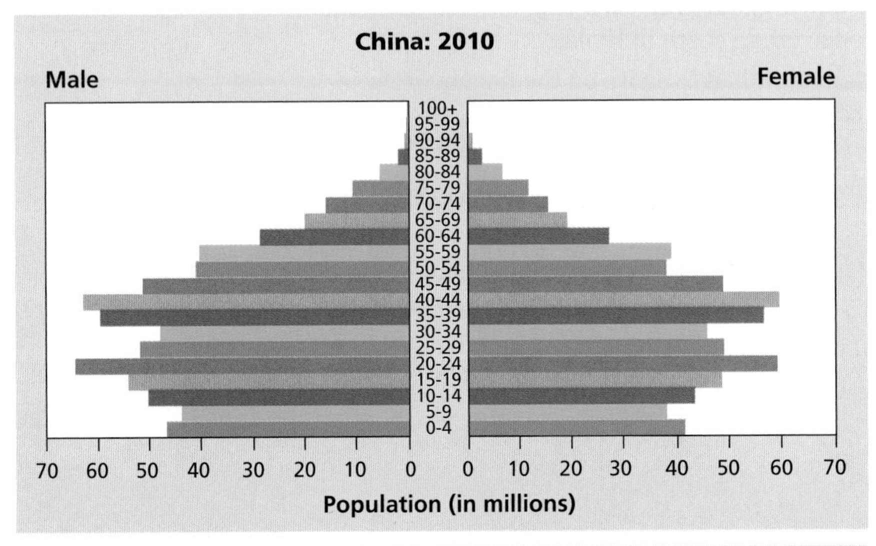

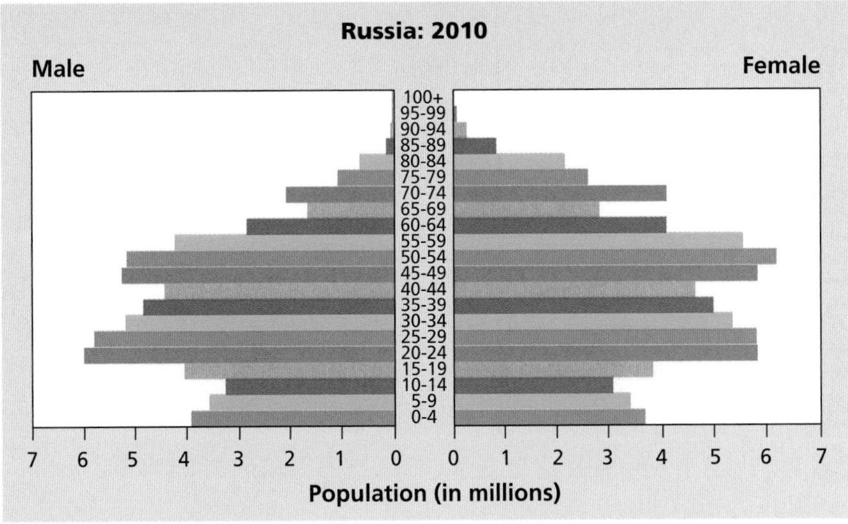

Source: U.S. Census Bureau, International Data Base

QY ANSWERS

no

Lesson 1-4 Box Plots

▶ **BIG IDEA** Box plots are a way to display important aspects of a numerical distribution.

In this lesson you will see how a *box plot* provides a visual summary of a distribution. One advantage of a box plot is that it readily shows how the data are spread. Here is a data set of scores on an algebra test with five key numbers identified.

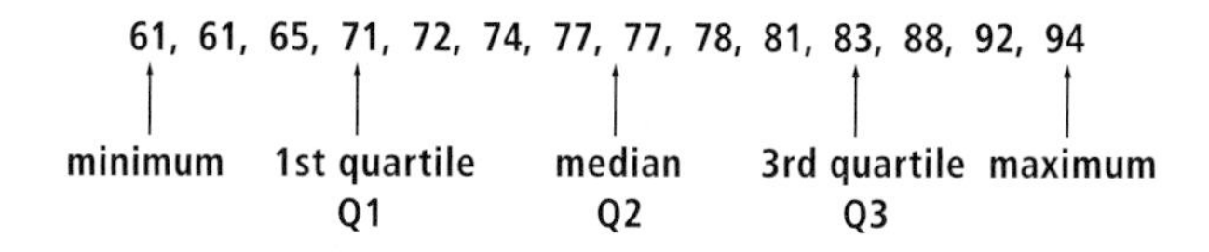

These are the five key numbers that are shown on a box plot. Below is a box plot of the data above.

Scores on an Algebra Test

60 65 70 75 80 85 90 95

What does this plot tell us about the data? To answer that question, we must first introduce some vocabulary.

Quartiles

Box plots were invented in the 1970s by the statistician John Tukey. Five key numbers are used to create a box plot:

1. the **minimum**, the least value of the variable;
2. the **first** (or **lower**) **quartile** Q_1, the median of the numbers below the median of the distribution;
3. the **second quartile** Q_2, the median of the full distribution;
4. the **third** (or **upper**) **quartile** Q_3, the median of the numbers above the median of the distribution; and
5. the **maximum**, the largest value of the variable.

These numbers are called the **five-number summary** of a distribution.

Vocabulary

box plot, box-and-whiskers plot
minimum
first (lower) quartile
second quartile
third (upper) quartile
maximum
five-number summary
interquartile range (IQR)
whiskers
outlier

Mental Math

Five students took a make-up test in a class, scoring 83, 90, 82, 70, and 85. The teacher felt the test was hard and added 3 points to each student's score. What was the average of the scores after the 3 points were added?

The diagram below shows how the numbers in a five-number summary are shown in a box plot. The difference $Q_3 - Q_1$ is called the **interquartile range**, or **IQR**. The IQR is the length of an interval in which you will find the middle 50% of the data. It is the length of the rectangle in the box plot.

Scores on an Algebra Test

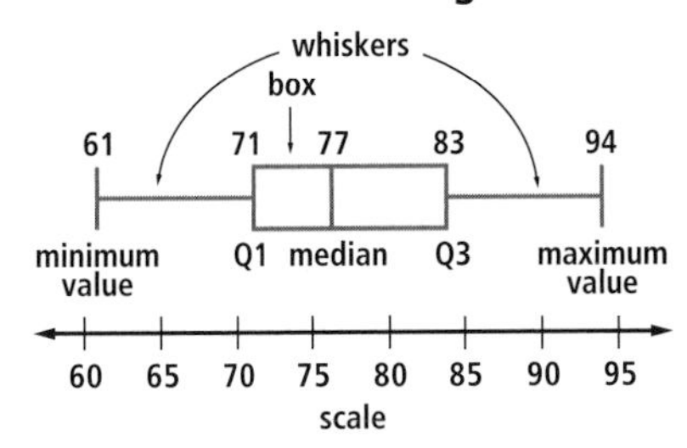

The vertical segment in the rectangle marks the median of the distribution. The horizontal segments from the rectangle to the maximum and minimum are called **whiskers**. For this reason, a box plot sometimes is called a **box-and-whiskers plot**. Some statistics utilities also display tick marks at the end of the whiskers, as shown above.

GUIDED

Example 1

The data 29, 36, 37, 38, 40, 43, 45, 46, 47, 50 are the ages of the best actor Oscar winners from 1997 to 2006.

a. Give the five-number summary of these ages.

b. Draw the box plot.

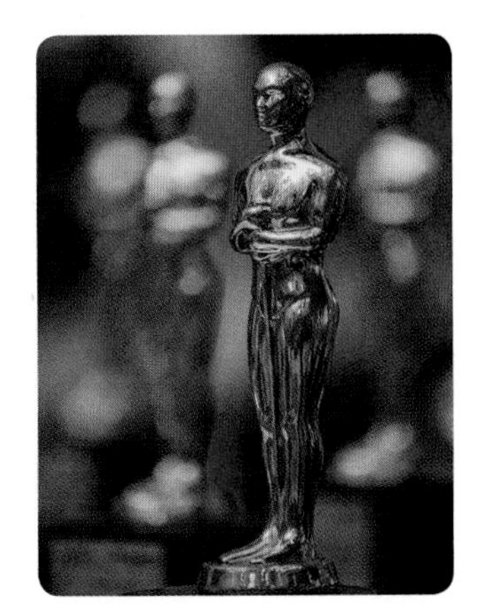

Solution

a. There are __?__ numbers in the list. The minimum value of the data is __?__. The maximum value of the data is __?__. Recall that the median for an even number of ordered data is the mean of the middle numbers. These data have a median of $\frac{40 + 43}{2} = 41.5$. The first quartile, Q_1 is the median of the numbers below the distribution median. The first quartile is __?__. The third quartile, Q_3 is the median of the numbers above the median. The third quartile is __?__.

b. The box goes from Q_1 to Q_3, that is, from __?__ to __?__. The whiskers go from the minimum value __?__ to the maximum value __?__. Here is a plot.

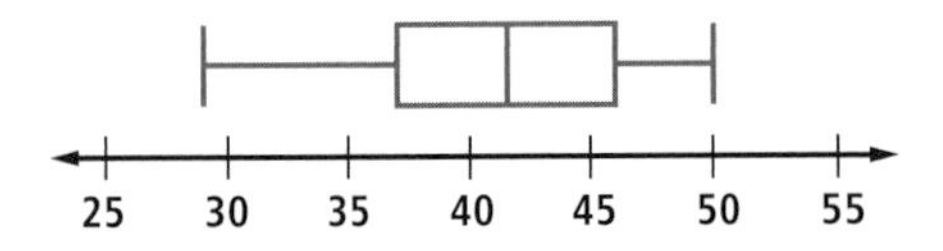

STOP See Quiz Yourself at the right.

QUIZ YOURSELF

Give the five number summary for the data set 12, 23, 23, 24, 34, 37.

Using Technology with Box Plots

You can use technology with a statistical application to find a five-number summary. Once a data set is entered into a spreadsheet, locate a menu choice for one-variable statistics calculations to show a variety of statistics, including the five-number summary. Most (but not all) technology with statistical applications can automatically generate a box plot. Below are screenshots showing the results using the age data from Guided Example 1. You will see the meanings of the symbols on the left screenshot below by the end of this chapter.

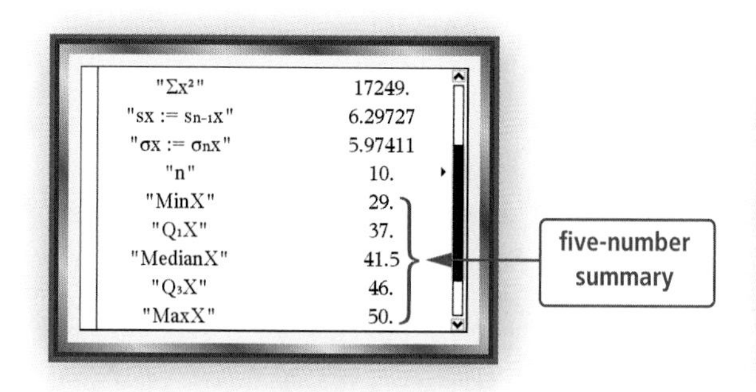

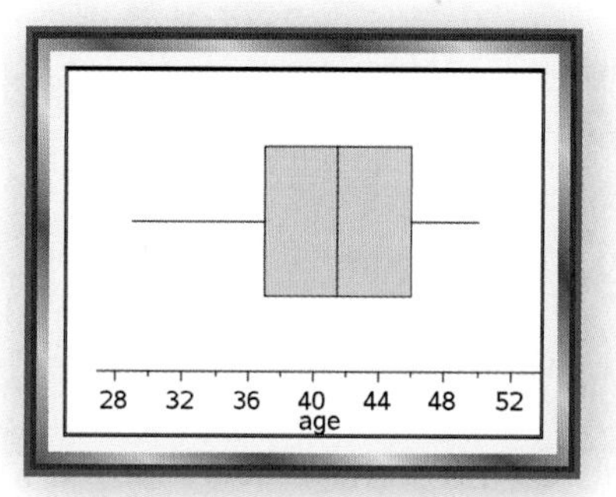

Using Graphs to See Outliers

Graphs can help you see data values that are extreme. The graph below is a histogram of the payrolls of the 30 Major League Baseball teams on opening day. Rounded to the nearest million dollars, total team payrolls were (in order) 24, 31, 37, 39, 52, 54, 58, 62, 67, 68, 69, 71, 71, 79, 82, 87, 88, 89, 90, 90, 94, 95, 100, 106, 108, 109, 109, 115, 143, and 190.

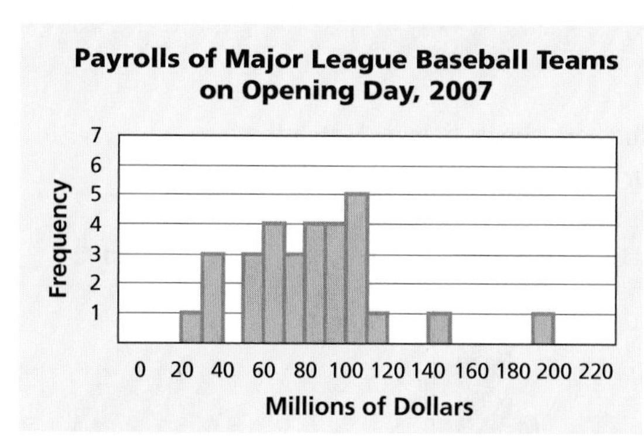

Source: *USA Today*

The histogram displays an extreme observation on the upper end (the payroll of the New York Yankees). Observations that are extreme are called ***outliers***. But what do we mean by extreme? Is \$190 million dollars an outlier? What about \$143 million? A common criterion for identifying outliers is the *1.5 × IQR criterion*. Any number larger than $Q_3 + 1.5(\text{IQR})$ or smaller than $Q_1 - 1.5(\text{IQR})$ is considered an **outlier**.

GUIDED

Example 2

The five-number summary for the Major League Baseball payroll data (in millions of dollars) is: minimum = 24, Q_1 = 62, median = 84.5, Q_3 =100, maximum = 190. Use the 1.5 × IQR criterion to determine if there are any outliers. If there are, identify them.

Solution $IQR = Q_3 - Q_1 =$?

$Q_3 + 1.5\,(IQR) =$?

$Q_1 - 1.5\,(IQR) =$?

So, any payroll greater than ? is an outlier.

Similarly, any payroll less than ? is an outlier.

The New York Yankees payroll of $190 million is the only outlier.

At the right is a box plot of the Major League Baseball payroll data created by software that uses the 1.5 × IQR criterion. Typically, when the whiskers are drawn, they represent all the values except the outliers. Here the right whisker ends at $143 million, and the outlier $190 million is shown by a dot. However, when the five-number summary is calculated, all data (including outliers) are used.

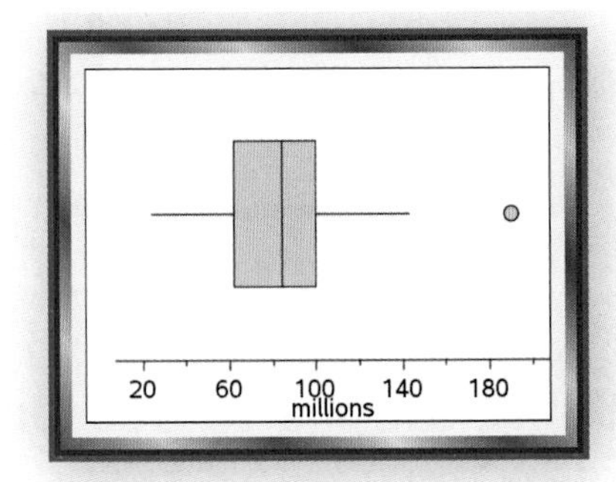

Comparing Histograms and Box Plots

A well-made histogram or a box plot can give information about the shape of a distribution and about extreme values. You can determine the general shape of one of these graphs from the shape of the other.

Example 3

Histograms and box plots of several distributions are shown below. Match the histograms with the box plots and explain your reasoning.

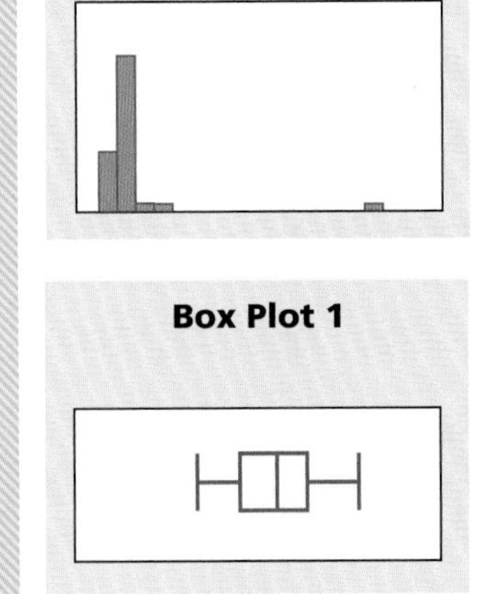

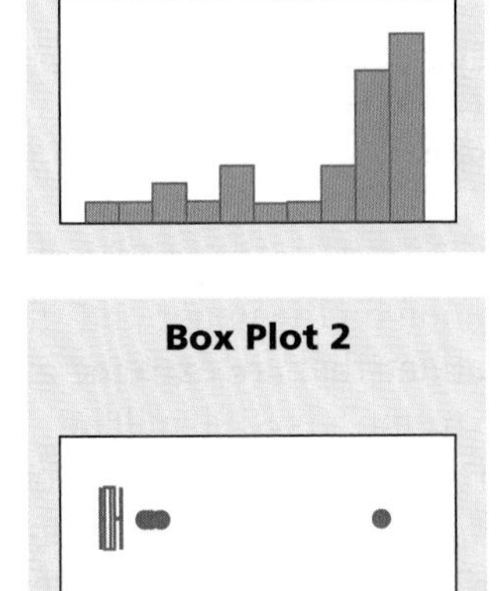

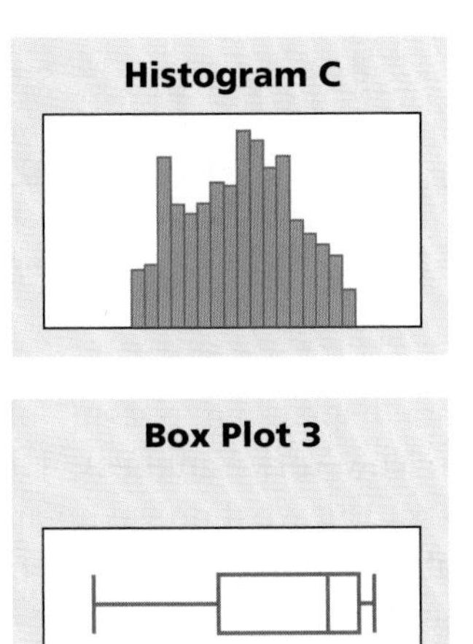

Solution In Histogram A, the extreme values on the upper end are outliers. So the histogram matches Box Plot 2.

In Histogram B, the long left tail on the histogram appears as a stretch of the lower 50% in the box plot. So it matches Box Plot 3.

Histogram C matches Box Plot 1. The histogram and box plot are both nearly symmetric.

Questions

COVERING THE IDEAS

1. **Multiple Choice** Which is not one of the numbers in the five-number summary for a set of data?

   A mean B median C maximum D minimum

2. The table at the right shows the mean student cost for room, board, and tuition in a public college for ten states in 2006-2007.

   a. Calculate the five-number summary for the data.

   b. Calculate the IQR for the data.

   c. Use the 1.5 × IQR criterion to determine if any of the costs are outliers.

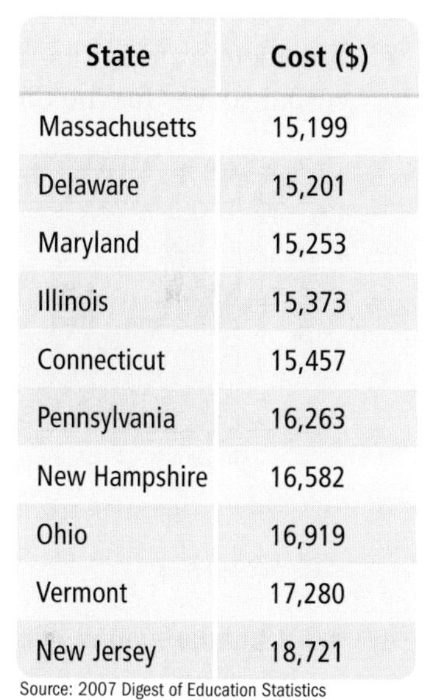

State	Cost ($)
Massachusetts	15,199
Delaware	15,201
Maryland	15,253
Illinois	15,373
Connecticut	15,457
Pennsylvania	16,263
New Hampshire	16,582
Ohio	16,919
Vermont	17,280
New Jersey	18,721

Source: 2007 Digest of Education Statistics

3. Create a box plot to fit this five-number summary.

   min = 45 $Q_1 = 56$ median = 66 $Q_3 = 71$ max = 87

4. In a box plot,

   a. about what percent of data should be less than Q_1?

   b. about what percent of data should fall between Q_1 and Q_3?

   c. about what percent of data should be greater than Q_3?

5. **Multiple Choice** A histogram of a distribution is shown at the right. Which of the three box plots below is a graph of this distribution? Justify your answer.

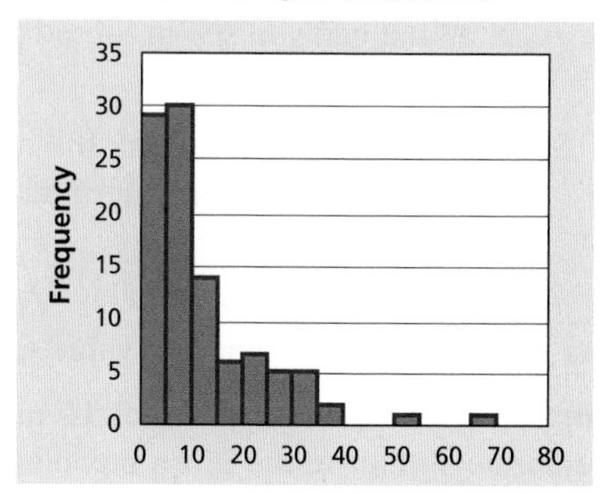

A
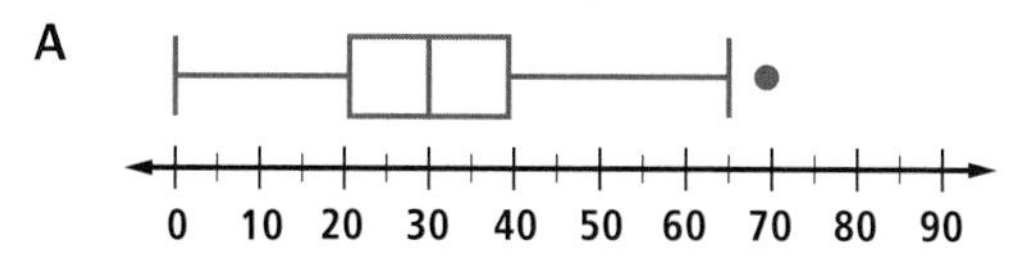

B
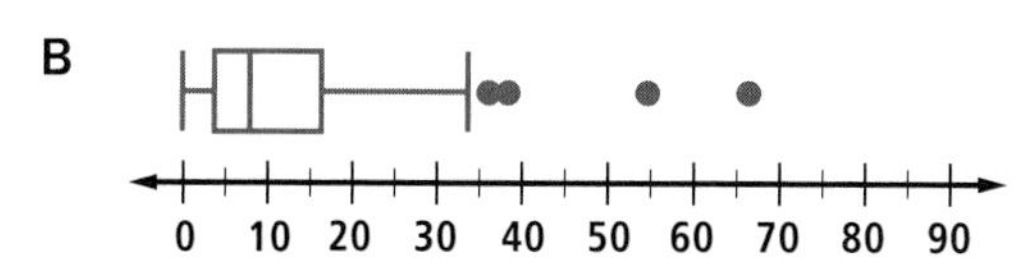

C
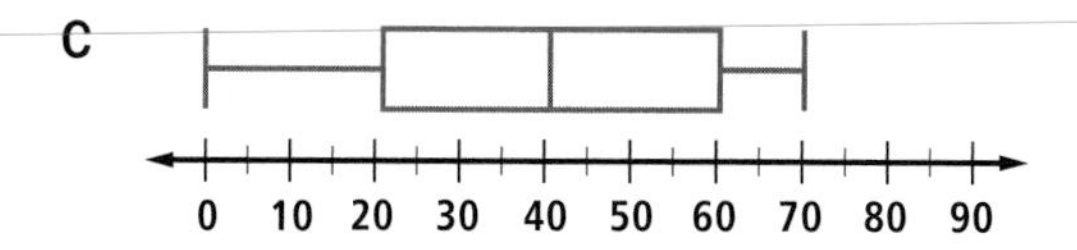

6. The following box plot displays data for the average monthly rainfall (measured in millimeters) at London Heathrow Airport between 1981 and 1991. Write the five number summary for the data. Identify outliers, if any exist.

Average Rainfall at London Heathrow Airport

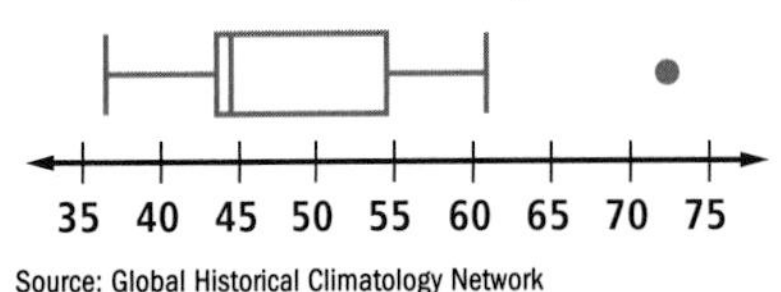

Source: Global Historical Climatology Network

APPLYING THE MATHEMATICS

7. The histogram and box plot below show the average 2003 SAT verbal scores for the 50 states plus Washington D.C.

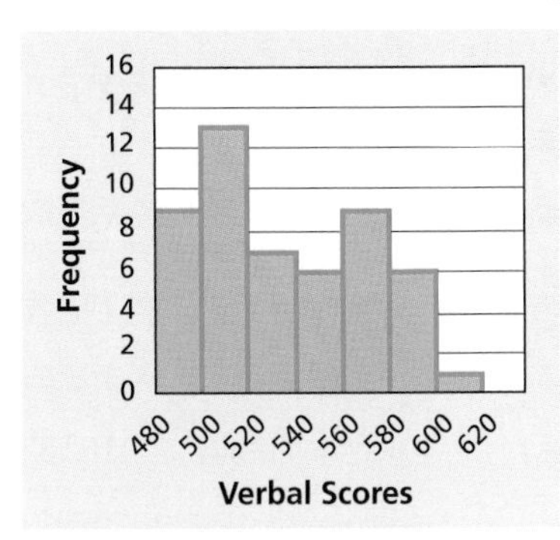

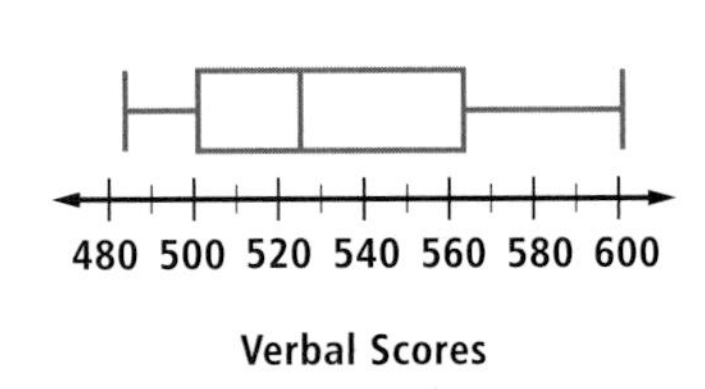

Source: The College Board

a. What can you see in the histogram that you cannot see in the box plot?

b. What can you see in the box plot that you cannot see in the histogram?

8. On a 50-point history test, the median was 34 and Q_1 was 28. One student got a perfect paper and qualified as an outlier under the $1.5 \times$ IQR criterion.

a. What is the largest value Q_3 could be?

b. What is the smallest value Q_3 could be?

In 9 and 10, construct a set of 10 numbers whose box plot would have a shape similar to the given box plot. There are no outliers.

9. 

10. 

REVIEW

11. The graph at the right shows the distribution of the maximum waiting times in minutes of 500 customers to a telephone call center.

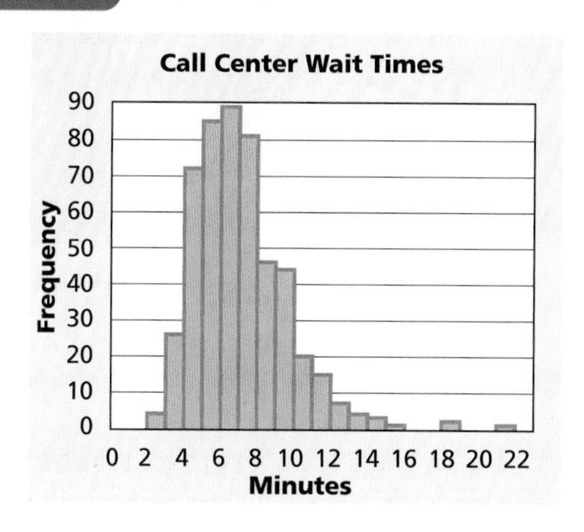

a. How many customers had to wait at least 10 minutes before their call was answered?

b. What proportion of people were answered within 5 minutes?

c. What percent of people were answered after waiting between 5 and 10 minutes?

d. Describe the shape of the distribution and relate that to the situation. **(Lesson 1-3)**

12. Nancy made a monthly investment in the shares of Mathco stock as shown below. That is, she bought 100 shares at \$6.75, 300 shares at \$7.25, and so on. **(Lesson 1-2)**

Shares	100	300	200	100	300	400	200	300	300
Cost per Share (\$)	6.75	7.25	8.25	9.00	6.25	5.75	7.75	8.50	9.75

What is Nancy's average cost per share of Mathco stock?

13. At the college Grace attends, she receives grade points for each hour of class. An A+ is worth 4.3, an A is worth 4, an A- is worth 3.7, a B+ is worth 3.3, a B is worth 3, and so on. Grace has 5 hours of A, 8 hours of A-, and 3 hours of C+. **(Lesson 1-2)**

a. What is Grace's grade point average without the 3 hours of C+?

b. What is Grace's grade point average with the 3 hours of C+?

c. How much lower is Grace's grade point average with the 3 hours of C+?

14. Written documents can sometimes be distinguished by the length of words used. A student, interested in how many letters are in the words English literature authors use, opened an Ernest Hemmingway novel to a random page and counted the frequency of word length for the first 250 words. Identify

a. the population.

b. the sample.

c. the variable of interest. **(Lesson 1-1)**

EXPLORATION

15. Suppose a data set has n elements with $n \geq 6$. As you know, if n is even, then the median of the set does not have to be an element of the set. For what values of n does the first quartile Q_1 not have to be an element of the set?

QY ANSWER

min $= 12$, $Q_1 = 23$,
median $= 23.5$,
$Q_3 = 34$, max $= 37$

Lesson

1-5 Cumulative Distributions

Vocabulary

cumulative data

cumulative distribution

percentile, *p*th percentile

BIG IDEA Cumulative distributions give the total number of values less than or equal to a particular value.

Meteorologists report rainfall in two different ways. On a day-to-day basis, rainfall affects our activities and even our mood, so the weather reports show how much rain falls each day. But in the long run, we also care about the *total* amount of rain to be sure that the soil and vegetation get enough moisture. So today's weather reports often include the total amount of precipitation that has fallen from January 1st through today. This total is an example of **cumulative data.** A function whose values are cumulative frequencies or cumulative relative frequencies is called a **cumulative distribution**.

Mental Math

Two sides of a triangle have lengths 80 and 55. What are all possible lengths of the third side?

Example 1

The table at the right and graph below show the normal precipitation in Los Angeles, California. The second column of the table shows the total, or cumulative, precipitation for the year up to and including the month. The graph is complete but some values in the table are missing.

a. **Determine the missing values in the table.**

b. **Draw a bar graph of the cumulative precipitation by month.**

c. **When has half the yearly precipitation fallen?**

d. **When has three fourths fallen?**

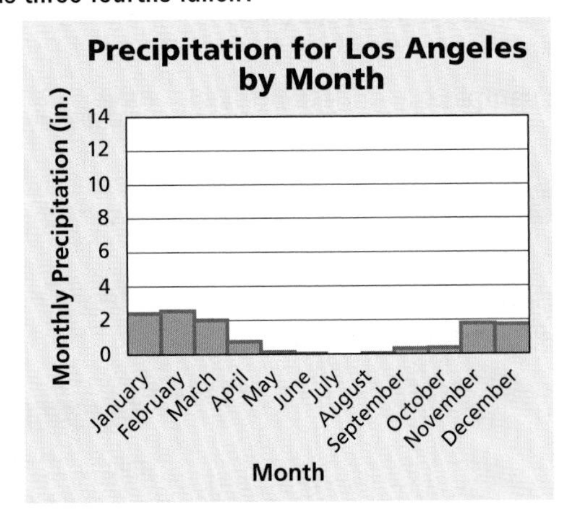

	Monthly Precipitation (inches)	Cumulative Precipitation (inches)
Jan	2.4	2.4
Feb	2.51	4.91
Mar	1.98	6.89
Apr	0.72	7.61
May	0.14	7.75
Jun	0.03	?
Jul	0.01	7.79
Aug	0.15	7.94
Sep	0.31	8.25
Oct	0.34	8.59
Nov	?	10.35
Dec	1.66	12.01

Source: National Climatic Data Center

Solution

a. The numbers in the cumulative column are a *running total*, that is, the cumulative total amount of rainfall at the end of each month is the cumulative total from the previous month increased by the current month.

Cumulative June = June + Cumulative May =
0.03 + 7.75 = 7.78 inches.

Work backwards to find the value for November rainfall. Use an equation.

Cumulative November	=	November + Cumulative October
10.35	=	November + 8.59
1.76 inches	=	November rainfall

b.

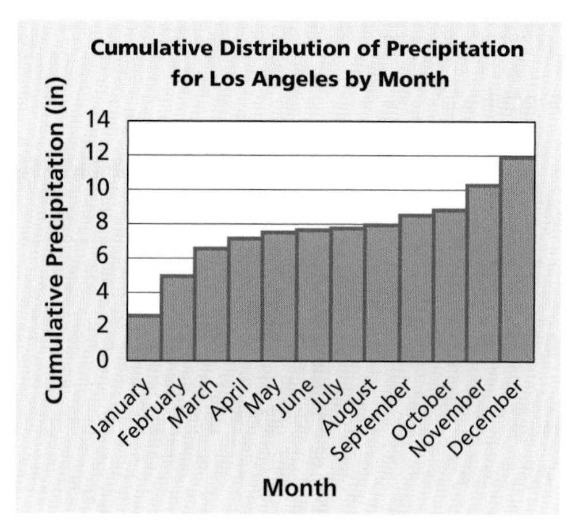

Notice that the bars get longer as the year progresses.

c. About 12 inches of precipitation falls in a year, so half the rain is 6 inches. This point is reached during March.

d. Three-quarters of 12 inches is 9 inches, which has usually fallen by the end of November.

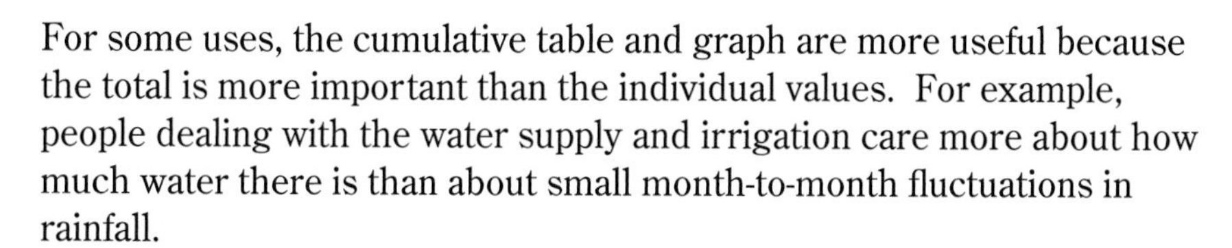

For some uses, the cumulative table and graph are more useful because the total is more important than the individual values. For example, people dealing with the water supply and irrigation care more about how much water there is than about small month-to-month fluctuations in rainfall.

In the previous example, there was no maximum amount of rain possible. But in other situations there is a natural upper bound.

GUIDED

Example 2

At 5:00 A.M. a sudden snowstorm caused Oak School to cancel classes. With no automated message-relay system, news had to spread by word-of-mouth. The table on the next page shows how many of the 170 families knew about the closing from 5 A.M. until 8 A.M.

a. Complete the table at the right to show how many families found out about the closing during each half hour.

b. Draw a line graph showing the cumulative number of families notified.

c. When had half the families been notified?

d. When had three quarters of the families been notified?

Time	Total Number of Families Notified during Preceding Half Hour	Total Number of Families who Know
5:30	6	6
6:00	?	14
6:30	?	48
7:00	104	152
7:30	?	168
8:00	2	170

Solution

a. The frequency for the half hour ending at 6:00 is Cumulative at 6:00 – Cumulative at 5:30.

 _?_ – _?_ = _?_

 The frequency for the half hour ending at 6:30 is _?_ – 14 = _?_.

 The frequency for the half hour ending at 7:30 is _?_.

b. Fill in the points and segments from 6:30 to 7:30 A.M.

c. Half the families equals _?_ families. This many families have been notified by _?_.

d. Three-quarters of the families equals _?_ families. This fraction of the families have been notified by _?_.

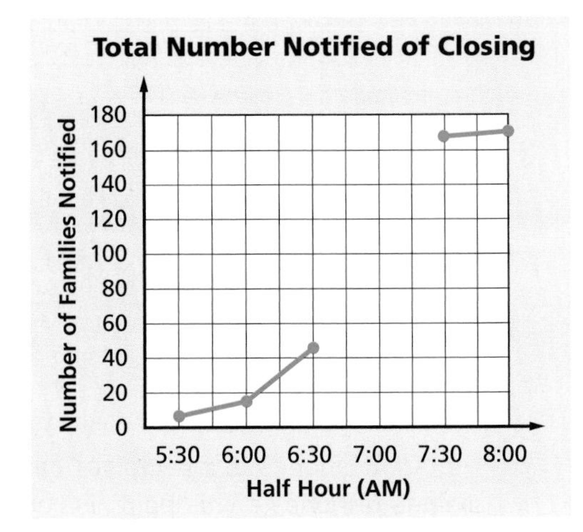

Percentiles

In the situation of Example 2, by 8:00 A.M., 100% of the families at Oak School were notified. In a situation like this, when a cumulative frequency has a maximum of 100%, it is common to speak of *percentiles*, the value at which a certain cumulative percent has been reached.

Definition of Percentile

The *p*th **percentile** of a set of numbers is a value in the set such that *p* percent of the numbers are less than or equal to that value.

For example, if the 85th family out of 170 had been notified at exactly 6:47 and the 86th family was notified at 6:49, then the 50th percentile would be at 6:47. Some statisticians define percentiles according to the percent of numbers strictly less than the value; according to their definition, the 50th percentile would be at 6:49.

Percentiles are commonly used for reporting results of standardized tests. You may have noticed that people are rarely said to be at the 100th percentile or 0th percentile. That is because the lowest percentile is typically reported as the 1st percentile, and the highest percentile as the 99th.

 **See Quiz Yourself at the right.**

QUIZ YOURSELF

What percentile corresponds to Q_3, the third quartile?

Activity

A school district administered a 30-point math test to all tenth graders in its three high schools. The student scores are in the table at the right. Construct cumulative frequency and cumulative relative frequency graphs.

Score Interval	Frequency
0	0
1–3	3
4–6	18
7–9	54
10–12	81
13–15	148
16–18	155
19–21	162
22–24	132
25–27	89
28–30	39

Step 1 Enter the data into columns A and B of a spreadsheet. Title column A "scores" and column B "freq". In column A, replace each interval with its midpoint. For example, enter "2" instead of "1–3".

Step 2 Make the title of column C "cumfreq". Then in cell C1, enter =b1. In cell C2, enter =c1+b2. This formula adds the frequency of the current interval to a running total. Copy that formula down through cell C11.

Step 3 Display the cumulative frequency in a line graph. Your display should look like the one shown at the right below.

Without knowing how many students took the test, it is hard to tell how meaningful a particular frequency or cumulative frequency value is. Relative frequencies would be more meaningful. The percentile score is a measure of *cumulative relative frequency.*

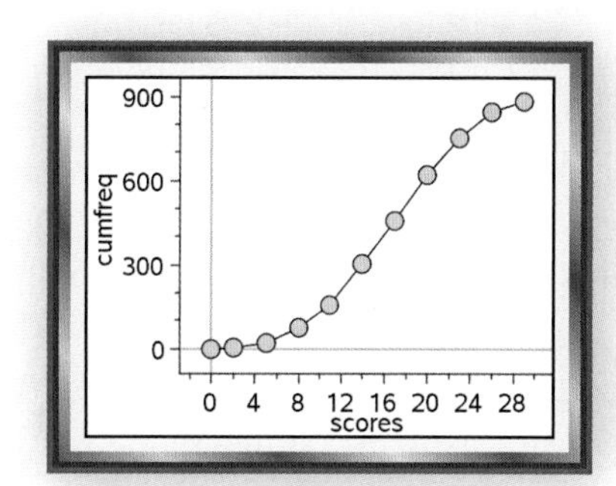

Step 4 Make the title of column D "relfreq" (for "relative frequency"). By scrolling down through column C of the spreadsheet, you can find the total number of students who took the test. Enter a formula in cell D1 that computes the relative frequency by dividing the value in B1 by that total. Copy it down through cell D11. Which score range has the highest relative frequency, and what is its relative frequency?

Step 5 Title column E "cumrelfreq" (for "cumulative relative frequency"). Use a procedure like the one in Step 2 to compute running totals of column D.

Step 6 Make a graph of cumulative relative frequency. Use it to find the quartiles.

Step 7 Construct a box plot.

Step 8 Using either your graph or the spreadsheet, estimate

a. the score corresponding to the 60th percentile.

b. the approximate percentile for a person with a score of 24.

Questions

COVERING THE IDEAS

1. Since 2003, the World Health Organization (WHO) has tracked new cases of Avian Influenza, or bird flu. The table at the right gives the number of people who have contracted the disease each year. Add columns to the table to show the total number of people who have contracted the disease at the end of each of the given years, relative frequency and cumulative relative frequency.

Year	Number of Cases
2003	4
2004	46
2005	98
2006	115
2007	88

Source: World Health Organization

2. A teacher gave a quiz which had a maximum score of 20 points. The chart below gives each score x and its frequency f.

x	7	9	12	13	14	15	16	17	18	19	20
f	1	1	2	1	3	9	16	9	4	3	1

   a. How many students took the quiz?
   b. How many students have scores less than or equal to the score at the 92nd percentile? What score is that?
   c. Use the data to draw a cumulative frequency graph.
   d. Use the data to draw a cumulative relative frequency graph.

3. a. Draw a bar graph showing the box office sales for each of the first 7 weeks of Spider-Man 3 (2007) given the tabulated cumulative values at the right. (In your graph, each week's value should only reflect the sales for that week.)
   b. Draw a cumulative relative frequency line graph for the total receipts.

Spider-Man 3 Box Office Sales

Week	Cumulative
1	$151,116,516
2	$240,236,828
3	$282,379,655
4	$307,754,583
5	$318,342,110
6	$325,585,149
7	$330,021,137

Source: The Numbers-Box Office Data

4. In a class of 30 students, the scores on a test worth 50 points were recorded as:

32	48	34	42	26	39	44	40	39	36
45	37	39	47	34	37	38	45	38	37
46	40	43	30	38	47	42	30	48	33

   a. Make a table of the data, showing the frequency and relative frequency of each score.
   b. Draw a cumulative frequency line graph and a cumulative relative frequency line graph for the data.
   c. Find the quartile values and construct a box plot.

5. **True or False** Consider a set of data.
   a. All data points less than or equal to the median correspond to the 50th percentile.
   b. All data points in the IQR correspond to the 50th percentile.
   c. If there are 12 different ordered numbers, the 3rd lowest value corresponds to the 25th percentile.

APPLYING THE MATHEMATICS

6. This chart gives percentile values for the heights of boys of different ages. For example, the bottom curve shows that the 3rd percentile for a 18-year-old boy is 162.5 cm, or about 64".
   a. What is the height of a 15-year-old boy at the 75th percentile?
   b. Give the median height of a 16-year-old boy.
   c. For what age is the 75th percentile 180 cm?
   d. Make a box plot of heights for 14-year-old boys. Use the 97th and 3rd percentiles as ends of the whiskers in your box plot.

Boys, Stature-for-age, Percentiles

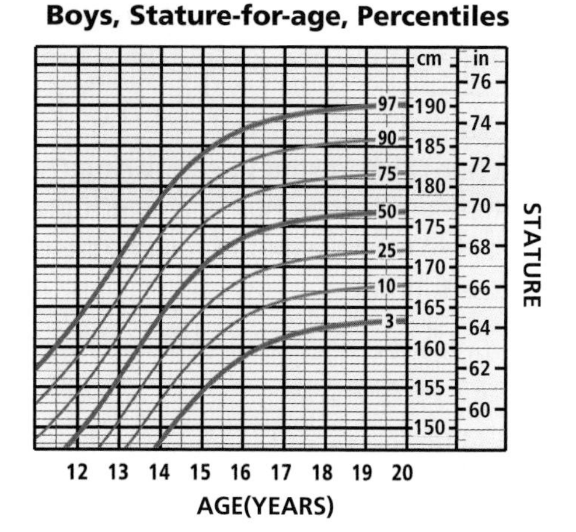

Source: National Center for Health Statistics

7. The table at the right shows the number of home runs that baseball player Roger Maris hit during his ten years in the American League.
   a. Describe a trend you see in the graph shown below.

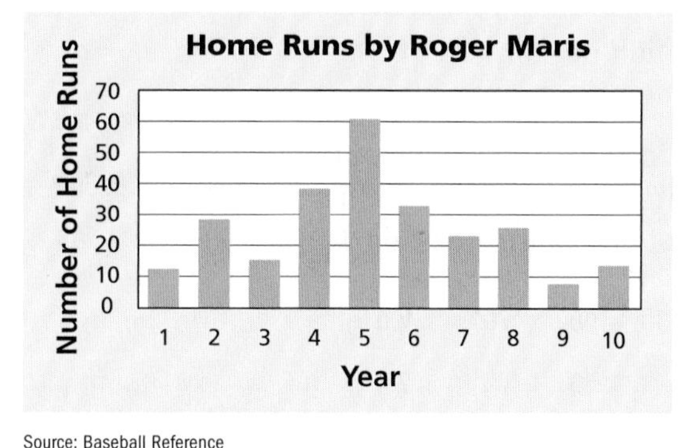

Source: Baseball Reference

Year	Home Runs
1	14
2	28
3	16
4	39
5	61
6	33
7	23
8	26
9	8
10	13

   b. Make a cumulative graph of the data. Describe how the same trend affects the cumulative graph.

8. Doctors studying a group of 240 elderly patients with Alzheimer's Disease noted six types of daily living skills in which they might be impaired. These are called impairments. The severity of Alzheimer's is measured by the total number n of impairments.
   a. For how many patients was $2 \leq n \leq 4$?
   b. Draw a cumulative frequency line graph for the data. Highlight the segments of the graph for which $2 \leq n \leq 4$.

Number of Impairments	Number of Patients
0	100
1	43
2	36
3	17
4	24
5	9
6	11

Source: *Journal of Gerontology*

9. The cumulative frequency graphs in this lesson are either steady or rising. Give an example of a situation in which such a graph would both rise and fall. Explain when this would happen.

REVIEW

In 10 and 11, construct a set of 12 numbers whose box plot would have a shape similar to the one given. There are no outliers. (Lesson 1-4)

10. 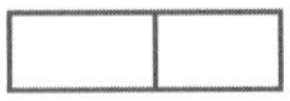

11. 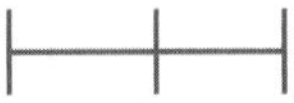

12. A teacher is interested in comparing the reading achievement of her 16 students to that of seventh graders nationwide. The nationwide average on the reading test is 7. Her 16 students score as follows:

6	9	6	10	9	6	7	8
6	5	9	4	14	5	8	7

a. Find the mean and median for these reading scores.

b. If the mean is used as the measure of central tendency, how does the class compare to the national norm?

c. If the median is used as the measure of central tendency, how does the class compare to the national norm?

d. Using the $1.5 \times$ IQR criterion, is the score of 14 an outlier?

e. Use a box plot to graph the distribution of scores. **(Lessons 1-4, 1-2)**

13. Prudence invested in shares of Large Co. Over time she bought different numbers of shares at different prices as shown in this table. Compute Prudence's average cost per share. **(Lesson 1-2)**

Number of Shares	100	300	200	1000	700	400
Price per Share ($)	58	62	65	59	72	75

14. ***Skill Sequence*** Solve each equation. **(Previous Course)**

a. $\frac{x}{3} = 9$ b. $\frac{x^2}{3} = 9$ c. $\frac{x^2}{3} = 9 + x$

15. Test your ability to enter a complex expression into a calculator. Evaluate $3\pi\sqrt{\frac{(1.1 + 2.9)^2}{1.1^2 + 2.9^2}}$ to the nearest thousandth. **(Previous Course)**

16. Without graphing, tell whether the point (-48, -103) is on the line with equation $y = -2x + 7$. Justify your answer. **(Previous Course)**

17. If $x = -3$ and $k = 10$, compute mentally: **(Previous Course)**

a. $|x - k|$ b. $|k - x|$ c. $(x - k)^2$

d. $(k - x)^2$ e. $(x - k)^3$ f. $(k - x)^3$

EXPLORATION

18. Pick out three characteristics about yourself that have a numerical value and about which you can find percentile information with respect to a larger group. Make your choices so that for one characteristic you have a low percentile, for one you have a high percentile, and for the third you are near the 50th percentile.

QY ANSWER

75th percentile

Lesson 1-6 Measures of Spread: Variance and Standard Deviation

Vocabulary

range

deviations

population variance, σ^2

population standard deviation, σ

sample variance, s^2

sample standard deviation, s

BIG IDEA Variance and standard deviation depend on the mean of a set of numbers. Calculating these measures of spread depends on whether the set is a sample or population.

Mental Math

Lenny's average score after 3 tests is 88. What score on the 4th test would bring Lenny's average up to exactly 90?

So far in this book, you have studied the mean, the mode, the five-number summary, and the IQR. These statistics help to describe the distribution of numbers in a data set. In the Activity below you will use these statistics to compare two given data sets.

Activity 1

The two dot plots below display frequency distributions of the height of the players on two hypothetical women's basketball teams.

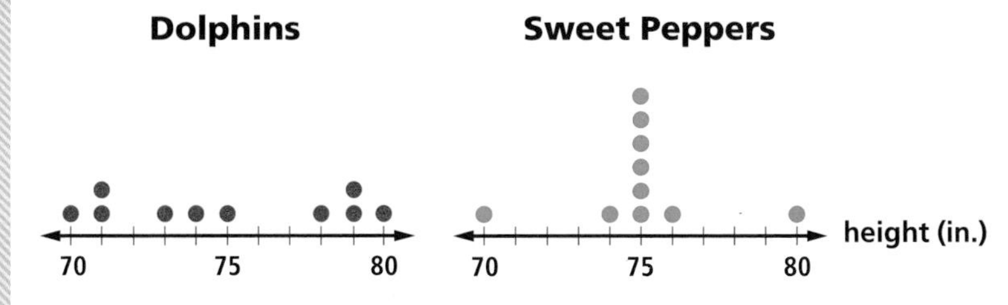

Step 1 Describe what you think is the main difference between the two dot plots from just looking at the graphs.

Step 2 Find the mean, median, mode, five-number summary, and IQR for each data set.

Step 3 Determine if any of the values from Step 2 are appropriate for distinguishing the main difference between the heights of the members of the two teams. Justify your conclusions.

In the Activity you may have concluded that the IQR helps to distinguish between the spreads of the two dot plots. It is one of the *measures of spread of a distribution*. Yet the IQR for the Sweet Peppers is 0 inches, which implies no spread. This number is not sensitive enough to indicate that there are Sweet Pepper players who are not 75" tall.

The simplest measure of the spread of a distribution is its *range*. The **range** is the difference of the maximum and minimum values of the variable. Each team has a mean height of 75" and a range of 80" – 70", or 10". In this way the distributions are quite similar. Yet the heights of the Dolphins seem more spread out than those of the Sweet Peppers.

So we need other statistical measures to better describe the spread of the heights of the players of each team. Two measures of spread that are influenced by every data point are *variance* and *standard deviation*.

The Variance and Standard Deviation of a Population

The Dolphins can be viewed as a population of ten women. When the set of data is a population, Greek letters are used for mean, variance, and standard deviation. The mean is labeled μ (mu), variance as σ^2 (sigma-squared), and standard deviation as σ (sigma). The variance for a population is calculated from the squares of **deviations**, or differences of each data value x_i from the mean μ. The shortest Dolphin player is 70" tall. The deviation $x_i - \mu$ for that player is $(70 - 75) = -5$ in. The square of her deviation is 25 square inches. Another of the Dolphin players is 78 inches tall. Her squared deviation is $(78 - 75)^2 = 9$ in^2.

The **population variance** is the mean of the squared deviations. That is, where n is the number of objects in a population, the variance is the sum of the squared deviations divided by n. The **population standard deviation** is the square root of the population variance. Example 1 shows how you can compute population variance and standard deviation by hand or by using a statistics utility.

Example 1

Find the variance and standard deviation for the heights of the Dolphins (treating them as a population).

Solution To find the variance and standard deviation it helps to organize the work step-by-step.

Step 1 Write the data x_i in a column. Find the mean by adding these numbers and dividing by n, the number of data points, which in this case is 10. Since the sum is 750, $\mu = 75$.

Step 2 In the next column record the result of subtracting the mean from each score, yielding $x_i - \mu$, which in this case is $x_i - 75$. Deviations are either positive, zero, or negative.

Step 3 Square each deviation and record each result in the next column.

Step 4 Add the squares of the deviations. Divide the sum of the squared deviations by n, in this case 10, to obtain the variance σ^2.

Step 5 Find the square root of the variance to get the standard deviation σ.

Results of these steps using technology are shown at the right and without using technology on the next page.

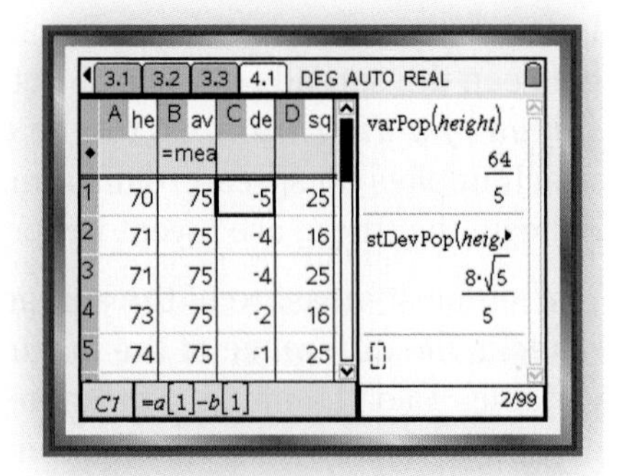

	Height (in.) x_i	Deviation (in.) $x_i - \mu$	Square of Deviation (in^2) $(x_i - \mu)^2$
	70	70 – 75 = -5	25
	71	71 – 75 = -4	16
	71	-4	16
	73	-2	4
	74	-1	1
	75	0	0
	78	3	9
	79	4	16
	79	4	16
	80	5	25
Sum	750	0	128

The mean μ is $\frac{750}{10} = 75$ in., the variance $\sigma^2 = \frac{128}{10} = \frac{64}{5} = 12.8$ in^2, and the standard deviation $\sigma = \sqrt{\frac{128}{10}} = \frac{8\sqrt{5}}{5} = \sqrt{12.8} \approx 3.58$ in.

Notice that the sum of the deviations equals 0. This is a great way to check your work. Also, notice that when the deviations are squared, values farther from the mean contribute more to the variance than values close to the mean. For instance, a height of 80 contributes $(80 - 75)^2 = 5^2 = 25$ to the sum of squared deviations, but 74 contributes only $(74 - 75)^2 = (-1)^2 = 1$. Because of this, groups with more data close to the mean generally have smaller standard deviations than groups with more data far from the mean.

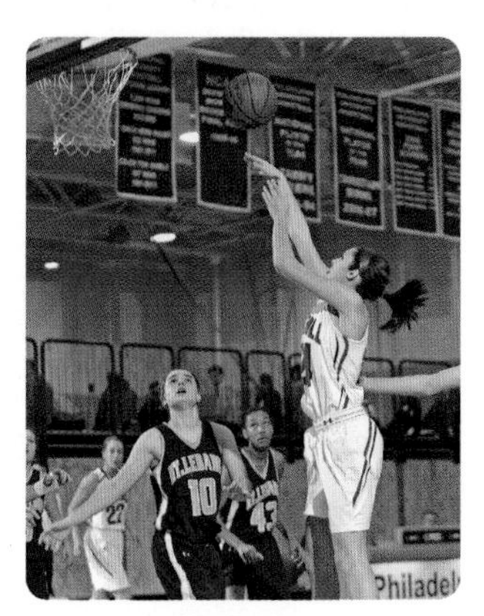

Basketball Beginnings

Basketball was introduced to women at Smith College in 1892, just one year after the game was invented.

Below is a picture of the Dolphins' data showing how the standard deviation of 3.58 relates to the distribution.

Dolphins

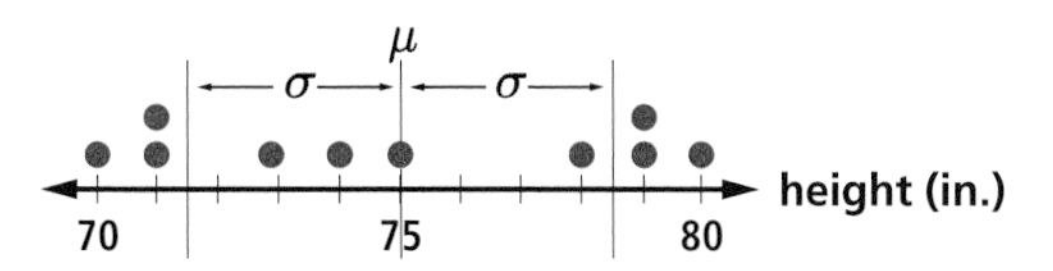

You are asked to calculate the variance and standard deviation for the Sweet Peppers in Question 3.

Formulas for the variance and standard deviation are usually written using Σ-notation. For a set of n numbers $x_1, x_2, x_3, \ldots, x_n$ each deviation from the mean can be written as $x_i - \mu$, and the square of the deviation as $(x_i - \mu)^2$. Because the definition of the variance and standard deviation are based on the mean, they can be used only when it makes sense to calculate a mean.

Definition of Variance and Standard Deviation of a Population

Let μ be the mean of the population data set $x_1, x_2, \ldots, x_n$. Then the **variance** σ^2 and **standard deviation** σ **of the population** are

$$\sigma^2 = \frac{\text{sum of squared deviations}}{n} = \frac{\sum_{i=1}^{n}(x_i - \mu)^2}{n}$$

and $$\sigma = \sqrt{\text{variance}} = \sqrt{\frac{\sum_{i=1}^{n}(x_i - \mu)^2}{n}}.$$

Variance and Standard Deviation of a Sample

For many years statisticians only used the population formulas. Over time, mathematicians and statisticians established that dividing by n for a sample variance did not produce the best estimate of the population variance. They showed that when using a sample, dividing by $n - 1$ rather than by n provided a better estimate of population variance.

When the data is a sample, Roman letters are used for mean, variance, and standard deviation. The mean is labeled $\bar{x}$ (read "x-bar"), variance is s^2, and standard deviation is s. The only difference in the formula is that $n - 1$ is used in place of n.

Definition of Variance and Standard Deviation of a Sample

Let $\bar{x}$ be the mean of the sample data set $x_1, x_2, \ldots x_n$.

Then the **variance** s^2 and **standard deviation** s **of the sample** are

$$s^2 = \frac{\text{sum of squared deviations}}{n-1} = \frac{\sum_{i=1}^{n}(x_i - \bar{x})^2}{n-1}$$

and $$s = \sqrt{\text{variance}} = \sqrt{\frac{\sum_{i=1}^{n}(x_i - \bar{x})^2}{n-1}}.$$

CAUTION: In this book most of the data come from samples, so unless directed otherwise, use the variance and standard deviation formulas for samples.

GUIDED

Example 2

According to the U.S. Department of Agriculture, ten to twenty earthworms per cubic foot is a sign of healthy soil. Mr. Green checked the soil in his garden by digging 7 one-cubic-foot holes and counting the earthworms. He found the following counts: 4, 23, 15, 10, 8, 12, 18.

Calculate the sample variance and sample standard deviation of the numbers of earthworms per cubic foot.

Solution Follow the same steps used in Example 1, but since this data represents a sample, use the variance and standard deviation formulas for samples. The symbols x_i, $x_i - \bar{x}$, and $(x_i - \bar{x})^2$ represent the earthworm count, deviation from the mean, and squared deviation, respectively.

	Count (worms) x_i	Deviation (worms) $x_i - \bar{x}$	Square of Deviation (worms squared) $(x_i - \bar{x})^2$
	4	-8.86	78.5
	23	?	102.82
	15	2.14	?
	10	-2.86	8.18
	8	?	23.62
	?	-0.86	0.74
	?	5.14	?
sum	$\sum_{i=1}^{7} x_i =$?	? ≈ 0	$\sum_{i=1}^{7} (x_i - \bar{x})^2 =$?

Wiggle, Squiggle, and Squirm In one acre of land, you can often find more than one million worms.

The mean $\bar{x} = \frac{\sum_{i=1}^{7} x_i}{7} = \frac{90}{7} \approx 12.86$ worms.

The variance $s^2 = \frac{\sum_{i=1}^{7} (x_i - \bar{x})^2}{?} = \frac{?}{?} = \underline{?}$ worms squared.

The standard deviation $s = \sqrt{\text{variance}} = \sqrt{\underline{?}} \approx \underline{?}$ worms, to two decimal places.

Because of these different formulas, some statistics utilities have two sets of symbols: s^2 and s, and σ and σ^2. Other calculators and programs use only one set of formulas for variance and standard deviation.

Activity 2

Use your calculator or statistics software to find (to the nearest tenth) the standard deviation of the following data set.

89, 79, 74, 67, 99, 91, 84, 81

If more than one standard deviation is given, record both. (You should find that the mean is 83 and both the sample and population standard deviation are between 9 and 11.)

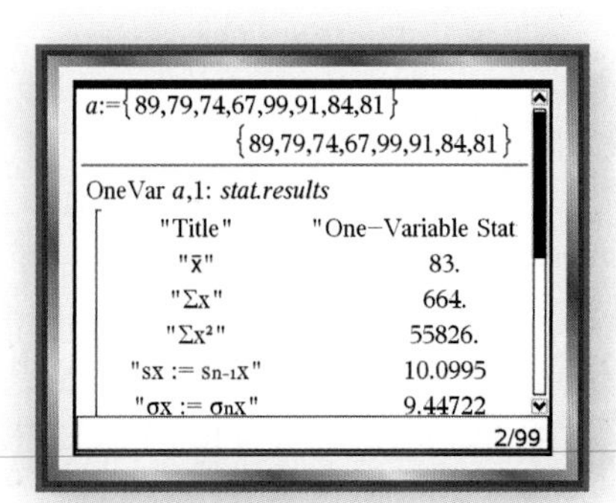

Questions

COVERING THE IDEAS

1. State whether the statistic is a measure of center or a measure of spread.

   a. mean b. range c. variance
   d. interquartile range e. standard deviation f. median

2. **Multiple Choice** The standard deviation of a set of scores is

   A the sum of the deviations.
   B the difference between the highest and lowest scores.
   C the score that occurs with the greatest frequency.
   D none of the above.

3. Use the heights of the Dolphins and Sweet Peppers.

   a. Calculate the variance and standard deviation for the heights of the Sweet Peppers using the formulas for populations.
   b. Find the difference between the means.
   c. Find the difference between the ranges.
   d. Find the difference between the standard deviations.
   e. Explain in your own words what the differences in Parts b–d tell you about the two data sets.

4. What statistics change if the Dolphins and Sweet Peppers are considered samples of all women basketball players?

5. If the standard deviation $s = 4.5$ cm, find the sample variance.

In 6 and 7, the measurements refer to pulse rates of two students while jogging, in beats per minute. Use a calculator or statistical software to find the mean and sample standard deviation for each situation.

6. student A with four measurements: 100, 120, 115, 133
7. student B with five measurements: 110, 120, 124, 116, 120

8. Suppose you used the formula for the sample standard deviation in Example 1. Would your answer be greater than, equal to, or less than the population standard deviation shown there?

9. Each of the following situations produces data that can be summarized with mean and standard deviation. Which would require population formulas for variance and standard deviation rather than sample formulas?

   A The Environmental Protection Agency measures carbon monoxide content of air at 15 locations of a metropolitan area.
   B An algebra teacher has scores from his students' final exams.
   C A consumer magazine tests four cars from each of three brands of hybrid vehicles to evaluate operating cost per mile.
   D A fan records the number of home runs per game for the Boston Red Sox in the 2008 season.

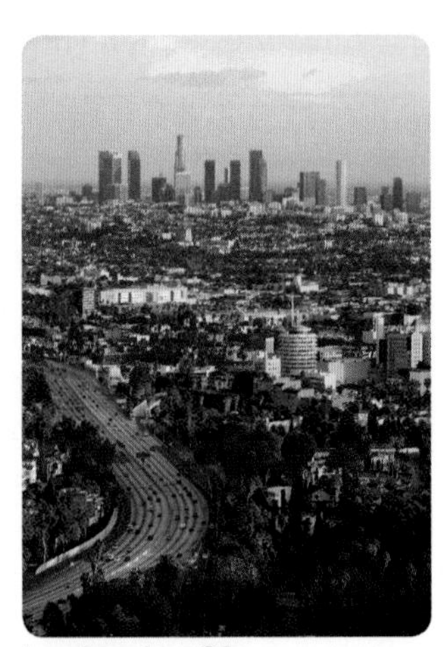

Los Angeles, CA

APPLYING THE MATHEMATICS

10. Suppose you know the distance in miles each student in a class lives from school. For this data set, state the unit for each statistic.

a. mean
b. range
c. variance
d. standard deviation

11. Beth found the variance of a data set to be –11. Why must her answer be wrong?

12. Suppose two samples have the same mean, but different standard deviations s_1 and s_2, with $s_1 < s_2$. Which sample shows more variability?

13. a. Consider the weights (in kilograms) of a group of deer. If the standard deviation is 7.8 kg, what is the variance?

b. If the variance is 19 kg^2, what is the standard deviation?

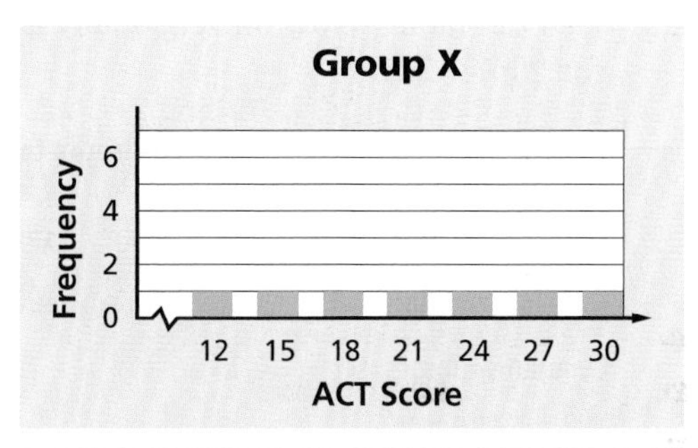

14. Use the hypothetical frequency distributions of ACT scores for groups X, Y, and Z at the right.

a. Match each group with its best description.

i. consistently near the mean
ii. very widely spread
iii. evenly distributed.

b. Without calculating, tell which group's ACT scores have the greatest standard deviation and which have the smallest.

c. Verify your answer to Part b with calculations.

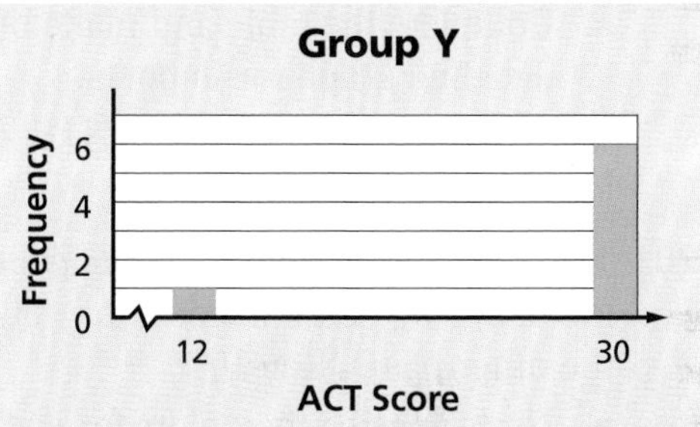

15. More than 1.3 million students in the class of 2007 took the ACT. On the mathematics section, $\mu = 21.0$ and $\sigma = 5.1$. Students receive scores rounded to the nearest whole number. What is the interval of student scores that lie within one standard deviation of the mean?

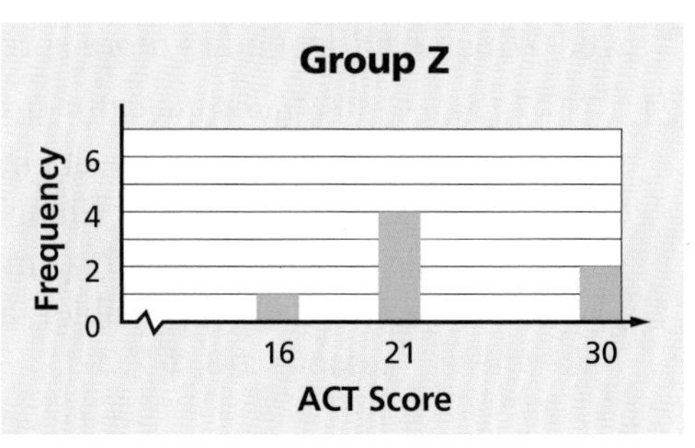

For 16 and 17, use the following data that represent the times (rounded to the nearest 5 seconds) for 20 sixth-graders to run 400 meters.

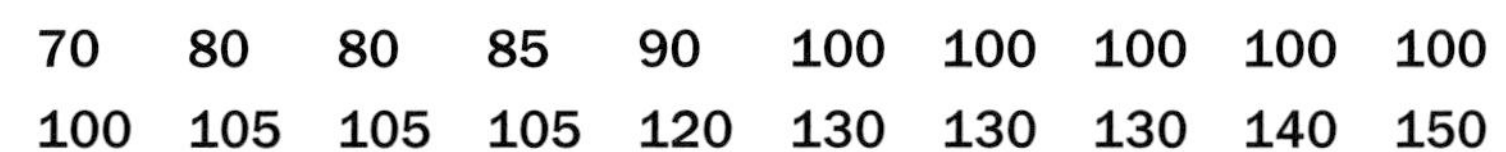

70	80	80	85	90	100	100	100	100	100
100	105	105	105	120	130	130	130	140	150

16. Find a sample of five students out of the 20 whose standard deviation for running time is as small as possible.

17. Find a sample of four running times whose standard deviation is larger than 25 seconds. Compute the standard deviation.

18. **Multiple Choice** A class of students is said to be *homogeneous* if the students in the class are very much alike on some measure. Here are four classes of students who were tested on a 20-point chemistry test. Which class is the most homogeneous with respect to scores on the test? Explain your answer.

A	$n = 20$	$\bar{x} = 15.3$	$s = 2.5$
B	$n = 25$	$\bar{x} = 12.1$	$s = 5.4$
C	$n = 18$	$\bar{x} = 11.3$	$s = 3.2$
D	$n = 30$	$\bar{x} = 10.4$	$s = 3.2$

REVIEW

19. Use the table below on seasonally adjusted U.S. domestic imports for 2007.

U.S. Total Imports in Goods and Services ($ billions)

TOTAL	JAN	FEB	MAR	APR	MAY	JUN	JUL	AUG	SEP	OCT	NOV	DEC
Total (per month)	187	186	192	191	193	195	197	197	198	200	205	203
Cumulative Total	?	?	?	?	?	?	?	?	?	?	?	?

Source: U.S. Census Bureau

a. Complete the table and make both a histogram and a cumulative data line graph for imports.

b. What was the total cost of U.S. domestic imports for 2007? **(Lessons 1-5, 1-3)**

20. Two data sets of heights of people each have minimum = 50", median = 67", and maximum = 80". One data set has IQR = 15"; the other has IQR = 10".

a. Draw possible box plots for each data set.

b. Which data set shows more spread? **(Lessons 1-4)**

21. The histogram below shows the number of states receiving the number of legal permanent residents specified in each interval in 2006. Write a paragraph describing immigration in 2006. Include both specific information such as maximum, minimum, mean, or median values (when possible), and general trends such as skewness. **(Lessons 1-3, 1-2, 1-1)**

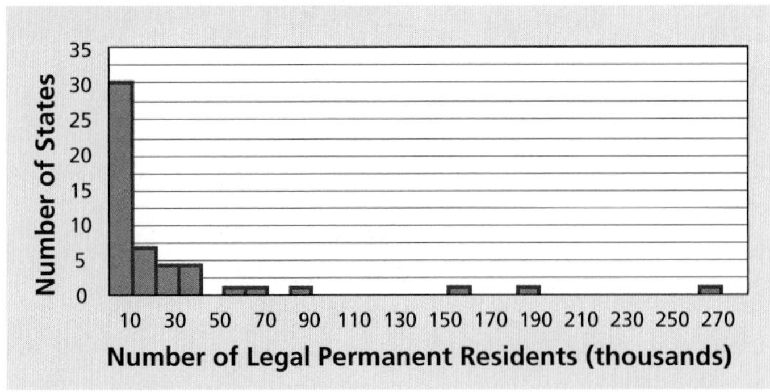

Source: U.S. Department of Homeland Security

In 22 and 23, use these data on the percent of Advanced Placement Examinations in Mathematics or Computer Science taken by female high school students.

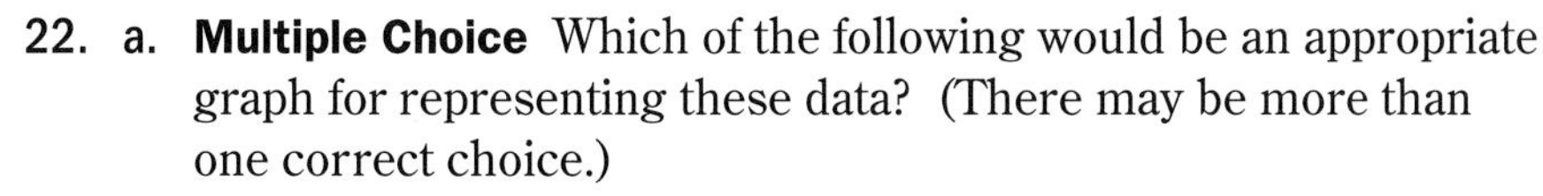

Year	1974	1979	1984	1989	1994	2004	2006	2007
Percent	26	32	35	36	43	44	45	46

Source: The College Board

22. a. **Multiple Choice** Which of the following would be an appropriate graph for representing these data? (There may be more than one correct choice.)
    A box plot
    B cumulative frequency graph
    C histogram
    D line graph
  b. Draw such a graph, and describe trend(s) in the data. **(Lessons 1-3, 1-1)**

23. The total number of students taking AP Exams in Mathematics or Computer Science was about 121,000 in 1994 and about 311,520 in 2004.
  a. What was the average annual change in the number of women taking AP Exams in these areas during this period?
  b. What was the average annual increase in the number of men taking these exams in this period? **(Lesson 1-2, Previous Course)**

24. **Multiple Choice** $\sum_{i=1}^{n} x_i$ equals **(Lesson 1-2)**
  A $\bar{x}$.
  B $\frac{\bar{x}}{n}$.
  C $\frac{n}{\bar{x}}$.
  D none of these.

EXPLORATION

25. The Russian mathematician Pafnuti L. Chebychev (1821 – 1894) proved a remarkable theorem called Chebychev's Inequality: In any data set, if p is the fraction of the data that lies within k standard deviations to either side of the mean, then $p \geq 1 - \frac{1}{k^2}$.
  a. According to Chebychev's Theorem, what percent of a data set must lie within 2 standard deviations of the mean?
  b. What percent must lie within 3 standard deviations?
  c. Test Chebychev's Theorem on a data set of your choice.

Lesson 1-7 Comparing Numerical Distributions

BIG IDEA A written comparison of data sets requires presentation of specific statistics as well as statements of similarities and differences.

The previous lessons used statistics to describe distributions of numerical variables and to develop cumulative distributions. This lesson asks you to compare two or more distributions. The emphasis is on clear, descriptive writing.

A written description of distributions compares and contrasts their *shapes*, *centers*, and *spreads*. The description should use the vocabulary and statistical measures you have learned in prior lessons. Although "a picture is worth a thousand words," you should not need that many words. Your description should be clear enough so that a reader who does not have graphs of the distributions can create mental images of them.

Mental Math

Multiple Choice

If an integer is divisible by 6 and by 9, then the integer must be divisible by which of the following?

A 12 **B** 15

C 18 **D** 21

Activity 1

Below and at the right are histograms, statistical summaries, and box plots of the distribution of heights (in inches) of all players in the National Basketball Association (NBA) and the National Football League (NFL) during the 2007–2008 season.

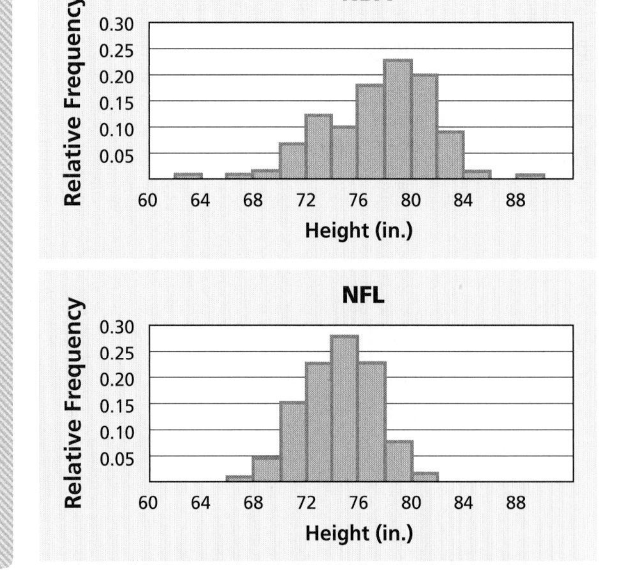

Statistics	Heights (in inches)	
	NBA	NFL
mean	79.2	74.0
median	80	74
interquartile range	5	4
standard deviation	3.63	2.58
range	25	15

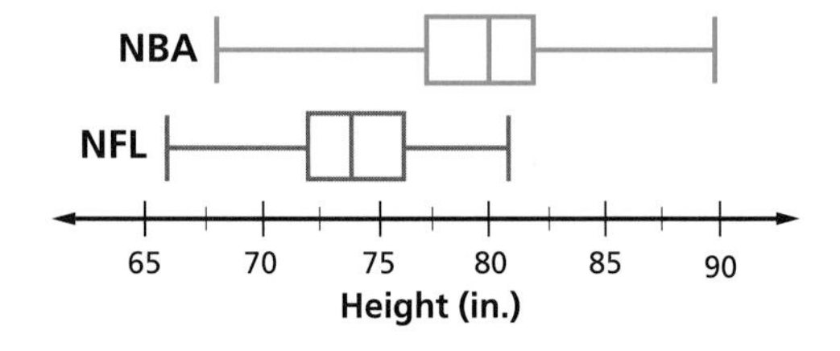

Step 1 Write a sentence about one piece of information conveyed by the statistical summaries table to compare NBA and NFL player heights.

Step 2 Write one sentence describing a comparison of NBA and NFL player heights that the box plots tell you that is not found in the table.

Step 3 Write one sentence comparing NFL and NBA player heights using a piece of information from the two histograms that is not found in either the box plots or the statistical table.

When comparing statistical summaries, notice how many times larger one number is than another. Note where the greatest relative differences are. Besides looking at the values of the 5-number summary in the box plots, examine where a quartile in one distribution might be in the other distribution. Discuss outliers, if possible. The histograms give a better picture of the shape of the distribution.

Information from several types of displays of two distributions can help you to give a more complete comparison of them.

Example 1

Compare the distributions of the heights of the players in the NBA and in the NFL given in Activity 1.

Solution This description begins by comparing the histograms.

The distribution of heights of players in the NBA in 2007-2008 is centered around 80 inches, while the NFL distribution of heights is symmetric about the 74–76 bar. There are tails at both ends of the NBA histogram indicating very few players shorter than 72 inches or taller than 86 inches. There are no tails on the NFL histogram.

This second part of the description compares the two box plots.

The minimum heights for both the NBA and the NFL players are virtually the same, but the maximum height of the NBA players is about 9 inches higher than that of the NFL, indicating a larger spread of heights in the NBA. About 25% of the NBA players are taller than all players in the NFL. The middle 50% of the heights of the NBA players are taller than the middle 50% of the NFL players.

This description ends by comparing the statistical summary tables.

The mean and the median for the NBA are nearly the same, as is the case for the NFL. So using either statistic to describe the average for the NFL or the NBA would be appropriate. The mean and median for the NBA players are about 7% higher than the respective NFL statistics. So, on average, the NBA players are taller than the NFL players.

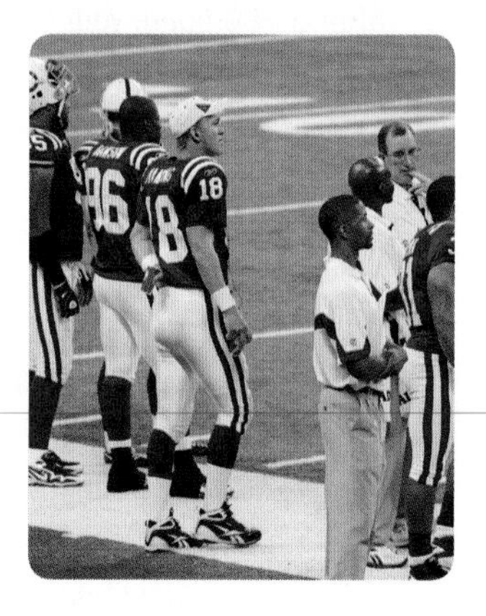

(continued on next page)

The standard deviation of the heights of the basketball players is 1 inch larger than the standard deviation for football players. This means overall there is more variability among the heights of the basketball players than of the football players. An overall difference in spread is also shown by the range of the heights of the NBA players being 10 points higher than that of the NFL players. The interquartile range for the NBA players is 5", which is slightly larger than the IQR of 4" for NFL players, indicating more spread in the heights of the middle 50% of the NBA players.

Activity 2 compares your class data with data from another sample.

Activity 2

The table, histogram, and box plot below describe the number of states visited by a sample of 20 adults in their lifetimes. A *visit* means that a person stayed for at least one night, drove or took a train through, or was in the airport. (Flying over does not count.) The count includes a person's home state.

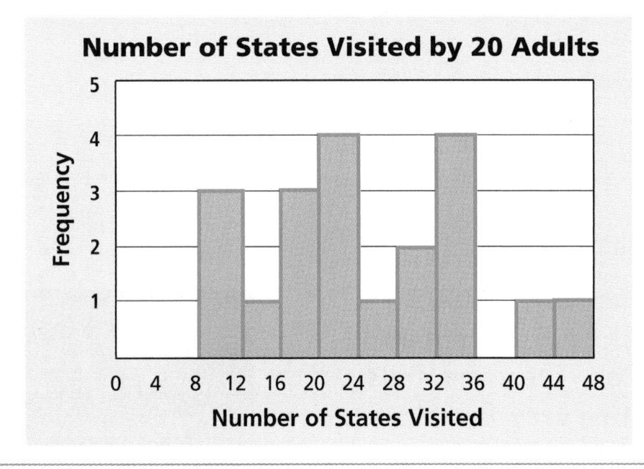

Number of States Visited				
mean	median	IQR	standard deviation	range
24.7	22.5	15.5	10.32	36

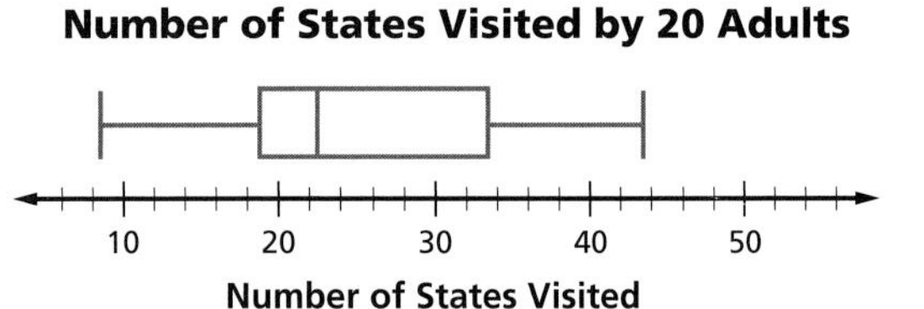

Step 1 Collect data concerning the total number of states each student in your class has visited.

Step 2 For your collected data, create a summary table, histogram, and box plot similar to those for the sample data.

Step 3 Compare your class data to the data from the adult sample. Use the outline below to guide your writing:

- Describe the variable involved. What is being measured in both sets of data?
- Use the five-number summary, histogram, and box plot to help you answer these questions:

  What is the shape of each data set? Are the data skewed? How can you tell?

  What is the center of each data set? Which measure is more appropriate to report: the mean or the median?

  What is the spread of each data set? Your choice of center statistic determines the spread you report.
- Explain the similarities or differences in the two data sets.

As you practice writing this kind of statistical summary, here is a checklist of things to remember.

- ✓ Describe the variable involved.
- ✓ Address all three major topics: shape, center, and spread.
- ✓ Give the appropriate units.
- ✓ Mention both data sets in your comparisons.
- ✓ Use words to give descriptions of your numbers. Use numbers to make your words more specific.

Questions

COVERING THE IDEAS

1. *Population density* is a measure of how crowded an area is. The information below compares Western and Eastern states in the United States in this regard. (The Mississippi River is the boundary between West and East.) Write a paragraph comparing the population densities of the Western and Eastern states.

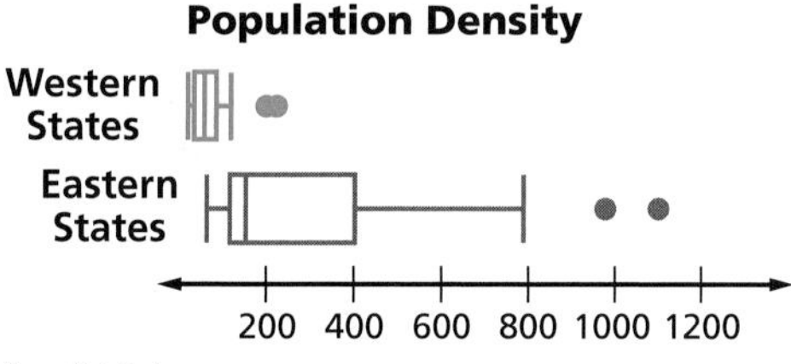

Source: State Master

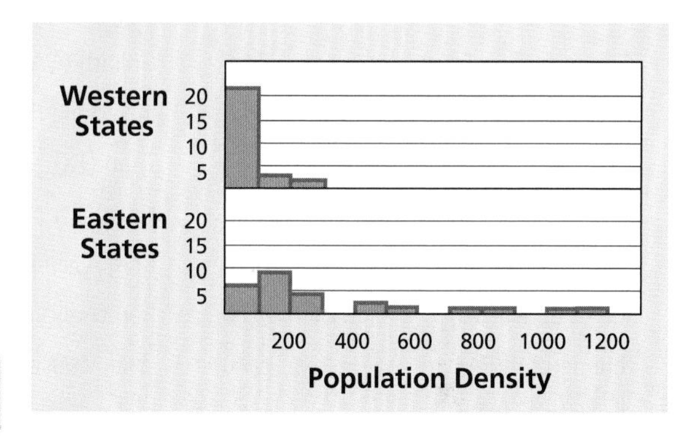

Western Density		Eastern Density	
mean	52.4	mean	301.4
median	38.6	median	172.3
standard deviation	54.4	standard deviation	298.3
interquartile range	55.4	interquartile range	299.4
Q_1	15.3	Q_1	101.7
Q_3	70.7	Q_3	401.1

2. Below is a list of terms you have seen in this chapter. Sort the terms into three groups: those dealing with shape, those with center, and those dealing with spread.

interquartile range	range	symmetric
standard deviation	skewed	variance
median	mean	
outlier	tail	

3. In clear, descriptive writing, what three characteristics of distributions should be discussed?

4. Refer to the information given in Activity 1. Suppose you are told a player is in the NBA or the NFL. What can you conclude about the league to which the player belongs, and what reasons can you give for your conclusion if:

   a. the player is 82 inches tall? b. the player is 71 inches tall?

5. The comparison in Activity 2 uses the median and IQR. Compare the same distributions using the mean and standard deviation.

APPLYING THE MATHEMATICS

6. Use these data to write a description comparing the maximum speeds on two samples of roller coasters—wood and steel.

Maximum Speeds of Roller Coasters

Wood		Steel	
Name (Location)	**Speed (mph)**	**Name (Location)**	**Speed (mph)**
The Beast (OH)	64.8	Wild Thing (MN)	74.0
Balder (Sweden)	55.9	Fujiyama (Japan)	80.8
Colossos (Germany)	74.6	Goliath (GA)	70.0
The Cyclone (NY)	60.0	Incredible Hulk (FL)	67.0
El Toro (NJ)	75.0	Titan (TX)	85.0
Megafobia (UK)	48.0	Nitro (NJ)	80.0
Screamin' Eagle (MO)	62.0	Borg Assimilator (NC)	51.0
Son of Beast (OH)	78.0	Iron Dragon (OH)	40.0
Thunderbolt (PA)	55.0	Wild Mouse (PA)	28.0
The Voyage (IN)	67.4	Hypersonic XLC (VA)	80.0
Mean Streak (OH)	65.0	Corkscrew (Canada)	40.0
Giant Dipper (CA)	55.0	Dragon Khan (Spain)	65.0
Minebuster (Canada)	55.9	The Great American Scream Machine (NJ)	68.0
Colossus (CA)	62.0	Wicked Twister (OH)	72.0
Le Monstre (Canada)	59.7	The Joker's Jinx (MD)	60.0
Thunderbird (Finland)	46.6		
Thunder Road (NC)	58.0		

Source: Roller Coaster DataBase

Astroland's Cyclone roller coaster, Coney Island, New York

7. The table below shows the length in meters of the 10 longest bridges in the United States and the 10 longest bridges outside of the United States. Write a descriptive paragraph comparing the length of the longest bridges in the United States to the longest bridges elsewhere in the world.

Longest Bridges

U.S.

Name	Length (m)
Lake Pontchartrain Causeway	38,422
Manchac Swamp Bridge	36,710
Atchafalaya Swamp Freeway	29,290
Chesapeake Bay Bridge Tunnel	24,140
Bonnet Carré Spillway	17,702
Jubilee Parkway	12,875
San Mateo-Hayward Bridge	11,265
Seven Mile Bridge	10,887
Sunshine Skyway Bridge	8,851
Twin Span	8,851

Outside U.S.

Name	Length (m)
Bang Na Expressway	54,000
Yangcun Bridge of Beijing-Tianjin Intercity Rail	35,812
Hangzhou Bridge	35,673
Runyang Bridge	35,660
Donghai Bridge	32,500
King Fahd Causeway	26,000
The No. 1 Bridge of Tianjin Binhai Mass Transit	25,800
Liangshui River Bridge of Beijing-Tianjin Intercity Rail	21,563
Yongding River Bridge of Beijing-Tianjin Intercity Rail	21,133
6th October Bridge	20,500

Source:Bridgemeister

Are We There Yet?

The Donghai Bridge, in China, is 32.5 km long and links Shanghai with the Yangshan Deep Water Port.

8. The graph at the right shows results from a test of coordination that was administered to both girls and boys. The results are described with percentiles. Compare boys with girls who
   a. score between 10 and 20.
   b. score between 30 and 40.
   c. score between 50 and 60.

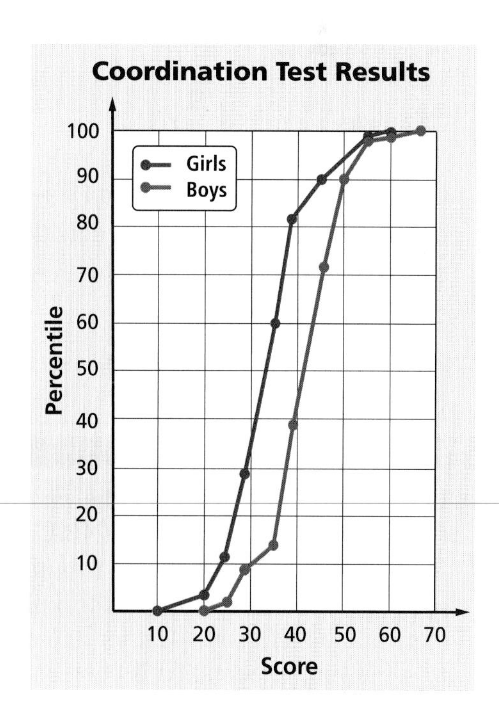

REVIEW

9. a. Consider the height in inches of a group of students. If the standard deviation is 4.3 in., what is the variance?
   b. If the variance is 9.6 in^2, what is the standard deviation? **(Lesson 1-6)**

10. The data in this table show how quickly 6 elevators can go from the ground floor to the top level. Let x = number of floors and y = the time in seconds. **(Lesson 1-6)**
    a. Calculate $\bar{x}$ and s_x.
    b. Calculate $\bar{y}$ and s_y.

Building	Number of Floors	Elevator Time (sec)
Taipei 101 (Taiwan)	101	30.2
Yokohama Landmark Tower (Japan)	73	23.7
Burj Dubai (UAE)	162	55.5
Sunshine 60 (Japan)	60	24.0
John Hancock Center (USA)	100	37.6
Freshwater Place (Australia)	63	22.8

Sources: Emporis.com; Forbes.com

11. A class of 30 students received the following scores on their 100-point mid-semester physics test. **(Lesson 1-5)**

63	96	68	84	52	79	88	80	75	72
90	75	78	97	68	74	76	91	76	74
92	80	85	60	77	94	81	60	96	65

    a. Make a table of the data, showing the frequency and relative frequency of each score.
    b. Draw a cumulative frequency line graph for the data.

In 12 and 13, consider the following data set: 13, 27, 11, 15, 9, 16, 18, 11, 6, 8, 11. (Lesson 1-2)

12. a. Calculate $\sum_{i=1}^{11} x_i$.
    b. Calculate $\bar{x}$.

13. Calculate $\sum_{i=1}^{11} (x_i)^2 - \left(\sum_{i=1}^{11} x_i\right)^2$.

14. Consider the line with equation $y + 7 = 3(x + 2)$. **(Previous Course)**
    a. Show that $y = 3x - 1$ is an equation for the same line.
    b. What are the slope and y-intercept of the line?
    c. Sketch a graph of the line.

15. Find an equation for the line passing through $(5, 0)$ and $(0, \pi)$. **(Previous Course)**

EXPLORATION

16. In Example 1 of this lesson, the variable studied is height, and the data come from two sets (NBA basketball players and NFL football players). The two distributions are quite different. Think of another pair of sets with distributions of a variable that you would expect to be different. Gather information about these distributions to see if your expectation is correct.

Lesson 1-8 Using Statistics to Solve a Mystery: The Case of *The Federalist* Papers

BIG IDEA Basic statistical analysis is used in many fields to investigate important questions.

Mental Math

What is the mean of $4x + 5$ and $-8x + 2$?

Disputes about authorship arise in the press, the legal system, and even among historians. One such dispute was a "mystery" concerning some famous documents in United States history, *The Federalist* papers. Frequency distributions were used to investigate this mystery.

Background

The Federalist papers were written between 1787 and 1788 under the pen name "Publius" to persuade the citizens of the State of New York to ratify the Constitution. Of the 85 *Federalist* papers, 14 were known to be written by James Madison, 51 by Alexander Hamilton, 5 by John Jay, and 3 were joint works. The remaining 12 papers were called "disputed" because historians were unsure whether they were written by Hamilton or Madison.

For many years historians tried to determine which author wrote each of the disputed papers. The dispute could not be settled by comparing the ideas in the papers because the philosophies of the men were so similar. In the 1960s two statisticians, Frederick Mosteller of Harvard University and David Wallace of the University of Chicago, used computers to count the frequency of key words in documents known to have been authored by the two men: 48 papers by Hamilton and 50 papers by Madison.

Key words were chosen that were not tied too closely to either author's writing style. The words were also chosen to be independent of the context of the paper being examined. For their study, they chose to use "by," "from," and "to." Mosteller and Wallace calculated the relative frequency of the occurrence of each word. If the key word "by" occurred 32 times in a paper of 2075 words, the rate of occurrence was $\frac{32}{2075} = 0.0154$. They reported each relative frequency as the number of occurrences per 1000 words: $(0.0154) \cdot 1000 = 15.4$ occurrences per 1000 words. Note that 15.4 per 1000 equals 1.54 per 100, or 1.54%.

James Madison (1751–1809)

The table on the next page shows the frequencies of the three key words in the 48 papers by Hamilton (H) and the 50 papers by Madison (M). For example, the word "from" was used by Hamilton 5–7 times in 21 of his 48 papers. Take a few minutes to read the table before going on.

Frequency Distribution of Rate per Thousand Words of the Words "by," "from," and "to" in 48 Hamilton and 50 Madison Papers								
"by"			"from"			"to"		
Rate	H	M	Rate	H	M	Rate	H	M
1-3*	2		1-3*	3	3	23-26*		3
3-5	7		3-5	15	19	26-29	2	2
5-7	12	5	5-7	21	17	29-32	2	11
7-9	18	7	7-9	9	6	32-35	4	11
9-11	4	8	9-11		1	35-38	6	7
11-13	5	16	11-13		3	38-41	10	7
13-15		6	13-15		1	41-44	8	6
15-17		5				44-47	10	1
17-19		3				47-50	3	2
						50-53	1	
						53-56	1	
						56-59	1	
Totals	48	50	Totals	48	50	Totals	48	50

***Each interval excludes its upper endpoint. Thus a paper with a rate of exactly 3 per 100 words would appear in the count for the 3-5 interval.**

 **See Quiz Yourself at the right.**

QUIZ YOURSELF

What word from the table does Madison use 13-15 times in 6 papers?

Activity 1

Below are the counts for the words "by," "from," and "to" in four *Federalist* papers known to be written by either Hamilton or Madison. Follow the steps to determine the author of each of the four papers.

Federalist Paper	Frequency			Number of Words	Rate per 1000 words		
	"by"	"from"	"to"		"by"	"from"	"to"
A	11	9	126	2714	?	?	?
B	42	12	75	2750	?	?	?
C	13	11	109	2269	?	?	?
D	29	9	102	2913	?	?	?

Step 1 Complete the table and find the rate per thousand words of "by," "from," and "to" for each unknown paper.

Step 2 Determine where each of the rates calculated in Step 1 fit in the table at the top of the page.

Step 3 Based on where the rates fit into the table, conjecture as to the authorship of each of the papers A, B, C, and D. Justify your answers.

Analyzing the Frequency of the Word "by"

The table at the top of the preceding page, which analyzes papers of known authorship shows that the word "by" is used much more frequently by Madison than it is by Hamilton. The rate of its occurrence seems to distinguish between the two authors. The word "to" seems more used by Hamilton. In contrast, the use of the word "from" does not distinguish one man's writing from the other.

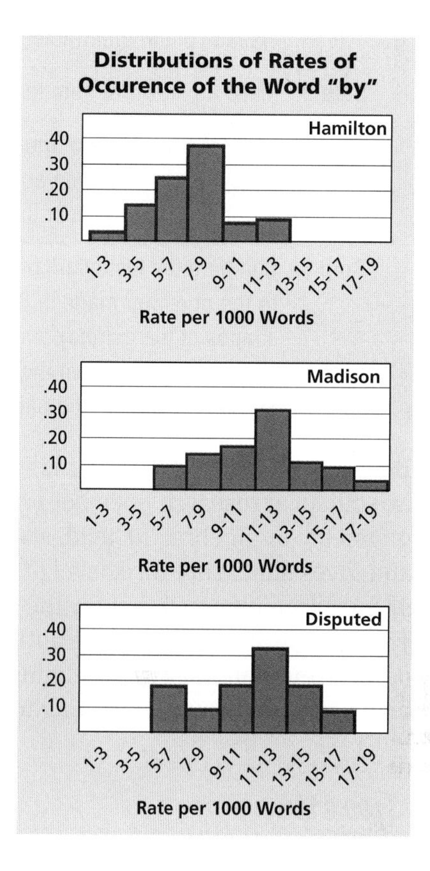

Based on this information, Mosteller and Wallace compared the use of the word "by" in the disputed *Federalist* papers to its use in the papers known to be authored by Hamilton and Madison. The relative frequency distributions comparing the use of the word "by" in the Hamilton papers, the Madison papers, and the disputed papers are shown at the right.

The shape of the graph for the disputed papers is clearly more like the shape of Madison's than that of Hamilton's. The median rate of occurrence of the word "by" is 11–13 for Madison's papers and for the disputed papers, and the median rate is only 7–9 for Hamilton's papers.

This evidence suggests that the disputed papers are Madison's. In fact, the full research study demonstrated that it is extremely likely that Madison authored 11 of the 12 disputed papers and probably the 12th as well.

Analyzing the Frequency of the Word "to"

Activity 2

Data about the use of the word "to" in the disputed *Federalist* papers are below.

Step 1 Complete this table and find the rate per thousand words for "to".

Federalist Paper	Frequency of "to"	Number of Words	Rate per Thousand Words
No. 49	58	1656	?
No. 50	28	1113	?
No. 51	50	1921	?
No. 52	72	1854	?
No. 53	73	2172	?
No. 54	61	2005	?
No. 55	78	2047	?
No. 56	39	1571	?
No. 57	74	2213	?
No. 58	61	2093	?
No. 62	80	2391	?
No. 63	88	3046	?

(continued on next page)

Step 2	Make a relative frequency table using the same bins that Mosteller and Wallace did in their table.
Step 3	Make a histogram from the data in the table you made in Step 2.
Step 4	Using the rate data of the word "to" from the table on page 64, make a histogram for Hamilton and one for Madison. Make sure the bin widths are the same as in your histogram from Step 3.
Step 5	Compare the histogram of the disputed *Federalist* papers from Step 3 to the ones you made in Step 4 for Hamilton and Madison. How do the shapes of the graphs compare? What about the median rates? Based on your evidence from the word "to", who do you believe to be the author of the disputed papers?

Beware! Statistical reasoning is ***inferential***, not ***deductive***. That is, statistical research does not *prove* findings with certainty as in geometry or algebra. Instead, statistics gives evidence for what is *likely*, and gives a measure for a level of confidence. For example, Mosteller and Wallace reported that for the *most* disputed paper, their evidence yields odds of 80 to 1 that the author was Madison. They called these odds strong but not overwhelming. The odds were much higher and considered to be overwhelming for the other 11 papers.

Questions

COVERING THE IDEAS

1. Why were *The Federalist* papers written?

In 2–4, refer to the table on page 62.

2. a. In how many of Hamilton's papers was the word "to" used at a rate of between 38 and 41 per 1000?
   b. In how many of Madison's papers was the word "to" used at a rate of between 38 and 41 per 1000?
3. How many papers of Hamilton and Madison were compared to produce the table?
4. If a Hamilton paper uses "by" with a rate of 11 words per 1000, which bin is it in?

In 5 and 6, refer to the histograms of the word "by" on page 63.

5. In what percent of the disputed papers was the word "by" used 11–13 times per thousand?
6. Would you expect fewer occurrences of the word "by" in a 1000-word essay by Hamilton or by Madison? Why?
7. What is meant by *inferential reasoning*?
8. Suppose a disputed paper has a rate of 6 uses of "from," 30 uses of "to," and 15 uses of "by," each per 1000 words. Who is the likely author, Hamilton or Madison? Explain your choice.

APPLYING THE MATHEMATICS

9. Suppose one of the disputed *Federalist* papers has 2107 words. The paper has 17 uses of "by," 8 uses of "from," and 97 uses of "to." Who is the likely author, Hamilton or Madison? Explain your choice.

10. Compute the weighted averages per 1000 words of the word "by" in all the papers known to be written by Hamilton and Madison. Use the median for each interval.

11. People diagnosed as ill with flu or a cold were asked questions about their illness and had their body temperatures measured. People were asked to "rate the pain in your muscles on a scale of 0–10 where 0 is no pain and 10 is very severe pain." Further, each was asked to rate their coughing on a scale of 0 (no cough) to 10 (constant coughing). CAUTION: These data are not to be used to diagnose cold or flu.

	Pain Rating	
Temp	**Flu**	**Cold**
96-97	1	5
97-98	1	6
98-99	3	15
99-100	4	14
100-101	10	9
101-102	11	1
102-103	9	0
103-104	5	0
104-105	1	0
Total	**45**	**50**

	Pain Rating	
Muscle	**Flu**	**Cold**
0-1	6	12
2-3	12	22
4-5	17	10
6-7	8	5
8-10	2	1
Total	**45**	**50**

	Rating	
Cough	**Flu**	**Cold**
0-1	7	10
2-3	9	9
4-5	18	23
6-7	8	7
8-10	3	1
Total	**45**	**50**

a. Marco has a temperature of 101.2. He rates his muscle pain as a 7, and cough as a 5. Which illness does Marco likely have? Explain your decision.

b. Annette rates her muscle pain as 3 and cough as 5. Her temperature is 98.3. Which illness does Annette likely have? Explain your answer.

REVIEW

In 12 and 13, describe a possible shape for the distribution of the variable. Explain your reasoning. (Lesson 1-3)

12. the prices of homes sold in the U. S. last year

13. the numbers of points scored by the winning team in WNBA (Women's National Basketball Association) games played last year

14. The table below includes the heights in feet of the 20 tallest buildings in the U.S. and the 20 tallest buildings outside of the U.S. Compute five-number summaries for each data set. Then write a descriptive paragraph comparing the data sets. **(Lesson 1-7)**

20 Tallest Buildings

U.S.				Outside U.S.			
Name	**Hgt (ft)**	**Name**	**Hgt (ft)**	**Name**	**Hgt (ft)**	**Name**	**Hgt (ft)**
Sears Tower	1451	Prudential Plaza	995	Taipei 101	1667	Tuntex Sky	1140
Empire State	1250	Wells Fargo	992	Petronas 1	1483	The Center	1127
Aon Center	1136	311 S Wacker	961	Petronas 2	1483	Rose Tower	1127
John Hancock	1127	AIB Building	952	Jin Mao Building	1380	Shimao Plaza	1093
Chrysler Building	1046	Key Tower	947	Two IFC	1362	Minsheng Bank	1087
NY Times Building	1046	1 Liberty Place	945	CITIC Plaza	1283	Ryugyong Hotel	1083
Bank of America, GA	1039	Columbia Center	933	Shun Hing Square	1260	Q1	1058
US Bank Tower	1018	Trump Building	927	Central Plaza	1227	Burj al Arab	1053
AT&T Corp	1007	Bank of America, TX	921	Bank of China	1205	Nina Tower	1046
Morgan/Chase	1002	Citigroup Center	915	Emirates Tower	1165	Menara Tel	1017

Source: Emporis Research

15. The frequency diagram at the right shows how many pairs of shoes each of 100 people owns. Let i = the number of pairs of shoes, and $f(i)$ = the frequency of that number. **(Lessons 1-3, 1-2)**

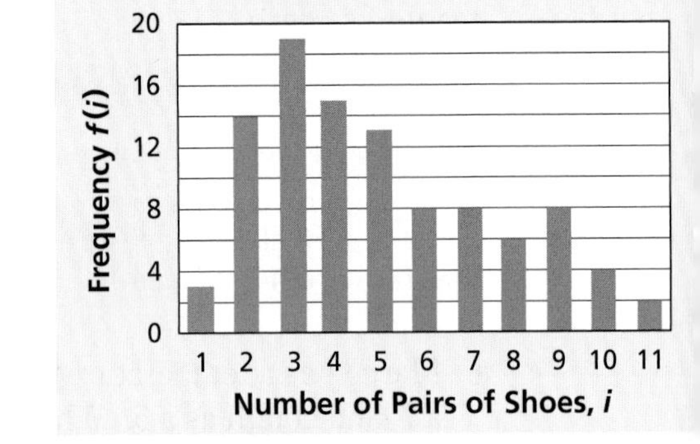

a. What is the mode?

b. What is $f(5)$?

c. Evaluate $\sum_{i=4}^{11} f(i)$.

d. What does the quantity $\dfrac{\sum_{i=1}^{11}(i \cdot f(i))}{\sum_{i=1}^{11} f(i)}$ represent?

16. If the interval from a to b is split into n equal parts the left most part is from a to $a + \frac{b-a}{n}$. Write the expression $a + \frac{b-a}{n}$ as a single fraction with denominator n. **(Previous Course)**

17. Consider the graphs of $y = -x^2$ and $y = -|x|$. **(Previous Course)**

a. Describe several ways the graphs are alike.

b. Describe several ways the graphs are different.

EXPLORATION

18. *Cliometrics* is the name given to the application of statistics to economic history. Find out how this word originated and discuss at least one thing that cliometricians have found.

QY ANSWERS

"by"

Chapter 1 Projects

A project presents an opportunity for you to extend your knowledge of a topic related to the material of this chapter. You should allow more time for a project than you do for typical homework questions.

1 Automobile Survey

What automobiles are most popular in your area?

a. Go to a large parking lot (near a shopping center, office building, or school) and classify at least 60 automobiles by the following criteria: style (van, truck, limousine, sports car, and so on), color, and manufacturer.

b. Report the results with at least three tables or displays.

c. Write a paragraph or two summarizing and interpreting your findings.

d. Describe the differences that might have resulted if you had collected your data at a different location (e.g., senior citizens center, executive office garage, used car lot) or at a different time (church lot on Sunday, movies on a budget night).

2 Sabermetrics: The Fastest Player

Statistical means are used to compare baseball players who played at different times. For instance, data on the number of bases stolen, times caught stealing, doubles, triples, and so on, can be used to compare the speeds of different players in order to determine who really was the fastest player. Using a baseball almanac or the Internet, investigate two players of your choosing. Devise a method and use it to determine the faster player.

3 Coin Circulation

It is common practice to collect pennies instead of spending them (thus putting them back in circulation). From a jar of pennies:

a. Choose a sample of 100 pennies. In a table, enter the frequency of pennies minted in a particular year and the relative frequencies.

b. Make a histogram of the data.

c. Do the same for nickels and compare the distributions.

d. Conjecture reasons for the differences in the distribution.

4 Graphing and Interpreting Statistical Data

Obtain a copy of a recent edition of either the *Statistical Abstract of the United States* or an almanac. Pick data that you find interesting or surprising (or both) that are presented in a table. Design a poster, at least 22" × 28", that interprets the data in the table, including displays to support your interpretation. Make sure that you choose a suitable headline for your poster that will attract people's attention.

5 Class Survey

Compile a database of information about the members of your mathematics class.

a. Ask each person to complete an information sheet for the following data. Where appropriate, measure in centimeters.
   - gender
   - height
   - hand span
   - foot length
   - time spent getting ready for school each day
   - travel time to school each morning
   - time at work each day
   - time spent on homework each day
   - number of people living at home
   - language most often spoken
   - time spent on the Internet each day
   - usual mode of travel to school (bus, walk, skateboard, bike, car, and so on)
   - time spent watching TV each day
   - time spent exercising each day

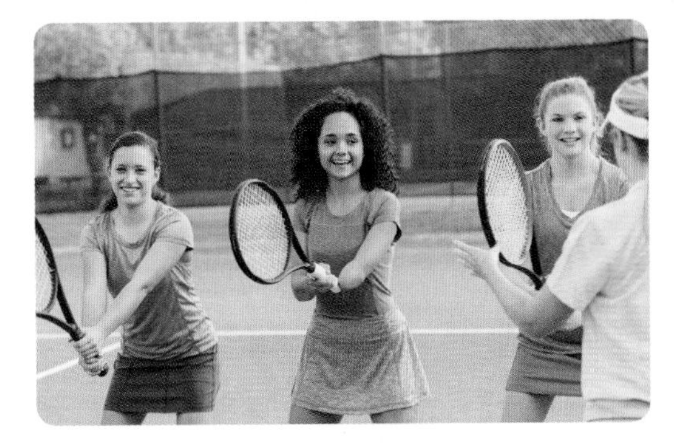

b. Construct a computer or calculator file with the class database. The database will be used again in later chapters.

c. Choose at least three variables. Use a statistics utility to display and summarize the data. For each variable, decide which type of display is most appropriate (box plot, stemplot, bar graph, and so on). Whenever appropriate, calculate statistics such as the mean, median, standard deviation, percentiles, and range.

d. Write a short paragraph describing a "typical" student in your class in terms of the variables you analyzed. For numerical variables, this will involve interpreting both the center and spread of the distributions.

e. Find a variable whose value differs quite a bit by gender. Find a variable whose value doesn't differ much by gender. Justify your conclusions with statistical values or displays.

6 The Disputed *Federalist* Papers

Stylistic words other than those discussed in Lesson 1-8 may be used to confirm or deny the results found by Mosteller and Wallace. In a manner comparable to that carried out in Lesson 1-8, choose a stylistic word such as "of," "in," "because," "that," "and," "but," and so on, and count its occurrences in *The Federalist* papers written by Hamilton and Madison, as well as the disputed papers. Confirm or deny Mosteller and Wallace's results using these data. The documents may all be found online, and words can be counted quickly using a word-processor search application.

Chapter 1 Summary and Vocabulary

This page lists the most important terms and phrases for this chapter. You should be able to give a general description and a specific example of each and a precise definition for those marked with an asterisk (*).

- Statistics is the science of the collection, organization, analysis, and interpretation of data. The set of *all* individuals or objects that could be studied is the **population**. A **sample** is a subset of the population.
- Data can be organized and displayed in tables and graphs. Bar graphs, line graphs, box plots, dotplots, cumulative frequency graphs, and histograms are kinds of graphs studied in this chapter.
- Summary statistics and graphs enable comparisons between data sets. Good written descriptions of data sets include discussion of shape, centers, and spreads. Measures such as the mean, median, and sometimes the mode indicate centers of the data. Measures of spread include the range, interquartile range, **variance**, and **standard deviation**. The **five-number summary** (minimum, first quartile, median (2nd quartile), third quartile, maximum) is one way to summarize the center and spread in a data set. The symbol for summation, Σ (sigma), provides a short way to write expressions for several of these measures.
- Comparisons depend on common units. Relative frequencies, percents, rates, and percentiles in data sets are statistics useful because comparisons are not affected by the size of samples.

Vocabulary

1-1
*statistic(s)
data, datum
variable
*population
*sample
survey
census
representative sample
categorical variable
numerical variable

1-2
*mean
*median
*measures of center, measures of central tendency
mode
subscripted variables
sigma, Σ
index, i
summation notation, sigma notation, Σ-notation
weighted average
relative frequency

1-3
distribution
histogram
bins
frequency histogram
relative frequency histogram
skewed
symmetric
population pyramid

1-4
box plot, box-and-whiskers plot
minimum
first (lower) quartile
second quartile
third (upper) quartile
maximum
*five-number summary
*interquartile range (IQR)
whiskers
outlier

1-5
cumulative data
cumulative distribution
*percentile, *p*th percentile

1-6
range
deviation
*population variance, σ^2
*population standard deviation, σ
*sample variance, s^2
*sample standard deviation, s

Chapter 1 Self-Test

Take this test as you would take a test in class. You will need a calculator. Then use the Selected Answers section in the back of the book to check your work.

In 1–5, the following data set is the number of points a high school football team scored in 14 games.

$x_1 = 14, x_2 = 21, x_3 = 0, x_4 = 45, x_5 = 20,$
$x_6 = 31, x_7 = 33, x_8 = 17, x_9 = 21, x_{10} = 7,$
$x_{11} = 6, x_{12} = 52, x_{13} = 19, x_{14} = 16.$

1. Find the variance and standard deviation of the data. Use population statistics.
2. Write an expression for the mean using Σ-notation.
3. Draw a box plot of the data.
4. Does the data set contain any outliers under the $1.5 \times$ IQR criterion? If so, name them. If not, explain why not.
5. If another data set of points scored in 14 games by a second high school football team had a smaller standard deviation than that found in Question 1, which team would have more variability in points scored?
6. Consider the table below of the number of staff positions in California public schools for 2003. Complete the table to find the relative frequency of each position.

Position	Frequency	Relative Frequency
Principals	13,340	?
Teachers	304,311	?
Aides	69,201	?
Office/Clerical	36,116	?
Counselors	6,640	?
Librarians	1,218	?
Other	109,381	?

Source: National Center for Education Statistics

7. **True or False** Extreme data affects the median of a data set less than it affects the mean.
8. Consider the box plot below.

2002–2003 Revenues for Public Schools by State (Billions of Dollars)

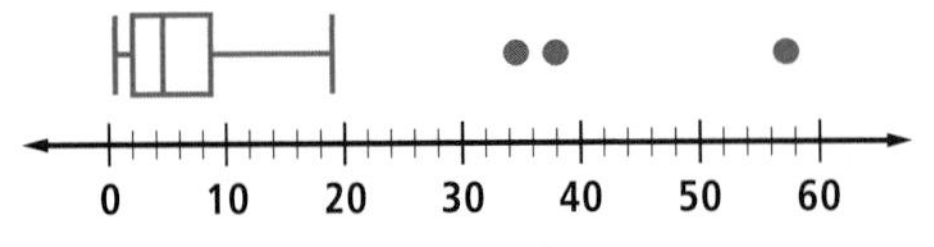

Source: National Center for Education Statistics

   a. Estimate a five-number summary for the box plot.

   b. Between which two values are the middle 50% of the data?

9. Tell whether each given measure of spread is directly calculated using the mean.

   a. range   b. variance

   c. IQR   d. standard deviation

10. All mountains in the world whose altitude is at least 8,000 meters are represented in the histogram below.

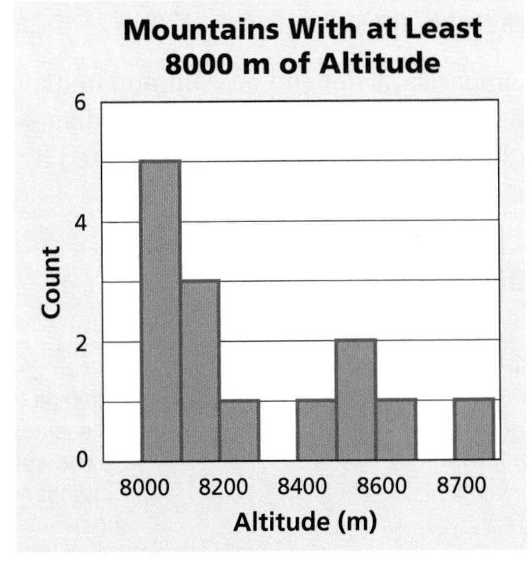

Source: The World Almanac, 2007

   a. How many mountains in the world are at least 8,000 meters high?

   b. Describe the shape of the distribution.

11. Consider the table below about the U.S. Budget in billions of dollars.

a. Which source was expected to contribute the greatest amount of additional receipts in 2007 as compared to 2006? Explain how you know.

b. Which source was expected to provide the greatest percent increase in revenues from 2007 to 2008?

	Actual	Estimate	
Category	2006	2007	2008
Personal Income Taxes	1,043.9	1,168.8	1,246.6
Corporate Income Taxes	353.9	342.1	314.9
Insurance and Retirement Receipts	837.8	837.4	974.2
Excise Taxes	74.0	57.1	68.1
Estate and Gift Taxes	27.9	25.3	25.7
Customs Duties	24.8	26.8	29.2
Miscellaneous Receipts	45.0	46.7	50.7
Total Receipts	2,407.3	2,540.1	2662.5

Source: Budget of the United States Government, 2008

12. The monthly average snowfall (in inches) in Boulder, Colorado, from a recent year are shown at the right.

a. Compute the cumulative snowfalls by month.

b. Draw a cumulative line graph of the annual snowfall.

c. What percent of the annual snowfall had fallen by the end of March?

Month	Snowfall (inches)
January	5.5
February	11.4
March	23.3
April	2.9
May	0.1
June	0
July	0
August	0
September	0
October	15.2
November	12.0
December	45.5

Source: National Oceanic and Atmospheric Administration

13. Carol's college computes grade point averages (GPAs) by assigning points for grades: A = 4, A– = 3.7, B+ = 3.3, B = 3, B– = 2.7, C+ = 2.3, C = 2, C– = 1.7, D+ = 1.3, D = 1, F = 0. It then weights the grade by the number of semester hours for each course. Compute Carol's first semester GPA based on the report below.

Course	Grade	Hours
Writing	B+	2
Calculus	A	4
Spanish I	C+	6
Am Lit	B	4

In 14 and 15, use the following histograms and box plots of test scores from two classes of high school students.

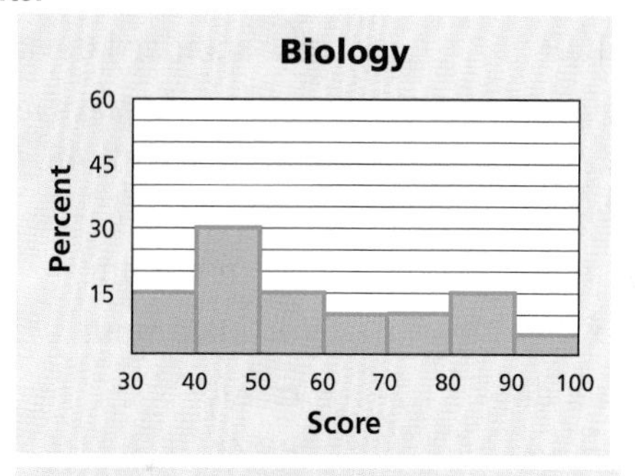

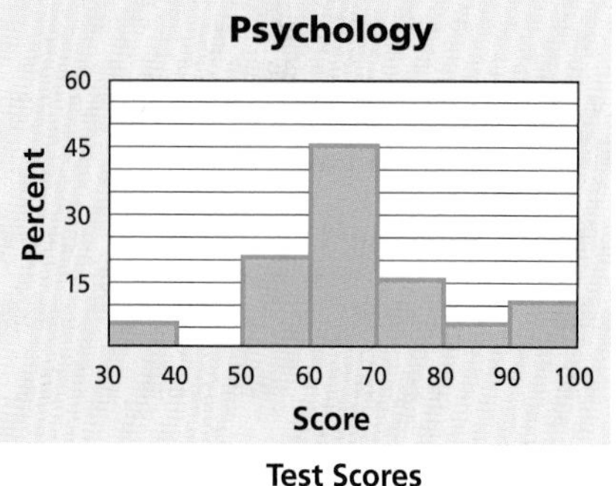

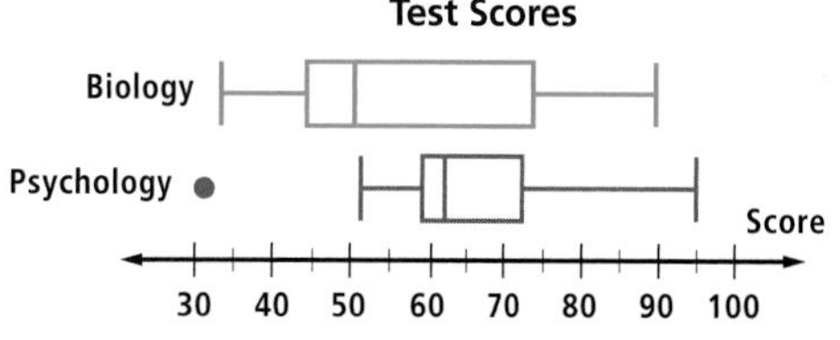

14. Use the histograms. What percent of students scored at least 80 on the psychology test?

15. Compare and contrast the two distributions using the information in the box plots.

Chapter 1 Chapter Review

SKILLS
PROPERTIES
USES
REPRESENTATIONS

The Chapter Review questions are grouped according to the SPUR Objectives in this chapter.

SKILLS Procedures used to get answers.

OBJECTIVE A Calculate measures of center and spread for data sets. (Lessons 1-2, 1-4, 1-6)

In 1 and 2, use the data set {4, 8, 12, 14, 14, 16, 16, 18, 18, 20} for a sample. Find each statistic.

1. a. range
   b. median
   c. IQR
2. a. mean
   b. variance
   c. standard deviation

In 3 and 4, the ages in years of some of Max's relatives are {2, 12, 24, 9, 14, 29, 33, 9, 15, 61, 42, 48, 42, 56}. Calculate each statistic.

3. a. median
   b. $\overline{x}$
   c. mode
   d. range
   e. s
4. a. minimum
   b. maximum
   c. Q_1
   d. Q_2
   e. Q_3
5. Give the five-number summary in square miles for the data set below.

State	Area (mi^2)	State	Area (mi^2)
Alaska	571,951	Montana	145,552
Utah	82,144	Wisconsin	54,310
California	155,959	Illinois	55,584
Rhode Island	1,045	Maine	30,862
Pennsylvania	44,817	Texas	261,797
Michigan	56,804	Florida	53,927

Source: U.S. Geological Survey

6. The median of a data set is 125, Q_1 is 99, and the IQR is 50. In this data set:
   a. what is Q_3?
   b. x is an outlier. What might x be?

OBJECTIVE B Calculate averages with weights, frequencies, and relative frequencies. (Lesson 1-2)

7. If a city averages 70 cm of rain per month during the 5-month rainy season and 10 cm per month during the rest of the year, what is the mean amount of rain per month for the year?
8. If an A is worth 4 points and a B+ is worth 3.3 points, what is the weighted average of a student who gets 24 credits of A and 16 credits of B+?
9. In Mrs. Clutz's physics class, 30% of girls have 3 notebooks, 20% have 5 notebooks and the rest have 7 notebooks each. On average, how many notebooks does a girl in Mrs. Clutz's class have?
10. In a school, 3 classrooms have no computers, 14 classrooms have 1 computer, 6 classrooms have 2 computers, and 1 classroom has 3 computers. What is the average number of computers per classroom?
11. Mr. Fisher counts tests as 60% of a student's grade, homework as 20%, and quizzes and classwork as 20%. If Mark has a test grade of 82 and a homework grade of 95, what classwork grade does he need to average 86?
12. One school gives a bonus weight to Advanced Placement (AP) courses. Brian received A's in all his classes. An A was worth 4 points in a non-AP class. He took 2 AP courses and 3 non-AP courses. His grade point average was 4.2 for the semester. What is an A in an AP course worth?

PROPERTIES The principles behind the mathematics.

OBJECTIVE C Use Σ-notation to represent a sum or mean. (Lesson 1-2)

In 13 and 14, suppose g_i equals the number of points Karen scored in the ith basketball game of this season.

$g_1 = 10 \quad g_2 = 24 \quad g_3 = 14 \quad g_4 = 12$

$g_5 = 27 \quad g_6 = 16 \quad g_7 = 20 \quad g_8 = 22$

13. a. Write an expression using Σ-notation that indicates the mean number of points Karen scored for the eight games.
    b. Evaluate the expression in Part a.
    c. Evaluate $\sum_{i=4}^{7} g_i$.

14. a. Evaluate $\sum_{i=1}^{4} g_i - \sum_{i=5}^{8} g_i$.
    b. Explain what Part a means in this situation.

15. Betty bought BigCo shares each month as follows.

Month	1	2	3	4	5	6	7	8
# of Shares	100	300	200	200	300	100	200	100
$ per Share	4.25	5.75	4.50	5.25	6.00	7.50	5.75	6.50

    a. Let n_i be the number of shares and p_i be the price per share in month i. Write an expression using Σ-notation for the average price per share of Betty's BigCo buys.
    b. Evaluate Part a.

OBJECTIVE D Describe relations between measures of center and spread. (Lessons 1-2, 1-5, 1-6)

16. Which is usually affected more by extreme values and outliers in the data set, the mean or the median?

17. a. **True or False** The mean and mode of a data set are always equal.
    b. If your answer to Part a is true, explain why. If false, give a counterexample.

18. **True or False** If there are 8 ordered points in a data set, the 4th lowest value corresponds to the 50th percentile.

Multiple Choice In 19-21, select the best answer.

19. The Q_1, Q_2, and Q_3 statistics refer to which of the following percentiles respectively?
    A 10%, 20%, 30%
    B 5%, 50%, 95%
    C 25%, 50%, 75%
    D 33%, 50%, 66%

20. Which statistic is used in the calculation of variance?
    A median
    B mean
    C interquartile range
    D mode

21. If the variance of a set of n numbers is y, what is the standard deviation of the set?
    A y^2
    B $\frac{y}{n-1}$
    C $\sqrt{y}$
    D $\frac{\sqrt{y}}{n}$

22. If the mean of 50 scores is exactly 26.4, and the mean of 49 of those scores is exactly 26, what is the 50th score?

USES Applications of mathematics in real-world situations

OBJECTIVE E Use statistics to draw conclusions about data. (Lessons 1-4, 1-6)

23. In a biology experiment, Jon recorded the number of days it took for a nest of 20 snapping turtle eggs to hatch. He obtained the sample data below.

58	63	74	60	51	51	83	48	66	50
94	53	60	74	60	78	83	82	60	83

    a. Find the five-number summary for these data.
    b. Identify all outliers using the 1.5 × IQR criterion.

24. Suppose the heights of boys in one physical education class had a mean of 65 inches and a standard deviation of 3.2. The heights of boys in another physical education class had a mean of 67 inches and a standard deviation of 2.1. Which class had more variation in height? Justify your answer.

25. The table below gives the median sales price of existing single family homes during 2006 for 12 metropolitan areas. Prices are in thousands of dollars.

Area	Price	Area	Price
Akron, OH	114.6	Charleston, WV	199.4
Albuquerque, NM	184.2	Gainesville, FL	213.2
Atlantic City, NJ	254.8	Honolulu, HI	630.0
Bismarck, ND	134.9	Indianapolis, IN	119.3
Boulder, CO	366.4	Jackson, MS	147.1
Buffalo-Niagara Falls, NY	97.9	Nassau-Suffolk, NY	474.7

Source: National Association of Realtors

a. Compute the five-number summary of these data.

b. What conclusions can you make from these data regarding the price of housing in metropolitan areas in the United States?

26. Fill in the Blanks A distribution has $Q_1 = 50$, $Q_2 = 60$, $Q_3 = 80$, and no outliers. The data in the distribution must be no smaller than __?__ and no larger than __?__.

27. A canvassing organization in Austin, Texas, sent out 20 canvassers to collect money to support renewable energy resources. The amount of money each canvasser collected (in dollars) during the day is: 100, 68, 212, 300, 0, 90, 180, 23, 80, 440, 140, 163, 117, 140, 445, 180, 200, 140, 250, 500.

a. Find the mean, median, and sample standard deviation of the amounts collected.

b. Which amount or amounts are the outliers to the collections?

c. Removing the outliers, find the new mean, median, and standard deviation of the amounts.

Voting-Age Population, Percent Reporting Registered, and Voted

Age, Gender, and Education	Voting-age population (millions)				Percent reporting they voted			
Year	1992	1996	2000	2004	1992	1996	2000	2004
Total	185.7	193.7	202.6	215.7	61.3	54.2	54.7	58.3
Age								
18 to 20 years old	9.7	10.8	11.9	11.5	38.5	31.2	28.4	41.0
21 to 24 years old	14.6	13.9	14.9	16.4	45.7	33.4	35.4	42.5
25 to 34 years old	41.6	40.1	37.3	39.0	53.2	43.1	43.7	46.9
35 to 44 years old	39.7	43.3	44.5	43.1	63.6	54.9	55.0	56.9
45 to 64 years old	49.1	53.7	61.4	71.0	70.0	64.4	64.1	66.6
65 years old and over	30.8	31.9	32.8	34.7	70.1	67.0	67.6	68.9
Gender								
Male	88.6	92.6	97.1	103.8	60.2	52.8	53.1	56.3
Female	97.1	101.0	105.5	111.9	62.3	55.5	56.2	60.1
Education Completed:								
8 years or less	15.4	14.1	12.9	12.6	35.1	28.1	26.8	23.6
Less than high school graduate	21.0	21.0	20.1	20.7	41.2	33.8	33.6	34.6
High school graduate or GED	65.3	65.2	66.3	68.5	57.5	49.1	49.4	52.4
Some college or associate's degree	46.7	50.9	55.3	58.9	68.7	60.5	60.3	66.1
Bachelor's or advanced degree	37.4	42.5	48.0	54.9	81.0	73.0	72.0	74.2

OBJECTIVE F Determine relationships and interpret data given in a table. (Lesson 1-1)

In 28-33, use the table on the previous page from the *Statistical Abstract of the United States 2007*.

28. Which numbers in the table total 215.7?

29. Identify the variables for which data are provided in this table.

30. Explain the meaning of the number 28.4 in the seventh column, titled "2000".

31. How many women reported that they voted in the United States in 2004?

32. How many high school graduates or people with a GED reported that they voted in the United States in 1992?

33. In which age group did the largest percent of its members report that they voted?

OBJECTIVE G Use statistics to compare and contrast two data sets in the context of a situation. (Lessons 1-7, 1-8)

34. Use the data below of normal daily mean temperatures by month in Juneau, Alaska, and Minneapolis-St. Paul, Minnesota. Data have been rounded to the nearest degree Fahrenheit.

On average, which city shows greater variability in temperature? Use a measure of spread to justify your answer.

Month	Juneau, AK	Minneapolis-St. Paul, MN
January	24	12
February	28	18
March	33	31
April	40	46
May	47	59
June	53	68
July	56	74
August	55	71
September	49	61
October	42	49
November	32	33
December	27	18

Source: U.S. National Oceanic and Atmospheric Administration

35. The tables, histograms, and box plots below contain the times, to the nearest tenth of a minute, it took to run 2 miles by a class of 12 male and 12 female high school students. Write a paragraph comparing and contrasting the two distributions.

Statistics	Time (in minutes)	
	female	male
mean	13.8	13.6
median	13.3	13.2
interquartile range	5.0	5.7
standard deviation	3.4	3.6
range	10.8	11.6

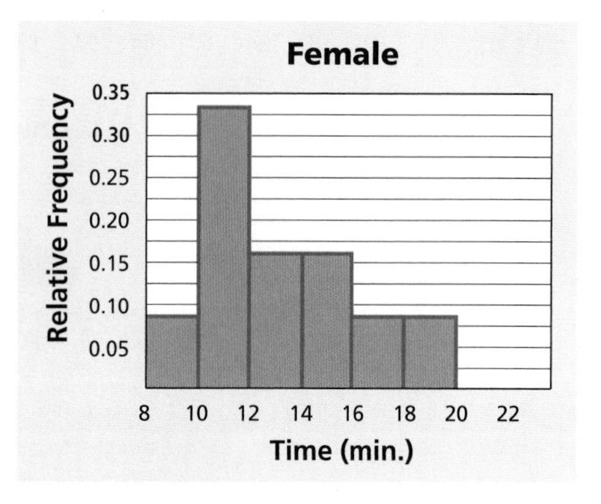

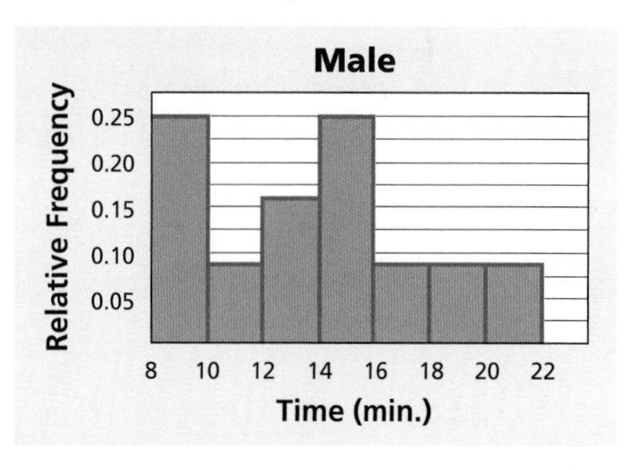

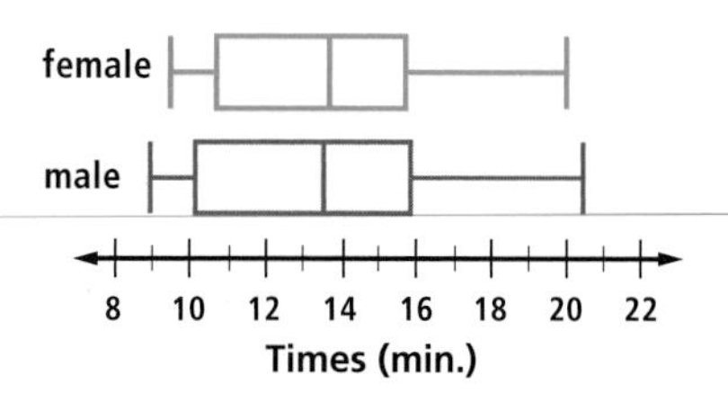

REPRESENTATIONS Pictures, graphs, or objects that illustrate concepts

OBJECTIVE H Read, interpret, and draw histograms and population pyramids from data. (Lesson 1-3)

36. Display the data on canvassing given in Question 27 in a histogram with bin width 50.

37. The data below represent the scores of a class of 20 students on a recent geometry test. Make a histogram with the data using a bin width of 5.

80	70	100	96	84	76	94	88	91	81
79	83	89	93	91	94	65	84	97	83

38. Use the displays below, which show population pyramids of the populations of the United States and India in the year 2007 in millions.

a. For the United States, estimate which age group had the greatest population. What is it?

b. For India, estimate which age group had the seventh greatest population. What is it?

c. Describe the difference between the populations of the U.S. and India in the age interval 0-9.

d. What factors might account for such different shapes of the age structure of these two cultures?

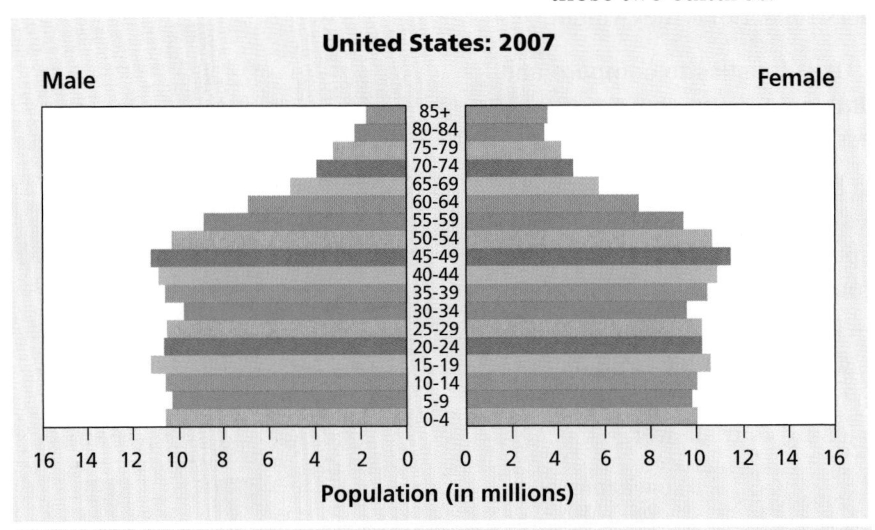

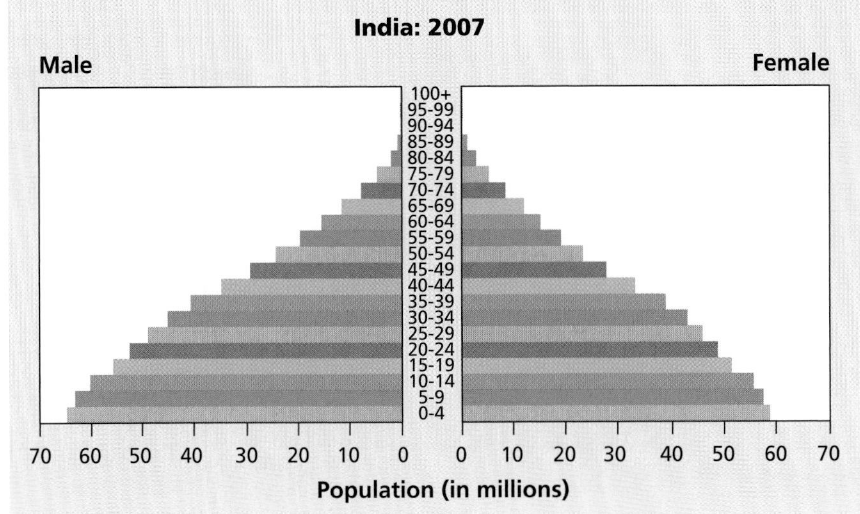

Source: U.S. Census Bureau, International Data Base

39. Ms. T. Chare made the following histogram of grades on her final exam in algebra.

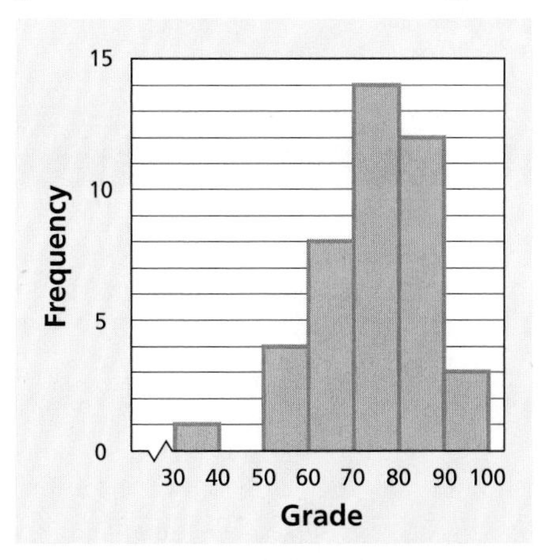

a. How many students took the exam?

b. True or False The median score is between 60 and 70 points.

c. What percent of her students scored 80 points or more?

OBJECTIVE I Read, interpret, and draw box plots from data. (Lesson 1-4)

40. Refer to this box plot of spending by a sample of baseball fans at a recent major league home game.

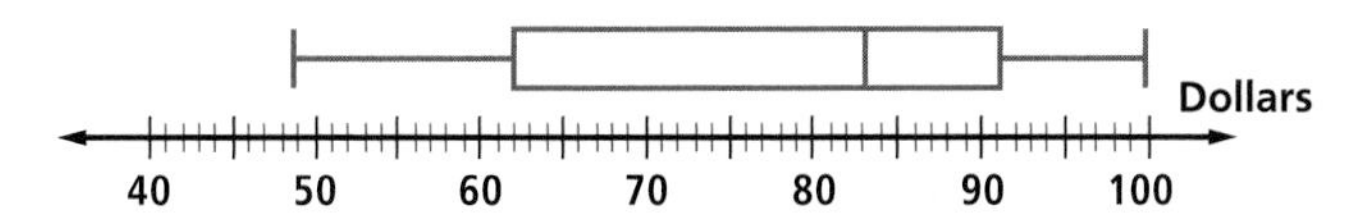

a. What is the median spending?

b. What is Q_3?

c. What percent of the fans spent between \$62 and \$91?

d. Describe the interval below Q_1.

e. Between which quartiles is there the greatest spread?

41. Use the data below of temperatures on the 25th day of the month in St. Louis, Missouri, to create a box plot.

Month	1	2	3	4	5	6	7	8	9	10	11	12
°F	10	36	41	68	60	78	72	68	78	45	31	29

Source: National Oceanic and Atmospheric Administration

42. Refer to the table below describing the estimated proven natural gas reserves in billions of cubic meters in 2007 for various countries. Draw a box plot of the data in the table, identifying any outliers with dots.

Country	Gas Reserves
Iran	28,080
Iraq	3,170
Former USSR	58,112
Algeria	4,504
Nigeria	5,215
Saudi Arabia	7,305
United Arab Emirates	6,072
United States	5,978
Venezuela	4,838

Source: Organization of Oil Exporting Countries

OBJECTIVE J Calculate and draw line graphs of cumulative frequencies and cumulative relative frequencies from tables of frequencies. (Lesson 1-5)

43. Glenda works on commission. For the first six months of this year she earned: \$3250, \$4125, \$2750, \$5375, \$4500, and \$3800.

a. Create a cumulative frequency table for Glenda's wages for the six months.

b. During which month did Glenda reach half her total sales?

c. Draw a cumulative relative frequency line graph for her total wages for the six months.

44. Below are the scores on a 10-point test in a class of 20 students.

3	5	3	4	7	5	6	8	10	5
6	8	7	1	6	7	8	7	8	7

a. Make a table showing the frequency and relative frequency of each score.

b. Draw a cumulative frequency line graph for the data.

Chapter

2 Functions and Models

Contents

In this chapter, you will study the modeling of data involving relations between two variables. For example, the data in the table at the left from the U.S. Department of the Treasury show the United States national debt at the end of selected fiscal years. These data, graphed in the scatterplot on the next page, show that the national debt has been increasing dramatically since 1965. The table shows that in 2005 (105 years after 1900), the national debt was almost 8 trillion dollars.

To describe patterns in data, *mathematical models* are used. A **mathematical model** is a mathematical description of a real situation, usually involving some simplification and assumptions concerning the situation. Mathematical models are often used to make estimates and predictions.

U.S. National Debt

Year	Amount (billions of dollars)
1965	317
1970	371
1975	533
1980	908
1985	1,823
1990	3,233
1995	4,974
2000	5,674
2005	7,933

Source: U.S. Department of the Treasury

If x is the number of years after 1900 and y is the debt in billions of dollars, then the relationship between x and y from 1965 to 2005 can be modeled by the exponential function with equation $y = 0.814(1.093)^x$, which is graphed on the scatterplot below.

Using this model, you might estimate the national debt in 2002 or predict its value in 2015, even though this information was not given.

In this chapter, you will study linear, quadratic, and exponential models, and see how they are used in business, government, and recreational pursuits.

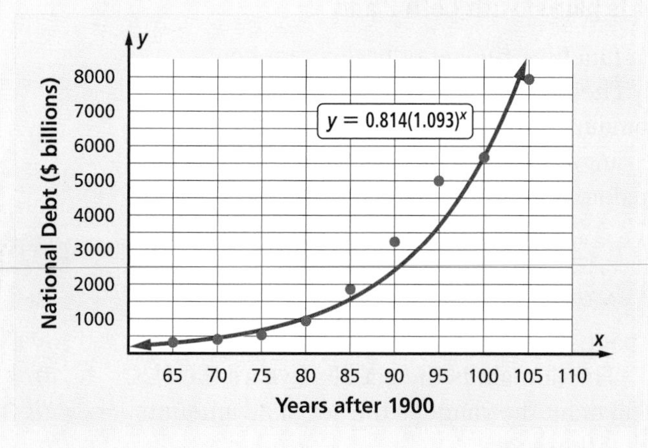

Lesson 2-1 The Language of Functions

Vocabulary

mathematical model
relation
independent variable
dependent variable
function, ordered pair definition
domain of a function
range of a function
function, correspondence definition
real function
member of a set, element of a set, $\in$
piecewise definition of a function
value of a function

▶ **BIG IDEA** A function is a special type of relation that can be described by ordered pairs, graphs, written rules or algebraic rules such as equations.

On pages 78 and 79, nine ordered pairs of numbers are listed and graphed. The first coordinate is the number of years after 1900; the second is the U.S. national debt (in billions of dollars) for that year. Any set of ordered pairs is a **relation**. In many contexts, the second number in each ordered pair depends in some way on the first number. When this happens, the first variable in a relation is called the **independent variable** and the second variable is called the **dependent variable**. For the national debt, the number x of years after 1900 is the independent variable and the debt y in that year is the dependent variable.

Mental Math

Write each expression as a power of x.

a. $(x^{50})^3$

b. $x^{50} \cdot x^{-22}$

c. $\frac{x^{50}}{x^{53}}$

What Is a Function?

When each value of the independent variable determines exactly one value of the dependent variable, the relation is called a *function*.

Ordered Pair Definition of Function

A **function** is a set of ordered pairs (x, y) in which each first component (x) is paired with exactly one second component (y).

For example, $f = \{(1, 2), (2, 4), (3, 7)\}$ is a function, but

$g = \{(1, 2), (2, 4), (1, 7)\}$ is not a function because 1 is paired with both 2 and 7.

In the set of ordered pairs of a function, the set of first components is the **domain** of the function. The set of second components is the **range** of the function. The domain consists of all allowable values of the independent variable; the range is the set of possible values for the dependent variable. The domain of the function f above is $\{1, 2, 3\}$, and the range is $\{2, 4, 7\}$.

QY1

▶ **QY1**

Give the domain and range of the function $h = \{(5, 1), (6, 3), (7, 1), (8, 3)\}$.

For the national debt data on page 78, we can say, "The U.S. national debt is a function of the year." The domain is the set of all years the U.S. has had and will have a national debt; the range is the set of all amounts of the national debt at the end of those years.

Another definition of *function* stresses the correspondence between the independent and dependent variables.

Correspondence Definition of Function

A **function** is a correspondence between two sets A and B in which each element of A corresponds to exactly one element of B.

The domain is the set A of values of the *independent variable.* The range is the set of only those elements of B that correspond to elements in A; these are the values of the *dependent variable.*

For most functions studied in this course, A and B are sets of real numbers. Functions whose domain and range are sets of real numbers are called **real functions.** Unless the domain of a function is explicitly stated, you may assume that it is the set of all real numbers for which the function is defined. Real functions have the useful characteristic that they can be pictured by a coordinate graph.

We use the symbols in the table at the right to represent sets of numbers. "z" stands for the German verb *zahlen,* "to count," and "q" stands for "quotient." These symbols can also be used in descriptions of other sets. If x is in a set A, then x is said to be a **member** or **element** of A, written $x \in A$. For instance, $3\frac{1}{2} \in \mathbb{Q}$. Similarly, every even integer can be written as $2 \cdot n$, where n is in the set of all integers. So we can write the set of even integers as $\{2n \mid n \in \mathbb{Z}\}$, read "the set of $2n$ such that n is an integer."

The symbol	represents the set of all
$\mathbb{Z}$	integers.
$\mathbb{R}$	real numbers.
$\mathbb{Q}$	rational numbers.
$\mathbb{N}$	natural numbers.

Example 1

A bakery charges \$2.00 per muffin. Customers get a \$2.00 discount for every 6 muffins purchased.

a Which statement is true: "the cost c is a function of the number m of muffins" or "the number m of muffins is a function of the cost c?"

b. Identify the independent and dependent variables of the function.

c. State the domain and range of the function.

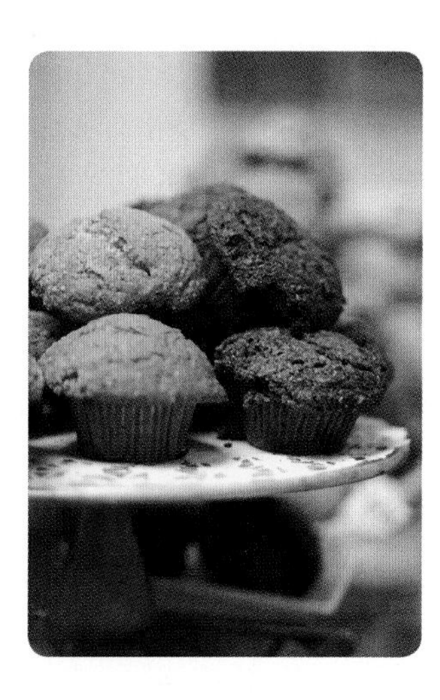

Solution

a. Because there is exactly one cost c for a given number of muffins, the cost is a function of the number of muffins. A customer who buys 5 muffins pays the same amount as one who buys 6 muffins, so the number of muffins is not a function of the cost.

b. Because c depends on m, m is the independent variable and c is the dependent variable.

c. The domain is the set of all possible values of m. Because "negative muffins" does not make sense and you cannot buy part of a muffin, the domain is the set of nonnegative integers, which can be written $\{m \mid m \in \mathbb{Z} \text{ and } m \geq 0\}$. The range is the set of all possible values of c. Any even-number cost in dollars is possible, so the range is the set of nonnegative even integers.

Descriptions of Functions

Functions can be described in many ways. Some frequently-used descriptions are (1) tables or lists of ordered pairs, (2) rules expressed in words or equations, and (3) coordinate graphs. You should know how to recognize functions described in each of these ways, and how to convert from one description to another.

GUIDED

Example 2

Let n be an integer with $n \geq 2$. Graph the function that shows how many elements of an ordered set of n different integers are greater than the median of that set.

Solution Copy and complete a table similar to the one below with possible sets for the values of n from 2 to 10. Circle and count the elements greater than the median. Call this count C. Make a scatterplot of the points (n, C) for n from 2 to 10.

n	Sample set (ordered)	Count C of numbers greater than the median
2	{2, (5)}	1
3	{24, _?_, (68)}	1
4	{13, _?_, (35, _?_)}	_?_
5	{_?_, _?_, _?_, (_?_, _?_)}	_?_
⋮	⋮	⋮

Each ordered pair of this function is of the form (n, C) where n is the number of elements in the set and C is the number of elements greater than the median. So for integer values of n from 2 to 10, the ordered pairs of the function were found in the table: (2, 1), (3, 1), (4, _?_), (5, _?_), (6, _?_), (7, _?_), (8, _?_), (9, _?_), and (10, _?_). The ordered pairs are graphed at the right. Notice that we do not connect the dots, because the values of n are *discrete*. Also notice that pairs of n-values share the same count or frequency, C.

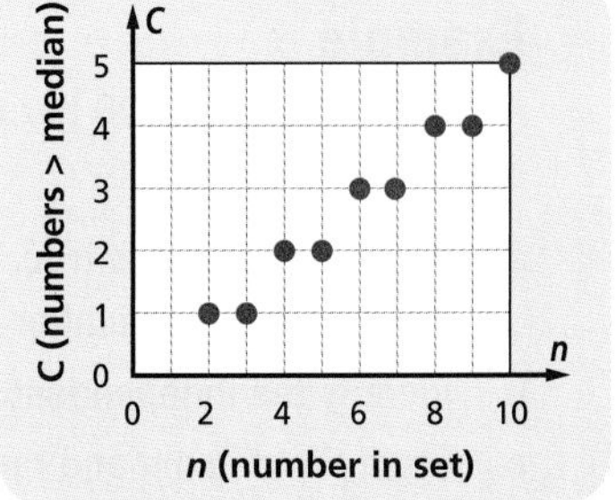

STOP QY2

In Example 2, the function is described in a table and a graph. Another description of the function in Example 2 combines two rules, one when n is even, and the other when n is odd.

$$C = \begin{cases} \frac{n}{2}, & \text{when } n \text{ is even} \\ \frac{n-1}{2}, & \text{when } n \text{ is odd} \end{cases}$$

For instance, when $n = 13$, n is odd, so $C = \frac{13-1}{2} = 6$. This type of description of a function is called a **piecewise definition** because it breaks the domain into pieces, and there is a rule for each piece.

QY2

In Example 2, for what value(s) of n does $C = 7$?

Often the domain and range of a function can be determined solely from a graph or an equation.

Example 3

A rule for the function graphed at the right is $y = 2^x - 4$. Find the domain and range of the function.

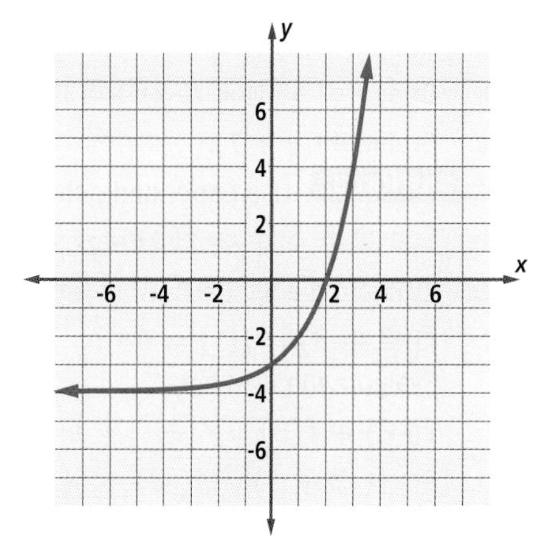

Solution The domain is the set of x-values for which $2^x - 4$ is defined, which is the set $\mathbb{R}$ of all real numbers. From the graph, the range appears to be the set of all real numbers greater than -4, which can be written as $\{y \mid y > -4\}$.

In a function, there is only one member of the range paired with each member of the domain. So, in a graph using rectangular coordinates, if y is a function of x, no vertical line will intersect the graph at more than one point. This is often referred to as the *vertical line test* for determining whether a relation is a function. You can see how this works on the two graphs of relations shown below. Only the relation graphed at the left is a function.

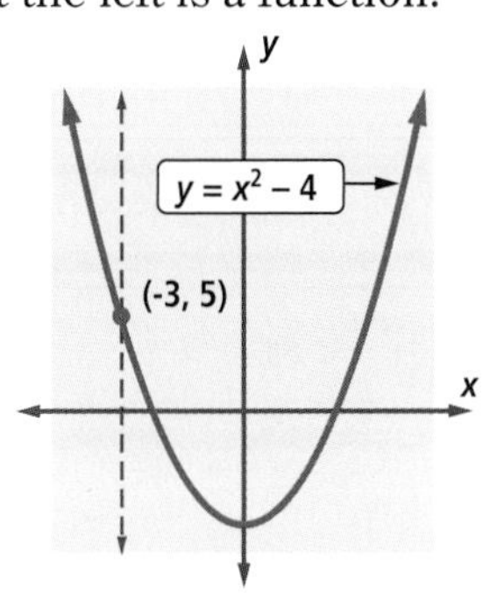

y is a function of x: no vertical line intersects the graph more than once.

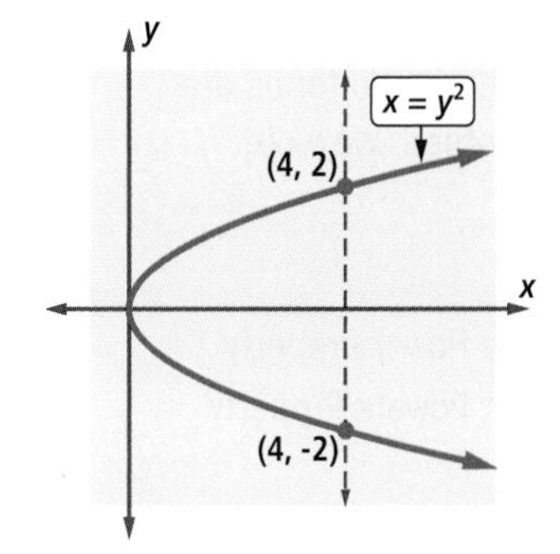

y is not a function of x: there is at least one vertical line that intersects the graph more than once.

Naming Functions and Their Values

A function is usually named by a single letter such as f or g. For a function f, the symbol $f(x)$, which is read "f of x," indicates the value of the dependent variable when the independent variable is x. $f(x)$ is also called the **value** of the function at x. This symbol was first used by the mathematician Leonhard Euler (pronounced "oiler") in the 18th century. Euler's notation is particularly useful when evaluating functions at specific values of the independent variable. This notation is also used when defining functions on a computer algebra system (CAS).

Example 4

Suppose f is the function defined by the rule $f(x) = 4 \cdot \left(\frac{1}{2}\right)^x$ for all real numbers x.

a. **Evaluate $f(5)$.**

b. **Does $f(-2 + 3) = f(-2) + f(3)$?**

c. **Evaluate $f(q + 1)$.**

Solution 1 Use paper and pencil.

a. Substitute 5 for x: $f(5) = 4 \cdot \left(\frac{1}{2}\right)^5 = 4 \cdot \frac{1}{32} = \frac{1}{8}$.

b. Evaluate the left side. Work witin parentheses first.

$f(-2 + 3) = f(1) = 4 \cdot \left(\frac{1}{2}\right)^1 = 2$

Evaluate the right side.

$f(-2) + f(3) = 4\left(\frac{1}{2}\right)^{-2} + 4\left(\frac{1}{2}\right)^3 = 16 + \frac{1}{2} = 16\frac{1}{2}$

So, $f(-2 + 3) \neq f(-2) + f(3)$.

c. Substitute $q + 1$ for x in the rule.

$f(q + 1) = 4 \cdot \left(\frac{1}{2}\right)^{q+1}$

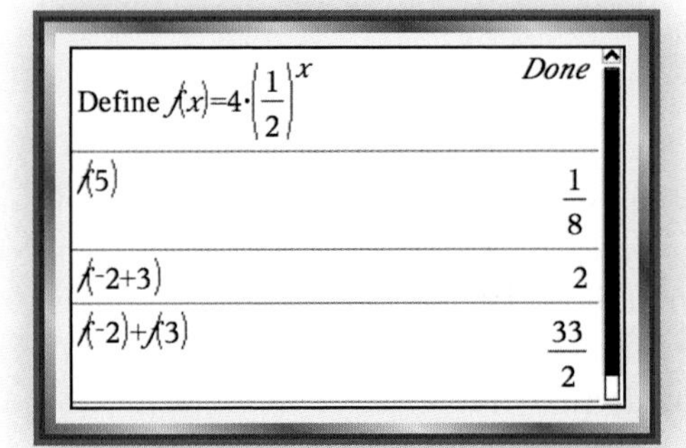

Solution 2 Define the function f on a CAS.

a. Input $f(5)$.

b. Input $f(-2 + 3)$. Input $f(-2) + f(3)$. The screenshot verifies the calculations in Solution 1.

c. The screenshot at the right indicates that $f(q + 1) = 2 \cdot 2^{-q}$. Yet our answer in Solution 1 was $f(q + 1) = 4 \cdot \left(\frac{1}{2}\right)^{q+1}$. To show that these two forms are equivalent, convert one answer to the other. We write 4 and $\frac{1}{2}$ as powers of 2.

$4 \cdot \left(\frac{1}{2}\right)^{q+1} = 2^2 \cdot (2^{-1})^{q+1} x^{-n} = \frac{1}{x^n}$

$= 2^2 \cdot 2^{-q-1}$ Power of a Power Property

$= 2^1 \cdot 2^{-q}$ Product of Powers Property

The result can be verified using a CAS.

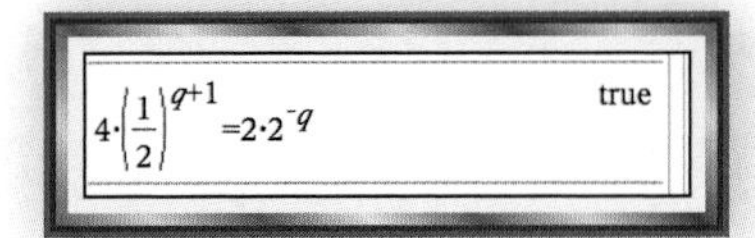

Part b of Example 4 illustrates that, in general, $f(a + b) \neq f(a) + f(b)$. That is, there is no general distributive property for functions over addition.

Questions

COVERING THE IDEAS

In 1–3, identify the independent variable and the dependent variable.

1. A parent bases a child's allowance on the number of chores completed by the child.
2. The participation grade in a class is calculated, in part, by the student's attendance.
3. Trees grow from sunlight and water.

In 4 and 5, give a definition of the term.

4. domain

5. range

6. An online photo lab usually charges \$0.25 per print to make a color print from a digital file. During a special promotion, customers receive a \$0.50 discount if 12 or more prints are made.
   a. Which is true, "the cost c is a function of the number n of prints made" or "the number n of prints made is a function of the cost c?"
   b. What is the cost of making 20 color prints from digital files?
   c. List all the ordered pairs (n, c) for $0 \le n \le 16$.
   d. Graph the relation in Part c for $0 \le n \le 16$.
   e. Write a piecewise formula for c in terms of n.

7. Consider the function defined by $y = f(x)$. What symbol represents each of the following?
   a. the function
   b. the dependent variable
   c. the value of the function
   d. the independent variable

8. Let g be the function defined by $g(t) = t^2 - 5$.
   a. Compute $g(-7)$.
   b. Find the value(s) of t such that $g(t) = 12$.
   c. Find the domain of g.
   d. Find the range of g.
   e. Evaluate $g(p + 3)$.

9. Consider $h(x) = \sqrt{x + 3}$.
   a. Evaluate $h(q - 1)$.
   b. For what value of q does $h(q - 1) = 3$?

In 10–12, a relation is graphed on a rectangular coordinate grid. Tell whether the relation is a function.

10.

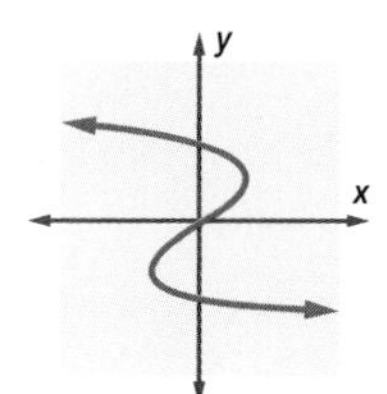

11.

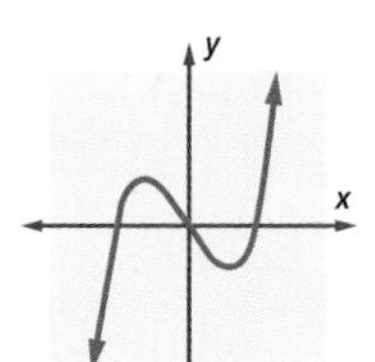

12.

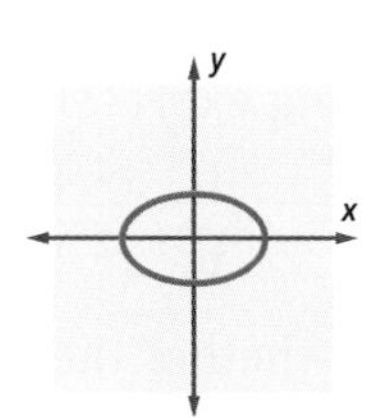

APPLYING THE MATHEMATICS

13. Refer to the graph of $y = f(x)$ at the right.
    a. Determine the domain and range of f.
    b. Find $f(2)$.
    c. When does $f(x) = 1.2$?

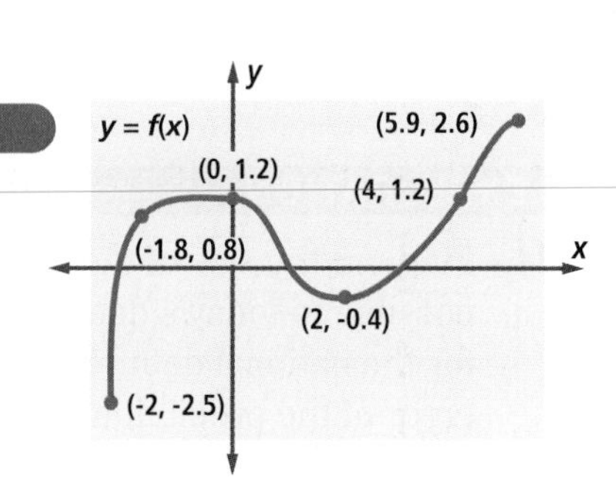

14. Refer to the national debt model, $y = 0.814(1.093)^x$, on page 79.
   a. According to the model, what was the national debt in 1995? How close is this to the actual value?
   b. According to the graph of the model, in what year was the national debt first one trillion dollars?

In 15–17, consider the relation defined by each sentence. Sketch a graph. Tell if the sentence defines a function. If so, give its domain and range.

15. $y = 3 \cdot 2^x$ 16. $y = -\sqrt{x}$ 17. $y < x + 1$

18. At the right is the graph of a function whose equation is $f(t) = 3 + 2 \cdot (3)^{-t}$.
   a. Find the domain and range of the function.
   b. Use the graph to approximate $f\left(\frac{1}{2}\right)$.
   c. Use the graph to estimate the value of t such that $f(t) = 3.5$.

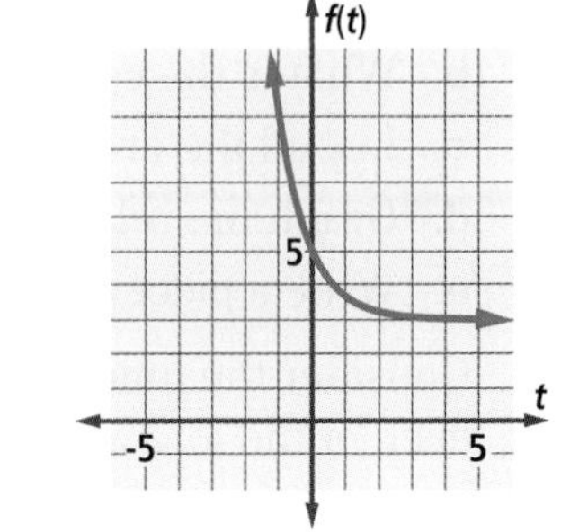

REVIEW

In 19 and 20, suppose that six used cars of a particular make and model are advertised in the newspaper for the following prices: $14,950; $15,250; $14,500; $14,700; $14,250; $14,900. (Lessons 1-6, 1-2)

19. Let c_i = the cost of the ith used car advertised.
   a. Evaluate $\frac{\sum_{i=1}^{6} c_i}{6}$.
   b. Which statistical measure is represented by the quantity in Part a?

20. a. Without calculating, tell why the standard deviation of this data set will be less than 500.
   b. Calculate the standard deviation to verify your answer to Part a.

21. ***Skill Sequence*** Find the missing expression. **(Previous Course)**
   a. $x^2 - 12x + \underline{\ ?\ } = (x - 6)^2$ b. $x^2 + \underline{\ ?\ } + 25 = (x + 5)^2$
   c. $x^2 + 22x + 121 = (\underline{\ ?\ })^2$ d. $x^2 + 2ax + a^2 = (\underline{\ ?\ })^2$

22. Without graphing, determine whether the point (3, -4) is on the line with equation $y = 3x - 5$. **(Previous Course)**

23. Write an equation for the line with slope $-\frac{3}{2}$ that passes through (-6, 4). **(Previous Course)**

EXPLORATION

24. Find out the size of the U.S. national debt for a date as close as possible to today's date. What does the equation on page 79 predict for the national debt on the date you have? Calculate the percent error of the prediction.

QY ANSWERS

1. domain {5, 6, 7, 8}; range {1, 3}.
2. $n = 14$ and $n = 15$

Lesson 2-2 Linear Models

Vocabulary

linear function
linear model
interpolation
extrapolation
observed values
predicted values
residual
sum of squared residuals

▶ **BIG IDEA** The sum of squared deviations is a statistic for determining which of two lines fits the data better.

A **linear function** is a set of ordered pairs (x, y) satisfying an equation of the form $y = mx + b$, where m and b are constants. Recall that the graph of every such function is a line with slope m and y-intercept b.

Mental Math

What is the sum of a set 25 elements whose mean is 300?

Fitting a Line to Data

When data in a scatterplot lie near a line, we can create a **linear model** for the data, that is, a model of one variable as a linear function of the other. Even if the linear function does not contain all of the data points, it may still be useful in describing the overall trend of the data or in predicting values of either the dependent or independent variable.

In this lesson and the next, you will learn several techniques for constructing linear models. One technique is to fit a line to data "by eye," that is, to draw a line that seems close to the data.

GUIDED

Example 1

Jewelers emphasize that the price of a diamond is determined by cut, carat weight, color and clarity. The table at the right gives carat weights and approximate prices in U.S. dollars for twenty diamond rings sold at a recent auction in Singapore. All rings are of the same quality gold and contain a single diamond. The data are graphed below.

Prices of Diamond Rings Sold in Singapore

y
1800
1600
1400
1200
1000
800
600
400
200
0
Price (U.S. $)
x
0
0.1
0.2
0.3
0.4
Weight (carats of diamonds)

Weight x	Price y (U.S. $)
0.18	702.00
0.17	517.50
0.25	963.00
0.29	1290.00
0.27	1080.00
0.15	484.50
0.20	747.00
0.25	1017.00
0.21	724.50
0.17	529.50
0.35	1629.00
0.33	1417.50
0.26	994.50
0.16	513.00
0.12	334.50
0.18	664.50
0.15	430.50
0.16	507.00
0.16	498.00
0.23	829.50

Source: Journal of Statistics Education

The linear model is based on the weight of the diamond used. Although the data are not collinear, the line through (0.18, 600) and (0.32, 1400) seems close to the points. It has been added to the graph. An equation of this graphed line is one linear model for these data. Is the size alone a good predictor of price?

a. Find an equation of the graphed line which relates weight and price.

b. Interpret the slope of the line in the context of the problem.

c. Use the model to estimate the price of a 0.3-carat diamond ring.

d. Why is the model not good for predicting the cost of a 0.05-carat diamond ring?

e. Why is the set of data not a function?

Hard Rock
The largest rough diamond ever found weighed 3106 carats.

Solution

a. We use the points (0.18, 600) and (0.32, 1400) to find an equation of the line:

$$m = \frac{y_2 - y_1}{x_2 - x_1} = \underline{\ ?\ } \approx \underline{\ ?\ }$$

Substitute $m = 5714.29$ and (0.18, 600) into the point-slope form of the equation for a line.

$$\begin{aligned} y - y_1 &= m(x - x_1) \\ y - \underline{\ ?\ } &= 5714.29(x - \underline{\ ?\ }) \\ y &= \underline{\ ?\ } \end{aligned}$$

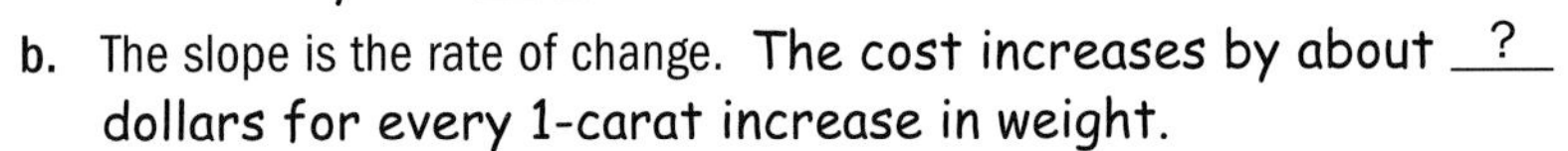

b. The slope is the rate of change. The cost increases by about __?__ dollars for every 1-carat increase in weight.

c. To predict the cost from the weight, substitute $x = 0.3$ and solve for y. When $x = 0.3$, $y =$ __?__. According to the model, the cost of a 0.3-carat diamond ring will be about __?__.

d. Substitute 0.05 for x and solve for y. The model predicts that a 0.05-carat diamond ring will cost __?__. This means that the seller would pay you to take the ring! This is not plausible, so the model is not a good predictor.

e. There is at least one x-value that does not have a unique y-value. There are two diamonds at 0.17 carats that have different prices, so the data set is not a function.

In this example, known diamond weights range from 0.12 to 0.35 carats. If you use the model to predict the price of a 0.3-carat diamond, you are making a prediction *between* known values. Prediction between known values is called **interpolation.** If you calculate the price of a diamond weighing 0.05 carats, you are making a prediction *beyond* known values. Prediction beyond known values is called **extrapolation**. Extrapolation is usually more hazardous than interpolation, because it depends on an assumption that a relationship will continue past the known data. In this case, a diamond much smaller than those in the sample might be inexpensive, but it will not be free.

Measuring How Well a Line Models Data

Diamond Ring Prices by Weight of Diamond	
Weight	Price (U.S. dollars)
0.15	484.50
0.16	507.00
0.18	702.00
0.25	963.00
0.27	1080.00
0.33	1417.50
0.23	829.50

In Chapter 1, you studied the sample variance, which was computed as the sum of the squared deviations from the mean divided by $n - 1$. A similar statistic, the *residual*, tells how far away data are from your chosen model. At the right is a table showing a smaller sample of the diamond ring data on page 87. These data, collected from sources such as experiments or surveys, are called **observed values**. Below, the scatterplot of these data is shown together with the graph of the line with equation $y = 2400x + 400$. This equation seems to be a pretty good model for the data. However, is it the best model? To compare models, we calculate *residuals*. The values predicted by a model are called **predicted values**. The observed value minus the predicted value is the **residual**. A residual is positive when the observed value is higher than what is predicted by the model. A residual is negative when the observed value is lower than what the model predicts.

For instance, a 0.16-carat diamond ring sold for \$507.00. This is lower than the price predicted by the model.

$$\begin{aligned}\text{predicted price} &= 2400 \cdot (0.16) + 400 \\ &= \$784\end{aligned}$$

The residual, or error, is

$$\begin{aligned}\text{residual} &= \text{observed value} - \text{predicted value} \\ &= 507 - 784 \\ &= -277.\end{aligned}$$

The actual cost was \$277 less than predicted. On the graph, we see that the observed value is 277 units below the linear model $y = 2400x + 400$. That is why the residual is negative. Every data point has a residual, so n data points provide n residuals. The absolute value of a residual is the length of the vertical segment from the data point to the corresponding point on the linear model.

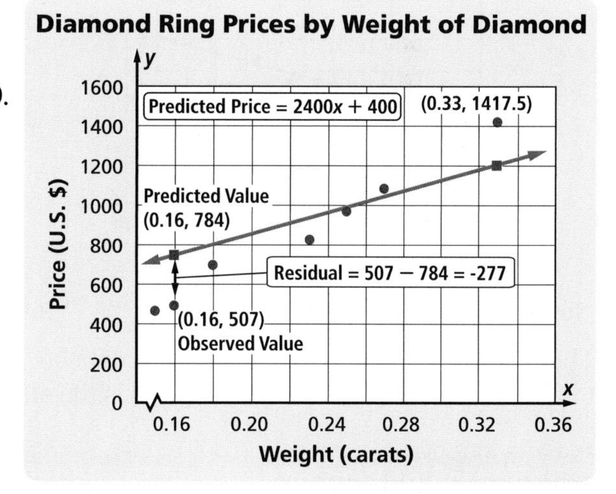

QY

QY

What is the residual for the 0.33-carat diamond?

On the graphs below, this line and another linear model are drawn to fit the seven points. Recall that when you compute variance, you add the squared deviations from the mean. Similarly, when modeling data with a line, we can measure the variation from the line by adding the squares of the residuals. You can use spreadsheets to calculate residuals and the sum of squared residuals, as shown below.

	A weight	B price	C predicted	D residuals
◆			=2400*a[]+400	=b[]-c[]
1	0.15	484.5	760.	-275.5
2	0.16	507	784.	-277.
3	0.18	702	832.	-130.
4	0.25	963	1000.	-37.
5	0.27	1080	1048.	32.

D residuals:=b[]-c[]

	C predicted	D residuals	E sqresid	F sumsq...
◆	=2400*a[]+4	=b[]-c[]	=d[]^2	
1	760.	-275.5	75900.3	237779.
2	784.	-277.	76729.	
3	832.	-130.	16900.	
4	1000.	-37.	1369.	
5	1048.	32.	1024.	

F1 =sum(e[])

You can also think of (residual)^2 as the area of a geometric square whose side has length equal to the absolute value of the residual. These geometric squares are drawn on the graphs below.

Linear Model 1

Squares are shown for a line that does not go through any data points.

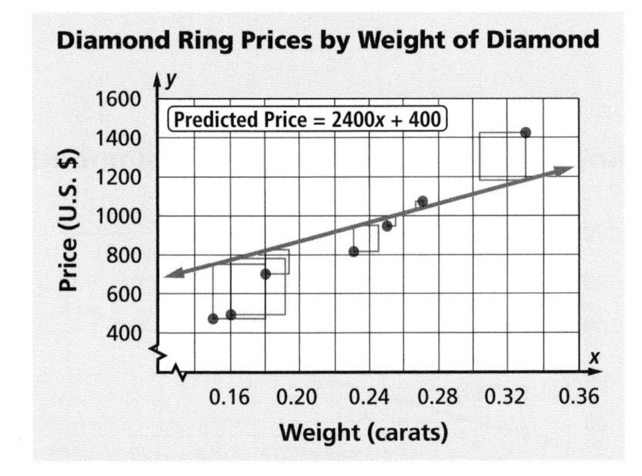

Total area of the squares ≈ 237,800

Linear Model 2

Squares are shown for a line through two of the data points.

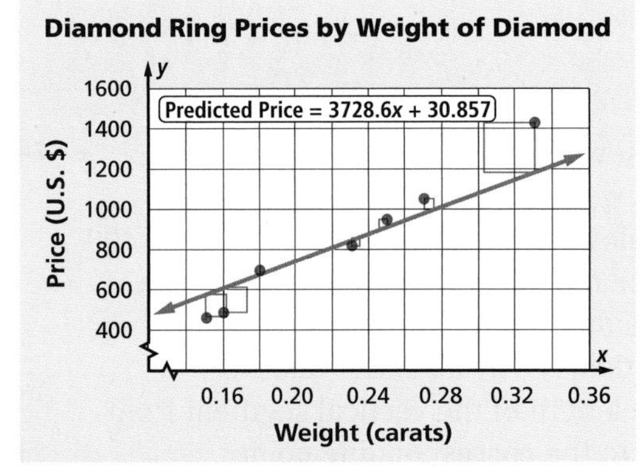

Total area of the squares ≈ 59,870

The second line is a better model of the data because it has a smaller total area of the squares. The total area is the **sum of squared residuals.**

Definition of Sum of Squared Residuals

Sum of squared residuals $= \sum_{i=1}^{n} (\text{observed } y_i - \text{predicted } y_i)^2$

The sum of squared residuals is a statistic that measures lack of fit. If you compare two lines, the one with the larger sum of squared residuals is not as good a model as the one with the smaller sum of squared residuals. If you have many possible lines you could use to model data, compute the sum of squared residuals for each model. The line that gives the smallest value provides the best fit to the data.

Activity

This table shows the number of televisions per 100 people in 1997 and the number of unemployed per 100 people in 2008 for nine countries from around the world. Some people suggest that increased TV viewing leads to a less productive workforce. Are these statistics related?

Country	TVs per 100	Unemployed per 100
Argentina	22.3	7.8
Bulgaria	40.0	6.3
India	6.5	6.8
Israel	29.9	6.1
Netherlands	51.8	4.5
New Zealand	52.3	4.0
Poland	33.7	9.7
South Africa	12.3	21.7
South Korea	34.7	3.2

Source: UNESCO Institute for Statistics, CIA World Fact Book

Step 1 Use a statistics utility to create a scatterplot with number of TVs on the x-axis and number of unemployed people on the y-axis.

Step 2 Add a movable line to the scatterplot.

Step 3 Move the line as close to $y = -0.3x + 17$ as you can. Your screen should look similar to the one at the right. Record the sum of squared residuals.

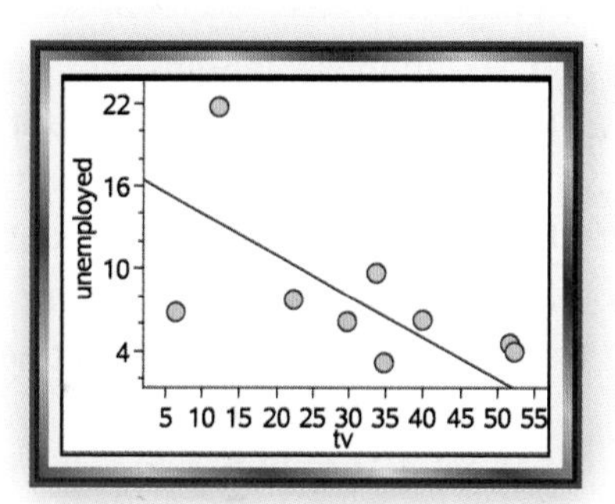

Step 4 Move the line so that it goes through the points for Bulgaria and the Netherlands. Record the equation. Record the sum of squared residuals.

Step 5 Tell which line is the better model of the data. Explain your choice.

Step 6 Move the line until you get the smallest sum of squares that you can. Record the equation and the sum of squared residuals.

Questions

COVERING THE IDEAS

1. Define *linear function*.
2. Refer to the diamond price data in Example 1.
   a. Find an equation for the line that passes through the data points (0.35, 1629) and (0.12, 334.50).
   b. Write a sentence that states how cost increases for every carat increase in weight according to your equation.
   c. Use your line to predict the price of a 0.17-carat diamond.
   d. Is this interpolation or extrapolation?
3. **Fill in the Blanks** For a data set, residuals are the differences between __?__ values and __?__ values.

4. When a residual is positive, is the observed value higher or lower than the predicted value?

In 5 and 6, a diamond speculator used the line with equation $y = 5000x - 250$ to estimate the price of diamond rings.

5. a. What would the speculator predict for the price of the 0.29-carat diamond ring?
   b. According to the data in Example 1, what is the residual for the 0.29-carat diamond ring?
6. a. What would the speculator predict as a price for the 0.21-carat diamond ring?
   b. What is the residual for this diamond ring?
7. Does the phrase "sum of squared residuals" mean "first sum the residuals and then square the sum" or "first square each residual and then sum the squares?"
8. Consider the linear model $f(x) = 5.2x - 3$.
   a. Given a residual of -0.2 at $x = 1$, find the observed value.
   b. Given a residual of 2.1 at $x = 4$, find the observed value.
   c. Given a residual of 0 at $x = 3$, find the observed value.

APPLYING THE MATHEMATICS

In 9 and 10, suppose Jane asked members of her family how many states they had visited. Her data are in the table at the right.

Family Member	Age	States Visited
Jane	16	15
Cousin Xia	16	18
Dad	40	27
Grandma	70	42
Brother Ed	10	9
Uncle Ralph	45	19

9. Jane plotted the data and drew a line through the points for Ed and Grandma. She found its equation to be $y = 0.55x + 3.5$.
   a. What is the slope of this line and what does it represent?
   b. Complete the spreadsheet below to calculate the residuals for this model.
   c. What is the sum of squared residuals for Jane's model?
   d. Jane used her equation to predict how many states she will have visited at age 30. Is this interpolation or extrapolation?
   e. What is Jane's prediction from Part d?

◇	A	B	C	D
1	Age	States Visited	Predicted	Residual^2
2	16	15	12.3	7.29
3	16	18		
4	40	27	25.5	2.25
5	70	42	42.0	0
6	10	9		
7	45	19	28.25	85.56

10. Xia thought that drawing the line through the points for Ed and Dad would look better because then there would be two points above the line and two below the line. Her line has equation $y = 0.6x + 3$.

a. What is the residual for Grandma using Xia's model?
b. What is the squared residual for Dad using Xia's model?
c. Calculate the sum of squared residuals for Xia's model.
d. Xia uses the equation to predict how many states Grandma will have visited when she reaches 100. Is this interpolation or extrapolation?
e. What is Xia's prediction from Part d?

11. The table at the right gives expected life spans and gestation periods for selected animals. *Gestation* for mammals is development time in the mother's uterus until birth. For humans, the mean time is 266 days.

Animal	Expected Life Span (years)	Mean Gestation (days)
Black Bear	20	215
Camel	40	390
Elephant	55	660
Gorilla	50	251
Gray Squirrel	9	42
Hippopotamus	45	240
Lion	12	100
Moose	27	240
Wolf	10	63

Source: BBC

a. Make a scatterplot with Life Span as the independent variable.
b. Draw the line through the points for Lion and Elephant. Do you think this line fits the data well?
c. Find an equation for the line in Part b in slope-intercept form.
d. Interpret the slope of the line in Part b.
e. What is the residual for the Moose under this model?
f. Use the line in Part b to predict the gestation period for the Blue Whale, which has an estimated life span of 80 years. (The actual gestation period is estimated at 11-12 months.) Is this interpolation or extrapolation?
g. What is the sum of squared residuals for the line?

REVIEW

12. Consider the function {(–2, 8), (–1, 8), (0, 8), (3, 8), (8, 8)}. **(Lesson 2-1)**
a. State its domain.
b. Give its range.

In 13 and 14, a sample of five math test scores is given. (Lesson 1-6)
a. Find the standard deviation of the data set.
b. Explain a conclusion you can draw about the spread of the data set by just looking at it.

13. 63, 77, 81, 83, 92

14. 94, 95, 95, 96, 100

EXPLORATION

15. Use statistical software to draw a line that has the smallest sum of squared residuals for the data in the table in Question 11.
a. What is the equation?
b. What is the sum of squared residuals?

QY ANSWER

$225.50

Lesson 2-3 Linear Regression and Correlation

Vocabulary

method of least squares
line of best fit, least squares line, regression line
center of mass
correlation coefficient
perfect correlation
strong correlation
weak correlation

BIG IDEA The regression line is the line of best fit to data. The correlation coefficient measures the strength and direction of a linear pattern in data.

The sum of squared residuals can be used to determine which of two lines is the better fit to a specific set of data. When data are fairly linear, like the diamond weights and prices in Lesson 2-2, you are likely to find a good model quickly. When there is a large spread of data points on a scatterplot, you might have to try several equations. Even if you pick the equation with the smallest sum of squared residuals, there is likely another line with an even smaller sum of squared residuals that you did not try. How can you find the best model?

Mental Math

Suppose you thought the price of a new car was going to be \$19,000. Instead, it was \$20,000. By what percent of the actual price were you off?

The Line of Best Fit

The problem of finding the best linear model for a set of data emerged from studies of astronomy, geography, and navigation in the late 1700s and early 1800s. A method for finding the line of best fit was first published by the French mathematician Adrien Legendre in 1805. His approach to the problem is called the **method of least squares**, because he used the Sum of Squared Residuals to determine the linear equation of best fit.

A sextant is an instrument used to determine latitude and longitude, using angle measures.

The **line of best fit**, also known as the **least squares line** or **regression line**, has three important properties:

1. It is the line that minimizes the sum of squared residuals, and it is unique. There is only one line of best fit for a set of data.
2. It contains the **center of mass** of the data, that is, the point $(\bar{x}, \bar{y})$ whose coordinates are the mean of the x-values and the mean of the y-values.
3. Its slope and intercept can be computed directly from the coordinates of the given data points.

Although the formula for the slope of the least squares line uses only addition, subtraction, squaring, and division, it is too complex for computation by hand. Every statistics utility contains a *regression* routine that will take the coordinates of a set of data points and compute the slope and y-intercept of the line of best fit.

Example

a. Use a statistics utility to find a line of best fit for the data about the weight in carats and the price of diamond rings from Lesson 2-2.

b. According to the regression line, how much will a 0.5-carat diamond ring cost?

c. Verify that the center of mass (0.212, $793.65) is on the line.

d. Find the sum of squared residuals for the linear regression.

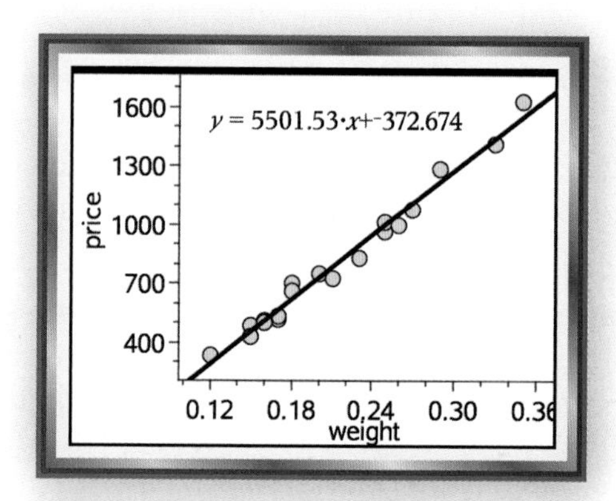

Solution

a. Enter the entire data set from Lesson 2-2 into a statistics utility. Label the weight as x and the price as y. Use the linear regression feature to find an equation for y as a function of x. The linear regression model is $y = 5501.53x - 372.67$.

b. Substitute 0.5 for x and calculate y.

$y = 5501.53(0.5) - 372.67 \approx 2378.1$

According to the linear regression model, the price would be $2378.10.

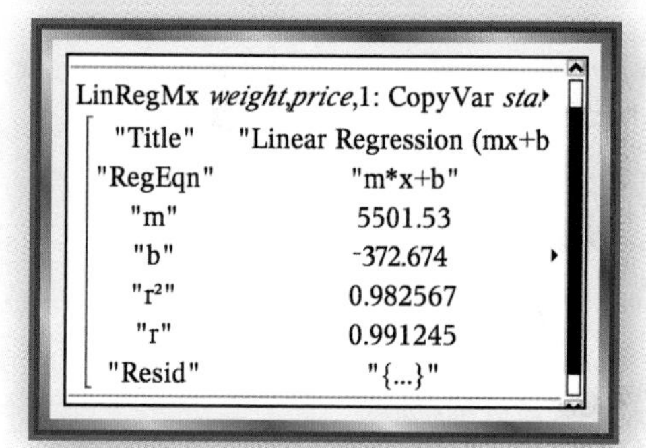

c. Substitute the coordinates of the center of mass for x and y in the regression equation.

$y = 5501.53x - 372.67$

Does $793.65 = 5501.53(0.212) - 372.67$?

Yes. It checks.

d. On many statistics utilities, once the regression equation is given, the residuals are calculated and stored until the next regression is found. In the screenshot at the right, the residuals are stored under the name Resid. So the sum of the squared residuals is 41842.9.

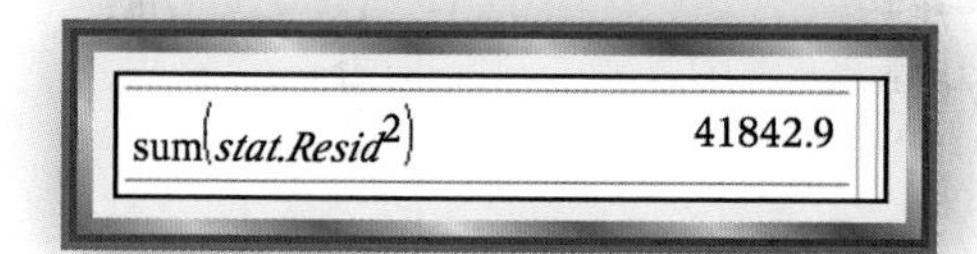

QY1

> **QY1**
> Use the least-squares regression line to estimate the price of a diamond ring weighing 0.17 carats.

Correlation

For the diamond ring data, the regression line is the best linear model for these data. But how good is "best?"

To measure the strength of the linear relation between two variables, a *correlation coefficient* is used. The correlation coefficient is often denoted by the letter r. The terms "co-relation" and "regression" were introduced by the English researcher Sir Francis Galton in the 1880s in a study comparing the heights of children with the heights of their parents. The statistic we now use for correlation was given a mathematical foundation by the English statistician Karl Pearson in 1896. Statisticians today use Pearson's formula, which can be evaluated automatically with a statistics utility.

Definition of Correlation Coefficient (Pearson's Formula)

The **correlation coefficient** for a population with n elements is

$$r = \frac{1}{n}\sum_{i=1}^{n}\left(\frac{x_i - \bar{x}}{s_x}\right)\left(\frac{y_i - \bar{y}}{s_y}\right).$$

$\frac{1}{n}$ is replaced by $\frac{1}{n-1}$ for data from a sample.

Unlike the sum of squared residuals, which can be measured for any line modeling a set of data, the correlation coefficient describes fit and direction for the regression line only.

It can be proved that, regardless of the data set, the correlation coefficient r is always a number between -1 and 1. Some data sets and the corresponding values of r are shown in the scatterplots below.

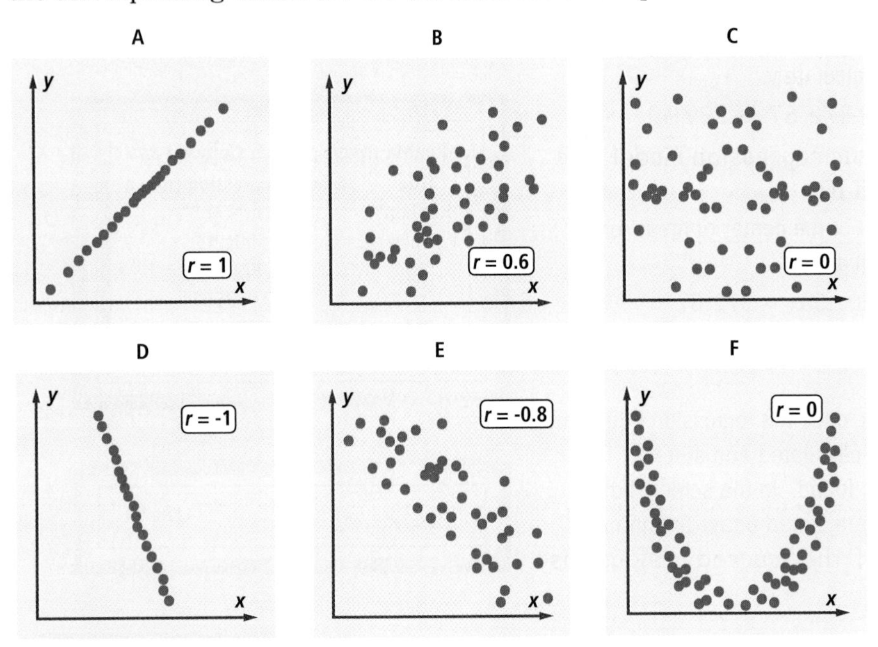

You should use your statistics utility to calculate the value of r. However, you should be able to interpret the value of r in the context of your data. In general, the sign of r indicates the *direction* of the relation between the variables, and its magnitude indicates the *strength* of the relation. Positive values of r indicate a ***positive association*** between the variables. That is, larger values of one variable are associated with larger values of the other. Negative values of r indicate a ***negative association*** between the variables. That is, larger values of one variable are associated with smaller values of the other.

The extreme values of 1 and -1 indicate a perfect linear relation, as in scatterplots A and D. That is, all data points lie on a line. Thus, a situation in which $r = \pm 1$ is sometimes called a **perfect correlation**.

A relation for which most of the data fall close to a line (scatterplot E) is called a **strong correlation**. A **weak correlation** is one for which, although a linear trend can be seen, many points are not very close to the line (scatterplot B). A value of r close or equal to 0 (scatterplots C and F) indicates that the variables are not related by a linear model. Note, however, that as indicated in scatterplot F, if $r = 0$, the variables might be strongly related in some other way as denoted by the pattern. A number line below summarizes these relations.

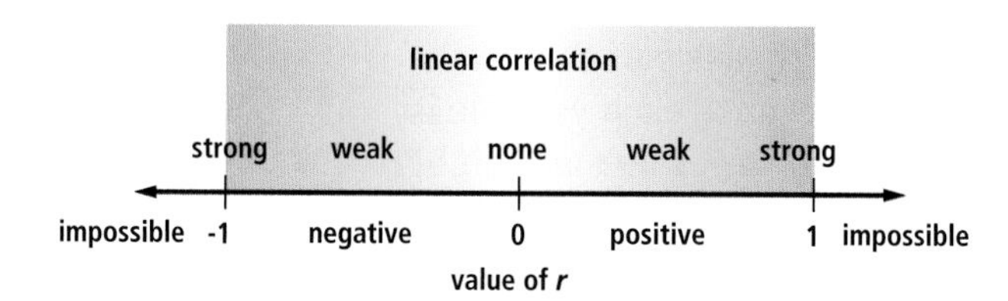

There are no strict rules about how large a correlation must be to be considered strong. In some cases, $|r| = 0.5$ is considered fairly strong, and in others it might be considered moderate or weak.

QY2

Without calculating the correlation coefficient, you can get a sense of its value by looking at the numerical data or at a scatterplot of the data.

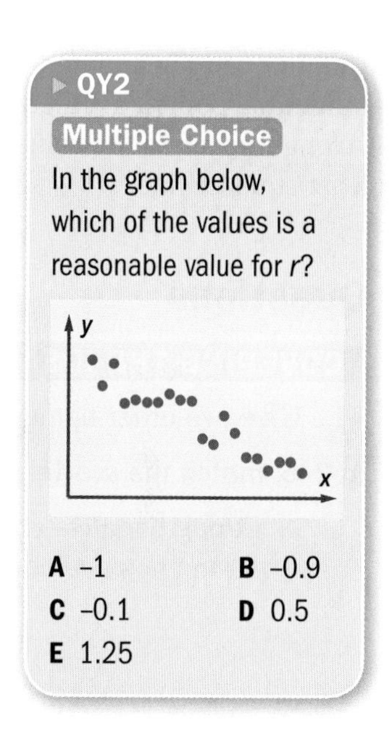
▸ QY2

Multiple Choice

In the graph below, which of the values is a reasonable value for r?

A -1 B -0.9
C -0.1 D 0.5
E 1.25

Activity

Set 1		Set 2		Set 3		Set 4		Set 5	
x	y	x	y	x	y	x	y	x	y
10	30	4	10	250	3	3	250	-3	9
11	40	8	9	300	9	9	300	-2	4
15	80	13	2	500	11	11	500	0	0
12	50	11	5	750	10	10	750	2	4
14	70	8	4	600	12	12	600	3	9

Step 1 Look at the pattern in each data table and predict what the correlation coefficient might be for that data set.

Step 2 Draw a scatterplot of each data set. Use the scatterplot to predict the correlation coefficient, altering your prediction from Step 1 if necessary.

Step 3 Use a statistics utility to determine the regression line and correlation coefficient for each data set. Compare your prediction from Step 2 to the actual correlation coefficient.

Some statistics utilities give values of r^2 rather than of r. This is because r^2 is used in advanced statistical techniques. You can calculate r by taking the square root of r^2, and then determine the sign of correlation according to the direction in the scatterplot.

QY3

▸ QY3

An r^2 value of 0.6 yields what possible values for r, to the nearest tenth?

Cautions about Correlation

It is important to note that while r provides a mathematical measure of *linearity*, it does not provide information about *cause and effect*. For instance, there is a large positive correlation between shoe size and reading level of children. But this does not mean that learning to read better causes your feet to grow or that wearing bigger shoes improves your reading.

It is up to the people who analyze and interpret the data to determine why two variables might be related. In the case of shoe size and reading level, the correlation is strong because each variable is related to age. Older children generally have both larger feet and higher reading skills than younger children. Similarly, the data in the Activity in Lesson 2-2 give a relationship between TVs and unemployment. The correlation is -0.8, which is strong, but it does not imply that unemployment can be reduced by providing a country with more TVs. This idea is sometimes summarized as *correlation does not imply causation.*

Another caution about correlation and regression: watch out for influential points. Outliers in either the x- or y-coordinate can have a strong impact on the values of slope, y-intercept, and correlation of the least squares line.

Questions

COVERING THE IDEAS

1. Give two other names for the regression line.

In 2–5, match the scatterplot with the best description.

A strong negative correlation
B weak negative correlation
C strong positive correlation
D almost no correlation

2.

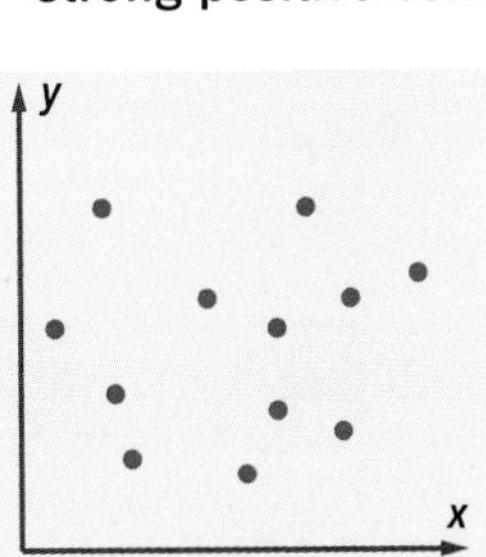

3.

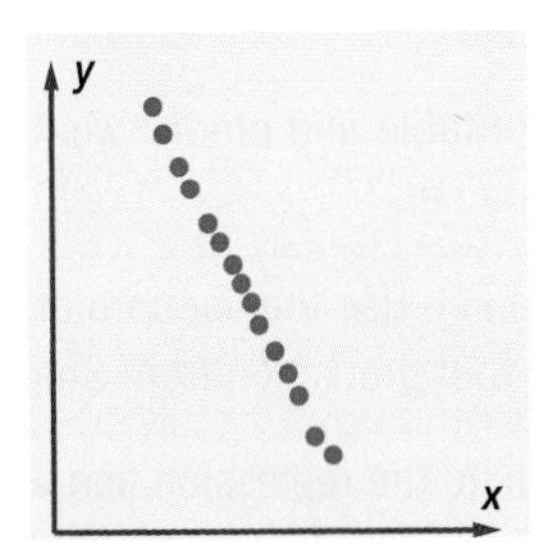

4.

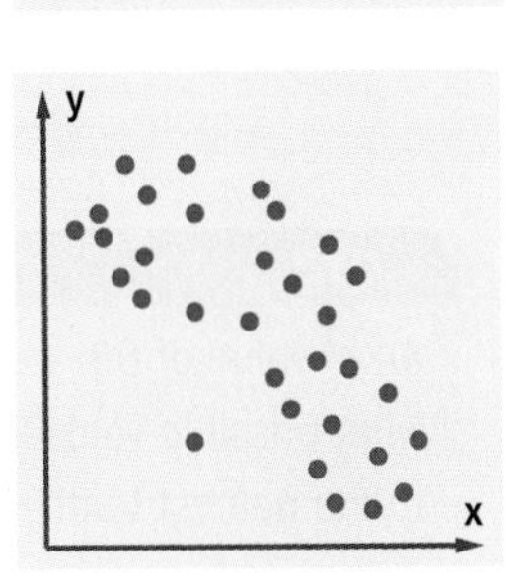

5.

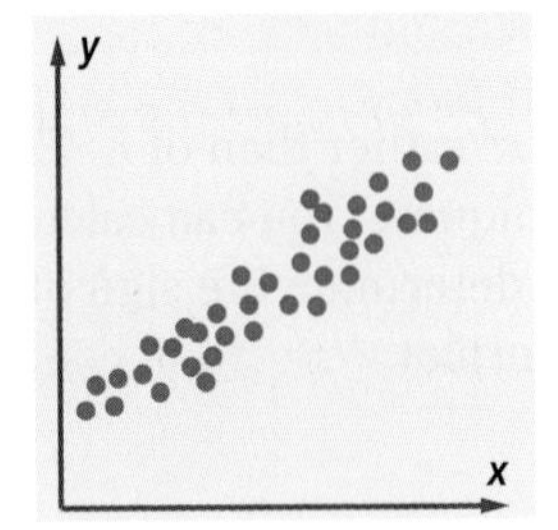

In 6–8, refer to the Activity.

6. Why is the correlation coefficient 1 for Set 1?
7. The x- and y-values in Sets 3 and 4 are swapped. How does this affect r? Does this make sense?
8. The line of best fit for Set 5 has a slope of 0 and r of 0. However, there is a definitive pattern. What other type of model would fit these data?
9. What point must be on the line of best fit?
10. Draw a scatterplot showing perfect positive correlation.
11. Suppose for some data set, $r^2 = 0.4$. Find all possible values of r.
12. **True or False** A negative value of r implies a negative slope for a linear regression line.

APPLYING THE MATHEMATICS

13. Make up a data set in which all the data lie on a single horizontal line, as shown at the right. Calculate the correlation coefficient for your set. Explain the result that you get.

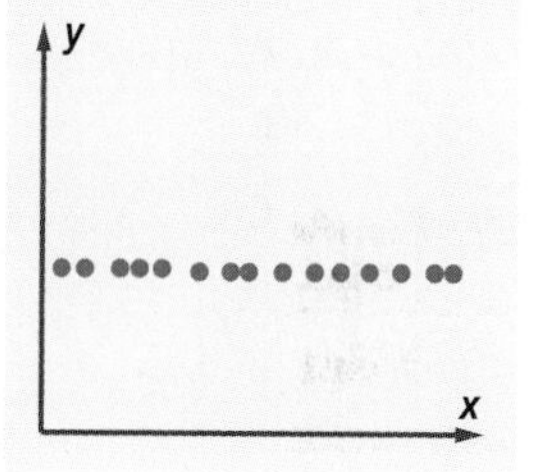

In 14 and 15, state whether you think the correlation coefficient is positive, negative, or almost zero. Explain your answer.

14. number of putts sunk in golf and the distance of ball from hole
15. a person's height and the distance he/she lives from school
16. Heavy metals can enter the food chain when metal-rich discharges from mines contaminate streams, rivers, and lakes. The table shows the lead and zinc contents in milligrams of metal per kilogram of fish (mg/kg) for 10 whole fish (4 rainbow trout, 4 large scale suckers, and 2 mountain whitefish) taken from the Spokane River during July, August, and October of 1999.

	Rainbow Trout				Large Scale Sucker				Mountain Whitefish	
Lead (mg/kg)	0.73	1.14	0.60	1.59	4.34	1.98	3.12	1.80	0.65	0.56
Zinc (mg/kg)	45.3	50.8	40.2	64.0	150.0	106.0	90.8	58.8	35.4	28.4

Source: Quantitative Environmental Learning Project, Seattle Central Community College

a. Use a statistics utility to find an equation of the least squares line to predict the amount of zinc from the amount of lead.
b. Use a statistics utility to find an equation of the least squares line to predict the amount of lead from the amount of zinc.
c. Are your answers to Parts a and b the same?
d. Use an appropriate equation to predict the amount of zinc in a fish that has $2\,\frac{\text{mg}}{\text{kg}}$ of lead.

17. A regression line for a set of data is $y = 10x + 4$. If the sum of squared residuals is 0, what is the correlation coefficient?

18. The scatterplot at the right shows the engine size (in liters) of 14 models of cars and their respective fuel economies (in miles per gallon).

    a. Is the association positive or negative? Explain your answer.

    b. Is the correlation coefficient positive or negative? Explain your answer.

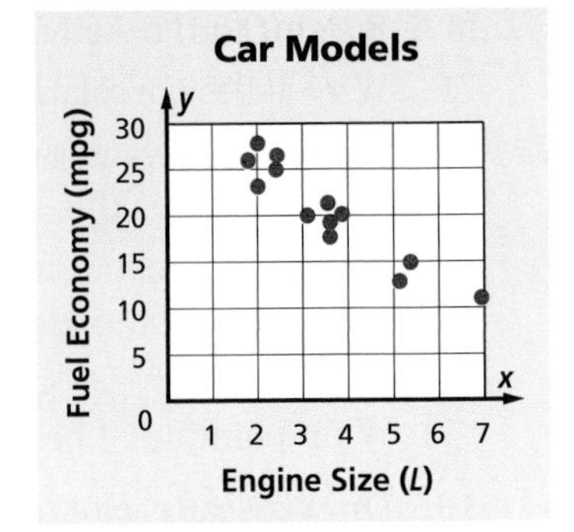

Source: United States Department of Energy

19. There is a 0.8 correlation between the total sales tax collected in Florida each year from 1960–2000 and the numbers of shark attacks in those years. Does this mean the more sales tax, the more shark attacks? Explain why or why not.

20. The data set below gives the greenhouse emissions and fuel economy for five car models.

Car Model	City MPG	Greenhouse Emissions (tons per year)
A	21	8.0
B	11	14.1
C	19	8.7
D	16	9.6
E	24	6.8

a. Use a statistics utility to calculate the regression line and the correlation coefficient. Use city mpg as the independent variable.

b. It can be proven that the slope of the regression line is given by the formula slope $= r \cdot \frac{s_y}{s_x}$, where r is the correlation coefficient, s_y is the standard deviation of the dependent variable n, and s_x is the standard deviation of the independent variable. Confirm that this formula gives you the same answer as you found in Part a.

c. Compute the values of $\bar{x}$ and $\bar{y}$.

d. Show that the point $(\bar{x}, \bar{y})$ lies on the least squares line.

REVIEW

21. A rule of thumb on body measurement is "arm span is equal to height." Selena and Luis used data on 15 adult men to write an equation expressing arm span y as a function of height x. Selena used the rule of thumb. She found the sum of squared residuals for the model $y = x$ was 3050. Luis fit a line on the scatterplot. His line had equation $y = 0.9x + 0.2$. His sum of squared residuals was 4109. Which line had the better fit? **(Lesson 2-2)**

22. Suppose $h(x) = \sqrt{x + 9}$. **(Lesson 2-1)**

    a. Find $h(16)$.

    b. What is the domain of h?

    c. Give the range of h.

23. Suppose $g(t) = t^2 - 6t - 6$. For what value(s) of t does $g(t) = 0$? **(Lesson 2-1)**

24. Let $y = 5x^2 + 2x$. Is y a function of x? Why or why not? **(Previous Course)**

25. ***Skill Sequence*** Rewrite each expression without fractions. **(Previous Course)**

a. $\frac{t}{\frac{1}{5t}}$ b. $\frac{4x}{\frac{x}{y}}$ c. $\left(\frac{r}{\frac{1}{5}}\right)^2$

EXPLORATION

26. In Parts a–d, use or modify the data as indicated. Then use a statistics utility to compute the regression line and the correlation coefficient for the data. Record your results.

a. Use the unemployment and TV data from Lesson 2-2.

b. Use the data in Part a but replace the data for South Africa with that of Nicaragua: 6.8 TVs per 100 people and 3.9 unemployed per 100 people.

c. Leave Nicaragua in the data set. Replace the Netherlands with Spain, which has 40.2 TVs per 100 people and 13.9 unemployed per 100 people.

d. Change the Spain data to the extreme situation of 80 TVs per 100 people and unemployment of 50 per 100 people. (No country has these statistics.)

e. Write a paragraph summarizing what you have found.

QY ANSWERS

1. \$562.59

2. B

3. $r \approx 0.8$ or $r \approx -0.8$

Lesson

2-4 Exponential Functions

Vocabulary

growth factor
exponential function with base *b*
exponential growth function
exponential growth curve
exponential decay function
asymptote

▶ **BIG IDEA** Exponential functions describe quantities that grow or decay by a constant factor.

Linear functions exhibit an additive pattern of change. As the value of x increases by 1, the value of $f(x) = mx + b$ increases by a constant amount m, which is the slope of the graph of the function f. *Exponential functions* are based on multiplication rather than addition. In an exponential function f with equation $f(x) = ab^x$, increasing x by 1 causes the function value to be multiplied by a constant amount b, the *growth factor*. Sometimes the growth factor is not given directly, but instead is described as a percent of change.

Mental Math

Suppose you earn $1000 on a job and 10% is taken out for income tax. Then your employer gives you a 10% raise. How much will you have after the raise and tax?

Exponential Growth and Decay

One common application of exponential functions is to describe changes in population.

Example 1

The town of Centerburg is suffering from a decline in population. A demographer has predicted that, under current conditions, the current population of 28,500 will decrease by 2% each year over the next decade. However, a manufacturer claims that a new factory will reverse the trend and cause the population to grow by 4% annually instead.

a. Create equations to describe the population of Centerburg as a function of time under these two conditions:

Function *f*:	**the population of 28,500 increases by 4% each year (with the new factory)**
Function *g*:	**the population of 28,500 decreases by 2% each year (with no new factory)**

b. Compare the projected populations after 10 years.

Solution

a. An annual growth rate of 4% means that each year the population is 1.04 times larger than the previous year's population. In contrast, if the population shrinks 2% each year, then each year it is multiplied by 0.98. Equations for the two functions can be written as
$f(x) = 28500(1.04)^x$ and $g(x) = 28500(0.98)^x$.

b. Evaluate these two functions when $x = 10$ years.

$f(10) = 28500(1.04)^{10} \approx 42{,}190$
$g(10) = 28500(0.98)^{10} \approx 23{,}290$

One way to compare is to subtract to find out how much the two projections $f(10)$ and $g(10)$ differ.

$$f(10) - g(10) = 42{,}190 - 23{,}290 = 18{,}900$$

Under the assumption that the factory is built and there is an annual growth rate of 4%, the population in ten years will be about 42,190. If, however, there is no change and there is an annual decrease of 2%, then the population in ten years will be about 23,290. The manufacturer claims that building the factory will lead to 18,900 more people living in Centerburg than would otherwise be there.

A second way to compare is by graphing the functions.

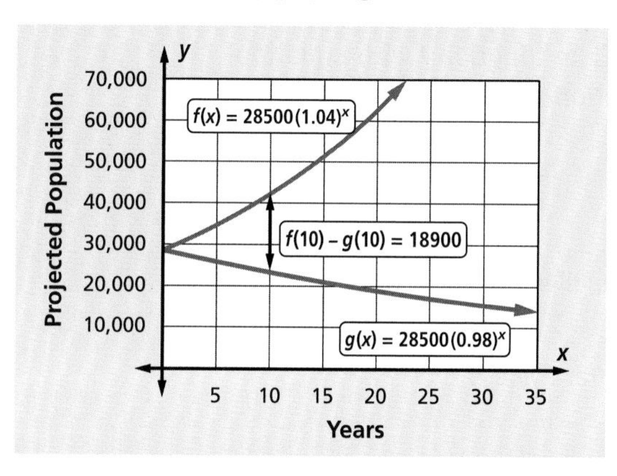

The graphs show that the populations differ by over 15,000 in 10 years and differ by about 30,000 in 15 years.

Each function in Example 1 has an equation of the form $y = ab^x$ in which the initial population a is repeatedly multiplied by the **growth factor** b. This is true of all exponential functions.

Definition of Exponential Function

An **exponential function with base *b*** and initial value a is a function with an equation of the form

$$f(x) = ab^x,$$

where $a \neq 0$, $b > 0$, and $b \neq 1$.

When an exponential function has a growth factor that is greater than 1, as in the situation in which the factory is built in Centerburg, the function is called an **exponential growth function**, and its graph is called an **exponential growth curve**.

In contrast, in a situation like Centerburg's population without the factory, the growth factor is between 0 and 1 and the function is called an **exponential decay function**. The word "decay" comes from the fact that these exponential functions model situations, such as radioactive decay, in which a quantity is diminishing by a constant factor. (Note that we still call b a growth factor, even though the function is decreasing.)

QY1

Note that in an exponential function, the independent variable is in the exponent. Thus, the function f with equation $f(x) = x^2$ is *not* exponential, even though it involves an exponent.

Another example of exponential growth is the value of an investment earning compound interest. Suppose you deposit P dollars in an account which pays an annual percentage yield r. If you make no deposits or withdrawals, each year your balance is multiplied by $1 + r$. After t years, your balance A is given by $A = P(1 + r)^t$. The function f with equation $A = f(r) = P(1 + r)^t$ is an exponential function with base $1 + r$.

QY1

When $a = 400$ and $b = 2.5$, tell whether $f(x) = ab^x$ describes a growth or decay function, and why.

GUIDED

Example 2

\$2500 is invested in an account with a 5.3% annual yield.

a. What is the yearly growth factor?

b. What is the equation for the balance A after t years?

c. What is the balance after 7 years?

d. What value is associated with $t = -2$? What does $t = -2$ mean?

Solution

a. $r = 0.053$. So the yearly growth factor is $(1 + r) =$ _?_.

b. The initial value is _?_ and the growth factor is _?_, so $A =$ _?_.

c. $A = 2500(\_?\_)^7 =$ _?_.

d. $A = 2500(\_?\_)^{-2} =$ _?_. If the account had been started 2 years ago, an investment of _?_ would have produced \$2500 this year.

The Roles of the Constants a and b in $f(x) = ab^x$

Examples 1 and 2 show that the value of a in $f(x) = ab^x$ represents the initial value, and the value of b determines whether f models exponential growth or exponential decay. But how do the values of a and b affect the graph of the function?

Activity

Experiment with your graphing technology to find values of a and b in $f(x) = ab^x$ that produce each graph below. Write an equation for each function. Be sure that your graphs contain the points marked on the graphs shown. Use trace or a table to check values of each function, as shown at the right.

	A	B
◆		=f1(a[])
1	-2	1/9
2	-1	1/3
3	0	1
4	1	3
5	2	9

B3 =1

13.89, -13.89, -10, 2, 10

a.

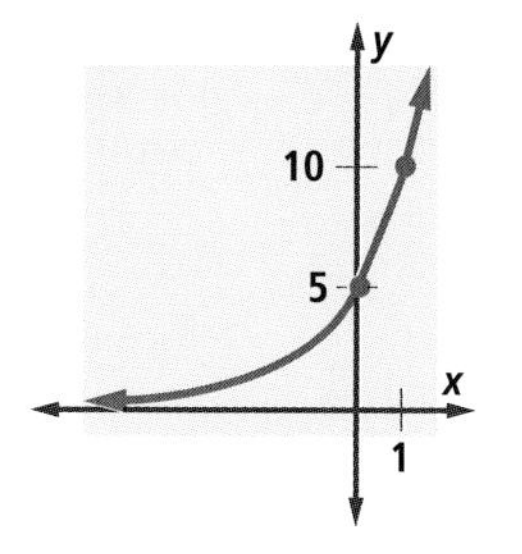

b.

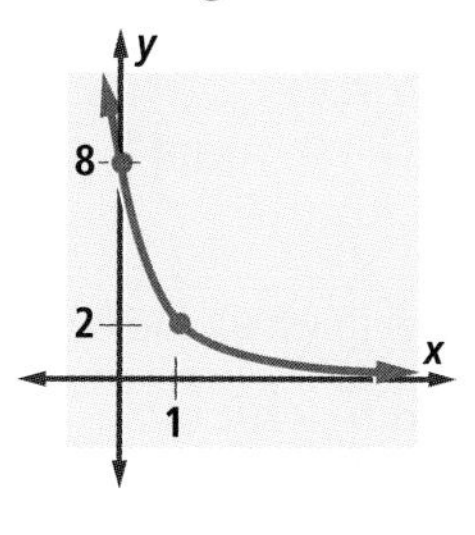

c.

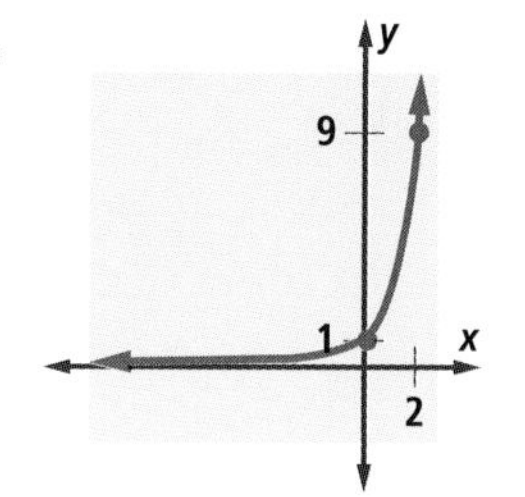

d.

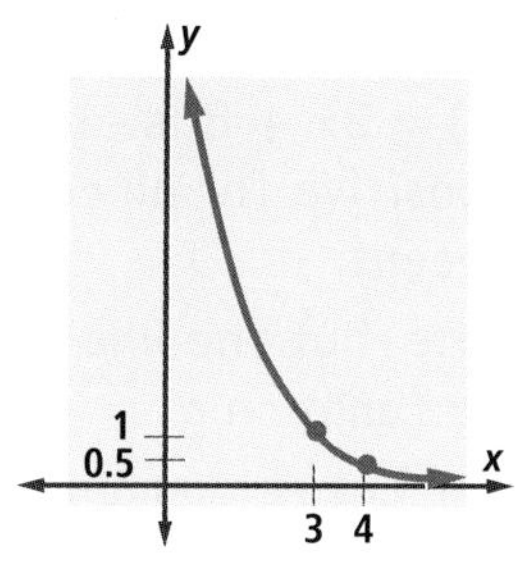

e.

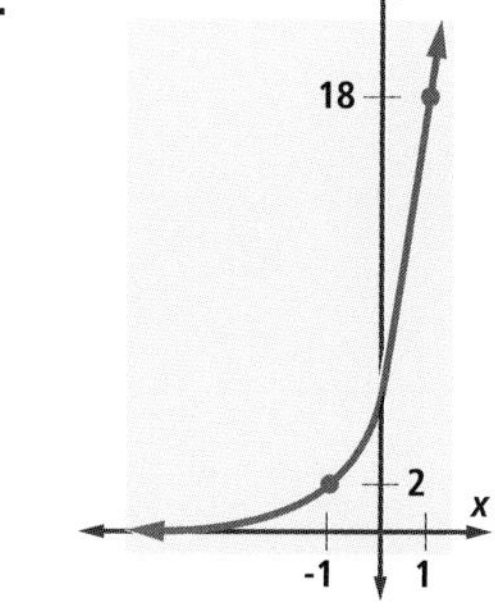

f.

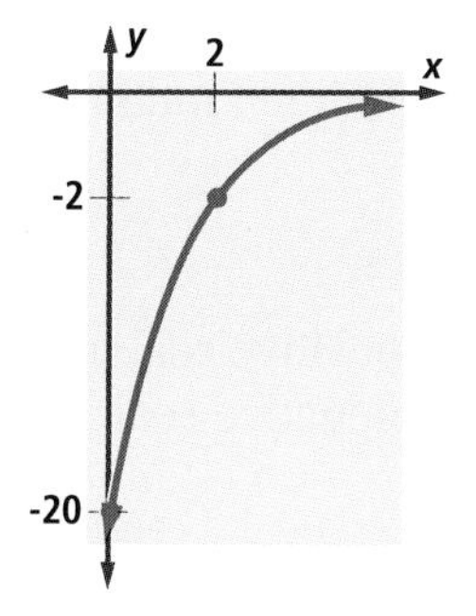

Common Features of Exponential Functions

The six graphs in the Activity have much in common. For Parts a, c, and e, as you move to the left on the graph, the values of y get closer and closer to 0 and the curve approaches the x-axis. On the other hand, for Parts b, d, and f, moving to the right produces y-values that get closer and closer to 0. In all cases, the x-axis is a horizontal *asymptote* of the function. An **asymptote** is a line that the graph of a function $y = f(x)$ approaches as x approaches a fixed value or increases or decreases without bound. Note that the asymptote is not actually part of the graph.

QY2

> **QY2**
>
> Use the graphs in the Activity.
>
> a. Which functions are increasing?
>
> b. Which functions have a range that is the set of positive real numbers?

GUIDED

Example 3

For each function characteristic, compare and contrast the linear function $f(x) = 2x + 8$ with the exponential function $g(x) = 8 \cdot 4^x$.

a. domain and range
b. y-intercept and x-intercept
c. asymptote
d. increasing or decreasing

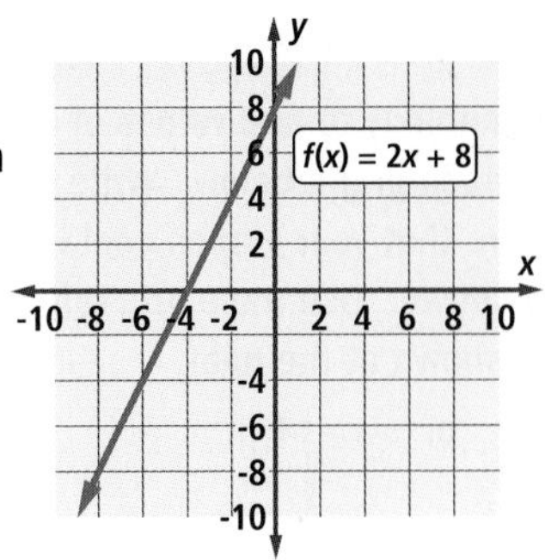

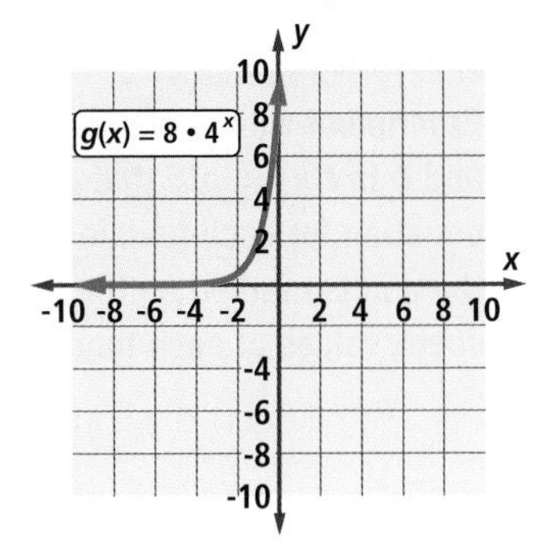

Solution

Look for similarities between the functions and ways in which they differ.

a. The two functions have the same domain, which is __?__. But the range of f is the set of all real numbers, while the range of g is __?__.

b. Both functions have a y-intercept of __?__.
The x-intercept of the graph of $f(x) = 2x + 8$ is –4 since $f(-4) = 0$. However, since $8 \cdot 4^x$ is positive for all real numbers x, function g has no x-intercept.

c. Function f does not have an asymptote, but the line with equation __?__ is an asymptote for function g.

d. Both functions are __?__ over their entire domain.

The definition of an exponential function f places restrictions on the values allowed for a and b in $f(x) = ab^x$. The base b must be a positive number other than 1, and $a \neq 0$. All these exponential functions have the following properties:

(1) The domain is the set of real numbers.

(2) If $a > 0$, then the range is the set of positive real numbers. If $a < 0$, then the range is the set of negative real numbers.

(3) The graph contains the point $(0, a)$.

(4) The x-axis is an asymptote for the graph.

(5) For positive values of a, if $b > 1$, then the function is increasing. If $0 < b < 1$, then the function is decreasing. For negative values of a, which are used less often, this situation is reversed.

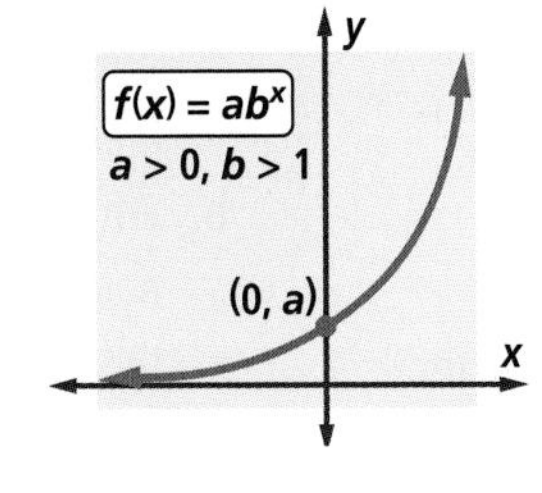

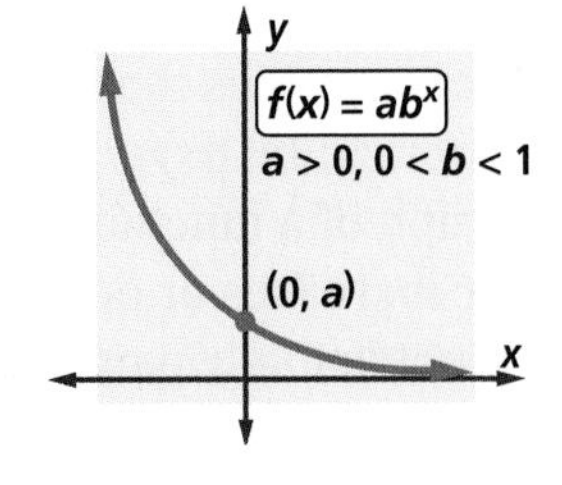

Questions

COVERING THE IDEAS

1. In 2008, the population of Ireland was 4,156,119, with an average annual growth rate of 1.13%. Assume this growth rate continues.
   a. Estimate, to the nearest thousand, the population of Ireland in 2009.
   b. Express the population P as a function of n, the number of years after 2008.
   c. Estimate, to the nearest thousand, Ireland's population in the year 2020.

Cobh, a seaport on the south coast of Ireland

2. Tell whether the equation describes an exponential function. If it does, give the values of a and b in $f(x) = ab^x$.
   a. $g(m) = 11 \cdot 4^m$ b. $s(t) = 6$ c. $j(z) = z^2$ d. $w(x) = 0.6^x$

3. On your 21st birthday, $4000 is invested for you at an annual yield of 8%. You do not withdraw any money from the account.
   a. Write a formula for A, the balance in the account after t years.
   b. What will be the balance in the account when you are 65?
   c. How much money would have to have been invested when you were 17 at the same yield to get the same result as in Part b?

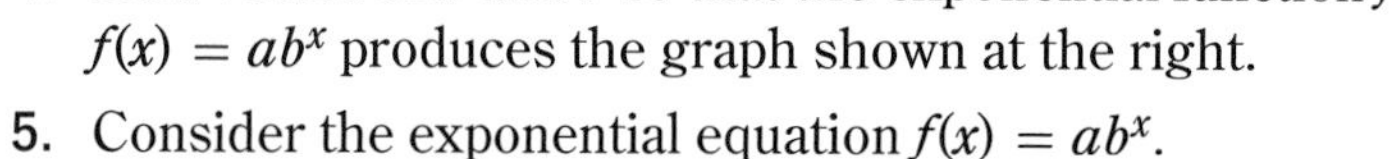

4. Find values of a and b so that the exponential function f with $f(x) = ab^x$ produces the graph shown at the right.

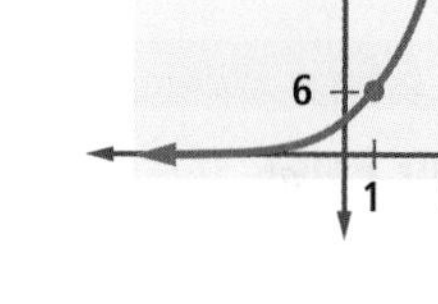

5. Consider the exponential equation $f(x) = ab^x$.
   a. **True or False** The initial value of the function is $f(1)$.
   b. If $0 < b < 1$, what type of exponential function is f?

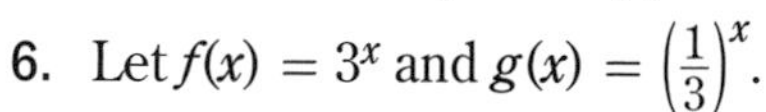

6. Let $f(x) = 3^x$ and $g(x) = \left(\frac{1}{3}\right)^x$.
   a. Graph the functions f and g on one set of axes for $-3 < x < 3$.
   b. Compare and contrast these two functions.
   c. Which function, f or g, represents exponential decay?

APPLYING THE MATHEMATICS

7. Consider $f(x) = 4^x$ and $g(x) = 5^x$.
   a. Without graphing, tell which function has greater values when $x > 0$.
   b. Without graphing, tell which function has greater values when $x < 0$.
   c. Check your answers to Parts a and b by graphing f and g on the same set of axes.

8. On the same axes, graph $f(x) = 2 \cdot 3^x$ and $g(x) = 3 \cdot 2^x$.
   a. Which function has the greater y-intercept?
   b. For what value of x is $f(x) = g(x)$? Explain how you found your answer.
   c. For what values of x is $f(x) < g(x)$?

9. a. **Multiple Choice** Which equation describes the function defined by the table given at the right?

   A $y = 48\left(\frac{1}{2}\right)^x$ B $y = 12\left(\frac{1}{2}\right)^x$

   C $y = 3(2)^x$ D $y = 12(2)^x$

   b. Explain how the initial value and growth factor from your answer to Part a are seen in the table.

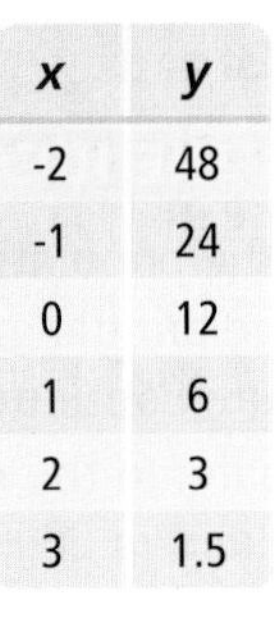

x	y
-2	48
-1	24
0	12
1	6
2	3
3	1.5

10. a. In 2007, China's population was estimated to be about 1,322,000,000. Its average annual growth rate was about 0.6%. If this growth rate continues, what will be the population of China in the year 2015?

    b. In the year 2000, the annual growth rate of China was 0.9%. Using this growth rate and the 2007 population of China, what would be the projected population in 2015? How much greater is this than your answer to Part a?

11. A gardener recycles yard waste in a compost bin in which each month 90% of the previous month's material is still present.

    a. How much of an initial 20 cubic feet of material would remain after 1, 2, and 3 months?

    b. After n months, how much of an initial 20 cubic feet of material would remain?

    c. **True or False** After six months, more than half the material will have decayed.

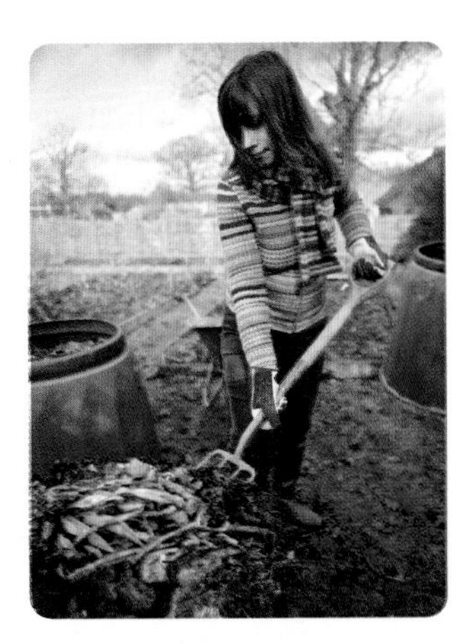

12. Compare the functions $y = x^2$ and $y = 2^x$. Mention each feature.

    a. type of function b. range c. y-intercept d. asymptote

REVIEW

13. **True or False** If a and b have a correlation coefficient of 1, then either a causes b or b causes a. Explain your answer. **(Lesson 2-3)**

14. **True or False** The correlation coefficient $r = -0.85$ indicates a weaker relation than the correlation coefficient $r = 0.65$. Explain your answer. **(Lesson 2-3)**

15. Let $f(x) = \frac{3x - 6}{x + 2}$. **(Lesson 2-1)**

    a. Evaluate $f(1)$.

    b. If $f(x) = 0$, find x.

    c. **True or False** $f(1) + f(2) = f(1 + 2)$. Justify your answer.

    d. **True or False** The domain and range of f are equal. Justify your answer.

    e. Evaluate $f(p - 3)$.

16. The box plot at the right pictures a data set that has no outliers.

a. What is the percentile of A?

b. What is the percentile of B?

c. What is the percentile of C?

d. In what percentile range does D fall?
(Lessons 1-5, 1-4)

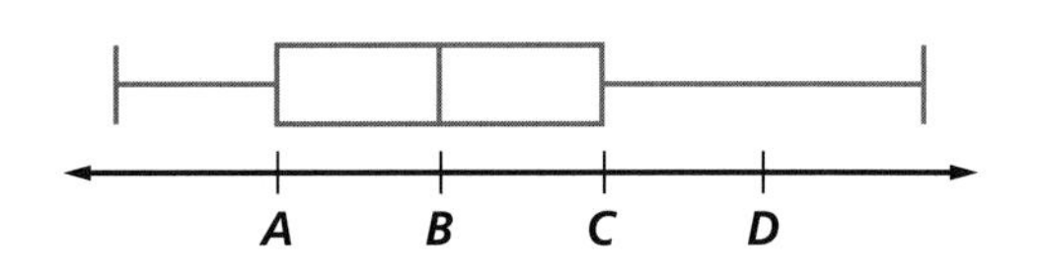

17. Solve the system $\begin{cases} 180 = rs^5 \\ 20 = rs^3 \end{cases}$. **(Previous Course)**

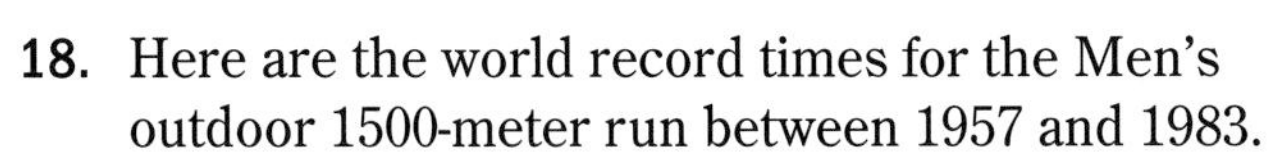

18. Here are the world record times for the Men's outdoor 1500-meter run between 1957 and 1983.

a. Enter the data into a statistics utility. Let the number of years after 1957 be the independent variable. Find the correlation coefficient between the number of years after 1957 and the record time.

b. What does the value of the correlation coefficient tell you about the relation between the two variables?

c. Find an equation for the line of best fit predicting world records for this event.

d. Using your model, predict what the record might have been in 2008. The actual record was 3.43 minutes. What does the comparison between your prediction and the actual time indicate about the strength of your model in extrapolation?
(Lessons 2-3, 2-2)

Year	Runner and Country	Time (seconds)
1957	Jungwirth (Czech)	218.1
1958	Elliott (Australia)	216.0
1960	Elliott (Australia)	215.6
1967	Ryun (U.S.)	213.1
1974	Bayi (Tanzania)	212.2
1979	Coe (U.K.)	212.1
1979	Ovett (U.K.)	212.1
1980	Ovett (U.K.)	211.36
1983	Maree (U.S.)	211.23
1983	Ovett (U.K.)	210.28

Source: *The Fascination of Statistics*; IAAF

Sebastian Coe (#254), shown here in the 1980 Olympics 1500-meter run, co-held the world record of 3 min 31.36 sec at the time.

EXPLORATION

19. a. Explain why a city whose population is growing at 5% per year does not grow twice as fast as a city whose population is growing at 2.5% per year.

b. City A grows 10% in 2011 and 2% in each year from 2012 to 2020. City B grows 2% in each year from 2011 to 2019 and then 10% in 2020. Compare the total percentage increase from 2011 to 2020 for the populations of these cities.

c. Generalize the result of Part b.

QY ANSWERS

1. $f(x) = 400 \cdot 2.5^x$ describes an exponential growth function because $2.5 > 1$.

2. **a.** a, c, e, f
b. a, b, c, d, e

Lesson

2-5 Exponential Models

Vocabulary

exponential regression
half-life

▶ **BIG IDEA** Exponential models are used in many fields in which data follow some kind of natural growth or decay. Exponential regression is used to fit exponential models to data.

Mental Math

Veronica is preparing a study plan for a biology test. She plans to study for four days, and each day she plans to study 50% longer than the previous day. On the first day she plans to study for 24 minutes. How many hours and minutes total will she study?

Finding an Exponential Function Using a System of Equations

Exponential models describe situations in diverse fields such as biology, paleontology, sociology, physics, and economics.

Populations very often grow at a constant rate, at least in the short run. Therefore, it is natural to fit an exponential model to population data. The average population growth rate in the U.S. is about 0.883%, but in areas that are growing quickly, the rate can be much higher.

Example 1

Huntley, Illinois had been a small farming town. But when a large housing development was built, the population growth pattern changed. Two special censuses gave village planners the data in the table at the right.

Year	Population
2003	12,270
2005	16,719

Source: The Village of Huntley

a. Find an exponential model for the data. Let $p(t)$ be the population t years after 2000.

b. Predict the population of Huntley in the year 2015.

Huntley, IL

Solution

a. A general equation for the model is $p(t) = ab^t$. From the table, $p(3) = 12{,}270$ and $p(5) = 16{,}719$. Substitute these values into the general equation to get a system.

$$\begin{cases} 12{,}270 = ab^3 \\ 16{,}719 = ab^5 \end{cases}$$

Divide the second equation by the first. This can be done because $a \neq 0$ and $b \neq 0$.

$$\frac{16{,}719}{12{,}270} = b^2$$

Because the base b must be positive, take the positive square root.

$$b = \sqrt{\frac{16{,}719}{12{,}270}} \approx 1.1673$$

This is the growth factor. The population was growing about 17% per year.

To find a, substitute this value of b into one of the two equations in the system. Using the first equation,

$$12{,}270 = a(1.1673)^3$$
$$a \approx \frac{12{,}270}{(1.1673)^3} \approx 7714.$$

According to this model, the initial population of Huntley in 2000 was 7714. The full exponential model is $p(t) = 7714(1.1673)^t$.

Check Use a CAS to solve the system of equations. A partial solution is shown at the right. The negative answers are not valid in this context.

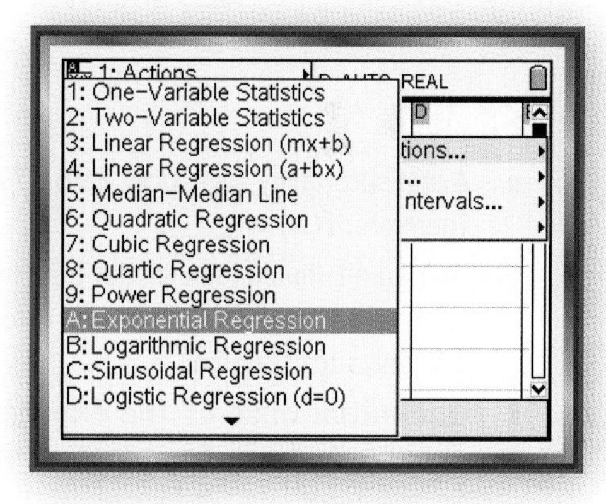

b. The model predicts the population in 2015 to be $p(15) = a \cdot b^{15} = 7714(1.1673)^{15} \approx 78{,}500$. Actually, demographers expect this growth rate to slow before then, but the model describes what would happen based on the two special census years.

Exponential Regression

The model for the population of Huntley was derived from just two data points. Two points are enough to algebraically determine an exponential function that is a perfect fit for those two points. If there are more than two data points, there may not be a model that fits all the data perfectly. However, if many data points follow an approximately exponential pattern, a statistics utility will find an **exponential regression** curve that models the pattern well.

GUIDED

Example 2

Bald eagles were once threatened with extinction. In the 48 contiguous states, their numbers were at an all-time low of 417 in 1963. But protection programs helped them rebound. In 2007, they were removed from the list of endangered species kept by the U.S. Fish and Wildlife Services.

a. **Use a statistics utility to fit an exponential model of the form $f(x) = ab^x$ to the data. Let x be "years after 1960." Report the values of a and b in the table on the next page to the nearest thousandth.**

b. **Superimpose the graph of the exponential model on the scatterplot and describe how well the exponential curve fits the data.**

c. **Identify the initial amount and the growth factor and explain their meanings.**

d. **Find the residuals for the model's predicted values for 2000 and 2005.**

(continued on next page)

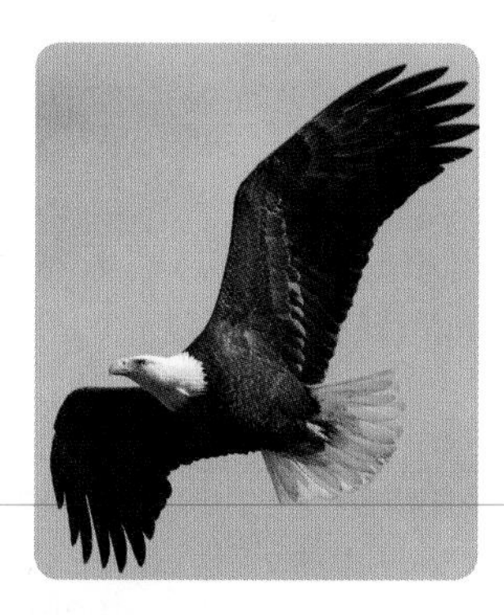

Bald eagles have a wing span of 6 to 8 feet.

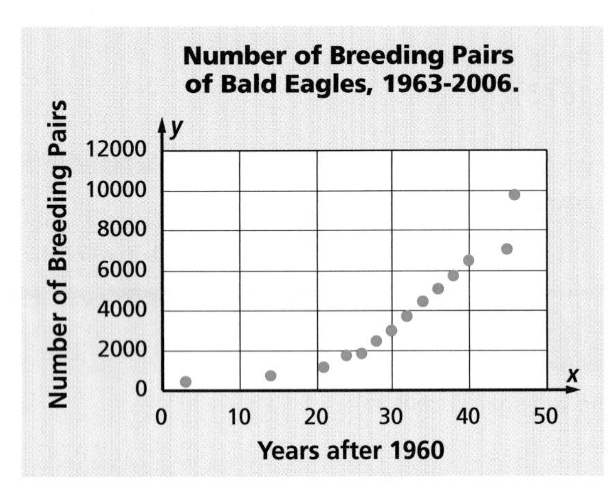

Year	Number of Breeding Pairs
1963	417
1974	791
1981	1188
1984	1757
1986	1875
1988	2475
1990	3035
1992	3749
1994	4449
1996	5094
1998	5748
2000	6471
2005	7066
2006	9789

Source: U.S. Fish and Wildlife Services

Solution

a. A statistics utility gives $a \approx 296.177$ and $b \approx 1.079$. Therefore, $f(x) =$ _?_.

b. Your graph should look similar to the screenshot at the right. The model is close to most of the points of the scatterplot.

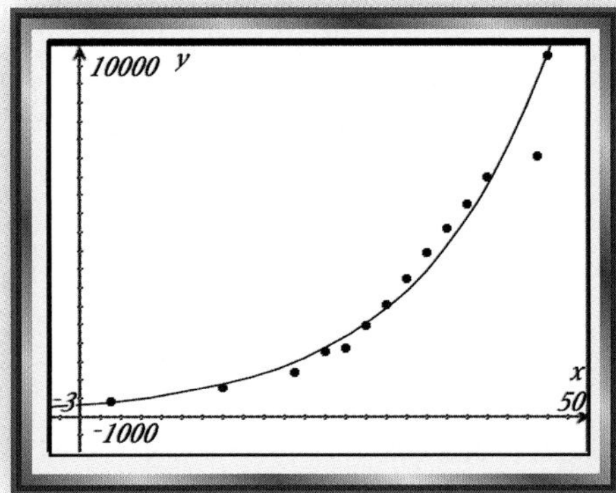

c. The initial value of the exponential model is about _?_ breeding pairs, which corresponds to the year _?_. The growth factor is _?_. This means that during the years 1963 through 2006, the eagle population had a growth rate of about _?_ % per year.

d. In 2000, the predicted number of breeding pairs is $f(40) \approx$ _?_ pairs. So for 2000, residual = observed value − predicted value = _?_ − _?_ = _?_. In 2005, the predicted number of breeding pairs is ≈ _?_ pairs. For 2005, error = _?_ − _?_ = _?_.

Letting x = years after 1960 makes a difference in the model. If the actual year is used as the independent variable, the statistics utility model is $f(\text{year}) = (9.181 \cdot 10^{-63}) \cdot (1.079)^{\text{year}}$. The exponents are so large that a small change in the growth factor due to rounding, for example, would create a large difference in the predicted values. This is a consideration for exponential models. However, with linear and quadratic models, it often does not make a difference whether you use the year as it is or use the number of years from a given point in time.

QY

QY

a. Without a calculator, estimate the value of $1.077^{1963} - 1.076^{1963}$.

b. Estimate the value in Part a with a calculator.

c. How close was your estimate to the actual value?

Half-Life and Exponential Decay

Radioactive elements are useful in situations involving detective work, such as diagnosing health problems with barium x-rays or finding the age of archeological artifacts with carbon dating.

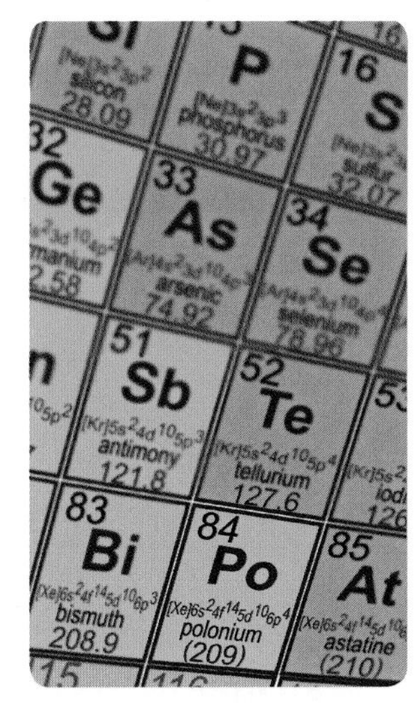

The **half-life** of a radioactive element is the amount of time it takes an original quantity to decay to half that amount. If you know the half-life of a radioactive element and the amount of the substance at one point in time, you can find the original amount.

In 2007, the element polonium was in the news when London police detectives investigated the poisoning of former Russian KGB agent Alexander Litvinenko. Since polonium had never been known to be used in a poisoning, the authorities did not look for evidence of it until weeks after the crime had taken place. As a consequence, they had to work backwards from the evidence to calculate the amount of polonium used on the victim. They made use of the fact that the half-life of polonium is 138 days.

Example 3

Detectives in the Litvinenko investigation found polonium on a cup in a hotel that he had visited. Suppose that 4 micrograms were found, and it had been 30 days since Litvinenko was there.

a. Find how much polonium was on the cup originally.

b. Derive a model for this situation.

Solution

a. Let t represent the number of days since a micrograms of polonium were placed on the cup. Then $f(t) = ab^t$ is the amount of polonium remaining. First, find the daily decay factor b. The half-life tells us that it takes 138 days for a micrograms of polonium to decay to $\frac{1}{2}a$ micrograms. So when $t = 138$, we know that $f(t) = \frac{1}{2}a$. Substitute these values into $f(t) = ab^t$.

$$\frac{1}{2}a = ab^{138}$$

$$\frac{1}{2} = b^{138} \quad \text{Divide both sides by } a.$$

$$\left(\frac{1}{2}\right)^{\frac{1}{138}} = \left(b^{138}\right)^{\frac{1}{138}} \quad \text{Raise each side to the } \frac{1}{138}\text{th power.}$$

$$b \approx 0.995$$

The decay factor is about 0.995. The polonium decay function is $f(t) = a(0.995)^t$. Now use the information that there were 4 micrograms after 30 days. Again, substitute.

$$4 \approx a(0.995)^{30}$$

$$a \approx 4.649$$

There were about 4.65 micrograms of polonium originally.

b. Substitute the values of a and b into $f(t) = ab^t$.
After t days, there are $f(t) = 4.65(0.995)^t$ micrograms of polonium remaining.

Check

Graph $f(x) = 4.65(0.995)^t$. Trace to see that (0, 4.65) and (30, 4.00) are on the graph.

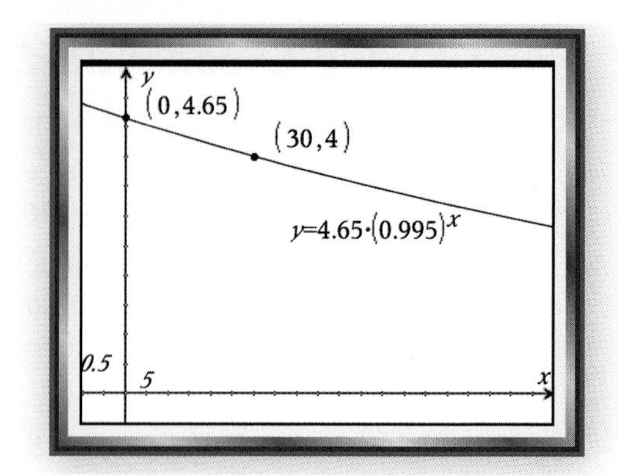

Questions

COVERING THE IDEAS

1. Suppose an exponential function with equation $f(t) = ab^t$ contains the two points (3, 20) and (10, 156).
   a. Write the system of equations that results from substituting the two points into the equation.
   b. Solve the system to yield an equation for the function. Round your values for a and b to the nearest thousandth.
   c. Check your equation for the two points (3, 20) and (10, 156).

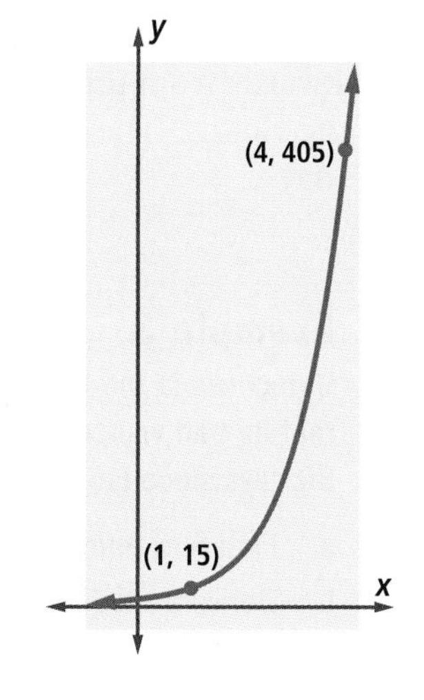

2. The graph at the right shows two ordered pairs that lie on the graph of an exponential function. Find an equation for the exponential function that contains the two points.
3. In a study of the change in an insect population, there were about 170 insects four weeks after the study began, and about 320 after two more weeks. Assume an exponential model of growth.
   a. Find an equation relating the population to the number of weeks after the study began.
   b. Estimate the initial number of insects.
   c. Predict the number of insects five weeks after the study began.
4. The prices of some diamonds of different sizes are given in the table at the right.
   a. Find an exponential regression model for this data.
   b. According to your model, what would be the price of a 2-carat diamond?

Weight (carats)	Price
0.25	$504
0.40	$1,040
0.55	$1,925
0.80	$3,680
1.25	$10,000
1.65	$18,150

5. The half-life of barium, which is used in CAT scans for medical diagnosis, is 2.6 minutes. Suppose a patient swallows a drink containing 10 units of barium prior to getting a CAT scan.
   a. Find an exponential model for the amount of barium left in the patient's system as a function of the number of minutes that have passed since drinking the barium.
   b. Find the amount of barium left after an hour.

6. Safety engineers monitor workplaces to see that workers are not exposed to unsafe levels of hazardous chemicals. Suppose that one chemical has a half-life of 7 days. If a worker currently has 18 units of this chemical in his body and was exposed 5 days ago, how much was the initial dose in his body?

7. The table at the right shows how the number of U.S. cell-phone subscribers has grown since 1985.
   a. Create an exponential regression model for the number of subscribers t years after 1985. Report your value of a rounded to the nearest integer and b rounded to the nearest thousandth.
   b. Describe how well the model fits the data.
   c. Determine the year in which the value predicted by the model differs by the greatest *percent* from the actual value.

Year	Number of Cell-Phone Subscribers (thousands)
1985	340
1987	1,231
1989	3,509
1991	7,557
1993	16,009
1995	33,786
1997	55,312
1999	86,047
2001	128,375
2003	158,722

Source: CTIA - The Wireless Association

APPLYING THE MATHEMATICS

8. Benjamin Franklin specified in his will that "1000 pounds sterling" were to be given to the town of Boston for the purposes of providing loans at interest to apprentices. He expected the loans to be repaid and unused money to be well invested. He predicted, "If this plan is executed, and succeeds as projected without interruption for one hundred years, the sum will then be one hundred and thirty-one thousand pounds…." What annual yield did Franklin expect on his gift to Boston?

9. Radium has a half-life of 1620 years. Suppose 3 g of radium is present initially.
   a. Complete the table below for this situation.

Number of Half-Lives	0	1	2	3
t = Number of Years After Start	0	1620		
$f(t)$ = Amount of Radium Present (grams)	3			

   b. Give an exponential model for the amount of radium left as a function of time t.
   c. How much radium would you expect to find after 4000 years?

10. A tour guide noticed that larger groups took more time to assemble for an event. The guide collected the data below.

Number of People	2	3	4	5	6	7	8	9	10
Time to Assemble (minutes)	2	2.6	3.4	4.4	5.7	7.4	9.7	12.5	16.3

   a. Make a scatterplot of the data.
   b. Find the linear regression model and graph it.
   c. Find the exponential regression model and graph it.
   d. Which of the two models seems to fit the data better? Why?

REVIEW

11. Consider $f(x) = 7^x$ and $g(x) = 8^x$. Without graphing, which function has greater values when:

    a. $x > 0$ b. $x < 0$ **(Lesson 2-4)**

In 12 and 13, Fill in the Blank. (Lesson 2-4)

12. All functions with equations of the form $y = a \bullet b^x$ contain the point ___?___.

13. The graph of every function in the form $y = b^x$ where $b > 0$ and $b \neq 1$ is always above the ___?___.

14. Use the Huntley data from Example 1. **(Lesson 2-2)**

    a. Find a linear model determined by the data for 2003 and 2005.

    b. What does the linear model predict for the 2015 population?

In 15 and 16, give an example of a function satisfying the given condition. In each case, give a domain and a rule. (Lesson 2-1)

15. The range is the set of all negative real numbers.

16. The range is the set of all nonnegative real numbers.

17. In 2004, of the 50 countries in the Western Hemisphere that consume petroleum, the mean consumption was about 608 thousand barrels per day per country. **(Lesson 1-2)**

    a. Find the total number of barrels consumed per day by all 50 countries.

    b. The United States consumed 20,731 thousand barrels per day in 2004. What would the mean be if the U.S. was not included?

    c. What percentage of the Western Hemisphere consumption is due to the U.S. consumption?

    d. Explain why mean consumption for all 50 countries is misleading given the U.S. consumption.

18. Consider an isosceles right triangle. **(Previous Course)**

    a. If the legs have length s, how long is the hypotenuse?

    b. If the hypotenuse has length h, how long are the legs?

EXPLORATION

19. Find out what happened to Alexander Litvinenko.

Over a barrel?
In 2007, the U.S. imported 3,437,000 barrels of petroleum per day.

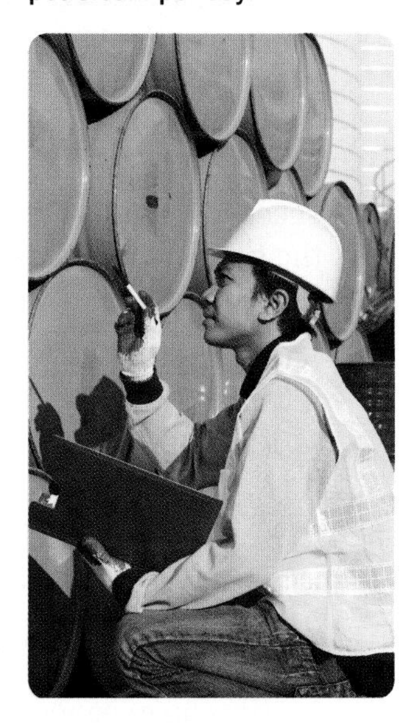

QY ANSWER

a. Answers vary.

b. about $1.455 \cdot 10^{63}$, a number with 64 digits in base 10

c. Most people estimate a much smaller difference.

Lesson 2-6 Quadratic Models

Vocabulary

quadratic model
quadratic regression

▸ BIG IDEA Quadratic models are appropriate to consider when you think data will increase to a peak and then drop, or when data will decrease to a low point and then come back.

Linear functions are appropriate for modeling situations involving a constant amount of change. Exponential functions model situations involving a constant percent of change. In this lesson, we focus on **quadratic models**, that is, models based on quadratic functions.

Mental Math

The identity $(x + y)^2 = x^2 + 2xy + y^2$ shows how to calculate $(x + 1)^2$ from x^2. Use it to find:

a. 41^2, since $40^2 = 1600$.

b. 101^2.

Properties of Quadratic Functions

A *quadratic function* is a function with an equation that can be put into the form $f(x) = ax^2 + bx + c$, where $a \neq 0$. Recall that the graph of a quadratic function is a *parabola*. If $a > 0$, the parabola opens up and has a *minimum point*, as shown at the left below. If $a < 0$, the parabola opens down and has a *maximum point*, as shown at the right below. The existence of these extrema distinguishes quadratic functions from linear and exponential functions.

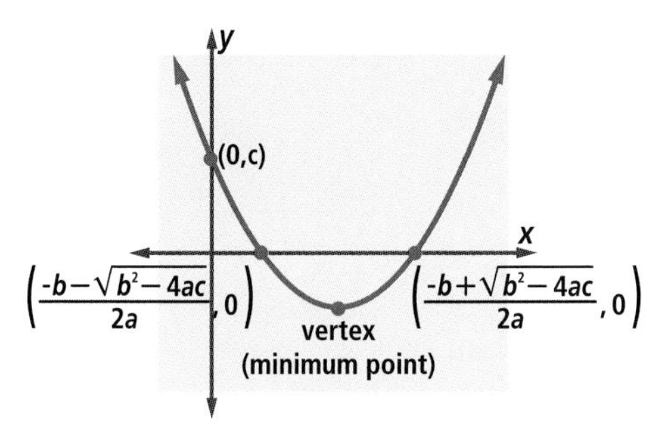

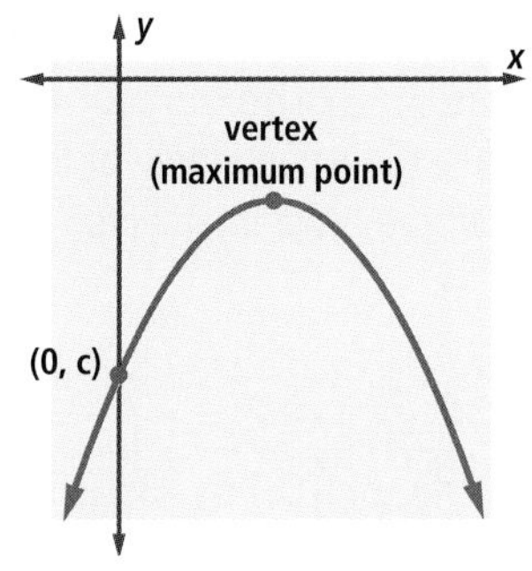

$f(x) = ax^2 + bx + c$
$a > 0$
$b^2 - 4ac > 0$

$f(x) = ax^2 + bx + c$
$a < 0$
$b^2 - 4ac < 0$

The domain of a quadratic function is the set of all real numbers. When $a < 0$, the range is the set of all real numbers less than or equal to the maximum value. When $a > 0$, the range is the set of all real numbers greater than or equal to the minimum value. The *y*-intercept is the *y*-coordinate of the point where $x = 0$:

$$f(0) = a \cdot 0^2 + b \cdot 0 + c = c.$$

So, regardless of the value of a, c is the *y*-intercept.

The x-intercepts are the x-coordinates of the points where $y = 0$. The x-intercepts exist only when $b^2 - 4ac \geq 0$, and then can be found by solving the quadratic equation $ax^2 + bx + c = 0$. From the Quadratic Formula, the x-intercepts are

$$\frac{-b + \sqrt{b^2 - 4ac}}{2a} \text{ and } \frac{-b - \sqrt{b^2 - 4ac}}{2a}.$$

The maximum or minimum point of any quadratic function occurs at the x-value that is the mean of the solutions to the equation $f(x) = 0$, that is, when $x = -\frac{b}{2a}$.

GUIDED

Example 1

Consider the function f with equation $f(x) = 2x^2 - 3x - 2$.

a. Find the x- and y-intercepts of its graph.

b. Tell whether the parabola has a maximum or minimum point, and find its coordinates.

Solution 1

a. Since $f(0) = \underline{\ ?\ }$, the y-intercept is -2.

To find the x-intercepts, let $f(x) = 0$ and solve for x.

$$2x^2 - 3x - 2 = 0$$

$$x = \frac{\underline{\ ?\ } \pm \sqrt{\underline{\ ?\ }^2 - 4 \cdot \underline{\ ?\ } \cdot \underline{\ ?\ }}}{2 \cdot \underline{\ ?\ }}$$

$$= \frac{\underline{\ ?\ } \pm \sqrt{\underline{\ ?\ }}}{?}$$

$$= 2 \text{ or } -\frac{1}{2}$$

The x-intercepts are 2 and $-\frac{1}{2}$.

b. Because the coefficient of x^2 is $\underline{\ ?\ }$, the vertex is a minimum point. Since parabolas are symmetric, the x-coordinate of the minimum (or maximum) point occurs at the mean of the two x-intercepts.

$$\frac{2 + \left(-\frac{1}{2}\right)}{2} = \underline{\ ?\ }$$

$$f(\underline{\ ?\ }) = 2(\underline{\ ?\ })^2 - 3(\underline{\ ?\ }) - 2 = -\frac{50}{16} = -\frac{25}{8}$$

So the minimum point is $\left(\frac{?}{?}, -\frac{25}{8}\right)$, or (0.75, -3.125).

Solution 2

Use a CAS to find the x- and y-intercepts.

Define $f1(x)=2 \cdot x^2-3 \cdot x-2$	Done
$f1(0)$	-2
solve$(f1(x)=0,x)$	$x=\frac{-1}{2}$ or $x=2$

fMin$(f1(x),x)$	$x=\frac{3}{4}$
$f1\left(\frac{3}{4}\right)$	$\frac{-25}{8}$

QY

> **QY**
> What is the domain of the function in Example 1?

Using Known Quadratic Models

In the 17th century, extending earlier work of Galileo, Isaac Newton showed that the height h of an object at time t after it has been thrown with an initial velocity v_0 from an initial height h_0 satisfies the formula

$$h = -\frac{1}{2}gt^2 + v_0t + h_0,$$

where g is the acceleration due to gravity. Recall that velocity is the rate of change of distance with respect to time; it is measured in units such as miles per hour or meters per second. Acceleration is the rate at which velocity changes, so it is measured in units such as miles per hour *per hour* or meters per second^2. Near the surface of Earth, g is approximately $32\ \frac{\text{ft}}{\text{sec}^2}$ or $9.8\ \frac{\text{m}}{\text{sec}^2}$.

Taking a dive
The world record for the highest dive is 53.90 meters. It took place in Villers-le-Lac, France.

GUIDED

Example 2

A ball is thrown upward from a height of 15 m with initial velocity 20 $\frac{\text{m}}{\text{sec}}$.

a. Find the relation between height h and time t after the ball is released.

b. How high is the ball after 3 seconds?

c. When will the ball hit the ground?

Solution

a. The conditions satisfy Newton's equation. Here v_0 = __?__ $\frac{\text{m}}{\text{sec}}$, and h_0 = __?__ m. Use the metric system value $g = 9.8$.

$$h = -\frac{1}{2}(\underline{\ ?\ })t^2 + \underline{\ ?\ }t + 15$$

$$h = -4.9t^2 + 20t + 15$$

b. Here $t = 3$ and you are asked to find h.

$$h = -4.9(3)^2 + 20 \cdot 3 + 15 = \underline{\ ?\ }$$

After 3 seconds, the ball is __?__ meters high.

c. At ground level, h = __?__. Solve $0 = -4.9t^2 + 20t + 15$ for t. Use a CAS to get $t \approx$ __?__ or $t \approx$ __?__. The negative value of t does not make sense in this situation, so we use the positive value. The ball will hit the ground after __?__ seconds.

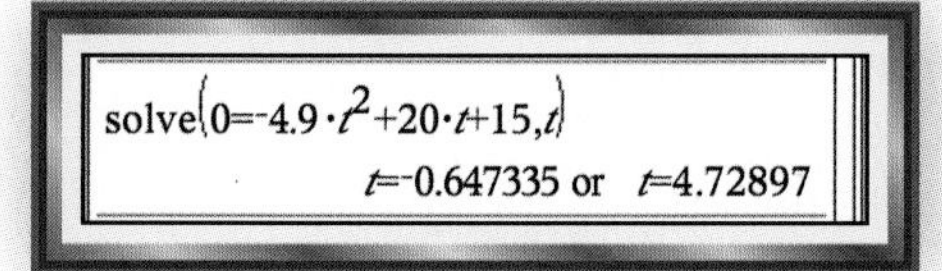

Finding the Quadratic Model through Three Points

The picture at the right shows that two points do not determine a parabola. To compute a unique quadratic model, you need a minimum of three noncollinear points.

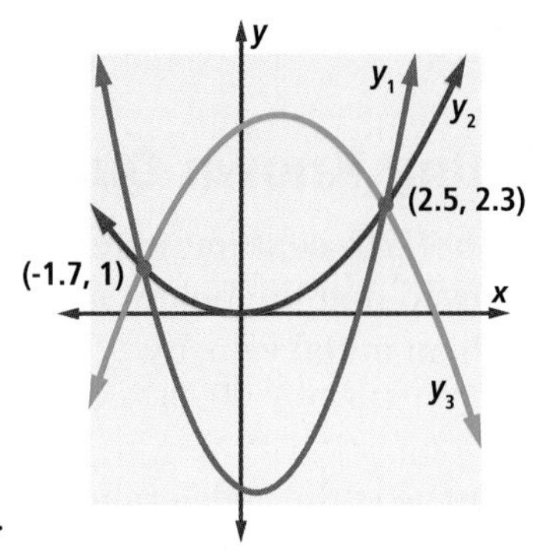

There are many parabolas that pass through the two points (-1.7, 1) and (2.5, 2.3).

One way to fit a quadratic model to data is to identify specific points on the model and set up a system of equations. The system must allow you to solve for the values of a, b, and c in the equation $y = ax^2 + bx + c$.

Alternately, you can use *quadratic regression*, a calculation available in many statistics packages. **Quadratic regression** is a technique, similar to the method of least squares, that finds an equation for the best-fitting parabola through a set of points.

Example 3

The parabola at the right contains points (1, –9), (6, –4), and (–0.2, 12.12). Find its equation.

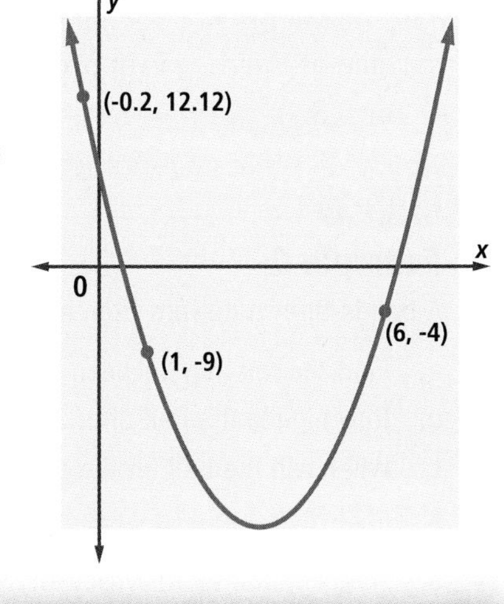

Solution 1 Because the ordered pairs (x, y) are solutions of the equation $y = ax^2 + bx + c$, substitute to get 3 linear equations, each with a, b, and c as unknowns.

$$\begin{aligned} f(1) &= -9 = a(1)^2 + b(1) + c \\ f(6) &= -4 = a(6)^2 + b(6) + c \\ f(-0.2) &= 12.12 = a(-0.2)^2 + b(-0.2) + c \end{aligned}$$

This produces a system of three equations.

$$\begin{cases} -9 = a + b + c \\ -4 = 36a + 6b + c \\ 12.12 = 0.04a - 0.2b + c \end{cases}$$

Use the `solve` command on a CAS. An equation for the parabola that contains these three points is $f(x) = 3x^2 - 20x + 8$.

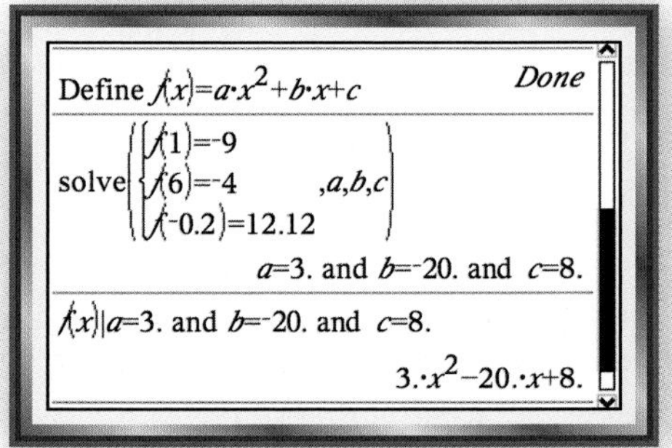

Solution 2 Use quadratic regression. You are asked to do this in Question 8.

Check Substitute the points into the equation.

Does $3(1)^2 - 20(1) + 8 = -9$? Yes.
Does $3(6)^2 - 20(6) + 8 = -4$? Yes.
Does $3(-0.2)^2 - 20(-0.2) + 8 = 12.12$? Yes, it checks.

When data points all lie on a single parabola, as in Example 3, the system strategy will yield an exact model. The model found by solving a system will be identical to the model formed by using quadratic regression. However, if the data show a quadratic trend, but not an exact quadratic fit, then the two solution strategies may yield slightly different equations.

Fitting a Quadratic Model through More Than Three Points

The following table contains data that might be collected by farmers interested in increasing the weight of their pigs. Suppose twenty-four randomly selected pigs were each given a daily dosage (in pellets) of a food supplement. Each group of three pigs received a dosage from 0 to 7 pellets, and the average percent weight gain for each group was recorded. The table below shows the average percent weight gain for each group of three pigs in relation to the number of pellets they were given daily.

Dosage (pellets)	0	1	2	3	4	5	6	7
Percent Weight Gain	10	13	21	24	22	20	16	13

The data show that more is not necessarily better. The pigs' bodies start rejecting the supplement when the dosage is too high. So there is a peak in the data and a quadratic model might be appropriate. A scatterplot of the data and the graph of the quadratic regression model $y = -1.0x^2 + 7.2x + 9.3$ are shown below. With the exception of the point (3, 24), the data points lie fairly close to the parabola.

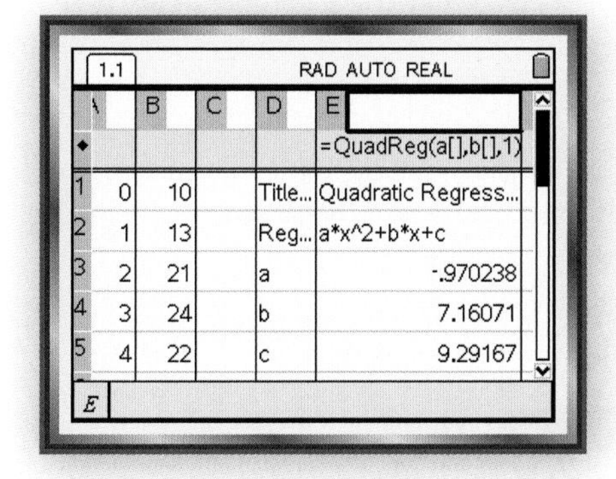

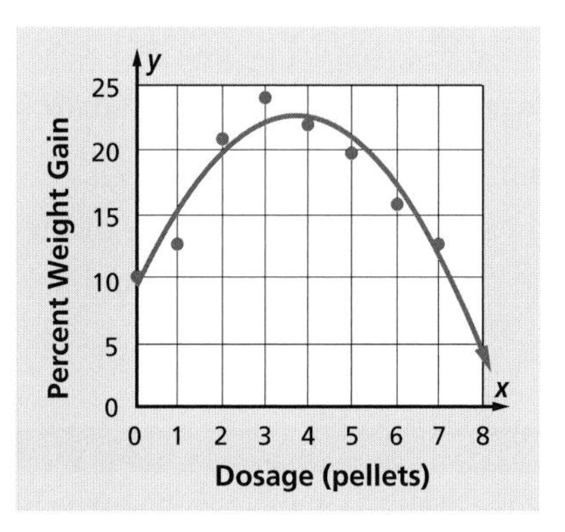

There is something quite different about this application when compared to the application in Example 2. There is no theory that links dosage with percent weight gain as there is with projectile motion. Models such as this one are called *impressionistic models* or *non-theory-based models*, because no theory exists that explains why the model fits the data. This is different from Example 2, where the well-established theory of gravity and all sorts of real data have verified that the height of a projectile is a quadratic function of time.

Questions

COVERING THE IDEAS

1. What is the general form of an equation of a quadratic function?
2. What values of x are the solutions to $ax^2 + bx + c = 0$?
3. What is the range of the function f in Example 1?
4. Consider the graph of the function f with $f(x) = 2x^2 - x - 4$.
   a. Give its y- and x-intercepts.
   b. Sketch the part of the graph where $-3 \le x \le 3$.
   c. Give the coordinates of the minimum point.
5. Tell whether the graph of the equation has a maximum point, a minimum point, or neither.
   a. $y = 8x^2 - 3x - 7$
   b. $y = 2x + 4x^2$
   c. $y = 6 - 2x^2$
   d. $y = -x^2 + 5x + 177$
6. Repeat Example 2 as if the ball were on the moon. Acceleration due to gravity on the moon is $1.6\ \frac{\text{m}}{\text{s}^2}$.

In 7 and 8, refer to Example 3.

7. Using the quadratic model, show that $f(4) = -24$.
8. Find the quadratic model by using quadratic regression.
9. Refer to the data about weight gain in pigs on page 121.
   a. What does the model predict for the percent weight gain for pigs fed 4.5 pellets daily?
   b. Is the prediction in Part a extrapolation or interpolation?
10. A parabola contains the points (0, 1), (4, 5), and (8, 7).
    a. Graph these points and estimate the coordinates of the vertex.
    b. Find an equation for the parabola by setting up and solving a system of equations.
    c. Check your estimate in Part a.

APPLYING THE MATHEMATICS

11. The table below shows the largest number of pieces $f(n)$ into which a pizza can be cut by n straight cuts.

 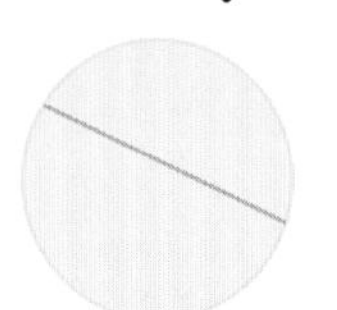 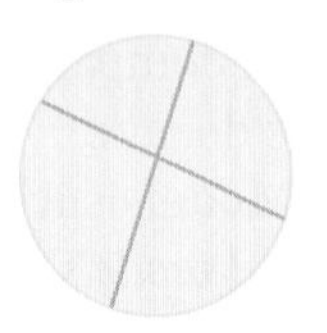 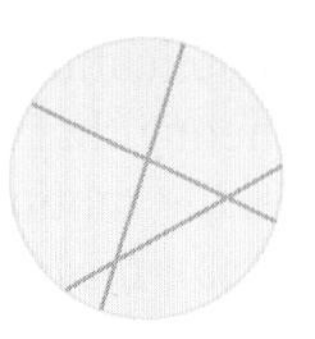 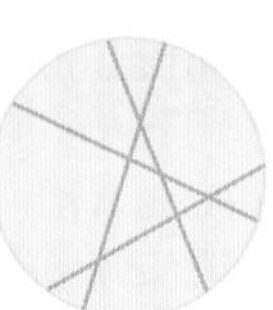

n	0	1	2	3	4
$f(n)$	1	2	4	7	11

   a. Fit a quadratic model to these data using regression.
   b. Use your model to find the greatest number of pieces produced by 5 straight cuts. Check your answer by drawing a diagram.

12. A piece of an artery or a vein is approximately the shape of a cylinder. The French physiologist and physician Jean Louis Poiseuille (1799–1869) discovered experimentally that the velocity v at which blood travels through arteries or veins is a function of the distance r of the blood from the axis of symmetry of the cylinder. For example, for a wide arterial capillary, the following formula might apply: $v = 1.185 - (185 \cdot 10^4)r^2$, where r is measured in cm and v in $\frac{\text{cm}}{\text{sec}}$.
    a. Find the velocity of blood traveling on the axis of symmetry of this capillary.
    b. Find the velocity of blood traveling $6 \cdot 10^{-4}$ cm from the axis of symmetry.
    c. According to this model, where in the capillary is the velocity of the blood 0?
    d. What is the domain of the function mapping r onto v?
    e. Sketch a graph of this function.

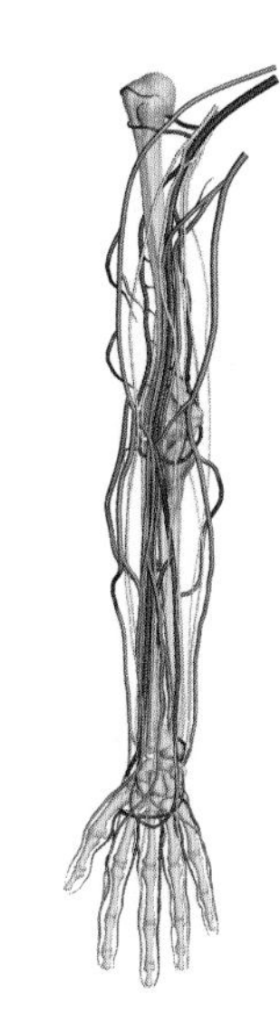

13. Use the table at the right showing the amount of bar iron exported to England from the American Colonies from 1762 to 1774. Bar iron is measured in "old" tons of 2240 pounds.
    a. Construct a scatterplot for these data with the independent variable as years after 1762.
    b. Find the quadratic regression model for these data.
    c. Use your quadratic model to predict the amount of bar iron exported in 1776. (The actual value was 28 old tons.)
    d. Why is extrapolation to 1776 inappropriate?

Year	Bar Iron Exported (old tons)
1762	110
1763	310
1765	1079
1768	1990
1770	1716
1771	2222
1773	838
1774	639

Source: U.S. Census Bureau

14. The Center for Disease Control studies trends in high school smoking. The percent of students in grades 9 through 12 who reported smoking cigarettes on 20 of the 30 days preceding the administration of the National Youth Risk Behavior Survey (frequent cigarette use) increased in the 1990s, but decreased after 1999.

Years after 1990	1	3	5	7	9	11	13	15
% of Frequent Cigarette Use $= y$	12.7	13.8	16.1	16.7	16.8	13.8	9.7	9.4

    a. Construct a scatterplot for these data.
    b. Calculate the sum of squared residuals for the quadratic model $y = -0.2x^2 + 2x + 11$.
    c. Find the quadratic regression model for the data. Calculate the sum of squared residuals for this model.
    d. Which model has a smaller sum of squared residuals?
    e. Using the regression model, predict the cigarette use in 2006 and 2010. Do you think the predictions are reasonable?

REVIEW

15. The half-life of Th-232 (Thorium) is 14.05 billion years. Suppose a sample contains 100 grams of pure thorium. **(Lesson 2-5)**
    a. Find an equation for the amount of Th-232 left as a function of the number of billions of years that have passed.
    b. Find the amount left after 13.7 billion years, the estimated age of our atmosphere.

In 16 and 17, Sri Lanka and Madagascar are two island nations in the Indian Ocean. In 2007, the population of Sri Lanka was about 20.9 million people and the population of Madagascar was about 19.4 million people. In 2007, the population of Sri Lanka was growing at a rate of about 0.98% annually, while the population of Madagascar was growing at a rate of about 3.01% annually. Let the function $S(x) = ab^x$ represent the population (in millions) of Sri Lanka x years after 2007, and let the function $M(x) = cd^x$ represent the population (in millions) of Madagascar x years after 2007.

Kandy
A town in the center of Sri Lanka

16. a. Determine a, b, c, and d and write formulas for $S(x)$ and $M(x)$.
    b. Assuming constant growth rates, use your formulas from Part a to estimate the populations of Sri Lanka and Madagascar in 1997.
    c. Graph $y = S(x)$ and $y = M(x)$ on the same set of axes for $0 \le x \le 50$.
    d. Make a prediction about how the future populations of Sri Lanka and Madagascar will compare if current trends continue. **(Lesson 2-4)**

17. When the function S is used to predict the population of Sri Lanka in 2008, the residual is 2.848. What was the observed population? **(Lesson 2-2)**

18. What are the maximum and minimum possible values for a correlation coefficient? **(Lesson 2-3)**

19. ***Skill Sequence*** Solve for x. **(Previous Course)**
    a. $(5x - 10)(x - 3) = 0$ b. $3x^2 = 1 - 2x$ c. $3x^4 = 1 - 2x^2$

20. Sketch the graph of $y = \frac{1}{x}$. **(Previous Course)**

EXPLORATION

21. a. Find the vertices for the family of quadratic functions $y = x^2 + bx + 1$ for $b = \{-4, -3, -2, -1, 0, 1, 2, 3, 4\}$.
    b. Graph all of your collected vertices in a scatterplot.
    c. Find an exact quadratic model for the data.
    d. If you had started this question with $y = 2x^2 + bx + 1$, what would you predict the quadratic model of the vertex data to be? Why?

QY ANSWER

$\mathbb{R}$

Lesson 2-7 Inverse Variation Models

Vocabulary

varies inversely as, is inversely proportional to
constant of variation, constant of proportionality
inverse-square relationship
varies inversely as the square of, is inversely proportional to the square of
power function

BIG IDEA Inverse and inverse square functions model many physical situations.

Mental Math

Rewrite the expression as a power of a single variable, if possible.

a. $\frac{1}{x}$

b. $\frac{1}{y^2}$

c. $(z^{-3})^5$

d. $(\sqrt{w})^{24}$

Inverse Variation

Suppose you have 6 pounds of ground meat to make into hamburger patties of equal weight. The more patties you make, the less each patty will weigh. More specifically, the weight W of each patty is related to the number N of patties produced by the equation

$$W = \frac{6}{N}.$$

In this situation, the weight per patty *varies inversely as* (or *is inversely proportional to*) the number of patties.

QY1

QY1

Using the relationship above, make a table of the weight W per patty for $N = 12, 24, 36,$ and 48 patties. Will the weight ever equal zero?

In general, we say that y **varies inversely as** x (or y **is inversely proportional to** x) whenever $y = \frac{k}{x}$. The parameter k is called the **constant of variation** (or **constant of proportionality**), and cannot be 0. In the hamburger situation, it is impossible for either N or W to be negative, but in other situations the domain and range may include negative values. Graphing $y = \frac{k}{x}$ for both positive and negative values of x shows that there are two basic types of graphs, depending on the value of k. Sometimes you will see the expression $\frac{k}{x}$ written as kx^{-1}.

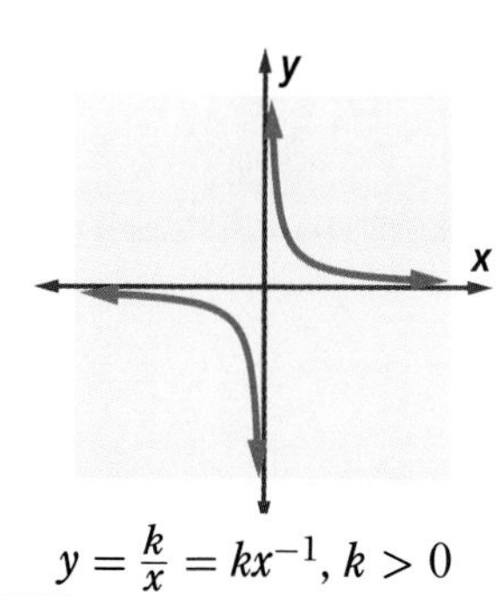

$y = \frac{k}{x} = kx^{-1}, k > 0$

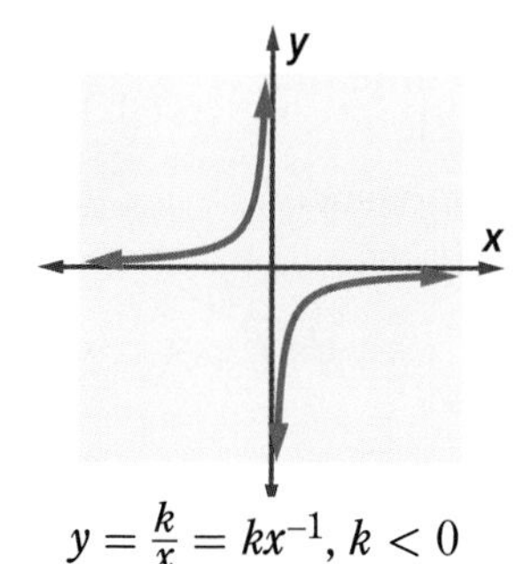

$y = \frac{k}{x} = kx^{-1}, k < 0$

Activity

Step 1 Graph $y = \frac{k}{x}$ for different positive values of k. On some graphing utilities, you can use a slider for k as shown on the next page. How does increasing the value of k affect the graph?

(continued on next page)

Step 2 Graph $y = \frac{k}{x}$ for different negative values of k. How does the sign of k affect the graph? Which features of the graph are invariant, that is, which features do not depend on the value of k?

Step 3 Fix a particular positive value of k and pick a point on the graph with $x > 0$. Then trace the graph, moving to the left. As x gets closer to 0, what happens to the y-coordinate?

Step 4 Now pick to a point on the graph with $x < 0$. Then trace the graph, moving to the right. As x gets closer to 0, what happens to the y-coordinate?

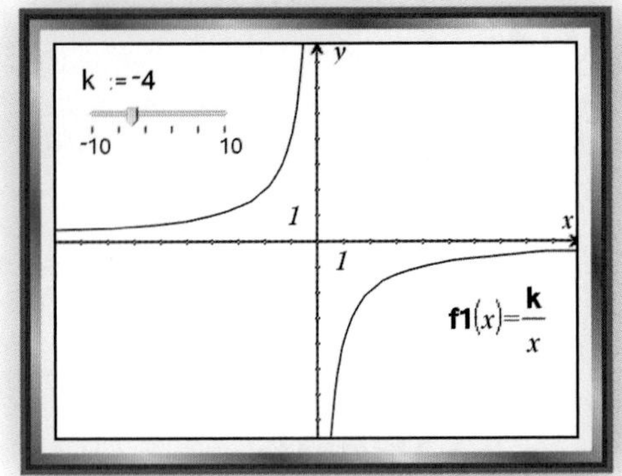

The graphs suggest that there are points on the graph of $y = \frac{k}{x}$ corresponding to all real values of x except $x = 0$, and to all real values of y except $y = 0$. The domain of the function with equation $y = \frac{k}{x}$ is therefore $\{x|\ x \neq 0\}$ and the range is therefore $\{y|\ y \neq 0\}$.

The graphs in the activity are *hyperbolas*. Both hyperbolas have a *horizontal asymptote* at $y = 0$. In numerical terms, as x gets larger, y gets closer and closer to 0. In the hamburger situation, this means that with a fixed amount of meat, as the number of hamburgers increases, the weight of each burger decreases until, with enough burgers, the weight of each can be as small as you wish.

In addition, both graphs have a *vertical asymptote* at $x = 0$, because as x gets closer to 0, y gets larger and larger (or more and more negative) without ever reaching a bound.

Determining the Constant of Variation

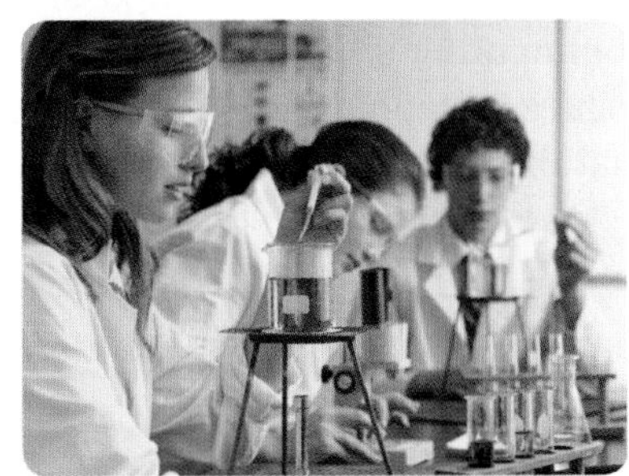

When one quantity varies inversely as another, you can determine the constant of variation from one data point. Inverse variation is common in the physical world. For instance, according to Boyle's Law, the volume of a gas varies inversely as the pressure. Pressure is measured in kilopascals (kPa) and volume is measured in milliliters (mL).

GUIDED

Example 1

In a chemistry lab, you collect data on the pressure and volume of a gas.

Volume (mL)	20	30	40	50	60	70	80	90	100
Pressure (kPa)	253.3	160.5	120.9	101.6	84.6	70.9	64.2	53.8	49.3

a. **Find a formula relating the pressure and volume of the gas sample you studied in the lab.**

b. **Graph both the data and the model on a single set of axes.**

c. **Use residuals to assess the quality of your model.**

d. **What volume corresponds to a pressure of 40 kPa?**

Solution

a. The model is of the form $P = \frac{k}{V}$. We start by selecting a "typical" point in the middle of the data set: $(V, P) = (60, 84.6)$.

Set $84.6 = \frac{k}{?}$ and solve for k.

$k = (\underline{\ ?\ })(\underline{\ ?\ }) = 5076$.

Therefore, $P = \underline{\ ?\ }$.

b. Enter the data into two columns of a spreadsheet. Generate a scatterplot and superimpose a graph of the model. The graph is shown at the right.

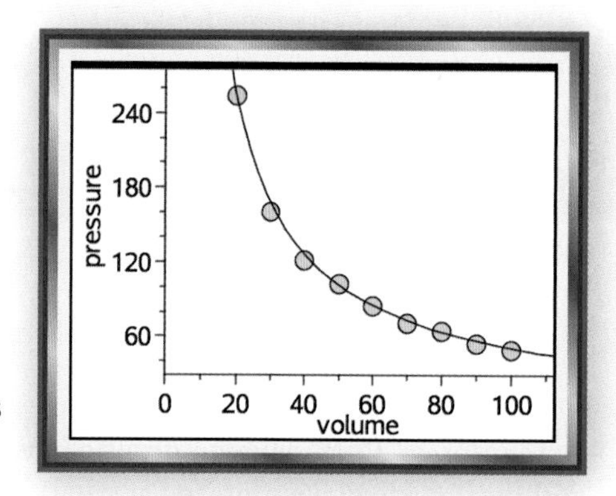

c. Add a third column that computes the predicted values and a fourth that computes the difference between the observed values and the predicted values as shown at the right.

There are more negative than positive residuals, so we might consider ___?___ (increasing/decreasing) the value of k we derived, but the residuals are not large and do not show a clear pattern that would suggest a different model shape. The low absolute values of the residuals and the lack of a clear pattern suggest that the model is fairly accurate.

	A vol...	B pres...	C model	D residual
◆			=5076./volu	
1	20	253.3	253.8	-0.5
2	30	160.5	169.2	-8.7
3	40	120.9	126.9	-6.
4	50	101.6	101.52	0.08
5	60	84.6	84.6	0.
6	70	70.9	72.5143	-1.61429

D1 =*b1*−*c1*

d. Substitute 40 for P in the equation from Part a: $40 = \frac{5076}{V}$. Then solve for V: $V = \underline{\ ?\ }$ mL.

Inverse-Square Relationships

In science contexts, the relationship $y = \frac{k}{x^2} = kx^{-2}$ is very common. Such relationships are called **inverse-square relationships**, and we write ***y* varies inversely as the square of *x*** (or ***y* is inversely proportional to the square of *x***). In many respects, inverse-square relationships are similar to inverse-variation situations.

Example 2

The force exerted by the electrical field between two charged objects is inversely proportional to the square of the distance between them. At a distance of 1.5 meters, you measure a force of 30 newtons. What will be the force at a distance of 3.0 meters? At 15 meters?

Solution First compute the constant of variation, then substitute the known distances to find the unknown force.

Because $F = \frac{k}{d^2}$, we can write $30 \text{ N} = \frac{k}{(1.5 \text{ m})^2}$ and solve for k.

$k = 30 \text{ N} \cdot 1.5^2 \text{ m}^2 = 67.5 \text{ N} \cdot \text{m}^2$.

Then use the equation $F = \frac{67.5 \text{ N} \cdot \text{m}^2}{d^2}$, substituting 3.0 m and 15 m for d.

When $d = 3.0$ m, the force is 7.5 N; when $d = 15$ m, the force is 0.3 N.

Graphs of $y = \frac{k}{x^2}$ share many of the features of the graphs of $y = \frac{k}{x}$.

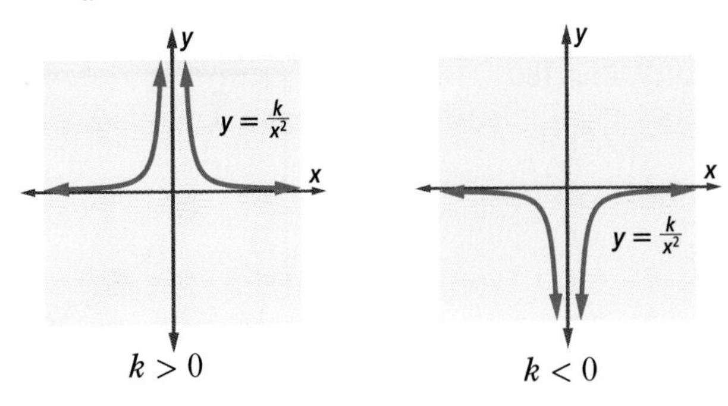

Asymptotes: Both types of graphs have horizontal asymptotes at $y = 0$ and have vertical asymptotes at $x = 0$.

Domain: Because $\frac{k}{x^2}$ is defined for all values of x except when $x = 0$, the domain of $y = \frac{k}{x^2}$ is $\{x|\ x \neq 0\}$. $y = \frac{k}{x}$ has the same domain.
Range: The graphs above suggest that the range of $y = \frac{k}{x^2}$ depends on the value of k: for $k > 0$, the range is $\{y|\ y > 0\}$, while for $k < 0$, the range is $\{y|\ y < 0\}$. The inverse variation function has range $\{y|\ y \neq 0\}$.
As with $y = \frac{k}{x}$, making the absolute value of k larger moves the graph of $y = \frac{k}{x^2}$ further away from the origin.
Note also that as $|x|$ increases, y gets close to zero much more quickly for $y = \frac{k}{x^2}$ than for $y = \frac{k}{x}$.

 QY2

> **▶ QY2**
>
> Consider $f(x) = \frac{6}{x}$ and $g(x) = \frac{6}{x^2}$. Compare $f(10)$ and $g(10)$, $f(100)$ and $g(100)$, $f(1000)$ and $g(1000)$. For each pair, which value is closer to zero?

Reciprocals of Power Functions

Recall that a **power function** is a function with an equation of the form $y = ax^n$, where n is an integer greater than 1. In this case we say that y varies directly as x^n. The reciprocal of a power function is a function of the form $y = \frac{1}{ax^n}$, or alternatively, $y = bx^{-n}$, where the coefficient $b = \frac{1}{a}$. In these cases, we say that y varies inversely as x^n. Inverse-variation functions are the reciprocals of direct-variation functions. Properties of these functions are summarized in the tables below.

Power Functions	Reciprocal Power Functions
$y = ax^n$	$y = \frac{a}{x^n}$
pass through origin	have vertical asymptote at $x = 0$
domain is R	domain is $\{x \mid x \neq 0\}$
range is R (odd exponents) or $\{y \mid y \geq 0\}$ (even exponents)	range is $\{y \mid y \neq 0\}$ (odd exponents) or $\{y \mid y > 0\}$ (even exponents)
rise sharply as x gets larger and larger	approach 0 as x gets larger and larger
rise or fall sharply as x gets smaller and smaller (depending on whether n is even or odd)	approach 0 as x gets smaller and smaller

Questions

COVERING THE IDEAS

1. **Multiple Choice** In which equation does W vary inversely as the square of g?

   A $W = \frac{k}{g}$ B $W = kg^2$ C $W = \frac{k}{g^2}$ D $W = k\sqrt{g}$

2. **Multiple Choice** Which of the following is *not* a characteristic of the function $y = \frac{k}{x}$ or its graph?

   A domain is the set of all real numbers

   B horizontal asymptote at $y = 0$

   C vertical asymptote at $x = 0$

   D shape is a hyperbola

3. Suppose that y varies inversely as x, and that $y = 45$ when $x = 10$.

   a. Compute the constant of variation.

   b. Find y when $x = 2$.

4. Suppose 240 hot dogs were ordered for a picnic, and x people finished them all.

   a. Write a formula for y, the mean number of hot dogs each person ate.

   b. Graph the relation you found in Part a.

   c. Your graph has a horizontal asymptote. Find its equation, and explain its meaning in the context of the problem.

5. What kind of variation is described by $y = 11.1x^{-2}$?

6. Refer to Example 2. When two electrically-charged particles are 0.4 m apart, the force between them is 12 N. What will the force be when they are 0.2 m apart?

7. In a chemistry experiment, the data in the table below were collected on the pressure and volume of a sample of gas. According to Boyle's Law, the pressure varies inversely as the volume.

***V* (mL)**	200	220	240	260	280	300	320
***P* (kPa)**	142.5	131.2	119.6	112.9	101.7	103.2	95.4

   a. Use the data point (240, 119.6) to compute the constant of proportionality.

   b. Write a formula for P in terms of V using your value from Part a.

   c. Graph the data and the formula you found in Part b on the same set of axes.

   d. Compute the sum of squared residuals for this model.

APPLYING THE MATHEMATICS

8. **True or False** The time it takes you to walk a certain distance is inversely proportional to your average speed.

9. The acceleration of a falling object due to Earth's gravity varies inversely as the square of the object's distance from the center of Earth.
   a. Earth's radius at sea level is about 6378 km and the acceleration due to gravity on Earth's surface is about 9.8 $\frac{\text{m}}{\text{s}^2}$. Compute the constant of proportionality.
   b. Compute the acceleration due to Earth's gravity for an object in orbit 10,000 km above Earth's surface.
   c. Compute the acceleration due to Earth's gravity for an object as far away as the Moon, 384,400 km from Earth's center.

REVIEW

10. Data records show the distance a ball traveled when hit at various angles at a constant bat velocity of 100 mph. **(Lesson 2-6)**

Angle (degrees)	35	40	55	65	70	75
Distance (reached feet)	294	308	294	239	201	156

   a. Construct a scatterplot for these data.
   b. Find a quadratic model for these data.

11. A wrecking ball swings to knock down a building. The following data were collected giving the height of the ball during its swing at various distances from the building. **(Lesson 2-6)**

Distance (feet)	100	75	50	25	0	-25
Height reached (feet)	601	536	499	501	510	554

   a. Find a quadratic regression model for these data.
   b. Determine the sum of squared residuals for your model.

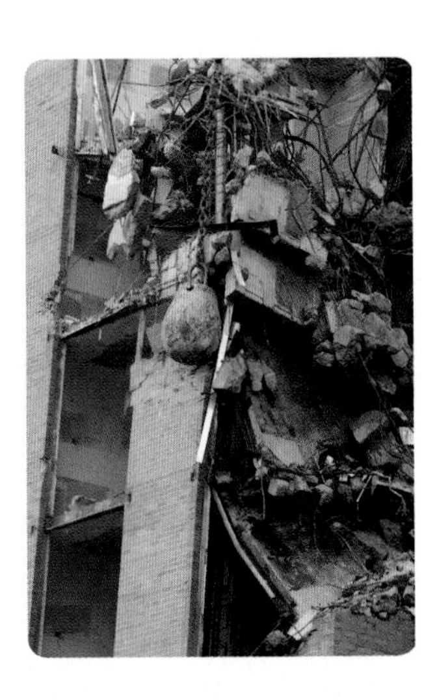

12. Consider the function $r: t \to 5(0.83)^t$. **(Lesson 2-4)**
   a. Give the domain of r.
   b. Give the range of r.
   c. State equations for any asymptotes of the graph of r.

13. Henry looked at the census data for Iowa City, Iowa, for the years 1960, 1970, 1980, 1990 and 2000 and used it to create a linear model of population growth. You do not need to have the data to answer these questions. **(Lesson 2-2)**
   a. Henry used his model to predict the population of Iowa City in 1984. Was his prediction a result of interpolation or extrapolation?
   b. Henry also used his model to predict the population of Iowa City in 2016. Would you expect his prediction for 1984 or his prediction for 2016 to have a larger error? Explain your answer.

14. The table at the right shows the winning jumps in the men's long jump event at the Olympic games. **(Lesson 2-3)**
    a. Make a scatterplot of these data.
    b. Find the line of best fit predicting the winning jump for a given year.
    c. What does the slope of the line tell you about the average rate of change in the length of the winning long jump from 1896 to 2004?
    d. Use the line of best fit to predict the winning jump for the Beijing Olympics in 2008.
    e. What is the residual of the prediction? (The actual jump by Irving Jahir Saladino Aranda was 8.34 m.)
    f. Which data points seem to be outliers here? Give a plausible explanation for why those outliers might have occurred.

Year	Gold Medalist	Jump
1896	Ellery Clark, United States	6.34 m
1900	Alvin Kraenzlein, United States	7.19 m
1904	Myer Prinstein, United States	7.34 m
1908	Francis Irons, United States	7.48 m
1912	Albert Gutterson, United States	7.60 m
1920	William Pettersson, Sweden	7.15 m
1924	DeHart Hubbard, United States	7.45 m
1928	Edward B. Hamm, United States	7.74 m
1932	Edward Gordon, United States	7.64 m
1936	Jesse Owens, United States	8.06 m
1948	Willie Steele, United States	7.82 m
1952	Jerome Biffle, United States	7.57 m
1956	Gregory Bell, United States	7.83 m
1960	Ralph Boston, United States	8.12 m
1964	Lynn Davies, Great Britain	8.07 m
1968	Robert Beamon, United States	8.90 m
1972	Randy Williams, United States	8.24 m
1976	Arnie Robinson, United States	8.35 m
1980	Lutz Dombrowski, East Germany	8.54 m
1984	Carl Lewis, United States	8.54 m
1988	Carl Lewis, United States	8.72 m
1992	Carl Lewis, United States	8.67 m
1996	Carl Lewis, United States	8.50 m
2000	Ivan Pedroso, Cuba	8.55 m
2004	Dwight Phillips, United States	8.59 m

Source: Athletics Weekly 2008

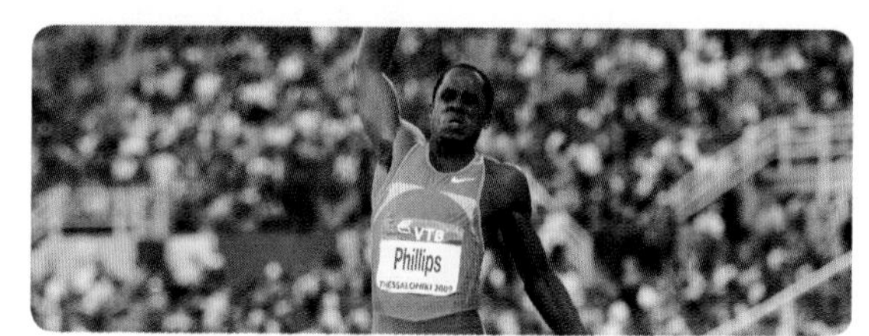

Dwight Phillips at the World Athletics Final, 2009

15. a. Which is generally least affected by outliers in the data set, the mean or median?
    b. If a data set is skewed with a tail to the left, then which is larger: the mean or median? **(Lessons 1-3, 1-2)**

EXPLORATION

16. According to the mathematics in the lesson, it is never possible to entirely escape Earth's gravity: since gravitational force varies inversely as the square of the distance, no matter how large the distance gets, the force of gravity from Earth never actually equals zero. Yet astronomical missions such as the Voyager continue to travel away from Earth, without burning rockets continually. Research the concept of escape velocity and explain why such missions are possible.

17. In Question 7, you used a data point to determine an inverse-variation model for the data. Experiment with different data points to see if you can find a model with a lower sum of squared residuals.

QY ANSWERS

1. no;

N	12	24	36	48
W	$\frac{1}{2}$	$\frac{1}{4}$	$\frac{1}{6}$	$\frac{1}{8}$

2. $g(x)$ is closer to zero than $f(x)$ for $x = 10, 100$, and 1000.

Lesson 2-8 Choosing a Good Model

Vocabulary

residual plot

▶ **BIG IDEA** The residual plot can help you choose a model for a given data set.

You have seen how to fit linear, exponential, quadratic, and inverse variation models to data. One aspect of modeling, however, is worthy of further discussion: *How do you know you have found a good model?* If two models appear to fit the data equally well, it is usually wise to pick the simpler of the two, but how can you tell how well data fit a model? One measure of how well a model fits the data is the correlation coefficient, but this applies only to linear models.

Another method to determine how well a model fits data is to analyze the residuals. When you evaluate a model at the x-value for a particular data point, you are likely to get a predicted y-value that is different from the observed y-value. Recall that the residual is the difference.

residual = error = observed y-value − predicted y-value

Mental Math

Name the figure that is the graph of the equation.

a. $3xy = 4$

b. $3x + y = 4$

c. $3x^2 + y = 4$

d. $y = 3 \cdot 4^x$

Residual Plots

The graph at the left below shows a scatterplot and linear regression model for a data set. The graph at the right below shows the *residual plot* for that model. The point (9, 5.22) marked on the scatterplot has its corresponding point marked on the residual plot, (9, 0.96). This means that the data point $x = 9$ has an observed value of 5.22 and a residual value of 0.96 under the linear regression model.

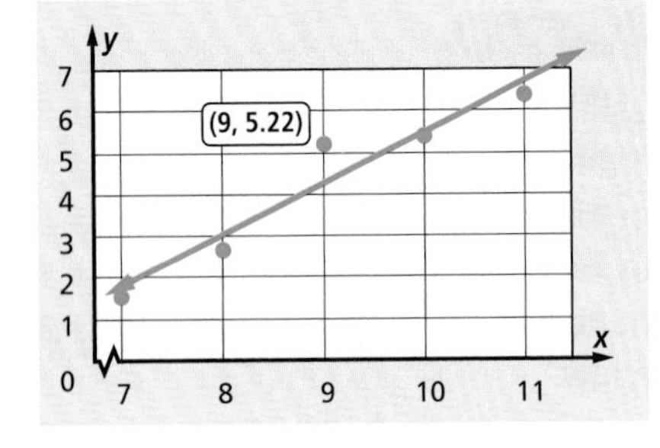

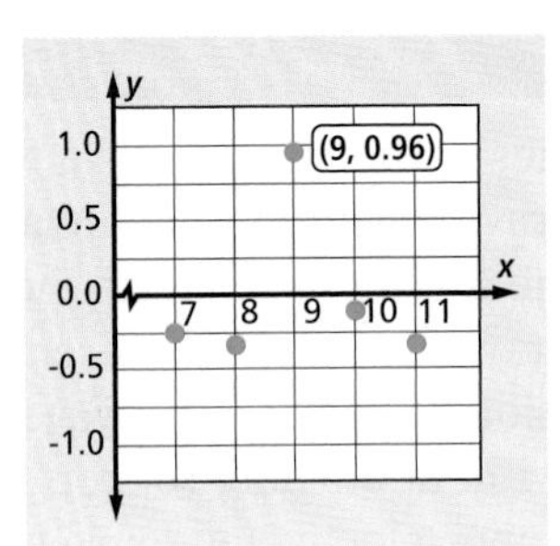

STOP QY

▶ **QY**

What is the predicted value for $x = 9$?

A **residual plot** pairs each x-value from the data set with its residual. If the residuals are clustered around the x-axis, as shown in the leftmost diagram below, the model is likely to be a good fit for the data. If, however, the residuals have a different pattern, then a better model can probably be found. The second and third graphs below show patterns of residuals indicating that both models need improvement.

Residual Pattern:	clustered around x-axis	same side of axis at ends	shaped like funnel
Residual Plot:		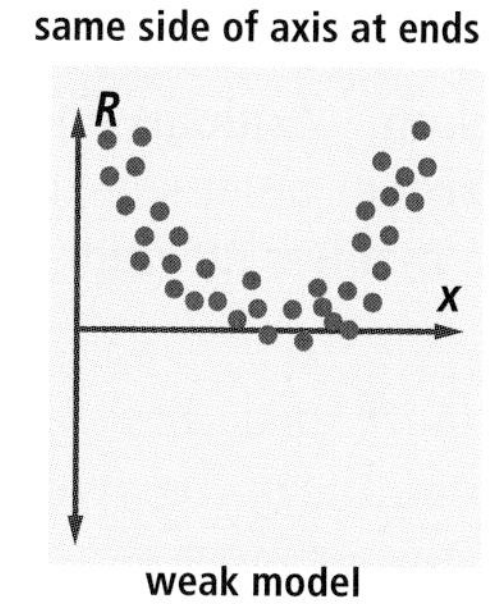	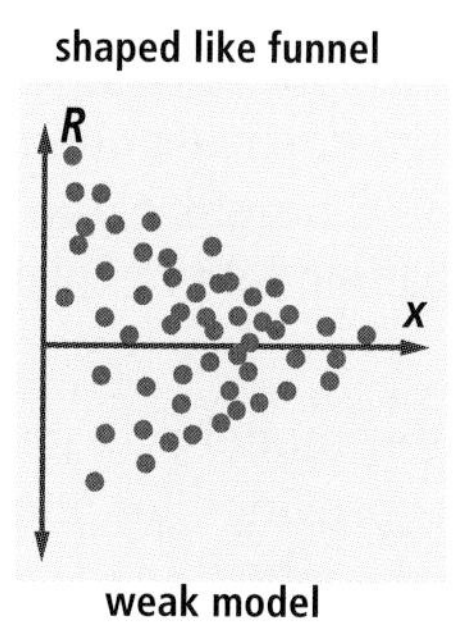
	good model	weak model	weak model

Analyzing Residuals for a Linear Model

Activity 1

The table at the right shows the length *L* of each day (sunrise to sunset) observed at a city in the Northern Hemisphere every 10 days for 100 days beginning on August 31st. *D* is the number of days after this date.

D (days)	*L* (minutes)
0	793
10	766
20	739
30	711
40	684
50	657
60	631
70	607
80	586
90	568
100	556

Step 1 Use a statistics utility to create the scatterplot of the data.

Step 2 Find the regression line and correlation coefficient for these data. What does the correlation coefficient indicate about a linear model for these data?

Step 3 Reproduce the spreadsheet below. Then use your calculated regression line to fill in both the predicted day lengths and the associated residuals. A few entries have already been filled in.

◇	A	B	C	D
1	Days (D)	Observed Day Length (L)	Predicted Day Length (p)	Residual (R = L – p)
2	0	793	786.05	6.95
3	10	766	761.53	4.47
4	20	739		
5	30	711		-1.49
6	40	684		
7	50	657		
8	60	631		
9	70	607		
10	80	586		
11	90	568		
12	100	556		

(continued on next page)

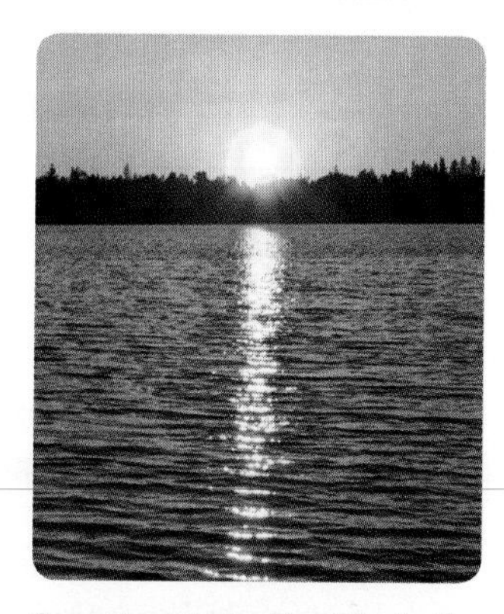

Sunset on Lake Franklin in the Chequamegon-Nicolet Forest in Northern Wisconsin

Step 4 Plot the residual set of points (D, R) on the same axes as the scatterplot in Step 1.

Step 5 Based on the residual plot, what do you conclude about a linear model for these data?

Activity 1 tested the theory that the number of minutes of daylight decreases in a linear fashion. Even though the correlation coefficient in Step 2 indicates that a line is a good model, the residuals show that there is a better model. Therefore, the researcher must seek another theory or more realistic model to explain the manner in which daylight decreases.

One way to seek a better model is to gather more data. In Activity 1, the hours of sunlight were provided for about 3 months of the year. This is not the full domain of the situation. Data for two or more years would show a periodic wavy pattern that requires functions you will study in later chapters of these book.

Analyzing Residuals for Other Models
Exponential Model

Activity 2

Step 1 Use a statistics utility to create a scatterplot of the U.S. Census data shown at the right. Use years after 1790 as the independent variable.

Step 2 Compute the exponential regression model for these data.

Step 3 Using the exponential regression model, compute the predicted populations and residuals. Organize the data in a spreadsheet like the one shown below.

Step 4 Plot the set of residual points (Y, R).

Step 5 Based on the residual plot, why is the exponential model not a good fit for these data?

◇	A	B	C	D	E
1	Year	Year – 1790 = (Y)	Population (m)	Predicted Population (p)	Residual (R = m – p)
2	1790	0	4	6.03	-2.03
3	1800	10	5	7.39	-2.39
4	1810	20	7	9.04	-2.04

Year	Population (millions)	Year	Population (millions)
1790	4	1900	76
1800	5	1910	92
1810	7	1920	106
1820	10	1930	123
1830	13	1940	132
1840	17	1950	151
1850	23	1960	179
1860	31	1970	203
1870	40	1980	227
1880	50	1990	249
1890	63	2000	281

Source: U.S. Census Bureau

Activity 2 suggests that although exponential regression models are useful for modeling population growth, they frequently break down over time. Limited resources and other factors prohibit populations from growing indefinitely.

A graph of the quadratic regression model of the U.S. population data, $y = 0.007x^2 - 0.1138x + 6.1097$, is shown at the left below. It turns out that this quadratic model is a much better fit than the exponential model. The residuals are relatively small and they cluster in a horizontal band centered around the x-axis. However, the quadratic model is an impressionistic model of population growth because there is no theory that supports a quadratic relationship between year and population.

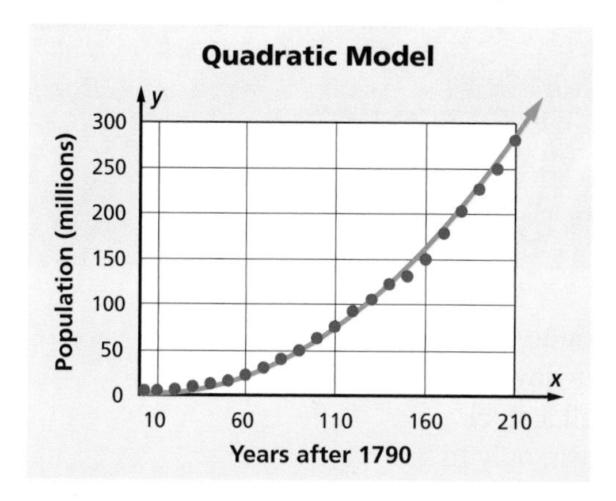

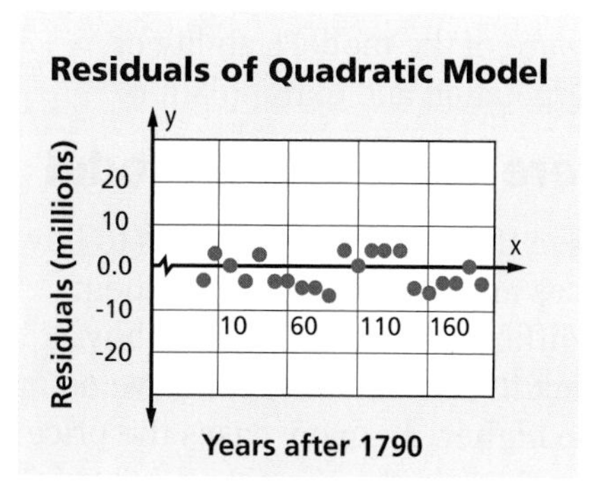

As mentioned in earlier lessons, extrapolation is risky business. This is particularly true when there is no theory to support the model. The quadratic model is a very good model for the population of the U.S. from 1790 to 2000, but there is no assurance that the model will make accurate predictions for years outside the data set. An even better model than the quadratic model might be a piecewise function, with each piece chosen to best fit a portion of the domain.

Careful Modeling

You should consider at least five things when building a model from data. Assuming that your data are a representative sample of the population of interest, you should:

1. Build a model from theory, if possible. Some real-world situations suggest certain models.
2. Graph the data on a scatterplot. Draw the model on the same graph. A good model should follow any pattern or trend in the data.

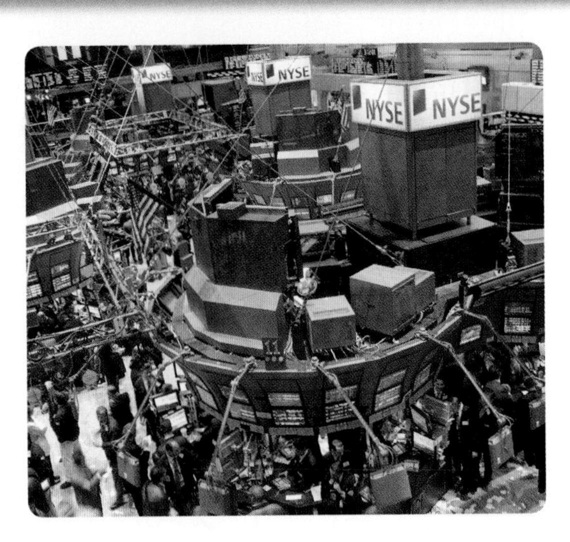

3. Graph the residuals. If the residual plot does not fall within a relatively narrow horizontal band centered around the x-axis or if there is a pattern to the residuals, you may have to change your theory or look for a better model.
4. Use the correlation coefficient to check whether a linear model is appropriate.
5. In all cases, be aware of the model's ability or limitation for interpolation and extrapolation.

Sometimes There Is No Good Model

Everyone who invests in the stock market wants to buy stocks when their prices are low and sell when their prices are high. The difficulty is that when you buy today because you think the price is low, you have no guarantee that the price tomorrow will be higher. In many cases the price goes down and your investment loses value. In the same way, you might sell a stock today because you think prices are high and likely to go down, only to find out that the prices are even higher tomorrow. No model has been developed that can accurately predict changes in the stock market, but many people make a living by developing models that work for a short time or in special situations.

The graph below shows the Dow Jones industrial average from July, 2001 to March, 2009.

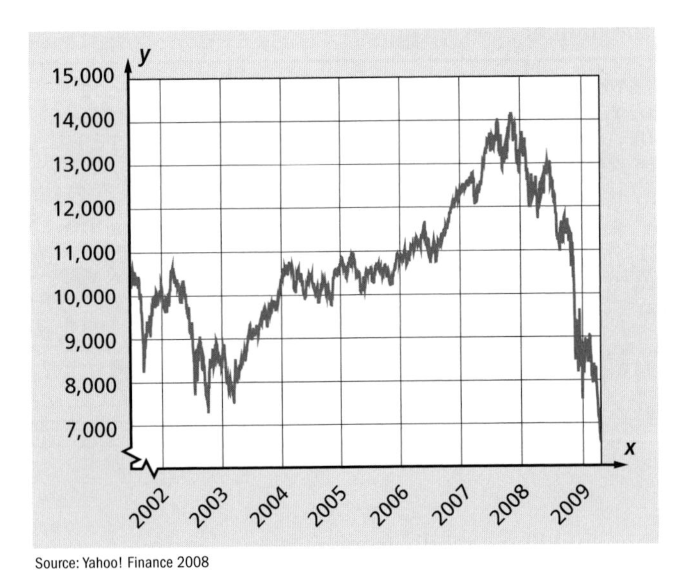

Source: Yahoo! Finance 2008

The graph shows many individual trends, but no overall, consistent trend. There are many fluctuations. Catastrophic events often cause significant changes in the graph.

One example of this was the destruction of the World Trade Center on September 11, 2001, which initiated a drop in the stock market. (See if you can locate the drop in the Dow-Jones average due to September 11 on the graph.) Even the most sophisticated stock traders with powerful statistical models were caught off guard by this event.

Questions

COVERING THE IDEAS

1. What characterizes a residual plot of a good model?

2. **Multiple Choice** Look at the residual plots below. Which indicates the best model? Explain your answer.

A B C

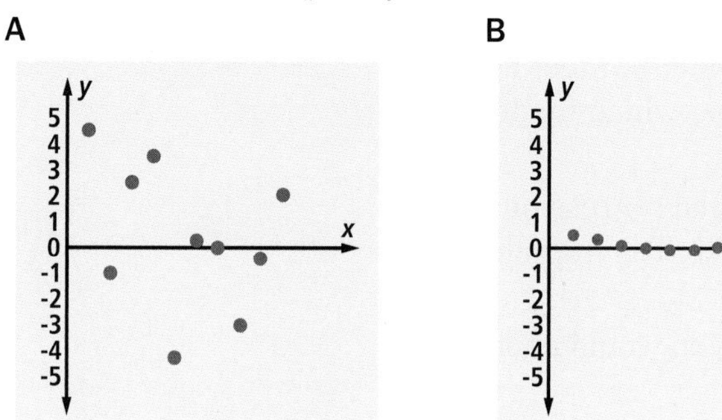

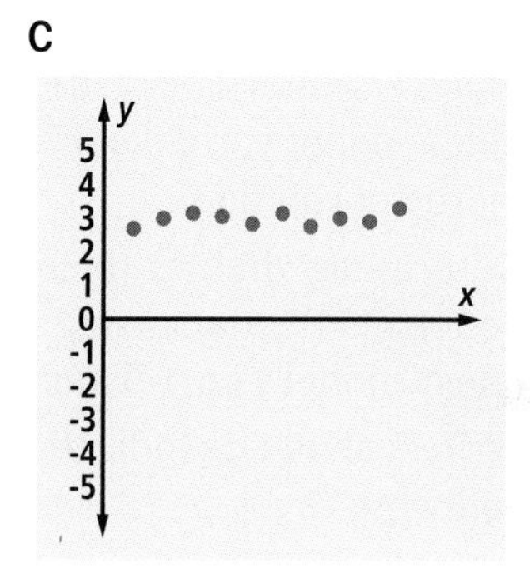

3. a. An exponential regression model for the length L of the Dth day in Activity 1 is $L = 792.63(0.996312)^D$. Compute a table of predicted values using this model.

   b. Compute a table of residual values using the results from Part a.

   c. Create a plot of the residuals from Part b and explain if the exponential regression model is a good model for this data.

In 4–6, refer to Activity 2.

4. Why is it reasonable to expect that an exponential model is an appropriate theoretical model for the population data?

5. a. Find an exponential regression model for the population data using only the years from 1790 to 1880.

   b. Make a spreadsheet with year, census population, predicted population, and residual.

   c. Plot the residuals.

   d. Is an exponential model appropriate for the 1790–1880 time period? Explain your decision.

6. a. Use the quadratic regression equation to predict the 1900 population to the nearest million.

   b. Use the value in Part a to calculate the residual for the 1900 population.

7. Why do you think no good model has been found for predicting the future prices of stocks?

APPLYING THE MATHEMATICS

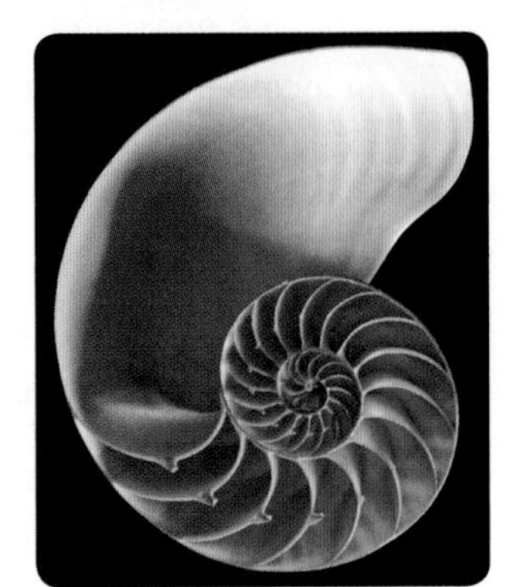

8. The chambered nautilus is a cephalopod mollusk, a relative of octopuses and squids. It creates a spiral shell that has inspired mathematicians and poets for centuries. As it grows, the animal partitions off increasingly larger cells to inhabit. The table below gives the volume (in cubic centimeters) of chambers 10 through 20 of a nautilus.

Chamber number	10	11	12	13	14	15	16	17	18	19	20
Volume (cm^3)	0.2	0.5	0.6	0.6	0.8	0.9	1	1.5	1.7	2.1	2.5

Source: Paleobiology

   a. Is a linear or exponential function the more appropriate theoretical model for the data? Give a reason for your answer.

   b. Make a scatterplot of the data, using chamber number as the independent variable. From the scatterplot, which model seems more appropriate?

   c. Determine whether the linear or exponential regression model better fits the data using residual plots as well as the value of the linear model's correlation coefficient.

   d. Write a sentence indicating which model you would choose, and why.

9. The table at the right shows how many millions of miles motor vehicles drove in the U.S. in certain years.

   a. Is any theoretical model appropriate? If so, what kind and why?

   b. Make a scatterplot of the data. What kind of model seems appropriate?

   c. Sketch the residual plots for the linear, exponential, and quadratic regression models.

   d. Write a sentence indicating which model you would choose and why.

Year	Miles (millions)
1920	47,600
1930	206,320
1940	302,188
1950	458,246
1960	718,762
1970	1,109,724
1980	1,527,295
1990	2,144,362
2000	2,746,925

Source: Historical Statistical Abstract

10. Seasonally-adjusted U.S. unemployment figures for June of each year are given in the table below.

Year	1997	1998	1999	2000	2001	2002	2003	2004	2005	2006	2007	2008
Unemployed (thousands)	6799	6212	5951	5651	6484	8393	9266	8280	7536	7017	6997	8499

Source: Bureau of Labor Statistics

   a. Make a scatterplot of the data, plotting the number of years after 1997 as the independent variable.

   b. Which, if any, of a linear, exponential, or quadratic function seems to model these data? Justify your answer.

REVIEW

11. Recall that the force between electrical fields is inversely proportional to the square of the distance between them. When two electrically charged particles are 0.8 m apart, the force between them is 5 N. What will the force between them be when they are 0.1 m apart? **(Lesson 2-7)**

12. Multiple Choice If grades g in school vary inversely as the number t of hours spent watching TV every day, then which of the following describes the relationship between g and t? **(Lesson 2-7)**

A $g = kt^2$ **B** $g = \frac{k}{t}$

C $g^2 = \frac{k}{t}$ **D** $g = \frac{k}{t^2}$

13. The population of Florida from 1950 to 2000 is given in the table below. **(Lesson 2-6)**

Year	1950	1960	1970	1980	1990	2000
Population (thousands)	2,771	4,952	6,791	9,747	12,938	15,982

Source: Historical Statistical Abstract

a. Make a scatterplot of the data, plotting number of years after 1950 as the independent variable.

b. Find a quadratic regression model for the data.

c. Use your model to predict the population of Florida in 2015.

Tampa, Florida

14. A 6" diameter frozen personal cheese pizza lists its total calories as 590. About how many calories are in a single 2-square-inch bite of pizza? **(Previous Course)**

EXPLORATION

15. Find population data over a number of years for your city or state. Find the linear, quadratic, or exponential model that best fits the data. Use your model to predict the population of your city or state in the year 2050. Do you believe this a reasonable prediction? Explain your response.

QY ANSWER

4.26

Chapter 2 Projects

1 Temperature vs. Latitude

At the right are the latitude (in degrees North) and the average daily maximum temperature in October for various cities in North America.

a. Use a statistics utility to draw a scatterplot with latitude on the x-axis and average daily maximum temperature on the y-axis.

b. Find a linear regression model for these data. Interpret the sign and magnitude of the slope of the regression line.

c. Over what domain do you expect the regression line to fit the data well?

d. Predict the average daily October maximum temperature for these cities:

Detroit, MI	$42°22'$ N
Tampa, FL	$27°49'$ N

e. The actual average daily high temperature in October is 62°F for Detroit and 84°F for Tampa. Find the residual for each of your predicted values in Part d.

f. Which cities appear to be outliers? Give plausible reasons why these cities might have a different relation between latitude and temperature than the others.

g. Find the latitude and average daily maximum temperatures in cities in other parts of the world, e.g., in Africa, Asia, or Europe. Explain any big differences between the regression lines you find for these areas and the one found in Part b.

Place	Latitude	Temperature (°F)
Caribou, ME	46.87	51
Chicago, IL	41.98	64
Denver, CO	39.77	68
Great Falls, MT	47.48	58
Juneau, AK	58.3	47
Kansas City, MO	39.32	69
Mexico City, Mexico	19.42	72
New Orleans, LA	29.98	80
Ottawa, Canada	45.43	55
Salt Lake City, UT	40.78	65
San Francisco, CA	37.62	70
Seattle, WA	47.45	60
Washington, DC	38.85	68

Source: National Climatic Data Center

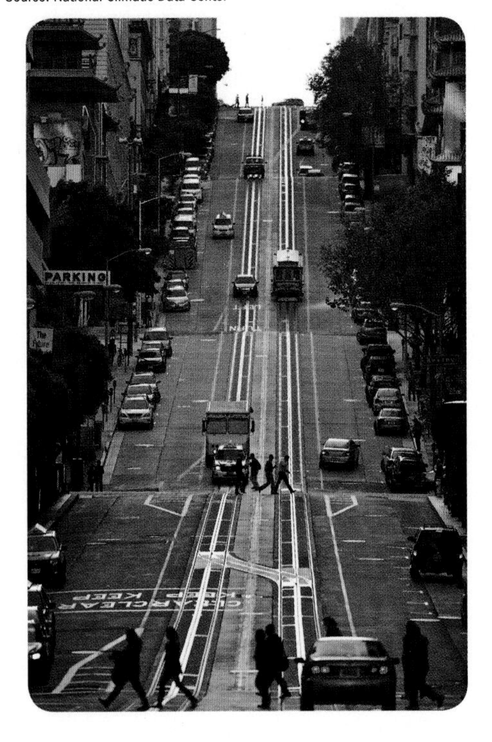

2 Class Survey Data Revisited

Use the class database constructed as a project at the end of Chapter 1. Some people claim that many high school students spend too much time watching TV or surfing the Internet.

a. Make a scatterplot of time spent watching TV versus time spent doing homework. Make a second scatterplot of time spent surfing the Internet versus time spent doing homework. Find the lines of best fit for each scatterplot.

b. Are there any outliers? Remove these data from the set and obtain an equation for a new regression line.

c. From the data, do you think students generally spend too much time watching TV or surfing the Internet? Why or why not?

3 Five Years From Now

The U.S. Bureau of Labor Statistics computes the Consumer Price Index "CPI" as a gauge of inflation.

a. Visit the BLS website and find data for the CPI in January for each year since 1982.

b. Plot the data on a scatterplot. Use a linear model to find an equation for the line of best fit for the data.

c. Find an exponential regression model for the same data.

d. Which model better fits the data?

e. Pick a new car you would like to own and find out how much it costs. Use the model you selected to calculate the cost of the car five years from now.

4 Light Intensity

The amount of energy given off by a flashlight is constant. If you aim a flashlight at a wall and then step back, the area illuminated by the flashlight will get larger. Since the intensity of the light is measured by the amount of light energy that strikes a given area, the intensity decreases as you step back. In this project, assume that your flashlight is the standard of measure, so its energy coefficient is 1.

a. Enter a dark room and place the flashlight 1 foot from a wall. Measure the diameter of the region that is illuminated by the light and calculate the area of illumination. Repeat the process, moving the flashlight back at one-foot intervals until you have gathered at least ten data points.

b. Make a scatterplot of your distance and illumination data. Does the scatterplot suggest how distance from the wall is related to the area of illumination?

c. Use a statistics utility to find the equation for a line of best fit for your data. Plot this regression line on your scatterplot. How well does the regression line seem to model the data? Does this mathematical model support the theory in Part b? If not, try another model to see if you can improve the fit using the sum of squared residuals.

Chapter 2 Summary and Vocabulary

- Any set of ordered pairs is a relation. **Functions** are those relations in which no two ordered pairs have the same first component. A function can also be viewed as a correspondence between two sets A and B, which relates each element of A (the function's **domain**) to exactly one element of B. Functions can be defined by giving a rule for the correspondence, a graph, a table, or a description in words, and by indicating the domain of the function.

- In this chapter, sets of data are modeled by linear, quadratic, and exponential functions. Scatterplots can be used to represent the data and to determine the type of relationship and the feasibility of a particular model. Some models are theory-based, as when a known law of physics is behind the mathematics. Other models are impressionistic in that there is no known theory to explain why the model should fit. Even if the model looks like a good fit, a plot of residuals may reveal that the model needs improvement.

- Linear functions model constant increase or constant decrease. A linear model can be approximated by drawing a line close to all the data points. The **line of best fit** minimizes the sum of the squared residuals between observed and predicted values, and can be found with a statistics utility.

- The strength of a linear relation between two variables is measured by the **correlation coefficient**, r. The sign of the correlation coefficient indicates the direction of the relation between the variables, and its magnitude indicates the strength of the linearity. Although perfect correlations ($r = \pm 1$) are rare, an r with an absolute value close to 1 indicates a strong linear relation. Strong correlations indicate a linear relationship between variables, but correlation does not necessarily imply causation. An r-value close to zero indicates that the variables are not related linearly.

- **Exponential functions with base *b***, which are of the general form $y = ab^x$ with $a > 0$, $b > 0$, and $b \neq 1$ model exponential growth (when $b > 1$) and decay (when $0 < b < 1$). Quadratic models of the form $y = ax^2 + bx + c$ include theory-based models for projectile motion. Most statistics utilities can find exponential and quadratic models of best fit. Many physical situations are modeled by functions of inverse variation with equations of the form $y = \frac{k}{x}$ or by functions of inverse-square variation with equations of the form $y = \frac{k}{x^2}$. The graphs of these functions have vertical and horizontal asymptotes.

Vocabulary

2-1
mathematical model
relation
independent variable
dependent variable
function, ordered pair definition
domain of a function
range of a function
function, correspondence definition
real function
member of a set, element of a set, $\in$
piecewise definition of a function
value of a function

2-2
linear function
linear model
interpolation
extrapolation
observed values
predicted values
residual
sum of squared residuals

2-3
method of least squares
line of best fit, least squares line, regression line
center of mass
correlation coefficient
perfect correlation
strong correlation
weak correlation

- Choosing a good model requires both an analysis of the data set and an understanding of how the data were obtained. **Residuals**, the differences between the observed values and the values predicted by the model, can be used to judge the quality of a model. If the residuals are large or if there is a pattern to the residuals, the model may be inadequate and another should be sought. Piecewise functions may provide a composite model that fits better than any function based on a single rule.

- Function models can produce very reasonable predictions for intervals over which the original data were observed. However, one should use caution when making predictions outside the domain of observed values. Interpolation is safer than extrapolation.

Vocabulary

2-4
growth factor
exponential function with base *b*
exponential growth function
exponential growth curve
exponential decay function
asymptote

2-5
exponential regression
half-life

2-6
quadratic model
quadratic regression

2-7
varies inversely as, is inversely proportional to
constant of variation, constant of proportionality
inverse-square relationship
varies inversely as the square of, is inversely proportional to the square of
power function

2-8
residual plot

Properties and Theorems

Pearson's Formula for Linear Correlation (p. 96)

Chapter 2 Self-Test

Take this test as you would take a test in class. You will need a calculator. Then use the Selected Answers section in the back of the book to check your work.

1. a. Evaluate $f(2)$ to the nearest tenth when $f(x) = -9.8x^2 - 5.2x + 12.7$.
   b. Evaluate $g(2n)$ if $g(x) = 5x^2 - 3$.

2. The line with equation $y = 1$ is a horizontal asymptote to the graph below. Give the domain and range of the graphed function.

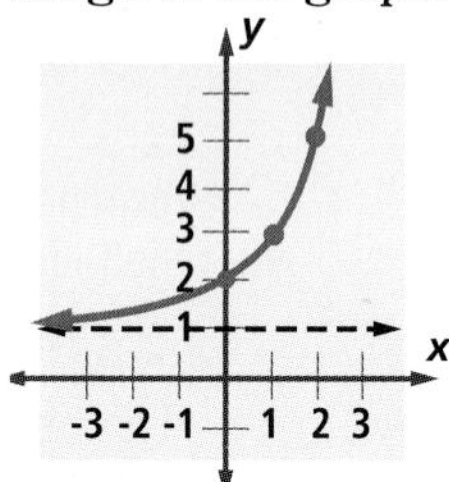

3. What are the domain and range of p when $p(x) = -\frac{2}{x}$?

4.

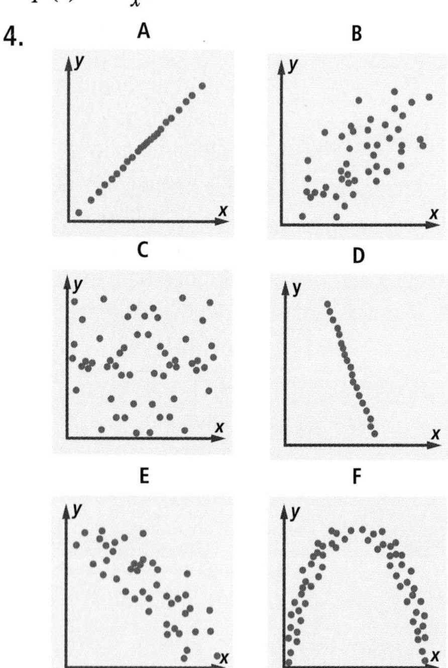

   a. Which scatterplots indicate a negative association between variables?
   b. Which scatterplots indicate a correlation coefficient close to zero?
   c. Which correlation coefficient(s) equal 1?

5. The equation $P = 3{,}424{,}000(1.013)^x$ can be used to model the population P of New Zealand x years after 1993. In 2008, this model produced a residual of 17,000 rounded to the nearest thousand. What was the actual population in 2008?

6. Stephen and Chris fit different lines to a scatterplot by eye. The sum of squared residuals was 34 for Stephen's line and was 576 for Chris's line. Use the sum of squared residuals to explain which of their models is a better fit to the data.

7. Suppose a ball is thrown upward at a velocity of 44 $\frac{\text{ft}}{\text{sec}}$ from a cliff 200 feet above a dry riverbed. Use the formula $h = -\frac{1}{2}gt^2 + v_0t + h_0$ where $g = 32\ \frac{\text{ft}}{\text{sec}^2}$.
   a. Write an equation for the height h (in feet above the riverbed) of the ball after t seconds.
   b. At what time will the ball hit the riverbed?

In 8 and 9, suppose that $y = 60$ when $x = 10$. For each situation,

   a. compute the constant of variation.
   b. find y when $x = 3$.

8. when y varies inversely as x

9. when y varies inversely as the square of x

10. Does the function f with $f(x) = 105(1.2)^x$ model exponential decay or exponential growth? How do you know?

11. A movie studio uses the regression equation $y = 2.51x + 471.10$ to predict how much money a movie will earn based on the cost of making the movie. Here, y = world revenues in millions of dollars and x = the movie's budget in millions of dollars. The movie "Regression without a Cause" had a budget of \$115 million and earned \$524 million. What is the residual?

12. Cesium-137 has a half life of 30 years. How much of a 10-gram sample will be left after 20 years?

13. The residuals for a linear and inverse-square model are graphed below. What do these graphs tell you about the appropriateness of each model?

Linear Model

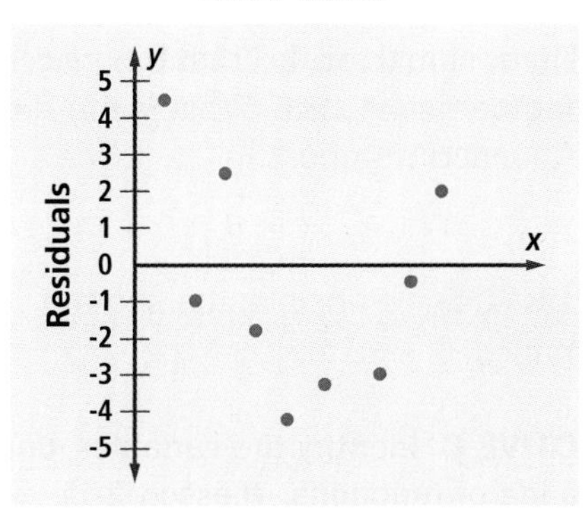

Inverse-Square Model

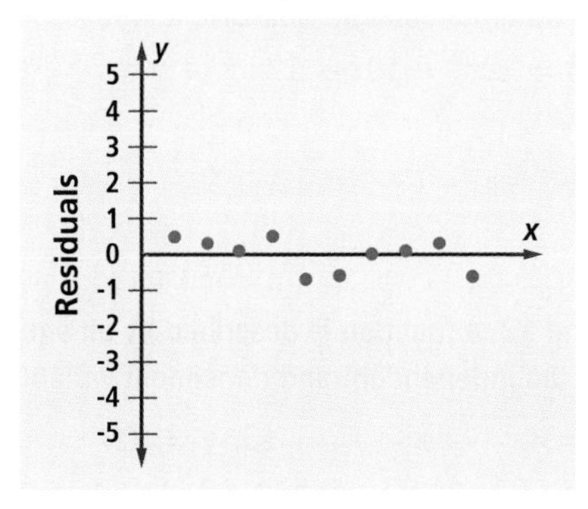

14. The table below shows the height h in feet of a ball above ground level t seconds after being thrown off the top of a building.

t	1	2	3	4	5	6
h	298	302	277	219	128	8

a. Fit a quadratic model to the data.

b. Is this a theory-based model or is it an impressionistic model?

c. From your model, what is the height of the ball 4 seconds after it was thrown?

15. The table below lists the percent of U.S. citizens aged 18–24 who voted in midterm elections from 1974 to 2002.

Year	Percent 18-24 Year-Olds Who Voted
1974	25.4
1978	25.1
1982	26.6
1986	23.9
1990	22.9
1994	22.2
1998	18.5
2002	19.4

Source: Center for Information and Research on Civic Learning and Engagement

a. Using the line of best fit with the number of years after 1974 as the independent variable, what does the slope tell you about voter turnout in midterm elections among 18–24 year olds?

b. Find the correlation coefficient. What does the correlation coefficient tell you about the relationship between the year and voter turnout in midterm elections among 18–24 year olds?

16. Sonia researched the number of fast-food restaurants in her city in several years and recorded the data in the table below.

Years after 2000	Number of Restaurants
1	31
2	36
3	39
4	48
5	53
6	63
7	71

a. Make a scatterplot of the data. Which model appears appropriate for the data?

b. Plot the residuals for the regression equations for various models.

c. Write a sentence or two explaining your choice of best fit model.

Chapter 2 Chapter Review

SKILLS
PROPERTIES
USES
REPRESENTATIONS

SKILLS Procedures used to get answers

OBJECTIVE A Work with $f(x)$ notation for function values. (Lesson 2-1).

In 1 and 2, let $f(x) = 4^x$.

1. Evaluate.
   a. $f(1)$ b. $f(-1)$ c. $\frac{f(6)}{f(3)}$
2. **True or False** Justify your answer.
   a. $f(2) + f(-2) = 0$
   b. $f(2) \cdot f(3) = f(5)$
3. Let $g(x) = x^3 + 2$.
   a. If $g(x) = 3$, find x.
   b. Does $g(4) - g(2) = g(2)$? Justify your answer.
4. Suppose $p(x) = 5x - 2$.
   Evaluate $p(n + 1) - p(n)$.

OBJECTIVE B Compute residuals from observed and predicted values. (Lessons 2-2, 2-8)

In 5–7, let $F(n)$ be the nth Fibonacci number as in the following table for $1 < n < 9$.

n	1	2	3	4	5	6	7	8	9
$F(n)$	1	1	2	3	5	8	13	21	34

5. a. The line of best fit for these nine ordered pairs is $F(n) = 3.65n - 8.47$. Calculate the residuals for the first 9 Fibonacci numbers.
   b. The 16th Fibonacci number is 987. Calculate the residual for this number.
   c. Calculate the sum of squared residuals for the first 9 Fibonacci numbers for the linear model.
6. An exponential model for these 9 Fibonacci numbers is $F(n) = 0.4935(1.594)^x$. Calculate the residual for the 16th Fibonacci number for the exponential model.
7. a. Calculate the residuals for the quadratic model for these 9 Fibonacci numbers, $F(n) = 0.787x^2 - 4.218x + 5.952$.
   b. The residual for the 15th Fibonacci number is 490.288. What is the 15th Fibonacci number?

PROPERTIES Principles behind the mathematics

OBJECTIVE C Identify the variables, domain, and range of functions. (Lesson 2-1)

In 8–11, a function is described by an equation.

a. State its domain. b. Give its range.

8. $f(x) = -2x^2 + 10x - 12$
9. $r(t) = 4 \cdot 3^t$
10. $j(x) = \frac{3}{-x}$
11. $I = \frac{1}{d^2}$

In 12 and 13, a function is described by an equation. Identify the independent and dependent variables.

12. $P = (x - 3)^4$ 13. $y = f(t)$

14. Ruth has \$23, and n friends join her for coffee. Coffee cost \$2 per cup. The rule $R(n) = 2(n + 1)$ describes the amount $R(n)$ in dollars that Ruth will spend on coffee. What is a reasonable domain and range for the function R?

In 15–17, tell if the statement is true or false.

15. The line of best fit is the linear model with the least sum of squared residuals.
16. The correlation coefficient r is measured in the same unit as the slope of the regression line.
17. Correlation does not imply causation.

OBJECTIVE D Identify properties of regression lines and of the correlation coefficient. (Lesson 2-3)

In 18–21, r represents a correlation coefficient.

18. **Multiple Choice** $r = 0.05$ indicates what?

   A a strong positive association

   B a weak positive association

   C a weak negative association

   D a strong negative association

19. What value of r indicates a perfect positive association?

20. For a set of data, the line of best fit is given by $y = 8.4 - 3.6x$ and $r^2 = 0.60$. What is r?

21. Explain what is meant by a strong positive correlation.

OBJECTIVE E Describe properties of quadratic, exponential, and inverse variation functions. (Lessons 2-4, 2-6, 2-7)

22. Consider the function f with $f(x) = ab^x$ with $b > 0$ and $a > 0$.

   a. Under what conditions is f decreasing?

   b. Under what conditions is f increasing?

23. Which of the functions, $a: t \rightarrow 0.8(1.1)^t$ or $b: t \rightarrow 1.1(0.8)^t$, models exponential growth and which models exponential decay?

24. Without graphing, how can you tell whether the graph of $f(x) = 5x^2 + 2x + 1$ has a maximum or minimum point?

In 25 and 26, identify the quadrants in which the graph of each equation appears.

25. a. $y = \frac{1}{x}$ b. $y = \frac{1}{-x}$

26. a. $y = \frac{1}{x^2}$ b. $y = \frac{1}{-x^2}$

In 27–30, suppose $f(x) = \frac{k}{x}$ and $g(x) = \frac{k}{x^2}$, with $k > 0$.

27. In Quadrant I, what happens to $f(x)$ as x increases?

28. In Quadrant II, what happens to $g(x)$ as x increases?

29. Suppose $f(x) = 40$ when $x = 10$. Compute k.

30. Suppose $g(x) = 40$ when $x = 10$. Compute k.

USES Applications of mathematics in real-world situations

OBJECTIVE F Find and interpret linear, quadratic, and exponential regressions and models. (Lessons 2-2, 2-3, 2-5, 2-6)

31. As part of a biology lab, Ben recorded the population of Drosophila (fruit flies) over a five-week period as follows: 2, 13, 74, 482, and 2793.

   a. Find the exponential regression equation of best fit for these data.

   b. Calculate the residual for the week the population was 74.

32. The table below shows the height h in feet of a ball above ground level t seconds after being thrown off the top of a building.

t	1	2	3	5	6
h	279	291	271	135	19

   a. Find the quadratic model of best fit for these data.

   b. Use your model to predict the height of the ball after 4 seconds.

   c. Is your prediction in Part b extrapolation or interpolation?

In 33 and 34, the heights (in inches) and shoes sizes of seven boys are recorded below.

Height (in.)	75	66	72	68	71	69	70
Shoe Size	13	$8\frac{1}{2}$	11	9	12	$10\frac{1}{2}$	10

33. a. Using the maximum and minimum height points, fit a linear model to the data by eye.

   b. Calculate the sum of squared residuals.

34. a. Find the line of best fit for predicting shoe size from height.

   b. Find the sum of squared residuals for the line of best fit in Part a.

   c. Why should the sum in Part b be less than your answer for Question 33b?

35. Actinium-226 has a half-life of 29 hours. If 100 mg of actinium-226 remain after 34 hours, how many mg were there originally?

36. The intensity I of light striking an object is inversely proportional to the square of the distance d from the source of the light. The intensity of light on a movie screen is 75 candles when the projector is 20 feet away.

a. Create a model for this situation.

b. Use your model to predict the intensity of light striking the screen when the projector is 25 feet away.

OBJECTIVE G Evaluate which type of model is more appropriate for data. (Lesson 2-8)

In 37–40, a situation is described. State whether a linear, exponential, or quadratic model is most appropriate for the situation. If no model is appropriate, say so. Explain your answer.

37. In 1859, Thomas Austin released 24 gray rabbits in Australia. The rabbits multiplied so quickly that within 20 years they were referred to as the "gray carpet."

38. In a physics experiment, Tim rolled a marble down an inclined plane and timed how long it took the marble to cover a given distance. Tim observed that the marble kept going faster and faster due to gravitational acceleration.

39. A tailor wanted to reduce the amount of time it took to make measurements, so he decided to make a model to predict the length of inseams based on sleeve length.

40. Jim knew that if he could accurately predict the price of gold, he would become rich.

41. Susie is developing a model for cell division in bacteria. She uses a microscope to count the number of bacteria present in a Petri dish at the end of each hour for eight hours. Her data are shown below.

End of Hour	1	2	3	4	5	6	7	8
Bacteria	2	4	8	16	30	61	124	241

a. Is any theoretical model appropriate for the data? If so, what theory and why?

b. Make a scatterplot. Which model appears appropriate for the data? Why?

c. Confirm your choice from Parts a and b by plotting the residuals.

42. Estimates of the population of New Orleans county from 2000 to 2005 are listed below.

a. Make a scatterplot. Which model appears appropriate for the data?

b. Plot the residual plots for the linear, quadratic, and exponential regression models. Do the plots give a clear choice of the best model?

c. Calculate the correlation coefficient for the linear model.

Year	Estimated Population
2000	484,674
2001	477,548
2002	472,085
2003	466,767
2004	460,556
2005	453,726

Source: U.S. Census Bureau

d. Write a paragraph explaining your choice of best fit model.

e. According to the model, what population would you expect in 2006?

f. The actual population in 2006 was 210,198. What is the residual?

g. Explain why your model is so far off.

REPRESENTATIONS Pictures, graphs, or objects that illustrate concepts

OBJECTIVE H Interpret properties of relations from graphs. (Lesson 2-1)

In 43–46, state whether the graph represents a function mapping x to y. If the relation is a function, determine its domain and range.

43. 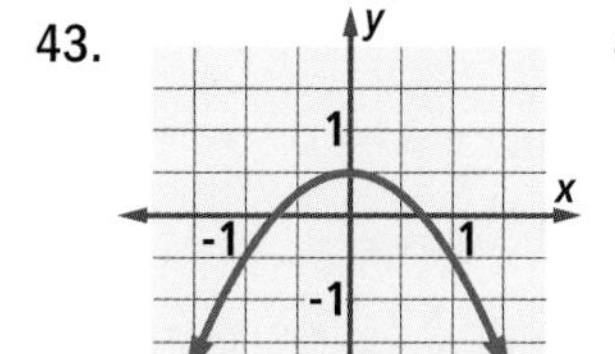

44. 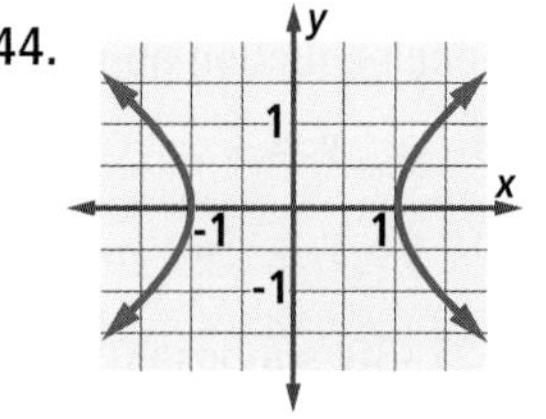

45. 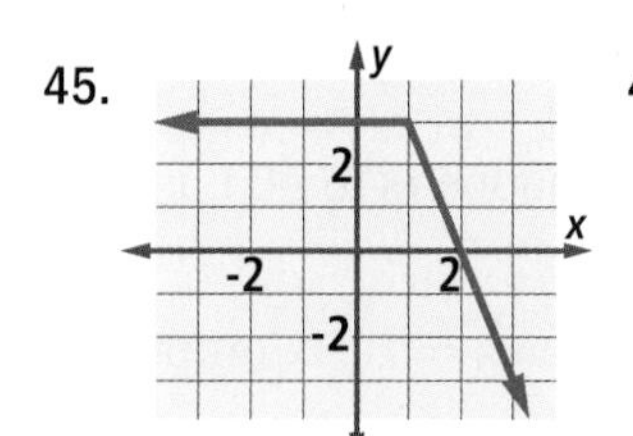

46. 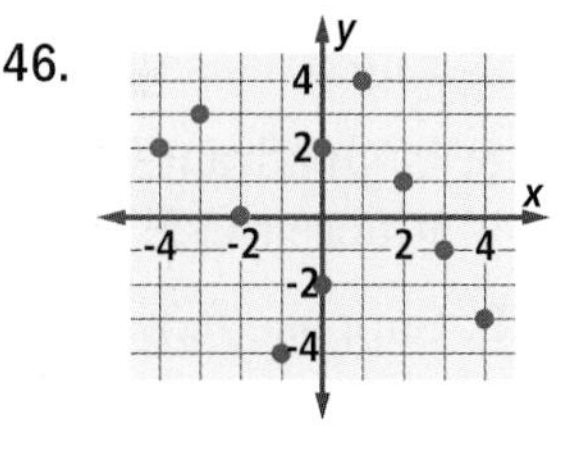

OBJECTIVE I **Use scatterplots and residual plots to draw conclusions about models for data. (Lessons 2-2, 2-3, 2-8)**

In 47–50, use the displays below. They show the number of hours 12 students spent studying for a test and their scores on the test. Both linear and quadratic models have been fit to the data and the regression equations are given.

$y = 1.887x + 60.762$

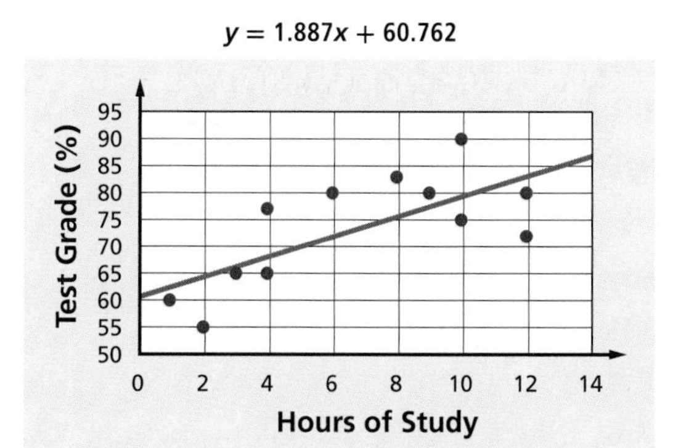

$y = -0.455x^2 + 7.979x + 46.775$

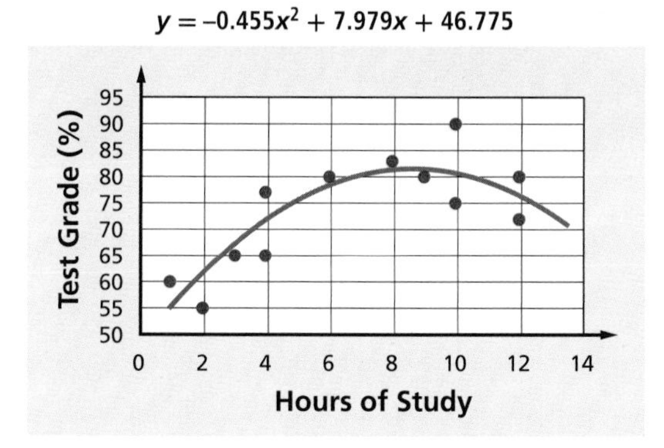

47. Consider the quadratic model and the student who studied eight hours for the test.
    a. What is the observed y-value?
    b. What is the predicted y-value?
    c. What is the residual?

48. Repeat Question 47 for the linear model and the student who studied eight hours for the test.

49. According to the quadratic model, what is the ideal amount of time to study in order to achieve a maximum score? Does this make sense? Explain why or why not.

50. The residuals for each model are graphed below. What do these graphs tell you about the appropriateness of each model?

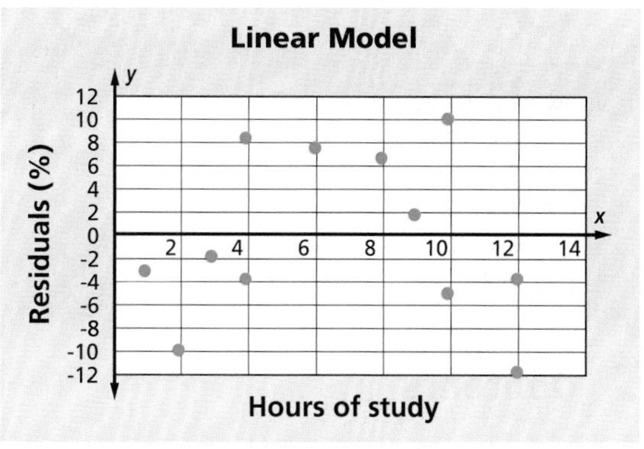

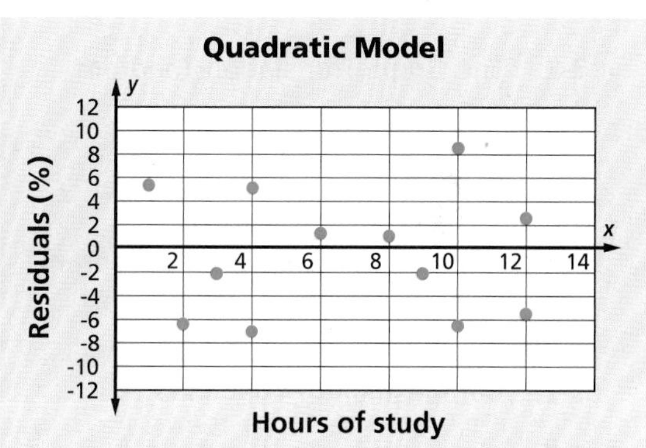

In 51–54, a scatterplot is given. State whether the correlation coefficient is likely to be positive, negative, or approximately zero.

51. 52.

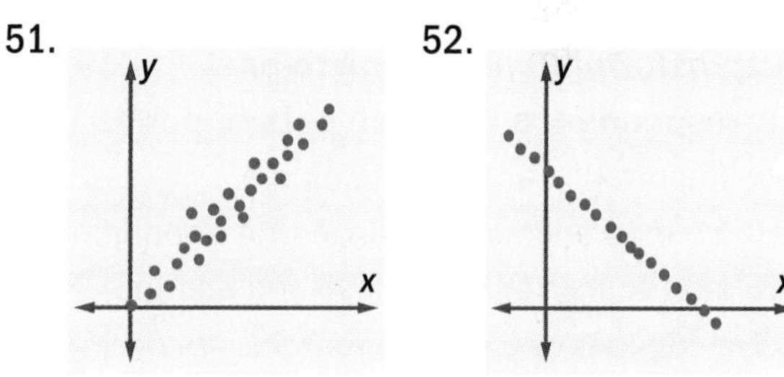

53. 54.

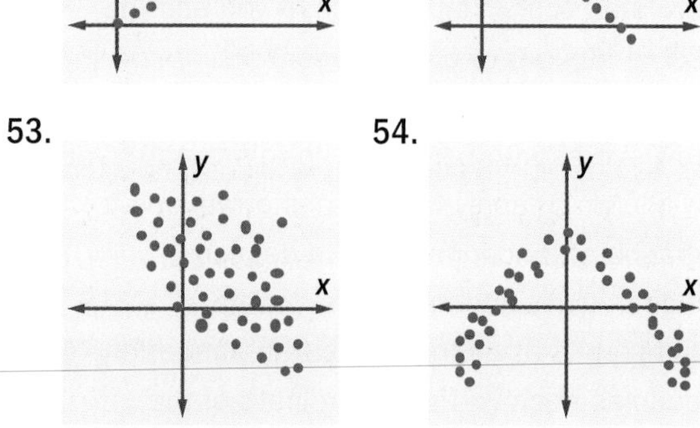

Chapter

3 Transformations of Graphs and Data

Contents

A *transformation* is a one-to-one correspondence between sets of points. Two important transformations are translations and scale changes. A translation of data occurs, for example, if you add the same number to every student's score on a test. A scale change occurs when you change raw scores into percents. In this chapter you will learn to apply these transformations to graphs of functions that are defined algebraically and to graphs of data sets. In the graphs at the right, the transformation doubles each vertical coordinate of the preimage to create the image. It is an example of a scale change.

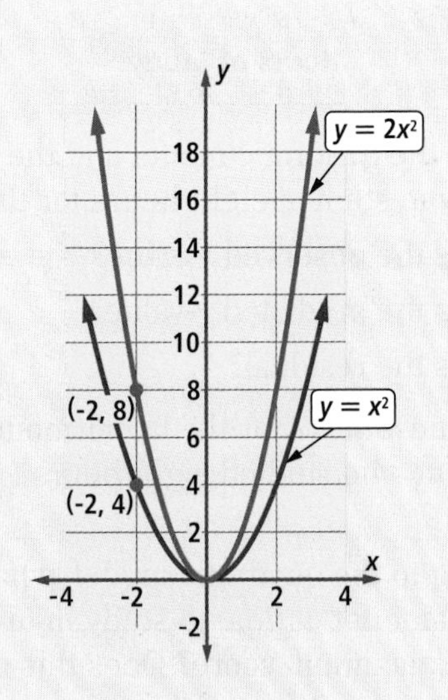

An example of a figure and its *translation* image is shown below. As mentioned in Lesson 2-5, it is common to let the independent variable be years after a starting year in order to minimize differences in residuals. In this case, the original data is translated so that time is measured in *years after 1960*.

Translations and scale changes can be described by algebraic formulas. In this chapter you will study these descriptions and the effects of these transformations on equations of functions and on statistical measures.

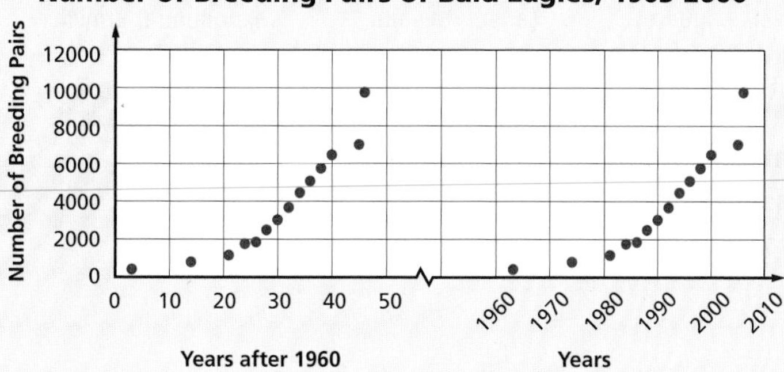

Lesson 3-1 Graphs of Parent Functions

Vocabulary

parent function

window

▸ **BIG IDEA** Knowledge of the features and graphs of parent functions helps in analyzing more complex functions.

Mental Math

Describe the shape of the graph of each equation.

a. $3x - 4y = 7$

b. $3x^2 - 4y = 7$

c. $3x \cdot 4y = 7$

Parent Functions

Many useful functions cluster in families. The **parent function** of the family is usually the member that has the simplest equation. For example, the parent of all linear functions has equation $f(x) = x$. The parent of the quadratic function g with $g(x) = -4.9x^2 + 31x + 5$ has equation $f(x) = x^2$. Eight important parent functions are graphed below.

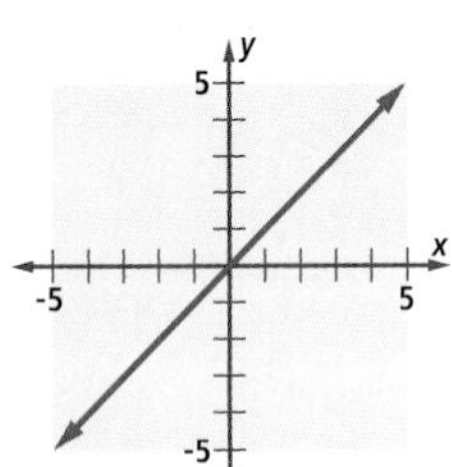

$f(x) = x$
(linear)

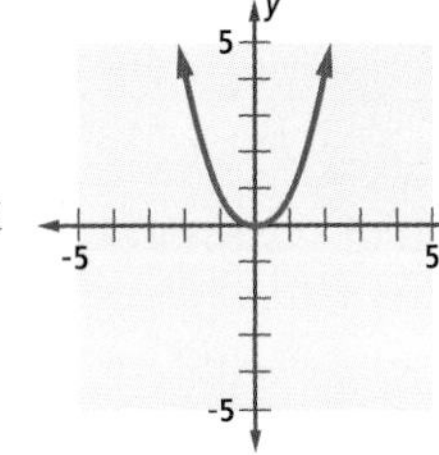

$f(x) = x^2$
(quadratic)

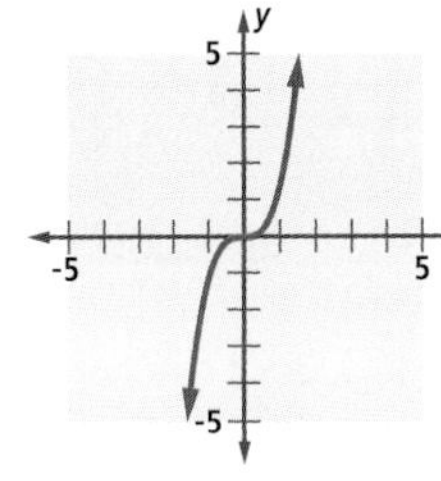

$f(x) = x^3$
(cubic)

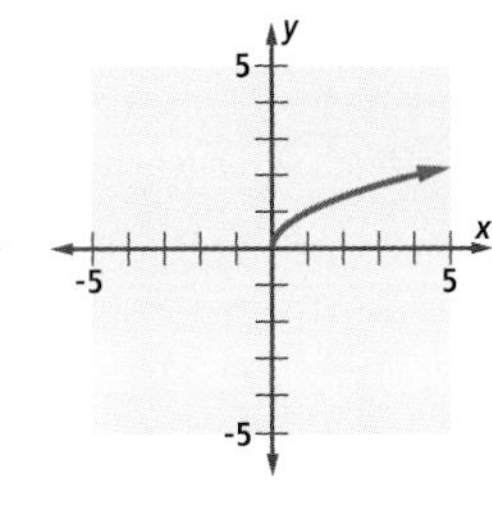

$f(x) = \sqrt{x}$
(square root)

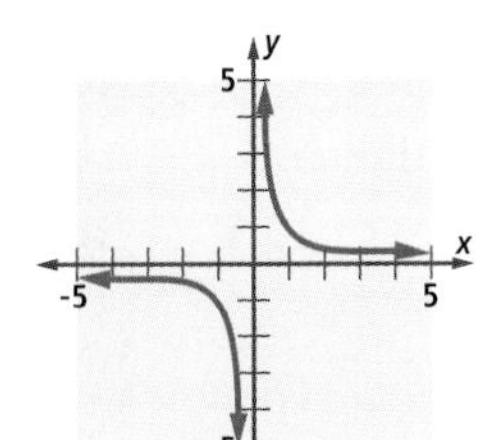

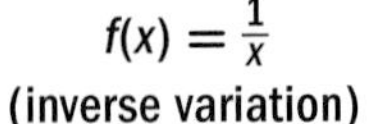

$f(x) = \frac{1}{x}$
(inverse variation)

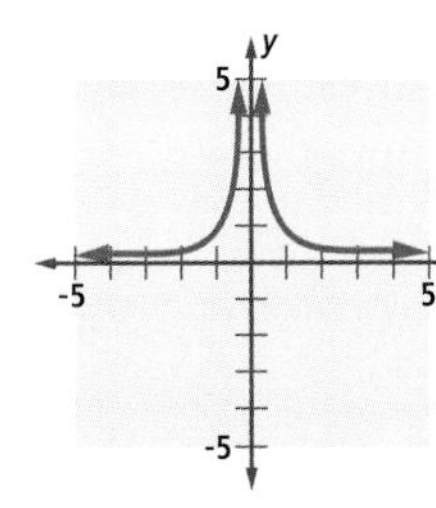

$f(x) = \frac{1}{x^2}$
(inverse square)

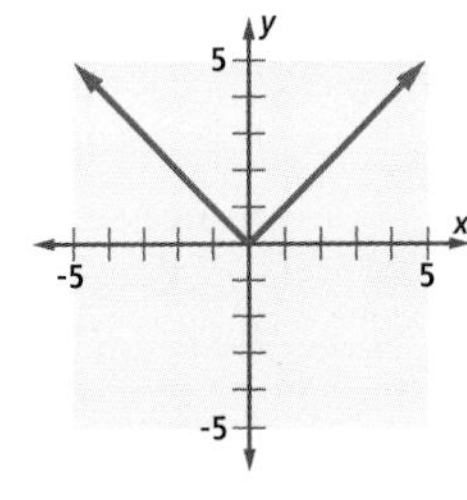

$f(x) = |x|$
(absolute value)

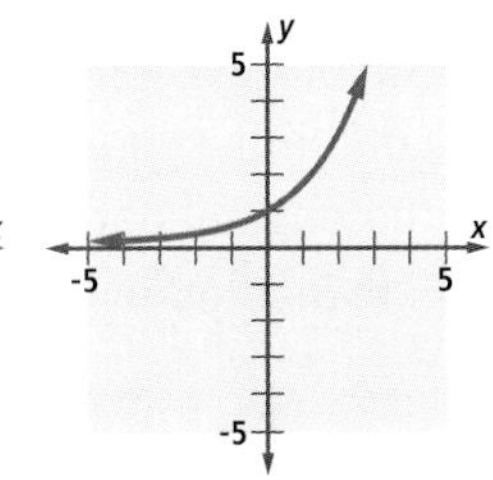

$f(x) = 2^x$
(exponential growth)

QY1

Tell which parent functions have each characteristic.

a. The domain does not include 0.

b. The graph contains the origin.

Choosing Windows of Graphs

When you display an image on a computer screen, its appearance can be changed by zooming in and out and by scrolling left and right or up and down. These changes affect what is displayed on the screen. At the left below is a document as it appears when it is opened. If you zoom out, then scroll up and left, you might see the picture at the right below.

lish Justice, insure dome
uility, provide for the co
e, promote the general
re, and secure the Blessin
y to ourselves and our Po
ordain and establish thi

We the People of the United States,
in Order to form a more perfect Union,
establish Justice, insure domestic
Tranquility, provide for the common
defence, promote the general
Welfare, and secure the Blessings of
Liberty to ourselves and our Poster-
ity, do ordain and establish this
Constitution for the United States
of America.

Similarly, when you plot a function on graph paper or with a graphing utility, you want to choose the viewing **window** that shows important aspects of the function. Graphing utilities have a standard window that is used as a default for plotting functions. The standard window is usually appropriate for parent functions but often misses important features of graphs of their offspring. Your knowledge of the parent graphs can help you choose a good window. On graphing utilities, the window is described by identifying the least and greatest values of x and y that will be shown, `xmin`, `xmax`, `ymin`, and `ymax`.

Example 1

a. Display the graph of $g(x) = |x + 25| - 10$ in an appropriate window.

b. State the domain and range of the function g.

Solution

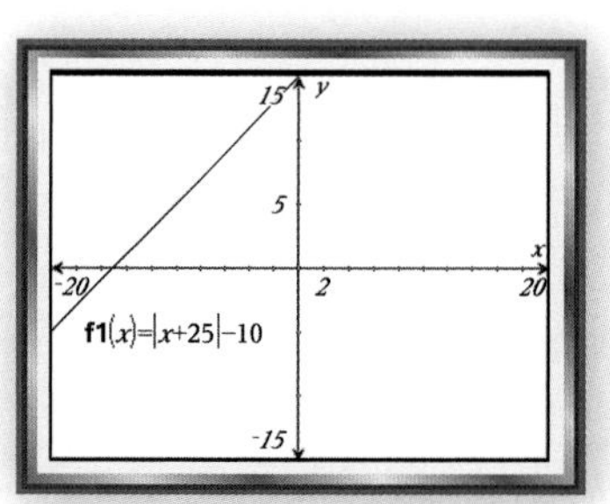

a. The graph of $g(x) = |x + 25| - 10$ in a standard window as shown at the right does not display important features of g. The parent of the function is the absolute value function, with equation $f(x) = |x|$. A key feature of the graph of the absolute value function is its vertex. The vertex is the minimum point of the graph.
The smallest value of $f(x) = |x|$ is when $x = 0$. For $f(x) = |x + 25| - 10$, the minimum value occurs when $|x + 25| = 0$, or when $x = -25$. Since $f(-25) = -10$, the vertex is $(-25, -10)$. Choose `xmin`, `xmax`, `ymin`, and `ymax` to include $(-25, -10)$. The window $-55 \leq x \leq 5$, $-20 \leq y \leq 30$ displays the important features of the graph, showing the vertex and the x- and y- intercepts and vertex.

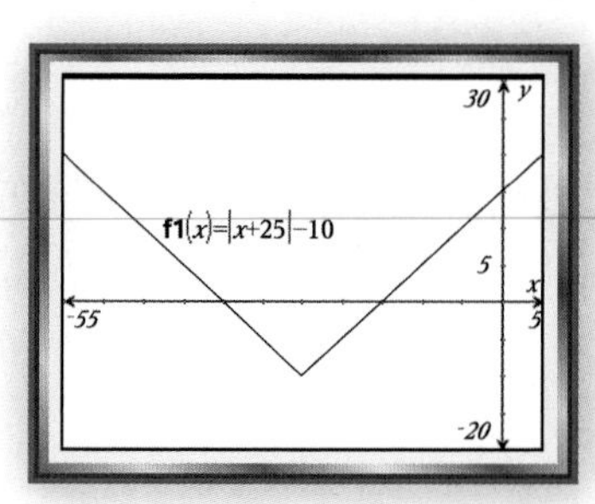

(continued on next page)

b. Any value of x produces a value of $f(x)$, so the domain is the set of all real numbers, $\mathbb{R}$. Since the minimum y-coordinate on the graph is -10, the range is $\{y| \ y \geq -10\}$.

STOP QY2

▶ QY2

Identify xmin, xmax, ymin, and ymax in the window of the second graph of Example 1.

Example 2

Graph the real function g with equation $g(x) = \sqrt{x - 8} + 5$ in a window that shows the graph's important features. State the domain and range.

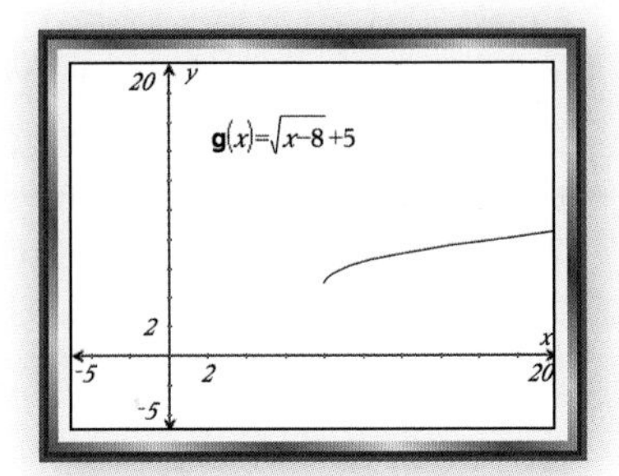

Solution In the real number system, square roots can only be evaluated for nonnegative numbers. Thus, $\sqrt{x - 8}$ is defined only when $x - 8 \geq 0$ or $x \geq 8$. When $x = 8$, $g(x) = 5$. When $x \geq 8$, $g(x) \geq 5$. So pick a window that shows the axes, the point (8, 5), and points to the right and above (8, 5). We chose xmin = -5, xmax = 20, ymin = -5, and ymax = 20.

The graph shows that the domain is $\{x| \ x \geq 8\}$ and the range is $\{y| \ y \geq 5\}$.

When functions are used to model real-world phenomena, the domain may be restricted. In Example 3, the quadratic for falling bodies works well while the object is in air, but not while the object is in water.

Example 3

Stephen, who is 1.7 m tall, dives off a 3 m springboard. An equation modeling his height $h(x)$ in meters above the water at time x in seconds is

$$h(x) = -4.9x^2 + 4.5x + 4.7.$$

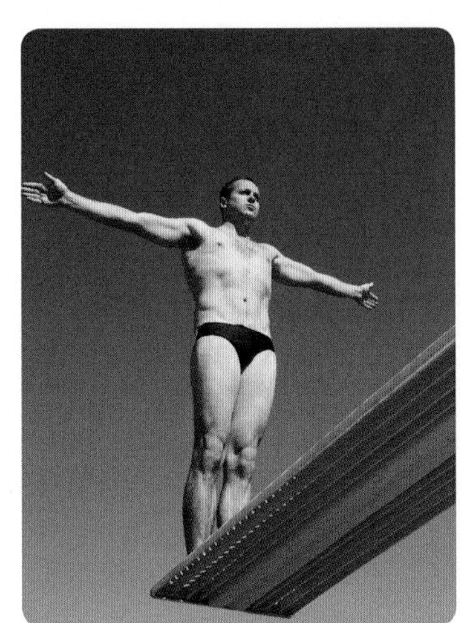

a. Create a graph that would be helpful for determining the maximum height of his dive and how long it lasts.

b. What is the domain and range of h within the context of this situation?

Solution

a. The parent function has equation $y = x^2$, so we know the graph of h is a parabola. In this situation, the equation only applies to the time from when Stephen leaves the board to when he enters the water, which from the standard window graph, shown at the left at the top of the next page, we see is between $x = 0$ and $x = 1.5$. Even though negative values of x do not have meaning in this situation, it is still useful to include negative values for the window settings in order to see the intercepts. You may need to modify the vertical window settings to include the vertex. The graph at the right on the next page is shown on the window $-1 \leq x \leq 5$, $-1 \leq y \leq 7$.

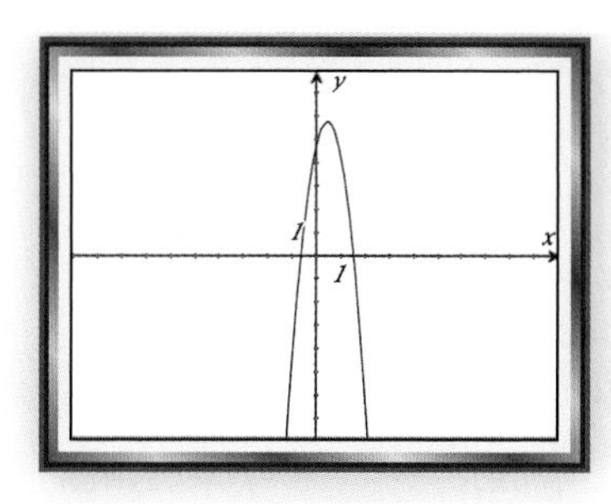

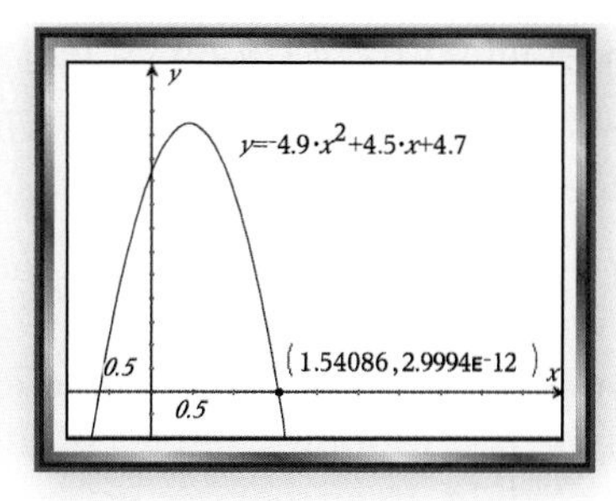

b. The maximum height is about 5.73 meters and the dive lasts about 1.5 seconds. The TRACE function on your calculator can help find these maximum x- and y- values. The diver's fall through air has domain close to $\{x \mid 0 \leq x \leq 1.5\}$. The range is about $\{h(x) \mid 0 \leq h(x) \leq 5.73\}$.

Asymptotes

The parent functions with equations $y = \frac{k}{x}$ and $y = \frac{k}{x^2}$, pictured on page 152, each have the x- and y-axes as asymptotes. As x moves closer and closer to 0, the curve approaches the y-axis, so the y-axis is a vertical asymptote. The curve approaches the horizontal asymptote, which is the x-axis, when x increases or decreases without bound. Asymptotes exist on all graphs of offspring of the parents $y = \frac{k}{x}$ and $y = \frac{k}{x^2}$.

How are asymptotes marked on a graph if they are not the axes? Consider the function f with equation $f(x) = \frac{1}{x-5}$. At the vertical asymptote, the graph appears to "jump" as shown at the right. This *discontinuity* occurs when $x = 5$, since $f(5) = \frac{1}{5-5} = \frac{1}{0}$ and division by 0 is undefined. The vertical asymptote of $x = 0$ (the y-axis) for the graph of the parent function $h(x) = \frac{1}{x}$ has moved to $x = 5$ for its offspring $f(x) = \frac{1}{x-5}$. While points on the vertical asymptote are not actually part of the graph, the asymptote helps to explain the behavior of the function. So we mark the asymptote, showing it with a dashed line. The x-axis is still a horizontal asymptote even though you will not necessarily see dashed line notation appear over the x-axis.

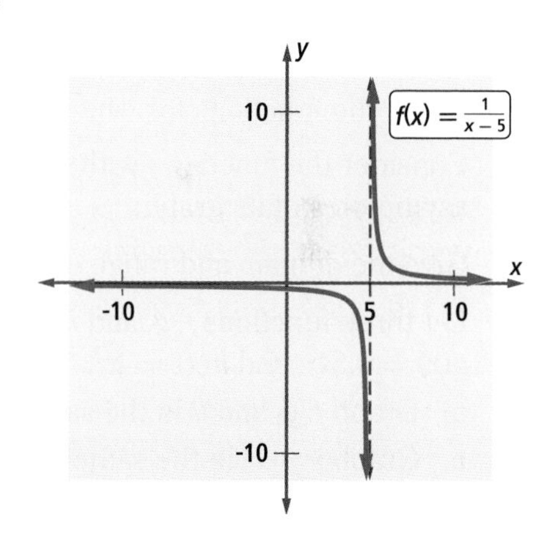

Some graphing utilities handle discontinuous graphs in misleading ways. A vertical line may appear where there should be a vertical asymptote. Therefore it is important to both know what a good graph looks like, and to know the limitations of whatever graphing technology you are using. A good graph meets the following criteria:

- Axes are labeled appropriately, with the scales shown.
- The window is chosen so that the characteristic shape of the graph can be seen.
- The intercepts are shown.

- Any endpoints are indicated as included (•) or excluded (◦).
- Asymptotes are shown with dotted lines. (An exception is when the x- or y-axis is an asymptote since the axes already appear.)

If you are not careful in choosing a window, what you see on your screen may appear to be in the family of one parent function when the function you are considering actually belongs to another family of functions. Consider the graph of $f(x) = x^3$ in the two windows below. At the left, the function appears to be linear. At the right, its true cubic nature is shown.

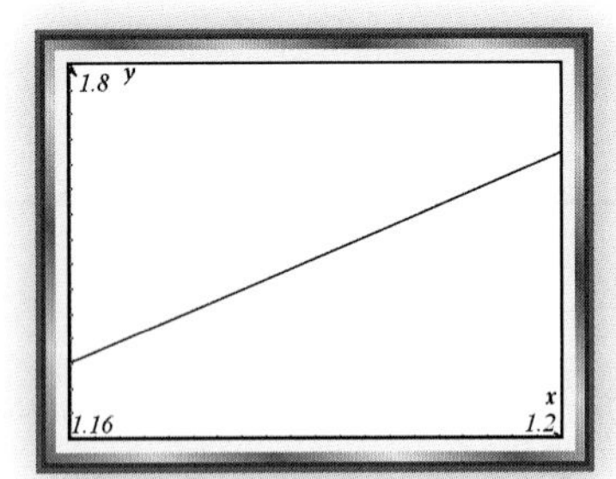

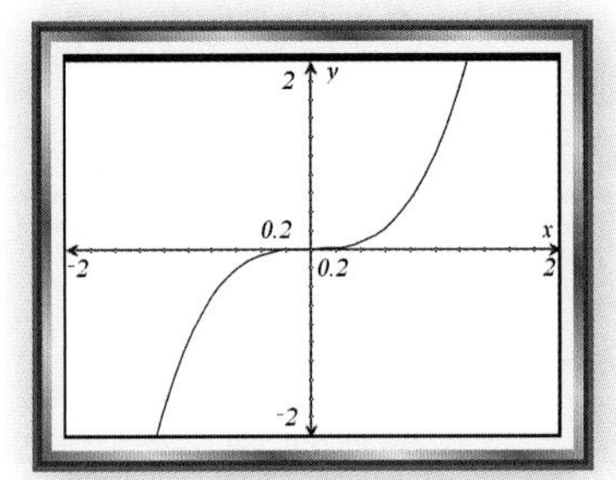

Questions

COVERING THE IDEAS

In 1–3, refer to the parent functions at the beginning of the lesson.

1. Name the functions whose graphs pass through Quadrant III.
2. Name the functions whose range is the set of all real numbers.
3. Name the functions for which the y-axis is a line of symmetry.
4. Consider the function f with $f(x) = 3^x - 5$. Give equations for any asymptotes of its graph.
5. Give the domain and range of h with $h(x) = |x + 30| + 12$.
6. Let three functions f, g, and h be defined by $f(x) = 3x^2$, $g(x) = 0.5x^2$, and $h(x) = -x^2$.
   a. Graph f, g, and h in the same window on a graphing utility.
   b. Graph $y = x^2$ in the same window. Describe how each graph in Part a is related to the graph of the parent function.

In 7 and 8, equations for two functions are given.

a. Sketch graphs of each pair of functions on the same set of axes.

b. How are the two graphs related?

7. $y = \sqrt{x}$, $y = \sqrt{x - 5}$
8. $y = |x|$, $y = |4x|$
9. State the dimensions of the default window of your graphing utility.

10. **a.** Graph the function f with $f(x) = x^3 - x$ in the given windows.

i. $-1 \le x \le 1, -1 \le y \le 1$ **ii.** $-5 \le x \le 5, -5 \le y \le 5$

iii. $-10 \le x \le 10, -10 \le y \le 10$ **iv.** $-100 \le x \le 100, -100 \le y \le 100$

b. Which window provides the most useful graph? Why?

APPLYING THE MATHEMATICS

11. **a.** Graph $f(x) = \frac{1}{x}$, $g(x) = \frac{1}{x+6}$, and $h(x) = \frac{1}{x} + 10$ on the same set of axes.

b. At what value(s) of x is each of f, g, and h discontinuous?

c. Give an equation of the vertical asymptote of each curve.

d. How is each of g and h related to f?

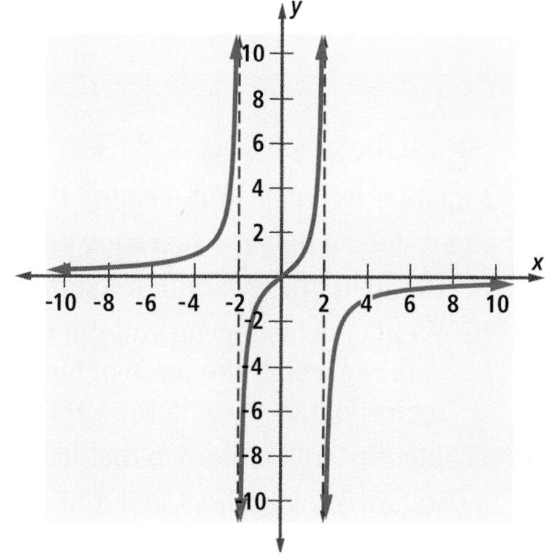

12. The graph at the right shows the function f with $f(x) = \frac{3x}{4 - x^2}$. Find a window for which the visible portion of the graph looks like the graph of

a. a cubic function.

b. a square root function.

13. Let f be the function with $f(x) = \sqrt{x^2 + 30x + 200}$.

a. What is the domain of f?

b. Choose an appropriate window and sketch the graph.

14. The relationship between the wrist circumference and neck circumference in humans is modeled by $y = 1.92x + 4.23$, where x is the wrist circumference (cm) and y is the neck circumference (cm). What is a reasonable domain in this context?

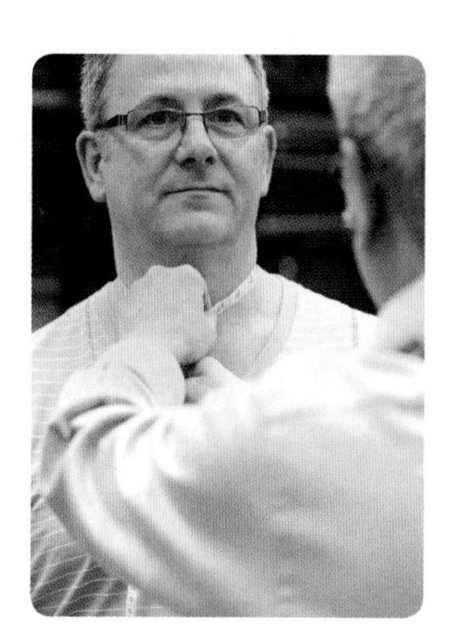

15. **a.** Use your graphing utility to sketch the graphs of $f(x) = \frac{1}{x^2}$ and $g(x) = \frac{4}{x^2}$ on the same set of axes. Choose a window that allows you to compare the two functions.

b. **True or False** For all x, $f(x) < g(x)$.

c. Justify your response to Part b both algebraically (by using the formulas) and geometrically (by using the graphs).

REVIEW

16. Consider the function $r : t \rightarrow 5(0.83)^t$. **(Lesson 2-4)**

a. Give the domain of r.

b. Give the range of r.

c. State equations for any asymptotes of the graph of r.

In 17–19, state whether the graph represents a function. (Lesson 2-1)

17.

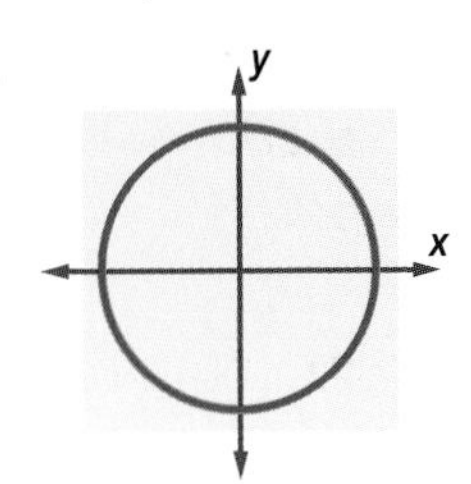

18.

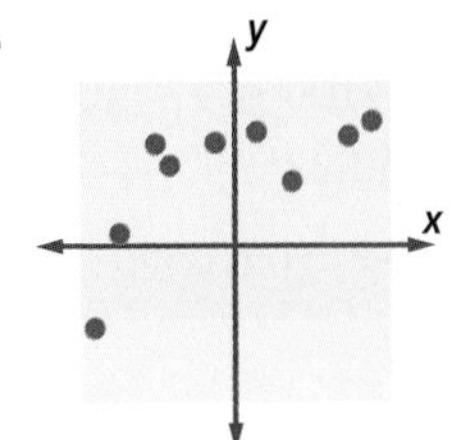

19.

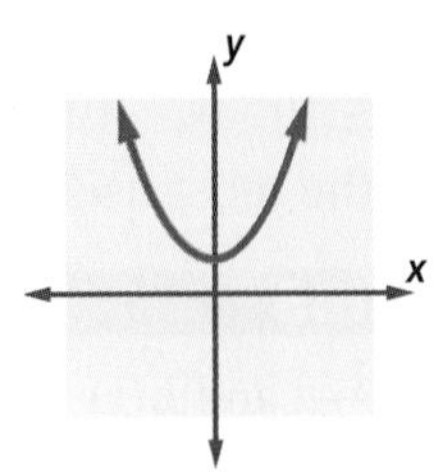

20. ***Skill Sequence*** Simplify. **(Previous Course)**

a. $\frac{1}{\frac{1}{6}}$ b. $\frac{1}{\frac{1}{n}}$ c. $\left(\frac{1}{n}\right)^{-1}$

21. Sketch the graph of $(x - 1)^2 + (y - 1)^2 = 4$. **(Previous Course)**

22. Consider the table at the right. Prices are in thousands of dollars. **(Lessons 1-6, 1-4)**

a. Compute the five-number summary.

b. What conclusion can you draw from these data regarding the price of housing in large metropolitan areas in the United States?

c. Identify any outliers in the data.

d. Remove any outliers and find the mean and standard deviation for the remaining data.

Median Sales Price
Existing Single-Family Homes, 2006

Atlanta	171.8	New York	469.3
Chicago	273.5	Philadelphia	230.2
Cleveland	134.4	St. Louis	148.4
Dallas	149.5	San Francisco	736.8
Detroit	151.7	Seattle	361.2
Houston	149.1	Washington, D.C.	431.0
Miami	371.2		

Source: National Association of Realtors

23. a. Draw $\triangle ABC$ with $A = (5, 3)$, $B = (-3, -4)$, and $C = (-2, 2)$.

b. Draw $\triangle A'B'C'$, the triangle whose x-coordinates are 6 more than the corresponding coordinates in $\triangle ABC$.

c. How are $\triangle ABC$ and $\triangle A'B'C'$ related? **(Previous Course)**

EXPLORATION

24. a. Graph the function f with equation $f(x) = \frac{5x^2 - 271x + 3600}{(x - 24)(x - 30)} + 20$. Set your window to $10 \le x \le 40$, $10 \le y \le 40$. A window-change applet may be provided by your teacher.

b. Find a window that shows only a portion of the graph that looks like a parabola. Record the viewing window.

c. Find a window that shows only a part of the graph that looks like a line.

d. Find a window that shows a graph that may have inverse-square as a parent.

e. See how many of the other parent functions you can mimic by zooming in on part of the graph of f.

QY ANSWERS

1. a. inverse variation, inverse square

b. all except inverse variation, inverse square, and exponential growth

2. -55; 5; -20; 30

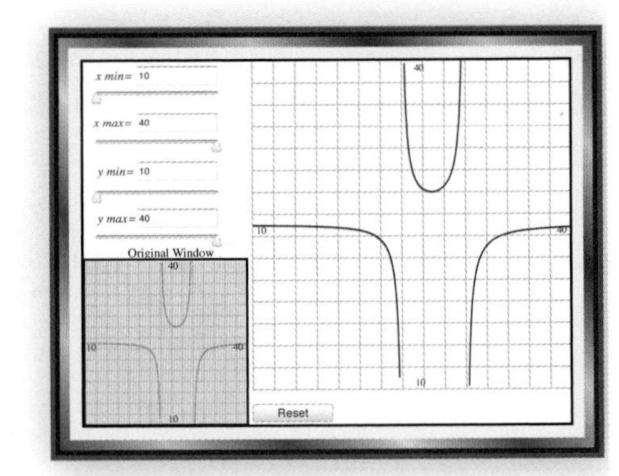

Lesson 3-2 The Graph-Translation Theorem

Vocabulary

transformation
preimage
image
translation

▶ **BIG IDEA** When the graph of a function is translated vertically or horizontally, its equation changes in a related way.

The Translation Image of a Graph

A **transformation** is a one-to-one correspondence between sets of points. Another term for "correspondence" is "mapping." We say that one set, the **preimage**, is mapped onto the other set, the **image**. One type of transformation is a **translation**, which shifts the graph horizontally or vertically. You can quickly describe and interpret translation images of functions of common parent functions. Below, the graph of the parent function f is the preimage. The graph of g is its image under a *vertical translation*.

Mental Math

Consider the data set {2, 2, 3, 4, 6}.

a. Find the mean, median, and mode of the data set.

b. Which measure is least descriptive of the center of the data set?

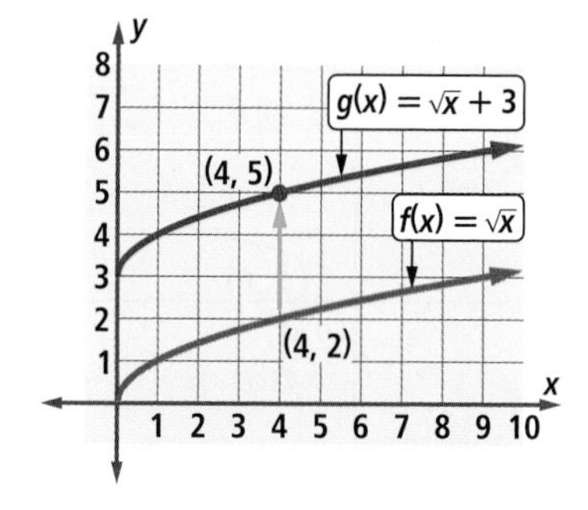

x	$f(x)$	$g(x)$
1	1	4
2	$\sqrt{2}$	$\sqrt{2} + 3$
3	$\sqrt{3}$	$\sqrt{3} + 3$
4	2	5
5	$\sqrt{5}$	$\sqrt{5} + 3$

This translation can also be written as $T(x, y) = (x, y + 3)$ or $T: (x, y) \rightarrow (x, y + 3)$, which is read "$(x, y)$ is mapped onto $(x, y + 3)$." The graph of $g(x) = \sqrt{x} + 3$ is the translation image of the graph $f(x) = \sqrt{x}$.

The general translation of the plane translates figures horizontally and vertically at the same time.

Definition of Translation

A **translation** in the plane is a transformation that maps each point (x, y) onto $(x + h, y + k)$, where h and k are constant.

Example 1

The graph of $y = x^2$ is shown at the right, together with its image under a translation T. The point (0, 0), which is the vertex of the preimage, maps onto the vertex (-3, 1) of the image.

a. Find a rule for the translation T.

b. Find the image of (2, 4) under this translation.

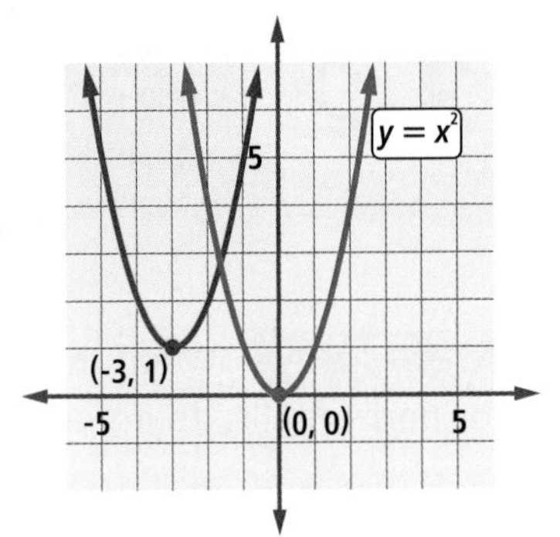

Solution

a. The second graph has been obtained from the graph of $y = x^2$ by a translation 3 units to the left and 1 unit up. Thus,

$T(x, y) = (x - 3, y + 1)$.

b. $T(2, 4) = (2 - 3, 4 + 1) = (-1, 5)$

QY1

QY1

What is the range of the image function in Example 1?

You can use technology to explore translations.

Activity

MATERIALS Graph variation application supplied by your teacher or the Internet

Consider $f1(x) = \sqrt{x}$. The graph of $f2$ is the image of the graph of $f1$ under a translation of h units horizontally and k units vertically. Points (0, 0) and (1, 1) are on the preimage graph, and P and Q are their images.

Step 1 The screen at the right shows the instance in which h is 5 and k is 4. Give the coordinates of P and Q.

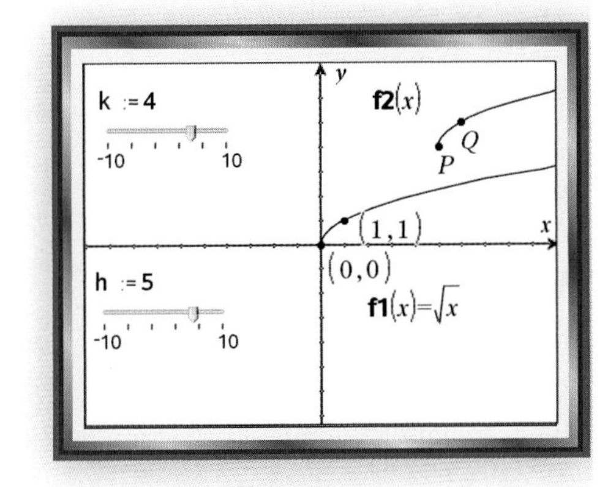

Step 2 Adjust the sliders so that $h = -3$ and $k = 2$. Give the coordinates of P and Q on the image.

Step 3 Find values of h and k so that the endpoint of the image is at (8, -6).

The Graph-Translation Theorem

Graphed in blue at the right is the circle with equation $x^2 + y^2 = 25$ and one point that lies on it, (3, -4). When they are translated 10 units right and 2 units down, the images are the circle with equation $(x - 10)^2 + (y + 2)^2 = 25$ and the point (13, -6), which are shown in red. The preimage point (3, -4) and its image (13, -6) satisfy the equations of their respective circles.

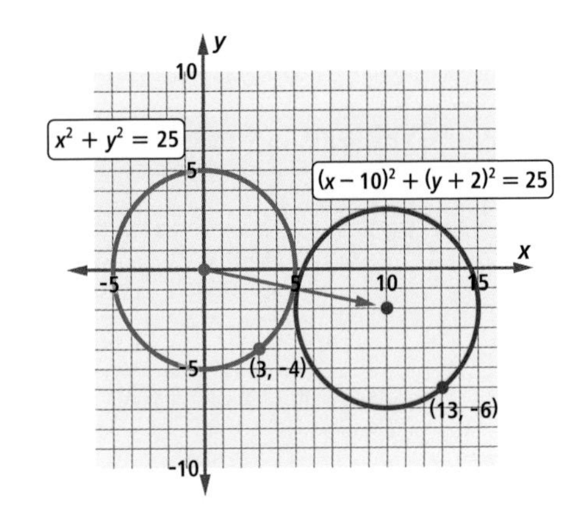

If we let T stand for this translation, then T: $(x, y) \rightarrow (x + 10, y - 2)$. Name the image point (x', y'). Then $x' = x + 10$ and $y' = y - 2$. Solving for x and y gives $x' - 10 = x$ and $y' + 2 = y$. Substitute in the equation $x^2 + y^2 = 25$ and you get $(x' - 10)^2 + (y' + 2)^2 = 25$. It is customary to write the image circle equation without the primes.

There is a direct relationship between replacing a variable expression in an equation and finding the image of a graph under a transformation. Consider the graphs of $f(x) = |x|$ and $g(x) = |x - 4|$ at the right. As the arrow from the point (8, 8) to (12, 8) indicates, the graph of $g(x) = |x - 4|$ is the image of the graph of $f(x) = |x|$ under the translation 4 units to the right, or $(x, y) \rightarrow (x + 4, y)$. Note that adding 4 to each x-coordinate corresponds to replacing x by $x - 4$ in the equation of the preimage. This leads to an important generalization.

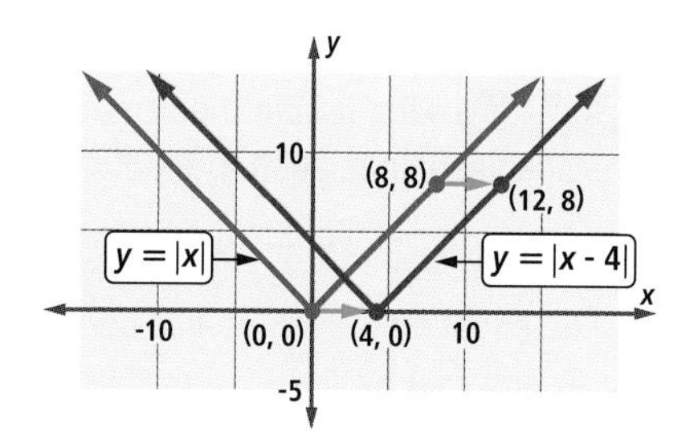

Graph-Translation Theorem

Given a preimage graph described by a sentence in x and y, the following two processes yield the same image:

(1) replacing x by $x - h$ and y by $y - k$ in the sentence;

(2) applying the translation $(x, y) \rightarrow (x + h, y + k)$ to the preimage graph.

That is, for $T(x, y) = (x + h, y + k)$, an equation of the image of $y = f(x)$ is $y - k = f(x - h)$.

The Graph-Translation Theorem can be applied to write an equation if a graph is given, and to sketch a graph if an equation is given. Translations also occur in situations that might not be described with an equation.

GUIDED

Example 2

At the right are graphs of the function $y = C(x) = \sqrt{25 - x^2}$ and its image $y = D(x)$ under the translation $(x, y) \rightarrow (x + 5, y - 4)$. Both are semicircles. Find an equation for the image.

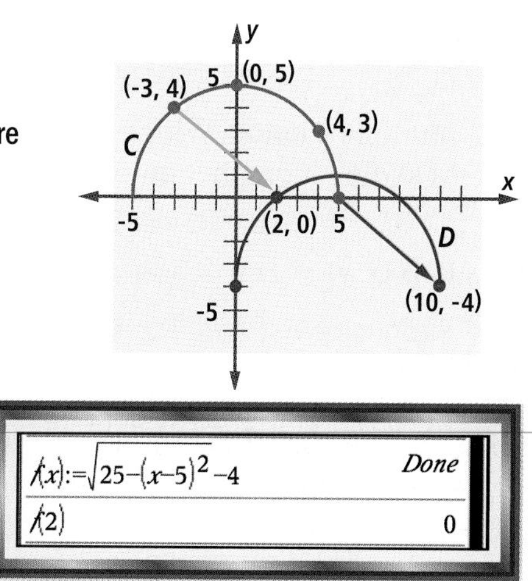

Solution Because the translation is 4 units down, replace y with _?_. Because the translation is 5 units to the right, replace x with _?_. This gives the equation _?_. Solve for y. $y =$ _?_

Check 1 The point (–3, 4) is on C. According to the translation, the image of (–3, 4) is (_?_, _?_). Show that this image point checks in your equation for D.

Check 2 Use a graphing utility to plot C and D on the same axes.

 QY2

QY2

What are the domains of C and D?

Example 3

Sketch a graph of $y = \frac{1}{(x+1)^2} - 4$.

Solution First rewrite the sentence in the form $y - k = \frac{1}{(x-h)^2}$ to see the replacements in relation to the function $y = \frac{1}{x^2}$.

$$y - -4 = \frac{1}{(x+1)^2} = \frac{1}{(x - -1)^2}$$

In the equation $y = \frac{1}{x^2}$, y has been replaced by $y - -4$ and x has been replaced by $x - -1$. Thus, by the Graph-Translation Theorem, the graph of $y - -4 = \frac{1}{(x - -1)^2}$ is the image of the graph of $y = \frac{1}{x^2}$ under the translation $T(x, y) = (x - 1, y - 4)$. Therefore, its graph is translated 1 unit to the left and 4 units down from the graph of the parent inverse-square function. In particular, its asymptotes are $x = -1$ and $y = -4$. With this knowledge, its graph can be sketched. The graphs of the parent function and the given equation are drawn at the right.

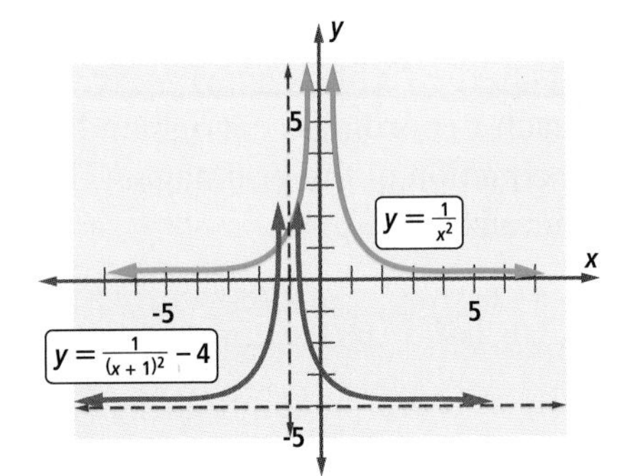

Questions

COVERING THE IDEAS

In 1–3, find the image of each point under the given translation.

1. (2, -5), move left 3 units and up 6 units
2. (-14, -7), horizontal translation of 2 units and a vertical translation of -3 units
3. (p, q), horizontal translation of a units and vertical translation of b units

In 4 and 5, find the image of the point under T: $(x, y) \rightarrow (x - 3, y + 4)$.

4. (1, -2)
5. (r, s)
6. **Multiple Choice** Which rule is for a translation T that has the effect of sliding a graph 3 units down and 7 units to the left?
   A $T(x, y) = (x - 3, y - 7)$
   B $T(x, y) = (x + 3, y + 7)$
   C $T(x, y) = (x - 7, y + 3)$
   D $T(x, y) = (x - 7, y - 3)$
7. Suppose $y = \frac{1}{x}$ and $T(x, y) = (x + 3, y + 4)$.
   a. Find the images of (1, 1), (-1, -1), and (0.5, 2) under T.
   b. Verify that the three images under T satisfy $y - 4 = \frac{1}{x - 3}$.

8. The graph of $f(x) = \sqrt{x}$ is shown at the right, together with its image under a translation T. The image of (0, 0) on the graph is the point (-1, 3).
   a. Find a rule for the translation T.
   b. Find the image of (9, 3) under this translation.
   c. Find an equation for the image g.

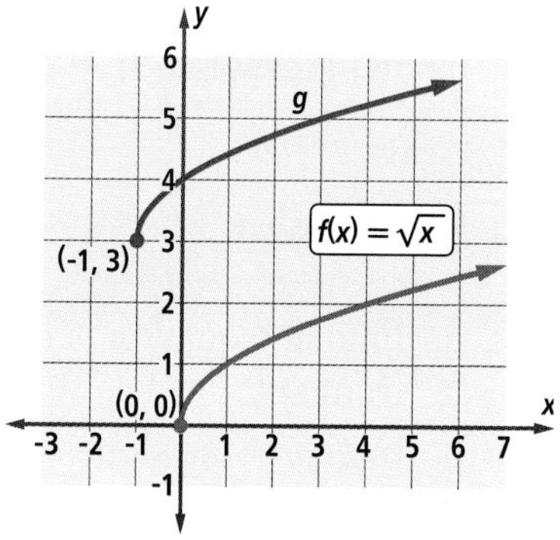

9. Use the Graph-Translation Theorem to find an equation of the image of $y = |x|$ under $T(x, y) = (x - 4, y + 6)$.

In 10 and 11, rules for two functions are given.
a. State a rule for a translation that maps *f* onto *g*.
b. Graph *f* and *g* on the same set of axes.

10. $f(x) = \frac{1}{x}$; $g(x) = \frac{1}{x-2}$

11. $f(x) = x^2$; $g(x) = (x + 2)^2 + 1$

APPLYING THE MATHEMATICS

12. a. Graph f and g, where $f(x) = \frac{1}{x}$ and $g(x) = \frac{1}{x+3} - 2$, on the same set of axes.
    b. Find equations for the asymptotes of g. How are they related to the asymptotes of f?
    c. Give the domain and range of f and g.

13. The parabola $y = (x - 3)^2 + 5$ is the image of the parabola $y = (x + 2)^2 + 7$ under a translation $T: (x, y) \rightarrow (x + h, y + k)$. What are the values of h and k?

14. **Multiple Choice** A circle has radius 4 and center (2, -3). Which of the following might be an equation for the circle?
    A $(x + 2)^2 + (y + 3)^2 = 16$
    B $(x - 2)^2 + (y - 3)^2 = 16$
    C $(x - 2)^2 + (y + 3)^2 = 16$
    D $(x + 2)^2 + (y - 3)^2 = 16$

15. The formula $N = 0.82(1.09)^t$ gives the approximate current U.S. national debt (in billions of dollars), t years after 1900.
    a. Compute the estimated national debt for 2007.
    b. Convert the formula to one that maps the actual year y onto the debt.

REVIEW

16. Consider the equation $y = \frac{1}{x-1}$. **(Lesson 3-1)**
    a. Identify the parent equation.
    b. Graph the equation and its parent on the same axis.

17. Use the mean lengths of largemouth bass of various ages given below. **(Lesson 2-3)**

Age (years)	1	2	3	4	5	6	7	8	9	10
Length (inches)	6.3	9.0	11.6	13.5	15.8	17.4	18.9	19.8	20.3	20.7

Source: Illinois Department of Natural Resources

a. Find an equation of the line of best fit for predicting length from age.

b. Interpret the slope of the line.

c. Use the line to predict the length of a twelve-year-old largemouth bass in Illinois.

d. Suggest a reason for being cautious about your prediction in Part c.

e. Find the correlation coefficient r between the age of the fish and its mean length.

f. Interpret the sign of r.

The world-record largemouth bass, caught in 1932 in Georgia, was 32.5 inches long and 28.5 inches in girth.

18. ***Skill Sequence*** Factor. **(Previous Course)**

a. $k^2 - 9$ b. $9 - 16t^2$ c. $(p + 4)^2 - 25$

EXPLORATION

19. a. Consider the linear equation $f(x) = 3x - 5$. Find an equation for the image of the graph of f under the following transformations.

i. $T(x, y) = (x + 1, y + 3)$

ii. $T(x, y) = (x + 2, y + 6)$

iii. $T(x, y) = (x - 4, y - 12)$

iv. $T(x, y) = (x + 1, y + 5)$

b. Make a conjecture based on the results of Part a.

c. Prove your conjecture in Part b.

d. Generalize this problem to any line of the form $y = mx + b$.

QY ANSWERS

1. $\{y \mid y \geq 1\}$

2. For C, $-5 \leq x \leq 5$. For D, $0 \leq x \leq 10$.

Lesson 3-3 Translations of Data

Vocabulary

invariant

▶ **BIG IDEA** Adding (or subtracting) the same value to every number in a data set adds (or subtracts) that value to measures of center but does not affect measures of spread.

Translating Data

The 2007 United States Open Golf Championship was played at the Oakmont Country Club in Pennsylvania. In golf, *par* is a predetermined number of strokes that a good golfer should require to complete a hole. For the Oakmont course, pars on the 18 holes totaled 70 strokes, so the course was considered to be a par-70 course. A golfer's progress through the 18 holes of a golf course is tracked by how many strokes the golfer is above or below par. For example, in his first round of the Open, Tiger Woods took 71 strokes, which was one over par, or +1. Angel Cabrera, the eventual champion, took 69 strokes, which was one under par, or -1, for the round.

The dimples on a golf ball make the ball travel farther.

Mental Math

Calculate the average score for five basketball games that had scores of 92, 93, 94, 98, and 98.

News reports give both the raw scores, 71 and 69, and the scores relative to a par, 1 and -1. Notice that if a player's score is s, then his score in relation to par is $s - 70$. This is an example of a *translation of data*. Translations of data produce distributions that have predictable shapes, centers, and spreads.

 QY

▶ QY

A golf course is rated at par 72. If a player's stroke score is x, what is his score relative to par?

A translation of a set of data $\{x_1, x_2, ..., x_n\}$ is a transformation that maps each x_i to $x_i + h$, where h is some constant. If T is the translation, then this transformation can be described as

$$T: x \rightarrow x + h \text{ or } T(x) = x + h.$$

The number $x + h$, or the point it represents, is the translation image of x. In the U.S. Open, the transformation T mapping each number of strokes onto its image has the rule $T(x) = x - 70$.

Activity

In the United States, the passage of the 19th amendment to the Constitution in 1920 gave women the right to vote. This activity compares the year in which women earned the right to vote in the U.S. to the year women achieved that right in other countries.

Step 1 Enter the years below into a statistics utility. Label the column `year`.

1893	New Zealand	1920	United States	1949	China	1974	Jordan
1902	Australia	1921	Sweden	1950	India	1976	Portugal
1906	Finland	1928	Britain	1954	Colombia	1989	Namibia
1913	Norway	1928	Ireland	1957	Malaysia	1990	Western Samoa
1915	Denmark	1931	Spain	1962	Algeria	1993	Kazakhstan
1917	Canada	1944	France	1963	Iran	1993	Moldova
1918	Austria	1945	Italy	1963	Morocco	1994	South Africa
1918	Germany	1947	Argentina	1964	Libya	2005	Kuwait
1918	Poland	1947	Japan	1967	Ecuador		
1918	Russia	1947	Mexico	1971	Switzerland		
1919	Netherlands	1947	Pakistan	1972	Bangladesh		

Source: New York Times

Step 2 Make a histogram of the data. Use bin size 10. Compute the mean, median, mode, range, IQR, and standard deviation for the data.

Step 3 To compare the year in which women got the right to vote in different countries to the U.S., subtract 1920, the U.S. or *baseline* value, from all data points. To do this, first create a slider called "baseline" that takes on values from 0 to 2000. Title a second column `newyear`. In the formula line, enter = `year - baseline`.

Step 4 Adjust your slider so that baseline = 1920. What is the adjusted value for Libya? For Finland? Interpret each in a sentence.

Step 5 Compute the mean, median, mode, range, IQR, and standard deviation for the adjusted data. Which values are the same as the original data? Which are different?

Step 6 Now make a histogram of the adjusted data. Use bin size 10. How does it compare to the histogram for the original data? What happens to the histogram bars when you move the slider?

Step 7 Make box plots of both the original and the adjusted data. Again, move the slider around. Describe how the box plot changes as you change the data.

Step 8 Which numbers in the five-number summary are affected by translating the data, and which ones are unaffected by the translation?

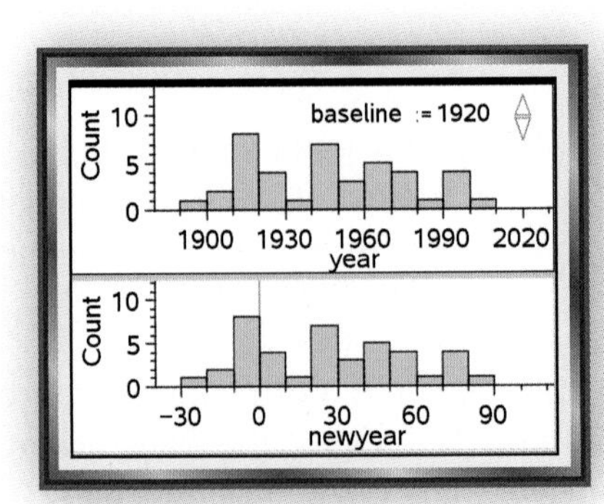

Measures of Center of Translated Data

Some of the results in the Activity are generalized in this theorem.

Theorem (Centers of Translated Data)

Adding h to each number in a data set adds h to each of the mean, median, and mode.

Proof Let $\{x_1, x_2, \ldots, x_n\}$ be a data set. Consider the translation in which x_i is mapped to $x_i + h$. To find the mean of the image set, you must evaluate

$$\frac{\sum_{i=1}^{n} (x_i + h)}{n}.$$

By definition of Σ, this expression represents

$$\frac{\overbrace{(x_1 + h) + (x_2 + h) + (x_3 + h) + \cdots + (x_n + h)}^{n \text{ terms}}}{n}.$$

Using the Associative and Commutative Properties of Addition, we rewrite the expression as

$$\frac{(x_1 + x_2 + x_3 + \cdots + x_n) + \overbrace{(h + h + h + \cdots + h)}^{n \text{ terms}}}{n}$$

$$= \frac{\left(\sum_{i=1}^{n} x_i\right) + nh}{n}$$

$$= \frac{\sum_{i=1}^{n} x_i}{n} + \frac{nh}{n}$$

$$= \frac{\sum_{i=1}^{n} x_i}{n} + h = \bar{x} + h.$$

Thus under a translation by h, the mean of the image set of data is h units more than the mean of the original set of data. It also can be shown that after a translation of h units, the median and mode of the image set are also increased by h units.

Measures of Spread for Translated Data

In the Activity, you saw that the range of the distributions, the IQR, and the standard deviation before and after the translation remain unchanged. To see that this is true in general for a data set $\{x_1, x_2, \ldots, x_n\}$ under a translation by h, first recall that range = maximum − minimum. Because a translation does not change the relative positions of the data points, the minimum value of the translated data is the image of the original minimum, and the maximum value of the translated data is the image of the original maximum.

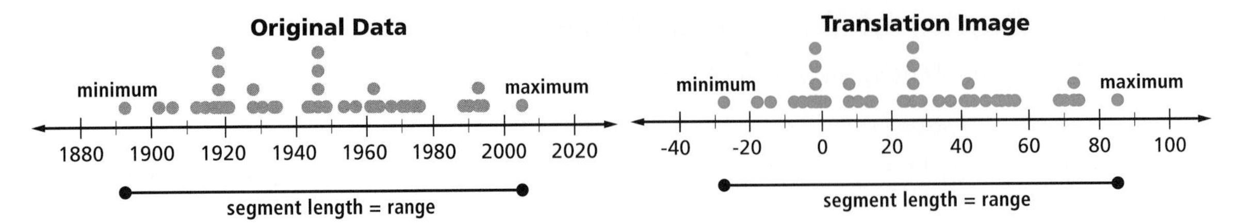

Under a translation of h units, the minimum m is mapped to $m + h$ and the maximum M is mapped to $M + h$. The range of the translated data is

$$(M + h) - (m + h) = M - m,$$

which is the range of the original data. Therefore the range remains unchanged after the translation.

In the calculation of the variance and standard deviation of image data under a translation by h, the mean $\bar{x}$ becomes $\bar{x} + h$. So, each new deviation equals $(x_i + h) - (\bar{x} + h) = x_i - \bar{x}$, which is the original deviation. Because each individual deviation stays the same under a translation, the variance and standard deviation also stay the same.

Theorem (Spreads of Translated Data)

Adding h to each number in a data set does not change the range, interquartile range, variance, or standard deviation of the data.

Because the measures of spread of a data set do not vary under a translation, they are said to be **invariant** under a translation.

The preceding theorems can be used to compute measures of center or spread when data are increased or decreased.

GUIDED

Example

In a local produce store, cantaloupes sell for 99¢ per pound. A clerk weighed 30 melons and computed the following statistics: mean = 3.5 lb, standard deviation = 8 oz, median = 3.4 lb, and IQR = 1 lb. After finishing his task, the clerk noticed that the scales were not correctly calibrated. The scale was set at 3 oz as its starting weight, not 0. Find the correct values for the mean, standard deviation, median, and interquartile range for the 30 melons.

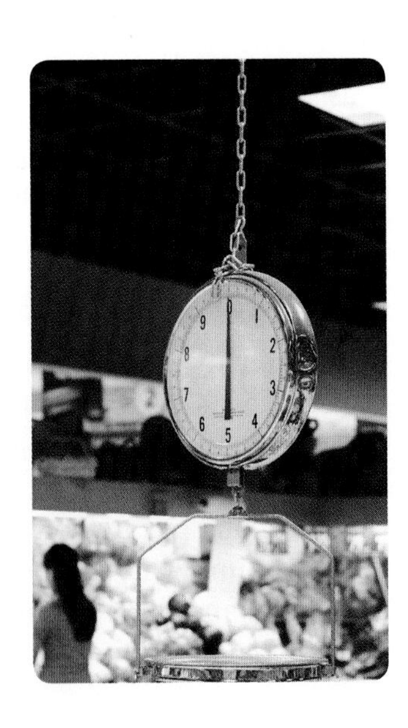

Solution The clerk recorded the weight of each cantaloupe as 3 ounces too heavy. The data need to be translated 3 ounces smaller. The mean and the median will therefore be decreased by 3 ounces.

Mean = 3.5 lb – __?__ = 3 lb 8 oz – __?__ = __?__

Median = 3.4 lb – __?__ = 3 lb 6.4 oz – __?__ = __?__

But the standard deviation and IQR will be unchanged.

Standard deviation = __?__ and IQR = __?__

Questions

COVERING THE IDEAS

1. A transformation that maps a number x to $x + h$ is called a(n) __?__.

2. The box plots at the right show the distribution of December salaries for 10 employees in a start-up company before and after their year-end bonus was added. How much was the bonus?

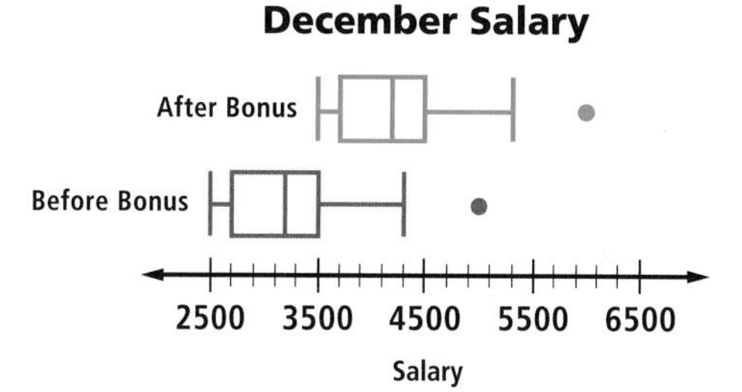

3. Kayla withdrew the 5 amounts shown below from an ATM while visiting her sister in another town.
   Withdrawal amounts:

   \$50 \$100 \$80 \$120 \$100.
   a. Find the following statistics for the five withdrawals.
      i. range
      ii. mode
      iii. median
      iv. mean
      v. variance
      vi. standard deviation
   b. Since Kayla was not using her own bank's ATM, she was charged a fee of \$1.50 for each transaction. Use your answers from Part a to compute the same statistics for the amounts withdrawn including the transaction fees.

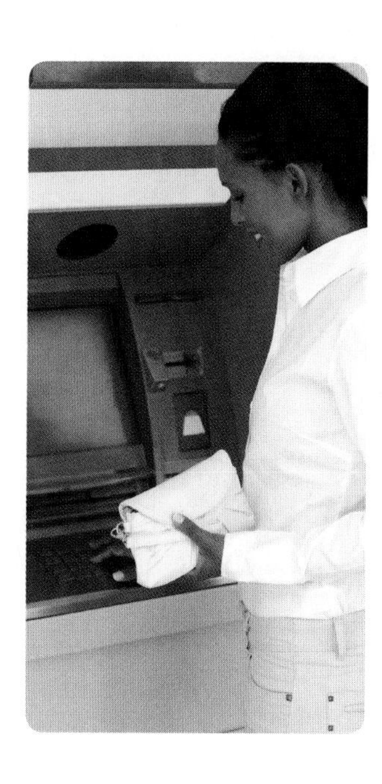

4. Suppose $x_1 = -3$, $x_2 = 5$, $x_3 = 6.1$, $x_4 = 2.4$, and $x_5 = 3.2$. Evaluate the given expression.
   a. $\sum_{i=1}^{5} (x_i + 8)$
   b. $\sum_{i=1}^{5} x_i + 8$

5. For a set of n test scores, the mean was $\overline{x}$ and the standard deviation was s. Later, every score was increased by b bonus points.
   a. What is the mean of the translated scores?
   b. What is the standard deviation of the translated scores?

6. Name four statistical measures that are invariant under a translation.

7. A set of data is translated. Find the missing values in the table.

	Original Data	Transformed Data
cases	10	?
mean	?	53
standard deviation	8.03	?
median	70	59.5
range	23	?
IQR	12	12

8. Consider the two frequency distributions below.

Original Scores	Frequency
2	1
3	2
7	3
8	2
10	6

Transformed Scores	Frequency
10	1
11	2
15	3
16	2
18	6

a. Make a dot-frequency diagram showing the two sets of scores.
b. Identify the transformation used to get the transformed scores.
c. Find the interquartile range, mode, mean, and median for the original scores.
d. Use the theorems of the lesson and your answers from Parts b and c to give the IQR, mode, mean, and median for the transformed data.

APPLYING THE MATHEMATICS

In 9 and 10, consider the data at the right, which give the scores from the first two tests in a class of 15 students taking *FST*.

Test 1	Test 2
84	79
84	81
66	61
98	98
70	67
86	83
74	48
60	61
88	88
93	91
81	81
90	96
82	87
57	53
92	95

9. a. Enter these data into a statistics utility, naming the lists *score 1* and *score 2*.
b. Find the means and standard deviations of *score 1* and *score 2*.
c. Draw a scatterplot with *score 1* on the horizontal axis and *score 2* on the vertical axis.
d. Find the correlation coefficient.
e. Find the line of best fit for predicting *score 2* from *score 1*.
f. Suppose the teacher decided to add 5 bonus points to each score on the first test. Add 5 to each *score 1*. Draw a new scatterplot. How is this scatterplot different from the one in Part c?
g. Compute the correlation coefficient and the least squares regression equation for the new scores on test 1 and the scores on test 2. Compare your answers to those in Parts d and e. What value(s) are invariant under this transformation? What value(s) have changed?
h. Suppose the teacher made an error when computing the scores for the second test, and to correct his error he subtracted 3 points from each score. Draw a scatterplot with the *score 1 plus bonus* on the horizontal axis and *score 2 minus 3* on the vertical axis. How is this scatterplot different from those drawn in Parts c and f?
i. Compute the correlation coefficient and the least squares regression equation for the transformed scores.
j. Compare the correlation coefficients, slopes and intercepts. What value(s) are invariant under these transformations? What value(s) have changed?

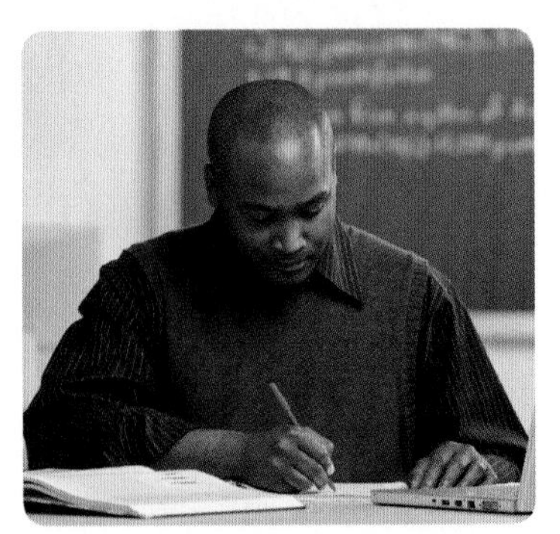

10. **Multiple Choice** Generalize the results of Question 9. If bivariate data are translated, which (if any) of the following are invariant?

A means of the two variables

B standard deviations of the two variables

C intercept of the line of best fit

D slope of the line of best fit

E correlation coefficient between the two variables

11. Mentally calculate the mean of the heights of the starters on a basketball team: 6′, 6′2″, 6′3″, 6′4″, 6′6″.

12. Let $\{x_1, x_2, x_3, ..., x_n\}$ be a data set with median m, and h a constant.

a. Suppose that n is odd. Explain why the median of the set $\{x_1 + h, x_2 + h, ..., x_n + h\}$ is $m + h$.

b. Explain why the median of the translated set is also $m + h$ when n is even.

REVIEW

13. Consider the equation $y + 3 = \frac{1}{x + 2}$. **(Lesson 3-2)**

a. Graph the equation using paper and pencil.

b. Check your work with a graphing utility.

c. Give a rule for the translation that maps the graph of $y = \frac{1}{x}$ onto the graph in Parts a and b.

14. Suppose the translation $T: (x, y) \longrightarrow (x - 8, y + 13)$ is applied to the graph of the function with equation $y = \frac{1}{x^2}$. **(Lesson 3-2)**

a. Find an equation for the image.

b. Sketch graphs of the image and preimage.

15. Let a real function f be defined by $f: x \longrightarrow \sqrt{x + 99}$. What is the domain of f? **(Lesson 3-1)**

16. ***Skill Sequence*** Solve for t. **(Previous Course)**

a. $rt = 18$ b. $rt = 7 + r$ c. $r = \frac{5}{t}$ d. $r + 4 = \frac{1}{t}$

17. a. Draw $\triangle ABC$ with $A = (0, 2)$, $B = (4, 4)$, and $C = (3, 0)$.

b. Let r be the transformation that maps each point (x, y) onto $(2x, 3y)$. Draw $A'B'C' = r(\triangle ABC)$.

c. What transformation is r? **(Previous Course)**

EXPLORATION

18. Give an example of a situation other than an athletic event where data are often translated, and explain why the translated data are used.

QY ANSWER

$x - 72$

Lesson 3-4 Symmetries of Graphs

Vocabulary

reflection-symmetric
axis of symmetry
line of symmetry
symmetric about a point
point symmetry
center of symmetry
even function
odd function

BIG IDEA Rules for functions can be used to determine symmetries of graphs and vice versa.

Recall from your previous courses that a figure is **reflection-symmetric** if and only if the figure can be mapped onto itself by a reflection over some line *l*. The reflecting line *l* is called the **axis** or **line of symmetry** of the figure. A line of symmetry can be any line in the plane. Similarly, a figure is **symmetric about point** *p* or has **point symmetry** if and only if the figure can be mapped onto itself under a rotation of 180° around *P*. The point *P* is called a **center of symmetry**.

Because graphs are sets of points, these definitions apply to graphs. Any symmetry of a figure implies that one part of the figure is congruent to another part. So when a graph has symmetry, the symmetry shortens the time in drawing the graph and helps in studying it.

Mental Math

What is the least number of symmetry lines the figure can have?

a. parabola

b. rectangle

c. equilateral triangle

d. circle

Activity 1

The diagram at the right shows half of a graph.

Step 1 Copy the diagram. Draw the other half of the graph so that the result is point-symmetric about the origin. Label this half *A*.

Step 2 Draw the other half of the original graph so that the result is symmetric with respect to the *y*-axis. Label this half *B*.

Step 3 Draw the other half of the original graph so that it is symmetric over the *x*-axis. Label the graph *C*.

Step 4 What symmetries does the union of graphs *A*, *B*, and *C* and the original graph possess?

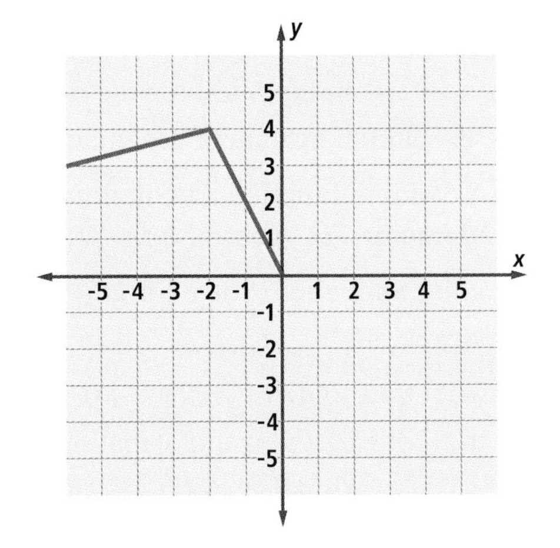

Recall also that the reflection image of (x, y) over the x-axis is $(x, -y)$, and the reflection image of (x, y) over the y-axis is $(-x, y)$. Also the image of (x, y) under a rotation of 180° about the origin is $(-x, -y)$. Combining these facts with the definitions of reflection, symmetry, and point symmetry yields three theorems.

Theorem (Symmetry over *y*-axis)

A graph is symmetric with respect to the *y*-axis if and only if for every point (x, y) on the graph, $(-x, y)$ is also on the graph.

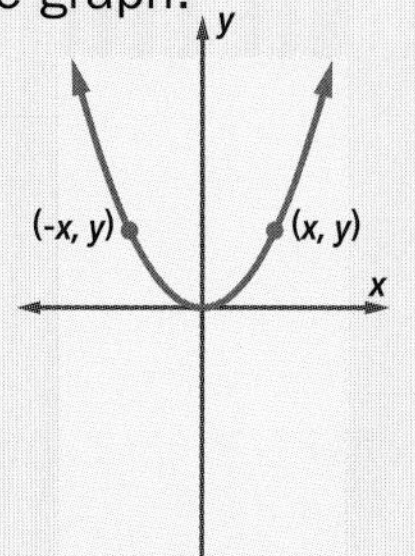

Theorem (Symmetry over *x*-axis)

A graph is symmetric with respect to the *x*-axis if and only if for every point (x, y) on the graph, $(x, -y)$ is also on the graph.

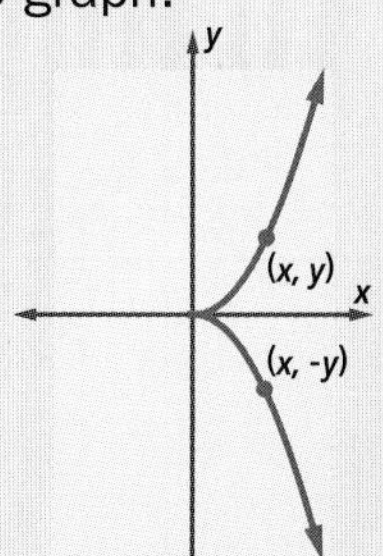

Theorem (Symmetry about the Origin)

A graph is symmetric to the origin if and only if for every point (x, y) on the graph, $(-x, -y)$ is also on the graph.

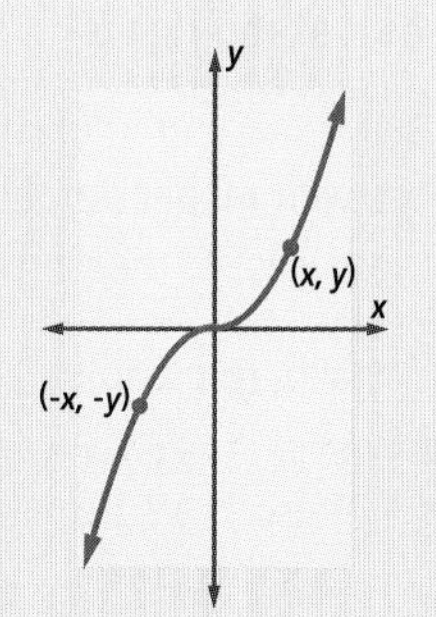

Proving That a Graph Has Symmetry

One point can be a counterexample that shows a graph does not have a certain type of symmetry. But one point cannot determine that a general relationship holds. The graphs of $f(x) = x^2$ and $g(x) = \dfrac{10}{3x^2 + 1}$ seem symmetric with respect to the y-axis. But this is not a proof.

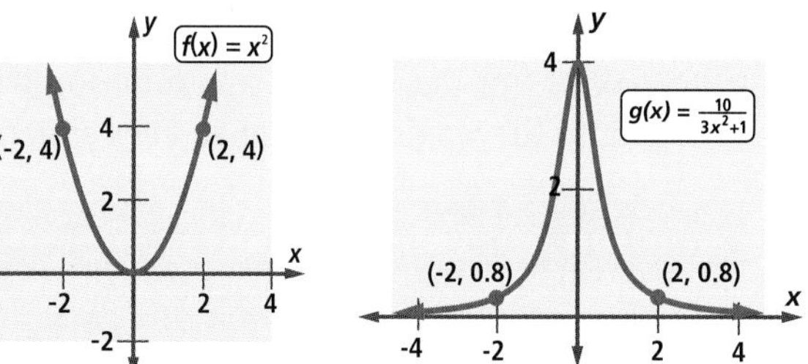

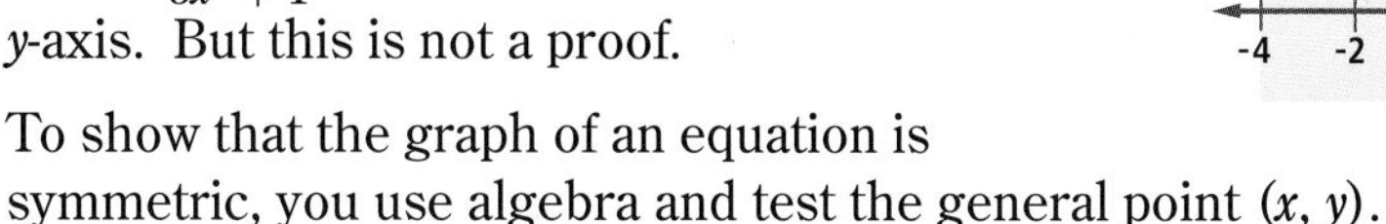

To show that the graph of an equation is symmetric, you use algebra and test the general point (x, y).

Example 1

Prove that the graph of $f(x) = \dfrac{1}{x^2 + 1}$ is symmetric with respect to the *y*-axis.

Solution To show that the graph of f is symmetric with respect to the y-axis, you need to show that for all (x, y) on the graph, $(-x, y)$ is also on the graph. In other words, you need to prove $f(x) = f(-x)$ for all x in the domain of f.

$$f(x) = \frac{10}{3x^2 + 1}$$

$$f(-x) = \frac{10}{3(-x)^2 + 1} = \frac{10}{3x^2 + 1}$$

So $f(x) = f(-x)$. Therefore the graph of $f(x) = \dfrac{10}{3x^2 + 1}$ is symmetric with respect to the y-axis.

Using Symmetry to Aid in Graphing

The ideas of symmetry apply to graphs of relations that are not functions. For instance, the graph of $x = |y|$ is symmetric over the x-axis, since if $x = |y|$, then $x = |-y|$.

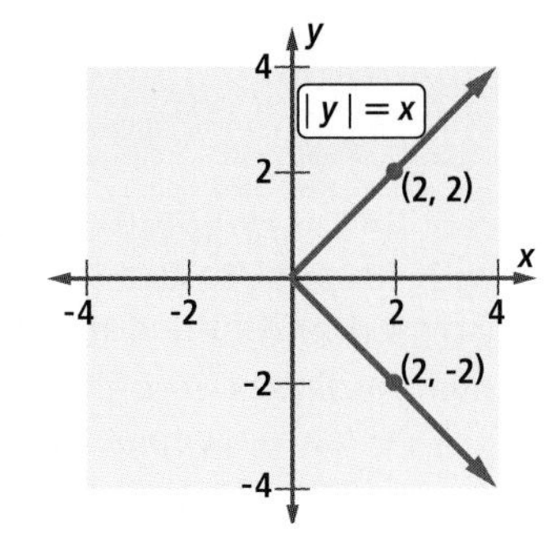

Activity 2

The part of the graph of $xy = 12$ that is in Quadrant I is shown at the right.

Step 1 Test to see if the equation is symmetric with respect to the y-axis, the x-axis, or the origin.

Step 2 Use the results to complete the graph.

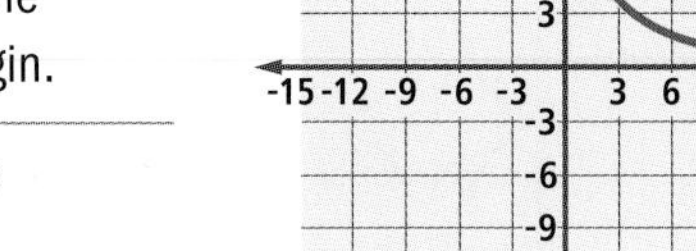

Even and Odd Functions

In previous courses and Lesson 2-7, you studied power functions with equations such as $y = x^2$, $y = x^3$, or $y = x^4$. The power functions f with $f(x) = ax^n$, where $a \neq 0$ and n is even, can all be proved to be symmetric over the y-axis. For this reason, any function whose graph is symmetric with respect to the y-axis is called an *even function*.

Definition of Even Function

A function is an **even function** if and only if for all values of x in its domain, $f(-x) = f(x)$.

By the method of Example 1, the graphs of the power functions f with $f(x) = ax^n$, where $a \neq 0$ and n is odd, can all be proved to be symmetric about the origin. For this reason, a function whose graph is symmetric about the origin is called an *odd function*.

Definition of Odd Function

A function f is an **odd function** if and only if for all values of x in its domain, $f(-x) = -f(x)$.

QY

If you are not sure if a function is even, odd, or neither, then a graphing utility or CAS may help you decide.

QY

Look at the graphs for the theorems on page 173 of this lesson.

a. Which graph shows an even function?

b. Which graph shows an odd function?

Example 2

Determine whether the function $f: x \rightarrow 4x - 2x^3$ is odd, even, or neither. If it appears to be even or odd, prove it.

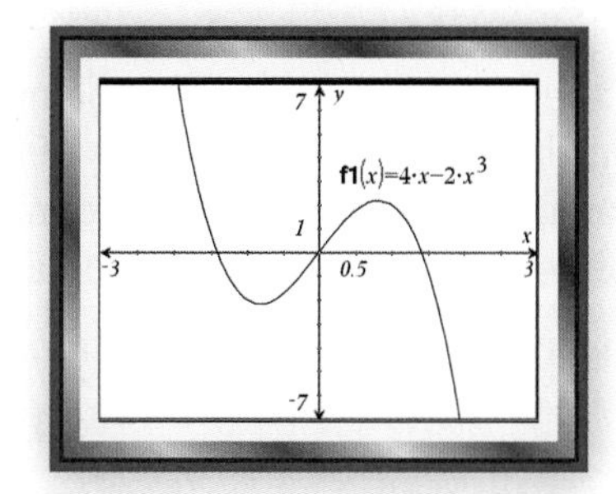

Solution 1 Draw a graph to see if f appears to be even or odd. A graph of f is shown at the right.

The graph appears to be symmetric about the origin, so f seems to be an odd function. To prove this, suppose (x, y) is on the graph. That is, $y = f(x) = 4x - 2x^3$. Now consider $f(-x)$.

$$\begin{aligned} f(-x) &= 4(-x) - 2(-x)^3 \\ &= -4x + 2x^3 \\ &= -(4x - 2x^3) \end{aligned}$$

Thus, $f(-x) = -f(x)$ for all x, so f is an odd function.

Solution 2 Define $f(x) = 4x - 2x^3$ on a CAS. Evaluate $f(-x)$ and $-f(x)$, as shown at the right. The expression for $f(-x)$ is the same as for $-f(x)$. For all x, $f(-x) = -f(x)$, so f is odd.

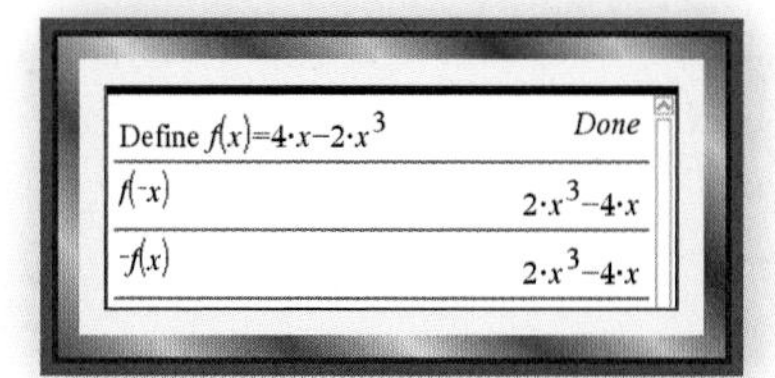

Using the Graph-Translation Theorem to Find Symmetries

You have seen graphs with symmetry related to the x-axis, the y-axis, and the origin. However, a line or point of symmetry may be located in other positions. If the graph is the translation image of a familiar graph, the symmetry of the known graph can give information about symmetry in the image.

Example 3

Consider the function F with $y = F(x) = \frac{1}{(x+3)^2} - 7$.

a. Give equations for the asymptotes of its graph.

b. Describe any lines or points of symmetry.

Solution

a. Rewrite the equation as $y + 7 = \frac{1}{(x+3)^2}$. This shows that, by the Graph-Translation Theorem, the graph of F is the image of the graph of $y = \frac{1}{x^2}$ under the translation $T(x, y) = (x - 3, y - 7)$. The graph of the parent function has asymptotes $x = 0$ and $y = 0$. Each asymptote is translated 3 units to the left and 7 units down. So, the asymptotes of F are $x = -3$ and $y = -7$.

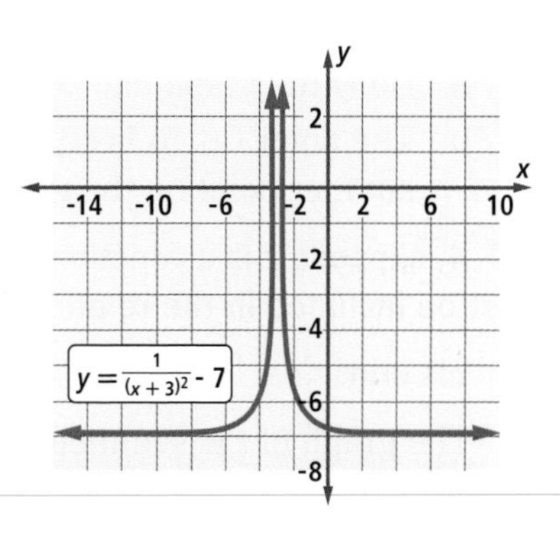

b. Since the graph of $y = \frac{1}{x^2}$ is symmetric over the y-axis, the graph of F is symmetric over the vertical line $x = -3$. Indeed, this line is the translation image of $x = 0$ under $T: (x, y) \rightarrow (x - 3, y - 7)$.

Check Sketch a graph of F.

Properties of Graphs of Translation Images

In general, if f is a function and each point (x, y) on its graph is mapped to $(x + h, y + k)$, then the graph of the image is congruent to the graph of the preimage, and all key points and lines are also mapped under this translation. Specifically, lines of symmetry map to lines of symmetry, maxima to maxima, minima to minima, vertices to vertices, symmetry points to symmetry points, and asymptotes to asymptotes.

This chart summarizes the characteristics of even and odd functions.

	Even Function	Odd Function
Symmetry	symmetric over y-axis	symmetric about the origin
Transformation	$(x, y) \rightarrow (-x, y)$	$(x, y) \rightarrow (-x, -y)$
Function Notation	$f(-x) = f(x)$	$f(-x) = -f(x)$
Sample Graph and Equation	$y = x^2$	$y = x^3$

Questions

COVERING THE IDEAS

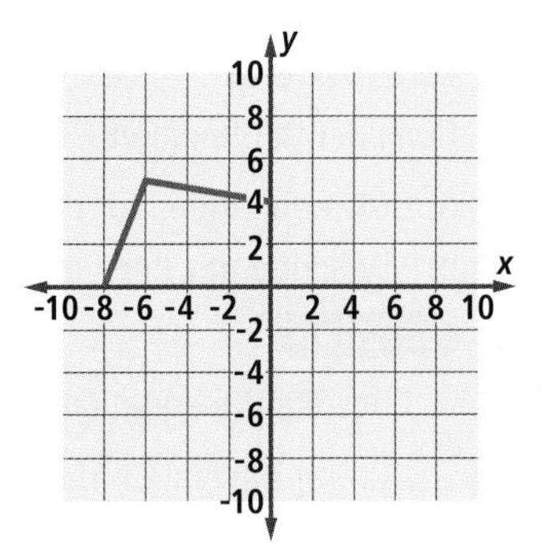

1. The part of a graph that is in Quadrant II is shown. The graph is symmetric over the x-axis and passes through only two quadrants.
   a. Copy the graph and complete it.
   b. The point (-6, 5) is on the graph. Use symmetry to find another point on the graph.

In 2–4, suppose z is a function that includes (4, -2). What other point must be included in the relation if z has the stated property?

2. z is odd.
3. z is even.
4. The graph of z is symmetric with respect to the x-axis.
5. The point (-3, -9) satisfies the equation $y = x \cdot |x|$. Use this point to test whether the graph of $y = x \cdot |x|$ appears to be symmetric
   a. over the y-axis. b. over the x-axis. c. about the origin.

6. Prove that the function f defined by $f\colon x \rightarrow 6x^{-1}$ is an odd function.

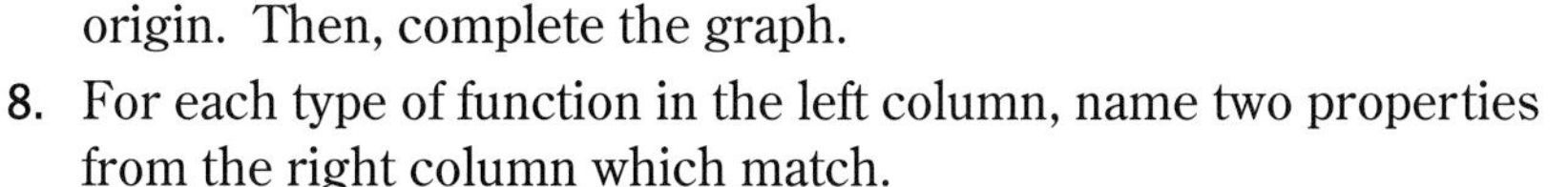

7. The part of the graph of $y = \frac{8}{2 + x^2}$ that lies in Quadrant I is at the right. Test for symmetry with respect to the y-axis, x-axis, and origin. Then, complete the graph.

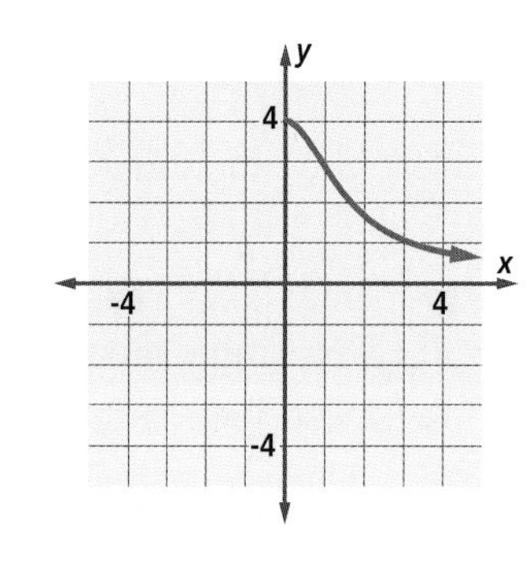

8. For each type of function in the left column, name two properties from the right column which match.

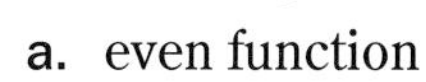

a. even function
b. odd function

i. The graph is symmetric about the origin.
ii. If (x, y) is in the function, so is $(-x, y)$.
iii. If (x, y) is in the function, so is $(-x, -y)$.
iv. The graph is symmetric over the y-axis.
v. The graph is symmetric over the x-axis.
vi. If (x, y) is in the function, so is $(x, -y)$.

In 9 and 10, an equation of a function is given. Tell if the function is odd, even, or neither.

9. $f(x) = -246x$

10. $g(x) = \frac{1}{2}x^3 + 3$

11. Consider the function $f: x \rightarrow \frac{1}{x - 6} + 1$.
   a. Sketch a graph of $y = f(x)$.
   b. Give equations for the asymptotes of the graph.
   c. How are these asymptotes related to the asymptotes of the parent function?

APPLYING THE MATHEMATICS

12. At the right is the graph of a function f that is point-symmetric with respect to the origin. Sketch the graph that is the translation image of the graph of f with $(-6, 4)$ for a center of symmetry.

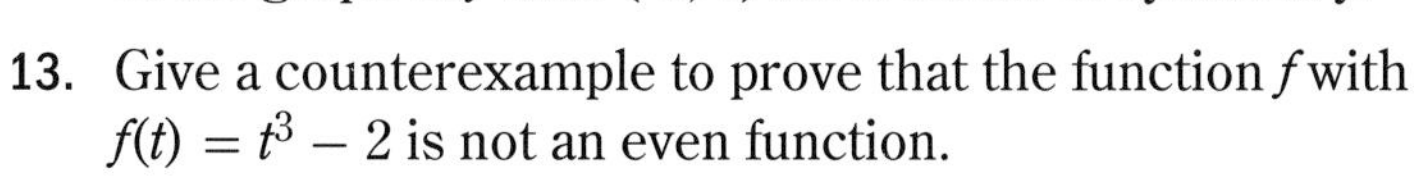

13. Give a counterexample to prove that the function f with $f(t) = t^3 - 2$ is not an even function.

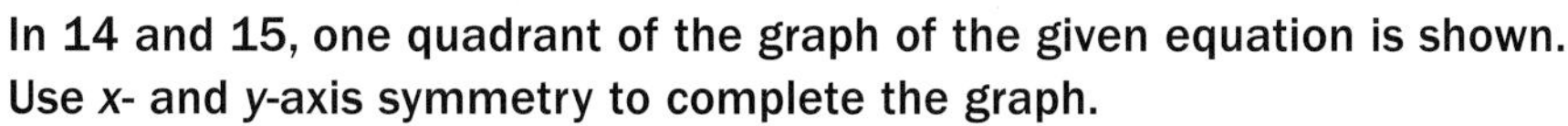

In 14 and 15, one quadrant of the graph of the given equation is shown. Use x- and y-axis symmetry to complete the graph.

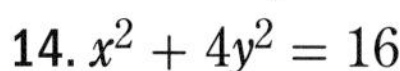

14. $x^2 + 4y^2 = 16$

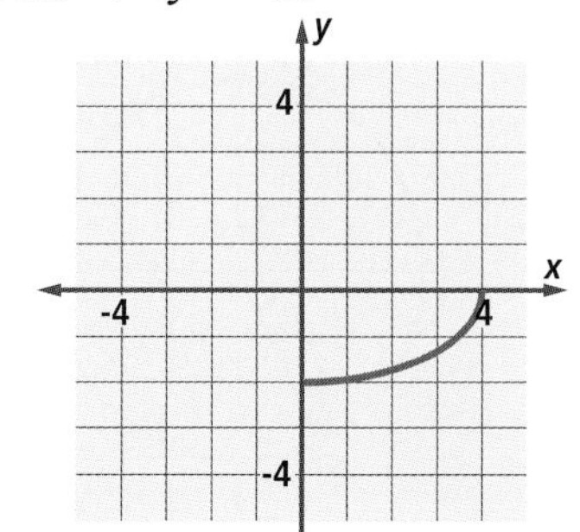

15. $|x| + |y| = 3$

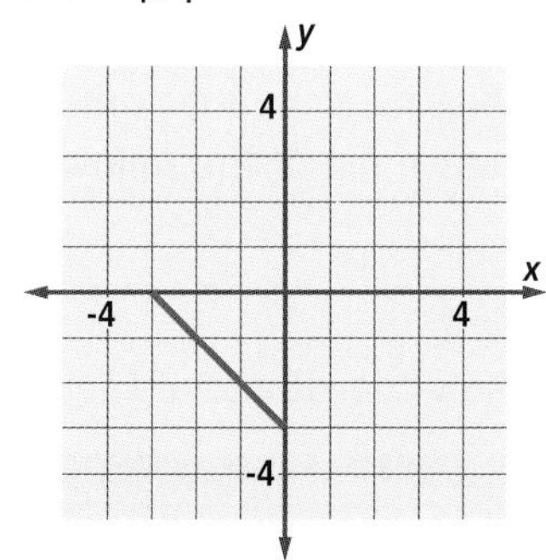

16. Recall from geometry that a circle whose center is at the origin has an equation of the form $x^2 + y^2 = r^2$. Prove algebraically that the circle with equation $x^2 + y^2 = 9$ has all three types of symmetry discussed in this lesson.

17. Prove that if f and g are odd functions, so is $f + g$. *Hint*: Consider $(f + g)(x)$ and $(f + g)(-x)$.

18. Use the function g with $g(x) = x^3$ as a parent function. The functions h and j with $h(x) = -x^3$ and $j(x) = x^3 + 2$ are related to it.
    a. Sketch the graphs of all three functions.
    b. Determine if each function is odd, even, or neither.
    c. Describe a transformation that maps g onto h.
    d. Describe a transformation that maps g onto j.
    e. Describe the symmetries of the three graphs. If there is reflection symmetry, give the equation of the line of symmetry. If there is point symmetry, give the coordinates of the center of symmetry.

REVIEW

19. The class results of a test on *Moby Dick* in American Literature were $\bar{x} = 67$ and $s = 13$. The teacher decides to add 4 points to each score. **(Lesson 3-3)**
    a. What is the class mean after this transformation?
    b. What is the class standard deviation after this transformation?

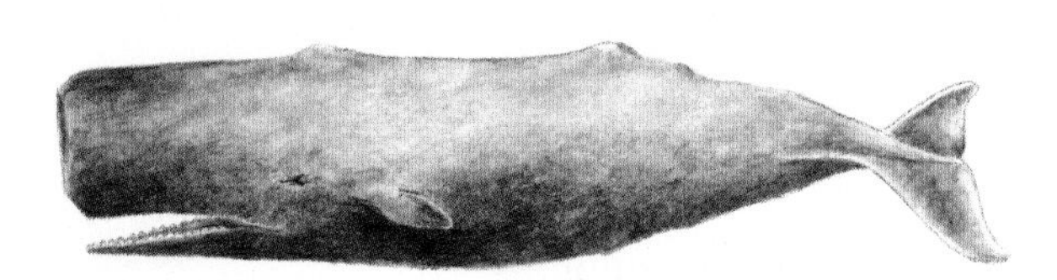

A whale of a tale
Moby Dick was a sperm whale.

20. ***Skill Sequence*** If $t_1 = 10$, $t_2 = -1$, $t_3 = 16$, and $t_4 = 6$, evaluate the expression. **(Lessons 3-3, 1-2)**

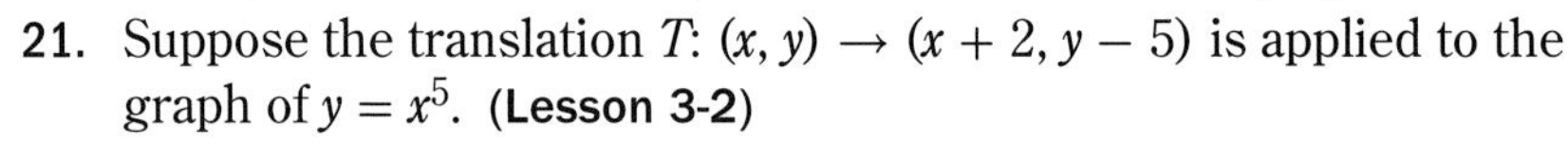

    a. $\sum_{i=1}^{4} t_i - 2$ b. $\sum_{i=1}^{4} (t_i - 2)$ c. $-2\sum_{i=1}^{4} t_i$

21. Suppose the translation $T: (x, y) \rightarrow (x + 2, y - 5)$ is applied to the graph of $y = x^5$. **(Lesson 3-2)**
    a. Find an equation for the image.
    b. Are the graphs of the preimage and image congruent? Explain why or why not.

In 22–25, use the circle at the right. $P = (a, b)$ and Q is the image of P under a 180° rotation around O. Express each of the following in terms of a and b. (Previous Course)

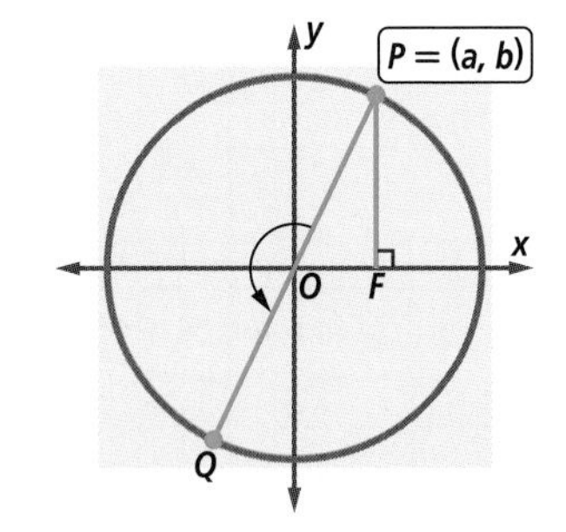

22. OF 23. OP

24. x-coordinate of Q 25. y-coordinate of Q

EXPLORATION

26. a. Can the graph of a function be mapped onto itself by a rotation whose magnitude is not 180° or 360°? If so, give some examples. If not, explain why not.
    b. Consider the question of Part a for graphs of relations which are not functions.

QY ANSWER

a. the graph with symmetry over the y-axis

b. the graph with symmetry about the origin

Lesson 3-5 The Graph Scale-Change Theorem

Vocabulary

horizontal and vertical scale change, scale factor

size change

▶ **BIG IDEA** The graph of a function can be scaled horizontally, vertically, or in both directions at the same time.

Vertical Scale Changes

Mental Math

What is $R_{270°}(1, 0)$?

Consider the graph of $y = f1(x) = x^3 + 3x^2 - 4x$ shown both in the graph and function table below. What happens when you multiply all the y-values of the graph by 2? What would the resulting graph and table look like? Activity 1 will help you answer these questions.

Activity 1

MATERIALS Graphing utility or slider graph application from your teacher

Step 1 Graph $f1(x) = x^3 + 3x^2 - 4x$ with window $-11 \le x \le 13$ and $-10 \le y \le 40$.

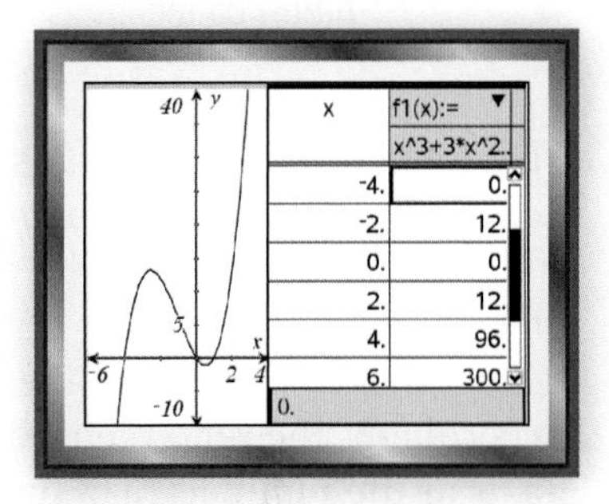

Step 2 Graph $f2(x) = 3(x^3 + 3x^2 - 4x) = 3 \cdot f1(x)$ on the same axes. Fill in the table of values for $f1(x)$ and $f2(x)$ only. Describe how the $f2(x)$ values relate to the $f1(x)$ values.

x	f1(x)	f2(x)	0.5 · f1(x)	1.5 · f1(x)	2 · f1(x)
–4	?	?	?	?	?
2	?	?	?	?	?
7	?	?	?	?	?
10	?	?	?	?	?

Step 3 Repeat Step 2 with $f3(x) = b(x^3 + 3x^2 - 4x)$ and use a slider to vary the value of b. Set the slider to 0.5, 1.5, and then 2. For each of these b-values, use a function table to fill in a column of the table in Step 2. Describe how the $f3(x)$ values relate to the corresponding values of $f1(x)$.

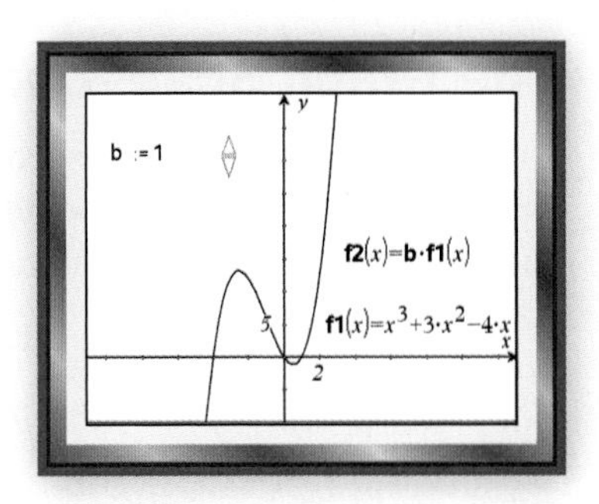

In Step 2 of Activity 1, we say that the graph of $f2$ is the image of the graph of $f1$ under a *vertical scale change* of magnitude 3. Each point on the graph of $f2$ is the image of a point on the graph of f under the mapping $(x, y) \rightarrow (x, 3y)$. You can create the same change by replacing y with $\frac{y}{3}$ in the equation for $f1$, because $\frac{y}{3} = x^3 + 3x^2 - 4x$ is equivalent to $y = 3(x^3 + 3x^2 - 4x)$, or $y = 3f1(x)$ in function notation. Similarly, if you replace y with $2y$ in the original equation, you obtain $2y = x^3 + 3x^2 - 4x$, which is equivalent to $y = \frac{1}{2}(x^3 + 3x^2 - 4x)$ or $y = \frac{1}{2}f(x)$.

 QY

Horizontal Scale Changes

Replacing the variable y by ky in an equation results in a vertical scale change. What happens when the variable x is replaced by $\frac{x}{2}$? By $4x$?

> **QY**
> Replacing y by $\frac{y}{4}$ in the equation for $f1$ yields an equation equivalent to $y =$ ___?___. What is the effect on the graph of f?

Activity 2

MATERIALS Graphing utility or slider graph application provided by your teacher.

Step 1 Consider the graph of $f1$ in Activity 1. Complete the table at the right.

x	$f1(x)$
-4	?
-1	?
2	?

Step 2 If $f4(x) = f1\left(\frac{x}{a}\right)$, then $f4(x) = \left(\frac{x}{a}\right)^3 + 3\left(\frac{x}{a}\right)^2 - 4\left(\frac{x}{a}\right)$. Graph $f4$ and use a slider to vary the value of a.

Step 3 a. What are the x- and y-intercepts of $f1$?
b. How do the intercepts of $f4$ change as a changes?

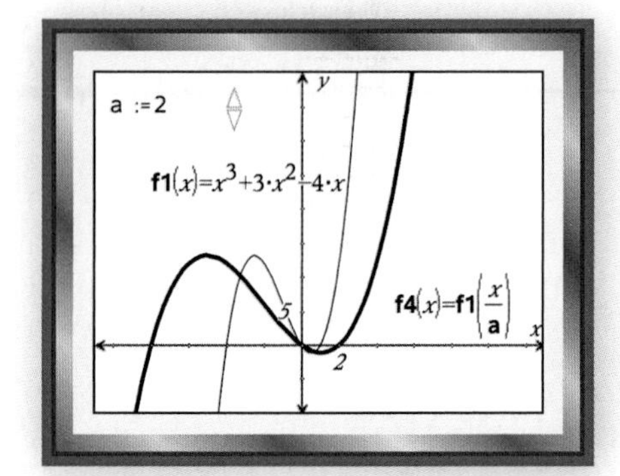

The graph of $f4$ in Activity 2 is the image of the graph of $f1$ under a *horizontal scale change* of magnitude a. Each point on the graph of $f4$ is the image of a point on the graph of $f1$ under the mapping $(x, y) \rightarrow (ax, y)$. Replacing x by $\frac{x}{2}$ in the equation doubles the x-values of the preimage points while the corresponding y-values remain the same. Accordingly, the x-intercepts of the image are two times as far from the y-axis as the x-intercepts of the preimage.

The Graph Scale-Change Theorem

In general, a **scale change** centered at the origin with **horizontal scale factor** $a \neq 0$ and **vertical scale factor** $b \neq 0$ is a transformation that maps (x, y) to (ax, by). The scale change S can be described by

$$S: (x, y) \rightarrow (ax, by) \quad \text{or} \quad S(x, y) = (ax, by).$$

If $a = 1$ and $b \neq 1$, then the scale change is a **vertical scale change**. If $b = 1$ and $a \neq 1$, then the scale change is a **horizontal scale change**. When $a = b$, the scale change is called a **size change**. Notice that in the preceding instances, replacing x by $\frac{x}{2}$ in an equation for a function results in the scale change $S: (x, y) \rightarrow (2x, y)$; and replacing y by $\frac{y}{3}$ leads to the scale change $S: (x, y) \rightarrow (x, 3y)$. These results generalize.

Graph Scale-Change Theorem

Given a preimage graph described by a sentence in x and y, the following two processes yield the same image graph:
(1) replacing x by $\frac{x}{a}$ and y by $\frac{y}{b}$ in the sentence;
(2) applying the scale change $(x, y) \rightarrow (ax, by)$ to the preimage graph.

Proof Name the image point (x', y'). So $x' = ax$ and $y' = by$. Solving for x and y gives $\frac{x'}{a} = x$ and $\frac{y'}{b} = y$. The image of $y = f(x)$ will be $\frac{y'}{a} = f\left(\frac{x'}{a}\right)$. The image equation is written without the primes.

Unlike translations, scale changes do not produce congruent images unless $a = b = 1$. Notice also that multiplication in the scale change corresponds to division in the equation of the image. This is analogous to the Graph-Translation Theorem in Lesson 3-2, where addition in the translation $(x, y) \rightarrow (x + h, y + k)$ corresponds to subtraction in the image equation $y - k = f(x - h)$.

GUIDED

Example 1

The relation described by $x^2 + y^2 = 25$ is graphed at the right.

a. Find images of points labeled A–F on the graph under S: $(x, y) \rightarrow (2x, y)$.

b. Copy the circle onto graph paper; then graph the image on the same axes.

c. Write an equation for the image relation.

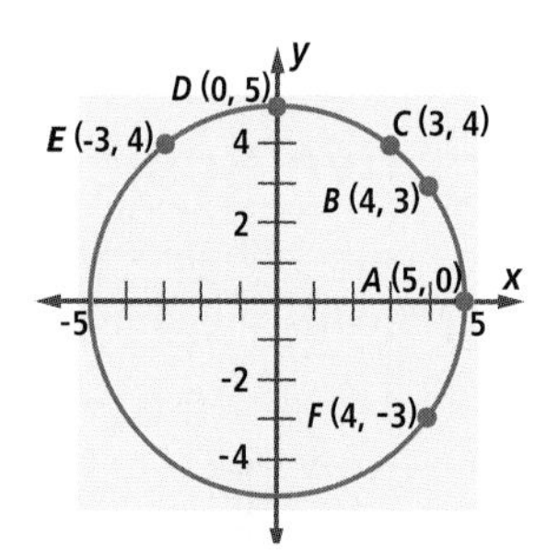

Solution

a. Copy and complete the table below.

b. Plot the preimage and image points on graph paper and draw a smooth curve connecting the image points. A partial graph is drawn below.

c. According to the Graph Scale-Change Theorem, an equation for an image under S: $(x, y) \rightarrow (2x, y)$ can be found by replacing x by _?_ in the equation for the preimage. The result is the equation _?_ .

	Preimage		Image	
Point	x	y	$2x$	y
A	5	0	10	0
B	?	?	?	?
C	?	?	?	?
D	?	?	?	?
E	?	?	?	?
F	?	?	?	?

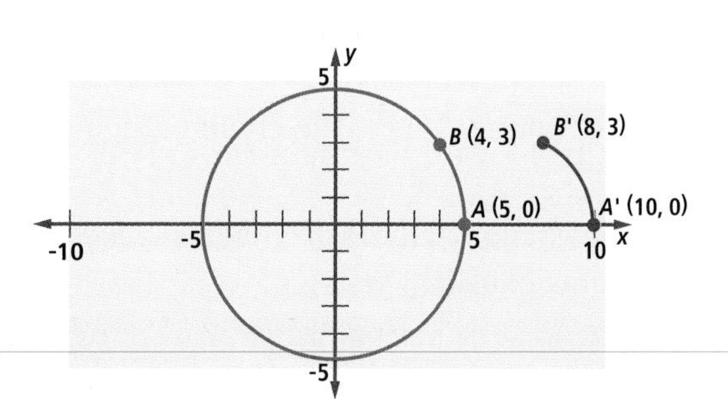

Example 2

A graph and table for $y = f(x)$ are given at the right. Draw the graph of $\frac{y}{3} = f(2x)$.

x	f(x)
-6	2
-3	-1
0	-1
2	3
6	0

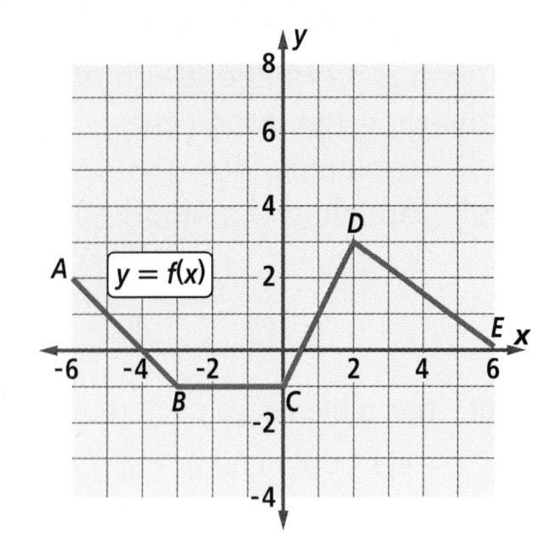

Solution Rewrite $\frac{y}{3} = f(2x)$ as $\frac{y}{3} = f\left(\frac{x}{\frac{1}{2}}\right)$. By the Graph Scale-Change Theorem, replacing x by $\frac{x}{\frac{1}{2}}$ and y by $\frac{y}{3}$ is the same as applying the scale change $(x, y) \rightarrow \left(\frac{1}{2}x, 3y\right)$.

So, $A = (-6, 2) \Rightarrow (-3, 6) = A'$

$B = (-3, -1) \Rightarrow (-3/2, -3) = B'$

$C = (0, -1) \Rightarrow (0, -3) = C'$

$D = (2, 3) \Rightarrow (1, 9) = D'$

and $E = (6, 0) \Rightarrow (3, 0) = E'$.

The graph of the image is shown at the right.

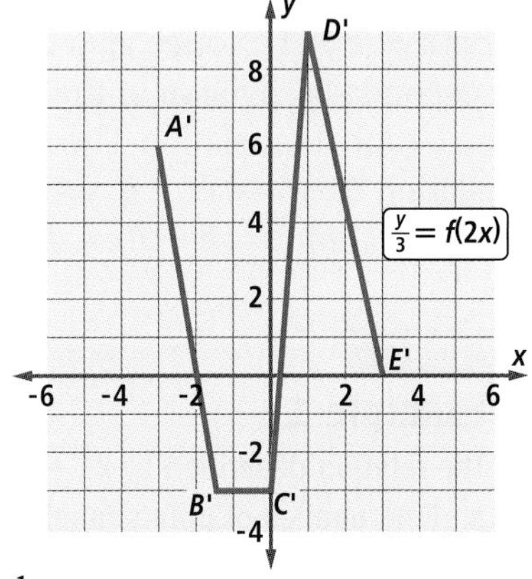

Negative Scale Factors

Notice what happens when a scale factor is -1. Consider the horizontal and vertical scale changes H and V with scale factors equal to -1.

$$H: (x, y) \rightarrow (-x, y) \text{ and } V: (x, y) \rightarrow (x, -y)$$

In H, each x-value is replaced by its opposite, which produces a reflection over the y-axis. Similarly, in V, replacing y by $-y$ produces a reflection over the x-axis. More generally, a scale factor of $-k$ combines the effect of a scale factor of k and a reflection over the appropriate axis.

Questions

COVERING THE IDEAS

1. **True or False** Under every scale change, the preimage and image are congruent.
2. Under a scale change with horizontal scale factor a and vertical factor b, the image of (x, y) is __?__.
3. Refer to the Graph Scale-Change Theorem. Why are the restrictions $a \neq 0$ and $b \neq 0$ necessary?
4. If S maps each point (x, y) to $\left(\frac{x}{2}, 6y\right)$, give an equation for the image of $y = f(x)$ under S.

5. Consider the function $f1$ used in Activities 1 and 2.
   a. Write a formula for $f1\left(\frac{x}{3}\right)$.
   b. How is the graph of $y = f1\left(\frac{x}{3}\right)$ related to the graph of $y = f1(x)$?

6. **Multiple Choice** Which of these transformations is a size change?
   A $(x, y) \rightarrow (3x, 3y)$ B $(x, y) \rightarrow (3x, y)$
   C $(x, y) \rightarrow (x + 3, y + 3)$ D $(x, y) \rightarrow \left(\frac{x}{3}, y\right)$

7. Functions f and g with $f(x) = -2x^2 + 5x + 3$ and $g(x) = \frac{1}{2}f(x)$ are graphed at the right.
   a. What scale change maps the graph of f to the graph of g?
   b. The x-intercepts of f are at $x = -\frac{1}{2}$ and $x = 3$. Where are the x-intercepts of g?
   c. How do the y-intercepts of f and g compare?
   d. The vertex of the graph of f is (1.25, 6.125). What is the vertex of the graph of g?

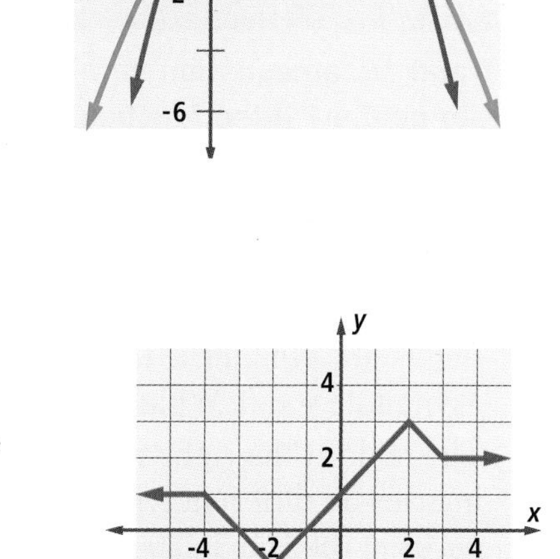

8. Consider the parabola with equation $y = x^2$. Let $S(x, y) = \left(2x, \frac{y}{7}\right)$.
   a. Find the images of (-3, 9), (0, 0), and $\left(\frac{1}{2}, \frac{1}{4}\right)$ under S.
   b. Write an equation for the image of the parabola under S.

9. The graph of a function f is shown at the right.
   a. Graph the image of f under $S(x, y) = \left(\frac{1}{2}x, 3y\right)$.
   b. Find the x- and y-intercepts of the image.
   c. Find the coordinates of the point where the y-value of the image of f reaches its maximum.

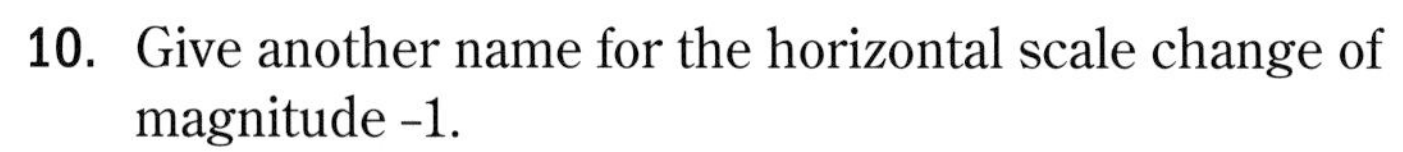

10. Give another name for the horizontal scale change of magnitude -1.

11. Describe the scale change that maps the graph of $y = \sqrt{x}$ onto the graph of $y = \sqrt{\frac{x}{12}}$.

APPLYING THE MATHEMATICS

12. Refer to the parabolas at the right. The graph of g is the image of the graph of f under what
    a. horizontal scale change?   b. vertical scale change?
    c. size change?

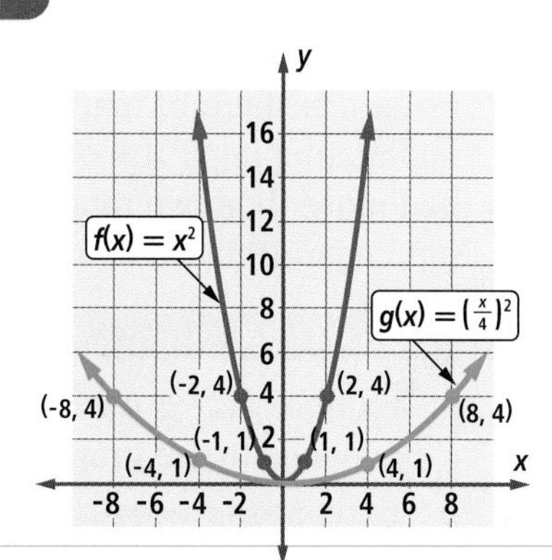

13. Write an equation for the image of the graph of $y = x + \frac{1}{x}$ under each transformation.
    a. $S(x, y) = (2x, 2y)$ b. $S(x, y) = \left(\frac{x}{3}, -y\right)$

14. A scale change maps (10, 0) onto (2, 0) and (-5, 8) onto (-1, 2). What is the equation of the image of the graph of $f(x) = x^3 - 8$ under the scale change?

In 15 and 16, give a rule for a scale change that maps the graph of f onto the graph of g.

15.

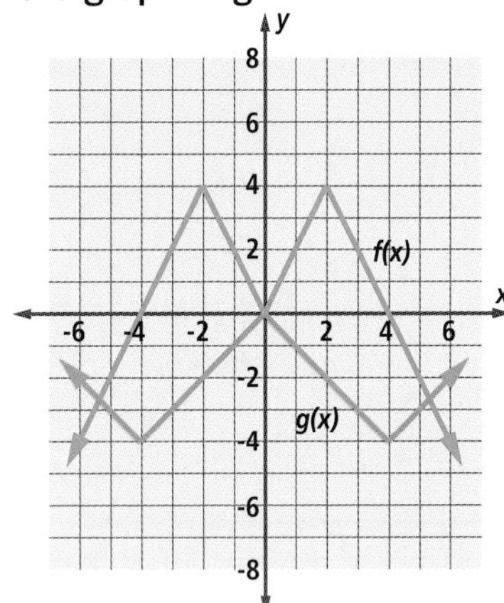

16.

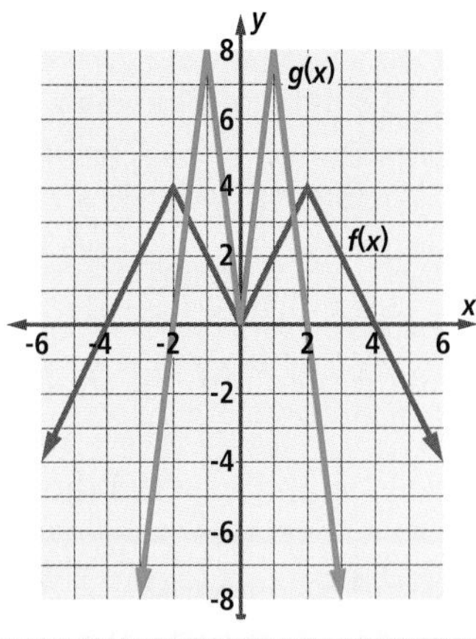

REVIEW

In 17 and 18, an equation for a function is given. Is the function odd, even, or neither? If the function is odd or even, prove it. (Lesson 3-4)

17. $f(x) = (5x + 4)^3$

18. $g(x) = 5x^4 + 4$

19. If $f(x) = -g(x)$ for all x in the common domain of f and g, how are the graphs of f and g related? **(Lesson 3-4)**

20. One of the parent functions presented in Lesson 3-1 has a graph that is not symmetric to the x-axis, y-axis, or origin. It has the asymptote $y = 0$. Which is it? **(Lesson 3-1)**

21. The table at the right shows the number of injuries on different types of rides in amusement parks in the U.S. in 2003, 2004, and 2005. Use the table to explain whether each statement is supported by the data. **(Lesson 1-1)**

	2003	2004	2005
Total	1954	1648	1713
Children's Rides	277	219	192
Family and Adult Rides	1173	806	1131
Roller Coasters	504	613	390

Source: National Safety Council

a. Injuries on children's rides decreased slightly each year.

b. The number of injuries on roller coasters decreased from 2003 to 2004.

c. Roller coasters are not as safe as children's rides.

22. Pizza π restaurant made 300 pizzas yesterday; 64% of the pizzas had no toppings, 10% of the pizzas had two toppings, and 26% had more than two toppings. How many pizzas had at least two toppings? **(Previous Course)**

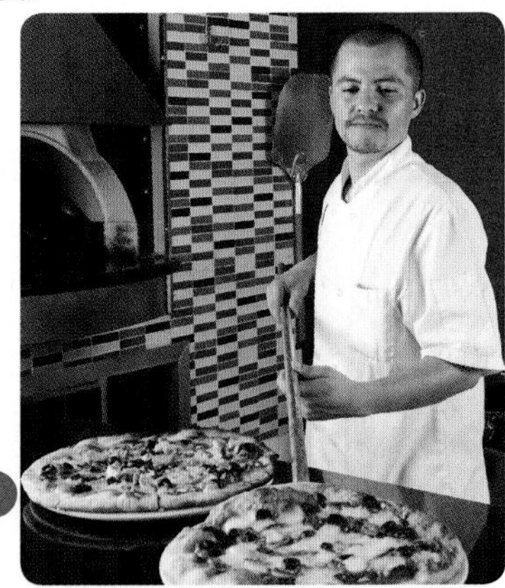

EXPLORATION

23. a. Explore $g(x) = b(x^3 + 3x^2 - 4x)$ for $b < 0$. Explain what happens to the graph of g as b changes.

b. Explore $h(x) = \left(\frac{x}{a}\right)^3 + 3\left(\frac{x}{a}\right)^2 - 4\left(\frac{x}{a}\right)$ for $a < 0$. Explain what happens to the graph of h as a changes.

QY ANSWER

$4(x^3 + 3x^2 - 4x)$; a vertical scale change of magnitude 4

Lesson 3-6 Scale Changes of Data

Vocabulary

scale change of a data set
scale factor
scale image

▶ **BIG IDEA** Multiplying every number in a data set by k multiplies all measures of center and the standard deviation and range by k, while the variance is multiplied by k^2.

Scale changes can also be applied to data sets. A useful example to consider is the Consumer Price Index (CPI), a measure of inflation. To calculate the CPI, the cost of a specified basket of goods is totaled in a particular base year, then scaled so the cost equals 100. Costs in later years are then compared to the base year cost.

The table at the right gives the CPI at five-year intervals beginning at 1950, with 1967 as the base year.

Scaling allows prices in any year to be compared to prices in any other year. Just solve a proportion.

Year	CPI
1950	72.1
1955	80.2
1960	88.7
1965	94.5
1970	116.3
1975	161.2
1980	298.8
1985	322.2
1990	391.4
1995	456.5
2000	515.8
2005	585.0
June 2008	655.5

Mental Math

Find the x-intercept(s) of the graph of the equation.

a. $y = |x + \pi|$

b. $y = |x - \pi|$

c. $y = |x| + \pi$

d. $y = |x| - \pi$

Example 1

Suppose a refrigerator cost $800 in 1995. What might you expect the cost of a similar refrigerator to be in June 2008?

Solution Set up a proportion.

$$\frac{\text{cost in June 2008}}{\text{cost in 1995}} = \frac{\text{CPI in June 2008}}{\text{CPI in 1995}}$$

Let x = cost of a refrigerator in June 2008. Substitute, using the CPI values given in the table.

$$\frac{x}{\$800} = \frac{655.5}{456.5}$$

Solve the proportion.

$$x = 800 \cdot \frac{655.5}{456.5} = \$1148.74... \approx \$1150$$

The cost of a similar refrigerator in June 2008 was probably about $1150.

The ratio $\frac{655.5}{456.5}$ in Example 1 is a *scale factor*. In this case, $\frac{655.5}{456.5} \approx 1.436$, indicating there was about a 43.6% increase in prices from 1995 to June 2008. By multiplying the 1995 price by 1.436, you can estimate what the June 2008 price of an item would be if the cost kept pace with inflation.

QY

If a person earned \$2,500 a month in 1995, what would the person need to have earned in June 2008 to keep up with inflation?

A **scale change** of a set of data $\{x_1, x_2, \ldots, x_n\}$ is a transformation that maps each x_i to ax_i, where a is a nonzero constant. That is, S is a scale change if and only if there is a nonzero constant a with

$$S: x \rightarrow ax, \text{ or } S(x) = ax.$$

The number a is called the **scale factor** of the scale change. The number ax or the point it represents is called the **scale image** of x.

In the situation above, the 1995 cost x of an item can be mapped onto the estimated June 2008 cost via the scale change

$$S: x \rightarrow 1.436x, \text{ or } S(x) = 1.436x.$$

Scaling and Measures of Center

Scale changes of data, like translations, affect statistical measures derived from the data.

Activity

The CPI in 1998 was about 496. Here are average prices of some grocery items in that year.

Year	Coffee 1 pound	Eggs 1 dozen	Gasoline 1 gallon	Orange Juice 12-oz can	Ground Beef 1 pound	Chicken 1 pound
1998	4.03	1.12	1.13	1.60	1.82	1.02

Source: Bureau of Labor Statistics

Step 1 Calculate the scale factor needed to predict costs of items in 2008 from 1998 prices.

Step 2 Enter the price data for 1998 into a spreadsheet like the one on page 187.

Step 3 Use the scale factor from Step 1 to compute the predicted June 2008 prices of the same items. Record these costs in a new column.

Step 4 Calculate the mean, median, range, variance, and standard deviation of each set of prices. Record the results to the nearest hundredth in another column.

Step 5 Multiply the 1998 statistics by the scale factor from Step 1 and place the results in an additional column. Your spreadsheet should look similar to the one on the next page.

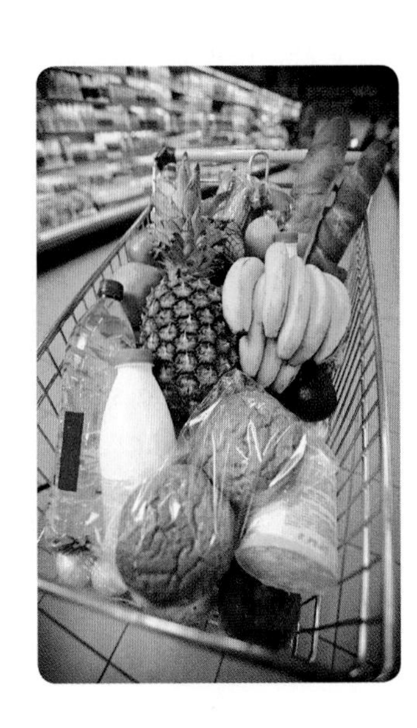

Step 6 Compare the results of Step 5 to those of Step 4. Which of the 2008 statistics can be found by scaling the corresponding 1998 statistics, rather than by calculating from the costs of the groceries?

The spreadsheet evaluates the formula B2 × (655.5/496).

E2 × (655.5/496)

◇	A	B	C	D	E	F	G
1	Items	1998 costs	2008 costs (predicted)		1998 statistics	2008 statistics	1998 statistics x scale factor
2	Coffee 1 Pound	$4.03	5.33	Mean			
3	Eggs 1 dozen	$1.12		Median			
4	Gasoline 1 gallon	$1.13		Range			
5	Orange Juice 12-oz can	$1.60		Variance			
6	Ground Beef 1 pound	$1.82		Standard Deviation			
7	Chicken 1 pound	$1.02					
8							

The Activity shows how scaling data affects statistics for measures of center and spread. These ideas are applied in Example 2.

Example 2

As a fund-raiser, club members sell candy for $2.50 per box. The number of boxes each of the 17 members sold is given below.

27, 30, 32, 32, 34, 35, 35, 37, 38, 39, 40, 41, 41, 43, 44, 44, 50

a. **Compute the mean, median, standard deviation, and IQR of the numbers of boxes sold.**

b. **Find the amount of money each member collected.**

c. **Compute the mean, median, standard deviation, and IQR of the amounts of money collected by scaling the values in Part a.**

Solution

a. For the number of boxes, mean = 37.8, median = 38, standard deviation = 5.89, and IQR = 9.

b. Apply the transformation $x_i \longrightarrow 2.5x_i$ to each value.
The number of dollars each member collected are 67.50, 75, 80, 80, 85, 87.50, 87.50, 92.50, 95, 97.50, 100, 102.50, 102.50, 107.50, 110, 110, and 125.

c. For the money collected, mean $= 37.8 \cdot 2.5 = 94.5$, median $= 38 \cdot 2.5 = 95$, standard deviation $= 5.89 \cdot 2.5 = 14.73$, and IQR $= 9 \cdot 2.5 = 22.5$.

Box plots of the numbers of candy boxes sold and amounts collected illustrate the effects of scaling on the data values, the center (mean and median), and the spread (range, IQR, and standard deviation).

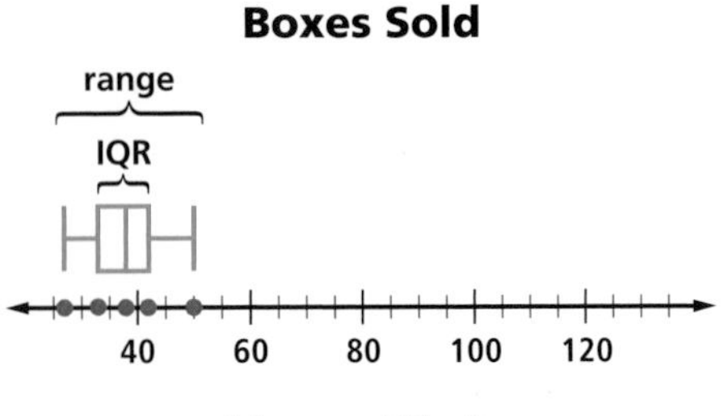

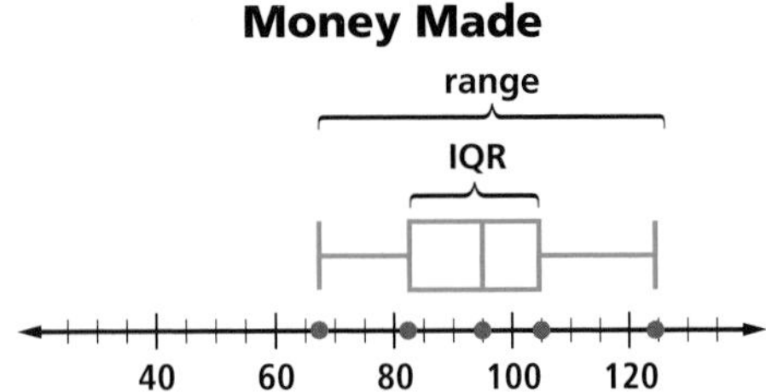

As you saw in the Activity and Examples, scale changes not only affect the data, they affect measures of center as well. This effect is stated in the theorem below.

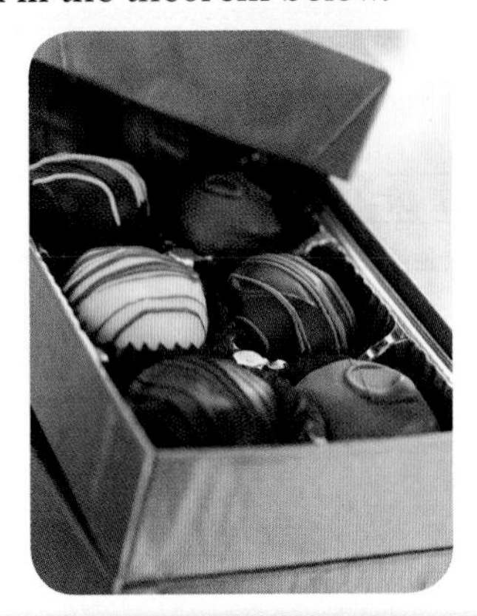

Theorem (Centers of Scaled Data)

Multiplying each element of a data set by the factor *a* multiplies each of the mode, mean, and median by the factor *a*.

Proof We prove the mean part of the theorem here; it can also be proved that the median and mode are multiplied by *a* (see Question 9 for the median). To describe the effect of a scale change on statistical measures for a general data set, represent the set as $\{x_1, x_2, x_3, ..., x_n\}$. Under a scale change with scale factor *a*, the image data set is $\{ax_1, ax_2, ax_3, ..., ax_n\} = \{x_1', x_2', ..., x_n'\}$. Let $\bar{x}$ be the mean of the original data set and $\overline{x'}$ be the mean of the image set.

The mean $\overline{x'} = \dfrac{\sum_{i=1}^{n} (ax_i)}{n}$.

By definition of Σ,

$$\overline{x'} = \frac{ax_1 + ax_2 + ax_3 + \cdots + ax_n}{n}$$

$$= \frac{a(x_1 + x_2 + x_3 + \cdots + x_n)}{n}$$

$$= \left(\frac{a\sum_{i=1}^{n} x_i}{n}\right) = a\left(\frac{\sum_{i=1}^{n} x_i}{n}\right).$$

Thus $\overline{x'} = a\bar{x}$. So, under a scale change, the mean of a set of data is mapped to the mean of the image set of data.

Scaling and Measures of Spread

The Activity and Example 2 show that measures of spread in a scale-change image data set are predictable. Both the range and standard deviation can be found by multiplying by the scale factor.

The predicted variance can also be found using the scale factor in a different way. These effects are described in the following theorem.

Theorem (Spreads of Scaled Data)

If each element of a data set is multiplied by $a > 0$, then the variance is a^2 times the original variance, the standard deviation is a times the original standard deviation, and the range is a times the original range.

Proof Consider the data set $\{x_1, x_2, ..., x_n\}$ and its image $\{ax_1, ax_2, ..., ax_n\}$ under a scale change of magnitude a. By the Centers of Scaled Data Theorem, the mean of the image data set is $a\bar{x}$, where $\bar{x}$ is the mean of the original data set. So the variance of the image data is given by

$$\frac{\sum_{i=1}^{n}(ax_i - a\bar{x})^2}{n-1} = \frac{\sum_{i=1}^{n}[a(x_i - \bar{x})]^2}{n-1} \quad \text{Distributive Property}$$

$$= \frac{\sum_{i=1}^{n}[a^2(x_i - \bar{x})^2]}{n-1} \quad \text{Power of a Product Property}$$

Applying the Distributive Property,

$$\sum_{i=1}^{n}[a^2(x_i - \bar{x})^2] = a^2\sum_{i=1}^{n}(x_i - \bar{x})^2.$$

Hence, the variance of the image data is given by

$$\frac{a^2\sum_{i=1}^{n}(x_i - \bar{x})^2}{n-1} = a^2\left(\frac{\sum_{i=1}^{n}(x_i - \bar{x})^2}{n-1}\right) = a^2s^2,$$

where s^2 is the variance of the original data set. Thus, the variance of the image data is a^2 times the variance of the original data.

To get the standard deviation of the image data, take the square root of the variance. Thus, the standard deviation of the image data is $|a| \cdot s$, which is $|a|$ times the standard deviation of the original data set. It can also be proved that the range and IQR of the image data are $|a|$ times the range and IQR, respectively, of the original data set.

Questions

COVERING THE IDEAS

1. Define *scale change* of a set of data.
2. A Rambler, a small car, cost about \$2000 in 1965. What would a comparable car have cost in 2005?
3. In what 5-year period from 1950 to 2005 was there the greatest percent increase in CPI, and what was that percent?

4. Consider the candy sales in Example 2. Suppose that a box costs \$5 instead of \$2.50.
   a. For the money collected, calculate each statistic.
      i. range ii. mode iii. median
      iv. mean v. variance vi. standard deviation
   b. Draw a box plot of the amounts of money collected.

5. A restaurant employs 11 workers whose individual earnings for an 8-hour day are summarized by this box plot. Suppose that each employee begins working 6 hours instead of 8. Assume employees earn the same hourly wage, regardless of the number of hours worked.

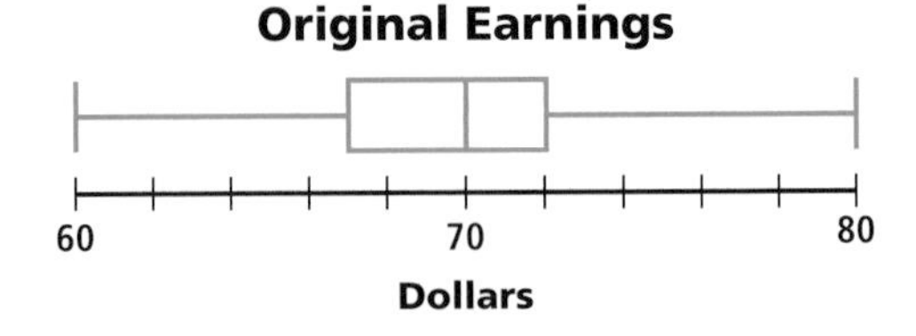

   a. What scale factor would be used to find each person's new daily earnings?
   b. Draw a box plot of the earnings for the 6-hour work day.

6. Suppose all elements of a data set are multiplied by x. Explain why the variance is multiplied by x^2.

7. **Multiple Choice** The box plot at the right represents a data set D. Which box plot below represents the image of D under the transformation $T: x \rightarrow \frac{1}{3}x$?

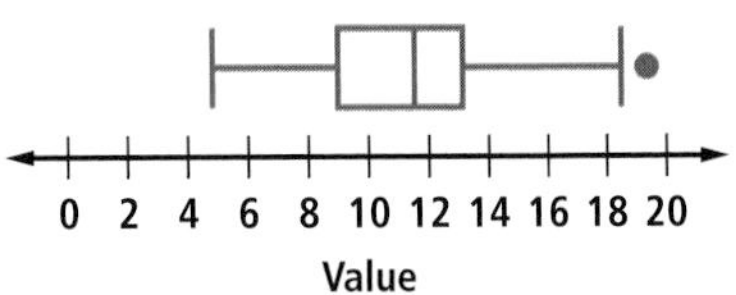

A
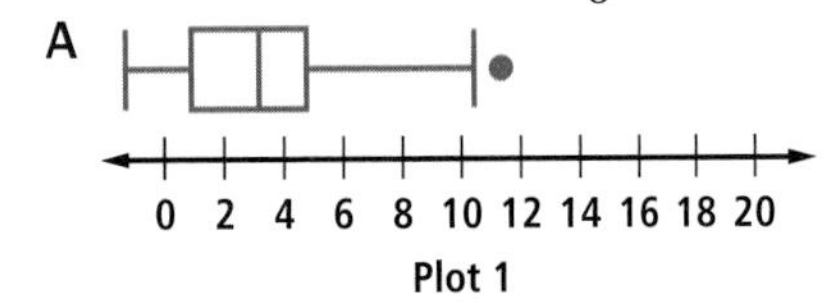

Plot 1

B
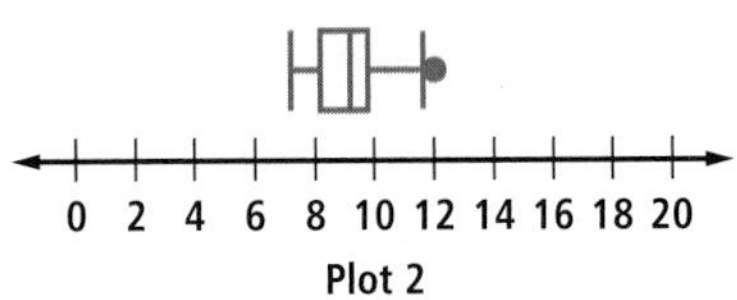

Plot 2

C
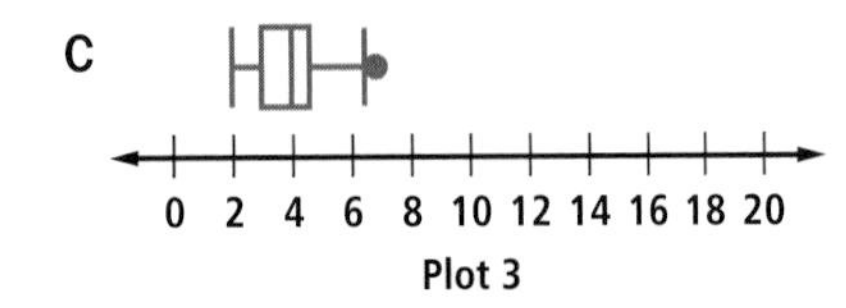

Plot 3

D
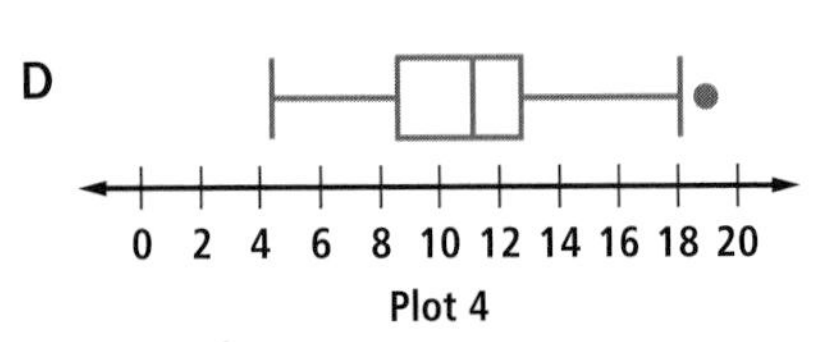

Plot 4

APPLYING THE MATHEMATICS

8. Suppose $Y_1 = 11$, $Y_2 = -3$, $Y_3 = -2$, $Y_4 = 7$, $Y_5 = 5$, $Y_6 = 4$. Evaluate each expression.
   a. $\sum_{i=1}^{6} 10Y_i$
   b. $\sum_{i=1}^{6} rY_i$
   c. $\sum_{i=1}^{6} \left(\frac{Y_i}{m}\right)$
   d. $\sum_{i=1}^{6} (Y_i + 2)$

9. Prove the Centers of Scale Changes of Data Theorem for medians.

10. For a large city, the median house price one year was \$258,200 with an interquartile range of \$81,000. Assume that house prices rise 2% over the next year. What would be the median and IQR for house prices in the next year?

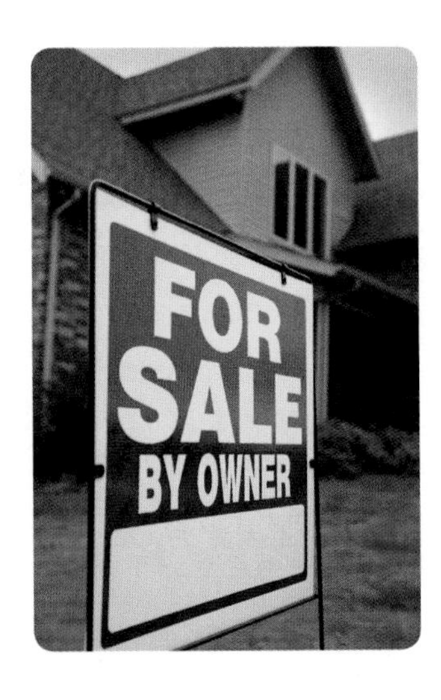

11. Let $\bar{x}$ = the mean and s = the standard deviation of scores on a test for a class of n students. Suppose everyone's score is multiplied by r, and then increased by a bonus b. For the new scores, find the

a. mean. b. variance. c. standard deviation.

12. Let M represent the maximum value of a data set and let m represent the minimum value.

a. Write an expression for the range r of the data set.

b. After a scale change with scale factor $d > 0$, what are the maximum and minimum values of the image data set?

c. Write and simplify an expression for the range of the image data.

d. How would your answers to Parts a–c change if $d < 0$?

13. Consider the following data, which give the height h in cm and weight w in kg of twelve students.

h (cm)	144	168	140	157	153	162	160	160	166	166	173	165
w (kg)	68	86	69	77	85	74	78	84	84	83	82	82

a. Enter these data into a statistics utility. Create a scatterplot with h on the horizontal axis.

b. Find the line of best fit for predicting weight from height.

c. Compute the correlation coefficient.

d. Use a statistics utility to convert the height to inches (1 in. = 2.54 cm) and weight to pounds (1 lb = 0.454 kg). Draw a new scatterplot. How is the scatterplot different from that in Part a?

e. Compute the correlation coefficient and the regression equation for predicting weight from height in Part d.

f. Which of the following statistics remain invariant under scale changes of the data?

i. correlation coefficient

ii. slope of the regression line

iii. y-intercept of the regression line

REVIEW

14. Consider the functions f and g with $f(x) = x$ and $g(x) = 7x$.

a. Describe a scale change that maps the graph of f onto the graph of g.

b. Describe a scale change that maps the graph of g onto the graph of f. **(Lesson 3-5)**

In 15–18, match the equation with its graph. (Lessons 3-5, 3-2)

15. $f(x) = |x + 5|$
16. $g(x) = |3x|$
17. $h(x) = \frac{1}{3}|x|$
18. $j(x) = |x - 5|$

A

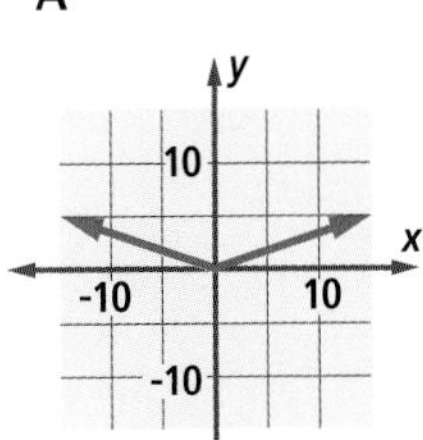

B

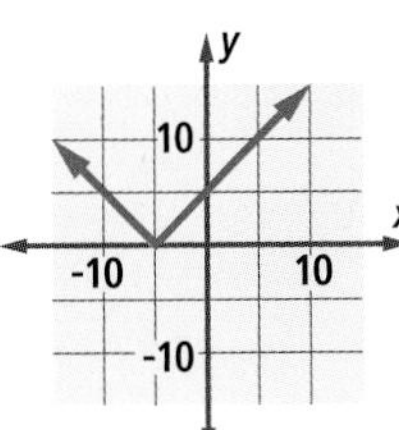

C

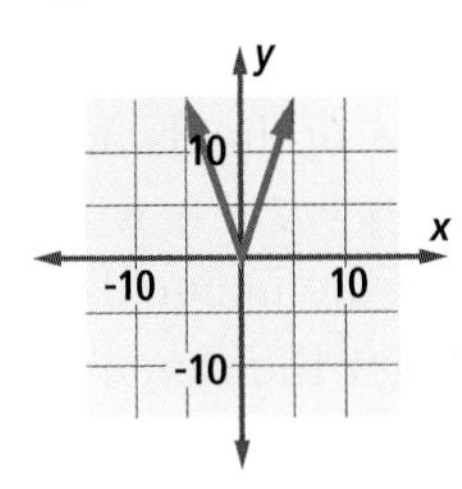

D

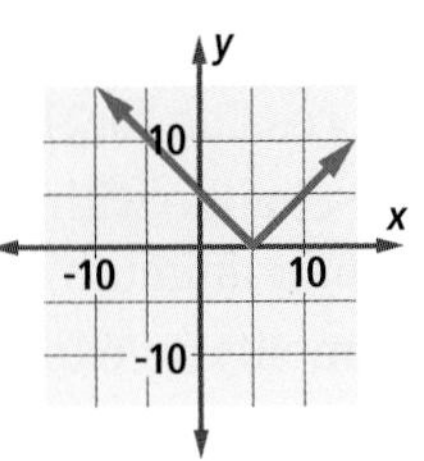

19. Name all the functions in Questions 15–18 that are even functions. **(Lesson 3-4)**

20. A certain hyperbola H is a translation image of the graph of $y = \frac{1}{x}$ and has asymptotes $x = 2$ and $y = -5$. Give an equation for H. **(Lessons 3-2, 3-1)**

In 21–23, if possible, factor the given expression. (Previous Course)

21. $6x^3 - 18x^2$
22. $9m^2n^2 - 49$
23. $3r^2 - 2r - 5$

24. ***Skill Sequence*** Rewrite each of the following in the form $a(x - h)$. **(Previous Course)**
    a. $7x - 21$
    b. $21x - 7\pi$
    c. $7x + 5\pi$

EXPLORATION

25. Take a data set of positive values and take the square root of each value. Which statistics, if any, are invariant? Which are affected in predictable ways?

QY ANSWER

$3,590

Lesson 3-7 Composition of Functions

Vocabulary

composite
function composition

BIG IDEA Following one process by another process creates a composite that itself can be viewed as a single process.

Mental Math

Start with a number n.

a. Subtract 4 from it. Then multiply the difference by 3. What number results?

b. Multiply it by 3. Then subtract 4 from the product. What number results?

The presidents of the United States from 1989–1992 and 2001–2008 were George H.W. Bush and his son. This was not the first time that a father and son had been presidents. The 2nd and 6th presidents, John Adams and John Quincy Adams, were also father and son. Here is part of John Quincy Adams's family tree. It shows that John Adams and Abigail Smith are John Quincy Adams's parents. Because President John Adams had the same name as his father, we call the father "Sr." and the son "Jr." here even though they never used those names.

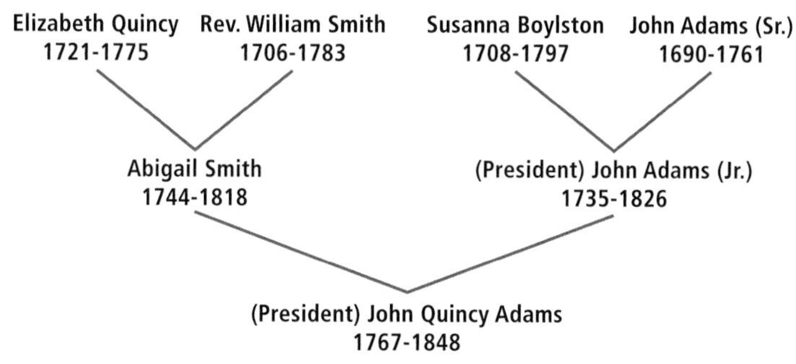

John Adams

John Quincy Adams

Functions can be used to describe the relationships between members of this tree. For example, suppose m is the function defined by

$$m(x) \text{ is the mother of } x,$$

and f is the function defined by

$$f(x) \text{ is the father of } x.$$

Then m(John Quincy Adams) = Abigail Smith, m(Abigail Smith) = Elizabeth Quincy, f(John Quincy Adams) = John Adams Jr., and so on.

Functions can be combined so that the value of one function becomes the argument of another. For example, since

m(John Quincy Adams) = Abigail Smith,

and f(Abigail Smith) = Rev. William Smith,

this combination can be written

$f(m$(John Quincy Adams)) = f(Abigail Smith) = Rev. William Smith.

In words, this equation says that the father of the mother of John Quincy Adams is Rev. William Smith.

We say that m and f have been *composed* to make a new function, which we could call the "maternal grandfather" function. We denote this function by the symbol $f \circ m$, read as "the *composite* of f with m."

Definition of Composite Function

Suppose f and g are functions. The **composite** of g with f, written $\mathbf{g \circ f}$, is the function defined by

$$(g \circ f)(x) = g(f(x)).$$

The domain of $g \circ f$ is the set of values of x in the domain of f for which $f(x)$ is in the domain of g.

The composite $g \circ f$ can be written without parentheses when applied to an argument, as in $g \circ f(x)$. Parentheses make it easier to see that there is one composite function applied to the argument, as in the following.

$(f \circ m)$(John Quincy Adams) = $f(m$(John Quincy Adams))
= f(Abigail Smith)
= Rev. William Smith

Composition of Functions Is Not Commutative

The operation that yields the composite of two functions is called **function composition.** Order makes a difference in function composition.

$(m \circ f)$(John Quincy Adams) = $m(f$(John Quincy Adams))
= m(John Adams Jr.)
= Susanna Boylston

The mother of the father of John Quincy Adams is Susanna Boylston. The function $m \circ f$ might be called the "paternal grandmother" function.

Notice that the two functions $f \circ m$ and $m \circ f$ are different functions. The range of $f \circ m$ contains only men, while the range of $m \circ f$ contains only women. This illustrates that *composition of functions is not commutative.* The next two examples show this with functions of real numbers.

Example 1

Let f and g be defined by $f(x) = 2x^2 + 3x$ and $g(x) = x - 7$. Evaluate.

a. $(f \circ g)(-2)$ b. $(g \circ f)(-2)$

Solution

a. To evaluate $(f \circ g)(-2)$, first evaluate $g(-2)$.

$$g(-2) = -2 - 7 = -9$$

Then use this output as the input to f. So

$$f(g(-2)) = f(-9)$$
$$= 2(-9)^2 + 3(-9) = 135.$$

b. $g \circ f(-2) = g(f(-2)) = g(2(-2)^2 + 3(-2))$
$= g(2)$
$= 2 - 7 = -5$

It can be tedious to evaluate points for composites of functions. An alternative is to find a formula for the composite.

GUIDED

Example 2

Let $f(x) = 2x^2 + 3x$ and $g(x) = x - 7$.

a. Derive a formula for $(f \circ g)(x)$.

b. Give a simplified formula for $(g \circ f)(x)$.

c. Verify that $f \circ g \neq g \circ f$ by graphing.

Solution

a. In $(f \circ g)(x) = f(g(x))$, first substitute $x - 7$ for $g(x)$.

$$f(g(x)) = f(x - 7)$$

Now use $x - 7$ as the input to function f.

$$f(x - 7) = 2(\underline{\ ?\ })^2 + 3(\underline{\ ?\ })$$
$$= 2(\underline{\ ?\ } - \underline{\ ?\ } + 49) + 3x - 21$$
$$= \underline{\ ?\ }$$

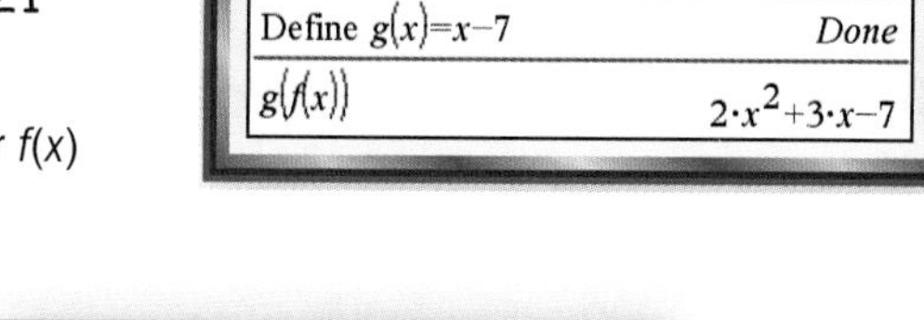

b. To find a formula for $(g \circ f)(x)$, substitute the expression for $f(x)$ first, or use a CAS. Define the functions on your CAS and evaluate the composite directly.

c. Graph the two functions, $f \circ g$ and $g \circ f$, using a graphing utility. You can see that the graphs are different. So the functions are different.

Notice that although $f \circ g$ and $g \circ f$ are ***not*** the same function, there is at least one value of x at which they have the same y-value. This is the x-value at the point of intersection of the two parabolas.

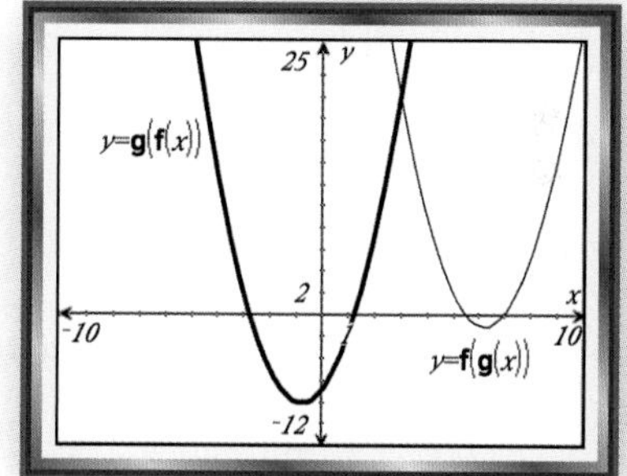

QY

Finding the Domain of a Composite Function

The domain of a composite function $g \circ f$ is the set of all elements for which $g \circ f$ is defined. So, to be in the domain of $g \circ f$, a number x must be in the domain of f, and the corresponding $f(x)$ value must be in the domain of g.

▶ QY

True or False $f \circ g$ is the image of the graph of f under a translation 7 units to the right.

Example 3

Let f and g be real functions defined by $f(m) = \sqrt{m}$ and $g(m) = \frac{2}{m-3}$. Find the domain of $g \circ f$.

Solution 1 Because f is a real function, the domain of f is the set of all nonnegative real numbers. The domain of g is the set of all real numbers but 3, so all values of $f(m)$ except when $\sqrt{m} = 3$ are in the domain of g. Thus, the domain of $g \circ f$ is the set of real numbers m with $m \geq 0$ and $m \neq 9$.

Solution 2 Find a formula for $g \circ f$ and analyze the domain.

$(g \circ f)(m) = g(f(m)) = g(\sqrt{m}) = \frac{1}{\sqrt{m} - 3}$

$\sqrt{m}$ is defined in the real number system only for $m \geq 0$.

$\sqrt{m} - 3 = 0$ when $m = 9$. So, the domain of $g \circ f$ is $\{m: m \geq 0 \text{ and } m \neq 9\}$.

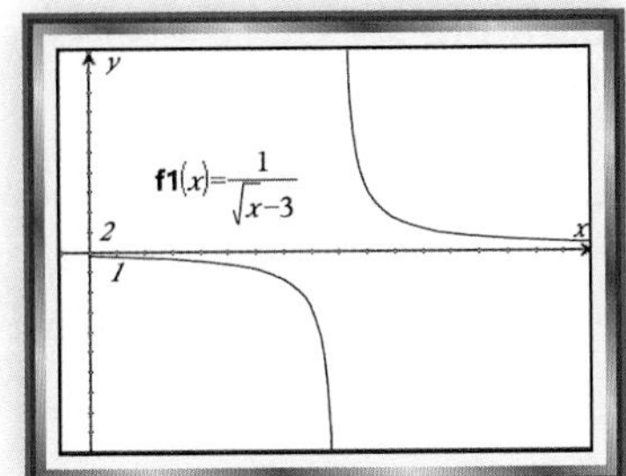

Check Use a graphing utility to graph $g(f(x)) = \frac{1}{\sqrt{x} - 3}$. You should get a graph like the one at the right with an asymptote at $x = 9$.

Example 3 shows that the domain of the composite can be different from the domain of either of the component functions.

Composition of Transformations

Because transformations are functions, they can be composed. Like other functions, composition of transformations is not commutative.

GUIDED

Example 4

Let $S:(x, y) \rightarrow (2x, y)$ and let $T:(x, y) \rightarrow (x + 4, y - 3)$.

a. Describe S and T in words.

b. Write a formula for the composite $(T \circ S)(x, y)$ and describe it in words.

c. Write a formula for the composite $(S \circ T)(x, y)$ and describe it in words.

Solution

a. S is a horizontal scale change of magnitude __?__ and T is a __?__.

b. $(T \circ S)(x, y) = T(S(x, y)) = T(2x, y) = (2x + 4, y - 3)$.
$T \circ S$ is a horizontal scale change of magnitude 2, followed by a translation 4 units right and 3 units down.

c. $(S \circ T)(x, y) = S(T(x, y)) = S($__?__, __?__$) = (2($__?__$)$, __?__$) =$ (__?__, __?__). $T \circ S$ is a translation 4 units right and 3 units down, followed by a horizontal scale change of magnitude 2.

Questions

COVERING THE IDEAS

In 1 and 2, consider John Quincy Adams's family tree.

1. What biological relation does the composite $m \circ m$ represent?
2. John Quincy Adams married Louisa Catherine Johnson. Suppose they have a child, x.
   a. Evaluate $(m \circ f)(x)$. b. Explain why $(f \circ m)(x) \neq (m \circ f)(x)$.

In 3 and 4, refer to the functions *f* and *g* of Examples 1 and 2.

3. Verify that $f(g(0)) \neq g(f(0))$.
4. **True or False** a. $f(g(6)) = g(f(6))$ b. $f(g(3)) = g(f(3))$
5. Let $M(t) = 2t - 1$ and $N(t) = \frac{3}{t+1}$.
   a. Find a formula for $(M \circ N)(t)$. b. State the domain of $M \circ N$.
6. **True or False** Composition of functions is commutative.

In 7 and 8, let $f(x) = (x + 1)^2$ and $g(x) = x - 2$.

7. Evaluate $f(g(\text{-}5))$ and $g(f(\text{-}5))$.
8. Show that $g \circ f \neq f \circ g$ by graphing.

In 9 and 10, $f(x) = 2x^3 - 1$ and $g(x) = 3x$.

9. Evaluate each expression.
   a. $f(g(\text{-}1))$ b. $(f \circ g)(0)$
10. a. Find a formula for $(f \circ g)(x)$.
    b. State the domain of $f \circ g$.
    c. For what value of x does $(f \circ g)(x) = (g \circ f)(x)$?
11. Let T be a transformation that translates each point right 3 and up 1; let S be a vertical scale change of magnitude $\frac{1}{4}$.
    a. **Fill in the Blanks** T: $(x, y) \to$ (__?__, __?__) and S: $(x, y) \to$ (__?__, __?__).
    b. Find a formulas for $T \circ S$ and $S \circ T$.

APPLYING THE MATHEMATICS

12. Consider the sets A, B, and C at the right.
    a. Evaluate $g(f(3))$.
    b. The composite $g \circ f$ maps 4 to what number?
    c. If $f(x) = x + 1$ and $g(x) = \sqrt{x}$, write a formula for $g(f(x))$.
    d. If the domain of f is extended to the set of all reals, what is the domain of $g \circ f$?

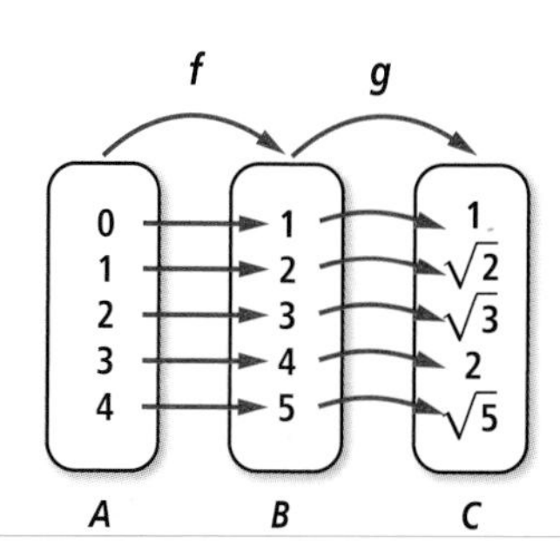

13. If $S(x, y) = (\text{-}y, x)$, find a formula for $S \circ S$.
14. If $T(x, y) = (x + 6, 2y)$, find a formula for $T \circ T$.

In 15–17, suppose $D(x) = 0.9x$ and $R(x) = x - 100$.

15. Explain why it is appropriate to call D a discount function and R a rebate function.

16. a. Evaluate $D(R(1200))$ and $R(D(1200))$.
    b. If you are buying a flat screen TV for \$1200, is it better to apply the discount after the rebate or before the rebate?

17. Find rules for $D \circ R(x)$ and for $R \circ D(x)$. Prove that $D \circ R \neq R \circ D$.

18. From these tables, evaluate each expression, if possible.

x	-5	-3	-1	0	1	2	3	4	5
$f(x)$	-5	-4	-3	0	3	2	1	0	-1

x	0	1	4	9	16	25
$g(x)$	0	1	2	3	4	5

a. $f(g(4))$ b. $g(f(4))$ c. $f(g(1))$ d. $g(f(1))$

19. a. One mile is 5280 feet. Write a formula for a function m that converts number of feet to number of miles.
    b. One mile is exactly 1.609344 kilometers. Write a formula for a function k that converts number of miles to number of kilometers.
    c. Write a rule for a composite function that converts feet to kilometers.
    d. How many feet is 5 kilometers?

20. Let f and g be real functions with $f(x) = \frac{1}{x}$ and $g(x) = x + 4$.
    a. Give equations for all asymptotes of the graph of $f \circ g$.
    b. Give equations for all asymptotes of the graph of $g \circ f$.

REVIEW

21. The graph of $y = f(x)$ is drawn at the right. Draw the graph of $y = 2f(3x)$. **(Lesson 3-5)**

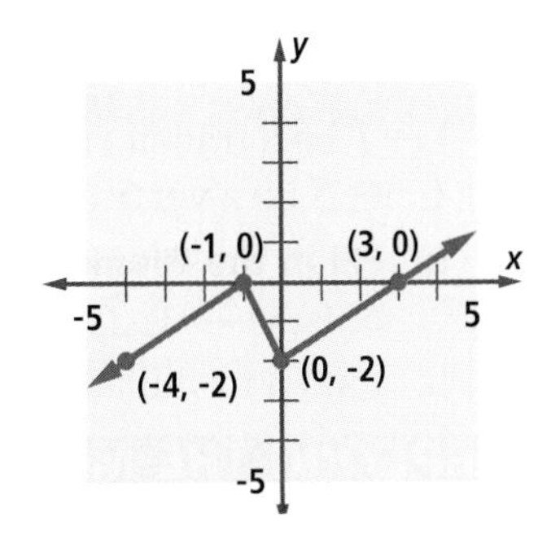

22. Prices of pies at Benny's Bakery have a mean of \$13.58 and a variance of 2.18. Assuming that customers do not change their purchasing pattern, what will be the effect on the mean and standard deviation under each circumstance? **(Lessons 3-6, 3-3)**
    a. The price per pie increases by 50 cents.
    b. The price per pie increases by 5%.

23. Find an equation for the image of the graph of $f(x) = x^2$ under the transformation $S: (x, y) \rightarrow \left(\frac{x}{4}, y\right)$. **(Lesson 3-5)**

24. ***Skill Sequence*** Find all real solutions. **(Previous Course)**
    a. $(5x - 10)(x - 3) = 0$ b. $3x^2 = 1 - 2x$ c. $3x^4 = 1 - 2x^2$

EXPLORATION

25. Find a function f such that $f(f(x)) = x$ yet $f(x) \neq x$.

QY ANSWER

true

Lesson 3-8 Inverses of Functions

Vocabulary

inverse of a function
identity function

▶ **BIG IDEA** From a function's description as a set of ordered pairs, by an equation, or by a graph, its inverse can be similarly described.

Mental Math

What operation undoes each action?

a. adding $\frac{2}{3}$ to a number

b. multiplying a number by $\frac{\pi}{2}$

c. squaring a positive number

The volume of a cube is $V = e^3$, where e is the length of an edge. Considering V as a function of e, we can write $V = f(e)$, and f is the *cubing* function. For example, the volume of a cube with 7-cm edges is $f(7) = 7^3 = 343 \text{ cm}^3$. However, sometimes you know the volume of a cube and need to find the edge length. Suppose a cube has a volume of 250 cm^3. Solve for the edge length by finding the cube root of 250; $e = \sqrt[3]{250} \approx 6.3$ cm. In general, $e = \sqrt[3]{V}$. Considering e as a function of V, we can write $e = g(V) = \sqrt[3]{V}$. The functions $f: e \rightarrow e^3$ and $g: V \rightarrow \sqrt[3]{V}$ are examples of *inverse functions*.

Finding the Inverse of a Function

Recall that a function can be considered as a set of ordered pairs in which each first element is paired with exactly one second element. If you switch coordinates in the pairs, the resulting set of ordered pairs is called the **inverse of the function**.

Example 1

Let $f = \{(-3, -5), (-2, 0), (-1, 3), (0, 4), (1, 3), (2, 0), (3, -5)\}$.
Describe the inverse of f. Is the inverse a function?

Solution

Let g be the inverse of f. The ordered pairs in g are found by switching the x- and y-coordinates of each pair in f.

$g = \{(-5, -3), (0, -2), (3, -1), (4, 0), (3, 1), (0, 2), (-5, 3)\}$

The inverse is not a function because there are ordered pairs in which the same first element is paired with different second elements, such as $(-5, -3)$ and $(-5, 3)$.

If the original function is described by an equation, then switching the variables in the equation gives an equation for its inverse. This has the same effect as switching the coordinates of every ordered pair.

Example 2

a. Give an equation for the inverse of the function described by $y = -x^2 + 4$.

b. Sketch a graph of $y = -x^2 + 4$ and the inverse on the same set of axes.

c. Is the inverse a function?

Solution

a. To form an equation for the inverse, switch x and y. The inverse is described by the equation $x = -y^2 + 4$.

b. The graphs of $y = -x^2 + 4$ and its inverse, $x = -y^2 + 4$, are drawn at the right.

c. The graph of the inverse contains ordered pairs with the same first coordinate but different second coordinate and fails the vertical-line test. So $x = -y^2 + 4$ is not an equation of a function.

function $y = -x^2 + 4$; $y = x$; inverse $x = -y^2 + 4$

At the right are tables with some ordered pairs of each relation in Example 2. Notice that the ordered pairs, from the function to the inverse, follow the mapping $(x, y) \rightarrow (y, x)$. This mapping can be seen graphically as a reflection over the line $y = x$. So, the graphs of a function and its inverse are reflection images of each other over the line $y = x$.

function $y = -x^2 + 4$

x	y
-2	0
-1	3
0	4
1	3
2	0
3	-5
4	-12

inverse $x = -y^2 + 4$

x	y
0	-2
3	-1
4	0
3	1
0	2
-5	3
-12	4

QY1

QY1

If (3,12.5) is a point on the graph of a function, what point is on the graph of its inverse?

Activity

Step 1 Let $a(x) = \frac{1}{x - 3} + 4$. Describe in words what the function does to a number x according to the order of operations.

Step 2 Now describe in words how to "undo" the process you described in Step 1. Call this function b and write a formula for $b(x)$.

Step 3 Check your answers to Steps 1 and 2 by choosing a value for x, inputting that x-value into the formula for $a(x)$, and then substituting that output into the formula for $b(x)$. What was your result?

The Activity shows that when one function undoes the effects of another function, the original input x results. When two functions are comprised of the operations of each other in reverse order, they are inverse functions of each other. Sometimes it is more convenient to switch the x- and y-coordinates of the original function first and then perform the inverse operations to arrive at the inverse of the original function.

Example 3

Consider the function f with $f(x) = \frac{1}{x-3} + 4$.

a. Give an equation for the inverse of f.

b. Graph f and its inverse on the same set of axes.

Solution 1

a. Let $y = \frac{1}{x-3} + 4$. Switch x and y to find an equation for the inverse.

$$x = \frac{1}{y-3} + 4$$

This equation answers the question, but we usually solve for y.

$$x - 4 = \frac{1}{y-3} \quad \text{Subtract 4.}$$

$$y - 3 = \frac{1}{x-4} \quad \text{Take reciprocals.}$$

From this equation, by the Graph Translation Theorem, you can see that the graph of the inverse is the image of the graph of $y = \frac{1}{x}$ under the translation $(x, y) \rightarrow (x + 4, y + 3)$.

$$y = \frac{1}{x-4} + 3 \quad \text{Add 3.}$$

b. The graphs of $y = \frac{1}{x-3} + 4$ and of $y = \frac{1}{x-4} + 3$ are shown at the right. To check that they are inverses, we also graphed $y = x$. Notice that each branch of the inverse is the image of one of the branches of the original hyperbola under a reflection over $y = x$.

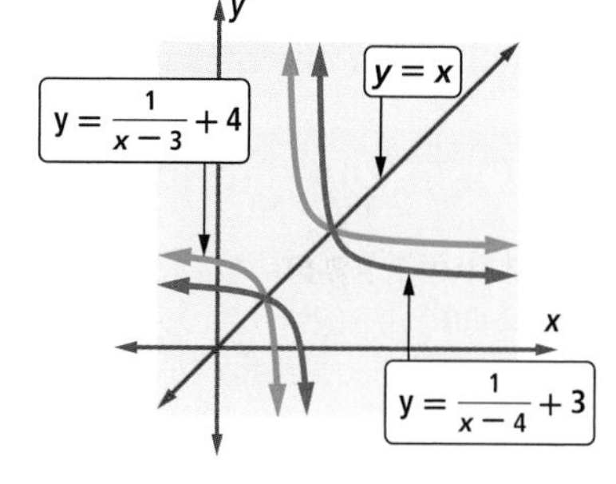

Solution 2 Use the `solve` command on a CAS. Notice that the solution given here is in a different form than Solution 1. The CAS rewrote $\frac{1}{x-4} + 3$ using common denominators.

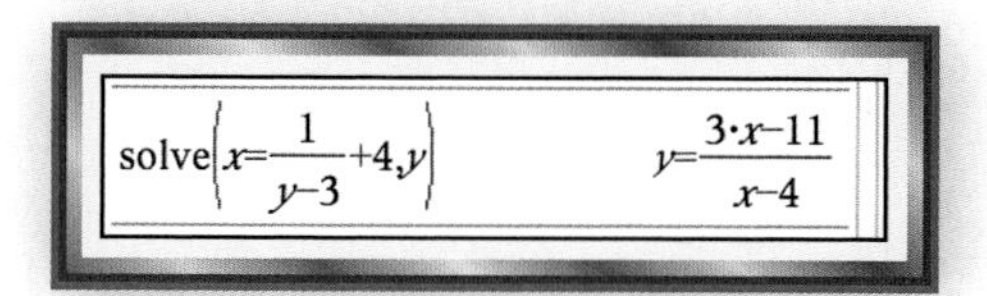

QY2

Examples 2 and 3 demonstrate a significant feature of inverses. That is, not all inverses are functions. Looking at the graph and applying the definition of function reveals that the inverse from Example 2 is not a function, while the inverse from Example 3 is a function.

When the inverse of a function f is a function, it is denoted by the symbol f^{-1}, read "f inverse." With this notation, the rule for the inverse of f in Example 3 can be written $f^{-1}(x) = \frac{3x - 11}{x - 4}$. *Caution*: Note that f^{-1} does not denote the reciprocal of f, which is $\frac{1}{f}$.

QY3

QY2

Show that $\frac{1}{x-4} + 3 = \frac{3x - 11}{x - 4}$.

QY3

a. What is the vertical asymptote of the graph of f in Example 3?

b. What is the vertical asymptote of the graph of f^{-1}?

Inverse Functions and Composition of Functions

Because the inverse of a function is found by switching the x- and y-coordinates, the domain and range of the inverse are found by switching the domain and range of the original function. Thus, the domain of f^{-1} is the range of f, and the range of f^{-1} is the domain of f. Hence, if f is a function whose inverse is also a function, $f(f^{-1}(x))$ and $f^{-1}(f(x))$ can always be calculated.

For example, for the function in Example 3,

$$f(f^{-1}(2)) = f\left(\frac{1}{2-4} + 3\right) = f(2.5) = \left(\frac{1}{2.5-3} + 4\right) = 2,$$

and $$f^{-1}(f(2)) = f^{-1}\left(\frac{1}{2-3} + 4\right) = f^{-1}(3) = \left(\frac{1}{3-4} + 3\right) = 2.$$

As you will see in Example 4, for these functions, $f \circ f^{-1}(x) = f^{-1} \circ f(x) = x$ for all values of x for which the composites are defined. This is why f^{-1} is called the inverse of f; f^{-1} undoes the effect of f. This property of the composition of inverses is an instance of the following theorem.

Inverses of Functions Theorem

Given any two functions f and g, f and g are inverse functions if and only if $f(g(x)) = x$ for all x in the domain of g, and $g(f(x)) = x$ for all x in the domain of f.

When f and g are inverse functions, then $f = g^{-1}$ and $g = f^{-1}$. The theorem states that two functions are inverses of each other if and only if $f \circ g$ and $g \circ f$ are the function I with $I(x) = x$, a function which is called an **identity function**. The Inverses of Functions Theorem enables you to test whether two functions are inverse functions even if you have not derived one from the other.

Example 4

a. Use the Inverses of Functions Theorem to determine whether f and g, defined by $f(x) = \frac{2x-4}{x+1}$ and $g(x) = \frac{x-1}{2x+4}$, are inverses.

b. Verify your result in Part a by graphing f and g.

Solution

a. Define both relations on a CAS. Use the CAS to evaluate $f(g(x))$

Because $(f \circ g)(x) \neq x$, the functions are not inverses.

Define $f(x)=\frac{2\cdot x-4}{x+1}$	*Done*
Define $g(x)=\frac{x-1}{2\cdot x+4}$	*Done*
$f(g(x))$	$\frac{-2\cdot(x+3)}{x+1}$

b. Graph f, g, and $y = x$. The graph of g is shown using a bold line. Note that the graphs of f and g are not reflection images of each other over $y = x$.

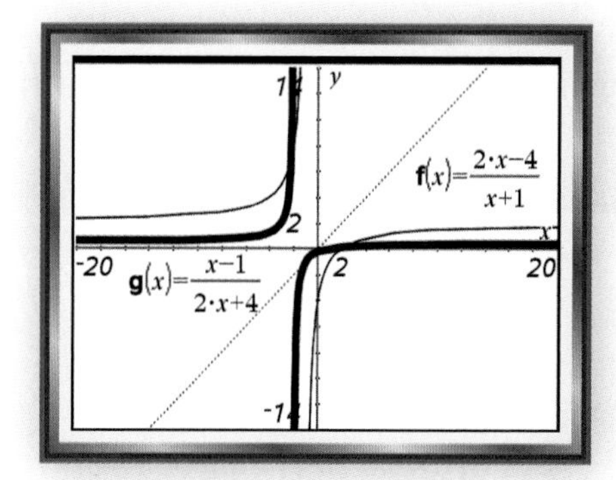

Questions

COVERING THE IDEAS

1. Define *inverse of a function*.

2. Let $f = \{(1, -4), (2, -6), (3, -8), (4, -10)\}$.
   a. Find g, the inverse of f.
   b. Graph f and g on the same set of axes.
   c. What transformation relates the graphs of f and g?

3. a. Suppose $f(x) = -2x^2$. Graph f and its inverse on the same axes.
   b. Explain why the inverse of f is not a function.

4. Give an example, different from those in the lesson, of a function whose inverse is not a function.

In 5–7, write an equation for the inverse of the function with the given equation. Solve your equation for y. Is the inverse a function?

5. $y = 2x - 4$ 6. $y = -x^3$ 7. $y = \sqrt{x}$

8. A rule for a function h is given. Is the inverse of h a function?
   a. $h(x) = |x + 2|$ b. $h(x) = x + 2$

In 9 and 10, determine if f and g are inverses by finding $g \circ f(x)$ and $f \circ g(x)$. Then check your conclusion by graphing the functions.

9. $f(x) = x^3$; $g(x) = \sqrt[3]{x}$ 10. $f(x) = \frac{2}{x} - 5$; $g(x) = \frac{2}{x + 5}$

APPLYING THE MATHEMATICS

11. At one point in the summer of 2008, one U.S. dollar was worth 10.033 Mexican pesos. Let $M(x)$ be the cost in pesos of an item priced at x U.S. dollars and $U(x)$ be the cost in dollars of an item priced at x Mexican pesos.

   a. Write expressions for $M(x)$ and $U(x)$.
   b. What was the U.S. price of an item which cost 20,000 pesos?
   c. Are M and U inverses of each other? Justify your answer.

12. The rule for converting from degrees Fahrenheit to degrees Celsius is "subtract 32, then multiply by $\frac{5}{9}$."

 a. Determine the rule for converting Celsius to Fahrenheit.

 b. Do the two rules represent inverse functions? Why or why not?

In 13 and 14, a graph is given.

 a. Sketch the graph of the inverse of the function.

 b. State whether or not the inverse is a function.

13.

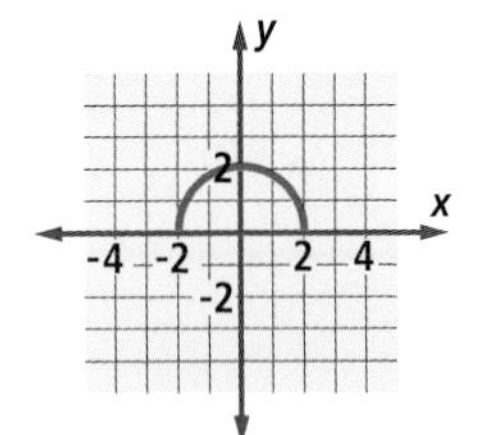

14.

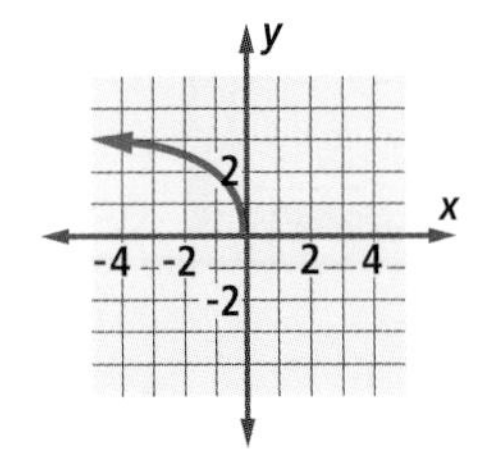

15. If h is the reciprocal function defined by $h(p) = \frac{1}{p}$, show that $h(p) = h^{-1}(p)$ for all $p \neq 0$. That is, h is its own inverse.

16. a. Let $f(x) = mx + b$, where $m \neq 0$. Find a formula for $f^{-1}(x)$.

 b. **True or False** The inverse of every function whose graph is a line is a function whose graph is a line. If false, give a counterexample.

17. An empty box weighs 11 ounces. It is filled with bolts weighing 8 oz each. The total weight in ounces of the box is given by $w(n) = 11 + 8n$ where n is the number of bolts. Write a formula for the inverse function which finds the number of bolts given the total weight.

REVIEW

In 18 and 19, let $v(t) = 39t$ and $r(t) = t + 17$. (Lesson 3-7)

18. Evaluate each composite.

 a. $(r \circ v)(-1)$ b. $r \circ r(-1)$

19. Find a formula for each function.

 a. $v \circ r$ b. $v \circ v$

20. Consider the Fahrenheit to Celsius rule in Question 12 and the rule to convert from Celsius to Kelvin of "add 273.15." **(Lesson 3-7)**

 a. Write equations for functions that convert from Fahrenheit to Celsius and from Celsius to Kelvin.

 b. Find the composite function. What does this function do?

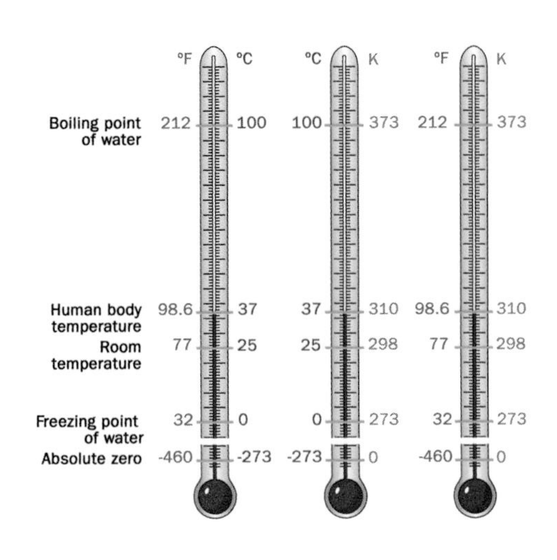

21. a. Prove that the function p defined by $p(x) = -|x| + 7$ is an even function.

 b. What type of symmetry does the graph of $y = p(x)$ have? **(Lesson 3-4)**

In 22 and 23, use the data in the table at the right that contains the number of cell phone subscribers (in thousands) and the average length of local calls.

	Subscribers	Min/Call
1990	5,283	2.20
1995	33,786	2.15
2000	109,478	2.56
2005	207,896	3.00

Source: CTIA - The Wireless Association

22. a. Graph the data for subscribers for a given year.
    b. Find the best exponential model relating the year to the number of subscribers.
    c. Use the model to predict the number of cell phone subscribers in 2010. **(Lesson 2-5)**

23. a. Graph the data for the average call length for a given year.
    b. Find a good model relating the year to the call length.
    c. Use the model to predict the average length of a cell phone call in 2010. **(Lesson 2-3)**

24. The table below contains average hourly earnings of production workers in the U.S. by month in 2006. **(Lesson 2-3)**
    The line of best fit for this data set is $y = 0.06x + 16.38$.

Month	1	2	3	4	5	6	7	8	9	10	11	12
Hourly Wage	16.43	16.49	16.55	16.63	16.66	16.73	16.79	16.84	16.88	16.94	16.99	17.07

Source: Bureau of Labor Statistics

    a. Use the line of best fit to predict the hourly wage for April 2008.
    b. Use a spreadsheet to find the residual and squared residual for each month.
    c. What is the sum of the squared residuals? What conclusion can you draw about that number?

25. Rewrite $p_1q_1 + p_2q_2 + \ldots + p_nq_n$ using Σ-notation. **(Lesson 1-2)**

EXPLORATION

26. The field of cryptology relies on inverses. Research the Caesar cipher to find out what it is and how it relates to inverse functions.

QY ANSWERS

1. (12.5, 3)
2. $\frac{1}{x-4} + 3$
$= \frac{1}{x-4} + \frac{3(x-4)}{x-4} = \frac{1 + 3x - 12}{x-4}$
$= \frac{3x - 11}{x-4}$
3. **a.** $x = 3$ **b.** $x = 4$

Lesson

3-9 z-Scores

Vocabulary

z-score
raw data
standardized data

▶ **BIG IDEA** *z*-scores enable a score to be compared to other scores in the same data set.

What is a *z*-Score?

Sometimes a person wants to know how his or her score or salary compares to a group as a whole. One way to do this is to compute a percentile. Another way is to analyze how many standard deviations the score or salary is above or below the mean.

The U.S. Bureau of Labor Statistics tracks the median hourly wages for 22 general occupational categories. The 2007 median wages for certain occupations are listed in the table below. The mean of the 7 median hourly wages is $\bar{x} = \$22.90$ and the standard deviation is $s = \$10.70$.

Food Production	Protective Services	Construction	Education	Healthcare	Architecture/ Engineering	Management
\$8.24	\$16.11	\$17.57	\$20.47	\$26.17	\$31.14	\$40.60

In general, the transformation that maps each wage x to the score $\frac{x - \bar{x}}{s} = \frac{x - 22.90}{10.70}$ tells you how many standard deviations that wage is above or below the mean. For example, the median wage for Management is $\frac{40.60 - 22.90}{10.70} \approx 1.7$ standard deviations above the mean, while the median wage for Food Production is $\frac{8.24 - 22.90}{10.70} \approx -1.4$ standard deviations above, or 1.4 standard deviations below, the mean.

In the table below, row L_1 is the original set of wages, L_2 is the image of L_1 under the translation T where $T(x) = x - \bar{x} = x - 22.90$, and L_3 is the image of L_2 under the scale change S where $S(x) = \frac{x}{s} = \frac{x}{10.70}$.

L_1	8.24	16.11	17.57	20.47	26.17	31.14	40.60
L_2	-14.66	-6.79	-5.31	-2.43	3.27	8.24	17.70
L_3	-1.4	-0.6	-0.5	-0.2	0.3	0.8	1.7

The transformation that maps the original data set L_1 onto L_3 is the composite $S \circ T$, where $S \circ T(x) = S(T(x)) = S(x - 22.90) = \frac{x - 22.90}{10.70}$. $S \circ T(x)$ is called the *z*-score for the value x. The *z*-score for Protective Services is -0.6; this means that \$16.11 is 0.6 standard deviations below the mean. In the same way, the 0.8 *z*-score for Architecture/Engineering means that \$31.14 is 0.8 standard deviations above the mean.

Mental Math

Suppose the mean of a set of scores is 486 and the standard deviation is 37. What number is:

a. 1 standard deviation above the mean? (remember - this is mental math!)

b. 1 standard deviation below the mean?

c. 2 standard deviations above the mean?

d. 2 standard deviations below the mean?

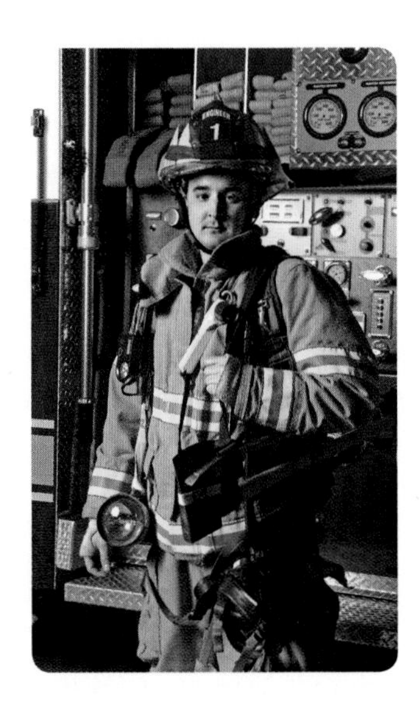

The preceding discussion can be generalized in the following definition.

Definition of z-Score

Suppose a data set has mean $\bar{x}$ and standard deviation s. The **z-score** for a member x of this data set is

$$z = \frac{\text{deviation}}{\text{standard deviation}} = \frac{x - \bar{x}}{s}.$$

A positive *z*-score tells how many standard deviations the score is above the mean. A negative *z*-score tells how far below the mean the score is.

 QY

QY

Find the *z*-score for \$20.47 in the table and describe what it means.

Example 1

Nancy scored 87 on a math quiz on which the mean score was 70 and the standard deviation of the scores was 8. Find her *z*-score and tell how far her score was from the mean.

Solution Her z-score is

$$z = \frac{87 - 70}{8} \approx 2.1,$$

so her score was 2.1 standard deviations above the mean.

Sometimes the original data are called **raw data** or raw scores, and the results of the transformation are called **standardized data** or standardized scores. In Example 1, a raw score of 87 corresponds to a standardized score of 2.1.

Properties of z-Scores

What happens to the mean and standard deviation of a data set if each score is converted to a *z*-score? Refer again to the median wages shown in row L_1 of the table on page 206. Because adding (or subtracting) the number h to every number in a data set adds (or subtracts) h to the mean, the mean of the data set in L_2 is $22.90 - 22.90 = 0$. Under a translation of a data set, the standard deviation is invariant. Thus, the standard deviation of the data set in L_2 is still 10.70. Because the scale change S with $S(x) = ax$ multiplies both the mean and standard deviation by a, the mean of the data set in L_3 is $\frac{1}{10.70} \cdot 0 = 0$ and the standard deviation is $\frac{1}{10.70} \cdot 10.70 = 1$. Thus, the *z*-scores in L_3 have mean 0 and standard deviation 1. In general, we have the following theorem.

Theorem (Mean and Standard Deviation of z-Scores)

If a data set has mean $\bar{x}$ and standard deviation s, the mean of its *z*-scores will be 0, and the standard deviation of its *z*-scores will be 1.

Using *z*-Scores to Make Comparisons

Standardized scores, or *z*-scores, make it easier to compare different sets of numbers.

GUIDED

Example 2

Mark scored 78 on a history test on which the mean was 71 and the standard deviation was 10. He scored 68 on a chemistry test on which the mean was 62 and the standard deviation was 6. Use z-scores to determine on which test he performed better compared to his classmates.

Solution His z-scores are:

history: $\frac{? - 71}{10} = \underline{\ ?\ }$

chemistry: $\frac{68 - ?}{?} = \underline{\ ?\ }$.

Because his z-score on the __?__ test is higher, Mark performed better on that test compared to his classmates.

In Example 2 you should notice that Mark scored above the mean on both tests, but the *z*-scores provide more information. The *z*-scores are sensitive to the fact that the scores on the history test are more spread out than those on the chemistry test.

Questions

COVERING THE IDEAS

In 1–3, refer to the data sets in rows L_2 and L_3 on page 206.

1. By computing directly, find the mean and standard deviation of each data set.
2. What does the *z*-score of -0.2 mean?
3. Which median wage is 0.5 standard deviations below the mean?
4. Find the *z*-score for a test score of 84 for each situation.
   a. mean = 89; standard deviation = 6.1
   b. mean = 67; standard deviation = 8.8
5. In which situation, 4a or 4b, is the test score better compared to others who took the test?
6. What is a standardized score?
7. A data set has a mean of 19 and a standard deviation of 4.3. How can the data set be transformed so that the mean is 0 and the standard deviation is 1?
8. Refer to Example 2. Mary scored 54 on the history test and 47 on the chemistry test. Use *z*-scores to determine on which test she did better compared to her classmates.

9. In 2007, Colorado high school students' average ACT scores were: English, 19.7; Math, 20.1; Reading, 20.8; and Science, 20.4. Jennie's scores were: English, 24; Math, 21; Reading, 18; and Science, 20. Find Jennie's *z*-score for each section assuming that each section had a standard deviation of 1.8.

APPLYING THE MATHEMATICS

In 10 and 11, the mean boys' time for a one-mile race was 6 minutes 6 seconds with a standard deviation of 16 seconds. The mean girls' time was 8 minutes 15 seconds with a standard deviation of 24 seconds.

10. Who is faster relative to others of his or her gender, a boy who runs a mile in 5 minutes 34 seconds, or a girl who runs a mile in 7 minutes 35 seconds?

11. Suppose a girl runs the race in 8 minutes 30 seconds. What boy's time would have the same *z*-score?

12. A student got a *z*-score of 1.33 on a test with a mean of 73 and a standard deviation of 9. What was the student's raw score?

13. The graph at the right shows how much customers spend at a grocery store. The costs have mean \$125 and standard deviation \$65. Copy the horizontal axis as shown.
    a. Change each value labeled on the axis to a *z*-score. Values for 100 and 250 are shown.
    b. Locate the points on the axis that correspond to *z*-scores of 0, 1, and -1.

Frequency
10
8
6
4
2
0 50 100 150 200 250 300 350
Grocery Costs

-0.4 1.9 *z*-score
0 50 100 150 200 250 300 350 cost

14. A teacher tells a class that the mean raw score on a test was 58. Alex has a raw score of 73 and a *z*-score of 1.25. What was the standard deviation on the test?

15. Considering the data sets in L_1 and L_3 at the start of the lesson, use the translation $T(x) = x - \bar{x}$ and the scale change $S(x) = \frac{x}{s}$.
    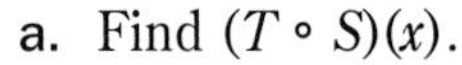
    a. Find $(T \circ S)(x)$.
    b. Apply $T \circ S$ to the data set in L_1. What are the mean and standard deviation of this new data set?
    c. Are these the same mean and standard deviation as for the data set in L_3? Explain why or why not.

In 16–19, use the table at the right with the scores from two tests.

16. How many standard deviations above the mean is Fiona's score on the physics test?

17. On which test did Raj do better compared to others who took the test?

	Physics	Mathematics
Fiona	84	64
Raj	70	76
Test Statistics	mean: 60 standard deviation: 6.9	mean: 70 standard deviation: 10.7

18. Simon had a 58 on the physics test. He scored equally well (in terms of z-score) on the mathematics test. What was his raw score on the mathematics test?

19. Andrea had the same raw score on each test, and she also had the same z-score on each test. What raw score did she get?

REVIEW

20. a. Write an equation for the inverse of g where $g(t) = \sqrt{t + 4}$.
    b. Is the inverse of g a function? **(Lesson 3-8)**

21. **True or False** Let $f(x) = x^3$ and $g(x) = x^{-3}$. **(Lessons 3-8, 3-7)**
    a. For all x, $f(g(x)) = g(f(x))$. b. f and g are inverses.

In 22 and 23, tell whether the statement is *true* or *false*. If it is false, give a counterexample. (Lessons 3-8, 3-4)

22. The inverse of an even function is a function.

23. The inverse of an odd function is a function.

24. Consider a 15% discount function D where $D(x) = 0.85x$ and a 7% total-with-tax function T where $T(x) = 1.07x$. **(Lesson 3-7)**

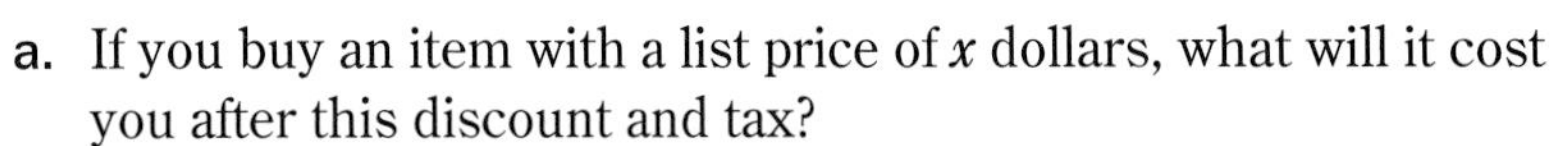

    a. If you buy an item with a list price of x dollars, what will it cost you after this discount and tax?
    b. Which is better to take first, the discount or the tax? Explain.

25. **Multiple Choice** A transformed set of data has a variance three times that of the original set. How were the data transformed? **(Lesson 3-6)**
    A translated by 3 B multiplied by $\sqrt{3}$
    C multiplied by 3 D multiplied by 9

26. **Multiple Choice** The graph of which relation has point symmetry? **(Lesson 3-4)**
    A $y = |x|$ B $y = x^2$
    C $y = x^3$ D $y = \frac{1}{x^2}$

EXPLORATION

27. In each of the 22 general occupation categories tracked by the U.S. Bureau of Labor Statistics (BLS), there are a number of more specific occupation descriptions. Visit the BLS website and identify a specific occupation description that interests you.
    a. Calculate a mean and standard deviation for the median wages of the specific occupations within the general category.
    b. Calculate the z-scores for the highest and lowest median wages in the general occupation category. What do these z-scores tell you about the range and distribution of wages within this group?
    c. Calculate the z-score for the occupation you have chosen. Describe how the wage for your occupation compares to the other wages within the category.

QY ANSWER

The z-score is -0.2, so the median wage for Education is 0.2 standard deviations below the average median wage for the seven occupations.

Chapter 3 Projects

1 Sports Handicaps

To enable a wide variety of players to play against one another in golf and bowling, players are given handicaps which are taken from or added to their scores.

a. Obtain the full data set of raw scores for a golfer or bowler for a season. Use these to determine the handicap for this athlete.

b. Calculate the descriptive statistics you studied in Chapter 1 for this data set.

c. Apply the handicap of this person to these scores and calculate the same descriptive statistics for the image set.

d. Show how the results that you found in Parts b and c agree with theorems stated in this chapter.

e. Discuss whether you think the handicap was too small, just about right, or too large. In your discussion, you might wish to consider the following questions: Would this handicap enable the person to compete with a professional? With a good high school athlete? What handicap do you think you would need?

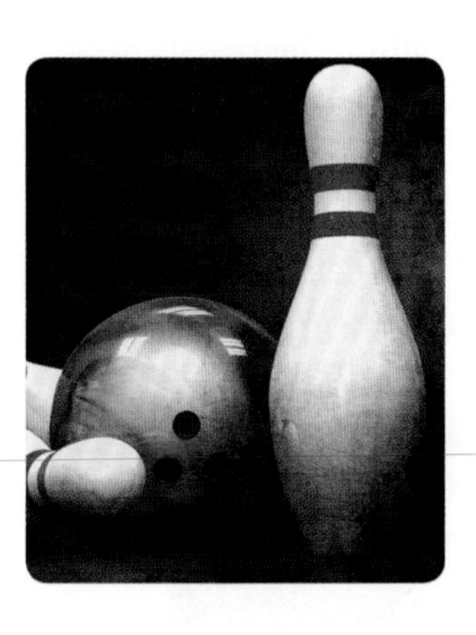

Got a moment to spare? Bowling pins are 15 inches high, the same as the circumference of the pin.

2 Facial Symmetry and Asymmetry

According to researchers such as Paul Ekman, spontaneous emotional reactions elicit more symmetrical facial expressions than deliberate emotional reactions. According to Ekman, of the 10,000 facial expressions the 42 facial muscles can form, only about 3000 are relevant to emotion.

a. Using a program such as Photo Booth, taking digital photos, or using already developed pictures, obtain and study pictures of your face while you are deliberately trying to smile, frown, squint, and scowl.

b. Find already taken pictures of yourself smiling, frowning, squinting, or scowling in which you were naturally doing so. These might have been taken when you were unaware of the camera.

c. Compare the pictures. Do you agree with Ekman that spontaneous expressions are more symmetrical?

d. Take a straight-on picture of your face and then use Photo Booth's Mirror tool and the left half of your face to create a picture of you that is vertically symmetrical. Compare this picture to the original. What do you notice?

3 Transformation Groups

The set of all scale changes under composition forms a mathematical structure called a *commutative group*. The properties of this commutative group are listed here.

i. Closure: If S_1 and S_2 are scale changes, then so is $S_2 \circ S_1$.

ii. Commutativity: $S_1 \circ S_2 = S_2 \circ S_1$.

iii. Associativity: $(S_3 \circ S_2) \circ S_1 = S_3 \circ (S_2 \circ S_1)$

iv. Identity: There is a scale change I such that $S \circ I = I \circ S = S$.

v. Inverses: For every scale change S there is an inverse scale change S^{-1} such that $S \circ S^{-1} = S^{-1} \circ S = I$.

Each of these properties can be proven. For instance, for property ii, let $S_1(x, y) = (ax, by)$ and $S_2(x, y) = (cx, dy)$. Then

$$(S_2 \circ S_1)(x) = S_2(S_1(x, y)) = S_2(ax, by) = (cax, dby).$$

$$(S_1 \circ S_2)(x) = S_1(S_2(x, y)) = S_1(cx, dy) = (acx, bdy).$$

Because multiplication of real numbers is commutative, $ac = ca$ and $bd = db$. So $S_1 \circ S_2 = S_2 \circ S_1$.

a. Prove the remaining properties for scale changes.

b. Does the set of all translations form a commutative group under composition?

4 CPI/Rate of Inflation

The U.S. Bureau of Labor Statistics (BLS) is responsible for gathering the data for the Consumer Price Index (CPI). Visit the BLS database website at www.bls.gov/data. Search the "Average Price Data" for your region or type of city and create a basket of 10 goods that includes gasoline, electricity, and various foods and beverages. Make sure that the goods you have selected have data available for a ten-year period.

a. Use your basket of goods to create a CPI for your area for the ten-year period.

b. Ask someone who is at least ten years older than you to look at your list of goods and estimate the percentage increase in cost of each item over the ten-year period. Use this information to create a "subjective CPI."

c. Compare the actual CPI for the ten-year period to your CPI and the subjective CPI. Which is a more accurate estimation of the actual CPI, your CPI or the subjective CPI? Why do you think one is better than the other?

d. Research the items that are included in the basket of goods used to calculate the actual CPI. Argue why this basket is or is not a good indication of the cost of living for a family in your neighborhood.

5 Class Survey Revisited, Yet Again

Use the class survey database constructed as a project in Chapter 1. Consider the variables which involve units, for example, height, hand span, foot length, etc.

a. Convert these measurements to other units (for example, if the data were originally in centimeters, change to inches). Apply transformations to create new data sets from the original data.

b. Drop the units from both the original and the image data. Compare the descriptive statistics of the image with those of the original data to confirm the results of Lesson 3-3.

c. Examine some relations between pairs of original variables (e.g., between height and foot length) and between their images under these transformations. What statistics are invariant under these transformations?

Chapter 3 Summary and Vocabulary

- Equations, graphs of functions, and data can be transformed in similar ways. Two such **transformations** – **translations** and **scale changes** – are studied in this chapter. When graphs of functions are translated or scaled, the **images** resemble the graphs of the **preimages**, or original **parent functions**. Translations slide graphs, whereas scale changes stretch or shrink them horizontally and vertically.

- Connections among transformations, function equations, and graphs are given by the Graph-Translation and Graph Scale-Change Theorems. By the Graph-Translation Theorem, a translation $(x, y) \to (x + h, y + k)$ transforms the graph of $y = f(x)$ to a graph of $y - k = f(x - h)$. The Graph Scale-Change Theorem asserts that scaling the graph of $y = f(x)$ by **horizontal scale factor** a and **vertical scale factor** b, $(x, y) \to (ax, by)$, produces the graph of $\frac{y}{b} = f\left(\frac{x}{a}\right)$. Some features of graphs can be predicted if just the equation is given, and equations can be written if just the graph is given.

- Symmetries of graphs can be determined from their equations. **Odd functions** have **point symmetry** around the origin; **even functions** are **reflection-symmetric** over the y-axis.

- Asymptotes, discontinuities, and other features of functions can be identified from a graph of the function. Knowing the parent function of a transformed function can assist in this identification.

- Under a translation of magnitude h, measures of center for a data set are translated by h, while measures of spread are unaffected. In contrast, when data are multiplied by a factor $a > 0$, measures of center, the standard deviation, and the range are multiplied by a, and the variance is multiplied by a^2.

- **Composites** of functions are formed by letting one function g operate on those outputs of another function f that are in g's domain. The composite of f followed by g is written as $g \circ f$, and is defined by $(g \circ f)(x) = g(f(x))$. In general, **function composition** is not commutative. That is, $f \circ g$ and $g \circ f$ are usually different functions.

- The **inverse** of a function $f: x \to y$ can be obtained by switching x's and y's in its equation or by switching x- and y-coordinates in a set of ordered pairs. When the resulting relation is a function, it is denoted by f^{-1}. The graphs of a function and its inverse are reflection images of each other with respect to the line $y = x$. Another characteristic property of inverse functions is that the composite of a function f and its inverse f^{-1} is the **identity function** I. That is, for all x in the domain of the composite, $f(f^{-1}(x)) = f^{-1}(f(x)) = I(x) = x$.

Vocabulary

3-1
window
parent function

3-2
transformation
preimage
image
translation

3-3
invariant

3-4
reflection-symmetric
axis of symmetry
line of symmetry
symmetry about a point
point symmetry
center of symmetry
even function
odd function

3-5
horizontal and vertical scale change
horizontal and vertical scale factor
size change

3-6
scale change of a data set
scale factor
scale image

3-7
composite
function composition

- Suppose a data set has mean $\bar{x}$ and standard deviation s. A **z-score** is the result of a composite of a specific translation $(T(x) = x - \bar{x})$ followed by a scale change $(S(x) = \frac{x}{s})$ of the translated data. The z-score, $z = \frac{x - \bar{x}}{s}$, corresponding to the raw score x, tells how many standard deviations x is above or below the mean. A data set transformed in this way has mean 0 and standard deviation 1. Using z-scores makes it possible to compare scores from different data sets.

Vocabulary

3-8
inverse of a function
identity function

3-9
z-score
raw data
standardized data

Properties and Theorems

Graph-Translation Theorem (p. 161)
Theorem (Centers of Translated Data) (p. 167)
Theorem (Spreads of Translated Data) (p. 168)
Theorem (Symmetry over the *y*-axis) (p. 173)
Theorem (Symmetry over the *x*-axis) (p. 173)
Theorem (Symmetry about the Origin) (p. 173)
Graph Scale-Change Theorem (p. 181)
Theorem (Centers of Scaled Data) (p. 188)
Theorem (Spreads of Scaled Data) (p. 189)
Inverses of Functions Theorem (p. 202)
Theorem (Mean and Standard Deviation of *z*-scores) (p. 207)

Chapter 3 Self-Test

Take this test as you would take a test in class. You will need a calculator. Then use the Selected Answers section in the back of the book to check your work.

1. Give equations for the asymptotes of the graph of the function $h: x \to \frac{1}{x-7} - 9$.

In 2 and 3, let the translation $T: (x, y) \to (x + 4, y - 2)$ be applied to the graph of $y = x^2$.

2. Write an equation for the image.

3. What are the coordinates of the vertex of the image?

4. The graph of a function f is below. Sketch a graph of its image under the transformation S when $S(x, y) = (2x, -y)$.

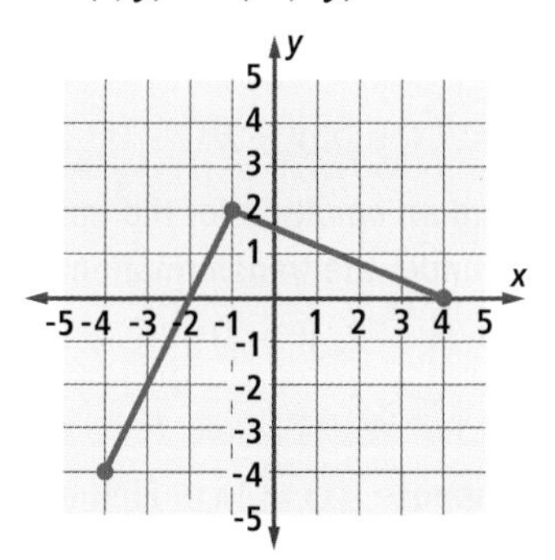

In 5 and 6, suppose a chemistry student finds masses of different samples of potassium chloride (KCl) and subtracts 150 from each value before computing the statistics below.

mean: 4.2 g — standard deviation: 1.3 g
minimum: 0.4 g — maximum: 13.6 g
median: 3.8 g — Q_1: 3.6 g — Q_3: 4.6 g

5. For the actual sample masses, give the
   a. median. b. range. c. IQR.

6. For the actual sample masses, give each statistic from Question 5 in ounces using 1 gram $\approx$ 0.035 ounces.

In 7–9, let f be a real function with $f(x) = 4\sqrt[3]{x + 2}$.

7. Find an equation for the inverse of f.

8. Is the inverse of f a function? Support your answer.

9. Sketch graphs of f and its inverse.

In 10 and 11, let m and n be real functions with $m(x) = 16 - 5x$ and $n(x) = x + \sqrt{x}$.

10. Write an expression for $n(m(x))$.

11. Give the domain of $n \circ m$.

12. Suppose the scale change $S(x, y) = \left(3x, \frac{y}{2}\right)$ is applied to the graph of $y = \frac{1}{x^2}$. Write an equation for the image.

13. A data set has a mean of 83 and a standard deviation of 7. What transformation should be applied so that the image set has a mean of 0 and a standard deviation of 1?

14. Tell whether the function j with equation $j(x) = x^2 - 5$ is even, odd, or neither. Support your answer with algebra.

15. The graph below of a function g is an angle.

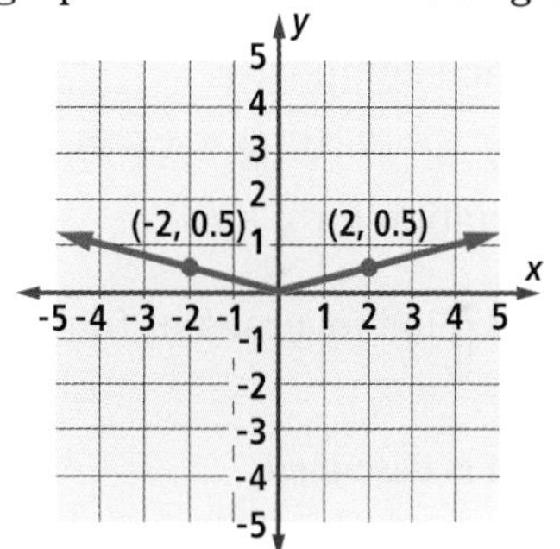

   a. What is the parent function of g?
   b. Find an equation for g.
   c. What symmetries does the graph of g have?
   d. Give the range of g.

16. Give a rule for the reflection over the line $y = x$.

17. Suppose a population of mice has a mean weight of 30 g and a standard deviation of 3.4 g. A population of moose has a mean weight of 910 lb and a standard deviation of 185 lb. Which animal is heavier relative to its population, a 37-g mouse or a 1260-lb moose? Explain your answer.

Chapter 3 Chapter Review

SKILLS
PROPERTIES
USES
REPRESENTATIONS

SKILLS Procedures used to get answers

OBJECTIVE A Find equations for and values of composites of functions. (Lesson 3-7)

In 1 and 2, consider the functions m mapping A to B, and n mapping B to C, as shown at the right.

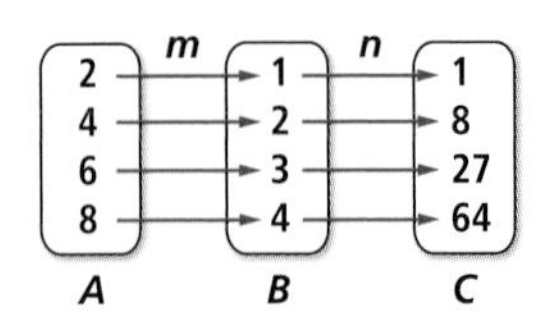

1. What is $n(m(6))$?
2. Find $(n \circ m)(2)$.

In 3 and 4, let $f(t) = 7t - 9$ and $g(t) = t^2 - t$.

3. Evaluate each composite.
   a. $f(g(-2))$ b. $g(f(-2))$
4. Find a formula for $f(f(t))$.

In 5 and 6, let $f(x) = \frac{4}{x}$ and $g(x) = x - 8$.

5. Evaluate each composite.
   a. $(f \circ g)(-8)$ b. $(g \circ g)(-8)$
6. Find an equation for $g \circ f$.

OBJECTIVE B Find inverses of functions. (Lesson 3-8).

In 7–9, a function is described.

a. Describe the inverse using a set of ordered pairs or an equation.

b. State whether the inverse is a function.

7. $f(x) = |x|$ 8. $\ell(x) = \frac{4}{x - 3}$
9. $g = \{(3, 4), (4, 12), (5, 9), (6, 0), (7, 11)\}$

PROPERTIES Principles behind the mathematics

OBJECTIVE C Use the Graph-Translation Theorem or the Graph Scale-Change Theorem to find transformation images. (Lessons 3-2, 3-5)

10. **Multiple Choice** Which scale change stretches a graph horizontally by a factor of 4 and shrinks it vertically by a factor of 11?
    A $S(x, y) = (4x, 11y)$ B $S(x, y) = \left(\frac{x}{4}, 11y\right)$
    C $S(x, y) = \left(\frac{x}{4}, \frac{y}{11}\right)$ D $S(x, y) = \left(4x, \frac{y}{11}\right)$
11. **Multiple Choice** Which translation has the effect on a graph of moving each point 5 units down and 9 units to the right?
    A $T(x, y) = (x - 5, y + 9)$
    B $T(x, y) = (x + 9, y - 5)$
    C $T(x, y) = (x - 9, y + 5)$
    D $T(x, y) = (x + 5, y - 9)$

In 12 and 13, find an equation for the image of the graph of $y = x^2$ under the transformation.

12. $T: (x, y) \rightarrow (x - 2, y + 7)$
13. $S: (x, y) \rightarrow \left(\frac{x}{2}, 3y\right)$

In 14 and 15, suppose $f(x) = |x|$. Find an equation for the image of the graph of f under the transformation.

14. $S(x, y) = \left(5x, \frac{y}{4}\right)$
15. $T(x, y) = (x - 1, y)$
16. Describe a transformation that maps the graph of $\sqrt{x}$ onto the graph of $\sqrt{5x}$.
17. Describe a transformation that maps the graph of $y = 8^x$ onto the graph of $y = 8^x + 4$.

OBJECTIVE D Describe the effects of translations and scale changes on functions and their graphs. (Lessons 3-2, 3-4, 3-5)

In 18–21, describe how the graph of the image is related to the graph of the preimage when the given change is made in the equation for a function or relation.

18. x is replaced by $x - 90$.
19. y is replaced by $y + 15.3$.
20. x is replaced by $3x$.

21. y is replaced by $\frac{y}{8}$.

22. Match each general transformation with one of Questions 18–21 above.

a. $T: (x, y) = (x, y + k)$ b. $T: (x, y) \to (x, by)$

c. $T: (x, y) = (ax, y)$ d. $T: (x, y) \to (x + h, y)$

23. **True or False** Under a translation, asymptotes of the preimage are mapped to asymptotes of the image.

24. Give a rule for a scale change that has the effect of reflecting the graph over the x-axis.

OBJECTIVE E Describe and identify symmetries and asymptotes of graphs. (Lessons 3-1, 3-4)

In 25–28, a function is described by an equation. Determine if the function is odd, even, or neither.

25. $k(t) = |3t + 4|$ 26. $j(m) = 10m^3$

27. $f(x) = 3|x| + 1$ 28. $s(y) = 7y^2 - 3y^4$

29. Consider the function f with $f(x) = \frac{1}{2x - 1}$. Give equations for any asymptotes of its graph.

30. If a graph has a horizontal asymptote at $y = 3$ and a vertical asymptote at $x = -2$, around what point might you center the screen to see the graph on your calculator?

OBJECTIVE F Identify properties of composites and inverses of functions. (Lessons 3-7, 3-8)

31. If f and g are real functions with $f(x) = \sqrt{x}$ and $g(x) = x - 2$, what is the domain of $f \circ g$?

32. If two relations are inverses of each other, what transformation maps the graph of one onto the graph of the other?

In 33-37, true or false.

33. If f is a function with an inverse f^{-1}, then $f^{-1}(x) = \frac{1}{f(x)}$.

34. If $f(x) = x^2$ and $g(x) = \sqrt{x} + 2$, for all x, $f \circ g(x) = g \circ f(x)$.

35. If f is a function, then $f(f^{-1}(x)) = x$ for all x in the domain of f^{-1}.

36. For all real functions f and g, $f \circ g = g \circ f$.

37. A scale change with magnitude a followed by a translation of h units is the same as a translation of h units followed by a scale change with magnitude a.

OBJECTIVE G Identify properties of z-scores. (Lesson 3-9)

In 38 and 39, a new data set is formed by taking the z-scores from raw scores with mean $\bar{x}$ and standard deviation s.

38. What is the standard deviation of the standardized data set?

39. What is the mean of the standardized data set?

In 40 and 41, a z-score is given. Explain what it tells you about the original data point in terms of the mean and standard deviation of the original data set.

40. $z = 0.04$ 41. $z = -1.3$

USES Applications of mathematics in real-world situations

OBJECTIVE H Use translations, scale changes, or z-scores to describe and analyze data and statistics. (Lessons 3-3, 3-6, 3-9)

In 42 and 43, consider the table below.

Raw Scores	Scaled Scores	Frequency
4	16	1
5	20	3
6	24	2
8	32	7
9	36	2

42. a. Find the mode, mean, and median of the raw data.

b. Identify the transformation used to scale the scores.

c. Find the mode, mean, and median of the scaled scores.

d. What theorem of translated data is shown in Parts b and c?

43. a. What is the range of the raw scores?

 b. What is the range of the scaled scores?

 c. What theorem of translated data is shown in Parts a and b?

44. Use a translation to mentally calculate the average of these bowling scores: 103, 114, 107, 101, 105.

45. For a sample of a certain butterfly species, a scientist found a mean length of 1.82 inches with a range of 2.71 inches and a standard deviation of 0.46 inches. If the data are converted to millimeters (1 inch = 25.4 mm), give the following statistics of the resulting data set.

 a. mean b. range c. variance

46. Ken scored 83 on an *FST* test on which the mean was 77 and the standard deviation was 4.1. His score on an English test was 63. On the English test, the mean was 54 and the standard deviation was 5.3. On which test did he do better compared to his classmates?

47. A pod of porpoises has a mean weight of 163 kg and a standard deviation of 29 kg. A school of yellow fin tuna has a mean weight of 83 lb and a standard deviation of 17 lb. Which animal is heavier relative to its group, a porpoise which weighs 207 kg or a tuna which weighs 107 lb?

REPRESENTATIONS Pictures, graphs, or objects that illustrate concepts

OBJECTIVE I Recognize functions and their properties from their graphs. (Lessons 3-1, 3-4, 3-8)

In 48 and 49, determine whether the function graphed seems to have an inverse which is a function. Give a reason for your answer.

48.

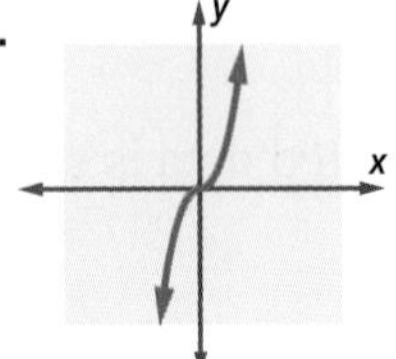

49.

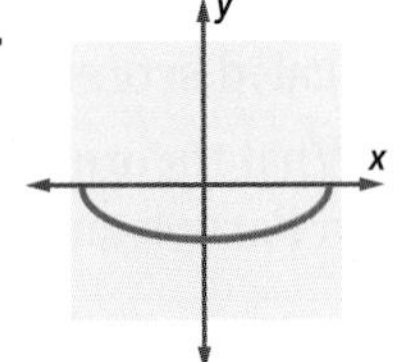

In 50–55, match each graph with the equation of its parent function.

A $y = |x|$ B $y = \sqrt{x}$ C $y = x$

D $y = \frac{1}{x}$ E $y = x^3$ F $y = x^2$

50.

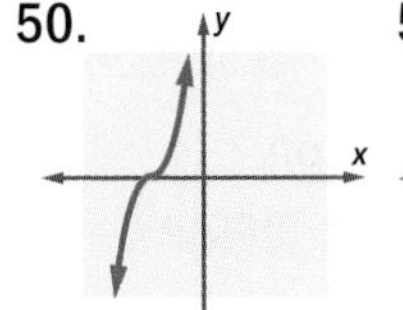

51.

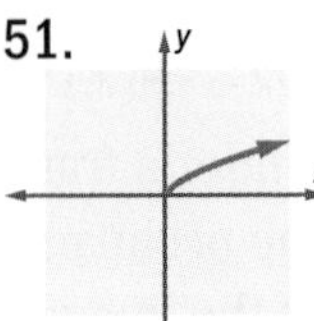

52.

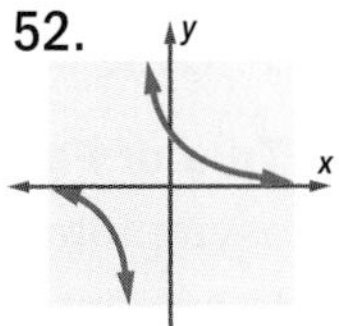

53. 54.

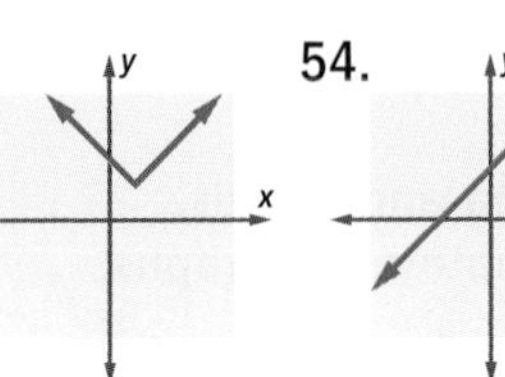

55.

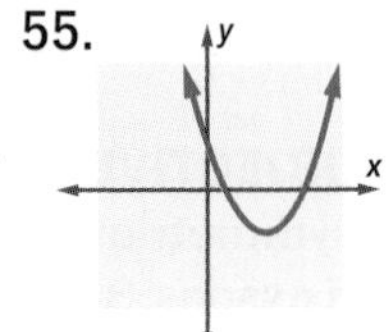

In 56 and 57, classify the function graphed as possibly odd, possibly even, or neither.

56.

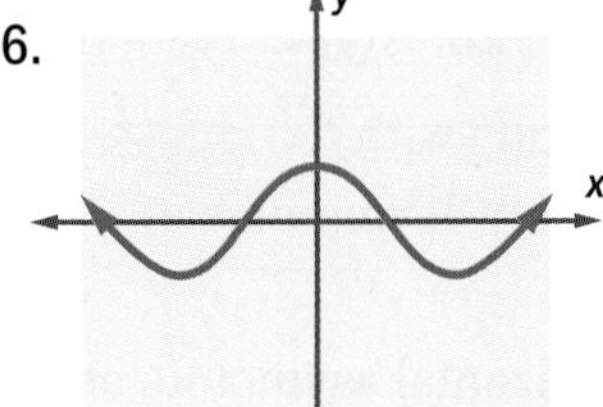

57.

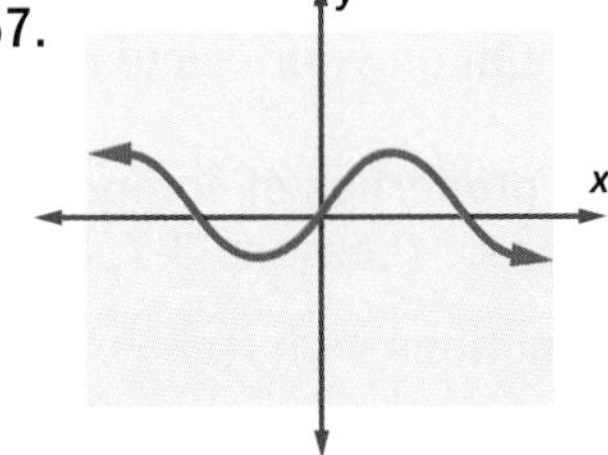

OBJECTIVE J Apply the Graph-Translation Theorem or the Graph Scale-Change Theorem to make or identify graphs. (Lessons 3-2, 3-5)

58. Sketch the graph of $y = |2x| + 10$ and its parent function on the same set of axes.

59. Let $k(x) = 12 - \frac{2}{x + 4}$.

 a. Sketch a graph of the function k.

 b. Give equations of any asymptotes.

 c. Give the coordinates of the x-intercepts.

60. The graph of $y = f(x)$ is drawn below. Draw the graph of $y = 4f(-x)$.

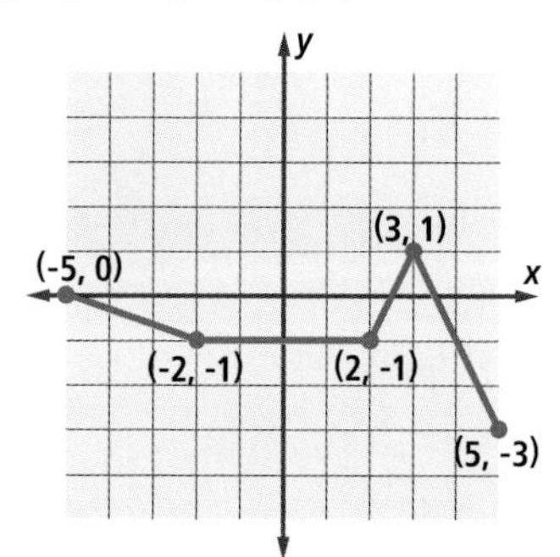

61. Give an equation for the transformation that maps the graph of f onto the graph of g.

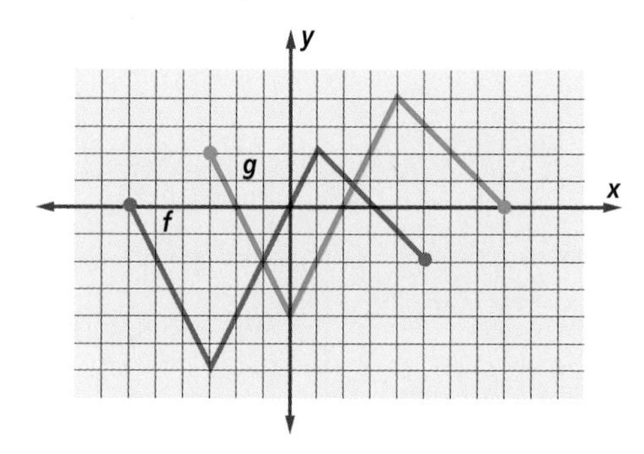

62. a. Graph $f(x) = x^2$ and $g(x) = 2x^2$ on the same set of axes.

b. True or False g is the image of f under the transformation $S: (x, y) \rightarrow (2x, y)$.

In 63–65, the graph is a translation or a scale-change image of the graph of the given parent function. Write an equation for the function that is graphed.

63. parent: $y = x^2$

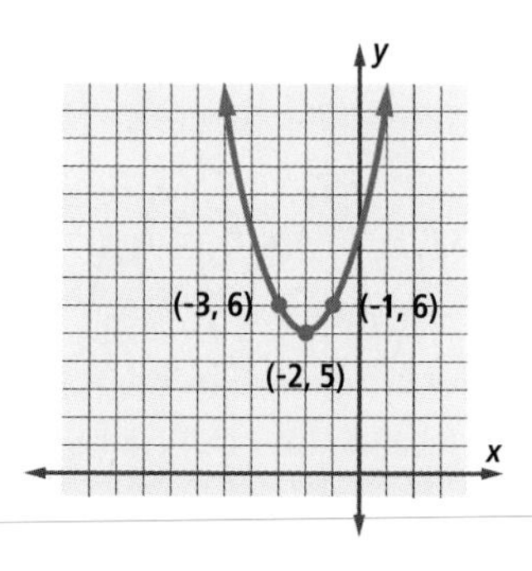

64. parent: $y = \sqrt{x}$

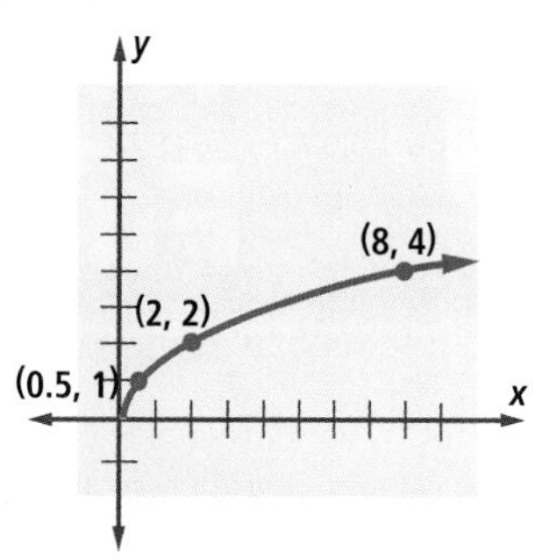

65. parent: $y = \frac{1}{x^2}$

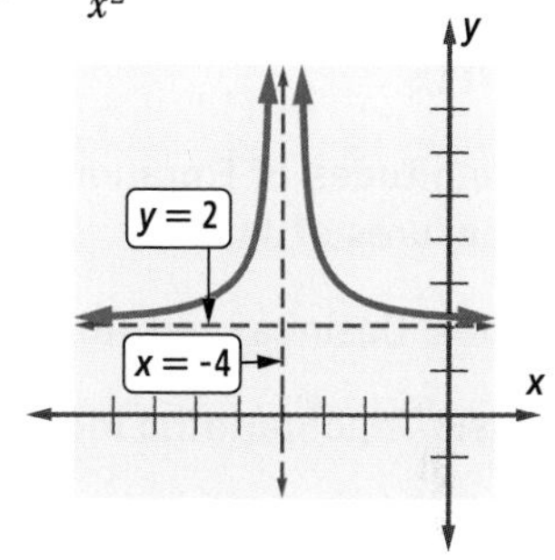

OBJECTIVE K Graph inverses of functions. (Lesson 3-8)

In 66–68, a rule for a function is given.
a. Find an equation for its inverse.
b. Graph the function and its inverse.
c. Determine if the inverse is a function.

66. $y = -2x^3$ **67.** $h(x) = \frac{2}{x}$ **68.** $j(x) = -x^2 + 4$

69. Tell whether or not f and g graphed below are inverses of each other. Explain your reasoning.

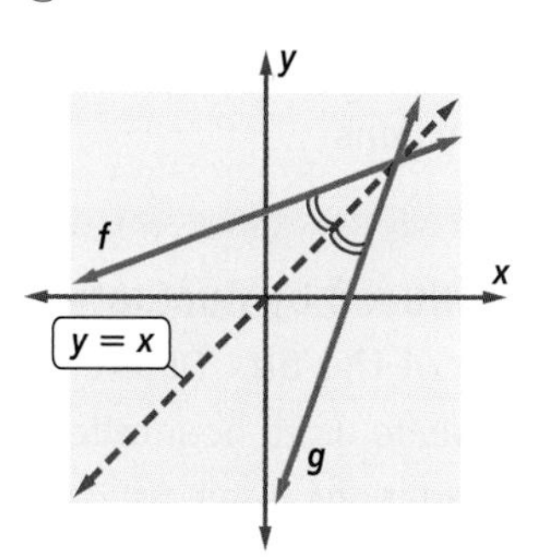

Chapter

4 Trigonometric Functions

Contents

Sound is produced by fluctuations in the pressure of the air. Different kinds of fluctuations cause us to hear different kinds of sounds. Variations in air pressure can be picked up by a microphone and can be displayed by an oscilloscope as a graph of air pressure versus time.

Sound Waves Produced by a Violin and a Flute

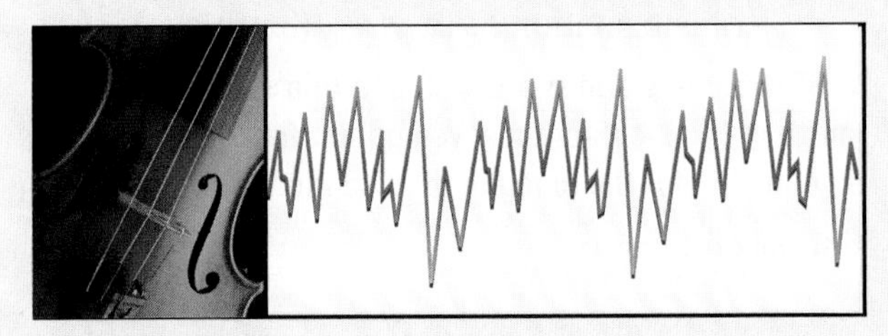

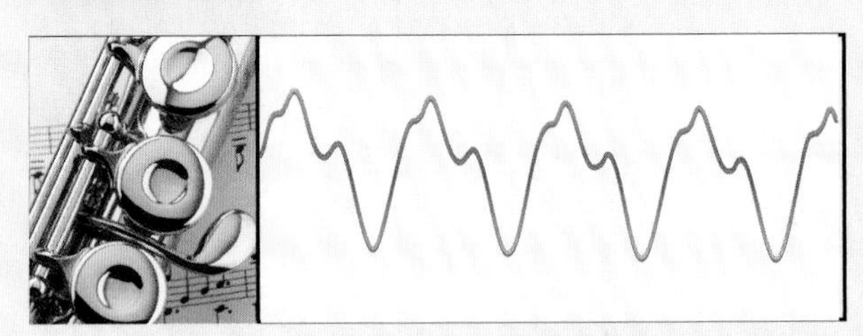

Graphs of different sounds produced by various instruments are on page 220. From such graphs you can see why sound is said to travel in waves.

A pure tone is a tone in which air pressure varies *sinusoidally* with time; that is, as a sine wave, which is the graph of the *sine function*. Pure tones seldom occur in nature, but they can be produced by certain tuning forks and electronic music synthesizers. Mathematically and physiologically, pure tones or sine waves are the foundation of all musical sound.

The sine function, and the related functions called the cosine and tangent functions, are examples of *trigonometric functions*. In this chapter, you will study some of the properties of these functions and study the effects of transformations on the graphs of the parent trigonometric functions. You will also learn how the trigonometric functions are used to model sound, electricity, and other periodic phenomena.

Lesson 4-1 Magnitudes of Rotations and Measures of Arcs

Vocabulary

rotation
center of the rotation
rotation image
magnitude of a rotation
revolution, full turn
half turn
quarter turn
radian

▶ **BIG IDEA** Magnitudes of rotations are described in revolutions, degrees, and radians.

Mental Math

A pie chart is constructed to represent the following ice cream preferences: vanilla, 50%; chocolate, 30%; strawberry, 20%. Find the angle measure of each sector of the pie chart.

A **rotation** is a transformation under which each point in the plane turns a fixed magnitude around a fixed point called the **center of the rotation**. In the figure at the right, A and B are on the same circle with center Q. Point A has been rotated counterclockwise about the center Q to the position of B, its **rotation image**.

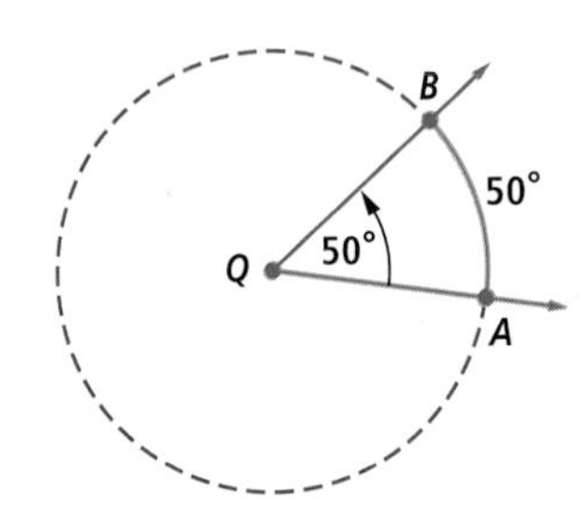

Revolutions and Degrees

There is a way to describe how much A has been rotated to get to B. Use the measure of the *central angle*, $\angle AQB$. Since $m\angle AQB = 50°$, the **magnitude** of the rotation is 50°. When rotations are measured in this way, the rotation of 360° is called one **revolution**, or a **full turn**. A rotation of 180° or $\frac{1}{2}$ revolution is called a **half turn**, and a rotation of 90° or $\frac{1}{4}$ revolution is called a **quarter turn**.

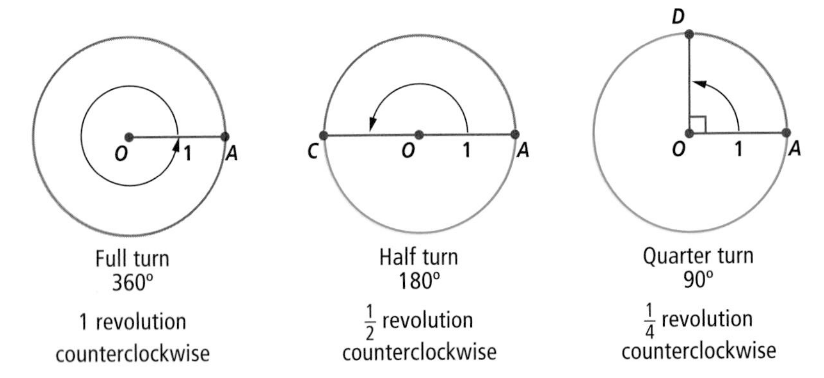

Above, you could also rotate B 50° clockwise to get to A. The clockwise direction is the negative direction in trigonometry because the four quadrants are numbered in a counterclockwise order. So the rotation that maps B onto A is said to be a -50° rotation, or to have magnitude -50°.

You can multiply both sides of the conversion equation

$$1 \text{ revolution} = 360°$$

to find how many degrees there are in any multiple of a revolution.

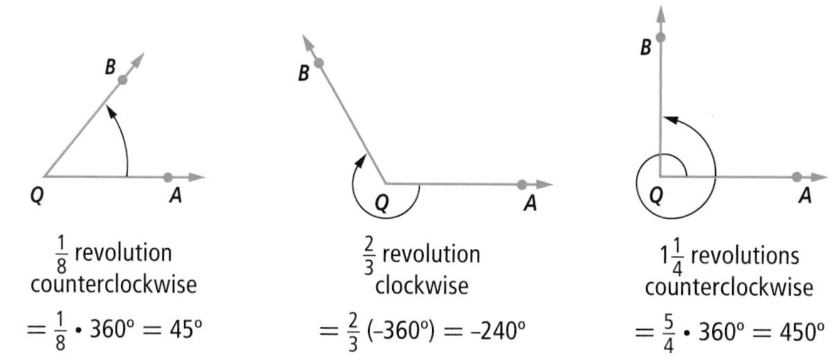

In skateboarding, snowboarding, and many other sports, you may see rotations called 360s, 540s, 720s, and so on. In gymnastics and figure skating, rotations are measured in revolutions or turns. A 720 means 2 revolutions. In these *physical rotations*, adding 360° or 1 revolution to a turn creates a different movement. But in *mathematical rotations*, all that matters is where you begin and where you end. Physically turning 360° means you turned all the way around, but a mathematical rotation of 360° is the same as a mathematical rotation of -360°, and both are the same as if you did nothing at all!

Let's roll Since 2004, June 21 has been designated "Go Skateboarding Day."

For this reason, the same rotation can have many different magnitudes. For instance, a rotation of $\frac{1}{8}$ revolution counterclockwise also has magnitude $1\frac{1}{8}$ revolutions counterclockwise or $\frac{7}{8}$ revolution clockwise. In degrees, that rotation has magnitude 45°, 405°, or -315°, respectively. Adding or subtracting 1 revolution (or 360°) to the magnitude of a rotation does not change the rotation.

Radian Measure

In a rotation, points that are farther from the center "move" or "turn" a greater distance along the arc of a circle than points that are closer to the center. For example, at the right, S moves farther to get to E than G moves to get to C. Measuring a rotation in degrees or revolutions does not tell anything about how far a point has moved. To solve this problem, a unit is needed that is related to the length of the arc. That unit is the *radian*. Radians have been in use for about 125 years and are important in the study of calculus and other advanced mathematics.

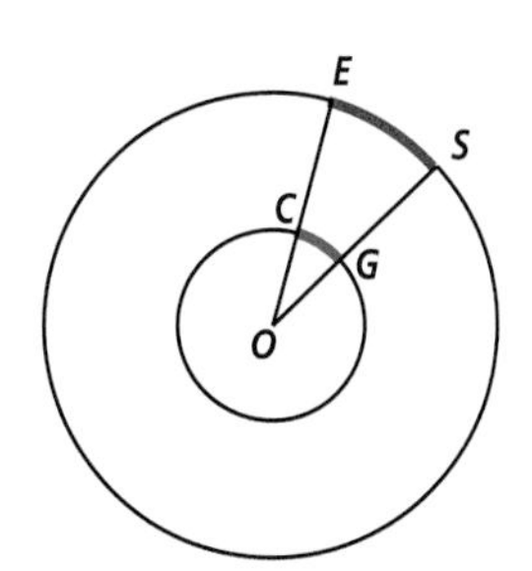

$\overset{\frown}{SE}$ is longer than $\overset{\frown}{GC}$.

Radian measure is based on arc lengths in the unit circle. As you know, if a circle has radius r, its circumference is $2\pi r$. We distinguish between 360° and $2\pi r$ because 360° is *arc measure* and $2\pi r$ is *arc length*.

 QY1

QY1

A point moves halfway around a circle of radius 5.

a. What is the length of the arc traversed by the point?

b. How many degrees is the rotation?

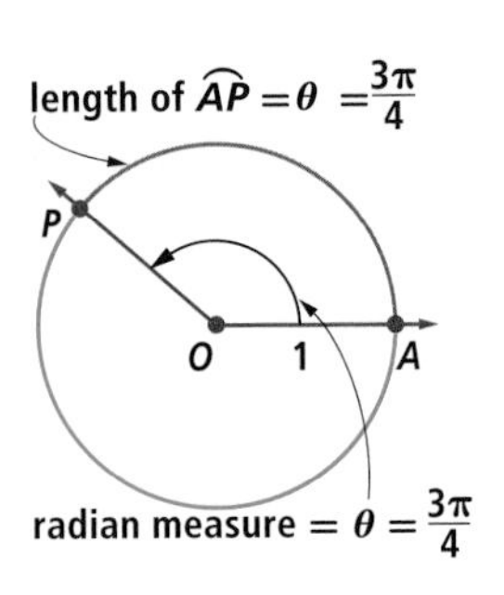

A *unit circle* (radius = 1) has circumference $C = 2\pi \cdot 1 = 2\pi$, which is approximately 6.28. Consider point A on a unit circle O, as pictured at the right. The magnitude in **radians** of the rotation with center O that maps A onto P is defined as the numerical length of $\widehat{AP}$. In this drawing, $\widehat{AP}$ is $\frac{3}{8}$ of a circle, so the length of $\widehat{AP}$ is $\frac{3}{8}$ of the circumference, or $\frac{3}{8} \cdot 2\pi = \frac{3\pi}{4} \approx 2.356$. Notice that $m\angle AOP = 135°$. So 135° corresponds to a rotation of $\frac{3\pi}{4}$ radians.

The radian measure of a full turn is 2π because if point A was physically turned through one complete revolution, it would travel over an arc of length 2π. Similarly, the radian measure of a half-turn is π and of a quarter turn is $\frac{\pi}{2}$.

As when rotations are measured in degrees, a clockwise rotation gives rise to a negative radian magnitude. Also, just as adding or subtracting multiples of 360° to the magnitude of a rotation gives rise to the same rotation, so does adding or subtracting multiples of 2π radians. The three circles below show that a rotation of magnitude $\frac{7\pi}{6}$ radians is the same as a rotation of $\frac{7\pi}{6} - 2\pi = -\frac{5\pi}{6}$ radians, and is the same as a rotation of $2\pi + \frac{7\pi}{6} = \frac{19\pi}{6}$ radians.

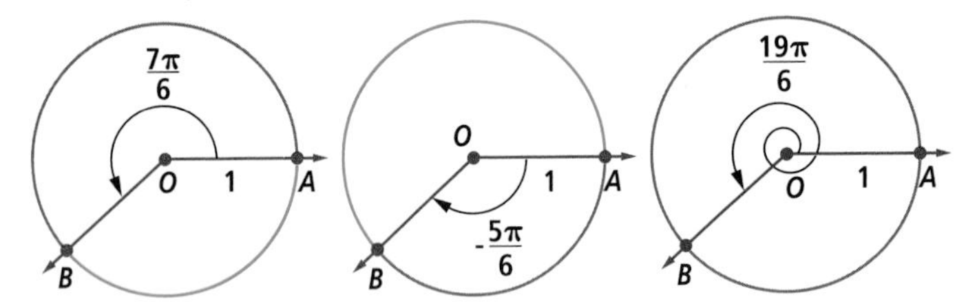

Converting between Radians and Degrees

Some degree measures are easy to convert to radians. For example, 30° is $\frac{1}{12}$ of 360°. Use the conversion formula

$$360° = 1 \text{ revolution} = 2\pi \text{ radians.}$$

Divide by 12 to obtain

$$30° = \frac{1}{12} \text{ revolution} = \frac{\pi}{6} \text{ radians.}$$

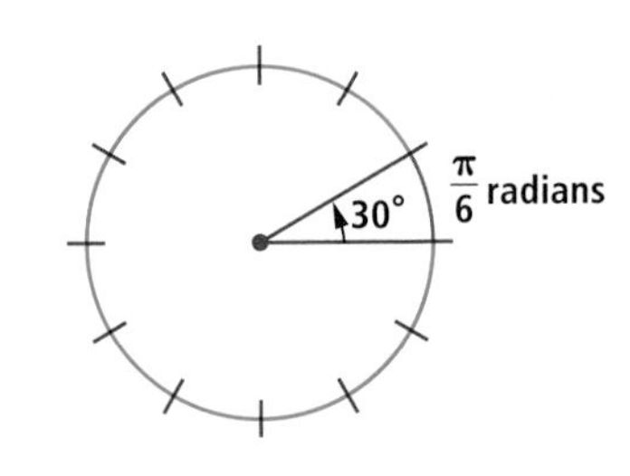

Activity

The table below lists some of the equivalent measures that result from the basic relationship 2π radians = 360°. Copy and fill in the rest of the table.

Degrees	0°	30°	?	?	90°	120°	?	150°	180°	360°
Radians	0	$\frac{\pi}{6}$	$\frac{\pi}{4}$	$\frac{\pi}{3}$	$\frac{\pi}{2}$	?	$\frac{3\pi}{4}$	?	π	?
Revolutions	?	?	$\frac{1}{8}$	?	?	?	$\frac{3}{8}$	?	$\frac{1}{2}$	1

The values in the Activity are common radian values. You should know them without having to do any paper-and-pencil or calculator work. To convert any rotation measure from one unit to another, use the conversion formula below, since 2π radians = 360°.

$$\pi \text{ radians} = 180°$$

Dividing both sides of the conversion formula by the quantity on one side gives rise to two conversion factors, each equal to 1.

$$1 = \frac{180°}{\pi \text{ radians}} \quad \text{or} \quad 1 = \frac{\pi \text{ radians}}{180°}$$

Example 1

a. Convert 1000° to radians exactly.

b. Convert 1000° to radians approximately.

Solution

a. Multiply by the appropriate conversion factor.

$1000° = 1000° \cdot \frac{\pi \text{ radians}}{180°} = \frac{1000}{180}\pi = \frac{50}{9}\pi$ radians

b. Use a calculator to get a decimal approximation for $\frac{50\pi}{9}$.

$\frac{50\pi}{9} \approx 17.453$, so 1000° is about 17.5 radians.

Caution: Computer algebra systems and calculators work in both radians and degrees. Be sure to set the mode to the unit you want.

How large in degrees is an angle or rotation of magnitude 1 radian? Think: If point B on a unit circle is rotated 1 unit around the center to Q, as shown at the right, what is $m\angle BOQ$? The circumference of a unit circle is $2\pi \approx 6.28$. So there are about 6.28 radians in one revolution, and one radian is slightly less than $\frac{1}{6}$ revolution. But $\frac{1}{6}$ revolution is equivalent to 60°. So one radian should be slightly less than 60°.

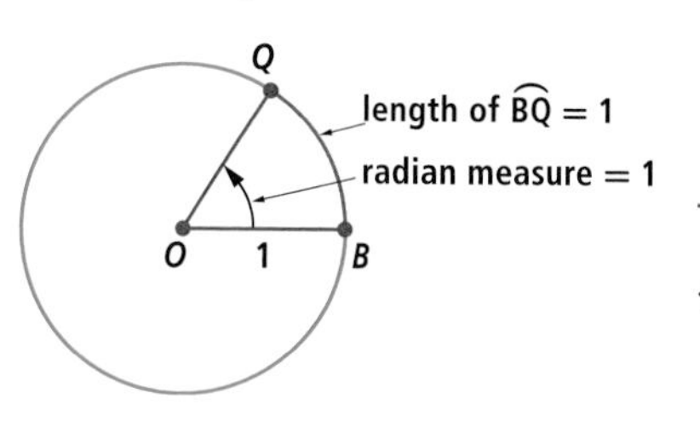

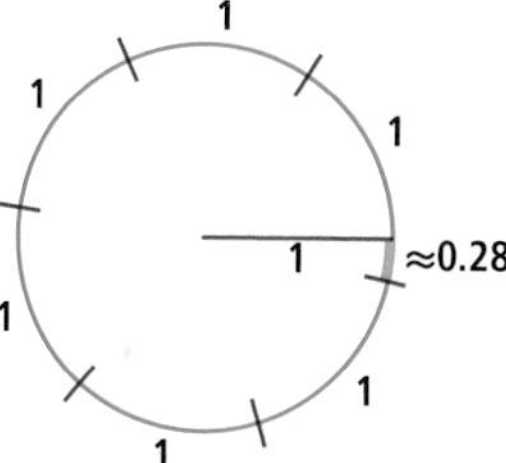

Example 2

Convert 1 radian to degrees.

Solution Use the conversion factor $\frac{180°}{\pi \text{ radians}}$.

So 1 radian = $1 \cancel{\text{radian}} \cdot \frac{180°}{\pi \cancel{\text{radians}}} = \frac{180°}{\pi} \approx 57.3°$.

Why Are Radians Used?

You may have studied angles and rotations for years and never used radians. You may be wondering why radians are used and if they are ever needed. One advantage of radians over degrees is that certain formulas are simpler when written with radians.

GUIDED

Example 3

Find the length of an arc of a 50° central angle in a circle of radius 6 feet.

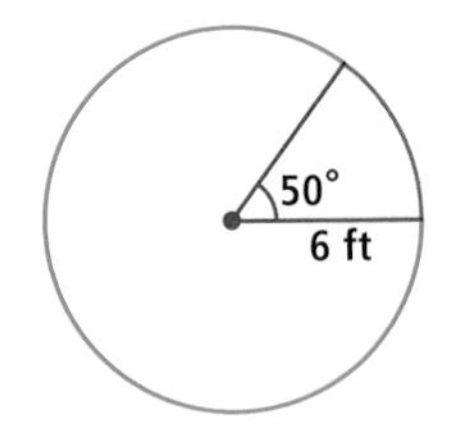

Solution The 50° arc is $\frac{50}{360}$ of the circumference of the circle. The circumference has length $2\pi r$, or __?__ ft. So, the length of the arc is $\frac{50}{360} \cdot$ __?__, which simplifies to __?__ ft exactly, or __?__ ft, to the nearest hundredth.

Example 3 is easily generalized. Notice how much simpler the formula is if the central angle is measured in radians.

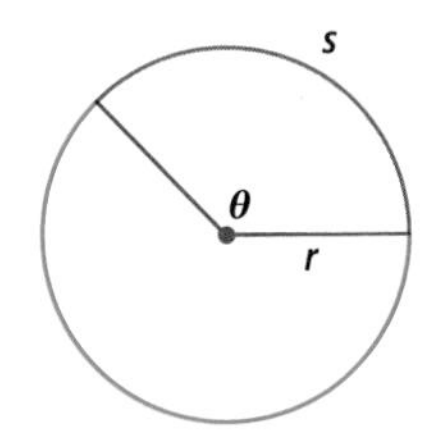

Circle Arc Length Formula

If s is the length of the arc of a central angle of θ radians in a circle of radius r, then $s = r\theta$.

Proof The central angle is $\frac{\theta}{2\pi}$ of a revolution. So the length s of the arc is $\frac{\theta}{2\pi}$ of the circumference. The circumference of the circle is $2\pi r$.
Thus, $s = \frac{\theta}{2\pi} \cdot 2\pi r = r\theta$.

QY2

QY2
Find the length of a $\frac{5}{18}\pi$ radian arc in a circle of radius 6 feet.

Example 4

A swing hangs from chains that are 8 ft long. How far does the seat of the swing travel if it moves through an angle of 1.25 radians?

Solution 1 Since the angle is 1.25 radians, the length of the intercepted arc on the unit circle is 1.25. The arc length on the 8-foot circle is 8 times the length of the arc on the unit circle. The distance traveled is $8 \cdot 1.25 = 10$ feet.

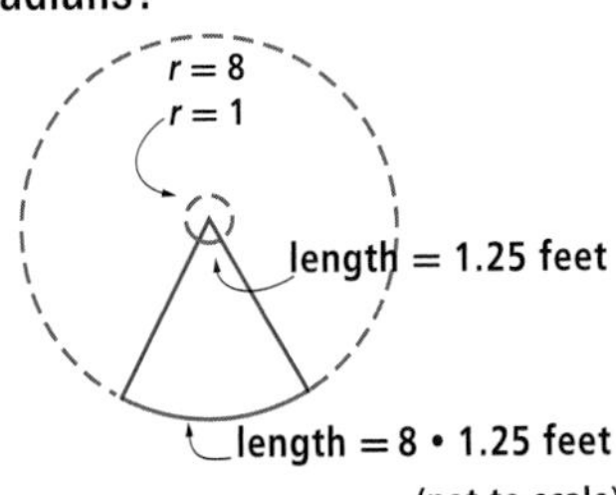

Solution 2 Use the Circle Arc Length Formula.
$s = r\theta$. The swing travels $8 \cdot 1.25$, or 10 feet.

Radians are so commonly used in mathematics that when no unit is given in a problem that could be in degrees or radians, it is understood that the measure is in radians.

Questions

COVERING THE IDEAS

1. Convert $\frac{9}{10}$ revolution to degrees.
2. Convert -805° to revolutions, rounding to the nearest tenth.

3. At the right is a graph of a circle with radius 1.
   a. Give the length of $\overset{\frown}{AC}$.
   b. What is the smallest positive magnitude in radians of the rotation with center O that maps A to C?

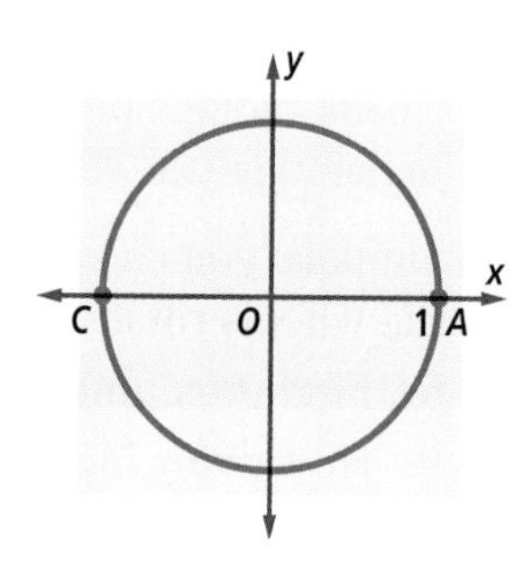

In 4 and 5, draw a circle with radius 1.
a. On this circle, heavily shade an arc with the given length.
b. Give the degree measure of the central angle of this arc.

4. $\frac{2\pi}{3}$ 5. 1

In 6 and 7, draw a unit circle and an arc with the given radian measure.

6. $\frac{3\pi}{2}$ 7. 2

8. Convert the measure to degrees. Round to the nearest thousandth.
   a. -0.2 radians b. -0.2π radians
9. If a skateboarder does a "540," what is the magnitude of the rotation
   a. in revolutions? b. in radians?

In 10 and 11, convert to radians exactly without using a calculator.

10. 225° 11. -80°

12. a. Draw an angle representing a rotation with measure $\frac{7\pi}{12}$ radians.
    b. Give two other radian measures of the same rotation.
13. Use the circle at the right. If $m\angle ABC = \frac{5\pi}{6}$ radians and the radius of the circle is 33, compute the length of $\overset{\frown}{AC}$.

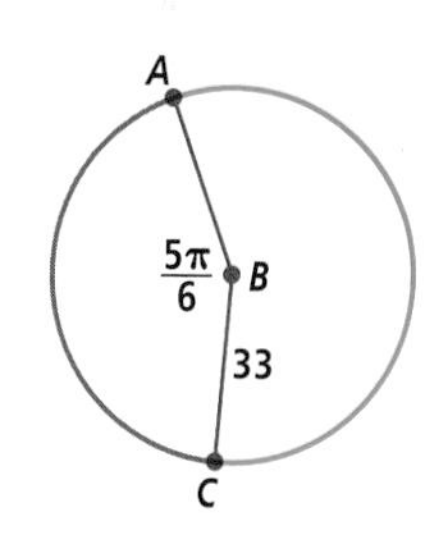

In 14 and 15, use a circle with diameter 8 cm.

14. Find the length of the arc intercepted by an angle of $\frac{3\pi}{4}$.
15. Find the length of the arc intercepted by an angle of 63°.
16. Suppose the blades of a wind turbine are 16' long. What is the distance traveled by a point on its tip as the blade rotates $\frac{3}{5}$ of a revolution?

APPLYING THE MATHEMATICS

17. On the clock tower on the Houses of Parliament in London, England, the minute hand is about 14 feet long. How many feet does the end of the minute hand move in 5 minutes?
18. An angle whose measure is $\frac{\pi}{2}$ is about __?__ times as large as an angle whose measure is $\frac{\pi}{2}^{\circ}$.
19. Musicians use a metronome to produce a steady beat as they practice. Many mechanical metronomes have a swinging arm with a weight to control the tempo. Suppose that a metronome arm is 4 inches long, moves through an angle of $\frac{\pi}{3}$, and beats at a rate of 160 beats per minute. How far does the tip of the arm travel in
    a. 1 beat? b. 1 hour?

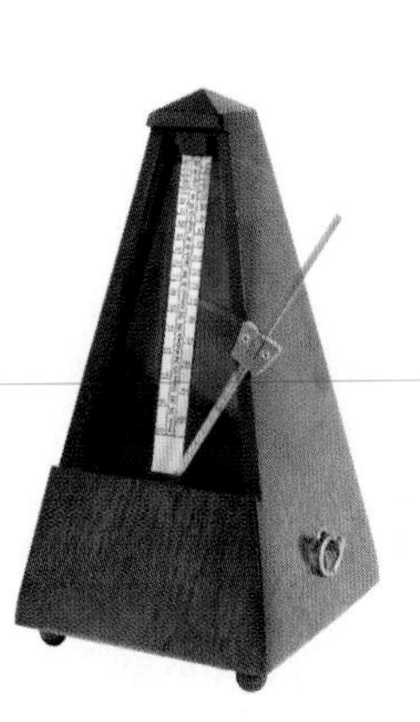

20. The planet Jupiter rotates on its axis at a rate of approximately 0.6334 radians per hour. What is the approximate length of the Jovian day (the time it takes Jupiter to make a complete revolution)?

21. Suppose you can ride a bike with 22" wheels (in diameter) so that the wheels rotate 150 revolutions per minute.
    a. Find the number of inches traveled during each revolution.
    b. How many inches are traveled each minute?
    c. Use your answer from Part b to find the speed, in miles per hour, that you are traveling.

22. Recall that when greater precision is desired, a degree is split into 60 *minutes* (abbreviated '). The diagram at the right shows a cross section of Earth. G represents Grand Rapids, MI and M represents Montgomery, AL. Assume that Grand Rapids is directly north of Montgomery. If the radius of Earth is about 3960 miles, estimate the air distance from Grand Rapids to Montgomery.

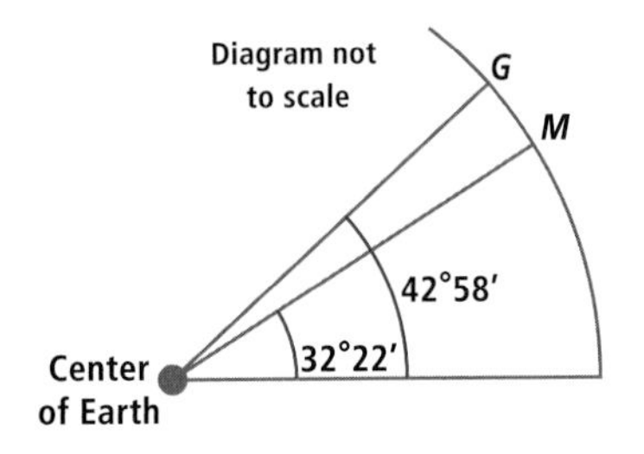

REVIEW

23. Use the Graph-Translation Theorem to find the equation of the image of $y = x^2$ under T: $(x, y) \rightarrow (x + 3, y - 2)$. **(Lesson 3-2)**

24. You discovered a new element, Yournameium, which has a half-life of 15 hours. Suppose the initial amount is A_0. How much will remain after each number of hours? **(Lesson 2-4)**
    a. 45 b. 8 c. 42 d. t

25. The *New York Times* held a contest pitting professionals' stock choices with stocks chosen by throwing darts at a dartboard every six months. The values in the table and box plots below represent the points gained or lost by the stocks. **(Lessons 1-7, 1-6, 1-4, 1-2)**

Period #	Pros	Darts
1	51.2	11.7
2	25.2	1.1
3	-3.3	-3.1
4	7.7	-1.4
5	-21.0	7.7
6	-13.0	15.4
7	-2.5	3.6
8	-19.6	5.7
9	6.3	-5.7
10	-5.1	6.9
11	14.1	1.8

Change in Stock Value

Pros

Darts

-30 -20 -10 0 10 20 30 40 50 60

    a. Find the mean, median, and standard deviation for each data set.
    b. Compare and contrast the distributions using the box plots.

26. Given the point P on the circle with center (0, 0) at the right, fill in the coordinates of the remaining reflection images in the various quadrants. **(Previous Course)**

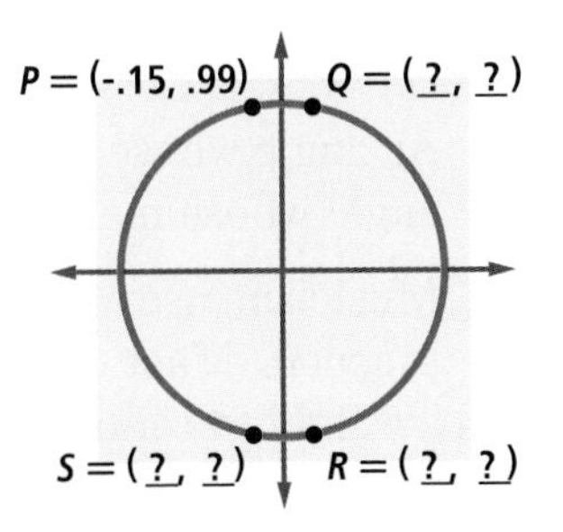

EXPLORATION

27. Derive a formula for the area of a sector in terms of the radius r of the circle and the length x of its boundary arc x in radians.

QY ANSWERS

1. a. 5π b. 180°
2. $\frac{5\pi}{3} \approx 5.24$ ft

Lesson

4-2 Sines, Cosines, and Tangents

Vocabulary

unit circle
cosine, cos
sine, sin
circular function
trigonometric function
tangent, tan

BIG IDEA The sine and cosine functions relate magnitudes of rotations to coordinates of points on the unit circle.

Mental Math

The following regular polygons are inscribed in a circle. Find the measure of the angle formed by two rays from the center of the circle that contain adjacent vertices of the polygon.

a. triangle **b.** square

c. pentagon **d.** octagon

The Sine, Cosine, and Tangent Functions

The **unit circle** is the circle with center at the origin and radius 1, as shown at the right.

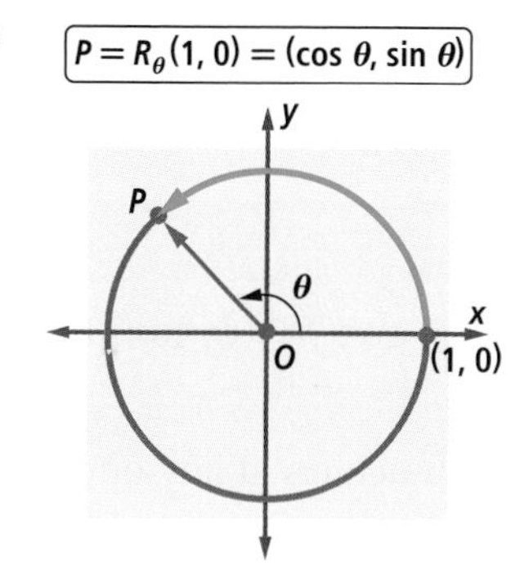

Consider the rotation of magnitude θ with center at the origin. We call this R_θ. Regardless of the value of θ, the image of (1, 0) under R_θ is on the unit circle. Call this point P. We associate two numbers with each value of θ. The **cosine** of θ (abbreviated $\cos\theta$) is the x-coordinate of P; the **sine** of θ (abbreviated $\sin\theta$) is the y-coordinate of P.

Definition of cos θ and sin θ

For all real numbers θ, **($\cos\theta$, $\sin\theta$)** is the image of the point (1, 0) under a rotation of magnitude θ about the origin. That is, $(\cos\theta, \sin\theta) = R_\theta(1, 0)$.

Because their definitions are based on a circle, the sine and cosine functions are sometimes called **circular functions** of θ. They are also called **trigonometric functions**, from the Greek word meaning "triangle measure," as you will see in the applications for triangles presented in Chapter 5. To find values of trigonometric functions when θ is a multiple of $\frac{\pi}{2}$ or 90°, you can use the above definition and mentally rotate (1, 0).

Example 1

Evaluate $\cos\pi$ and $\sin\pi$.

Solution Because no degree sign is given, π is measured in radians. Think of $R_\pi(1, 0)$, the image of (1, 0) under a rotation of π. $R_\pi(1, 0) = (-1, 0)$. So, by definition, $(\cos\pi, \sin\pi) = (-1, 0)$. Thus, $\cos\pi = -1$ and $\sin\pi = 0$.

Check Use a calculator. Make sure it is in radian mode.

cos(π)	-1
sin(π)	0

Because π radians = 180°, Example 1 shows that cos 180° = -1 and sin 180° = 0. Cosines and sines of other multiples of $\frac{\pi}{2}$ or 90° are shown on the unit circle below.

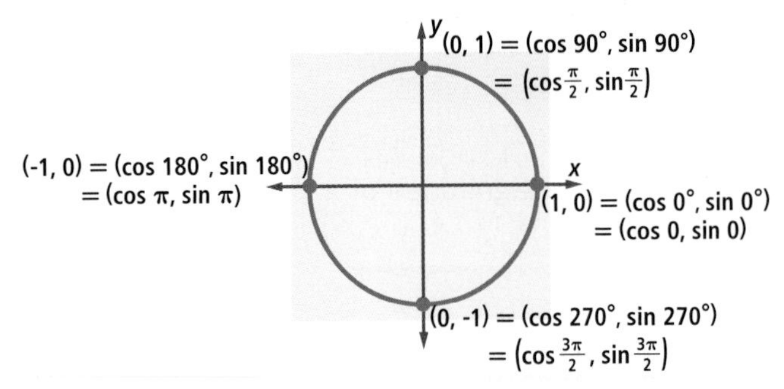

The third most common circular function is defined in terms of the sine and cosine functions. The **tangent** of θ (abbreviated tan θ) equals the ratio of sin θ to cos θ.

Definition of Tangent

For all real numbers θ, provided $\cos \theta \neq 0$, $\mathbf{tan}\ \boldsymbol{\theta} = \frac{\sin \theta}{\cos \theta}$.

When cos θ *does* equal zero, which occurs at any odd multiple of 90°, tan θ is undefined.

GUIDED

Example 2

a. Evaluate tan π.

b. Evaluate tan (–270°).

Solution

a. From Example 1, cos π = _?_ and sin π = _?_. So tan π = _?_.

b. cos (–270°) = _?_, sin (–270°) = _?_, so tan (–270°) is _?_.

For any value of θ, you can approximate sin θ, cos θ, and tan θ to the nearest tenth with a good drawing.

Activity

MATERIALS compass, protractor, graph paper, and calculator

Step 1 Work with a partner. Draw a set of coordinate axes on graph paper. Let each square on your grid have side length 0.1 unit. With the origin as center, draw a circle of radius 1. Label the figure as at the right.

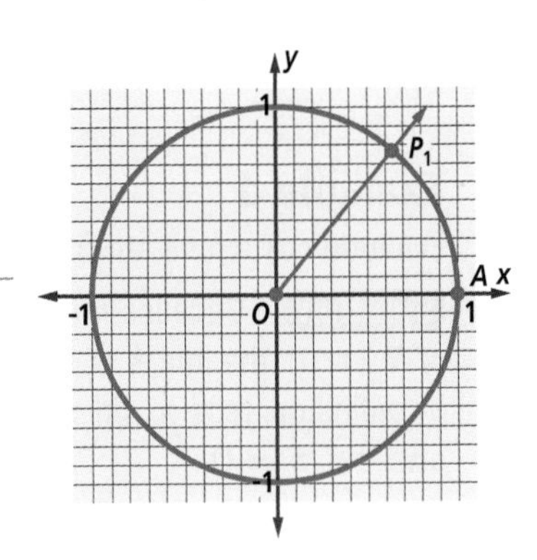

Step 2 a. Use a protractor to mark the image of $A = (1, 0)$ under a rotation of 50°. Label this point P_1, as shown at the right.

b. Use the grid to estimate the x- and y-coordinates of P_1.

c. Estimate the slope of $\overrightarrow{OP_1}$.

d. Use your calculator to find cos 50°, sin 50°, and tan 50°. Make sure your calculator is set to degree mode.

Step 3 a. Use a protractor to mark the image of A under $R_{155°}$. Label this point P_2.

b. Use the grid to estimate the x- and y-coordinates of P_2. Estimate the slope of $\overrightarrow{OP_2}$.

c. With your calculator, find cos 155°, sin 155°, and tan 155°.

Step 4 Repeat Step 3 if the rotation has magnitude –100°. Call the image P_3.

Step 5 How is tan θ related to the slope of the line through the origin and $R_\theta(A)$?

You can find better approximations to other values of sin θ, cos θ, or tan θ using a calculator.

Example 3

Suppose the tips of the arms of a starfish determine the vertices of a regular pentagon. The point $A = (1, 0)$ is at the tip of one arm, and so is one vertex of a regular pentagon *ABCDE* inscribed in the unit circle, as shown at the right. Find the coordinates of B to the nearest thousandth.

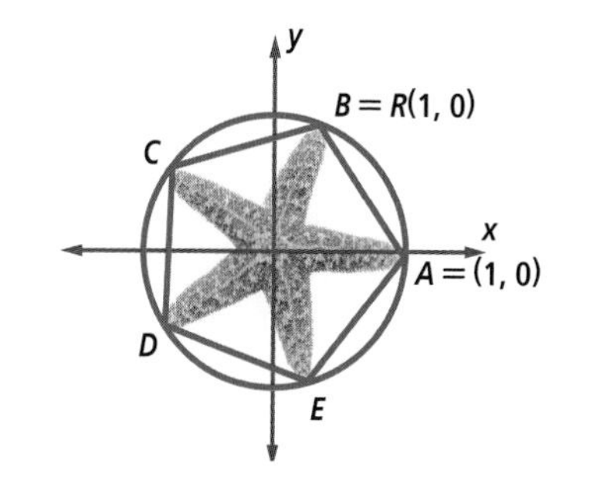

Solution Since the full circle measures 2π around, the measure of arc AB is $\frac{2\pi}{5}$. So $B = R_{\frac{2\pi}{5}}(1, 0)$. Consequently, $B = \left(\cos \frac{2\pi}{5}, \sin \frac{2\pi}{5}\right)$. A calculator shows $B \approx (0.309, 0.951)$.

For a given value of θ, you can determine whether sin θ, cos θ, and tan θ are positive or negative without using a calculator by using coordinate geometry and the unit circle. The cosine is positive when $R_\theta(1, 0)$ is in the first or fourth quadrant. The sine is positive when the image is in the first or second quadrant. The tangent is positive when the sine and cosine have the same sign and negative when they have opposite signs. The following table summarizes this information for values of θ between 0 and 360° or 2π.

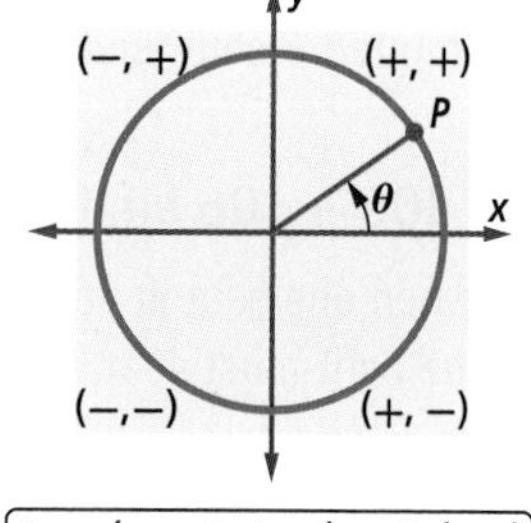

$P = (\cos \theta, \sin \theta) = R_\theta(1, 0)$

θ (radians)	θ (degrees)	quadrant of $R_\theta(1, 0)$	cos θ	sin θ	tan θ
$0 < \theta < \frac{\pi}{2}$	$0° < \theta < 90°$	first	+	+	+
$\frac{\pi}{2} < \theta < \pi$	$90° < \theta < 180°$	second	–	+	–
$\pi < \theta < \frac{3\pi}{2}$	$180° < \theta < 270°$	third	–	–	+
$\frac{3\pi}{2} < \theta < 2\pi$	$270° < \theta < 360°$	fourth	+	–	–

The applications of sines, cosines, and tangents are many and diverse, including the location of points in the plane and the calculation of certain distances.

Example 4

As of 2008, the largest Ferris wheel in North America is the Texas Star at Fair Park in Dallas, Texas. Its seats hang from 44 spokes. This Ferris wheel is 212 feet tall. How high is the seat off the ground as you travel around the wheel?

Solution We need to make some assumptions. Assume that you get on the Ferris wheel when the seat is at the wheel's lowest point and that this is at ground level. Also assume the seat is the same distance directly below the end of the spoke the entire way around.

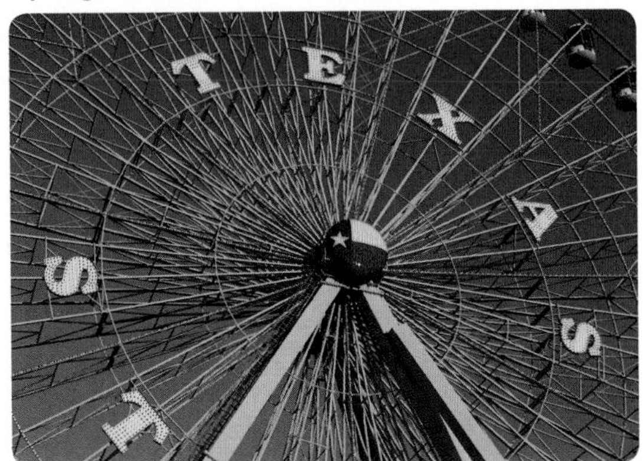

The key to answering the question is to realize that the height of the seat is determined by the magnitude of rotation of the spoke from the horizontal. To see this, imagine the Ferris wheel on a coordinate system whose origin is the center of the wheel. Think of the circle centered at the origin with radius 106 feet. By the definition of the sine, when the spoke has turned θ counterclockwise from the horizontal, the height of the end of the spoke *above the center of the wheel* is given by $106 \sin \theta$.

Add the radius 106 to get the height of the seat *above the ground*. Thus, in general, a seat that has been rotated θ counterclockwise from the horizontal is at a height

$$106 + 106 \sin \theta$$

feet above the ground. Thus, when one seat is at the bottom, going counterclockwise from the right-most seat, the 44 seats on the Ferris wheel are at heights

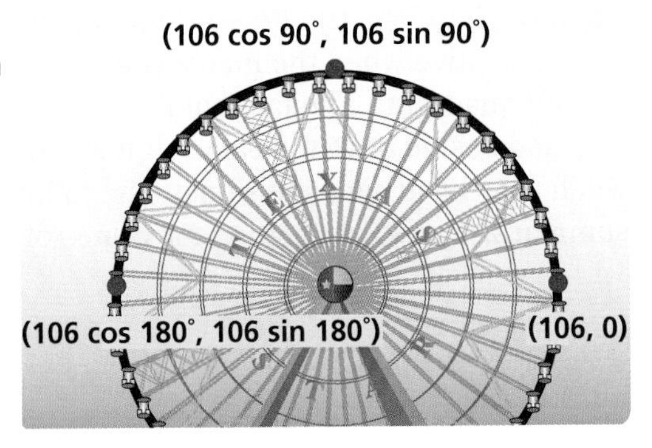

$$106 + 106 \sin 0 = 106 \text{ feet}$$

$$106 + 106 \sin\left(\frac{2\pi}{44}\right) \approx 121 \text{ feet}$$

$$106 + 106 \sin\left(2 \cdot \frac{2\pi}{44}\right) \approx 136 \text{ feet}$$

$$106 + 106 \sin\left(3 \cdot \frac{2\pi}{44}\right) \approx 150 \text{ feet}$$

$$106 + 106 \sin\left(4 \cdot \frac{2\pi}{44}\right) \approx 163 \text{ feet}$$

and so on.

Questions

COVERING THE IDEAS

1. Suppose the point $A = (1, 0)$ is rotated a magnitude θ around the point $O = (0, 0)$.
   a. $\cos \theta$ is the __?__ of $R_\theta(A)$.
   b. $\sin \theta$ is the __?__ of $R_\theta(A)$.

In 2–4, use the figure at the right. Which point is $R_\theta(1, 0)$ for the given value of θ?

y
B = (0, 1)
x
C = (-1, 0)
A = (1, 0)
D = (0, -1)

2. 3π
3. -50π
4. $-450°$
5. How is $\tan \theta$ related to $\cos \theta$ and $\sin \theta$?

In 6–8, give exact values without a calculator.

6. a. $\sin(-270°)$ b. $\cos(-270°)$ c. $\tan(-270°)$
7. a. $\sin 3\pi$ b. $\cos 3\pi$ c. $\tan 3\pi$
8. a. $\sin 0$ b. $\cos 0$ c. $\tan 0$
9. a. Give two values of θ in degrees for which $\tan \theta$ is undefined.
   b. Give two values of θ in radians for which $\tan \theta$ is undefined.

In 10 and 11, find the coordinates of the indicated image to the nearest thousandth.

10. $R_{67°}$
11. $R_{1 \text{ (radian)}}$
12. a. Use a calculator to approximate tan 200° to three decimal places.
    b. Use a picture to explain how you could have found the sign of tan 200° without using a calculator.

In 13 and 14, let $P = R_\theta(1, 0)$.

13. If P is in the fourth quadrant, state the sign of the following.
    a. $\cos \theta$ b. $\sin \theta$ c. $\tan \theta$
14. If $\cos \theta < 0$ and $\sin \theta < 0$, in what quadrant is P?

In 15–17, refer to Example 4.

15. How high is the seat above the ground when it is at the top of the Ferris wheel?
16. How high is the seat above the ground when it has been rotated $\frac{\pi}{3}$ from the horizontal?
17. Suppose the seat next to you is at ground level. How high are you off the ground?

APPLYING THE MATHEMATICS

18. a. In the pentagon of Example 3, find the coordinates of C, D, and E to the nearest thousandth.
    b. Why do you only need to use a calculator for one of the points?
19. Find three values of θ for which $\cos \theta = -1$.

20. For what values of θ between 0 and 2π is $\sin \theta$ positive?

21. As θ increases from 0 to 90°, tell whether $\cos \theta$ increases or decreases.

22. The name "tangent function" is derived from the use of the word "tangent" in geometry. Here is how. At the right, line ℓ is tangent to the unit circle at $A = (1, 0)$, P is the image of a rotation of A with magnitude θ and center O, and $\overrightarrow{OP}$ intersects ℓ at Q.

   a. When $0 < \theta < \frac{\pi}{2}$, prove that $QA = \tan \theta$.

   b. Draw a diagram similar to the one at the right for the case of $\frac{\pi}{2} < \theta < \pi$. Explain how to find $\tan \theta$ from your diagram.

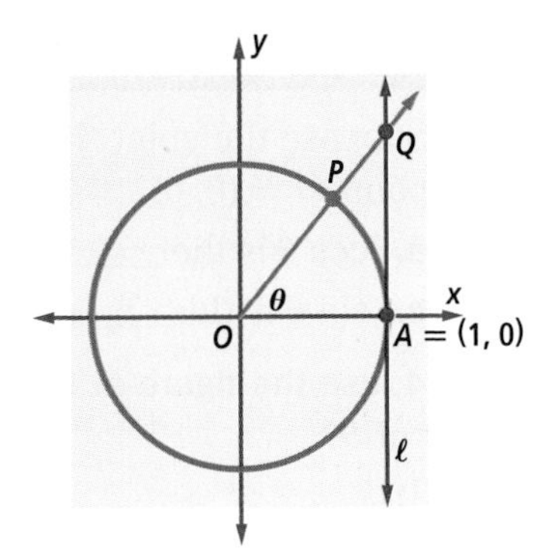

REVIEW

23. Convert $\frac{5}{6}$ revolution clockwise to degrees. **(Lesson 4-1)**

24. Let A' be the image of $A = (1, 0)$ under the rotation of $-\frac{2\pi}{3}$ with center (0, 0). Give two other magnitudes of the rotation with center (0, 0) such that the image of A is A'. **(Lesson 4-1)**

25. In isosceles $\triangle ABC$ at the right, $AB = 1$. What is the length of $\overline{BC}$? **(Previous Course)**

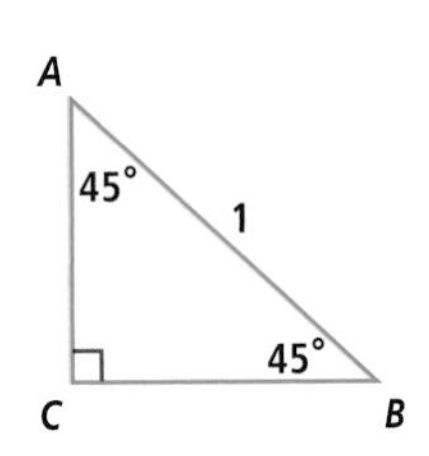

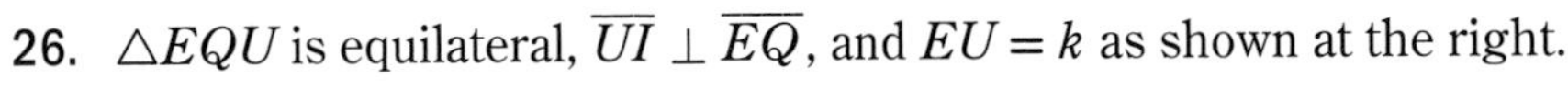

26. $\triangle EQU$ is equilateral, $\overline{UI} \perp \overline{EQ}$, and $EU = k$ as shown at the right.

   a. Find EI in terms of k.

   b. Find UI in terms of k. **(Previous Course)**

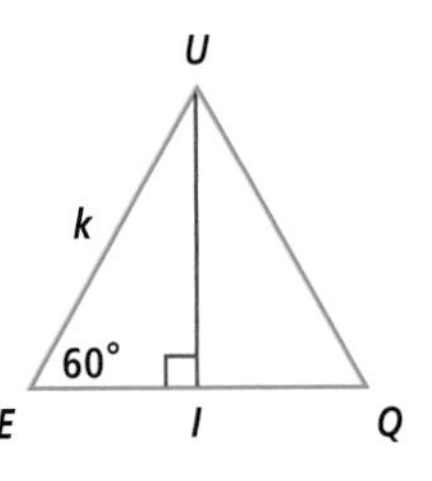

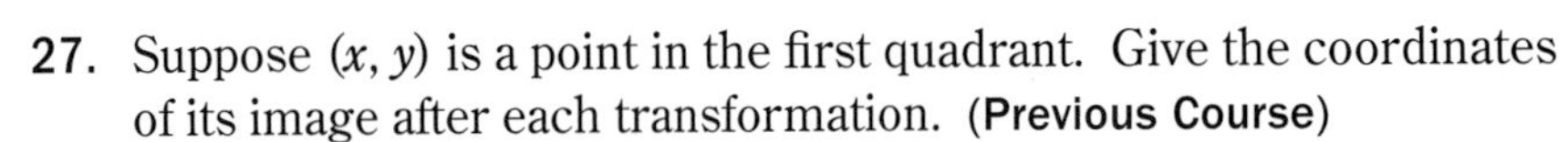

27. Suppose (x, y) is a point in the first quadrant. Give the coordinates of its image after each transformation. **(Previous Course)**

   a. reflection over the y-axis

   b. reflection over the x-axis

   c. rotation of 180° around (0, 0)

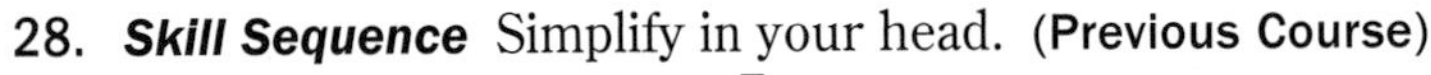

28. ***Skill Sequence*** Simplify in your head. **(Previous Course)**

   a. $\dfrac{\frac{1}{13}}{\frac{7}{13}}$ b. $\dfrac{\frac{\sqrt{5}}{13}}{\frac{7}{13}}$ c. $\dfrac{\frac{1}{3}}{\frac{\sqrt{5}}{3}}$

EXPLORATION

29. The first Ferris wheel was designed by George Washington Gale Ferris, Jr., a Pittsburgh bridge builder, for the World's Columbian Exposition in Chicago in 1892–1893. It was also the largest Ferris wheel ever built. It could seat 2160 people at one time. Research this Ferris wheel for the additional information needed to answer Questions 15–17. Then answer the questions.

The World's Columbian Exposition was a celebration of the 400th anniversary of Columbus arriving in the new world.

Lesson 4-3 Basic Trigonometric Identities

Vocabulary

identity

▶ **BIG IDEA** If you know cos θ, you can easily find cos($-\theta$), cos($90° - \theta$), cos($180° - \theta$), and cos($180° + \theta$) without a calculator, and similarly for sin θ and tan θ.

An **identity** is an equation that is true for all values of the variables for which the expressions on each side are defined. There are five theorems in this lesson; all are identities.

Mental Math

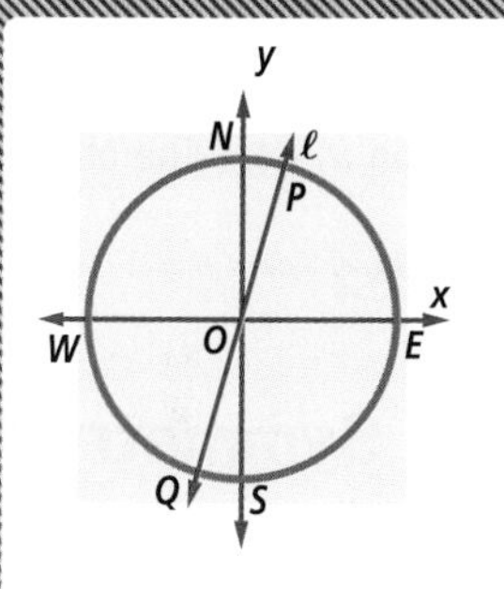

True or False

a. $\angle POE$ and $\angle POW$ are complementary.

b. $\angle POE$ and $\angle PON$ are supplementary.

c. $m\angle POE = m\angle QOW$

d. $m\angle POW = \pi - m\angle POE$

The Pythagorean Identity

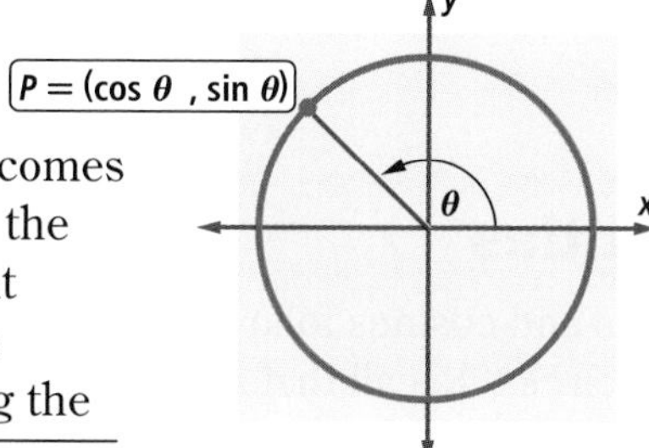

The first identity we derive in this lesson comes directly from the equation $x^2 + y^2 = 1$ for the unit circle. Because, for every θ, the point $P = (\cos \theta, \sin \theta)$ is on the unit circle, the distance from P to (0, 0) must be 1. Using the Distance Formula, $\sqrt{(\cos \theta - 0)^2 + (\sin \theta - 0)^2} = 1$. Squaring both sides of the equation gives $(\cos \theta)^2 + (\sin \theta)^2 = 1$. This argument proves a theorem called the *Pythagorean Identity.*

Pythagorean Identity Theorem

For every θ, $\cos^2\theta + \sin^2\theta = 1$.

An abbreviated version of $(\cos \theta)^2$ is $\cos^2\theta$, the square of the cosine of θ. Similarly, $(\sin \theta)^2$ is written $\sin^2\theta$ and $(\tan \theta)^2$ is written $\tan^2\theta$. Notice that we do *not* write $\cos \theta^2$ for $(\cos \theta)^2$.

QY1

The name of the above identity comes from the Pythagorean Theorem because in the first quadrant, as shown at the right, cos θ and sin θ are the sides of a right triangle with hypotenuse 1. Among other things, the Pythagorean Identity enables you to obtain either cos θ or sin θ if you know the other.

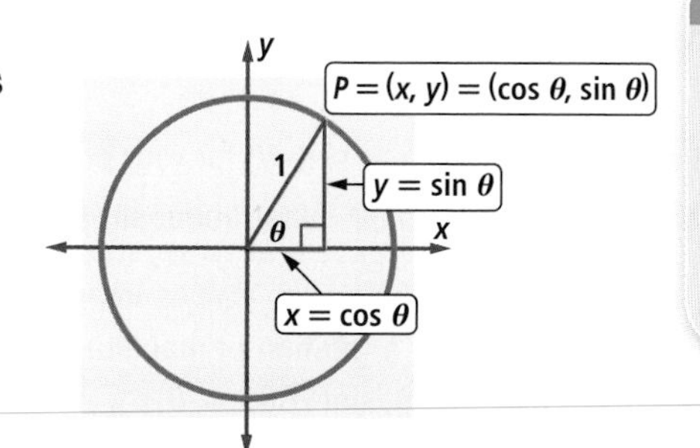

▶ **QY1**

Which two expressions are equal?

A $\tan^2\theta$

B $\tan \theta^2$

C $(\tan \theta)^2$

Example 1

If $\cos \theta = \frac{3}{5}$, find $\sin \theta$.

Solution Substitute into the Pythagorean Identity.

$$\left(\frac{3}{5}\right)^2 + \sin^2 \theta = 1$$

$$\frac{9}{25} + \sin^2 \theta = 1$$

$$\sin^2 \theta = \frac{16}{25}$$

$$\sin \theta = \pm \frac{4}{5}$$

Thus, $\sin \theta = \frac{4}{5}$ or $\sin \theta = -\frac{4}{5}$.

Check Refer to the unit circle. The vertical line $x = \frac{3}{5}$ intersects the unit circle in two points. One is in the first quadrant, in which case the y-coordinate ($\sin \theta$) is $\frac{4}{5}$. The other is in the fourth quadrant, where $\sin \theta$ is $-\frac{4}{5}$.

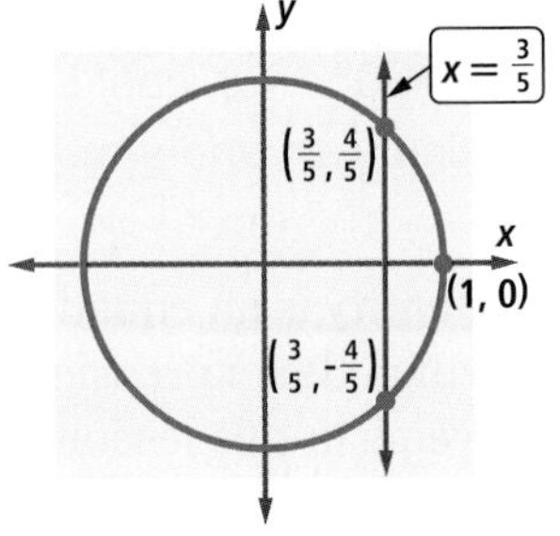

STOP QY2

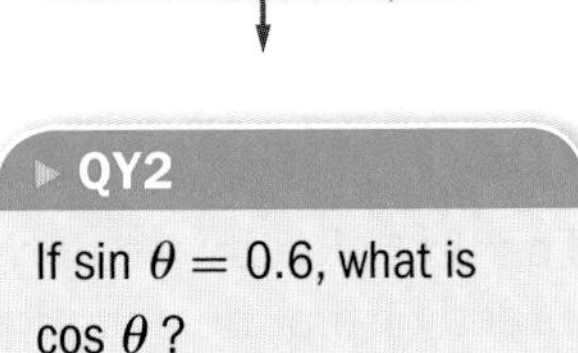

QY2

If $\sin \theta = 0.6$, what is $\cos \theta$?

The Symmetry Identities

Many other properties of sines and cosines follow from their definitions and the symmetry of the unit circle. Recall that a circle is symmetric to any line through its center. This means that the reflection image of any point over one of these lines also lies on the circle.

Activity 1

MATERIALS DGS or graph paper, compass, and protractor

Step 1 Draw a unit circle on a coordinate grid. Plot the point $A = (1, 0)$. Pick a value of θ between 0° and 90°. Let a point P in the first quadrant be the image of A under the rotation R_θ. Find the values of $\cos \theta$ and $\sin \theta$ from the coordinates of P. A sample is shown at the right.

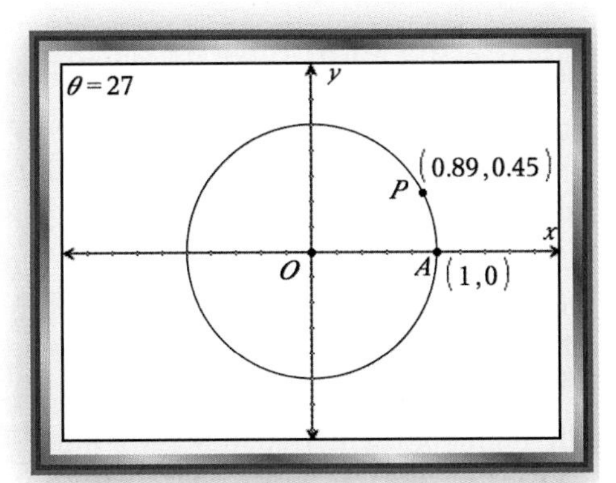

Step 2 Reflect P over the x-axis. Call its image Q. Notice that Q is the image of $(1, 0)$ under a rotation of magnitude $-\theta$. Consequently, $Q = (\cos(-\theta), \sin(-\theta))$.

a. What are the values of $\cos(-\theta)$ and $\sin(-\theta)$ for your point Q?

b. How are $\cos \theta$ and $\cos(-\theta)$ related? What about $\sin \theta$ and $\sin(-\theta)$?

Step 3 Rotate your point P 180° around the circle. Call its image H. Notice that H is the image of $(1, 0)$ under a rotation of magnitude $(180° + \theta)$. Consequently, $H = (\cos(180° + \theta), \sin(180° + \theta))$.

a. What are the values of $\cos(180° + \theta)$ and $\sin(180° + \theta)$ for your point H?

b. How are $\cos \theta$ and $\cos(180° + \theta)$ related? How are $\sin \theta$ and $\sin(180° + \theta)$ related?

Step 4 Use a calculator to find cos θ and sin θ for your value of θ in Step 1. Then find cos(-θ) and sin(-θ), and also cos(180° + θ) and sin(180° + θ). Explain any differences between the values displayed by the calculator and what you found in Steps 2 and 3.

Save your work for Activity 2.

Activity 1 is based on the following ideas: When a point *P* on the unit circle is reflected over either axis, or when it is rotated through a half turn, either the coordinates of the three images are equal to the coordinates of *P* or they are opposites of the coodinates of *P*. The magnitudes of the rotations that map (1, 0) onto these points are θ (for *P* at the right), -θ (for *Q*), 180° + θ (for *H*), and 180° – θ (for *J*). So the sines and cosines of these magnitudes are either equal or opposites.

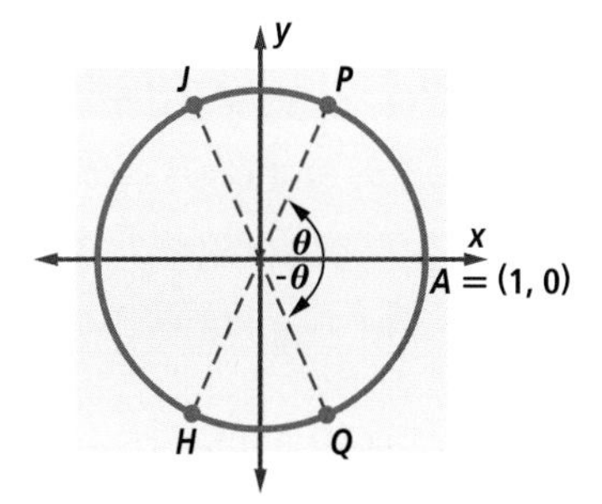

Sines and Cosines of Opposites

Rotations of magnitude θ and -θ go in opposite directions. The two rotation images are reflection images of each other over the *x*-axis. Thus they have the same first coordinates (cosines) but opposite second coordinates (sines). It follows that the ratios of the *y*-coordinates to the *x*-coordinate are opposites. This argument proves the following theorem.

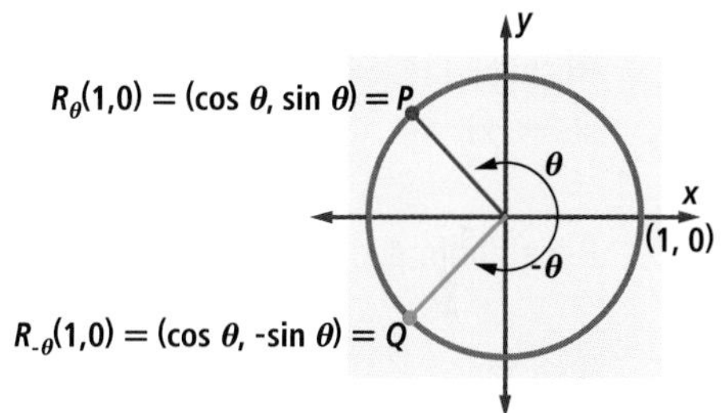

Opposites Theorem

For all θ,

$$\cos(-\theta) = \cos\theta, \quad \sin(-\theta) = -\sin\theta, \quad \text{and } \tan(-\theta) = -\tan\theta.$$

Sines and Cosines of θ + 180° or $\theta + \pi$

Adding 180° or π to the argument θ of a trigonometric function is equivalent to rotating halfway around the unit circle.

Half-Turn Theorem

For all θ, $\cos(180° + \theta) = -\cos\theta = \cos(\pi + \theta)$

$\sin(180° + \theta) = -\sin\theta = \sin(\pi + \theta)$

and $\tan(180° + \theta) = \tan\theta = \tan(\pi + \theta)$.

Proof Let $A = (1, 0)$ and let $P = R_\theta(A) = R_\theta(1, 0) = (\cos\theta, \sin\theta)$. Now let Q be the image of P under $R_{180°}$. Because $R_{180°}$ maps (a, b) to $(-a, -b)$, Q has coordinates $(-\cos\theta, -\sin\theta)$. But Q is also the image of A under a rotation of magnitude $180° + \theta$. So Q also has coordinates $(\cos(180° + \theta), \sin(180° + \theta))$. Equating the two ordered pairs for Q proves the first two parts of the theorem. The third part follows by dividing the second equation by the first.

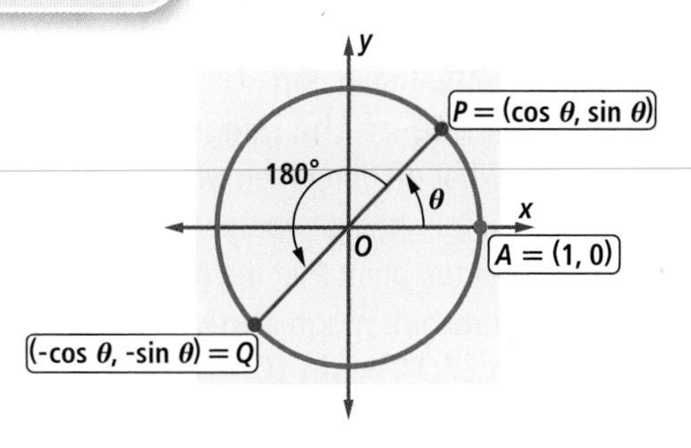

Sines and Cosines of Supplements

Recall that if an angle has measure θ, then its supplement has measure $180° - \theta$, that is, $\pi - \theta$. Activity 1 shows that the values of the trigonometric functions of θ and $180° - \theta$ are related, as stated in the following theorem.

> **Supplements Theorem**
>
> For all θ, $\sin(180° - \theta) = \sin\theta = \sin(\pi - \theta)$
>
> $\cos(180° - \theta) = -\cos\theta = \cos(\pi - \theta)$
>
> and $\tan(180° - \theta) = -\tan\theta = \tan(\pi - \theta)$.

Proof Let $P = (\cos\theta, \sin\theta)$. Let Q be the reflection image of P over the y-axis, as in the diagram at the right. Because the reflection image of (x, y) over the y-axis is $(-x, y)$,

$(\cos(180° - \theta), \sin(180° - \theta)) = (-\cos\theta, \sin\theta) = Q$

$P = (\cos\theta, \sin\theta)$

$180° - \theta$

$(-1, 0) = B$

$A = (1, 0)$

$$Q = (-\cos\theta, \sin\theta).$$

Recall from geometry that reflections preserve angle measure, so

$$m\angle QOB = m\angle POA = \theta.$$

Also, since $\angle AOQ$ and $\angle QOB$ are a linear pair,

$$m\angle AOQ = 180° - \theta.$$

So, by the definitions of cosine and sine,

$$Q = (\cos(180° - \theta), \sin(180° - \theta)).$$

Thus, $(\cos(180° - \theta), \sin(180° - \theta)) = (-\cos\theta, \sin\theta)$.

The x-coordinates are equal, so

$$\cos(180° - \theta) = -\cos\theta.$$

Likewise, the y-coordinates are equal, so

$$\sin(180° - \theta) = \sin\theta.$$

Dividing the latter of these equations by the former gives the third part of the Supplements Theorem,

$$\tan(180° - \theta) = -\tan\theta.$$

QY3

> **QY3**
>
> Suppose $\sin\theta = 0.496$ and $\cos\theta = 0.868$. Without using a calculator, find
>
> **a.** $\sin(\pi - \theta)$.
>
> **b.** $\cos(180° - \theta)$.

Example 2

Given that $\sin 10° \approx 0.1736$, find a value of x other than $10°$ and between $0°$ and $360°$ for which $\sin x = 0.1736$.

Solution Think: $\sin 10°$ is the second coordinate of the image of $(1, 0)$ under $R_{10°}$. What other rotation will give the same second coordinate? It is the rotation that gives the reflection image of the point P in the diagram at the right. That rotation has magnitude $180° - 10°$, or $170°$. So $\sin 170° = \sin 10° = 0.1736$, and $x = 170°$.

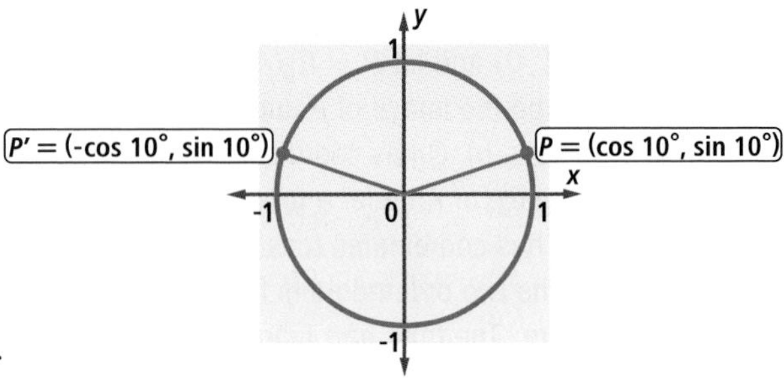

If the requirement that $0° < x < 360°$ in Example 2 is relaxed, there are other answers. Because you can add or subtract 360° to the magnitude of any rotation and get the same rotation, $\sin 10° = \sin 170° = \sin 530° = \sin(-190°)$. Also, in radians, $\sin\left(\frac{\pi}{18}\right) = \sin\left(\frac{17\pi}{18}\right) = \sin\left(\frac{53\pi}{18}\right) = \sin\left(\frac{-19\pi}{18}\right)$.

QY4

▶ QY4

Given that $\cos 10° \approx 0.9848$, find a value of x other than 10° for which $\cos x = 0.9848$.

Sines and Cosines of Complements

Activity 2

MATERIALS DGS or graph paper

Step 1 Begin with the graph from Step 3 of Activity 1. Hide points H and Q. Draw the line $y = x$. Again pick a value of θ between 0° and 90° and let $P = R_\theta(1, 0)$. Find $\cos \theta$ and $\sin \theta$ for your value of θ.

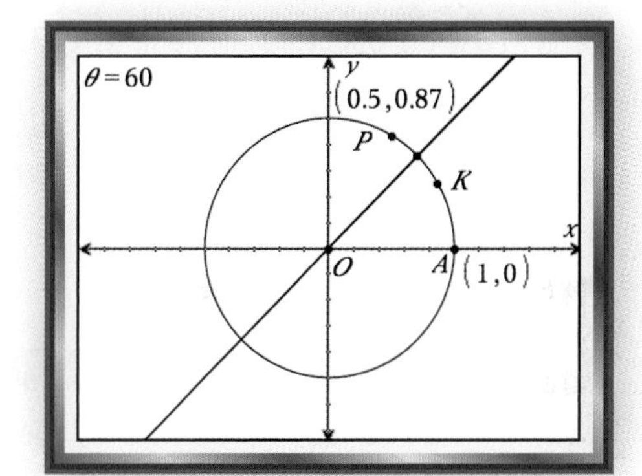

Step 2 Reflect point P over $y = x$ and call its image K. From your knowledge of reflections, what are the coordinates of K?

Step 3 In terms of θ, what is the magnitude of the rotation that maps (1, 0) onto K? (Hint: K is as far from A along the circle as P is from the point (0, 1).) Answer in both degrees and radians.

Step 4 Develop an identity that relates the sine and cosine of your answers to Step 3 to the sine and cosine of θ.

If an angle has measure θ, then its complement has measure $90° - \theta$ or $\frac{\pi}{2} - \theta$. Activity 2 shows that the sines and cosines of θ and $90° - \theta$ are related.

Complements Theorem

For all θ,

$$\sin(90° - \theta) = \cos \theta = \sin\left(\frac{\pi}{2} - \theta\right)$$

and $\cos(90° - \theta) = \sin \theta = \cos\left(\frac{\pi}{2} - \theta\right)$.

These theorems can help extend your knowledge of circular functions.

Example 3

Given that $\sin 30° = \frac{1}{2}$, compute the exact value of each function below.

a. $\cos 60°$ **b.** $\cos 30°$ **c.** $\sin 150°$

d. $\cos 210°$ **e.** $\sin(-30°)$

(continued on next page)

Solution

a. Use the Complements Theorem.
$\cos 60° = \sin(90° - 60°) = \sin 30°$. So $\cos 60° = \frac{1}{2}$.

b. Use the Pythagorean Identity Theorem. $\sin^2 30° + \cos^2 30° = 1$.
So $\cos^2 30° = 1 - \left(\frac{1}{2}\right)^2 = \frac{3}{4}$. Thus, $\cos 30° = \pm\sqrt{\frac{3}{4}} = \pm\frac{\sqrt{3}}{2}$.
However, we know $\cos 30°$ is positive, so $\cos 30° = \frac{\sqrt{3}}{2}$.

c. Use the Supplements Theorem. $\sin 150° = \sin(180° - 150°) = \sin 30°$. So $\sin 150° = \frac{1}{2}$.

d. Use the Half-Turn Theorem. $\cos 210° = \cos(180° + 30°) = -\cos 30° = -\frac{\sqrt{3}}{2}$.

e. Use the Opposites Theorem. $\sin(-30°) = -\sin 30° = -\frac{1}{2}$.

In using these identities, you should also be able to use the unit circle to do a visual check of your answers or to derive a property if you forget one.

Questions

COVERING THE IDEAS

1. **True or False** When $\theta = 180°$, $\cos^2\theta + \sin^2\theta = 1$.

2. a. If $\sin\theta = \frac{24}{25}$, what are two possible values of $\cos\theta$?
   b. Draw a picture to justify your answers to Part a.

3. If $\tan\theta = 3$, what is $\tan(-\theta)$?

4. a. **True or False** $\cos 14° = \cos(-14°)$
   b. Justify your answer to Part a with a unit circle diagram.

In 5 and 6, refer to the figure at the right. $P = R_\theta(1, 0)$, $P' = r_{y\text{-axis}}(P)$, $P'' = R_{180°}(P)$, and $P''' = r_{x\text{-axis}}(P)$.

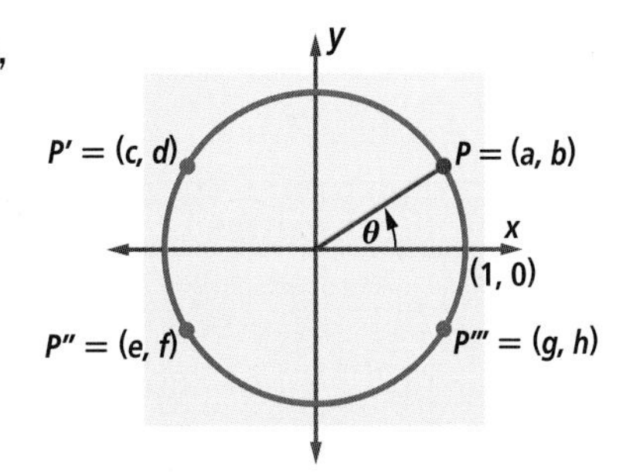

5. Which coordinates equal $\cos(180° - \theta)$?

6. Which coordinates equal $\sin(180° + \theta)$?

7. **True or False** $\sin(-\theta) = \sin\theta$

In 8 and 9, $\sin\theta = \frac{1}{3}$. Evaluate without using a calculator.

8. $\sin(-\theta)$

9. $\sin(180° - \theta)$

10. Using what you know about $\sin(180° - \theta)$ and $\cos(180° - \theta)$, explain why $\tan(180° - \theta) = -\tan\theta$.

11. Use a calculator to verify the three parts of the Supplements Theorem when $\theta = 146.5°$.

In 12 and 13, suppose $\cos x = \frac{5}{13}$. Evaluate without using a calculator.

12. $\cos(180° + x)$

13. $\sin(90° - x)$

In 14 and 15, $\tan y = k$. Evaluate.

14. $\tan(-y)$

15. $\tan(180° - y)$

16. Copy the table below, filling in the blank entries and completing the diagrams, to summarize the theorems in this lesson.

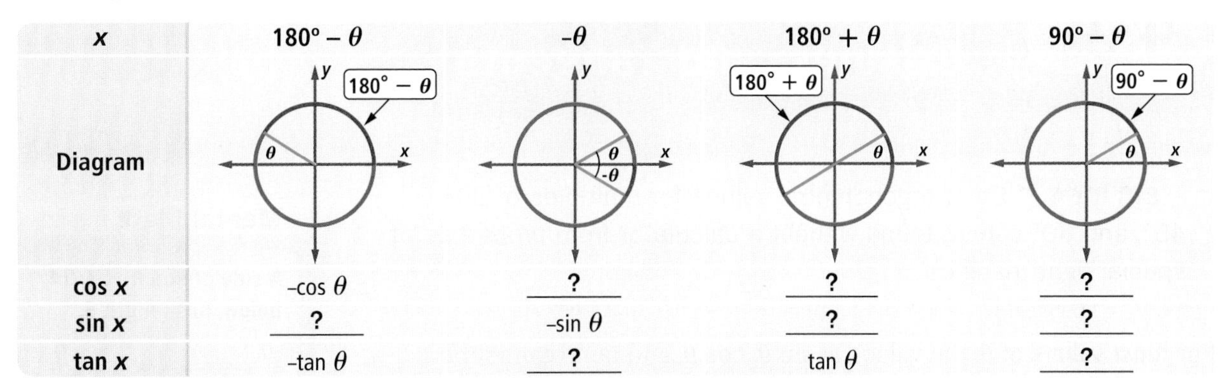

x	$180° - \theta$	$-\theta$	$180° + \theta$	$90° - \theta$
Diagram				
$\cos x$	$-\cos\theta$	?	?	?
$\sin x$	?	$-\sin\theta$	?	?
$\tan x$	$-\tan\theta$	?	$\tan\theta$	?

APPLYING THE MATHEMATICS

In 17–21, the display below shows inputs and outputs of a CAS in degree mode. What theorem justifies each statement?

17.	cos(350)	cos(10)
18.	sin(160)	sin(20)
19.	sin(202)	-sin(22)
20.	cos(84)	sin(6)
21.	tan(187)	tan(7)

22. Prove that $\sin(\pi - \theta) = \sin\pi - \sin\theta$ is *not* an identity.

In 23–26, from the fact that $\sin 18° = \frac{\sqrt{5}-1}{4}$, find each value.

23. $\sin 162°$ 24. $\sin(-18°)$ 25. $\sin\frac{11\pi}{10}$ 26. $\cos\frac{2\pi}{5}$

REVIEW

In 27–29, without using a calculator, give exact values. (Lesson 4-2)

27. $\sin 90°$ 28. $\cos 810°$ 29. $\tan(90° + 90°)$

30. Convert $\frac{11}{6}$ clockwise revolutions to degrees. **(Lesson 4-1)**

31. **a.** What is the magnitude of the rotation of the minute hand of a clock in 6 minutes?

b. What is the measure of the angle between the minute hand and the second hand of a clock at exactly 12:06 A.M.? **(Lesson 4-1)**

32. Find an equation for the image of the graph of $y = x^2$ under the scale change $(x, y) \to \left(\frac{1}{2}x, 5y\right)$. **(Lesson 3-5)**

EXPLORATION

33. Use a calculator to investigate whether $\frac{\cos^2\theta}{1 - \sin\theta} = 1 + \sin\theta$ is an identity. Try to prove your conclusion, either by providing a counterexample or by using definitions and properties.

QY ANSWERS

1. A and C
2. $\cos\theta = \pm\sqrt{1 - 0.6^2}$
$= \pm 0.8$
3. **a.** 0.496 **b.** -0.868
4. Answers vary. Samples: -10°, 370°, -350°

Lesson 4-4 Exact Values of Sines, Cosines, and Tangents

BIG IDEA Exact trigonometric values for multiples of 30°, 45°, and 60° can be found without a calculator from properties of special right triangles.

For most values of θ, the values of $\sin \theta$, $\cos \theta$, and $\tan \theta$ cannot be found exactly and must be approximated. For this reason, you used approximate values found with a calculator in previous lessons.

In this lesson, you will apply what you know about 45°-45°-90° and 30°-60°-90° triangles to obtain exact values of $\cos \theta$, $\sin \theta$, and $\tan \theta$ when θ is a multiple of 30°, 45°, or 60°.

Mental Math

A side of square *SQUA*, below, has length 5.

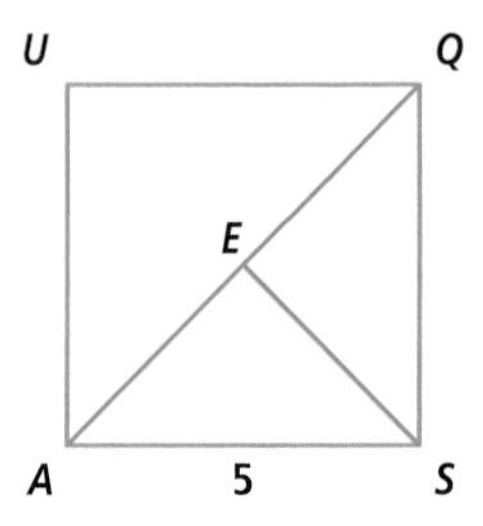

a. What is the length of $\overline{AQ}$?

b. If E is the midpoint of $\overline{AQ}$, what is the length of $\overline{SE}$?

Exact Values of Trigonometric Functions for $\theta = 45°$

You can use the properties of isosceles right triangles to find cos 45° and sin 45°.

GUIDED

Example 1

Use $\triangle OPF$ at the right to compute the exact values of cos 45° and sin 45°. Justify your answer.

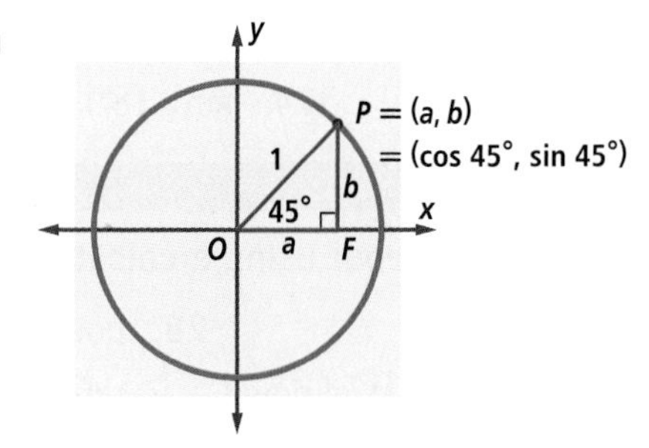

Solution Because $m\angle FOP = 45°$, $m\angle P = 45°$. So $\triangle OPF$ is isosceles with legs $\overline{OF}$ and __?__. a and b are the lengths of the legs, so $a = b$. By the Pythagorean Theorem, $a^2 + b^2 = 1$, so $2a^2 = 1$, and $a^2 =$ __?__. Therefore, $a = b = \pm\frac{1}{\sqrt{2}}$. Because a and b are lengths, $a = b = \frac{1}{\sqrt{2}}$.

But $\cos 45° = a$ and $\sin 45° =$ __?__.

Thus, $\cos 45° = \sin 45° = \frac{1}{\sqrt{2}} = \frac{\sqrt{2}}{2}$.

 QY1

QY1

Explain why $\tan 45° = 1$.

Exact Values of Trigonometric Functions for $\theta = 30°$ and $\theta = 60°$

In Example 3 of Lesson 4-3, you were told that $\sin 30° = \frac{1}{2}$. You can verify this by using properties of equilateral triangles.

GUIDED

Example 2

Derive the exact values of cos 30° and sin 30°.

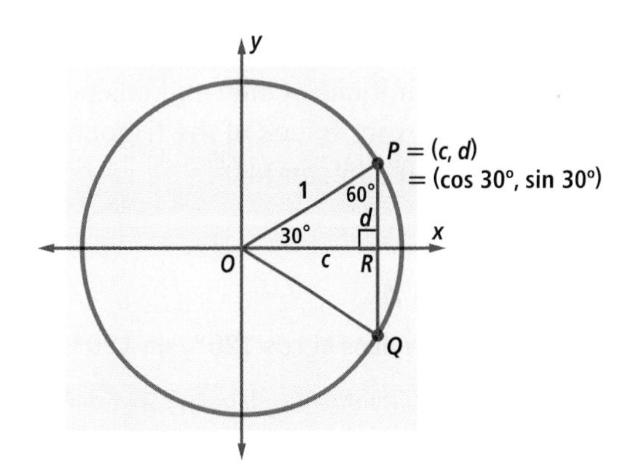

Solution In equilateral $\triangle OPQ$ at the right, since $OP = 1$, $PQ = \underline{\ ?\ }$. Consequently, $PR = d = \underline{\ ?\ }$.

By the Pythagorean Theorem, $c^2 + d^2 = 1$.

$$\text{So } c^2 + \underline{\ ?\ } = 1.$$

$$c^2 = \underline{\ ?\ }$$

$$c = \underline{\ ?\ }$$

Thus, $(\cos 30°, \sin 30°) = (c, d) = (\underline{\ ?\ }, \underline{\ ?\ })$.

So, $\cos 30° = \underline{\ ?\ }$ and $\sin 30° = \underline{\ ?\ }$.

To obtain the exact values of cos 60° and sin 60°, use the Complements Theorem: $\cos 60° = \sin 30° = \frac{1}{2}$ and $\sin 60° = \cos 30° = \frac{\sqrt{3}}{2}$.

Example 3

Find the exact value of tan 30°.

Solution Use $\tan\theta = \frac{\sin\theta}{\cos\theta}$.

$$\tan 30° = \frac{\sin 30°}{\cos 30°} = \frac{\frac{1}{2}}{\frac{\sqrt{3}}{2}} = \frac{1}{2} \cdot \frac{2}{\sqrt{3}} = \frac{1}{\sqrt{3}} = \frac{\sqrt{3}}{3}.$$

QY2

> **QY2**
>
> Find the exact value of tan 60°.

You should memorize the exact values of cos θ, sin θ, and tan θ for $\theta = 30°, 45°$, and 60°. They are important tools in mathematics and science because they are exact. To help you learn them, they are summarized below.

$\sin 45° = \frac{\sqrt{2}}{2} = \sin\frac{\pi}{4}$ $\quad \sin 30° = \frac{1}{2} = \sin\frac{\pi}{6}$ $\quad \sin 60° = \frac{\sqrt{3}}{2} = \sin\frac{\pi}{3}$

$\cos 45° = \frac{\sqrt{2}}{2} = \cos\frac{\pi}{4}$ $\quad \cos 30° = \frac{\sqrt{3}}{2} = \cos\frac{\pi}{6}$ $\quad \cos 60° = \frac{1}{2} = \cos\frac{\pi}{3}$

$\tan 45° = 1 = \tan\frac{\pi}{4}$ $\quad \tan 30° = \frac{\sqrt{3}}{3} = \tan\frac{\pi}{6}$ $\quad \tan 60° = \sqrt{3} = \tan\frac{\pi}{3}$

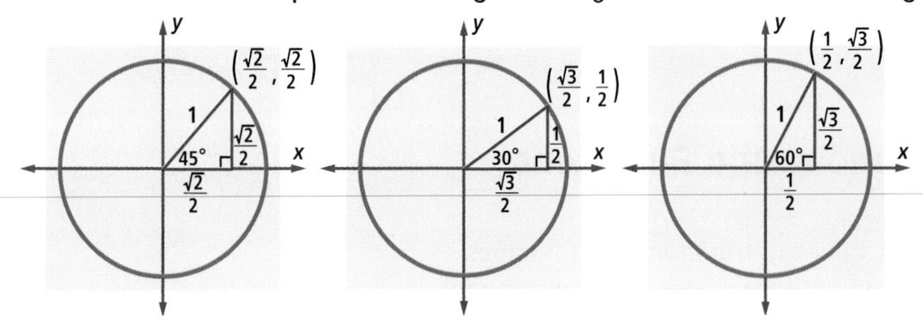

STOP QY3

> **QY3**
>
> Which theorem verifies that sin 30° = cos 60°?

Exact Values for Sines and Cosines of Multiples of 30°, 45°, and 60°

Using the definitions of sine and cosine and the Symmetry Identities, you can find exact values of the trigonometric functions for all integer multiples of 30°, 45°, and 60°.

GUIDED

Example 4

Find exact values of cos 120°, sin 120°, and tan 120°.

Solution By the Supplements Theorem,

$\cos 120° = \underline{\ \ ?\ \ } = -\frac{1}{2}$ and $\sin 120° = \underline{\ \ ?\ \ } = \frac{\sqrt{3}}{2}$.

$\tan 120° = \frac{\sin 120°}{\cos 120°} = \frac{\frac{\sqrt{3}}{2}}{-\frac{1}{2}} = -\sqrt{3}$.

Check Use a calculator.

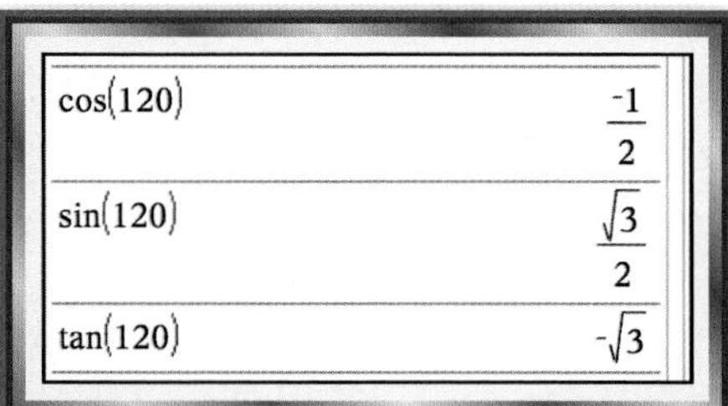

On the unit circle below are the images of (1, 0) under rotations of integer multiples of 30° or 45° between 0° and 360°. You should be able to calculate exact values of the sine, cosine, and tangent functions for all pictured values of θ by relating them to one of the points in the first quadrant or on the axes.

Activity

Copy the unit circle and the exact values of (cos θ, sin θ) given at the right. Use your knowledge of reflections and symmetries to add the exact values of trigonometric functions for multiples of 30°, 45° and 60° in Quadrants II, III, and IV.

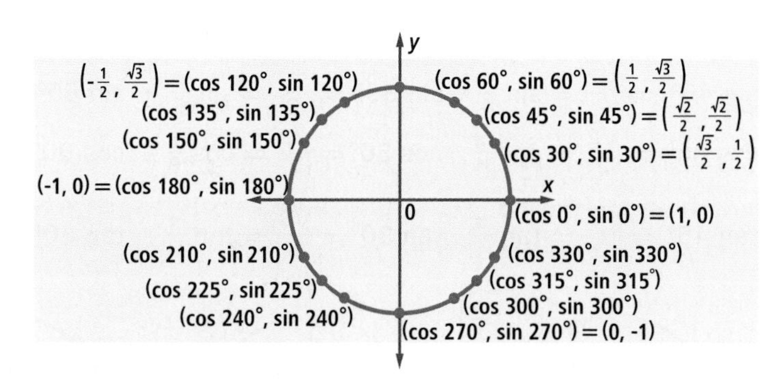

Exact Values for Trigonometric Functions of Radians

It is important to know the exact values of trigonometric functions for certain radians. You can compute those values by converting to degrees, but in the long run, it is helpful to learn to "think radian."

GUIDED

Example 5

Without using technology, compute the exact value of each trigonometric function below.

a. $\sin \frac{\pi}{4}$ b. $\cos \frac{5\pi}{6}$ c. $\tan \pi$

Solution

a. Convert to degrees: $\frac{\pi}{4} \cdot \frac{180^\circ}{\pi} = 45^\circ$. $\sin 45^\circ = \frac{\sqrt{2}}{2} = \sin \frac{\pi}{4}$

b. $\frac{5\pi}{6} = \underline{\ ?\ }^\circ$, so $\cos \frac{5\pi}{6} = \underline{\ ?\ }$.

c. $\tan \pi = \frac{\sin \pi}{\cos \pi} = \frac{?}{?} = \underline{\ ?\ }$

Questions

COVERING THE IDEAS

In 1–3, refer to the unit circle at the right in which $m\angle POA = 30^\circ$, $m\angle QOA = 45^\circ$, and $m\angle ROA = 60^\circ$. Name a segment whose length equals the following.

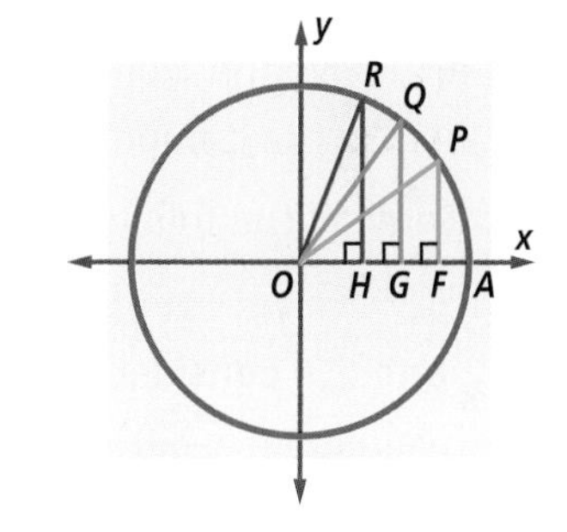

1. $\cos 30^\circ$ 2. $\sin 45^\circ$ 3. $\sin 60^\circ$

4. Evaluate.
   a. $\cos \frac{\pi}{3}$ b. $\tan \frac{\pi}{4}$ c. $\sin \frac{\pi}{6}$

In 5–10, find the exact value.

5. a. $\sin 240^\circ$ b. $\cos 240^\circ$ c. $\tan \frac{4\pi}{3}$
6. a. $\sin \frac{3\pi}{4}$ b. $\cos \frac{3\pi}{4}$ c. $\tan 135^\circ$
7. a. $\sin \frac{11\pi}{6}$ b. $\cos(-30^\circ)$ c. $\tan \frac{11\pi}{6}$
8. $\sin 210^\circ$ 9. $\cos \frac{5\pi}{3}$ 10. $\tan(-405^\circ)$
11. Draw a unit circle as in the Activity, labeling the angles in *radians* and filling in all the values of the trigonometric functions.

APPLYING THE MATHEMATICS

12. a. Find two values of θ between -90° and 90° for which $\cos \theta = \frac{1}{2}$.
    b. Find two values of θ between 270° and 450° for which $\cos \theta = \frac{1}{2}$.
    c. What is the relationship between the two pairs of angles formed in Parts a and b?
13. Consider the equation $\sin \theta = -\frac{1}{2}$.
    a. Draw a unit circle and mark the two points for which $\sin \theta = -\frac{1}{2}$.
    b. Give two values of θ between 0° and 360° that satisfy the equation.
    c. Give two values of θ between 0 and 2π radians that satisfy the equation.
14. a. Find two values of θ between 0 and 2π such that $\cos \theta = \sin \theta$.
    b. What is the value of $\tan \theta$ for each value of θ in Part a?

15. **True or False** If $\tan \theta = \pm 1$, then $\theta = (45n)^\circ$ and n is an odd integer. Justify your answer.

16. The regular nonagon $ABCDEFGHI$ pictured here is inscribed in the unit circle.
    a. Give the exact coordinates of point B in terms of θ.
    b. Give the value of θ in radians.
    c. Estimate AB to the nearest thousandth.

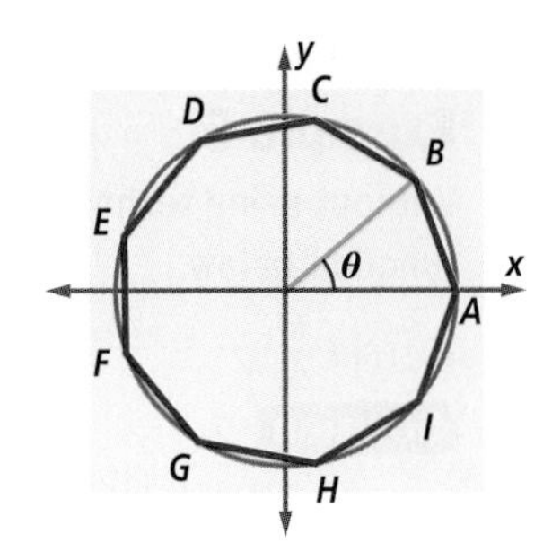

REVIEW

17. Without using a calculator, given that $\sin 52^\circ \approx 0.788$, estimate each value. **(Lesson 4-3)**
    a. $\sin(-52^\circ)$ b. $\sin 128^\circ$ c. $\sin 232^\circ$ d. $\cos 38^\circ$

18. **True or False** For all θ, $\cos(\theta + 90^\circ) = \sin \theta$. **(Lesson 4-3)**

19. Without using a calculator, give the exact value for $\sin\left(-\frac{\pi}{2}\right)$. **(Lesson 4-2)**

20. a. Prove that $\cos \theta \cdot \tan \theta = \sin \theta$ for all $\cos \theta \neq 0$.
    b. Why is it impossible to have $\cos \theta = 0$ in Part a? **(Lesson 4-2)**

21. Convert the following measures to radians. **(Lesson 4-1)**
    a. 135° b. 390° c. -215° d. -270°

In 22 and 23, consider $g(t) = t^2 + 1$ and $f(t) = 3t - 1$. (Lesson 3-7)

22. Evaluate $g(f(-80))$.

23. Find a formula for $(f \circ g)(t)$.

24. When a certain drug enters the blood stream, its potency decreases exponentially with a half-life of 8 hours. Suppose the initial amount of drug present is A. How much of the drug will be present after each number of hours? **(Lesson 2-5)**

    a. 8 b. 24 c. t

EXPLORATION

25. A regular triangle, hexagon, and dodecagon have been inscribed in the unit circle. Find the exact perimeter of each polygon. You may find a CAS useful.

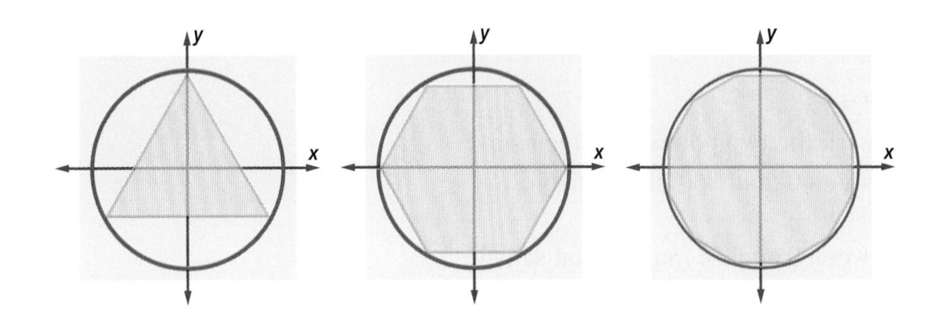

QY ANSWERS

1. $\tan 45^\circ = \frac{\sin 45^\circ}{\cos 45^\circ} = \frac{\frac{\sqrt{2}}{2}}{\frac{\sqrt{2}}{2}} = 1$

2. $\tan 60^\circ = \frac{\sin 60^\circ}{\cos 60^\circ} = \frac{\frac{\sqrt{3}}{2}}{\frac{1}{2}} = \sqrt{3}$

3. Complements Theorem

Lesson 4-5 The Sine and Cosine Functions

Vocabulary

sine function

cosine function

▶ **BIG IDEA** The values of cos θ and sin θ determine functions with equations $y = \sin x$ and $y = \cos x$ whose domain is the set of all real numbers.

From the exact values of sines, cosines, and tangents you calculated in Lesson 4-4, you can see the shape of a function called the *sine function*.

Mental Math

If gasoline costs $4.00 a gallon and a car gets 25 miles to the gallon, what does it cost for gas per mile?

Activity 1

Step 1 The table below contains some exact values of sin θ. It also shows decimal approximations to those values. Complete the table, using a unit circle to help you.

θ (degrees)	0°	30°	45°	60°	90°	120°	135°	150°	180°
θ (radians)	0	$\frac{\pi}{6}$	$\frac{\pi}{4}$	$\frac{\pi}{3}$	$\frac{\pi}{2}$	$\frac{2\pi}{3}$	$\frac{3\pi}{4}$	$\frac{5\pi}{6}$	π
sin θ (exact)	0	$\frac{1}{2}$	?	?	?	$\frac{\sqrt{3}}{2}$	?	?	?
sin θ (approx.)	0	0.5	?	?	?	?	0.707	?	?

θ (degrees)	210°	225°	240°	270°	300°	315°	330°	360°
θ (radians)	$\frac{7\pi}{6}$	$\frac{5\pi}{4}$	$\frac{4\pi}{3}$	$\frac{3\pi}{2}$	?	?	?	2π
sin θ (exact)	?	$-\frac{\sqrt{2}}{2}$	?	?	?	?	?	?
sin θ (approx.)	?	?	?	?	-0.866	?	?	?

Step 2 Here is a graph of the first five points in the first part of the table. Copy this graph, and on it plot the points you found in Step 1. Then draw a smooth curve through the points.

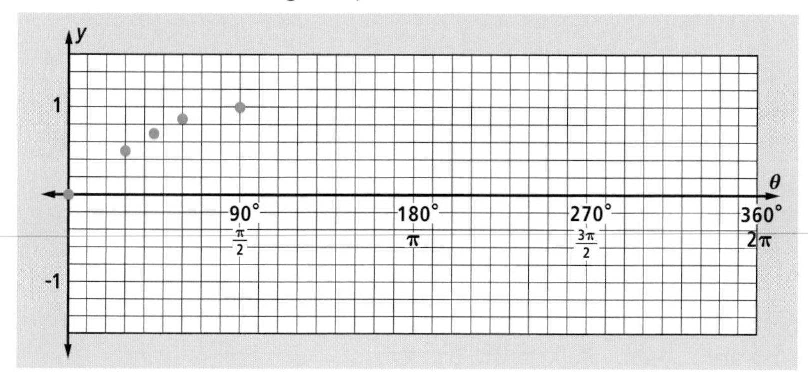

(continued on next page)

Step 3 Check Step 2 by using a graphing utility to plot $y = \sin\theta$ for $0° \leq \theta \leq 360°$ and for $0 \leq \theta \leq 2\pi$.

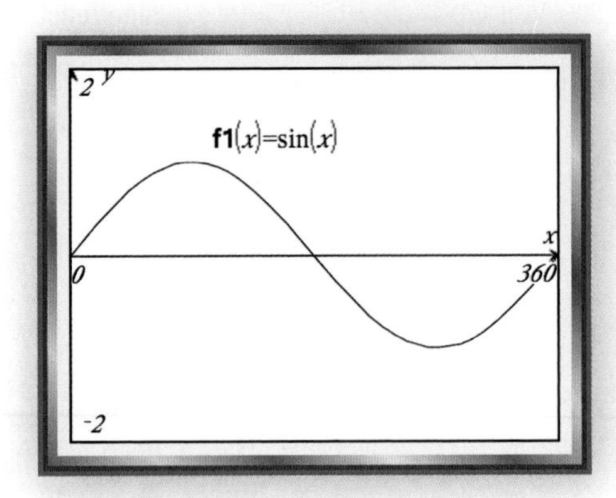

The Graph of the Sine Function

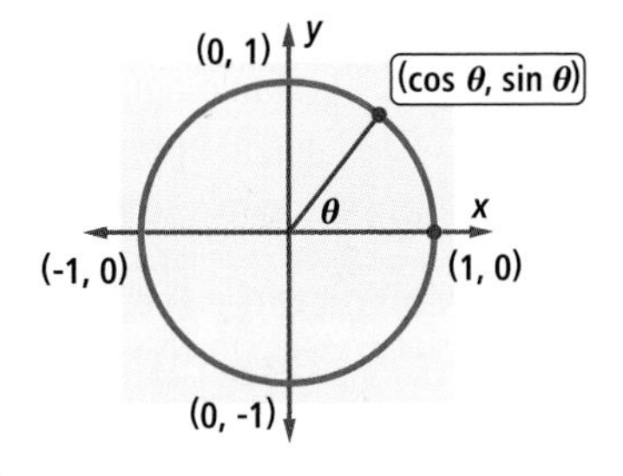

The function that maps each real number θ to the y-coordinate of the image of (1, 0) under a rotation of θ is called the **sine function**. From the unit circle, you can tell that $\sin\theta$ is positive when $0° < \theta < 180°$ and negative when $180° < \theta < 360°$. The maximum value is 1, when $\theta = 90°$, and the minimum value is -1, when $\theta = 270°$.

QY

Restate the preceding paragraph for θ in radians.

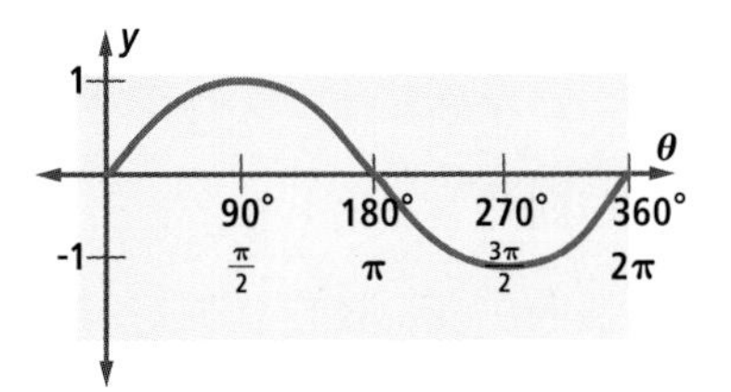

A graph of the sine function, for $0° \leq \theta \leq 360°$, is shown at the right. To make it easier to locate zeros, maxima, and minima, the scale on the horizontal axis is in multiples of $\frac{\pi}{2}$ and 90°.

This is one *cycle* of the graph of the sine function. Because the image of (1, 0) under a rotation of θ repeats itself every 2π radians, the y-coordinates in the ordered pairs of the function f with equation $f(\theta) = \sin\theta$ repeat every 2π. Thus, the graph above can be easily extended both to the right and left without calculating any new sine values. The graph of the entire sine function has infinitely many cycles. A graph showing three complete cycles of the sine function appears below.

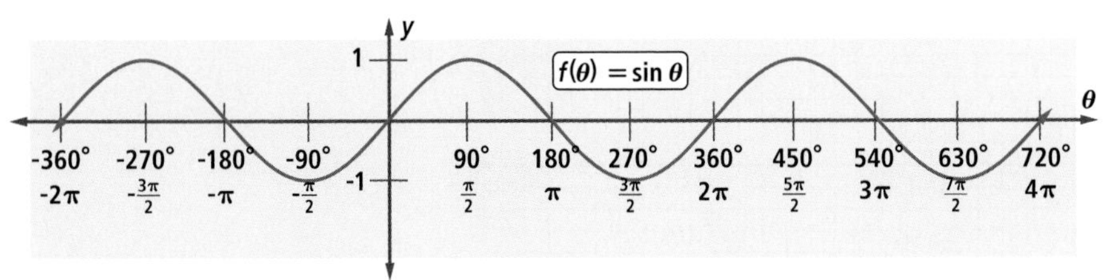

Notice from the graph that the y-intercept of the sine function is 0. The sine function's x-intercepts (zeros) are ..., -2π, $-\pi$, 0, π, 2π, 3π, 4π, ..., that is, the integer multiples of π.

As the graph on the previous page makes clear, the domain of the sine function is the set of real numbers. Because the maximum and minimum values of the sine function are 1 and -1 (the y-intercepts of the unit circle) the range is the interval $-1 \leq y \leq 1$. Also notice that the graph of the sine function is point-symmetric about the origin. Thus, the sine function is an odd function. This is because of the Opposites Theorem that states for all θ, $\sin(-\theta) = -\sin \theta$.

The Graph of the Cosine Function

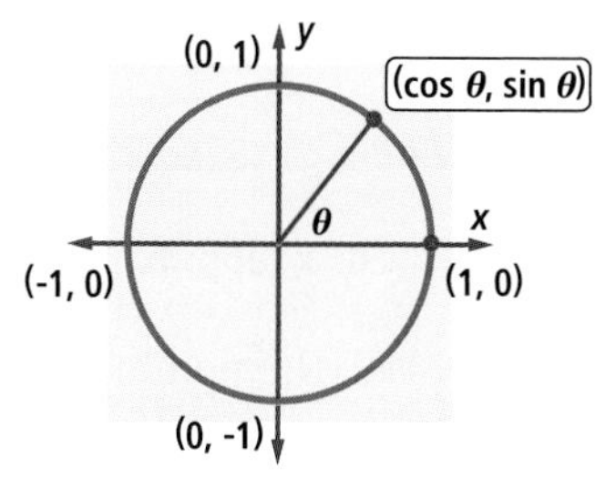

Remember that the image of (1, 0) under a rotation of magnitude θ is $(\cos \theta, \sin \theta)$. The function that maps each real number θ to the *first* coordinate of the image of (1, 0) under a rotation of θ is called the **cosine function**. The cosine function has many characteristics like those of the sine function. A graph of the cosine function is shown below.

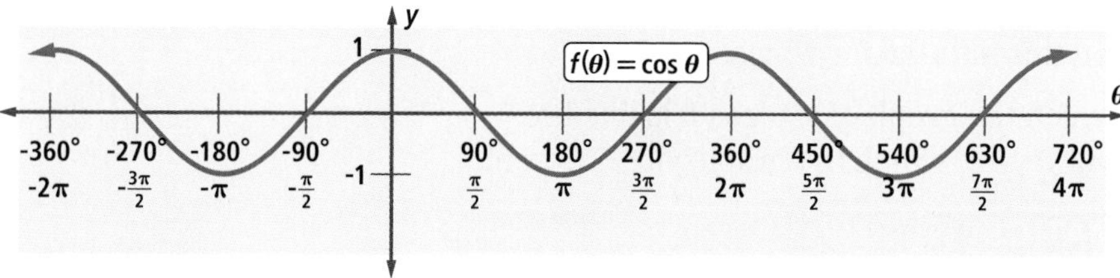

Activity 2

Use the definitions and graphs of the sine and cosine functions to fill in the table.

	sine function (degrees)	sine function (radians)	cosine function (degrees)	cosine function (radians)
Domain	?	?	?	?
Range	?	?	?	?
Zeros	?	?	?	?
Maxima	$\sin \theta = 1$ when $\theta = 90°, 450°, 810°, \ldots$	?	?	?
Minima	?	?	?	?

Questions

COVERING THE IDEAS

1. a. Identify the domain and the range of the sine function.
   b. Find five values of x such that $\sin x = 0$.

2. a. Sketch a graph of $y = \sin x$ for $0 \leq x \leq 2\pi$.
   b. Find all values of x on this interval such that $\sin x = 1$.
   c. Find all values of x on this interval for which $\sin x = 0.5$.

3. a. Copy the table below. Fill in exact and approximate values (rounded to three decimal places) for some of the coordinates of points on the graph of the cosine function.

x	0	$\frac{\pi}{6}$	$\frac{\pi}{4}$	$\frac{\pi}{3}$	$\frac{\pi}{2}$	$\frac{2\pi}{3}$	$\frac{3\pi}{4}$	$\frac{5\pi}{6}$
cos x (exact)	1	$\frac{\sqrt{3}}{2}$	$\frac{\sqrt{2}}{2}$	$\frac{1}{2}$	0	?	?	$-\frac{\sqrt{3}}{2}$
cos x (approx.)	1	0.866	0.707	0.5	0	?	?	-0.866
x	π	$\frac{7\pi}{6}$	$\frac{5\pi}{4}$	$\frac{4\pi}{3}$	$\frac{3\pi}{2}$	$\frac{5\pi}{3}$	$\frac{7\pi}{4}$	$\frac{11\pi}{6}$	2π
cos x (exact)	-1	?	?	?	?	?	?	?	?
cos x (approx.)	-1	?	?	?	?	?	?	?	?

b. Use the points from the table to graph $y = \cos x$.

4. a. Sketch a graph of $y = \cos \theta$ for $-360° \leq \theta \leq 720°$.
   b. Find five values of θ on this interval for which $\cos \theta = 0$.

5. Describe three ways in which the graph of $y = \cos \theta$ is like the graph of $y = \sin \theta$ and two ways in which the graphs are different.

APPLYING THE MATHEMATICS

6. Describe the translation with the smallest positive magnitude that maps the graph of $g(x) = \cos x$ onto that of $y = \sin x$.

7. The graph of the sine function is reflection-symmetric over the line with equation $x = \frac{\pi}{2}$.
   a. What property of sines is a result of this symmetry?
   b. Name two other lines of symmetry for the graph.

8. In a stable environment, predator-prey populations can be modeled by sine waves. Refer to the graph below.

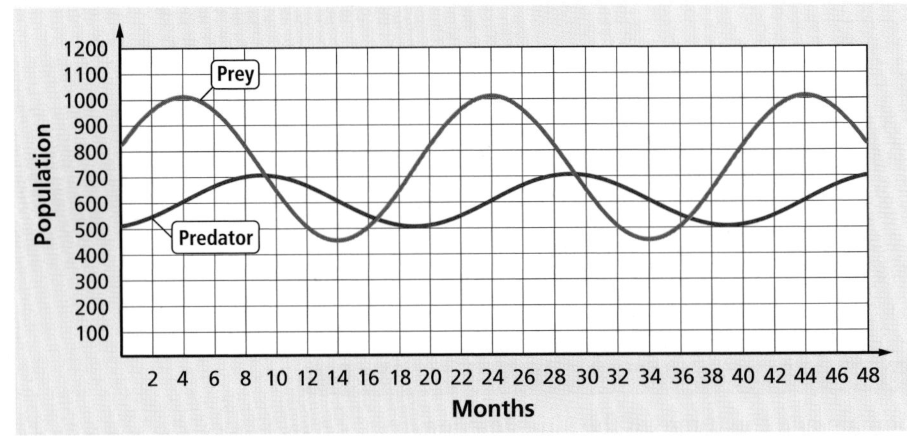

a. Describe what is happening with the prey population when the predator population is at its peak.
b. Describe what is happening with the prey population when the predators are the fewest.

9. Use the graph of $y = f(x)$ at the right. Suppose f is known to be either the cosine function or the sine function.

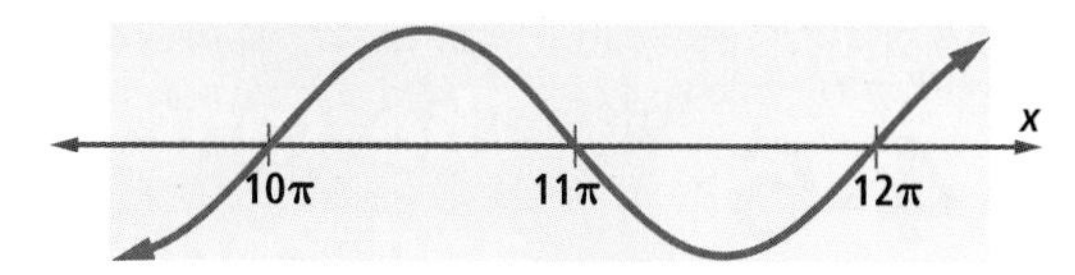

a. Evaluate $f\left(\frac{23\pi}{2}\right)$.

b. For what value of x, in the interval from 10π to 12π, does $f(x) = 1$?

c. Tell whether f is the cosine function or sine function. Justify your answer.

10. The graph of the cosine function is reflection-symmetric to the y-axis. What property of cosines is a result of this symmetry?

REVIEW

In 11 and 12, *A* is a point on a circle with center at the origin. Find the coordinates of *A* for the given value of θ. (Lesson 4-4)

11.

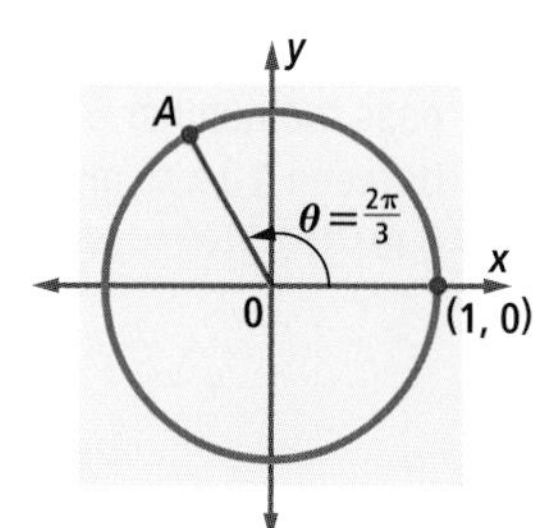

12.

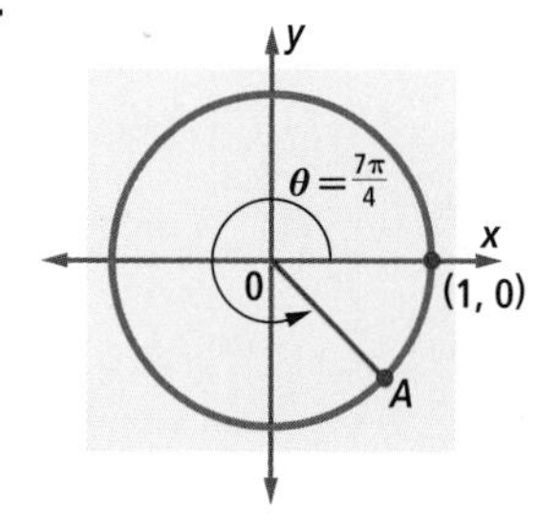

13. In radians, what is the sum of the measures of the angles of a pentagon? **(Lesson 4-1)**

14. An old 78 RPM record revolves through 78 revolutions in a minute. How many radians is this per second? **(Lesson 4-1)**

15. The measure of an angle is k radians. Convert this measure to degrees. **(Lesson 4-1)**

16. The students in Ms. T. Chare's 1st period geometry class measured their heights h in centimeters and recorded the following five-number summary of their data: $\overline{h} = 165$; min $= 137$; $Q_1 = 154$; median $= 168$; $Q_3 = 174$; max $= 188$. Are there any outliers in the data set? Explain your answer. **(Lesson 1-4)**

EXPLORATION

17. At what angle to the x-axis does the graph of $y = \sin x$ pass through $(0, 0)$? Give numerical and visual evidence supporting your answer.

QY ANSWER

From the unit circle, you can tell that $\sin\theta$ is positive when $0 < \theta < \pi$ and negative when $\pi < \theta < 2\pi$. The maximum value is 1, when $\theta = \frac{\pi}{2}$, and the minimum value is –1, when $\theta = \frac{3\pi}{2}$.

Lesson

4-6 The Tangent Function and Periodicity

Vocabulary

tangent function
periodic function
period of a function

BIG IDEA The sine and cosine functions are periodic, repeating every 2π or 360°. The tangent function is periodic, repeating every π or 180°.

A frieze pattern is a visual design that repeats over and over along a line. The frieze pattern at the right appears on the Chan Chan ruins in Trujillo, Peru.

In Lesson 4-5, you used values of sine and cosine to graph trigonometric functions. You also observed that, like frieze patterns, their graphs repeat as you move horizontally. This lesson extends those ideas to the tangent function.

Mental Math

How many times does the minute hand of a clock pass the number 6 between 10 A.M. and 6 P.M.?

The Tangent Function

The correspondence $\theta \rightarrow \tan \theta$, when θ is a real number, defines the **tangent function**. From the definition $\tan \theta = \frac{\sin \theta}{\cos \theta}$, values for the tangent function can be generated.

Activity

Step 1 The table below contains some exact values of $\tan \theta$. It also shows decimal equivalents of those values. Fill in the missing values.

θ	0	$30° = \frac{\pi}{6}$	$45° = \frac{\pi}{4}$	$60° = \frac{\pi}{3}$	$90° = \frac{\pi}{2}$	$120° = \frac{2\pi}{3}$	$135° = \frac{3\pi}{4}$	$150° = \frac{5\pi}{6}$	$180° = \pi$
$\tan \theta$ (exact)	0	$\frac{\sqrt{3}}{3}$	1	$\sqrt{3}$	undefined	?	?	?	0
$\tan \theta$ (approx.)	0	0.577	1	1.732	undefined	?	?	?	0

θ	$210° = \frac{7\pi}{6}$	$225° = \frac{5\pi}{4}$	$240° = \frac{4\pi}{3}$	$270° = \frac{3\pi}{2}$	$300° = \frac{5\pi}{3}$	$315° = \frac{7\pi}{4}$	$330° = \frac{11\pi}{6}$	$360° = 2\pi$
$\tan \theta$ (exact)	?	?	?	?	?	?	?	?
$\tan \theta$ (approx.)	?	?	?	?	?	?	?	?

Step 2 At the right, a graph of the values of the tangent function given in the first part of the table from Step 1 is shown. Copy this graph, and add the points you found in Step 1 to the graph.

Step 3 Draw a smooth curve through these points, to show the graph of $y = \tan \theta$ for all θ, $0° \leq \theta \leq 360°, 0 \leq \theta \leq 2\pi$ where $\tan \theta$ is defined.

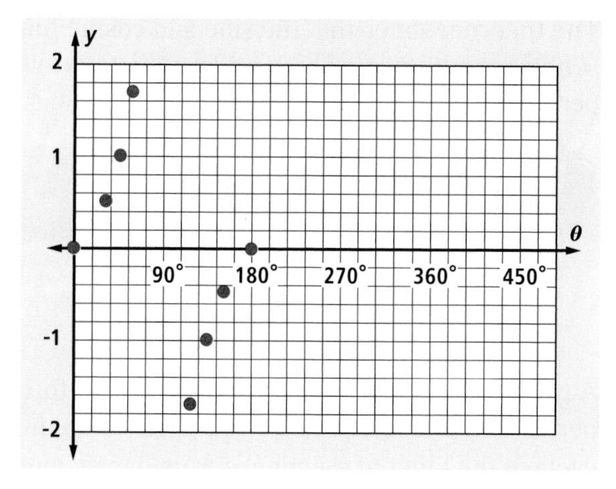

The Graph of the Tangent Function

At the right is a graph of $y = \tan x$ for $-\frac{3\pi}{2} \leq x \leq \frac{5\pi}{2}$. Notice that this graph looks strikingly different from the graphs of both the sine and cosine functions. The tangent function has asymptotes and does not have a maximum or minimum value.

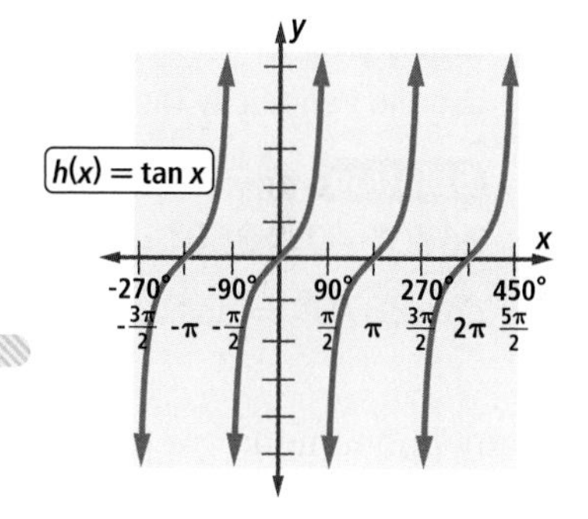

Example 1

Consider $f(x) = \tan x$.

a. **Give the domain and range of the function *f*.**

b. **Is *f* an odd function, an even function, or neither? Justify your answer.**

Solution

a. Because the tangent function has multiple vertical asymptotes, the domain of the tangent function is the set of all real numbers except odd multiples of 90° or $\frac{\pi}{2}$. Notice that the tangent function has no minimum or maximum values. Therefore, its range is the set of all real numbers.

b. From the Opposites Theorem, $\tan(-x) = -\tan x$ for all x. Thus, the tangent function is an odd function.

Periodicity and the Trigonometric Functions

The periodic nature displayed by sine, cosine, and tangent is summarized in the following theorem.

Periodicity Theorem

For any real number x and any integer n,

$$\sin x = \sin(x + n \cdot 2\pi) = \sin(x + n \cdot 360°)$$
$$\cos x = \cos(x + n \cdot 2\pi) = \cos(x + n \cdot 360°)$$
$$\tan x = \tan(x + n \cdot \pi) = \tan(x + n \cdot 180°).$$

The theorem states that the sine and cosine functions are *periodic functions* with *period* 360° or 2π radians, while the tangent function is a periodic function with period 180° or π radians.

Definitions of Periodic and Period

A function f is **periodic** if there is a positive number p such that $f(x + p) = f(x)$ for all x in the domain of f. The smallest such p, if it exists, is called the **period** of f.

A part of the function from any particular x to $x + p$, where p is the period of the function, is called a *cycle* of the function. For instance, one cycle of the tangent function is from 0 to π; another is from $-\frac{\pi}{2}$ to $\frac{\pi}{2}$.

Example 2

Use the Periodicity Theorem to find cos 2670°.

Solution $\frac{2670°}{360°} \approx 7.4$, so $2670° - 7 \cdot 360°$ will be less than 360°.
$2670° - 7 \cdot 360° = 150°$, so $R_{2670°} = R_{150°}$. Therefore,
$\cos 2670° = \cos 150° = -\frac{\sqrt{3}}{2}$.

Many phenomena are periodic, including tides, calendars, heart beats, actions of circular gears, phases of the moon, and seasons of the year.

GUIDED

Example 3

The graph at the right shows normal human blood pressure as a function of time. Blood pressure is *systolic* when the heart is contracting and *diastolic* when the heart is expanding. The changes from systolic to diastolic blood pressure create the pulse. For this function, determine each.

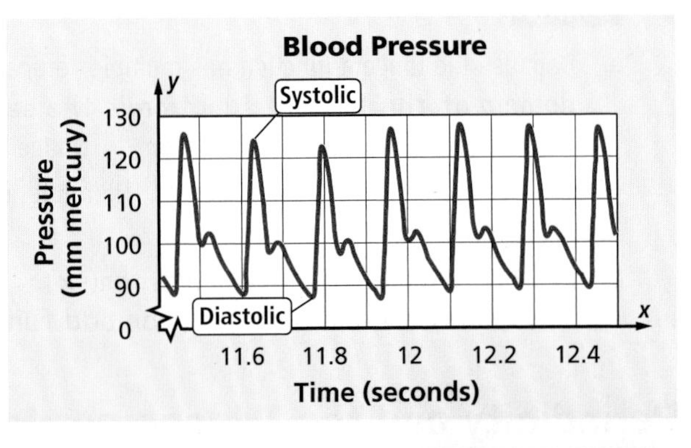

a. the maximum and minimum values

b. the range

c. the period

Solution

a. The maximum and minimum values of the graph are those values in which the graph obtains a highest and lowest point, respectively. The maximum value on this graph is _?_, the minimum value is _?_.

b. The range is the maximum value minus the minimum value. From Part a, the range shown on this graph is _?_.

c. The period is the range of x-values for the smallest section of the graph that can be translated horizontally onto itself. The period shown on this graph is _?_ seconds.

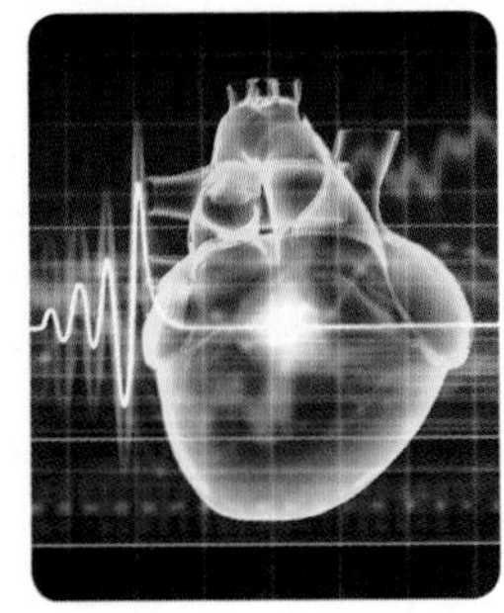

Questions

COVERING THE IDEAS

1. a. List all values of θ between 0° and 360° such that $\cos\theta = 0$.
   b. What is $f(\theta) = \tan\theta$ for these θ-values?
   c. What do these values of θ mean for the graph of the tangent function?
2. List all of the values of x from 0 to 2π for which $\sin x = 0$. What do these x-values indicate for the graph of the tangent function?

In 3–5, use the Periodicity Theorem to evaluate.

3. $\sin 495°$ 4. $\cos 810°$ 5. $\tan 3570°$
6. Given that $\tan\frac{4\pi}{9} \approx 5.671$, use the Periodicity Theorem to evaluate.
   a. $\tan\frac{13\pi}{9}$ b. $\tan -\frac{5\pi}{9}$ c. $\tan\frac{22\pi}{9}$
7. What is the period of the function with the given equation?
   a. $y = \sin x$ b. $y = \cos x$ c. $y = \tan x$

APPLYING THE MATHEMATICS

8. Suppose that f is a periodic function whose domain is the real numbers. One cycle of f is graphed at the right.
   a. What is the period of f?
   b. Graph f on the interval $-15 \leq x \leq 15$.
   c. Find $f(51)$.
   d. Find four integer values of x such that $f(x) = 0$.

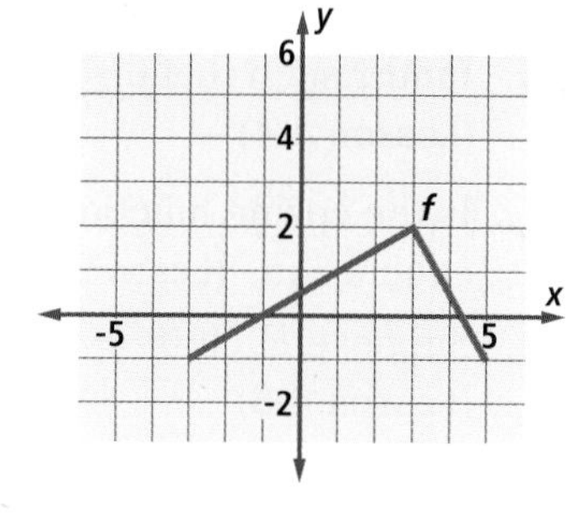

9. If one endpoint of a cycle of the cosine function is 90°, where is the other endpoint?
10. If one endpoint of a cycle of the tangent function is $\frac{\pi}{2}$, where is the other endpoint?
11. State equations for two of the asymptotes of the tangent function
    a. in radians. b. in degrees.
12. Let $f(n)$ be the number in the nth decimal place of $\frac{1}{7}$.
    a. Give the values of $f(1)$, $f(2)$, $f(3)$, and $f(4)$.
    b. f is a periodic function. What is its period?
13. The table at the right contains hourly data for the height of tide relative to the mean low water level in Pago Pago, American Samoa on October 5, 2008.
    a. Create a scatterplot of the data.
    b. Determine the range of the data.
    c. From the scatterplot, estimate the period of the data.

Hour	Height	Hour	Height
0	1.59	12	2.03
1	1.29	13	1.65
2	0.91	14	1.14
3	0.61	15	0.83
4	0.65	16	0.62
5	0.79	17	0.49
6	1.04	18	0.53
7	1.32	19	0.87
8	1.75	20	1.15
9	2.01	21	1.49
10	2.20	22	1.77
11	2.21	23	1.87

Source: National Oceanographic and Atmospheric Administration

REVIEW

14. Fill in the blanks for the graph of the sine function at the right. **(Lesson 4-5)**

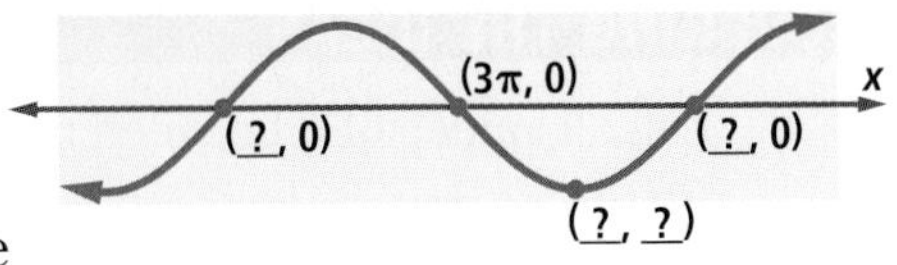

15. Refer to the predator-prey graph below. **(Lesson 4-5)** For both the predator and the prey functions, determine the
    a. domain. b. maximum and minimum. c. period.

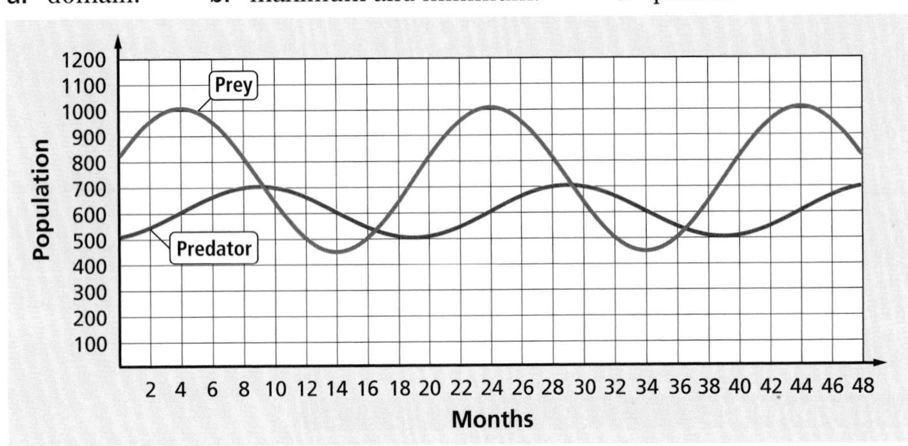

16. Find x such that $0 \le x \le 2\pi$, if $\cos x = \frac{1}{2}$ and $\sin x = -\frac{\sqrt{3}}{2}$. **(Lesson 4-4)**

17. Is the cosine function odd, even, or neither? Justify your conclusion. **(Lessons 4-5, 3-4)**

18. Suppose $R_\theta(1, 0) = (-0.75, y)$ and is a point in Quadrant II. Find y. **(Lesson 4-3)**

19. Under some translation T, the point (-6, 2) is mapped to (0, 7).
    a. State the rule for T. b. Find $T(9, 9)$. **(Lesson 3-2)**

20. The graph at the right is from *Weather on the Planets*, by George Ohring. It shows how temperature is a function of latitude on Earth and Mars when it is spring in one hemisphere on each planet. Let L be the latitude on each planet, $E(L)$ be the average temperature at latitude L on Earth, and $M(L)$ be the average temperature at latitude L on Mars.
    a. Is L the dependent or independent variable?
    b. Estimate $E(60)$.
    c. Estimate $E(0) - M(0)$, and state what quantity this expression represents.
    d. What is the range of M? **(Lesson 2-1)**

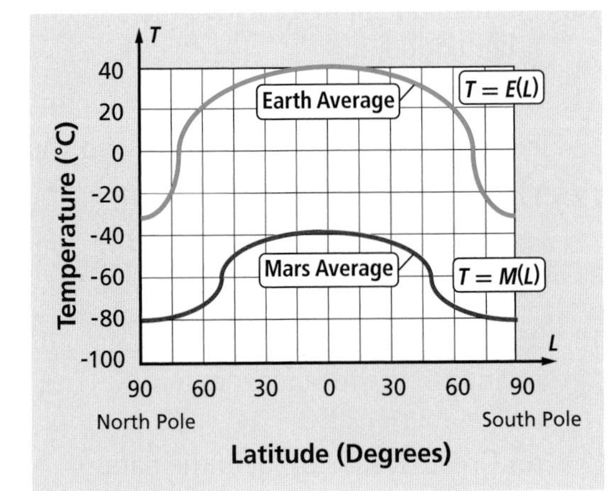

EXPLORATION

21. The word *tangent* has another meaning in geometry. It also has another meaning in English. What are these meanings?

Lesson 4-7 Scale-Change Images of Trigonometric Functions

Vocabulary

sine wave
amplitude
frequency

▶ **BIG IDEA** The Graph Scale-Change Theorem can be applied to obtain the equation, amplitude, and period of scale-change images of the graph of a parent trigonometric function.

Sine Waves

A pure tone, such as that produced by a tuning fork, travels in a *sine wave*. A **sine wave** is the image of the graph of the sine or cosine function under a composite of translations and scale changes. The pitch of the tone is related to the period of the wave; the longer the period, the lower the pitch. The intensity of the tone is related to the *amplitude* of the wave. The **amplitude** of a sine wave is half the distance between its maximum and minimum values.

Mental Math

The expression $3r - 8$ is in the form $ax - h$. Rewrite it in the form $\frac{x - \frac{h}{a}}{\frac{1}{a}}$.

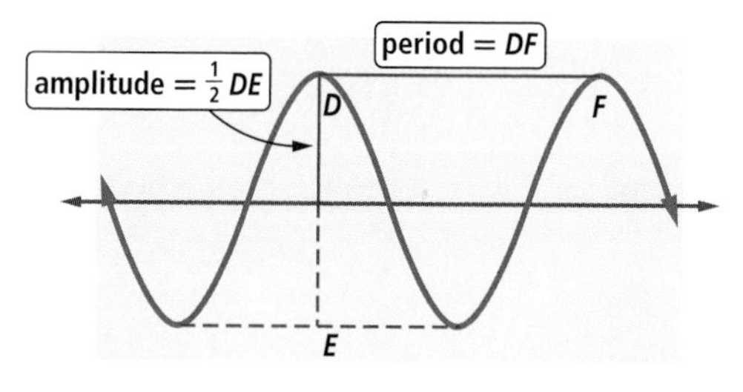

An oscilloscope is used to record changes in the voltage of an electric current.

STOP QY1

Stretching a sound sine wave vertically changes the intensity of the tone. Stretching a sound wave horizontally changes its pitch. Recall from Lesson 3-5 that a scale change is a mapping $S: (x, y) \rightarrow (ax, by)$ centered at the origin with $a \neq 0$ and $b \neq 0$. Under this scale change, an equation for the image of $y = f(x)$ is $\frac{y}{b} = f\left(\frac{x}{a}\right)$.

▶ **QY1**

The range of the sine function is $\{y \mid -1 \leq y \leq 1\}$, so the amplitude of the graph of $y = \sin x$ is __?__.

Example 1 shows how a scale change affects both the amplitude and period of a sine wave.

Example 1

Consider the function with equation $y = 6\cos\left(\frac{x}{3}\right)$.

a. **Explain how this function is related to its parent function, the cosine function.**

b. **Identify its amplitude and its period.**

(continued on next page)

Solution

a. Divide each side of the given equation by 6. This rewrites the function rule in a form that can be analyzed using the Graph Scale-Change Theorem.

$$\frac{y}{6} = \cos\left(\frac{x}{3}\right)$$

In the equation $y = \cos x$, y has been replaced by $\frac{y}{6}$ and x by $\frac{x}{3}$. The graph of the parent function is stretched vertically by a factor of 6 and horizontally by a factor of 3.

b. The vertical stretch means that the maximum and minimum values of the parent function are multiplied by 6. Hence, the given function has amplitude $\frac{1}{2}(6 - -6) = 6$. The horizontal stretch means that the period 2π of the parent function is also stretched by a factor of 3. So the given function has a period of $3(2\pi) = 6\pi$.

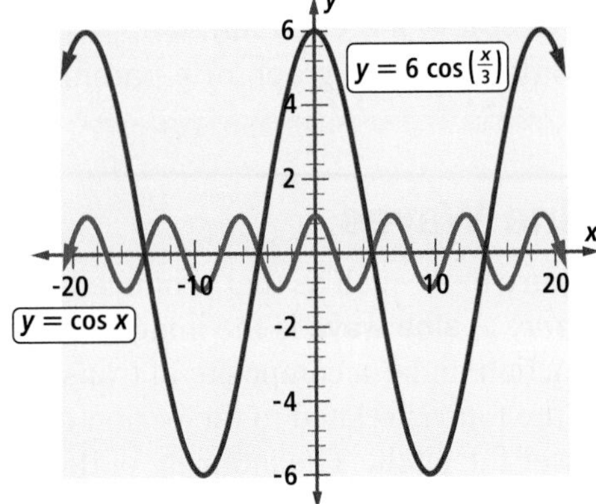

Check Graphs of parts of $y = \cos x$ and $y = 6\cos\left(\frac{x}{3}\right)$ are shown at the right. Only a little more than two cycles of $y = 6\cos\left(\frac{x}{3}\right)$ are shown, but from this you can see that the amplitude and period found above were correct.

If the graphs of $y = \cos x$ and $y = 6\cos\left(\frac{x}{3}\right)$ in Example 1 represented sound waves, the sound of $y = 6\cos\left(\frac{x}{3}\right)$ would be 6 times as loud and have a lower pitch than that of $y = \cos x$.

In general, the functions defined by $\frac{y}{b} = \sin\left(\frac{x}{a}\right)$ and $\frac{y}{b} = \cos\left(\frac{x}{a}\right)$ are equivalent to the functions defined by

$$y = b\sin\left(\frac{x}{a}\right) \quad \text{and} \quad y = b\cos\left(\frac{x}{a}\right),$$

where $a \neq 0$ and $b \neq 0$, and their graphs are images of the graphs of the parent functions

$$y = \sin x \qquad \text{and} \quad y = \cos x$$

under the scale change that maps (x, y) to (ax, by). The theorem below indicates the relationship of the constants a and b to the properties of the sine waves.

Theorem (Properties of Sine Waves)

The graphs of the functions defined by $y = b\sin\left(\frac{x}{a}\right)$ and $y = b\cos\left(\frac{x}{a}\right)$ have amplitude $= |b|$ and period $= 2\pi|a|$.

 **QY2**

There is a corresponding theorem for the graph of the function with equation $\frac{y}{b} = \tan\left(\frac{x}{a}\right)$. However, the parent tangent function does not have an amplitude and the period of the parent tangent function is π, so the period of $\frac{y}{b} = \tan\left(\frac{x}{a}\right)$ is $\pi|a|$.

▶ QY2

In the theorem, why is absolute value used in the calculation of both amplitude and period?

Example 2

The graph at the right shows an image of the graph of $y = \sin x$ under a scale change. Find an equation for the image.

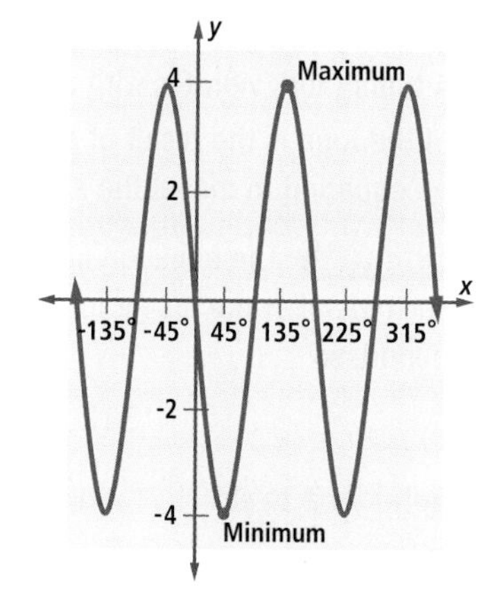

Solution An equation for the image is of the form $y = b \sin\left(\frac{x}{a}\right)$. From the graph, there is a minimum at (45°, –4) and a maximum at (135°, 4). The difference between the maximum and minimum values of y is 8, so the amplitude is 4. The graph shows a cycle from 0° to 180°, so the period is 180°. Therefore, $|b| = 4$ and $360°|a| = 180°$. Thus, $b = 4$ or -4 and $a = \frac{1}{2}$ or $-\frac{1}{2}$. Consider the four possibilities.

$$y = 4 \sin(2x)$$
$$y = -4 \sin(2x)$$
$$y = 4 \sin(-2x)$$
$$y = -4 \sin(-2x)$$

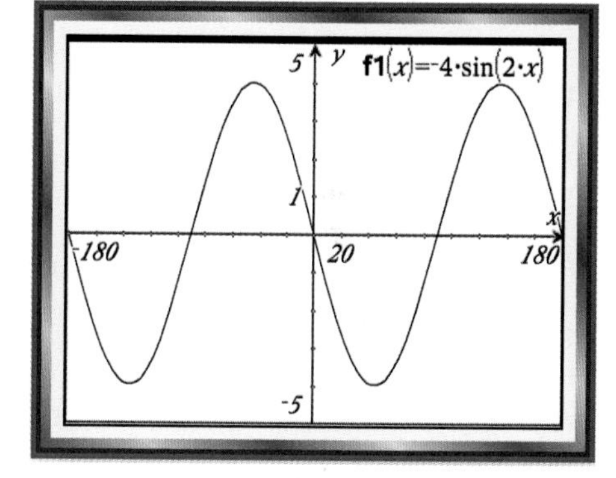

Notice that the graph pictured must be a reflection image of the graph of the parent sine function since, for example, starting at zero, the graph decreases as you move towards 45°. One equation that will produce the graph is $y = -4 \sin(2x)$.

Check Use a graphing utility to check that your equation produces the given graph.

STOP QY3

QY3

Which of the other choices in the solution will produce the graph?

The Frequency of a Sine Wave

Notice that the graph of $y = \cos x$ in Example 1 completes 3 cycles for every one completed by the graph of $y = 6\cos\left(\frac{x}{3}\right)$. We say that $y = \cos x$ has three times the *frequency* of $y = 6\cos\left(\frac{x}{3}\right)$. In general, the **frequency** of a periodic function is the reciprocal of the period, and represents the number of cycles the curve completes per unit of the independent variable. Thus, the frequency of the cosine function is $\frac{1}{2\pi}$, and the frequency of the function $y = 6\cos\left(\frac{x}{3}\right)$ is $\frac{1}{6\pi}$.

When a sine wave represents sound, doubling the frequency results in a pitch one octave higher. So the graph of $y = 6\cos\left(\frac{x}{3}\right)$ represents a sound with pitch between one and two octaves lower than the pitch represented by $y = \cos x$ and with 6 times the intensity. It is common in these situations to view the x-axis as representing time. In sound waves, the y-axis represents pressure, typically measured in newtons (abbreviated N) per square meter, $\frac{\text{N}}{\text{m}^2}$.

Example 3

A tuning fork vibrates with a frequency of 512 cycles per second. The intensity of the tone is the result of a vibration whose maximum pressure is 22 $\frac{N}{m^2}$. Find an equation to model the sound wave produced by the tuning fork.

Solution The equation has the form $y = b\sin\left(\frac{x}{a}\right)$ where *x* is the time in seconds after the tuning fork is struck. The frequency is the reciprocal of the period, so

$$512 = \frac{1}{2\pi|a|}.$$

Solve for *a* to get $a = \pm\frac{1}{1024\pi}$.

The maximum pressure of the air gives the amplitude $b = 22$. Choosing the positive value of *a*, the equation is $y = 22\sin(1024\pi x)$.

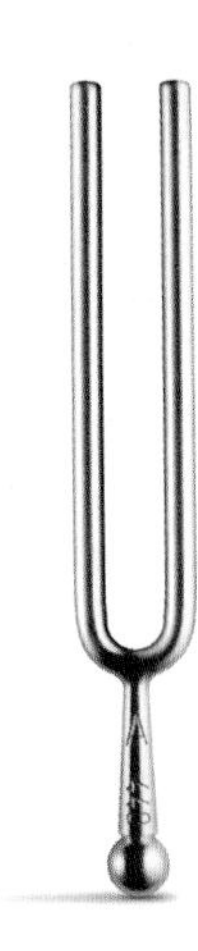

Tuning In Tuning forks are most commonly used to tune musical instruments to the note "A."

Knowing the number of cycles per unit of independent variable can help you solve trigonometric equations.

Example 4

Without using technology, determine how many solutions each equation below has on the interval $0 \le x \le 2\pi$. Confirm your answer with a graph.

a. $\cos(3x) = 0.8$ b. $5\tan\left(\frac{x}{2}\right) = 3$

Solution

a. The parent cosine function has two solutions in the interval $0 \le x \le 2\pi$ because this domain represents one cycle. $y = \cos(3x)$ is the image of $y = \cos x$ under a horizontal shrink of magnitude $\frac{1}{3}$. That means that each cycle of $y = \cos(3x)$ is one-third as long as a cycle of $y = \cos x$, so there will be three cycles on the interval $0 \le x \le 2\pi$ and six solutions. A graph confirms that there are six points of intersection of $y = \cos(3x)$ and $y = 0.8$ on $0 \le x \le 2\pi$.

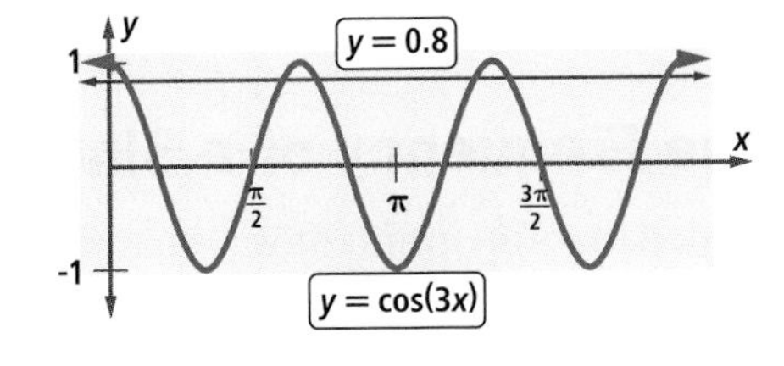

b. The graph of $y = 5\tan\left(\frac{x}{2}\right)$ is the image of the graph of $y = \tan x$ under a horizontal stretch by a factor of 2. So each cycle is twice as long, and there are half as many points of intersection on the interval $0 \le x \le 2\pi$. Therefore, there is only one solution. The graph at the right confirms a single solution.

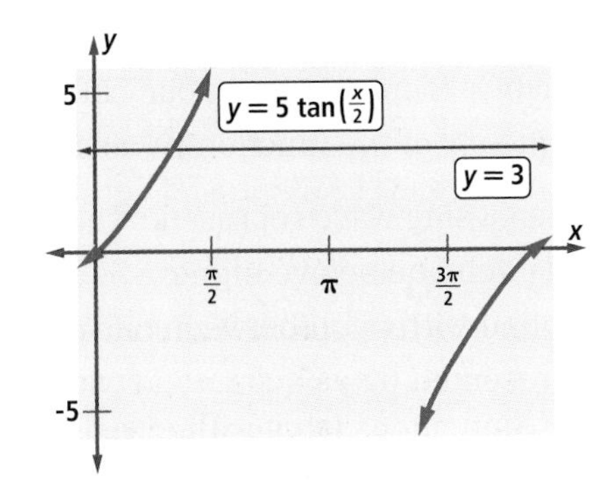

Questions

COVERING THE IDEAS

1. Consider the function with equation $y = \frac{1}{5}\sin x$.
   a. **True or False** The graph of this function is a sine wave.
   b. What is its period?
   c. What is its amplitude?
   d. Sketch graphs of $y = \frac{1}{5}\sin x$ and $y = \sin x$ on the same set of axes for $-\pi \le x \le 2\pi$.
   e. Describe how the two graphs in Part d are related.

In 2 and 3, an equation for a sine wave is given.
a. Find its amplitude. b. Find its period.

2. $y = 3\cos x$
3. $\frac{y}{4} = \cos\left(\frac{x}{3}\right)$

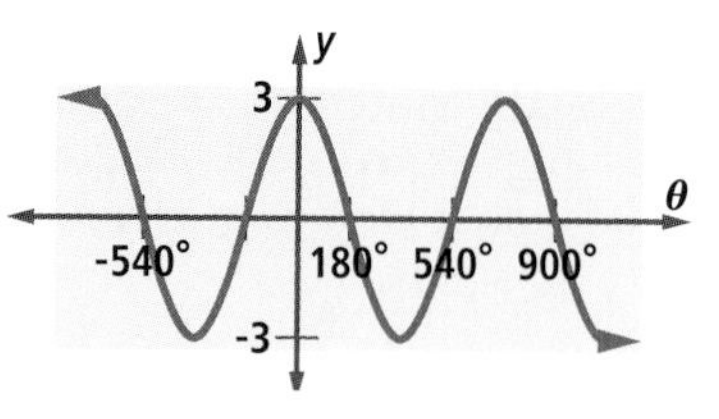

4. **Multiple Choice** Which equation could yield the graph at the right?

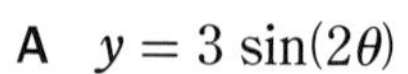

A $y = 3\sin(2\theta)$
B $y = 3\cos(2\theta)$
C $y = 3\cos\left(\frac{\theta}{2}\right)$
D $y = 3\sin\left(\frac{\theta}{2}\right)$

5. a. Find an equation of the image of the graph of $y = \sin x$ under the transformation $(x, y) \rightarrow (5x, y)$.
   b. Find the amplitude and period of the image.
6. a. Give the period and amplitude of $y = \frac{1}{5}\sin(3\theta)$.
   b. Check using a graphing utility.

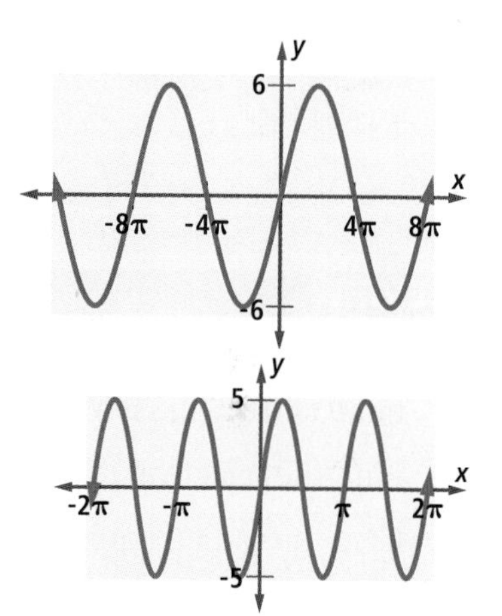

7. The graph at the right is an image of the graph of $y = \sin x$ under a scale change. Find an equation for this curve.
8. Refer to the graph sketched at the right.
   a. Identify the amplitude, period, and frequency.
   b. **Fill in the Blanks** If this graph represents a sound wave, then that sound is __?__ times as loud and has __?__ times the frequency of the parent sound wave with equation $y = \sin x$.

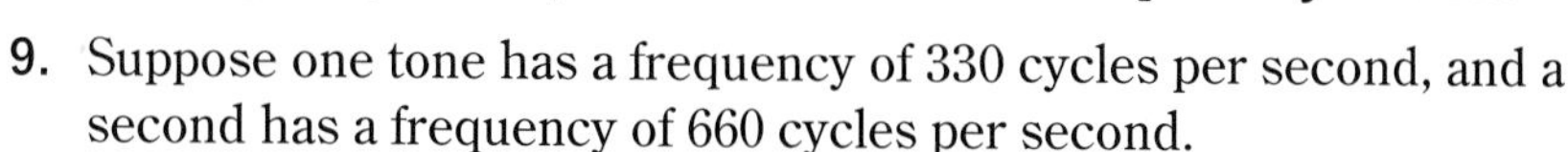

9. Suppose one tone has a frequency of 330 cycles per second, and a second has a frequency of 660 cycles per second.
   a. Which has the higher pitch?
   b. How many octaves higher is that pitch?
10. How many solutions does $\sin(5x) = 0.65$ have for $0 \le x \le 2\pi$?
11. Sketch one cycle of $6y = \cos\left(\frac{x}{4}\right)$, and label the zeros of the function.
12. **Multiple Choice** A sound wave whose parent is the graph of $y = \sin x$ has 3 times the frequency and 7 times the amplitude of the parent. What is a possible equation for this sound wave?

A $y = 7\sin(3x)$
B $y = 7\sin\left(\frac{1}{3}x\right)$

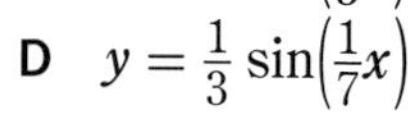

C $y = 3\sin\left(\frac{1}{7}x\right)$
D $y = \frac{1}{3}\sin\left(\frac{1}{7}x\right)$

Most modern full-size pianos have 88 keys spanning $7\frac{1}{4}$ octaves.

13. Consider a tuning fork vibrating at 440 cycles per second and displacing air molecules by a maximum of 32 $\frac{\text{N}}{\text{m}^2}$. Give a possible equation for the sound wave that is produced.

APPLYING THE MATHEMATICS

14. Residential electricity is called AC for "alternating current," because the direction of current flow alternates through a circuit. The current (measured in amperes) is a sine function of time. The graph at the right models an AC situation.

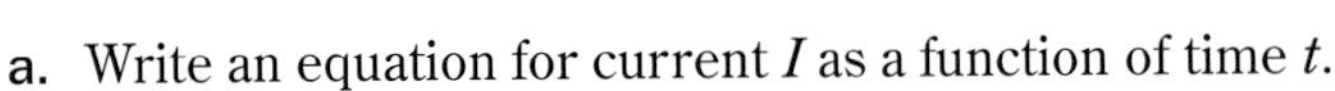

Current (amperes)
Time (seconds)

a. Write an equation for current I as a function of time t.

b. Find the current produced at 0.04 seconds.

15. Which of the functions f, g, and h, defined by $f(x) = \tan x$, $g(x) = \tan(3x)$, and $h(x) = 3\sin(2x)$, have the same period?

In 16–18, match each equation with its graph below.

16. $\frac{y}{2} = \sin\left(\frac{x}{2}\right)$ 17. $2y = \sin\left(\frac{x}{2}\right)$ 18. $\frac{y}{2} = \sin(2x)$

A

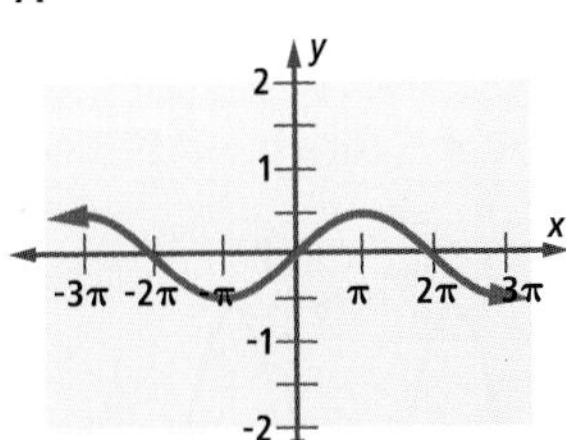

B

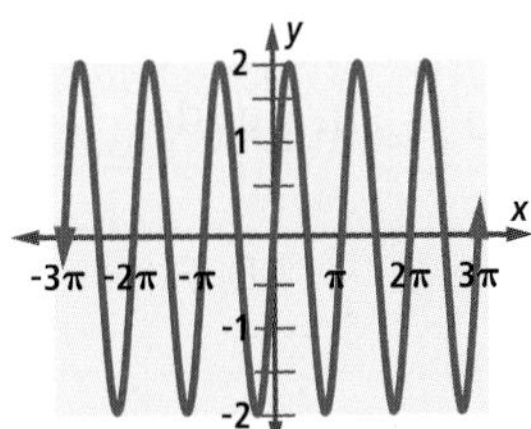

C

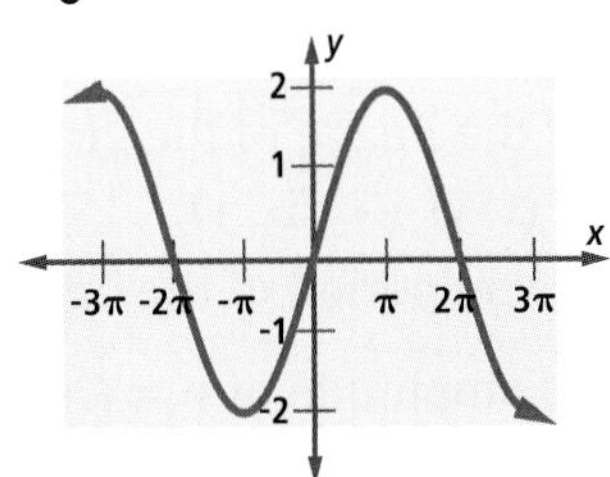

REVIEW

19. Given that $\tan 0.675 \approx 0.8$, find three other values of θ with $\tan\theta \approx 0.8$. **(Lesson 4-6)**

20. Find the exact value of $\tan(-120°)$. **(Lesson 4-4)**

21. Give the radian equivalent to each. **(Lesson 4-1)**

a. -720° b. 225° c. 315°

22. State the Graph-Translation Theorem. **(Lesson 3-2)**

23. a. Graph $f(x) = x^3$ and its image under the translation $T: (x, y) \to (x + 3, y - 1)$.

b. Find an equation for the image. **(Lesson 3-2)**

EXPLORATION

24. *Pitch* and *loudness* are common words for the frequency and amplitude of sound. Light waves are also modeled with trigonometric functions.

a. What properties of light do the frequency and amplitude of light waves represent?

b. Name some other characteristics that sound waves and light waves share.

QY ANSWERS

1. 1
2. Both are distances and cannot be negative.
3. $y = 4\sin(-2x)$

Lesson

4-8 Translation Images of Trigonometric Functions

Vocabulary

phase shift

▶ **BIG IDEA** Translations do not affect the period or amplitude of parent trigonometric functions.

Phase Shifts

The graph of $y = \cos x$ is shown in blue along with a graph of a translation image.

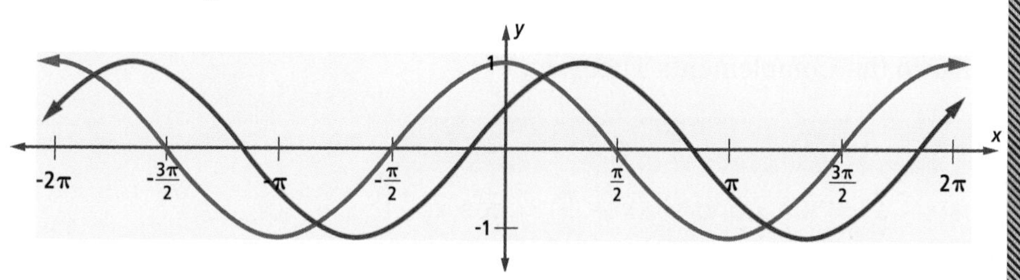

Recall from Lesson 3-2 how to translate graphs of functions horizontally and vertically. Horizontal translations of trigonometric functions have a special name: *phase shifts*. This name comes from the study of sound and electricity, where trigonometric functions are often applied. In general, the **phase shift** of a sine wave is the least positive or the greatest negative magnitude of a horizontal translation that maps the graph of $\left(\frac{y}{b}\right) = \cos\left(\frac{x}{a}\right)$ or $\left(\frac{y}{b}\right) = \sin\left(\frac{x}{a}\right)$ onto the wave.

Mental Math

Give the vertex of the parabola described by each equation.

a. $y = -x^2$

b. $y = (x - 5)^2 + 2$

c. $y = 4(x + 50)^2$

d. $y = ax^2 - c$

GUIDED

Example 1

Consider the function h with $h(x) = \sin(x + 60°)$. Identify the phase shift.

Solution The expression $x + 60°$ is equivalent to $x -$ _?_.
Rewrite the equation as $h(x) = \sin(x - -60°)$. The graph of h is the image of the graph $y = \sin x$ under a horizontal translation of _?_ (left/right). Thus the phase shift is _?_.

People who work with electricity, such as electrical engineers and electricians, use phase shifts. In an alternating current circuit, for example, two waves—the voltage and the current flow—are involved. If these waves coincide, then they are said to be *in phase*. If the current flow lags behind the voltage, then the circuit is *out of phase* and an *inductance* is created. Inductance helps to keep the current flow stable.

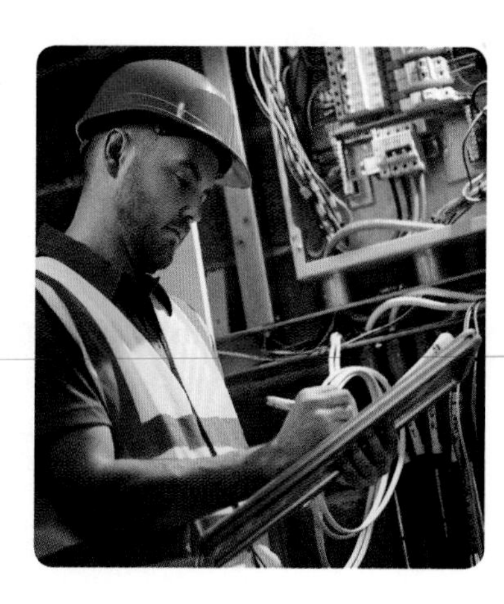

Example 2

Maximum inductance in an alternating current occurs when the current flow lags behind the voltage by $\frac{\pi}{2}$. In a situation of maximum inductance, find an equation for the current, and sketch the two waves. Assume that the two waves have the same amplitude and period, and that the voltage is modeled by the equation $y = \cos x$.

Solution Maximum inductance occurs when the current has a phase shift of $\frac{\pi}{2}$. There is no vertical shift. If the equation for the current is of the form $y - k = \cos(x - h)$, then $k = 0$ and $h = \frac{\pi}{2}$. An equation for the current is $y = \cos(x - \frac{\pi}{2})$. The waves are graphed at the right.

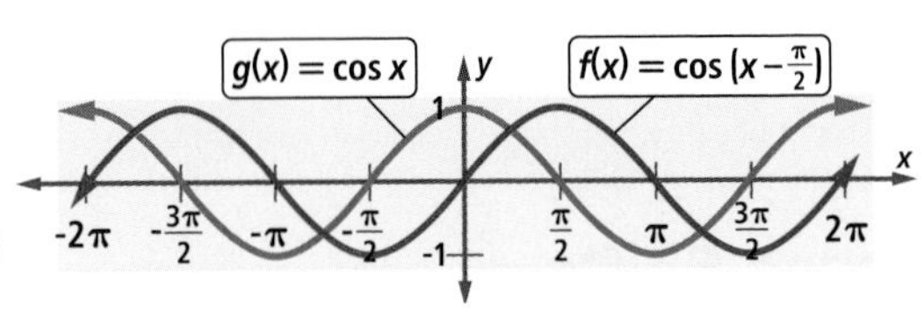

Examine the graphs carefully. Notice that the graph of $y = \cos(x - \frac{\pi}{2})$ seems to coincide with the graph of $y = \sin x$. This relationship gives rise to an identity that is similar to the Complements Theorem.

Theorem (Phase Shift Identity)

For all real numbers x, $\cos(x - \frac{\pi}{2}) = \sin x$ and $\sin(x + \frac{\pi}{2}) = \cos x$.

Because the graph of the cosine function is a translation image of the graph of the sine function, these graphs are congruent.

Translation Images of Sine Waves

Example 3

Consider the graph of the function f, where $f(x) = \cos x$.

a. Find an equation for its image g under the translation $(x, y) \rightarrow (x, y - 2)$, and sketch a graph of $y = g(x)$.

b. Find the amplitude and period of the function g.

Solution

a. By the Graph-Translation Theorem, an equation for the image is $y = \cos x - 2$. This results in a vertical shift two units down from the parent function. Graphs of f and g are on the right.

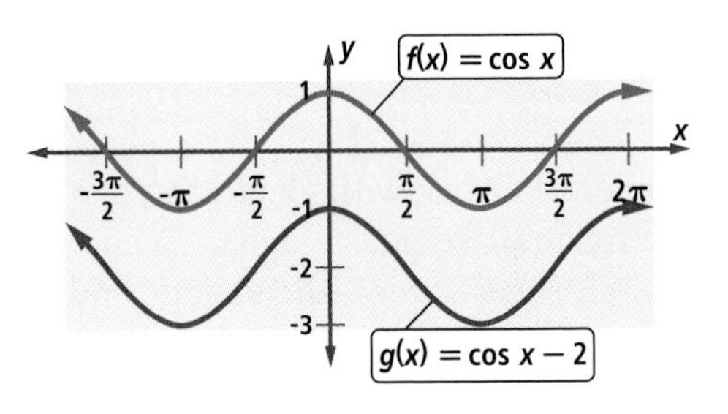

b. The maximum and minimum values of the cosine function are 1 and −1, respectively. So the maximum and minimum values of g are −1 and −3, respectively. Thus, the amplitude of g is $\frac{1}{2}|(-1 - -3)| = \frac{1}{2} \cdot 2 = 1$, the same as the amplitude of the cosine function. Similarly, the period of g is 2π, the same as the period of the parent function.

In general, if two curves are translation images of each other, then they are congruent. Thus, a translation of a sine or cosine wave preserves both its amplitude and its period.

From an analysis of a graph showing a translation image of any of the parent trigonometric functions, you can determine an equation for a translation image.

Example 4

Find an equation for the translation image of the graph of the cosine function shown below.

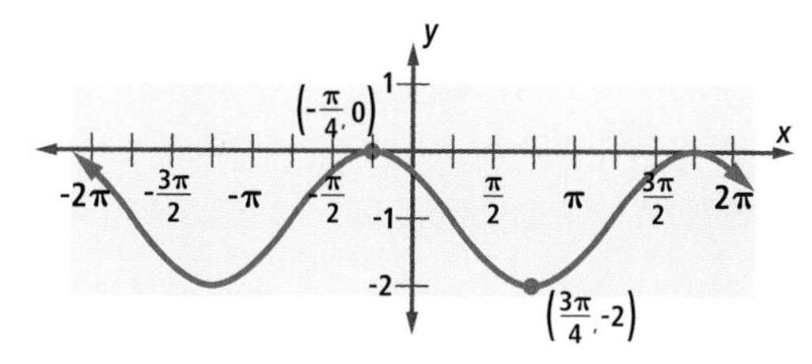

Solution There is both a phase shift and a vertical shift. Since the coordinates of a relative maximum are given, it is convenient to consider $\left(-\frac{\pi}{4}, 0\right)$ as the image of the maximum point (0, 1) of the cosine function. You can consider the point (0, 1) as having been translated $\frac{\pi}{4}$ units to the left and 1 unit down. So the phase shift is $= -\frac{\pi}{4}$.

The vertical shift is 1 unit down. Thus, an equation for this graph is $y = \cos\left(x + \frac{\pi}{4}\right) - 1$.

The two graphs are congruent, so the amplitude and period remain the same.

Check A graph of $y = \cos\left(x + \frac{\pi}{4}\right) - 1$ shows that the graph corresponds to the given graph.

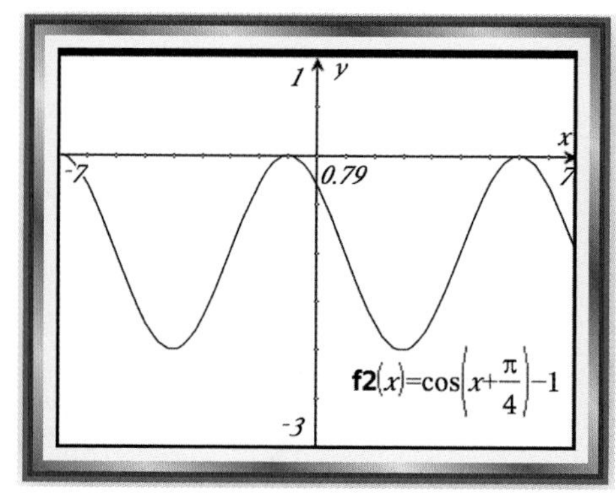

QY

In Example 4, you could have thought of the graph as a translation image of the sine function. (See Question 7.) In general, when using a sine or cosine function as a model (as you will in Lesson 4-10), the choice of which function is a matter of convenience. Sines and cosines can be used interchangeably if you pay attention to the phase shift.

▸ QY

Identify the phase shift and vertical translation of the graph of $y = \sin x$ so that its image is the graph of $y = \sin\left(x - \frac{\pi}{3}\right) + 4$.

Questions

COVERING THE IDEAS

1. **True or False** The graph of any function and its image under a translation are congruent.

2. Consider the function with equation $y = \cos\left(x + \frac{\pi}{3}\right)$.
   a. Identify the phase shift from $y = \cos x$.
   b. Copy and complete the table at the right, which shows the translation images of the points $(0, 1)$, $\left(\frac{\pi}{2}, 0\right)$, $(\pi, -1)$, $\left(\frac{3\pi}{2}, 0\right)$, and $(2\pi, 1)$.
   c. Use the points found in Part b to help you sketch two cycles of the graph of $y = \cos\left(x + \frac{\pi}{3}\right)$.

Preimage on $= \cos$	Image on $= \cos\left(+ \frac{\pi}{3}\right)$
$(0, 1)$	?
$\left(\frac{\pi}{2}, 0\right)$	?
$(\pi, -1)$	?
$\left(\frac{3\pi}{2}, 0\right)$	?
$(2\pi, 1)$	?

3. Consider the function with equation $y = \sin\left(x + \frac{3\pi}{4}\right) + 5$
   a. Identify the phase shift from $y = \sin x$.
   b. Identify the vertical shift.

4. Consider the translation $T: (x, y) \rightarrow \left(x + \frac{\pi}{6}, y + 0.5\right)$.
   a. Find an equation for the image of the graph of the sine function under T.
   b. Find the amplitude, the period, and the phase shift of the image.

In 5 and 6, write an equation for the function that is graphed.

5.

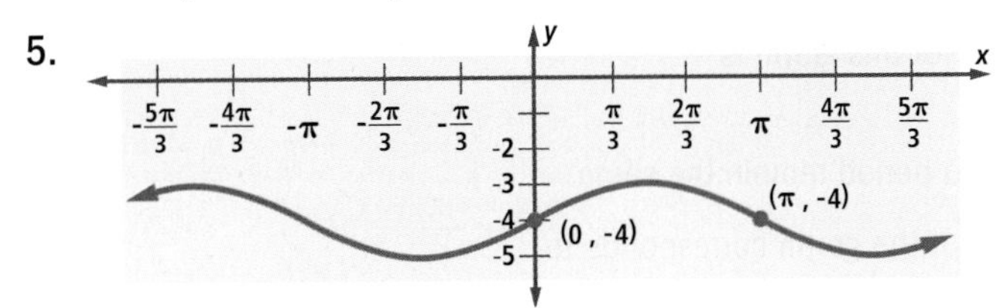

6.

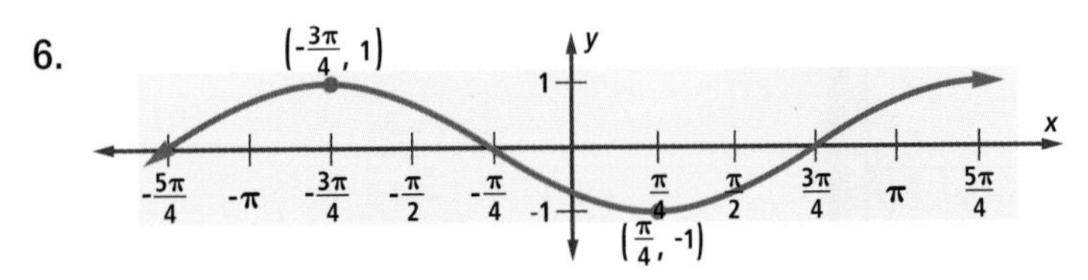

7. Refer to Example 4. Find an equation for the graph, thinking of it as a translation image of the graph of the sine function.

8. Consider the graph at the right.
   a. Write an equation using the sine function to describe the graph.
   b. Write an equation using the cosine function to describe the graph.

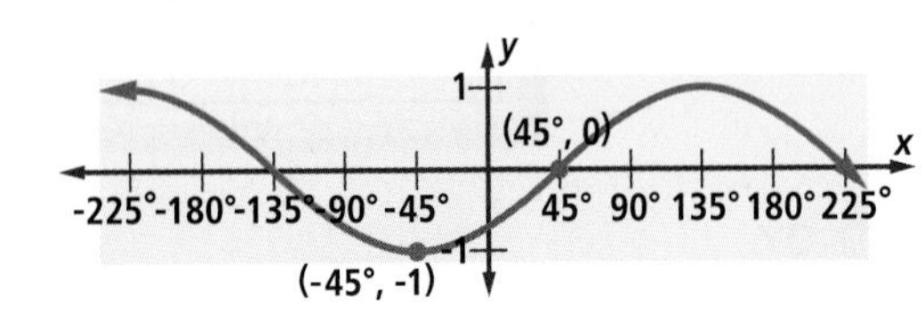

9. a. Give an equation for a translation that could be used to map the graph of $f(x) = \sin x$ onto the graph of $g(x) = \cos x$.
   b. Use your answer to Part a to write an expression for $\cos x$ in terms of $\sin x$.

APPLYING THE MATHEMATICS

10. When two waves from the same source travel different distances to reach an object—for example, when radio waves are reflected off of buildings in a city—they may arrive out of phase. Suppose that two such waves are shown in the graph at the right.

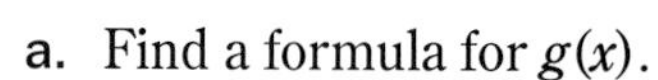

- **a.** Find a formula for $g(x)$.
- **b.** The signal you would receive is represented by the sum $f + g$. Graph that function.
- **c.** Find the period, amplitude, and phase shift from the sine function of the function you graphed in Part b.
- **d.** Identify a value of x for which $f(x) + g(x) = 0$. How can that value be found using just the graphs of f and g?

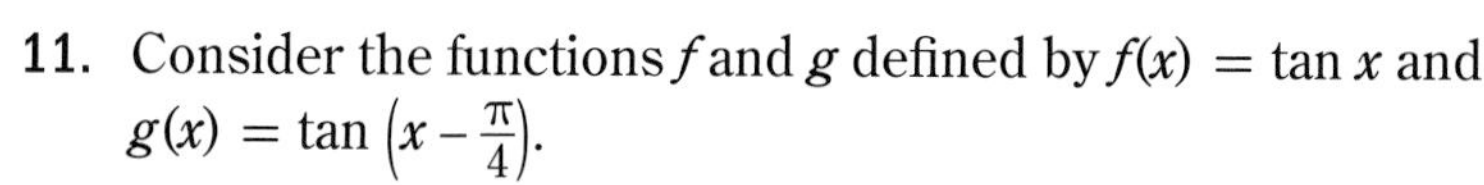

11. Consider the functions f and g defined by $f(x) = \tan x$ and $g(x) = \tan\left(x - \frac{\pi}{4}\right)$.

- **a.** Give the period for both f and g.
- **b.** Describe the transformation that maps the graph of f onto the graph of g.
- **c.** What is the image of the point $\left(\frac{\pi}{4}, 1\right)$ under this transformation?
- **d.** Write the equations of the asymptotes of the graph of g on the interval $\frac{\pi}{4} \le x \le \frac{9\pi}{4}$.
- **e.** Sketch two cycles of the graph of g on the interval $\frac{\pi}{4} \le x \le \frac{9\pi}{4}$.

12. The height h in meters of the tide in a harbor is given by $h = 0.8\cos\left(\frac{\pi}{6}t\right) + 6.5$, where t is the time in hours after high tide.

- **a.** Calculate h at $t = 0$, $t = 1$, and $t = 2$.
- **b.** Sketch a graph of this function for $0 \le t \le 24$.
- **c.** What is the minimum height of the tide during this 24-hour time period?
- **d.** At what times during the 24 hours after high tide does the minimum height occur?
- **e.** What is the period of this sine wave and what does the period mean in terms of the tide?

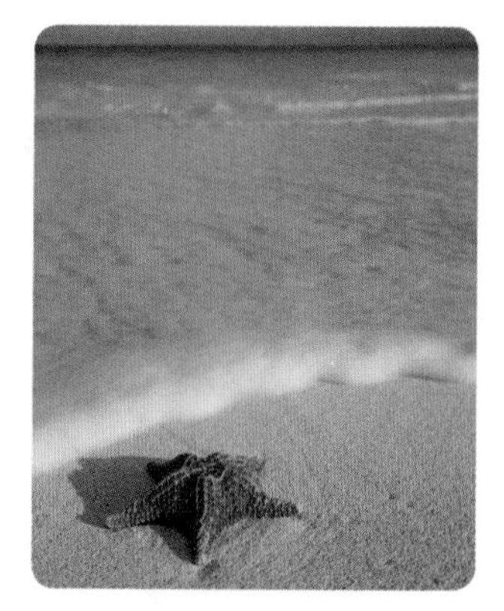

Tides describe the rise and fall in sea level relative to the land due to gravitational pull of the moon and Sun.

REVIEW

In 13 and 14, an equation for a function is given.
a. Find its amplitude. b. Find its period. (Lesson 4-7)

13. $y = 4\sin x$

14. $y = 5\cos\left(\frac{x}{2}\right)$

15. Sketch one complete cycle of the graph of $3y = \sin\left(\frac{x}{2}\right)$, and label the zeros of the function shown on the graph. **(Lesson 4-7)**

In 16 and 17, give the degree equivalent. (Lesson 4-1)

16. 7π

17. $\frac{5\pi}{4}$

18. Let f and g be functions whose equations are $f(x) = x^2 - 3$ and $g(x) = 2x + 1$. Find a formula for $f(g(x))$. **(Lesson 3-7)**

19. a. Draw $\triangle NOW$, where $N = (2, 6)$, $O = (-1, 4)$, and $W = (1, -4)$.
    b. Draw $\triangle N'O'W'$, the reflection image of $\triangle NOW$ over the y-axis.
    c. If r_y represents reflection over the y-axis, then $r_y : (x, y) \rightarrow$ __?__.
    (Lesson 3-4)

20. Match each transformation with the *best* description.
    (Lessons 3-5, 3-4, 3-2)

a. $M(x, y) = (x - 7, y + 5)$	(i) reflection over the x-axis
b. $N(x, y) = (y, x)$	(ii) reflection over the line $y = x$
c. $P(x, y) = (x, -y)$	(iii) scale change
d. $Q(x, y) = (0.2x, 20y)$	(iv) size change
e. $V(x, y) = \left(\frac{x}{2}, \frac{y}{2}\right)$	(v) translation

EXPLORATION

21. Noise canceling headphones were developed using the concept of "antiphase."
    a. Find out what antiphase means and how it is used in noise canceling headphones.
    b. If a sound wave is modeled by the equation $y = 12\cos\left(\frac{x}{4}\right)$, write an equation for the antiphase that would cancel the sound.
    c. Find other applications where antiphasing is used.

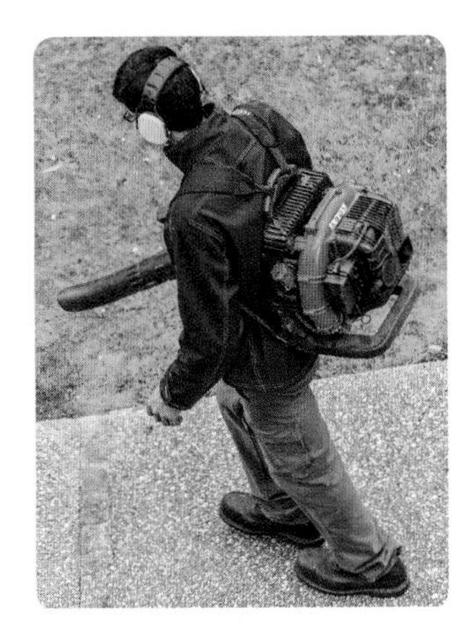

QY ANSWER

phase shift: $\frac{\pi}{3}$;

vertical translation: 4

Lesson

4-9 The Graph-Standardization Theorem

BIG IDEA The image of the graph of a parent trigonometric function under the composite of a scale change followed by a translation has a predictable equation, amplitude, phase shift, and period.

Recall that a sine wave is the image of the graph of $y = \sin x$ under a composite of translations and scale changes.

In Lessons 4-7 and 4-8, you saw how scale changes affect the amplitude and frequency of sine waves and how translations introduce phase shifts and vertical shifts. In this lesson, you will see how composites of scale changes and translations affect sine waves.

Mental Math

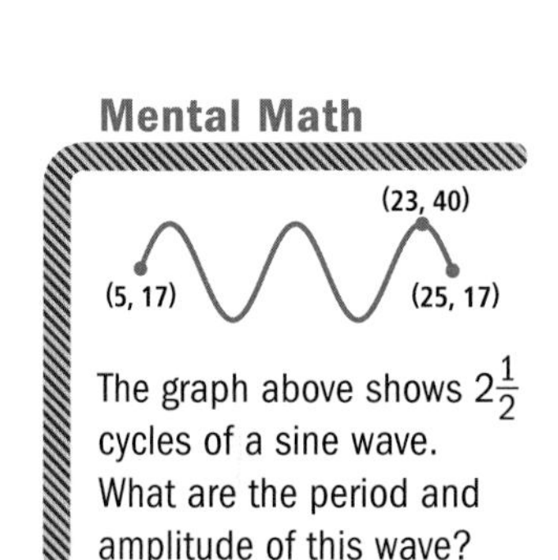

The graph above shows $2\frac{1}{2}$ cycles of a sine wave. What are the period and amplitude of this wave?

A Specific Example

We first apply a composite to a graph that is not a sine wave. Consider the parabola with equation $y = x^2$. Apply the scale change S: $(x, y) = (2x, 3y)$. This stretches the parent graph by a factor of 2 in the x-direction and by a factor of 3 in the y-direction. To this image apply the translation T: $(x, y) \rightarrow (x + 1, y - 2)$. This translates the image 1 unit to the right and 2 units down. The graphs below show the order of transformations.

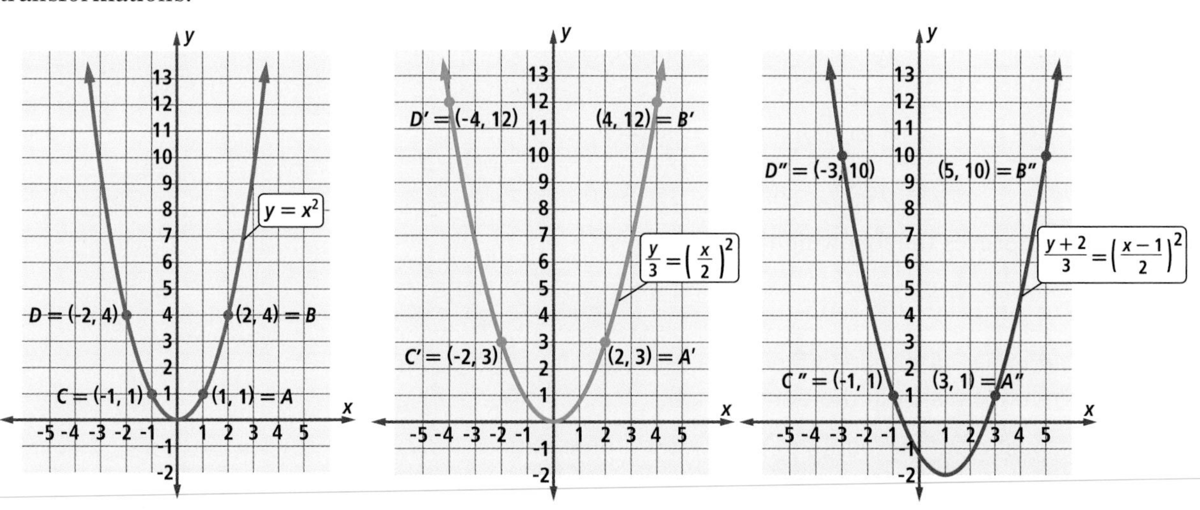

By the Graph Scale-Change Theorem, the image of the graph of $y = x^2$ under the scale change S: $(x, y) \rightarrow (2x, 3y)$ has equation

$$\frac{y}{3} = \left(\frac{x}{2}\right)^2.$$

By the Graph-Translation Theorem, the image of the graph of this new equation under T: $(x, y) \rightarrow (x + 1, y - 2)$ is

$$\frac{y + 2}{3} = \left(\frac{x - 1}{2}\right)^2.$$

This is an equation for the image of the graph of $y = x^2$ under the composite $T \circ S$. For instance, the point $D'' = (-3, 10)$ is the image of $D = (-2, 4)$ under $T \circ S$. When $(-3, 10)$ is substituted into the equation $\frac{y + 2}{3} = \left(\frac{x - 1}{2}\right)^2$, we get $\frac{10 + 2}{3} = \left(\frac{-3 - 1}{2}\right)^2$. It checks.

 QY

▶ QY

Find the coordinates of the vertex of each parabola on the previous page.

The General Idea

Suppose S is a scale change and T a translation with

$$S(x, y) = (ax, by), \text{ where } a \neq 0 \text{ and } b \neq 0,$$
$$\text{and } T(x, y) = (x + h, y + k).$$

Then the translation image of the scale-change image of (x, y) is

$$\begin{aligned} T \circ S(x, y) &= T(S(x, y)) \\ &= T(ax, by) \\ &= (ax + h, by + k). \end{aligned}$$

Thus, $T \circ S$ maps (x, y) to $(ax + h, by + k)$. That is, if (x', y') is the image of (x, y) under $T \circ S$, then

$$T \circ S(x, y) = (x', y') = (ax + h, by + k).$$

This equation for $T \circ S$ helps to determine how this transformation affects equations. Since

$$x' = ax + h \text{ and } y' = by + k,$$

it follows that $\frac{x' - h}{a} = x$ and $\frac{y' - k}{b} = y$.

This argument proves the following theorem.

Graph-Standardization Theorem

Given a preimage graph described by a sentence in x and y, the following processes yield the same graph:

(1) replacing x by $\frac{x - h}{a}$ and y by $\frac{y - k}{b}$ in the sentence;

(2) applying the scale change $(x, y) \rightarrow (ax, by)$ followed by the translation $(x, y) \rightarrow (x + h, y + k)$ to the preimage graph.

The Graph-Standardization Theorem applies to all relations graphed on a rectangular coordinate grid.

The Graph-Standardization Theorem and Trigonometric Functions

The scale change $(x, y) \to (ax, by)$ multiplies the period of a sine wave by $|a|$ and its amplitude by $|b|$. The translation $(x, y) \to (x + h, y + k)$ shifts the image h units horizontally and k units vertically. Combining these two transformations produces the following theorem.

Theorem (Characteristics of a Sine Wave)

The graphs of the functions with equations $\frac{y-k}{b} = \sin\left(\frac{x-h}{a}\right)$ and $\frac{y-k}{b} = \cos\left(\frac{x-h}{a}\right)$, with $a \neq 0$ and $b \neq 0$, have

amplitude $= |b|$, period $= 2\pi|a|$,

phase shift $= h$, and vertical shift $= k$.

Forms of Equations

We call $\frac{y-k}{b} = \sin\left(\frac{x-h}{a}\right)$ or $\frac{y-k}{b} = \cos\left(\frac{x-h}{a}\right)$ the *graph-standardized form* of the equation for a sine or cosine function. When an equation is in this form, you can use the above theorem to determine characteristics of the sine wave.

Example 1

a. Explain how the graph of $\frac{y-1}{2} = \cos\left(\frac{x+\pi}{3}\right)$ is related to the graph of $y = \cos x$.

b. Identify the amplitude, period, vertical shift, and phase shift of this function.

Solution

a. The given equation results from the equation $y = \cos x$ by replacing x by $\frac{x-(-\pi)}{3}$ and y by $\frac{y-1}{2}$. Thus the graph of $\frac{y-1}{2} = \cos\left(\frac{x+\pi}{3}\right)$ is the image of the graph of $y = \cos x$ under the scale change $(x, y) \to (3x, 2y)$ followed by the translation $(x, y) \to (x - \pi, y + 1)$.

b. From the Characteristics Theorem above, you can determine that the amplitude of $\frac{y-1}{2} = \cos\left(\frac{x+\pi}{3}\right)$ is 2; its period is 6π, the vertical shift is up 1 and the phase shift is $-\pi$.

Check Graph $y = \cos x$, as shown in blue. Next, graph its image under the scale change as shown by the curve drawn in red. Then graph the translation image of the red curve to get the green curve.

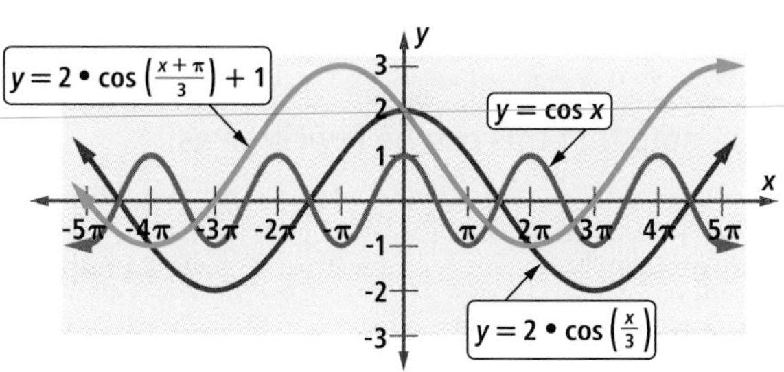

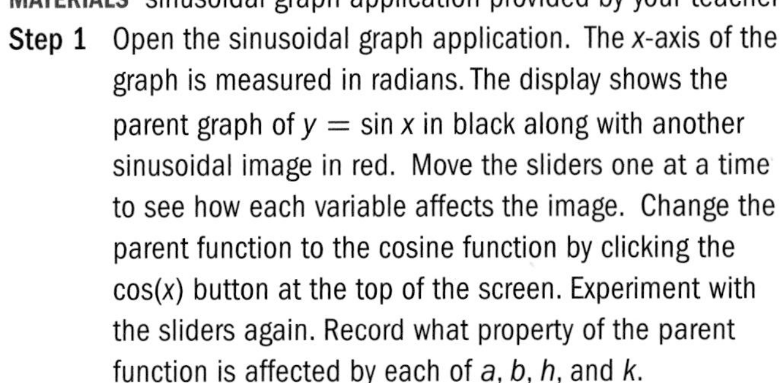

Activity

MATERIALS sinusoidal graph application provided by your teacher

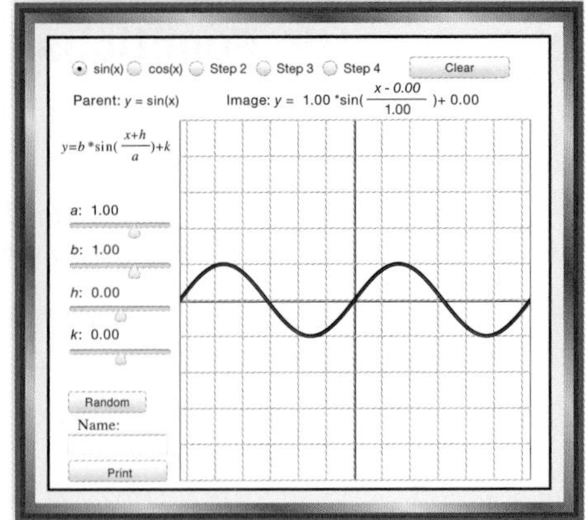

Step 1 Open the sinusoidal graph application. The x-axis of the graph is measured in radians. The display shows the parent graph of $y = \sin x$ in black along with another sinusoidal image in red. Move the sliders one at a time to see how each variable affects the image. Change the parent function to the cosine function by clicking the cos(x) button at the top of the screen. Experiment with the sliders again. Record what property of the parent function is affected by each of a, b, h, and k.

Step 2 Click the Step 2 check box button at the top of the screen to create a new sinusoidal image in blue. Note that the parent function is the sine function. Move one slider to scale the red graph to have the same amplitude as the blue graph. Then move another slider to have the red graph coincide with the blue graph. Write the equation of the red graph in $\frac{y-k}{b} = \sin\left(\frac{x-h}{a}\right)$ form, and then simplify $\frac{y-k}{b}$ and $\frac{x-h}{a}$ if possible.

Step 3 Click the Step 3 check box. Note that the parent function has changed to the cosine function. Move a slider to make the red graph have the same period as the blue graph. Then move a slider to translate the red graph to match the blue graph. Write the equation of the red graph in $\frac{y-k}{b} = \cos\left(\frac{x-h}{a}\right)$ form, and then simplify.

Step 4 Click the Step 4 check box. What is the parent function? Move the sliders to make the red graph completely match the blue graph. Write an equation for the red graph in graph-standardized form.

Step 5 Choose a parent function by clicking the sin(x) or cos(x) button at the top of the screen. Press the Random button to generate the graph of a random sinusoidal function. Adjust the sliders for a and b to transform the red graph to have the same shape as the generated graph, then adjust the sliders for k and h to translate your graph to coincide with the generated graph. Record your values of a, b, h, and k, and write the equation in graph-standardized form.

The form $y = A\sin(Bx + C) + D$ or $y = A\cos(Bx + C) + D$ is used in many applications. To convert $\left(\frac{y-1}{2}\right) = \cos\left(\frac{x-\pi}{3}\right)$ to this form, multiply the equation by 2 and add 1 to both sides to get

$$y = 2\cos\left(\frac{x-\pi}{3}\right) + 1.$$

Then, note that this can be rewritten as

$$y = 2\cos\left(\frac{x}{3} - \frac{\pi}{3}\right) + 1,$$

or equivalently,

$$y = 2\cos\left(\frac{1}{3}x - \frac{\pi}{3}\right) + 1.$$

In this form, $A = 2$, $B = \frac{1}{3}$, $C = -\frac{\pi}{3}$, and $D = 1$.

It is useful to be able to convert between forms. To find the characteristics of the sine wave, it helps to rewrite an equation in graph-standardized form.

Example 2

Consider the graph of $y = 2\sin(3x + \pi)$.

a. Describe this graph as the image of the graph of $y = \sin x$ under a composite of transformations.

b. Without graphing, determine the amplitude, period, vertical shift, and phase shift of the sine wave.

Solution

a. Convert the equation into graph-standardized form.

$$\frac{y}{2} = \sin(3x + \pi) = \sin\left(3\left(x + \frac{\pi}{3}\right)\right) = \sin\left(\frac{x - \left(\frac{-\pi}{3}\right)}{\frac{1}{3}}\right)$$

Thus the graph of $y = 2\sin(3x + \pi)$ is the image of the graph of $y = \sin x$ under the scale change $(x, y) \rightarrow \left(\frac{1}{3}x, 2y\right)$ followed by the translation $(x, y) \rightarrow \left(x - \frac{\pi}{3}, y\right)$.

b. The amplitude of the sine wave is $|2| = 2$. The period is $2\pi\left(\frac{1}{3}\right) = \frac{2\pi}{3}$. There is no vertical shift. The phase shift is $\frac{-\pi}{3}$.

Circular Motion

Trigonometric functions of the forms studied in this lesson arise naturally from situations of circular motion.

Example 3

The first Ferris wheel was built in 1893 for the Columbian Exposition. The radius of the wheel was about 131 feet and its center was about 140 feet above the ground. Find an equation that models the height above ground level of a car at time *x* (in minutes) assuming the wheel is continually rotating at 9 minutes per revolution, and that the car is initially at the maximum height.

Solution Let y be the height of the car that starts at the top. Since the car is initially at maximum height, there is no horizontal translation (phase shift) if the cosine function is used as the parent. So model this situation with the equation $\frac{y - k}{b} = \cos\left(\frac{x - h}{a}\right)$.

Because there is no translation, $h = 0$.

Reinventing the wheel

The original Ferris wheel had 36 cars that could hold 60 passengers each.

(*continued on next page*)

The amplitude is the radius of the wheel, so the amplitude b is 131. Since the center was 140 feet above ground, there is a vertical translation of $k = 140$. The period (one complete revolution or 2π) takes 9 minutes, so if we want time in minutes, we need to shrink the period from 2π to 9. A horizontal shrink with magnitude $\frac{9}{2\pi}$ will do this. Thus $a = \frac{9}{2\pi}$.

Substituting $a = \frac{9}{2\pi}$, $b = 131$, $h = 0$, and $k = 140$ into the equation, we obtain $\frac{y - 140}{131} = \cos\left(\frac{x - 0}{\frac{9}{2\pi}}\right)$, or $y = 131 \cos\left(\frac{2\pi}{9}x\right) + 140$.

Check After $2\frac{1}{4}$ minutes, the car should be at the level of the center of the wheel. Is this predicted by the equation? Substitute $2\frac{1}{4}$ for x.

$y = 131 \cdot \cos\left(\frac{2\pi}{9} \cdot 2\frac{1}{4}\right) + 140 = 140$. It checks.

Questions

COVERING THE IDEAS

1. On page 269, the graph of $\frac{y+2}{3} = \left(\frac{x-1}{2}\right)^2$ is shown to be the image of the graph of $y = x^2$ under the composite of a translation and a scale change. Find the image of each point under that composite.
   a. (0, 0) b. (10, 100) c. (-10, 100)
2. What transformations and in what order can be applied to the graph of $y = x^2$ to obtain the graph of $\frac{y-7}{4} = \left(\frac{x+8}{3}\right)^2$?
3. Consider the graph of the function with equation $y = \sin\left(\frac{x-\pi}{3}\right)$.
   a. Give the amplitude, period, and phase shift of the graph.
   b. The graph is the image of the graph of $y = \sin x$ under the composite of what two transformations?
4. Consider the graph of $y = \frac{1}{5}\cos\left(\frac{x+\pi}{6}\right)$.
   a. Give the amplitude, period, and phase shift of the graph.
   b. The graph is the image of the graph of $y = \cos x$ under the composite of what two transformations?

In 5 and 6, an equation of a sine wave is given.
a. Write the equation in graph-standardized form.
b. Find its amplitude, period, phase shift, and vertical shift from the graph of the parent function.

5. $y = 7\sin(\pi x) - 5$
6. $s = 6 + 2\cos(3t + 4)$

In 7–9, refer to Example 3.

7. What are the minimum and maximum heights of the cars?
8. Check that after $4\frac{1}{2}$ minutes you are at the minimum height.
9. Model the height y of the wheel at time x by an equation in the form $\frac{y-k}{b} = \sin\left(\frac{x-h}{a}\right)$, using the sine function as the parent.

APPLYING THE MATHEMATICS

10. Suppose $f(x) = \tan\left(\frac{x+\pi}{4}\right)$.
   a. Describe a scale change and translation whose composite maps the graph of $y = \tan x$ onto the graph of $y = f(x)$.
   b. Draw a graph of $y = f(x)$ and state its period and phase shift.
11. Create a version of the Graph-Standardization Theorem for the tangent function.

In 12–14, write an equation for the function whose graph has the given characteristics.

12. Parent $y = \cos x$, phase shift $-\frac{\pi}{3}$, period $\frac{\pi}{2}$, and amplitude 9.
13. Parent $y = \sin x$, amplitude 5, period 6π, phase shift $\frac{\pi}{4}$, and vertical shift -1.
14. Parent $y = x^3$, scaled by S: $(x, y) \rightarrow (2x, 3y)$, then translated by T: $(x, y) \rightarrow (x - 1, y + 5)$.
15. The function graphed at the right has maximum value 15, minimum value -5, and period 2π. Write an equation for it.
16. For the sine wave modeled by $y = A \sin(Bx + C) + D$, give formulas in terms of A, B, C, and D for the
   a. amplitude. b. period. c. phase shift.

REVIEW

17. Consider the translation T: $(x, y) \rightarrow \left(x + \frac{\pi}{3}, y + 2\right)$. Find an equation for the image of the graph of $y = |x|$ under T. **(Lesson 3-2)**
18. Identify the following characteristics for the function g defined by $g(x) = A \sin(Bx)$. **(Lesson 4-7)**
   a. period b. amplitude c. domain d. range
19. Solve $\sin c = 0$ for $0 \leq c \leq 6\pi$. **(Lesson 4-5)**
20. A teacher made two forms of a test. The test scores are below.

Form A	93	62	89	77	68	94	73	82	85	76	83	79
Form B	65	87	71	76	67	87	76	81	77	82	62	78

   Compared to the other students who took the same test, who performed better, the student with an 85 on Form A or the student with an 82 on Form B? **(Lesson 3-9)**
21. Simplify. **(Previous Course)**
   a. $\frac{10^8}{10^3}$ b. $\frac{6 \cdot 10^5}{2 \cdot 10^{-3}}$ c. $\frac{2.8 \cdot 10^{-2}}{1.4 \cdot 10^3}$ d. $\frac{a \cdot 10^m}{b \cdot 10^n}$

EXPLORATION

22. Does it make a difference whether you translate a graph first and then scale the image, or scale the graph first and then translate the image? Defend your answer with at least two examples.

QY ANSWER

(0, 0), (0, 0), (1, –2)

Lesson 4-10 Modeling with Trigonometric Functions

Vocabulary

simple harmonic motion

▶ **BIG IDEA** The Graph-Standardization Theorem can be used to build an equation that models real-world periodic data.

Mental Math

Find an equation for $f \circ g$ if f and g have the given equations.

a. $f(x) = x^2$ and $g(x) = x^5$

b. $f(x) = x^5$ and $g(x) = x^2$

c. $f(x) = \sqrt{x}$ and $g(x) = 4x - 17$

d. $f(x) = 4x - 17$ and $g(x) = \sqrt{x}$

Many natural phenomena involving periodic behavior can be modeled by a single sine or cosine function.

Activity

The tables below show the number N of hours of daylight in Jacksonville, FL as a function of the number D of days after December 31st of a certain year. The data were collected on the first day of each month over a two-year period. Neither year was a leap year.

D	1	32	60	91	121	152	182	213	244	274	305	335
N	10.23	10.77	11.55	12.5	13.37	13.98	14.08	13.62	12.77	11.88	10.98	10.35

D	366	397	425	456	486	517	547	578	609	639	670	700
N	10.23	10.77	11.53	12.48	13.35	13.98	14.08	13.62	12.78	11.88	10.98	10.35

Source: The Old Farmer's Almanac, 2009

Step 1 Work with a partner and use a statistics utility to create a scatterplot of the data. The data should appear periodic.

Step 2 Assuming the data fit a sine function, estimate the following from the graph and interpret its meaning in this situation.

a. amplitude **b.** period **c.** vertical translation **d.** phase shift

Step 3 Use the information from Step 2 to create a model of the form $\frac{y-k}{b} = \sin\left(\frac{x-h}{a}\right)$. Graph your model on the same grid as the scatterplot. You may need to convert your equation to a form that your technology can graph. How well does your model seem to fit the data?

Step 4 Check your model by using the sine regression function on your calculator.

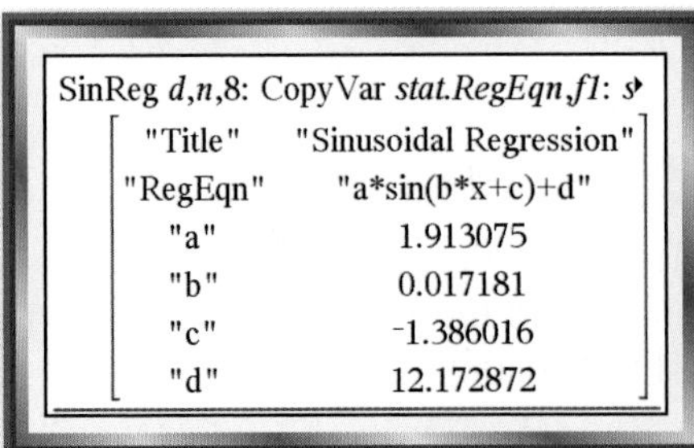

U.S. Census for 2007 ranked Jacksonville, FL as the 12th largest city in the United States with a population of 805,605.

Simple Harmonic Motion

The motion of a pendulum swinging back and forth in a vacuum and the motion of a weight bobbing on a spring are examples of *harmonic motion*. The graphs of these motions appear to be sine waves. Motion that can be described using a sine or cosine function is called **simple harmonic motion**. Each point on the graph corresponds to a location of the pendulum or weight at a particular time.

Example 1

A pendulum swings back and forth in a vacuum. Its distance from an object is captured using a motion detector for the object. The setup is pictured below at the left; at the right is a graph of the pendulum's distance from the motion detector. Write an equation for the distance from the pendulum to the object as a function of time.

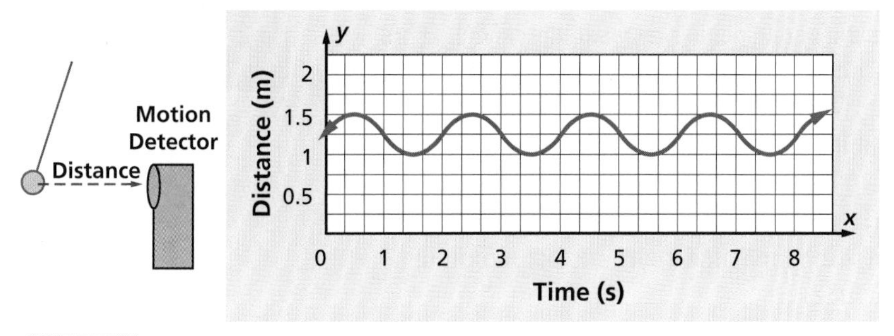

Solution We want to find the values of a, b, h, and k in the equation $\frac{y-k}{b} = \sin\left(\frac{x-h}{a}\right)$. First, identify the period P, because $a = \frac{P}{2\pi}$. The pendulum makes a round-trip every 2 seconds, so the period is 2 seconds. Thus $a = \frac{2}{2\pi} \approx 0.3183$. Next, calculate the amplitude b. The amplitude is half of the difference between the maximum and minimum y-values. So $b = \frac{1.5 - 1.0}{2} = 0.25$. Then find the vertical shift k, the distance from the x-axis to the center line of the graph. This distance is 1.25, so $k = 1.25$. Finally, compare the graph to a graph of the sine or cosine function. We pick the sine function. The phase shift is 0 with respect to $y = \sin x$, so $h = 0$.

Now substitute in the equation to write a formula for the function.
$\frac{y - 1.25}{0.25} = \sin\left(\frac{x}{0.3183}\right)$ or $y = 0.25 \sin\left(\frac{x}{0.3183}\right) + 1.25$, where y is the distance (in meters) of the pendulum from the object and x is time (in seconds).

Interpreting a Given Model

When you are given the equation of a sine wave model, you can interpret the coefficients to describe the situation.

GUIDED

Example 2

When an oven is set to a particular temperature, the heat level rises and falls, actually fluctuating slightly above and below that level as time passes. Assume that when a particular oven is set to 425°F, the oven temperature t in degrees Fahrenheit m minutes after the burner first shuts off satisfies $t = 425 + 6\cos(0.9m)$.

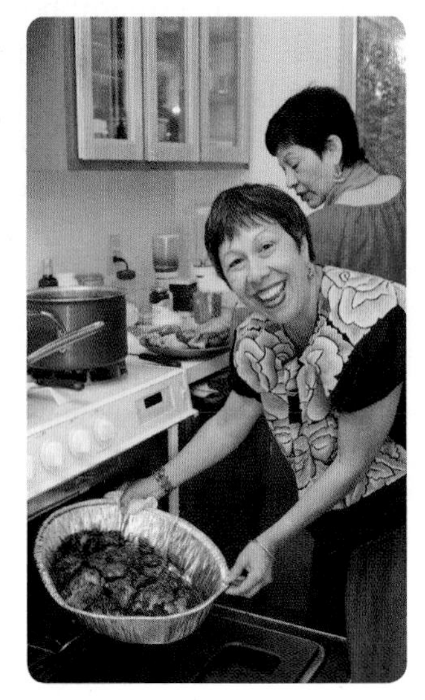

a. What are the maximum and minimum temperatures of the oven at this setting?

b. What is the period of this sine wave? What does the period represent?

Solution First, transform the equation into the form $\frac{t-k}{b} = \cos\left(\frac{m-h}{a}\right)$ to identify the coefficients of the model.

$$\frac{t - 425}{?} = \cos\left(\frac{m}{?}\right)$$

a. The sum and difference of the amplitude and vertical translation of the sine wave tell us the maximum and minimum values of the model. This function has an amplitude of __?__ and a vertical translation of __?__. Using these values, the maximum temperature of the oven is __?__ while the minimum temperature is __?__.

b. The period $= 2\pi|a| =$ __?__. For this sine wave, the period represents how long it takes the oven to go from one maximum (or minimum) temperature to the next maximum (or minimum) temperature.

Modeling an arbitrary sine wave, especially one with a nonzero phase shift, can be an involved process. It is easier if you focus your attention on one graph characteristic at a time. By asking a question like "What's the period?" or "What's the amplitude?" you can deal with one piece of the model at a time instead of trying to figure out everything at once.

Questions

COVERING THE IDEAS

In 1 and 2, use the sine regression model from Step 4 of the Activity.

1. What is the calendar day on which the number of hours of daylight is the greatest (the summer solstice) for the first year? How many hours of daylight are there on that day?
2. How many hours of daylight does your model predict for January 1st for the next year after the data shown?
3. Bernie has just purchased an oven. The manufacturer claims that less heat escapes from this oven than from the oven in Example 2. Assume that Bernie's oven takes twice as long to cool from the maximum to the minimum temperature. Write an equation to model Bernie's oven temperature as he cooks a roast at 425°.

In 4 and 5, a pendulum's motion is captured and graphed below.

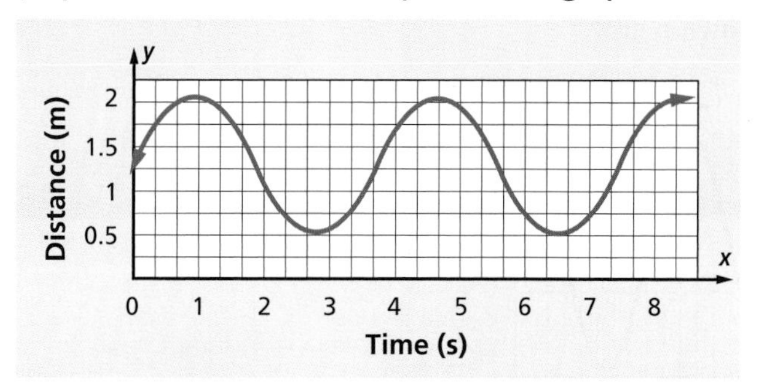

4. a. To the nearest 0.1 second, identify the time(s) at which the pendulum is furthest from the motion detector.

   b. To the nearest 0.1 second, identify the time(s) at which the pendulum is closest to the motion detector.

5. a. For each attribute below, describe its meaning in this situation, and find its value.

   i. amplitude ii. period iii. vertical shift

   b. Write a formula for the pendulum's motion in terms of time.

6. Write a formula describing the graph below.

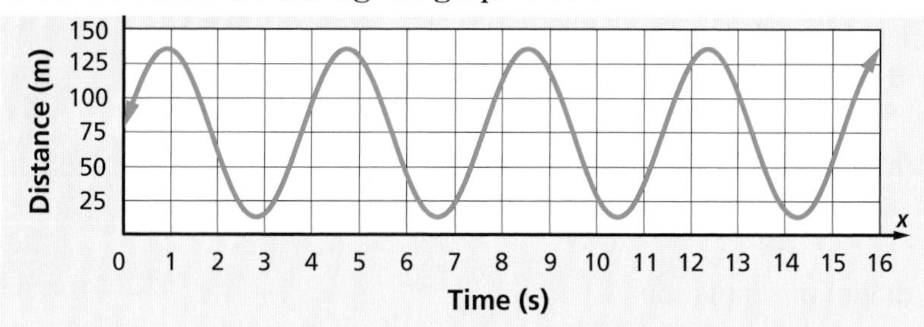

7. A model for an earthquake tremor is $f(t) = 1.6 \sin(188.5t)$. Find the period and amplitude of the tremor.

APPLYING THE MATHEMATICS

8. A pendulum swings for 100 seconds; its motion is captured and graphed below. Considering only period, vertical shift, and amplitude, which properties seem to be changing over time? Which seem to be constant? Why might this be so?

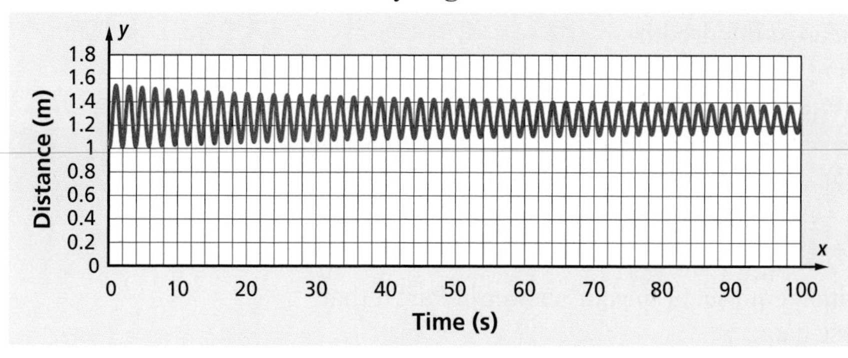

9. Rose Tate's height above ground level was tracked as she traveled on a Ferris wheel. The graph is shown below.

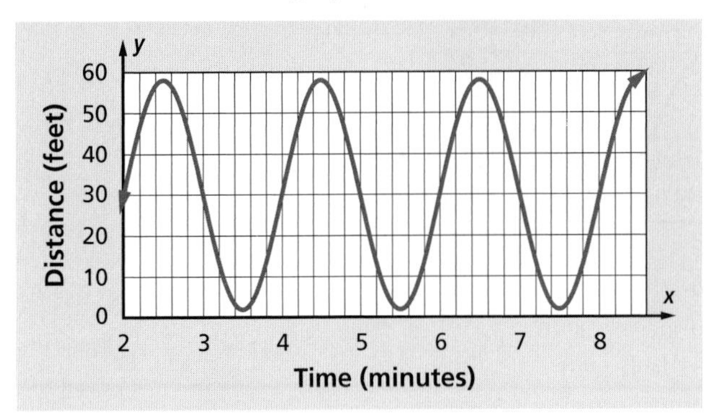

a. What is the radius of the Ferris wheel?

b. How high was Rose above the ground at her lowest point?

c. How high was she when she boarded the Ferris wheel?

d. How long does it take the Ferris wheel to make one revolution?

e. Write an equation that approximates the graph.

10. The average adult normally breathes in and out every 5 seconds. The amount taken in and expelled from the lungs in a single breath is called the person's *tidal volume*. In adults, the tidal volume is about 16 fluid ounces. The maximum amount usually held in the lungs is 48 ounces.

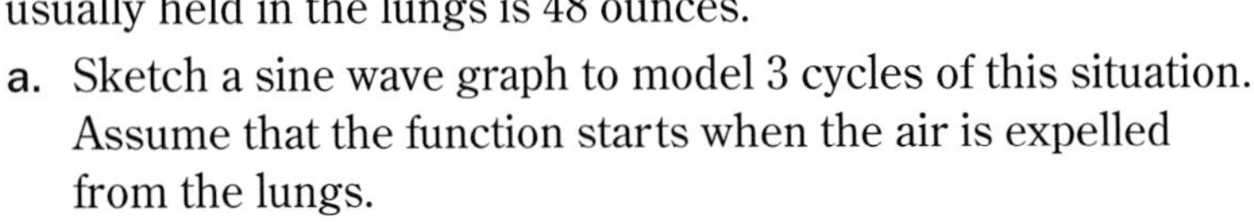

a. Sketch a sine wave graph to model 3 cycles of this situation. Assume that the function starts when the air is expelled from the lungs.

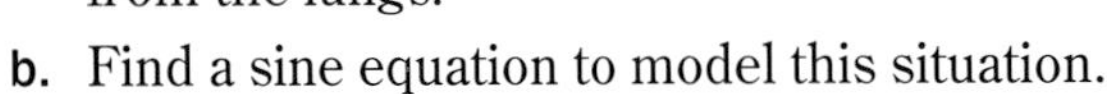

b. Find a sine equation to model this situation.

11. Average monthly temperatures in degrees Fahrenheit for Chicago for the years 1971–2000 are given below.

month	Jan	Feb	Mar	Apr	May	Jun	Jul	Aug	Sep	Oct	Nov	Dec
avg. temp (°F)	22.0	27.0	37.3	47.8	58.7	68.2	73.3	71.7	63.8	52.1	39.3	27.4

Source: Statistical Abstract of the United States, 2009

a. Draw a scatterplot of these data. Plot $x =$ the number of the month on the horizontal axis, and $y =$ the temperature on the vertical axis.

b. Carefully sketch a sine curve to fit the data and estimate its amplitude to the nearest hundredth.

c. What is the period of this function?

d. **Multiple Choice** Which of these four models is best for these data?

A $\frac{y}{a} = \cos\left(\frac{\pi x - 3.5}{6}\right)$ B $\frac{y - 49}{-a} = \cos\left(\frac{\pi x - 3.5}{6}\right)$

C $\frac{y - 49}{a} = \cos\left(\frac{\pi x - 3.5}{6}\right)$ D $\frac{y}{-a} = \cos\left(\frac{\pi x - 3.5}{6}\right)$

e. Find another equation equivalent to your answer in Part d that also describes these data.

12. Listed below are the hours of daylight in International Falls, Minnesota, for ten days of the year.

Date	1/1	2/28	3/21	4/27	5/6	6/21	8/14	9/23	10/25	12/21
Hours	8.38	10.95	12.27	14.40	14.87	16.15	14.40	12.12	10.27	8.30

Source: The Old Farmer's Almanac 2009

a. Draw a scatterplot of the data indicating dates as days of the year. (1/1 would be 1, 2/28 would be 59, and so on.) Assume this year was not a leap year.

b. Using paper and pencil, fit a sine wave to the scatterplot.

c. Determine an equation of a cosine function that models the data.

d. Using the model in Part c, estimate the hours of daylight on July 3.

e. Estimate what days International Falls had at least 10 hours of daylight.

REVIEW

13. Consider the function with equation $y = \cos\left(\frac{x + \pi}{4}\right)$.

a. Sketch a graph of the function.

b. Give the amplitude, period, and phase shift of its graph.

c. The graph is the image of the graph of $y = \cos x$ under the composite of which two transformations? **(Lesson 4-9)**

14. Given $\cos \theta = m$, find each value. **(Lesson 4-3)**

a. $\cos(-\theta)$ b. $\cos(\pi - \theta)$ c. $\sin(\theta - \pi)$

In 15 and 16, refer to the unit circle at the right. (Lesson 4-3)

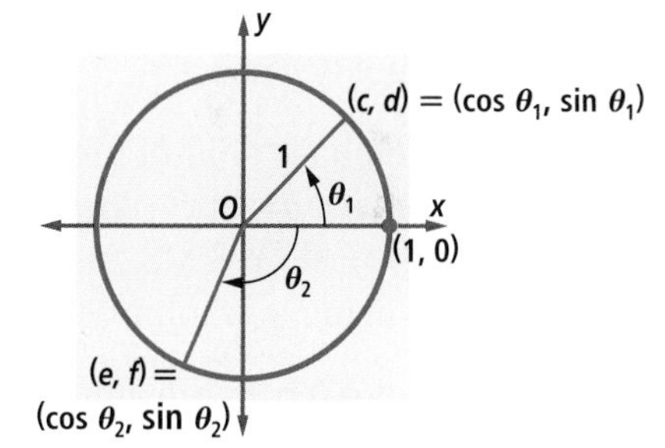

15. a. If $c = d$, find the exact value of c.

b. What is the value of θ_1?

16. If $\theta_2 = \frac{-2\pi}{3}$, find e and f.

In 17 and 18, let $f(x) = x^2 - 2$ and $g(x) = \frac{1}{3x - 1}$. (Lesson 3-7)

17. a. Calculate $g \circ f \circ g(-2)$.

b. Calculate $f \circ g \circ f(-2)$.

18. a. Give the domain of $f \circ g \circ f$.

b. Give the domain of $g \circ f \circ g$.

EXPLORATION

19. The French physicist Léon Foucault used a large pendulum to demonstrate the fact that Earth rotates on its axis. The original Foucault Pendulum hangs in the Panthéon in Paris, France; copies are exhibited in science museums around the world. How does the pendulum's motion show Earth's rotation? Research Foucault to find the answer and to find out about his other inventions.

Chapter 4 Projects

1 Sums of Sine Waves with the Same Period

Consider functions with equations of the form $f(x) = a \sin x + b \cos x$. Here are some examples:

$$a = 1, b = 1: f(x) = \sin x + \cos x$$
$$a = \sqrt{3}, b = 1: f(x) = \sqrt{3} \sin x + \cos x$$
$$a = 1, b = -1: f(x) = \sin x - \cos x$$
$$a = 5, b = 12: f(x) = 5 \sin x + 12 \cos x$$

a. Can you predict the shapes of the graphs of such functions from the values of a and b? Graph the four examples above. Then choose some other values for a and b. What shapes are the graphs?

b. Can you predict the amplitude of a graph before you draw it? For example, what is the amplitude of $f(x) = 3 \sin x + 4 \cos x$? Check your prediction using a graphing utility. What is the period of the graph of $f(x) = a \sin x + b \cos x$?

c. What is the phase shift of the graph of $f(x) = a \sin x + b \cos x$? Predict the phase shift of $g(x) = \sin x + \sqrt{3} \cos x$. Check your prediction using a graphing utility.

d. For $a > 0$ and $b > 0$, the amplitude and the phase shift are related to a right triangle with legs of lengths a and b. Find out what the relationship is.

e. How are the graphs of $y = a \sin x - b \cos x$ and $y = a \sin x + b \cos x$ related if a and b are positive? How does your response to this question change if a and b are not both positive?

2 Beat Matching and DJing 101

Beat matching, once an art mastered only by DJs (disc jockeys) with vinyl records and turntables, can now be mastered by anyone using digital mp3 files, software, and special equipment such as Scratch Live. Some software allows you to see the graphs of files in order to match songs with similar frequencies, or beats, in order to mix different songs together as DJs do. Pick two songs, parts of which share a beat or rhythm which you can identify by ear using the mp3s and audio software.

a. Compare the period of the beat you hear in both songs. Are they similar?

b. What about the amplitude?

c. If you were to translate one of the songs at the cusp of the beat you are looking at and start it at the same time as the associated cusp in the second song, what would happen if you let both songs play at the same time?

d. Beat juggling can be performed when a new rhythmical composition is created using two songs and manipulating the arrangement and order of what is being played. A simple beat juggle might play one or more bars of the beat in song A, and then cross over to the beatmatched song B for several bars before going back to song A, and so on. Using audio software, copy and paste portions of the beat you analyzed in Parts a–c from both songs and compose a beat juggling session by putting them next to each other and letting them play. If your teacher allows, play the remix for your class.

3 "Noncircular" Functions

The trigonometric functions are defined in terms of a unit circle. Imagine using a different shape centered at the origin O and passing through $A = (1, 0)$. Below are three possible alternatives: two squares (Figures A and B) and a regular octagon (Figure C).

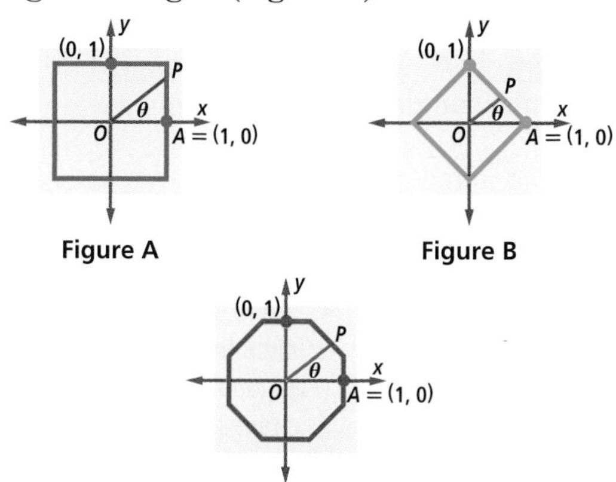

Figure A

Figure B

Figure C

In each case, define the functions side, coside, and tide (to distinguish them from sine, cosine, and tangent) as follows. For any rotation θ in radians, let P be the point on the figure such that $m\angle POA = \theta$. Then define

$$P = (\text{coside }\theta, \text{side }\theta)$$
$$\text{and tide }\theta = \frac{\text{side }\theta}{\text{coside }\theta}, \text{ if coside }\theta \neq 0.$$

a. Choose either Figure A or Figure B. Draw (separate) graphs of $s(x) = \text{side } x$, $c(x) = \text{coside } x$, and $t(x) = \text{tide } x$. Compare and contrast these with the graphs of the trigonometric functions.

b. Suggest theorems about the side, coside, and tide functions analogous to those studied in this chapter, such as the Pythagorean Identity, the Opposites Theorem, or the Supplements Theorem. Prove your results.

c. Choose one of the two other figures and repeat Parts a and b.

4 Square Waves

a. Graph the functions

$$F_1(x) = \sin x,$$
$$F_2(x) = \sin x + \tfrac{1}{3}\sin(3x),$$
$$F_3(x) = \sin x + \tfrac{1}{3}\sin(3x) + \tfrac{1}{5}\sin(5x), \text{ and}$$
$$F_{10}(x) = \sin x + \tfrac{1}{3}\sin(3x) + \tfrac{1}{5}\sin(5x) + \cdots + \tfrac{1}{19}\sin(19x).$$

Describe the patterns you observe in the general shape, amplitude, and period. Predict the behavior of the function

$$F_n(x) = \sum_{i=1}^{n} \frac{\sin(2i-1)}{(2i-1)x} \text{ for large } n.$$

b. These kinds of waves are important in understanding alternating current in electricity. Find out how and why.

5 Sunrise and Sunset Times

From an almanac or other reference, find the times of sunrise and sunset for various dates in a particular city in a particular year.

a. Make scatterplots of date versus time of sunrise and date versus time of sunset. Sketch curves of good fit through the data. Find an equation to model each set of data.

b. How does Daylight Saving Time affect a graph of time of sunset or sunrise?

c. From the equations of sunset and sunrise in relation to date, how could you determine an equation for the number of hours of daylight in relation to the date?

d. How does the latitude or longitude of a city affect the time of sunset or sunrise? Collect other data and supply theoretical evidence to support your answer.

Chapter 4 Summary and Vocabulary

- Degrees, **radians**, and **revolutions** measure **magnitudes** of rotations. Radians are often preferred for their convenience. By convention, counterclockwise rotations have positive measure and clockwise rotations have negative measure. The formula $360^\circ = 1$ revolution $= 2\pi$ radians can be used to convert from one unit to another. In a circle with radius r, the length of an arc determined by a central angle of measure θ radians is $r\theta$.

- The image of (1, 0) under the rotation R_θ, with center (0, 0) and magnitude θ, is defined to be **($\cos \theta$, $\sin \theta$)**. For all θ, since $R_{\theta+2\pi} = R_\theta$, $\cos(\theta + 2\pi) = \cos \theta$ and $\sin(\theta + 2\pi) = \sin \theta$. The **sine function** $\theta \to \sin \theta$ and **cosine function** $\theta \to \cos \theta$ each have domain the set of real numbers and range $\{y| -1 \le y \le 1\}$. For all θ except odd integer multiples of $\frac{\pi}{2}$, **$\tan \theta$** $= \frac{\sin \theta}{\cos \theta}$. The range of the **tangent function** $\theta \to \tan \theta$ is the set of all real numbers. Graphs of these **trigonometric functions** are found on pages 248, 249, and 253.

- **Sine waves** are images of graphs of the parent **circular functions** under composites of translations and scale changes. The sine wave with equation $\frac{y-k}{b} = \sin\left(\frac{x-h}{a}\right)$ or $\frac{y-k}{b} = \cos\left(\frac{x-h}{a}\right)$ has **amplitude** $|b|$, **period** $2\pi|a|$, **phase shift** h, and vertical shift k from its parent. Periodic phenomena and heights of objects that travel in circles on a vertical plane can be described by these equations.

- The Pythagorean Identity, $\sin^2\theta + \cos^2\theta = 1$, comes from the definition of sine and cosine and the equation of a **unit circle**. Symmetries of the unit circle give rise to properties of the sine, cosine, and tangent of all x for which they are defined. These properties are described in the table at the top of the next page.

Vocabulary

4-1
*rotation, rotation image
center of the rotation
magnitude of a rotation
revolution
full turn
half turn
quarter turn
*radian

4-2
*unit circle
*cosine, cos
*sine, sin
circular function
trigonometric function
*tangent, tan

4-3
*identity

4-5
*sine function
*cosine function

4-6
tangent function
*periodic function
period of a function

4-7
*sine wave
*amplitude
*frequency

4-8
*phase shift

4-10
simple harmonic motion

Theorem	Symmetry of circle	Properties
Opposites Theorem	over the x-axis	$\cos(-\theta) = \cos\theta$ $\sin(-\theta) = -\sin\theta$ $\tan(-\theta) = -\tan\theta$
Supplements Theorem	over the y-axis	$\cos(\pi - \theta) = -\cos\theta$ $\sin(\pi - \theta) = \sin\theta$ $\tan(\pi - \theta) = -\tan\theta$
Complements Theorem	with respect to the line $y = x$	$\cos\left(\frac{\pi}{2} - \theta\right) = \sin\theta$ $\sin\left(\frac{\pi}{2} - \theta\right) = \cos\theta$
Half-Turn Theorem	about the origin	$\cos(\pi + \theta) = -\cos\theta$ $\sin(\pi + \theta) = -\sin\theta$ $\tan(\pi + \theta) = \tan\theta$

- From the properties above and plane geometry, you can determine exact values of the sine, cosine, and tangent functions for arguments that are multiples of $\frac{\pi}{4}$ or $\frac{\pi}{6}$.

Theorems and Properties

Chapter 4 Self-Test

Take this test as you would take a test in class. You will need a calculator. Then use the Selected Answers section in the back of the book to check your work.

In 1–5, give exact values.

1. Convert $\frac{7}{9}$ of a revolution to radians.
2. How many degrees equal $-\frac{5\pi}{2}$ radians?
3. Convert 120° to radians. Give your answer in terms of π.
4. Find $\sin\left(\frac{3\pi}{4}\right)$.
5. What is tan 60°?
6. Approximate $\cos\left(\frac{7\pi}{5}\right)$ to four decimal places.
7. Identify each of the following for the function with equation $y = \sin(x - \pi) + 3$.
   a. domain b. range
   c. amplitude d. period
   e. phase shift f. vertical shift
8. a. For what values of θ from 0 to 2π are $\cos\theta$ and $\tan\theta$ both positive?
   b. For what values in the same range are they both negative?

In 9 and 10, consider the unit circle below.

9. Find $\sin\theta$.
10. Find $\cos(\theta + \pi)$.

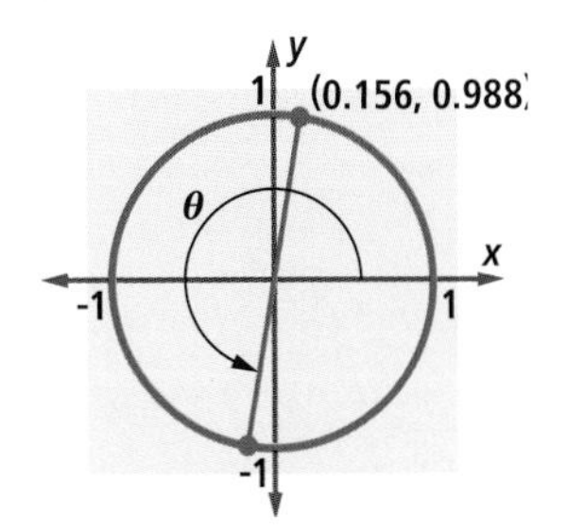

11. **True or False** Justify your answer. For all θ, $\cos(\theta + 3\pi) = \cos\theta$.

12. Below is part of the graph of a function f. Which could be an equation for f: $f(x) = \sin x$ or $f(x) = \cos x$? Justify your answer.

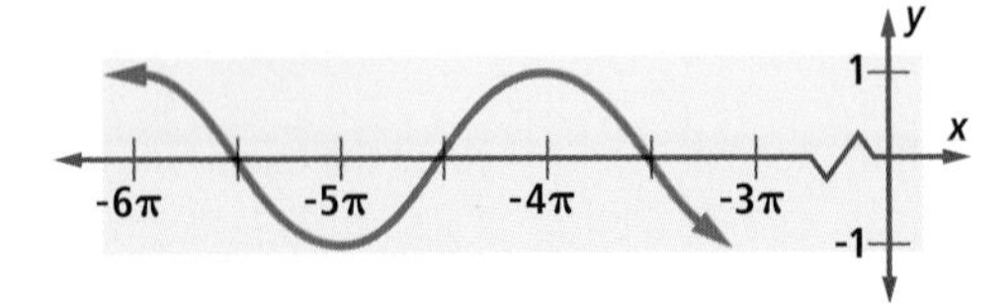

13. Below is a graph of a sine wave together with a general form of an equation for it. Find the values of a, b, h, and k in the equation.

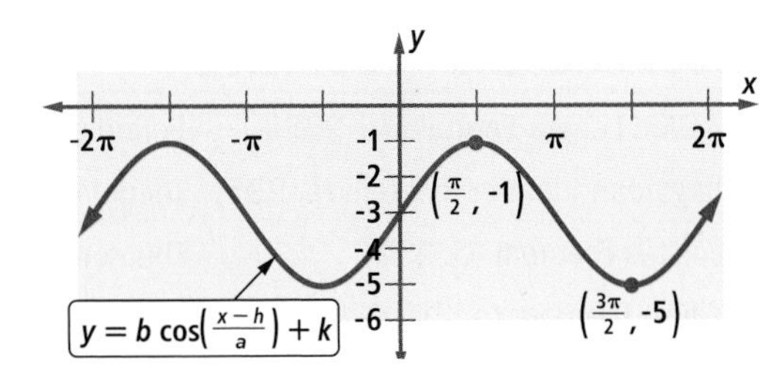

14. Give an exact value of $\cos(-\theta)$ if $\sin\theta = \frac{5}{13}$ and $\frac{\pi}{2} < \theta < \pi$.
15. Suppose the transformation $(x, y) \rightarrow \left(\frac{x}{3} + \pi, y\right)$ is applied to the graph of $y = \tan x$. Write an equation for the image.
16. **Multiple Choice** Which property of a trigonometric function changes under a vertical scale change of its graph?
   A phase shift B amplitude
   C frequency D period

17. A sine wave model for the average temperature T for Grand Rapids, Michigan, during month n can be found using the data and the graph below.

Month	0	3	6	9	12
Temp. (°F)	22	46	71	51	22

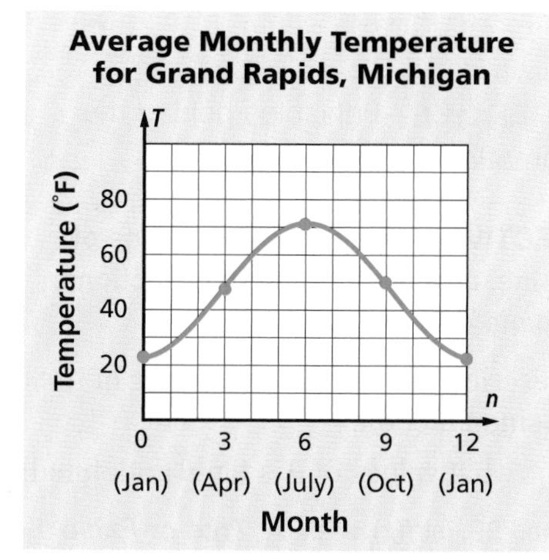

a. What is the amplitude of the wave?

b. What is the period of the wave?

c. Write an equation to model these data.

d. Use the model to estimate the average temperature for Grand Rapids in February.

18. The voltage E in volts in a circuit after t seconds $(t > 0)$ is given by $E = 12\cos(14\pi t)$.

a. What is the maximum voltage achieved in the circuit?

b. Find three times at which the voltage achieves its maximum value.

19. **Multiple Choice** Which equation could yield the graph below?

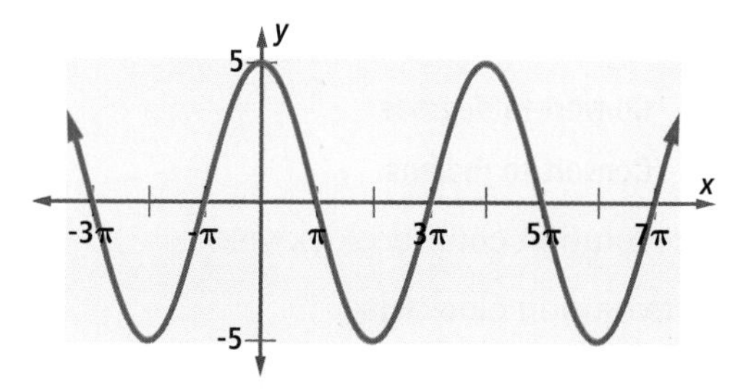

A $y = 5\sin(2x)$ B $y = 5\cos(2x)$

C $y = 5\sin\left(\frac{x}{2}\right)$ D $y = 5\cos\left(\frac{x}{2}\right)$

20. The hum you hear on a radio when it is not functioning properly is a sinusoidal sound wave with a frequency of 60 vibrations per second. If the amplitude of this wave is 0.1, write an equation for the displacement y as a function of time t.

Chapter 4 Chapter Review

SKILLS
PROPERTIES
USES
REPRESENTATIONS

SKILLS Procedures used to get answers

OBJECTIVE A Convert between degrees, radians, and revolutions. (Lesson 4-1)

In 1 and 2, a rotation is given.

a. Convert to degrees.

b. Convert to radians.

1. $\frac{2}{5}$ revolution counterclockwise
2. $\frac{1}{3}$ revolution clockwise

In 3 and 4, convert to degrees without using a calculator.

3. $-\frac{5\pi}{6}$ radians
4. $\frac{\pi}{12}$ radians

In 5 and 6, convert to radians without using a calculator.

5. 225°
6. 330°

In 7 and 8, tell how many revolutions are represented by a rotation of the given magnitude.

7. $\frac{2\pi}{3}$
8. 630°

OBJECTIVE B Find values of sines, cosines, and tangents. (Lessons 4-4, 4-5)

In 9–11, give exact values without using a calculator.

9. $\sin\left(\frac{\pi}{4}\right)$
10. $\cos\left(\frac{\pi}{3}\right)$
11. $\tan\left(\frac{\pi}{6}\right)$

In 12–14, approximate to the nearest thousandth.

12. $\tan 1.3$
13. $\cos 0.0926$
14. $\sin 0.4563$

In 15–17, evaluate to the nearest hundredth.

15. $\cos 3$
16. $\sin(-4.2)$
17. $\tan 251^\circ$

In 18–21, give exact values without using a calculator.

18. $\sin\left(\frac{5\pi}{4}\right)$
19. $\cos 210^\circ$
20. $\tan\left(-\frac{\pi}{4}\right)$
21. $\cos\left(\frac{9\pi}{2}\right)$
22. Give two values of θ between -2π and 2π such that $\sin \theta = 1$.

In 23 and 24, let $P = R_\theta(1, 0)$. Find the coordinates of P when θ is the following.

23. 5π
24. $\frac{2}{5}$ of a revolution clockwise
25. Solve $\cos x = -\frac{1}{2}$ exactly for x in the interval $-\frac{\pi}{2} \le x \le \frac{5\pi}{2}$.

PROPERTIES Principles behind the mathematics

OBJECTIVE C Apply the definitions of the sine, cosine, and tangent functions. (Lessons 4-2, 4-3, 4-5, 4-6)

26. Describe the domain and range of the cosine function.
27. For what values of θ is $\tan \theta$ undefined?

In 28 and 29, let $f(x) = \sin x$. **True or False** Explain your answer.

28. f is an even function.
29. The maximum value of f is 1.
30. In what domain interval(s) between 0 and 2π are the values of both the sine and tangent functions negative?
31. **Multiple Choice** For which values of θ is $\sin \theta < 0$ and $\cos \theta > 0$?

A $0 < \theta < \frac{\pi}{2}$

B $\frac{\pi}{2} < \theta < \pi$

C $\pi < \theta < \frac{3\pi}{2}$

D $\frac{3\pi}{2} < \theta < 2\pi$

OBJECTIVE D Apply theorems about sines, cosines, and tangents. (Lessons 4-3, 4-6)

32. Why is the statement that for all θ, $\sin(\pi - \theta) = \sin \theta$ called the Supplements Theorem?
33. Use theorems about sines and cosines to prove that $-\sin\left(\frac{\pi}{2} - \theta\right) = \cos(\pi - \theta)$.

In 34–37, given $\sin \theta = k$ for $0 < \theta < \frac{\pi}{2}$, without using a calculator, find each.

34. $\sin\left(\frac{\pi}{2} - \theta\right)$
35. $\cos(\pi - \theta)$
36. $\sin(\theta - \pi)$
37. $\cos(\pi + \theta)$

In 38–40, **True or False** Justify your answer.

38. For all θ, $\cos(\theta + 5\pi) = \cos\theta$.
39. For all θ, $\sin(\theta + 4\pi) = \sin\theta$.
40. For all θ, $\cos^2\theta + \sin^2\theta = \tan^2\theta$.
41. If $\cos\theta = -\frac{1}{4}$, without using a calculator, find all possible values of
    a. $\sin\theta$. b. $\tan\theta$.

OBJECTIVE E **Identify the amplitude, period, frequency, phase shift, and other properties of trigonometric functions. (Lessons 4-7, 4-8, 4-9)**

In 42-45, an equation is given. If they exist, find the following features of the graph of the equation.

a. the period b. the amplitude c. the phase shift

42. $\frac{y}{5} = \cos\left(\frac{x}{2}\right)$
43. $y = 2\sin(3\pi x)$
44. $y = 2\sin\left(x - \frac{\pi}{3}\right)$
45. $h(\theta) = \frac{1}{2}\tan(3\theta)$
46. Identify each of the following for the function defined by the equation $y = -4\sin\left(\frac{x}{3}\right)$.
    a. amplitude b. period c. frequency
47. State the maximum and minimum values of the function f with $f(t) = 10 + 5\cos(2t)$.
48. Suppose the transformation $(x, y) \rightarrow (3x, y - 1)$ is applied to the graph of $y = \sin x$.
    a. State an equation for the image.
    b. Find the amplitude, period, phase shift, and vertical shift of the image.
49. Let $S(x, y) = \left(\frac{x}{4}, -3y\right)$ and $T(x, y) = (x + 7, y)$.
    a. Find the image of the graph of $y = \cos x$ under the composite $T \circ S$.
    b. Find a single transformation that maps $y = \cos x$ to the graph in Part a.

USES Applications of mathematics in real-world situations

OBJECTIVE F **Write and solve equations for phenomena described by trigonometric functions. (Lessons 4-7, 4-8, 4-9)**

50. An alternating current I in amps of a circuit at time t in seconds is given by the formula $I = 35\cos(72\pi t)$.
    a. Find the maximum and minimum values of the current.
    b. How many times per second does the current reach its maximum value?
51. The voltage V in volts in a circuit after t seconds is given by $V = 110\cos(50\pi t)$.
    a. Find the first time the voltage is 55 volts.
    b. Find three times at which the voltage is maximized.
52. A certain sound wave has equation $y = 34\cos(15\pi t)$. Give an equation of a sound wave with twice the frequency and half the intensity of this one.
53. A pendulum is shown at the right. The angular displacement from vertical (in radians) as a function of time t in seconds is given by $f(t) = \frac{1}{2}\sin\left(3t + \frac{\pi}{2}\right)$.

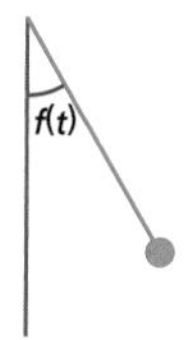

    a. What is the initial angular displacement from the pendulum's rest position?
    b. What is the frequency of f?
    c. How long will it take for the pendulum to make 5 complete swings?

OBJECTIVE G **Find equations of trigonometric functions to model periodic phenomena. (Lesson 4-10)**

54. Suppose the height h in meters of a tide at time t is given in the table below.

t	0	5.1	6.2	11.9	18.2	23.7
h	3.7	0.2	0.8	3.5	0.5	3.7

    a. Fit a sine wave to these data.
    b. What is the period of the sine wave?
    c. What is the amplitude of the sine wave?

55. The length of the day from sunrise to sunset in a city at 34° N latitude (such as Los Angeles, California) is given in the table below.

Date	Days after Jan 1	Length (hours)
1/1	0	9.93
2/1	31	10.55
3/1	59	11.45
4/1	90	12.55
5/1	120	13.55
6/1	151	14.27
7/1	181	14.38
8/1	212	13.83
9/1	243	12.88
10/1	273	11.83
11/1	304	10.8
12/1	334	10.07
1/1	365	9.93

Source: National Oceanic and Atmospheric Administration

a. Find an equation using the sine function to model these data.

b. What is the period of any function used to model these data?

c. Use your model from Part a to predict the length of the day on December 21, the winter solstice, at 34° N latitude.

56. The figure below shows a water wheel rotating at 5 revolutions per minute. The distance d of point P from the surface of the water as a function of time t in seconds can be modeled by a sine wave with an equation of the form $\frac{d-k}{b} = \sin\left(\frac{t-h}{a}\right)$.

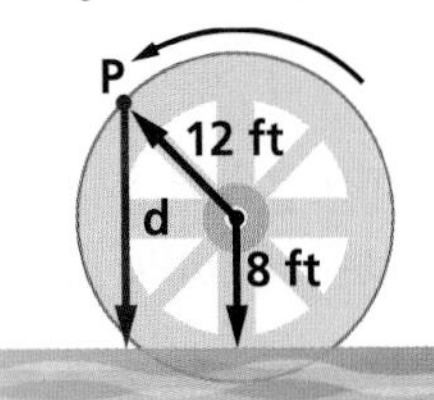

a. What is the amplitude of the sine wave?

b. What is the period of the sine wave?

c. After point P emerges from the water, you start a stopwatch when $d = 8$. Write an equation for the sine wave.

d. Approximately when does point P first reach its highest point?

57. A gear with a 3-cm radius rotates counterclockwise at a rate of 150 revolutions per minute. The gear starts with the tooth labeled A level with the center of the wheel as shown below.

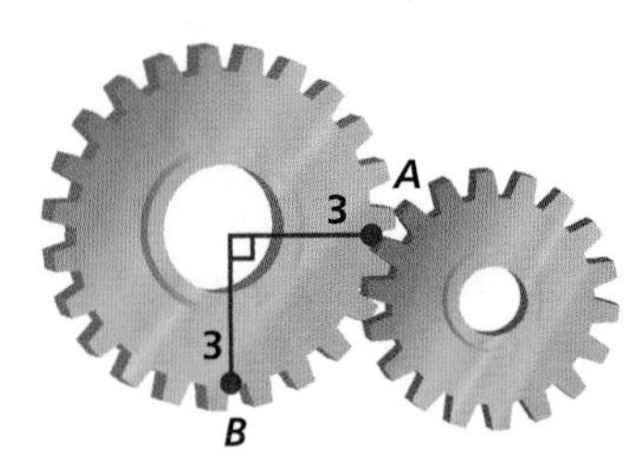

a. Write an equation to give the vertical distance y that point A is above or below its starting position at time t in seconds.

b. How far above or below the starting position will point A be after 4 minutes?

c. Write an equation for the vertical distance h that point A is above point B at time t.

REPRESENTATIONS Pictures, graphs, or objects that illustrate concepts

OBJECTIVE H Use the unit circle to find values of sines, cosines, and tangents. (Lessons 4-2, 4-3)

In 58–61, refer to the unit circle below.

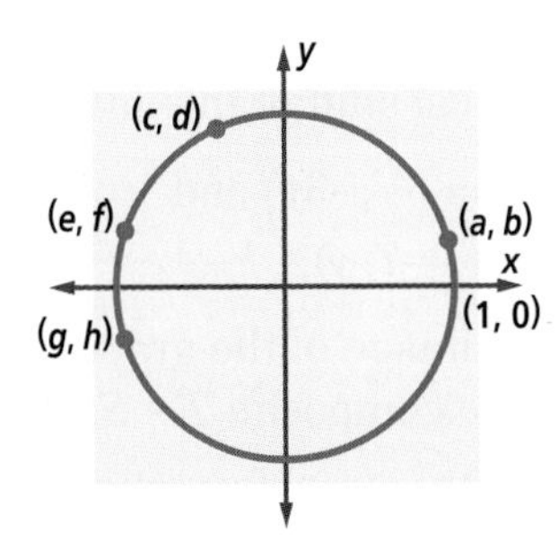

a. Identify the letter that best represents the value given.

b. Estimate the value to the nearest thousandth.

58. $\sin 20°$

59. $\cos(520°)$

60. $\cos\left(\frac{8\pi}{9}\right)$

61. $\sin\left(-\frac{28\pi}{9}\right)$

62. Use a unit circle to show why the cosine function is an even function.

In 63–65, let P' be the reflection image of P over the y-axis, as shown in the unit circle below. State the value of the following.

63. $\sin(180° - \theta)$

64. $\tan(-\theta)$

65. $\cos\left(\frac{\pi}{2} - \theta\right)$

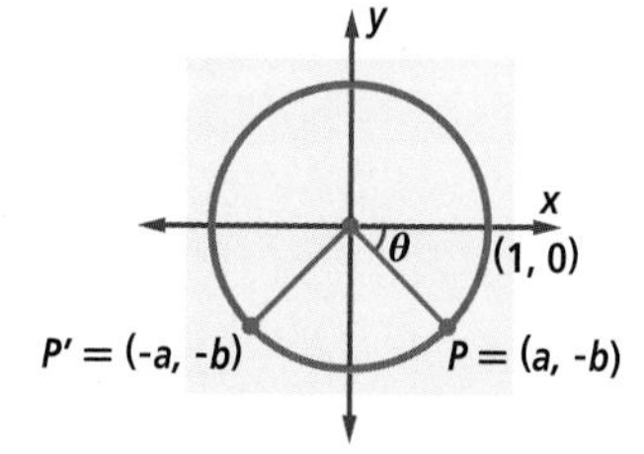

OBJECTIVE I Draw or interpret graphs of the parent sine, cosine, and tangent functions in degrees or radians. (Lessons 4-5, 4-6)

66. Sketch a graph of the cosine function without using technology, indicating key points and the period.

67. Below is part of the graph of a function f. Which of the following could be an equation for f: $f(x) = -\cos x$ or $f(x) = -\sin x$? Justify your answer.

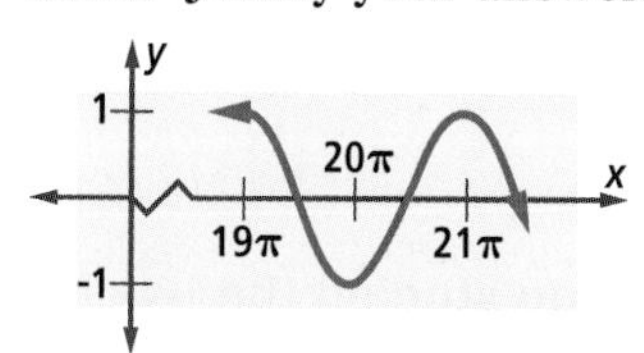

68. Consider the graphs of $f(x) = 4 \sin x$ and $g(x) = 4 \cos x + 8$. What translation maps the graph of f onto the graph of g?

69. a. Use a graphing utility to graph $y = \sin x$ and $y = \sin(\pi - x)$ on the same set of axes.

b. Describe the relationship between the graphs.

c. What theorem does the relationship between these two graphs represent?

70. a. Sketch a graph of the tangent function.

b. Write equations for two of the asymptotes of the graph of the tangent function.

OBJECTIVE J Graph and describe transformation images of graphs of trigonometric functions. (Lessons 4-7, 4-8, 4-9)

In 71–73, match each graph with its equation.

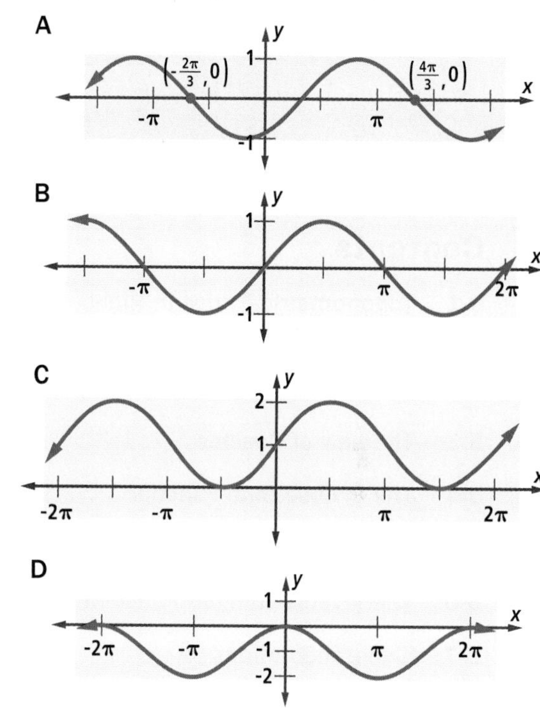

71. $y = 1 + \sin x$

72. $y = \cos\left(x - \frac{\pi}{2}\right)$

73. $y = \sin\left(x - \frac{\pi}{3}\right)$

In 74 and 75, sketch one cycle of the graph without using a graphing utility.

74. $y = \frac{1}{3}\sin(2x)$

75. $y = 7\cos\left(\frac{\pi}{2}x\right)$

76. Write an equation for the image of the graph of $y = \sin x$ under a phase shift of $\frac{3\pi}{4}$.

77. Write an equation for the image of the graph of $y = \cos x$ under a phase shift of $-\frac{\pi}{3}$ and with a period of 4π.

78. Consider the equation $y = 4 - 7\sin(2x + 1)$.

a. Sketch a graph of the equation.

b. State the period and maximum value of the function with this equation, if they exist.

Chapter

5 Trigonometry

Contents

In Chapter 4, you studied the use of the sine, cosine, and tangent functions in describing circular motion and other periodic phenomena. The first known uses of these functions arose from the need to solve practical, everyday problems in which lengths or angles in triangles had to be found. This is why the study of the sine, cosine, and tangent functions is called *trigonometry*, a word derived from Greek words meaning "triangle measurement."

The origins of trigonometry have been traced back to the Egyptians of the 13th century B.C.E., whose tables of shadow lengths correspond to today's tangent and cotangent functions. The Babylonians and Greeks used trigonometry to study the heavens, and more recent travelers used it to navigate.

In this chapter, you will study additional properties of these functions and you will use trigonometry to solve problems, such as the following, that are based on triangle measurement.

In baseball, how long is a throw from 10 feet behind second base to first base?

What is the diameter of the Moon?

How far is it from New York to New Delhi, India?

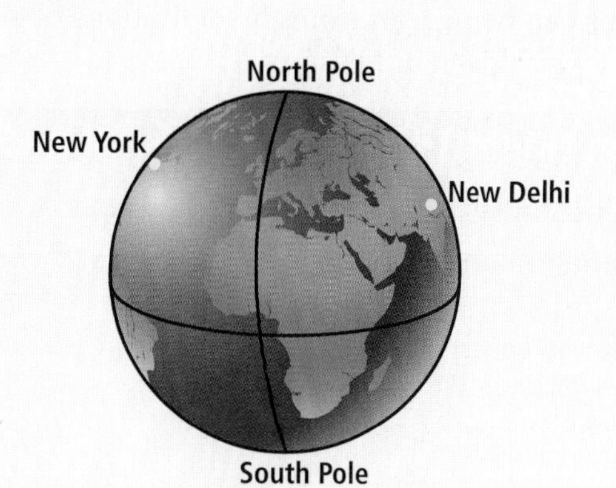

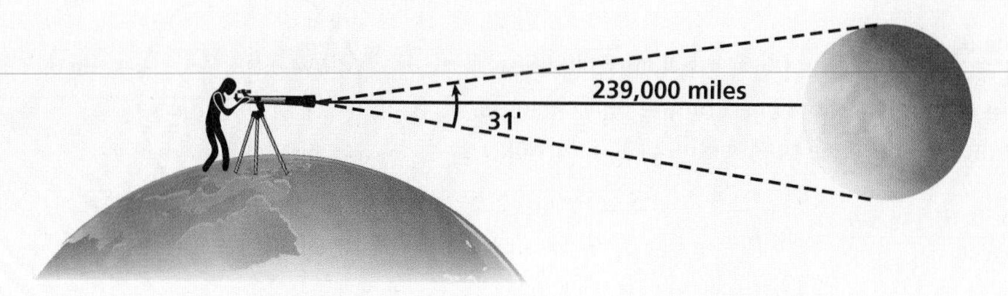

Lesson 5-1 Trigonometric Ratios in Right Triangles

Vocabulary

angle of depression
angle of elevation

▶ **BIG IDEA** The sine, cosine, and tangent of magnitudes between 0° and 90° each equal a ratio of sides of a right triangle.

Mental Math

Consider the four numbers 0.5, 0.7, 0.8 and 0.9.

a. Which is the best approximation to $\frac{\sqrt{2}}{2}$?

b. Which is the best approximation to $\frac{\sqrt{3}}{3}$?

c. Which is the best approximation to $\frac{\sqrt{4}}{4}$?

The picture below reviews some terminology about right triangles. Examine it closely. Each leg is opposite one acute angle and adjacent to the other. Recall that the side opposite a vertex is named by the small letter of that vertex. For example, opposite vertex B is side b.

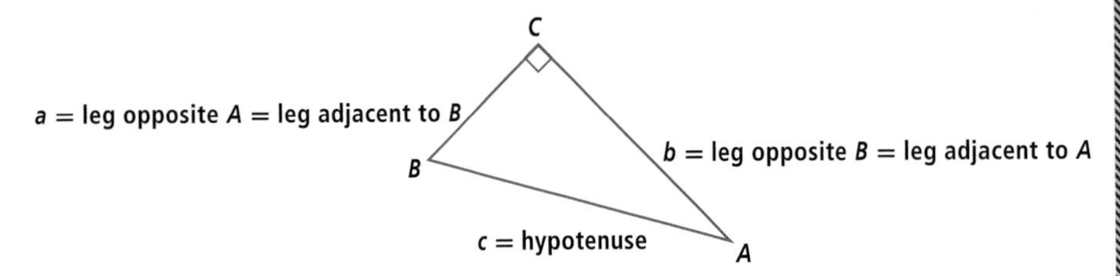

Relating Sines, Cosines, and Tangents to Sides in Right Triangles

In an earlier course, you may have seen sines, cosines, and tangents defined in terms of the lengths of sides in right triangles. These relationships can be *proved* from the definitions of sine, cosine, and tangent that were given in Lesson 4-2.

Theorem (Right Triangle Ratios for Sine, Cosine, and Tangent)

If θ is the measure of an acute angle in a right triangle, then

$$\frac{\text{leg opposite } \theta}{\text{hypotenuse}} = \sin \theta,$$
$$\frac{\text{leg adjacent to } \theta}{\text{hypotenuse}} = \cos \theta, \text{ and}$$
$$\frac{\text{leg opposite } \theta}{\text{leg adjacent to } \theta} = \tan \theta.$$

Proof Here we prove that $\frac{\text{leg opposite } \theta}{\text{hypotenuse}} = \sin \theta$. The other proofs are similar. A right triangle with hypotenuse 1 can be positioned on the unit circle as shown at the right. The coordinates of P give the lengths of the legs of $\triangle PAE$, so $PE = \sin \theta$. Thus,

$$\frac{\text{leg opposite } \theta}{\text{hypotenuse}} = \frac{\sin \theta}{1} = \sin \theta.$$

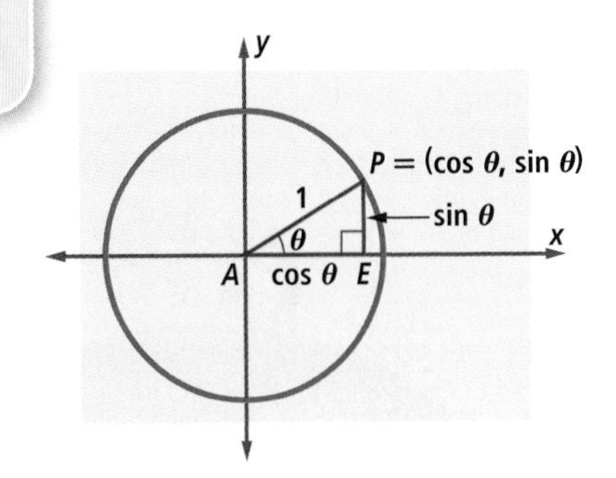

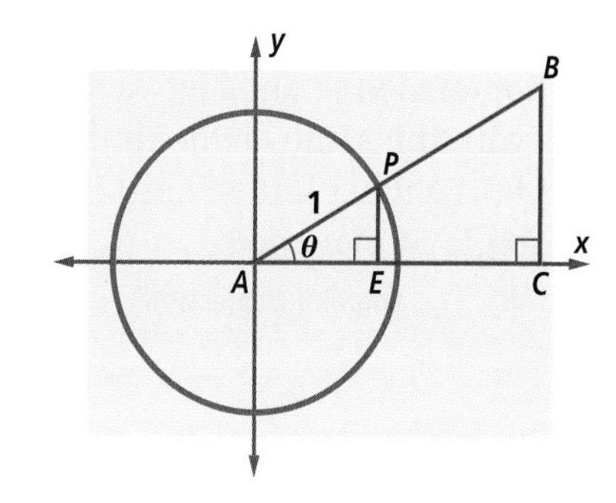

If the hypotenuse of $\triangle ABC$ is not 1, place the triangle as in the diagram at the right, where both $\triangle PAE$ and $\triangle BAC$ contain $\angle A$ and a right angle. The triangles are similar, so

$$\frac{BC}{BA} = \frac{PE}{PA}.$$

$PE = \sin\theta$ and $PA = 1$, so substitute.

$$\frac{BC}{BA} = \frac{\sin\theta}{1} = \sin\theta.$$

In $\triangle BAC$, $\overline{BC}$ is opposite θ and $\overline{BA}$ is the hypotenuse. So

$$\frac{\text{leg opposite }\theta}{\text{hypotenuse}} = \sin\theta.$$

In the above proof, $\theta = \text{m}\angle A$, so you could write $\sin(\text{m}\angle A) = \frac{\text{leg opposite } A}{\text{hypotenuse}}$. However, it is customary to write $\sin A$, $\cos A$, and $\tan A$ rather than $\sin(\text{m}\angle A)$, $\cos(\text{m}\angle A)$, and $\tan(\text{m}\angle A)$.

GUIDED

Example 1

In $\triangle ABC$, the leg opposite A has length 20, the leg adjacent to A has length 21, and the hypotenuse has length 29. Calculate $\sin A$, $\cos A$, and $\tan A$.

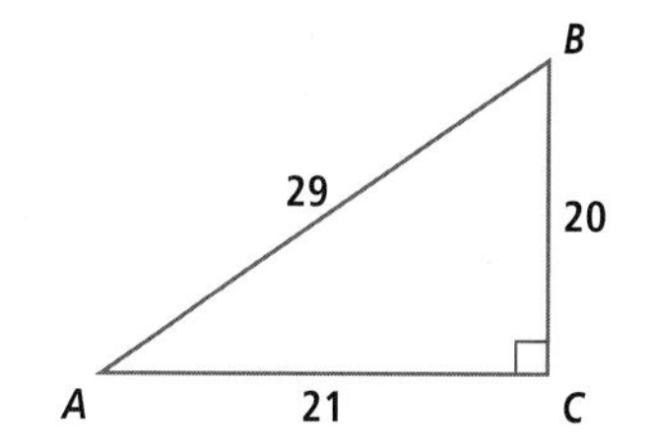

Solution $\sin A = \frac{\text{leg opposite } A}{?} = \underline{\ ?\ }$

$\cos A = \frac{?}{\text{hypotenuse}} = \underline{\ ?\ }$

$\tan A = \frac{?}{?} = \underline{\ ?\ }$

Finding Sides in Right Triangles

In many situations, you know only the length of one side and the measure of one acute angle in a right triangle. Using the appropriate trigonometric ratio, you can determine the lengths of the other sides.

Example 2

Sonar on a salvage boat locates an object at a downward angle of 50° and a distance of 510 meters, as shown in the drawing. How far below the level of the water is the object?

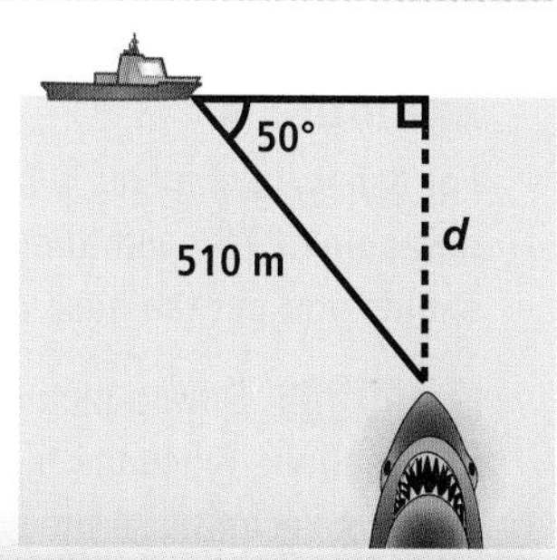

Solution In this situation, the measure of an acute angle and the length of the hypotenuse in a right triangle are known. The problem asks for the length of the leg opposite the known acute angle. The sine ratio involves these two lengths, so it is the appropriate ratio to use.

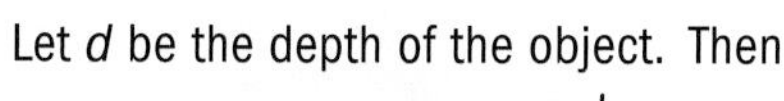
Let d be the depth of the object. Then

$$\sin 50° = \frac{d}{510},$$

so $510 \sin 50° = d$.

A calculator shows that $d = 390.68... \approx 391$ meters.

510·sin(50°)	390.683
$\frac{391}{510}$	0.766667
sin(50°)	0.766044

Check Compare $\frac{391}{510}$ with $\sin 50°$. They should be very close.

The 50° angle in Example 1 is called an *angle of depression*. The **angle of depression** to an object is the angle formed by a horizontal ray and a ray with the same endpoint that contains the object if the object is below the horizontal. If the object is above the horizontal ray, then the angle is an **angle of elevation**.

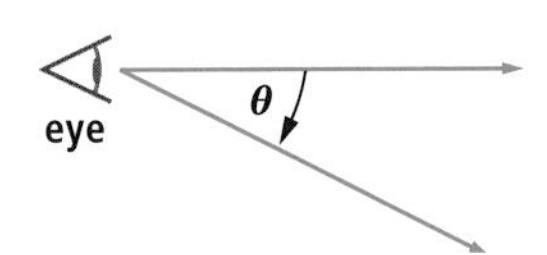

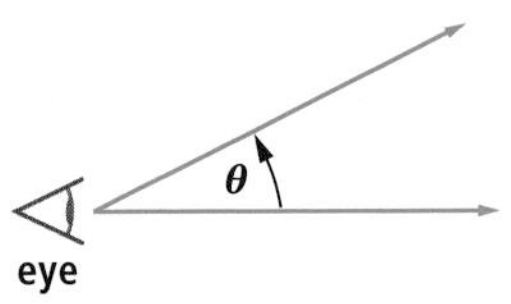

GUIDED

Example 3

You wonder how tall a grain silo is. You walk 60′ from the silo and find that you have to look up at a 38° angle to see the top. If your eyes are 5′3″ off the ground, how tall is the silo without the roof?

Solution Draw a diagram showing the right triangle formed by the top *T* of the silo, your eye *E*, and the point *P* below *T* at your eye level. Angle *TEP* is an angle of elevation. Given are m∠*TEP* and *EP*, the length of the leg adjacent to it. You must find *TP*, the length of the leg opposite ∠*E*. Use the tangent ratio and solve for *TP*.

$$\tan 38° = \frac{TP}{?}$$

$$TP = \underline{\ ?\ } \tan 38°$$

So $TP \approx \underline{\ ?\ }$ feet.

Add 5.25 feet to *TP* to obtain the height of the silo.

The silo is about ___?___ feet tall.

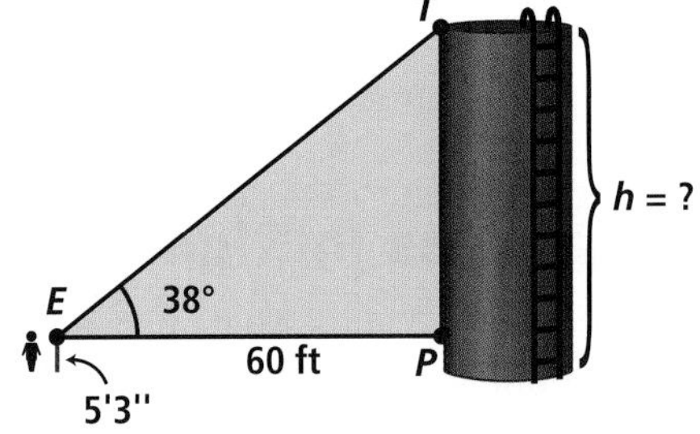

Check Divide your value of *TP* by 60. The quotient should be very close to tan 38°.

Example 4

For safety, in climbing a ladder against a wall, the ladder should ideally make an angle of 75° with the ground. If a ladder is 12′ long, how far from the wall should you put the base of the ladder?

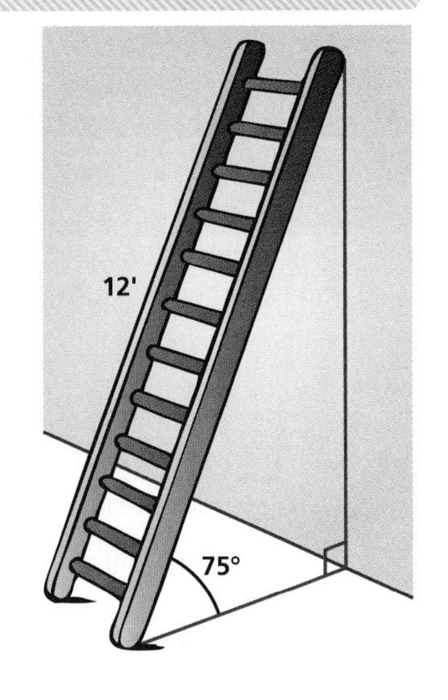

Solution Draw a picture, as shown at the right, including the known information. Since the hypotenuse is known and the leg adjacent to a known angle is desired, use the cosine function.

$$\cos 75° = \frac{\ell}{12}$$

$$\ell = 12 \cos 75° \approx 3.11 \text{ feet}$$

You should put the ladder about 3.1 feet from the wall.

Questions

COVERING THE IDEAS

1. Prove the theorem in this lesson for tan θ.

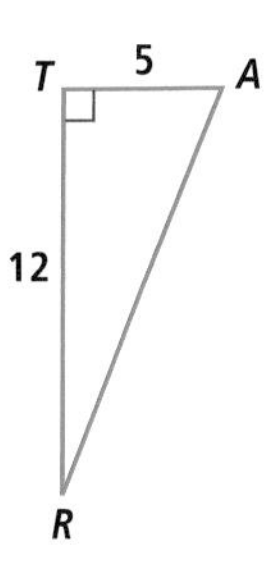

2. Consider $\triangle ART$ as shown at the right. Give the exact value of
   a. $\sin A$.   b. $\cos A$.   c. $\tan A$.

In 3 and 4, find the unknown length to the nearest tenth.

3. 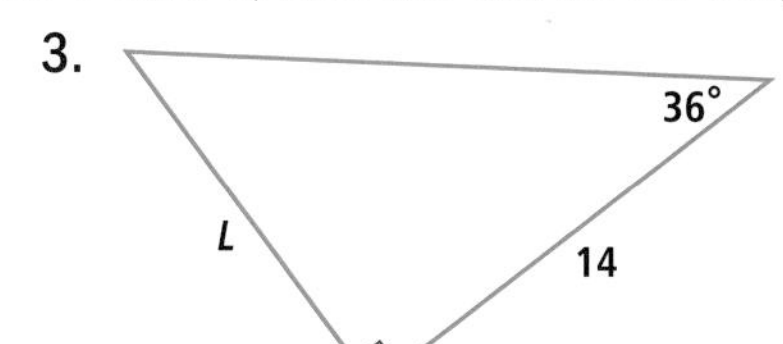

4. 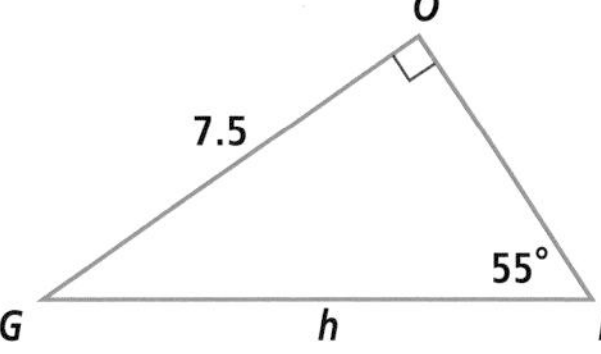

5. Refer to Example 2. Suppose sonar on the surface of the sea locates an object at an angle of depression of 65° and a distance of 320 feet. How deep is the object?

6. Refer to Example 3. To determine the height of a flagpole, you walk 30′ away from it and measure the angle of elevation to its top to be about 50°. How high is the flagpole?

7. Consider $\triangle BED$ with a right angle at E. $BE = 3$ and $DE = 4$.
   a. Find exact values for $\sin D$ and $\cos B$.
   b. Explain why these two values are equal.

APPLYING THE MATHEMATICS

8. Suppose you are looking at Taipei 101, a building in Taipei, Taiwan, which has a height of about 509 meters. From where you are, you estimate the angle of elevation to the top of this building to be about 10°.
   a. How far are you from the building, to the nearest ten meters?
   b. If the angle of elevation is only correct to the nearest three degrees, give the maximum and minimum values for your distance from Taipei 101.

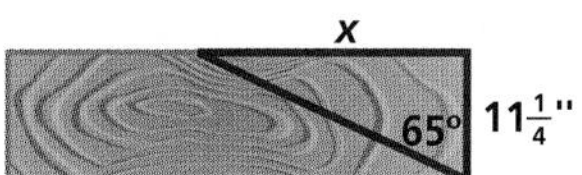

9. Lumber sold in the U.S. as 12″ is actually only $11\frac{1}{4}$″ wide. Suppose you need to cut a 65° angle for a woodwork project, but you don't have a protractor. To the nearest $\frac{1}{8}$ inch, how many inches should you mark off on the other side of a 12″ board?

10. A triangle is drawn in the coordinate plane as shown at the right.
    a. What is $m\angle EDF$?
    b. Find the coordinates of E to the nearest tenth.

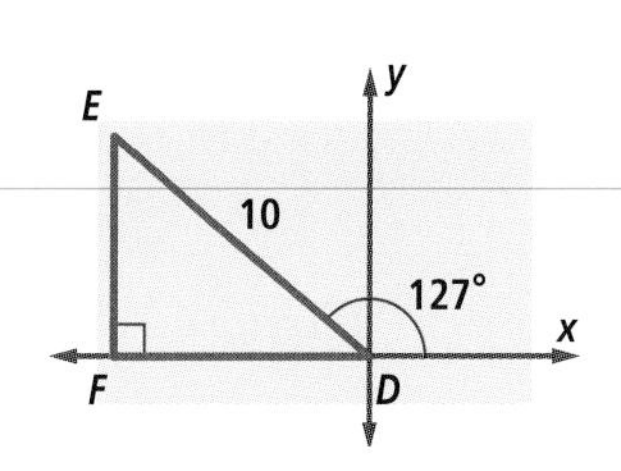

11. A right triangle ABC is drawn in the coordinate plane as shown at the right. Use the given information to find the coordinates of point B. Round lengths to the nearest tenth.

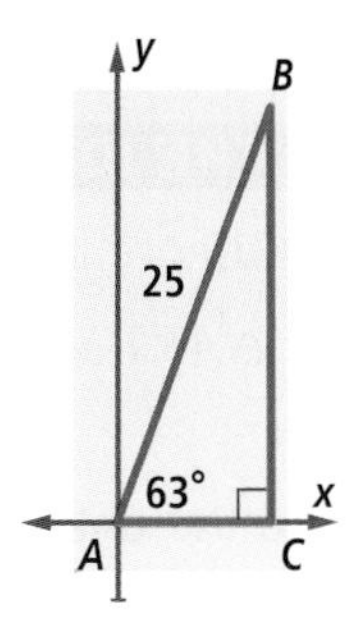

12. A ship is on a bearing of 47° east of north as shown at the right.
    a. How far has the ship sailed if it is 185 km north of its original position?
    b. How far east of its original position is the ship?
    c. If the ship's average speed is 12 km per hour, for how long has it been sailing?

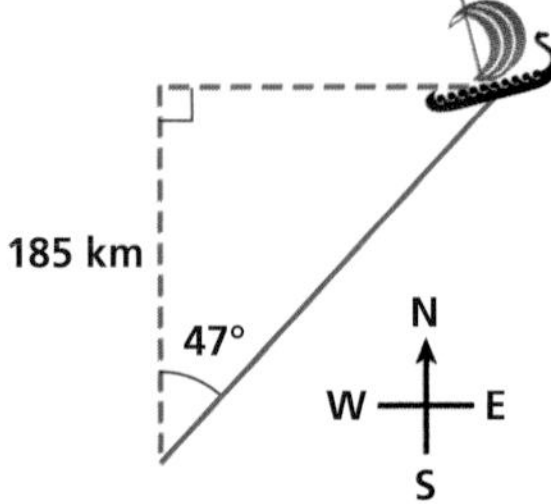

REVIEW

In 13 and 14, use identities to simplify. (Lesson 4-3)

13. $\sin(90° - \theta)$

14. $\tan(180° - \theta)$

15. If $\sin \theta = 0.28$ and $90° < \theta < 180°$, compute. **(Lessons 4-3, 4-2)**
    a. $\sin(180° - \theta)$
    b. $\tan \theta$
    c. $\cos(180° + \theta)$

16. Find the distance between the points. **(Previous Course)**
    a. (x_1, y_1) and (x_2, y_2)
    b. $(0, 0)$ and $(\cos \theta, \sin \theta)$
    c. $(6, -7)$ and $(7, -6)$

EXPLORATION

17. For small angles, the length of an arc and the length of the associated chord of a circle are approximately equal. Astronomical measurements are commonly made by measuring the small angle formed by the extremities of a distant object. The Moon covers an angle of 31 minutes (31′) when it is about 239,000 miles away.

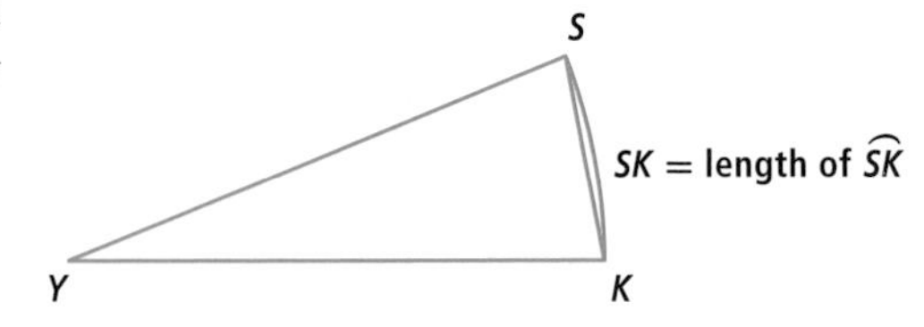

    a. From this information, estimate the diameter of the Moon by finding the length of $\overset{\frown}{SK}$.
    b. Estimate the diameter of the Moon by finding the length of $\overline{SK}$ using a trigonometric function.

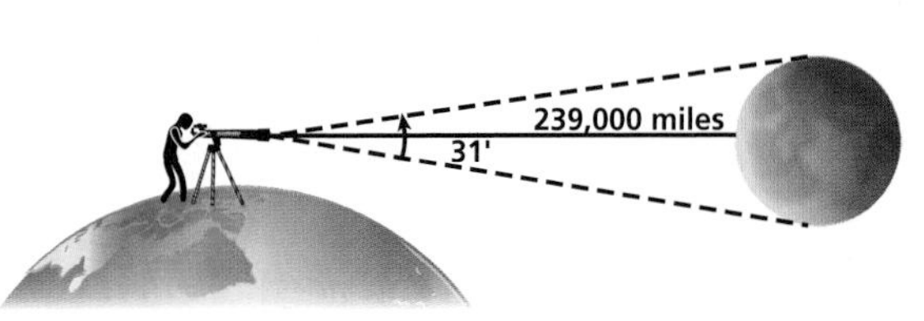

    c. Compare your estimates to the value found in an almanac or other reference book. What do you think accounts for the differences?

Lesson 5-2 The Inverse Cosine Function

Vocabulary

inverse cosine function, $\cos^{-1}$, Arccos

BIG IDEA If you know $\cos \theta$ and that $0 \leq \theta \leq 180°$, then θ is uniquely determined.

In Lesson 5-1, you used trigonometric functions to find the length of a side in a right triangle. By using inverse trigonometric functions, you can find measures of angles of triangles. In this lesson, properties of the inverse cosine function, $\cos^{-1}$, are examined.

Mental Math

Find the exact value.

a. $\cos 60°$

b. $\cos \frac{\pi}{4}$

c. $\cos -225°$

Recall the ladder problem in Example 4 of Lesson 5-1. The ladder is unsafe to climb if the angle it makes with the ground is less than 75°. Suppose you place a 16′ ladder so that its base is 5′ from a wall. Is this safe? In this case,

$$\cos \theta = \frac{5}{16} = 0.3125.$$

A problem in solving this equation for θ is that there are infinitely many values of θ that satisfy the equation. We need to restrict the domain of the cosine function to obtain a solution between 0° and 90°.

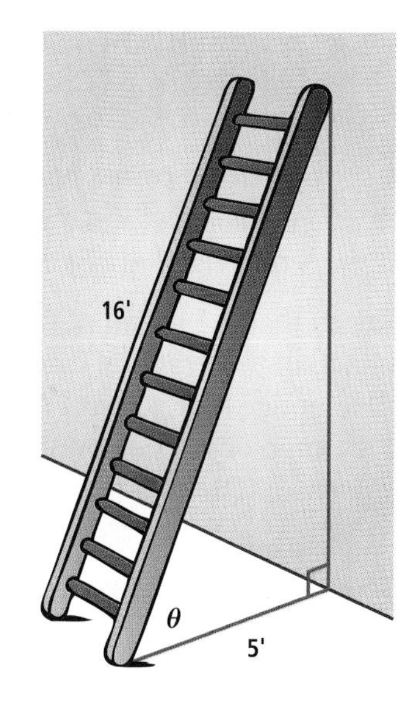

The Domain of the Restricted Cosine Function

Recall the graph of $y = \cos x$.

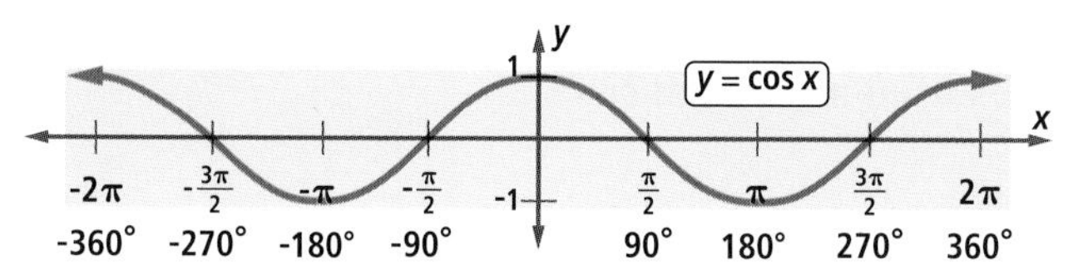

Many values of x yield the same value of y. For this reason, in order for the inverse of the cosine function to be a function, the domain must be restricted. There are many possible restricted domains. Here are three criteria for an appropriate domain:

1. The domain should include the angles between 0 and $\frac{\pi}{2}$ (or 0° and 90°), because they are the measures of the acute angles of a right triangle.
2. On the restricted domain, the function should take on all the values of the range, that is, all real numbers between –1 and 1.
3. The function should be continuous on the restricted domain.

Only one domain fits all three criteria. It is $\{x \mid 0 \leq x \leq \pi\}$ or $\{x \mid 0° \leq x \leq 180°\}$. The function that is the inverse of this restricted cosine function is denoted **cos^{-1}** and called the **inverse cosine** function. You can read $\cos^{-1} x$ as "the number whose cosine is x."

Definition of Inverse Cosine Function

$y = \cos^{-1} x$ if and only if $x = \cos y$ and $0 \leq y \leq \pi$.

Calculators are programmed to give approximations to values of the $\cos^{-1}$ function. To solve $\cos \theta = 0.3125$ from the ladder problem (on the previous page), note that

$$\theta = \cos^{-1}(0.3125) \approx 71.8°.$$

So the angle is a little smaller than what would be considered safe.

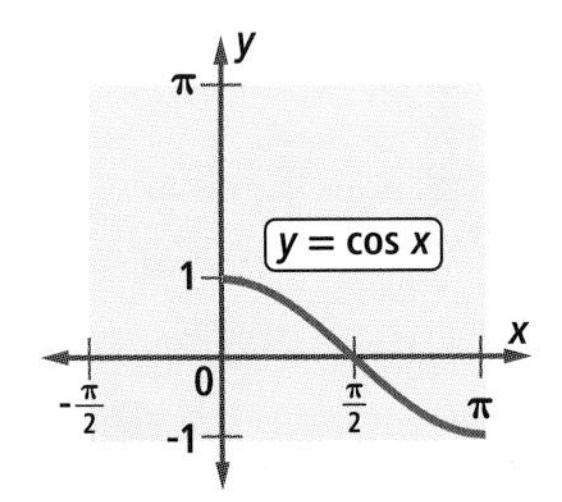

restricted cosine function
domain: $\{x \mid 0 \leq x \leq \pi\}$
range: $\{y \mid -1 \leq y \leq 1\}$

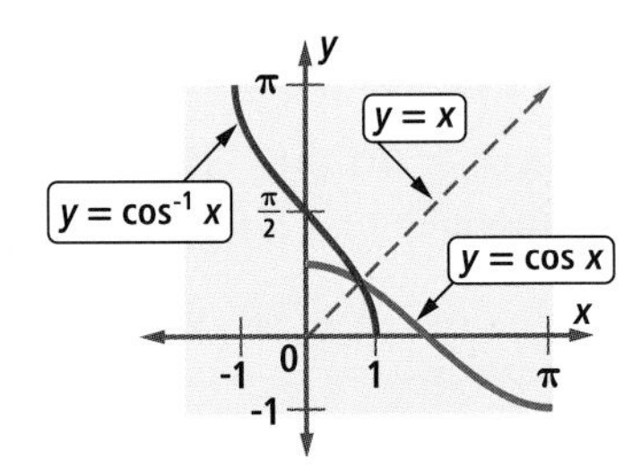

inverse cosine function
domain: $\{x \mid -1 \leq x \leq 1\}$
range: $\{y \mid 0 \leq y \leq \pi\}$

Recall that the graph of the inverse of any function is the reflection image of the graph of that function over the line with equation $y = x$. Also notice that, as with all inverse functions, the domain of $y = \cos^{-1} x$ is the range of $y = \cos x$ and the range of $y = \cos^{-1} x$ is the domain of $y = \cos x$. The notation **Arccos** is sometimes used in place of $\cos^{-1}$.

Evaluating Inverse Cosines

The values of the inverse cosine function must lie in the range of $y = \cos^{-1} x$, that is, from 0 to π or from 0° to 180°.

Example 1

Evaluate $\cos^{-1}\left(\frac{\sqrt{3}}{2}\right)$ without a calculator. Give an exact answer in radians.

Solution Recall the unit circle. If $y = \cos^{-1}\left(\frac{\sqrt{3}}{2}\right)$, then, by definition of $\cos^{-1}$, y is the unique number on the interval $0 \leq y \leq \pi$ whose cosine is $\frac{\sqrt{3}}{2}$. Because $\cos \frac{\pi}{6} = \frac{\sqrt{3}}{2}$ and $0 \leq \frac{\pi}{6} \leq \pi$, $\cos^{-1}\left(\frac{\sqrt{3}}{2}\right) = \frac{\pi}{6}$.

Check Enter $\cos^{-1}\left(\frac{\sqrt{3}}{2}\right)$ on a calculator set to radian mode. Multiply this value by 6 and compare it to π.

$\cos^{-1}\left(\frac{\sqrt{3}}{2}\right)$	0.523599
0.52359877559828·6	3.14159

Example 2

Evaluate $\cos^{-1}(-0.33333)$. Give your answer in degrees.

Solution According to the definition, $y = \cos^{-1}(-0.33333)$ if and only if $\cos y = -0.33333$ and $0° \leq y \leq 180°$. Use a calculator. The answer is approximately 109.47°.

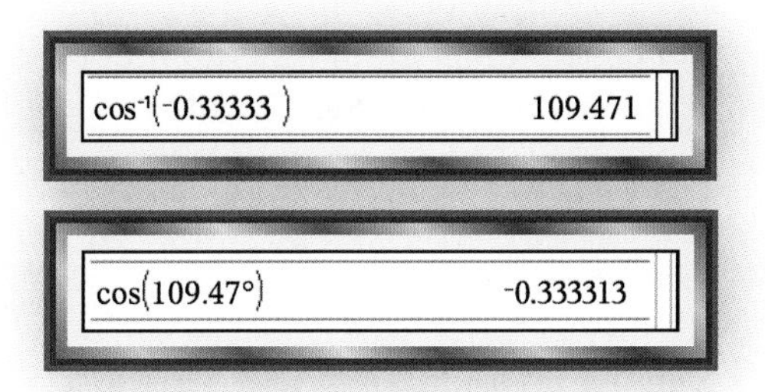

Check Is $0° \leq 109.47° \leq 180°$? Yes, so the answer from the calculator is in the correct range. Now evaluate cos(109.47°). Our calculator gives the answer as –0.333313, which is about –0.33333.

Applications of the Inverse Cosine

Wherever there are applications of the cosine function, applications of the inverse cosine function may appear.

Example 3

The Landsat 7 satellite orbits Earth at an altitude of 705 km (438 miles) above the equator. It can only see a portion of Earth's surface (bounded by a horizon circle as shown below at the left) at any given time. Imagine looking at a cross section of the satellite and Earth, as shown below at the right. Point *C* is the center of Earth, *H* is a point on the horizon circle, and *S* is the location of the satellite. Let $\theta = m\angle HCS$. The altitude *a* of the satellite is the distance in kilometers of the satellite above Earth. Notice that $\angle CHS$ is a right angle, because the tangent to a circle is perpendicular to the point of tangency.

a. Write a formula for θ in terms of *r* and *a*.

b. The radius of Earth is about 6378 km. To the nearest degree, what is θ?

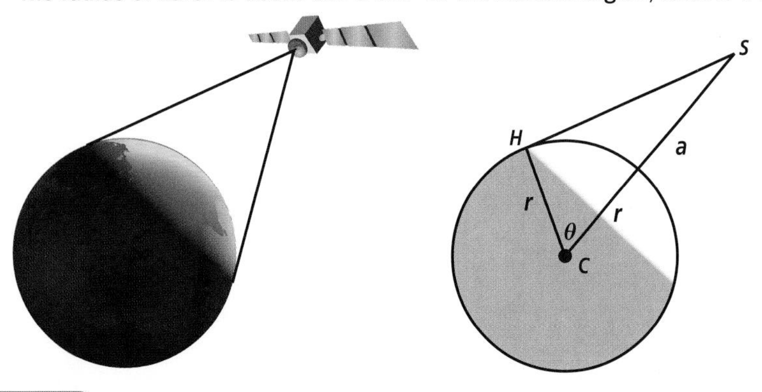

Solution

a. Since the hypotenuse of $\triangle CHS$ and the side adjacent to θ involve *r* and *a*, use the cosine of θ.

$$\cos \theta = \frac{r}{r + a}$$

Solve for θ. $\theta = \cos^{-1}\left(\frac{r}{r + a}\right)$

b. For $r = 6378$ and $a = 705$,

$$\theta = \cos^{-1}\left(\frac{6378}{6378 + 705}\right) \approx \cos^{-1}(0.900466) \approx 26°.$$

Questions

COVERING THE IDEAS

In 1–3, the graphs below show several ways to restrict the domain of $y = \cos x$. Each graph fails to meet the criteria for appropriate domain in some way. How does it fail?

1. 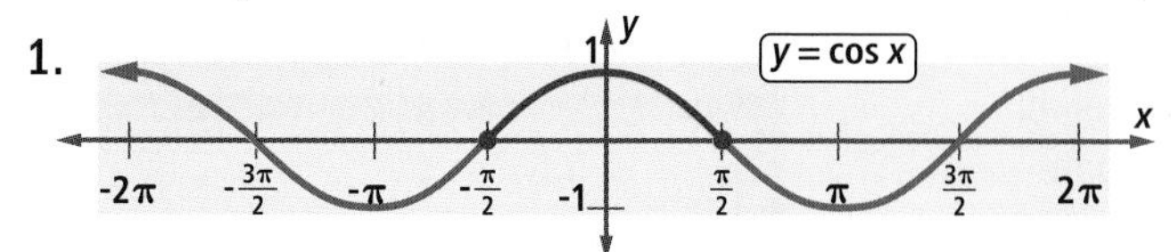

   domain: $\{x \mid -\frac{\pi}{2} \le x \le \frac{\pi}{2}\}$

2. 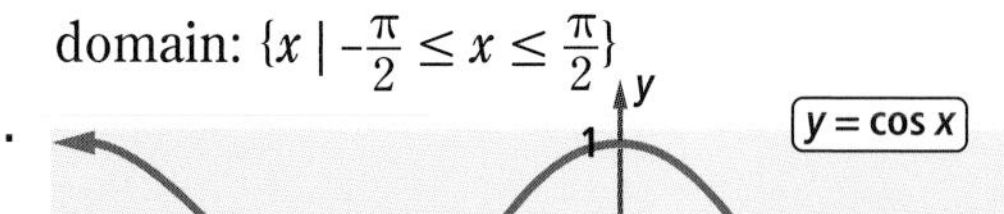
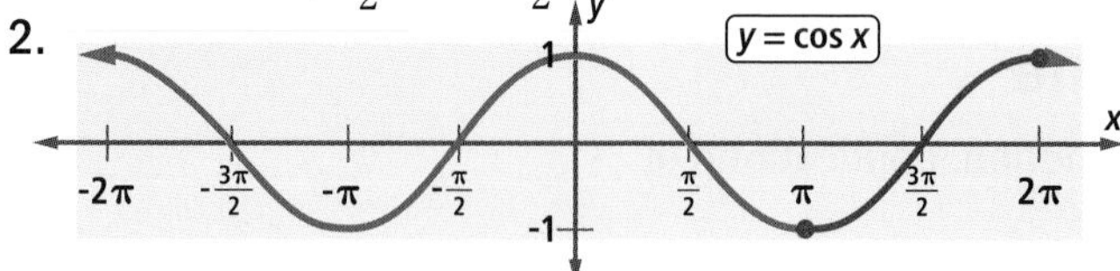

   domain: $\{x \mid \pi \le x \le 2\pi\}$

3. 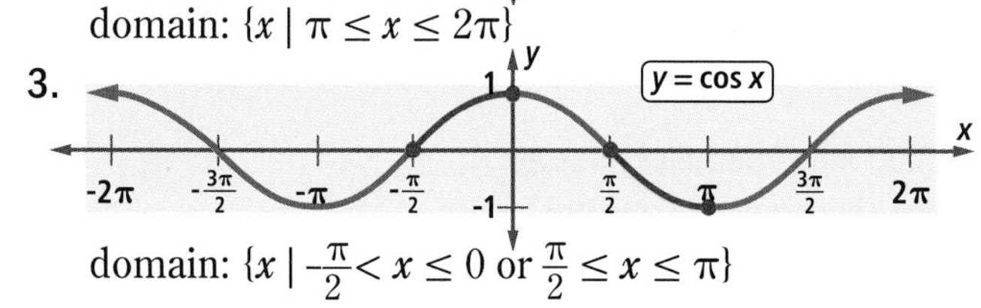

   domain: $\{x \mid -\frac{\pi}{2} < x \le 0 \text{ or } \frac{\pi}{2} \le x \le \pi\}$

4. a. What transformation maps the graph of the restricted cosine function onto the graph of the inverse cosine function?

   b. State the domain and range of the inverse cosine function.

In 5–7, find the exact value of the expression in radians and degrees without a calculator.

5. $\cos^{-1}\left(\frac{\sqrt{2}}{2}\right)$

6. Arccos 0

7. $\cos^{-1}\left(-\frac{\sqrt{3}}{2}\right)$

8. Refer to the diagram at the right. For safety, many people recommend no more than a 1:4 ratio between distance from a wall and the length of a ladder. What is the measure of the angle the ladder makes with the ground when this ratio is used?

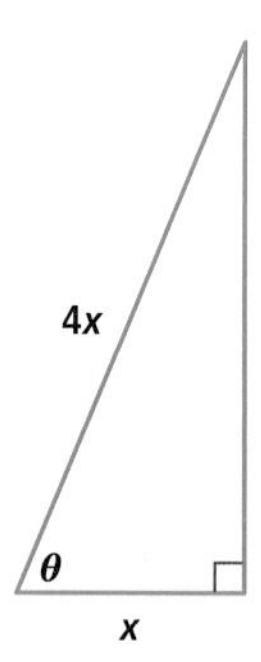

9. Refer to Example 3. Give the measure of θ when the height of a satellite above Earth is about 402 km (250 miles), as is the case with the International Space Station.

APPLYING THE MATHEMATICS

In 10 and 11, compute without using your calculator.

10. $\cos\left(\cos^{-1}\left(-\frac{\sqrt{2}}{2}\right)\right)$

11. $\cos^{-1}(\cos 410°)$

In 12 and 13, find five solutions to each equation to the nearest degree.

12. $\cos x = \frac{1}{2}$ **13.** $\cos y = 0.9263$

14. A 26″ bicycle tire has a chalk mark at its top. As the bike moves forward, the height h of the chalk mark in terms of distance d traveled can be modeled by $h = f(d) = 13 + 13\cos\left(\frac{d}{13}\right)$, with d measured in inches.

a. What is the period of the function f? How do you know?

b. Graph one period of f.

c. How far has the bike traveled the first time the chalk mark is 20″ above ground? The second time?

REVIEW

15. A fishing boat sails at a steady speed of 10 mph on a bearing of 55° (from north). If the boat set sail at 2:20 A.M., describe its location east and north of its port at 6:20 A.M. **(Lesson 5-1)**

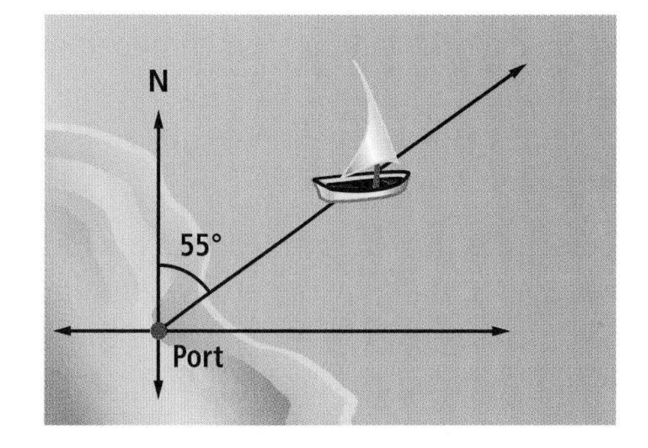

16. Consider $\triangle ABC$ below.

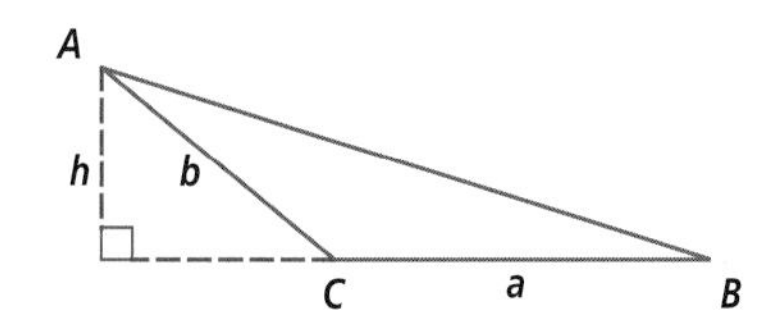

a. Find h, the altitude to side $\overline{BC}$, in terms of b and $m\angle ACB$.

b. Derive a formula for the area of $\triangle ABC$ in terms of a, b, and $m\angle ACB$. **(Lessons 5-1, 4-3)**

17. Find images of each of the following under a scale change $S: (x, y) \rightarrow \left(80x, \frac{y}{8}\right)$.

a. $(5, -4)$ **b.** $(0, 1)$ **c.** $\left(\frac{3}{4}, \frac{1}{2}\right)$ **(Lesson 3-5)**

18. A data set consists of exactly four numbers. The mode of the data set is 5, the mean and median are 10 and 15. Which number is the mean? Explain how you can tell. **(Lesson 1-2)**

EXPLORATION

19. a. Find $\sin(\cos^{-1} 0.8)$.

b. Find $\sin\left(\cos^{-1}\left(\frac{21}{29}\right)\right)$.

c. Generalize the results above. That is, find $\sin(\cos^{-1} a)$ when $0° \leq \cos^{-1} a \leq 90°$.

Lesson 5-3 The Law of Cosines

BIG IDEA Given SAS or SSS in a triangle, the Law of Cosines enables you to find the remaining sides or measures of angles of the triangle.

Mental Math

Simplify

a. $\cos(-t)$

b. $\cos(180° - t)$

c. $\cos(90° - t)$

Trigonometry enables lengths of sides and angle measures to be found in *any* triangle, not just right triangles. One important application is in finding distances that are difficult to measure directly. Here is an example.

A discus is thrown by means of a whirling movement made by the athlete within a circle 2.5 m in diameter. The winner is the one who makes the longest throw without stepping outside of the throwing circle before the discus touches the ground.

Jim Thorpe (1887-1953)
Named by one group the greatest athlete of the 20th century

The length of a throw is the distance from the edge of the circle to where the discus lands. But how is the length measured? In the past a referee rolled out a long measuring tape, but now lasers and the *Law of Cosines* are used.

This desired length is W in the diagram at the right. Because $\triangle MZL$ is not a right triangle, the methods of Lesson 5-1 cannot be used directly. However, the problem can be solved using a powerful theorem called the *Law of Cosines*.

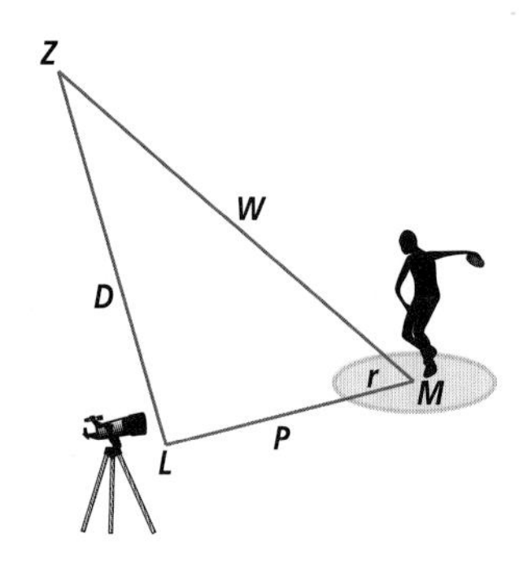

The Law of Cosines is proved using the definitions of cosine and sine and the formula for the distance between two points in the coordinate plane. Also, remember that for any θ, $\cos^2 \theta + \sin^2 \theta = 1$.

Theorem (Law of Cosines)

In any $\triangle ABC$, $c^2 = a^2 + b^2 - 2ab \cos C$.

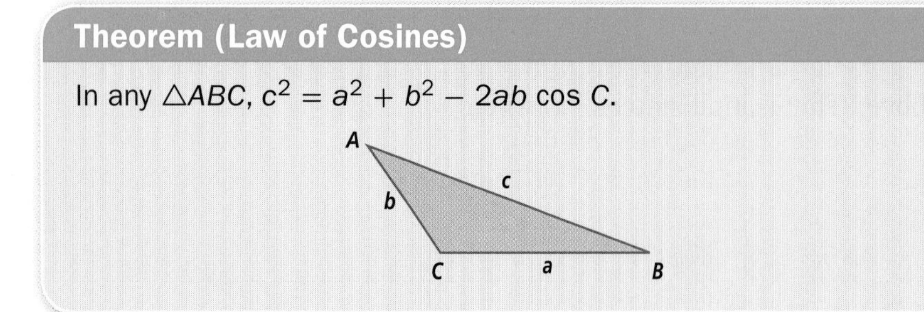

Proof Impose a coordinate plane on $\triangle ABC$ so that $C = (0, 0)$ and $\overline{BC}$ lies on the x-axis. Since $BC = a$, the coordinates of B are $(a, 0)$. To find the coordinates of A, draw a unit circle that intersects $\overrightarrow{CA}$ at D. By the definition of the cosine and sine, $D = (\cos C, \sin C)$. Since $AC = b$, A can be considered the image of D under the size change $(x, y) \rightarrow (bx, by)$.

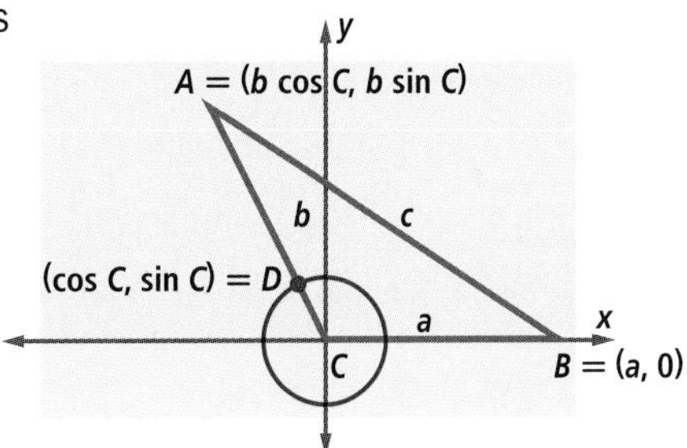

So, $A = (b \cos C, b \sin C)$.

By the Pythagorean distance formula, $c = \sqrt{(b \cos C - a)^2 + (b \sin C - 0)^2}$.

Now we rewrite the equation into the form in which it appears in the statement of the theorem.

Square both sides. $c^2 = (b \cos C - a)^2 + (b \sin C - 0)^2$

Expand the binomials. $c^2 = b^2\cos^2 C - 2ab \cos C + a^2 + b^2\sin^2 C$

Apply the Commutative Property of Addition. $c^2 = a^2 + b^2\sin^2 C + b^2\cos^2 C - 2ab \cos C$

Factor. $c^2 = a^2 + b^2 (\sin^2 C + \cos^2 C) - 2ab \cos C$

Use the Pythagorean Identity. $c^2 = a^2 + b^2 - 2ab \cos C$

QY

QY

Suppose $a = 12$, $b = 5$, and $\cos C = \frac{1}{2}$. What is c?

Using the Law of Cosines to Find a Side

Note the power of the Law of Cosines. Given two sides and the included angle of *any* triangle (the SAS condition), you can find the third side.

Example 1

Suppose that the distance of a laser telescope to the center of the throwing circle in a discus competition is 30.10 meters, from the telescope to the target is 51.02 meters, and the angle between the telescope's paths to the target and to the center of the throwing circle is 95°. Find the length of a throw from the circle to the target.

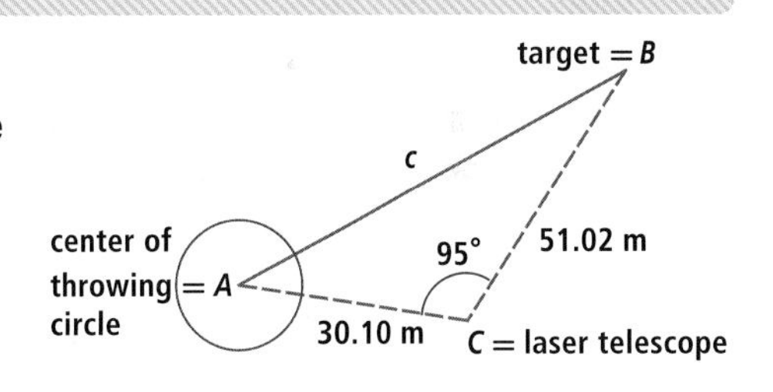

Solution 1 Use the Law of Cosines since two sides and the included angle are given. For this situation, $a = 51.02$ m, $b = 30.10$ m, and $C = 95°$.

$$c^2 = a^2 + b^2 - 2ab \cos C$$
$$= 51.02^2 + 30.10^2 - 2(51.02)(30.10)\cos 95° \approx 3776.7409$$

So, $c \approx \sqrt{3776.7409} \approx 61.46$ meters

The length c is measured from the center of the throwing circle, but discus throws are measured from the edge of the circle. So the radius of the 2.5-m diameter throwing circle must be subtracted.

length of throw $= c -$ radius of throwing circle
$= 61.46 - 1.25 = 60.21$ m

(continued on next page)

Solution 2 Use a CAS to substitute values into the Law of Cosines using the "such that" feature. Since most technology will not distinguish between C and c, we renamed C as cc.

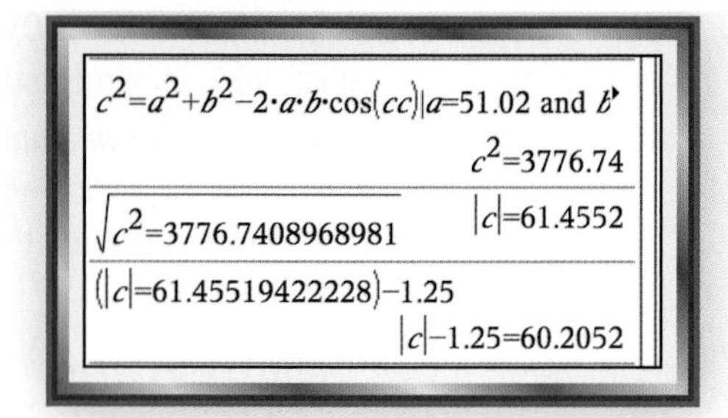

Check Since $\angle C$ is obtuse, c is the longest side of $\triangle ABC$ and 61.46 is larger than the lengths of the other two sides. Also, by the Triangle Inequality, c must be less than $30.10 + 51.02$, and it is. So the answer is reasonable.

The first known appearance of the Law of Cosines is in Book II of Euclid's *Elements,* written around 300 B.C.E. The Law of Cosines is a generalization of the Pythagorean Theorem, since if $m\angle C = 90°$, then

$$\begin{aligned} c^2 &= a^2 + b^2 - 2ab\cos 90° \\ &= a^2 + b^2 - 2ab \cdot 0, \\ \text{so } c^2 &= a^2 + b^2. \end{aligned}$$

Using the Law of Cosines to Find an Angle

Example 1 shows how the Law of Cosines can be used given SAS. Because each angle between 0° and 180° has a unique cosine, the Law of Cosines can also be used to find the measure of any angle of a triangle if the lengths of all three sides are known (the SSS condition). The Law of Cosines can be rewritten to give a formula for the measure of the angle.

Example 2

$\triangle ABC$ has side lengths of 5.3, 6, and 8 as shown at the right. Find the measure of its largest angle.

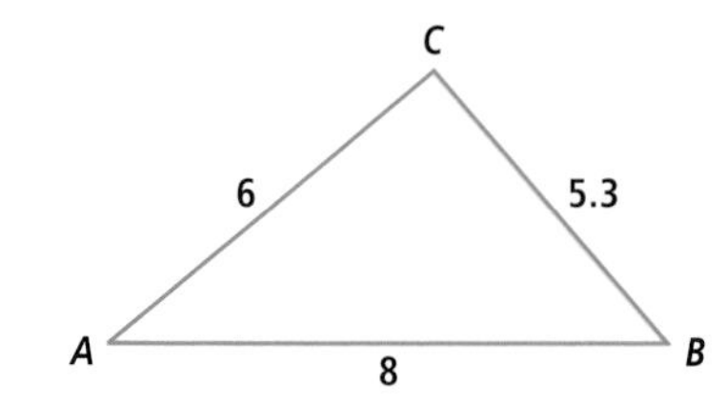

Solution 1 In any triangle, the largest angle is opposite the longest side.
The longest side is $\overline{AB}$ and so angle C has the largest measure.
Rewrite the Law of Cosines in terms of a, b, and c.

$$c^2 = a^2 + b^2 - 2ab \cdot \cos C$$

$$c^2 - a^2 - b^2 = -2ab \cdot \cos C$$

$$\cos C = \frac{c^2 - a^2 - b^2}{-2ab}$$

Making appropriate substitutions,

$$\cos C = \frac{8^2 - 5.3^2 - 6^2}{-2 \cdot 5.3 \cdot 6} = 0.00141....$$

Use a calculator and the inverse cosine function to calculate C.

$\cos^{-1}(0.00141...) \approx 89.9°$.

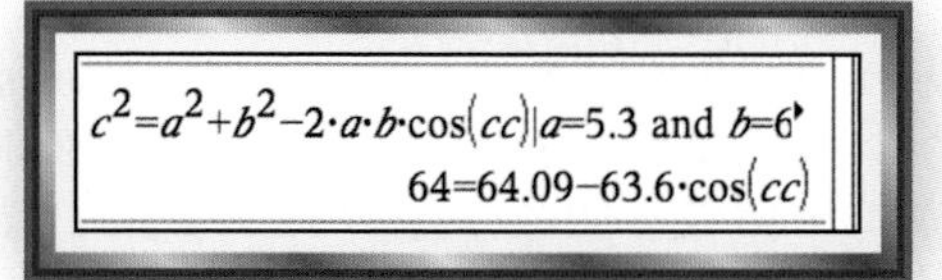

Solution 2 Use a CAS to substitute values into the Law of Cosines and solve the equation step-by-step as shown at the right. Note that the CAS does not simplify $\cos^{-1}(\cos(cc))$ due to domain restrictions.

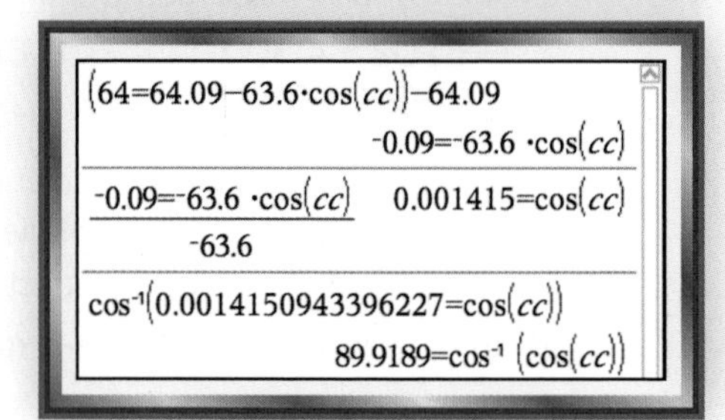

Using the Law of Cosines for finding the included angle is straightforward, but the algebra can be tricky. Fortunately, with a CAS you can substitute directly into the Law of Cosines formula for a, b, and c and solve for C.

Questions

COVERING THE IDEAS

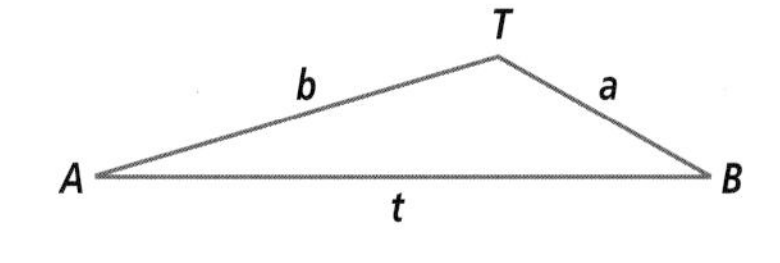

1. Refer to triangle TAB at the right.
   a. Write a formula for t in terms of $m\angle T$, a, and b.
   b. Suppose $a = 4$ cm, $b = 7$ cm, and $m\angle T = 135°$. Find the exact length t.

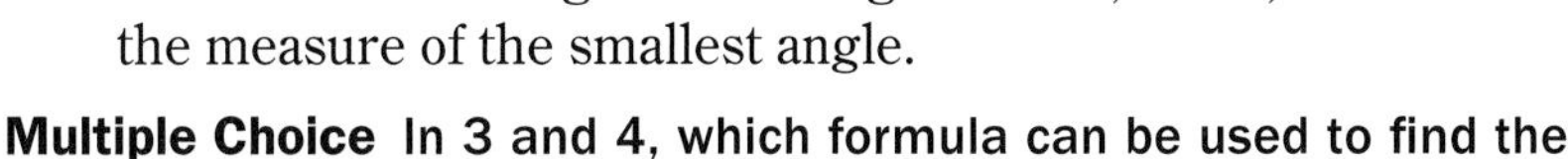

2. The sides of a triangle are of lengths 7.1 m, 3.9 m, and 7.1 m. Find the measure of the smallest angle.

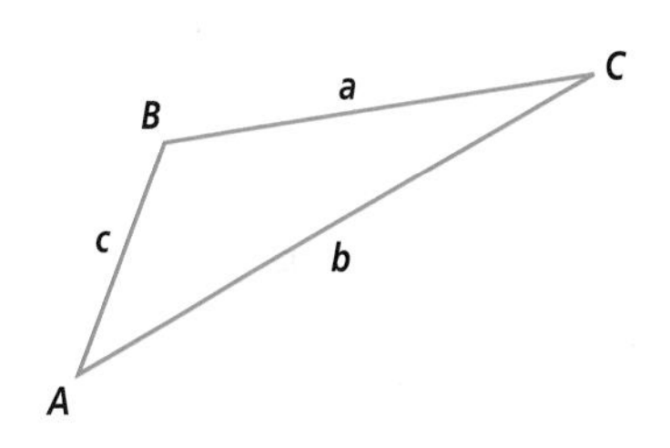

Multiple Choice In 3 and 4, which formula can be used to find the required measure directly?

A $a^2 = b^2 + c^2 - 2bc \cos A$ B $b^2 = a^2 + c^2 - 2ac \cos B$

C $c^2 = a^2 + b^2 - 2ab \cos C$ D none of the above

3. Given: a, b, $m\angle A$. Find c.
4. Given: a, b, c. Find $m\angle B$.

In 5 and 6, refer to the proof of the Law of Cosines.

5. a. Find the slopes of $\overline{CD}$ and $\overline{CA}$.
   b. Use the results of Part a to show that A, D, and C are collinear.
6. Use the distance formula to verify that $AC = b$.
7. $\triangle ABC$ is isosceles, with $AB = AC = 15$ cm, and $m\angle A = 150°$. Use the Law of Cosines to find the exact length of $\overline{BC}$.
8. It is 1.9 miles from a house to a restaurant and 1.2 miles to a pier, as shown at the right. The angle between the two lines of sight is 87°. How far is it from the pier across the water to the restaurant?

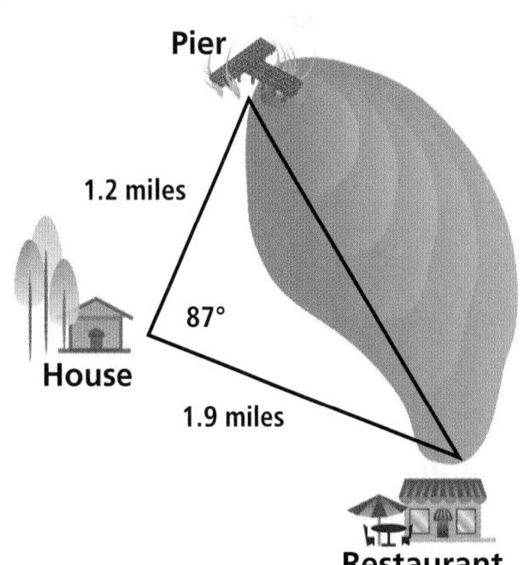

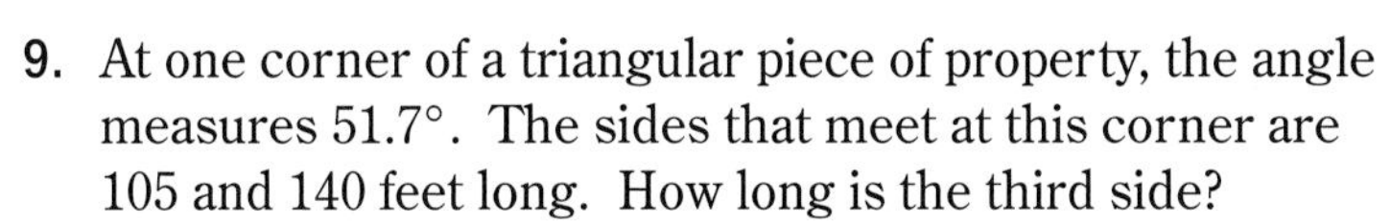

9. At one corner of a triangular piece of property, the angle measures 51.7°. The sides that meet at this corner are 105 and 140 feet long. How long is the third side?
10. In $\triangle XYZ$, $x = 12''$, $y = 14''$, and $z = 16''$. Find $m\angle Z$ and $m\angle X$.

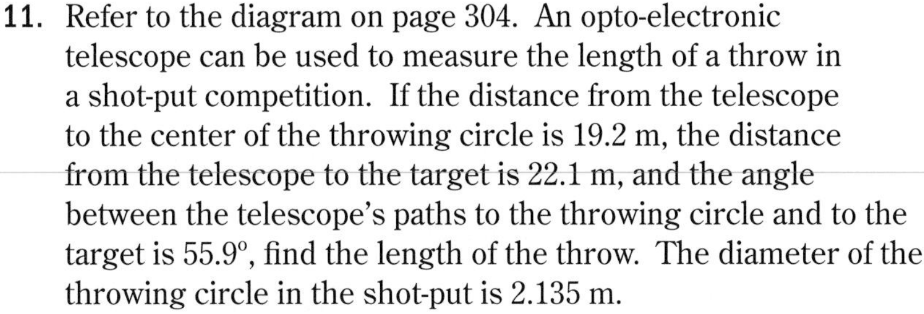

11. Refer to the diagram on page 304. An opto-electronic telescope can be used to measure the length of a throw in a shot-put competition. If the distance from the telescope to the center of the throwing circle is 19.2 m, the distance from the telescope to the target is 22.1 m, and the angle between the telescope's paths to the throwing circle and to the target is 55.9°, find the length of the throw. The diameter of the throwing circle in the shot-put is 2.135 m.

APPLYING THE MATHEMATICS

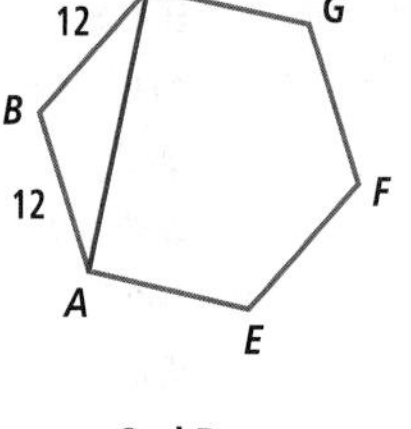

12. Find the exact length of $\overline{AD}$ of the regular hexagon shown.

13. Using the Law of Cosines, show that there can be no triangle with sides 3 cm, 2 cm, and 7 cm long.

14. A gannet (a sea bird) has a nest on an island and flies in a horizontal line for 160 m over the sea looking for fish, then flies 150 m downward, diving to catch a fish. The direct flight back to the nest is 250 m. At what angle did the bird turn to dive into the water?

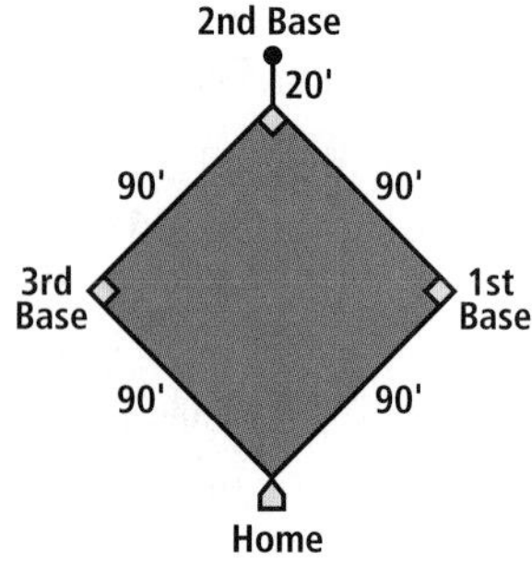

15. A baseball player is 20 feet behind 2nd base. Use the figure at the right to determine how far it is to throw from there to 1st base.

16. Refer to Question 8. A canoe located at the restaurant and a runner decide to race. The runner takes the path from the restaurant to the house to the pier. If the runner averages 9.3 mph and the boat averages 5.9 mph, who wins and by how much time?

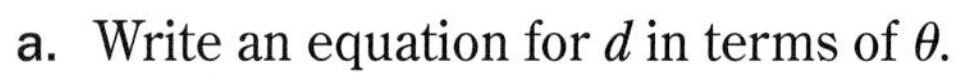

17. Suppose a parallelogram is made with bars of 8″ and 15″ that are hinged so the included angle θ between its sides can be changed as shown at the right.

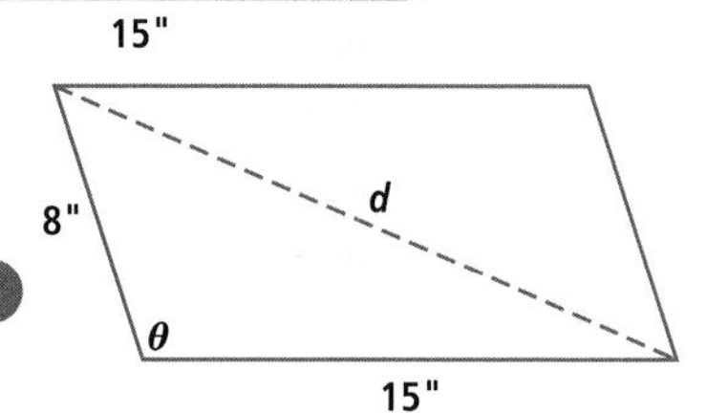

a. Write an equation for d in terms of θ.

b. Solve the equation for θ.

REVIEW

18. a. State the domain and range of the inverse cosine function.

b. How are its domain and range related to the domain and range of the restricted cosine function? **(Lesson 5-2)**

19. Compute without using your calculator. **(Lesson 5-2)**

a. $\cos^{-1} 1$ b. $\cos^{-1}\left(-\frac{\sqrt{3}}{2}\right)$

20. To estimate the height of a 2-story building, Howard walks 40 steps away from the building. The angle of elevation at this point to the top edge of the building is about 60°. About how tall is the building if Howard's steps are about 9 inches long? **(Lesson 5-1)**

21. Consider the equation $\cos \theta = -\frac{1}{2}$. **(Lesson 4-4)**

a. Draw a unit circle and mark the two points for which $\cos \theta = -\frac{1}{2}$.

b. Give two values of θ between 530° and 890° that make the equation true.

c. Give a negative value of θ that makes the equation true.

22. **Multiple Choice** A parabola has its vertex at (–5, 6). Which of the following might be an equation for the parabola? **(Lesson 3-2)**

A $f(x) = (x - 5)^2 - 6$ B $f(x) = (x - 5)^2 + 6$

C $f(x) = (x + 5)^2 + 6$ D $f(x) = (x + 5)^2 - 6$

EXPLORATION

23. Find BC in Question 7 without using trigonometry.

QY ANSWER

$c = \sqrt{109}$

Lesson 5-4 The Inverse Sine Function

Vocabulary

inverse sine function, $\sin^{-1}$, Arcsin

▶ **BIG IDEA** If you know $\sin \theta$ and that $0° \le \theta \le 180°$, two values of θ are possible unless $\theta = 90°$.

In Lesson 5-2, you studied the inverse cosine function. This lesson is about the *inverse sine function.*

In Lesson 4-3, you saw that for all x, $\sin x = \sin(\pi - x)$, so an angle and its supplement have the same sine. Thus, if you know the sine of an angle, the angle is usually not uniquely determined. For instance, in $\triangle ABC$, if $\sin C \approx 0.84787$, the measure of $\angle C$ could be about 58° or about 122°, as shown below.

Mental Math

Find θ if θ is an acute angle and

a. $\sin \theta = \frac{\sqrt{2}}{2}$.

b. $\sin \theta = \frac{1}{2}$.

c. $\sin \theta = \frac{\sqrt{3}}{2}$.

If you evaluate $\sin^{-1}(0.84787)$ on a calculator, it will return the degree or equivalent radian measure shown at the right. The calculator ignores the value between 90° and 180°. Here is why.

$\sin^{-1}(0.84787)$	57.980749
$\sin^{-1}(0.84787)$	1.011955

The Domain of the Restricted Sine Function

At the right is a graph of the sine function $y = \sin x$ (in blue), and its reflection image $x = \sin y$ (in red) over the line with equation $y = x$. Notice that the inverse of the sine function is not a function.

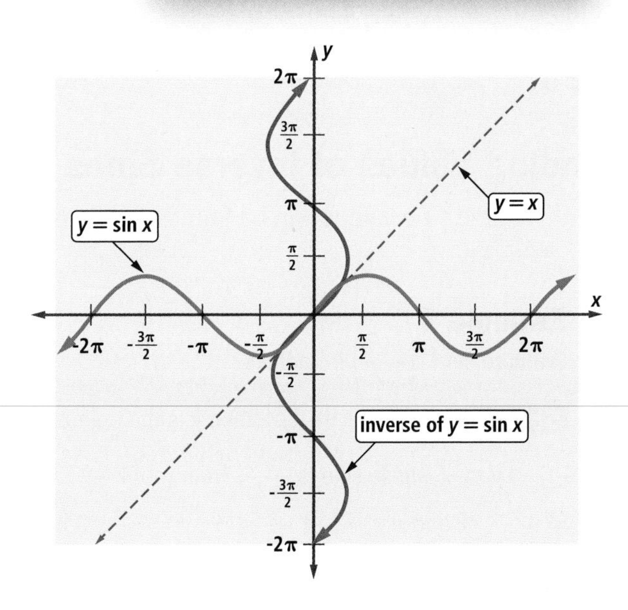

However, as with the cosine function, it is possible to restrict the domain of the sine function so that its inverse is a function. The following three criteria stated in Lesson 5-2 can again be met.

1. The domain should include the angles between 0 and $\frac{\pi}{2}$ (or 0° and 90°), because they are the measures of the acute angles of a right triangle.
2. On the restricted domain, the function should take on all the values of the range, that is, all real numbers between -1 and 1.
3. The function should be continuous on the restricted domain.

These criteria force the restriction of the domain of $y = \sin x$ to $\{x \mid -\frac{\pi}{2} \leq x \leq \frac{\pi}{2}\}$ or $\{x \mid -90° \leq x \leq 90°\}$. The **inverse sine function**, denoted **$\sin^{-1}$**, is the inverse of this restricted sine function. You can read $\sin^{-1} x$ as "the number (or angle) whose sine is x", just as you did with $\cos^{-1} x$.

Definition of Inverse Sine Function

$y = \sin^{-1} x$ if and only if $x = \sin y$ and $-\frac{\pi}{2} \leq y \leq \frac{\pi}{2}$.

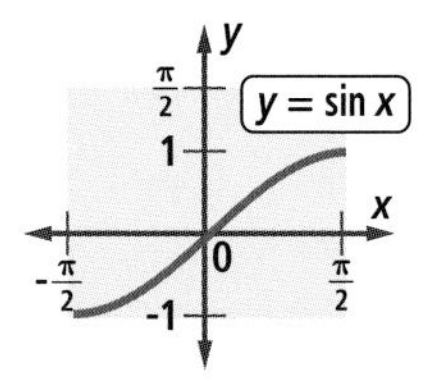

restricted sine function
domain: $\{x \mid -\frac{\pi}{2} \leq x \leq \frac{\pi}{2}\}$
range: $\{y \mid -1 \leq y \leq 1\}$

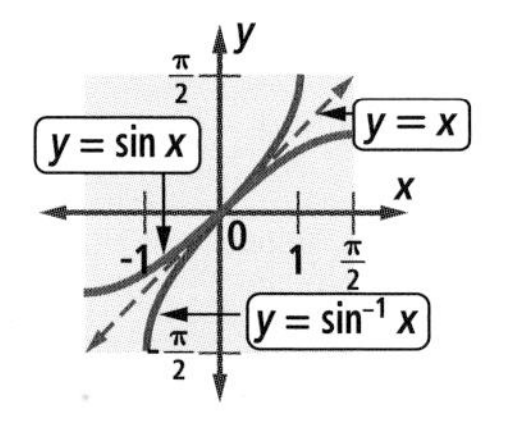

inverse sine function
domain: $\{x \mid -1 \leq x \leq 1\}$
range: $\{y \mid -\frac{\pi}{2} \leq y \leq \frac{\pi}{2}\}$

As always, the domain and range of a function and its inverse are switched. Thus, the domain of $y = \sin^{-1} x$ is $-1 \leq x \leq 1$, and the range is $-\frac{\pi}{2} \leq y \leq \frac{\pi}{2}$ (or $-90° \leq y \leq 90°$). That is, when $y = \sin^{-1} x$, y is the number from $-\frac{\pi}{2}$ to $\frac{\pi}{2}$ whose sine is x. The notation **Arcsin** is sometimes used in place of $\sin^{-1}$.

QY

> **QY**
> What is Arcsin 1?

Finding Values of Inverse Sines

Some inverse sine values can be found exactly without a calculator.

Example 1

Evaluate $\sin^{-1}\left(-\frac{\sqrt{3}}{2}\right)$ in radians.

Solution If $y = \sin^{-1}\left(-\frac{\sqrt{3}}{2}\right)$, then y is the unique number in the interval $-\frac{\pi}{2} \leq y \leq \frac{\pi}{2}$ whose sine is $-\frac{\sqrt{3}}{2}$. From your knowledge of exact values, $\sin\left(-\frac{\pi}{3}\right) = -\frac{\sqrt{3}}{2}$. Because $-\frac{\pi}{2} \leq -\frac{\pi}{3} \leq \frac{\pi}{2}$, $y = -\frac{\pi}{3}$. Thus, $\sin^{-1}\left(-\frac{\sqrt{3}}{2}\right) = -\frac{\pi}{3}$.

All values of the inverse sine function can be estimated using a calculator.

GUIDED

Example 2

Suppose θ is an angle in a triangle and $\sin \theta = 0.7214$. Find θ in degrees.

Solution One solution is $\sin^{-1}(0.7214) \approx$ __?__. The second solution is the measure of a supplement of this angle, __?__.

Check Evaluate sin __?__ with a calculator. The display shows 0.7213977... It checks.

Recall that if f and f^{-1} are inverse functions, then $f \circ f^{-1}(x) = x$ for all x in the domain of f^{-1}, and $f^{-1} \circ f(x) = x$ for all x in the domain of f. Thus, $\sin(\sin^{-1} x) = x$ for all x in the domain of $\sin^{-1}$, and $\sin^{-1}(\sin x) = x$ for all x in the restricted domain of the sine function.

Example 3

Explain why $\sin^{-1}\left(\sin \frac{5\pi}{4}\right) \neq \frac{5\pi}{4}$.

Solution 1 Evaluate directly.

$$\sin \frac{5\pi}{4} = -\frac{\sqrt{2}}{2}$$

$$\sin^{-1}\left(\sin \frac{5\pi}{4}\right) = \sin^{-1}\left(-\frac{\sqrt{2}}{2}\right) = -\frac{\pi}{4}$$

Thus, $\sin^{-1}\left(\sin \frac{5\pi}{4}\right) \neq \frac{5\pi}{4}$.

Solution 2 Since $\frac{5\pi}{4}$ is not in the restricted domain of the sine function necessary for the inverse to be a function, it is not in the range of the inverse sine function. So it is impossible for $\frac{5\pi}{4}$ to be a value of the inverse sine function.

Applying the Inverse Sine Function

The inverse sine function has applications in all situations involving the sine function.

Activity

An 18-ft ladder leans against a building as shown at the right.

Step 1 Express θ, the measure of the angle the ladder makes with the ground, as a function of h, the height of the top of the ladder.

Step 2 Pick three possible values of h and find the three corresponding values of θ.

Step 3 a. Describe the set of all possible values of h.

b. Describe the set of all possible values of θ.

(continued on next page)

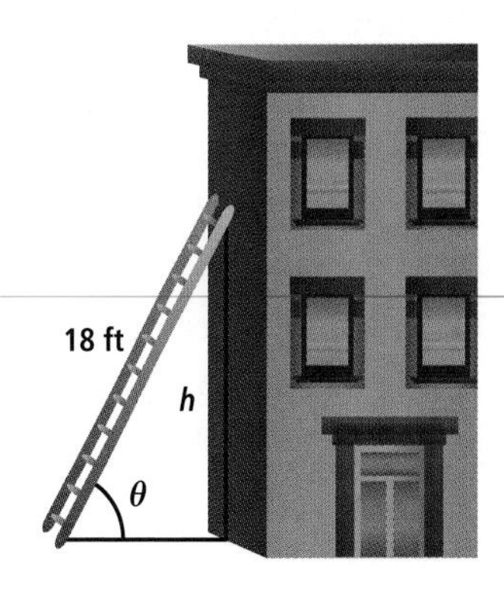

Step 4 a. Graph the function in Step 1 over the domain in Step 3a, so that the entire range in Step 3b shows.

b. On your graph, identify the 3 points corresponding to your answers to Step 2.

Questions

COVERING THE IDEAS

1. Explain why the inverse of the sine function is not a function.
2. How is the expression $\theta = \sin^{-1} k$ read?
3. a. Complete the table of values below.

point on = sin	$\left(-\frac{\pi}{2}, -1\right)$	$\left(-\frac{\pi}{3}, ?\right)$	$\left(-\frac{\pi}{4}, ?\right)$	$\left(-\frac{\pi}{6}, ?\right)$	(0, ?)	$\left(\frac{\pi}{6}, ?\right)$	$\left(\frac{\pi}{4}, ?\right)$	$\left(\frac{\pi}{3}, ?\right)$	$\left(\frac{\pi}{2}, ?\right)$
corresponding point on = $\sin^{-1}$	$\left(-1, -\frac{\pi}{2}\right)$,	(?, ?)	(?, ?)	(?, ?)	(?, ?)	(?, ?)	(?, ?)	(?, ?)	(?, ?)

b. Graph $y = \sin x$ and $y = \sin^{-1} x$ on the same set of axes.

c. State the domain and range of the inverse sine function.

d. What transformation maps the graph of $y = \sin x$ to the graph of $y = \sin^{-1} x$?

In 4–6, find an exact value in radians and degrees without a calculator.

4. $\sin^{-1}\left(\frac{1}{2}\right)$
5. Arcsin 1
6. $\sin^{-1}\left(-\frac{\sqrt{2}}{2}\right)$
7. A plane flying at an altitude of 32,000 feet (about 6 miles) descends to an airport at a constant angle θ in radians on a path m miles long.

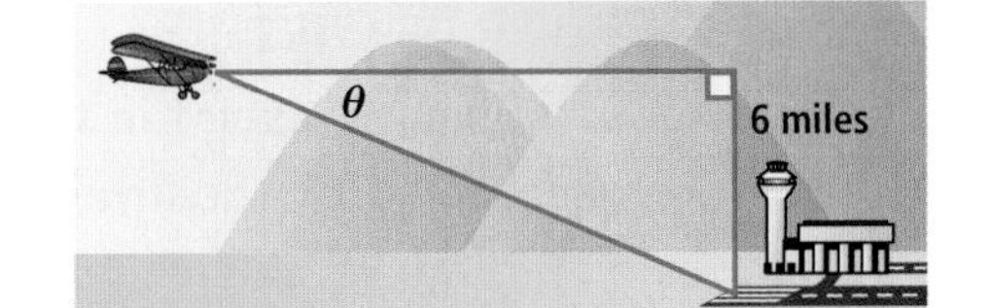

a. Write an equation for θ as a function of m.

b. Graph your function from Part a using $0 \le m \le 100$.

c. If $m = 30$, find θ to the nearest tenth. Explain the meaning of your answer in the context of the problem.

8. Suppose $0° < x < 180°$ and $\sin x = 0.1$. Find all possible values of x to the nearest tenth of a degree.

APPLYING THE MATHEMATICS

9. One statement below is true; the other is false. Which one is false, and why?

A If $\theta = \sin^{-1} n$, then $n = \sin \theta$. B If $n = \sin \theta$, then $\theta = \sin^{-1} n$.

10. Explain why $\{x \mid 0 \le x \le \pi\}$ is not used as the domain of the restricted sine function.
11. Explain why $\sin(\sin^{-1} x) = x$ for all x such that $-1 \le x \le 1$.
12. a. Evaluate $\cos(\sin^{-1} 0.8)$ using a calculator.

b. Draw an appropriate triangle to show how the answer to Part a could have been found without a calculator.

c. Evaluate $\cos\left(\sin^{-1}\left(\frac{b}{c}\right)\right)$, where $b \neq 0$ and $c \neq 0$.

13. The equation $E = 4\sin(60\pi t)$ describes the electrical voltage E in a circuit at a time t.

 a. Solve for t in terms of E.

 b. How is the graph of t as a function of E related to the graph of $E = 4\sin(60\pi t)$?

14. A satellite orbits Earth. At time $t = 0$ hours it is at its farthest distance of 600 miles above Earth. At $t = 1$ hour, it is at its closest distance of 450 miles above Earth.

 a. Assume that the distance d varies sinusoidally with time. Write an equation using the sine function to model this situation.

 b. Solve this equation for t in terms of d.

 c. What are the first four times the satellite is 500 miles above Earth?

15. Find the smallest positive value of x in radians with $2\sin x + 6 = 5$.

16. Example 3 in Lesson 5-2 shows the diagram at the right representing Earth and the Landsat 7 satellite.

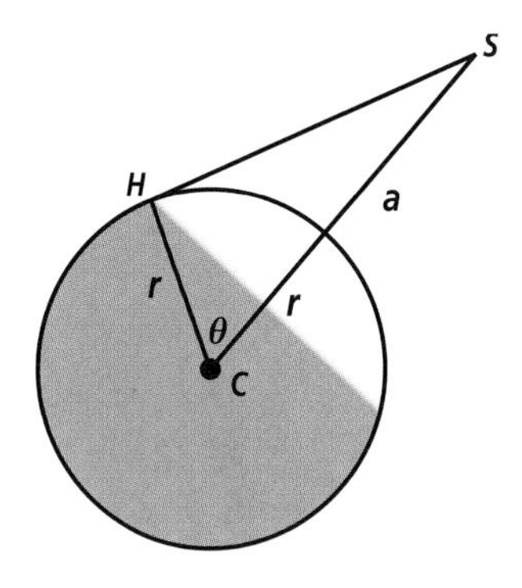

 a. Give a formula for $m\angle HSC$ in terms of r and a. ($\angle HSC$ is called the *angular separation* of the horizon.)

 b. Give a relationship between $\cos\theta$ and $\sin(m\angle HSC)$.

REVIEW

17. The figure at the right represents the way a golfer played a hole to avoid two water hazards. To the nearest yard, how far is it from the tee to the green directly over the water? **(Lesson 5-3)**

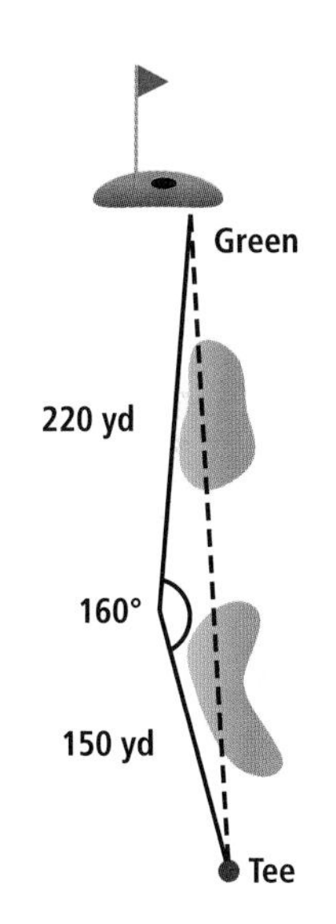

18. In $\triangle PQR$, $p = 19$, $q = 5$, and $r = 17$. Find $m\angle P$. **(Lesson 5-3)**

In 19 and 20, use a calculator only to check answers. (Lessons 4-6, 4-5, 4-3)

19. Given $\sin 144° \approx 0.588$,

 a. find θ such that $\theta > 360°$ and $\sin\theta \approx 0.588$.

 b. find θ such that $\theta < 0°$ and $\sin\theta \approx 0.588$.

 c. find θ such that $\theta \neq 144°$, $0° \le \theta < 360°$, and $\sin\theta \approx 0.588$.

20. Given $\cos 80° \approx 0.174$, find each value.

 a. $\cos 440°$ b. $\cos -440°$ c. $\cos 460°$

21. Consider the graph of the function f with $f(x) = 2x^2 + x - 15$.

 a. Find its y-intercept and x-intercepts.

 b. Tell whether the graph has a maximum or minimum point and find its coordinates. **(Lesson 2-6)**

EXPLORATION

22. Graph the function with equation $y = \sin^{-1}(\sin x)$ on a calculator on the domain $-720° \le x \le 720°$. Describe and explain the result.

QY ANSWER

$\frac{\pi}{2}$ or 90°

Lesson 5-5 The Law of Sines

▸ **BIG IDEA** The Law of Sines allows you to find unknown lengths in a triangle given the ASA, AAS, or SSA conditions.

The Law of Cosines is not always helpful in finding sides and angles in triangles. Consider this situation. Some campers want to find the distance f across a lake. They measure the distance from the flagpole to the headquarters and measure the angles as indicated. This provides them with the measures of two angles and the included side of $\triangle FBH$. The ASA condition is met, so this is enough information to find f.

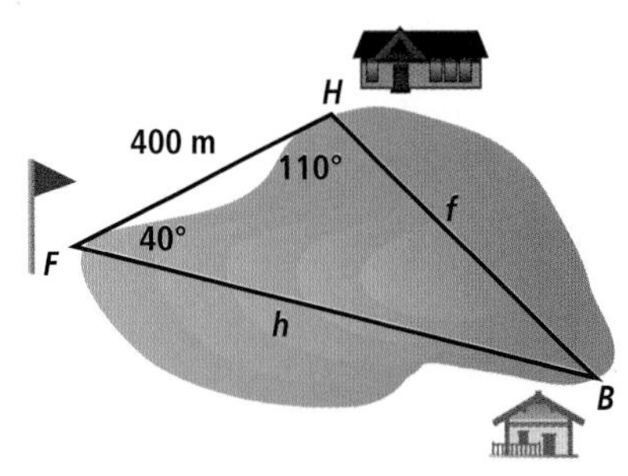

However, when the campers apply the Law of Cosines to find f, they get

$$f^2 = b^2 + h^2 - 2bh \cos F$$
$$f^2 = 400^2 + h^2 - 2 \cdot 400 \cdot h \cos 40°$$
$$f^2 = 160{,}000 + h^2 - 800h \cos 40°.$$

The result is a single equation with two unknowns, so the campers are unable to proceed any further with this solution.

Fortunately, there is another way to find f, using a theorem known as the *Law of Sines*. The proof of the Law of Sines involves the area of a triangle, so we first review area.

Mental Math

1. Which of these sets of numbers cannot be lengths of sides of a right triangle?

2. Which of these sets of numbers cannot be lengths of sides of any triangle?

A 30, 10, 40

B 30, 50, 40

C 30, 60, 40

D 30, 80, 40

The SAS Area Formula for a Triangle

The area K of triangle ABC as shown at the right is given by the familiar formula $K = \frac{1}{2}bh$.

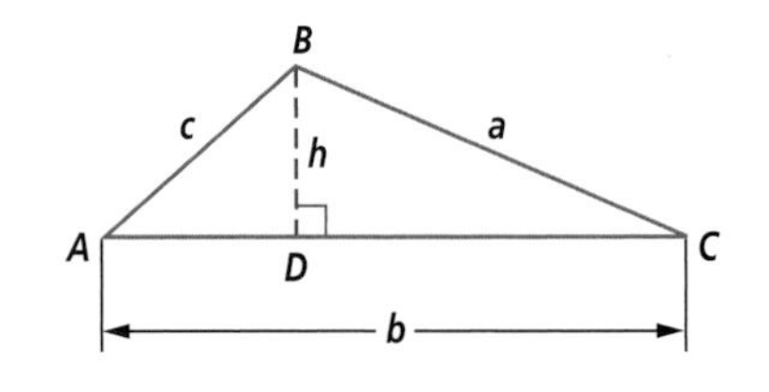

If h is not known, you can find h using right triangle BCD.

$$\sin C = \frac{h}{a}, \text{ so } h = a \sin C.$$

Substituting this value in the area formula gives

$$K = \frac{1}{2} ab \sin C.$$

Similarly, $\sin A = \frac{h}{c}$. So $h = c \sin A$, and so another formula is

$$K = \frac{1}{2} bc \sin A.$$

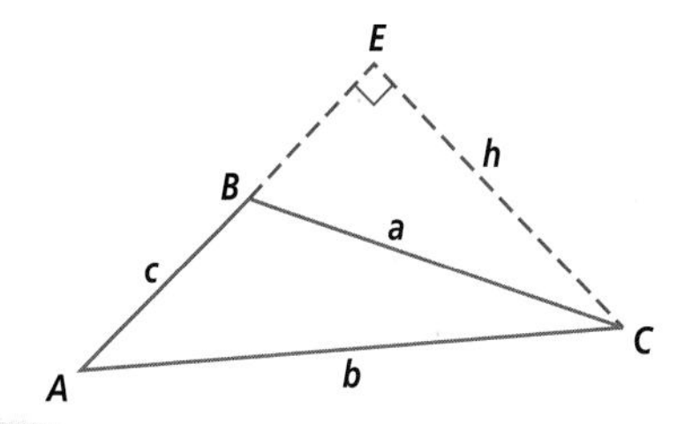

Activity

For $\triangle ABC$ at the right, use $\overline{AB}$ as the base and h as the altitude. Derive yet a third formula for the area of $\triangle ABC$.

This argument proves the following theorem.

Theorem (SAS Area Formula for a Triangle)

In any triangle, the area is one-half the product of the lengths of any two sides and the sine of their included angle.

A Proof of the Law of Sines

The Law of Sines is one of the most beautiful results in all of mathematics—simple and elegant.

Theorem (Law of Sines)

In any triangle ABC, $\frac{\sin A}{a} = \frac{\sin B}{b} = \frac{\sin C}{c}$.

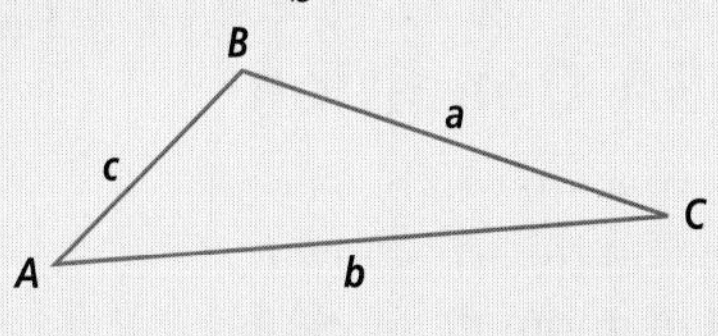

Proof Let $\triangle ABC$ be any triangle. To find the area of $\triangle ABC$ with the SAS Formula, any two sides and their included angle may be used. Because the area of a given triangle is constant,

$$\frac{1}{2} bc \sin A = \frac{1}{2} ac \sin B = \frac{1}{2} ab \sin C.$$

Multiply by 2. $bc \sin A = ac \sin B = ab \sin C$

Divide by abc. $\frac{bc \sin A}{abc} = \frac{ac \sin B}{abc} = \frac{ab \sin C}{abc}$

Rewrite the fractions in lowest terms. $\frac{\sin A}{a} = \frac{\sin B}{b} = \frac{\sin C}{c}$

Using the Law of Sines with the ASA Condition

You can use the Law of Sines to find the length of a second side of a triangle given the measures of two angles and one side (the ASA or AAS conditions).

Example 1

Refer to the camp situation on page 314. Find the distance *f* across the lake.

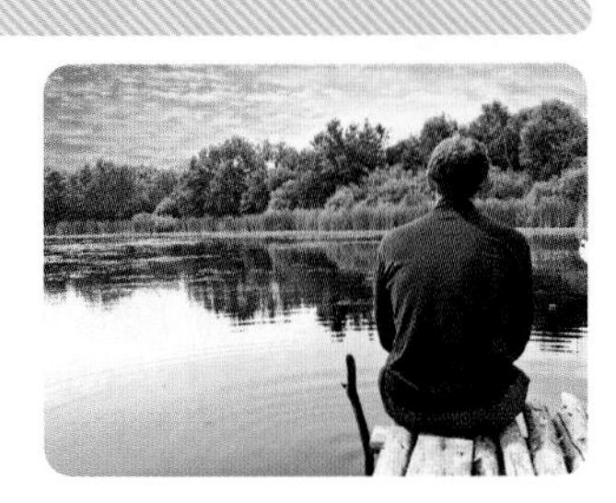

Solution Since the sum of the measures of the angles of a triangle is 180°,

$$m\angle B = 180° - 40° - 110° = 30°.$$

Use the Law of Sines to find *f*. First write the Law of Sines for this triangle.

$$\frac{\sin F}{f} = \frac{\sin B}{b} = \frac{\sin H}{h}$$

$$\frac{\sin 40°}{f} = \frac{\sin 30°}{400} = \frac{\sin 110°}{h}$$

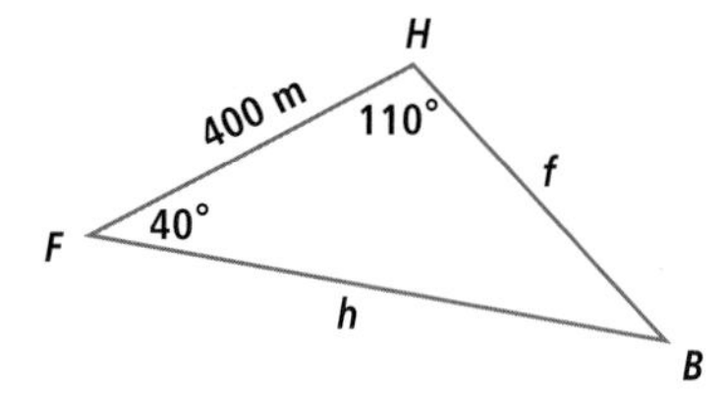

Use the left and middle fractions to solve for *f*.

$$f \sin 30° = 400 \sin 40°$$

$$f = \frac{400 \sin 40°}{\sin 30°} \approx 514$$

It is about 514 meters across the lake from the beach house to headquarters.

Check Recall that in a triangle the longer sides are opposite the larger angles. The 514-meter side is opposite the 40° angle, and the shorter 400-meter side is opposite the (smaller) 30° angle, as it should be.

QY

QY

Find the value of *h* in Example 1.

Using the Law of Sines with the SSA Condition

The Law of Sines can also be used to determine the measure of a second angle of a triangle when two sides and a nonincluded angle are known. This is the SSA condition. However, as you studied in geometry, SSA is not a condition that, in general, guarantees congruence. Thus, when the Law of Sines is used in an SSA situation, it is possible to get no solution, one solution, or two solutions.

When the side opposite the given angle is larger than the other given side, the SSA condition leads to a unique triangle. We then call it the *SsA condition*.

Example 2

In $\triangle XYZ$, $y = 5$, $z = 3$, and $m\angle Y = 60°$. Find $m\angle Z$.

Solution Make a rough sketch. Use the Law of Sines to get

$$\frac{\sin 60°}{5} = \frac{\sin Z}{3}.$$

$$\sin Z = \frac{3}{5}\sin 60° = \frac{3\sqrt{3}}{10} \approx 0.52$$

$$m\angle Z = \sin^{-1}(0.52) \approx 31.3°$$

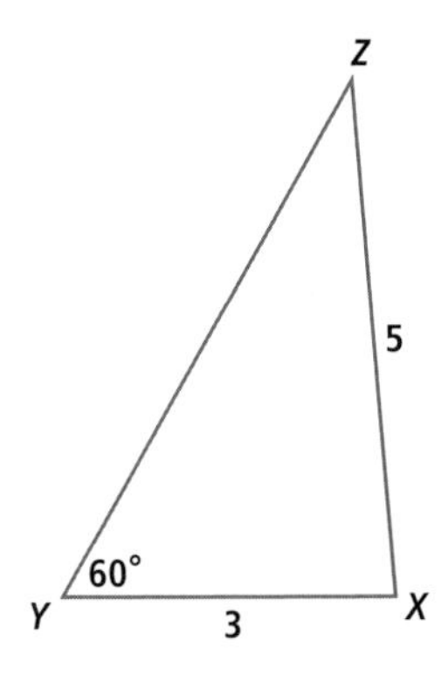

Test whether there is a second solution. From the Supplements Theorem,

$\sin(180° - 31.3°) = \sin 148.7° \approx 0.52$ also. But if 148.7° were a solution to this problem, then the sum of the angles of $\triangle XYZ$ would be more than 180°. So $m\angle Z = 31.3°$ is the only solution.

Example 3 illustrates when two solutions are possible.

GUIDED

Example 3

In $\triangle ABC$, $m\angle A = 35°$, $AB = 8$, and $BC = 6$. Find all possible values of $m\angle C$ to the nearest degree.

Solution Make a rough sketch. Use the Law of Sines.

By the Law of Sines, $\frac{?}{6} = \frac{\sin C}{?}$.

Solve for $m\angle C$. $\sin C = \frac{?}{6} \approx$ _?_

$m\angle C = \sin^{-1}($_?_$) \approx$ _?_

A triangle determined by this value is pictured at the right.

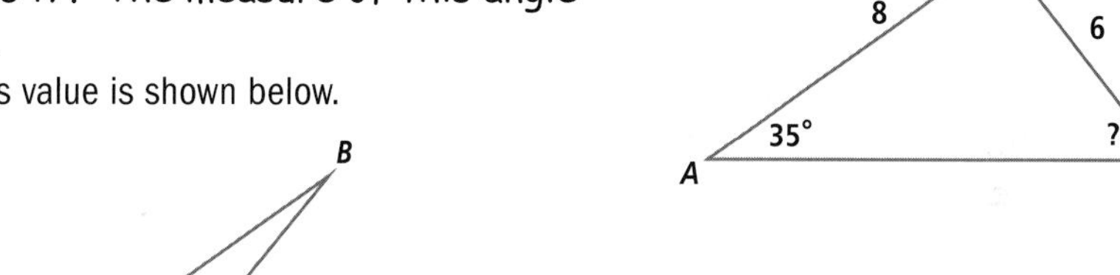

However, from the Supplements Theorem, a second angle C has $\sin C = 0.7647$. The measure of this angle is $180° -$ _?_ $\approx$ _?_ °.

A triangle determined by this value is shown below.

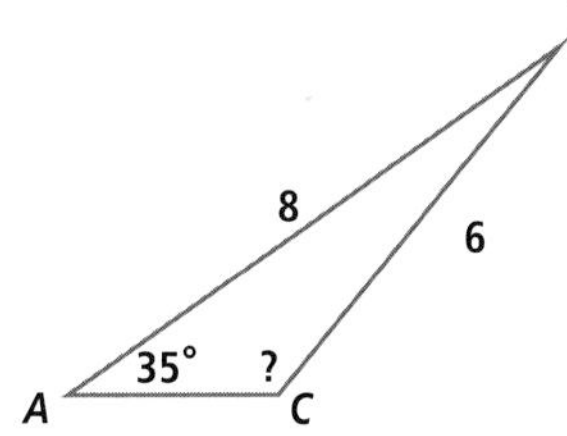

So there are two possible values of $m\angle C$, _?_ and _?_, and the two triangles are not congruent.

As you will see in the Questions, there are also SSA cases when no triangle is possible.

In general, when looking for measures of angles or sides in triangles, try methods involving simpler computations first. If these methods do not work, use the following.

1. If given two angles in a triangle, find the third angle by subtracting the total of the given angles from 180° or π radians, depending on the given units.
2. If a triangle is a right triangle, use right triangle trigonometric ratios.
3. If a triangle is not a right triangle, consider the Law of Sines. It is useful for the ASA, AAS, and SSA conditions.
4. If the Law of Sines is not helpful, use the Law of Cosines. The Law of Cosines is most directly applicable to the SAS and SSS conditions.

Questions

COVERING THE IDEAS

1. Find the area of $\triangle ABC$ where $a = 35$, $b = 26$, and $m\angle C = 117°$.
2. Without a calculator, find the area of $\triangle DEF$ where $d = 40, f = 20$, and $m\angle E = 30°$.
3. In any triangle, the ratio of the sine of any angle to __?__ is constant.
4. Tell whether the statement is equivalent to the Law of Sines for $\triangle ABC$.
   a. $\frac{A}{a} = \frac{B}{b} = \frac{C}{c}$
   b. $\frac{a}{\sin A} = \frac{b}{\sin B} = \frac{c}{\sin C}$
   c. $\frac{a}{b} = \frac{\sin A}{\sin B}$, $\frac{b}{c} = \frac{\sin B}{\sin C}$, $\frac{a}{c} = \frac{\sin A}{\sin C}$
   d. $ab \sin C = bc \sin A = ac \sin B$
5. In $\triangle XYZ$ at the right, find x to the nearest tenth.

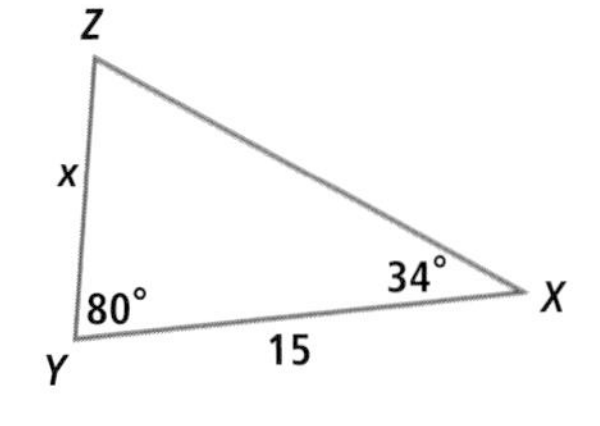

6. In $\triangle EFG$ below, find $m\angle F$ to the nearest tenth of a degree.

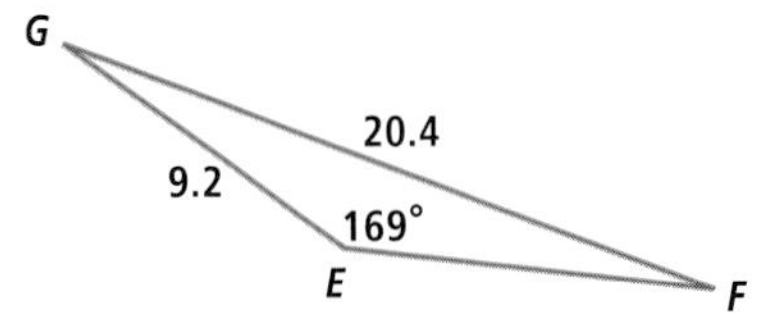

7. In $\triangle ABC$, suppose $m\angle C = 45°$, $m\angle A = 30°$, and $c = 10$. Find the lengths of the other two sides to the nearest hundredth.
8. The Costas want to check the survey of a plot of land they are thinking of buying. The plot is triangular with one side on the lakefront, with dimensions as shown at the right.
   a. Find the measure of the angle between the lakeshore and the 452-foot side.
   b. About how many feet of lake frontage does the plot have?

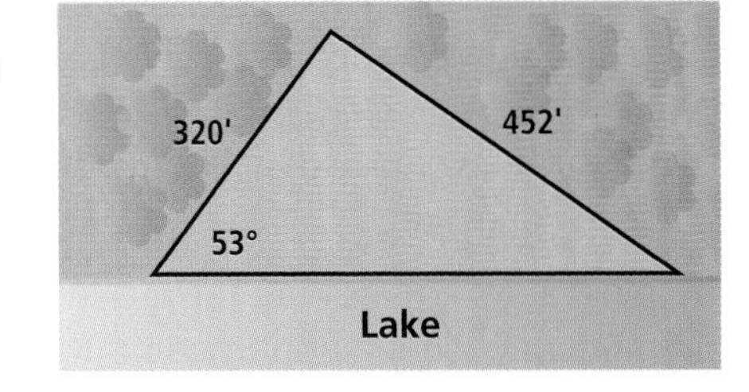

9. In $\triangle XYZ$, $x = 2$ cm, $z = 3$ cm, and $m\angle X = 40°$.
   a. Find the two possible measures of $\angle Z$.
   b. Draw two noncongruent triangles satisfying these conditions.

In 10 and 11, **True or False.**

10. When given SSA conditions, it is possible for two different triangles to be determined.
11. When given ASA conditions, it is possible for two different triangles to be determined.

APPLYING THE MATHEMATICS

12. Use the SAS Triangle Formula to find the area of an equilateral triangle with sides of length 20.
13. A piston driven by a 3-cm radial arm has length 9 cm. As the radial arm moves from 80° to 70° off its axis, how much does x change?

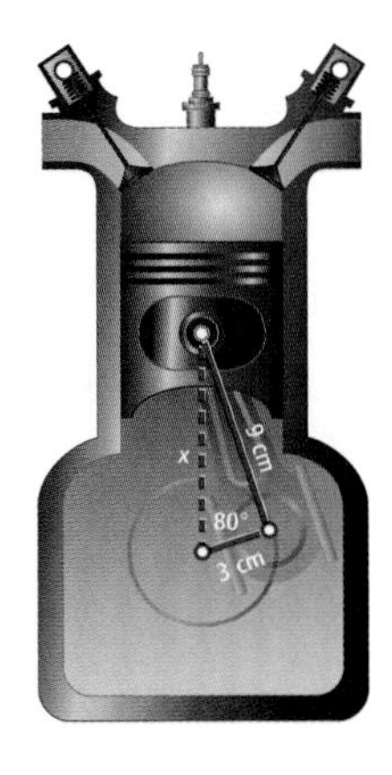

14. Richard spots a tornado 40° S of W. He calls Joanne who lives 1.4 miles due south of his house. She sees the cloud at 25° S of W. How close are Richard and Joanne to the tornado?

15. In $\triangle WXY$, $WX = 100$ and $m\angle W = 45°$. What values of XY will yield the following number of possible measures for $\angle Y$?

a. exactly two **b.** exactly one **c.** none

REVIEW

In 16 and 17, evaluate in degrees and radians without a calculator. (Lesson 5-4)

16. $\sin^{-1}(0)$ **17.** $\sin^{-1}\left(\frac{1}{2}\right)$

In 18–20, find the exact value. (Lesson 4-4)

18. $\cos 45°$ **19.** $\sin\left(\frac{2\pi}{3}\right)$ **20.** $\tan \pi$

In 21 and 22, $\sin \theta = t$. Evaluate without a calculator. (Lesson 4-3)

21. $\sin(-\theta)$ **22.** $\sin(180° - \theta)$

23. The graph at the right is a translation or scale-change image of the graph of $y = |x|$. Write an equation for the graph. **(Lessons 3-5, 3-2)**

y
x
(-5, -2)
(-1, -2)
(-3, -4)

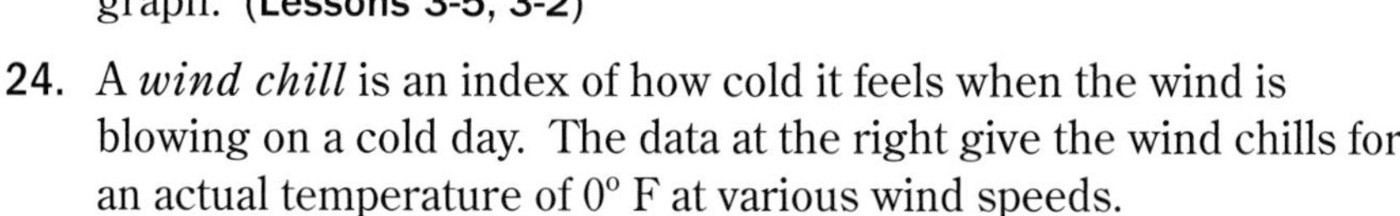

24. A *wind chill* is an index of how cold it feels when the wind is blowing on a cold day. The data at the right give the wind chills for an actual temperature of 0° F at various wind speeds.

a. Identify the independent and dependent variables in this case.

b. Construct a scatterplot of these data.

c. Find a suitable quadratic model for the data. **(Lessons 2-6, 2-2)**

Wind Chill (°F)	Wind Speed (mph)
-11	5
-16	10
-19	15
-22	20
-24	25
-26	30
-27	35
-29	40
-30	45

Source: National Oceanic and Atmospheric Administration

25. ***Skill Sequence*** Suppose $X_1 = 56, X_2 = 73, X_3 = 68, X_4 = 65, X_5 = 58, X_6 = 91$, and $X_7 = 79$. Evaluate. **(Lesson 1-2)**

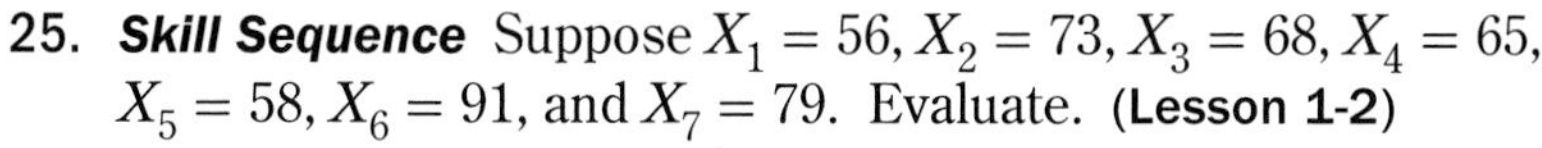

a. $\sum_{i=1}^{7} 8X_i$ **b.** $\sum_{i=1}^{7} kX_i$ **c.** $\sum_{i=1}^{7} \left(\frac{X_i}{k}\right)$ **d.** $\sum_{i=1}^{7} (X_i - 8)$

EXPLORATION

26. $\triangle ABC$ is inscribed in $\odot O$. Prove that the diameter of $\odot O$ is $\frac{a}{\sin A}$.

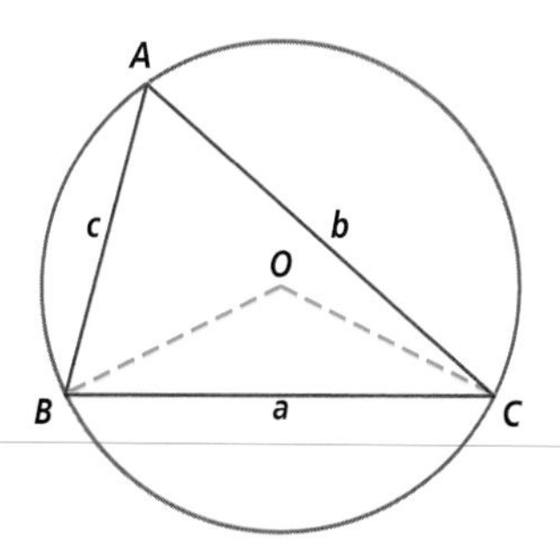

QY ANSWER

$h \approx 752$ m

Lesson

5-6 The Inverse Tangent Function

Vocabulary

inverse tangent function, $\tan^{-1}$, Arctan

▸ **BIG IDEA** If you know $\tan \theta$, and that $-90° \leq \theta \leq 90°$, then θ is uniquely determined.

You have seen situations in which the tangent function is used. For instance, if a plane flying at an altitude of h miles begins its descent d miles from an airport, then the angle θ of the path of descent satisfies

$$\tan \theta = \frac{h}{d}.$$

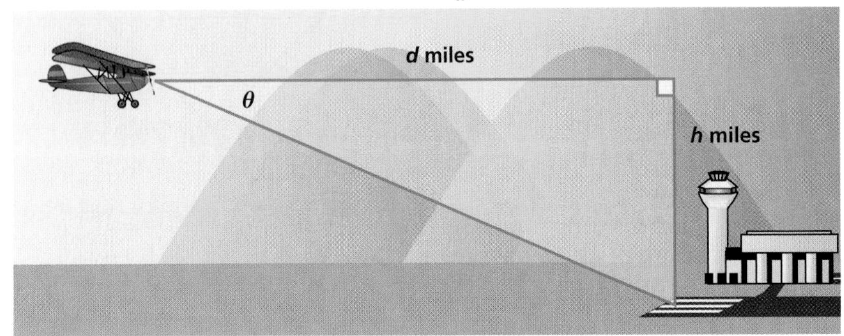

Mental Math

Find θ if $0° \leq \theta \leq 90°$ and

a. $\tan \theta = 0$.

b. $\tan \theta = 1$.

c. $\tan \theta = \sqrt{3}$.

To find θ, you can use the [TAN-1] key on a calculator. This key returns a single value of the function called the *inverse tangent function*.

Defining the Inverse Tangent Function

At the right is a graph of the inverse of the tangent function, $x = \tan y$. You may notice that it is a reflection image of the graph $\tan x$ over the line with equation $y = x$. Like the sine and cosine functions, the inverse of the tangent function is not a function, but it is possible to restrict the domain of the tangent function so that its inverse is a function.

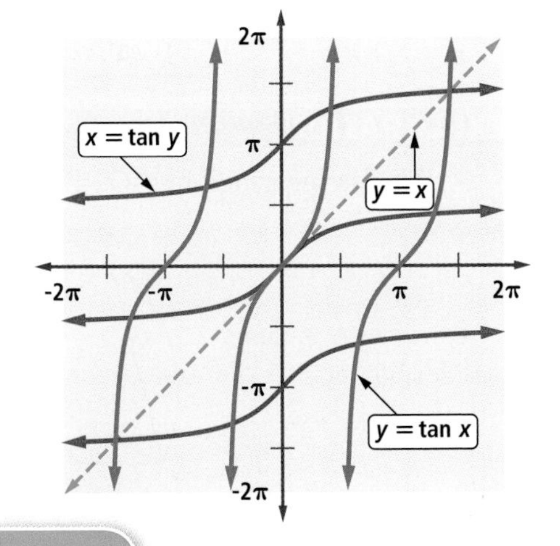

Using the criteria for choosing an appropriate domain used in Lessons 5-2 and 5-4 leads to restricting the domain of the tangent function to $-\frac{\pi}{2} < x < \frac{\pi}{2}$. This **inverse tangent function**, denoted as **$\tan^{-1}$**, is the inverse of this restricted function.

Definition of Inverse Tangent Function

$y = \tan^{-1} x$ if and only if $x = \tan y$ and $-\frac{\pi}{2} < y < \frac{\pi}{2}$.

The restricted tangent function and inverse tangent function are graphed below. Notice that, unlike the $\sin^{-1}$ and $\cos^{-1}$ functions, the $\tan^{-1}$ function has the set of all real numbers as its domain.

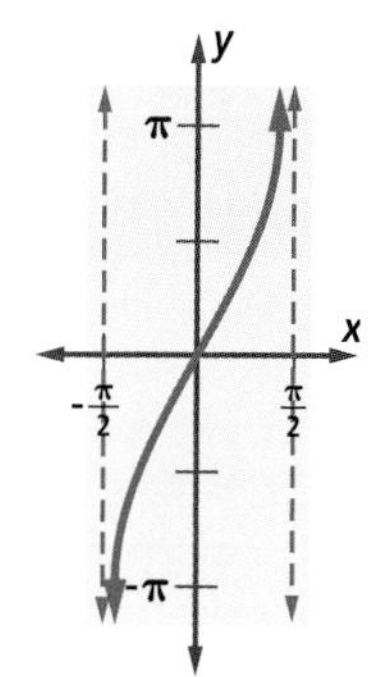

restricted tangent function
domain: $\{x \mid -\frac{\pi}{2} < x < \frac{\pi}{2}\}$
range: $\mathbb{R}$

inverse tangent function
domain: $\mathbb{R}$
range: $\{y \mid -\frac{\pi}{2} < y < \frac{\pi}{2}\}$

QY

You can read $y = \tan^{-1} x$ as "y is the number (or angle) whose tangent is x." The notation **Arctan** is sometimes used in place of $\tan^{-1}$.

QY

Draw a right triangle with an angle θ such that $\theta = \tan^{-1} 4$.

Finding Values of the Inverse Tangent Function

Some values of the inverse tangent function can be found exactly without using a calculator.

Example 1

Evaluate $\tan^{-1}(-1)$ in radians.

Solution If $y = \tan^{-1}(-1)$, then by definition of $\tan^{-1}$, y is the unique number in the interval $-\frac{\pi}{2} < y < \frac{\pi}{2}$ with $\tan y = -1$. So $\sin y$ and $\cos y$ are opposites. On the unit circle, this occurs in the interval desired only when $y = -\frac{\pi}{4}$.

All values of the inverse tangent function can be estimated using a calculator.

Example 2

Evaluate Arctan 4 in degrees.

Solution Apply the definition of the inverse tangent function: $y = \text{Arctan } 4$ if and only if $\tan y = 4$ and $-90° < y < 90°$. Set your calculator to degree mode. Use it to get

Arctan 4 $\approx$ 75.96°.

Check Evaluate tan 75.96° with a calculator. This results in a display of about 3.9988... . It checks. Also, look at the drawing for the QY. The measure of angle θ is about 76°.

Applying the Inverse Tangent Function

Example 3

A plane is flying at an altitude of 32,000 feet (about 6 miles) and wants to descend to an airport runway x miles away at a constant angle θ.

a. **Write an equation for θ as a function of x.**

b. **If $x = 30$, find θ to the nearest degree.**

Solution

a. Draw a diagram. Let P = the position of the plane before descent, A = the point of contact on the airport runway, and T = the point above A needed to form a right triangle. Then $\theta = m\angle APT$.

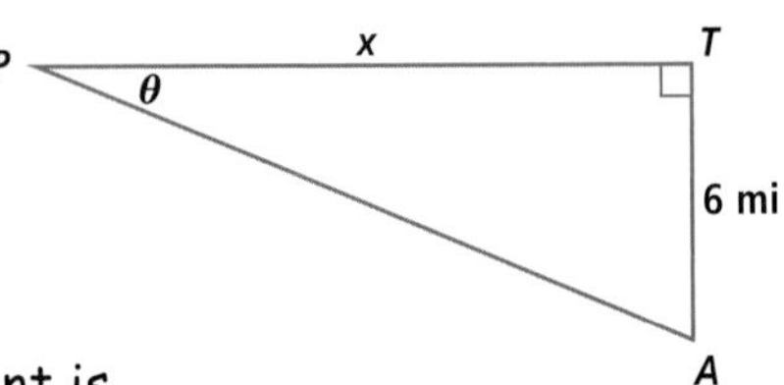

Then $\tan \theta = \frac{6}{x}$ and $\theta = \tan^{-1}\left(\frac{6}{x}\right)$.

b. When $x = 30$, $\theta = \tan^{-1}\left(\frac{6}{30}\right) \approx 11.31°$. The angle of descent is about 11°.

The angle of descent in Example 3 is quite steep. This is why planes flying higher than 30,000 feet often begin their descent as far away as 180 miles from their destination.

Example 4

By federal law, every new public building must have an access ramp with slope $\frac{1}{12}$ or less. What is the maximum angle of elevation for a ramp?

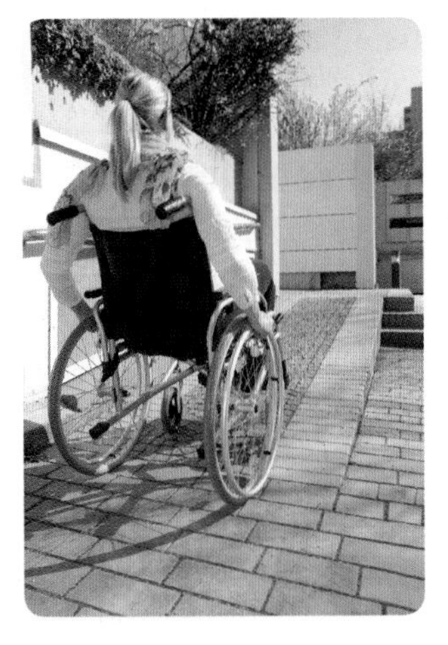

Solution First, draw a picture.

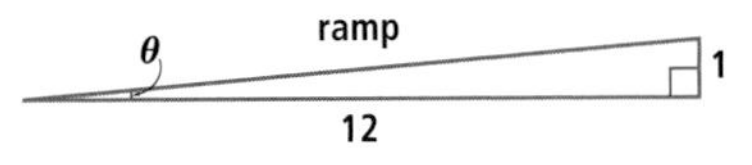

In this situation, $\tan \theta = \frac{1}{12}$. Because the ratio is composed of nonnegative numbers, the angle will be in Quadrant I.

$$\theta = \tan^{-1}\left(\frac{1}{12}\right) \approx 4.7636°$$

The maximum angle of elevation for an access ramp is 4.8°.

Check Does $\tan 4.7636° = \frac{1}{12}$? Since $\tan 4.7636° = 0.083332...$ and $\frac{1}{12} = 0.08\overline{3}$, the answer checks.

Questions

COVERING THE IDEAS

1. a. Find θ to the nearest tenth of a degree if $\tan \theta = 0.07$.

   b. Interpret your answer to Part a in the situation of a plane flying at an altitude of 5.5 miles, 78.6 miles from an airport.

2. How is the expression "$\theta = \tan^{-1} k$" read?

3. **a.** Complete the table of values below.

point on $y = \tan x$	$(-1.5, ?)$	$\left(-\frac{\pi}{3}, ?\right)$	$\left(-\frac{\pi}{4}, ?\right)$	$\left(-\frac{\pi}{6}, ?\right)$	$(0, 0)$	$\left(\frac{\pi}{6}, ?\right)$	$\left(\frac{\pi}{4}, ?\right)$	$\left(\frac{\pi}{3}, ?\right)$	$(1.5, ?)$
corresponding point on $y = \tan^{-1}x$	$(?, ?)$	$(?, ?)$	$(?, ?)$	$(?, ?)$	$(0, 0)$	$(?, ?)$	$(?, ?)$	$(?, ?)$	$(?, ?)$

b. Plot the points above, to graph $y = \tan x$ and $y = \tan^{-1} x$ on the same coordinate system. Use the same scale on each axis.

c. State the domain and range of the inverse tangent function.

d. What transformation maps the graph of $y = \tan x$ to the graph of $y = \tan^{-1}x$?

In 4–6, find the exact value of the expression without a calculator.

4. $\tan^{-1}(\sqrt{3})$
5. Arctan 1
6. $\tan^{-1}\left(-\frac{\sqrt{3}}{3}\right)$

In 7 and 8, refer to Example 3.

7. **a.** Graph $y = \tan^{-1}\left(\frac{6}{x}\right)$ in the window $-5 \le x \le 100$, $-5 \le y \le 2$.

   b. Trace the graph to estimate y when $x = 70$.

   c. Explain what the answer to Part b means in context.

8. If a plane flying at 37,000 feet (about 7 miles) begins a descent 180 miles from the airport, find the angle of descent to the nearest tenth of a degree.

APPLYING THE MATHEMATICS

9. **a.** Let θ be the measure of the acute angle between the line with equation $y = 2x - 5$ and the x-axis. Prove that $\theta = \tan^{-1} 2$.

   b. Generalize Part a.

10. A radar tracking station is located 2 miles from a rocket launch pad. If a rocket is launched straight up, express the angle of elevation θ of the rocket from the tracking station as a function of the altitude a (in miles) of the rocket.

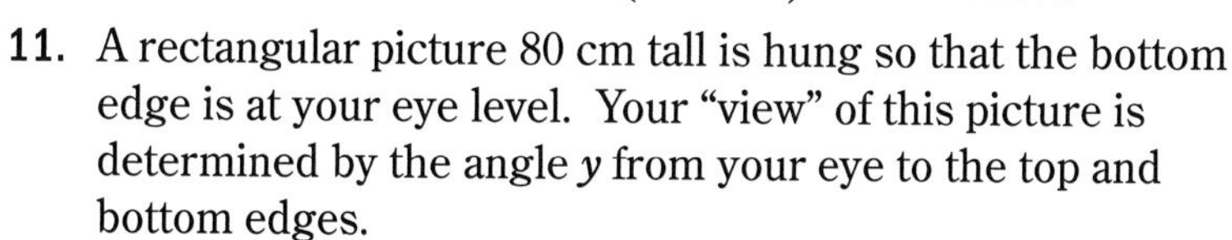

11. A rectangular picture 80 cm tall is hung so that the bottom edge is at your eye level. Your "view" of this picture is determined by the angle y from your eye to the top and bottom edges.

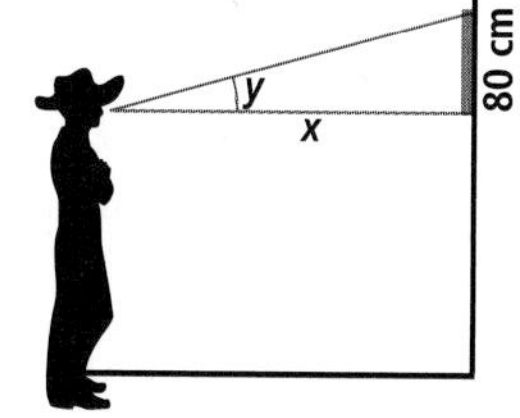

   a. Write an equation for y as a function of the distance x in cm between your eye and the picture.

   b. Suppose that the picture is raised so that the bottom edge is 30 cm above your eye level. Express y as a function of x.

In 12 and 13, compute without a calculator.

12. $\tan(\tan^{-1}(-19))$
13. $\tan^{-1}\left(\tan\left(\frac{2\pi}{3}\right)\right)$

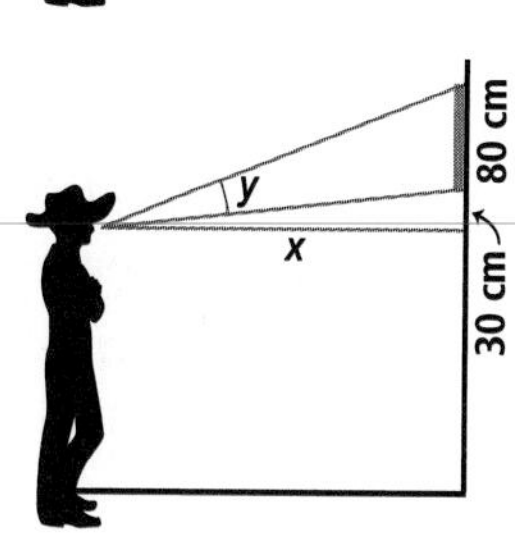

14. Consider the expression $\tan^{-1}\left(\frac{a}{b}\right) + \tan^{-1}\left(\frac{b}{a}\right)$.

   a. Choose several values for a and b and evaluate this expression.

   b. Make and prove a conjecture about the value of this expression in terms of a and b.

   c. Explain how this expression can be related to right triangles.

REVIEW

15. Owners of some land wished to determine the distance from A to B. They walked 250 feet from A directly towards B but encountered a swamp. So they turned left $50°$ and walked 160 feet, then turned right $63°$ and walked directly to B. How far is it from A to B? **(Lesson 5-5)**

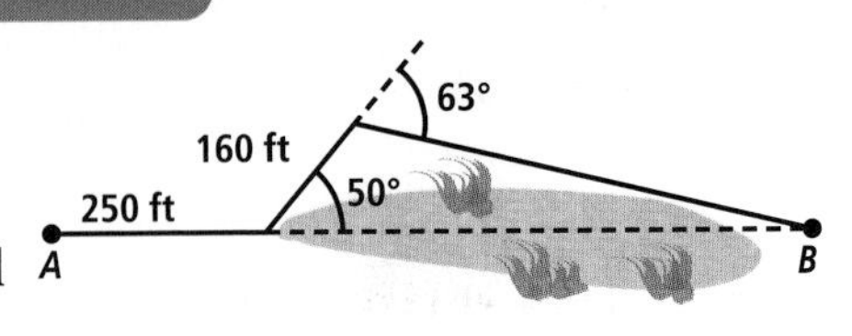

Multiple Choice In 16-21, a triangle is given with an unknown side or angle measure x. Which strategy for finding x is computationally the easiest? **(Lessons 5-5, 5-3, 5-1)**

A definitions of right triangle trigonometric ratios

B Law of Sines

C Law of Cosines

16. 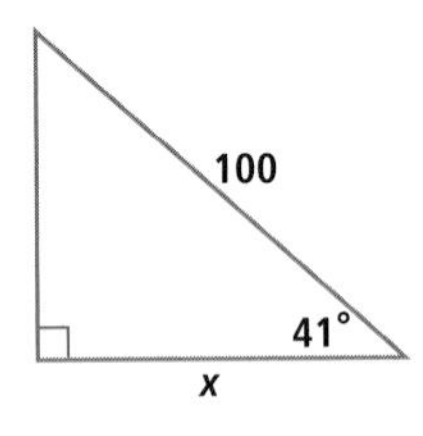

17. 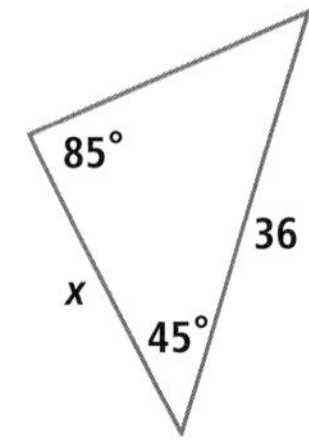

18. 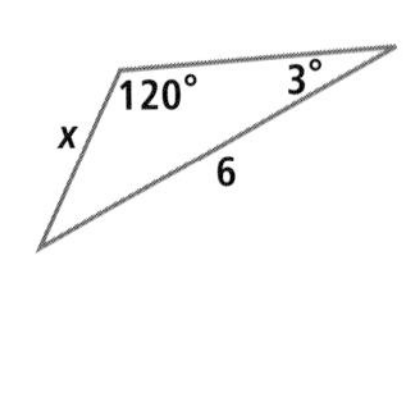

19. 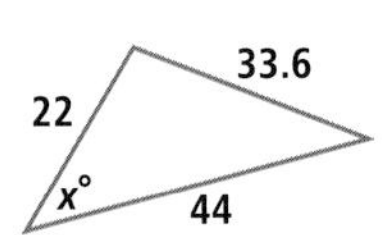

20. 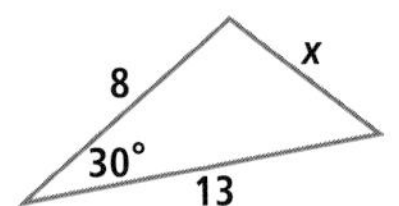

21. 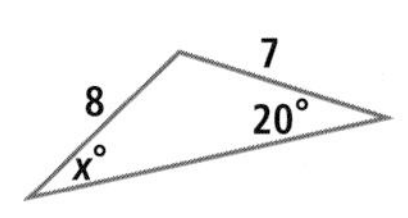

22. How many values of x between 0 and 2π are there such that $\tan x = -\sqrt{3}$? **(Lesson 4-4)**

23. **Skill Sequence** Solve for x. **(Previous Course)**

a. $(7x - 7)(x - 3) = 0$ b. $6x^2 = 5 - 7x$ c. $6x^4 = -26x^2 - 24$

EXPLORATION

24. In 1671, James Gregory discovered that when $-1 \le x \le 1$, $\tan^{-1} x = \sum_{i=0}^{\infty} \frac{(-1)^i x^{(2i+1)}}{2i+1} = x - \frac{x^3}{3} + \frac{x^5}{5} - \frac{x^7}{7} + \frac{x^9}{9} - \ldots$. In 2002, Yasumasa Kanada of Tokyo University calculated π to 1,241,100,000,000 decimal places using Gregory's series and the formula (proved from geometry) that
$\pi = 48\tan^{-1}\left(\frac{1}{49}\right) + 128\tan^{-1}\left(\frac{1}{57}\right) - 20\tan^{-1}\left(\frac{1}{239}\right) + 48\tan^{-1}\left(\frac{1}{110{,}443}\right)$.

a. Use a calculator to verify that the formula for π used by Kanada seems to be correct.

b. Use the first three terms of Gregory's series to estimate each term in Kanada's formula and thus find an estimate for π. To how many decimal places is this estimate accurate?

c. Use more terms of Gregory's series to find an estimate accurate to more decimal places.

QY ANSWER

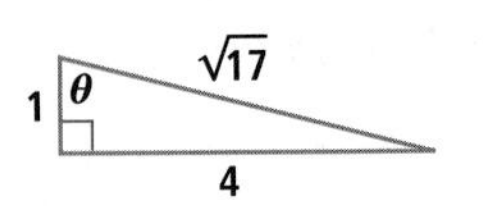

Lesson 5-7 General Solutions to Trigonometric Equations

Vocabulary

trigonometric equation

general solution to a trigonometric equation

BIG IDEA You can find all the solutions of a trigonometric equation using inverse trigonometric functions and the Periodicity Theorem.

A **trigonometric equation** is an equation in which the variable to be found is an argument of the sine, cosine or tangent function. Examples are $\cos\theta = 0.6$, $9 = 12\sin(2\pi t)$, $\tan(2x) = 3$, and $2\sin^2\theta = 1 - \sin\theta$.

Mental Math

Describe all solutions to $\sin x = 0$ over the given domain for x.

a. $0° \le x \le 90°$

b. $0° \le x \le 360°$

c. $-720° \le x \le 720°$

d. set of all real numbers

Activity

Step 1 Graph $y = \tan(2x)$ on the interval $-\frac{\pi}{2} < x < \frac{\pi}{2}$. Use the graph to tell how many solutions the equation $\tan(2x) = 3$ has on the interval $-\frac{\pi}{2} < x < \frac{\pi}{2}$.

Step 2 Use your graph to estimate these solutions to the nearest hundredth and check each solution by substituting into the original equation.

Step 3 Use a graph to determine all solutions to the equation $\tan(2x) = 3$ on the interval $-2\pi < x < 2\pi$. Explain how you could have found all these solutions without graphing.

Step 4 Describe *all* real solutions to the equation $\tan(2x) = 3$, when x can be any real number.

The Activity shows that the number of solutions to a trigonometric equation can vary significantly depending on the domain of the variable. Three domains commonly arise:

(1) the restricted domains of the sine, cosine, and tangent functions that are used in obtaining their inverse functions;

(2) an interval equal in size to the period of the function under study;

(3) the set of all real numbers for which the function is defined.

Using the inverse trigonometric functions $\cos^{-1}$, $\sin^{-1}$, or $\tan^{-1}$ on your calculator will help you find solutions for domain (1). To find the solutions for domains (2) and (3), you may need to use the properties of the trigonometric functions studied in Chapter 4, and you will find it helpful to graph the function. For domain (3), the periodic nature of the functions generates infinitely many solutions on the set of real numbers.

The solution that arises from domain (3), considering all real values of the variable, is called the **general solution** to the trigonometric equation.

The Simplest Trigonometric Equations

The simplest trigonometric equations are of the form $a \cdot f(x) = b$, where f is the sine, cosine, or tangent function.

Example 1

Consider the equation $\cos\theta = 0.6$. Round all solutions to the nearest thousandth.

a. **Find all solutions between 0 and π.**

b. **Solve the equation when $0 \le \theta \le 2\pi$.**

c. **Describe the general solution.**

Solution

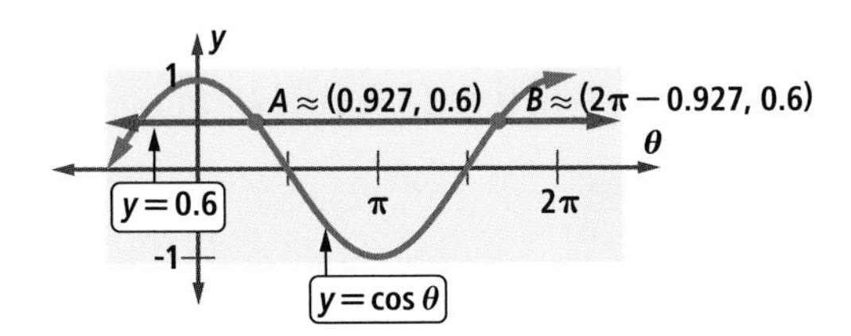

a. By definition of the inverse cosine function, $\theta = \cos^{-1} 0.6$. In radians, $\theta \approx 0.927$.

b. On the same set of axes, graph $y = \cos\theta$ and $y = 0.6$ over the interval $0 \le \theta \le 2\pi$. Each point of intersection corresponds to a solution to the equation $\cos\theta = 0.6$.

Point A represents the solution found in Part a. Point B shows another solution in the interval $0 \le \theta \le 2\pi$. The symmetry of the cosine graph shows this solution to be approximately $2\pi - 0.927$, or about 5.356. Thus, when $\cos\theta = 0.6$ and $0 \le \theta \le 2\pi$, $\theta \approx 0.927$ or $\theta \approx 5.356$.

c. The general solution follows from Part b and the Periodicity Theorem. Adding or subtracting multiples of 2π to the two solutions in Part b generates all solutions to $\cos\theta = 0.6$.

$$\begin{array}{ll} \theta \approx 0.927 + 2\pi & \theta \approx 5.356 + 2\pi \\ \theta \approx 0.927 + 4\pi & \theta \approx 5.356 + 4\pi \\ \vdots & \vdots \\ \theta \approx 0.927 - 2\pi & \theta \approx 5.356 - 2\pi \\ \theta \approx 0.927 - 4\pi & \theta \approx 5.356 - 4\pi \\ \vdots & \vdots \end{array}$$

Thus, the general solution to $\cos\theta = 0.6$ is $\theta \approx 0.927 + 2\pi n$ or $\theta \approx 5.356 + 2\pi n$ for all integers n. These solutions represent all points of intersection of the line $y = 0.6$ and the graph of $y = \cos\theta$.

Check A CAS solution of $\cos x = 0.6$ is shown at the right. Note the CAS outputs the general solution using $\pm$ 0.9273.

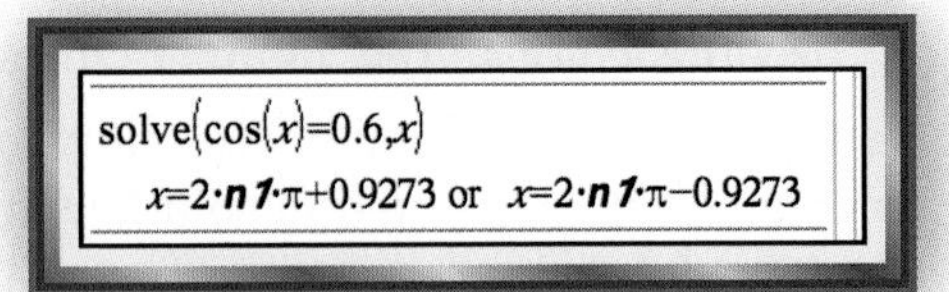

The process used in Example 1 can be employed to solve many trigonometric equations. That is, first find one solution using an inverse trigonometric function, and then use other properties of these functions, particularly the Periodicity Theorem, to find all the solutions.

GUIDED

Example 2

The output voltage E (in volts) of a circuit after t seconds ($t > 0$) is given by $E = 12\sin(2\pi t)$. To the nearest 0.01 second, at which times in the first three seconds does $E = 9$?

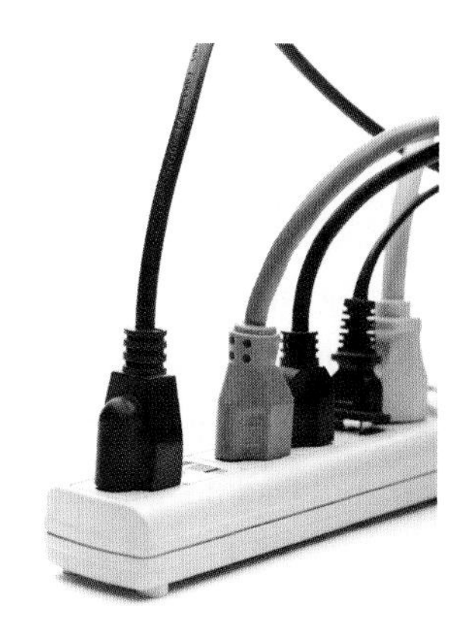

Solution Substitute 9 for E in the given equation.

$$\underline{\ ?\ } = 12\sin(2\pi t)$$

Solve for $\sin(2\pi t)$. $\quad \underline{\ ?\ } = \sin(2\pi t)$

Think of $2\pi t$ as a "chunk" representing θ and solve for $2\pi t$ to find a first quadrant solution.

$$2\pi t = \sin^{-1}\left(\frac{9}{12}\right) \approx 0.8481$$

So, a first solution is $\theta_1 \approx 0.8481$.

Another solution can be found using the Supplements Theorem, $\sin\theta = \sin(\pi - \theta)$. Thus,

$$\sin(2\pi t) = \sin(\pi - 2\pi t) \approx \sin(\pi - 0.8481) \approx \sin \underline{\ ?\ }.$$

So, $\sin \underline{\ ?\ } \approx \frac{9}{12}$, from which a second solution is $\theta_2 \approx 2.2935$.

Because the period of the sine function is 2π, if multiples of 2π are added to or subtracted from 0.8481 and 2.2935, other solutions are generated. So the general solution is

$$2\pi t \approx 0.8481 + 2\pi n \text{ or } 2\pi t \approx 2.2935 + 2\pi n, \text{ for all integers } n.$$

Divide both sides by 2π to find that

$$t \approx 0.1350 + n \quad \text{or} \quad t \approx 0.3650 + n, \text{ for all integers } n.$$

Specifically, on the interval $0 \le t \le 3$,

$$\begin{aligned} t &\approx 0.1350 + 0 \quad \text{or} \quad & t &\approx 0.3650 + 0 \\ &\approx 0.1350 + 1 & &\approx 0.3650 + 1 \\ &\approx 0.1350 + 2 & &\approx 0.3650 + 2 \end{aligned}$$

There are six solutions: 0.1350, $\underline{\ ?\ }$, $\underline{\ ?\ }$, $\underline{\ ?\ }$, $\underline{\ ?\ }$, $\underline{\ ?\ }$.

Check 1 Substitute each of these six values of x in the original equation, and verify that $E \approx 9$ for each value. For instance, for the first two values,

$$12\sin(2\pi(0.1350)) \approx \underline{\ ?\ } \text{ and } 12\sin(2\pi(0.3650)) \approx \underline{\ ?\ }.$$

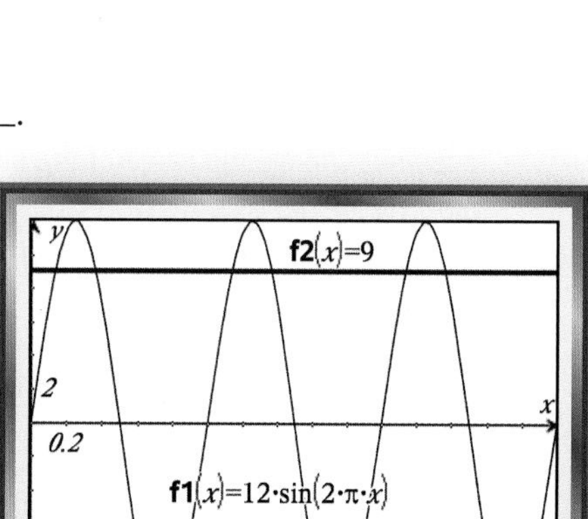

Check 2 Graph $y = 12\sin(2\pi x)$ and $y = 9$. Use a window that includes the maximum and minimum values of this function, 12 and –12, respectively. Notice also that $E = 12\sin(2\pi x)$ has period equal to $\frac{2\pi}{2\pi} = 1$. There are two solutions per period for a total of 6 solutions in the interval $0 \le x \le 3$.

Trigonometric Equations with Quadratic Form

In the quadratic equation $ax^2 + bx + c = 0$, where $a \neq 0$, if x is replaced by an unknown value of a trigonometric function, the resulting equation is a trigonometric equation with quadratic form.

Example 3

Consider $2 \sin^2 \theta = 1 - \sin \theta$.

a. **Find all solutions in the interval $0 \leq \theta \leq 2\pi$.**

b. **Find the general solution.**

Solution 1

a. Think of $\sin \theta$ as a chunk. The equation has quadratic form, so rewrite the equation with one side equal to 0, and solve by factoring or by using the quadratic formula.

$$2 \sin^2\theta = 1 - \sin \theta$$

$$2 \sin^2\theta + \sin \theta - 1 = 0$$

$$(2 \sin \theta - 1)(\sin \theta + 1) = 0$$

So $\quad 2 \sin \theta - 1 = 0 \quad \text{or} \quad \sin \theta + 1 = 0$

$$\sin \theta = \frac{1}{2} \quad \text{or} \quad \sin \theta = -1$$

One solution to each of these equations can be found using the inverse sine function.

$$\sin^{-1}\left(\frac{1}{2}\right) = \frac{\pi}{6} \quad \text{or} \quad \sin^{-1}(-1) = -\frac{\pi}{2}$$

So $\quad \theta = \frac{\pi}{6} \quad \text{or} \quad \theta = -\frac{\pi}{2}.$

By the Supplements Theorem, $\sin \theta = \sin (\pi - \theta)$. Thus, $\pi - \frac{\pi}{6} = \frac{5\pi}{6}$ and $\pi - \left(-\frac{\pi}{2}\right) = \frac{3\pi}{2}$ also satisfy the given equation. Even though $-\frac{\pi}{2}$ is not in the target interval, $\frac{3\pi}{2}$ is. So the solutions to $2 \sin^2 \theta = 1 - \sin \theta$ for $0 \leq \theta \leq 2\pi$ are $\frac{\pi}{6}$, $\frac{5\pi}{6}$, and $\frac{3\pi}{2}$.

b. By the Periodicity Theorem, adding multiples of the period, 2π, does not change the value of the sine.
So the general solution is
$\theta = \frac{\pi}{6} + 2\pi n$, $\theta = \frac{5\pi}{6} + 2\pi n$, or $\theta = \frac{3\pi}{2} + 2\pi n$, for any integer n.

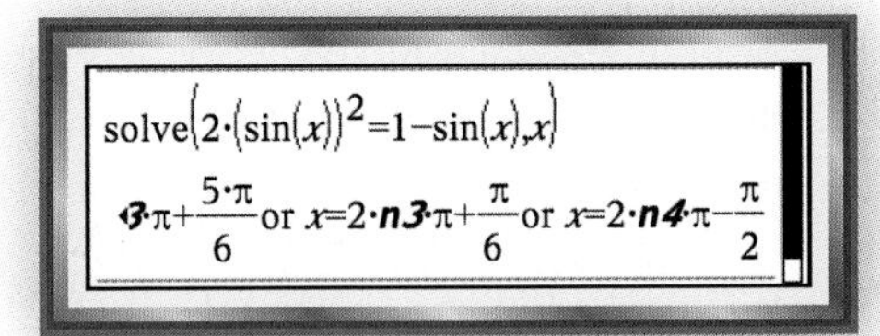

Solution 2 Use a CAS. One CAS gives the solutions $2n\pi + \frac{5\pi}{6}$ or $2n\pi + \frac{\pi}{6}$ or $2n\pi - \frac{\pi}{2}$. This looks like the Solution 1 answer except for $2n\pi - \frac{\pi}{2}$. But notice that you can add 2π to this solution without affecting its being a solution.

$2n\pi - \frac{\pi}{2} + 2\pi = 2n\pi + \frac{3\pi}{2}$, which is found in Solution 1.

Check Graph $y = (\sin x)^2$ and $y = 1 - \sin x$ on the same set of axes. Notice that there are three points of intersection on $0 \leq x \leq 2\pi$. The x-coordinates of these points are the solutions found in Part a.

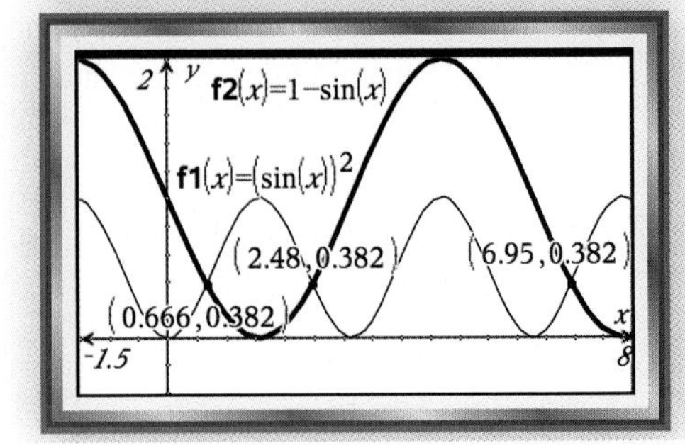

Questions

COVERING THE IDEAS

1. Give the three commonly used domains for solutions to trigonometric equations.

2. The CAS solution to Example 1 does not look the same as the paper solution. Explain why it shows that the paper solution is correct.

3. Consider the equation $\cos \theta = 0.72$.
   a. Find all solutions between 0 and 2π.
   b. Draw a graph to illustrate the solutions in Part b.
   c. Give a general solution to the equation.

In 4 and 5, refer to Example 2.

4. Show that 2.1349 and 2.3650 are solutions to $12 \sin(2\pi t) = 9$.

5. a. How many solutions are there to the equation $12 \sin(2\pi t) = 9$ for $0 < t < 5$?
   b. Use a graph to justify your answer to Part a.
   c. Give all solutions to $12 \sin(2\pi t) = 9$ for $3 < t < 5$.

6. Suppose the output voltage E (in volts) of a circuit after t seconds $(t > 0)$ is given by $E = 20 \cos(4\pi t)$. To the nearest 0.01 second, at which times in the first 2 seconds is E equal to 10?

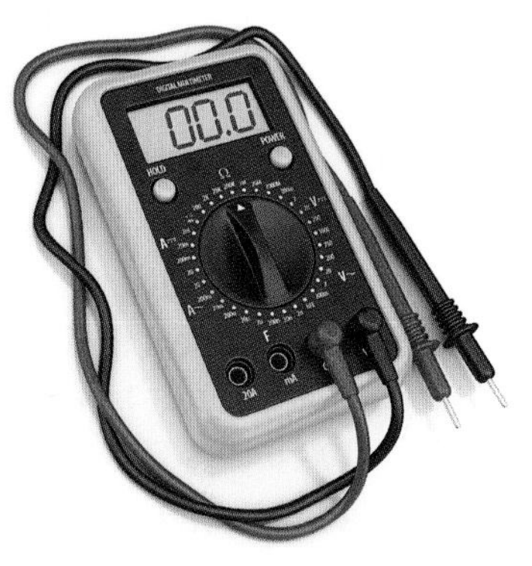

A battery tester is a voltage meter.

In 7 and 8, find all values of θ on the interval $0 \leq \theta < 2\pi$ that satisfy the equation.

7. $5 \sin \theta + 1 = 0$

8. $4 \cos^2 \theta - 1 = 0$

9. Solve $\cos \theta = 0.132$ to the nearest degree for the domain indicated.
   a. $\{\theta \mid 0° \leq \theta < 90°\}$
   b. $\{\theta \mid 0° \leq \theta < 360°\}$
   c. set of all real numbers

10. Solve $2 \cos^2 x - 3 \cos x + 1 = 0$ for the indicated domain.
   a. $\{x \mid 0 \leq x < 2\pi\}$
   b. all real numbers

APPLYING THE MATHEMATICS

11. The number of hours y of daylight in Seattle x days after March 21 can be modeled by the equation $y = 12.25 + 3.75 \sin\left(\frac{2\pi x}{365}\right)$.
   a. Use the equation to find the first two times after March 21 that there will be 11.5 hours of daylight.
   b. Convert your answers in Part a to dates.

The Space Needle in Seattle is approximately 605 feet tall.

12. a. Give a general solution to $3 \cos \theta = 7$.
   b. Under what conditions will $a \cos \theta = b$ have solutions?

In 13 and 14, solve for $0 \leq \theta \leq 2\pi$.

13. $\tan \theta = 3$

14. $\tan \theta - \sqrt{3} = 2 \tan \theta$

REVIEW

15. Mariah sits in the center seat of the first row of the Palace Movie Theater, as shown at the right. She is 8.5 feet away from the screen. When she is seated, her eyes are 3.1 feet above the floor. The bottom of the screen is 6.2 feet above the floor, and the top of the screen is 15.6 feet above the floor. Her angle of vision is formed by the bottom and top of the screen and her eyes. Find the measure of her angle of vision. **(Lesson 5-6)**

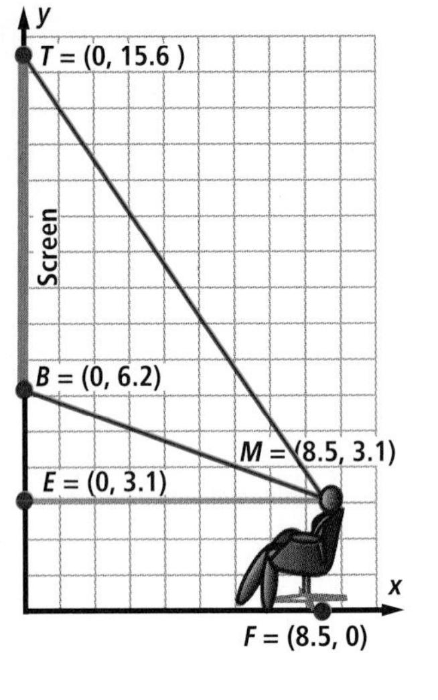

16. In $\triangle XYZ$, $x = 12$, $z = 10$, and $m\angle Z = 43°$. Find all possible values for $m\angle X$, $m\angle Y$, and XZ. **(Lesson 5-5)**

17. A portion of a roller-coaster track is to be sinusoidal as illustrated below. The high and low points of the track are separated by 60 meters horizontally and 35 meters vertically. The low point is 5 meters below ground level.

y
Track
35 m
x
Ground
} 5 m
60 m

   a. Write an equation for the height y in meters of a point on the track at a distance x meters from the high point.

   b. The vertical timbers are to be placed every 5 meters horizontally from the foot of the high point. How long should the timbers be? **(Lesson 4-10)**

In 18–23, tell whether the function described by the equation is odd, even, or neither. (Lessons 4-6, 4-5, 3-4)

18. $y = \lvert x \rvert$	19. $y = \sqrt{x}$	20. $y = x^{10}$
21. $y = 10x$	22. $y = \cos x$	23. $y = \tan x$

EXPLORATION

24. a. How many solutions are there to the given equation on the interval $0 \le \theta < 2\pi$?

   i. $\cos \theta = 0.5$ ii. $\cos(2\theta) = 0.5$

   iii. $\cos(3\theta) = 0.5$ iv. $\cos(4\theta) = 0.5$

   b. Generalize your results in Part a.

   c. How many solutions are there to the equation $\sin(n\theta) = a$, where $\lvert a \rvert < 1$, and n is a positive integer, on the interval $0 \le \theta < 2\pi$?

Lesson 5-8 Parametric Equations for Circles and Ellipses

Vocabulary

parameter
parametric equations
equation for the unit circle in standard form

BIG IDEA Parametric equations use separate functions to define coordinates x and y and to produce graphs.

A circle cannot be the graph of a function with equation $y = f(x)$ because there exist many pairs of points with the same first coordinate. However, you can write each coordinate as a function of a third variable. We call the variable t. A variable that determines other variables is called a **parameter.** When the coordinates of points on a curve (or line) are each expressed with an equation written in terms of a parameter, the equations are called **parametric equations**.

Mental Math

Tell whether the number is positive, zero, or negative.

a. $\sin 1000°$

b. $-|-\sin 1000°|$

c. $\cos 73° - \sin 73°$

d. $\sqrt{10}\sin(10\pi)$

Activity 1

Set your graphing utility to degree mode.

Step 1 Find out how you can enter parametric equations into your technology.

Enter $x_t = \cos t$ and $y_t = \sin t$.

Step 2 Choose a suitable window and graph the two equations. Describe what happens.

Step 3 If possible, animate the graph and run your animation from 0° to 1080° or from 0 to 6π.

The equations in Activity 1 are parametric equations for the unit circle. This is because any point P on the unit circle can be considered as the image of (1, 0) under a rotation of magnitude θ, and by definition, $R_\theta(1, 0) = (\cos \theta, \sin \theta)$. In this case, θ is the parameter.

Recall the Pythagorean Identity $\cos^2 \theta + \sin^2 \theta = 1$. Substituting x for $\cos \theta$ and y for $\sin \theta$, we get $x^2 + y^2 = 1$, the **equation for the unit circle in standard form**.

Activity 2

Set your graphing utility to degree mode and choose parametric for graph type.

Step 1 Graph the equations $\begin{cases} x = 2 \cos t \\ y = 2 \sin t \end{cases}$ and $\begin{cases} x = 3 \cos t \\ y = 3 \sin t \end{cases}$, both for $0° \le t \le 360°$. What is the effect of the 2 or the 3 on the graph?

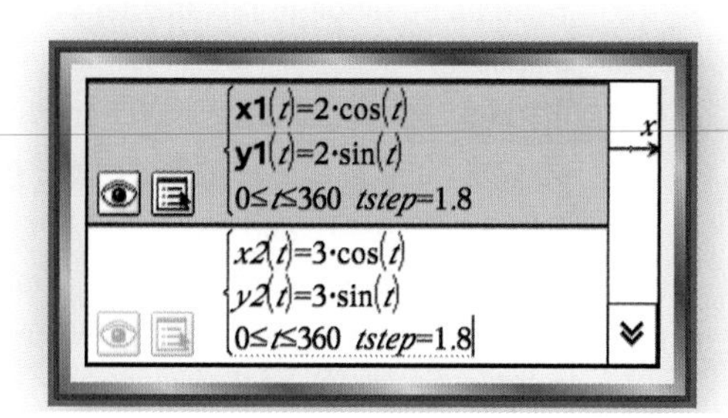

(continued on next page)

Step 2 Generalize your observations in Step 1 to describe the graph of $\begin{cases} x = r\cos t \\ y = r\sin t \end{cases}$ for $0° \le t \le 360°$.

Activity 2 suggests that the image of the unit circle under the size change with center (0, 0) and magnitude r is given by the parametric equations $\begin{cases} x = r\cos t \\ y = r\sin t \end{cases}$. This result can be proved using the Pythagorean Identity.

Theorem (Parametric Equation for a Circle)

The circle with center (0, 0) and radius r has parametric equations $\begin{cases} x = r\cos t \\ y = r\sin t \end{cases}$, $0° \le t \le 360°$ or $0 \le t \le 2\pi$.

Proof Rewrite the parametric equations as $\begin{cases} \frac{x}{r} = \cos t \\ \frac{y}{r} = \sin t \end{cases}$.

We know $\cos^2 t + \sin^2 t = 1$, because of the Pythagorean Identity.

Substitute $\frac{x}{r}$ for $\cos t$ and $\frac{y}{r}$ for $\sin t$. $\left(\frac{x}{r}\right)^2 + \left(\frac{y}{r}\right)^2 = 1$

Use the Power of a Quotient Property. $\left(\frac{x^2}{r^2}\right) + \left(\frac{y^2}{r^2}\right) = 1$

Multiply both sides of the equation by r^2. $x^2 + y^2 = r^2$

This is an equation for the circle centered at the origin with radius r.

Notice that multiplying $\cos t$ and $\sin t$ by r makes them r times as large; however, this transformation is equivalent to replacing x with $\frac{x}{r}$ and y with $\frac{y}{r}$. This substitution is exactly what the Graph Scale-Change Theorem states: the unit circle $x^2 + y^2 = 1$ is transformed by a size change of magnitude r.

QY1

> **QY1**
> a. $x^2 + y^2 = 64$ is the image of the unit circle under a size change of what magnitude?
> b. Write parametric equations for this circle.

Scale Changes and Parametric Equations

By multiplying x- and y-coordinates by constants, you produce a scale-change image. When the constants are not equal, the image of the unit circle under such a transformation is not a circle, but an ellipse.

GUIDED

Example 1

a. Graph the ellipse $\begin{cases} x = 2\cos t \\ y = 5\sin t \end{cases}$, $0° \le t \le 360°$.

b. Write an equation in rectangular coordinates for the ellipse.

Solution

a. Make a table of values for $0° \le t \le 90°$. Also include $t = 180°$ and $t = 270°$ in the table. Some values have been filled in for you. Plot the points on a rectangular grid. Use symmetries over the axes to complete a sketch of the ellipse.

t	$x = 2\cos t$	$y = 5\sin t$
0°	2.00	0
30°	?	2.50
60°	?	?
90°	0	?
180°	?	?
270°	?	-5.00

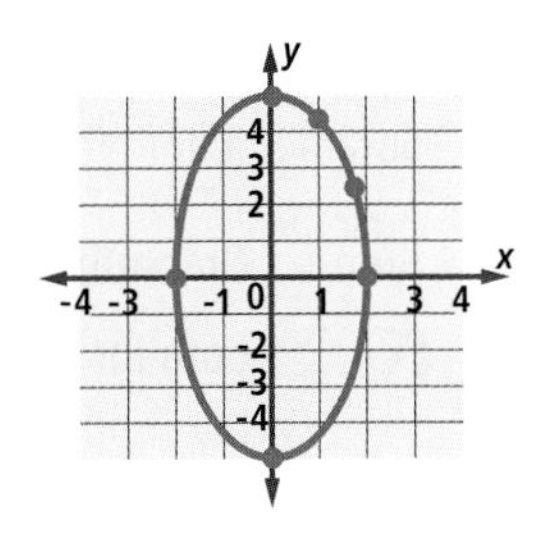

b. $\cos^2 t + \sin^2 t = 1$ Pythagorean Identity

$\underline{\ ?\ } + \underline{\ ?\ } = 1$ Substitute $\frac{x}{2}$ for cos t and $\underline{\ ?\ }$ for sin t.

$\underline{\ ?\ } + \underline{\ ?\ } = 1$ Apply the Power of a Quotient Property.

Activity 3

Step 1 Graph the ellipse $\begin{cases} x = 2\cos t \\ y = 5\sin t \end{cases}$, $0° \le t \le 360°$, from Example 1.

Step 2 On the same grid, graph $\begin{cases} x = 2\cos t + 4 \\ y = 5\sin t - 3 \end{cases}$, for $0° \le t \le 360°$.
Describe the differences between the two graphs.

The ellipse of Step 1 of Activity 3 can be mapped onto the ellipse of Step 2 by the translation $(x, y) \rightarrow (x + 4, y - 3)$. This result suggests the following theorem.

Theorem (Parametric Equation for a Circle with Center (*h*, *k*))

The circle with center (h, k) and radius r has parametric equations $\begin{cases} x = h + r\cos t \\ y = k + r\sin t \end{cases}$, $0° \le t \le 360°$ or $0 \le t \le 2\pi$.

Proof From the parametric equations,

$$\begin{cases} x - h = r\cos t \\ y - k = r\sin t \end{cases}.$$

Thus,

$$\begin{aligned}(x - h)^2 + (y - k)^2 &= r^2\cos^2 t + r^2\sin^2 t \\ &= r^2(\cos^2 t + \sin^2 t) \\ &= r^2 \cdot 1 \\ &= r^2.\end{aligned}$$

GUIDED

Example 2

Write parametric equations for the circle with center (–4, 5) and radius 2.

Solution The circle with radius 2 and center (–4, 5) is the image of the graph of $\begin{cases} x = \cos t \\ y = \sin t \end{cases}$ first under S: $(x, y) \rightarrow (2x, 2y)$ and then the translation T: $(x, y) \rightarrow (x - 4, y + 5)$.

Under S, the equations for the image of the circle are $\begin{cases} x = \underline{\ ?\ } \\ y = \underline{\ ?\ } \end{cases}$.

To move the center to (–4, 5), add $\underline{\ ?\ }$ to the x-coordinates and $\underline{\ ?\ }$ to the y-coordinates. Therefore, parametric equations for this circle are $\begin{cases} x = \underline{\ ?\ } + 2\cos t \\ y = 5 + \underline{\ ?\ } \end{cases}$.

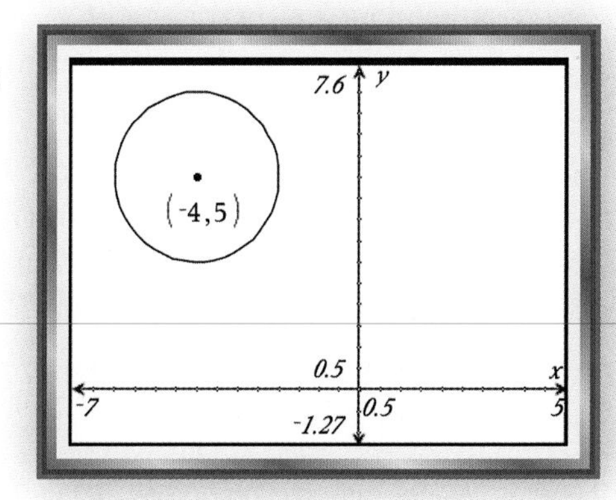

Check Graph the parametric equations using a graphing utility.

 QY2

> ▸ **QY2**
>
> What is an equation in standard form for the circle of Example 2?

Questions

COVERING THE IDEAS

In 1 and 2, write parametric equations for the circle described.

1. center (3, -2) and radius 25
2. $(x - 8)^2 + (y + 4)^2 = 9$
3. Write an equation for the ellipse $\begin{cases} x = 4 \cos t \\ y = \sin t \end{cases}$ in rectangular form.
4. Write equations of the circle graphed at the right in standard rectangular and parametric form.

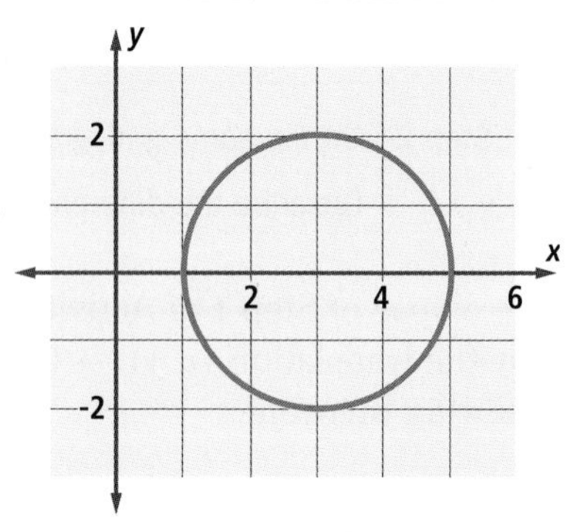

5. Write an equation of the circle $\begin{cases} x = 3 + 5 \cos t \\ y = -2 + 5 \sin t \end{cases}$, $0 \le t \le 2\pi$ in standard rectangular form.
6. Write parametric equations for the lower half of the circle with center (-4, 3) and radius 6.
7. Write parametric equations for the circles at the right.

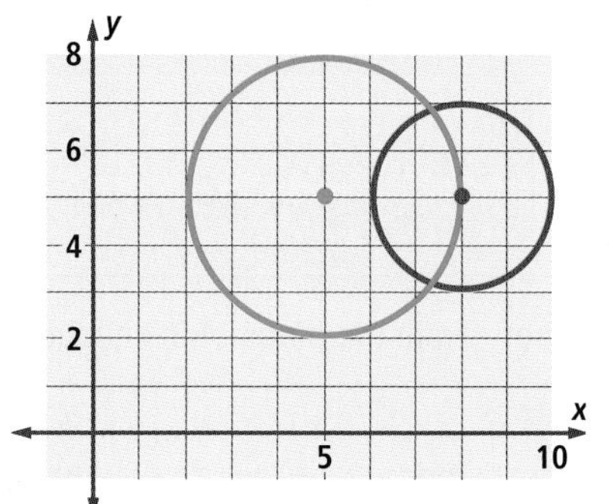

8. a. Graph $\begin{cases} x = \cos t \\ y = \sin t \end{cases}$, $360° \le t \le 720°$.
   b. Compare this with the graph of $\begin{cases} x = \cos t \\ y = \sin t \end{cases}$, $0° \le t \le 360°$.
9. As t increases from 0° to 360°, what happens to the corresponding point on the graph of $\begin{cases} x = 8 \cos t \\ y = 7 \sin t \end{cases}$?
10. The unit circle is translated 6 units to the left and 3 units up.
    a. Write an equation in rectangular form for the transformed circle.
    b. Write parametric equations for the original circle and its image.

APPLYING THE MATHEMATICS

11. A circle with center at (8, -3) has a radius of 0.5. This circle is the image of the unit circle under what transformation?
12. Let $S(x, y) = (4x, 4y)$ and $T(x, y) = (x - 2, y + 5)$. Find equations for the image of the graph of $\begin{cases} x = \cos t \\ y = \sin t \end{cases}$ under
    a. $S \circ T$.
    b. $T \circ S$.
13. a. Graph the parametric equations $\begin{cases} x = 5 \cos t \\ y = 3 \sin t \end{cases}$.
    b. Describe the shape of the graph.
    c. What transformation has been applied to the unit circle in the horizontal direction? In the vertical direction?
    d. Describe how the graph differs from the graph of $\begin{cases} x = 5 \cos t \\ y = 5 \sin t \end{cases}$.
    e. Write an equation in standard form that has the same graph as the equation in Part d.

14. The unit circle is transformed with the mapping $S: (x, y) \rightarrow (9x, 9y)$. Find a mapping that will transform the image back to the unit circle.

15. Consider the sets of parametric equations below. Compare and contrast the curves they trace out, and how those curves are traced. (Many graphing utilities have an animation mode that shows a point moving along a parametric curve.)

   a. $\begin{cases} x = \cos t \\ y = \sin t \end{cases}$ b. $\begin{cases} x = \sin t \\ y = \cos t \end{cases}$ c. $\begin{cases} x = \cos(2t) \\ y = \sin(2t) \end{cases}$

REVIEW

16. Consider $8 \cos^2\theta = 3 - 2 \cos \theta$. **(Lesson 5-7)**

   a. Find all solutions in the interval $0 \leq \theta \leq 2\pi$.

   b. Find the general solution.

17. How many solutions are there to the equation $6 \sin(3\pi t) = 2$ when $0 < t < 8$? **(Lesson 5-7)**

18. A hill slopes upward at an angle of 6° with the horizontal. A tree grows vertically on the hill. When the angle of elevation of the Sun is 24°, the tree casts a shadow 41 m long. If the shadow is entirely on the hill, how tall is the tree? **(Lesson 5-5)**

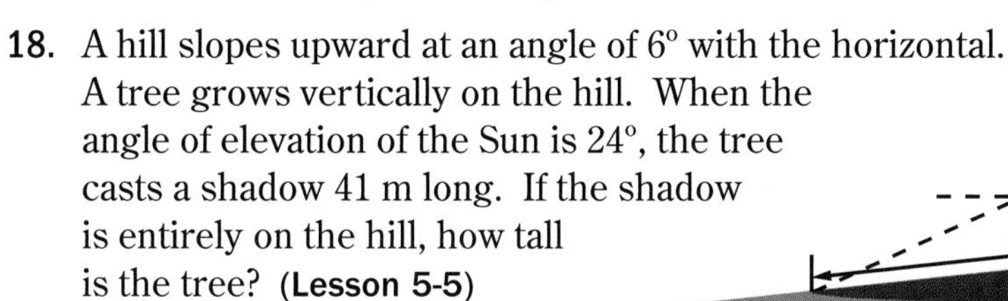

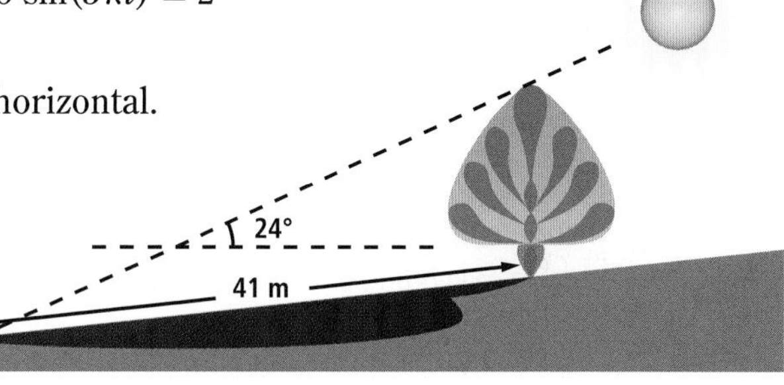

19. Consider the function with equation $y = -3 \cos\left(\frac{x - \pi}{6}\right) + 7$. Give the amplitude, period, vertical shift, and phase shift of the function. **(Lesson 4-9)**

20. For the function with equation $y = \tan\left(x + \frac{\pi}{4}\right)$, determine the **(Lesson 4-8)**

   a. domain. b. range. c. period.

21. Given $f(x) = |x|$ and $g(x) = 1 - x^2$, let $h(x) = f(g(x))$. **(Lesson 3-7)**

   a. Write an expression for $h(x)$.

   b. State the domain and range of h.

EXPLORATION

22. Use parametric equations to construct the picture at the right.

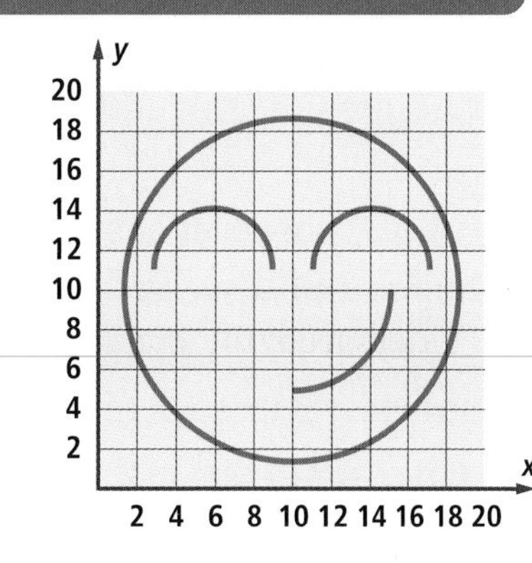

QY ANSWERS

1. **a.** 8

   b. $\begin{cases} x = 8 \cos t \\ y = 8 \sin t \end{cases}$

2. $(x + 4)^2 + (y - 5)^2 = 4$

Lesson

5-9 The Secant, Cosecant, and Cotangent Functions

Vocabulary

secant, sec
cosecant, csc
cotangent, cot
reciprocal trigonometric functions

BIG IDEA Reciprocals of the sine, cosine, and tangent functions have special names.

Activity 1

Step 1 Six ratios can be created using two of the sides of $\triangle ABC$. Fill in the blanks with three of the ratios.

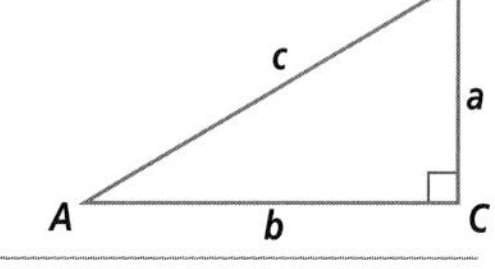

$\sin A = \underline{\ ?\ }$ $\cos A = \underline{\ ?\ }$

$\tan A = \underline{\ ?\ }$

Step 2 There are three other ratios that can be created from the sides of the triangle. Each one of these ratios is the reciprocal of a ratio in Step 1.

reciprocal of $\sin A = \frac{1}{\sin A} = \underline{\ ?\ }$

reciprocal of $\cos A = \frac{1}{\cos A} = \underline{\ ?\ }$

reciprocal of $\tan A = \frac{1}{\tan A} = \underline{\ ?\ }$

Step 3 Pick three values of a, b, and c that can be sides of a right triangle. Evaluate the three reciprocals for these values to 4 decimal places.

Step 4 Square the numbers in Step 3. Do you notice anything unusual? Compare your result with others in your class.

Mental Math

Give the reciprocal of each number as a fraction, decimal, or integer.

a. $-\frac{37}{99}$ **b.** 0.02

c. –1 **d.** 0.6

Naming the Reciprocal Functions

The reciprocals of the sine, cosine, and tangent functions have special names.

Definition of Secant, Cosecant, and Cotangent

Let x be any real number. Then

the **secant** of x = **sec** $x = \frac{1}{\cos x}$, whenever $\cos x \neq 0$;

the **cosecant** of x = **csc** $x = \frac{1}{\sin x}$, whenever $\sin x \neq 0$;

the **cotangent** of x = **cot** $x = \frac{1}{\tan x} = \frac{\cos x}{\sin x}$, whenever $\sin x \neq 0$.

For example, $\cos 60° = \frac{1}{2}$, so $\sec 60° = \frac{1}{\cos 60°} = \frac{1}{\frac{1}{2}} = 2$.

 QY1

QY1

What is the value of csc 60°?

Because each of the secant, cosecant, and cotangent functions can be expressed as the reciprocal of a parent trigonometric function, these functions are sometimes called **reciprocal trigonometric functions**. Notice that since the denominator in each of the definitions can be 0, there are real numbers for which each of these functions is not defined. For instance, $\cot x = \frac{1}{\tan x}$ is not defined when $\tan x = 0$; that is, when $x = k\pi$ for any integer k.

Example 1

Find $\sec\left(\frac{7\pi}{6}\right)$ exactly.

Solution By definition of secant,

$$\sec\left(\frac{7\pi}{6}\right) = \frac{1}{\cos\left(\frac{7\pi}{6}\right)} = \frac{1}{-\frac{\sqrt{3}}{2}} = \frac{-2}{\sqrt{3}} = \frac{-2\sqrt{3}}{3}.$$

Check Use a CAS to compute $\sec\left(\frac{7\pi}{6}\right)$ or $\frac{1}{\cos\left(\frac{7\pi}{6}\right)}$.

Caution: The reciprocal of a function and the inverse of a function are different functions. For example, the reciprocal of the cosine function (the secant function) and the inverse of the cosine function ($\cos^{-1} x$) are quite different functions.

Activity 2

MATERIALS graph paper

Step 1 Copy and complete the table below.

	0	$\frac{\pi}{6}$	$\frac{\pi}{4}$	$\frac{\pi}{3}$	$\frac{\pi}{2}$	$\frac{2\pi}{3}$	$\frac{3\pi}{4}$	$\frac{5\pi}{6}$	π	$\frac{4\pi}{3}$	$\frac{3\pi}{2}$	$\frac{5\pi}{3}$
$\sin x$	?	?	?	?	1	?	?	?	0	?	?	?
$\csc x$	?	?	?	?	?	?	?	?	undefined	?	?	?

Step 2 On graph paper, draw a horizontal axis marked in radians from $-\pi$ to 3π. Mark the vertical axis from –4 to 4.

Step 3 Graph $y = \sin x$ for $0 \le x \le 2\pi$. Use coordinates from your table. Connect the points with a smooth curve.

Step 4 Draw dotted vertical lines where $\csc x$ is undefined.

Step 5 Plot the values of the cosecant function for $0 \le x \le 2\pi$.

Step 6 Within each pair of asymptotes, connect the points with a smooth curve. You should see "U" and upside-down "U" shapes.

Step 7 Extend the graph of $y = \sin x$ to $-\pi \le x \le 3\pi$.

Step 8 Complete the graph of $y = \csc x$, $-\pi \le x \le 3\pi$. Draw asymptotes first.

STOP QY2

QY2

What is the range of the cosecant function?

Graphs of the Reciprocal Functions

You made a graph of the cosecant function in Activity 2 by connecting points. Because the expressions csc x and sin x are reciprocals, you could also use the properties of the sine function to determine characteristics of the cosecant function.

Function Properties	Graph Characteristics
When sin x is positive, csc x is positive. When sin x is negative, csc x is negative.	For a given value of x, the graphs of $y = \sin x$ and $y = \csc x$ are on the same side of the x-axis.
When $\sin x = 0$, csc is undefined.	If $\sin k = 0$, then there is a vertical asymptote of $y = \csc x$ at $x = k$.
Because $\lvert \sin x \rvert \leq 1$ for all x, $\lvert \csc x \rvert \geq 1$ for all x. The smaller $\lvert \sin x \rvert$ is, the larger $\lvert \csc x \rvert$ is.	The closer the graph of $y = \sin x$ is to the x-axis, the farther the graph of $y = \csc x$ is.
$\csc x = \sin x$ when $\sin x = \pm 1$.	The graphs intersect when $\sin x = \pm 1$.

These properties are exhibited in the graphs below. The graph of $y = \sin x$ is in blue; the graph of $y = \csc x$ is in red.

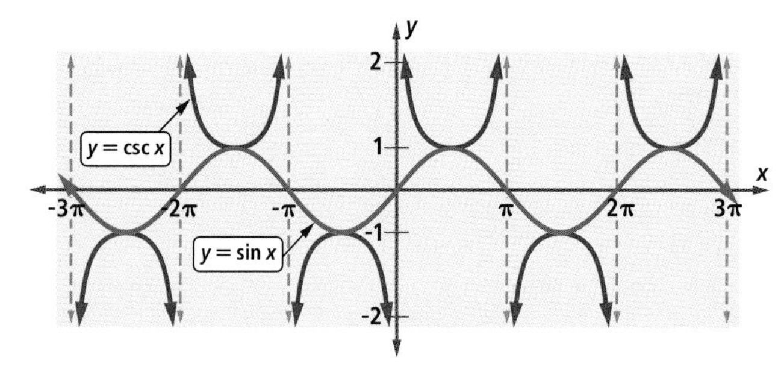

Example 2

Consider the function with equation $y = \csc x$.

a. **Identify its domain and range.**

b. **Find its period.**

c. **Identify any minimum or maximum values.**

Solution

a. The domain consists of all real numbers except $x = n\pi$, for any integer n. From both the definition and the graph above, you can see that the range is $\{y \mid y \leq -1 \text{ or } y \geq 1\}$.

b. The period of $y = \sin x$ is 2π. Consequently, the graph of $y = \csc x$ also has period 2π.

c. There are no maximum or minimum values of csc x. However, 1 can be a relative minimum and -1 can be a relative maximum of the cosecant function.

Because the graph of the cosine function is a translation image of the graph of the sine function, the graph of the secant function is a translation image of the graph of the cosecant function.

The graph of $y = \cot x$ is shown below in red. It is a reflection image of the graph of $y = \tan x$ over any vertical line with equation $x = \frac{\pi}{4} + n\pi$, where n is an integer. The graph of $y = \tan x$ is drawn in blue.

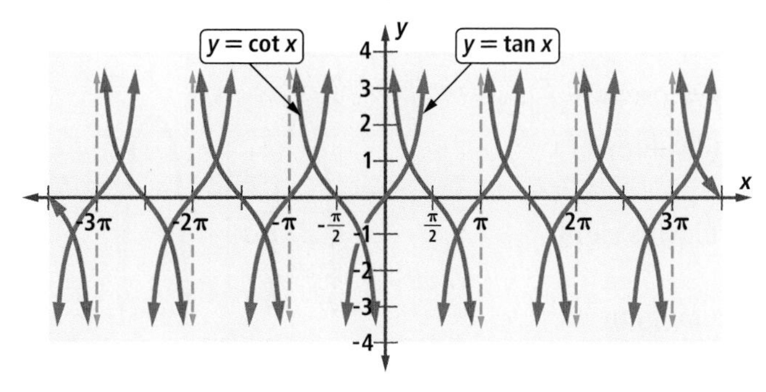

Questions

COVERING THE IDEAS

Fill in the Blank In 1 and 2, use sin, cos, tan, cot, sec, or csc.

1. $\frac{1}{\tan 36°} = \underline{\ ?\ }\ 36°$

2. $\csc\left(\frac{\pi}{3}\right) = \frac{1}{\underline{\ ?\ }\left(\frac{\pi}{3}\right)}$

3. a. Graph $y = \cos x$ for $-2\pi \le x \le 2\pi$.
   b. On the same axes, graph $y = \sec x$. Use dashed lines to identify asymptotes.

In 4–9, evaluate without technology.

4. $\csc\left(\frac{\pi}{2}\right)$
5. $\sec 210°$
6. $\cot(-180°)$
7. $\sec\left(-\frac{7\pi}{6}\right)$
8. $\cot\left(\frac{25\pi}{4}\right)$
9. $\csc\left(\frac{\pi}{4}\right)$
10. Describe the asymptotes of the cotangent function in radians.
11. Give the domain, range, period, and asymptotes of $y = \csc x$.
12. Consider the triangle at the right. In terms of x, y, and z, find
    a. $\sec \theta$.
    b. $\cot \theta$.
    c. $\csc \theta$.

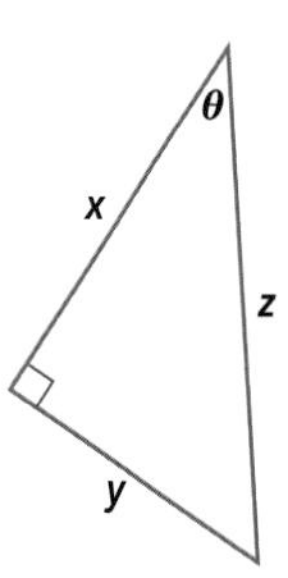

APPLYING THE MATHEMATICS

13. Tell whether the equation describes an odd function.
    a. $y = \sin x$
    b. $y = \cos x$
    c. $y = \cot x$
    d. $y = \csc x$
14. Explain why the graph of the secant function has no points between the horizontal lines with equations $y = 1$ and $y = -1$.
15. If $\sin x = 0.70$, compute.
    a. $\csc x$
    b. $\csc(-x)$
    c. $\csc(\pi - x)$
    d. $\cos x$
    e. $\sec x$
    f. $\cot x$

16. The line with equation $y = \frac{2}{\sqrt{2}}$ intersects the graph of $y = \sec x$ as shown at the right. One point of intersection is $\left(-\frac{\pi}{4}, \frac{2}{\sqrt{2}}\right)$. Find the coordinates of the five other points of intersection shown here.

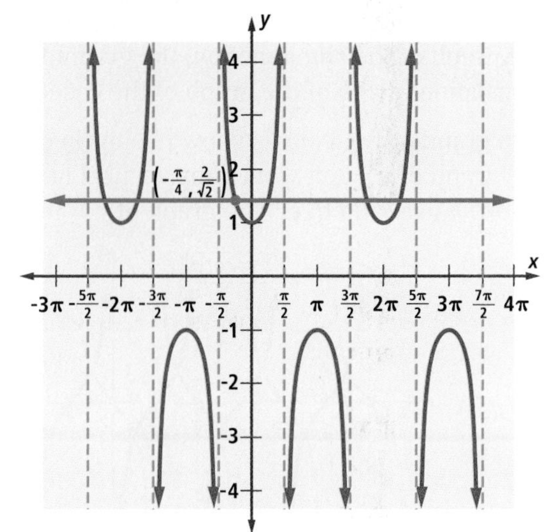

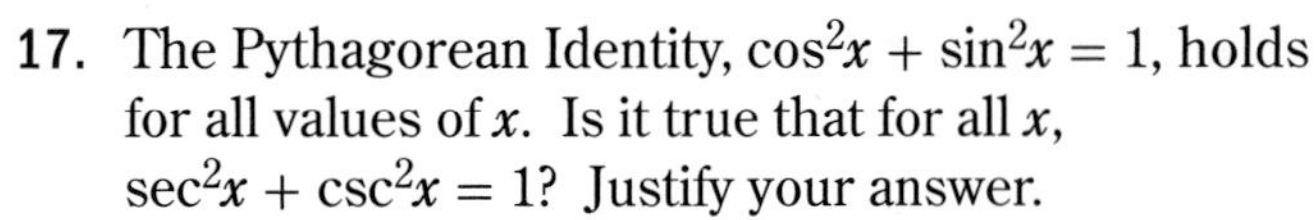

17. The Pythagorean Identity, $\cos^2 x + \sin^2 x = 1$, holds for all values of x. Is it true that for all x, $\sec^2 x + \csc^2 x = 1$? Justify your answer.

18. Give the asymptotes of the graph of $y = \csc\left(x + \frac{\pi}{3}\right)$.

19. Explain why $\cot(\pi - x) = -\cot x$ for all values of x for which $\cot x$ is defined.

20. a. Explain why, in right triangle ABC with right angle C, $\tan A = \cot B$.
    b. What property of the graph of the tangent function does this explain?

21. Simplify the product $(\sin x)(\cos x)(\tan x)(\csc x)(\sec x)(\cot x)$.

REVIEW

22. Consider the circle whose parametric equations are $\begin{cases} x = -4 + 4\cos t \\ y = 2 + 4\sin t \end{cases}$, $0 \le t \le 360°$. **(Lesson 5-8)**
    a. Find the circle's center and radius.
    b. Write an equation for the circle in rectangular coordinates.

23. Consider the circle whose equation is $x^2 + (y + 6)^2 = 17$. Find parametric equations for this circle. **(Lesson 5-8)**

24. The figure at the right shows a waterwheel rotating at 4 revolutions per minute. The distance d of point P from the surface of the water as a function of time t in seconds can be modeled by a sine wave with equation of the form $\frac{d - k}{b} = \sin\left(\frac{t - h}{a}\right)$. **(Lessons 4-10, 4-7)**
    a. What are the amplitude and period of the distance function?
    b. After point P emerges from the water, you start a stopwatch when $d = 7$. Write an equation for the distance function.
    c. Approximately when does P first reach its highest point?

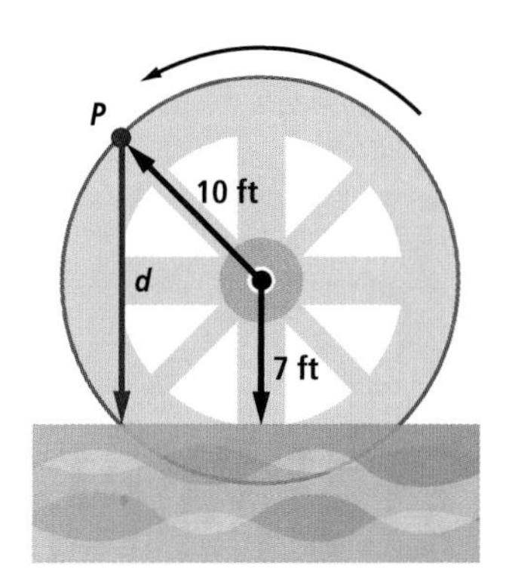

25. **Fill in the Blank** Under a scale change with horizontal factor 3 and vertical factor $\frac{1}{5}$, the image of (x, y) is ___?___. **(Lesson 3-5)**

26. The graph of $y = (x + 5)^4 + 7$ is the image of the graph of $y = (x - 7)^4 + 2$ under what translation? **(Lesson 3-2)**

EXPLORATION

27. In the drawing of the first quadrant of the unit circle at the right, $\overline{OB}$ has been extended to intersect the horizontal line $y = 1$ at G. Tell which segment lengths represent each of the trigonometric functions of θ. For example, $AB = \sin \theta$.

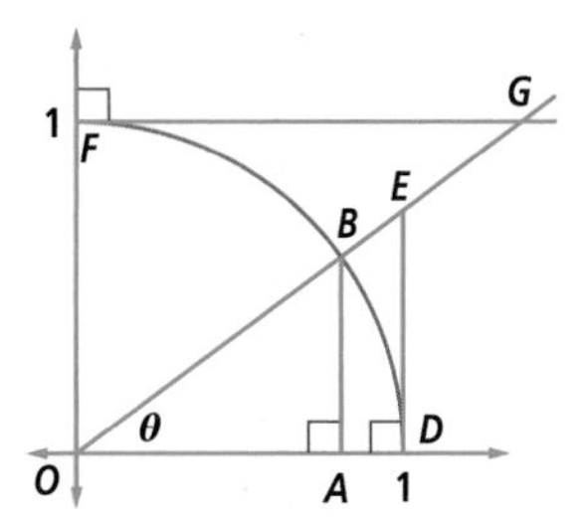

QY ANSWERS

1. $\frac{2}{\sqrt{3}}$ or $\frac{2\sqrt{3}}{3}$
2. $\{y \mid |y| \ge 1\}$

Lesson 5-10 From New York to New Delhi

Vocabulary

great circle

meridian

longitude

latitude

▶ **BIG IDEA** The Spherical Law of Cosines can be used to find the distance between any two points on a sphere.

In this lesson we review some geography and define "shortest distances" on a sphere. Then we use these ideas to find the shortest distance between two cities having the same longitude or latitude. Finally, we calculate the shortest distance between New York and New Delhi to illustrate how you can use trigonometry to find the shortest distance between any two cities on Earth.

Mental Math

A measure in degrees and minutes is given. Convert to degrees in decimal form.

a. $23°30'$

b. $48°20'$

c. $92°15'$

d. $54°45'$

Activity

MATERIALS Internet access

Step 1 Below is a Mollweide equal-area projection of Earth with New York and New Delhi's positions marked. Find the route that appears shortest to you. Identify a landmark along your route so you can use it later in this Activity.

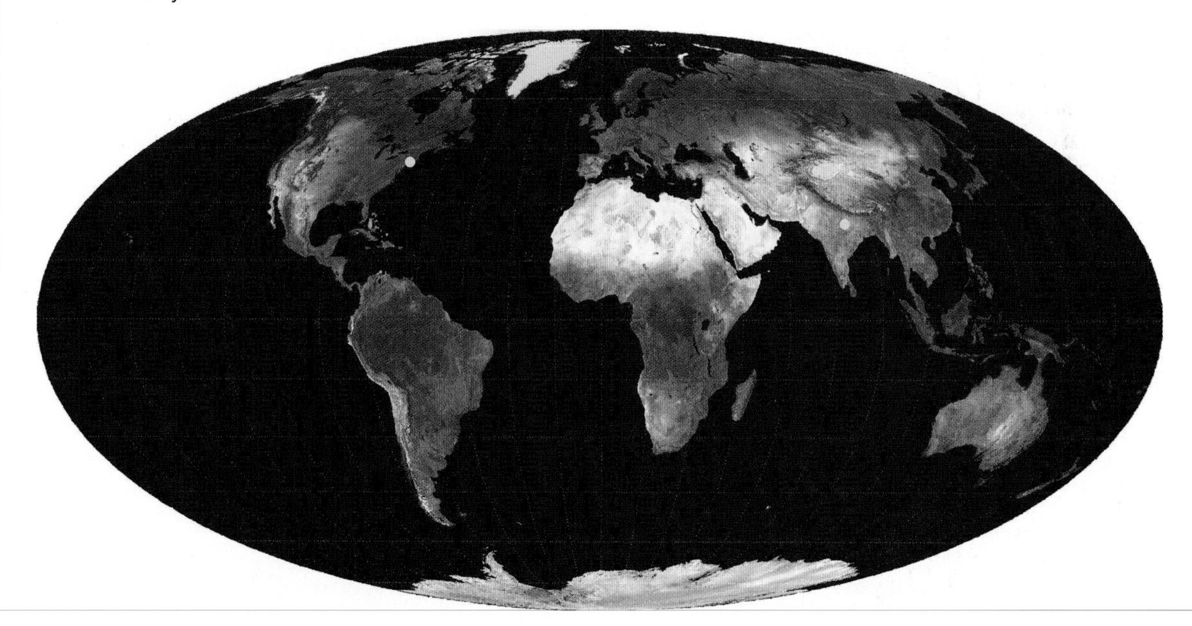

(continued on next page)

Step 2 Open a dynamic geography utility. Zoom and rotate the image until you can see New York City marked on the globe. Put a placemark on New York City so you can find it easily later.

Step 3 Find New Delhi and put a placemark there.

Step 4 Because New York City and New Delhi are so far apart, it is hard to see both on the same view of the globe. Zoom around and spin the globe to locate your landmark from Step 1. Put a placemark there and name it "Flyover."

Step 5 Use the measurement tool to find the length of the path that goes from New York to your "Flyover" location. Measure the distance from "Flyover" to New Delhi. Add the results to find the total length of your path from New York to New Delhi.

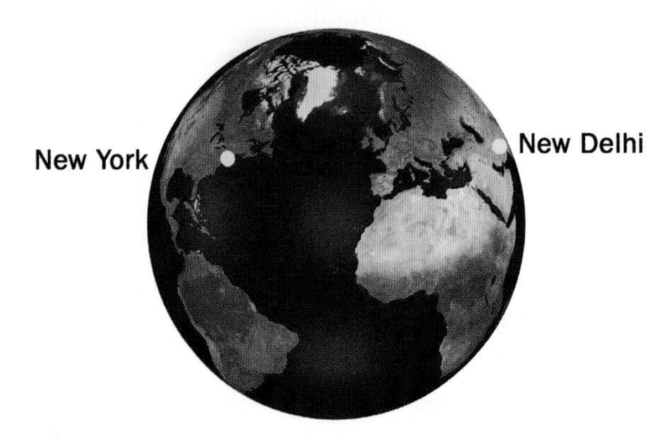

Step 6 When rotating Earth along your path, you probably noticed that the path doesn't look as straight as it did on the Mollweide projection. Because Earth is a sphere, any flat map will distort shapes on Earth's surface. Adjust your "Flyover" point to make your entire path as straight as possible. How has the total distance changed?

Circles on a Sphere

The Activity shows that the shortest distance between two points on Earth does not always appear as a straight line on a map. If you fly from New York City to New Delhi, you might be surprised to see snowbound fjords from your window. Even though New Delhi is *south* of New York City, the shortest flight path passes over Greenland and Finland. This occurs because, on a sphere, the shortest distance between two points is the arc of a *great circle*.

On a sphere, any intersection with a plane forms a circle. A **great circle** is created when that plane passes through the center of the sphere. This circle has the same center as the sphere. Thus, a great circle has the same radius and circumference as the sphere.

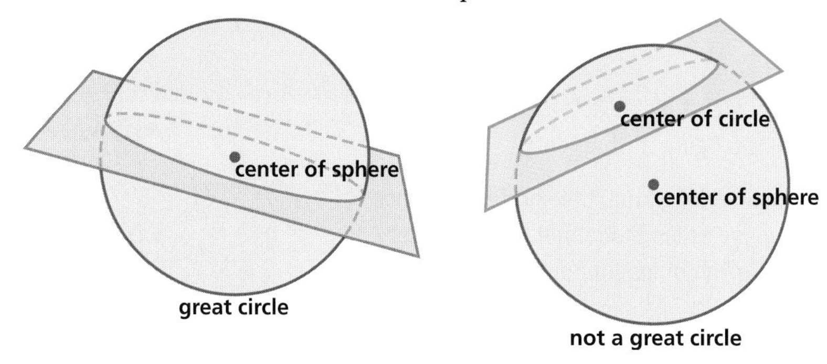

On Earth, which is approximately a sphere of radius 3960 miles, the equator is a great circle. There are infinitely many other great circles containing the north pole N and the south pole S. Each semicircle with endpoints N and S is called a "line" of longitude, or **meridian**. The meridian through Greenwich, England, is called the *Greenwich meridian* or *prime meridian*. **Longitudes** are measured using angles east or west of Greenwich, so all longitudes are between 0° and 180°. In the figure at the right, the longitude of P is ϕ (the Greek letter phi). Because ϕ is east of Greenwich, ϕ measures longitude east. The meridian that is 180°W (and 180°E) is called the *International Date Line*.

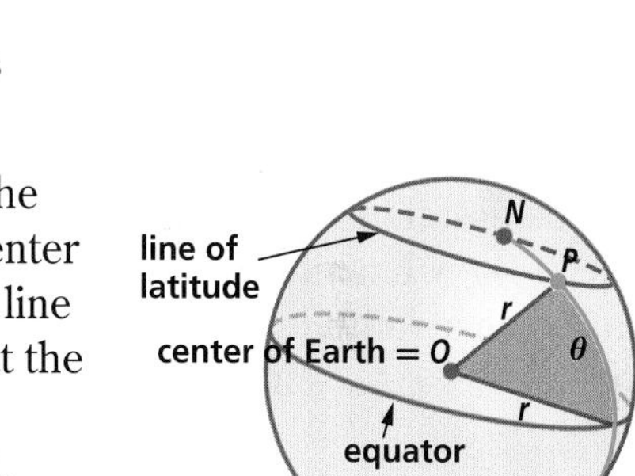

Latitudes measure the extent to which a point is north or south of the equator. They are determined by an angle whose vertex is at the center of Earth and whose sides pass through the endpoints of an arc on a line of longitude. So all latitudes are between 0° and 90°. In the figure at the right, the latitude of P is θ. Because P is north of the equator, θ measures latitude north. Thus the location of P is described as ϕ°E, θ°N. Notice that each "line" of latitude is a circle, but, except for the equator, lines of latitude are *not* great circles.

Distances between Points with the Same Longitude

The position of any point on Earth can be determined by its longitude and latitude. In the figure at the right, Y represents New York City. Its location is about 74°00'W, 40°43'N. This means that New York is on a line of longitude 74° west of the Greenwich meridian and on a line of latitude 40°43', or 40.72° north of the equator. (The symbol ' stands for "minutes of arc"; one minute of arc is $\frac{1}{60}$ of a degree.) Since these coordinates identify a specific point on Earth, and cities cover a large area, the locations are approximate.

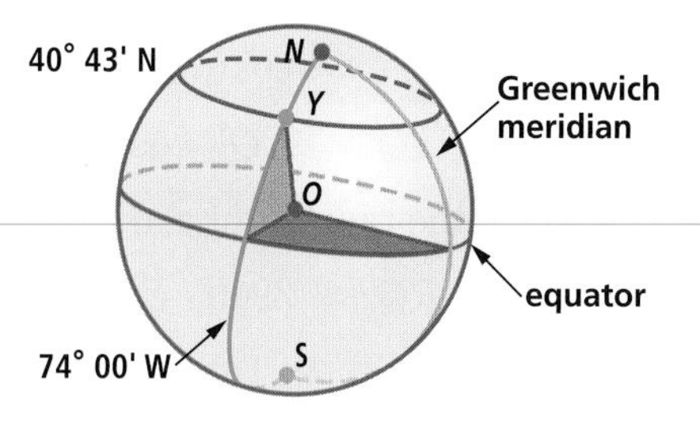

Because lines of longitude are arcs of great circles and all great circles on Earth are roughly congruent, the distance between two points on the same line of longitude can be found easily.

Example 1

Find the distance between New York City and Bogotá, Colombia, located at about 74°W and 4°37'N (labeled *B* in the figure at the right).

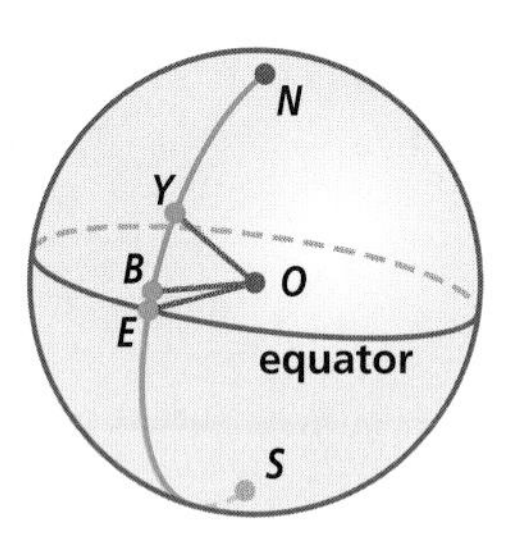

Solution Because *Y* and *B* have the same longitude, the shortest distance between them is the length of $\overset{\frown}{YB}$ on the line of longitude that is 74°W. Let *E* be the point where the equator intersects this line of longitude.

$$\begin{aligned} m\angle YOB &= m\angle YOE - m\angle BOE \\ &= 40\tfrac{43}{60}^{\circ} - 4\tfrac{37}{60}^{\circ} \\ &= 36\tfrac{6}{60}^{\circ} = 36.1^{\circ} \end{aligned}$$

So, the distance between New York City and Bogotá, Colombia is $\frac{36.1^{\circ}}{360^{\circ}}$ of the circumference of a great circle, or $\frac{36.1^{\circ}}{360^{\circ}} \cdot 2 \cdot \pi \cdot 3960 \text{ mi} \approx 2495 \text{ mi}$.

Distances Between Points at the Same Latitude

Consider Tehran, Iran (35°41'N, 51°25'E), and Tokyo, Japan (35°41'N, 139°46'E). Many people think that the shortest distance between Tehran and Tokyo is along the 35°41'N line of latitude, but this is not the case, because lines of latitude are not great circles.

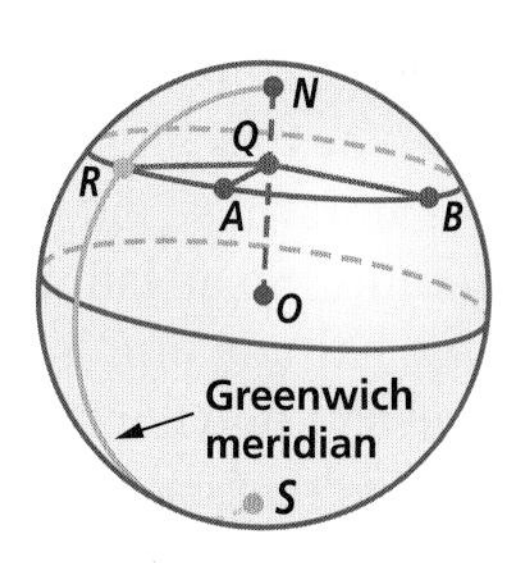

Let us first find the distance between Tehran (*A*) and Tokyo (*B*) along the 35°41'N line of latitude. Let *R* be the point on the Greenwich meridian at latitude 35°41'. If *Q* is the center of the circle that is this line of latitude, $m\overset{\frown}{AB} = m\angle AQB$

So,
$$\begin{aligned} m\overset{\frown}{AB} &= m\angle RQB - m\angle RQA \\ &= 139^{\circ}46' - 51^{\circ}25' = 88^{\circ}21' \end{aligned}$$

The distance between Tehran and Tokyo going due east is thus $\frac{88^{\circ}21'}{360^{\circ}}$ of the circumference of the circle of latitude they are on.

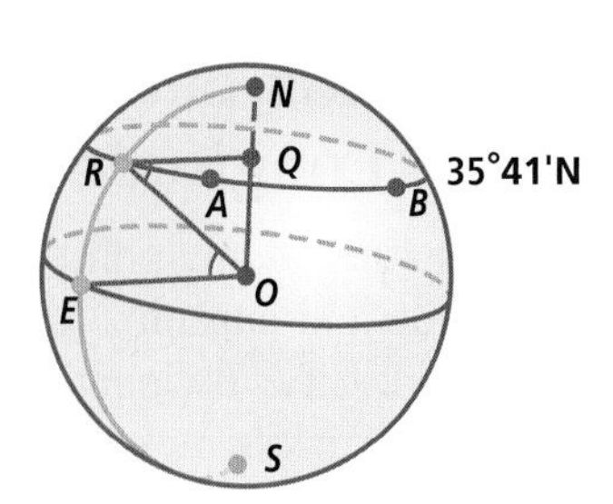

Refer to the figure at the right. To find the circumference of circle *Q*, we need the radius of circle *Q*. One radius of this circle is $\overline{RQ}$. Because $\overline{RQ} \parallel \overline{EO}$, $m\angle QRO = 35^{\circ}41'$. Also,

$$\frac{RQ}{RO} = \cos(m\angle QRO).$$

Hence,
$$RQ = RO \cdot \cos(m\angle QRO).$$

But *RO* is the radius of Earth, about 3960 miles. Therefore,

$$RQ \approx 3960 \cdot \cos 35^{\circ}41' \approx 3217 \text{ mi}.$$

So the distance from Tehran to Tokyo along the line of latitude is about

$$\frac{88^{\circ}21'}{360^{\circ}} \cdot 2\pi \cdot 3217 \text{ mi} \approx 4961 \text{ mi}.$$

To find the *shortest* distance between cities that have different longitudes, we need to use *spherical triangles,* that is, triangles whose sides are arcs of great circles measured in degrees, and a *Spherical Law of Cosines*, which is presented here without proof.

Theorem (Spherical Law of Cosines)

If ABC is a spherical triangle with arcs a, b, and c, then $\cos c = \cos a \cos b + \sin a \sin b \cos C$.

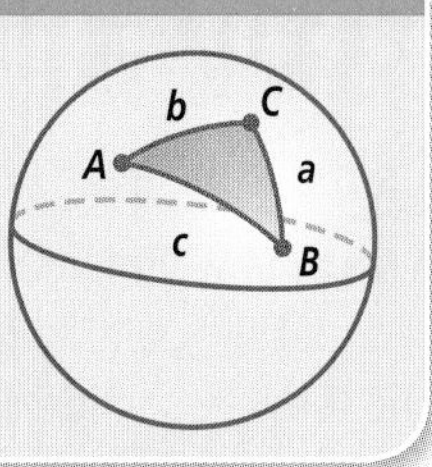

You can use the Spherical Law of Cosines to find the great circle distance between Tehran and Tokyo. The key is to let Tehran and Tokyo be two vertices of the spherical triangle and to let one of the poles be the third vertex. In Example 2, the North Pole is used.

Example 2

Find the length of n, the great circle arc from Tehran to Tokyo.

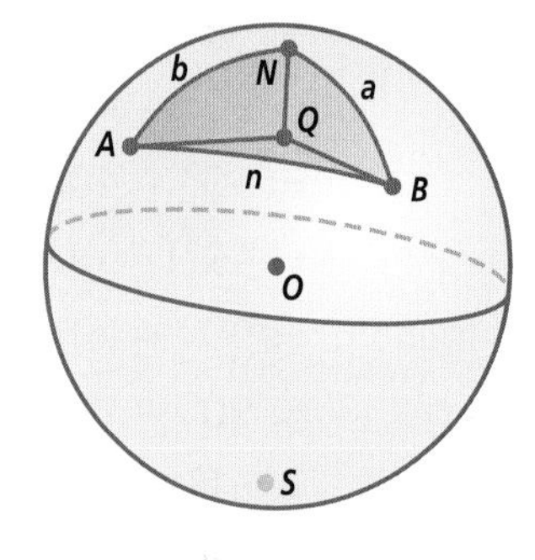

Solution Let A = Tehran, B = Tokyo, and N = the North Pole. In spherical $\triangle ABN$, by the Spherical Law of Cosines,

$$\cos n = \cos a \cos b + \sin a \sin b \cos N.$$

$m\angle N$ is $88°21'$, the same as $m\angle AQB$ that was found before in calculating the distance from Tehran to Tokyo along the line of latitude.

Arc measures a and b are easy to find.

$$b = 90° - \text{latitude of Tehran}$$
$$= 90° - 35°41' = 54°19'$$

Because Tehran and Tokyo have the same latitude, $a = 54°19'$ also. Now substitute into the Spherical Law of Cosines.

$$\cos n = \cos 54°19' \cos 54°19' + \sin 54°19' \sin 54°19' \cos 88°21'$$
$$\approx 0.3592$$
$$n = \cos^{-1}(0.3592) \approx 68.9°$$

So, the length of the great circle arc from Tehran to Tokyo is about $\frac{68.9°}{360°} \cdot 2\pi \cdot 3960 \text{ mi} \approx 4762 \text{ mi}$.

Notice that this is about 200 mi shorter than the path along the line of latitude.

Tehran, Iran

Distance between Any Two Points on a Sphere

The most general and most common problem of this type is to find the shortest distance between two cities not on the same latitude or longitude, such as New York and New Delhi.

Example 3

Find the length of *n*, the great circle arc from New York to New Delhi. New Delhi is located at 72°12'E, 28°38'N.

New Delhi, India

Solution Let A = New York, B = New Delhi, and N = the North Pole. In spherical $\triangle ABN$, by the Spherical Law of Cosines:

$$\cos n = \cos a \cos b + \sin a \sin b \cos N.$$

Here n is $m\widehat{AB}$, $a = m\widehat{BN}$, and $b = m\widehat{AN}$. To find $m\angle N$, extend the meridians through New York and New Delhi to points C and D on the equator. Let point E be the intersection of the prime meridian and the equator. Then

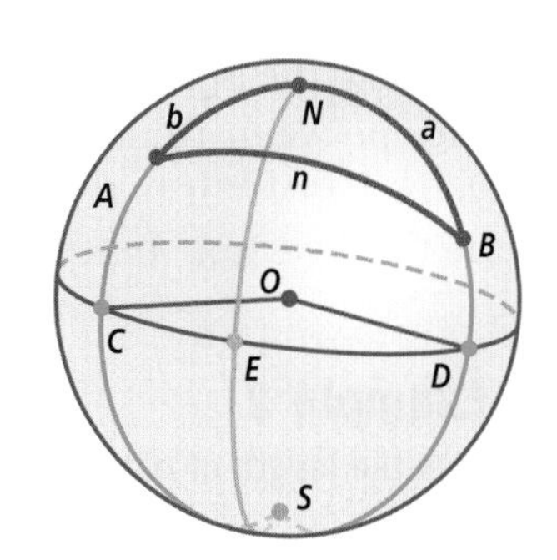

$$\begin{aligned} m\angle N &= m\angle COD \\ &= m\angle COE + m\angle EOD \\ &= 74°00' + 77°12' \\ &= 151°12'. \end{aligned}$$

Arc measures b and a are again easy to find.

$$\begin{aligned} b &= 90° - \text{latitude of New York} \\ &= 90° - 40°43' \\ &= 49°17' \\ \text{and } a &= 90° - \text{latitude of New Delhi} \\ &= 90° - 28°38' \\ &= 61°22'. \end{aligned}$$

Now substitute into the Spherical Law of Cosines.

$$\begin{aligned} \cos n &= \cos 61°22' \cos 49°17' + \sin 61°22' \sin 49°17' \cos 151°12' \\ \cos n &\approx -0.2704 \\ n &= \cos^{-1}(-0.2704) \approx 105.7° \end{aligned}$$

So, the length of the great circle path from New York to New Delhi is about $\frac{105.7°}{360°} \cdot 2\pi \cdot 3960 \text{ mi} \approx 7305 \text{ mi}$.

Problems involving distances in navigation and astronomy have been important for millennia and led to the development of trigonometry. Spherical trigonometry, in fact, began in ancient Greece before plane trigonometry. Euclid knew some of its fundamentals, and by the time of Menelaus (about 100 A.D.), Greek trigonometry reached its peak. Nasir-Eddin (1201–1274), an Arabian mathematician, systematized both plane and spherical trigonometry, but his work was unknown in Europe until the middle of the fifteenth century. Johannes Muller (1436–1476), also known as Regiomontanus, first enunciated the Spherical Law of Cosines given in this lesson.

Questions

COVERING THE IDEAS

In 1 and 2, tell whether the figure is *always*, *sometimes but not always*, or *never* a great circle.

1. line of longitude
2. line of latitude

In 3 and 4, state the common name of the meridian.

3. 0°W
4. 180°E

5. What is the name of the great circle that is 0°N latitude?
6. Find the distance from Capetown, South Africa (18°31'E, 33°58'S) to the South Pole.
7. Find the distance between Paris, France (2°20'E, 48°52'N) and Barcelona, Spain (2°11'E, 41°23'N). Assume that the cities are on the same meridian.

Casa Mila, designed by Gaudi, in Barcelona

8. **True or False** An arc of a line of latitude (other than the equator) can be the side of a spherical triangle.
9. In step 1 of the Activity, an apparently straight path from New York to New Delhi would take the plane over west Africa; one such path is about 8600 miles long. Using the actual distance between New York and New Delhi, estimate how much time you would save by travelling the shortest route in an airplane flying at 550 mph.
10. Find the great circle distance between Paris, Texas (95°33'W, 33°39'N), and Paris, France (2°20'E, 48°52'N).

APPLYING THE MATHEMATICS

In 11 and 12, consider Chicago, Illinois (88°W, 42°N), Providence, Rhode Island (71°W, 42°N), and Rome, Italy (12°W, 42°N).

11. a. Find the distance from Chicago to Providence along the line of latitude.
    b. Find the great circle distance from Chicago to Providence.
    c. How much longer is the line of latitude distance?
12. a. Find the distance from Chicago to Rome along the line of latitude.
    b. Find the great circle distance from Chicago to Rome.
    c. How much longer is the line of latitude distance?
    d. To the nearest percent, how much longer is the line of latitude distance?
13. Find the great circle distance between Prague, Czech Republic (14°26'E, 50°5'N), and Rio de Janeiro, Brazil (43°12'W, 22°57'S).
14. What is the largest possible great circle distance between two points on Earth?

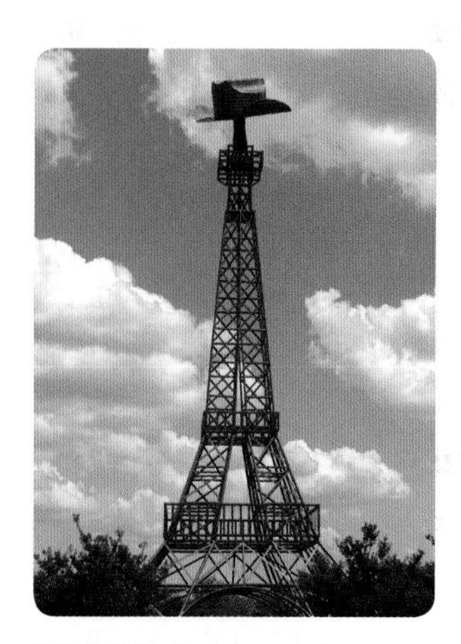

"Paris" is the name of 15 municipalities in the U.S. Only Paris, TX, has a replica of the Eiffel Tower with a cowboy hat.

REVIEW

15. Suppose $f(x) = \cot\left(\frac{x + \pi}{3}\right)$. **(Lessons 5-9, 4-9)**

a. Describe a scale change and translation whose composite maps the graph of $y = \cot x$ onto the graph of $y = f(x)$.

b. Draw a graph of $y = f(x)$.

c. State the period and phase shift of $f(x)$.

16. If $\sec x = 2.9$, compute the five other trigonometric functions of x. Assume $0 \le x \le \frac{\pi}{2}$. **(Lesson 5-9)**

17. As t increases from 0 to 2π, what happens to the corresponding point on the graph of $\begin{cases} x = 5 \sin t \\ y = 2 \cos t \end{cases}$? **(Lesson 5-8)**

18. In $\triangle CAT$ below, find x to the nearest tenth. **(Lesson 5-5)**

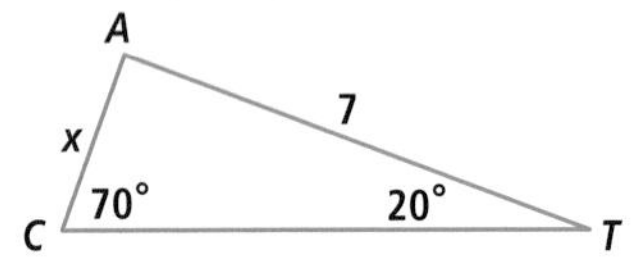

19. In $\triangle DOG$ at the right, find $m\angle G$ to the nearest tenth of a degree. **(Lesson 5-5)**

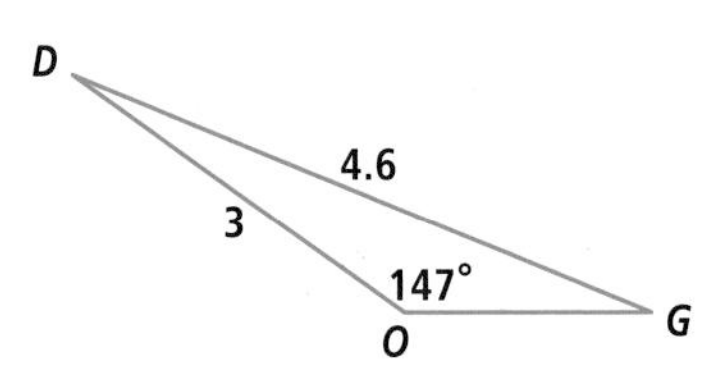

20. In $\triangle ABC$, $a = 23$, $b = 17$, and $c = 33$. Find $m\angle A$. **(Lesson 5-3)**

21. Micky lives on a lake. He wants to go from his house to Lookout Point. He can drive on County Road A for 0.9 miles, then turn onto County Road D and drive for another 1.4 miles, or he can take his power boat directly across the lake. The angle between County Roads A and B measures 57°, as shown at the right. If he drives, his average speed will be 40 mph, and if he takes the power boat, his average speed will be 20 mph. Which way will be quicker? **(Lesson 5-3)**

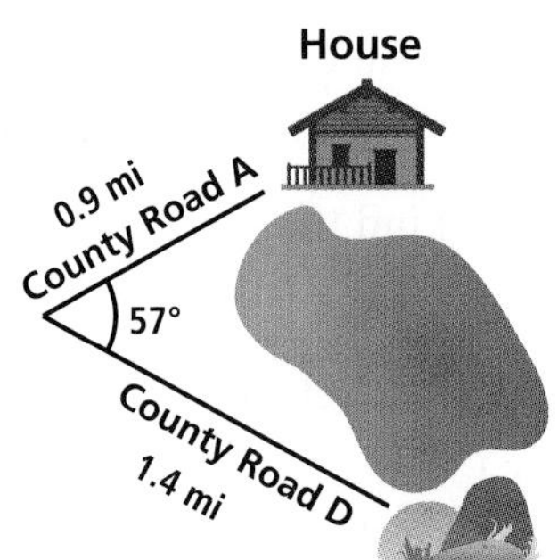

22. Consider these data:

12 15 14 13 17 18 30 14 19 18 15 14

a. Find the median and the IQR.

b. Compute the mean and standard deviation.

c. Which of these measures of central tendency and variability provides a better description of the sample? Explain your answer. **(Lessons 1-6, 1-2)**

23. ***Skill Sequence*** Solve. **(Previous Course)**

a. $x(x - 6) = 0$

b. $(y + 23)(3y - 21) = 0$

c. $(z + 3)(z - 7)(z + 15) = 0$

EXPLORATION

24. Some 3D geometry software allows you to construct points on the same line of latitude, the arc between the points along that line, and the great circle arc. Explore how the difference between the great circle and latitude path is affected by the distance between the points.

Chapter 5 Projects

1 Determining Maximum Altitude

On its web site, the National Aeronautics and Space Administration (NASA) describes how two observers can calculate the altitude reached by a model rocket by measuring the distance L between the observers, the angles a and d of elevation to the rocket, and the angles b and c between the observers and the launch pad.

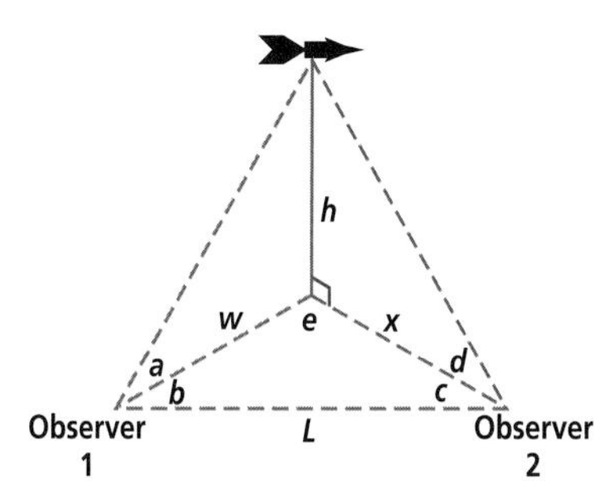

a. One of the equations NASA provides is $h = \frac{L \tan a \tan c}{\cos b(\tan b + \tan c)}$. Explore the NASA site to find out how NASA derived this formula.

b. Use the Law of Sines to derive another formula for h in terms of a, c, e, and L.

c. Verify that the formulas in Parts a and b give similar results when $L = 200$, $a = 78°$, $d = 70°$, $b = 80°$, and $c = 35°$.

2 Spherical Triangles

When three great circular arcs intersect, they form a spherical triangle. Research the history, geometry, and 3-dimensional spherical analogs of 2-dimensional right triangle trigonometry. Make a poster displaying your findings and share these findings with your class.

3 Bond Angles in Molecules

In the methane molecule (CH_4), a single carbon atom lies at the center C of a regular tetrahedron $ABDE$ whose vertices are four hydrogen atoms, as shown both chemically and geometrically below. In this way, the hydrogen atoms achieve their maximum separation.

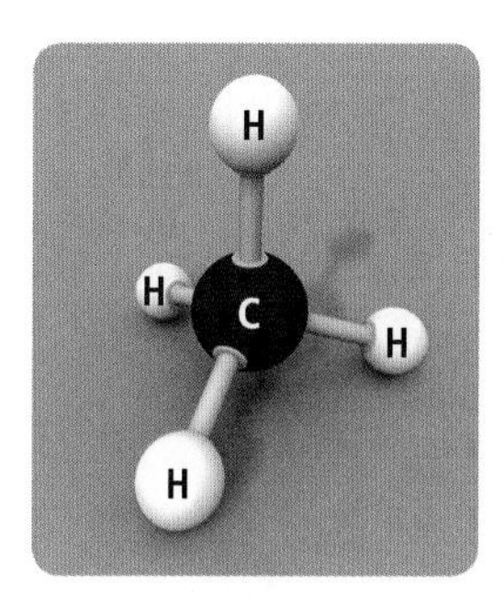

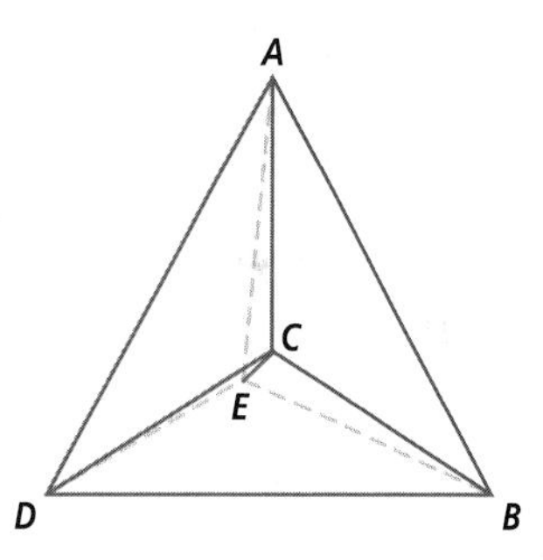

The angles with vertex at C are *bond angles*. One such bond angle is $\angle ACB$. These bond angles all have the same measure, a measure characteristic of many organic molecules.

a. Find the measure of this bond angle using geometry and trigonometry.

b. Consult a chemistry textbook to check your answer to Part a and also to find the carbon-hydrogen bond length in the methane molecule. Find the bond lengths and bond angles in the molecules ethane (C_2H_6), ethylene (C_2H_4), acetylene (C_2H_2), and cyclopropane (C_3H_6), and write a report on why bond angles are important to chemists.

4 The Noon Day Project

This annual Internet event invites classrooms around the world to calculate the circumference of Earth from measurements of the shadow cast by a meter stick at the same time at different locations. This method was first used by Eratosthenes over 2000 years ago!

The project works by using the shadow lengths of a meter stick from two different locations to calculate a *central angle*. A sample observation is shown below.

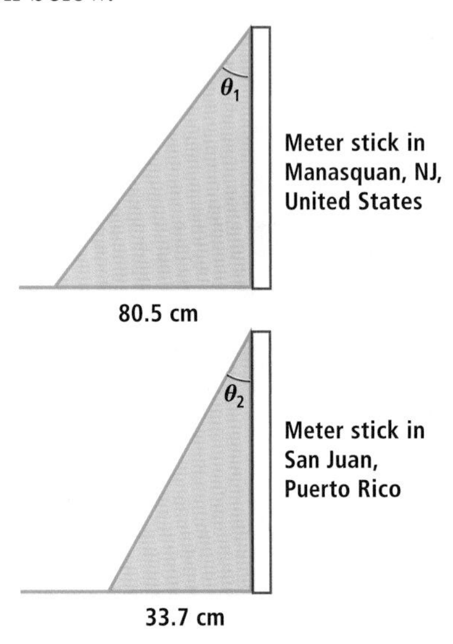

The central angle equals $|\theta_1 - \theta_2|$.

Old San Juan, Puerto Rico
Castillo de San Felipe del Morro

a. Use trigonometry to calculate the central angle between Manasquan and San Juan as well as between your home and a friend or family member's home somewhere else across the United States or the world.

b. How many copies of each central angle fit in a complete 360° revolution of Earth?

c. Each central angle measures the distance between the latitudes of each pair of locations. Using a dynamic geography utility, atlas, or the fact that 1 degree of latitude $\approx$ 111 km, approximate the circumference of Earth using each central angle calculated in Part a and each number of copies calculated in Part b.

d. Calculate the error in each of your estimates, given that the average circumference of Earth is 40,008 km.

5 Application of the Law of Sines

Use the Law of Sines to prove five of the following statements for every triangle ABC.

a. $\frac{a}{b} = \frac{\sin A}{\sin B}$

b. $\frac{a - c}{c} = \frac{\sin A - \sin C}{\sin C}$

c. $\frac{b + c}{b - c} = \frac{\sin B + \sin C}{\sin B - \sin C}$

d. The bisector of an interior angle of a triangle divides the opposite side into parts whose ratio is equal to the ratio of the sides adjacent to the angle bisected. That is, if $\overrightarrow{AD}$ bisects $\angle BAC$, then $\frac{x}{y} = \frac{c}{b}$.

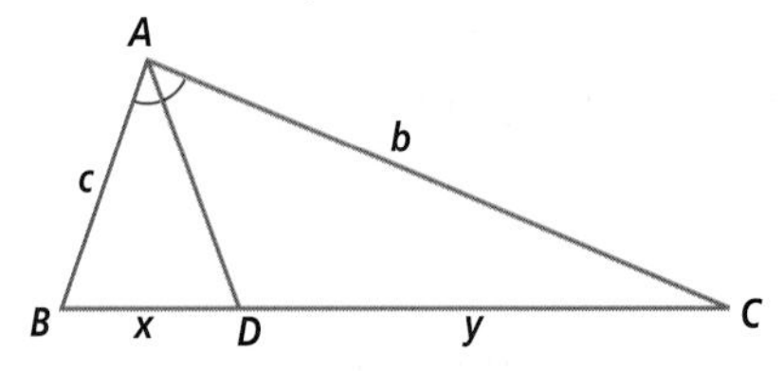

e. In an acute scalene triangle, the largest angle is opposite the longest side.

f. A triangle is equilateral if and only if it is equiangular.

Chapter 5 Summary and Vocabulary

- The sine, cosine, and tangent of θ, when $0° < \theta < 90°$ or $0 < \theta < \frac{\pi}{2}$, are ratios of side lengths of right triangles. In a right triangle,

$$\cos\theta = \frac{\text{adjacent leg}}{\text{hypotenuse}},$$
$$\sin\theta = \frac{\text{opposite leg}}{\text{hypotenuse}},$$
$$\tan\theta = \frac{\text{opposite leg}}{\text{adjacent leg}}.$$

These ratios can be evaluated exactly for some angles and estimated from a good drawing. Better approximations to all values can be found using a calculator or computer. These values can be used to find lengths of sides and measures of angles of right triangles.

- To find unknown sides and angles in all triangles, the following two theorems are helpful.

Law of Cosines: In any $\triangle ABC$, $c^2 = a^2 + b^2 - 2ab\cos C$.

Law of Sines: In any $\triangle ABC$, $\frac{a}{\sin A} = \frac{b}{\sin B} = \frac{c}{\sin C}$.

The Law of Cosines is helpful when the given information is SAS or SSS. The Law of Sines is helpful when the given information is ASA or AAS. In the case of SSA, zero, one, or two triangles may fit the given information.

- Sines and cosines may also be used to obtain arc lengths and arc measures on a sphere. One application of this idea is in finding the distance between two points on Earth given their **latitudes** and **longitudes**.

- The inverses of the parent sine, cosine, and tangent functions are not functions. However, if the domains of the parent functions (and, equivalently, the ranges of their inverses) are restricted as noted below, the inverses are functions.

$y = \cos^{-1} x$ if and only if $x = \cos y$ and is defined when $0 \le y \le \pi$.

$y = \sin^{-1} x$ if and only if $x = \sin y$ and is defined when $-\frac{\pi}{2} \le y \le \frac{\pi}{2}$.

$y = \tan^{-1} x$ if and only if $x = \tan y$ and is defined when $-\frac{\pi}{2} < y < \frac{\pi}{2}$.

The inverse trigonometric function keys on a calculator typically give values only for arguments in the above intervals.

Vocabulary

5-1
angle of depression
angle of elevation

5-2
*inverse cosine function
*$\cos^{-1}$, Arccos

5-4
*inverse sine function
*$\sin^{-1}$, Arcsin

5-6
*inverse tangent function
*$\tan^{-1}$, Arctan

5-7
trigonometric equation
general solution to a trigonometric equation

5-8
parameter
parametric equations
equation for the unit circle in standard form

5-9
secant, sec
cosecant, csc
cotangent, cot
reciprocal trigonometric functions

5-10
great circle
meridian
longitude
latitude

- Equations involving the trigonometric functions are called **trigonometric equations**. Solving trigonometric equations involves the use of inverses of the trigonometric functions. This leads to solutions in a restricted domain. Due to the periodic nature of trigonometric functions, a **general solution** can be obtained by using the period of the parent function and the properties of sines, cosines, and tangents.

- Three other trigonometric functions are the **reciprocal trigonometric functions**: For all θ,

$$\textbf{cosecant } \theta = \textbf{csc } \theta = \frac{1}{\sin \theta},$$
$$\textbf{secant } \theta = \textbf{sec } \theta = \frac{1}{\cos \theta},$$
$$\textbf{cotangent } \theta = \textbf{cot } \theta = \frac{1}{\tan \theta}.$$

When two or more variables are defined in terms of a third variable t, t is a **parameter** and the defining equations are called **parametric equations**. When x and y are defined in terms of a parameter, the points (x, y) describe a curve. The set of parametric equations $\begin{cases} x = \cos \theta \\ y = \sin \theta \end{cases}$ defines the unit circle, while $\begin{cases} x = a \cos \theta \\ y = b \sin \theta \end{cases}$ defines an ellipse.

Theorems and Properties

Right Triangle Ratios for Sine, Cosine, and Tangent (p. 294)
Law of Cosines (p. 304)
SAS Area Formula for a Triangle (p. 315)
Law of Sines (p. 315)
Parametric Equation for a Circle (p. 332)
Parametric Equation for a Circle with Center (h, k) (p. 333)
Spherical Law of Cosines (p. 345)

Chapter 5 Self-Test

Take this test as you would take a test in class. You will need a calculator. Then use the Selected Answers section in the back of the book to check your work.

In 1–3, consider $\triangle ABC$ below.

1. Find sin B exactly.
2. Find tan C exactly.
3. Find m$\angle B$ to the nearest tenth of a degree using $\cos^{-1}$.

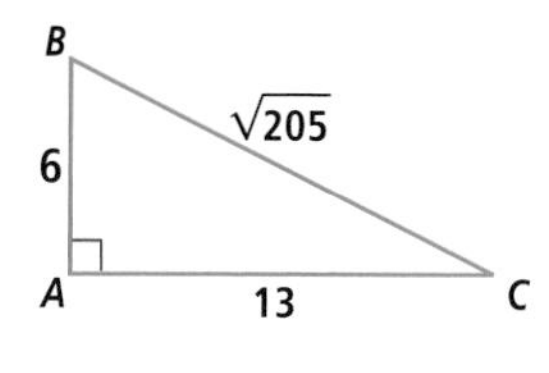

In 4–6, evaluate without a calculator.

4. $\sin^{-1}\left(\frac{1}{2}\right)$ 5. $\tan^{-1} 1$ 6. $\cot\left(\frac{3\pi}{4}\right)$
7. Explain why $\sin^{-1}\left(\sin\left(-\frac{2\pi}{3}\right)\right) \neq -\frac{2\pi}{3}$.
8. a. Draw a graph of $y = \tan^{-1} x$.
   b. Give the domain and range of the inverse tangent function.
9. Solve $\cos \theta = -0.125$ over each domain.
   a. $\{\theta \mid 0 \le \theta \le \pi\}$
   b. $\{\theta \mid 0 \le \theta \le 2\pi\}$
   c. the set of all real numbers
10. Describe the general solution to $2 \cos \theta + \sqrt{3} = 0$.
11. Find x in the triangle below. Round your answer to the nearest tenth.

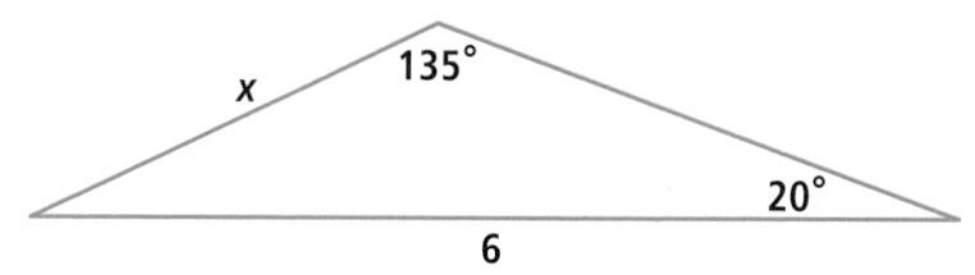

12. Explain why applying the Law of Cosines always leads to a situation with only one answer.
13. Find all values of θ between $-\pi$ and π such that $\cos^2\theta = 3 - 3\cos\theta$.
14. Suppose a 12-m ladder makes an angle with measure θ with the ground. Find an expression for the height h at which the ladder touches the wall in terms of θ.

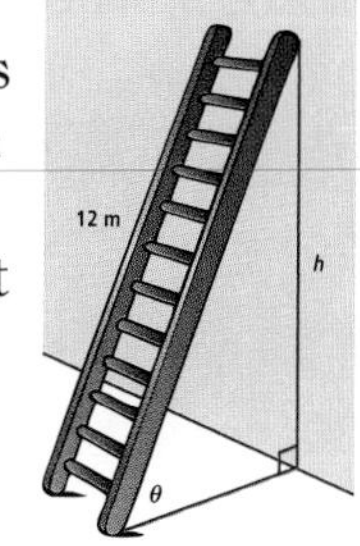

15. New Orleans has about 14 hours of daylight at the summer solstice and 9.3 hours at the winter solstice. An equation for number h of hours of daylight as a function of the number d of days after March 21 is given by $h = 11.65 + 2.35 \sin\left(\frac{d}{\frac{365}{2\pi}}\right)$. Find the next four times New Orleans has 13.6 hours of daylight. Give your answers in days after March 21.
16. The diagram below shows a triangular playground that is bounded by Main Street, High Street, and Central Avenue. The city wants to erect a fence around the playground. Assume that fencing is sold by the foot. What is the minimum amount of fencing the city must buy?

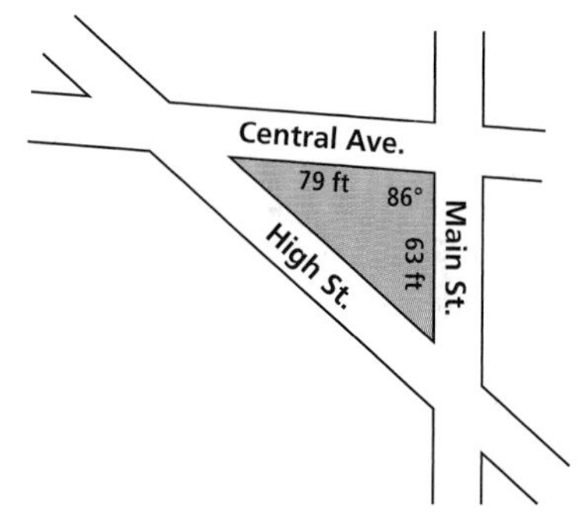

17. Write equations for the circle graphed at the right in rectangular and parametric form.
18. Give two exact values of x for which sec x is undefined.

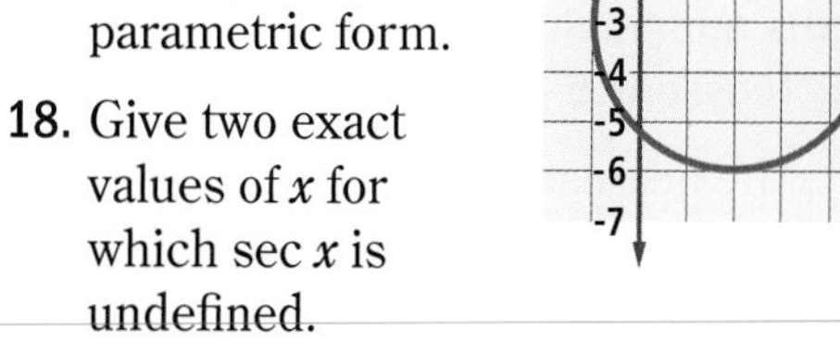

19. Estimate to the nearest mile the distance between New London, Connecticut (41.35 N, 72.11 W) and London, England (51.32 N, 0.5 W). Assume Earth is a sphere with radius 3960 mi.

Chapter 5 Chapter Review

SKILLS
PROPERTIES
USES
REPRESENTATIONS

SKILLS Procedures used to get answers

OBJECTIVE A Find sines, cosines, and tangents of angles. (Lesson 5-1)

In 1 and 2, find the sine, cosine, and tangent of $\angle A$.

1.

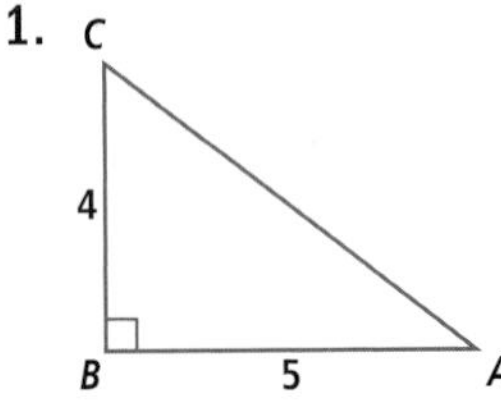

2.

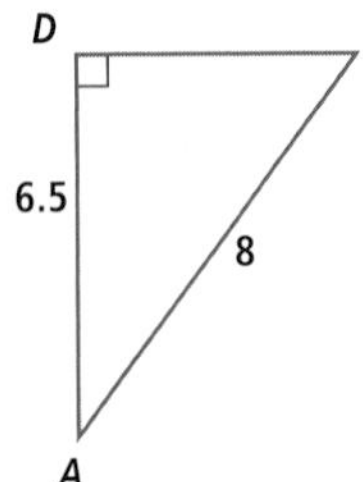

In 3–5, find the exact value without using a calculator.

3. $\sin 60^\circ$ 4. $\cos\left(\frac{\pi}{4}\right)$ 5. $\cos\left(\frac{5\pi}{6}\right)$

In 6 and 7, approximate to the nearest thousandth.

6. $\tan\left(\frac{\pi}{13}\right)$ 7. $\sin 23^\circ 42'$

OBJECTIVE B Evaluate inverse trigonometric functions. (Lessons 5-2, 5-4, 5-6)

In 8–11, evaluate without a calculator, giving answers in degrees and radians.

8. $\sin^{-1}\left(\frac{\sqrt{3}}{2}\right)$ 9. $\text{Arctan}(-1)$

10. $\cos^{-1} -\frac{\sqrt{2}}{2}$ 11. $\tan^{-1}\left(\sqrt{3}\right)$

In 12 and 13, evaluate without a calculator.

12. $\cos\left(\text{Arccos}\left(\frac{7}{11}\right)\right)$ 13. $\sin^{-1}\left(\sin\left(\frac{\pi}{3}\right)\right)$

OBJECTIVE C Use trigonometry to find lengths and angle measures in triangles. (Lessons 5-1, 5-3, 5-5)

14. Find, to the nearest tenth of a degree, the measures of the acute angles of an 8-15-17 right triangle.

In 15 and 16, find x.

15.

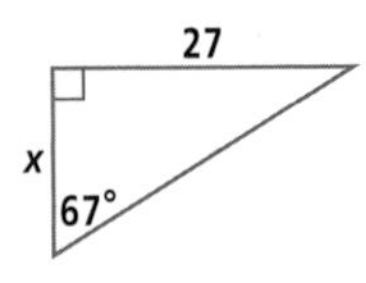

16.

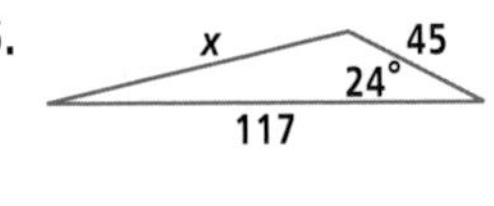

17. In $\triangle ABC$, $m\angle A = 37^\circ$, $m\angle B = 72^\circ$, and $b = 8$. Find a.

18. In the diagram at the right, $\overline{AB} \parallel \overline{CD}$. The circle has diameter 35 mm and $AB = CD = 30$ mm. Find θ to the nearest tenth of a degree.

19. In $\triangle KLM$, $m\angle K = 22.3^\circ$, $k = 32$, and $\ell = 53$. Find all possible values of $m\angle L$ to the nearest tenth of a degree.

20. $\triangle ABC$ is isosceles with $AB = AC = 43$ cm and $m\angle A = 130^\circ$. Find BC.

21. $\triangle ABC$ has sides $AB = 31$, $BC = 43$, and $CA = 17$. Find all interior angles of $\triangle ABC$.

OBJECTIVE D Solve trigonometric equations. (Lessons 5-2, 5-4, 5-6, 5-7)

In 22–24, find θ to the nearest hundredth, where $0 < \theta < \frac{\pi}{2}$.

22. $\tan\theta = 1.9$

23. $\cos\theta = 0.43$

24. $\sin\theta = 0.9876$

25. Find the exact radian value of θ between 0 and $\frac{\pi}{2}$ such that $\sin\theta = \frac{\sqrt{3}}{2}$.

26. Give the number of solutions to the equation, given $0 \le x \le 2\pi$. Justify your reasoning.

a. $7\sin x = 3$ b. $3\sin x = 7$

c. $3\tan x = 7$

In 27–29, solve, given that $0 \le \theta \le 2\pi$.

27. $\cos\theta = -0.81$

28. $\tan\theta = \sqrt{2}$

29. $\sin\theta = \frac{4}{7}$

In 30 and 31, solve, given that $2\pi \le \theta \le 4\pi$.

30. $\cos^2\theta + 2\cos\theta + 1 = 0$

31. $\tan\theta = 2$

In 32–37, describe the general solution in radians.

32. $\cos x = 0.68$

33. $2 \sin y = -\sqrt{3}$

34. $6 \tan z - 9 = 0$

35. $\tan(3w) = 3.12$

36. $\cos(2\pi t) = -0.341$

37. $2 \sin^2(3\theta) - \sin(3\theta) = 1$

OBJECTIVE E Find secants, cosecants, and cotangents. (Lesson 5-9)

38. In $\triangle URN$ at the right, identify
a. $\cot U$.
b. $\csc R$.

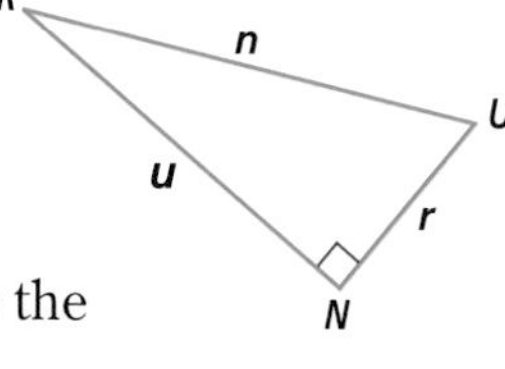

39. If $\cot \theta = \frac{3}{\sqrt{10}}$, what is the value of $\sin \theta$?

40. Give the exact value of $\sec\left(\frac{\pi}{3}\right)$.

41. Give the exact value of $\cot(-765^\circ)$.

42. If $\cos x = 0.70$, compute
a. $\sec x$. b. $\sec(\pi - x)$.

43. Write $\frac{1}{\sin x \cot x}$ as the value of a single trigonometric function.

PROPERTIES Principles behind the mathematics

OBJECTIVE F Interpret the Law of Sines, Law of Cosines, and related theorems. (Lessons 5-3, 5-5)

44. Explain why the two triangles below have the same area.

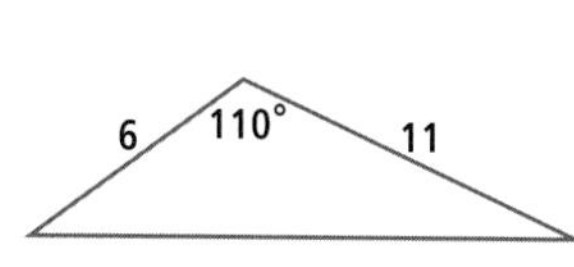

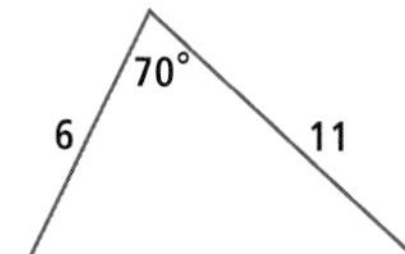

45. Explain how the Pythagorean Theorem is a special case of the Law of Cosines.

46. In $\triangle EFG$, $m\angle E = 30^\circ$, $EF = 5$, and $FG = 8$. Alex claims that there is exactly one triangle satisfying these conditions. Keesha claims that there are two. Who is correct? Why?

47. Explain why the Law of Sines shows that no triangle ABC can have $AB = 40$, $m\angle BAC = 70^\circ$, and $BC = 20$.

OBJECTIVE G State and use properties of inverse trigonometric functions. (Lessons 5-2, 5-4, 5-6)

48. For what values of θ is this statement true? If $k = \cos \theta$, then $\theta = \cos^{-1} k$.

49. **Multiple Choice** If $\tan 3x = b$ and $-\frac{\pi}{2} < 3x < \frac{\pi}{2}$, then x equals ___?___.

A $3 \tan^{-1} b$
B $\tan^{-1}\left(\frac{b}{3}\right)$
C $\tan^{-1}(3b)$
D $\frac{1}{3} \tan^{-1} b$

50. Why must the domain of the sine function be restricted in order to define $y = \sin^{-1} x$?

51. State the domain and range of the inverse cosine function.

52. **Fill in the Blank** The function with equation $y = \text{Arctan } x$ has range ___?___.

53. Explain why $\cos^{-1}\cos\left(-\frac{\pi}{4}\right) \neq -\frac{\pi}{4}$.

In 54 and 55, find two positive and two negative solutions.

54. Use degrees and round θ to the nearest tenth: $\sin \theta = 0.63$.

55. Use radians and round t to the nearest thousandth: $\cos t = -0.38$.

USES Applications of mathematics in real-world situations

OBJECTIVE H Use trigonometry to solve problems involving right triangles. (Lesson 5-1)

56. A ladder against a wall makes a 75° angle with the ground. If the base of the ladder is 4 feet from the wall, find the ladder's length.

57. A building casts a shadow 22 m long when the elevation of the Sun is at 51°, as shown at the right. How high is the building?

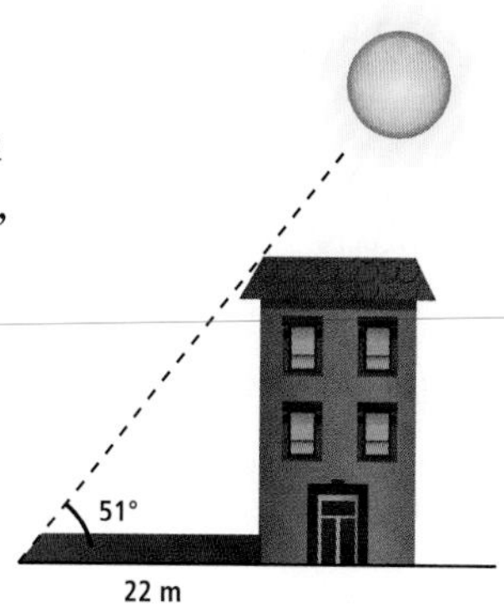

58. Jack is steadying a 50-foot flagpole with two 60-foot-long guy wires. What acute angle θ will the wires make with the ground?

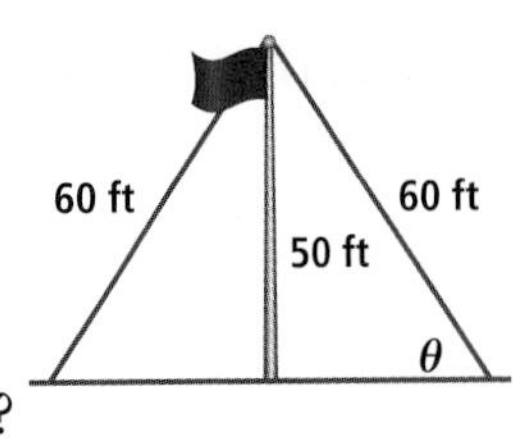

59. A sailboat sails due west from a port at 1 P.M. At 3 P.M., the bearing of a lighthouse is 42° clockwise from north.

  a. If the lighthouse is 20 miles due north of the port, how far out to sea is the sailboat?

  b. What is the speed of the sailboat?

  c. At what time will the lighthouse lie on a bearing of 60° (clockwise from north) to the sailboat, assuming the speed of the ship remains the same?

OBJECTIVE I Solve real-world problems using the Law of Sines and the Law of Cosines. (Lessons 5-3, 5-5)

60. A team of surveyors measuring from A to B across a pond places a stake at S, a point from which they measure distances to A and B. The measures are shown in the diagram. Find AB.

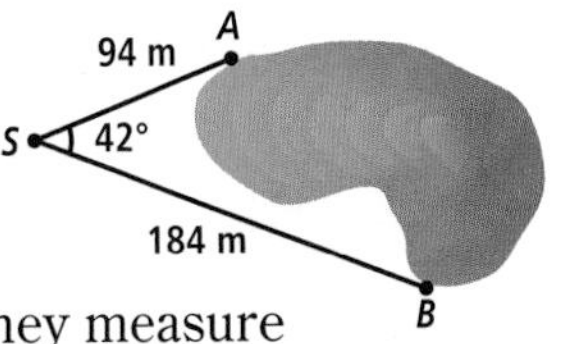

61. When the wind is directly behind a sail boat, the boat will sail faster if it tacks, or sails at an angle to the desired direction. Suppose a boat sails a course beginning at point A, going to C where it tacks, and then to D, where it tacks again, ending at B. $AC = 350'$ and $BD = 55'$. How far is point A from point B? Round your answer to the nearest ten feet.

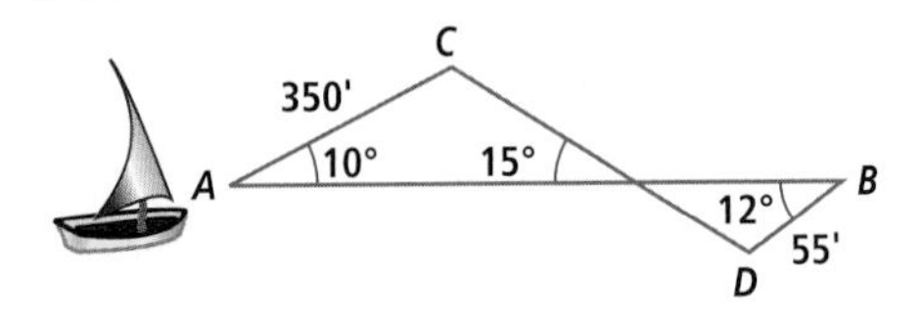

62. An airport controller notes that a plane 14° west of south and 20 miles from the airport is flying toward another plane that is 18 miles directly south of the airport. If the planes are at the same altitude, how far apart are the planes to the nearest tenth of a mile?

63. Forest lookout stations at A and B are 18 miles apart. The ranger in A spots a fire 13° east of north. The ranger at B locates the fire 27° north of west with respect to B.

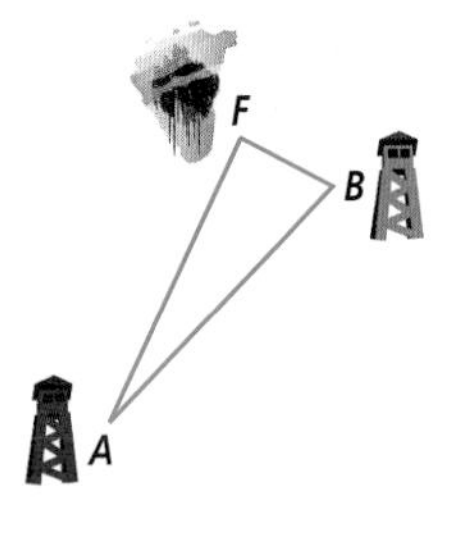

  a. If B is directly northeast of A, find the distance of the fire from A.

  b. Find the distance of the fire from B.

OBJECTIVE J Write and solve equations for phenomena described by trigonometric functions. (Lessons 5-2, 5-4, 5-7)

64. The length c of chord $\overline{AB}$ in $\odot O$ depends on the magnitude of central angle θ.

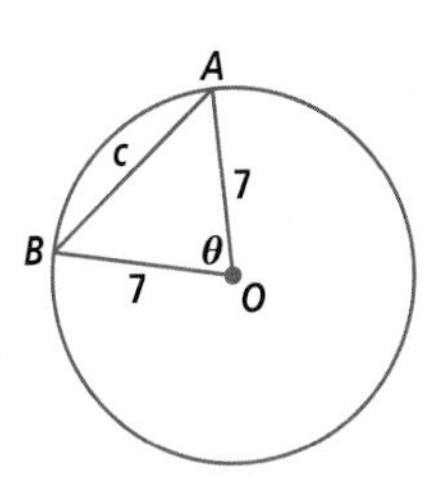

  a. Use the Law of Cosines to write an equation for c in terms of θ.

  b. Solve the equation from Part a for θ.

  c. What is an appropriate domain for c?

  d. Use your equation to find the value of θ when $c = 11$.

65. The distance d in feet from the ground of a paddle on a mill wheel after t minutes can be modeled by $d = 11 + 15\sin(\pi(t - 3))$.

  a. Graph the model for $t > 0$. Why does it dip below the t-axis?

  b. On your graph, mark the first two times that the paddle is $5'$ above the ground.

  c. Find the first two times the paddle is $5'$ above the ground by solving $5 = 11 + 15\sin(\pi(t - 3))$.

66. On a compass used to draw arcs, the leg that holds the pencil measures 5.9″. Find a formula for the angle θ at the top of the compass in terms of the radius of the desired circle.

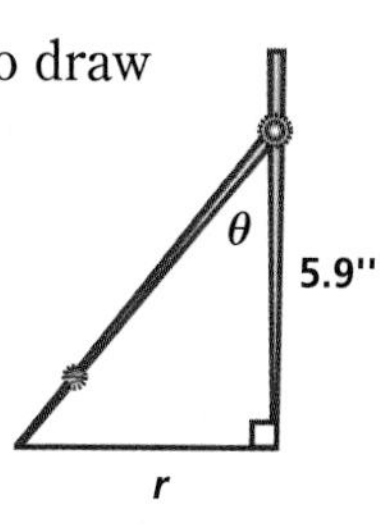

67. The vertical displacement d of a mass oscillating at the end of a spring, measured in cm, is given by the equation $d = 6\sin(\pi t)$ where t is the time in seconds.

 a. Solve this equation for t.

 b. At what time does d first equal 4.5 cm?

68. The measured voltage E in a circuit after t seconds ($t > 0$) is given by $E = 15\cos(3\pi t)$. Find, to the nearest 0.01 second, the first five times that $E = 13$.

OBJECTIVE K **Find the shortest distance between two points on Earth given their longitude and latitude. (Lesson 5-10)**

In 69–71, assume Earth is a sphere with radius 3960 mi.

69. Jackson, Mississippi (90°12'W, 32°22'N), and St. Louis, Missouri (90°12'W, 38°35'N), are on the same meridian. Estimate the distance from Jackson to St. Louis.

70. Find the shortest distance between Ankara, Turkey (32°55'W, 39°55'N), and Beijing, China (116°25'W, 39°55'N).

71. Find the shortest distance between Chicago, Illinois (87°39'W, 41°51'N) and Sydney, Australia (151°13'E, 33°51'S).

REPRESENTATIONS Pictures, graphs, or objects that illustrate concepts

OBJECTIVE L **Graph or identify graphs of inverse trigonometric functions. (Lessons 5-2, 5-4, 5-6)**

72. Graph $y = \cos x$ and $y = \cos^{-1} x$ for $0 \le x \le \pi$ on the same set of axes.

73. **Multiple Choice** Which equation is graphed at the right?

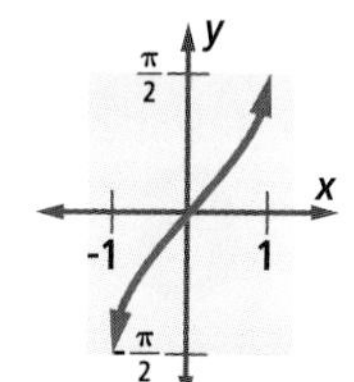

 A $y = \tan^{-1} x$ **B** $y = \cos^{-1} x$

 C $y = \sin x$ **D** $y = \sin^{-1} x$

74. Sarah graphed $y = \cos^{-1} x$ on her calculator in degree mode. Her screen looked like the one drawn below.

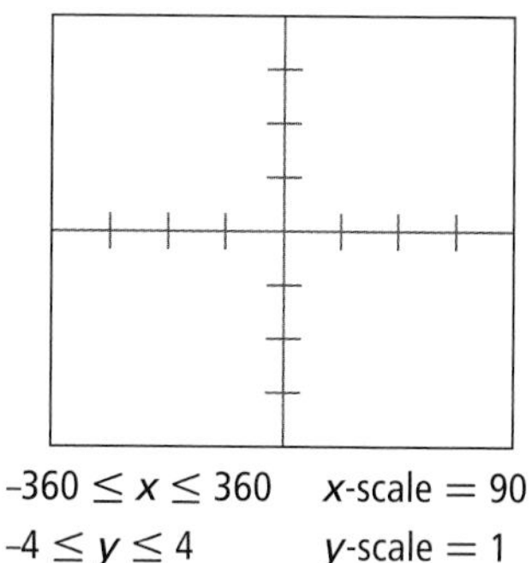

$-360 \le x \le 360$ x-scale $= 90$

$-4 \le y \le 4$ y-scale $= 1$

 a. What should Sarah do to display the graph?

 b. What would the display look like?

OBJECTIVE M **Graph parametric equations of circles and ellipses. (Lesson 5-8)**

75. Graph the parametric equations $\begin{cases} x = 3\cos t \\ y = 5\sin t \end{cases}$.

76. Write parametric equations for the circles graphed below.

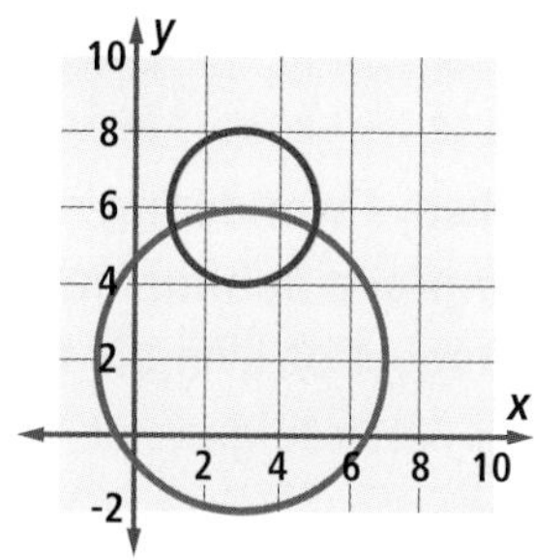

77. The unit circle is translated 7 units to the right and 5 units down.

 a. Write an equation in rectangular form for the image circle.

 b. Write parametric equations for the original circle and its image.

78. Find parametric equations for the circle with standard equation $(x + 3)^2 + (y - 6)^2 = 81$.

Chapter

6 Counting, Probability, and Inference

Contents

People are always wondering whether one way to improve health, learning, or some other aspect of living is better than another. Is it better to exercise daily, or is every other day enough? Does eating a good breakfast help a person score higher on a test? Does taking a particular vitamin lower your risk of getting a particular disease?

In these kinds of situations, the treatment in question rarely works all the time. It will help some people but not everyone. Those people who benefit will say that their treatment worked, while others will say that another method worked. Which should you choose?

This is a typical problem in which *inferential statistics* plays an important role. Inferential statistics works from real data to determine whether events are different from what might be expected randomly.

One important historical example of the use of inferential statistics is the research and development process that led to the first polio vaccine. In the first half of the 20th century, polio crippled large numbers of people annually in the United States. The president at the time, Franklin D. Roosevelt, survived an attack of polio in 1921 that left him permanently paralyzed from the waist down. He set a national policy to fight the disease.

Scientists knew that a virus caused polio, but they had no cure. In the early 1950s, Jonas Salk of the University of Pittsburgh developed a vaccine that seemed to prevent polio. But how could one know? There were tens of thousands of polio cases, but they represented only a small percentage of the population. The study to show the vaccine worked had to be quite large. The table below summarizes the data that Salk and his team collected. All the children were 6–9 years old.

From these data, would you conclude that the vaccine was successful? The scientists did, and despite vocal criticism, the Salk vaccine was put into widespread production. After some false starts, polio vaccinations became standard for children in the U.S., and after a few years, the more effective Sabin vaccine replaced Salk's. The number of polio cases declined precipitously, and today polio has been nearly eradicated in the U.S. and many other countries world wide.

The analysis of this and other similar experiments is based on randomization. In this chapter, you will study the concepts that underlie this kind of analysis and work with an inferential statistic that allows you to draw conclusions from certain kinds of data.

	Total	Developed Paralytic Polio	Did Not Develop Paralytic Polio
Vaccinated Children	200,745	33	200,712
Unvaccinated Children	201,229	115	201,114

Source: Journal of the American Statstical Association, December 1965

Lesson

6-1 Introduction to Probability

Vocabulary

probability theory
experiment
outcome
sample space
event
probability of an event, $P(E)$
fair, unbiased
randomly, at random
empty set, null set

▸ **BIG IDEA** When outcomes of a situation are equally likely, the probability of an event is the percent or fraction of outcomes that fit the event.

Probability theory is the branch of mathematics that studies situations in which there is an element of chance. In this lesson we review the basic ideas of probability you have likely seen in previous courses.

The Outcomes and Sample Space of an Experiment

A situation with several possible results is called an **experiment.** Probabilities are a measure of how relatively often the results occur. Each result of an experiment is called an **outcome**. The set of all possible outcomes of an experiment is the **sample space** for the experiment. Here are a few experiments and their sample spaces.

Experiment	Sample Space
flipping a coin	{heads, tails}
tossing a six-sided die	{1, 2, 3, 4, 5 6}
taking an antibiotic for a sore throat	{sore throat cured, sore throat continues}
picking an integer from 1 to 100	$\{n \in \mathbb{Z} \mid 1 \leq n \leq 100\}$

Notice that an experiment does not have to take place in a laboratory!

Mental Math

At the intersection of Large Avenue and Busy Street, the lights are timed as follows.

Large Ave	Busy Street	Time (s)
Green	Red	20
Yellow	Red	5
Red	Green	30
Red	Yellow	5

a. For a driver on Busy Street, what fraction of the time is the light red?

b. For a driver on Large Avenue, what percent of the time is the light green?

c. If you are a person who drives every day on Large Avenue through this intersection, what are the chances of the light being green today and tomorrow?

Example 1

Two six-sided dice, one red and one green, are thrown, and both numbers are recorded.

a. List all possible outcomes for the experiment.

b. How many outcomes are in the sample space for this experiment?

Solution

a. Each one of the six possible outcomes for the red die can occur with any one of the six possible outcomes for the green die. The outcomes are pictured on the next page. Instead of drawing the dice, you might list the outcomes using ordered pairs (red, green), from right to left, top to bottom: (1, 1), (2, 1), (3, 1), (4, 1), (5, 1), (6, 1), (1, 2), (2, 2), and so on, until (6, 6).

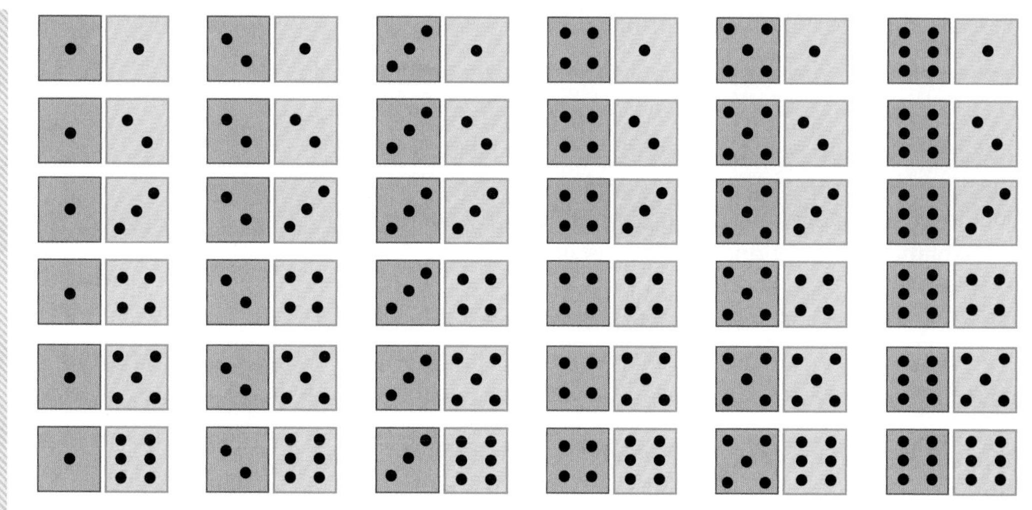

b. The list and picture show that there are 36 outcomes in the sample space.

 QY1

▶ QY1

A person tosses a penny, nickel, and dime and records if each is heads or tails. List all possible outcomes.

Events and Probabilities

When two dice are thrown, people are often concerned with the sum of the dots that show on the dice. For example, they may want to get a sum of 7. This is an *event*. In general, an **event** is any subset of the sample space of an experiment. Here are some possible events in the experiment of tossing the two dice from Example 1.

Event description	Outcomes in the event
tossing "doubles"	(1, 1), (2, 2), (3, 3), (4, 4), (5, 5), (6,6)
tossing a sum of 10	(4, 6), (5, 5), (6, 4)
tossing a 3 on the red die	(3, 1), (3, 2), (3, 3), (3, 4) (3, 5), (3, 6)
tossing a sum of 1	none

 QY2

▶ QY2

In the experiment of tossing two six-sided dice from Example 1, list the outcomes in the event "tossing a difference of 4".

The ***probability of an event*** is a number from 0 to 1 that measures the likelihood, or chance, that the event will occur. Probabilities may be written as fractions, decimals, percents, or relative frequencies. An event that is expected to happen one-quarter of the time has a probability that may be written as $\frac{1}{4}$, $\frac{2}{8}$, 0.25, 25%, or any other way equivalent to these.

Let $N(E)$ be the number of outcomes in an event E, and $N(S)$ be the number of outcomes in the sample space S. For the experiment "tossing two dice," $N(S) = 36$. If E_1 = "tossing a sum less than 4," then $E_1 = \{(1, 1), (1, 2), (2, 1)\}$ and $N(E_1) = 3$. So if all outcomes in S were *equally likely*, you would expect to roll a sum less than 4 $\frac{3}{36}$ or $\frac{1}{12}$ of the time. "Equally likely" means each outcome has an equal chance of happening.

This idea is generalized in the definition of probability.

Definition of Probability of an Event

Let E be an event in a finite sample space S. Let $N(E)$ and $N(S)$ be the numbers of elements in E and S, respectively. If each outcome in S is equally likely, then the probability that E occurs, called the **probability of *E*** and denoted ***P*(*E*)**, is given by

$$P(E) = \frac{N(E)}{N(S)} = \frac{\text{number of outcomes in the event}}{\text{number of outcomes in the sample space}}.$$

QY3

When a sample space has n outcomes all of which are equally likely, then the experiment is called **fair** or **unbiased** and each outcome has probability $\frac{1}{n}$. The outcomes are said to occur **randomly** or **at random**.

> **QY3**
>
> In the tossing of two fair dice from Example 1, what is the probability of a sum of 4?

GUIDED

Example 2

An experiment consists of tossing two fair coins and counting the number of heads. Consider the events 1 head in all, 2 heads in all, and 0 heads in all. Are these events equally likely?

Solution Begin by listing all possible outcomes using a tree diagram. Let H = heads, T = tails.

First Coin Toss	Second Coin Toss	Outcome
H	H	H H
	T	?
T	H	?
	T	?

The sample space is {_?_, _?_, _?_, _?_}.

There is/are _?_ outcome(s) in the event 0 heads, _?_ outcome(s) in the event 1 head, and _?_ outcome(s) in the event 2 heads. These events _?_ (are/are not) equally likely.

Caution: You must know or be told that outcomes are equally likely before you can apply the definition of probability.

GUIDED

Example 3

A researcher is studying the number of boys and girls in families with three children. Assume that the birth of a boy or a girl is equally likely.

a. List the sample space.

b. Find the probability that a family of three children has exactly one boy.

Solution Let B represent a boy, G represent a girl, and a triple of letters stand for the genders of the children from oldest to youngest.

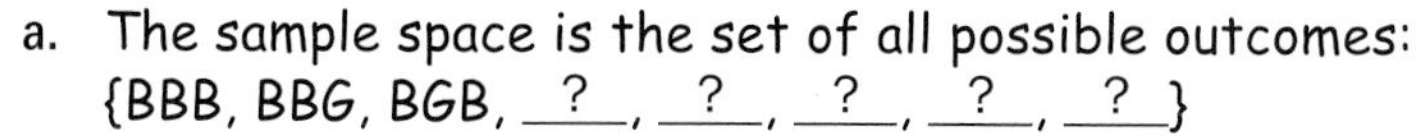

a. The sample space is the set of all possible outcomes:
{BBB, BBG, BGB, __?__, __?__, __?__, __?__, __?__}

b. There are __?__ outcomes in the sample space. __?__ outcomes are in the event "exactly one boy." So P(one boy) $= \frac{?}{?}$.

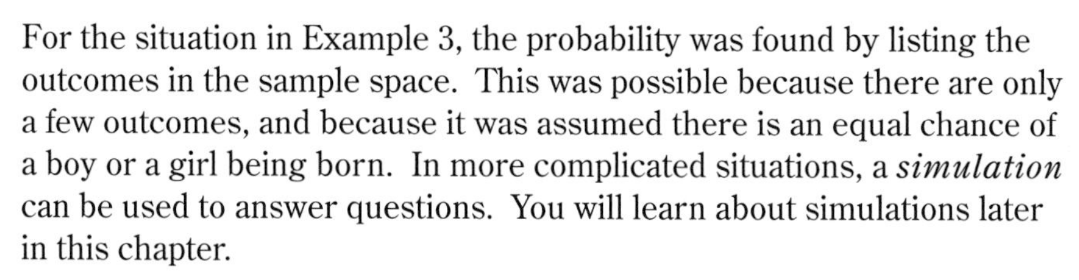

For the situation in Example 3, the probability was found by listing the outcomes in the sample space. This was possible because there are only a few outcomes, and because it was assumed there is an equal chance of a boy or a girl being born. In more complicated situations, a *simulation* can be used to answer questions. You will learn about simulations later in this chapter.

If an event contains no possible outcomes, then it cannot occur and it has probability 0. For instance, in Example 3, the event "four boys" has probability 0. Recall from earlier courses that a set with no elements is called the **empty set** or the **null set** and is denoted either by the symbol { } or Ø. With this notation, we write $P(\{\ \}) = P(\emptyset) = 0$. If an event is certain to happen, then it contains all the possible outcomes in the sample space, and has probability 1. That is, $N(E) = N(S)$, so $P(E) = \frac{N(E)}{N(S)} = 1$. These properties are summarized in the theorem below.

Theorem (Basic Properties of Probability)

Let S be the sample space associated with an experiment. Then, for any outcome or event E in S,

(i) $0 \le P(E) \le 1$.

(ii) if $E = S$, then $P(E) = 1$.

(iii) if $E = \emptyset$, then $P(E) = 0$.

Relative Frequencies and Probabilities

When tossing a coin, you might *assume* that heads and tails are equally likely. From this, it follows that the probability of a head is $\frac{1}{2}$ and the probability of a tail is $\frac{1}{2}$. In 2500 tosses of the coin, you would expect about 1250 heads. If you have some reason to expect that a particular coin is biased, you might toss the coin a large number of times and take the long-term *relative frequency* of heads as an estimate of the probability.

For instance, if 1205 heads occur in 2500 tosses, then the relative frequency of heads is $\frac{1205}{2500} = 48.2\%$. Then you are faced with a decision: Is the coin a fair coin and did the 1205 heads just occur by random chance, or is the coin slightly biased towards heads? Later in this chapter, you will see some mathematics that can help you decide.

While relative frequencies and probabilities are related, the meanings of the two values differ. The values of both range from 0 to 1. However, a relative frequency of 0 means an event *has not occurred*; for example, the relative frequency of snow on July 4th in Florida since 1900 is 0. In contrast, a probability of 0 in a finite sample space means the event *cannot occur.* For example, the probability of tossing a sum of 17 with two fair six-sided dice is 0. We say that the outcome "tossing a sum of 17" is not in the sample space.

Now consider a value of 1. A relative frequency is 1 when an event *has occurred* in each known trial. A probability is 1 when an event *must always occur.* The probability is 1 that the toss of a pair of dice gives a number less than 13. Therefore if something has a probability of 1, then the relative frequency of the event will also be 1. However, if a relative frequency is 1, this does not guarantee that the probability is 1. The same holds for 0.

Questions

COVERING THE IDEAS

1. A person guesses randomly on an "Always, Sometimes, or Never" question. The correct answer to the question is "Always."
   a. List the sample space.
   b. Describe the event "getting a wrong answer".
   c. What is N("getting a wrong answer")?
   d. What is P("getting a wrong answer")?

In 2–5, two fair dice are rolled as in Example 1.
a. List the outcomes of the named event.
b. Give the probability of the event.

2. The sum of the numbers shown is 9.
3. The absolute value of the difference of the numbers shown is 2.
4. The sum of the numbers shown is either even or odd.
5. The product of the numbers shown is 11.
6. An egg is dropped onto a cement floor from a height of two feet. You wonder whether it will break or not.
   a. List all the outcomes of the experiment.
   b. Are the outcomes equally likely? Explain why or why not.

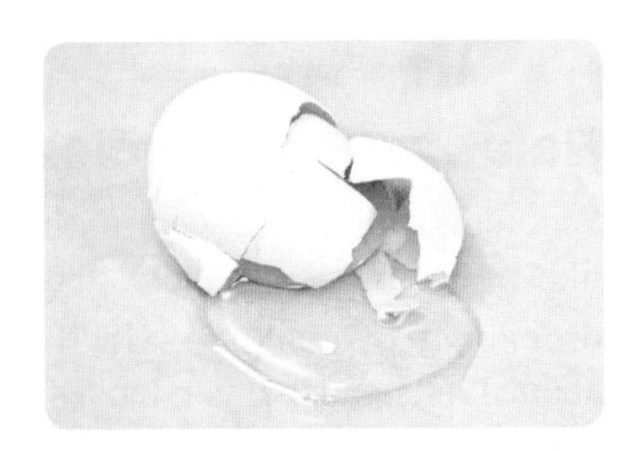

Oops!

7. Two fair coins are flipped as in Example 2.
   a. List the outcomes of the event "At least one tail appears."
   b. Give the probability of the event "At least one tail appears."

8. A closet contains w white shirts, b black shirts, and g grey shirts. One shirt is chosen at random.
   a. What is the probability a grey shirt is chosen?
   b. What is the probability a black shirt is chosen?
   c. What is the probability a white shirt is not chosen?

9. A family has four children.
   a. List the sample space for the genders of the children, using B for boy and G for girl.
   b. What is the probability that the family has 3 boys and 1 girl?

APPLYING THE MATHEMATICS

10. Three fair coins are flipped.
    a. List the outcomes of the events "0 heads," "1 head," "2 heads," and "3 heads."
    b. Give the probability of each event.

11. Suppose n fair coins are flipped.
    a. Make a table showing the probability of getting all heads for $n = 1, 2, 3$, and 4.
    b. Based on your table, predict the probability of getting all heads for $n = 5$.

12. Which of (1) or (2) is possible, and why?
    (1) The relative frequency of an event is 0, but the probability of the event is not 0.
    (2) The probability of an event is 0, but the relative frequency of the event is not 0.

13. Let A be an event. If $P(A) = 0.9$ and there are 460 equally likely outcomes in the sample space, how many outcomes are in A?

14. A whole number between 1 and 366 (inclusive) is chosen at random. Find the probability of each event.
    a. The number is divisible by 5.
    b. The number is divisible by 6.

REVIEW

15. Consider the equation $\cos \theta = -\frac{1}{2}$. **(Lesson 4-4)**
    a. Draw a unit circle and mark the two points $(\cos \theta, \sin \theta)$ for which $\cos \theta = -\frac{1}{2}$.
    b. Give two values of θ between 540° and 900° that make the equation true.
    c. Give a negative value of θ that makes the equation true.

16. The table at the right gives the names of fourteen models of cars from the 2007 model year, their engine sizes measured in liters, and their fuel economy in miles per gallon of gas. **(Lesson 2-3)**

a. Make a scatterplot of the data, using engine size as the independent variable.

b. Describe the direction of the association between fuel economy and engine size.

c. Find the equation of the least squares regression line.

d. What is the value of the slope of your regression line? What are its units?

e. Explain how the direction of the association is indicated in the value of the slope.

f. What is the correlation coefficient for the model? What does it tell you about the relationship between engine size and fuel economy?

Car Model	Engine Size (liters)	Fuel Economy (M.P.G.)
Acura RL	3.5	19
Audi S8	5.2	15
Bentley Azure	6.7	11
BMW M6	5	13
Buick Lucerne	3.8	20
Chevrolet Optra5	2	23
Chrysler 300A WD	3.5	18
Dodge Caliber	1.8	26
Ford Taurus	3	20
Honda Accord	2.4	26
Hyundai Elantra	2	28
Kia Optima	2.4	25
Saturn Aura	3.5	21
Toyota Camry	2.4	25

Source: fueleconomy.gov

17. An algebra teacher has 15 students who took a test in 2nd hour and 25 students each in 3rd and 7th hour who took the test. The following table summarizes the test results. **(Lesson 1-2)**

	2nd hour	3rd hour	7th hour
$\overline{x}$	72	80	83

a. What is the mean of the scores in all three classes combined?

b. A student in 3rd hour missed the quiz because of a field trip. When his score was averaged in, the mean for this class fell to 79.2. What was his score?

18. x% of the people who make reservations at a restaurant show up. What percent does not show up? **(Previous Course)**

19. **Fill in the Blanks** with "union" or "intersection."

a. An angle is the ___?___ of two rays with the same endpoint.

b. The set of all triangles is the ___?___ of the set of isosceles triangles and the set of scalene triangles.

c. The set of multiples of 35 is the ___?___ of the set of multiples of 7 and the set of multiples of 5. **(Previous Course)**

20. If $A = \{\text{bacon, eggs, cheese}\}$ and $B = \{\text{eggs, cheese, ham}\}$,

a. What is $A \cap B$?

b. What is $A \cup B$? **(Previous Course)**

EXPLORATION

21. The digits of the decimal expansion of π have been conjectured as constituting a set of random digits in the sense that each digit might occur with probability 0.1; each two-digit pair might occur with probability 0.01, etc. A number with this property is called a *normal number*. Find and summarize an article about this possible property of π or other irrational numbers.

QY ANSWERS

1. HHH, HHT, HTH, HTT, THH, THT, TTH, TTT

2. (5, 1); (6, 2); (1, 5); (2, 6)

3. $P(\text{sum of 4}) = \frac{4}{36} = \frac{1}{12}$

Lesson 6-2 Principles of Probability

Vocabulary

union of sets
disjoint sets
mutually exclusive sets
intersection of sets
complementary events
complement of *A*, not *A*

BIG IDEA The fundamental principles of probability are derived from basic principles of counting.

Because probabilities are often calculated by dividing one count by another, effective counting is essential to probability.

Mental Math

State as a simple fraction.

a. 2^{-5}

b. 5^{-2}

c. $3 \cdot 10^{-4}$

d. $\left(\frac{1}{2}\right)^{-6}$

The Addition Counting Principle

The Addition Counting Principle is the most basic principle of counting. Addition finds the number of elements in the *union* of two *disjoint* sets. $A \cup B$, the **union of sets** A and B, contains all elements that are either in A or in B. If A and B have no elements in common, they are called **disjoint sets** or **mutually exclusive sets**.

Addition Counting Principle (Mutually Exclusive Form)

If two finite sets A and B are mutually exclusive, then $N(A \cup B) = N(A) + N(B)$.

Activity 1

MATERIALS a pair of fair six-sided dice

Step 1 Toss a pair of fair six-sided dice 20 times, recording the sum of face-up numbers each time.

Step 2 Make a frequency table of each sum using the combined data for your class.

Step 3 a. What was the class's relative frequency of having a sum of 7?
b. What was the class's relative frequency of having a sum of 11?
c. What was the class's relative frequency of having a sum of 7 or 11?

Step 4 Which of the events in Step 3 are mutually exclusive?

In the Activity, you computed relative frequencies of several events as well as the union of mutually exclusive events. This idea extends to probability in the following way. Think of A and B as events in a finite sample space S. Divide both sides of the equation in the Addition Counting Principle by $N(S)$.

$$\frac{N(A \cup B)}{N(S)} = \frac{N(A)}{N(S)} + \frac{N(B)}{N(S)}$$

The fractions all stand for probabilities, identified in the next theorem.

Theorem (Probability of the Union of Mutually Exclusive Events)

If A and B are mutually exclusive events in the same finite sample space, then $P(A \cup B) = P(A) + P(B)$.

Activity 2

Step 1 Write the possible sums of the face-up numbers when two fair six-sided dice are tossed. Which sums are prime?

Step 2 Let each possible sum be an event. Find the probability of each event.

Step 3 Find the probability of each of the prime number events. Add them to find the probability that, when two fair six-sided dice are tossed, the sum of the face-up numbers is a prime number.

Overlapping Events

When two events are not mutually exclusive, then they have outcomes in common and are said to *overlap*. The set of overlapping events is the set of elements in both A and B, called the **intersection** of the sets A and B, denoted by $A \cap B$. The table below summarizes the language of set intersections.

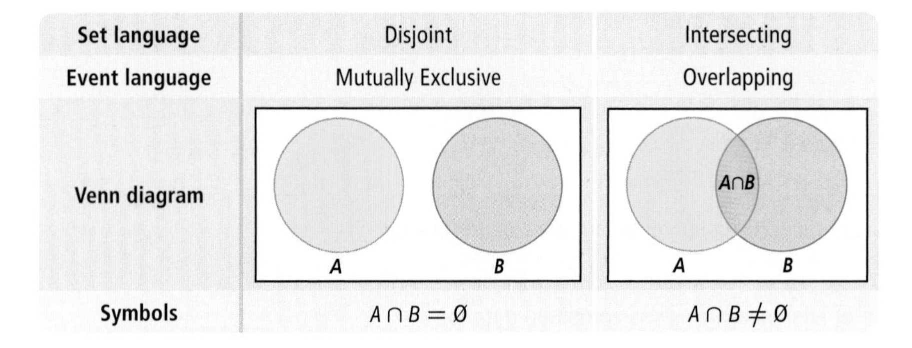

Set language	Disjoint	Intersecting
Event language	Mutually Exclusive	Overlapping
Venn diagram	A B	A∩B A B
Symbols	$A \cap B = \emptyset$	$A \cap B \neq \emptyset$

Example 1

Suppose at a high school, 298 students study only French, only Spanish, or both languages. The school reports 115 students study French and 209 study Spanish, but because 115 + 209 > 298, there must be students who study both languages. How many students study both?

Solution Let F be the set of French students and S be the set of Spanish students. Then $N(F) = 115$, $N(S) = 209$, and $N(F \cup S) = 298$. Let x be the number of students who study both languages. That is, $x = N(F \cap S)$. Then the number of students who only study French is $115 - x$ and the number of students who only study Spanish is $209 - x$. The Venn diagram at the right shows this.

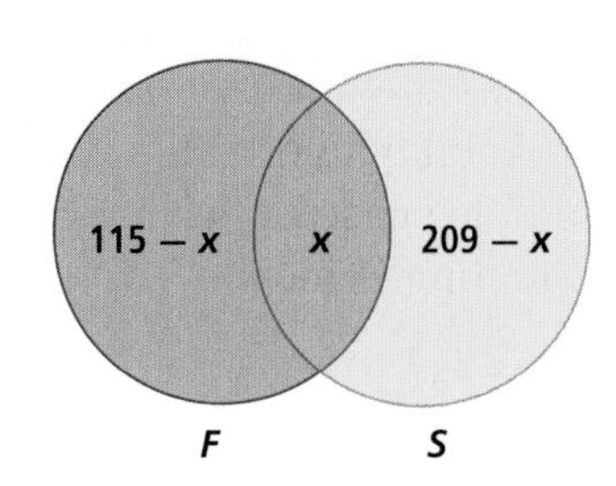

Studying only French, only Spanish, or both languages are mutually exclusive events. By the Addition Counting Principle,

$$N(F \cup S) = (115 - x) + x + (209 - x) = 324 - x.$$

Thus $N(F \cup S)$ can be found by subtracting the overlap from the total.

So $\quad N(F \cup S) = 298 = 324 - x.$

Solve the equation. $\quad x = 26$

Thus, there are 26 students studying both languages.

Check

$115 - 26 = 89$ students are studying only French, and $209 - 26 = 183$ are studying only Spanish. $89 + 183 + 26 = 298$, so it checks.

The flag of Spain

STOP **QY**

QY

What is the probability that a randomly selected student in the high school of Example 1 studies both French and Spanish?

This situation is a special case of a more general result.

Addition Counting Principle (General Form)

For any finite sets A and B, $N(A \cup B) = N(A) + N(B) - N(A \cap B)$.

Proof If $N(A \cap B) = x$, then the number of elements in A which are not in the intersection is $N(A) - x$. Similarly, there are $N(B) - x$ elements in B which are not in the intersection, as shown at the right.

Then $N(A \cup B) = (N(A) - x) + x + (N(B) - x)$

$= N(A) + N(B) - x$

$= N(A) + N(B) - N(A \cap B)$ because $N(A \cap B) = x$.

A B

$N(A) - x$ $\quad x \quad$ $N(B) - x$

$N(A \cup B)$

In Question 11, you are asked to show how the general form of the Addition Counting Principle can be used to derive the theorem below.

Theorem (Probability of the Union of Events, General Form)

If A and B are any events in the same finite sample space, then $P(A \text{ or } B) = P(A \cup B) = P(A) + P(B) - P(A \cap B)$.

Example 2

A pair of six-sided dice is thrown. If the dice are fair, what is the probability that the dice show doubles or a sum less than 10?

Solution Refer to the dice diagram in Example 1 of Lesson 6-1. Find the number of doubles. There are 6 doubles, so $P(\text{doubles}) = \frac{6}{36}$.

(continued on next page)

Find the number of sums less than 10. There are 30 pairs that sum less than 10, so P(sum under 10) = $\frac{30}{36}$. Find the number of doubles that sum less than 10. There are 4 doubles that sum less than 10, so P(doubles and sum under 10) = $\frac{4}{36}$. Use the Probability of the Union of Events theorem.

P(double or sum under 10)

= P(doubles) + P(sum under 10) − P(doubles and sum under 10)

$= \frac{6}{36} + \frac{30}{36} - \frac{4}{36} = \frac{32}{36}$ or about 88.9%.

GUIDED

Example 3

A pair of fair six-sided dice is thrown. What is the probability that exactly one die shows a 3 or the sum of the numbers is greater than 9?

Solution Again, refer to the dice diagram in Example 1 of Lesson 6-1.

Count outcomes to find *P*(one die shows 3), *P*(sum > 9), and *P*(one die shows 3 and sum > 9). P(one die shows 3) = $\frac{?}{36}$, P(sum > 9) = $\frac{?}{36}$, and P(one die shows 3 and sum > 9) = $\frac{?}{36}$.

Use the Probability of the Union of Events theorem.

P(one die shows 3 or sum > 9) = P(one die shows 3) + ___?___ − ___?___

$= \frac{?}{36} + \frac{6}{36} - \frac{?}{36} = \frac{?}{36}$.

Example 3 shows that the Probability of the Union of Mutually Exclusive Events Theorem is a special case of the Probability of the Union of Events Theorem. When A and B are mutually exclusive events, $A \cap B = \emptyset$ and so $P(A \cap B) = 0$. Then,

$$P(A \cup B) = P(A) + P(B) - P(A \cap B)$$

reduces to $$P(A \cup B) = P(A) + P(B).$$

Complementary Events

Sometimes events are mutually exclusive and their union is the entire sample space. Such events are called **complementary events.** The **complement of an event *A*** is called **not *A*.** Here are examples of events and their complements.

Experiment	Sample Space	Event	Complement
tossing a coin	{heads, tails}	{tails}	{heads}
tossing two coins	{HH, HT, TH, TT}	getting no heads {TT}	getting 1 or 2 heads {HH, HT, TH}
picking an integer from 1 to 100	$\{n \in \mathbb{Z} : 1 \leq n \leq 100\}$	picking a prime number	picking 1 or a composite number

By the definition of complementary events, the following theorem holds.

Theorem (Probability of Complements)

If A is any event, then $P(\text{not } A) = 1 - P(A)$.

Example 4

Refer to the 298 students studying languages in Example 1. If a student is selected at random, what is the probability that he is not studying both languages at the same time?

Solution 1 Let event B = studying both languages, and event not B = not studying both languages. We know there are 26 students studying both languages, so $N(B) = 26$ and $P(B) = \frac{26}{298}$.

Apply the Probability of Complements Theorem.

$P(\text{not } B) = 1 - P(B) = 1 - \frac{26}{298} = \frac{272}{298}$ or about 91.3%

Check $P(B) + P(\text{not } B) = \frac{26}{298} + \frac{272}{298} = \frac{298}{298} = 1$

Solution 2 Use the symbols from Solution 1. Since $N(B) = 26$ and $N(S) = 298$, $N(\text{not } B) = 298 - 26 = 272$.

Thus $P(\text{not } B) = \frac{N(\text{not } B)}{N(S)} = \frac{272}{298}$.

Questions

COVERING THE IDEAS

In 1–3, suppose two fair six-sided dice are rolled.

a. List the outcomes of the named event.

b. Give the probability of the event.

1. The sum is a multiple of 3.
2. Doubles or a sum over 7
3. The sum is a multiple of 3, but not a multiple of 2.
4. Consider the experiment of tossing 3 fair coins.
   a. Give a sample space for the experiment.
   b. Find $P(2 \text{ tails})$.
   c. Find $P(\text{at least } 2 \text{ tails})$.
   d. **True or False** $P(3 \text{ tails}) = P(0 \text{ tails})$
5. Assume that events A and B are in the same sample space. Under what conditions is $P(A \cup B) = P(A) + P(B)$ true?
6. Suppose two fair six-sided dice are rolled. Let X = the first die shows 1 and Y = the sum of the numbers on the dice is 5.
   a. Find $N(X \cup Y)$.
   b. Find $N(X \cap Y)$.
   c. State whether X and Y are mutually exclusive.

7. **True or False** Complementary events are a special case of mutually exclusive events. Explain your answer.

8. Explain why, for any event E, $P(\text{not } E) = 1 - P(E)$.

9. It is estimated that about 10% of males in the U.S. are left-handed. Identify a complement of this event and find its probability.

Six out of the last 12 U.S. presidents, from Truman through Obama, have been left-handed.

APPLYING THE MATHEMATICS

10. Suppose the probability that the next test will be on a Tuesday is 0.35 and the probability that it will be on a Thursday is 0.25.
    a. What is the probability that the next test is on a Tuesday or Thursday?
    b. What is the probability that it is not on a Tuesday?
    c. What is the probability that the test is on neither Tuesday nor Thursday?

11. Use the General Form of the Addition Counting Principle to derive the General Form of the Probability of the Union of Events Theorem.

In 12–15, consider tossing two fair six-sided dice and dividing the number on the first die by the number on the second to get a quotient *q*. Find each probability.

12. $P(q < 1)$

13. $P(q \leq 1)$

14. $P(q = \frac{1}{2} \text{ or } q = \frac{1}{3})$

15. $P(q \text{ is an integer})$

16. All human blood can be typed as one of O, A, B, or AB, and is either Rh positive or Rh negative. Refer to the table below.
    a. What is probability that a person is Rh negative?
    b. What is the proportion of Rh negative within each blood type?
    c. Which of the four blood types has the highest proportion of Rh negative?

ABO and Rh Blood Type Relative Frequencies in the United States

ABO Type	Rh Type	Relative Frequency
O	positive	38%
O	negative	7%
A	positive	34%
A	negative	6%
B	positive	9%
B	negative	2%
AB	positive	3%
AB	negative	1%

Source: American Association of Blood Banks

In 17 and 18, refer to the Venn diagram at the right.

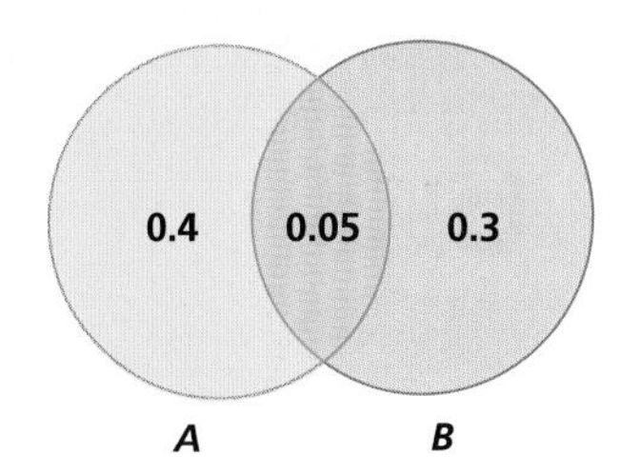

17. Are A and B mutually exclusive? Why or why not?

18. Calculate the following probabilities.
 a. $P(A \cap B)$
 b. $P(A)$
 c. $P(A \cup B)$
 d. $P(\text{not } B)$

REVIEW

19. In a game of chance, after rolling one fair six-sided die and flipping one fair coin, a player is said to win if heads and a prime number appear or if tails and the number 5 appear. **(Lesson 6-1)**
 a. Give the sample space for this experiment.
 b. Find $N(\text{loss})$ and $P(\text{loss})$
 c. Name an event that has three outcomes in this sample space.

20. If an experiment has only three outcomes X, Y, and Z, and $P(X) = a$ and $P(Y) = b$, what is $P(Z)$? **(Lesson 6-1)**

21. When $0 < x < 2\pi$, $\sin x = -\frac{48}{73}$, and $\cos x = -\frac{55}{73}$, find
 a. $\tan x$.
 b. $\sec x$.
 c. $\csc x$.
 d. $\cot x$.
 e. the value of x to the nearest hundredth. **(Lessons 5-9, 4-2)**

22. Give the amplitude, period, vertical shift, and phase shift of the graph of $y = 3 \cos\left(\frac{\pi - x}{6}\right)$. **(Lesson 4-9)**

23. During one day at a gas station, 60% of the gasoline sold was regular at \$3.25 per gallon; 30% was premium at \$3.55 per gallon; and 10% was diesel at \$3.70 per gallon. What was the weighted average cost per gallon for the day? **(Lesson 1-2)**

EXPLORATION

24. a. Draw a Venn diagram for $A \cup B \cup C$, where A, B, and C overlap.
 b. Extend the Probability of a Union of Events Theorem to cover any three events in the same sample space. That is, give a formula for $P(A \cup B \cup C)$.
 c. Give an example of the use of the formula you found in Part b.

QY ANSWER

$\frac{26}{298} \approx 8.7\%$

Lesson 6-3 Counting Strings with Replacement

Vocabulary

string
length of a string
independent events
dependent events

BIG IDEA If two events are independent, the probability that both events occur is the product of the probabilities of the events.

The Multiplication Counting Principle

The Blu Yonder travel company offers package vacations with a choice of economy class or business class flights and three options for accommodation (3-star hotel, 4-star hotel or 5-star hotel). The tree diagram below shows the different possibilities.

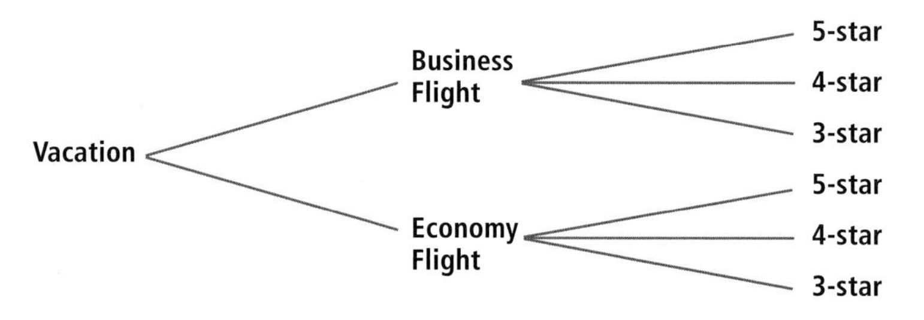

Counting the paths shows that there are six different possible options. Blu Yonder felt that there were not enough options. To try to increase sales, they decided to advertise that each vacation comes in one of five themes (adventure, sports, beaches, shopping, or sights). They knew that for each of the 6 possible vacations, there were 5 possible themes. The following tree diagram shows the 5 theme options for a vacation with a business class flight and 4-star accommodations.

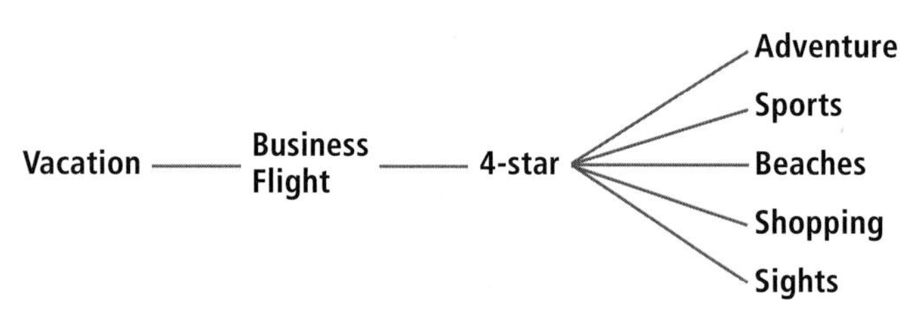

If we consider each possible vacation, there are $6 \cdot 5 = 30$ total choices. This is an instance of a basic use of multiplication.

A ski resort in Savoy, France.

Multiplication Counting Principle

Let A and B be any finite sets. The number of ways to choose one element from A and then one element from B is $N(A) \cdot N(B)$.

Mental Math

Suppose that $f(xy) = f(x) + f(y)$ and that $f(2) = 0.472$ and $f(3) = 0.295$.

a. Find $f(6)$.

b. Find $f(12)$.

The Multiplication Counting Principle extends to choices made from more than two sets. In the travel situation, we chose from three sets: flight classes, hotel accommodations, and themes. There were 2 flight classes, 3 hotels, and 5 themes, and the total number of possible vacations was $2 \cdot 3 \cdot 5 = 30$. In general, the number of ordered selections with one element from set A_1, one element from set A_2, ..., and one element from set A_k is $N(A_1) \cdot N(A_2) \cdot \dots \cdot N(A_k)$.

Ordered Symbols with Replacement

Here is a typical application of the Multiplication Counting Principle.

Example 1

A popular game show features a spinner divided into twenty-four congruent sectors and numbered something like the wheel shown at the right. The spinner cannot stop on a boundary line. You spin it twice. Describe a sample space *S* for this experiment, and determine the number of elements in *S*.

Solution There are two spins. Counting the sectors shows that each spin has twenty-four outcomes. Let x be the outcome of the first spin, and let y be the outcome of the second spin. So the sample space S consists of ordered pairs (x, y).

There are twenty-four choices for x and twenty-four for y. By the Multiplication Counting Principle, there are $24 \cdot 24 = 576$ elements in S.

In Example 1, we selected twice from the same set and found that there were 24^2 possible outcomes. If all outcomes are equally likely, the probability of each outcome is $\frac{1}{24^2}$.

 QY1

> **QY1**
>
> Suppose you spin the spinner in Example 1 three times.
>
> a. How many outcomes are there?
>
> b. What is the probability of each outcome?

Strings

When the symbols in a problem must be ordered, it is common to refer to the ordered list of symbols as a **string**. The number of symbols in a string is the **length** of the string.

Example 2

a. On a 28-question multiple-choice mathematics test, each question has 5 choices. How many possible completed answer sheets are there?

b. If you guess randomly on each question, what is the probability of answering all 28 questions correctly?

(*continued on next page*)

Solution

a. There are five ways to respond to each of the 28 questions. Think of each response as a choice from a set with 5 elements in it: a, b, c, d, or e. Let A_n be the set of possible responses from Question n. So $A_1 = A_2 = A_3 = \dots = A_{27} = A_{28} = \{a, b, c, d, e\}$.
The number of possible answer sheets is
$N(A_1) \cdot N(A_2) \cdot N(A_3) \cdot \dots \cdot N(A_{27}) \cdot N(A_{28})$
$= \underbrace{5 \cdot 5 \cdot 5 \cdot \dots \cdot 5 \cdot 5}_{28 \text{ factors}} = 5^{28}$
$= 37{,}252{,}902{,}984{,}619{,}140{,}625.$

b. When you guess randomly, all outcomes are equally likely. Only one of those 5^{28} outcomes is "all answers correct," so the probability of answering all questions correctly is $\frac{1}{5^{28}} = 5^{-28} \approx 0.0000000000000000000268$, which is very close to zero.

In Example 2, because there are 28 questions, each with 5 choices for an answer, finding the number of answer sheets depends on counting the number of strings of length 28. The strings are comprised of symbols from the set {a, b, c, d, e}. Each string corresponds to one possible answer sheet, assuming a student answers every question with one choice of answer. Three of the possible strings for answer sheets are listed below.

Answer Sheet Possibility: cabdcceabcdadcbbcdacdcabecab

Answer Sheet Possibility: bbbbbbbbbbbbbbbbbbbbbbbbbbbb

Answer Sheet Possibility: abcdabcdabcdabcdabcdabcdabcd

You saw in Example 2 that there are 5^{28} possible strings. These are *strings with replacement* because the symbols can be used over and over again.

Theorem (Strings with Replacement)

Let S be a set with n elements. Then there are n^k possible strings with replacement of length k with elements from S.

Proof Use the Multiplication Counting Principle. Here $N(S) = n$, so the number of possible ways to choose one element from S, each of k times, is

$$\underbrace{N(S) \cdot N(S) \cdot \dots \cdot N(S)}_{k \text{ factors}} = \underbrace{n \cdot n \cdot \dots \cdot n}_{k \text{ factors}} = n^k.$$

QY2

Example 3 uses both the Strings with Replacement Theorem and the Multiplication Counting Principle.

QY2

How many strings with replacement are there of length 2 with 3 symbols a, b, and c?

GUIDED

Example 3

In a certain state, license plates have two letters followed by 4 digits from 0 through 9. How many license plates are possible?

Solution First use the Strings with Replacement Theorem to compute the number of possibilities separately for letters and for numbers.

There are _?_ letters, so there are _?_ strings of 2 letters.

There are _?_ digits, so there are _?_ strings of 4 digits.

Now use the Multiplication Counting Principle to combine the counts for letters and numbers.

So there are _?_ • _?_ = 6,760,000 possible license plates.

QY3

> **QY3**
>
> Refer to Example 3. Suppose each license plate is equally likely and that your family has license plate QY 1234. What is the probability that the next car you see has license plate QY 1233 or QY 1235?

Independent Events

We think of events A and B as *independent* if the probability of A does not affect the probability of B. Selections with replacement are independent events because each selection is not affected by what has been selected previously. To find the probability that a number of independent events all occur, you can apply the Multiplication Counting Principle.

Example 4

The spinner shown here is used in a carnival game. It is assumed to be fair, so the spinner has the same probability of landing in each sector. If the spinner lands on an edge, the spin does not count. The game consists of two spins. You win if the first spin stops on an even number and the second spin stops on a multiple of 3. What is the probability of winning?

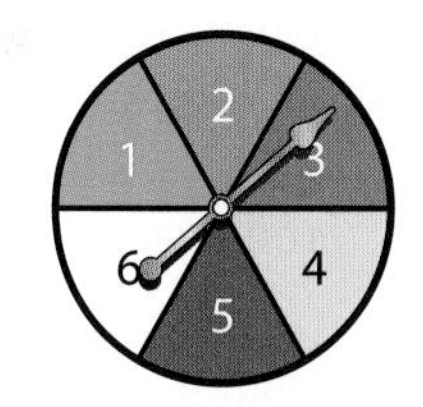

Solution The sample space S for spinning twice is illustrated at the right by the 36 dots. Let event A be that the first spin stops on an even number. Let event B be that the second spin stops on a multiple of 3. One outcome that satisfies both events is (2, 3). The event $A \cap B$ consists of the six circled ordered pairs in which the first component is even and the second component is a multiple of 3. The six circled ordered pairs represent those for which both A and B are true.

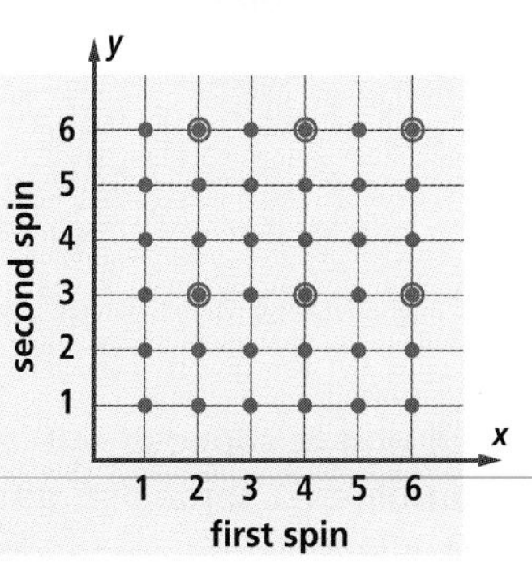

So $$P(A \cap B) = \frac{N(A \cap B)}{N(S)} = \frac{6}{36} = \frac{1}{6}.$$

In Example 4, notice that $P(A) = \frac{1}{2}$ and $P(B) = \frac{1}{3}$. Thus $P(A \cap B)$ is the product of $P(A)$ and $P(B)$. This is the defining characteristic of independent events.

Definition of Independent Events

Events A and B are **independent events** if and only if $P(A \cap B) = P(A) \cdot P(B)$.

Events that are not independent are called **dependent events**.

GUIDED

Example 5

Refer to Example 4. If event *B* were changed to be "the sum of both spins is greater than 8," show that the events *A* and *B* are dependent.

Solution The sample space and $P(A)$ are the same as in Example 4.

There are 10 outcomes in which the sum of both spins is greater than 8. Thus $N(B) =$ _?_ and $P(B) =$ _?_.

There are _?_ outcomes in which both the first spin is even and the sum of the spins is greater than 8. Thus $N(A \cap B) =$ _?_ and $P(A \cap B) =$ _?_. The events A and B are dependent because $P(A) \cdot P(B) = \frac{5}{36} \neq P(A \cap B)$.

Questions

COVERING THE IDEAS

1. The Blu Yonder travel company described on page 374 decides that in addition to themes, they will offer a choice of souvenir. Customers can pick a key chain, t-shirt, or canvas bag. How many different vacation packages are now offered?

2. a. How many different shirt-pants-sweater outfits are possible if you have six shirts, five pairs of pants, and two sweaters, assuming all items can go with each other?

   b. How many different shirt-pants-sweater outfits are possible if you have h shirts, ℓ pairs of pants, and w sweaters, assuming all items can go with each other?

3. Suppose a 5-character ID number consists of a letter followed by 4 digits from 0 through 9. How many ID numbers are possible?

In 4 and 5, suppose you have a spinner with five congruent sectors numbered 1 through 5 and a six-sided die, both fair. Decide whether the given events are independent or dependent. Justify your answer by computing appropriate probabilities.

4. $A =$ spinner shows 4; $B =$ sum of spinner and die is over 6

5. $A =$ spinner shows 5; $B =$ die and spinner show the same number

6. A test has 12 true-false questions.
   a. How many completed answer sheets with all questions answered are possible?
   b. If you guess on all the questions, what is the probability that you will get them all correct?
   c. If you know the answers to five questions and guess on the rest, what is the probability that you will get them all correct?

7. How many ways are there of answering a test with 6 true-false and 12 multiple-choice questions with 4 choices each?

8. A fair six-sided die is tossed 10 times, with results written in a string. The first trial gave a string with only 5s and 6s. What is the probability of the 10 tosses giving a string that has only numbers greater than 4?

APPLYING THE MATHEMATICS

9. An ice cream shop offers 34 flavors, 5 sauces, and 23 toppings. A sundae consists of one scoop of ice cream, a ladle of sauce, and a choice of one topping.
   a. How many different sundaes could be ordered?
   b. The shop offers 12 of its flavors in three varieties: regular, fat free, and sugar free. How many sundaes could be ordered?
   c. Suppose that a customer can omit the topping. How does that change the answer to Part a?

We all scream for ice cream! The average American consumes 48 pints of ice cream a year.

10. The teacher said there was about a one in a million chance of obtaining a perfect score by guessing on a test where all questions were true or false. Assuming the teacher was not exaggerating, about how many true-false questions were on this test?

11. A fair coin is tossed three times. Let event A be that at most one head occurs. Let event B be that both heads and tails occur at least once. Are A and B independent? Justify your answer using the definition of independence.

12. Two types of blood clotting disorders are ***thrombophilia*** (excessive blood clotting) and ***hemophilia*** (insufficient blood clotting). Suppose 51% of the age group under study is male. Assume that gender and thrombophilia are independent.
    a. One estimate is that 6.5% of the population has thrombophilia. What is the probability that a randomly selected person is male with thrombophilia?
    b. What is the probability that a person has thrombophilia or is male?
    c. Given that the relative frequency of hemophilia in the U.S. is about $\frac{1}{17{,}000}$ and the relative frequency of males with hemophilia is $\frac{1}{85{,}000}$, are these events independent? Justify your answer.

REVIEW

13. In Mrs. William's class of 35 students, 17 students came to school by bus today, and 13 came by car. The remaining students walked.
    a. What is the probability that a randomly selected student in Mrs. William's class walked to school today?
    b. What is the probability that a randomly selected student in Mrs. William's class who did not walk came by bus? **(Lesson 6-2)**

14. Give an example of two mutually exclusive events that are not complementary. **(Lesson 6-2)**

15. Suppose you roll two fair six-sided dice. What is the probability that the sum is a perfect square? **(Lesson 6-1)**

16. Consider the function f with $f(x) = x^2 - 1$. **(Lesson 3-5)**
    a. Sketch the graph of f under the transformation $S: (x, y) \rightarrow (3x, -2y)$.
    b. Give an equation for f under S.

17. Factor $8x^3 - 10x^2 - 3x$. **(Previous Course)**

In 18–21, recall that for any positive integer *n*, *n*!, read "*n* factorial," is the product of the integers from 1 to *n*. (Previous Course)

18. Give the values of 1!, 2!, 3!, 4!, and 5!.

19. Evaluate 10! with a calculator.

20. Evaluate $\frac{2010!}{2009!}$ without using a calculator.

21. Evaluate $\frac{12!}{3!9!}$.

EXPLORATION

22. License plates can have both letters and digits as seen in Example 3. Different states have different restrictions on the numbers of digits and letters per license plate. Research the rules and limitations of your state's license plates and determine the number of possible license plates in your state.

QY ANSWERS

1. **a.** $24^3 = 13824$
   b. $\frac{1}{24^3} = \frac{1}{13824} \approx 0.000072$
2. 9
3. $\frac{2}{26^2 \cdot 10^4 - 1} = \frac{2}{6760000 - 1}$

Lesson 6-4 Counting Strings without Replacement

Vocabulary

permutation

permutation of n objects taken r at a time, ${}_nP_r$

▶ **BIG IDEA** The number of permutations of length r from a set of n symbols can be determined by applying the Multiplication Counting Principle.

Mental Math

Suppose $x = 24$. Calculate:

a. $5x$.

b. $6 \cdot 5x$.

c. $\frac{6 \cdot 5x}{3 \cdot 2 \cdot 1}$.

d. $\frac{7 \cdot 6 \cdot 5 \cdot 4 \cdot 3 \cdot 2 \cdot 1}{3 \cdot 2 \cdot 1}$.

Consider five symbols a, b, c, d, and e that will be in a string. If the symbols can be used and replaced, then the string can be as long as you wish. But if the symbols cannot be replaced, then as you use each symbol, there is one less symbol available. So the maximum length of the string is 5. That is the idea of the counting problem in Example 1.

Example 1

The "Big 10" athletic conference consists of 11 teams: Illinois, Indiana, Iowa, Michigan, Michigan State, Minnesota, Northwestern, Ohio State, Penn State, Purdue, and Wisconsin. You want to predict which team will finish first in a particular sport, which second, and which third. How many different predictions are possible?

Solution First, be sure that you understand the problem. You are looking for predictions, so identify one. One prediction might be that Northwestern will finish first, followed by Penn State and then Wisconsin. To find the total number of different predictions, use the Multiplication Counting Principle.

	1st		2nd		3rd
11 teams can be predicted to finish first.	11				
For each of these 11, 10 teams can be second.	11	•	10		
For each of these 11 • 10 pairs of teams, 9 teams can be third.	11	•	10	•	9

So there are $11 \cdot 10 \cdot 9 = 990$ possible predictions.

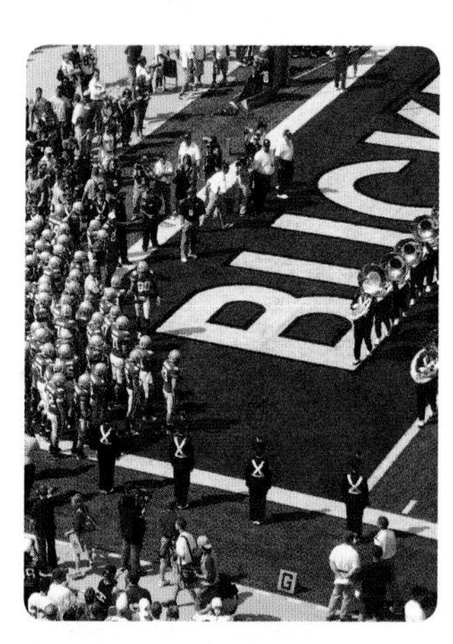

Ohio State, with over 51,000 students, is the largest university in the Big Ten.

Example 1 illustrates that there are 990 different strings of 11 different symbols with length 3 *without replacement*. If you picked the teams blindly, the probability of predicting the first, second, and third place finishers correctly is $\frac{1}{990}$.

 QY1

▶ **QY1**

What is the number of strings of 11 different symbols with length 3 *with* replacement?

An arrangement of teams, objects, or symbols without replacement is called a **permutation** of those objects. We say that the solution of Example 1 calculates the number of *permutations of 11 objects taken 3 at a time.* In general, for any positive integers n and r with $n \geq r$, the number of **permutations of *n* objects taken *r* at a time** is the number of strings of length r of n symbols without replacement. This number is denoted $_nP_r$. Generalizing the process in Example 1, we can prove the following theorem.

Theorem (Formula for $_nP_r$)

For any positive integers n and r with $n \geq r$, the number of permutations of n objects taken r at a time is

$$_nP_r = n(n-1)(n-2) \cdot \cdots \cdot (n-r+1).$$

Proof We want to know the number of strings of length r when n symbols are available but no symbol can be repeated. There are n possible choices for the first element of the string. Then there are $n - 1$ possible choices for the second element, $n - 2$ choices for the third element, and so on, until there are $n - (r - 1)$, or $n - r + 1$ choices for the rth element. By the Multiplication Counting Principle, the product of these numbers is the total number of possible strings.

QY2

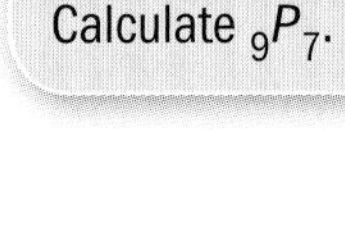
QY2

Calculate $_9P_7$.

Caution: The letter P in $_nP_r$ stands for permutation, not probability.

Example 1 shows that $_{11}P_3 = 11 \cdot (11-1) \cdot (11-2) = 11 \cdot 10 \cdot 9 = 990$. Notice that $11 \cdot 10 \cdot 9 = \frac{11 \cdot 10 \cdot 9 \cdot 8 \cdot 7 \cdot 6 \cdot 5 \cdot 4 \cdot 3 \cdot 2 \cdot 1}{8 \cdot 7 \cdot 6 \cdot 5 \cdot 4 \cdot 3 \cdot 2 \cdot 1} = \frac{11!}{8!}$, and $8 = 11 - 3$. In a similar way, any product of consecutive integers can be written as a quotient of factorials. This provides an alternate way of calculating $_nP_r$.

Theorem (Alternate Formula for $_nP_r$)

$$_nP_r = \frac{n!}{(n-r)!}$$

GUIDED

Example 2

How many different six-letter strings can be formed from six letters in the word PALINDROME without replacement?

Solution 1 Again first make sure that you understand the question. One possible string is DOMAIN. Other possible strings are PEORIA and DRLNPM. So the question asks for the number of permutations of __?__ objects taken __?__ at a time. Use the first formula for $_nP_r$, with $n =$ __?__ and $r =$ __?__.

$_{10}P_? = \underbrace{10 \cdot 9 \cdot 8 \cdot \text{\_?\_}}_{\text{\_?\_ factors}} =$ __?__ , so there are 151,200 strings.

Solution 2 Use the Alternate Formula for $_nP_r$.

$$_{10}P_? = \frac{10!}{(10 - \underline{\ ?\ })!} = 151{,}200 \text{ strings}$$

You can solve equations involving $_nP_r$ for n or r using either formula.

Example 3

Solve $_nP_6 = 17 \cdot {_nP_5}$.

Solution 1 Use the Formula for $_nP_r$. Translate $_nP_6$ and $_nP_5$ into their polynomial equivalents.

$$n(n-1)(n-2)(n-3)(n-4)(n-5)$$
$$= 17n(n-1)(n-2)(n-3)(n-4)$$

Divide each side of the equation by $n(n-1)(n-2)(n-3)(n-4)$.

$$n - 5 = 17$$
$$n = 22$$

Solution 2 Use the Alternate Formula for $_nP_r$. Translate $_nP_6$ and $_nP_5$ into their factorial equivalents.

$$\frac{n!}{(n-6)!} = 17 \cdot \frac{n!}{(n-5)!}$$

Divide both sides by $n!$ and use the Means-Extremes Property.

$$(n-5)! = 17(n-6)!$$

Use the fact that $n! = n(n-1)!$, or, more specifically, that $(n-a)! = (n-a)(n-a-1)!$.

$$(n-5)(n-6)! = 17(n-6)!$$
$$n - 5 = 17$$
$$n = 22$$

Check Find the permutation function on your calculator. Then compute $_{22}P_6$ and $17 \cdot {_{22}P_5}$. It checks.

nPr(22,6)	53721360
17·nPr(22,5)	53721360

Permutations of *n* Objects Taken *n* at a Time

In many situations, the number of permutations of *all* the objects in a set is desired. For instance, in Example 1, you might want to know the number of possible rankings of all 11 teams. You know there are $11 \cdot 10 \cdot 9$ ways of selecting the first three teams. As a result, there are $11 \cdot 10 \cdot 9 \cdot 8$ ways of selecting the first four teams. Continuing the process, there will be 11 factors in the product for selecting all 11 teams, so the number of strings with *length* 11 of 11 different objects, is

$$11 \cdot 10 \cdot 9 \cdot 8 \cdot 7 \cdot 6 \cdot 5 \cdot 4 \cdot 3 \cdot 2 \cdot 1 = 11!.$$

QY3

QY3

Use technology to find 11!.

The general formula for the number of permutations of n objects taken n at a time is a special case of the ${}_nP_r$ formula.

Corollary (Formula for ${}_nP_n$)

There are $n!$ permutations of n different elements. That is,

$${}_nP_n = n!.$$

Proof For all positive integers n and r with $n \geq r$,

$${}_nP_r = n(n-1)(n-2) \cdot \cdots \cdot (n-r+1).$$

Substitute n for r.

$${}_nP_n = n(n-1)(n-2) \cdot \cdots \cdot (n-n+1)$$
$$= n(n-1)(n-2) \cdot \cdots \cdot 1 = n!$$

 QY4

When $n = r$, the alternate formula for ${}_nP_r$ gives ${}_nP_n = \frac{n!}{(n-n)!} = \frac{n!}{0!}$. Because ${}_nP_n = n!$, in order to have the alternate formula for ${}_nP_r$ work for ${}_nP_n$, we must have $\frac{n!}{0!} = n!$. So we must define 0! to equal 1.

Definition

$$0! = 1$$

QY4

How many ways are there for a teacher to assign a group of 3 students the jobs of erasing the boards, wiping the tables, and turning off the lights?

Example 4

Tired of hearing his 17 students argue about who gets to be first in line, first-grade teacher Kerry Okie decides to put the students in a different order every time they get in line. He decides to list all the possible orders so that he will not duplicate any of them. If it takes him 20 seconds to write each order, how long will it take him to finish his list?

Solution Calculate ${}_nP_n$ with $n = 17$.

$17! = 355{,}687{,}428{,}096{,}000$.

So it would take $17! \cdot 20$ seconds to list all of the possible orderings. This is over 200,000,000 years.

Questions

COVERING THE IDEAS

1. List all the permutations of the letters in BAT.
2. Bashful, Doc, Dopey, Grumpy, Happy, Sleepy, and Sneezy go to work whistling in a different order each day. How many days can they go without repeating an order?
3. Write ${}_{117}P_4$ in each way.
   a. as a product of integers
   b. as a ratio of two factorials

In 4 and 5, evaluate.

4. ${}_{14}P_8$
5. ${}_{317}P_2$
6. Refer to the eleven teams in the "Big Ten" football conference. In how many ways can the first 5 positions in the standings be filled?
7. a. How many 3-letter permutations are there of the letters in TRIANGLE?
   b. How many of these permutations contain only consonants?
8. A student's schedule consists of six class periods (in Math, English, History, Language, Science, and Physical Education), two of which meet before lunch.
   a. How many ways can those first two classes be chosen?
   b. How many of these contain Math as one of the two classes?

In 9 and 10, evaluate the expression and explain your answer in terms of permutations of items from a set.

9. ${}_nP_1$
10. ${}_nP_n$
11. Explain why 0! is defined to equal 1.
12. Solve for ${}_nP_7 = 5 \cdot {}_nP_6$ for n.

APPLYING THE MATHEMATICS

13. a. Some automobile door locks use five buttons numbered 1–5. A combination consists of four different buttons pressed one at a time. How many such combinations are there?
    b. If a particular lock's combination is 2354 but you have forgotten it, what is the probability that you would guess it on the first try?

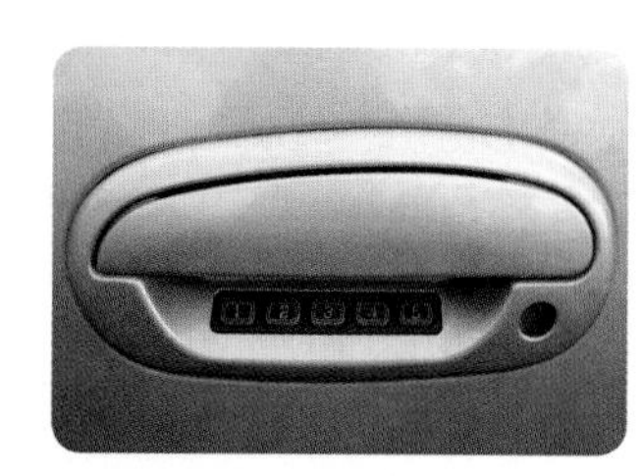

14. A student has textbooks in algebra, geometry, biology, chemistry, FST, and physics. These books are to be arranged on two shelves.
    a. How many different arrangements are possible if three books are on each shelf?
    b. How many different arrangements are possible if the three math books are on one shelf and the three science books are on another?
    c. How many different arrangements are possible if each shelf has at least one book?

15. Each row of an aircraft has three seats on one side of the aisle and two seats on the other. In how many different ways can a couple and their three children occupy a row of seats, if the two parents sit in the window seats?

16. a. Write ${}_nP_3$ as a polynomial. b. Write ${}_nP_3$ using factorials.

17. a. Use algebra to show that ${}_nP_0 = 1$.
    b. Explain why ${}_nP_0 = 1$ using a counting argument.

REVIEW

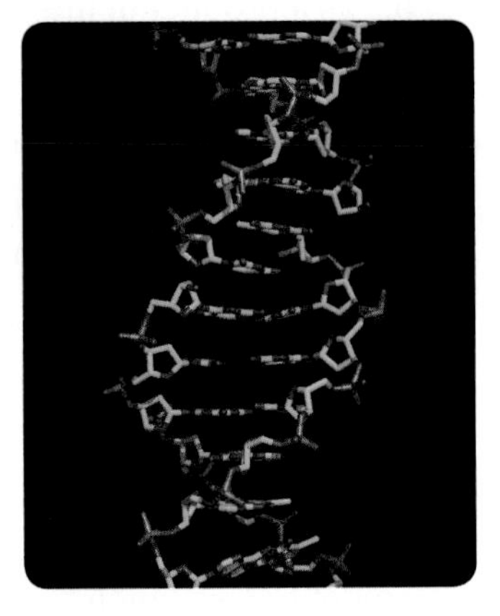

DNA strand

18. Every strand of human DNA consists of millions of nucleotides linked together to form a chain. Each nucleotide contains one of four nitrogenous bases: adenine, guanine, thymine, or cytosine. Sequences of these bases determine our genetic code. How many different possible sequences are there for a segment of DNA that is 50 nucleotides long? **(Lesson 6-3)**

19. Suppose a 7-character ID number consists of a letter followed by 5 digits from 0 to 9, followed by one more letter. How many ID numbers are possible? **(Lesson 6-3)**

20. A pair of fair six-sided dice are thrown. What is the probability that one die shows a 5 or the sum is greater than 8? **(Lesson 6-2)**

21. Give the exact value. **(Lesson 4-4)**
    a. $\tan \frac{\pi}{6}$ b. $\sin 150°$

22. Convert $4\frac{1}{2}$ revolutions
    a. to degrees. b. to radians. **(Lesson 4-1)**

23. Let $f(x) = x^2 - 5$ and $g(x) = \frac{1}{3x - 1}$. **(Lessons 3-7, 3-1)**
    a. Find a value of x that is not in the domain of g.
    b. Find a value of x that is not in the domain of $g \circ f$.

24. If all the numbers in a data set are tripled, what happens to the standard deviation of the set? **(Lesson 3-6, 1-6)**

EXPLORATION

25. For how many positive integers n does $n!$ end in exactly five 0s?

26. During winter, the first-grade class from Example 4 has a problem: when they put their hats on the radiator to dry out after recess, no student ever gets back his or her correct hat.
    a. If there are three students, how many ways are there to assign hats so that no student gets back his or her own hat? If all permutations are equally likely, what is the probability of this occurring?
    b. Repeat Part a if there are four students.
    c. In general, a derangement of a set of objects is a permutation in which no object is in its original position. Research derangements and find a method to generalize parts a and b.

QY ANSWERS

1. $11^3 = 1{,}331$
2. 181,440
3. 39,916,800
4. 6

Lesson 6-5 Contingency Tables

Vocabulary

contingency tables
Simpson's Paradox

▶ **BIG IDEA** A careful reading of contingency tables can uncover relationships in data.

Mental Math

Mr. Fisher has two FST classes.

	Boys	Girls
Class I	20	10
Class II	8	16

What is the probability a student is a boy if the student is randomly selected from all students

a. in Class I?

b. in Class II?

c. in either class?

The Titanic was a luxury liner that hit an iceberg and sunk on its maiden voyage in 1912. In the movie *Titanic,* Leonardo DiCaprio plays a heroic third-class passenger who gives his life to save that of a first-class woman (played by Kate Winslet). This image agrees with a popular conception of the disaster: the rich survived, and the middle-class and poor passengers drowned. How accurate is that conception?

Titanic Table 1 below lists the number of passengers and crew who survived and died (the possible outcomes) in the sinking of the Titanic, categorized by status (first-class, second-class, third-class, and crew).

Titanic Table 1: Status and Survival

	First	Second	Third	Crew
Survived	203	118	178	212
Died	122	167	528	673

Source: British Wreck Commissioner's Inquiry Report

STOP QY1

▶ **QY1**

Calculate the total number of people who survived and the total number who died.

Tables that divide outcomes among two or more categorical variables are called **contingency tables**. They help in analyzing complex situations.

Example 1

Refer to Titanic Table 1. Round to the nearest tenth of a percent.

a. Out of all people on the Titanic, what percent survived?

b. Find the percent of passengers in first class who survived.

c. Find the percent of passengers who survived that were in first class.

Solution

a. Add the numbers in each row. $203 + 118 + 178 + 212 = 711$ people survived and $122 + 167 + 528 + 673 = 1490$ people died. The total number of passengers was $711 + 1490 = 2201$. So $\frac{711}{2201} \approx 32.3\%$ survived.

(continued on next page)

The Titanic

b. There were 203 survivors in first class. The total number of passengers in first class was 203 + 122 = 325. Therefore, the percent of people in first class who survived is $\frac{203}{325} \approx 62.5\%$.

c. Again, there were 203 survivors in first class. The total number of people who survived was 711. Therefore, the percent of people who survived that were in first class is $\frac{203}{711} \approx 28.6\%$.

Many people find the discrepancy between the answers to Parts b and c confusing, because the questions seem so similar. As simple fractions, both answers have the same numerator, 203, which is the number of first-class survivors. But the denominators are different, because the two questions ask about different populations. Part b asks about the population of *passengers in first class*, of which there were 325. Part c asks about the population of *all passengers who survived*, of which there were 711. In forming a $\frac{part}{whole}$ ratio, you must be clear about the answer to the question "What is the whole?". For questions about ratios based on a contingency table, you must be careful to correctly identify the whole, that is, the *population* to which the question refers.

Tree diagrams can help to clarify the difference between Parts b and c above. In the diagram below at the left, the people are sorted first by status and then by who survived and who did not. In the tree diagram below at the right, the sorting is first by survival and then by status. Each of Parts b and c of Example 1 matches a different tree diagram, which isolates the population in question.

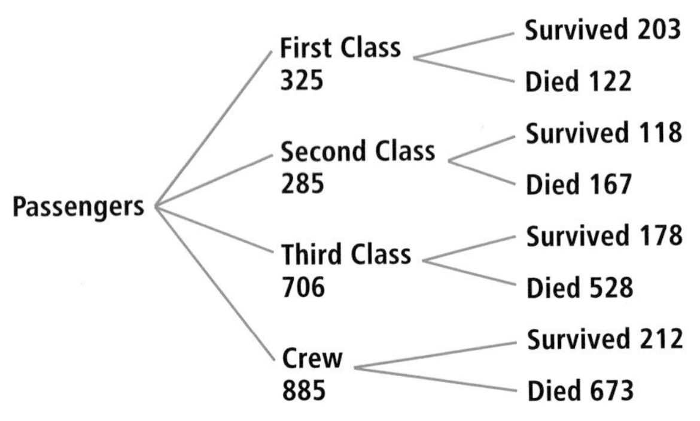

First branch separates by status: Of the 325 first class passengers, 203 survived, or $\frac{203}{325} \approx 62.5\%$.

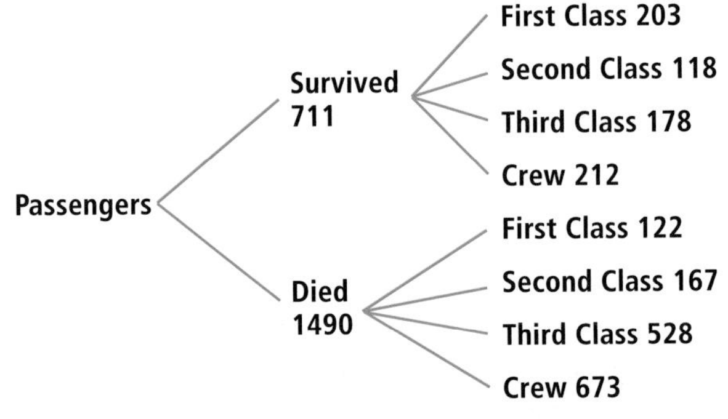

First branch separates survivors from deaths: Of the 711 people who survived, 203 were in first class, or $\frac{203}{711} \approx 28.6\%$.

QY2

▶ QY2

Which is greater, the chance that a survivor was a second-class passenger, or the chance that a second-class passenger survived?

Other Contingency Table Formats

Because it is common to ask questions that involve totals of rows or columns, those totals are frequently included in contingency tables.

GUIDED

Example 2

A 2001 study by the University of Texas Southwestern Medical Center examined 626 patients to see if there was a connection between getting a tattoo and infection with Hepatitis C (HCV). The results are in the contingency table below.

	Tattoo Done in Commercial Tattoo Parlor	Tattoo Done Elsewhere	No Tattoo
Has Hepatitis C	17	8	18
No Hepatitis C	35	53	495

Source: Haley RW, Fischer RP in *Medicine*, March 2001

a. Add row and column totals to the table.

b. What does the total of the third column represent?

c. What does the total of the second row represent?

d. According to this data, is someone with a tattoo done in a commercial parlor more or less likely to have HCV than someone with a tattoo done elsewhere?

e. Give at least one reason why the result in Part d might not reflect the safety of each kind of tattoo.

Solution

a.

	Tattoo Done in Commercial Tattoo Parlor	Tattoo Done Elsewhere	No tattoo	Totals
Has Hepatitis C	17	8	18	43
No Hepatitis C	35	53	495	?
Totals	52	?	?	626

b. The total of the third column represents all the patients who ___?___.

c. The total of the second row represents all the patients who ___?___.

d. The chance that someone with a tattoo from a commercial parlor had HCV is $\frac{17}{52} \approx 33\%$. The chance that someone with a tattoo done elsewhere has HCV is ≈ ___?___%. Therefore, someone with a tattoo done in a commercial parlor is ___?___ likely to have HCV than someone with a tattoo done elsewhere.

e. Here is a possible explanation. People with tattoos done in commercial tattoo parlors might have engaged in other activities that put them at risk for HCV.

It is common to express entries in a contingency table as percents of a row or a column total. Consider the table of Titanic survivor data below.

Titanic Table 2: Column Percents

	First	Second	Third	Crew
Survived % of column	203 62%	118 41%	178 25%	212 24%
Died % of column	122 38%	167 59%	528 75%	673 76%
Totals	325	285	706	885

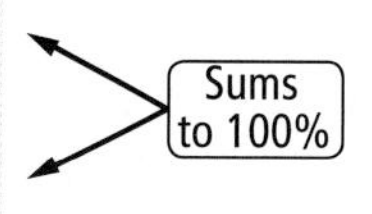

In Titanic Table 2, each percentage represents the percent of people of a given status who survived or died. Therefore, the percentages along each column add to 100%. The percentages along each row do not add to 100% because each value is a percent of a different total: 62% represents a fraction of the 325 first-class passengers, while 41% represents a fraction of the 285 second-class passengers. If you were interested in the class breakdown of survivors, you would compute percentages of row totals instead, as in Titanic Table 3.

Titanic Table 3: Row Percents

	First	Second	Third	Crew	Totals
Survived % of row	203 29%	118 17%	178 25%	212 30%	711
Died % of row	122 8%	167 11%	528 35%	673 45%	1490

 **QY3**

QY3

In Titanic Table 3, what does 17% represent?

Notice that while the percentages make some comparisons easier, they make other comparisons more difficult. For example, although the 17% of survivors who were in second class is larger than the 8% of deaths in first class, the number of second-class passengers who survived is smaller than the number of first-class passengers who died. The discrepancy arises because many more people died than lived, so a smaller percentage of deaths can represent a larger number of people.

When a contingency table gives percentages but does not specify whether they are percentages of rows or columns, you should check whether rows or columns add to 100% to find out what the percentages represent.

Example 3

Fifth-grade students in a school were surveyed about their favorite book series. The results are reported in the contingency table below.

	Harry Potter	Animorphs	Lemony Snicket	Lord of the Rings	Other
Boys	42%	15%	23%	11%	9%
Girls	51%	8%	28%	5%	8%

a. **What does the 23% in the first row, third column represent?**

b. **Which is larger: the number of boys who prefer Animorphs, or the number of girls who prefer Lemony Snicket?**

Solution

a. In the first column, 42% + 51% ≠ 100%. But the total of each row is 100%. The number in the first row, third column means that 23% of all boys surveyed chose Lemony Snicket as their favorite book series.

(continued on next page)

b. The contingency table gives percentages of boys and percentages of girls respectively. It does not give the number of boys or girls in each cell. Nor does it give the total number of boys and of girls, from which you could compute these figures yourself. Not enough information is given to answer this question.

 QY4

QY4

Answer Example 3b, if you know that 500 boys and 400 girls were surveyed.

Averaging and Simpson's Paradox

When data are combined or separated, the totals may not look like the parts. Here is a famous real-world example. In the 1970s, the University of California at Berkeley was sued for gender discrimination in graduate school admissions. As a result, it examined its admissions closely. In 1973, a typical year, 35% of 4321 female applicants were admitted, while 44% of 8442 male applicants were admitted. The women's acceptance rate was noticeably smaller than the men's.

The situation was actually more complicated than it looked. Applicants for graduate school apply to specific programs, not to the university as a whole, and acceptance rates vary widely from program to program. The contingency table below shows the men's and women's acceptance rates within each program.

Program	Men		Women		Overall	
	Applicants	% Admitted	Applicants	% Admitted	Applicants	% Admitted
A	825	62%	108	82%	933	64%
B	560	63%	25	68%	585	63%
C	325	37%	593	34%	918	35%
D	417	33%	375	35%	792	34%
E	191	28%	393	24%	584	25%
F	373	6%	341	7%	714	6%
Total	2691	44.5%	1835	30.4%	4526	38.8%

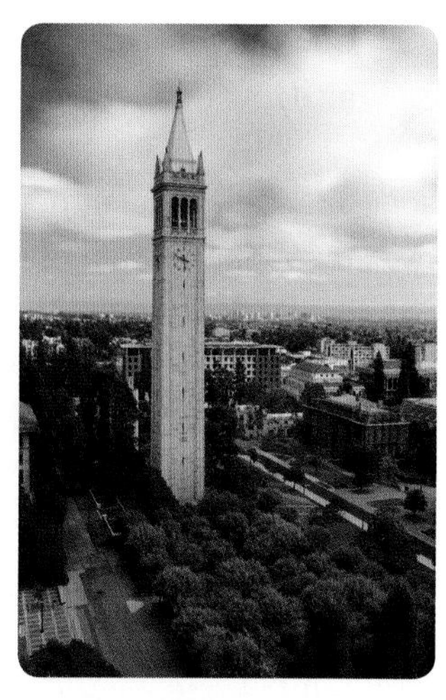

Jane K. Sather Tower, University of California, Berkeley

Within each program, men were accepted at a comparable or equal rate to women. The discrepancy arises because the programs to which the most men applied (programs A and B) had higher overall acceptance rates than the two most popular women's programs (C and E). In this case, the choice of program is a *confounding variable,* that is, an extra variable associated with one of the variables in the study (gender), that causes part of the observed effect.

The discrepancy between the overall acceptance rates and the acceptance rates within each program is an example of **Simpson's Paradox**, named after Edward H. Simpson, who wrote about the paradox in 1951. Simpson's Paradox can arise whenever different categories of data are combined (or *aggregated*) to compute an average or percent. Although Simpson's Paradox arises in a variety of situations, the best way to avoid being confused by it is to separate (or *disaggregate*) the data into the original categories.

Questions

COVERING THE IDEAS

1. A sample of 70 students at a high school was examined to investigate patterns of enrollment in language classes.

	Freshman	Sophomore	Junior	Senior
Spanish	8	11	5	2
French	2	5	0	1
No language	5	8	10	13

   a. What percent of students who took no language were seniors?
   b. What percent of students who were seniors took no language?
   c. Copy and extend the table to include row and column totals.

2. Use Titanic Table 1 on page 387.
   a. How many first- and second-class passengers were aboard the ship when it set sail?
   b. What percent of the first- and second-class passengers survived?
   c. What percent of third-class passengers survived?
   d. Do you think it is accurate to say poor passengers drowned at a disproportionate rate? Back up your response with evidence.

3. Use the table from the polio study on page 359.
   a. Give the row percents for vaccinated children.
   b. Give the column percents for those who developed paralytic polio.
   c. What percent of children who developed paralytic polio were vaccinated?

4. The contingency table for the graduate school admissions on page 391 shows that a higher percent of women than men were admitted to Program A. Find the number of men and number of women accepted to Program A. Explain how it is possible that the number of women accepted is lower when compared to men, even though the percentage of women accepted is higher.

5. The table below gives the number of motor vehicles involved in accidents in the U.S. in 2004, classified by vehicle type and severity.

	Fatality (1000s)	Injury (1000s)	No Injury (1000s)
Passenger Car	25.5	1,989.8	4,216.3
Pickup Truck	10.8	482.0	1,161.8
SUV	7.8	475.0	1085.8
Van	3.7	259.5	563.7
Other Light Truck	0.1	29.3	74.3

Source: Statistical Abstract of the United States, 2007-2008

   a. What is the total of the first row, and what does it represent?
   b. What is the total of the first column, and what does it represent?
   c. What is the chance that an SUV accident had a fatality?
   d. What is the chance that a vehicle in which there was a fatality was an SUV?

6. An electronics store does free repairs for the first year on everything it sells. The store sells three brands of DVD players: 50% of those sold are Brand A, 30% are Brand B, and 20% are Brand C. During the first year, 25% of Brand A machines need service, 10% of Brand B need service, and 5% of Brand C need service.
   a. Make a tree diagram for this situation
   b. What percent of all the machines sold need service?
   c. What percent of the machines needing service are Brand A?

APPLYING THE MATHEMATICS

In 7–9, use the data below. The British agency that conducted the inquiry into the Titanic disaster separated their findings according to gender and age, as given in the table below.

		1st Class	2nd Class	3rd Class	Steerage and crew	Total
Survived	Men	57	14	75	192	338
	Women	140	80	76	20	316
	Children	6	24	27	0	57
Died	Men	118	154	387	670	1329
	Women	4	13	89	3	109
	Children	0	0	52	0	52
Total	Men	175	168	462	862	1667
	Women	144	93	165	23	425
	Children	6	24	79	0	109

Source: British Wreck Commissioner's Inquiry Report

7. a. What percent of men in second class survived?
   b. What percent of survivors were women?
   c. What percent of survivors were in third class?

8. a. Who was more likely to survive: a man in first class, or a child in third class? Support your answer numerically.
   b. If someone survived, how likely is it that the person was either in first class or a woman?

9. Describe the difference between the three questions below, and give each value.
   a. What is the chance that a child in first class survived?
   b. What is the chance that a survivor was a child in first class?
   c. What is the chance that a person on the Titanic was a child who survived in first class?

10. The batting statistics in the table below for professional baseball players Derek Jeter and David Justice in the years 1995 and 1996 illustrate Simpson's Paradox.

	1995		1996		totals	
	at-bats	hits	at-bats	hits	at-bats	hits
Jeter	48	12	582	183		
Justice	411	104	140	45		

Source: Major League Baseball

a. Calculate the batting averages for Jeter and Justice in 1995 and in 1996.

b. Calculate the totals and batting averages for Jeter and Justice for the two years.

c. What is surprising about these results? What is the confounding variable?

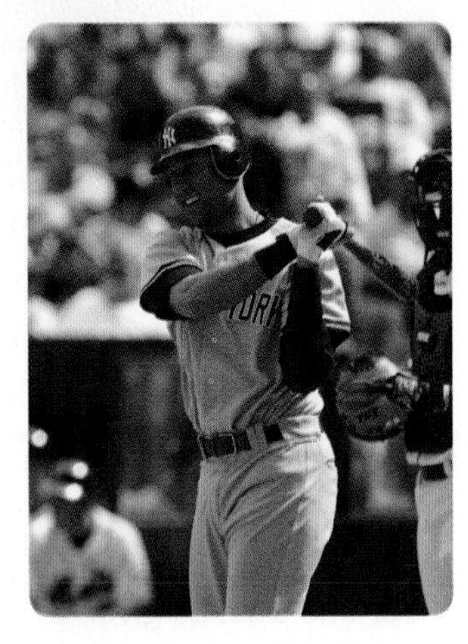

Derek Jeter of the New York Yankees

REVIEW

11. Write ${}_{98}P_3$ in each way. **(Lesson 6-4)**

a. as a product of integers

b. as a ratio of two factorials

In 12 and 13, evaluate. (Lesson 6-4)

12. ${}_{13}P_9$

13. ${}_{212}P_3$

14. If you must answer every question on a test, how many ways are there of answering a test with 12 true-false and 6 multiple-choice questions with 5 choices each? **(Lesson 6-3)**

15. Suppose two fair six-sided dice are rolled and the sum is a multiple of 2, but not a multiple of 3. **(Lesson 6-2)**

a. List the outcomes of the event.

b. Give the probability of the event.

16. In 2007, the population of Hungary was reported as 9,956,108 with an average annual growth rate of -0.253%. Assume this growth rate is constant. **(Lesson 2-5)**

a. Express the population P in terms of the number of years n after 2007.

b. Estimate the population of Hungary in 2000.

EXPLORATION

17. Many people believe that SUVs are safer than passenger cars because they believe that when an accident occurs, occupants of an SUV are safer than occupants of passenger cars. Use the accident data given in Question 5 to either support or refute this claim.

QY ANSWERS

1. 711 survived; 1490 died

2. The chance that a second-class passenger survived.

3. the percent of survivors who were second-class passengers

4. the number of girls who prefer Lemony Snicket

Lesson

6-6 Conditional Probability

Vocabulary

conditional probability of an event, $P(B|A)$

BIG IDEA The idea of conditional probability helps to explain and compute results from contingency tables.

Lesson 6-5 began with the following table about the numbers of people on the Titanic who survived and who died.

	First	Second	Third	Crew
Survived	203	118	178	212
Died	122	167	528	673

Source: British Wreck Commissioner's Inquiry Report

From this contingency table, you can determine probabilities that a person was in a particular row given that they were in a particular column, or vice-versa. For instance, you can determine the probability that a randomly-selected passenger survived (the first row) given that the passenger was in second class (the second column). Finding probabilities like this are the topic of this lesson.

Mental Math

Write each expression as a simple fraction in lowest terms.

a. $\frac{30\% + 15\%}{50\%}$

b. $\frac{0.9 \cdot 0.3}{0.3 \cdot 0.4 + 0.7 \cdot 0.6}$

c. $\frac{2x \cdot 4x}{5x \cdot 3x + 6x \cdot 4x}$

What Is Conditional Probability?

Let A and B be events with A = "the passenger was in second class" and B = "the passenger survived." We want the probability of B given A. Such a probability is called a *conditional probability*. The probability is $\frac{118}{118 + 167} = \frac{118}{285}$, or about 41%. Notice that $118 = N(A \cap B)$ and $285 = N(A)$. So, the conditional probability that a randomly-selected second-class passenger survived is $\frac{N(A \cap B)}{N(A)}$, in this case $\frac{118}{285}$.

In this situation, the sample space S is the set of all passengers and crew, and $N(S) = 2201$. If we divide both numerator and denominator by $N(S)$, we obtain

$$P(B \text{ given } A) = \frac{N(A \cap B)}{N(A)} = \frac{\frac{N(A \cap B)}{N(S)}}{\frac{N(A)}{N(S)}} = \frac{P(A \cap B)}{P(A)}.$$

Thus, conditional probabilities can be defined in terms of probabilities.

Definition of Conditional Probability

The **conditional probability of an event** B given an event A, written $\mathbf{P(B \mid A)}$, is $\frac{P(A \cap B)}{P(A)}$.

Conditional Probability and Medical Tests

You could calculate the conditional probability from the Titanic data without having a formula, by thinking clearly about the situation. A similar and very important situation involving conditional probability is in medicine. It is complex enough to require a systematic procedure for the calculation.

When a person takes a test to see if they have a particular allergy, disease, or other condition, the result is called *negative* if the test indicates the person does *not* have the condition. It is called *positive* if the test indicates the person *does* have the condition. Usually, the negative result is the one you want.

Few medical tests are 100% accurate. Taking this into account, there are two ways a test can be accurate and two ways it can be wrong. They are shown in the tree diagram below. The two ways it can be wrong are called a *false positive* (the test says you have the condition, but you don't) and *false negative* (the test says you don't have the condition, but you do.)

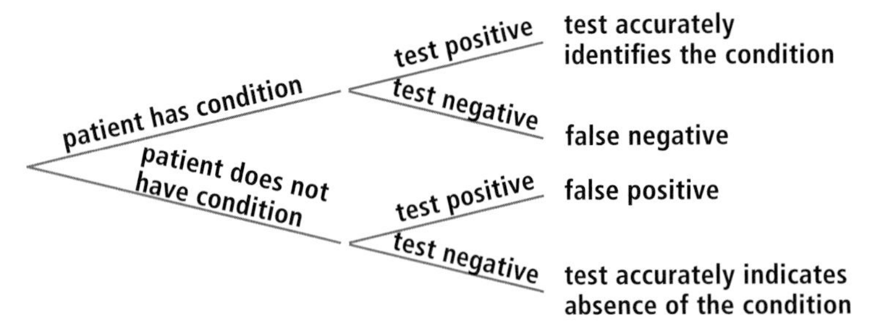

From a treatment point of view, the worst result is the false negative. Then you have a condition but the test shows you don't, so you are falsely led to believe that you do not need treatment. But false positives are also bad because they can be traumatic. A person can think they have a disease yet not have it. This is why it is useful to know how likely a false negative and false positive might be.

Example 1

An article in the *Journal of the American Medical Association* in 1997 reported that, when people go to their doctor's office with a sore throat and think they might have strep throat, 30% actually have strep throat. It noted that a current test for strep throat was 80% accurate if you have strep throat and 90% accurate if you do not. What is the probability that a person who receives a positive result from this test does not have the disease?

Solution First, organize the given information into a tree diagram. We let A = "tests positive" and B = "has strep throat."

We want to calculate $P((\text{not } B) \mid A)$, the probability that a person does not have strep throat even though the test indicates that the person does. Suppose T people are tested. Then $0.30T$ have strep throat and $0.70T$ do not.

Of the 0.30T who have strep throat:

$$0.80 \cdot 0.30T = 0.24T \text{ test positive.}$$

$$0.20 \cdot 0.30T = 0.06T \text{ test negative.}$$

Of the 0.70T who do not have strep throat:

$$0.10 \cdot 0.70T = 0.07T \text{ test positive.}$$

$$0.90 \cdot 0.70T = 0.63T \text{ test negative.}$$

You can now write these percents of T on a tree diagram. Each leaf is labeled with its probability, the product of the branches that lead to it.

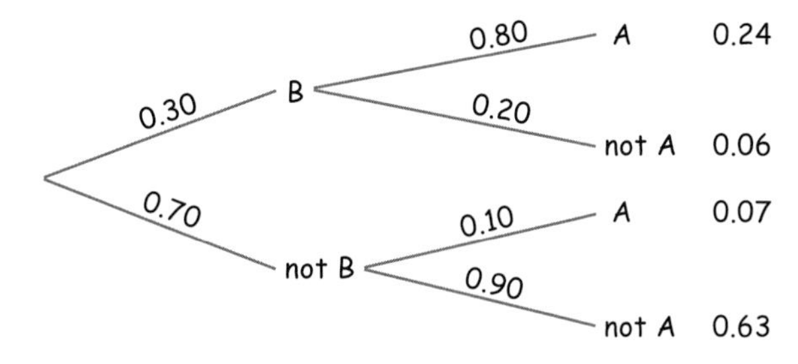

Another way to summarize these probabilities is with a table. Notice that the columns add to the totals in the bottom row.

	B = Patient has strep	not B = Patient does not have strep	Total
A = Test +	$P(A \cap B) = 0.24$	$P(A \cap \text{not } B) = 0.07$	$P(A) = 0.31$
not A = Test −	$P(\text{not } A \cap B) = 0.06$	$P(\text{not } A \cap \text{not } B) = 0.63$	$P(\text{not } A) = 0.69$
Total	$P(B) = 0.30$	$P(\text{not } B) = 0.70$	1.0

The probabilities in this table are descriptive of the population of patients who took the test.

From the definition of conditional probability,

$$P(\text{not } B \mid A) = \frac{P(A \text{ and not } B)}{P(A)} = \frac{0.07T}{0.07T + 0.24T} = \frac{0.07}{0.31} \approx 23\%.$$

Given that someone tests positive for strep, the chance he or she does not have the disease is therefore about 23%.

 QY

The answer 23% in Example 1 means that, in this situation, about 1 in 4 people for whom the test is positive do not have strep throat. In this case, a doctor may prescribe an antibiotic just to play it safe, but for a more serious disease, for which treatment may be painful and expensive, a second test may help to provide a more accurate result.

QY

In Example 1, what is the probability that a negative result is a false negative—that is, a person has strep but the test does not show it?

False Negatives

In cases such as cancer, where a disease can be fatal and prompt treatment is crucial to the patient's survival, false negatives are particularly troubling.

GUIDED

Example 2

Suppose all patients are tested for a serious disease that is estimated to be found in 0.5% of people. Suppose also that the test accurately spots the disease 98% of the time and accurately indicates no disease 95% of the time. What is the probability that a negative result is a false negative?

Solution Instead of a tree diagram, construct a contingency table with the given information. Let D = "has the disease" and A = "tests positive."

	D	not D
A	$0.98 \cdot 0.005 =$ ___?___	___?___ $\cdot$ ___?___ $=$ ___?___
not A	___?___ $\cdot\ 0.005 =$ ___?___	$0.95 \cdot$ ___?___ $=$ ___?___
Total	0.5% of total	___?___% of total

Thus, in this situation, the probability that someone who tests negative does in fact have the disease equals $P(\underline{\ ?\ } \mid \underline{\ ?\ }) = \dfrac{P(D \cap (\text{not } A))}{P(\text{not } A)} = \underline{\ ?\ }$.

The answer to Example 2 indicates that about 0.01% of the negatives are false negatives. It is important for tests to have a low rate of false negatives because in many cases, early detection and treatment is key to patient recovery or survival. An additional concern for doctors and hospitals is the potential liability from malpractice suits fueled by misdiagnoses. Question 3 asks for the false positive rate in the situation of Example 2, which may be quite high. This type of misdiagnosis is also troubling, considering its potential emotional, physical, and financial impact.

Questions

COVERING THE IDEAS

1. Refer to the table on page 395. What is the probability that a randomly selected passenger survived given that the passenger was in first class?
2. Refer to Example 1. For what percent of the patients is the strep throat test accurate?
3. a. Draw a tree diagram corresponding to Example 2.
   b. What is the probability that a positive result is a false positive?
   c. Your answer to Part b should surprise you. Explain why it makes sense.

4. Your house probably has a smoke detector whose purpose is to alert you to a fire.
   a. In this context, what is a typical instance of a false positive? Why might one occur?
   b. What is a typical instance of a false negative? Why might one occur?
   c. Which is riskier: a false positive or a false negative?

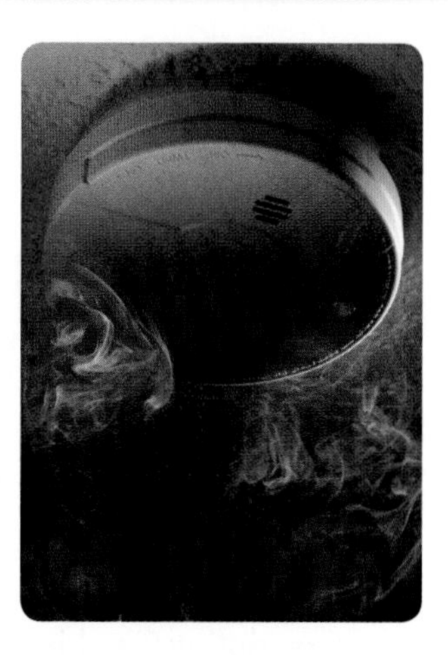

96% of U.S. households have a smoke detector. An estimated 890 lives could be saved each year if all homes had them.

In 5–7, use this information. Polygraph machines attempt to detect whether a statement is a lie by monitoring a person's skin temperature, blood pressure, breathing, and other physical symptoms believed to correlate with deception. In a lie detector test, a "positive" result means that a statement is detected as a lie. The most favorable estimates give a 90% rate of detecting lies, with 95% of truthful statements "passing" (i.e. not marked as lies). Although many studies have obtained far lower rates, true controlled experiments are both hard to create and rarely replicate "field conditions." Assume that a police department decides to use a polygraph and estimates that about 15% of people will lie at least once.

5. In this context, give the meaning of each term.
   a. false positive
   b. false negative
6. a. Make a table or a tree diagram to determine how likely it is that someone who tests positive really is lying.
   b. What is the likelihood that someone who tests positive is telling the truth?
   c. What is the likelihood that someone who tests negative is lying?
7. Use tree diagrams for Parts a and b.
   a. If the same test is used when the probability of a lie occurring is 40%, what is the likelihood a positive result is a false positive?
   b. If the test only has a 75% accuracy rate (for both true and false statements), but is used on the original applicant pool (15% lies), what is the likelihood a positive result is a false positive?
   c. Explain your result from Part b in words.

In 8–10, the standard HIV test accurately finds the condition 99% of the time and also 99% of the time accurately shows that a person does not have HIV. In the general population, the prevalence of HIV infection is about 0.5%.

8. If a test result comes back positive, what is the likelihood the person has the disease?
9. Suppose that all people who test positive are given the same test again. Also, assume that false positives are true "flukes," that is, there is no underlying condition that causes some people to receive false positive results more than others. What is the likelihood that a person who tests positive a second time actually has the disease?

10. Write a few sentences explaining the results in Questions 8 and 9. First, if a test is 99% accurate, how is it that only about $\frac{1}{3}$ of positive results are actually correct? Second, how can the same test be so much more accurate if used twice?

APPLYING THE MATHEMATICS

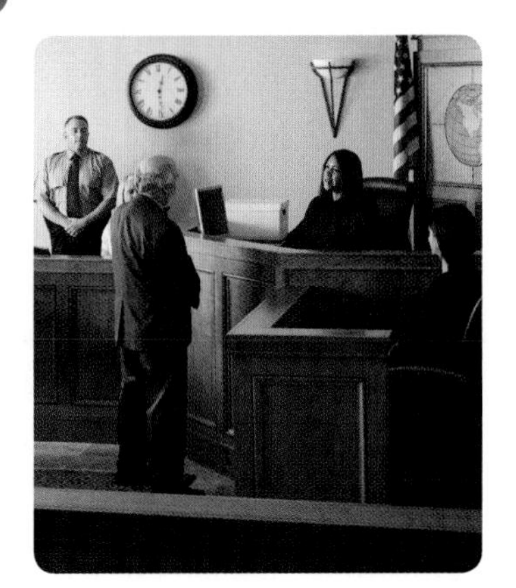

11. At a trial, a prosecutor argues that only one person in 12,000,000 matches the description of the perpetrator noted by witnesses at the scene. Therefore, the prosecutor concludes that a defendant who matches the description is guilty with probability $\frac{11{,}999{,}999}{12{,}000{,}000}$. Assume that the description is accurate, and that the accused person matches the description. Let M = "a person matches the description," and let C = "a person committed the crime."
    a. Which probability is represented by $\frac{1}{12{,}000{,}000}$: $P(C)$, $P(M)$, $P(C \mid M)$, or $P(M \mid C)$?
    b. Which expression in Part a represents the probability that the accused actually committed the crime, based on the description?
    c. In 2006, the population of California was about 36,000,000. If the crime was committed in California, estimate the number of people in California expected to match the description of the perpetrator.
    d. Use your answer to Part c to compute $P(C \mid M)$, assuming that the perpetrator was from California.
    e. This situation is referred to as the *prosecutor's fallacy.* Explain in ordinary terms the mistake that the prosecutor made.

REVIEW

In 12 and 13, the table below shows the numbers of flights that were on time and delayed for two airlines A and B at five airports in June 1991. (Lesson 6-5)

	Airline A			Airline B		
	On Time	Delayed	% Delayed	On Time	Delayed	% Delayed
Los Angeles	497	62	?	694	117	?
Phoenix	221	12	?	4840	415	?
San Diego	212	20	?	383	65	?
San Francisco	503	102	?	320	129	?
Seattle	1841	305	?	201	61	?
Total	?	?	?	?	?	?

12. a. Copy and complete the table.
    b. Identify the paradox contained in the table.
    c. Explain why your answer in Part b is an example of this paradox.

13. a. Calculate the probability that a plane of Airline A was delayed given that it landed in Seattle.
    b. Calculate the probability that a plane of Airline A that was delayed landed in Seattle.

14. To study the connection between car owner satisfaction and origin of production, car owners were surveyed. The results are summarized in the table below. **(Lesson 6-5)**

	Level of Satisfaction		
Owners of	High	Medium	Low
American	80	100	45
Japanese	40	30	20
European	25	35	25

a. Add appropriate row and column totals.
b. What percentage of European car owners reported a medium level of satisfaction?
c. What percentage of Japanese car owners were highly satisfied?

15. a. How many permutations consisting of four letters each can be formed from the letters of the word CONVEX?
    b. How many of the permutations in Part a begin with a vowel? **(Lesson 6-4)**

16. Consider the data in the table below:

i	1	2	3	4	5	6	7	8	9
a_i	1	1	2	3	5	8	13	21	34

a. Write and evaluate an expression in Σ-notation for $a_1 + a_2 + a_3 + a_4 + a_5 + a_6 + a_7$.
b. **Multiple Choice** Which expression below represents the variance of the data? **(Lessons 1-6, 1-2)**

A $\dfrac{\sum_{i=1}^{9} a_i}{9}$ B $\dfrac{\sum_{i=1}^{9} a_i^2}{9}$ C $\dfrac{\sum_{i=1}^{9} (a_i - \bar{a})^2}{8}$

D $\sqrt{\dfrac{\sum_{i=1}^{9} a_i^2}{8}}$ E $\dfrac{\sum_{i=1}^{9} (a_i - \bar{a})}{8}$

EXPLORATION

17. Look for information about the accuracy of a test for a medical condition you have heard about. Use that information, and information about the prevalence of the condition, to compute the likelihood of a false positive and of a false negative.

QY ANSWER

$\dfrac{0.06T}{0.06T + 0.63T} = \dfrac{6}{69} \approx 8.7\%$

Lesson 6-7 Designing Simulations

Vocabulary

simulation
random
trial
Monte Carlo method

BIG IDEA When you do not know how to calculate the probability of an event, you may be able to use a simulation to obtain an estimate of the probability.

A **simulation** of a real situation is an experimental model that attempts to capture all aspects of the original situation that affect the outcomes. A simulation allows people to plan and predict. Crash tests, pre-election polls, and fire drills are all examples of simulations for which it would be impractical or not feasible to wait for the real event to gather data. Even college entrance tests have "trial runs." The PSAT simulates the SAT; the PLAN simulates the ACT.

In this lesson we simulate an experiment by assigning numbers to the possible outcomes. You will determine outcomes by generating *random numbers*. A set of numbers is **random** if each number has the same probability of occurring, each pair of numbers has the same probability of occurring, each triple of numbers has the same probability of occurring, etc. When an experiment or simulation is repeated, each repetition is called a **trial**.

Mental Math

Calculate.

a. log 100

b. $\log \frac{1}{100}$

c. log 1

d. log 0

Using Random Numbers in a Simulation

The manufacturer of Sugar Oats cereal is promoting sales by putting a Sports Stars trading card in each box. The cards are randomly distributed so that the chance of obtaining each one is equally likely. Children are encouraged to collect a set of all four cards, which are shown below. Bobby wants his parents to buy the cereal, but they think they may need to buy many boxes in order to get all four cards.

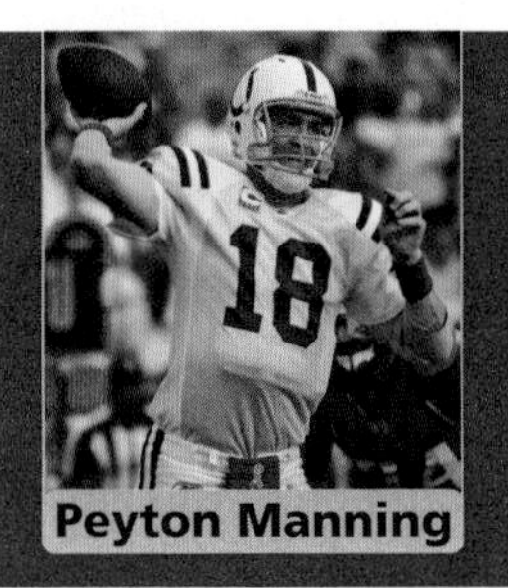

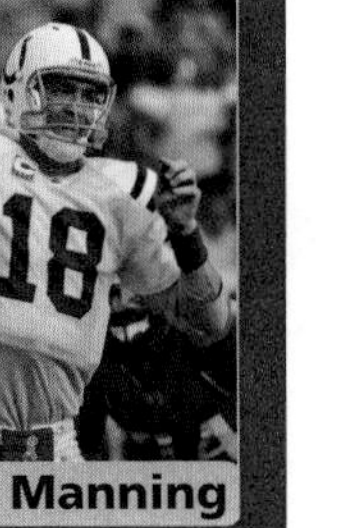

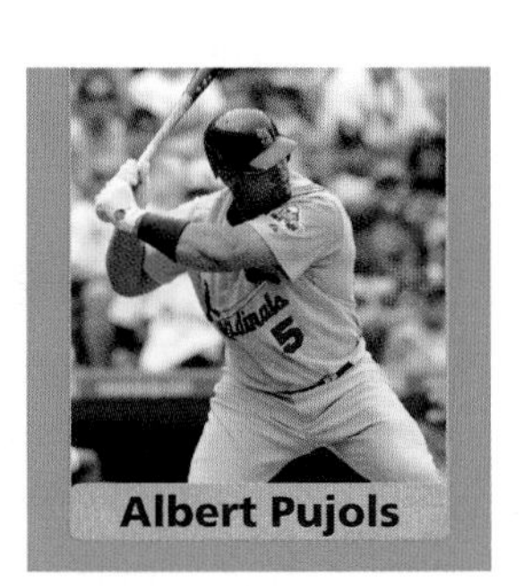

How many boxes might Bobby's parents have to buy in order to get all four cards? To examine this question, we do a simulation.

First, we assign the numbers 1, 2, 3, and 4 to the four outcomes from buying a single box. Each number indicates that the box contains a card of an athlete as follows.

1: Tiger Woods

2: Candace Parker

3: Peyton Manning

4: Albert Pujols

Then we use the random integer command on a calculator to simulate buying boxes of cereal, and keep buying until we have at least one of each type of card.

The screen at the right shows one trial, with each digit representing the card in one cereal box. This trial produced the digits ③,④, 4,②, 2,① with the circled numbers showing boxes that contained a card that Bobby did not yet have. The set was complete after 6 boxes, when the Tiger Woods card was found.

randInt(1,4)	3
randInt(1,4)	4.
randInt(1,4)	4.
randInt(1,4)	2.
randInt(1,4)	2.
randInt(1,4)	1.

Activity 1

Step 1 Perform one trial of the above simulation yourself. How many boxes did it take you to get all four Sports Stars cards? How does your result compare with those of your classmates?

Step 2 Bobby's parents refuse to buy dozens of boxes. Based on your results for Step 1, pick a number of boxes *B* that you think is likely to give Bobby a full set of Sports Stars cards and also satisfy his parents.

Step 3 Simulate buying B boxes ten times, noting each time whether you were successful in finding all four cards. Record your data in a table like the one at the right.

Trial	Card number for each of the *B* Boxes	Success?
1		
2		
3		
4		
5		
6		
7		
8		
9		
10		

Step 4 Combine your data with the rest of the class. Construct a graph showing the percent of times that buying *x* boxes resulted in having all 4 cards.

Step 5 Do you think it would be worth it to buy lots of boxes to attempt to obtain all four cards? Why or why not?

Monte Carlo Simulations

Using repeated trials to produce relative frequencies to estimate probabilities is often called a **Monte Carlo method**, named after the well-known Monte Carlo casino in the principality of Monaco. The process was developed and named by Stanislaw Ulam, a Polish-born mathematician who spent most of his career in the United States.

Ulam recognized that the power and speed of computers would allow for Monte Carlo experiments with huge numbers of trials, making this a valuable tool in science, business, and other fields. Some complex questions involving risk and decision-making that cannot be answered with other mathematical theories can be addressed with this method.

There are three steps to the Monte Carlo method.

1. Determine how the situation will be modeled.
2. Define what constitutes a trial and what will be recorded.
3. Specify the number of trials that will be run and how the estimated answer will be obtained.

A Monte Carlo method was used in Activity 1. Consider what made up a single trial. In the simulation, the numbers 1 to 4 represented the cards. A single trial involved generating B random numbers and recording whether all four Sports Stars cards had been obtained. You ran 10 trials and then calculated the percent of successful trials.

Designing a Monte Carlo Simulation

It is common for airlines to overbook, meaning that they sell more tickets for a flight than there are available seats. This compensates for the fact that many people do not show up for their scheduled flights.

But how many extra seats should be booked? If the airline sells too many tickets, it is likely that some passengers will have to be "bumped." If they sell too few tickets, many flights will depart with empty seats. Airlines use simulations to decide how many tickets to sell.

Suppose that an airline has a small commuter plane with 24 seats. Their past experience shows that 90% of passengers with tickets actually arrive at the airport. The airline must decide how many tickets to sell. To help in thinking about this decision, we use two simulations.

The first task is to decide how to assign random numbers to model the behavior of each passenger, who either shows up for the flight or does not. Unlike the simulation with the Sports Stars cards, here each outcome is *not* equally likely. So we must devise a method to assign show versus no-show to each passenger so that the outcomes are weighted 90%-10%. One way to do this is to generate random integers from 0 to 9 and call the appearance of one of the numbers a no-show. We choose to use 0 for no-show. Use either the Table of Random Numbers in the Appendix at the back of this book or technology to generate trials.

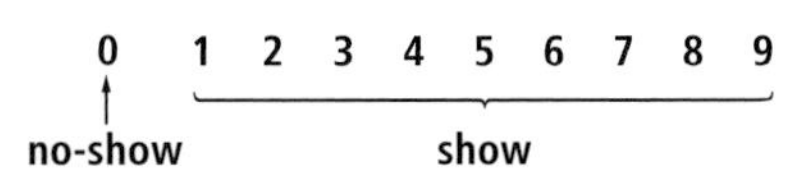

Example 1

Suppose that 24 tickets are sold for the small plane and that the probability of arrival for each passenger is 90%. Use a simulation to estimate how many of the 24 passengers will show up.

Solution 1 A trial will consist of 24 random digits, each representing one passenger. (Listing the "passengers" in blocks of five digits makes the list easier to read.) Six trials are listed below with the no-shows indicated in red. As technology is not always available, using a table of random digits is useful. In this case, we start from some digit in the table and sequentially generate trials.

Trial 1: 89404 63870 33212 74379 7135
Trial 2: 94595 56869 69014 60045 1842
Trial 3: 57740 84378 25331 12566 5867
Trial 4: 38867 62300 08158 17983 1643
Trial 5: 56865 05859 90106 31595 7154
Trial 6: 18663 72695 52180 20847 1223

For each trial, count the no-shows (0s) and those that arrived (digits 1-9).

Trial	no-shows	arrived
1	2	22
2	3	21
3	1	23
4	3	21
5	3	21
6	2	22

The average number of arrivals is $\frac{22 + 21 + 23 + 21 + 21 + 22}{6} \approx 21.7$. The simulation estimates that when 24 tickets are sold, about 22 passengers will show up.

Solution 2 The lists of results for six trials can be generated using technology. The first trial is shown at the right. It has two no-shows. You could generate five other trials and solve the problem as you did in Solution 1.

```
randInt(0,9,24)
{8,9,4,0,4,6,3,8,7,0,3,3,2,1,2,7,4,3,7,9,7,▸
```

The results in Example 1 indicate that it is reasonable for the airline to sell more tickets than there are seats. But how many more? To get a sense of the situation, change the simulation. Remove the stipulation that 24 tickets are sold, and ask a different question: *How many tickets must be sold to have a full flight?*

Example 2

Assuming a 90% probability that a given passenger shows up for a flight, estimate how many tickets should be sold in order to fill all 24 seat

Solution Use the same plan as in Example 1, except now generate random integers until all 24 seats are full. In each trial you must generate digits until your list contains 24 nonzero digits.

(*continued on next page*)

Trial	simulated reservations (0 means no-show)	tickets sold
1	24130 48360 22527 97265 76393 6	26
2	42167 93093 06243 61680 37856 16	27
3	37570 39975 81837 16656 06121 9	26
4	77921 06907 11008 42751 27756 534	28
5	09429 93969 52636 92737 88974	25
6	10365 61129 87529 85689 48237	25
7	07119 97336 71048 08178 77233 13	27
8	51085 12765 51821 51259 77452	25
9	82368 21382 52474 60268 89368	25
10	01001 54092 33362 94904 31273 04148	30

The last column shows the number of tickets that were sold for each flight in order to get 24 passengers who arrived. Averaging these, an estimate is that 26.4 tickets must be sold to have a full flight.

The results from Example 2 suggest that selling 25 tickets, which overbooks by 1, would not usually lead to passengers being bumped. Selling 26 tickets would lead to bumping approximately $\frac{4}{10} \approx 40\%$ of the time because of trials 5, 6, 8, and 9. However, it is important to note that these are rough estimates. While a computer would easily do millions of trials and give better estimates, such predictions only allow us to look at what usually happens. In real life, unusual situations do occur.

Using Simulations to Estimate Probabilities

Probabilities are determined in a variety of ways. In some situations, you might assume a probability, as when you assume a coin is fair. In other situations, you might conduct an experiment and use the long-term relative frequency of an event as an estimate of the probability of the event. In complicated situations, a simulation can be used to generate a relative frequency, thereby estimating the probability.

Activity 2

Two brown-eyed parents can have a blue-eyed child if they each carry a recessive gene for blue eyes. Assume that in these cases, the probability of blue eyes is $\frac{1}{4}$ and the probability of brown eyes is $\frac{3}{4}$. Design a simulation to estimate the probability of having exactly one of three children with brown eyes.

Step 1 Assign the digits 1, 2, 3, and 4 to represent the eye color when one child is born.

Step 2 Simulate forty trials of three-child families and record the results.

Step 3 Use the corresponding relative frequency to estimate the probability that a family with three children has exactly one with brown eyes.

Questions

COVERING THE IDEAS

1. Define *simulation.*
2. Use the random-number generator ten times on your calculator to add two random integers, each between 1 and 6, inclusive. Record the results. This simulates tossing two dice ten times.
3. A local store gives away a random coupon with each visit. Coupons come in three colors: red, blue, and green. If you collect a set of three coupons, one of each color, you get a free ice cream cone. You wish to estimate how many times you must visit the store in order to get a free ice cream cone.
   a. Describe how to assign random digits to model this situation.
   b. Perform 5 trials of your simulation and record the results.
   c. How many visits must you make in order to have a better than 50% chance to get a free ice cream cone?
4. Suppose that in Example 1, a no-show was represented by a 5 rather than by a 0. Using the same list of digits, record the number of no-shows and arrived passengers. Use the data to make another estimate of the number of passengers who will show up when 24 tickets are sold.
5. Suppose that 70% of adults are in favor of raising the driving age. Consider the question "If 10 adults are chosen at random. How often would all 10 be in favor of raising the driving age?" A simulation estimates the answer by assigning the digits 0-6 to "in favor" and the digits 7–9 to "against." Here are 6 trials of this simulation.

Trial 1:	09763	83473
Trial 2:	73577	88604
Trial 3:	67917	27354
Trial 4:	26575	36086
Trial 5:	15011	10536
Trial 6:	45766	96067

   a. Give the results that would be recorded for these trials.
   b. Use the results to answer the question "How often would all 10 adults be in favor of raising the driving age?"

APPLYING THE MATHEMATICS

6. A family has four children. Assume the birth of a boy is equally likely to the birth of a girl.
   a. List the sample space for the genders of the children in birth order, using B for boy and G for girl.
   b. What is the probability that the family has 3 boys and 1 girl?
   c. Run a simulation of 50 families of four children. How close is the relative frequency in the simulation to the actual probability?

7. A survey of a sample of people reveals the relative frequency of left-handedness to be 10% in men and 8% in women.
   a. Use a Monte Carlo simulation to estimate the probability that at least one person in a sample of one man and one woman is left-handed.
   b. Calculate the actual probability.

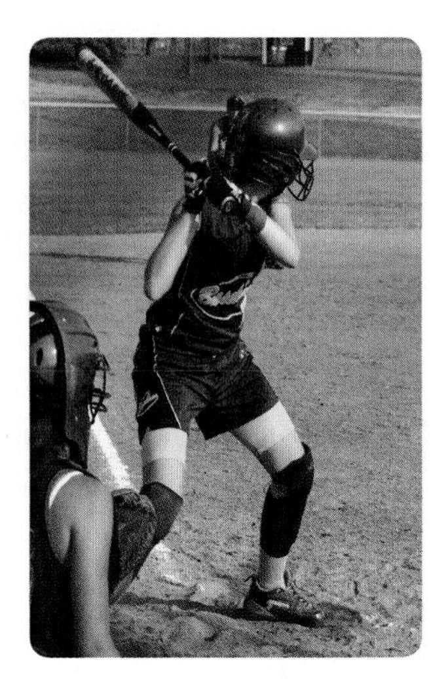

8. Suppose that a manufacturer knows that 2% of the nails produced are defective. Describe how you would design a simulation to determine each value.
   a. the probability that a box with 100 nails has four or more defective ones
   b. how many nails the company should put in each box in order to have at least 100 good nails in a box

9. A softball player has a batting average of 0.300. This means $\frac{\text{number of hits}}{\text{number of official at-bats}} = 0.300.$
   a. Design a simulation to illustrate the next five official at-bats of this player.
   b. Run at least ten trials of your simulation and record the results.
   c. Based on the data of your simulation, estimate the probability that the player gets no hits in the next five at-bats.

10. In 2006, most new cars in North America were a neutral color (gray, silver, black, or white). Select random integers from 1 to 100 to simulate the color of the next 3 cars you pass. Run 20 trials to estimate the probability that all 3 cars are a neutral color.

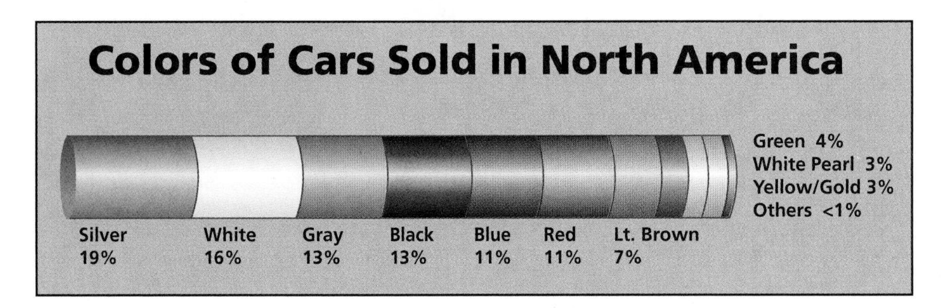

11. A diabetes test is accurate 90% of the time and gives a false result 10% of the time. In a population of ten people, one person is diabetic and the other nine are not. An accurate result for a diabetic means the test result is positive; an accurate result for a non-diabetic means the test result is negative.
    a. Design a simulation of giving the diabetes test to all ten people.
    b. According to fifty trials of your simulation, what is the likelihood that someone tests positive?
    c. According to fifty trials of your simulation, on average, how many non-diabetics test positive for diabetes?

REVIEW

12. A computer software security company estimated that 87.5% of email was spam. It claimed that its software was 98.3% successful in blocking spam while properly passing 99.63% of non-spam email.
    a. What is the probability that an email that is blocked is spam?
    b. Suppose that a sample of 100,000 emails is screened for spam. Use relative frequencies along with the given probabilities to calculate the relative frequency of blocked emails that are spam and confirm your answer to Part a.
    c. Which method of determining the statistic was easier to compute? Explain your answer. **(Lesson 6-6)**

13. A college scholarship program awards first, second and third prize cash awards to the winners. If 700 students applied this year, how many different award outcomes are possible? **(Lesson 6-4)**

In 14 and 15, evaluate. (Lesson 6-4)

14. $7!_{10}P_3$

15. $(n - r)!_nP_r$

16. How many different Thanksgiving dinners are possible if there is 1 turkey, 2 kinds of gravy, 5 side dishes, 4 desserts and 2 kinds of biscuits and if (unfortunately) there are no "seconds" and you can have only 1 serving from each category? **(Lesson 6-3)**

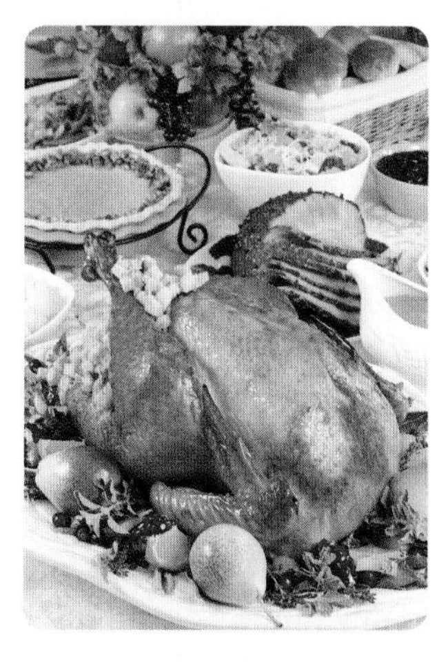

Benjamin Franklin proposed the turkey as the official U.S. bird. Today, 45 million turkeys are eaten each Thanksgiving.

17. a. Graph $f(x) = \frac{1}{x}$ and $g(x) = \frac{1}{x + 3} - 2$ on the same axes.
    b. Find equations for the asymptotes of the function g. How are they related to the asymptotes of f?
    c. Give the domain and range of f and g. **(Lessons 3-2, 3-1)**

EXPLORATION

18. The U.S. mint has produced quarters that each bear a state's name. Each coin has a P (for Philadelphia) or D (for Denver) on it to indicate where it was minted. Suppose you want a complete collection of the 100 coins and you already have 90. If the 100 coins appear at random, conduct a simulation to determine how many quarters you would need to examine before you had all 100.

19. Monty Hall was the host of the TV game show called *Let's Make a Deal!* At the end of each show, one contestant had a chance to win a car. The contestant was told that behind one door was a car and behind the other two doors was an item of little value (such as a goat). The contestant picked a door. Then, without opening the door the contestant picked, Monty would open another door that had a goat. Now the deal: Monty would allow the contestant to take what was behind the door they had chosen or switch doors.
    a. Should the contestant switch or not?
    b. Perform a simulation to determine whether the contestant should switch. Do enough trials to convince yourself which is the better strategy or whether it makes no difference.

Lesson

6-8 Two "Laws," but Only One Is Valid

Vocabulary

expected count of an outcome
expected count of an event

▸ **BIG IDEA** Expected counts and the Law of Large Numbers allow us to predict what happens when there are many trials of an experiment.

Expected vs. Observed Counts

In Jane Austen's famous novel *Pride and Prejudice*, the Bennet family has five daughters. Since, at that time in England, a family home could only be passed on to a male heir, the daughters faced the loss of their family home. Assuming that the probability a baby is a girl is 0.5, on average, we would expect that half of the children born would be girls, leading to an expectation of $5 \cdot \frac{1}{2} = 2.5$ girls. This is consistent with the following definition of *expected count*.

Definition of Expected Count

If an outcome in an experiment has probability p, then in n trials of the experiment, the **expected count** of the outcome is np.

Note that even though there cannot be a fraction of a child, the expected count does not have to be an integer because it measures what happens on average, not with any specific family.

Mental Math

Suppose class participation counts for 20% of a student's grade, homework counts 30%, and tests and quizzes count 50%.

a. What is the highest overall grade out of 100 that McKayla can get if she gets 75 on class participation and 85 on homework?

b. What is her lowest possible grade?

Example 1

A current estimate for the proportion of left-handers in the population is 9%. What is the expected count of left-handers in a classroom of 36 students?

Solution Since $p = 0.09$ and $n = 36$ students, the expected count of the outcome is $np = 36 \cdot 0.09 = 3.24$. About 3 left-handers would be expected in a class of 36 students.

If an event contains many outcomes, then the **expected count of an event** is the sum of the expected counts of the outcomes. For instance, in many places on Earth there is a dry season and a wet season. Suppose that the wet season is 125 days long and there is a 85% probability of rain on those days, while the dry season is the other 240 days of the year and there is a 10% chance of rain on those days. Then the expected number of days of rain in a year is the sum of the expected count for each season: $125 \cdot 0.85 + 240 \cdot 0.10$, or 130.25.

If you observed that it rained 144 days last year in this location, then the difference between observed and expected counts is 144 – 130.25, or 13.75. The difference between the relative frequency of rain you observed and the probability of rain is $\frac{144}{365} - \frac{130.25}{365} = \frac{13.75}{365}$, or about 0.0377 or 3.8%. The expected count does not tell what will happen in any particular case. It tells what we expect to happen ***in the long run***.

The Law of Large Numbers

Everyone expects a fair coin to come up heads 50% of the time, but does that happen in practice? During World War II, John Kerrich, a South African mathematician, was captured by the German army in Denmark and interned at a camp. To pass the time, Dr. Kerrich did experiments on chance processes. In one such experiment he tossed a coin 10,000 times and observed 5067 heads.

GUIDED

Example 2

In Kerrich's experiment, assume that the coin was fair.

a. What was the expected number of heads?

b. Find the difference between the observed and expected numbers of heads.

c. What was the relative frequency of heads?

d. What was the difference between the relative frequency and the probability of heads?

Solution

a. For a fair coin, P(heads) = __?__. Here, n = 10,000. So np = 10,000 • __?__ = __?__.

b. observed – expected = 5067 – __?__ = __?__

c. The relative frequency is $\frac{5067}{?}$ or, as a decimal, __?__.

d. relative frequency – probability = __?__ – __?__ = 0.0067

In Kerrich's experiment, he did not expect to get *exactly* 5000 heads and 5000 tails even if the coin were fair. The difference between the expected number, 5000, and the actual number, 5067, may seem like a lot. But remember that Kerrich tossed the coin 10,000 times! The relative frequency differed from the probability by only 0.0067. The relative frequency gets closer and closer to 0.5 as the number of trials increases. This idea is summarized by the following important statement about probability in the long run.

Law of Large Numbers

Suppose, in an experiment, an event E has probability p. Then if the experiment is repeated again and again, the relative frequency of the event will approach p.

The "Law of Averages"

President Ronald Reagan with baseball sportscaster Harry Caray

When a batter comes up late in the game without a hit, baseball sportscasters often say "He's due for a hit." A gambler who has had a run of bad luck will say, "My luck is due to change." Both of these statements are based on a principle often referred to as the "law of averages." But the "law of averages" is not valid.

The gambler who has had a run of bad luck may be correct that his luck has been bad, but he is no more likely to have a run of good luck now than when he began. The baseball player who has not gotten a hit is no more likely to get a hit now than before his slump.

How do the "Laws" Differ?

Returning to a coin situation, suppose you toss a coin 10 times and 8 heads appear. What is the probability of heads on the 11th toss? The "law of averages" implies that if a fair coin is tossed and more heads have occurred, you should expect more tails to occur going forward so that the number of heads and tails will be the same in the long run. Thus, the probability of heads on the 11th toss would be less than $\frac{1}{2}$. The Law of Large Numbers, by contrast, says that as the number of trials increases, the difference between the relative frequency of heads and the probability of heads will go to zero. The probability of heads, however, remains at $\frac{1}{2}$.

To understand why the Law of Large Numbers is true while the "law of averages" is faulty reasoning, perform the following experiment.

Activity

MATERIALS a coin, paper and a pencil

Step 1 Toss a coin once. Observe heads or tails. Record the outcome in a spreadsheet similar to the one shown below. Your data may be different.

◇	A	B	C	D	E	F
1	Toss *A*	Heads or Tails *B*	Cumulative Number of Heads *C*	Cumulative Number of Tails $D = A - C$	Relative Frequency of Heads $E = \frac{C}{A}$	Cumulative Number of Heads – Cumulative Number of Tails $F = C - D$
2	1	H	1	0	$\frac{1}{1}$	1
3	2	H	2	0	$\frac{2}{2}$	2
4	3	T	2	1	$\frac{2}{3}$	1
5	4	H	3	1	$\frac{3}{4}$	2
6	⋮	⋮	⋮	⋮	⋮	⋮

If B3 = "H"
THEN C3 = C2 + 1
ELSE C3 = C2

D[ ] = A [ ] – C [ ]

E[ ] = C [ ]/A [ ]

F[ ] = C [ ] – D [ ]

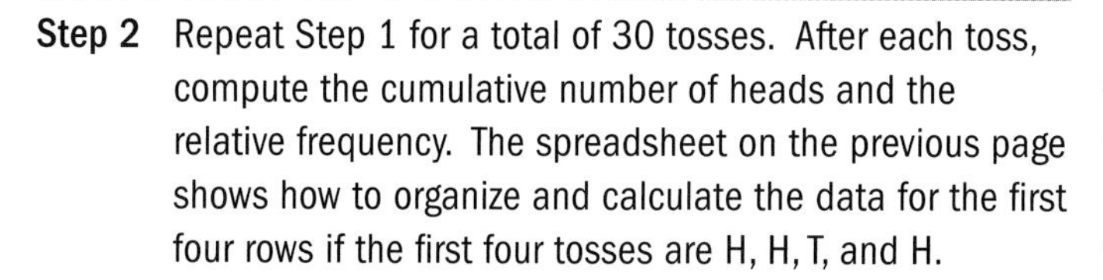

Step 2 Repeat Step 1 for a total of 30 tosses. After each toss, compute the cumulative number of heads and the relative frequency. The spreadsheet on the previous page shows how to organize and calculate the data for the first four rows if the first four tosses are H, H, T, and H.

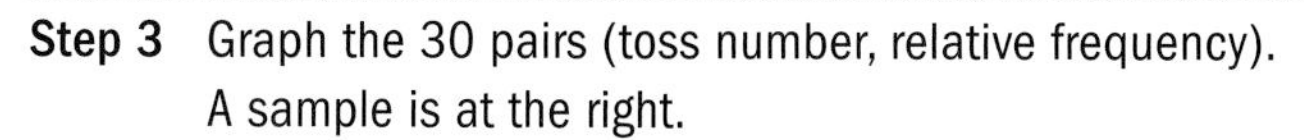

Step 3 Graph the 30 pairs (toss number, relative frequency). A sample is at the right.

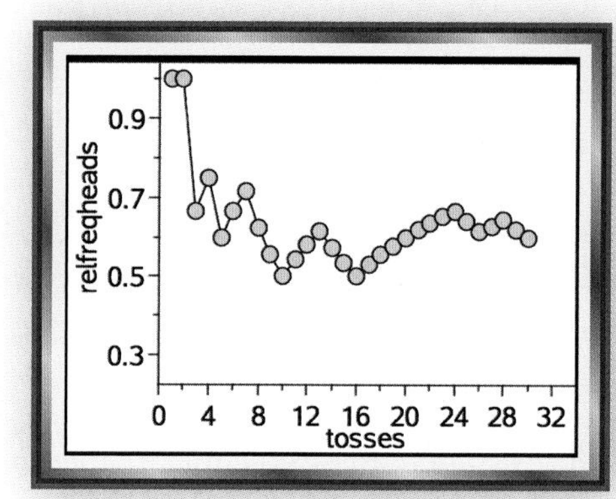

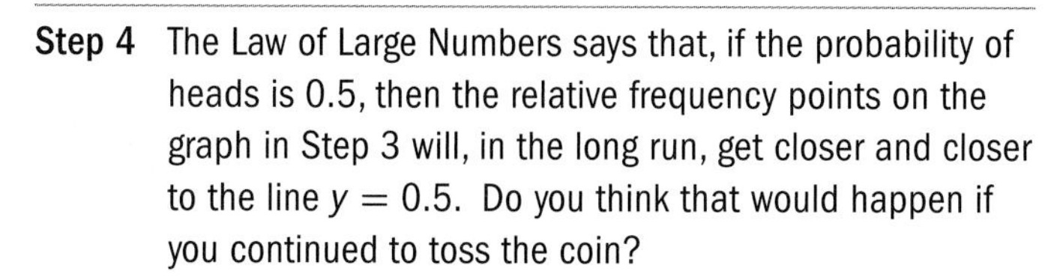

Step 4 The Law of Large Numbers says that, if the probability of heads is 0.5, then the relative frequency points on the graph in Step 3 will, in the long run, get closer and closer to the line $y = 0.5$. Do you think that would happen if you continued to toss the coin?

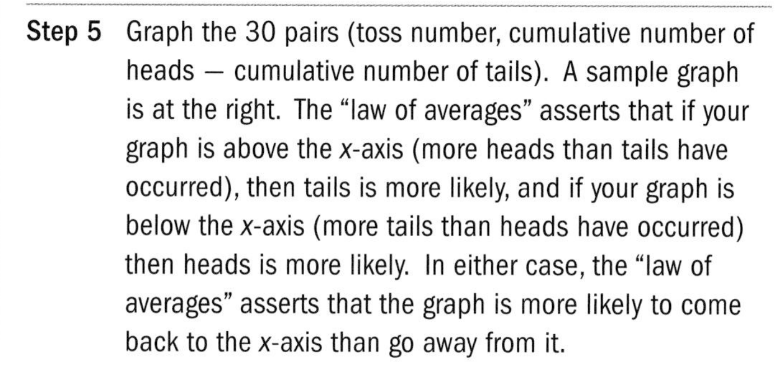

Step 5 Graph the 30 pairs (toss number, cumulative number of heads − cumulative number of tails). A sample graph is at the right. The "law of averages" asserts that if your graph is above the *x*-axis (more heads than tails have occurred), then tails is more likely, and if your graph is below the *x*-axis (more tails than heads have occurred) then heads is more likely. In either case, the "law of averages" asserts that the graph is more likely to come back to the *x*-axis than go away from it.

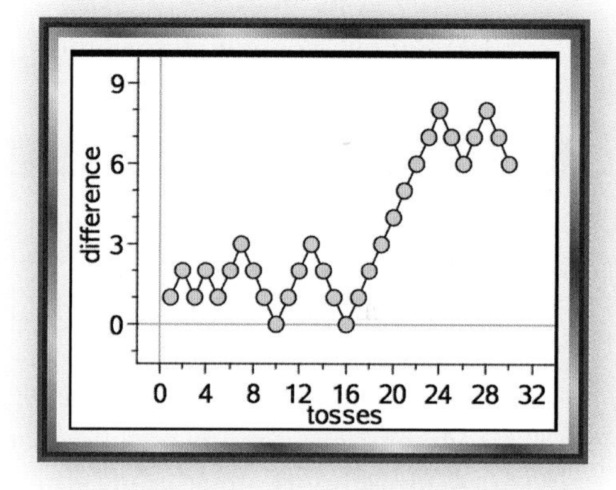

If you were to continue tossing coins, you might get results like those shown in the Activity.

In the language of expected and observed counts, as the number of trials increases, the "law of averages" incorrectly says that the difference between observed and expected values is more likely to go toward 0 than away from it.

Questions

COVERING THE IDEAS

1. Refer to the Bennet family page 410. If 72% of women are tongue curlers, what is the expected number of tongue curlers among the five Bennet daughters?

2. Estimates of the probability that a child is left-handed range from 0.07 to 0.15. Of the 480 children in an elementary school, how many are expected to be left-handed?

3. When tossing a pair of fair dice, the probability of rolling doubles is $\frac{1}{6}$. A class experimented and tossed pairs of dice 550 times, and got doubles 84 times.
   a. What is the difference between the expected count if the dice were fair and the observed count?
   b. What is the difference between the probability (assuming the dice are fair) and the relative frequency?
   c. Do you think the dice are fair? Why or why not?

It is estimated that from $\frac{2}{3}$ to $\frac{4}{5}$ of people can curl their tongue.

4. Since there was a tie between the top two players in a golf tournament, a play-off will be held. The winner will get the first prize of \$80,000, and the runner-up will get a consolation prize of \$35,000. The less-favored player is considered to have a 25% chance of winning. What is this player's expected winnings?

5. A student is playing a board game and needs to toss a 6 on a die to start. She has had five turns but still has not rolled a 6. She announces that she is due a 6 on the next couple of turns. Comment on the student's announcement.

6. A student tosses a coin 10 times and gets 8 tails.
   a. If the coin is fair, do these results contradict the Law of Large Numbers?
   b. What false conclusion would a person make following the "law of averages"?

APPLYING THE MATHEMATICS

In 7 and 8, use the fact that a field goal is worth 3 points and a touchdown is worth 6 points.

7. A football team is at the 28-yard line and it is 4th down, so there is only one more play that can be made. The coach feels that the team can successfully kick a field goal 35% of the time. The team could also try for a touchdown. The coach thinks that this would succeed 20% of the time. What should the team do?

8. When the team is at the 2-yard line, the coach feels that the team can make a touchdown 80% of the time. After the touchdown the team can get an extra point 90% of the time. What is the expected number of points the team will score in this situation?

9. A commuter drives to work. When the traffic is normal, about 75% of the time, the drive averages 30 minutes. When the traffic is heavy, about 25% of the time, the drive averages 40 minutes. What is the expected travel time?

10. In investing, a general rule is that the less risk of losing principal in an investment, the lower the guaranteed interest rate. Consider a \$10,000 bond that pays 10% interest annually, but with a 1 in 25 chance that the investment will lose all of its principal by the end of that year. What is the expected gain or loss in that year?

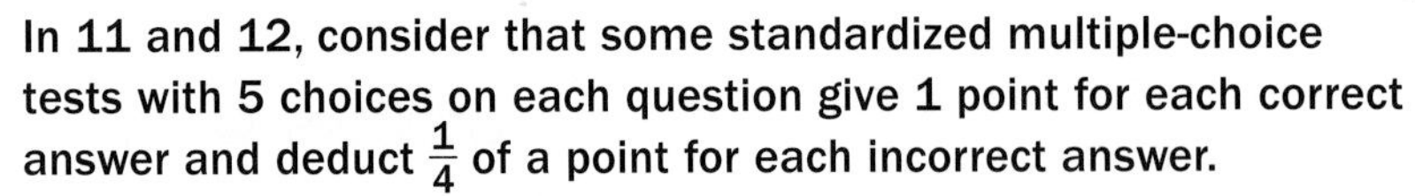

In 11 and 12, consider that some standardized multiple-choice tests with 5 choices on each question give 1 point for each correct answer and deduct $\frac{1}{4}$ of a point for each incorrect answer.

11. a. If you randomly guess on 10 questions, what is your expected point total? Explain your calculation.
    b. Generalize Part a to apply to a situation where you randomly guess on n questions.
    c. With this scoring, is it to your advantage in the long run to guess?

12. If on each question you can eliminate 1 of the choices, so you randomly guess from the other 4 choices, what is your expected point total on a 40 question test? Explain your calculations.

13. Leona rolls a die 10 times and gets the distribution described by the table below. She now suspects that the die is biased.

Outcome	1	2	3	4	5	6
Frequency	0	0	0	2	1	7

a. Make a relative frequency bar graph showing Leona's distribution.

b. Do the data provide evidence in support of Leona's assertion? Why or why not?

c. Leona decides to subject the die to a more complete test. She rolls it 100 times and gets the following distribution.

Outcome	1	2	3	4	5	6
Frequency	15	14	16	19	15	21

Do the new data provide evidence in support of Leona's assertion? Why or why not?

d. Which of the laws of this lesson is supported by Leona's results?

REVIEW

14. Suppose that a baseball player has a batting average of 0.300. Use your calculator to simulate his next 100 official at-bats, recording whether he gets a hit or not. What is the longest streak of hits that he gets in your simulation? **(Lesson 6-7)**

15. Suppose that 12% of women and 9% of men have attached ear lobes. **(Lesson 6-7)**

a. Use a Monte Carlo simulation to estimate the probability that at least one person in a sample of two women has attached ear lobes.

b. Calculate the actual probability

16. Consider the following situation. Your gym class will play football for the next 15 class days. Each class day, students are randomly assigned to one of the 11 field positions.
(Lessons 6-7, 6-1)

a. You decide to perform a simulation to see how likely it is that you will get to play quarterback over the 15 days. Which is the better simulation strategy, (1) or (2), and why?

(1) Drawing the numbers 0–10 from a box, assigning each position a number.

(2) Taking the sum of two dice and assigning each position a number between 2 and 12.

b. What is the number of trials in this experiment?

17. a. How many 3-letter permutations are there in the letters in WYOMING?

b. How many of the permutations in Part a contain only consonants? **(Lesson 6-4)**

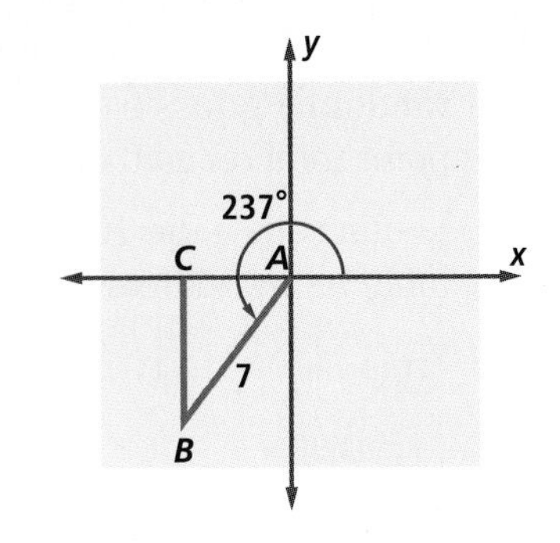

18. A triangle is drawn in the coordinate plane as shown.

a. Find m$\angle BAC$.

b. Find the coordinates of point B to the nearest tenth. **(Lessons 4-4, 4-1)**

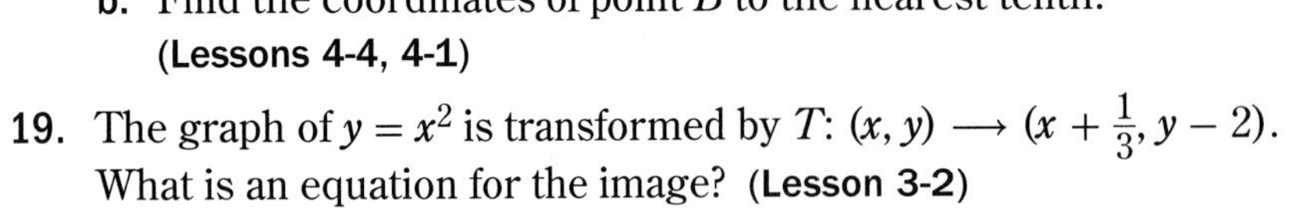

19. The graph of $y = x^2$ is transformed by T: $(x, y) \longrightarrow \left(x + \frac{1}{3}, y - 2\right)$. What is an equation for the image? **(Lesson 3-2)**

20. Put the following correlations in order from strongest correlation to weakest. **(Lesson 2-3)**

A $r = -0.92$ B $r^2 = 0.25$

C $r = 0.4$ D $r^2 = 0.8$

21. The bar graphs at the right show the distributions of 100 tosses for a six-sided green die with 1, 2, 3, 4, 5 and 6 on its faces, a six-sided red die with 2, 3, 4, 5, 6 and 6 on its faces, and an eight-sided blue die with 1, 2, 2, 3, 3, 4, 5 and 6 on its faces. Match each die with its graph. **(Lesson 1-1)**

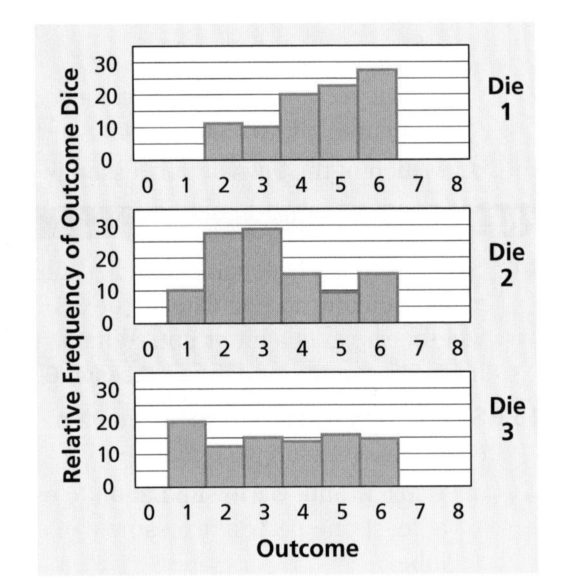

EXPLORATION

22. The Law of Large Numbers was first presented in the book *Ars Conjectandi* by Jakob Bernoulli (1654–1705). Research Bernoulli's contributions to the field of probability.

Lesson 6-9 The Chi-Square Test

Vocabulary

hypothesis
hypothesis testing
chi-square statistic, χ^2
degree of freedom
significance level

BIG IDEA The chi-square statistic measures the extent to which observed counts differ from expected counts.

Mental Math

Find a solution to the nearest 10.

a. $\frac{523}{1681} = \frac{x}{100}$

b. $\frac{74}{y} = \frac{26}{31}$

Suppose you toss a coin 50 times. Consider the following scenarios.

(1) You think the coin is fair and heads comes up 43 times. You would suspect that the coin is not fair. The event "43 or more heads in 50 tosses" is so unlikely with a fair coin that you would be right to think the coin is biased in favor of heads.

(2) You think the coin is fair and heads comes up 30 times. Most people would conclude that the coin is fair and only the normal variability of results in such an experiment has taken place.

(3) You think the probability of heads is 80%, and heads comes up 43 times. You might pat yourself on the back for being correct about the coin being weighted, but wonder if 80% is a reasonable probability.

QY1

QY1

What is the expected number of heads in each of scenarios (1), (2), (3)?

In all three scenarios, there is an expected number and an observed number of heads. The probability that has led to the expected number is a **hypothesis** about the situation. The tossing of the coin is an experiment to test that hypothesis. This use of statistics is called **hypothesis testing**.

Chi-Square Statistic

In hypothesis-testing situations like these, the *chi-square* or χ^2 statistic (χ is the Greek letter "chi", pronounced "kai") can be used. The chi-square statistic can help you determine whether certain results are due to chance variation or whether they indicate that the hypothesis is wrong. This statistic was introduced by the English statistician Karl Pearson in 1900.

Definition of Chi-square Statistic

If data is collected on a variable with k outcomes, the **chi-square statistic** $\chi^2 = \sum_{i=1}^{k} \frac{(a_i - e_i)^2}{e_i}$, where a_i is the observed frequency for outcome i, and e_i is the expected count of the outcome i.

The calculation of the chi-square statistic requires that you state hypotheses about probabilities in a situation so you can calculate the expected count. The value of the chi-square statistic is relatively small if data agrees with the model and relatively large if the data does not follow the model.

Example 1

To test if the coin is fair, compute the chi-square statistic for the coin that came up heads 43 times in 50 tosses.

Solution There are two outcomes for a coin toss, heads or tails, so $k = 2$. First determine the expected count of each: How many heads e_1 would be expected if the coin were fair? How many tails e_2 would be expected? If the coin is fair, P(heads) = 0.5 and P(tails) = 0.5. So the expected number of heads, $e_1 = 0.5 \cdot 50 = 25$. Similarly, $e_2 = 25$.

$$\chi^2 = \sum_{i=1}^{2} \frac{(a_i - e_i)^2}{e_i} = \frac{(a_1 - e_1)^2}{e_1} + \frac{(a_2 - e_2)^2}{e_2}$$

Scenario (1)	Outcome 1 HEADS	Outcome 2 TAILS
Observed Frequency	$a_1 = 43$	$a_2 = 7$
Expected Frequency	$e_1 = 25$	$e_2 = 25$
$\frac{(a_i - e_i)^2}{e_i}$	$\frac{(43 - 25)^2}{25} = 12.96$	$\frac{(7 - 25)^2}{25} = 12.96$

So, $\chi^2 = 12.96 + 12.96 = 25.92$

At this point, you have nothing with which to compare the chi-square value of 25.92. Complete Example 2 to get a comparison value and determine whether 25.92 is large or small.

GUIDED

Example 2

To test if the coin is fair, compute the chi-square statistic for the coin that came up heads 30 times in 50 tosses.

Solution Complete the table.

Scenario (2)	Outcome 1 HEADS	Outcome TAILS
Observed Frequency	$a_1 =$?	$a_2 =$?
Expected Frequency	$e_1 =$?	$e_2 =$?
$\frac{(a_i - e_i)^2}{e_i}$	$\frac{(30 - 25)^2}{25} =$?	$\frac{(? - ?)^2}{?} =$?

The chi-square statistic is $\chi^2 =$? + ? = 2.

Notice that the chi-square value for the coin in Example 2 is much smaller than the one for the coin in Example 1. This is because the observed frequencies are closer to the expected frequencies.

 QY2

Simulating the Likelihood of a Chi-square Value

We still have not determined how far the observed frequencies can be from expected frequencies to still be considered variation due to chance. One way to determine this is to simulate the coin-tossing experiment a large number of times, calculate the chi-square statistic for each experiment, and see how often a specific value occurs.

▶ **QY2**

Can the value of the chi-square statistic be negative? Why or why not?

Activity 1

MATERIALS chi-square application or technology file provided by your teacher

The following simulation shows how a particular χ^2 value ($\chi^2 = 2$ in Example 2) compares to 1000 generated χ^2 values. One experiment consists of a random generation of 50 heads or tails. The difference between these simulated counts of heads and tails with the expected numbers of heads and tails are used to calculate the chi-square statistic seen in Examples 1 and 2. This statistic is computed and recorded. One thousand experiments are run in one simulation.

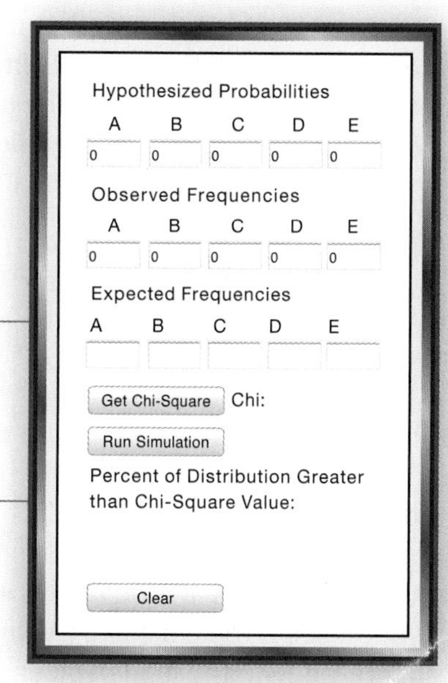

Step 1 Consider Example 2. Record 0.5 and 0.5 in the first two cells for probabilities. These are hypothesized probabilities because that is what χ^2 is testing: how well the observed frequencies under these probabilities match the expected count.

Step 2 Put the observed counts of heads and tails in the cells for observed frequencies. For Example 2, these are 30 and 20, respectively.

Step 3 Press the "Get Chi-Square" button and record the chi-square value.

Step 4 Run the simulation. When the simulation is finished, a distribution of the chi-square values will appear on your screen graphed as a histogram. Record the percent of experiments out of the 1000 that have χ^2 larger than the original chi-square value of 2. Compare your percent to that of others in your class.

Step 5 Summarize the results of the simulation in a sentence: If the hypothesized probabilities are accurate, a chi-square value larger than the original chi-square value would occur about __?__ of the time by chance.

Step 6 Repeat Steps 1–5 to simulate Example 1.

Step 7 Repeat Steps 1–5 to simulate scenario (3) on page 417.

Significance Level and the Chi-Square Table

The percent produced in the simulation of Example 2 in Activity 1 is usually between 11% and 12%. Thus, in the long run, you could expect about 12% of the experiments to have a χ^2 larger than 2 by chance. A relative frequency of 12% is about 1 in 8 and is not particularly unusual. This is not low enough to *reject* the hypothesis that the coin is fair.

It would be cumbersome to always calculate at least two chi-square values in order to compare the likelihood of one situation to another or to simulate 1000 experiments in order to estimate the likelihood of such a chi-square value. How do you decide when to reject a hypothesis? Luckily, you can compare a calculated chi-square value against a table like the one shown below. In this table, n is the number of outcomes and $n - 1$ is called the number of **degrees of freedom**. The probabilities 0.25, 0.10. 0.05, 0.01, and 0.001 indicate how likely a chi-square value that large is to occur. These are the critical probabilities we test hypotheses against, and are known as **significance levels.**

Critical Chi-Square Values

Degrees of Freedom $= n - 1$	Significance Level				
	0.25	0.10	0.05	0.01	0.001
1	1.32	2.71	3.84	6.63	10.83
2	2.77	4.61	5.99	9.21	13.82
3	4.11	6.25	7.81	11.34	16.27
4	5.39	7.78	9.49	13.28	18.47
5	6.63	9.24	11.07	15.09	20.51

Common practice is to use 0.01 or 0.05 as the significance level. If a chi-square value for the original experiment is *larger* than the chi-square value in the table at a particular significance level, then we reject the hypothesis. If the chi-square value you calculate is *less* than the chi-square value in the table, you cannot reject the hypothesis or conclude that the experiment is unlikely at the particular significance level. We say there is ***insufficient evidence*** to reject the hypothesis, not that the hypothesis is true.

For example, suppose we pick 0.01 as the significance level. In Example 1, $n = 2$, so look at row 1 in the table and the 0.01 column. It shows that a chi-square value as large as 6.63 would be expected to occur with a probability of 0.01. A fair coin flipped 50 times yielding 43 heads corresponds to a chi-square value of 25.92. Since $25.92 > 6.63$ at the 0.01 significance level, we reject the hypothesis that the coin is fair.

QY3

You are not expected to know how the values in the chi-square table were calculated. The mathematics needed to calculate them is usually studied in college-level statistics courses. In fact, instead of the table, you can use a chi-square function on a statistics utility that returns the probability for the experiment in question.

QY3

Test the hypothesis that the coin is fair in Example 1 using the chi-square table at the 0.001 significance level.

We did this for Example 2. We set the lower bound to the chi-square value 2 that was calculated, the upper bound to infinity, and input 1, the degrees of freedom. The calculator computation verifies that the result in Example 2 is not unlikely. A chi-square value of 2 or larger has a probability of 15.7%. We cannot reject the hypothesis that the coin is fair.

More than Two Outcomes

The probabilities of getting 30 or 43 heads in 50 tosses of a fair coin can be calculated exactly. You will learn how to calculate these probabilities in Chapter 10. So simulations are not needed for coin-tossing experiments. But when there are more than two outcomes to an experiment, calculating probabilities can be quite tedious or nearly impossible, and simulating the distribution of χ^2 values is very helpful.

This is the case with the Titanic data from Lesson 6-5. Activity 2 uses the chi-square statistic to answer the question, "Was the number of survivors so unequal by class that it is unlikely to have happened by chance?" There were more passengers in first-class than in second-class, so you might expect that there would be more survivors in first class than in second. About twice as many first-class passengers survived as second class. If chance of survival was evenly distributed, is this likely?

Activity 2

MATERIALS chi-square application or technology file provided by your teacher

Step 1 Here is the first table from Lesson 6-5, with row percent totals for the number of people in each class. We consider the "crew" as a class.

Passengers	First	Second	Third	Crew	Total
Survived	203	118	178	212	711
Died	122	167	528	673	1490
Totals (Percent)	325 14.8%	285 12.9%	706 32.1%	885 40.2%	2201

Describe a hypothesis to test by filling in the blanks: The relative frequency of survivors by class equals the relative frequencies of all passengers by class: 14.8% first class, __?__% second class, __?__% third class, and __?__% crew.

Step 2 Calculate expected counts. 14.8% of people on the Titanic were traveling first class, so under the hypothesis, 14.8% of the survivors should be first class. There were 711, survivors in all, so 14.8% of 711 or about 105 survivors are expected in first class. Fill in the table below. The total number of expected survivors should add to 711.

	First	Second	Third	Crew	Total
Actual (Observed) Number of Survivors	203	118	178	212	711
Expected Number of Survivors	105	$e_2 = ?$	$e_3 = ?$	$e_4 = ?$	711

(continued on next page)

Step 3 Compute the contributions to chi-square $\frac{(a_i - e_i)^2}{e_i}$ for each of the four classes using the values for e_2, e_3, and e_4 from Step 3. Place them in a third row of the table. The contribution for first class is given.

	First	Second	Third	Crew	Total
Actual (Observed) Number of Survivors	203	118	178	212	711
Expected Number of Survivors	105	e_2	e_3	e_4	711
Contribution to Chi-square	91.5	?	?	?	?

Step 4 Add the 4 numbers you found in Step 3 to get the chi-square statistic.

Step 5 Check by using a statistics utility to compute the chi-square statistic for the Titanic data. Some utilities use a chi square "goodness of fit" (GOF) function, which has as inputs the list of the observed values, the list of the expected values, and the degrees of freedom. The screenshot at the right shows a χ^2 value of 128.9. The "Pval" indicates that the probability that a value as high as about 128.9 would occur is about 9.21×10^{-28}, or almost 0. That means that a chi-square as large as 128.9 would occur by chance far less than 0.001 of the time. Therefore, we reject the hypothesis that the Titanic survivors were distributed proportionally to the number of passengers in each class when tested at the 0.001 significance level. There were significantly different percents of survivors in the classes than would be expected by chance.

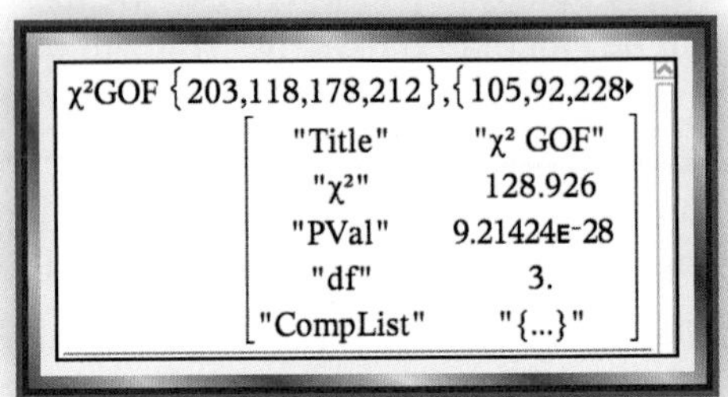

Hypothesized Probabilities

A	B	C	D	E
.148	.129	.321	.402	0

Observed Frequencies

A	B	C	D	E
203	118	178	212	0

Step 6 Analyze these data using the web applet provided by your teacher to verify that the probability of such a high χ^2 value is very close to 0.

Questions

COVERING THE IDEAS

In 1–4, refer to Activity 2.

1. **a.** What do the a_i stand for? **b.** What do the e_i stand for?
2. What equation needs to be solved to find e_3?
3. Show the calculation of $\frac{(a_3 - e_3)^2}{e_3}$.
4. Identify the degrees of freedom.
5. A coin is tossed 50 times and 33 heads occur.
   a. What value do you obtain for the chi-square statistic testing whether this coin is fair?
   b. What is the probability of chi-square value this large?
   c. Is this event so unusual that you would conclude that the coin is likely to be biased?

6. Recall from Lesson 6-3 that John Kerrich got 5067 heads in 10,000 tosses of a coin. Use a chi-square statistic to test the hypothesis that his coin was a fair coin at the 0.01 significance level.

7. A die is tossed 120 times and the values occur with the frequencies as in the table at the right. Accordingto a chi-square test, is there reason to believe the die is biased? Justify your answer.

Number Showing	1	2	3	4	5	6
Frequency	24	17	18	21	16	24

8. Here is part of a table from Lesson 6-5. Follow the steps to apply a chi-square test to determine whether there was a significant effect on having Hepatitis C depending on where the tattoo was done.

	Tattoo Done in Commercial Tattoo Parlor	Tattoo Done Elsewhere
Has Hepatitis C	17	8
No Hepatitis C	35	53

a. Copy the table and add totals in appropriate places.

b. What percent of people in this study had Hepatitis C?

c. Calculate the number of people in this study who got their tattoo in a commercial tattoo parlor who would be expected to have Hepatitis C if location had no effect.

d. Calculate the number of people in this study who did not get their tattoo in a commercial tattoo parlor who would be expected to have Hepatitis C if location had no effect.

e. Calculate the chi-square statistic using the actual numbers of people who had Hepatitis C and the expected counts you found in Parts c and d.

f. What is the probability of a chi-square value this large?

g. At the 0.01 significance level, does the chi-square test suggest that location had any effect on a person's chances of having Hepatitis C?

9. In the years 2001–2007, 111 storms in the Atlantic, Caribbean, and Gulf of Mexico were classified as cyclones by the National Hurricane Center. A cyclone is a closed, rotating weather system that may or may not have hurricane-strength winds. The data are presented in the table below.

Year	2001	2002	2003	2004	2005	2006	2007
Number of Cyclones	15	12	16	15	28	10	15

Source: National Hurricane Center

People wonder if some years have significantly more cyclones than other years, or if the differences among years are due to chance.

a. Calculate the numbers of expected storms in these years if each year had an equal number of storms. Copy the table above and add a row for the expected counts.

b. Find the chi-square statistic using these values.

c. Find the probability of a chi-square value this large.

d. Does the chi-square statistic provide evidence to reject the claim that cyclones occur in the same frequency over these years at the 0.05 significance level? Why or why not?

Aerial view of a well-defined cyclone

APPLYING THE MATHEMATICS

10. You toss a coin 100 times. At the 0.001 significance level, if the coin is fair, the number of heads should be between which two numbers?

In 11 and 12, consider that there are 4 quarters in a professional basketball game. Do teams tend to score more points in certain periods than others? Here are the points scored by each team in the 12 NBA basketball games played on November 16, 2007. Three games went into overtime, so teams were tied (T) at the end of the 4 quarters. Other teams are indicated by who won (W) and who lost (L).

Team won, lost, or tied	First Quarter	Second Quarter	Third Quarter	Fourth Quarter
L	30	26	18	27
W	22	31	28	29
L	24	22	19	26
W	24	28	24	16
T	26	32	26	23
T	23	22	37	25
T	32	30	19	33
T	23	35	26	30
W	32	15	24	34
L	24	27	23	15
L	20	26	22	23
W	29	19	14	41
L	21	24	28	15
W	14	13	29	36
W	25	20	24	26
L	10	22	27	11
L	24	19	31	20
W	19	20	29	31
L	24	22	18	20
W	25	24	20	21
T	31	16	26	24
T	28	29	25	15
L	26	29	32	18
W	34	30	32	26

11. Add the data from all 24 teams and perform a chi-square test to determine whether, in general, teams tend to score more points in certain periods than others. Use the 0.05 significance level.

12. Add the data from the 9 teams that won before overtime and perform a chi-square test to determine whether winning teams tend to score more points in certain periods than others. Use the 0.05 significance level.

REVIEW

13. A science experiment is expected to have a successful outcome 40% of the time. In 20 repetitions of the experiment, what is the expected count of the event of a failure? **(Lesson 6-8)**

14. **True or False** As Dr. Kerrich tossed his coin 100,000 times, according to the Law of Large Numbers, we would expect the relative frequency of heads to approach the probability of getting heads. **(Lesson 6-8)**

15. Suppose a person with a certain medical condition has a probability of 0.45 of recovering fully if the person undergoes surgery. Also suppose that 100 surgeries for this condition are performed. Design a simulation for this situation and explain how you would run the simulation using technology. **(Lesson 6-7)**

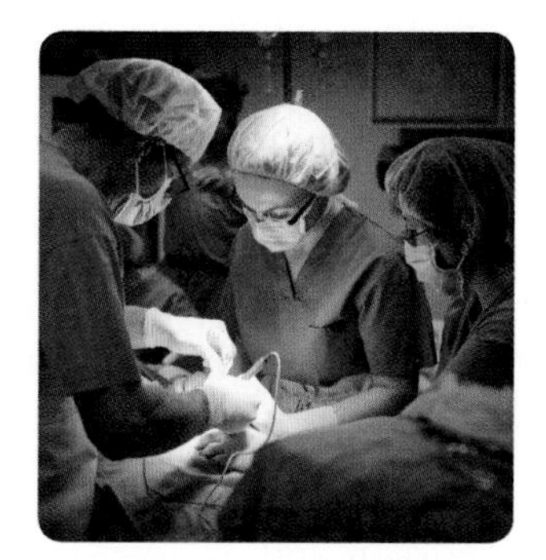

16. Is the equation an identity? If not, change the right side of the equation to make it an identity. **(Lesson 4-3)**
    a. $\cos(90° - \theta) = \cos\theta$
    b. $-\cos\theta = \cos(360° - \theta)$
    c. $\sin(-\theta) = -\sin\theta$
    d. $\tan(180° + \theta) = \tan\theta$

17. For a given set of data, $\bar{x} = 22$ and $s = 9$. Find the z-score for each number. **(Lesson 3-9)**
    a. 15
    b. 24
    c. 31

EXPLORATION

18. In *A Mathematician Reads the Newspaper*, John Paulos reports that if you spin a coin on its edge, heads will occur only about 30% of the time. Perform an experiment, spinning a penny on its edge at least 150 times. Then do a chi-square test to determine whether Paulos's report seems to be correct.

QY ANSWERS

1. 25; 25; 40
2. No. Each contribution to chi-square is nonnegative because each difference in the numerator is squared and the denominator is a count.
3. $10.83 < 25.92$, so the hypothesis is rejected even at the 0.001 level.

Chapter 6 Projects

1 The Sum of Many Dice

In Lesson 6-1, the 36 possible outcomes for a toss of two dice are listed and used to determine the probability of obtaining each possible sum. Since it becomes impractical to list all 216 outcomes for more than two dice, a simulation may be used to estimate the probabilities of getting various sums when three or more dice are tossed.

a. Simulate the sum of a toss of three fair dice 1000 times. Create a relative frequency graph of the sums obtained. Predict the probability of each.

b. Create and analyze the relative frequency graph of the sum of 10 fair dice tossed 1000 times.

c. Simulate 1000 tosses of two dice, one fair and one weighted such that a 6 occurs with probability $\frac{1}{2}$, and 1 through 5 each occur with probability $\frac{1}{10}$. Explain how you performed the simulation and display your results in a histogram.

2 Random Walk

A random walk is a trajectory generated by taking successive random steps.

a. Suppose a person stands on a north-south street at a random position marked "0." The person tosses a fair coin to determine which direction he walks. If a single toss of the coin yields heads, he takes one step north; otherwise, he takes one step south. After each step, he flips the coin again and takes one step, and so on.

i. If the man takes 10 steps, at what possible final positions might the man stand? Of the possible positions, which do you think are most likely and why?

ii. Simulate a walk of 10 steps 100 or more times using a spreadsheet application. Find and graph the relative frequencies of the final positions. Does this agree with your expectations?

b. The one-dimensional case of Part a can be extended to 2 or more dimensions. Describe a possible two-dimensional random walk. What are the possible final positions after 10 steps and what are the most probable positions?

3 The "Law of Averages"

Lesson 6-8 discussed the invalidity of the "law of averages." Here, we test this claim.

a. Generate a sequence of 500 tosses of a fair coin.

b. Look at the sequence to find "streaks" of 3 heads in a row. For each streak, record the result of the next toss, i.e. create the set of every toss following a streak of 3 heads. What are the relative frequencies of heads and tails in this set?

c. For the same 500 tosses, look at streaks of 3 tails and create the set of tosses following each streak. What are the relative frequencies of heads and tails in this set?

d. Is a streak of heads more likely followed by tails than heads or a streak of tails more likely followed by heads than tails? Explain your answer.

4 Probabilistic Analysis of Functions

Investigate the following situation. If a, b, and c are numbers from 0 to 9, what is the probability P that the function f with equation $f(x) = ax^2 + bx + c$ has real roots?

a. Consider the case where a, b, and c are integers from 0 to 9. Calculate the probability of having real roots. (Hint: There are 1000 ways to assign values to a, b, c. Also, consider the cases when a is zero and a is nonzero separately.)

b. Consider the case where a, b, and c are real numbers from 0 to 9. Use technology to simulate this case and estimate the probability that f has real roots.

Hideki Matsui of the Yankees and Ryan Howard of the Phillies during the 2009 World Series

5 Probability Trees

In Lesson 6-6, you used contingency tables and graphs to help you determine the probabilities involved with the diagnosis of strep throat. Graphs may be used in other situations to help determine probabilities.

A situation involving probabilities important to TV broadcasters is the World Series of Major League Baseball. Since there may be 4, 5, 6, or 7 games depending on the performance of the teams, broadcasters need to estimate the most likely length of a series in order to develop program and advertising schedules.

a. To help determine the most likely series length, a probability tree is helpful. Draw a probability tree of all possible outcomes of the World Series between two teams A and B. Assume each team has an equal chance of winning each game. (Hint: Since the series ends when one team wins 4 games, the various branches of the tree will terminate at different points.)

b. Use the tree to determine the most likely series length.

Chapter 6 Summary and Vocabulary

- Probabilities are calculated from assumptions about **outcomes** in a **sample space**. If all outcomes in a finite sample space are equally likely, the **probability of an event** is the ratio of the number of individual outcomes making up the **event** to the number of outcomes in the sample space. The probability of the **union** of two events A and B satisfies $P(A \cup B) = P(A) + P(B) - P(A \cap B)$. If A and B are **mutually exclusive,** $P(A \cup B) = P(A) + P(B)$. If A and B are **complementary**, $P(B) = 1 - P(A)$.

- The occurrence of one event may affect the probability that another event occurs. The probability of event B given that event A has occurred is written $P(B \mid A)$. If $P(B \mid A) = P(B)$, then events A and B are **independent**; this definition is equivalent to the statement $P(A \cap B) = P(A) \cdot P(B)$. **Contingency tables** help organize information in order to determine whether events are independent, to compute **conditional probabilities**, or to suggest patterns in data. It is important to be able to identify conditional probability situations clearly: for example, when a test comes back positive for a disease, the probability that you actually have the disease is a conditional probability different from the probability that an arbitrary test result is correct.

- The number of ways to choose one element each from two finite sets A and B is $N(A) \cdot N(B)$. An ordered list of symbols is a **string**. By the Multiplication Counting Principle, the number of strings of n different items, *without* replacement, is $n!$. The number of strings of **length** k of n items with replacement is n^k. The number of strings of r items out of a given set of n *without* replacement (also called a **permutation of *n* objects taken *r* at a time**) is denoted $_nP_r = n(n-1) \cdot \dots \cdot (n-r+1) = \frac{n!}{(n-r)!}$.

- Random numbers can be approximated manually (such as by throwing dice) or can be generated using a calculator or computer. Many real-life situations can be simulated by using **random** numbers to code experiments or events, for example, the number of people with reservations who actually show up for a flight. The use of randomness to generate relative frequencies obtained from simulated repeated trials of an experiment and then to estimate probabilities is called the **Monte Carlo method**.

Vocabulary

6-1
probability theory
experiment
*outcome
*sample space
*event
*probability of an event, $P(E)$
*fair, unbiased
randomly, at random
empty set, null set

6-2
union of sets
disjoint sets
mutually exclusive events
intersection of sets
complementary events
complement of A, not A

6-3
string
length of a string
independent events
dependent events

6-4
*permutation
*permutations of n objects taken r at a time, $_nP_r$

6-5
contingency table
Simpson's Paradox

6-6
*conditional probability of an event, $P(B \mid A)$

It is a commonly held belief that when an event has recently occurred that the complement event is now more likely. Though the Law of Large Numbers says that over a large number of trials the relative frequency of an event approaches its probability. A particular coin toss, for instance, is not any more likely to be heads if the previous five tosses have all been tails than if they were all heads. A **chi-square statistic** can suggest whether observed counts are close to **expected counts** and can be used to estimate the probability of an observed distribution of data agreeing with a hypothesis.

Vocabulary

6-7
simulation
random
trial
Monte Carlo method

6-8
expected count of an outcome
expected count of an event

6-9
hypothesis
hypothesis testing
chi-square statistic, χ^2
degrees of freedom
significance level

Properties and Theorems

Basic Properties of Probability (p. 363)
Addition Counting Principle (Mutually Exclusive Form) (p. 367)
Probability of the Union of Mutually Exclusive Events Theorem(p. 368)
Addition Counting Principle (General Form) (p. 369)
Probability of the Union of Events Theorem (General Form) (p.369)
Probability of Complements Theorem (p. 371)
Multiplication Counting Principle (p. 374)
Strings with Replacement Theorem (p. 376)
Formula for ${}_nP_r$ Theorem (p. 382)
Alternate Formula for ${}_nP_r$ Theorem (p. 382)
Formula for ${}_nP_n$ Corollary (p. 384)
Law of Large Numbers (p. 411)

Chapter 6 Self-Test

Take this test as you would take a test in class. You will need a calculator. Then use the Selected Answers section in the back of the book to check your work.

1. Consider the experiment of tossing three different coins.
   a. Write the sample space for the experiment.
   b. List the outcomes in the event "at least two tails show up."
2. a. Evaluate ${}_{15}P_6$.
   b. What is meant by ${}_{15}P_6$?
3. How many four-letter permutations can be made from the letters in SNOWFLAKE?
4. Consider the experiment of tossing a fair coin and a fair six-sided die. The sides of the coin are marked 1 and 2. The faces of the die are marked from 1 to 6. Find the probability that the sum is less than five.
5. A consumer group reports that 15% of Brand X dental floss packages contain less than the advertised length of floss. Four packages are chosen at random. Assume that the consumer group is correct.
   a. What is the probability that all four packages contain less than the advertised amount?
   b. What is the probability that all four packages contain at least the advertised amount?
6. If you have 4 pairs of jeans, 5 t-shirts, and 2 pairs of sneakers, how many different outfits consisting of a pair of jeans, a t-shirt, and a pair of sneakers can you make?
7. Suppose $P(A \cap B) = 0.6$, $P(A) = 0.7$, and $P(B) = 0.8$.
   a. Find $P(A \cup B)$. b. Find $P(B \mid A)$.
8. The criminal justice department at a local university estimates that in 85% of the cases brought to trial in the county court the defendants is guilty, that 10% of guilty defendants are found innocent, and that 5% of innocent defendants are found guilty.
   a. Make a table or tree diagram of the situation.
   b. What is the probability that a defendant who is found guilty is actually innocent?
   c. What is the probability that a defendant who is found innocent is actually guilty?

In 9 and 10, **True or False**. Explain your reasoning.

9. If A and B are complementary events and $P(A) = k$, then $P(B) = 1 - k$.
10. The expected count of the outcome 2 in 600 rolls of a fair six-sided die is 60.
11. The information booklet about a wildlife reservation park states that the probability of observing an eagle on a given day is 0.21, and the probability for a hawk is 0.17. The booklet also indicates that, based on records, 2% of the visitors observe both of these prey birds on the same day. In this park, are observing an eagle and observing a hawk independent events? Explain.

12. A high school guidance office states that 20% of past graduates majored in science, 45% in liberal arts, and 15% in engineering. The remaining 20% were spread out among a variety of other majors. A new survey of 780 current graduates turned up 125 science majors, 312 liberal arts majors, 138 engineering majors, and 205 others. Use the chi-square statistic to test the hypothesis that the proportions by major have remained the same.

In 13 and 14, suppose a student dormitory in a college shows the data in the table below:

	Freshmen	Sophomores	Juniors	Seniors
Own a car	70	60	70	90
Do not own a car	320	240	110	60

13. Estimate the probability that a student in the dormitory owns a car.
14. What percent of car owners are juniors?
15. John's batting average is 0.325. Describe a simulation to estimate the probability of his getting 4 or more hits out of 8 at bats.
16. Suppose a six-sided die is tossed 12,000 times.
    a. What is the expected number of times a 2 will appear if the die is fair?
    b. If the results of the experiment were as follows, what might you conclude?

Number	1	2	3	4	5	6
Occurrences	8271	1233	984	298	489	734

    c. Could you have this conclusion if after 10 trials you obtained three 1s, three 5s, and one each of the other outcomes? Explain why or why not.

Chapter 6 Chapter Review

SKILLS
PROPERTIES
USES
REPRESENTATIONS

SKILLS Procedures used to get answers

OBJECTIVE A Compute probabilites and expected counts of events in various contexts. (Lessons 6-1, 6-2, 6-8)

In 1 and 2, consider tossing a fair six-sided die and subtracting the number on the first toss from the number on the second, resulting in a difference *d*. Find each probability.

1. $P(d < 0)$
2. $P(d \text{ is an integer})$

In 3 and 4, two fair 6-sided dice are tossed. Find each.

3. $P(\text{doubles})$
4. $P(\text{sum is prime})$

In 5–7, three fair coins are tossed. Find each.

5. $P(\text{all heads})$
6. $P(2 \text{ heads})$
7. $P(\text{at least 1 tail})$

8. A current estimate of color blindness in the U.S. population is 1.3%. In a high school graduating class of 1200 students, how many would be expected to be color blind?

9. When tossing a fair die 230 times, what is the expected count of the outcome "3"?

10. According to a National Center for Health Statistics survey, about 4.7% of children between the ages of 5 and 17 missed 11 or more days of school because of illness or injury in the last year. What would be the expected count of students aged 5–17 in a school of 532 students to miss 11 or more days of school because of illness or injury?

OBJECTIVE B Find the number of strings with or without replacement. (Lessons 6-3, 6-4)

11. How many possible strings are there for the letters in the word TROPICAL?

12. How many three-letter permutations can be made from the letters in TROPICAL that contain no A?

13. How many five-letter permutations can be made from the letters in CABINET?

14. How many three-letter permutations can be made from CABINET that start and end with a vowel?

15. A coin is tossed eight times. How many possible outcomes are there?

OBJECTIVE C Evaluate expressions using factorials. (Lesson 6-4)

In 16–18, evaluate without using a calculator.

16. $6!$
17. $\frac{9!}{5!4!3!}$
18. $\frac{100!}{97!} \cdot \frac{48!}{50!} \cdot \frac{8!}{11!}$

19. Write $\frac{54!}{50!}$ as a product of integers.

20. Simplify $\frac{(a + 2)!}{(a + 1)!}$.

In 21–23, evaluate.

21. ${}_{12}P_3$
22. ${}_7P_5$
23. ${}_nP_1$

OBJECTIVE D Calculate probabilities using the General and Mutually Exclusive Forms of the Probability of the Union of Events, the Probability of Complements, and the Definition of Conditional Probability. (Lessons 6-2, 6-6)

24. If A and B are two events where $P(A) = 0.6$ and $P(A \cap B) = 0.05$, find $P(B \mid A)$.

25. If A and B are complementary events, find $P(A \cap B)$.

26. If $P(A) = 0.6$ and $P(B) = 0.7$, explain why A and B cannot be mutually exclusive.

27. A and B are two mutually exclusive events, $P(A) = 0.25$, and $P(B) = 0.3$. Find $P(\text{not } (A \text{ or } B))$.

28. What is the probability of a randomly chosen positive integer less than or equal to 100 being either a perfect square or a perfect cube?

PROPERTIES Principles behind the mathematics

OBJECTIVE E Determine whether events are independent or dependent. (Lesson 6-3)

In 29–31, determine whether the events *A* and *B* are independent or dependent.

29. $P(A) = 0.4$, $P(B) = 0.25$, $P(A \cap B) = 0.1$

30. A fair coin is tossed once; $A =$ heads occurs; $B =$ tails occurs.

31. A fair six-sided die is tossed twice; $A =$ first toss is a 5; $B =$ sum of the tosses is 7.

32. Determine if these two events are independent. Randomly choosing two socks, one after another, from a drawer, when
 a. the first sock is replaced.
 b. the first sock is not replaced.

OBJECTIVE F Discuss the Law of Large Numbers and the "law of averages." (Lesson 6-8)

33. The 2007–2008 NBA rookie of the year, Kevin Durant, finished the season with an 87.3% free-throw percentage during the regular season. After failing to make his first five free-throws, a radio announcer says, "Durant is due for a successful free-throw attempt." Comment on the announcer's claim.

34. A supercomputer randomly generates numbers between 0 and 9 every five seconds which are needed for an experiment. After 1 day (17280 digits) the frequency distribution of digits is as shown below.

Digit	0	1	2	3	4	5	6	7	8	9
Count	1711	1741	1699	1728	1745	1795	1733	1701	1689	1738

 What is the relative frequency of 2? What do you expect in the long run?

35. A state lottery costs $1 per ticket to play with a payoff of $700 and probability $\frac{1}{1000}$ of winning. Explain how the Law of Large Numbers assures the state of a profit using as an example an individual who plays this lottery every day for a year.

USES Applications of mathematics in real-world situations

OBJECTIVE G List sample spaces and events for experiments. (Lesson 6-1)

36. Consider the experiment of flipping a coin then rolling a die.
 a. Write the sample space.
 b. List the outcomes in the event "the coin shows heads or the die shows a non-prime even number."
 c. How many outcomes are in the event "the die shows a number less than 3"?

In 37–39, describe the sample space for the given experiment.

37. finding the direction the wind is blowing to the nearest degree clockwise from north at a particular time of day

38. taking a poll of the television programs people watch at a certain time of the week

39. finding the birthdays of people in a group

OBJECTIVE H Use a contingency table to compute percentages involving categorical variables. (Lesson 6-5)

In 40–42, a city recently opened a new park and planted many new trees of three different species. Shortly thereafter, a storm knocked over some of the trees. This contingency table gives a count of trees planted and knocked down by species.

Tree Species	Maple	Oak	Ash
Trees Planted	102	68	85
Trees Knocked Down	16	8	17

40. What percentage of each species was knocked down in the storm? Which species seems to be the sturdiest?

41. What percentage of the total trees were oaks before the storm? What percentage are oaks after the storm?

42. What percentage of the total trees in the park was knocked down in the storm?

OBJECTIVE I Calculate probabilities in real situations. (Lessons 6-3, 6-4, 6-6)

43. What is the probability of guessing correctly on exactly four of five multiple-choice questions with four options each?

44. What is the probability of four consecutive days without rain if the probability of rain on any particular day is 0.3, regardless of whether it rained the previous day?

45. Each morning, Elliot rides his bike to the train station, where he takes the train to work. Elliot is late getting to the station 8% of the time, and the train leaves early 9% of the time and it leaves late 12% of the time. If Elliot is late getting to the train station and the train is late leaving the station, he will still be able to get to work on time. What is the probability that he will be late getting to the station but still make it to work on time?

46. In tennis, each player is allowed two serves per point. Suppose Rafael Nadal, the 2008 Wimbeldon champion, gets his first serve in 85% of the time. When this happens, he wins the point 95% of the time. If he misses his first serve, he wins the point 60% of the time.
   a. Draw a tree diagram corresponding to the situation.
   b. Find the probability that Nadal does not win the point.
   c. Find the probability that a point that Nadal wins is from his first serve.
   d. Find the probability that he missed his first serve given he does not win the point.

47. Refer to the Titanic data in Lesson 6-5.
   a. What is the probability that someone who died on the Titanic was a crew member?
   b. What is the probability that a randomly-selected passenger survived, given that the passenger was a member of the crew?

OBJECTIVE J Use the Multiplication Counting Principle, the Strings with Replacement Theorem, and permutations to find the number of ways of arranging objects. (Lessons 6-3, 6-4).

48. What is the probability of correctly pairing 4 locks with their keys from a box with 9 keys in it on the first try?

49. Emma's school uniform allows for three different skirts, two pairs of shoes, and six blouses. How many different outfits does the dress code allow?

50. Phillip has five school subjects and he has eight different colored folders.
   a. How many ways are there for him to file his papers by subject in different folders?
   b. How many ways are there for Phillip to order his five subject folders?
   c. Using your answers to Parts a and b, how many total possibilities does Phillip have for organizing his papers considering both order and folder color?

51. How many different ways are there for a six-player volleyball team and their coach to line up in a row for a photo if the coach wants to stand at one of the ends?

OBJECTIVE K Design and conduct simulations with or without technology. (Lesson 6-7)

In 52 and 53, you are playing a game that involves six coins: three pennies, two nickels, and one dime. Each coin is flipped. If the coin shows heads, you get to keep it, but if the coin shows tails, you do not. Suppose you play the game 100 times.

52. Describe an experiment to estimate the probability of winning at least 18 cents.

53. Modify the experiment to determine the probability of winning at least 18 cents if one of the pennies is replaced with a nickel.

54. Describe how you could simulate this situation: You are going on a week-long vacation to a place where 10% of the days it rains. You want to know how many days it will rain, so you can pack accordingly.

OBJECTIVE L Use the chi-square statistic to determine whether or not an event is likely. (Lesson 6-9)

55. A sample of 6000 fruit flies contained 315, 1202, 1146, and 3337 flies of four species. A scientist claims that these four species of fruit flies should appear $\frac{1}{16}$, $\frac{3}{16}$, $\frac{3}{16}$, and $\frac{9}{16}$ of the time, respectively.

a. Calculate the expected counts.

b. What is the degree of freedom in this situation?

c. At the 0.01 significance level, is there sufficient evidence to reject the scientist's claim?

56. Below are the number of accidental injury hospitalizations in California in 2004:

Age	< 1	1-4	5-12	13-15	16-20
Injuries	1476	6017	8282	4563	9518

Is there sufficient evidence to say that the number of injuries in the five age groups appear in the following percentages 5%, 20%, 27%, 16%, and 32% respectively? Test at the 0.001 significiance level.

57. Three bridges into a city are hypothesized to be used by $\frac{1}{4}$, $\frac{5}{8}$, and $\frac{1}{8}$ of the cars during the morning rush hour. A highway study of a random sample of 790 cars indicated that 191, 481, and 118 cars use these bridges respectively. Run a simulation to determine whether the hypothesis should be rejected. Use a 0.05 significance level.

REPRESENTATIONS Pictures, graphs, or objects that illustrate concepts

OBJECTIVE M Represent information about relative frequencies or frequencies in a contingency table. (Lesson 6-5)

58. A poll is conducted in a local town to gauge the party affiliations of its registered voters.

Age	Democrat	Republican	Independent
18–35 (43%)	37%	31%	32%
> 35 (57%)	29%	37%	34%

a. What percentage of the town's voters are Democrats?

b. What percentage are Republicans?

c. Suppose that, at the next election, each voter must decide between a Republican or Democratic candidate for mayor. Assume everyone votes according to party line. If the over-35 Independents split exactly in half between the candidates, what percentage of the 18–35 Independents must vote Democratic for there to be a tie?

59. A study of SAT preparation courses asked students to identify whether they studied more than 3 hours per week during the six-week course. School district A used a private company to conduct its course while school district B conducted its own course. Students were tested before and after the course, and the number of students who had less than or more than a 25 point increase were recorded in the following table.

	Hours Studied	Performance Increase		Total
School District A		< 25 Points	> 25 Points	
	<3	27	179	206
	>3	42	293	335
	Hours Studied	Performance Increase		Total
School District B		< 25 Points	> 25 Points	
	<3	80	197	277
	>3	21	55	76

a. What is the relative frequency of students who went to school district A who studied more than 3 hours per week but whose scores rose less than 25 points? Studied less than 3 hours?

b. Repeat Part a for school district B.

c. Does number of hours of study seem to relate to test performance in each district?

d. Make a new contingency table combining the school districts. Recalculate the performance rates for students who studied for less than 3 hours.

e. What do these data exemplify?

Chapter

7 Polynomial Functions

Contents

A *polynomial* in one variable is a sum of multiples of nonnegative integer powers of that variable. The variety of polynomial functions makes them candidates for modeling many real-world situations, some of which they model exactly and others which they approximate.

Graphed on the next page are four polynomial functions. The quadratic function P_1 is like those that describe the path of a projectile. The cubic function P_2 gives the volume of a cube with side length x.

A quartic function like P_3 could be used to model a cross-section of the shape of the nose of the aircraft shown in the photo above. The polynomial function P_4, which is of degree 7, gives the amount you would have today if \$1000 had been invested 7 years ago, \$800 had been invested 5 years ago, and \$900 had been invested 4 years ago at an annual rate of $x - 1$.

In this chapter, you will see applications of polynomial functions and their graphs in physics, finance, and other fields, as well as see some of the beautiful mathematical properties these functions possess. The theory underlying these properties involves complex numbers, division, and factoring.

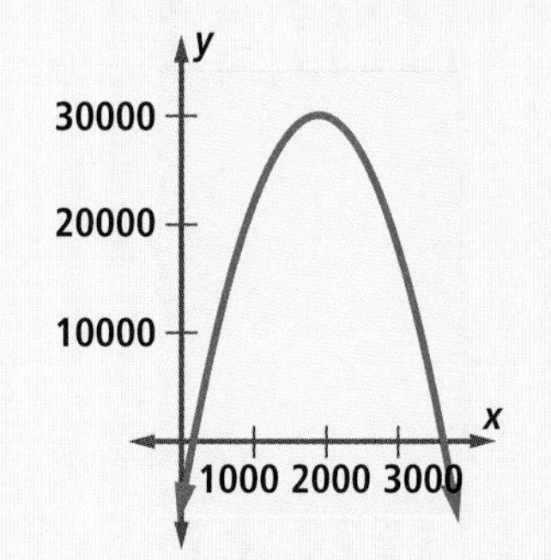

$P_1(x) = -0.01x^2 + 38x - 6000$

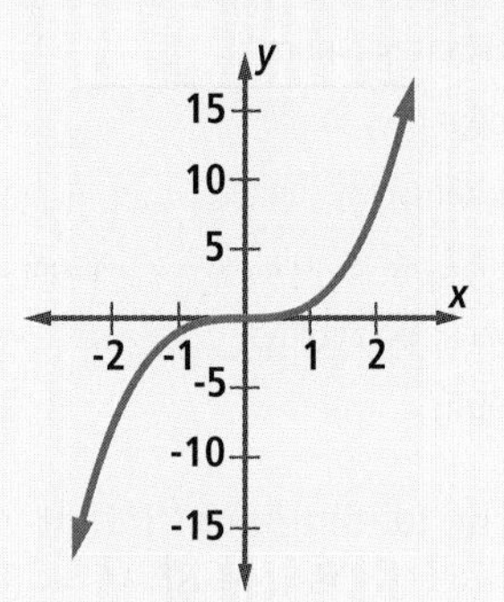

$P_2(x) = x^3$

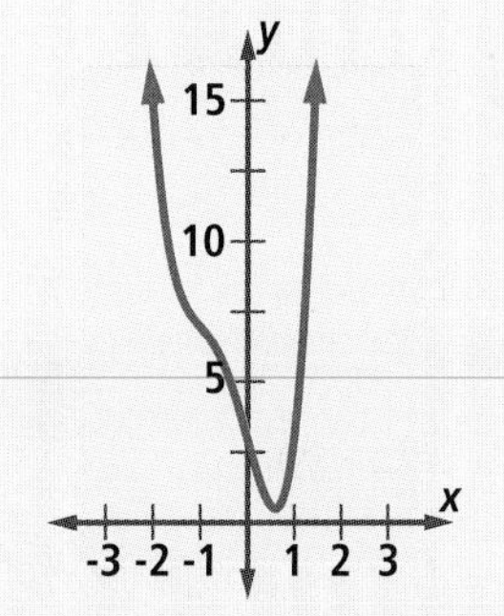

$P_3(x) = 2x^4 + 4x^3 - 6x + 3$

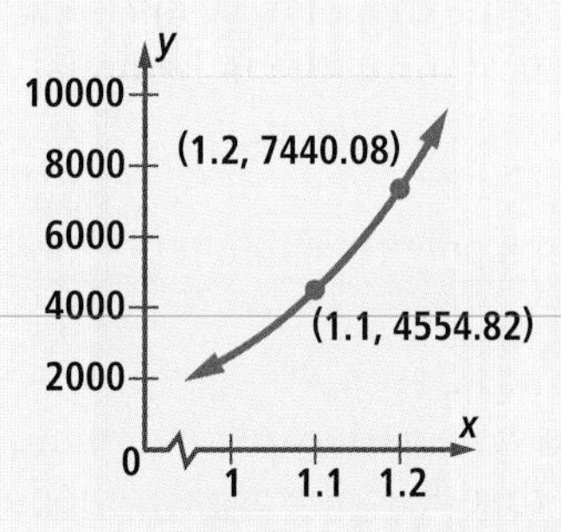

$P_4(x) = 1000x^7 + 800x^5 + 900x^4$

Lesson

7-1 Characteristics of Polynomial Functions

Vocabulary

polynomial in *x*
degree of a polynomial
coefficients of a polynomial
leading coefficient
standard form of a polynomial
polynomial function
maximum/minimum value of a function
extreme values, extrema
relative (local) maximum/minimum
increasing function
decreasing function
zeros of a polynomial

BIG IDEA Zeros, maxima, and minima are important features of the graphs of polynomial functions.

Mental Math

Expand.

a. $(x + 7)(x - 1)$

b. $(x^2 - xy)(x^2 + xy)$

The linear and quadratic expressions you studied in this and preceding courses are types of polynomial expressions. The general form of a polynomial in one variable is given in the following definition.

Definition of Polynomial

A **polynomial in *x*** is an expression of the form

$$a_nx^n + a_{n-1}x^{n-1} + a_{n-2}x^{n-2} + \dots + a_1x + a_0,$$

where n is a nonnegative integer and $a_n \neq 0$.

The number n is the **degree** of the polynomial and the numbers a_n, a_{n-1}, $a_{n-2}, \dots, a_0$ are its **coefficients.** Polynomials of degrees 0 through 5 have special names.

Degree	Name of Polynomial	Example
0	constant polynomial	2.38
1	linear polynomial	$4x + \pi$
2	quadratic polynomial	$y - y^2$
3	cubic polynomial	$2n^3 - 4n + 9$
4	quartic polynomial	$0.7 - 81t^2 + t^4$
5	quintic polynomial	$\frac{1}{30}x^5 + 2x^4 + 6x^3 + 2x$

The number a_n is called the **leading coefficient** of the polynomial. You might think that the leading coefficient of $0.7 - 81t^2 + t^4$ is 0.7, because it is the first coefficient. But for a polynomial in one variable, the leading coefficient comes first only when the terms of the polynomial are written in descending order of their exponents. This is called the **standard form of a polynomial**. In standard form, the polynomial above is $t^4 - 81t^2 + 0.7$. The leading coefficient is 1. The degree of a polynomial in one variable is the largest exponent of the variable, which in this case is 4. All the exponents in a polynomial must be nonnegative integers.

STOP QY1

QY1

What is the degree of $6t - \frac{1}{10}t^5 + 100t^2$?

Polynomials arise from a wide variety of situations. You have studied polynomials of degree 2 that describe area or model paths of projectiles, and polynomials of degree 3 that describe volume. Activity 1 explores a typical problem involving volume.

Activity 1

A company wants to produce an open-top box from a 60-cm by 45-cm piece of cardboard. They want the box to have the largest possible volume. The pictures below show the way the cardboard will be cut and what the box will look like after it is assembled.

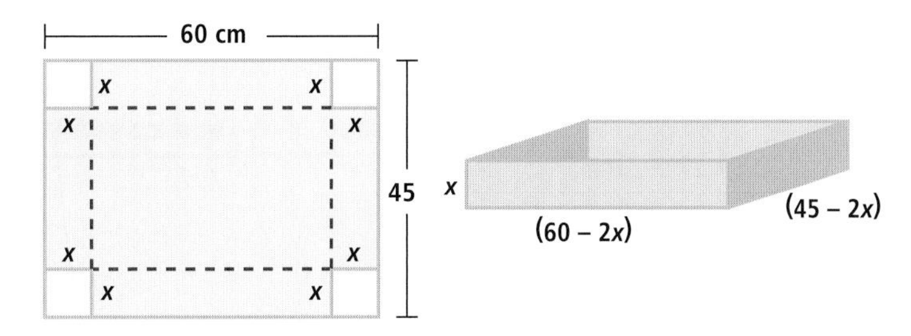

Step 1 Recall the formula for the volume of a rectangular solid is $V = \ell \cdot w \cdot h$. Write an equation for the volume V of the box in terms of x.

Step 2 For what values of x does V have meaning in this situation?

Step 3 Make a table of eight values of x and V on your interval from Step 2.

Step 4 What are the minimum and maximum values the volume took in your table?

Step 5 Use the results of Steps 2 and 4 to set the window on a graphing utility. Graph the function that maps x onto V.

Step 6 Find the maximum point of the graph.

Step 7 Explain the meaning of the coordinates of the maximum point in the context of the problem.

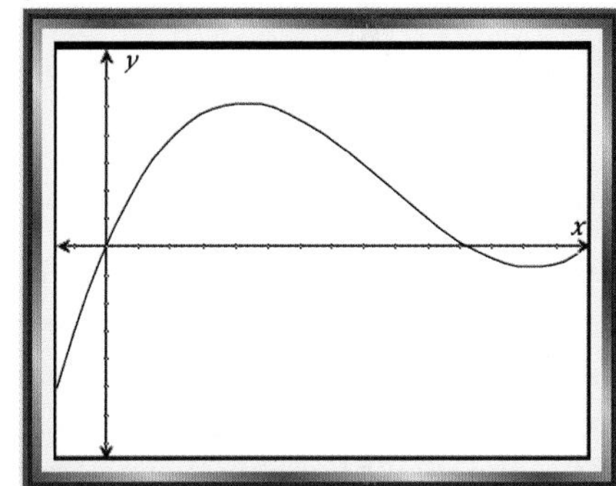

The volume of the box in Activity 1 depends on the length x of the side of the cut-out square, so the volume is considered as a function of x. You should have found that

$$V = 4x^3 - 210x^2 + 2700x.$$

This equation identifies the function as a polynomial function of degree 3. In general, a **polynomial function** is a function whose rule can be written as a polynomial.

Graphs of Polynomial Functions

You are familiar with graphs of polynomial functions of degree 1 or 2. The graph of every polynomial function of degree 1 is a line, and the graph of every polynomial function of degree 2 is a parabola. The graphs of polynomials of higher degree do not all have the same shape. For example, $y = x^3$ does not have the same graph as the function in Activity 1, but they both have degree 3. To describe the graphs of polynomial functions, it helps to know their maximum and minimum values and their intercepts.

Example 1

Let p be the polynomial function with equation $p(x) = (x - 1)^3(x + 4)^2$

a. **What is the degree of p?**

b. **Find the y-intercept of its graph.**

c. **Find all x-intercepts of its graph.**

Solution

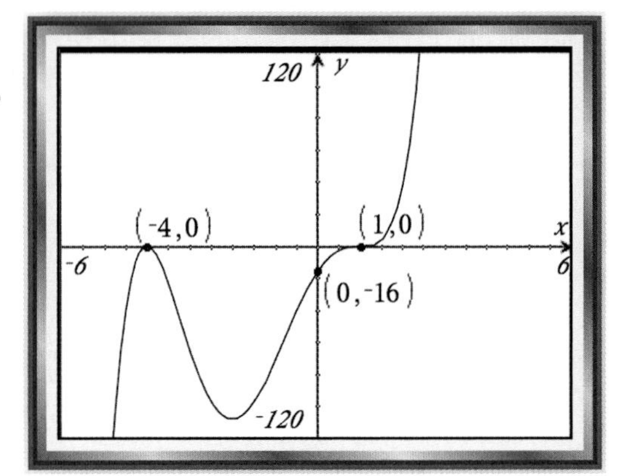

a. Expand $(x - 1)^3(x + 4)^2$.
$p(x) = x^5 + 5x^4 - 5x^3 - 25x^2 + 40x - 16$.
Its degree is the exponent of the highest power of x, so its degree is 5.

b. The y-intercept is equal to $p(0)$.
$p(0) = (0 - 1)^3(0 + 4)^2 = -16$.
So the y-intercept is -16.

c. The x-intercepts are the solutions to $p(x) = 0$.
$(x - 1)^3(x + 4)^2 = 0$ implies $x = 1$ or $x = -4$, so the x-intercepts are 1 and -4.

Check The graph of p at the right verifies the answers to Parts b and c.

Extrema of Functions

Recall that the range of a function is the set of possible values of the dependent variable. The **maximum value** of a function is the largest value in its range. Similarly, the **minimum value** of a function is the smallest value in its range. These are the **extreme values** or **extrema** of the function. Functions can also have *relative extrema*. A **relative** (or **local**) **maximum** or **relative** (or **local**) **minimum** is a value that is the maximum or the minimum on an interval $a < x < b$. It is also possible for a function to have no extrema at all.

Graphs I to IV below show some different combinations of extrema and relative extrema.

I

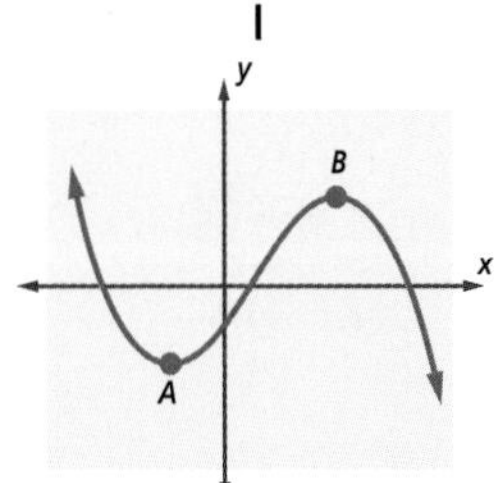

no minimum
no maximum
relative minimum at A
relative maximum at B

II

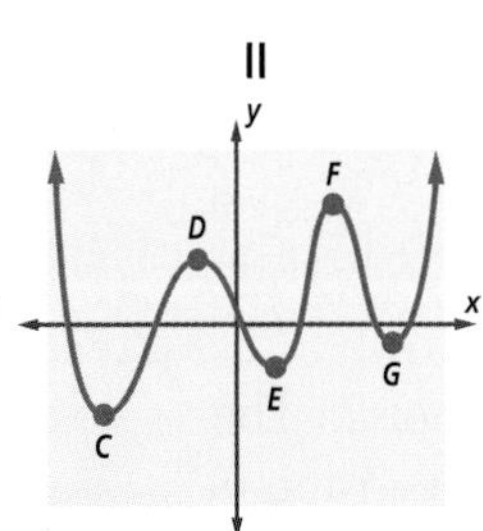

minimum at C
no maximum
relative minima at C, E, G
relative maxima at D, F

III

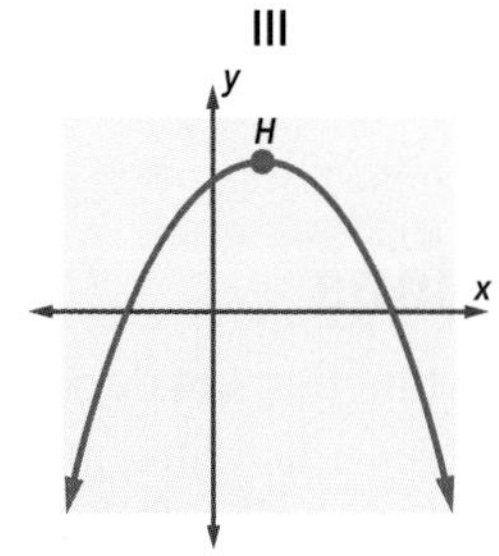

no minimum
maximum at H
no relative minimum
relative maximum at H

IV

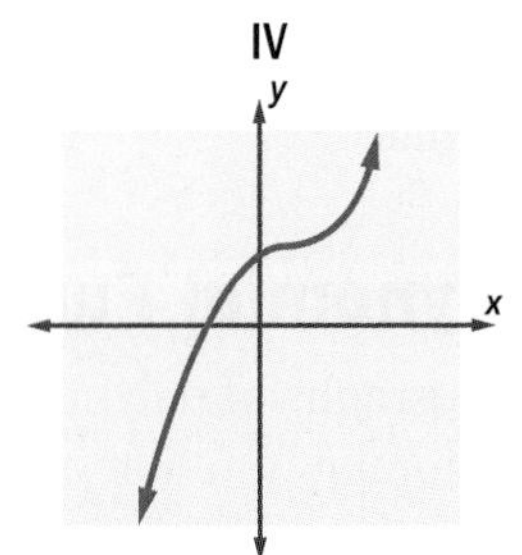

no minimum
no maximum
no relative minimum
no relative maximum

QY2

> **QY2**
> Using a calculator, graph the function p from Example 1 and determine its relative minimum value.

Intervals Where a Function is Increasing or Decreasing

A function is **increasing** on an interval if the slope of the line through any two points of the graph on that interval is positive. A function is **decreasing** on an interval if the slope of the line through any two points on that interval is negative. Points at which the function changes from increasing to decreasing (or vice versa) are relative extrema.

Example 2

Consider the graph of $f(x) = x^3 - 5x - 2$ shown at the right. A relative maximum occurs when $x \approx$ -1.3 and a relative minimum occurs when $x \approx 1.3$. Describe the intervals on which

a. f is increasing.

b. f is decreasing.

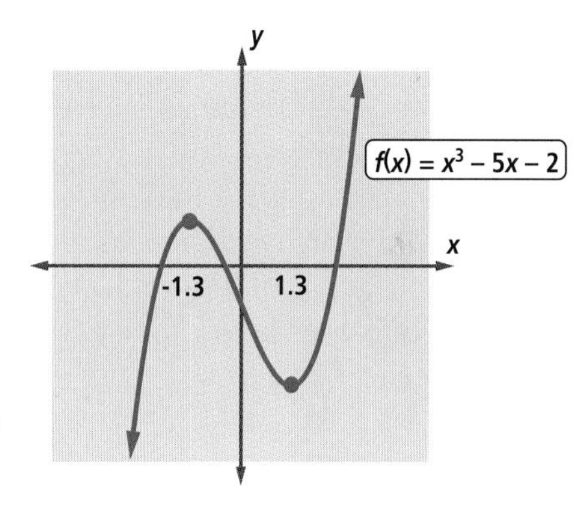

Solution The background shading separates the curve into three sections.

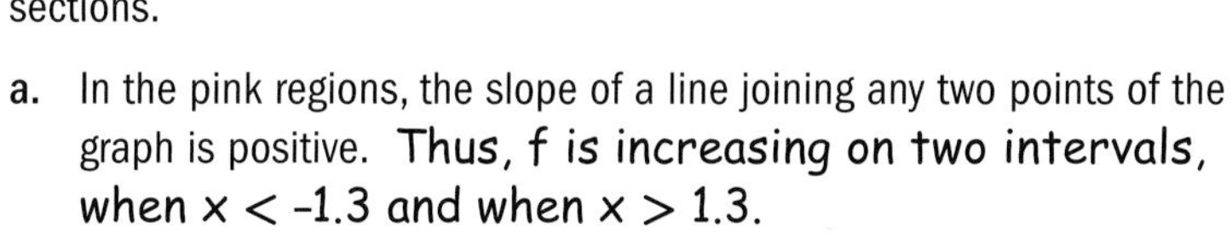

a. In the pink regions, the slope of a line joining any two points of the graph is positive. Thus, f is increasing on two intervals, when $x <$ -1.3 and when $x > 1.3$.

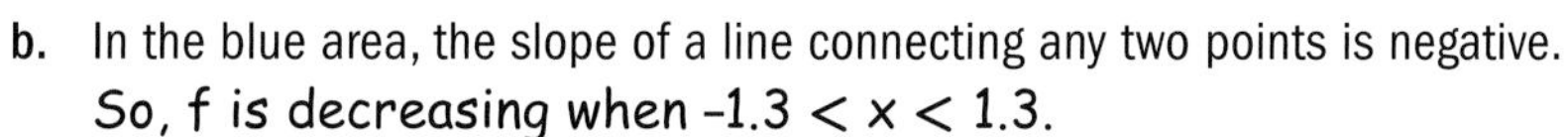

b. In the blue area, the slope of a line connecting any two points is negative. So, f is decreasing when -1.3 $< x < 1.3$.

When describing the behavior of a function on an interval, consider whether the values of the function are positive, negative, or zero and whether those function values are increasing, decreasing, or constant as the values of x increase.

GUIDED

Example 3

Consider the graph of $h(x) = -\frac{2}{3}x^4 + 3x^3 - 5x$ at the right. The x-coordinates of key points are labeled with letters.

a. Using the letter labels, in what interval(s) is h negative?

b. Using the letter labels, in what interval(s) is h decreasing?

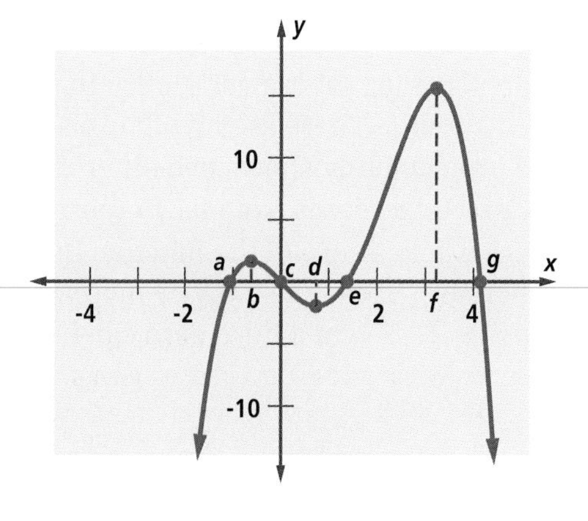

Solution

a. h is negative when the graph is below the x-axis. This occurs when __?__, $c < x < e$, or __?__.

b. h is decreasing when the graph goes down as x increases. This occurs when __?__ or __?__.

 QY3

> **QY3**
> In Example 3, what is the behavior of h on the interval $e < x < f$?

Zeros of Polynomial Functions

Recall that the x-intercepts of the graph of a function are the *zeros* of the function itself. Similarly given a polynomial function p with $p(x) = a_nx^n + a_{n-1}x^{n-1} + \dots + a_1x + a_0$, the zeros or *roots* of the polynomial function p are all values of x such that $p(x) = 0$. These are also called **zeros of the polynomial**.

You already know how to find the exact zero of a polynomial of degree 1. This polynomial function is of the form $p(x) = ax + b$, and its zero is $-\frac{b}{a}$. You can also find the zeros of any polynomial function of degree 2 because this is a quadratic polynomial function with an equation of the form $p(x) = ax^2 + bx + c$. The zeros are found using the Quadratic Formula.

Though there are formulas for exact values of zeros of polynomial functions of degrees 3 and 4, they are quite complicated, and there are no general formulas for finding exact zeros of all polynomial functions of degree higher than 4. The techniques you already know can be used to approximate zeros of higher-degree polynomials. Consider the graph and table of values below for $f(x) = x^3 - 5x - 2$. One of the zeros is labeled in red.

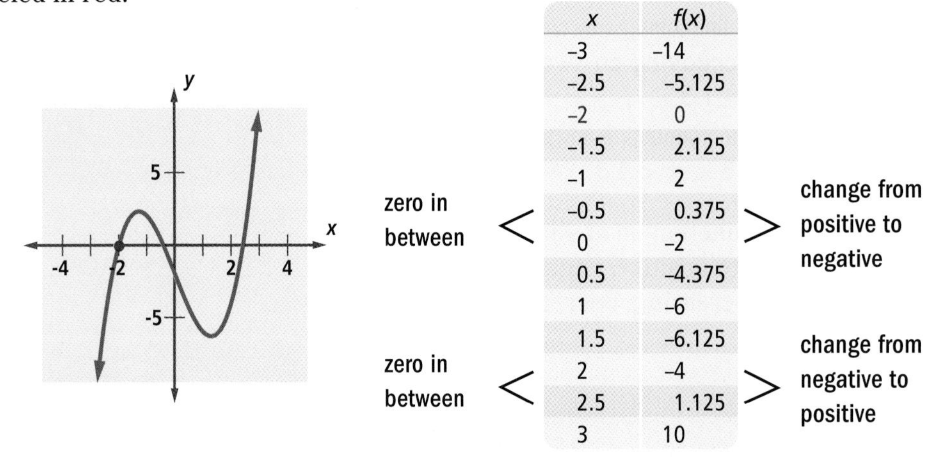

x	f(x)
-3	-14
-2.5	-5.125
-2	0
-1.5	2.125
-1	2
-0.5	0.375
0	-2
0.5	-4.375
1	-6
1.5	-6.125
2	-4
2.5	1.125
3	10

On the graph, you can see the x-intercepts by finding points where the graph crosses the x-axis. Notice that -2 is a value of x for which $f(x) = 0$ in the table. There are no other places in the table where $f(x) = 0$, yet the graph clearly shows two other x-intercepts. As x increases from -0.5 to 0, $f(x)$ decreases from 0.375 (positive) to -2 (negative). Therefore, there exists a zero of the function between -0.5 and 0. Likewise, as x increases from 2 to 2.5, $f(x)$ increases from -4 (negative) to 1.125 (positive). So yet a third zero can be found between 2 and 2.5.

Activity 2

Step 1 Use a spreadsheet to build a table of values for $f(x) = x^3 - 5x - 2$ from $x = 2$ to $x = 2.5$ in steps of 0.1. In what interval must there be a zero?

Step 2 Use the spreadsheet to evaluate $f(x)$ in 0.01 increments over the interval you found in Step 1. Estimate the zero of $f(x)$ accurate to the hundredths place.

Step 3 Repeat the procedure of Steps 1 and 2 to estimate the zero between -0.5 and 0 accurate to the hundredths place.

Finding Zeros and Extrema with Technology

CAS and graphing calculators have commands that will calculate zeros and extrema of functions. On one CAS the zeros of $f(x) = x^3 - 5x - 2$ are found by using the `zeros` command.

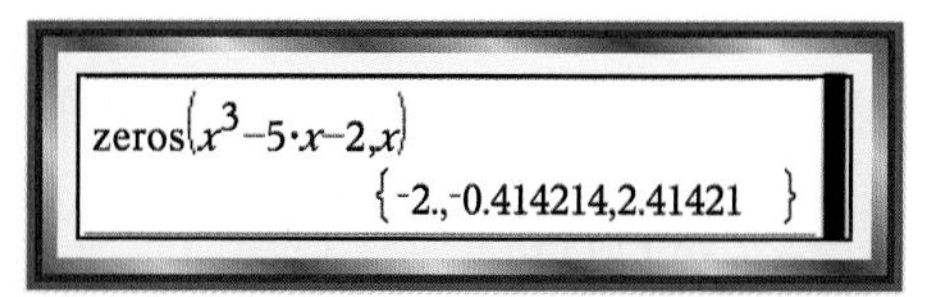

So the three zeros are approximately -2, -0.414, and 2.414. To find a relative minimum or relative maximum on a specified interval, the `fMax` or `fMin` command is used. For instance, the display at the right shows that on the interval $-0.414 < x < 2.414$, a relative minimum occurs when $x = 1.29$. This gives the x-coordinate of the relative minimum. The value of the relative minimum is $f(1.29) \approx -6.30$.

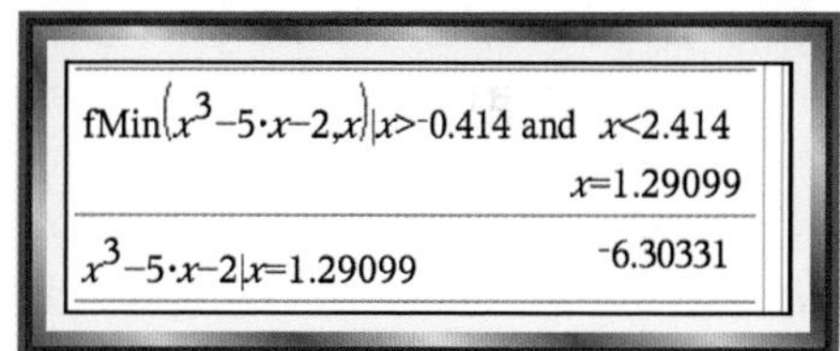

Questions

COVERING THE IDEAS

1. Why is $\frac{4}{5}x^3 - 2 + \frac{8}{x^3}$ not a polynomial?

In 2 and 3, refer to the volume equation from Activity 1.

2. a. Graph the equation on the window $-10 \le x \le 50$ and $-10{,}000 \le y \le 40{,}000$. Label all intercepts, zeros, and extrema. Use technology to find coordinates to the nearest integer.
   b. At what value of x does the relative maximum of V occur?
3. Explain what $V(30) = 0$ means in the context of the box.
4. An open box is constructed by cutting square corners with side x cm from a 150-cm by 100-cm sheet of cardboard as shown.
   a. Write a polynomial function for the volume of the box.
   b. Graph the function.
   c. When are the function values negative? What does this mean?
   d. Find the coordinates of the point at which the function has its maximum value.
   e. For what size square does the maximum volume occur?
   f. What is the maximum volume the box can have?

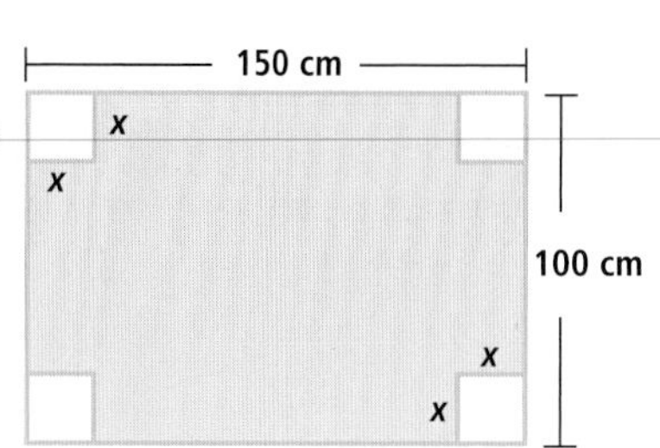

5. At the right is a graph of a function f with key x-coordinates labeled. In what interval(s)

   a. are the values of f positive?
   b. are the values of f negative?
   c. is f increasing?
   d. is f decreasing?

6. At the right is the graph of a 4th degree polynomial with key points labeled. Classify each named point as a minimum, maximum, local minimum, local maximum, or intercept. (Some points may be classified in more than one way.)

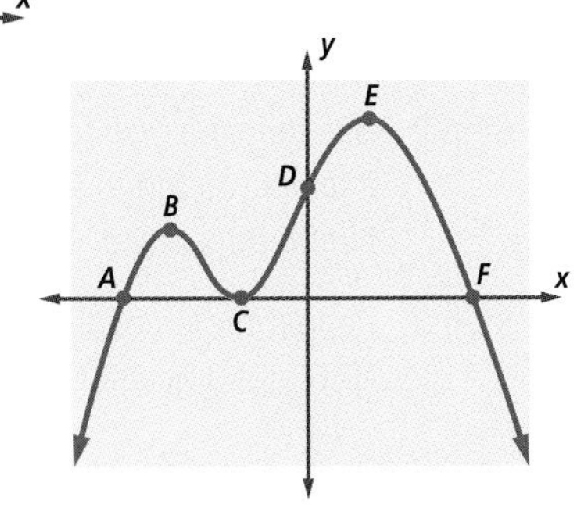

7. The table at the right gives values of a polynomial function.
   a. Plot these points and sketch a possible graph of $y = f(x)$.
   b. Between which pairs of consecutive integers must the zeros of f occur?

x	y
–2	9
–1	–2
0	5
1	9
2	–1
3	–29

8. Barry is saving his summer earnings for college. The table below shows the amount of money earned each summer.
   a. At the end of each summer, he puts his money in a savings account with annual rate r. Recall the compound interest formula, $A = P(1 + r)^t$, which gives the value of P dollars invested at an annual rate r after t years. Express Barry's total savings as a polynomial function of x, where $x = 1 + r$.

After Grade	Summer Earnings
8	\$800
9	\$850
10	\$1000
11	\$900
12	\$1200

   b. If the annual yield r is 7%, how much will be in his account when he goes to college?

APPLYING THE MATHEMATICS

9. Graph the function with the given equation. Then use the graph to match the function with a characteristic.

   a. $A(x) = 4 - x^2 - x$
   b. $B(x) = (x - 2)^3$
   c. $C(x) = 7x - 9$
   d. $D(x) = (x - 5)^2 - 6$
   e. $E(x) = x^3 - 5x$

   I. no relative extrema
   II. one maximum
   III. one minimum
   IV. both a relative maximum and a relative minimum

In 10-13, graph the function described by the given equation. Then estimate all extreme and relative extreme values to the nearest tenth.

10. $f(x) = x^4$
11. $g(t) = (t - 2)^4 + 3$
12. $h(x) = x^4 - x^2$
13. $j(x) = x^4 - 7x^3 - 1$

In 14 and 15, consider the function G, where $G(x) = x^3 + 3x^2 - 2x - 6$.

14. a. Sketch a graph of G using a graphing utility.
    b. Identify the y-intercept of G on your graph.
    c. Find the zero(s) of G to the nearest tenth.
    d. Find the coordinates of the relative extrema of G to the nearest hundredth.
    e. On what interval(s) are the values of G positive? On what interval(s) are the values of G negative?
    f. On what interval(s) is G increasing? On what interval(s) is G decreasing?

15. Use a CAS to find exact values of the zeros and relative extrema of G.

16. Rick saved his earnings for several summers just as Barry did in Question 8. A polynomial for his savings is

$$S(x) = 1200x^4 + 1300x^3 + 1400x^2 + 1900.$$

    a. What did Rick deposit in the bank the first summer he saved?
    b. One summer Rick did not save any money. Which summer was this?
    c. Evaluate $S(1.0575)$ and explain what it represents.

REVIEW

17. Estimate the distance from Cairo, Egypt (30° N, 31°E) to Cairo, Illinois (37° N, 89°W) to the nearest 20 miles. **(Lesson 5-10)**

18. **Fill in the Blank** A graph is symmetric with respect to the x-axis if for every point (x, y) on the graph, the reflection image of that point, __?__, is also on the graph. **(Lesson 3-4)**

19. If the graph of $y = (x + 3)^2 - 7$ is the translation image of the graph of $f(x) = x^2$ under T, find a rule for T. **(Lesson 3-2)**

20. A rocket follows a projectile path. It is fired from a nozzle 30 ft above the ground with an initial velocity of 525 feet per second. Its height h at time t seconds is given by $h = -16t^2 + 525t + 30$.
    a. After how many seconds will the rocket be 3000 ft above the ground?
    b. What will be the maximum height of the rocket?
    c. How many seconds after launching will the rocket hit the ground? **(Previous Course)**

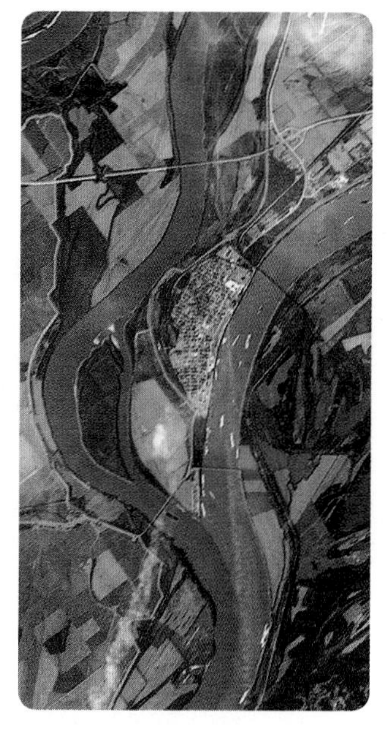

Cairo, IL is located where the Mississippi and Ohio Rivers meet.

In 21 and 22, factor each expression. (Previous Course)

21. $x^2 - 18x + 81$

22. $3x^2 - 2x - 8$

EXPLORATION

23. Create a quintic polynomial function that has exactly
    a. one real zero.
    b. three real zeros.
    c. five real zeros.
    d. four real zeros.

QY ANSWERS

1. 5
2. -108
3. $h(x)$ is positive and h is increasing.

Lesson

7-2 Polynomial Models

Vocabulary

figurate numbers

triangular numbers

tetrahedron

tetrahedral number

▶ **BIG IDEA** Given any finite set of points with different first coordinates, there is a polynomial function whose graph contains all the points.

In Lesson 7-1, you saw that polynomials can arise from volume situations. Polynomials also arise from many counting situations.

Mental Math

Solve.

a. $x(x - 7) = 0$

b. $(y + 9)(3y - 7) = 0$

c. $(z - 5)(z + 8) \cdot (z + 15) = 0$

Figurate Numbers

Figurate numbers are numbers of points in a patterned sequence of familiar geometric shapes. For example, the sum of the integers from 1 to n is given by the quadratic polynomial $t(n) = \frac{1}{2}n^2 + \frac{1}{2}n$. When n is a positive integer, the values of this polynomial are called **triangular numbers** because $t(n)$ is the number of points in the nth triangular array of dots.

1 3 6 10 15

QY1

▶ **QY1**

What is the value of $t(6)$?

Each array in the sequence of triangular numbers is 2-dimensional. Apples, oranges, and other fruit are often displayed in 3-dimensional layers as shown at the right. The outline of the array is a triangular pyramid, a four-sided polyhedron also known as a **tetrahedron**, so the result is a *tetrahedral* array. The total number of points in a 3-dimensional tetrahedral array is called a **tetrahedral number**.

Below are the four smallest tetrahedral numbers and the arrays that give rise to them.

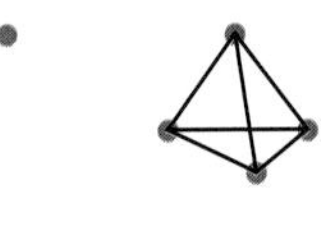

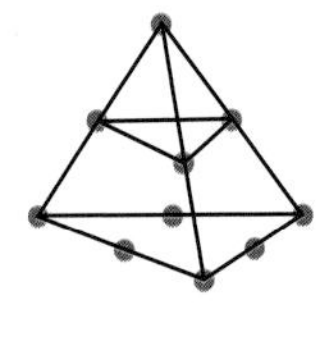

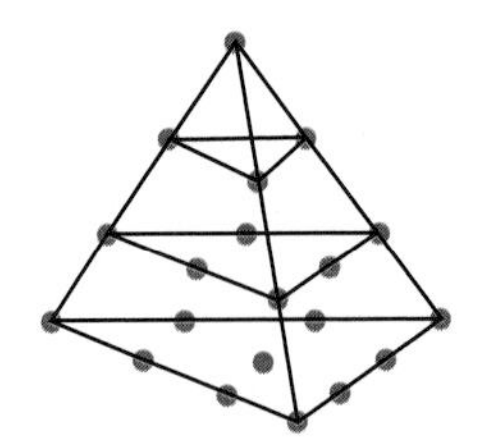

$T(1) = 1$ $T(2) = 4$ $T(3) = 10$ $T(4) = 20$

Notice that the number of dots in each layer is a triangular number. Consequently, each tetrahedral number $T(n)$ beyond the first is the sum of the previous tetrahedral number and a triangular number. For instance, the 3rd tetrahedral number is the sum of the 2nd tetrahedral number and the 3rd triangular number.

$$\begin{aligned} T(3) &= T(2) + t(3) \\ &= 4 + 6 \\ &= 10 \end{aligned}$$

QY2

Suppose that for the 25th anniversary of a store, the owner wants to arrange apples in a tetrahedral stack 25 rows high. The owner wants to know how many apples are needed.

QY2

What is the value of $T(5)$?

The owner would like an explicit formula for $T(n)$. She begins by graphing known values to see what kind of curve they suggest. At the right is a graph. This graph is clearly not linear. Is it a parabola? Is it part of a graph of a higher-degree polynomial? Or is it a part of an exponential curve?

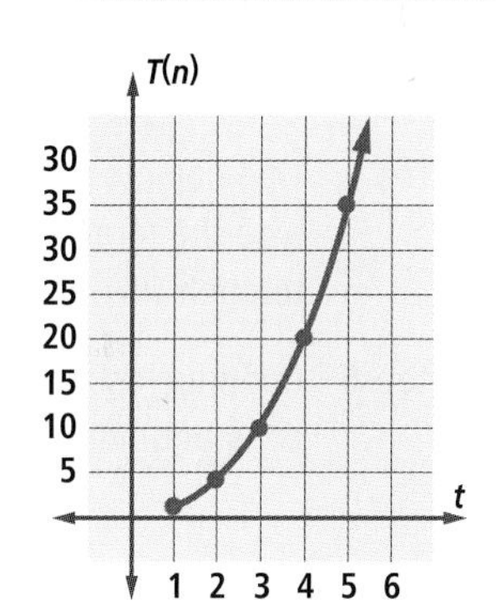

The family of polynomial functions in one variable has a special property that enables you to predict from a table of values whether or not a function is a polynomial. To examine this property, consider first some polynomials of small degree evaluated at consecutive integers.

For instance, consider the first degree polynomial $f(x) = 7x + 2$. Observe that the differences between consecutive values of $f(x)$, where the x-values are consecutive integers, are constant.

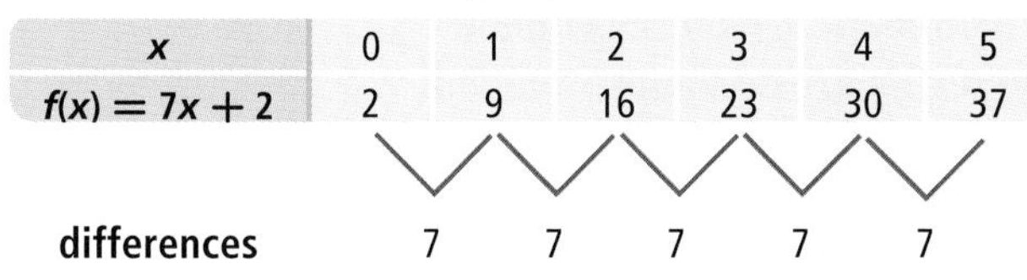

x	0	1	2	3	4	5
$f(x) = 7x + 2$	2	9	16	23	30	37

differences	7	7	7	7	7

The constant difference is equal to the slope of the graph of the function f where $f(x) = 7x + 2$.

For second degree polynomial functions, that is, quadratic functions, the differences between consecutive values are not equal, but differences of these differences are constant. For instance, if $g(x) = x^2 + x + 4$ is evaluated at the integers from -1 to 5 inclusive, the following results:

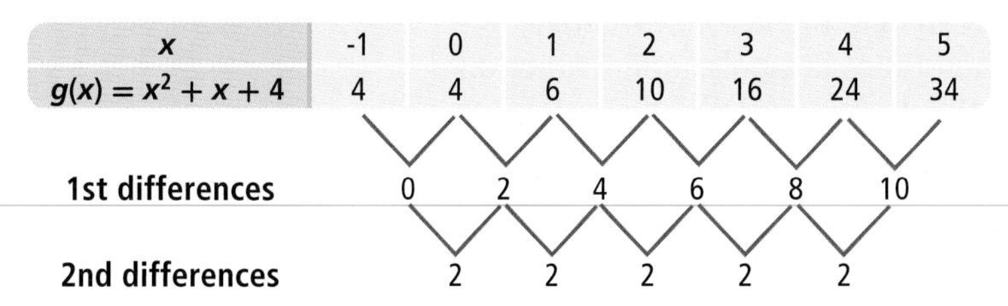

x	-1	0	1	2	3	4	5
$g(x) = x^2 + x + 4$	4	4	6	10	16	24	34

1st differences	0	2	4	6	8	10
2nd differences	2	2	2	2	2	

Consider now the cubic polynomial $h(x) = 3x^3 - 2x - 8$. You can use a spreadsheet to calculate its values, and the corresponding differences, for integers between 0 and 6 inclusive.

Activity

MATERIALS spreadsheet software

Step 1 In column A of a spreadsheet, type the numbers 0 through 6.

Step 2 In cell B1, type the formula
= 3A1^3 − 2 * A1 − 8
Fill down the formula from cell B1 to cell B7.

	A	B	C	D
1	0	-8		
2	1	-7		
3	2	12		
4	3	67		
5	4	176		
6	5	357		

B1 $=3 \cdot a1^3 - 2 \cdot a1 - 8$

Step 3 Title column C "first" for first difference. In cell C1, type the formula = B2 − B1. Fill down the formula through cell C6. What is the value in cell C6? Why shouldn't you copy the formula into cell C7?

	A	B	C first	D
1	0	-8	1	
2	1	-7		
3	2	12		
4	3	67		
5	4	176		
6	5	357		

C1 =b2−b1

Step 4 Title column D "second" for second difference. In cell D1, type the formula = C2 − C1. Fill down the formula through D5. What is the value in cell D5?

Step 5 Continue as in Steps 3 and 4 until you reach a column of constant differences. Which set of differences is constant?

The Polynomial Difference Theorem

Each of the preceding examples is an instance of the following theorem. Its proof is beyond the scope of this book and is therefore omitted. Recall from your previous courses that an *arithmetic sequence* is one in which the differences between consecutive terms are constant. You will study arithmetic sequences in Chapter 8.

Polynomial Difference Theorem

The function with equation $y = f(x)$ is a polynomial function of degree n if and only if, for any set of x-values that are consecutive terms of an arithmetic sequence, the nth differences of corresponding y-values are equal and nonzero.

The Polynomial Difference Theorem tells you that if values of a function produce constant differences (from x-values that are consecutive terms of an arithmetic sequence), then there is a polynomial function containing these points. Moreover, the degree of the polynomial is the n corresponding to the nth differences that are equal.

The x-values do not have to be consecutive integers. We use consecutive integer x-values in the examples here because consecutive integers are a convenient example of an arithmetic sequence.

QY3

QY3

The fifth differences of $f(1), f(2), f(3), f(4), \ldots$ all equal 3. What is the degree of f?

GUIDED

Example 1

Consider the tetrahedral number function T described on page 446. Is T a polynomial function? If so, what is the degree of the polynomial?

Solution Construct a table of consecutive tetrahedral numbers, and take differences between consecutive terms.

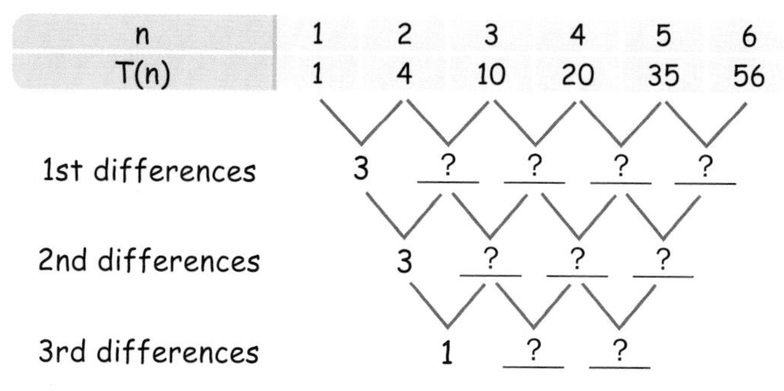

n	1	2	3	4	5	6
T(n)	1	4	10	20	35	56

1st differences 3 ? ? ? ?

2nd differences 3 ? ? ?

3rd differences 1 ? ?

The third differences appear to be constant. Thus, by the Polynomial Difference Theorem, the function $n \rightarrow T(n)$ appears to be a polynomial function of degree 3.

Solving a System to Find a Polynomial Model

In order to find an explicit formula for the cubic polynomial of Example 1, you can set up and solve a system of equations. You know the formula for $T(n)$ is of the form $T(n) = an^3 + bn^2 + cn + d$.

You must find four coefficients a, b, c, and d in order to determine $T(n)$. So use any four data points in the table of tetrahedral numbers, and substitute them into the cubic form above. It is easiest if the values of x are consecutive integers. Substituting 4, 3, 2, and 1 for n and the corresponding values of $T(n)$, we obtain the following four equations.

$$\begin{aligned} T(4) &= 20 = 64a + 16b + 4c + d \\ T(3) &= 10 = 27a + 9b + 3c + d \\ T(2) &= 4 \;\;= 8a + 4b + 2c + d \\ T(1) &= 1 \;\;= a + b + c + d \end{aligned}$$

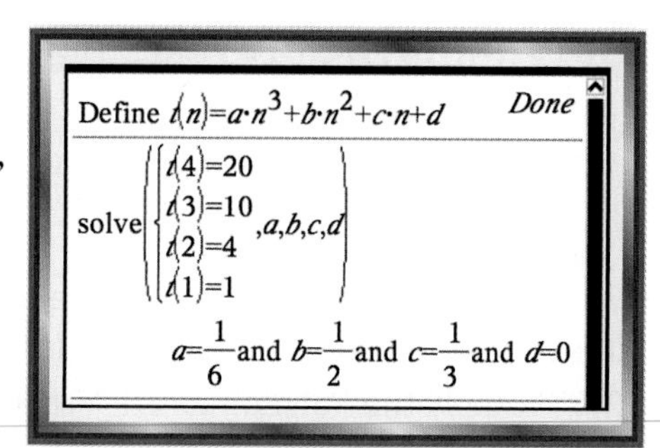

You can solve this system with a CAS using the `solve` command as was done in Lesson 2-6, and as shown at the right, but you might be able to solve it faster with pencil and paper. To eliminate d, subtract each successive pair of sentences above. This yields the following equivalent system:

$$\begin{aligned} 10 &= 37a + 7b + c \\ 6 &= 19a + 5b + c \\ 3 &= 7a + 3b + c \end{aligned}$$

Repeating the same procedure with the sentences above eliminates c.

$$\begin{aligned} 4 &= 18a + 2b \\ 3 &= 12a + 2b \end{aligned}$$

Subtracting these two equations gives $1 = 6a$; so $a = \frac{1}{6}$.

To find the other coefficients, substitute back. When $a = \frac{1}{6}$, $3 = 12\left(\frac{1}{6}\right) + 2b$. Thus, $b = \frac{1}{2}$. Similarly, $3 = 7a + 3b + c$, so when $a = \frac{1}{6}$ and $b = \frac{1}{2}$, $3 = 7\left(\frac{1}{6}\right) + 3\left(\frac{1}{2}\right) + c$. Thus, $c = \frac{1}{3}$. Finally, $1 = a + b + c + d$, so when $a = \frac{1}{6}$, $b = \frac{1}{2}$, and $c = \frac{1}{3}$, $d = 0$.

The resulting polynomial formula for the nth tetrahedral number is

$$T(n) = \frac{1}{6}n^3 + \frac{1}{2}n^2 + \frac{1}{3}n.$$

Using Regression to Determine a Polynomial

The cubic model can also be found by using *polynomial regression.*

Example 2

Use cubic regression to find a cubic model for the tetrahedral numbers.

Solution Enter the known values into a statistics utility.

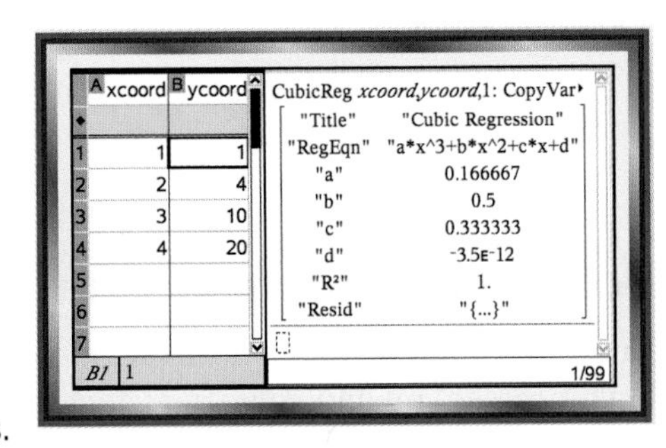

One statistics utility gives the cubic regression model $y = 0.1666667x^3 + 0.5x^2 + 0.3333333x + (-3.5E{-}12)$.

Each number except 0.5 is an approximation. 0.1666667 is about $\frac{1}{6}$, 0.3333333 is about $\frac{1}{3}$. Also, -3.5E–12 means -3.5×10^{-12}, which is approximately 0.

The cubic model is $y = \frac{1}{6}x^3 + \frac{1}{2}x^2 + \frac{1}{3}x$.

This agrees with the model found by solving the system of equations.

To find the cubic model for the tetrahedral numbers, four data points were used. Five points are needed to determine a quartic (4th degree) model. In general, to find the equation for a polynomial of degree n you need $n + 1$ points that are not modeled exactly by a polynomial of degree less than n.

Questions

COVERING THE IDEAS

1. Find the 7th tetrahedral number in two ways.
   a. Add the 7th triangular number to the 6th tetrahedral number.
   b. Use the explicit formula for $T(n)$ developed in the lesson.
2. If the 7th differences of the y-values for consecutive integral x-values all equal 4, then what is the degree of the polynomial?
3. To apply the Polynomial Difference Theorem, what must be true about the x-values?

In 4 and 5, determine if y is a polynomial function of x of degree ≤ 5. If the function is a polynomial function, find its degree.

4.

x	1	2	3	4	5	6	7	8	9
y	2	12	36	80	150	252	392	576	810

5.

x	1	3	5	7	9	11	13
y	2	8	32	128	512	2048	8192

6. The graph at the right contains the following points: (1, 216), (2, 125), (3, 64), (4, 27), (5, 8), (6, 1). Using what you have learned about finite differences, find an equation that fits the data points.

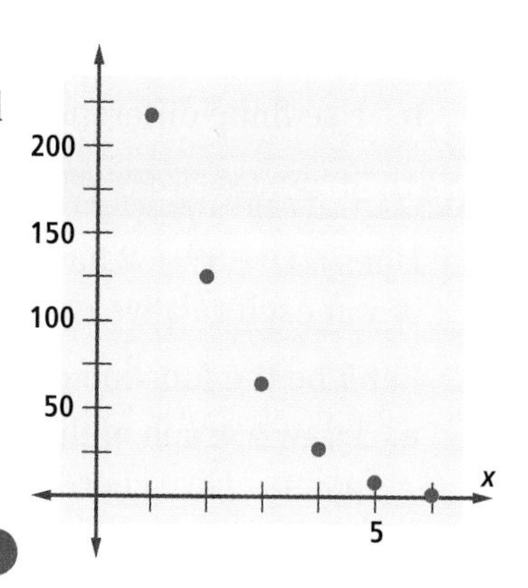

7. Consider a polynomial function f containing the points below.

x	0	1	2	3	4	5	6
y	0	9	28	69	144	265	444

a. Find the smallest possible degree of f.

b. Determine an equation for f. c. Find $f(7)$.

APPLYING THE MATHEMATICS

8. Polynomial functions are often used to approximate other functions on a limited domain. $f(x) = \frac{x^9}{362{,}880} - \frac{x^7}{5040} + \frac{x^5}{120} - \frac{x^3}{6} + x$ describes a polynomial function that approximates a function known to you.

a. Graph the polynomial on the window $-4 \leq x \leq 4$, $-2 \leq y \leq 2$ and determine what function it approximates.

b. On what domain is f a good approximation?

9. The partial table below shows values of a nth-degree polynomial function f. Find the missing differences and the values a, b, c, d, and n.

x	1	2	3	4	5	6	7
$f(x)$	-8.2	-11	33.4	a	b	c	d

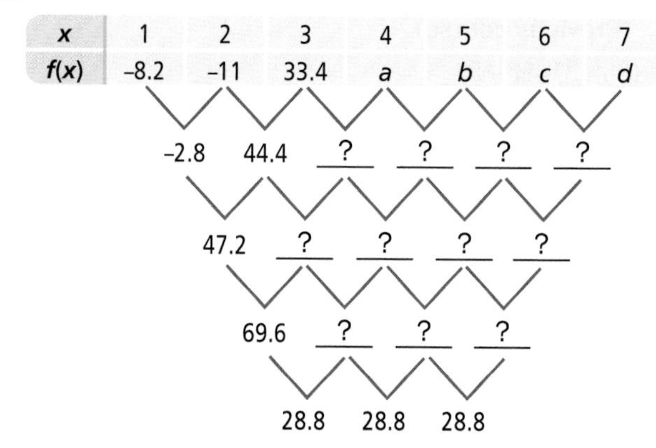

10. a. If $f(x) = ax^3 + bx^2 + cx + d$, find $f(1), f(2), f(3), f(4), f(5)$, and $f(6)$.

b. Prove that the third differences of these values are constant.

11. Refer again to the tetrahedral numbers in the lesson. Suppose you try to model the data with a quadratic function.

a. Use $T(1)$, $T(2)$, and $T(3)$ to find a quadratic model for the data, and verify that this formula works for the first three numbers in the sequence.

b. Verify that this quadratic model does ***not*** predict the 4th tetrahedral number in the sequence.

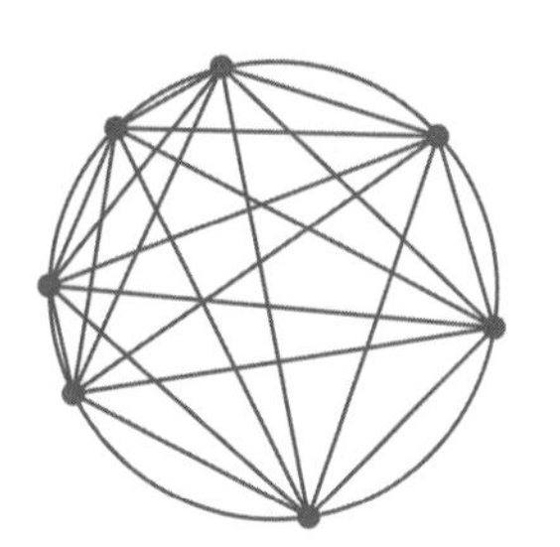

12. n points are arranged on a circle and all the chords drawn. Let $I(n)$ be the number of intersection points inside the circle if no 3 chords are concurrent.
    a. Compute $I(n)$ for $n = 4, 5, 6, 7, 8, 9$.
    b. Use finite differences to find a formula for $I(n)$.

REVIEW

13. Does $f(x) = x^3 + 2$ have any relative extrema? If so, state the value of x at each relative extremum. **(Lesson 7-1)**

14. Let h be the function defined by $h(x) = -25x^3 + 5x^2 + 75x + 125$.
    a. Draw a graph of the function. Label all intercepts and extrema points. Give coordinates accurate to the nearest tenth.
    b. In what interval(s) are the values of h positive?
    c. In what interval(s) are the values of h negative?
    d. In what interval(s) is h increasing?
    e. In what interval(s) is h decreasing? **(Lesson 7-1)**

15. The population of Russia in 2007 was about 141.4 million, with an annual growth rate of -0.5%. Assume the rate continues into the future. **(Lessons 2-5, 2-4)**
    a. Give a model for the population n years after 2007.
    b. Predict the population in 2020.
    c. Predict when the population will reach 125 million.

16. ***Skill Sequence*** Multiply and simplify. **(Previous course)**
    a. $(x + 3)(4x)$
    b. $(x + 3)(4x - 5)$
    c. $(x + 3)(2x^2 + 4x - 5)$
    d. $(x + 3)(2x^2 + 4x - 5)(x - 3)$

EXPLORATION

17. A complete set of *dominoes of order n* consists of L tiles with two halves each containing a number of dots according to the following criterion: Each whole number less than or equal to n (including the blank, representing 0) is paired with each other number exactly once. Note that this means that each number is paired with itself exactly once.
    a. Draw a complete set of dominoes of order 3.
    b. Make a table of the number of dominoes in a complete set for orders 1, 2, 3, 4, 5, and 6.
    c. Determine a formula for L in terms of n.

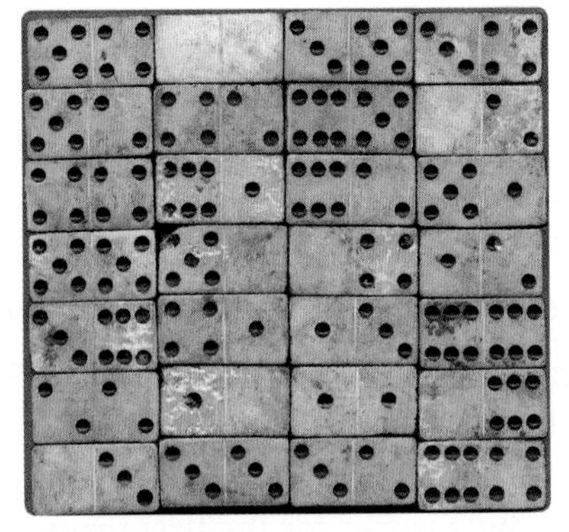

Dominoes, a game invented in China, dates back to 1120.

QY ANSWERS

1. 21
2. 35
3. 5

Lesson 7-3 Division and the Remainder Theorem

BIG IDEA When one polynomial is divided by another, the result can be written either as a rational expression or as a quotient polynomial with a remainder polynomial.

In earlier courses you learned how to add, subtract, and multiply polynomials. This lesson discusses division of polynomials. The procedure is similar to dividing integers and relies on the inverse relationship between multiplication and division.

$$\frac{\text{dividend}}{\text{divisor}} = \text{quotient or dividend} = \text{quotient} \cdot \text{divisor}$$

Suppose you wish to divide 84 by 12. If you know that $7 \cdot 12 = 84$, then from the definition of division you may conclude that $\frac{84}{12} = 7$. Recall that 84 is called the ***dividend***, 12 is called the ***divisor*** and 7 is the ***quotient***.

Similarly for polynomials, if you know $(x + 3)(x + 5) = x^2 + 8x + 15$, you may conclude that $\frac{x^2 + 8x + 15}{x + 5} = x + 3$, provided $x \neq -5$. Here the dividend is $x^2 + 8x + 15$, the divisor is $x + 5$, and the quotient is $x + 3$.

We also say that $x + 3$ is a *factor* of $x^2 + 8x + 15$. If you do not see how a polynomial can be factored, then you can use long division.

Mental Math

Rewrite the fraction in lowest terms.

a. $\frac{40}{84}$

b. $\frac{40x^3}{84x^6}$

c. $\frac{40(3y + 10)}{84(6y + 20)}$

Example 1

Divide $6x^2 + x - 2$ by $3x + 2$.

Solution Look at the first terms of both the dividend and divisor.

$$\begin{array}{r} 2x\phantom{{}+x-2} \\ 3x + 2\,\overline{)\,6x^2 + x - 2} \\ \underline{6x^2 + 4x}\phantom{{}-2} \\ -3x - 2 \end{array}$$

Think: $3x\overline{)6x^2} = 2x$. This is the first term in the quotient; write it above the dividend. Now multiply $3x + 2$ by $2x$ and subtract from $6x^2 + x - 2$.

Now look at the first term of the divisor and the new dividend.

$$\begin{array}{r} 2x - 1 \\ 3x + 2\,\overline{)\,6x^2 + x - 2} \\ \underline{6x^2 + 4x}\phantom{{}-2} \\ -3x - 2 \\ \underline{-3x - 2} \\ 0 \end{array}$$

Think: $3x\overline{)-3x} = -1$. This is the second term in the quotient; write it above the dividend to the right of $2x$. Multiply $3x + 2$ by -1 and subtract from $-3x - 2$. Since 0 is left, there is no remainder, and the division is finished.

Check Does $(3x + 2)(2x - 1) = 6x^2 + x - 2$? Yes. It checks.

If some of the coefficients in the dividend polynomial are zero, it helps to fill in all the missing powers of the variable, using zero coefficients.

Division of Polynomials Using a CAS

Most CAS can do all operations on polynomials. The input and output from two different machines are shown below.

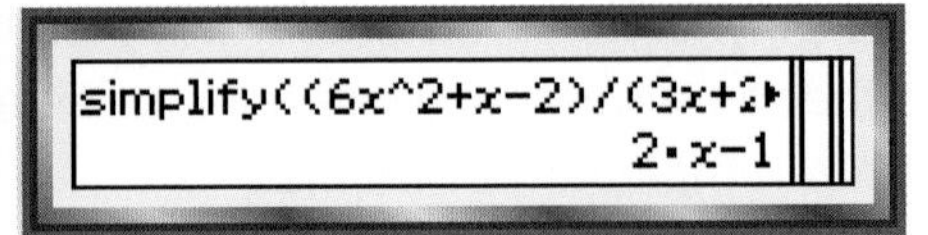

As you can see, some CAS perform some polynomial division without the use of a command. Test yours to see if you need the command.

Division with Remainders

When 1539 is divided by 8, the quotient is 192 and the remainder is 3. You can check that by multiplying 192 by 8 and adding 3.

$$1539 = 192 \cdot 8 + 3$$

In general, if an integer n is divided by a nonzero integer d, there is a unique quotient q and a unique remainder r with $0 \le r < d$, with

$$n = q \cdot d + r.$$

As in division of integers, not all polynomial division problems "come out even." In general, when a polynomial $p(x)$ is divided by a polynomial $d(x)$, it produces a quotient $q(x)$ and remainder $r(x)$. Either $r(x) = 0$ or $r(x)$ has degree less than the degree of the divisor, $d(x)$. In symbols,

$$p(x) = q(x) \cdot d(x) + r(x), \text{ or}$$
$$\frac{p(x)}{q(x)} = d(x) + \frac{r(x)}{q(x)}.$$

GUIDED

Example 2

Divide $3x^4 - 7x^3 + 5$ by $x^2 + 2$.

Solution The coefficients of x^2 and x in the dividend are zero. Insert $0x^2$ and $0x$ so that all powers of x appear.

$$\begin{array}{rl} & \underline{\ ?\ }x^2 - \underline{\ ?\ }x - \underline{\ ?\ } \\ x^2 + 2 \big) & 3x^4 - 7x^3 + 0x^2 + 0x + 5 \\ & \underline{3x^4 \qquad\quad + 6x^2} \\ & -7x^3 - 6x^2 + 0x \\ & \underline{-7x^3 \qquad\quad - 14x} \\ & -6x^2 + 14x + 5 \\ & \underline{-6x^2 \qquad\quad - 12} \\ & 14x + 17 \end{array}$$

Think: $x^2 \overline{)3x^4} = \underline{\ ?\ }x^2$

Multiply $x^2 + 2$ by $\underline{\ ?\ }x^2$ and subtract.

Think: $x^2 \overline{)-7x^3} = \underline{\ ?\ }x$

Multiply $x^2 + 2$ by $\underline{\ ?\ }x$ and subtract.

Think: $x^2 \overline{)-6x^2} = \underline{\ ?\ }$

Multiply $x^2 + 2$ by $\underline{\ ?\ }$ and subtract.

Because the degree of $14x + 17$ is less than that of $x^2 + 2$, the division is complete. You may write:

The quotient is __?__ x^2 – __?__ x – __?__ with remainder $14x + 17$, or

$$\frac{3x^4 - 7x^3 + 5}{x^2 - 2} = \underline{\ ?\ }x^2 - \underline{\ ?\ }x - \underline{\ ?\ } + \frac{14x + 17}{x^2 + 2}.$$

Check 1 Substitute a value for x and verify the last equation.

Check 2 Does dividend = quotient • divisor + remainder?

Does $3x^4 - 7x^3 + 5 = (3x^2 - 7x - 6)(x^2 + 2) + (14x + 17)$? Yes.

Dividing by $x - c$

Consider the special case when the divisor $d(x)$ is of the form $x - c$. For the remainder to be of a lower degree than the divisor, it must be a constant, possibly zero. Then

$p(x) = q(x) \cdot (x - c) + r(x)$, where $r(x)$ is a constant.

This equation is true for all x. In particular, it is true when $x = c$. Then

$p(c) = q(c) \cdot (c - c) + r(c)$.

This can be simplified.

$p(c) = q(c) \cdot 0 + r(c)$

$p(c) = r(c)$

This says the value of the polynomial at $x = c$ is precisely the remainder when p is divided by $x - c$. These steps prove the theorem below.

Remainder Theorem

If a polynomial $p(x)$ is divided by $x - c$, then the remainder is $p(c)$.

QY

QY

When $5x^4 - 3x + 2$ is divided by $x + 2$, what is the remainder?

Example 3

Find the remainder when $4x^5 - x^3 + 1$ is divided by $x - 5$.

Solution 1 Use the Remainder Theorem.
Let $p(x) = 4x^5 - x^3 + 1$ and find $p(5)$.

$p(5) = 4(5)^5 - (5)^3 + 1 = 12500 - 125 + 1 = 12{,}376$

So the remainder when $4x^5 - x^3 + 1$ is divided by $x - 5$ is 12,376.

This calculation could also be done on a CAS.

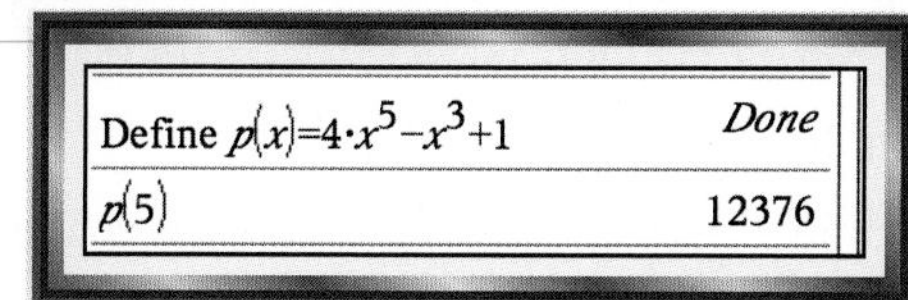

(continued on next page)

Solution 2 Use polynomial long division.

$$
\begin{array}{rl}
 & 4x^4 + 20x^3 + 99x^2 + 495x + 2475 \\
x - 5 \,) & \overline{4x^5 \qquad\qquad - x^3 \qquad\qquad\qquad + 1} \\
 & \underline{4x^5 - 20x^4} \\
 & 20x^4 - x^3 \\
 & \underline{20x^4 - 100x^3} \\
 & 99x^3 \\
 & \underline{99x^3 - 495x^2} \\
 & 495x^2 \\
 & \underline{495x^2 - 2475x} \\
 & 2475x + 1 \\
 & \underline{2475x - 12375} \\
 & 12376
\end{array}
$$

Solution 3 Use a CAS. The first time we tried this, we were surprised. Nothing happened!

$\frac{4 \cdot x^5 - x^3 + 1}{x-5} \qquad \frac{4 \cdot x^5 - x^3 + 1}{x-5}$

There are several CAS methods to check answers.

CAS Method 1: Instruct the CAS to rewrite the input as a proper fraction.

$\frac{\text{dividend}}{\text{divisor}} = \frac{\text{remainder}}{\text{divisor}} + \text{quotient}$

So the remainder is 12376.

$\text{propFrac}\left(\frac{4 \cdot x^5 - x^3 + 1}{x-5}\right)$

$\frac{12376}{x-5} + 4 \cdot x^4 + 20 \cdot x^3 + 99 \cdot x^2 + 495 \cdot x + 247\text{▸}$

CAS Method 2: Use the `expand` command.

$\text{expand}\left(\frac{4 \cdot x^5 - x^3 + 1}{x-5}\right)$

$\frac{12376}{x-5} + 4 \cdot x^4 + 20 \cdot x^3 + 99 \cdot x^2 + 495 \cdot x + 247\text{▸}$

CAS Method 3: On some CAS you can use two commands called `polyQuotient` and `polyRemainder` that take three inputs: the dividend, the divisor, and the independent variable.

$\text{polyQuotient}(4 \cdot x^5 - x^3 + 1, x-5, x)$

$4 \cdot x^4 + 20 \cdot x^3 + 99 \cdot x^2 + 495 \cdot x + 2475$

$\text{polyRemainder}(4 \cdot x^5 - x^3 + 1, x-5, x)$

12376

Notice that the Remainder Theorem has limitations. It only applies when dividing by a linear factor, and it does not produce a quotient. But it provides a quick way to find the remainder. And the Remainder Theorem has a very useful consequence, the Factor Theorem, which you will see in Lesson 7-4.

Questions

COVERING THE IDEAS

In 1 and 2, find the quotient when the first polynomial is divided by the second.

1. $2x^2 - 5x - 12, x - 4$
2. $8z^4 - 2z^3 - 3z - 6, 2z^2 + 1$
3. Given that $(x - 4)(x^2 + 7x - 1) = x^3 + 3x^2 - 29x + 4$, write each quotient as a polynomial.
   a. $\dfrac{x^3 + 3x^2 - 29x + 4}{x - 4}$
   b. $\dfrac{x^3 + 3x^2 - 29x + 4}{x^2 + 7x - 1}$
4. In a division problem, state two relationships between the dividend, divisor, quotient, and remainder.
5. Suppose $p(x) = d(x) \cdot q(x) + r(x)$, $p(x) = x^2 + 2x + 13$, and $d(x) = x + 2$. Find possible polynomials $q(x)$ and $r(x)$.
6. The input and output of a CAS are given at the right. Identify the dividend, divisor, quotient, and remainder.

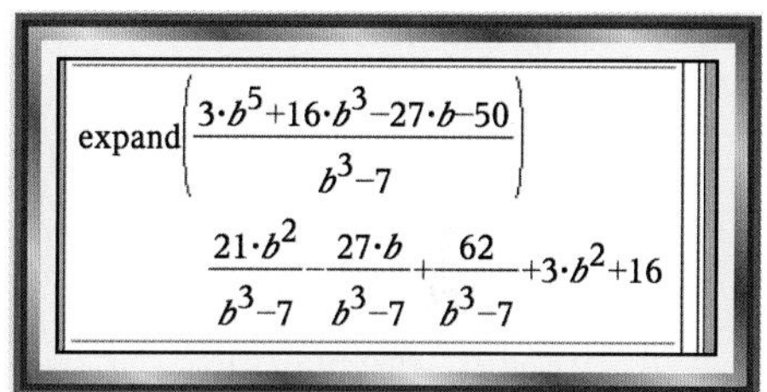

$$\text{expand}\left(\frac{3\cdot b^5+16\cdot b^3-27\cdot b-50}{b^3-7}\right)$$

$$\frac{21\cdot b^2}{b^3-7}-\frac{27\cdot b}{b^3-7}+\frac{62}{b^3-7}+3\cdot b^2+16$$

In 7 and 8, two polynomials are given.
   a. Use the Remainder Theorem to find the remainder when the first polynomial is divided by the second.
   b. Check by division.

7. $y^3 + 3y^2 - 9y - 40, y + 3$
8. $2x^3 - 9x^2 - x + 5, x + 5$
9. Suppose you know that when $p(x)$ is divided by $x - 76$, the remainder is zero. What is $p(76)$?

APPLYING THE MATHEMATICS

10. Let $f(x) = x^7 - 128$ and $g(x) = x - 2$. Express $\frac{f(x)}{g(x)}$ as a polynomial in simplest terms.

In 11 and 12, two polynomials are given.
   a. Find the quotient and remainder when the first polynomial is divided by the second.
   b. Show that the degree of the remainder is less than the degree of the divisor.
   c. Check by multiplying the divisor by the quotient and adding the remainder.

11. $3a^3 + 5a^2 - 10, a^2 - 4$
12. $x^4 - 6x^3 + 17x^2 - 2x - 3, x - 3$
13. What is the quotient if $x^3 + 2x^2y - 2xy^2 - y^3$ is divided by $x - y$?
14. Consider $g(x) = \dfrac{x^3 - 9x^2 + 27x - 27}{x^2 - 6x + 9}$.
    a. **True or False** For all $x \neq 3$, g is a linear function.
    b. Justify your answer to Part a.

15. a. Refer to Example 2. Use a graphing utility to plot on the same set of axes $f(x) = \frac{3x^4 - 7x^3 + 5}{x^2 + 2}$, $g(x) = 3x^2 - 7x - 6$, and $h(x) = 3x^2 - 7x - 6 + \frac{14x + 17}{x^2 + 2}$.
b. Write several sentences comparing and contrasting the graphs.

REVIEW

16. Consider the function h described by the data points below.

z	0	1	2	3	4	5	6
$h(z)$	3	2	7	24	59	118	207

a. Show that h may be a polynomial of degree less than 5, and find its degree.
b. Determine an equation for h.
c. Evaluate $h(7)$. **(Lesson 7-2)**

17. Let k be the function defined by $k(s) = s^3 + 4s^2 + s - 6$.
a. Sketch the graph of the function. Label all intercepts and extrema points. Give coordinates accurate to the nearest tenth.
b. In what interval(s) are the values of k positive?
c. In what interval(s) are the values of k negative?
d. In what interval(s) is k increasing?
e. In what interval(s) is k decreasing? **(Lesson 7-1)**

18. Does the function t defined by $t(x) = -x^3 + 3x^2 + x$ have any relative extrema? If so, state the coordinates of each one. **(Lesson 7-1)**

19. The classic 1954 Japanese film *The Seven Samurai* follows the story of a village of farmers who hire the seven to combat bandits who plan to return to the village to steal their crops. If the seven samurai stand in line formation, how many different orders are possible? **(Lesson 6-4)**

20. To draw a map, a cartographer needed to find the distances between point Z across a lake and each of points X and Y on the other side. The cartographer found $XY \approx 0.3$ miles, $m\angle X \approx 48°$, and $m\angle Y \approx 101°$. Find the distances from X to Z and from Y to Z. **(Lesson 5-5)**

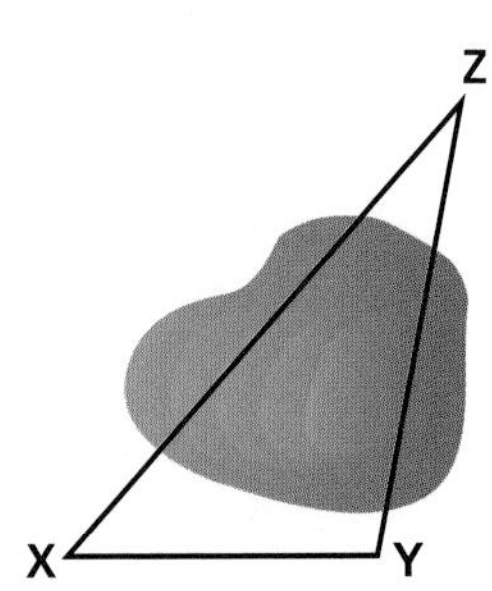

21. ***Skill Sequence*** Solve. **(Previous Course)**
a. $x(x + 8) = 0$
b. $(y - 4)(y - 7) = 0$
c. $(z - 6)(z + 2)(z + 11) = 0$

EXPLORATION

22. a. Divide each polynomial by $x - y$.
i. $x^3 - y^3$ ii. $x^5 - y^5$ iii. $x^7 - y^7$ iv. $x^9 - y^9$
b. Generalize the pattern in Part a.

QY ANSWER

88

Lesson 7-4 The Factor Theorem

▸ **BIG IDEA** The factors of a polynomial indicate the zeros of the corresponding polynomial function.

Activity

MATERIALS CAS

Step 1 Consider the polynomial function g described by $g(x) = x^2 - 3x - 10$. Graph the function using the window $-10 \leq x \leq 15$, $-250 \leq y \leq 100$. Identify the x-intercepts.

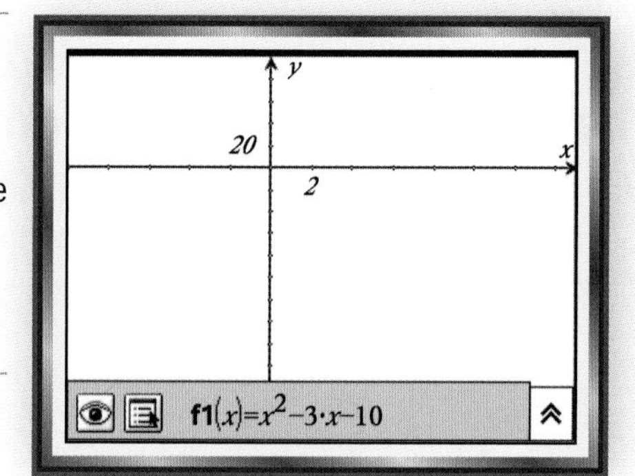

Step 2 Use a CAS to factor the polynomial.

Step 3 Use a CAS to find the zeros of the function.

Step 4 Repeat Steps 1-3 for the function $h: x \longrightarrow x^3 + x^2 - 12x$.

Step 5 Repeat Steps 1-3 for the function w with equation $w(x) = 2x^4 + 6x^3 - \frac{37}{2}x^2 + \frac{15}{2}x$.

Step 6 Describe the relationship among the factors of the polynomial, the zeros of the function, and the graph.

Mental Math

An equation of the form $p(x) = 0$ is given. Factor $p(x)$ and solve the equation.

a. $x^2 - 25x = 0$

b. $x^2 + x - 6 = 0$

How Are Zeros of a Polynomial Related to the Factors of the Polynomial?

The relations between the expanded form of a polynomial and its zeros are subtle. However, when looking at the *factored* form of a polynomial and its zeros, a pattern emerges. Each zero c of the polynomial corresponds with a linear factor $(x - c)$. These related facts are illustrated below for the polynomial graphed at the right.

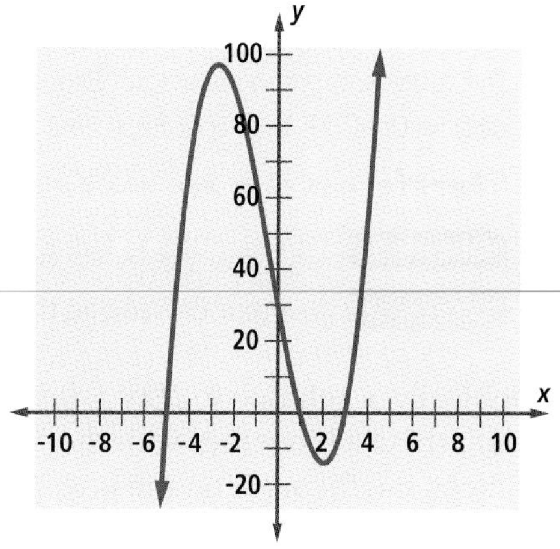

$$\text{Expanded form: } 2x^3 + 2x^2 - 34x + 30$$

$$\text{Factored form: } 2(x + 5)(x - 1)(x - 3)$$

$$\text{Zeros: } -5, 1, \text{ and } 3$$

This result holds for all polynomials and can be proved using the Remainder Theorem from Lesson 7-3.

Factor Theorem

For a polynomial $p(x)$, $(x - c)$ is a factor of $p(x)$ if and only if $p(c) = 0$.

Proof The theorem says "if and only if," so two statements must be proved:

(1) If $(x - c)$ is a factor of $p(x)$, then $p(c) = 0$.

(2) If $p(c) = 0$, then $(x - c)$ is a factor of $p(x)$.

In (1), $(x - c)$ is given as a factor of $p(x)$. So for all x, $p(x) = (x - c)\,q(x)$, where $q(x)$ is a polynomial. In particular, when $x = c$, $p(x) = p(c) = (c - c)q(c) = 0$. So $p(c) = 0$.

In (2), $p(c) = 0$. By the Remainder Theorem, $p(x) = (x - c)q(x) + p(c)$ for all x, where $q(x)$ is a polynomial. Since $p(c) = 0$, $p(x) = (x - c)q(x)$, making $(x - c)$ a factor of $p(x)$.

Using the Factor Theorem to Factor Polynomials

The Factor Theorem says that finding a zero of a polynomial will give a factor of the polynomial. So, if you need to factor a polynomial and cannot find a factor, look for a zero.

Example 1

Factor $p(x) = x^4 - 6x^3 - 4x^2 + 24x$.

Solution Examine a graph and function table of values for $p(x)$.

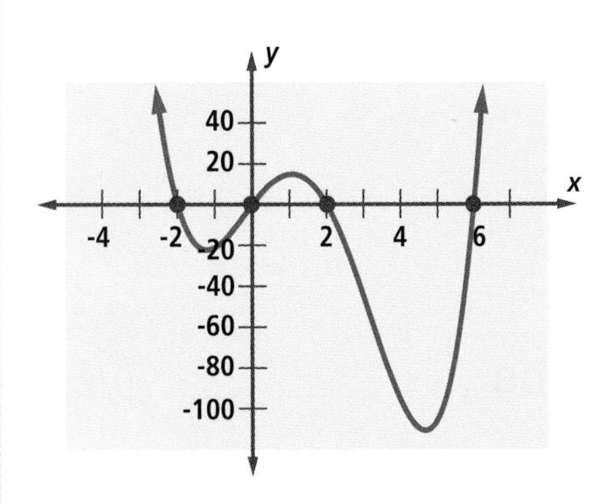

x	= ()
-2	0
-1	-21
0	0
1	15
2	0
3	-45
4	-96
5	-105
6	0
7	315

The table and graph show four values of x in red for which $p(x) = 0$: -2, 0, 2, and 6. Each zero determines a factor.

Therefore $p(x) = x(x + 2)(x - 2)(x - 6)$.

Check 1 Use your CAS to factor $p(x)$. It checks.

Check 2 Use your CAS to find the zeros of $p(x)$. It checks.

Define $p(x)=x^4-6\cdot x^3-4\cdot x^2+24\cdot x$	Done
factor$(p(x))$	$x\cdot(x-6)\cdot(x-2)\cdot(x+2)$
zeros$(p(x),x)$	$\{-2,0,2,6\}$

Graphically, a solution to $p(x) = 0$ is an x-intercept of the graph of p. Putting this fact together with the Remainder and Factor Theorems produces the theorem on the next page.

Factor-Solution-Intercept Equivalence Theorem

For any polynomial $p(x)$ with real coefficients and any real number c, the following are logically equivalent statements.

(1) $(x - c)$ is a factor of $p(x)$.
(2) $p(c) = 0$.
(3) c is an x-intercept of the graph of $y = p(x)$.
(4) c is a zero of $p(x)$.
(5) The remainder when $p(x)$ is divided by $(x - c)$ is 0.

Finding Polynomials from Known Zeros

The Factor Theorem can also be used to find polynomials when their zeros are given.

Example 2

Find an equation for the 3rd-degree polynomial function p graphed below.

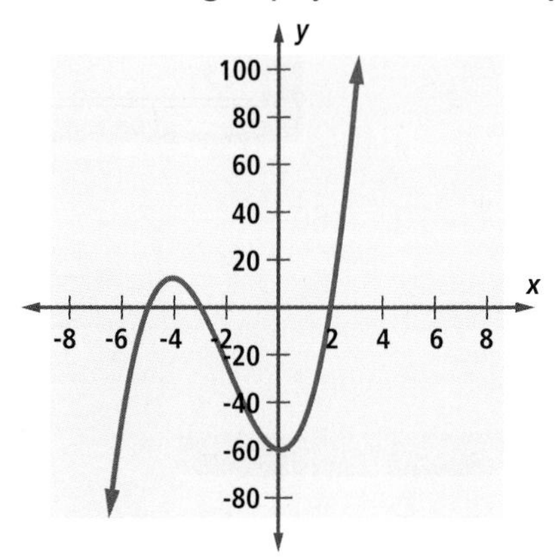

Solution Let $p(x)$ be the polynomial. The graph shows that $p(0) = -60$ and the zeros of p are –5, –3, and 2. By the Factor Theorem, three factors of $p(x)$ are $x + 5$, $x + 3$, and $x - 2$. The product of these factors is a 3rd-degree polynomial. There is also a constant factor a. So for some value of a, the polynomial is $p(x) = a(x + 5)(x + 3)(x - 2)$. The graph at the right shows three possible functions, each with a different value of a. None has the correct y-intercept. To find the correct value of a, substitute $(0, -60)$ into the equation.

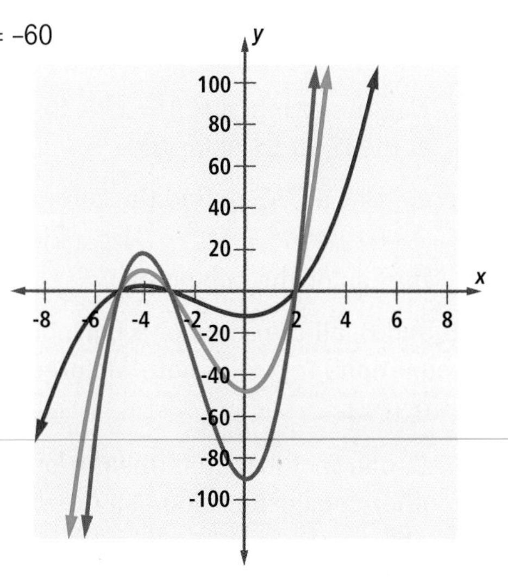

$$-60 = a(0 + 5)(0 + 3)(0 - 2)$$
$$-60 = -30a$$
$$a = 2$$

So, $p(x) = 2(x + 5)(x + 3)(x - 2)$.

In Example 3, no intercept is given, but the restriction that the polynomial must have integer coefficients limits possible answers.

Example 3

Find a 3rd-degree polynomial $p(x)$ with integer coefficients whose zeros are $1, \frac{3}{5}$, and $-\frac{2}{3}$.

Solution Each zero of $p(x)$ corresponds to a factor of the polynomial. From the given, by the Factor Theorem, $p(x)$ has factors $(x - 1)$, $\left(x - \frac{3}{5}\right)$, and $\left(x + \frac{2}{3}\right)$. So $p(x) = a(x - 1)\left(x - \frac{3}{5}\right)\left(x + \frac{2}{3}\right)$. To obtain integer coefficients, choose a value of a whose product with $\left(x - \frac{3}{5}\right)$ and $\left(x + \frac{2}{3}\right)$ will clear the fractions. We use the lowest common denominator, $a = 5 \cdot 3 = 15$.

$$p(x) = 5 \cdot 3(x - 1)\left(x - \frac{3}{5}\right)\left(x + \frac{2}{3}\right)$$

$$= (x - 1) \cdot 5\left(x - \frac{3}{5}\right) \cdot 3\left(x + \frac{2}{3}\right) \quad \text{Rearrange the factors.}$$

$$= (x - 1)(5x - 3)(3x + 2) \quad \text{Distribute.}$$

$$p(x) = 15x^3 - 14x^2 - 7x + 6$$

Define $p(x)=15\cdot x^3-14\cdot x^2-7\cdot x+6$	Done
$p(1)$	0
$p\left(\frac{3}{5}\right)$	0
$p\left(\frac{-2}{3}\right)$	0

Check Evaluate $p(x)$ at each zero. Use a CAS.

In Example 3, multiples of $p(x)$ such as $150x^3 - 140x^2 - 70x + 60$ or $-15x^3 + 14x^2 + 7x - 6$ are correct as well.

Questions

COVERING THE IDEAS

1. State the Factor Theorem.
2. If $f(x) = 2x^3 - 34x^2 + x - 17$ has a factor $(x - 17)$, what must be true about $f(17)$?
3. Use the graph of $f(x) = x^4 - 3x^3 - 39x^2 + 47x + 210$ at the right to factor $f(x)$.

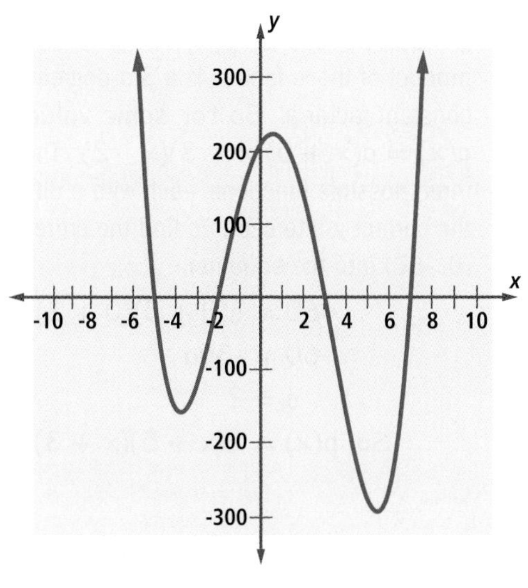

4. a. Use a CAS to find the zeros of the polynomial $t(n) = n^3 + 4n^2 - 31n - 70$.
   b. Factor the polynomial.

In 5 and 6, all the zeros of a polynomial $p(x)$ are given. Give an equation for $p(x)$ in both factored and standard form.

5. 0, 6, -4
6. $8, \frac{7}{2}$, and $-\frac{5}{3}$.

In 7–9, one fact is stated about a polynomial. State at least two other conclusions you can draw.

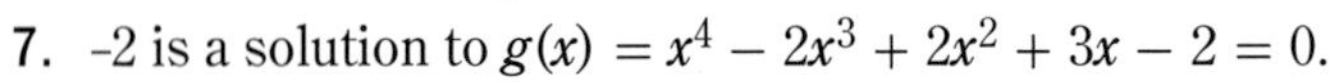

7. -2 is a solution to $g(x) = x^4 - 2x^3 + 2x^2 + 3x - 2 = 0$.
8. The graph of $h(x) = x^5 + x^4 - 18x^2$ crosses the x-axis at $x = 3$.

9. When $f(t) = 3t^5 - 46t - 4$ is divided by $t - 2$, there is no remainder.

10. The x-intercepts of a polynomial function are -6, -3, and 3 and its y-intercept is 27. Find the standard form of the polynomial.

APPLYING THE MATHEMATICS

11. On an exam, Grace factored $g(x) = 3x^3 + 4x^2 - 17x - 6$ as $(x - 3)(x - 2)(3x - 1)$.
    a. Graph $y = g(x)$ and $f(x) = (x - 3)(x - 2)(3x - 1)$ on the same set of axes.
    b. Is Grace's factorization correct? How can you tell?

12. Let $p(x) = (x - 2)^2(x + 3)k(x)$.
    a. What are two roots of the equation?
    b. If $p(3) = -1224$, what does $k(3)$ equal?
    c. If $p(x) = 9x^5 - 48x^4 - 201x^3 + 588x^2 + 876x - 2016$, find the zeros of k.

13. The factored form of a polynomial is given.
    a. Use a CAS to expand the polynomial.
    b. Explain how the zeros of the polynomial relate to the coefficients of the polynomial.
       i. $(x - a)(x - b)$
       ii. $(x - a)(x - b)(x - c)$
       iii. $(x - a)(x - b)(x - c)(x - d)$

In 14 and 15, a graph of a polynomial function with integer zeros is given. Find a possible equation of the given degree for the graph.

14. degree 3

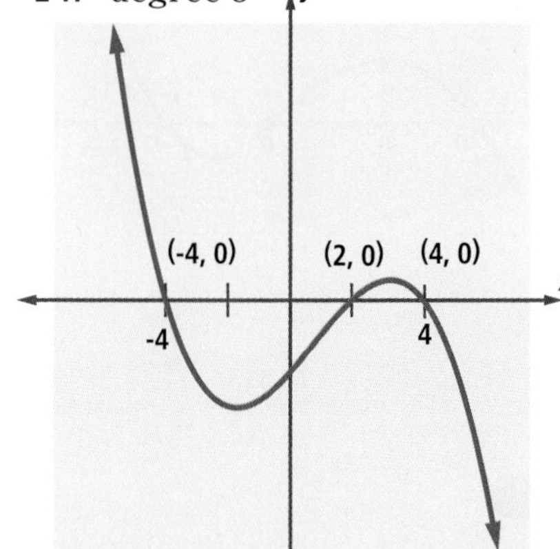

15. degree 4

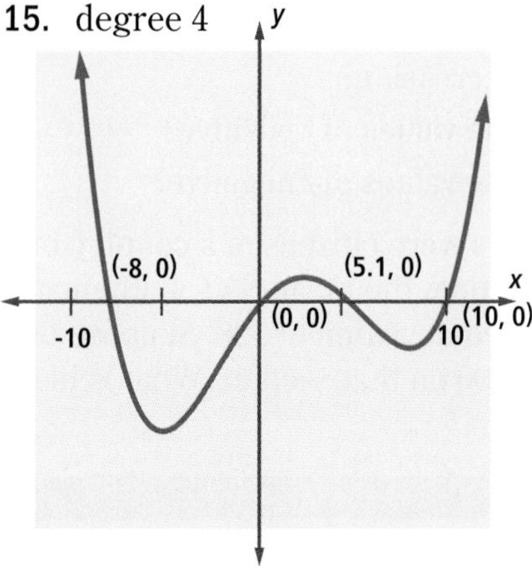

REVIEW

In 16 and 17, find the quotient when the first polynomial is divided by the second. (Lesson 7-3)

16. $3x^2 - 16x - 35, x - 7$

17. $6z^4 - 3z^3 - 3z - 6, 2z^2 + 2$

18. Use the data in the table below. **(Lesson 7-2)**

    a. Determine if v is a polynomial function of u of degree < 5.

u	1	2	3	4	5	6
v	0	–12	–20	–18	0	40

    b. If v is a polynomial function of u of degree < 5, determine the polynomial.

19. *Pentagonal numbers* $p(n)$ are the numbers of dots in the sequence of figures beginning with those pictured below. **(Lessons 7-2, 2-6)**

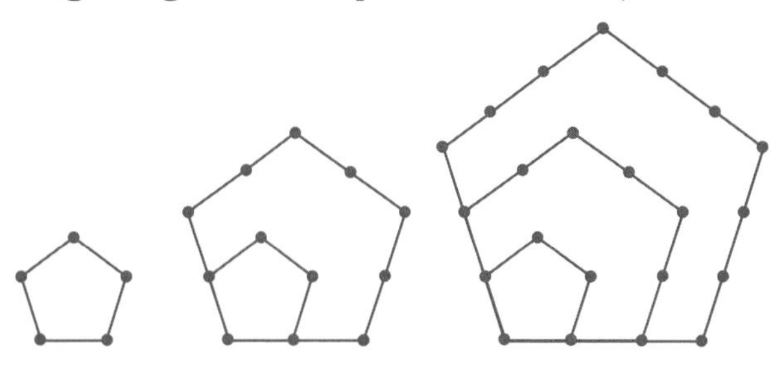

The Pentagon, near Washington, DC, houses approximately 23,000 employees.

    a. Show that there is a polynomial of degree 2 which generates these pentagonal numbers.

    b. Determine the quadratic polynomial $p(n)$.

    c. Use the polynomial of Part b to calculate $p(5)$ and $p(6)$, and verify your answer geometrically.

20. At the right is a graph of a polynomial $y = f(x)$. **(Lesson 7-1)**

    a. Identify the zeros of f.

    b. Identify the values of x where the relative extrema occur.

    c. In what interval(s) is f increasing?

    d. In what interval(s) is f decreasing?

    e. In what interval(s) are the values of f positive?

    f. In what interval(s) are the values of f negative?

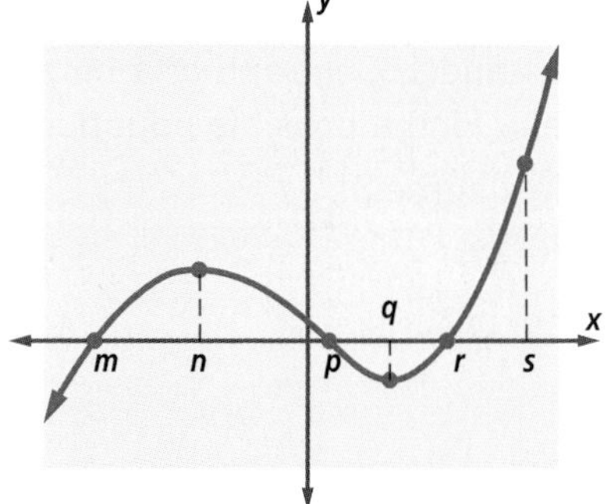

21. In 2007, 4% of SAT test-takers were citizens of a country other than the U.S. Of this population, the mean SAT mathematics score was 572 with a standard deviation of 128. Suppose one of these test-takers scored 700 on that section. What is his or her z-score? **(Lesson 3-9)**

EXPLORATION

22. The Factor-Solution-Intercept Equivalence Theorem lists five logically equivalent statements.

    a. State the command on your CAS for each of these statements.

    b. Use each command to explore the polynomial $21t^4 + 41t^3 - 115t^2 - 129t + 54$.

    c. Which command or commands was most useful in exploring the polynomial? Why?

Lesson 7-5 Complex Numbers

Vocabulary

i, imaginary numbers
complex number
real part
imaginary part
complex conjugates
discriminant

BIG IDEA Numbers of the form $a + bi$, where a and b are real and $i^2 = -1$, can be added, subtracted, multiplied, and divided much like polynomials, and they are the roots of many polynomial equations.

Consider the polynomial $x^2 + 1$. Its zeros are the solutions to $x^2 + 1 = 0$ or, more simply, $x^2 = -1$. Does this equation have any solutions?

The number of solutions depends on the domain of available numbers. If x is a real number, $x^2 = -1$ has no solution, since the square of any real number is nonnegative. The graph at the right verifies that $f(x) = x^2 + 1$ has no real zeros, because its graph does not cross the x-axis.

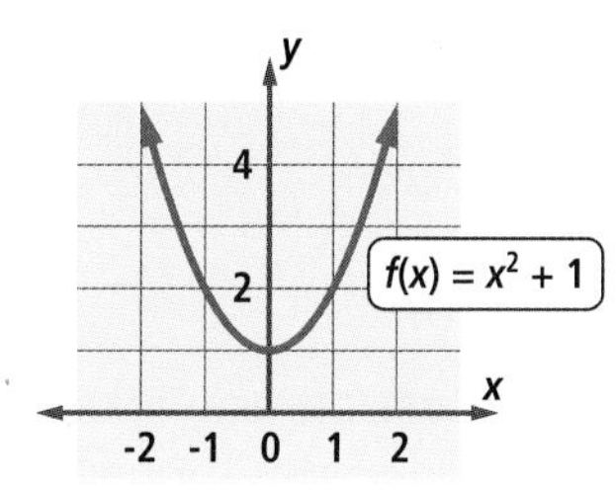

Mental Math

Between which two consecutive integers is the number?

a. $\sqrt{5}$

b. $\sqrt{15}$

c. $\sqrt{150}$

The Number *i*

It is possible to define solutions to $x^2 + 1 = 0$ that are not real numbers. Because $x^2 = -1$, these solutions are called *the square roots of negative 1*. One of these solutions is denoted as $\sqrt{-1}$. It is customary also to call this number i.

Definition of *i*

$i = \sqrt{-1}$

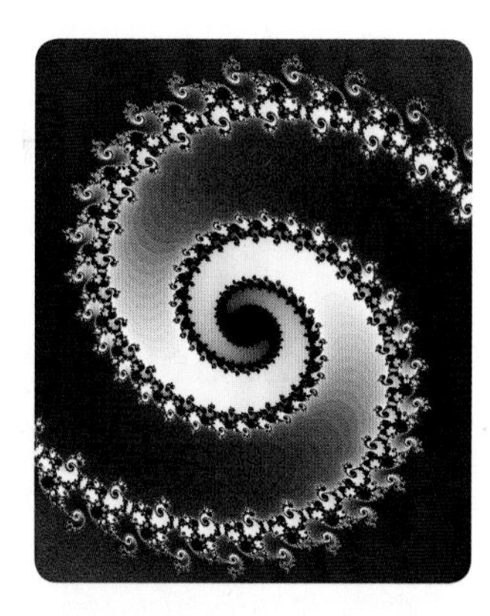

Fractals like the one pictured here are created using complex numbers.

The number i and its nonzero real number multiples bi are called **imaginary numbers** because when they were first used, mathematicians had no concrete way of representing them. They are, nevertheless, a reality in mathematics, and important applications have been found for them in fields ranging from control processing to quantum mechanics.

From its definition, $i \cdot i = i^2 = -1$. Assuming that properties of multiplication of real numbers can be extended to include i, then the following shows that $-i$ is another square root of -1.

$$(-i)^2 = (-i) \cdot (-i) = -1 \cdot i \cdot -1 \cdot i = -1 \cdot -1 \cdot i \cdot i = 1 \cdot i^2 = -1$$

Thus, in a domain including imaginary numbers, the polynomial $f(x) = x^2 + 1$ has two zeros, i and $-i$.

QY1

QY1

Verify that $f(x)$ has two zeros, i and $-i$, by evaluating $x^2 + 1$ for $x = i$ and $x = -i$.

Square Roots of Negative Numbers

If operations with imaginary numbers are assumed to satisfy the Commutative and Associative Properties of Multiplication, the square root of any negative number can be expressed as a multiple of i.

Example 1

Show that a square root of -5 is $i\sqrt{5}$.

Solution Multiply $i\sqrt{5}$ by itself.

$$i\sqrt{5} \cdot i\sqrt{5} = i \cdot i \cdot \sqrt{5}\sqrt{5}$$
$$= i^2 \cdot 5$$
$$= -1 \cdot 5 = -5$$

In general, if k is a positive real number, then $\sqrt{-k} = i\sqrt{k}$, so the square root of any negative number is the product of i and a real number. This means that all equations of the form $x^2 + k = 0$, with $k > 0$, have two solutions, $i\sqrt{k}$ and $-i\sqrt{k}$.

 QY2

One property of real numbers that does not extend to imaginary numbers concerns products of square roots. If a and b are positive real numbers, $\sqrt{a}\sqrt{b} = \sqrt{ab}$.

This is not the case when both a and b are negative. Notice that

$$\sqrt{-4}\sqrt{-16} = 2i \cdot 4i = -8,$$
$$\text{but } \sqrt{(-4)(-16)} = \sqrt{64} = 8.$$

For this reason, you should rewrite square roots of negative numbers in terms of i before performing any other operations.

QY2

What are the two solutions to the equation $x^2 + 7 = 0$?

Complex Solutions to Quadratic Equations

The sum of an imaginary number and a real number is called a *complex number.*

Definition of Complex Number

A **complex number** is a number of the form $a + bi$, where a and b are real numbers and $i = \sqrt{-1}$.

The number a is the **real part** and b is the **imaginary part** of $a + bi$. Notice both the real and imaginary parts of $a + bi$ are real numbers! For instance, in $3 - 4i$, the real part is 3 and the imaginary part is -4.

 QY3

QY3

Name the real part and the imaginary part of each complex number.

a. $-11 + 8i$

b. 5

All quadratic equations have solutions when the domain is the set of complex numbers.

Example 2

Solve $x^2 - 2x + 7 = 0$.

Solution Use the Quadratic Formula.

$$x = \frac{2 \pm \sqrt{(-2)^2 - 4(1)(7)}}{2 \cdot 1} = \frac{2 \pm \sqrt{-24}}{2} = \frac{2 \pm 2i\sqrt{6}}{2} = 1 \pm \sqrt{6}i$$

A CAS allows you to enter i and complex numbers in calculations. To obtain complex solutions when solving an equation like $x^2 - 2x + 7 = 0$, you must change the domain your CAS uses. This can be done for a single equation using a complex solve command, shown here as the abbreviation cSolve. Note the difference in results from solve.

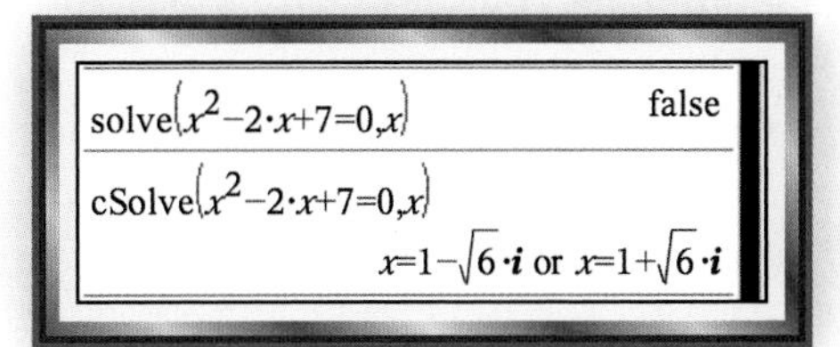

The two solutions of $x^2 - 2x + 7 = 0$ were found to be $1 + \sqrt{6}i$ and $1 - \sqrt{6}i$, complex numbers that are alike except for the sign of their imaginary part. A pair of complex numbers like these, of the form $a + bi$ and $a - bi$ where a and b are real numbers, are called **complex conjugates** of each other. When the coefficients of a quadratic with no real solutions are real numbers, the two nonreal solutions are complex conjugates. This is easily proved. Recall that $b^2 - 4ac$ is the **discriminant** of the quadratic equation $ax^2 + bx + c = 0$. If a quadratic equation with real coefficients has a negative discriminant, then the two solutions $\frac{-b + \sqrt{b^2 - 4ac}}{2a}$ and $\frac{-b - \sqrt{b^2 - 4ac}}{2a}$ equal $-\frac{b}{2a} + \frac{i\sqrt{4ac - b^2}}{2a}$ and $-\frac{b}{2a} - \frac{i\sqrt{4ac - b^2}}{2a}$, so they are complex conjugates of each other.

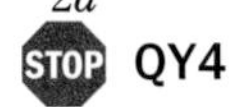 QY4

> **QY4**
> Use a CAS to find the complex conjugate solutions of $2x^2 - 12x + 19 = 0$.

Operations with Complex Numbers

The arithmetic of complex numbers is similar to the arithmetic of binomials, but with the additional fact that $i^2 = -1$.

GUIDED

Example 3

Let $z = 4 + 5i$ and $w = 2 - 3i$. Express each of the following in $a + bi$ form.

a. $z + w$ b. $z - w$ c. zw d. $\frac{z}{2}$

Solution

a. $z + w = (4 + 5i) + (\underline{\ ?\ })$
$= (4 + \underline{\ ?\ }) + (5i - \underline{\ ?\ }) = 6 + 2i$

b. $z - w = (4 + 5i) - (\underline{\ ?\ })$
$= 4 + 5i - \underline{\ ?\ } + \underline{\ ?\ }$
$= (4 - 2) + (\underline{\ ?\ }) = 2 + \underline{\ ?\ }$

(continued on next page)

c. $zw = (4 + 5i)(2 - 3i)$
$= 4(2 - 3i) + 5i(2 - 3i)$
$= 8 - 12i + \underline{\ ?\ } - \underline{\ ?\ }$
$= 8 - (\underline{\ ?\ }) + (\underline{\ ?\ })$ Use $i^2 = -1$.
$= 23 - 2i$

d. $\frac{z}{2} = \frac{4 + 5i}{2} = \frac{4}{2} + \frac{5}{2}i = (\underline{\ ?\ }) + (\underline{\ ?\ })i$

Check Use a calculator to check your solutions.

4+5·*i*+2−3·*i*	6+2·*i*
4+5·*i*−(2−3·*i*)	2+8·*i*
(4+5·*i*)·(2−3·*i*)	23−2·*i*
$\frac{4+5\cdot i}{2}$	$2+\frac{5}{2}\cdot i$

Example 3 illustrates the theorem below.

Theorem (Complex Number Operations)

Given two complex numbers $a + bi$ and $c + di$ and a real number r:

$(a + bi) + (c + di) = (a + c) + (b + d)i$ (complex number addition)

$(a + bi)(c + di) = (ac - bd) + (ad + bc)i$ (complex number multiplication)

$\frac{a + bi}{r} = \frac{a}{r} + \frac{b}{r}i.$ (division by any real number r)

It is not necessary to memorize this theorem. Simply remember to operate as you do with binomials: combine like terms when adding and use the Distributive Property when multiplying. It is important to know that when operating with complex numbers, even when doing powers and roots, the result is a complex number, and thus can be written in $a + bi$ form.

Conjugates and Division by a Complex Number

In Example 3, the calculation $\frac{4 + 5i}{2}$ involved dividing a complex number by a real number. The process of dividing by a complex number as in $\frac{4 - 3i}{5 + i}$ is more complicated, and requires a special result that comes from multiplying complex conjugates.

Theorem (Complex Conjugate Multiplication)

If a and b are real numbers, then $(a + bi)(a - bi) = a^2 + b^2$.

Proof Multiply the factors, using the Distributive Property.

$$\begin{aligned}(a + bi)(a - bi) &= a(a - bi) + bi(a - bi)\\ &= a^2 - abi + abi - (b^2i^2)\\ &= a^2 - b^2(-1)\\ &= a^2 + b^2\end{aligned}$$

The preceding proof confirms that the product of complex conjugates is a real number.

 QY5

QY5

Expand $(6 + 5i)(6 - 5i)$.

The complex conjugate of the divisor can be used to convert a quotient of complex numbers into a single number in $a + bi$ form.

Example 4

Express $\frac{4 - 3i}{5 + i}$ in $a + bi$ form.

Solution When the denominator is multiplied by its complex conjugate, the result is a real number. So multiply numerator and denominator by $5 - i$, the conjugate of the denominator.

$$\frac{4 - 3i}{5 + i} = \frac{4 - 3i}{5 + i} \cdot \frac{5 - i}{5 - i} = \frac{(4)(5 - i) - 3i(5 - i)}{25 + 1}$$

$$= \frac{20 - 4i - 15i + 3i^2}{26} = \frac{17 - 19i}{26} = \frac{17}{26} - \frac{19}{26}i$$

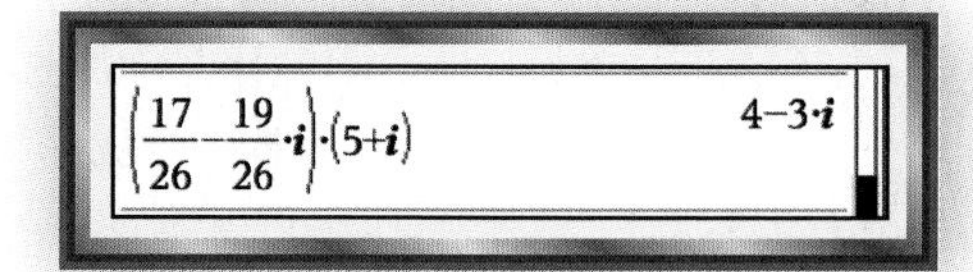

Check Multiply the quotient by the divisor. The result as shown on this calculator is $4 - 3i$, as it should be.

Complex Numbers and Factoring

Just as finding solutions to equations depends on the domain of allowable numbers, so does factoring. Notice how the domain determines whether a polynomial can be factored.

	Domain for the Coefficients of the Factors		
polynomial	set of integers	set of real numbers	set of complex numbers
$x^2 - 9$	$(x - 3)(x + 3)$	$(x - 3)(x + 3)$	$(x - 3)(x + 3)$
$x^2 - 5$	cannot factor	$(x - \sqrt{5})(x + \sqrt{5})$	$(x - \sqrt{5})(x + \sqrt{5})$
$x^2 + 36$	cannot factor	cannot factor	$(x - 6i)(x + 6i)$

A CAS can factor some polynomials over the set of polynomials with complex number coefficients. The complex factor command on one CAS is shown at the right.

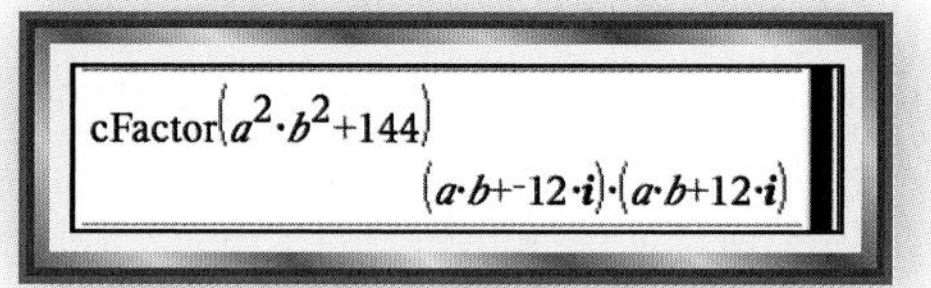

The Factor Theorem holds for complex numbers. For instance, $(x - 6i)(x + 6i)$ are factors of the polynomial $x^2 + 36$, while $6i$ and $-6i$ are solutions of the equation $x^2 + 36 = 0$.

How Are Various Types of Numbers Related?

For the complex number $a + bi$, if $a = 0$, then $a + bi = 0 + bi = bi$, so every imaginary number is also a complex number. Similarly, for any real number a, $a = a + 0i$. Thus, every real number is also a complex number. The diagram at the right shows how many types of numbers are related.

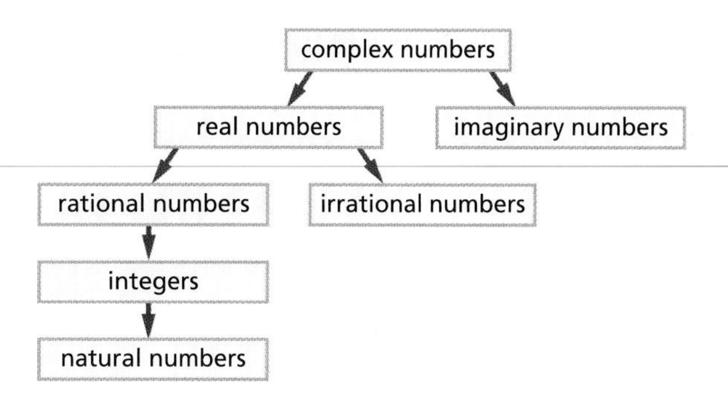

Questions

COVERING THE IDEAS

1. Show that $i\sqrt{13}$ and $-i\sqrt{13}$ are each square roots of -13.

In 2-4, simplify.

2. $\sqrt{-100}$
3. $\sqrt{-18}$
4. $(6i)^2$

In 5 and 6, a complex number is given.

a. Give the conjugate.

b. Give the product of the conjugates.

5. $3 - 2i$
6. $9i + 7$

7. Simplify each product.

   a. $\sqrt{-20}\sqrt{5}$

   b. $\sqrt{-20}\sqrt{-5}$

8. Factor $x^2 - 7$ over

   a. the set of polynomials with rational coefficients.

   b. the set of polynomials with real coefficients.

In 9 and 10, a quadratic equation is given.

a. Calculate its discriminant.

b. Solve and express the solutions in *a* + *bi* form.

9. $x^2 - 3x + 25 = 0$
10. $8x^2 + 215 = 11x$

11. In the equations of Questions 9 and 10, what relationship do the solutions have to one another?

In 12 and 13, a binomial is given.

a. Factor the binomial over the set of polynomials with complex coefficients.

b. Check by multiplying.

12. $x^2 + 64$
13. $7z^2 + 14$

14. a. Give a polynomial with factors $(x - 15i)$ and $(x + 15i)$.

    b. Give an equation where $15i$ and $-15i$ are zeros of the polynomial.

15. **True or False** Explain your answer.

    a. Every complex number is a real number.

    b. Every real number is a complex number.

16. What can you say about the graph of a quadratic polynomial function if its roots are not real?

17. By what number would you multiply $\frac{17 - 11i}{i + 12}$ to write the fraction in $a + bi$ form?

APPLYING THE MATHEMATICS

In 18–20, use this information. Impedance in an alternating-current (AC) circuit is the amount by which the circuit resists the flow of electricity. Commonly measured using complex numbers, the total impedance Z_T is calculated in a series circuit by adding the impedances ($Z_T = Z_1 + Z_2 + ...$) and in a parallel circuit by solving for Z_T using the form $\frac{1}{Z_T} = \frac{1}{Z_1} + \frac{1}{Z_2} + ...$ Find Z_T for each circuit and write in $a + bi$ form.

18. Series:

    $Z_1 = 3 + 7i$

    $Z_2 = -2 - 8i$

    $Z_3 = -6 + 4i$

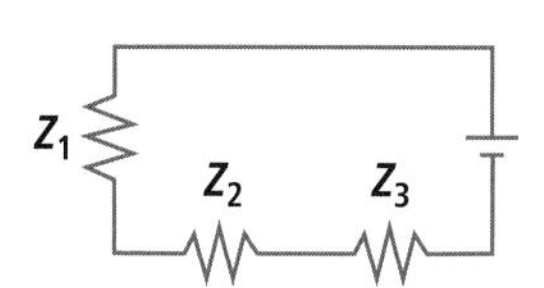

19. Parallel:

    $Z_1 = 2$

    $Z_2 = -4i$

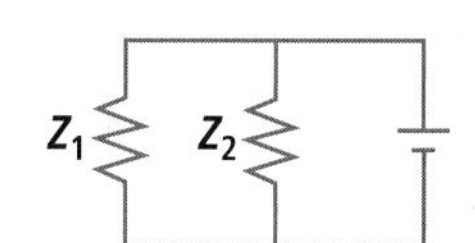

20. Parallel with series:

    $Z_1 = 9 - 2i$

    $Z_2 = 5 + 8i$

    $Z_3 = 14 + 6i$

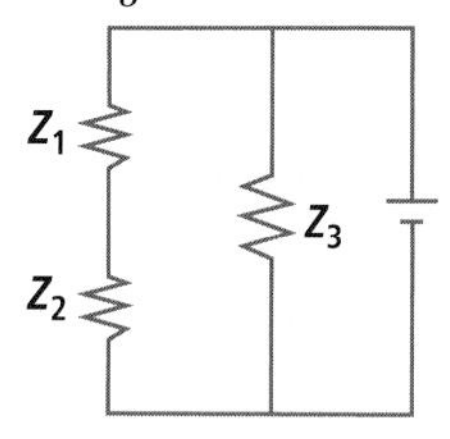

Nikola Tesla (1856-1943), who invented the alternating-current (AC) circuit used in the U.S.

In 21 and 22, expand and express in $a + bi$ form.

21. $(5 - 4i)^3$

22. $\frac{(1 + 3i)(2 + 5i)}{3 + 8i}$

23. Find i^2, i^3, i^4, and i^5.

24. Write a 3rd-degree polynomial equation in factored form and standard form that has solutions $1 + 5i$, $1 - 5i$, and 3.

25. Let $f(x) = x^2 - 6x + 10$.
    a. Evaluate $f(3 + i)$ and evaluate $f(3 - i)$.
    b. Use the results of Part a to factor $f(x)$ over the set of polynomials with complex coefficients.

26. Use the Quadratic Formula to factor $2x^2 - 3x + 4$ over the set of polynomials with complex coefficients.

27. Let $z = \frac{1}{2} + \frac{\sqrt{3}}{2}i$.
    a. Calculate z^3 and write the result in $a + bi$ form.
    b. Let w be the complex conjugate of z. Calculate w^3.
    c. **Fill in the Blank** Both z and w are cube roots of _?_.
    d. Find another cube root of the answer to Part c.

REVIEW

28. Let $h(x)$ be a polynomial function whose graph intersects the x-axis at (-1, 0). State at least two conclusions you can draw from this information. **(Lesson 7-4)**

29. Give rules for three different polynomial functions with zeros at -1, 3, and $-\frac{3}{4}$. **(Lesson 7-4)**

In 30–32, recall that the triangular numbers 1, 3, 6, 10, ... are the values of the function t where $t(n) = \frac{1}{2}n(n + 1)$, when n is a positive integer. The *square numbers* given by $s(n) = n^2$ are the numbers of dots in shapes like those pictured below. (Lesson 7-2)

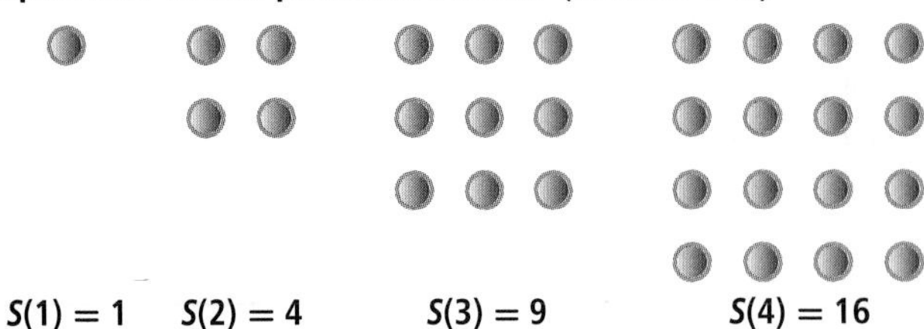

30. **Fill in the Blank** $s(n)$ is a polynomial function of n of degree __?__.
31. **Fill in the Blank** $t(n)$ is a polynomial function of n of degree __?__.
32. Prove: For all positive integers n, $s(n + 1) = t(n) + t(n + 1)$.
33. Do you remember the formula for the number of diagonals in a polygon? If you have forgotten the formula, you can derive it using quadratic regression. **(Lesson 2-6)**

 a. Count the number of diagonals in each polygon below and complete the table at the right.

Shape	Number of Sides	Number of Diagonals
Triangle		
Quadrilateral		
Pentagon		
Hexagon		

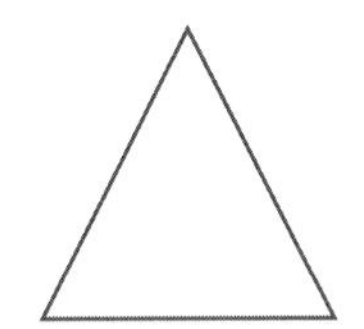
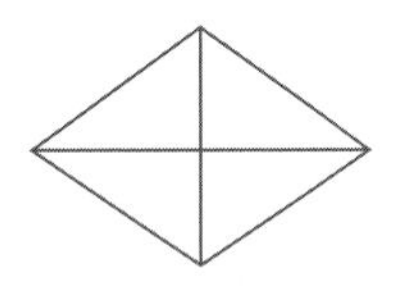
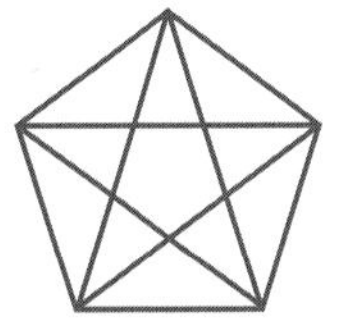
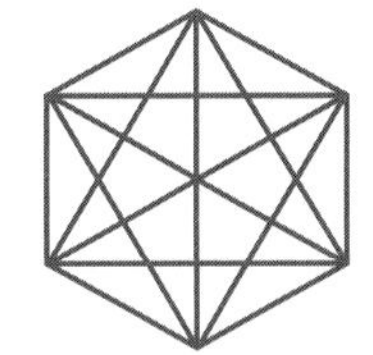

 b. Use quadratic regression to write an expression for the number d of diagonals in an n-gon.

 c. Is the model exact or approximate?

 d. Use the model to determine the number of diagonals in a 50-sided polygon.

EXPLORATION

34. Explore patterns in powers of i.

 a. Predict the value of each of i^{1992}, i^{1993}, and i^{2000}.

 b. Describe in words or in symbols how to evaluate a large power of i.

QY ANSWERS

1. $i^2 + 1 = -1 + 1 = 0$; $(-i)^2 + 1 = i^2 + 1 = -1 + 1 = 0$

2. $i\sqrt{7}, -i\sqrt{7}$

3a. real part $= -11$; imaginary part $= 8$

3b. real part $= 5$; imaginary part $= 0$

4. $3 + \frac{\sqrt{2}}{2}i, 3 - \frac{\sqrt{2}}{2}i$

5. 61

Lesson

7-6 The Fundamental Theorem of Algebra

Vocabulary

multiplicity

▶ **BIG IDEA** The zeros of a polynomial function with complex coefficients are complex numbers; the number of zeros depends only on the degree of the polynomial, not whether its coefficients are real or nonreal.

The Number of Zeros of a Quadratic Function

As you know, the solutions to any quadratic equation can be found using the Quadratic Formula: when $a \neq 0$, $ax^2 + bx + c = 0$ if and only if $x = \frac{-b \pm \sqrt{b^2 - 4ac}}{2a}$. When a, b, and c are real numbers, these values of x are the x-intercepts of the graph of $f(x) = ax^2 + bx + c$ if the solutions to the equation are real.

Below we have graphed three functions whose equations differ only in the value of c. This affects the value of $b^2 - 4ac$, the discriminant of the quadratic. The table below shows the relationships among the graph of a quadratic function, its zeros, and its factors as predicted by the Factor-Solution-Intercept Equivalence Theorem and further includes the effect of the discriminant.

Mental Math

Consider each expression of the form $ax^2 + bx + c$ and calculate $b^2 - 4ac$.

a. $x^2 - 6x + 7$

b. $3x^2 - 0.5$

c. $10x^2 + 5x + 4$

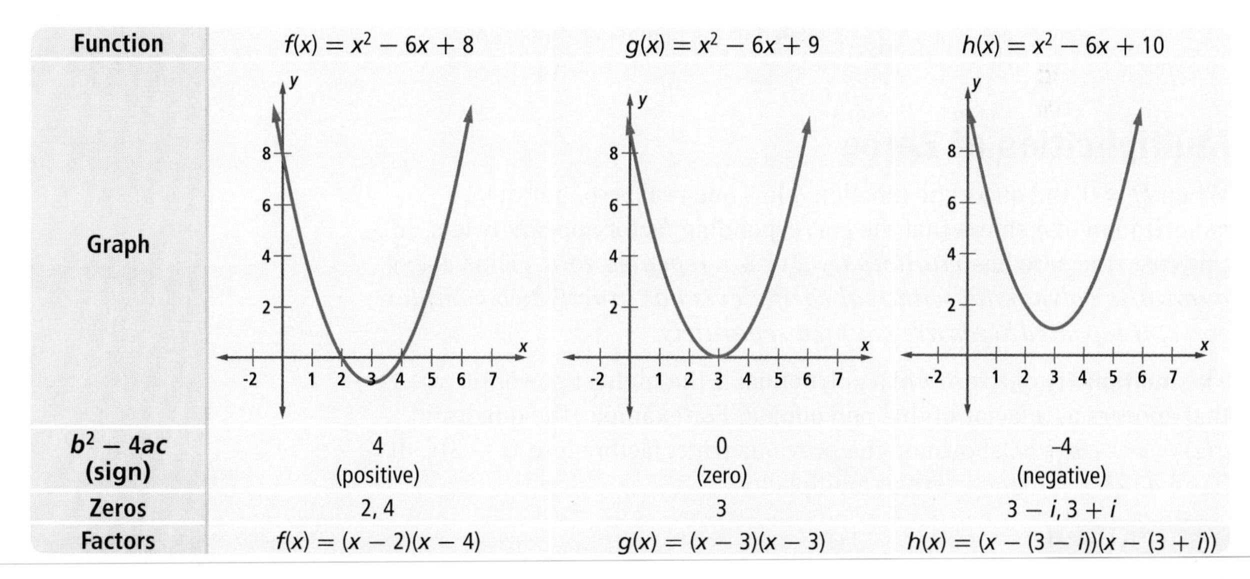

Function	$f(x) = x^2 - 6x + 8$	$g(x) = x^2 - 6x + 9$	$h(x) = x^2 - 6x + 10$
Graph			
$b^2 - 4ac$ (sign)	4 (positive)	0 (zero)	-4 (negative)
Zeros	2, 4	3	$3 - i, 3 + i$
Factors	$f(x) = (x - 2)(x - 4)$	$g(x) = (x - 3)(x - 3)$	$h(x) = (x - (3 - i))(x - (3 + i))$

In general, the graph of a real quadratic function q intersects the x-axis at two points when there are two real zeros, in exactly one point when there is a single real zero, and does not intersect the x-axis when its zeros are nonreal. If the discriminant $D = b^2 - 4ac$ is positive, q has two real zeros. If D is negative, q has two nonreal complex conjugate zeros.

Activity 1 expands the relationship among factors, the number of real zeros of a function, and the number of x-intercepts of the graph to include complex zeros of a polynomial function.

Activity 1

Work with a partner. Record your findings in a table like the one below.

Quartic Polynomial	Number of x-intercepts	Number of Real Zeros	Number of Complex Zeros	Thumbnail Sketch
	zero			
	one			
	two			
	three			
	four			

Step 1 Use a graphing utility to create a quartic polynomial function whose graph has no x-intercept.

Step 2 Use the CZEROS command on your CAS to find how many real and complex zeros your function has.

Step 3 Draw a thumbnail sketch of the graph of the function to complete the first row in the table above.

Step 4 Repeat Steps 1-3 with quartic polynomials with different numbers of x-intercepts to fill in the rest of the table.

Step 5 Compare your results with those of your classmates. How are the number and types of zeros related to the number of x-intercepts?

Multiplicities of Zeros

When $D = 0$, the quadratic function q has one real zero, but the factorization of q shows that the corresponding factor appears twice. In this case, the zero has *multiplicity 2* or is a *repeated* zero. Thus *every quadratic polynomial with real coefficients has exactly two complex zeros, if repeated zeros are counted separately.*

The **multiplicity** of a zero r of a polynomial is the highest power of $(x - r)$ that appears as a factor of that polynomial. For example, the quadratic $g(x) = x^2 - 6x + 9$, shown on the previous page, factors into $(x - 3)^2$. It has a *double zero*, or a zero of multiplicity 2.

Activity 2

Graph $y = (x + 1)(x - 2)^n$ for $n = 1, 2, 3, 4, 5$. Make a thumbnail sketch of each graph and describe any similarities and differences between the graphs in words.

As seen in Activity 2, changing the multiplicity of a zero does not affect the number or values of the x-intercepts of a graph. It does change the degree of the polynomial, and the shape of the graph.

The Number of Zeros of Higher-Degree Polynomial Functions

You may wonder how zeros and graphs of higher-degree polynomial functions are related. For example, are there formulas like the Quadratic Formula to provide zeros to all higher–degree polynomials? Can all polynomials be factored into linear factors as quadratics can? What is the greatest number of zeros a polynomial can have? These questions intrigued mathematicians for centuries.

Several Italian mathematicians of the 16th century studied cubics. The works of Scipione del Ferro (1465–1526) and Niccolo Tartaglia (1500–1557) were published by Girolamo Cardano (1501–1576) in 1545 in his treatise on algebra, *Ars Magna*. In that book, formulas for zeros for classes of cubic functions are shown and complex numbers are recognized as legitimate solutions to equations. Shortly after, Ludovico Ferrari (1522–1565), a student of Cardano, found a method for finding exact zeros of any polynomial of degree 4. In all these discoveries, no new numbers were needed beyond the complex numbers. There then became a search for a formula for zeros to all quintics and beyond.

In 1797, the following result connecting previous investigations was discovered by Karl Friedrich Gauss when he was 18 years old. It immediately made Gauss famous among mathematicians.

Fundamental Theorem of Algebra

If $p(x)$ is any polynomial of degree $n \geq 1$ with complex coefficients, then $p(x)$ has at least one complex zero.

This theorem, whose proof is beyond the scope of this book, is remarkable because it refers to all polynomials, whether their coefficients are real or nonreal. From the fact that there is one zero, we can prove there are no more than n zeros.

Let $p(x)$ be any polynomial of degree n. By the Fundamental Theorem of Algebra, $p(x)$ has at least one complex zero, call it c. Then, by the Factor and Remainder Theorems, when $p(x)$ is divided by $(x - c)$ the quotient $q(x)$ is a polynomial of degree $n - 1$. Now the quotient $q(x)$ has at least one complex zero, so divide $q(x)$ by the factor associated with that zero to get a new quotient $r(x)$ of degree $n - 2$, and so on. Each division reduces the degree of the previous polynomial by 1, so the process of repeated division can have at most n steps, each providing one zero.

Thus a consequence of the Fundamental Theorem of Algebra is that regardless of the degree of the polynomial, each of its zeros is a complex number. No new numbers are needed!

In general, the idea of multiplicities of zeros and the fact that no new numbers are needed to factor polynomials of any degree implies the following result.

Theorem (Number of Zeros of a Polynomial)

A polynomial of degree $n \geq 1$ with complex coefficients has exactly n complex zeros, if multiplicities are counted.

GUIDED

Example 1

Find the zeros of $p(x) = x^3 - 24x^2 + 144x$ and indicate their multiplicities.

Solution Notice that x is a factor of each term of $p(x)$, so x is also a factor of the sum.

$$p(x) = x(\underline{\ ?\ })$$

The quadratic factor is a perfect square trinomial. Thus $p(x) = x(x - 12)^2$. This means $p(x)$ has $\underline{\ ?\ }$ zeros counting multiplicities. The zero 0 has multiplicity $\underline{\ ?\ }$; the zero 12 has multiplicity $\underline{\ ?\ }$.

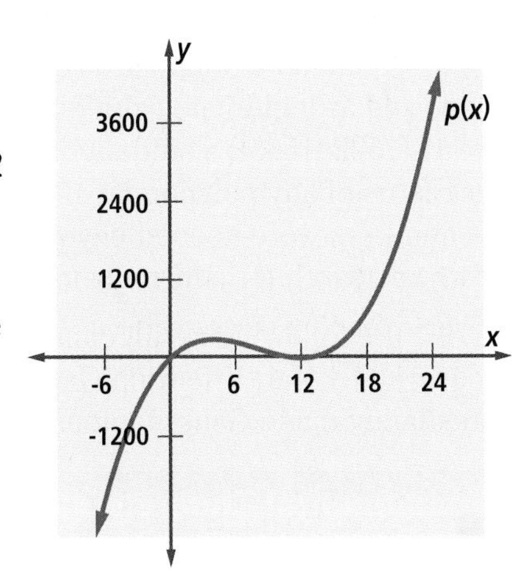

In Example 1, the polynomial $p(x)$ has degree 3 and has exactly 3 complex zeros counting multiplicities. The zeros (all of which are real in this case) are indicated in the graph at the right. So there is no need to look at a larger domain to find other zeros.

Determining the Possible Degrees of a Polynomial Function from Its Graph

From the graph of a polynomial function, you can determine information about its degree. Not only does the Number of Zeros of a Polynomial Theorem dictate the maximum number of intersections of a polynomial graph and the x-axis, it also determines the maximum number of intersections the graph may have with any horizontal line.

Polynomial Graph Wiggliness Theorem

Let $p(x)$ be a polynomial of degree $n \geq 1$. The graph of $p(x)$ can cross any horizontal line $y = d$ at most n times.

Proof Let $p(x)$ be a polynomial with degree $n \geq 1$. The x-coordinates of points of intersection of the graph of $y = p(x)$ and the horizontal line $y = d$ are the solutions of the equation $p(x) = d$. This equation is equivalent to $g(x) = p(x) - d = 0$. The degree of $g(x)$ is the same as the degree of $p(x)$ because the two polynomials differ only by a constant. Thus, $g(x)$ has at most n zeros. So the graph of $p(x)$ has at most n intersections with $y = d$.

Example 2

A polynomial equation $y = f(x)$ is graphed below. What is the lowest possible degree of $f(x)$?

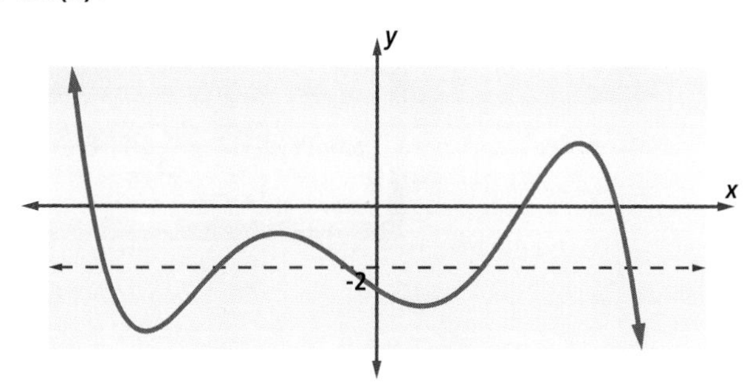

Solution The polynomial crosses the horizontal line y = -2 five times. So the degree of f(x) must be at least 5.

When one (or more) of the coefficients of the polynomial is nonreal, the solutions cannot be pictured on a standard graph. However, the Number of Zeros of a Polynomial Theorem still applies. For instance, the polynomial $g(x) = -2x^5 - ix$ has degree 5, so it has 5 zeros.

 QY

When a polynomial has real coefficients, its nonreal zeros always come in complex conjugate pairs. The proof of this theorem is long, so we omit it.

QY

How many zeros does the polynomial $f(x) = 6x^5 - 2x^3 + 7x^2 - 3$ have?

Conjugate Zeros Theorem

Let $p(x) = a_nx^n + a_{n-1}x^{n-1} + \cdots + a_1x + a_0$, where $a_n, a_{n-1}, \ldots a_1, a_0$ are all real numbers, and $a_n \neq 0$. If $z = a + bi$ is a zero of $p(x)$, then $a - bi$ is also a zero of $p(x)$.

Example 3

Let $p(x) = 3x^3 - 2x^2 + 12x - 8$.

a. Verify that $2i$ is a zero of $p(x)$.

b. Find the remaining zeros of $p(x)$ and their multiplicities.

Solution

a. $p(2i) = 3(2i)^3 - 2(2i)^2 + 12(2i) - 8$
$= 24i^3 - 8i^2 + 24i - 8$
$= -24i + 8 + 24i - 8$
$= 0$

(continued on next page)

b. Because $2i$ is a zero, then, by the Conjugate Zeros Theorem, so is its conjugate, $-2i$. The Factor Theorem implies that $(x - 2i)$ and $(x + 2i)$ are factors of $p(x)$. Thus their product $(x - 2i) \cdot (x + 2i) = x^2 + 4$ is a factor of $p(x)$.

Divide $p(x)$ by $x^2 + 4$ to find another factor:

$$\begin{array}{r} 3x - 2 \\ x^2 + 4\overline{)3x^3 - 2x^2 + 12x - 8} \\ \underline{3x^3 \qquad\quad + 12x \qquad} \\ -2x^2 \qquad\quad - 8 \\ \underline{-2x^2 \qquad\quad - 8} \\ 0 \end{array}$$

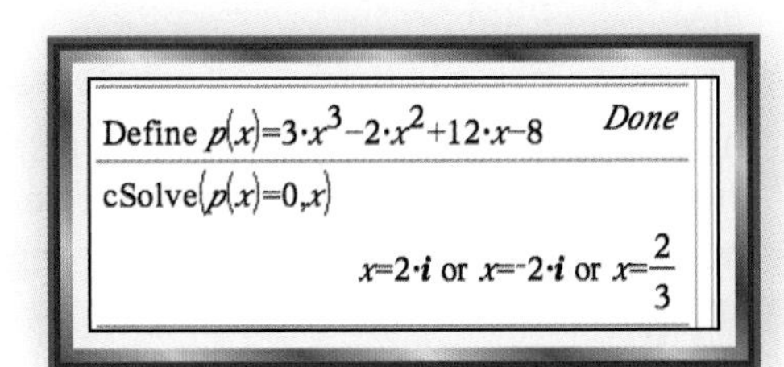

Thus $p(x) = (x^2 + 4)(3x - 2)$. So, the zeros of $p(x)$ are $2i$, $-2i$, and $\frac{2}{3}$.

Check 1 In Part a, $p(2i)$ was shown to equal 0. Similarly,

$$\begin{aligned} p(-2i) &= 3(-2i)^3 - 2(-2i)^2 + 12(-2i) - 8 \\ &= 24i + 8 - 24i - 8 = 0 \end{aligned}$$

$$\begin{aligned} \text{Also, } p\left(\tfrac{2}{3}\right) &= 3\left(\tfrac{2}{3}\right)^3 - 2\left(\tfrac{2}{3}\right)^2 + 12\left(\tfrac{2}{3}\right) - 8 \\ &= \tfrac{8}{9} - \tfrac{8}{9} + 8 - 8 = 0. \end{aligned}$$

Check 2 Solve $p(x) = 0$ using a CAS to find all its solutions.

Define $p(x)=3\cdot x^3-2\cdot x^2+12\cdot x-8$ *Done*

cSolve$(p(x)=0,x)$

$x=2\cdot i$ or $x=-2\cdot i$ or $x=\frac{2}{3}$

The question of finding a formula for exact zeros to all polynomials was not settled until the early 19th century. In 1824, Niels Abel, a Norwegian mathematician, wrote a conclusive proof that it is impossible to construct a general formula for zeros of any polynomial beyond degree 4. Abel's work had several effects. First, the theory he developed contributed to the foundation of another advanced branch of modern mathematics, *group theory*. Second, rather than searching for exact zeros to polynomials, mathematicians knew they had to rely on approximation techniques. These techniques are studied in an area of mathematics called *numerical analysis*.

Niels Abel (1802-1829)

Questions

COVERING THE IDEAS

1. What is the discriminant of the quadratic equation in y, with equation $wy^2 + xy + z = 0$?
2. Name three mathematicians who contributed to the analysis of zeros of cubic or quartic polynomials.
3. State the Fundamental Theorem of Algebra.

In 4–10, tell whether the statement is *true* or *false*.

4. Every polynomial has at least one real zero.
5. A polynomial of degree n has at most n real zeros.

6. A cubic polynomial may have 4 zeros.
7. The graph of a cubic polynomial function may intersect the x-axis exactly once.
8. A polynomial whose graph contains (2, 0), (-3, 0), and (1, 0) could be of degree greater than three.
9. All polynomials with real coefficients and degree $n \geq 1$ have zeros that are complex numbers.
10. If $p(x)$ is a polynomial with real coefficients and if $3 + 2i$ is a zero of $p(x)$, then $3 - 2i$ is a zero of $p(x)$.
11. Consider $q(x) = (x - 13)^2(x + 6)^3$.
    a. Identify the zeros of $q(x)$ and give their multiplicities.
    b. Verify your solution to Part a by graphing $y = q(x)$.
12. The polynomial $(x - 4)^3(x^2 + 1)$ has three zeros: 4, i, and $-i$. Tell why this is not a counterexample to the Number of Zeros of a Polynomial Theorem.
13. a. How many zeros does $r(n) = 8n^{15} - in$ have?
    b. How many of these zeros are real?
14. Consider the polynomial $p(x) = x^4 - 4x^3 + 30x^2 - 4x + 29$.
    a. Verify that i is a zero of $p(x)$.
    b. Find the remaining zeros of $p(x)$ and their multiplicities.
15. Suppose the graph at the right represents the polynomial function with equation $y = g(x)$. What is the lowest possible degree for $g(x)$?

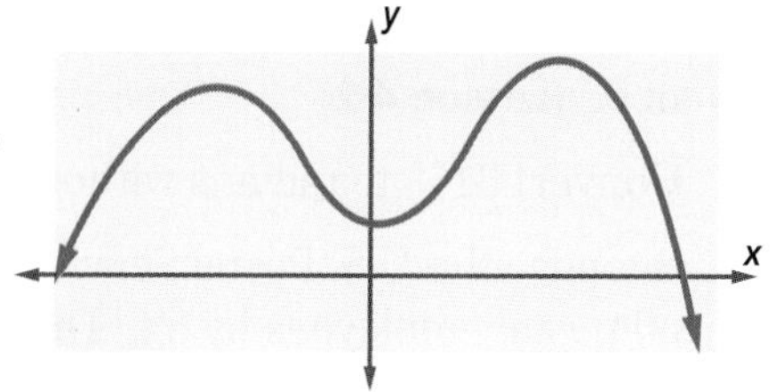

APPLYING THE MATHEMATICS

16. Find all zeros of $p(t) = 2t^5 - 72t$.
17. The zeros of $x^n - 1$ are called the *nth roots of unity.* Find the fourth roots of unity.
18. Suppose $p(x)$ is a polynomial with real coefficients and $p(3 + 7i) = 0$. What is $p(3 - 7i)$?
19. Find a polynomial with real coefficients, leading coefficient 1, and of the lowest degree possible that has the two zeros 2 and $1 - 3i$.
20. Find the zeros of $t(x) = 2x^2 + ix + 3$.
21. Tell whether each of the following could be part of the graph of a fourth degree polynomial function. Explain your answers.

a.

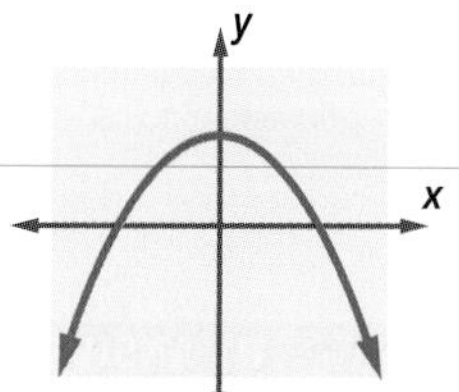

b.

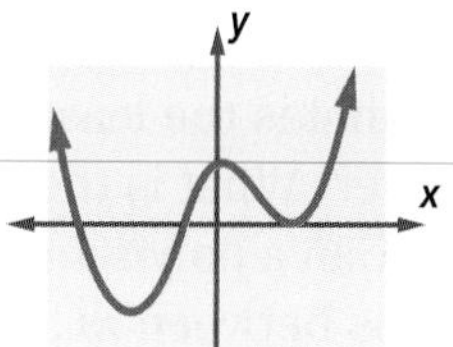

c.

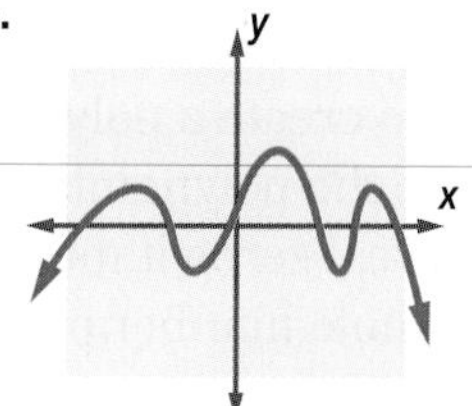

22. The curve pictured at the right is the graph of a polynomial function p of degree 5 with real coefficients. Copy the figure and insert a horizontal axis so that each condition is satisfied.

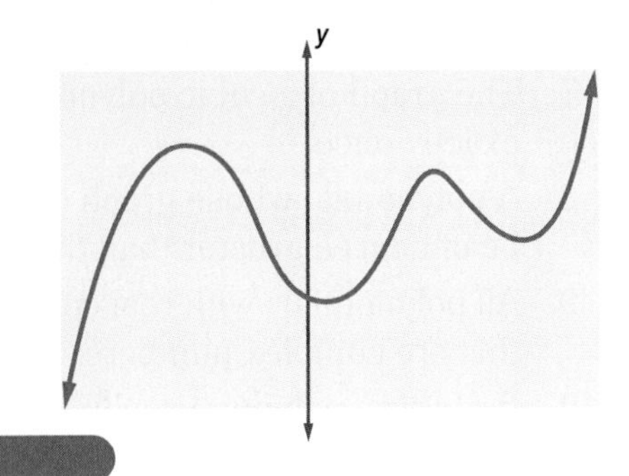

   a. p has one real root.
   b. p has three real roots.
   c. p has five real roots.

REVIEW

23. Let $z = 6 - 5i$. Write in $a + bi$ form. **(Lesson 7-5)**
   a. w, the complex conjugate of z
   b. $w + z$
   c. $w - z$
   d. wz
   e. $\frac{w}{z}$

24. How can you check a polynomial division problem when the remainder is not zero? **(Lesson 7-3)**

25. The side lengths of a triangle are 4 m, 6 m, and 9 m. Find the measure of its largest angle. **(Lesson 5-3)**

26. Suppose the point $A = (1, 0)$ is rotated a magnitude θ around the origin. Give the coordinates of the rotation image of A in terms of θ. **(Lesson 4-2)**

27. Convert 315° to radians without using a calculator. **(Lesson 4-1)**

28. Suppose the U.S. Postal Service decides to increase the salaries of all postal employees by 5%. Disregarding all other factors (such as promotions, resignations, or new hirings), how will this raise affect each statistic? **(Lessons 3-6, 1-6, 1-2)**
   a. the average salary of postal employees nationally
   b. the standard deviation of salaries

Wages and benefits for postal workers are negotiated by the National Association of Letter Carriers and the United States Postal Service.

EXPLORATION

29. Randomly pick the digits in a nine-digit number. Use the digits in order as the coefficients of a polynomial of degree 8 with alternating signs. For instance, if your number is 369,464,564, your polynomial is

$$y = 3x^8 - 6x^7 + 9x^6 - 4x^5 + 6x^4 - 4x^3 + 5x^2 - 6x + 4.$$

   a. Graph your polynomial equation. Tell how many real zeros it has.
   b. If the digits of a number may be any one of 0 through 9 (except the first digit cannot be 0), and the system of alternating signs of coefficients is used to create a polynomial, what is the least number m of real zeros the polynomial can have? What is the greatest number M of real zeros it may have? Can a polynomial of this type have any whole number of real zeros between m and M? Why or why not?

QY ANSWER

5

Lesson 7-7 Factoring Sums and Differences of Powers

BIG IDEA Expressions of the form $x^n \pm y^n$ can be factored into polynomials with real coefficients for any odd positive integer n.

Finding All Cube Roots of a Real Number

We usually think of 4 as being *the* cube root of 64, but are there others? Every solution to $x^3 = 64$ is a cube root of 64. According to the Number of Zeros Theorem, $x^3 - 64 = 0$ has three complex roots. If 4 is the only root, then its multiplicity must be three, so then $x^3 - 64$ would have to equal $(x - 4)^3$. These are not equal, so there must be other cube roots.

Mental Math

Evaluate.

a. $\sqrt{64}$ b. $\sqrt{-64}$

c. $\sqrt[3]{64}$ d. $\sqrt[3]{-64}$

Example 1

Find the cube roots of 64 other than 4.

Solution The desired cube roots are the solutions to $x^3 - 64 = 0$. Let $P(x) = x^3 - 64$. Since $P(4) = 0$, $(x - 4)$ is a factor of $x^3 - 64$. Use a CAS to divide and obtain the other factor, as shown at the right.

Thus $x^3 - 64 = (x - 4)(x^2 + 4x + 16)$. To complete the factoring, use the Quadratic Formula to find the zeros of $x^2 + 4x + 16 = 0$.

$$x = \frac{-4 \pm \sqrt{4^2 - 4 \cdot 16}}{2} = \frac{-4 \pm \sqrt{-48}}{2} = -2 \pm 2i\sqrt{3}$$

So the other cube roots of 64 are $-2 + 2i\sqrt{3}$ and $-2 - 2i\sqrt{3}$.

Check $(-2 + 2i\sqrt{3})^3 = (-2 + 2i\sqrt{3})(4 - 8i\sqrt{3} - 12)$

$= (-2 + 2i\sqrt{3})(-8 - 8i\sqrt{3})$

$= 16 + 16i\sqrt{3} - 16i\sqrt{3} + 48$

$= 64$

QY1

QY1

a. Give the complex conjugate of $-2 + 2i\sqrt{3}$.

b. **True or False** The complex conjugate from Part a is a cube root of 64.

Factoring Sums and Differences of Cubes

In Example 1, the difference of two cubes $x^3 - 64$ was factored. The method was based on knowing one zero, and thus one factor. More generally, consider the problem of factoring $x^3 - y^3$, the difference of any two cubes. To get a first factor, look for a solution to $x^3 - y^3 = 0$. Clearly $x = y$ is a solution, so $x - y$ is one factor. Divide $x^3 - y^3$ by $x - y$ to find the other factor.

The division is shown on a CAS at the right. Thus, $x^3 - y^3 = (x - y)(x^2 + xy + y^2)$. Neither of these factors can be factored further over the set of polynomials with real coefficients.

$\frac{x^3-y^3}{x-y}$ x^2+xy+y^2

This proves the second part of the following theorem. You are asked to prove the first part of the theorem in Question 10.

Sums and Differences of Cubes Theorem

For all x and y,

$$x^3 + y^3 = (x + y)(x^2 - xy + y^2) \text{ and}$$
$$x^3 - y^3 = (x - y)(x^2 + xy + y^2).$$

Any binomial that can be expressed in the form $x^3 + y^3$ or $x^3 - y^3$ can be factored using one of the formulas from the theorem.

 QY2

QY2

Expand $(x - 7) \cdot (x^2 + 7x + 49)$.

Example 2

Factor $27a^3 - 8b^6$ over the set of polynomials with integer coefficients.

Solution This is an instance of the difference of two cubes $x^3 - y^3$, where $x = 3a$ and $y = 2b^2$. Apply the previous theorem.

$$27a^3 - 8b^6 = (3a)^3 - (2b^2)^3$$
$$= (3a - 2b^2)((3a)^2 + (3a)(2b^2) + (2b^2)^2)$$
$$= (3a - 2b^2)(9a^2 + 6ab^2 + 4b^4)$$

You can use factorizations of polynomials to factor integers.

GUIDED

Example 3

Factor 1027 by recognizing that $1027 = 10^3 + 3^3$.

Solution Since $x^3 + y^3 = (x + y)(x^2 - xy + y^2)$,

$$10^3 + 3^3 = (\underline{\ ?\ } + \underline{\ ?\ })(\underline{\ ?\ } - \underline{\ ?\ } + \underline{\ ?\ })$$
$$= \underline{\ ?\ } \cdot \underline{\ ?\ }.$$

In previous courses you have factored a difference of two squares as $x^2 - y^2 = (x - y)(x + y)$. Is there a pattern for differences of powers with exponents greater than 2 or 3? The Activity explores this question.

Activity

Step 1 Use a CAS to factor each polynomial into polynomials with real coefficients.

a. $x^4 - y^4$ b. $x^5 - y^5$ c. $x^6 - y^6$

d. $x^7 - y^7$ e. $x^8 - y^8$

Step 2 a. For what values of n does $x^n - y^n$ have $x - y$ as a factor?

b. For what values of n does $x^n - y^n$ have exactly two factors (other than itself and 1)?

Step 3 Repeat Step 1 for the following polynomials.

a. $x^4 + y^4$ b. $x^5 + y^5$ c. $x^6 + y^6$

d. $x^7 + y^7$ e. $x^8 + y^8$

Step 4 Use the results of Step 3.

a. For what values of n does $x^n + y^n$ have $x + y$ as a factor?

b. For what values of n does $x^n + y^n$ have exactly two factors (other than itself and 1)?

Factoring Sums and Differences of Odd Powers

The method used to find factors of sums and differences of cubes generalizes to all *odd* powers.

Sums and Differences of Odd Powers Theorem

For all x and y and for all odd positive integers n,
$x^n + y^n = (x + y)(x^{n-1} - x^{n-2}y + x^{n-3}y^2 - \cdots - xy^{n-2} + y^{n-1})$ and
$x^n - y^n = (x - y)(x^{n-1} + x^{n-2}y + x^{n-3}y^2 + \cdots + xy^{n-2} + y^{n-1})$.

You are asked in Questions 5 and 6 to verify this theorem for specific values of n. Example 4 applies it to a sum of seventh powers.

Example 4

Factor $a^7 + b^7$.

Solution Since $a^7 + b^7$ is a sum, $a + b$ is a factor, and in the second factor the signs alternate. Thus,

$$a^7 + b^7 = (a + b)(a^6 - a^5b + a^4b^2 - a^3b^3 + a^2b^4 - ab^5 + b^6).$$

A similar theorem for the factorization of sums and differences of even powers does not exist. If n is a positive even integer, then $x^n + y^n$ does not have a linear factor with real coefficients. To factor the difference of two powers, $x^n - y^n$, when n is even, consider the even power as the square of some lower power, and reduce the problem to the difference of two squares. Example 5 illustrates a specific case.

Example 5

Factor $x^6 - 64$ completely over the set of polynomials with real coefficients.

Solution $x^6 = (x^3)^2$, and $64 = 2^6 = (2^3)^2$. So

$$x^6 - 64 = (x^3)^2 - (2^3)^2$$
$$= (x^3 - 2^3)(x^3 + 2^3).$$

(continued on next page)

Each of the factors is a sum or difference of cubes, so they can be factored further: $x^6 - 64 = (x - 2)(x^2 + 2x + 4)(x + 2)(x^2 - 2x + 4)$. Since $x^2 + 2x + 4$ and $x^2 - 2x + 4$ each have discriminant -12, they cannot be factored further into polynomials with real coefficients. So this factorization is complete.

Questions

COVERING THE IDEAS

1. Show that $-1 - i\sqrt{3}$ is a cube root of 8.

In 2 and 3, a polynomial is described.

a. Find the zeros of the polynomial by factoring.

b. Verify your result by drawing a graph.

2. $p(x) = 2x^3 - 32x$

3. $m(x) = x^3 - 125$

4. a. Show that $x - 5$ is a factor of $x^3 - 125$ by dividing.
   b. Check your answer by multiplying $x - 5$ by the quotient.
   c. Show that the quotient is not factorable over the set of polynomials with real coefficients by calculating its discriminant.
   d. Determine the three cube roots of 125.

In 5 and 6, a binomial is given.

a. Factor the binomial over integers using the Sums and Differences of Odd Powers Theorem.

b. Verify the factorization by multiplying.

5. $t^7 - 128$

6. $x^9 + y^9$

In 7–9, a binomial is given.

a. Describe the polynomial as a difference of squares, sum of squares, difference of cubes, or sum of cubes.

b. Factor the given polynomial without a CAS.

7. $a^2 - 64b^2$

8. $27c^3 + 1$

9. $125{,}000 - 64d^3$

10. Prove the first part of the Sums and Differences of Cubes Theorem.

In 11 and 12, write the number in base 10 and find its prime factorization.

11. $10^3 + 1$

12. $10^6 - 1$

In 13 and 14, True or False. Justify your answer.

13. $x + y$ is a factor of $x^4 + y^4$.

14. If n is an odd integer, $a - b$ is a factor of $a^n - b^n$.

APPLYING THE MATHEMATICS

In 15 and 16, factor completely over the set of polynomials with integer coefficients without using a CAS.

15. $24x^2y^2 - 6x^4$

16. $27a^3b^3 + 216c^3$

17. Refer to Example 5.
    a. Factor $x^6 - 64$ as a difference of cubes.
    b. Verify that the result you get in Part a can be factored further to get the solution shown in Example 5.

18. Let $p(x) = x^5 - 10x^3 + 9x$.
    a. Find the zeros of the polynomial function p by factoring.
    b. Verify your results to Part a by graphing the function p.

REVIEW

19. **True or False** The Fundamental Theorem of Algebra guarantees that the equation $5x^4 + 25x^3 + 4x^2 + 16 = 0$ has at least one real solution. Explain your answer. **(Lesson 7-6)**

20. **True or False** The polynomial $c(x) = x^3 - 1169x^2 + 1171x - 1170$ has at least one real zero. Explain how you know. **(Lesson 7-6)**

21. Calculate the total impedance Z_T for the parallel circuit at the right where $Z_1 = 5 + 3i$, $Z_2 = 5 - 2i$; and $Z_3 = 1 + i$. **(Lesson 7-5)**

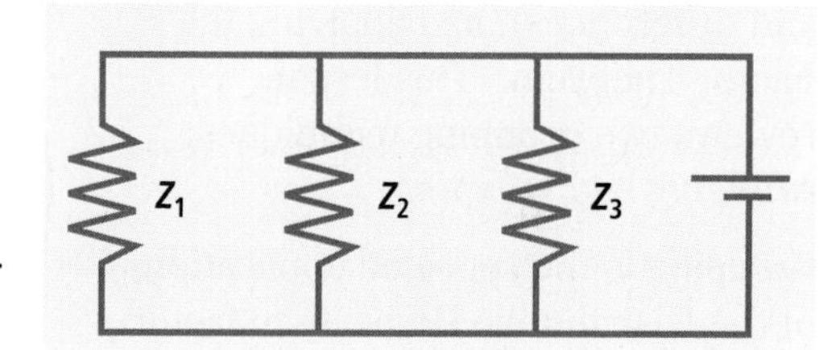

22. Consider the equation $V(t) = 2t^3 - 2t^2 - 4t$. Find the zeros of V by graphing, then factor $V(t)$ into linear factors. **(Lesson 7-4)**

23. Find a polynomial with zeros -2, $\frac{7}{6}$, and 9. **(Lesson 7-4)**

24. Consider the functions f and g with $f(x) = x^3 - 27x$ and $g(x) = (x - 2)^3 - 27(x - 2)$. **(Lessons 7-1, 3-4, 3-2)**
    a. Find the x- and y-intercepts of f.
    b. State whether f is odd, even, or neither.
    c. Find x- and y-intercepts of g.
    d. State whether g is odd, even, or neither.

EXPLORATION

25. Consider the polynomial which, when factored, is $(3x - 5)(2x + 9)(2x + 7)$.
    a. When it is expanded, what will the leading coefficient be?
    b. When it is expanded, what will the constant term be?
    c. Write the zeros of the polynomial as simple fractions.
    d. Ignoring the sign, what is the relationship of the numerators of the zeros to the linear factors?
    e. Ignoring the sign, what is the relationship of the denominators of the zeros to the linear factors?
    f. Consider the polynomial $(ax + b)(cx + d)(ex + f)(gx + h)$. When the polynomial is expanded, tell what the constant term will be, what the leading coefficient will be, and give the relationship of these products to the coefficients of the original polynomial in factored form.
    g. Generalize Parts a–f.

QY ANSWERS

1. a. $-2 - 2i\sqrt{3}$
   b. true
2. $x^3 - 343$

Lesson 7-8 Advanced Factoring Techniques

Vocabulary

grouping

chunking

*n*th roots of unity

▸ **BIG IDEA** By chunking, grouping, and repeated factoring, you can factor polynomial expressions that at first glance might seem too complicated to factor.

In previous lessons, polynomials were factored by using graphs, recognizing special patterns (sums and differences), and applying the Factor Theorem. This lesson reviews two grouping techniques: *chunking* and *grouping*.

Grouping is the repeated application of the Distributive Property to factor polynomials that contain groups of terms with common factors.

THE FAMILY CIRCUS. By Bil Keane

"Gee, Daddy, don'tcha know what a 'number sentence' is? How about 'chunking' and 'regrouping' and..."

Mental Math

Factor.

a. $a^2 - 4$

b. $b^2 + 6b + 9$

c. $c^2 + 5c + 6$

Example 1

Factor $t^3 - 3t^2 - 4t + 12$.

Solution Observe that the first two terms have a common factor of t^2 and the last two terms are each divisible by –4. Apply the Distributive Property. Factoring these pairs of terms yields

$$\underbrace{t^3 - 3t^2}_{\text{common factor of } t^2} \underbrace{- 4t + 12}_{\text{common factor of } -4} = t^2(t - 3) - 4(t - 3).$$

This shows that another common factor is $(t - 3)$. Now apply the Distributive Property again.

$$t^2(t - 3) - 4(t - 3) = (t^2 - 4)(t - 3)$$

Finally, factor the difference of squares.

$$= (t - 2)(t + 2)(t - 3)$$

Check Multiply the factors. The product shown at the right checks.

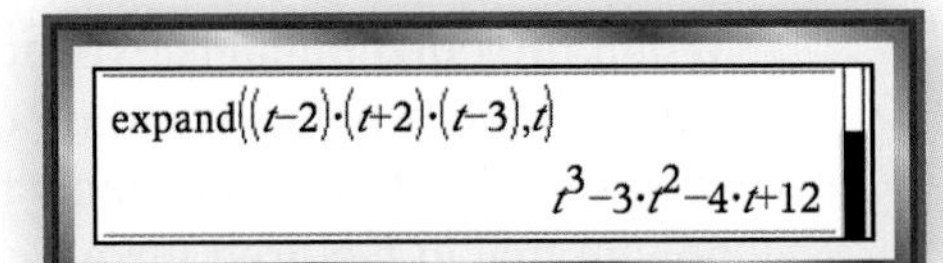

QY

> **QY**
> Use the result of Example 1 to solve $t^3 - 3t^2 - 4t + 12 = 0$.

Chunking

Chunking is the process of grouping some small bits of information into a single piece of information. For instance, when reading the word "store," you don't think "s, t, o, r, e." You chunk the five letters into one word. In algebra, chunking can be done by viewing an entire algebraic expression as one variable.

Example 2

Factor $(x^2 + 4x)^2 + 7(x^2 + 4x) + 12$.

Solution Notice that the expression $x^2 + 4x$ is used throughout the polynomial. Let $u = x^2 + 4x$. Then rewrite the polynomial as $u^2 + 7u + 12$. Now factor the quadratic:

$$u^2 + 7u + 12 = (u + 3)(u + 4)$$
$$= ((x^2 + 4x) + 3)((x^2 + 4x) + 4) \quad \text{Substitute for } u.$$
$$= (x^2 + 4x + 3)(x^2 + 4x + 4)$$

The factorization is not complete, because each of these quadratic factors can be factored further:

$$x^2 + 4x + 3 = (x + 3)(x + 1), \text{ and}$$
$$x^2 + 4x + 4 = (x + 2)^2.$$

Thus, $(x^2 + 4x)^2 + 7(x^2 + 4x) + 12 = (x + 3)(x + 1)(x + 2)^2$.

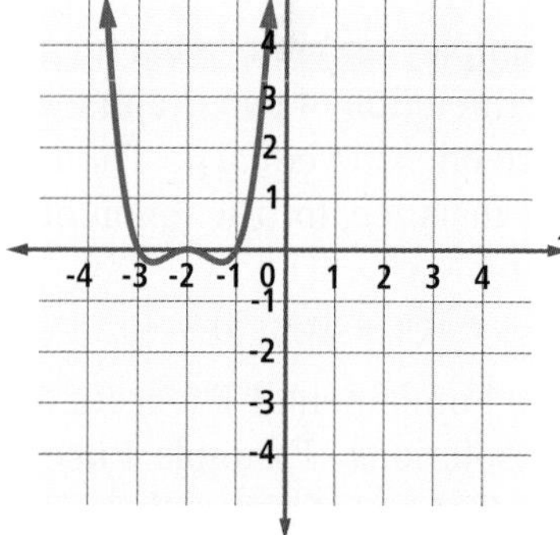

Check 1 Graph $p(x) = (x^2 + 4x)^2 + 7(x^2 + 4x) + 12$. The graph has x-intercepts at $x = -3, -2$, and -1. So, by the Factor Theorem, $p(x)$ has factors $(x + 3)$, $(x + 2)$ and $(x + 1)$. Using a CAS, divide $p(x)$ by the product of the factors. The quotient is $x + 2$, so the full factorization is $(x + 3)(x + 2)^2(x + 1)$.

Check 2 Use a CAS. Divide $p(x)$ by $(x + 3)(x + 1)(x + 2)^2$. The calculator shows 1, so the divisor and dividend are equal.

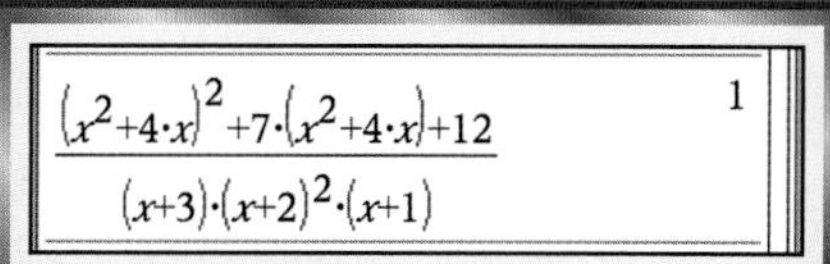

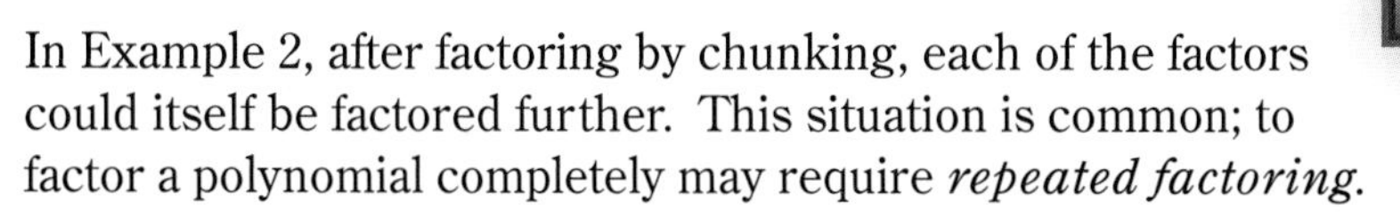

In Example 2, after factoring by chunking, each of the factors could itself be factored further. This situation is common; to factor a polynomial completely may require *repeated factoring*.

When a polynomial in two variables is equal to 0, grouping terms and chunking may help you draw the graph of the relation.

Example 3

Draw the graph of $x^2 + 2xy + y^2 + x + y - 2 = 0$.

Solution The form of the equation makes it difficult to evaluate numbers and plot by hand, and many graphing utilities cannot accept an equation if it is not solved for y.

(continued next page)

If you recognize the expression $x^2 + 2xy + y^2$ as a perfect square, that observation can help you group the terms.

$$0 = x^2 + 2xy + y^2 + x + y - 2$$
$$= (x + y)^2 + (x + y) - 2$$

Chunk $u = x + y$.

$$0 = u^2 + u - 2$$
$$= (u + 2)(u - 1)$$
$$= (x + y + 2)(x + y - 1)$$

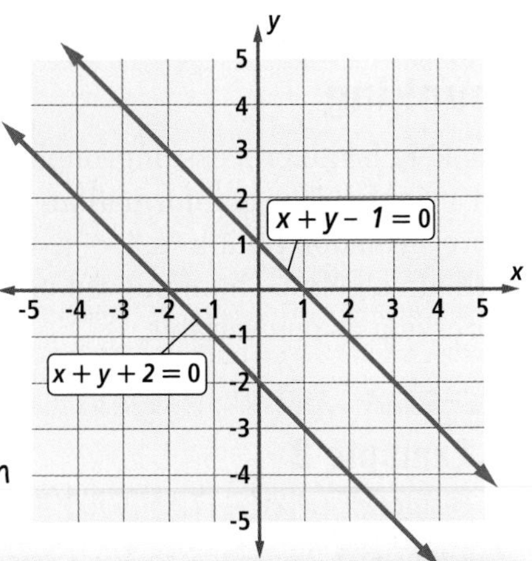

By the Zero Product Property, the original equation is true if and only if $x + y + 2 = 0$ or $x + y - 1 = 0$. The graphs of these equations are lines. Thus the graph of the original equation is the union of these two lines, as shown at the right.

Check 1 Check that the coordinates of 2 points on $x + y - 1 = 0$ and two points on $x + y + 2 = 0$ satisfy the original equation.

Check 2 Use a CAS to solve $x^2 + 2xy + y^2 + x + y - 2 = 0$ for y for graphing purposes.

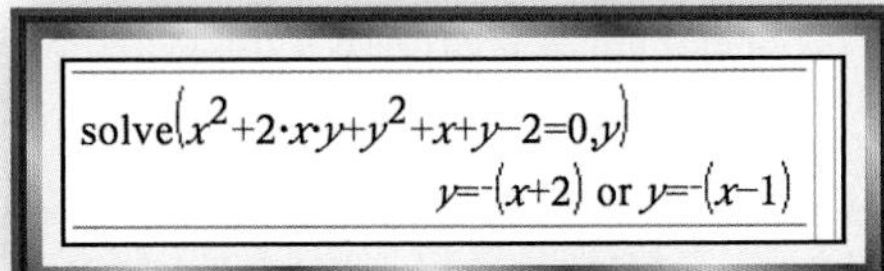

Example 3 exhibits a useful property of graphs. If you want an equation for the union of two graphs, work with the equation of each graph to have one side equal 0. Then multiply the two sides that are not zero. For instance, for the union of the graphs of $y = x^2$ and $y = 3x - 3$, notice that $y - x^2 = 0$ and $y - (3x - 3) = 0$. Thus, an equation for the union is $(y - x^2)(y - 3x + 3) = 0$.

The Fundamental Theorem of Algebra tells you that $x^6 - 1 = 0$ has six solutions. Example 4 uses factoring to find their values and uses the fact that factorizations into irreducible polynomials are unique to check the solution.

GUIDED

Example 4

a. Find all real solutions to $x^6 = 1$ by factoring.

b. Use the result of Part a to find the nonreal solutions to $x^6 = 1$.

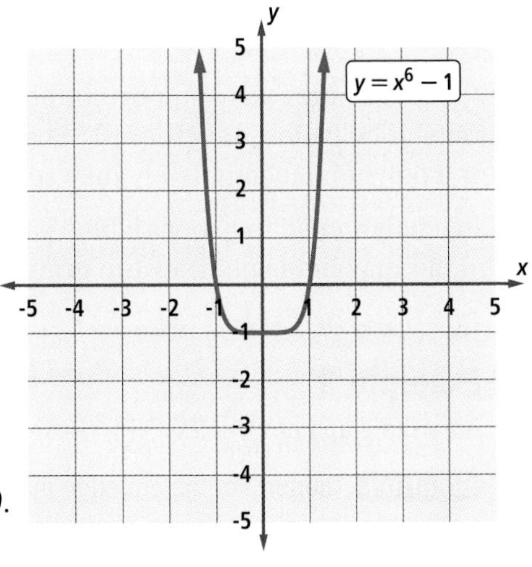

Solution

a. When $x^6 = 1$, $x^6 - 1 = 0$. Think of x^6 as $(x^3)^2$ and 1 as 1^2. Then $x^6 - 1$ is a difference of squares.

$$(x^3 - 1)(x^3 + 1) = 0$$

Each of these cubic polynomials can be factored further.

$$(x - 1)(\underline{\ ?\ })(\underline{\ ?\ })(x^2 - x + 1) = 0$$

By the Zero Product Property,

$x - 1 = 0$ or $\underline{\ ?\ } = 0$ or $\underline{\ ?\ } = 0$ or $x^2 - x + 1 = 0$.

There are two quadratic factors and the discriminant of each is negative, so 1 and -1 are the only real solutions to $x^6 = 1$.

b. Use the Quadratic Formula to find the zeros of the quadratic factors.

$x^2 - x + 1 = 0 \Rightarrow x = \frac{1 \pm \sqrt{(-1)^2 - 4(1)(1)}}{2(1)} = \frac{1}{2} \pm \frac{\sqrt{-3}}{2} = \frac{1}{2} \pm \frac{\sqrt{3}}{2}i$

$x^2 + x + 1 = 0 \Rightarrow x =$ ___?___ $=$ ___?___

Check From the Factor Theorem, you know that $x - 1$ is a factor of $x^6 - 1$. Dividing $x^6 - 1$ by $x - 1$ gives $x^5 + x^4 + x^3 + x^2 + x + 1$. Since factorizations into irreducible polynomials are unique, the product of the other three factors you found in Part a should be equal to this quotient. The screen at the right shows that this is the case.

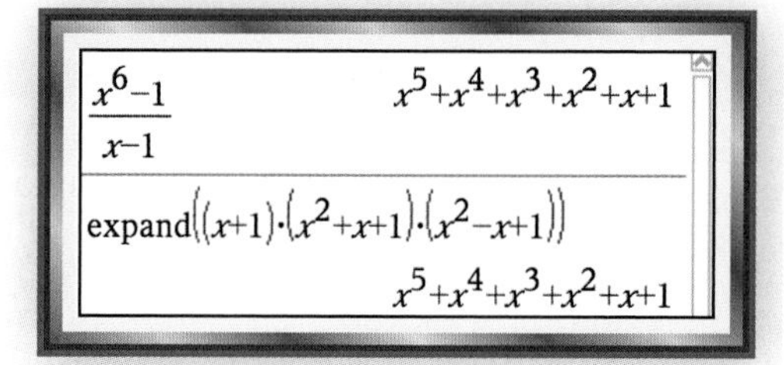

In Example 4, the six solutions to the equation $x^6 = 1$ are found. These are the 6 sixth *roots of unity*. The complex solutions to the equation $x^n = 1$ are called the ***n*th roots of unity**.

Questions

COVERING THE IDEAS

1. Refer to Example 1.
   a. Factor the polynomial by rewriting it as $t^3 - 4t + (-3t^2 + 12)$ and then applying the Distributive Property twice.
   b. Verify the factorization $(t - 2)(t + 2)(t - 3)$ by paper-and-pencil multiplication.
2. Factor the polynomial $x^3 + 5x^2 - 36x - 180$ by grouping the first and second terms and the third and fourth terms and then applying the Distributive Property twice.
3. Find the zeros of the function f with $f(x) = 2x^3 - 4x^2 - 2x + 4$
   a. by grouping and factoring.
   b. by drawing a graph.
4. Factor $(x^2 - 3x)^2 + 2(x^2 - 3x) - 24$ over the reals, without using technology.
5. a. Graph the set of ordered pairs (x, y) satisfying $y^3 + xy^2 - x^2y - x^3 = 0$.
   b. Describe the graph in words.
6. a. Factor the polynomial $x^6 - 1$ from Example 4 using the chunk $u = x^2$.
   b. Show that the solution to Example 4 is equivalent to your answer in Part a.

APPLYING THE MATHEMATICS

In 7 and 8, factor.

7. $ax - bx - by + ay + az - bz$
8. $z^2 + zx - yz - xy$

9. Use grouping to draw a graph of $\{(x, y) | y^2 - 2xy - 3y + 6x = 0\}$.

10. Find an equation describing the graph of the union of the two relations $y = x^4 - 2x^2 + 1$ and $x = y$ as shown at the right.

11. Let $f(x) = x^2 + 5x + 10$. Simplify $\frac{f(x) - f(b)}{x - b}$.

12. a. Factor $x^8 - 1$ over the set of polynomials with real coefficients by chunking.

 b. Now factor over the reals using a CAS. Which of the polynomial factors in Part a could be factored further over the reals?

 c. Use the factors from Part b to find the nonreal zeros of $x^8 - 1$.

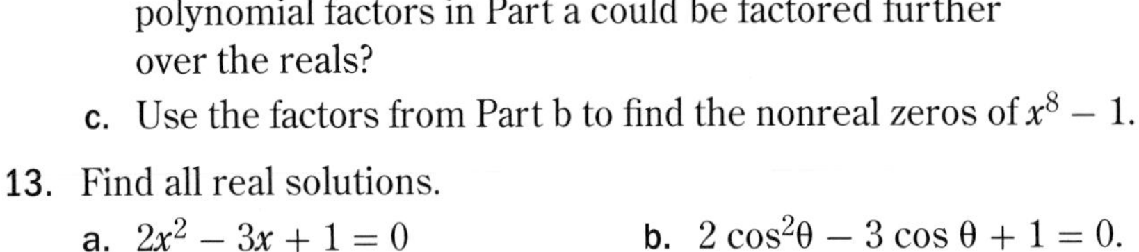

13. Find all real solutions.

 a. $2x^2 - 3x + 1 = 0$ b. $2 \cos^2\theta - 3 \cos \theta + 1 = 0$.

14. a. Find the sum of the 4th roots of unity.

 b. Find the product of the 4th roots of unity.

REVIEW

In 15–17, factor over the set of polynomials with real coefficients without using a CAS. (Lesson 7-7)

15. $t^3 + u^3$ 16. $16x^4 - y^4z^8$ 17. $y^6z^3 - 64x^6$

18. Determine the three cube roots of 125. **(Lesson 7-7)**

19. A meteorologist has predicted that the probability of rain in each of the two days of a weekend is 0.3. Explain how to code a set of random digits to simulate the weather on both days of the weekend. **(Lesson 6-7)**

20. a. How many permutations consisting of four letters each can be formed from the letters of FACTORING?

 b. How many of the permutations in Part a end in G? **(Lesson 6-4)**

21. **Fill in the Blanks** Prediction between known values of a data set is called __?__; prediction beyond known values is called __?__. **(Lesson 2-2)**

EXPLORATION

22. a. Consider the general quadratic equation in standard form, $ax^2 + bx + c = 0$. Let $a = 1$, and the roots be r_1 and r_2. Expand $(x - r_1)(x - r_2)$.

 b. Write both algebraic expressions for and descriptions in words of b and c in terms of r_1 and r_2.

 c. Consider the general cubic equation in standard form, $ax^3 + bx^2 + cx + d$. Let $a = 1$, and the roots be r_1, r_2, and r_3. Expand $(x - r_1)(x - r_2)(x - r_3)$.

 d. Write both algebraic expressions and descriptions in words for b, c, and d in terms of r_1, r_2, and r_3.

QY ANSWER

$t = 2, t = -2$, or $t = 3$

Chapter 7 Projects

1 Generating Functions: The Crazy Dice Problem

"A generating function is a clothesline on which we hang up a sequence of numbers for display." (Herbert S. Wilf) A *generating function* is a polynomial or infinite series in which the coefficient of the nth term is the value at n of some function that we are interested in. Generating functions allow you to transform a counting problem into an algebraic problem. An example of this is the following. Consider rolling two fair dice. As you know, if the sum of numbers on the dice is considered, there are 36 outcomes in this space. The sum 2 happens once, 3 happens twice, and so on.

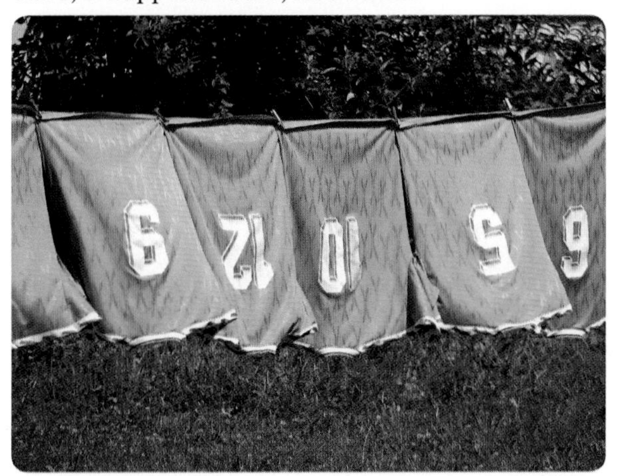

a. List the frequency of each outcome of the sum when two dice are rolled.

b. The polynomial $x^6 + x^5 + x^4 + x^3 + x^2 + x$ can represent one die with each coefficient representing the frequency of that exponent's face. Use a CAS to multiply $(x^6 + x^5 + x^4 + x^3 + x^2 + x)$ by itself.

c. What is the coefficient for the x^2 term? the x^3 term? What do these correspond to in terms of two dice?

d. Parts b and c show that rolling the dice together corresponds to the multiplication of polynomials. Using a CAS, determine the frequency distribution for the sums when 3 fair dice are rolled.

e. It is possible to put different positive integers on two fair dice such that the frequency distribution of sums is the same. That is, you can label the dice with different sets of numbers and still have the same likelihood of rolling any total as if the dice were fair and labeled 1 to 6. Given the following *incomplete* expression for the generating function for one of the dice, $(x^8 + x^6 + x^5 + x^4 + x^? + x)$, and with the help of a CAS, fill in the last side of the first die and determine the appropriate labels for the second die.

2 Graphing Complex Numbers

Complex numbers in the form $a + bi$ are graphed as points (a, b) in the coordinate plane. For example, $1 + 3i$ is graphed as the point $(1, 3)$.

a. Graph $1 + i$ in the coordinate plane.

b. Let $z = 1 + i$. Use a CAS to calculate the values of and plot $z^2, z^3, z^4, z^5, \ldots$.

c. Connect the points to form a smooth curve. What shape does the curve take?

d. Using the capabilities of a CAS, vary the value of the exponent m of z^m using a slider. What happens to the point on the coordinate plane as you increase m?

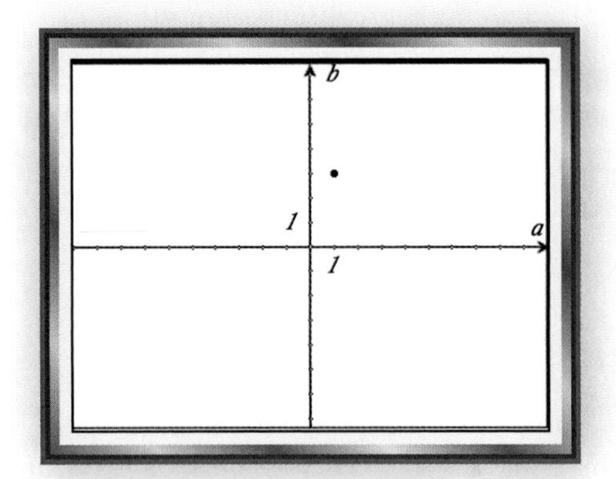

e. Repeat Parts a–d for the complex number $0.8 + 1.1i$ and for another complex number of your choosing.

3 Algebraic and Transcendental Numbers

Algebraic numbers and *transcendental numbers* are two types of numbers. Two transcendental numbers you have studied are π and e. What are the distinguishing characteristics of transcendental or algebraic numbers? How are they related to the rational and irrational numbers? Where are they used? What led to their discovery?

4 Properties of Cubics

Recall that the graph of every quadratic function q defined by $q(x) = ax^2 + bx + c$ is reflection-symmetric over the line $x = -\frac{b}{2a}$. The purpose of this project is to explore the symmetry of the graph of the cubic function u defined by $u(x) = ax^3 + bx^2 + cx + d$.

a. Graph $y = x^3$ and four other equations for cubic functions. Describe any line or rotation symmetry you observe.

b. Examine specifically the graphs of several functions with equations of the form $u(x) = ax^3 + px$. Prove that any function with an equation of this form is an odd function, and explain how your proof shows that the origin is a center of symmetry for the graph of $u(x) = ax^3 + px$.

c. Now, examine graphs of equations of the form $y = ax^3 + px + q$. What is the center of symmetry for these curves?

d. Consider equations of the form $u(x) = a(x - h)^3 + p(x - h) + q$. What are the coordinates of the center of symmetry for these curves?

e. When an equation of the form $u(x) = a(x - h)^3 + p(x - h) + q$ is expanded and like terms are combined, explain why it must be of the form $u(x) = ax^3 + bx^2 + cx + d$.

Chapter 7 Summary and Vocabulary

- A **polynomial in *x*** of **degree** n is an expression of the form $a_nx^n + a_{n-1}x^{n-1} + ... + a_1x + a_0$, where $a_n \neq 0$. Polynomials model situations such as volumes and can be fit to a wide variety of data.

- For any nth-degree polynomial, the nth differences of y-values corresponding to any arithmetic sequence of x-values are equal and nonzero. If a polynomial of degree n fits a given set of data, then $n + 1$ points give rise to $n + 1$ equations that determine its coefficients.

- The long division algorithm for dividing polynomials is very similar to that for dividing integers. If there is no remainder, the quotient is a factor of the dividend. If there is a remainder, its degree is less than the degree of the dividend.

- Let p be a **polynomial function**. Then the following are equivalent:

  c is a solution to $p(x) = 0$.

  c is an x-intercept of the graph of p.

  $x - c$ is a factor of $p(x)$.

  There is no remainder when $p(x)$ is divided by $x - c$.

- **Complex numbers**, or numbers of the form $a + bi$ where a and b are real numbers and $i = \sqrt{-1}$, are operated on in ways similar to those of polynomials. The Fundamental Theorem of Algebra guarantees that every polynomial of degree $n \geq 1$ with complex coefficients has exactly n complex zeros, if **multiplicities** are counted.

- Formulas exist for finding the exact solutions of $p(x) = 0$ when $p(x)$ is of degree 1, 2, 3, or 4. No general formula exists for finding exact solutions for any higher-degree polynomial equations. Even so, some polynomial equations can be solved by factoring. All sums and differences of the same odd power can be factored, as can differences of even powers. Other techniques include **grouping** of terms with common factors, **chunking**, and repeated factoring.

Vocabulary

7-1
polynomial in *x*
degree of a polynomial
coefficients of a polynomial
leading coefficient
standard form of a polynomial
polynomial function
maximum/minimum value of a function
extreme values, extrema
relative (local) maximum/minimum
increasing function
decreasing function
zeros of a polynomial

7-2
figurate numbers
triangular numbers
tetrahedron
tetrahedral number

7-5
i, imaginary numbers
complex number
real part
imaginary part
complex conjugates
discriminant

7-6
multiplicity

7-8
grouping
chunking
*n*th roots of unity

Properties and Theorems

Polynomial Difference Theorem (p. 448)
Remainder Theorem (p. 455)
Factor Theorem (p. 460)
Factor-Solution-Intercept Equivalence Theorem (p. 461)
Theorem (Complex Number Operations) (p. 468)
Theorem (Complex Conjugate Multiplication) (p. 468)
Fundamental Theorem of Algebra (p. 475)
Number of Zeros of a Polynomial Theorem (p. 476)
Polynomial Graph Wiggliness Theorem (p. 476)
Conjugate Zeros Theorem (p. 477)
Sums and Difference of Cubes Theorem (p. 482)
Sums and Differences of Odd Powers Theorem. (p. 483)

Chapter 7 Self-Test

Take this test as you would take a test in class. You will need a calculator. Then use the Selected Answers section in the back of the book to check your work.

In 1–5, consider the polynomial function g where $g(x) = -\frac{1}{2}x^4 + 2x^3 - 3x + 4$.

1. Graph g.
2. Find the coordinates of all maxima and minima, rounded to the nearest hundredth.
3. Determine all real zeros, rounded to the nearest hundredth.
4. Give the degree of the polynomial $g(x)$.
5. Find the number of nonreal zeros.
6. Determine the quotient and remainder when $3x^4 - 7x^3 + 3x^2 - x - 2$ is divided by $x + 2$.
7. **Multiple Choice** For the polynomial $f(x) = x^8 - 7x^6 + 6x^3 - 4x^2 + 12x - 8$, $f(1) = 0$. Which is a factor of $f(x)$?

   A $x - 1$ B $x + 1$

   C $x - 2$ D $x + 2$
8. Find a polynomial with real coefficients for which $2 - i$ and 5 are zeros.
9. a. By what complex number would you multiply $\frac{2 - 3i}{3 + 2i}$ to write the fraction in $a + bi$ form?

   b. Express $\frac{2 - 3i}{3 + 2i}$ in $a + bi$ form.
10. Let $k = 2 + 4i$ and $w = 3 + 7i$. Express each of the following in $a + bi$ form.

    a. $2k + w$ b. $k \cdot w$
11. The following are the first six pentagonal numbers, P_n: $P_1 = 1, P_2 = 5, P_3 = 12, P_4 = 22, P_5 = 35, P_6 = 51$.

    a. Determine a polynomial formula for P_n.

    b. Calculate P_{20}.
12. A rectangular plot of farmland will be bounded on one side by a river and on the other three sides by a connected electric fence 500 meters long. Call the length of the fence parallel to the river ℓ.

    a. Express the width of the plot as a function of ℓ.

    b. Express the area of the plot as a function of ℓ.

    c. State the domain for the situation.
13. Factor $9ab + 18b + 5a + 10$ by grouping.
14. Demonstrate how the Sum and Differences of Cubes Theorem can be used to factor $64m^6 + 27k^3$ over the set of polynomials with integer coefficients.
15. Factor $x^{10} - 1$ using chunking.
16. If the 5th differences of the values of a polynomial are all equal to 3 for consecutive integral x-values, what is the degree of the polynomial?

In 17–20, consider the cubic functions R, S, and T and their graphs below. All x-intercepts are shown.

$R(x) = x^3 + 4x^2 + x - 6$

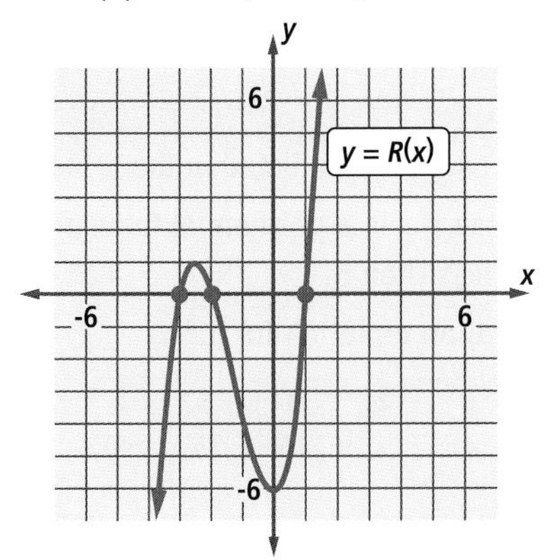

$S(x) = x^3 - 3x^2 + 3x - 1$

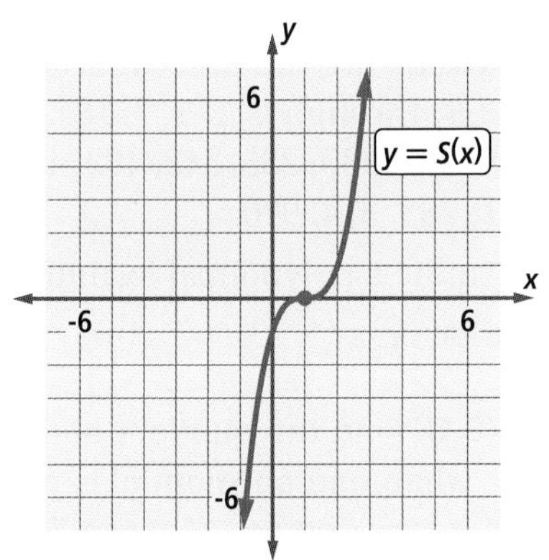

$T(x) = x^3 + 2.5x^2 + 0.5x - 1$

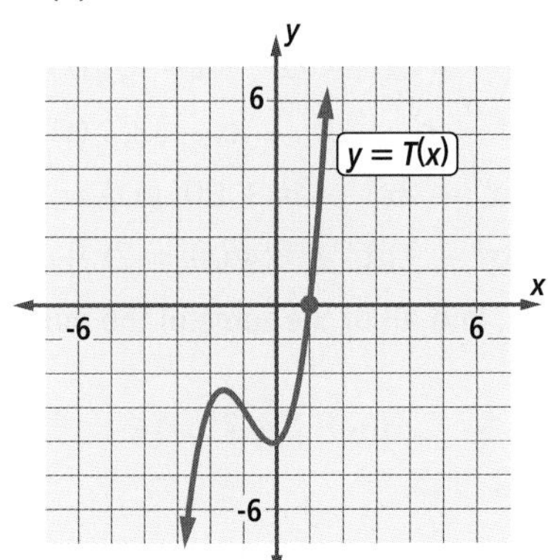

17. How many real zeros does each function appear to have?

18. **Fill in the Blank** These functions have $(x - 1)$ as a factor, therefore __?__ is a point on each graph.

19. **Multiple Choice** Function S has no nonreal zeros. Then 1 must be a zero of multiplicity

    A 1. **B** 2. **C** 3. **D** 4.

20. For function C, $x = 1$ is a zero of multiplicity 1. How many nonreal zeros does C have?

21. Part of the graph of a polynomial function p is shown below. What is the lowest degree the polynomial $p(x)$ can have?

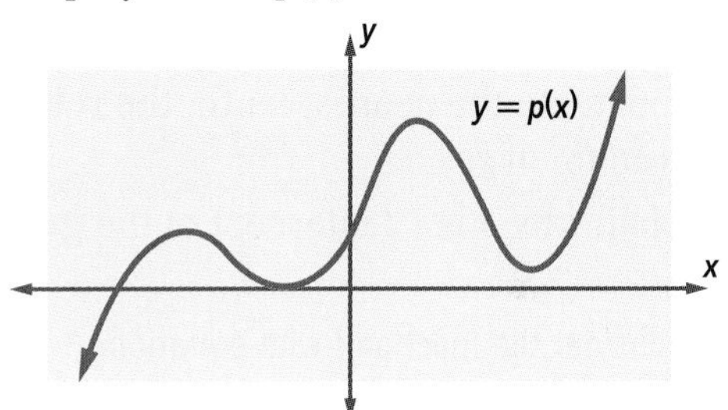

Chapter 7 Chapter Review

SKILLS
PROPERTIES
USES
REPRESENTATIONS

SKILLS Procedures used to get answers

OBJECTIVE A Calculate or approximate zeros and relative extrema of polynomial functions. (Lesson 7-1)

In 1–3, consider the function g with equation $g(t) = t^3 - 6t^2 + 7t + 4$.

1. Determine, to the nearest thousandth, the negative zero of g.
2. Estimate, to the nearest tenth, the relative maximum of g.
3. Explain why 4 is a t-intercept of the graph of g.

In 4–6, consider the function f with equation $f(x) = x^3 - x^2 - 10x + 10$.

4. Approximate, to the nearest hundredth, the largest x-intercept of the graph of f.
5. Estimate, to the nearest tenth, the relative minimum of f.
6. Explain why 1 is an x-intercept of the graph of f.

OBJECTIVE B Use finite differences and systems of equations to determine an equation for a polynomial function from data points. (Lesson 7-2)

In 7 and 8, suppose that in a set of data points, the x-values form an arithmetic sequence, the 4th set of differences are not constant, and the 5th differences of y-values are equal.

7. What is the smallest possible degree of a polynomial function that fits the set of data points?
8. What is the smallest number of data points that need to be used to determine the polynomial function of least degree?

In 9 and 10, use the data listed in the table.

a. Determine if y is a polynomial function of x of degree less than 5.

b. If so, find an equation for it.

9.

x	1	2	3	4	5	6	7
y	4	7	14	25	40	59	82

10.

x	2	4	6	8	10	12	14
y	6	13	26	45	70	101	138

11. The following are the first twenty *heptagonal numbers* h_1 through h_{20}: 1, 7, 18, 34, 55, 81, 112, 148, 189, 235, 286, 342, 403, 469, 540, 616, 697, 783, 874, 970.
    a. Determine a polynomial formula for h_n.
    b. Calculate h_{163}.

OBJECTIVE C Find the quotient and remainder when one polynomial is divided by another of lesser degree. (Lesson 7-3)

In 12–15, find the quotient and remainder when $f(x)$ is divided by $g(x)$.

12. $f(x) = 3x^4 + 5x^3 - 2x^2 + x - 7; g(x) = x - 1$
13. $f(x) = x^4 - x^3 - 2x^2 + 2x - 42; g(x) = x - 3$
14. $f(x) = 6x^3 + 2x^2 - 5x + 20; g(x) = 3x - 4$
15. $f(x) = 8x^4 + 16x^3 + 9; g(x) = 2x + 6$

In 16 and 17, find all of the zeros of the polynomial function.

16. $f(x) = 15x^5 - 40x^3 + 6x^2 - 16$
17. $f(x) = x^5 - y^5$
18. Using the complex factor command on a CAS, factor and find the zeros of $m(x) = x^4 + 4x^3 + 38x^2 + 100x + 325$.

OBJECTIVE D Factor polynomials and solve polynomial equations using the Factor Theorem, Sum and Difference of Cubes Theorem, Distributive Property, grouping terms, or chunking. (Lessons 7-4, 7-7, 7-8)

19. **Fill in the Blank** Since $\left(x + \sqrt{5}\right)$ is a factor of $x^4 - 3x^2 - 10$, __?__ is a solution to $x^4 - 3x^2 - 10 = 0$.

20. Solve $x^3 - x^2 + 4x - 4 = 0$.

21. Find an equation for a polynomial function with zeros $1 - i$, $1 + i$, and 2.

22. Factor $(x^2 - 3x)^2 - 2(x^2 - 3x) - 8$.

In 23-25, a polynomial is given.

a. Factor over the set of polynomials with real coefficients.

b. Factor over the set of polynomials with complex number coefficients.

23. $x^5 + x^2 - 6x^3 - 6$

24. $t^3 - 27$

25. $x^5 + 2x^4 - x - 2$

26. Find the three cube roots of 512.

27. Use the Quadratic Formula and the Factor Theorem to factor $x^2 - 2x + 3$.

OBJECTIVE E Perform operations with complex numbers. (Lesson 7-5)

In 28-31, let $m = 7 + 3i$ and $n = 11 - 8i$. Express the given number in $a + bi$ form.

28. $m + 3n$

29. $m - n$

30. $n^2 - 2m$

31. $\frac{mn}{3}$

32. Find $u^2 - v^2$ when $u = 2 + 6i$ and $v = -5 - 3i$.

In 33-35, rewrite in $a + bi$ form.

33. $(-3 + 3\sqrt{3}i)^3$

34. $\frac{29 + 2i}{4 + 7i}$

35. $\frac{6 - i}{2i - 3}$

PROPERTIES Principles behind the mathematics

OBJECTIVE F Apply the vocabulary of polynomials. (Lessons 7-1, 7-6)

36. Given the polynomial function P with $P(x) = 5x^9 + 6x^6 - 8x^3 + x^2 - 17$, state each.
    a. the degree of the polynomial
    b. the leading coefficient
    c. the coefficient of x^4
    d. the constant term
    e. the number of zeros P must have

37. Give an example of a cubic polynomial with three terms and a leading coefficient of 5.

True or False In 38–40, if the statement is false, change one word to make it true.

38. If a polynomial function p has a relative maximum at (-2, -5), then the graph of $y = p(x)$ never crosses the x-axis.

39. The roots of a polynomial function p refer to its y-intercepts.

40. If the graph of $y = p(x)$ has 3 x-intercepts, then $p(x)$ is of degree 3.

41. Consider the polynomial function q with $q(z) = (z - 4)(z + 6)^2(z - 8)^3$. Name a zero of q that has each multiplicity.
    a. one
    b. two
    c. three

OBJECTIVE G Apply the Remainder Theorem, Factor Theorem and Factor-Solution-Intercept Equivalence Theorem. (Lessons 7-3, 7-4)

42. If $f(x) = -x^3 + 20x^2 - 131x + 280$ has a factor of $(x - 7)$, what must be true about $f(7)$?

43. Find a 4th-degree polynomial $p(x)$ with integer coefficients whose zeros are 1, 2, 3, and 7.

In 44 and 45, one fact is stated about a polynomial. State at least two other conclusions you can draw.

44. $x = -11$ is a solution to the equation $0 = x^4 + 10x^3 - 28x^2 - 202x - 165$.

45. The graph of $k(x) = x^3 - 5x^2 + x - 5$ only crosses the x-axis once at $x = 5$.

In 46–48, $g(x) = (x + 2)^2(3x - 4)^2(5x + 1)^2$.

46. How many different zeros does g have?

47. What is the remainder when g is divided by $x - 1$?

48. Name one binomial that, when divided into g, returns a remainder of 0.

OBJECTIVE H Apply the Fundamental Theorem of Algebra and Conjugate Zeros Theorem. (Lesson 7-6)

49. Explain why every real number is equal to its complex conjugate.

50. Explain why every polynomial of odd degree with real coefficients must have at least one real zero.

51. If the graph of a fourth-degree polynomial does not cross the horizontal axis, what conclusions can be drawn about its zeros?

52. Let $f(x) = x^4 - 12x^2 - 64$. Confirm that $2i$ is a zero of $f(x)$. Then determine the remaining zeros.

53. **True or False** If $2 + i$ is a zero of $z^3 + 3z^2 - 23z + 35$, then $1 + i$ cannot be another zero of it. Explain your answer.

USES Applications of mathematics in real-world situations.

OBJECTIVE I Construct and interpret polynomials that model real situations. (Lessons 7-1, 7-2)

54. Waldo's Wonderfully Wacky Widgets are right cones with height h and radius r that sit on top of square boxes with side $s = 2r$.
    a. Write a formula for the volume of one of Waldo's Wonderfully Wacky Widgets.
    b. If the height of the cone is fixed at 10 inches, write a polynomial in r for the volume of a Wonderfully Wacky Widget.
    c. Identify the degree of the polynomial, and tell why it is reasonable in this situation.
    d. What is the leading coefficient of the polynomial?

55. A box with a top is constructed from a piece of 200-cm by 400-cm cardboard. The shaded regions are cut out and the flaps are folded to make a box, as in the diagram below. Write a polynomial in x for the volume of the box.

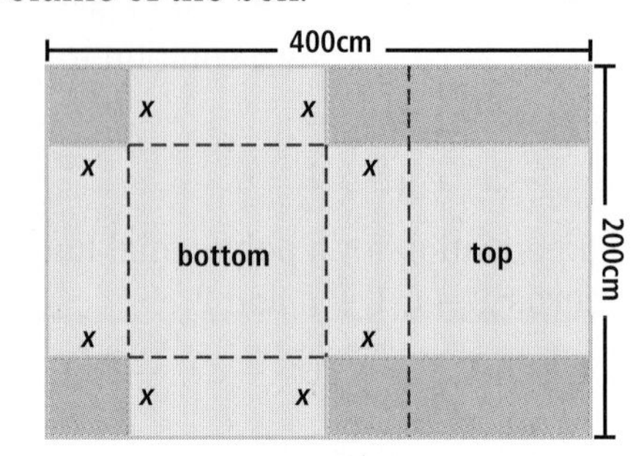

56. The number of games G needed for n checkers players to play each other twice is given in the table below.

n	2	3	4	5	6	...
G	2	6	12	20	30	...

    a. Find a polynomial for G in terms of n.
    b. Assume that there are 28 students in your class. How many games will your class need to play if each student plays every other student twice?

57. Garth's grandmother put \$350 in a savings account each year starting the day Garth was born. The account compounds annually at a rate of r.
    a. Write a polynomial in x, where $x = 1 + r$, that represents the value of the account on Garth's 13th birthday.
    b. If the annual rate on the account is 5.5%, how much money will be in the account on Garth's 13th birthday?

In 58–60, the government gives 3 farmers each 200 yards of fence and tells the farmers that they can claim any rectangular plot that is completely enclosed by their fence.

58. Mark starts with a width that is 20 yards longer than the length. Express the area as a function of the length ℓ.

59. Tommy starts with a length 25 yards longer than the width. Express the area as a function of the length ℓ.

60. Joe, the shrewdest of the trio, scribbles some numbers on a piece of paper before he sets the length of his plot. Given that farmer Joe gets the maximum area for his fence, what are the length ℓ, width w, area, and perimeter of his plot?

REPRESENTATIONS Pictures, graphs, or objects that illustrate concepts

OBJECTIVE J Relate properties of polynomial functions and their graphs. (Lessons 7-1, 7-4, 7-6)

In 61 and 62, an equation for a polynomial function is given.

a. Graph the function.

b. From the graph, estimate the coordinates of any relative extrema to the nearest tenth.

c. From the graph, estimate all real zeros.

61. $y = -2x^2 + 48x - 280$

62. $y = x^5 - x^3 + x^2 - 1$

63. Consider the part of the graph of $f(x) = 0.1x^5 - 2x^3 + x^2 - 27$ shown below.

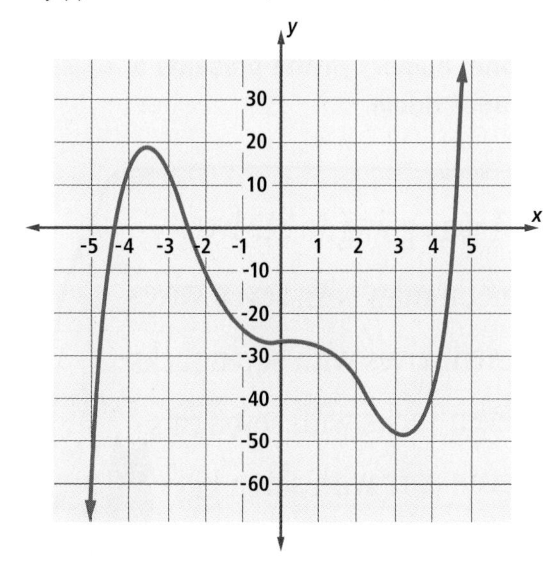

a. How many real zeros does f seem to have?

b. The number of real zeros and the degree of f are not equal. Give a possible explanation.

64. **True or False** The equation $0.1x^5 - 2x^3 + x^2 - 27 = 0$ has exactly three solutions. Explain your answer. If the statement is false, suggest a change to make it true.

65. **True or False** The graph of $y = p(x)$ can cross the x-axis at most n times if n is the degree of p.

66. Below is a graph of a function s with various x-coordinates labeled. Describe all intervals in which

a. the values of s are positive.

b. the values of s are negative.

c. the values of s are increasing.

d. the values of s are decreasing.

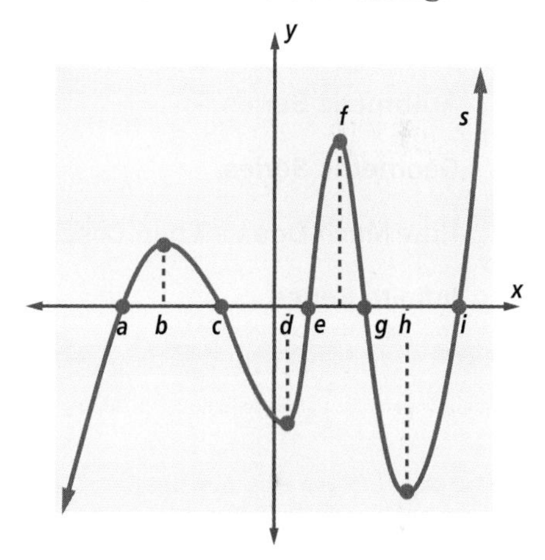

Chapter

8

Sequences and Series

Contents

A traditional nursery rhyme presents a mathematics riddle.

As I was going to St. Ives
I met a man with seven wives
Seven wives with seven sacks
Seven sacks with seven cats
Seven cats with seven kits
Kits, cats, sacks, wives
How many were going to St. Ives?'

Most people think about the solution as involving multiplication by sevens.

1 man
$1 \cdot 7 = 7$ wives
$7 \cdot 7 = 49$ sacks
$49 \cdot 7 = 343$ cats
$343 \cdot 7 = 2401$ kittens

The numbers 1, 7, 49, 343, 2401 form a *sequence*. It is called a *geometric sequence* because every term is 7 times the previous term. The kth term is 7^{k-1}. To find the total number of items, add up the numbers.

$$1 + 7 + 49 + 343 + 2401 = 2801$$

This addition is called a *series*. The series can be written in summation notation as

$$\sum_{k=1}^{5} 7^{k-1} = 1 + 7 + 7^2 + 7^3 + 7^4.$$

How many do you think were going to St. Ives?

If the seven kits had seven mice, and the seven mice had seven fleas, and so on forever, you could express the sum as the *infinite series* $\sum_{k=1}^{\infty} 7^{k-1}$. This infinite series has no finite sum, but you have seen many infinite series that do have finite sums. For example, the infinite series

$$\sum_{n=1}^{\infty} \frac{6}{10^n} = \frac{6}{10} + \frac{6}{100} + \frac{6}{1000} + \frac{6}{10{,}000} + \cdots = 0.\overline{6} = \frac{2}{3}.$$

In this chapter, you will study several kinds of sequences and series, and you will see how they are used in financial calculations and other situations.

Lesson 8-1 Arithmetic Sequences

Vocabulary

Fibonacci sequence
term of a sequence
harmonic sequence
explicit formula for a sequence
sequence
initial condition
recurrence relation
recursive formula for a sequence
arithmetic sequence, linear sequence

▶ **BIG IDEA** Arithmetic sequences are ordered lists of numbers with a common difference, making them analogous to linear functions.

You are familiar with many sequences. Here are some of them.

(1) the increasing sequence of all positive integers: 1, 2, 3, 4, 5, 6, ...

(2) the sequence of triangular numbers: 1, 3, 6, 10, 15, 21, 28, 36, ...

(3) the **Fibonacci sequence**: 1, 1, 2, 3, 5, 8, 13, 21, 34, ...

The numbers in a sequence are its **terms** and are identified by their position in the sequence as the 1st term, 2nd term, 3rd term, and so on. The terms of a sequence do not have to be integers, nor do they have to be in increasing order. Here are some examples.

(4) the sequence of positive integer powers of -7: -7, 49, -343, 2401, -16807, ...

(5) the sequence of reciprocals of the positive integers, called the **harmonic sequence**: $1, \frac{1}{2}, \frac{1}{3}, \frac{1}{4}, \frac{1}{5}, \frac{1}{6}, \dots$

A sequence does not have to follow a pattern and can be finite.

(6) a sequence of eight random 4-digit decimals between 0 and 1: 0.8378, 0.1148, 0.6181, 0.2311, 0.2050, 0.8095, 0.9846, 0.6344

(7) the sequence of prices of one ounce of gold (in dollars) on December 31st, in 10-year intervals from 1910 to 2000: 20.67, 20.67, 20.67, 34.50, 40.25, 36.50, 37.60, 641.20, 423.80, 272.15

Mental Math

How many stars are in the 999th term of the sequence below if each term has 3 more stars than the previous term?

Naming Sequences and Terms

We often denote a sequence by a single letter. For instance, we can call the sequence of triangular numbers T. Then the nth term of the sequence is identified as T_n. So, looking above, you can see that $T_1 = 1$, $T_2 = 3$, $T_3 = 6$, $T_4 = 10$, and so on. Some people name this sequence as $\{T_n\}$, with the braces and subscript signaling that it is a sequence. The most common variables used for subscripts are n, i, j, and k. When i is used as a subscript in a sequence, it does not stand for an imaginary number.

STOP QY1

▶ **QY1**

If F is the Fibonacci sequence, what is F_8?

Sequences Are Functions

Consider the sequence below. Let R_n be the number of dots in a rectangular array of dots with n columns and $n + 1$ rows.

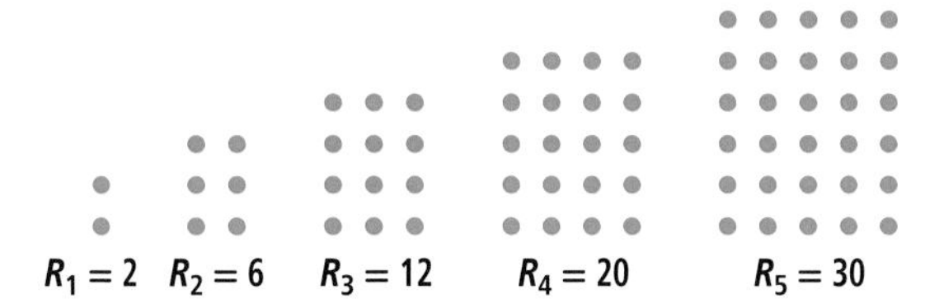

The nth array has n dots in each of $n + 1$ rows, so $R_n = n(n + 1)$.

$R_n = n(n + 1)$ is a formula that pairs each positive integer n with the corresponding number R_n. This is an example of an **explicit formula** because it shows the nth term of the sequence in terms of n. So you can think of the sequence R as containing the ordered pairs (1, 2), (2, 6), (3, 12), (4, 20), (5, 30), and so on. In this way, every sequence can be viewed as a function.

Definition

A **sequence** is a function whose domain is a set of consecutive integers greater than or equal to an integer k.

In ordinary function mode, a calculator or CAS uses a default domain of the real numbers. To limit the domain to positive integers, store consecutive integers in one list. Then use the equation of the sequence to generate the sequence values in another list, as shown below at the left.

Below at the right is a graph of the first 10 terms of the sequence R.

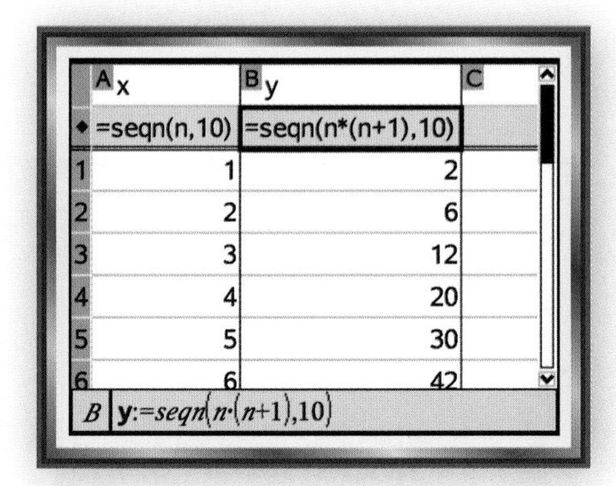

	A x	B y	C
◆	=seqn(n,10)	=seqn(n*(n+1),10)	
1	1	2	
2	2	6	
3	3	12	
4	4	20	
5	5	30	
6	6	42	

B y:=seqn(n·(n+1),10)

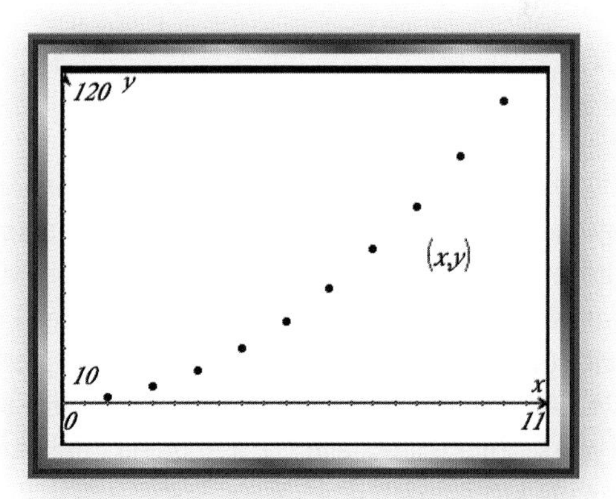

STOP QY2

QY2

Replicate the table and graph of the first 10 terms of the sequence R on your own technology.

Recursive Formulas for Sequences

Suppose you have \$250 in savings now and can save \$75 each week from a job. Then your total savings each week (the *balances* in your account) form the increasing sequence

$$250, 325, 400, 475, 550, \ldots.$$

You can translate the balances into a formula expressing how the terms of the sequence are related. Let B_n be the balance after n weeks.

$$B_1 = \$250$$
$$B_2 = B_1 + \$75 = \$325$$
$$B_3 = B_2 + \$75 = \$400$$
$$B_4 = B_3 + \$75 = \$475$$

In general, each week's balance is the previous week's balance plus \$75.

$$B_n = B_{n-1} + \$75$$

In this equation, B_n stands for the balance at week n, the previous week is week $n - 1$, and the balance of the previous week is B_{n-1}.

By itself, the equation $B_n = B_{n-1} + \$75$ does not allow you to determine the balances because it does not tell you the starting balance B_1. However, the two equations

$$\begin{cases} B_1 = 250 & \text{(the \textbf{initial condition})} \\ B_n = B_{n-1} + 75,\ n > 1 & \text{(the \textbf{recurrence relation})} \end{cases}$$

form a *recursive formula* that does define all terms of the sequence B. A **recursive formula** is one in which the first term or first few terms of a sequence are given, and the nth term is expressed in terms of the preceding term(s).

Activity

In 2006, the United States produced 251.3 million tons of garbage; over the period from 2000 to 2006, the amount of garbage produced annually increased by 2.2 million tons per year. Let G_n be the amount of garbage produced each year, with G_1 representing the amount of garbage produced in 2006. Assume the amount of increase stays the same for future years. Find G_2, G_3, and G_4, and predict the amount of garbage produced in 2015.

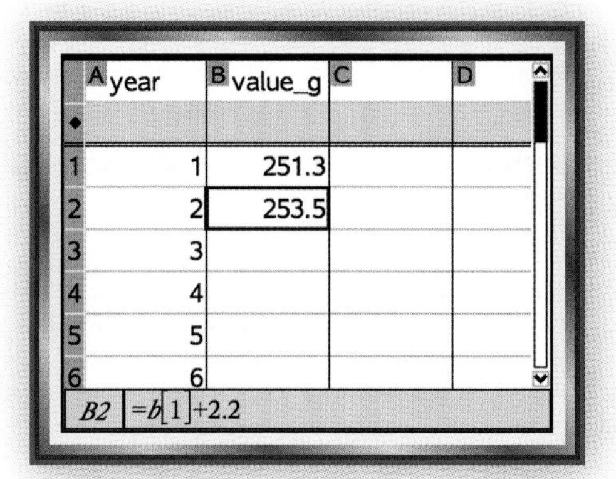

	A year	B value_g	C	D
◆				
1	1	251.3		
2	2	253.5		
3	3			
4	4			
5	5			
6	6			

B2 =b[1]+2.2

Step 1 Use a spreadsheet. If your spreadsheet allows for separate column titles, title column A "year" and column B "value_g". In cells A1-A10, enter the values 1-10, and in B1, enter the value of G_1 in millions of tons, 251.3.

Step 2 Because the value in B2 should be 2.2 million tons more than the value in B1, enter the formula =b1+2.2 in cell B2. Then fill this formula down through cell B10.

Step 3 Record the values that correspond to terms G_2, G_3, and G_4.

Step 4 2015 is __?__ years after 2006. So if G_1 represents the amount of garbage in 2006, the term representing the amount in 2015 is __?__. Find the value of the corresponding term from your spreadsheet.

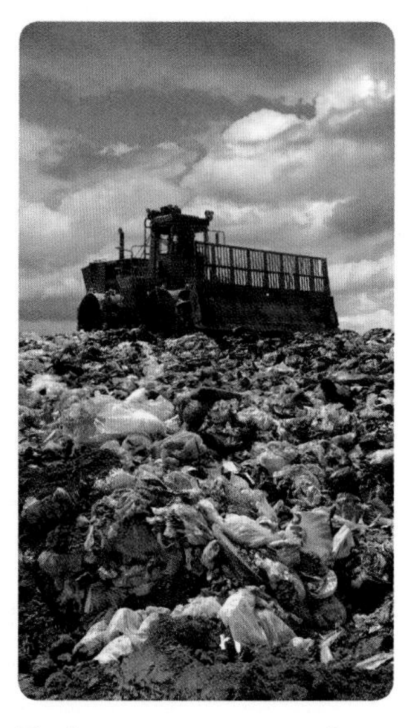

Reduce, reuse, recycle
In 2007, people in the U.S. recycled 33% of the trash they produced.

Arithmetic (Linear) Sequences

An **arithmetic** (air ith met′ ick) **sequence** or **linear sequence** a is one in which the difference $a_n - a_{n-1}$ between consecutive terms is constant. For example, the sequence with first term equal to -3 and a constant difference of 8 is an arithmetic sequence whose first six terms are

$$-3, 5, 13, 21, 29, 37.$$

Each term beyond the first is eight more than the previous term. Each term is also equal to -3 plus a number of 8s. For instance, if a_n is the nth term of this sequence, then $a_1 = -3$, $a_n = a_{n-1} + 8$, and in general, $a_n = -3 + 8 \cdot (n - 1)$. In this way, any arithmetic sequence can be described either explicitly or recursively.

Theorem (Formulas for Arithmetic Sequences)

Let n be a positive integer and a_1 and d be constants. Then the arithmetic sequence with first term a_1 and constant difference d can be described by the explicit formula

$$a_n = a_1 + (n - 1)d$$

and the recursive formula

$$\begin{cases} a_1 \\ a_n = a_{n-1} + d \quad \text{for all integers } n > 1. \end{cases}$$

The explicit formula in the theorem shows that arithmetic sequences are linear functions of n. This is why they are also called linear sequences. When given such a function as an explicit formula for a sequence, you can find a recursive formula for the sequence.

If you know two terms of an arithmetic sequence, you can find both an explicit formula and a recursive formula.

GUIDED

Example

Suppose a is an arithmetic sequence with $a_2 = 14$ and $a_6 = 4$. Find

a. an explicit formula for a.

b. a recursive formula for a.

(continued on next page)

Solution

a. Using the explicit formula in the theorem above, write an equation relating each term to a_1 and d.

$a_2 = a_1 + (2 - 1)d \Rightarrow 14 = a_1 + d$

$a_6 = a_1 + (6 - 1)d \Rightarrow \underline{\ ?\ } = a_1 + \underline{\ ?\ }\, d$

Solve this system. $d = \underline{\ ?\ }$ and $a_1 = 16.5$. Therefore,

$a_n = 16.5 + \underline{\ ?\ }\,(n - 1)$.

b. Since a_6 is the fourth term after a_2, $a_6 = 14 + 4d$.

n	2	3	4	5	6
a_n	14	$14 + d$	$14 + 2d$	$14 + 3d$	$14 + 4d = 4$

Then $14 + 4d = 4$, so $4d = -10$ and $d = -2.5$. Because $a_2 = a_1 + d$, $a_1 = \underline{\ ?\ }$. A recursive formula is

$$\begin{cases} a_1 = \underline{\ ?\ } \\ a_n = a_{n-1} + \underline{\ ?\ } \text{ for } n > 1. \end{cases}$$

QY3

QY3

In an arithmetic sequence with constant difference 11, if $a_3 = 17$, then $a_1 = \underline{\ ?\ }$.

In the Example, the explicit formula can be written in function notation:

$$a(n) = 16.5 - 2.5(n - 1)$$
$$= 19 - 2.5n.$$

This formula indicates that the graph of the sequence consists of points on the line with y-intercept 19 and slope -2.5. It also shows you why knowing two terms of the sequence determines the entire sequence.

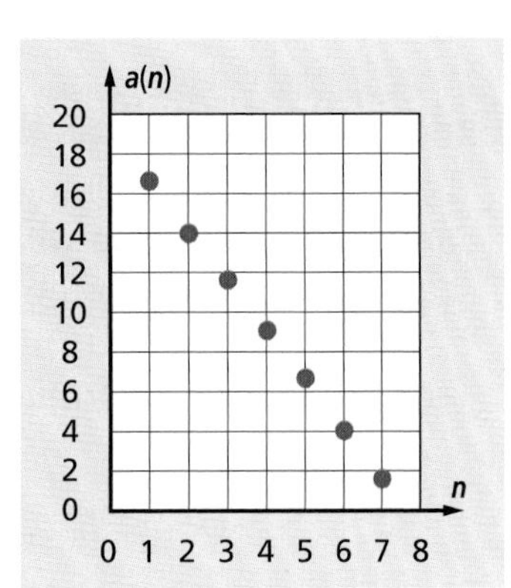

Questions

COVERING THE IDEAS

1. Suppose t_n is a term in a sequence.
   a. What expression stands for the preceding term?
   b. What expression stands for the next term?

2. Graph the first five terms of the sequence defined by

$$\begin{cases} s_1 = 10 \\ s_n = s_{n-1} - 4,\ n > 1. \end{cases}$$

In 3 and 4, **True or False.** If the statement is false, correct it to be true.

3. The domain of every sequence is the set of positive numbers.
4. The terms of a sequence are elements of its range.

In 5–8, an explicit or recursive formula for a sequence is given.

a. Determine the first three terms of the sequence.

b. Determine whether the sequence is an arithmetic sequence.

5. $\begin{cases} h_1 = 2 \\ h_n = h_{n-1} - 5, n \geq 2 \end{cases}$

6. $\begin{cases} g_1 = 38 \\ g_n = \frac{1}{4} + g_{n-1}, n \geq 2 \end{cases}$

7. $b_n = \frac{n^2(n+1)^2}{4}$

8. $d_n = (n+1)^2 - n^2$

9. a. Write an explicit formula for the increasing sequence of positive integers.
   b. Write a recursive formula for this sequence.
   c. Is this sequence arithmetic?

10. What term number is 1128 in the arithmetic sequence 8, 12, 16, ..., 1128, ...?

11. Suppose a major league baseball player has 1493 hits at this point in his career.
    a. If the player continues in baseball and gets 175 hits a year, how many years will it take to get 3000 hits?
    b. What is the position of the first term larger than 3000 in the arithmetic sequence 1493, 1668, ...?

12. An arithmetic sequence q has $q_1 = 40$ and $q_5 = 20$. Find
    a. a recursive formula for the sequence.
    b. an explicit formula for the sequence.

APPLYING THE MATHEMATICS

13. What is the constant difference of the arithmetic sequence $\frac{1}{2}, \frac{1}{3}, \frac{1}{6}, \ldots$?

14. You owe your parents \$650 for car insurance and you pay back \$75 per week from your job. Let w_n be the amount in dollars you owe at the beginning of week n.
    a. Compute w_1, w_2, w_3, and w_4.
    b. Find an explicit formula for w_n.
    c. Find the first value of n such that $w_n < 0$. Interpret your answer in the context of the situation.

15. A theater has 20 seats in the first row and five more seats in each subsequent row. If the last row has 125 seats, how many rows are there in the theater?

In 16–19, suppose that a is an arithmetic sequence with constant difference $d \neq 0$. Tell whether the sequence of terms defined must be an arithmetic sequence and why.

16. the sequence b with $b_n = a_n + 2$
17. the sequence q with $q_n = 2a_n$
18. the sequence s with $s_n = (a_n)^2$
19. the sequence h with $h_n = \frac{1}{a_n}$

REVIEW

20. Use a CAS to simulate rolling two fair six-sided dice 100 times. Did a sum of 7 occur more, less than, or equally as often as its expected value? **(Lesson 6-7)**

21. A bag contains 13 yellow, 8 blue, 7 red, 8 white, and 6 green marbles. Three marbles are drawn in succession, each marble being replaced before the next one is drawn. What is the probability of drawing a blue, then a yellow, and then a green marble? **(Lesson 6-3)**

22. Find the exact value of $\tan(-120°)$ without a calculator. **(Lesson 4-4)**

23. **True or False** The graphs of $f(x) = 3^x$ and $g(x) = x^3$ are reflection images of each other over the line $x = y$. **(Lesson 3-4)**

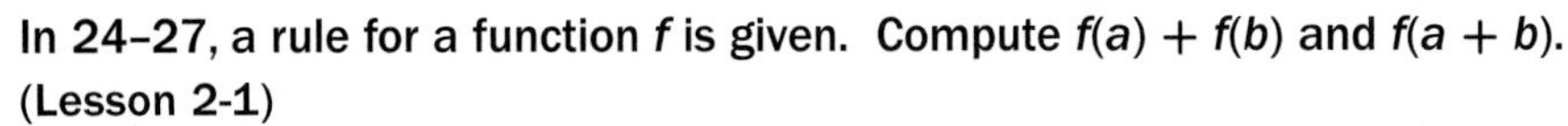

In 24–27, a rule for a function f is given. Compute $f(a) + f(b)$ and $f(a + b)$. (Lesson 2-1)

24. $f(x) = 3x$
25. $f(x) = x^2 - 2$
26. $f(x) = \frac{6}{x}$
27. $f(x) = \sqrt{x + 3}$

EXPLORATION

28. Consider the right triangle with sides 3, 4, and 5. Notice that the lengths of the sides form an arithmetic sequence.
    a. Determine three other right triangles with side lengths in arithmetic sequence.
    b. Let the side lengths of a right triangle be a, $a + d$, and $a + 2d$ units long. Use the Pythagorean Theorem to determine a relation between a and d.
    c. Based on the result in Part b, make a general statement about all the right triangles whose side lengths form an arithmetic sequence.

QY ANSWERS

1. 21

3. -5

Lesson 8-2 Geometric and Other Sequences

Vocabulary

geometric sequence, exponential sequence

arithmetic mean

geometric mean

▶ **BIG IDEA** Geometric sequences are exponential functions for which the domain is the set of positive integers; they have many properties analogous to those of arithmetic (linear) sequences.

Geometric (Exponential) Sequences

Recall that a **geometric sequence** or **exponential sequence** g is one in which the *ratio* $\frac{g_n}{g_{n-1}}$ between consecutive terms is constant. For instance, in the geometric sequence

$$10, \frac{5}{2}, \frac{5}{8}, \frac{5}{32}, \ldots,$$

the constant ratio is $\frac{1}{4}$. Below at the left is an explicit formula for the nth term of this sequence; at the right is a recursive formula.

$$g_n = 10\left(\frac{1}{4}\right)^{n-1} \qquad \begin{cases} g_1 = 10 \\ g_n = \frac{1}{4}g_{n-1} \text{ for all integers } n > 1 \end{cases}$$

Generalizing for any geometric sequence gives the following.

Mental Math

Find a simple rule that fits the first four terms of this sequence. Use your rule to predict the 5th and 6th terms of the sequence.

2, 4, 8, 14, ...

Theorem (Formulas for Geometric Sequences)

Let n be a positive integer and g_1 and r be constants. Then the geometric sequence with first term g_1 and constant ratio r can be described by the explicit formula

$$g_n = g_1 r^{n-1}$$

and the recursive formula

$$\begin{cases} g_1 \\ g_n = rg_{n-1} \text{ for all integers } n > 1. \end{cases}$$

QY1

Notice that geometric sequences are exponential functions for which the domain is the set of positive integers. This is why they are sometimes called exponential sequences.

▶ **QY1**

In the geometric sequence g: 5, 10, 20, 40, 80, ..., find g_1 and the constant ratio r.

GUIDED

Example 1

If you invest $10,000 in a 5-year CD with an annual yield of 4.83%, the amount you begin with and the amounts you have at the end of consecutive years form a finite geometric sequence.

a. Find an explicit formula for this sequence.

b. Find a recursive formula for this sequence.

Solution

a. Use the Compound Interest Formula $A = P(1 + r)^t$, where A is the amount you will have after t years at the rate r if you begin with P. From the given information,

$$A = \underline{\ ?\ }(1 + \underline{\ ?\ })^{\underline{\ ?\ }}$$

Using the notation of sequences, let A_t be the amount after t years. Then the explicit formula is

$$A_t = \underline{\ ?\ }(1.0483)^{\underline{\ ?\ }}.$$

b. Call the sequence A and let the variable subscript be t, for the number of years from now. Then $A_0 = \$10{,}000$ and the constant ratio is 1.0483. So a recursive formula for the sequence is

$$\begin{cases} A_0 = \underline{\ ?\ } \\ A_t = \underline{\ ?\ }\,A_{t-1}, \text{ for all integers } t \geq 1. \end{cases}$$

Notice how arithmetic and geometric sequences compare.

Sequence Type	Arithmetic	Geometric
Definition	Difference between consecutive terms is constant d	Ratio of consecutive terms is constant r
Terms	$a_1, a_1 + d, a_1 + 2d, \ldots$	$g_1, g_1r, g_1r^2, \ldots$
Explicit Formula	$a_n = a_1 + (n - 1)d$	$g_n = g_1 \cdot r^{n-1}$
Recurrence Relation	$a_n = a_{n-1} + d$	$g_n = g_{n-1} \cdot r$

Arithmetic and Geometric Means

Consider the arithmetic sequence 70, 77, 84, 91, 98, … . Pick three consecutive terms, such as 84, 91, and 98. Notice that the middle term is the mean of the other two terms: $91 = \frac{84 + 98}{2}$. This is why the mean is also called the **arithmetic mean.**

The following Activity asks you to explore a corresponding property of geometric sequences.

Activity

Step 1 Use a first term g_1 of your choosing and the constant ratio 0.8 of a geometric sequence g. Write the first seven terms of your sequence.

Step 2 Pick three consecutive terms of your sequence. Call them g_{k-1}, g_k, and g_{k+1}. Show that $g_k^2 = g_{k-1} \cdot g_{k+1}$.

Step 3 Repeat Step 1 using the same first term and the constant ratio –3. Does the property of Step 2 still hold?

Step 4 Use the formula for the nth term of a geometric sequence $g_n = g_1 r^{n-1}$ to prove that the property of Step 2 is a property of every geometric sequence.

The Activity demonstrates that if you pick three consecutive terms g_{k-1}, g_k, and g_{k+1} of any geometric sequence, then $g_k^2 = g_{k-1} \cdot g_{k+1}$. For instance, in the geometric sequence

$$5, 20, 80, 320, 1280, 5120, 20480, \ldots$$

suppose you pick 320, 1280, and 5120. Then

$$1280^2 = 320 \cdot 5120, \quad \text{and} \quad 1280 = \sqrt{320 \cdot 5120}.$$

You may recognize that 1280 is the *geometric mean* of 320 and 5120. This is why $\sqrt{xy}$ is called the **geometric mean** of x and y. The drawings of Questions 14–17 show some situations from geometry that gave rise to this name.

 QY2

QY2

What is the geometric mean of 2 and 32?

Finding Formulas for Arithmetic and Geometric Sequences

When you encounter a new sequence, the first question you should ask is whether it is one of the types of sequences you know. Identifying the type of sequence will help you determine a formula.

The distance between utility poles depends on the terrain, the type of utility, and the practices of the company that maintains them.

GUIDED

Example 2

For each sequence below, first determine whether it could be arithmetic, geometric, or neither; if it could be arithmetic or geometric, assume that it is and give the next two terms and an explicit formula.

a. $a_n = 18, 24, 30, 36, \ldots$

b. $b_n = 16, 12, 9, 6.75, \ldots$

c. $c_n = 1, 3, 6, 10, \ldots$

Solution First check to see whether the differences or ratios of consecutive terms are constant, then use the appropriate formula.

(continued on next page)

a. Each term of a is six more than the previous term. So a_n could be an arithmetic sequence with constant difference ___?___. Therefore, $a_5 = 42$; $a_6 =$ ___?___; $a_n =$ ___?___.

b. Since $b_2 - b_1 = -4$ and $b_3 - b_2 = -3$, b_n is not an arithmetic sequence. Compare ratios.

$\frac{b_2}{b_1} =$ ___?___ $\quad \frac{b_3}{b_2} =$ ___?___ $\quad \frac{b_4}{b_3} =$ ___?___

So b_n could be a geometric sequence with constant ratio ___?___. Then $b_5 =$ ___?___; $b_6 =$ ___?___; $b_n =$ ___?___.

c. $c_2 - c_1 = 2$; $c_3 - c_2 = 3$. The differences are not the same, so the sequence is not arithmetic.

$\frac{c_2}{c_1} = \frac{3}{1} = 3$; $\frac{c_3}{c_2} = \frac{6}{3} = 2$. The ratios are not the same. So c is neither an arithmetic nor a geometric sequence.

Unlike exponential functions, it is possible for a geometric sequence to have a *negative* ratio. For example, the sequence

$$8, -4, 2, -1, \frac{1}{2}, -\frac{1}{4}, \ldots$$

has $r = -\frac{1}{2}$. In such cases, the terms alternate from positive to negative.

Finding Formulas for Other Sequences

There is no single method you can use to find a formula for an arbitrary sequence. Instead, you have to look for patterns (perhaps using the method of finite differences from Lesson 7-2) and compare the sequence to ones you already know; even then, you may only be able to make an approximation. Examples 3 and 4 below present two unknown sequences to illustrate different strategies you can use.

Example 3

Find an explicit formula for the nth term of a sequence beginning with $a = \frac{1}{5}, \frac{4}{25}, \frac{9}{125}, \frac{16}{625}, \ldots$.

Solution The terms themselves do not form either an arithmetic or a geometric sequence. But the numerators 1, 4, 9, 16 are all perfect squares and the denominators 5, 25, 125, 625 are consecutive powers of 5. The numerator of a_n is n^2; the denominator of a_n is 5^n. So $a_n = \frac{n^2}{5^n}$.

Example 4

Let R_n be the number of dots in a rectangular array of dots with n columns and $n + 1$ rows. Find a formula for R_n.

Solution Looking at the situation geometrically provides an explicit formula. Each rectangle has dimensions n by $(n + 1)$ dots, so the total number of dots is $n(n + 1)$. Therefore, $R_n = n(n + 1) = n^2 + n$.

Check Draw the first several terms in the sequence.

$R_1 = 2 = 1 \cdot 2$ $R_2 = 6 = 2 \cdot 3$ $R_3 = 12 = 3 \cdot 4$ $R_4 = 20 = 4 \cdot 5$ $R_5 = 30 = 5 \cdot 6$

Questions

COVERING THE IDEAS

1. Consider the sequence $r_n = 14\left(\frac{6}{7}\right)^{n-1}$.
   a. Use the formula to find the first four terms of the sequence.
   b. Find the common ratio of the sequence.
   c. Use a table to find the first value of n such that $r_n < 0.5$.
2. Consider the sequence generated by the recursive formula
   $$\begin{cases} g_1 = 30 \\ g_n = \frac{2}{3}g_{n-1}, \text{ for all integers } n > 1. \end{cases}$$
   a. Write the first four terms of the sequence.
   b. Write an explicit formula for g_n.
   c. Find g_{32} using either the explicit or recursive formula.

In 3–6, two terms of a geometric sequence are given. Write a recursive and an explicit formula for each sequence.

3. $g_1 = 16, g_2 = \frac{1}{2}$
4. $g_1 = \frac{1}{3}, g_2 = -2$
5. $g_1 = 1000, g_3 = 250$
6. $g_2 = 30, g_4 = 750$
7. An investment of $1200 increases by 6.3% every year. Write an explicit formula for a geometric sequence v describing its value each year, with v_1 representing its initial value.
8. Find the explicit formula for the sequence q: $\frac{1}{2}, \frac{3}{4}, \frac{5}{8}, \frac{7}{16}, \ldots$.
9. The first two terms of a sequence are 15 and 33. Find an explicit formula for the sequence if it is
   a. arithmetic.
   b. geometric.
10. In a geometric sequence g, $g_8 = 9$ and $g_{10} = 10$. Find g_9 and g_{11}.

APPLYING THE MATHEMATICS

11. Several long distance runners are on a special ten-day exercise program. They are to run three miles on the first day, and on each successive day of the program they are to increase their distance by 10%. How far must they run on the sixth day?

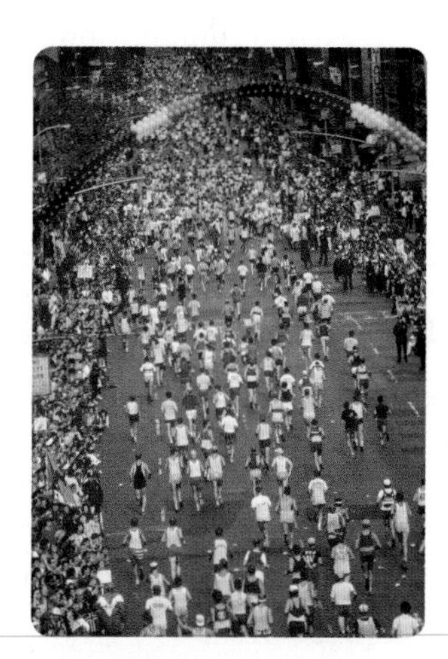

Off and running The first New York City Marathon in 1970 had 127 runners. Today, it attracts over 100,000.

12. Consider the sequence q: 12, 102, 1002, 10002, … .
    a. Could the sequence be arithmetic? Justify your answer.
    b. Could the sequence be geometric? Justify your answer.
    c. Find a formula that fits these terms of the sequence.

13. A particular car loses 17% of its value each year. Suppose the value in the first year is \$13,555.
    a. Find the value of the car in its second year.
    b. Write an explicit formula for the value of the car in its nth year.
    c. In how many years will the car be worth about \$1000?

In 14 and 15, $\overline{PQ}$ is a diameter of circle O. $\overline{RS}$ is a chord of the circle, $\overline{RS} \perp \overline{PQ}$, and $RT = ST$. Then it is known from geometry that ST is the geometric mean of PT and QT.

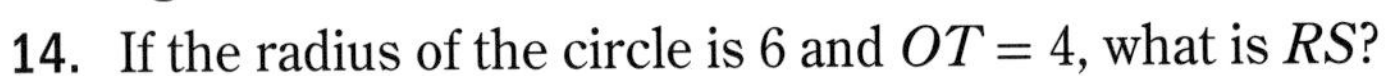

14. If the radius of the circle is 6 and $OT = 4$, what is RS?
15. If $RS = 8$ and $OT = 8$, what is PQ?

In 16 and 17, the midpoints of the sides of parallelogram $A_1B_1C_1D_1$ have been connected in order to form parallelogram $A_2B_2C_2D_2$, and so on to form new parallelograms.

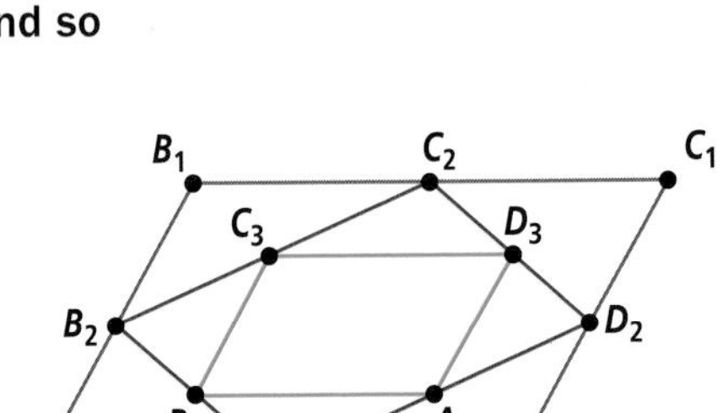

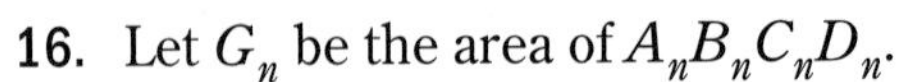

16. Let G_n be the area of $A_nB_nC_nD_n$.
    a. Use geometry to explain why $G_n = \frac{1}{2}G_{n-1}$.
    b. If the area of $A_1B_1C_1D_1$ is 100 square units, write an explicit formula for G_n.

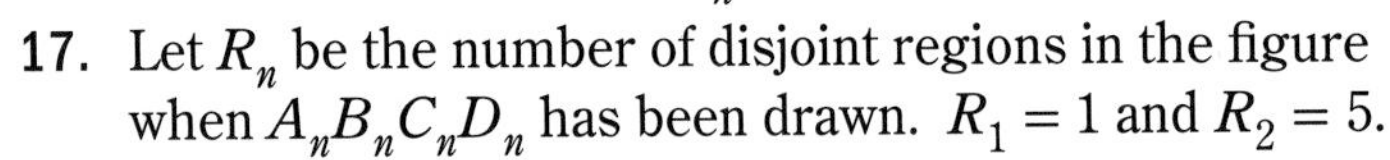

17. Let R_n be the number of disjoint regions in the figure when $A_nB_nC_nD_n$ has been drawn. $R_1 = 1$ and $R_2 = 5$.
    a. What are R_3 and R_4?
    b. Find a recursive formula and an explicit formula for R_n.

REVIEW

18. For the arithmetic sequence $4x + y, 3x, 2x - y, \ldots$, find
    a. the constant difference. b. the 50th term. **(Lesson 8-1)**

19. How many ways can a four-member family sit at their six-seat dining room table? **(Lesson 6-4)**

Some studies indicate having family dinners on a regular basis correlates positively with academic performance.

In 20 and 21, True or False. (Lesson 3-4)

20. If the graph of a function can be mapped onto itself under a rotation of 180° about the origin, then the function is even.
21. It is not possible for an even function to be one-to-one.

EXPLORATION

22. The numbers 1, 1, 2, 5, 14, 42, ... are called the *Catalan Numbers* after the Belgian mathematician Eugène Charles Catalan (1814–1894). They arise in many contexts, for example, as the number of ways of writing a product of n terms using parentheses to group terms and products into pairs.

2 factors: ab	1 way
3 factors: $(ab)c$ or $a(bc)$	2 ways
4 factors: $(ab)(cd)$, $((ab)c)d$, $(a(bc))d$, $a(b(cd))$, $a((bc)d)$	5 ways

    a. Find the next three Catalan numbers.
    b. Describe a geometric context in which they arise.

QY ANSWERS

1. $g_1 = 5$; $r = 2$

2. 8

Lesson

8-3 End Behavior of Sequences

Vocabulary

end behavior of a sequence

limit of a sequence c_n as n approaches infinity, $\lim_{n \to \infty} c_n$

diverge, divergent

converge, convergent

alternating harmonic sequence

▶ **BIG IDEA** For convergent sequences, as the term number approaches infinity, the value of the term approaches a fixed limit.

As with any function, what happens to the values t_n of a sequence t as n gets very large is called the **end behavior** of the sequence.

Mental Math

Suppose $c_n = \frac{a_n}{b_n}$ and a and b are either arithmetic or geometric sequences.

If $c_1 = \frac{2}{3}$, $c_2 = \frac{4}{9}$, and $c_3 = \frac{6}{27}$, what are c_4 and c_5?

Examples of End Behavior of Sequences

Examine the following two sequences and the graphs of their first 25 terms. Although both sequences decrease as n increases, there are important differences in their end behaviors.

$$b_n = -2n \qquad b: -2, -4, -6, -8, -10, \ldots$$

The sequence with terms $b_n = -2n$ is an arithmetic sequence. Each term is 2 less than its predecessor. Thus the terms of sequence b decrease without bound.

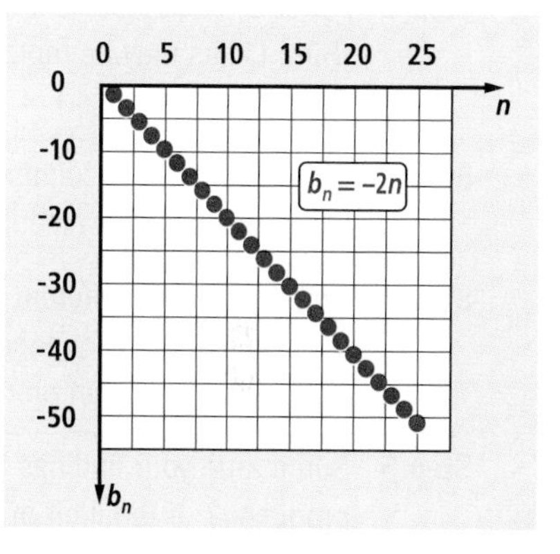

$$c_n = \frac{1}{n} \qquad c: 1, \frac{1}{2}, \frac{1}{3}, \frac{1}{4}, \frac{1}{5}, \ldots$$

In contrast, the terms of the harmonic sequence c get closer and closer to zero. No matter how large you make n, c_n never equals zero (although you may see 0 on a calculator because of rounding). We say that the terms of the sequence c approach 0 as n increases and that 0 is the **limit of c_n as n approaches infinity**, written $\lim_{n \to \infty} c_n$.

Because the terms b_n do not approach a constant value as n increases, the sequence b does not have a limit as n approaches infinity. So we say, "the limit does not exist." On the graph of b_n at the right, you can see that there is no horizontal asymptote.

A sequence that does not have a finite limit is said to **diverge** or to be **divergent**. Sequence b is divergent. A sequence that has a finite limit L is said to be **convergent** or to **converge** to L. The harmonic sequence c is convergent with limit 0. Also, a constant sequence is considered to be convergent. For instance, if for all n, $s_n = 5$, then each term is 5 and the sequence converges to 5.

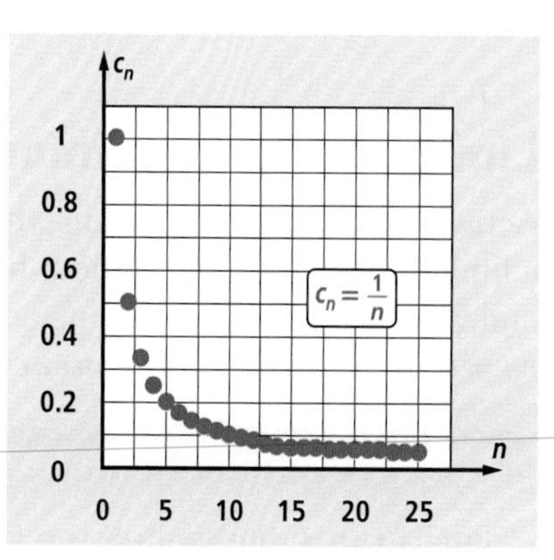

Arithmetic Operations on Sequences

You can perform arithmetic operations to combine terms of two sequences. Activity 1 makes use of the two sequences given above and three others given in the table on the next page.

nth Term	First Five Terms	50th Term	1000th Term
$b_n = -2n$	-2, -4, -6, -8, -10, ...	-100, ...	-2000, ...
$c_n = \frac{1}{n}$	$1, \frac{1}{2}, \frac{1}{3}, \frac{1}{4}, \frac{1}{5}, \ldots$	$\frac{1}{50}, \ldots$	$\frac{1}{1000}, \ldots$
$d_n = (-1)^n$	-1, 1, -1, 1, -1, ...	1, ...	1, ...
$e_n = \frac{1+4n}{n}$	$5, \frac{9}{2}, \frac{13}{3}, \frac{17}{4}, \frac{21}{5}, \ldots$	$\frac{201}{50}, \ldots$	$\frac{4001}{1000}, \ldots$
$f_n = 6$	6, 6, 6, 6, 6, ...	6, ...	6, ...

Activity 1

MATERIALS CAS

Step 1 Set column A of a CAS spreadsheet to display the term number n for $n = 1, 2, 3, 4, 5, 50$ and 1000. In columns B through F enter the formulas to display terms for the five sequences shown above.

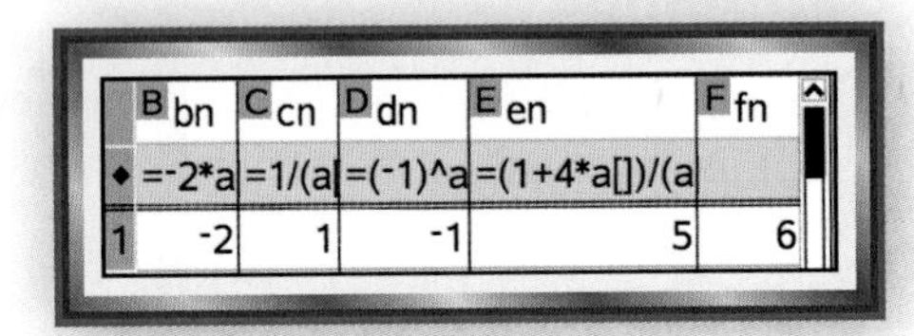

Step 2 For each sequence, is it convergent? If it is convergent, give its limit. (Hint: Limits may be more apparent if you use decimal approximations rather than fractions.)

Step 3 Enter a formula in column G for a new sequence, $g_n = b_n \cdot f_n$. Is this new sequence convergent? If so, give its limit.

Step 4 Experiment with different formulas for g_n formed by either adding k to the terms of the sequences b_n, c_n, d_n, and e_n or multiplying them by k. List the formulas that produce convergent sequences and give their limits.

Step 5 Summarize your findings from Step 4. When is a convergent sequence produced? If the limit of the original sequence is L, what is the limit of the new sequence? Keep your spreadsheet to use in Activity 2.

Limits and the Harmonic Sequence

Activity 1 shows sequences in which the terms get closer and closer to a limit L as n increases without bound. We say "n approaches infinity" and write it as $n \to \infty$. For instance, as n approaches infinity, the terms $c_n = \frac{1}{n}$ of the ***harmonic sequence*** are closer and closer to the limit 0.

Limit of Harmonic Sequence Property

$\lim_{n \to \infty} \left(\frac{1}{n}\right) = 0$; the limit of the harmonic sequence is 0.

The sentence $\lim_{n \to \infty} \left(\frac{1}{n}\right) = 0$ is read "the limit of $\frac{1}{n}$, as n approaches infinity, is 0."

Why does the harmonic sequence converge to 0? Graphically, the terms of the harmonic sequence lie on the branch of the graph of the inverse variation function $f(x) = \frac{1}{x}$ that is in the first quadrant. That is, the limit is geometrically represented by the fact that $y = 0$ is an asymptote to this graph. All graphs of convergent sequences have horizontal asymptotes.

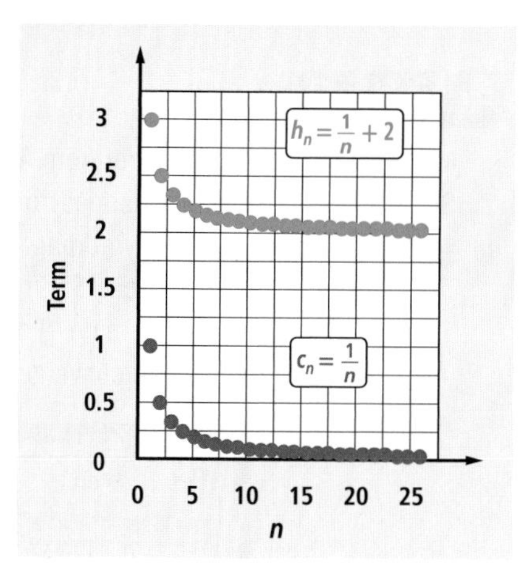

Activity 1 explored the results when arithmetic operations combine a constant with each term of a sequence. For instance, the lower graph at the right shows the first 25 terms of the harmonic sequence $c_n = \frac{1}{n}$. The upper graph shows the function $h_n = \frac{1}{n} + 2$ formed by adding 2 to each term of sequence c. You can see that $\lim_{n \to \infty} h_n = 2$. Generalized forms of these conclusions are given below.

Properties of Limits and Constants

For all nonzero constants k and sequences a for which $\lim a_n = L$:

(1) $\lim_{n \to \infty} k = k$

(2) $\lim_{n \to \infty} (a_n + k) = \lim_{n \to \infty} (a_n) + k = L + k$

(3) $\lim_{n \to \infty} (a_n - k) = \lim_{n \to \infty} (a_n) - k = L - k$

(4) $\lim_{n \to \infty} (a_n \cdot k) = \lim_{n \to \infty} (a_n) \cdot k = kL$

(5) $\lim_{n \to \infty} \left(\frac{a_n}{k}\right) = \frac{\lim_{n \to \infty} (a_n)}{k} = \frac{L}{k}$

 QY

> **QY**
> Find $\lim_{n \to \infty} w_n$ when $w_n = 16 + \frac{1}{n}$.

By knowing that the limit of the harmonic sequence is 0, you can often apply these properties to find the limit of a sequence with a more complicated formula. The idea is to use algebra to transform the expression into one that shows its relationship to $\frac{1}{n}$.

Example 1

Find the limit of the sequence P where $P_n = \frac{5n + 1}{n}$.

Solution Divide each term of the numerator by the denominator.

$\frac{5n + 1}{n} = \frac{5n}{n} + \frac{1}{n} = 5 + \left(\frac{1}{n}\right)$ — the formula for the harmonic sequence

Think of $5 + \frac{1}{n}$ as the constant 5 added to each term of the harmonic sequence given by $\frac{1}{n}$. Now consider limits. If every term of a sequence is 5, then its limit as $n \to \infty$ is also 5. We also know the limit of the harmonic sequence is 0. So

$$\lim_{n \to \infty} \left(\frac{5n + 1}{n}\right) = \lim_{n \to \infty} \left(5 + \frac{1}{n}\right)$$
$$= \lim_{n \to \infty} (5) + \lim_{n \to \infty} \left(\frac{1}{n}\right)$$
$$= 5 + 0 = 5.$$

Activity 2

MATERIALS CAS spreadsheet from Activity 1

Step 1 Open your spreadsheet from Activity 1. Enter formulas for g_n to create new sequences by adding, subtracting, multiplying, or dividing two of these sequences. For instance, $g_n = b_n + c_n$ might be the first sequence you examine. Form as many new sequences as you can that appear to be convergent.

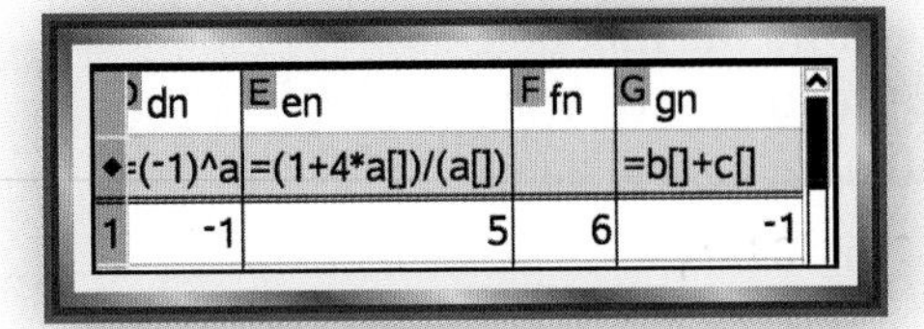

Step 2 For each sequence that seems convergent, give a formula for the nth term and the number that appears to be the limit.

One sequence that can be formed in Activity 2 is $g_n = c_n \cdot d_n = \frac{(-1)^n}{n}$. It is like the harmonic sequence, but every other term is negative. It is called the **alternating harmonic sequence**. Although the terms alternate between positive and negative, this sequence converges and has limit 0. Its graph has $y = 0$ as a horizontal asymptote, as shown at the right.

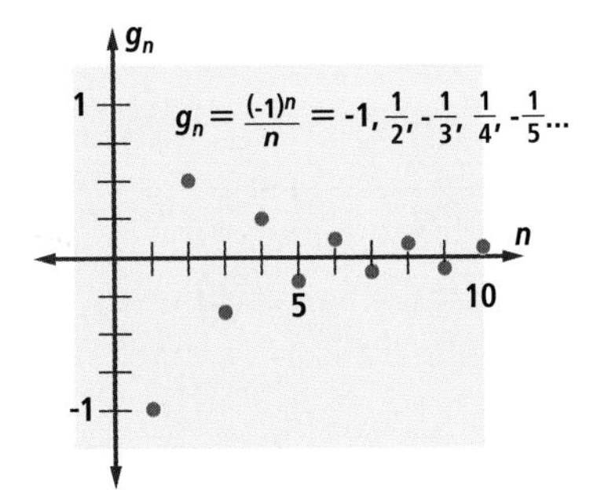

Limits of Results of Operations with Sequences

Activity 2 shows that when corresponding terms of two convergent sequences are combined with one of the four basic operations, their limits are combined in the same way.

Properties of Limits on Operations with Sequences

If a and b are two convergent sequences and $\lim_{n \to \infty} b_n \neq 0$, then

1. **Limit of a Sum Property** The limit of a sum of the sequences is the sum of their limits. $\lim_{n \to \infty} (a_n + b_n) = \lim_{n \to \infty} a_n + \lim_{n \to \infty} b_n$
2. **Limit of a Difference Property** The limit of a difference of the sequences is the difference of their limits. $\lim_{n \to \infty} (a_n - b_n) = \lim_{n \to \infty} a_n - \lim_{n \to \infty} b_n$
3. **Limit of a Product Property** The limit of a product of the sequences is the product of their limits. $\lim_{n \to \infty} (a_n \cdot b_n) = \lim_{n \to \infty} a_n \cdot \lim_{n \to \infty} b_n$
4. **Limit of a Quotient Property** The limit of a quotient of the sequences is the quotient of their limits. $\lim_{n \to \infty} \frac{a_n}{b_n} = \frac{\lim_{n \to \infty} a_n}{\lim_{n \to \infty} b_n}$

Note that the properties of limits on operations with sequences do not mention divergent sequences. Combining divergent sequences can produce another divergent sequence or, sometimes, a convergent sequence. Also notice that in the Limit of a Quotient Property, the limit of the divisor sequence b must be nonzero.

Using Properties of Limits

Tables and graphs can verify the end behavior of sequences.

Example 2

Consider $p_n = \frac{5n+1}{n}$ and $q_n = \frac{3n+2}{n}$. A new sequence $t_n = p_n \cdot q_n$ is formed by multiplying term values in the sequence p_n by corresponding term values in q_n.

a. Find $\lim_{n\to\infty} p_n$ and $\lim_{n\to\infty} q_n$.

b. Use the properties of limits to find $\lim_{n\to\infty} (p_n \cdot q_n)$.

c. Confirm your answer to Part b with a table and with a graph.

Solution

a. From Example 1, $\lim_{n\to\infty} p_n = 5$. $q_n = \frac{3n+2}{n} = \frac{3n}{n} + \frac{2}{n} = 3 + 2 \cdot \frac{1}{n}$. So, from the Properties of Limits and Constants, $\lim_{n\to\infty} q_n = 3 + 2 \cdot \lim_{n\to\infty} \frac{1}{n} = 3 + 2 \cdot 0 = 3$.

b. Use the Limit of a Product Property with the limits found in Part a.

$$\lim_{n\to\infty} (p_n \cdot q_n) = \lim_{n\to\infty} p_n \cdot \lim_{n\to\infty} q_n = 5 \cdot 3 = 15$$

c. A table of selected values of the new sequence $t_n = p_n \cdot q_n$ and a graph are shown below and at the right.

n	$p_n = \frac{5n+1}{n}$	$q_n = \frac{3n+2}{n}$	$t_n = p_n \cdot q_n$
1	6	5	30
2	5.5	4	22
3	5.333	3.667	19.556
100	5.01	3.02	15.130
1000	5.001	3.002	15.013

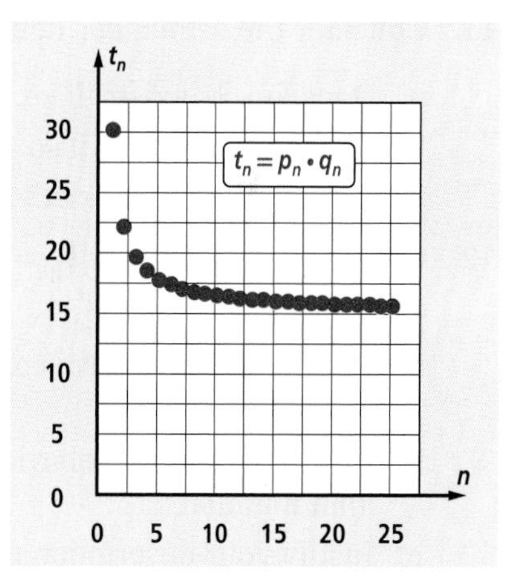

The table and graph show that as n increases, the terms get closer to 15.

Check Use the limit command on a CAS.

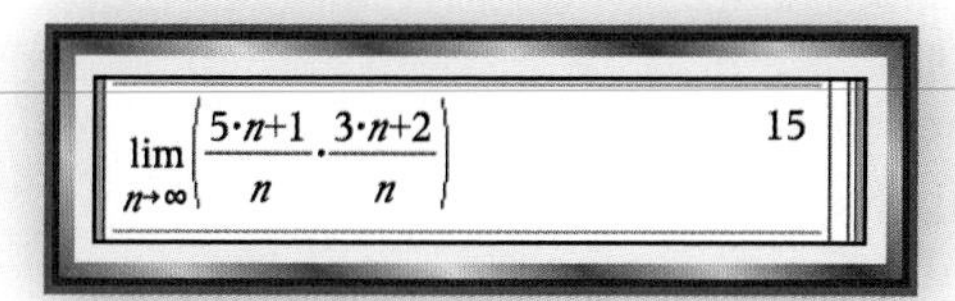

Questions

COVERING THE IDEAS

1. What is the *end behavior* of a sequence?

2. A graph of the sequence $t_n = 1.5, 1.5, 1.5, 1.5, \ldots$ is shown at the right. Does the sequence converge? If so, what is its limit?

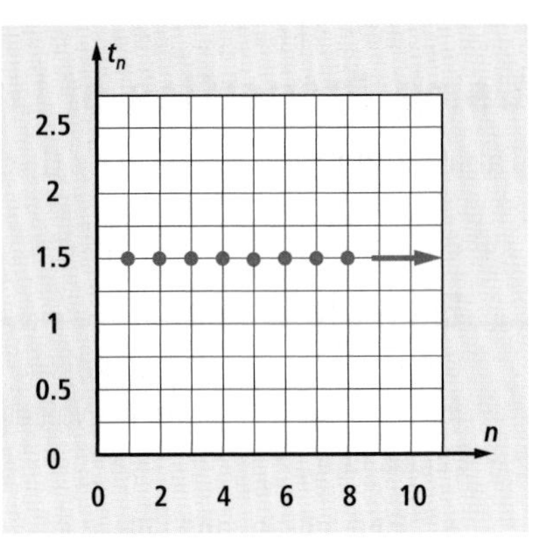

3. **True of False** If each term in a sequence is less than the preceding term, the sequence has a limit.

4. **True of False** If a sequence has a limit, then the graph of the sequence has a horizontal asymptote.

5. Is the sequence defined by $t_n = \frac{n}{n+1}$ convergent? Use a table and a graph to explain how you know.

In 6–8, a formula for the *n*th term of a sequence is given. Does the sequence converge? If it converges, give the limit.

6. $d_n = \frac{1}{n} + 6$

7. $b_n = -\frac{1}{10}n^2 + 1$

8. $f_n = 5 + \frac{3}{n}$

9. Consider the sequence with nth term $4 \cdot (-1)^n$.
   a. Write the first five terms of the sequence.
   b. Does this sequence converge? Explain.

10. Given that $\lim_{n \to \infty} \frac{3n-7}{4n} = \frac{3}{4}$, find $\lim_{n \to \infty}\left(2 \cdot \frac{3n-7}{4n} + 10\right)$.

APPLYING THE MATHEMATICS

11. Consider the sequences defined by $b_n = 5 + \frac{1}{n^2}$ and $c_n = n + 5$.
    a. Does $\lim_{n \to \infty} b_n$ exist? If so, give its value.
    b. Does $\lim_{n \to \infty} c_n$ exist? If so, give its value.
    c. Does $\lim_{n \to \infty} (b_n \cdot c_n)$ exist? If so, give its value.

12. Consider the sequence $b_n = \frac{4n+1}{n}$ and the reciprocal of that sequence $t_n = \frac{1}{b_n} = \frac{n}{4n+1}$.
    a. List the first five terms of b_n and t_n. Write your answers as fractions.
    b. Describe the end behavior of sequences b and t using limit notation.
    c. Justify your description using properties of limits.

13. Consider the sequence defined by the recursive formula

$$\begin{cases} h_1 = 6 \\ h_n = \frac{1}{3}h_{n-1} \text{ for } n > 1. \end{cases}$$

    a. Write the first five terms of the sequence in fraction form.
    b. Does the sequence converge? If so, find its limit.

14. If a radioactive element has half-life t and there are B mg of the element to start, then after n time periods of length t, there will be $a_n = B\left(\frac{1}{2}\right)^n$ mg of the element left.
    a. Does sequence a converge? If it converges, give its limit. If it diverges, explain how you know.
    b. Explain what your answer means in the real-world situation.

15. Repeat Part a of Question 14 for the sequence $c_n = \cos n$, where n is in degrees.

16. The sequence of decimals 0.8, 0.88, 0.888, 0.8888, ..., formed by an increasing number of 8s, converges to a rational number. Write that rational number as a fraction in lowest terms.

17. The sequences defined by $a_n = 5 + 2n$ and $b_n = 10 - 2n$ diverge. Examine $a_n + b_n$, $a_n - b_n$, $a_n \cdot b_n$, and $\frac{b_n}{a_n}$. If any converge, give their limits.

Marie Curie (1867–1934), best known for her work wih radioactivity, won two Nobel prizes.

REVIEW

18. Give an example of a sequence that is both geometric and arithmetic. **(Lessons 8-2, 8-1)**

19. Consider the geometric sequence 100, 70, 49, 34.3, ...
    a. Write an explicit formula for the sequence.
    b. Write a recursive formula for the sequence.
    c. What is the number of the first term that is less than 0.5? **(Lesson 8-2)**

20. Consider the sequence of squares formed by connecting midpoints of squares as shown. Let s_n equal the side length of the nth square.
    a. If $s_1 = 50$, find s_2 and s_3.
    b. Find an explicit formula for s_n.
    c. Is the sequence arithmetic, geometric, or neither? **(Lesson 8-2)**

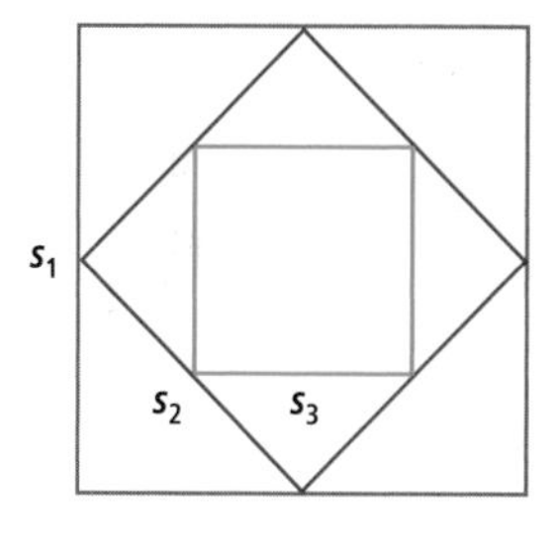

21. a. Draw one cycle of a sine wave with amplitude 6 and period π.
    b. Find an equation for the sine wave in Part a.
    c. State three values of the independent variable for which this sine wave achieves its minimum value. **(Lesson 4-5)**

22. ***Skill Sequence*** Represent using summation notation. **(Lesson 1-2)**
    a. $u_1 + u_2 + u_3 + u_4 + u_5$
    b. $v_7 + v_8 + v_9 + v_{10}$
    c. $w_1 + w_2 + w_3 + \ldots + w_n$

EXPLORATION

23. a. Identify two divergent sequences a and b whose product ab converges.
    b. Identify two different divergent sequences a and b whose difference $a - b$ converges.

QY ANSWER

16

Lesson

8-4 Arithmetic Series

Vocabulary

series
value, sum of a series
finite series
arithmetic series
infinite series

BIG IDEA Arithmetic series are sums of consecutive terms of arithmetic sequences and can be evaluated using any of several formulas.

Mental Math

Find the sum of the first 10 integers.

The Sum of Terms of a Sequence

Suppose that a virus enters a community and the number of new infections each week forms an arithmetic sequence. If there were 6 cases the first week and, in each week thereafter, 4 more new cases than the week before, then the sequence $a_n = 6, 10, 14, 18, 22, \ldots$ of new cases in the nth week can be described with the formulas below.

Recursive Formula

$$\begin{cases} a_1 = 6 \\ a_n = a_{n-1} + 4, n > 0 \end{cases}$$

Explicit Formula

$$a_n = 6 + (n - 1)4 = 6 + 4(n - 1)$$

It is important to keep track of the total number of cases, not just new cases. The table and graph below show the first five terms of the sequence and the cumulative totals.

Week	New Cases	Total Number Infected
1	6	6
2	10	$6 + 10 = 16$
3	14	$6 + 10 + 14 = 30$
4	18	$6 + 10 + 14 + 18 = 48$
5	22	$6 + 10 + 14 + 18 + 22 = 70$

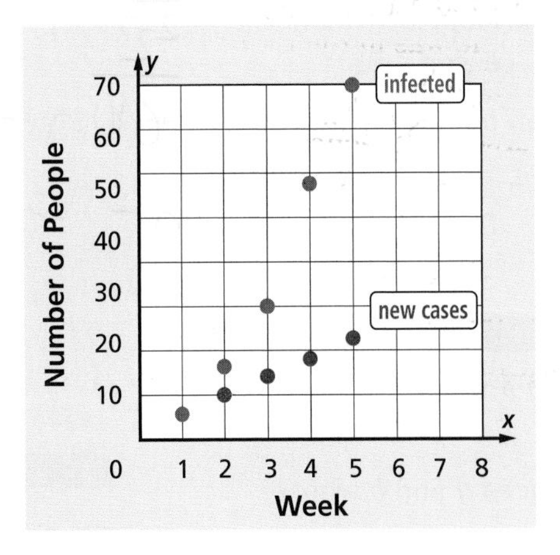

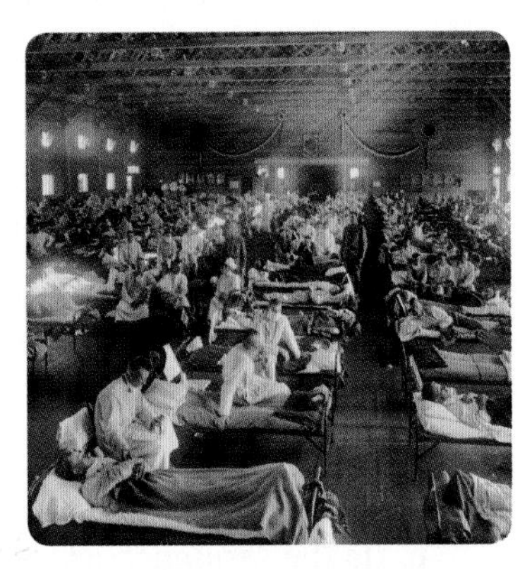

An emergency hospital in Funston, KS, during the influenza epidemic of 1918 that killed 50 to 100 million people worldwide

If a_n is the number of new cases in the nth week and S_n is the cumulative total number infected up to that week, then $S_n = \sum_{i=1}^{n} a_i$.

So, for example,

$$S_5 = \sum_{i=1}^{5} a_i = a_1 + a_2 + a_3 + a_4 + a_5 = 6 + 10 + 14 + 18 + 22 = 70.$$

To find a formula for S_n in terms of n, just substitute the explicit formula for a_i. Then $\quad S_n = \sum_{i=1}^{n} (6 + 4(i - 1))$.

For instance, $\quad S_5 = \sum_{i=1}^{5} (6 + 4(i - 1)) = 70$.

The indicated sum of the terms of a sequence is called a **series**. The numerical sum is called either the **value** or, simply, the **sum** of the series. If a finite number of terms are added, the resulting series is called a **finite series**. The series above is a finite series whose sum is 70. If a_1, a_2, a_3, ..., a_n are terms in a finite arithmetic sequence, then

$$S_n = a_1 + a_2 + a_3 + \dots + a_n = \sum_{i=1}^{n} a_i$$

is called a finite **arithmetic series**. If the number of terms added is infinite, the resulting series is an **infinite series**. Infinite series are discussed in Lesson 8-7.

A Story about Gauss

You can add every term in a finite arithmetic series to find the sum, but this can be tedious, even with a calculator or spreadsheet.

Activity

Work in a group without using technology. In no more than two minutes, find the sum of the integers from 1 to 100. Be ready to share with your classmates the method you used to find the sum.

Let's look at the way to add the integers from 1 to 100 used by the great mathematician Carl Friedrich Gauss when he was in the third grade. As punishment for misbehavior, everyone in his class was asked to add the integers from 1 to 100. Of course, his teacher thought the problem would take a long time. Here is how Gauss found the sum, according to his own reports.

Carl Friedrich Gauss (1777–1855) pictured on a German postal stamp

Suppose S_{100} represents the sum of the first 100 integers. So

$$S_{100} = 1 \quad + 2 \quad + 3 \quad + \cdots + 99 \quad + 100.$$

He rewrote the series beginning with the last term.

$$S_{100} = 100 + 99 + 98 + \cdots + 2 \quad + 1$$

Then he added the two equations.

$$2S_{100} = 101 + 101 + 101 + \cdots + 101 + 101$$

The right side of the equation has 100 terms, each equal to 101. Thus

$$2S_{100} = 100(101).$$

He divided each side of the equation by 2.

$$S_{100} = 5050$$

Gauss wrote nothing but the answer of 5050 on his slate, having done all the previous work in his head. You might think the teacher was angry at Gauss, but in fact this event led the teacher to realize that Gauss was very special, and he took steps to cultivate Gauss's talent.

QY1

> **QY1**
> Use Gauss's technique to find the sum of the first 50 integers.

Formulas for the Sum of Any Arithmetic Series

Gauss's strategy can be used to find a formula for the sum S_n of *any* arithmetic series. First, write S_n starting with the first term a_1 and successively add the constant difference d. Second, write S_n starting with the last term a_n and successively subtract the constant difference d.

$$S_n = a_1 + (a_1 + d) + (a_1 + 2d) + \cdots + (a_1 + (n - 1)d)$$

$$S_n = a_n + (a_n - d) + (a_n - 2d) + \cdots + (a_n - (n - 1)d)$$

Now add the two preceding equations.

$$2S_n = \underbrace{(a_1 + a_n) + (a_1 + a_n) + (a_1 + a_n) + \cdots + (a_1 + a_n)}_{n \text{ terms}}$$

The right side has n terms each equal to $a_1 + a_n$.

$$2S_n = n(a_1 + a_n)$$

Divide both sides by 2. $S_n = \frac{n}{2}(a_1 + a_n)$

This formula for S_n is useful if you know the first and nth terms of the series. If you do not know the nth term, you can find it using $a_n = a_1 + (n - 1)d$ from Lesson 8-1. This leads to a second formula which you are asked to prove in Question 17.

> **Theorem (Sum of an Arithmetic Series)**
>
> The sum $S_n = a_1 + a_2 + \ldots + a_n = \sum_{i=1}^{n} a_i$ of an arithmetic series with constant difference d is given by
>
> $$S_n = \frac{n}{2}(a_1 + a_n) \text{ or } S_n = \frac{n}{2}(2a_1 + (n - 1)d).$$

QY2

> **QY2**
> Use the theorem to find the sum of the arithmetic series 10 + 15 + 20 + 25 + 30 + 35 + 40 + 45 + 50.

The first formula for the sum of an arithmetic series indicates that you can find the sum if you know the first term, the last term, and the number of terms in the series.

Example 1

A student borrowed $8000 for college expenses. The loan was to be repaid over a 100-month period, with monthly payments as follows:

$120.00, $119.50, $119.00, ..., $70.50.

a. How much did the student pay over the life of the loan?

b. What was the total interest paid on the loan?

Solution

a. Find the sum 120.00 + 119.50 + 119.00 + ⋯ + 70.50. Because the terms show a constant difference ($d = 0.50$), this sum is an arithmetic series with $a_1 = 120.00$, $n = 100$, and $a_{100} = 70.50$.
Use the formula $S_n = \frac{n}{2}(a_1 + a_n)$.
$S_{100} = \frac{100}{2}(120.00 + 70.50) = 9525.$
The student paid back a total of $9525.

b. The interest was $9525 – $8000 = $1525.

The second formula for the sum of an arithmetic series enables you to calculate the sum if you know the first term, the constant difference, and the number of terms of the series.

GUIDED

Example 2

In training for a marathon, an athlete runs 7500 meters on the first day, 8000 meters the next day, 8500 meters the third day, and so on, each day running 500 meters more than on the previous day. What is the total distance the athlete will have run over 30 days?

Solution The distances form an arithmetic sequence with $a_1 =$ _?_ and $d =$ _?_.

Because the number of terms is known, use the second formula in the theorem.

$S_n = \frac{30}{2}(2a_1 + (30 - 1)d) = \frac{30}{2}$ _?_ = _?_

The athlete runs _?_ meters in thirty days.

The choice of formulas for S_n is useful if S_n is known and you must find one of a_1, a_n, n, or d.

Example 3

The main floor of a new auditorium is planned to seat 800 people, with seats arranged in 20 rows. Each row will have 2 more seats than the previous row. How many seats should there be in the first row?

(continued on next page)

The Walt Disney Concert Hall in Los Angeles, CA, home to the Los Angeles Philharmonic

Solution From the given information, the number of seats in each row forms an arithmetic sequence a with $n = 20$, $d = 2$, and $S_{20} = 800$. If you find a_1, you will be able to calculate the number of seats for each row. Substitute into the second formula for S_n to get an equation, then solve it.

$$S_{20} = \frac{20}{2}(2a_1 + (20 - 1)2)$$
$$800 = 10(2a_1 + 38)$$
$$800 = 20a_1 + 380$$

Thus, $a_1 = 21$.

The first row should have 21 seats. The other rows will have 23, 25, 27, ..., 59 seats.

$s=\frac{n\cdot(2\cdot a+(n-1)\cdot d)}{2}|n=20$ and $d=2$ and s ▸ $800=20\cdot(a+19)$

$\frac{800=20\cdot(a+19)}{20}$ $40=a+19$

$(40=a+19)-19$ $21=a$

Check Use a CAS. To enter the sigma notation, you need an explicit formula for the terms. Since the first term is 21 and the common difference is 2, $a_n = 21 + (n - 1)2$. The series is the sum of the first 20 terms of this sequence.

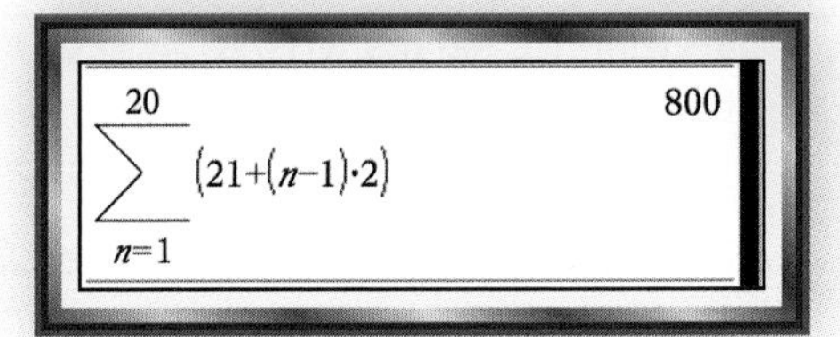

Questions

COVERING THE IDEAS

In 1 and 2, Multiple Choice. Consider the following choices.

A $16 + 19 + 22 + 25 + \cdots$ B $4 + 8 + 16 + 32 + \cdots$

C $-3, 4, -5, 6, \ldots$ D $8, -2, -12, -22, \ldots$

1. Which is an arithmetic sequence?
2. Which is an arithmetic series?
3. How are sequences and series related?
4. To support a charity, a club is planning a fund-raising drive that will last 7 days. The members hope to get \$50 in donations the first day and to increase donations by \$5 each day.
   a. Describe the meaning of a_n and S_n in this context.
   b. Write a formula for a_7 and evaluate it.
   c. Write a formula for S_7 and evaluate it.
5. A clock chimes once at 1 o'clock, twice at 2 o'clock, three times at 3 o'clock, and so on, with twelve chimes at noon and at midnight. How many times does the clock chime during the course of a 24-hour day?
6. Let $S_n = \sum_{k=1}^{n} (8k - 1)$.
   a. Write out the terms of S_5.
   b. What values of a_1, a_n, d, and n can be used to evaluate S_{12}?
   c. Find the value of S_{12}.

A pink ribbon is used worldwide to raise awareness for breast cancer.

In 7–10, copy the table below, which describes four arithmetic series, each one in a different way. For each series, fill in the remaining information.

	a_1	d	n	a_n	Sum (written out)	Sum (sigma notation)
7.	15	3	5	?	?	?
8.	62	–4	?	30	?	?
9.	?	?	?	?	$43 + 48 + 53 + \cdots + 78$	?
10.	?	?	?	?	?	$\sum_{n=1}^{6} (2n - 1)$

In 11 and 12, use the theorem in this lesson to calculate the sum.

11. $\sum_{i=1}^{1000} i$

12. $\sum_{j=1}^{50} (9j + 10)$

13. What sequence is being summed in Question 11?

APPLYING THE MATHEMATICS

In 14 and 15, follow the directions of Questions 7–10.

	a_1	d	n	a_n	Sum (written out)	Sum (sigma notation)	Sum
14.	12	-10	25	?	?	?	?
15.	2.5	1.5	?	20.5	?	?	?

16. Suppose a child builds a figure with colored blocks in stages as illustrated below.

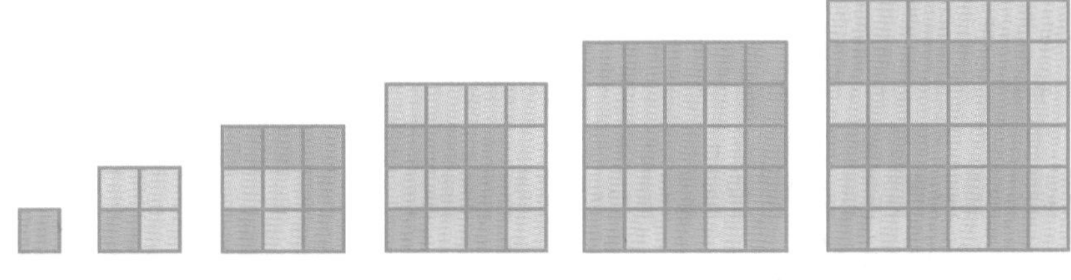

The number of blocks added to the figure at each step is a term in the arithmetic sequence of odd numbers 1, 3, 5, ... Suppose the pattern is continued to form a 10×10 square.

a. How many total blocks are there?

b. How many of the blocks are dark? How many are light?

17. Use the formulas $S_n = \frac{n}{2}(a_1 + a_n)$ and $a_n = a_1 + (n - 1)d$ to prove that the sum of the first n terms of an arithmetic series with first term a_1 and constant difference d is $S_n = \frac{n}{2}(2a_1 + (n - 1)d)$.

18. A sequence is described recursively as $\begin{cases} a_1 = 28 \\ a_n = a_{n-1} - 4, \text{ for } n > 1. \end{cases}$

a. Write an explicit formula for this sequence.

b. Find the sum of the first 30 terms of this sequence.

In 19 and 20, suppose a display of cans in a supermarket is built with one can on top, two cans in the next row, and one more can in each succeeding row.

19. If there are 12 rows of cans, how many cans are in the display?

20. If 200 cans are available to be displayed, how many rows are needed, and how many cans will be left over?

21. A series of k consecutive even integers begins with w.

$$\underbrace{w + (w + 2) + (w + 4) + (w + 6) + \cdots + \underline{\ ?\ }}_{k \text{ even numbers}}$$

a. Write an expression for the last term in the series.

b. Write an expression for the sum.

REVIEW

In 22 and 23, decide whether the sequence is convergent or divergent. If it is convergent, give its limit. (Lesson 8-3)

22. $-\frac{3}{4}, \frac{9}{5}, \frac{19}{20}, \ldots, \frac{2n^2+1}{3n^2-7}, \ldots$

23. $-\frac{3}{4}, -9, \frac{19}{2}, \ldots, \frac{2n^2+1}{3n-7}$

24. Let $t_n = \sin\left(\frac{\pi}{6}n\right)$.

a. Write the first five terms of the sequence t_n.

b. State whether the sequence is arithmetic, geometric, or neither.

c. Does $\lim\limits_{n \to \infty} t_n$ exist? If so, what is it? **(Lessons 8-3, 8-2, 8-1, 4-4)**

25. Consider the sequence of dots in rectangular arrays shown at the right. **(Lesson 8-2)**

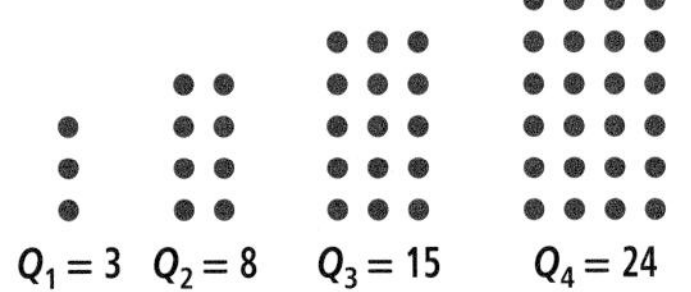

a. Write an explicit formula for Q_n.

b. What is Q_{99}?

26. Write the series $2^3 + 3^3 + 4^3 + \cdots + 100^3$ using sigma notation. Use a CAS to find the sum. **(Lesson 1-2)**

EXPLORATION

27. Question 16 pictures the arithmetic sequence of positive odd integers, 1, 3, 5, 7, ..., and the series formed by adding the terms of the sequence where

$$S_1 = a_1 = 1,$$

$$S_2 = \sum_{n=1}^{2} a_n = 1 + 3 = 4,$$

$$S_3 = \sum_{n=1}^{3} a_n = 1 + 3 + 5 = 9, \text{ and so on.}$$

a. Determine S_4, S_5, and S_6.

b. Based on the results in Part a, write a rule for the sum S_n of the first n odd integers.

c. Deduce your rule in Part b from the formula $S_n = \frac{n}{2}(a_1 + a_n)$.

QY ANSWERS

1. $2S_{50} = 50(51)$,
$S_{50} = 25(51) = 1275$

2. $\frac{9}{2}(10 + 50) = 270$

Lesson 8-5 Geometric Series

Vocabulary

geometric series
*n*th partial sum, S_n

BIG IDEA Geometric series are sums of consecutive terms of geometric sequences and can be evaluated if you know the first term and common ratio.

The story about Gauss in Lesson 8-4 is a true story. The example that begins this lesson is a legend.

Mental Math

If you kiss your mother goodbye once on Monday, twice on Tuesday, 4 times on Wednesday, 8 times on Thursday, and so on, how many kisses will you give your mother in the week?

An Example of a Geometric Series

The game of chess was invented 1500–2500 years ago. A story is told that the king of Persia enjoyed the game so much that he offered the game's inventor whatever he desired as a reward, as long as it was reasonable. Clearing a chessboard, the inventor asked for one single grain of wheat on the first square, twice that on the second square, twice that again on the third, twice that again on the fourth, and so on for the whole board. The king, ready to give jewelry, gold, and other riches, thought that this was a modest request, easily granted. Was it?

The numbers of grains on the squares are $2^0, 2^1, 2^2, ..., 2^{63}$. They form a geometric sequence with $g_1 = 1$ and $r = 2$, and so the *n*th term is $g_n = 2^{n-1}$. Let $S_n = \sum_{i=1}^{n} g_i$. Then the total award for the entire chessboard is S_{64}, where

$$S_{64} = \sum_{i=1}^{64} 2^{i-1} = 1 + 2 + 4 + 8 + \cdots + 2^{63}.$$

S_{64} is a finite geometric series.

World Chess Champion Elisabeth Paetz

Evaluating a Finite Geometric Series

To evaluate S_{64} without adding every term, you can use an approach similar to the one used by Gauss for arithmetic series. First, write the series. Then multiply each side of the equation by the constant ratio, 2.

$$S_{64} = 1 + 2 + 4 + \cdots + 2^{62} + 2^{63}$$
$$2S_{64} = \quad 2 + 4 + \cdots + 2^{62} + 2^{63} + 2^{64}$$

Then, subtract the second equation from the first.

$$S_{64} - 2S_{64} = 1 + 0 + 0 + \cdots + 0 + 0 - 2^{64}$$

Simplify. $\quad -S_{64} = 1 - 2^{64}$

So, $\quad S_{64} = 2^{64} - 1.$

The total number of grains of wheat on the chessboard would be $2^{64} - 1$, or 18,446,744,073,709,551,615. If each grain has volume 12.5 mm^3, about 230 cubic kilometers of wheat would be required, many times the amount of wheat in the world!

A Formula for the Sum of a Finite Geometric Series

In general, a **geometric series** is an indicated sum of terms of a geometric sequence. If a sequence has a first term of g_1 and constant ratio r, the ***n*th partial sum**, denoted, S_n can be written as the sum of terms or written in summation notation.

$$S_n = g_1 + g_1r + g_1r^2 + \cdots + g_1r^{n-1} = \sum_{i=1}^{n} g_1r^{i-1}$$

A formula for the sum S_n can be developed by generalizing the procedure used to find S_{64} from the chess legend.

$$\begin{aligned}
S_n &= g_1 + g_1r + g_1r^2 + \cdots + g_1r^{n-1} & \\
rS_n &= g_1r + g_1r^2 + \cdots + g_1r^{n-1} + g_1r^n & \text{Multiply } S_n \text{ by } r. \\
S_n - rS_n &= g_1 - g_1r^n & \text{Subtract the preceding equations.} \\
(1 - r)S_n &= g_1(1 - r^n) & \text{Factor.} \\
S_n &= \frac{g_1(1 - r^n)}{1 - r} & \text{Divide by } 1 - r.
\end{aligned}$$

This proves the following theorem.

Theorem (Sum of a Finite Geometric Series)

The sum $S_n = \sum_{i=1}^{n} g_1r^{i-1} = g_1 + g_1r + \cdots + g_1r^{n-1}$ of the finite geometric series with constant ratio $r \neq 1$ is given by

$$S_n = \frac{g_1(1 - r^n)}{1 - r}.$$

Example 1

A geometric series is given by $4 + (4 \cdot 3) + (4 \cdot 3^2) + (4 \cdot 3^3) + (4 \cdot 3^4)$. Evaluate S_5 by using

a. the formula above. b. sigma notation on a CAS.

Solution

a. The first term is 4 and the common ratio is 3. There are five terms.

$$S_5 = \frac{4(1 - 3^5)}{1 - 3} = \frac{4(-242)}{-2} = 484$$

b. First write an expression for the nth term: $t_n = 4 \cdot 3^{n-1}$. Sum from $n = 1$ to $n = 5$.

The CAS shows that $S_5 = 484$.

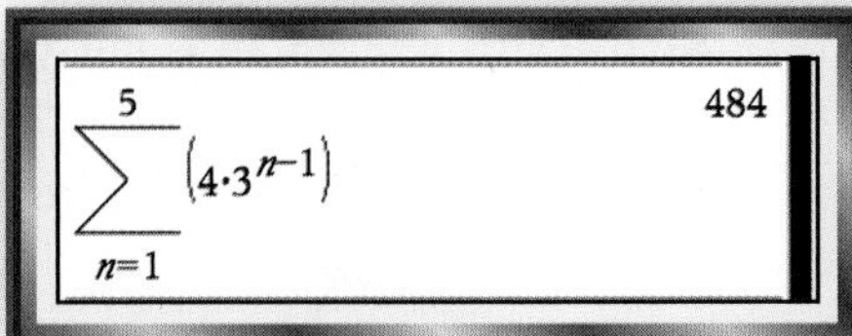

Check Write out the five terms and find the sum.

$S_5 = 4 + 12 + 36 + 108 + 324 = 484$

The constant ratio r in the formula for the sum of a finite geometric series can be negative. It also can be a fraction.

QY1

> **QY1**
>
> Find the sum of the first four terms of a geometric sequence with first term 6 and common ratio -5.

GUIDED

Example 2

Evaluate $\sum_{i=1}^{6} 24\left(\frac{1}{3}\right)^{i-1}$.

Solution To use the formula for the sum of a geometric series, you must determine g_1, n, and r.

$g_1 =$ _?_ $\quad n =$ _?_ $\quad r =$ _?_

$S_n = \frac{g_1(1 - r^n)}{1 - r} =$ _?_ $=$ _?_ as a mixed number.

Check Write each term in the sequence and add.

$24 + 8 + \frac{8}{3} + \frac{8}{9} + \frac{8}{27} + \frac{8}{81} =$ _?_

You can get an alternate formula for the sum of a geometric series by multiplying the numerator and denominator of the formula for S_n by -1. The alternate form avoids negative signs in cases where $r > 1$.

$$S_n = \frac{g_1(r^n - 1)}{r - 1}$$

Uses of Geometric Series

Example 3

Recall the disease outbreak from Lesson 8-4. This time assume that the first week there are 14 cases, and that each week thereafter the number of *new* infections increases by 25%.

a. Use sigma notation to express the number of people infected by the *n*th week.

b. How many people have the disease by the end of the 5th week?

Solution

a. First write out several terms of the series.

$$14 + 14(1.25) + 14(1.25)^2 + \cdots$$

A formula for the number of new cases in week n is $g_1 r^{n-1} = 14(1.25)^{n-1}$. The total number of people infected by the nth week is $S_n = \sum_{i=1}^{n} 14(1.25)^{i-1}$.

(continued on next page)

b. To calculate the total number of people that have the disease by the end of the 5th week, evaluate the series. Since $r > 1$, use the alternate formula for S_n.

$$S_n = \frac{g_1(r^n - 1)}{r - 1}$$

$$S_5 = \frac{14(1.25^5 - 1)}{1.25 - 1} \approx 114.8984$$

At the end of the 5th week, about 115 people have the disease.

QY2

In Examples 2 and 3, the number of terms of the series was known and you were asked to find the sum. In Example 4, this process is reversed. You are given a sum and are asked to find how many terms must be added to produce it. The tree structure used in Example 4 is common in many situations, such as the logic of computers, the branching structure in many natural shapes such as the passages in your lungs, and even family trees.

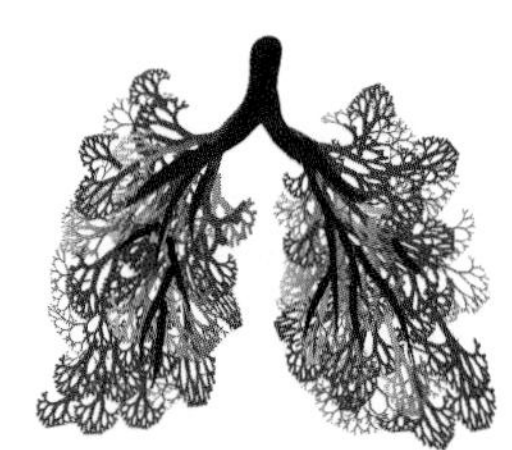

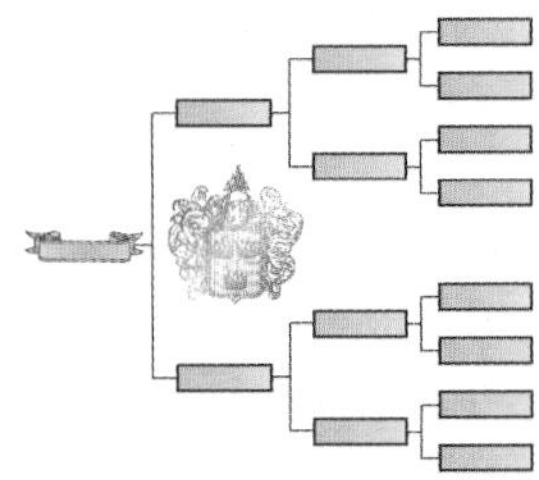

QY2

Suppose that a disease outbreak at a different place is modeled by $S_n = \sum_{i=1}^{n} 25(1.40)^{i-1}$.

a. How many new cases are there the first week? The second week?

b. What is the growth rate in the number of new cases?

Example 4

The maximum number of natural ancestors that you could have are 2 parents, 4 grandparents, 8 great-grandparents, and so on. Assuming that no one appears twice in your ancestral tree, in the past n generations you have S_n natural ancestors, where $S_n = 2 + 4 + 8 + \cdots + 2^n$. How many generations must you go back before you have a million natural ancestors, assuming that no one appears twice?

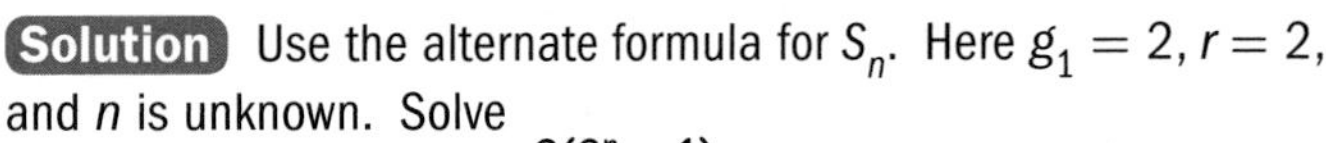

Solution Use the alternate formula for S_n. Here $g_1 = 2, r = 2$, and n is unknown. Solve

$$1{,}000{,}000 = \frac{2(2^n - 1)}{2 - 1}.$$

$$1{,}000{,}000 = 2(2^n - 1)$$

$$500{,}000 = 2^n - 1$$

$$2^n = 500{,}001$$

You can solve this equation with the help of logarithms (which you will review in the next chapter), by trial and error, or by using a solve command on a CAS, as shown at the right.

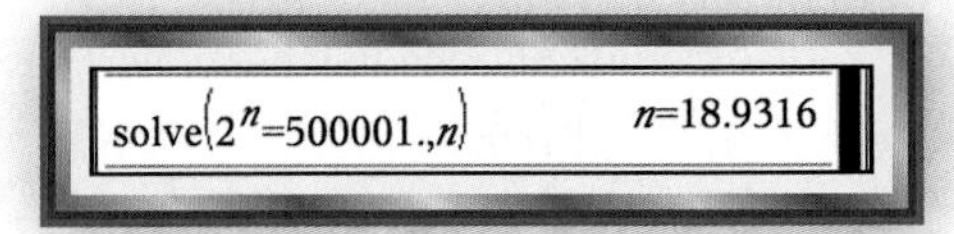

$$n \approx 18.9$$

Assuming no one appears twice on your tree, if you go back 19 generations, you will have had a million ancestors. Thus, it is almost certain that many people have appeared twice on your ancestral tree.

Check Make a spreadsheet with a column to show the number in each generation and a column for the partial sums. The total for generation n is the partial sum for $n - 1$ terms plus the number in generation n. For instance, the total for the third generation is found by adding the 8 third generation ancestors to the 2nd partial sum.

◇	A	B	C
1	Generations	# in generation	Total
2	1	2	2
3	2	4	6
4	3	8	14
5	4	16	30
⋮	⋮	⋮	⋮
19	18	262144	524286
20	19	524288	1048574

Here, 6 + 8 = 14. The formula for cell C4 is = C3 + B4.

Questions

COVERING THE IDEAS

1. Suppose that in the chess story on page 529, the king agreed to give the inventor 2^{n-1} grains of wheat for each of the n squares on a gameboard, but with the smaller board shown at the right.

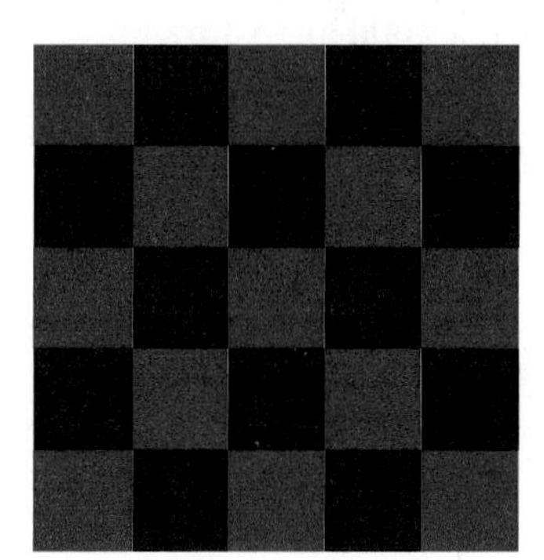

   a. Rewrite the equation $S_{64} = \sum_{i=1}^{64} 2^{i-1} = 1 + 2 + 4 + 8 + \cdots + 2^{63}$ to fit the situation with the smaller board.

   b. How many grains of wheat would be needed?

2. Consider the geometric sequence 4, 20, 100, 500, 2500, 12500, 62500, 312500.

   a. Give the values of g_1, r, and n.

   b. Use the theorem of this lesson to calculate the sum of the series $4 + 20 + 100 + 500 + 2500 + 12500 + 62500 + 312500$.

   c. Use the alternate formula $S_n = \frac{g_1(r^n - 1)}{r - 1}$ to calculate the sum. In what way does this process differ from your work in Part b?

3. a. Write the first 6 terms of the geometric sequence with first term -2 and constant ratio 4.

   b. Evaluate $\sum_{i=1}^{6} (-2)4^{i-1}$ by adding the numbers in Part a.

   c. Evaluate $\sum_{i=1}^{6} (-2)4^{i-1}$ using the theorem in the lesson.

4. Use a formula to find the sum of the geometric series $\frac{2}{3} + \frac{2}{9} + \frac{2}{27} + \frac{2}{81} + \frac{2}{243}$.

5. a. Express the geometric series $10 - 20 + 40 - 80 + 160 - 320 + 640 - 1280 + 2560$ using sigma notation.
   b. Find the sum of the series using a formula from this lesson.

In 6 and 7, evaluate.

6. $\sum_{j=1}^{7} 6(0.5)^{j-1}$

7. $\sum_{k=1}^{18} 10(1.5)^{k-1}$

8. As first prize winner in a lottery, you are offered a million dollars in cash or a prize consisting of one cent on June 1, two cents on June 2, four cents on June 3, and so on, with the amount doubling each day until the end of June. Which prize is more valuable? Justify your choice.

In 9 and 10, refer to Example 4.

9. Assuming no ancestor appears twice in a family tree, how many natural ancestors does a person have
   a. in the 5th generation in the past?
   b. in the past 5 generations in total?

10. How many generations must you go back before you have 10,000 total ancestors, assuming that no one appears twice?

APPLYING THE MATHEMATICS

11. Hector writes $\frac{6\left(1 - \left(\frac{1}{5}\right)^4\right)}{1 - \frac{1}{5}}$ as his first step in using the formula to find the sum of a given geometric series. Write the terms of the series Hector was given.

12. The set of a music show includes a backdrop of 8 nested triangles; four of the triangles are shown here. The smallest triangle (black) is the first built, having a perimeter of 0.75 feet. Each successive triangle has perimeter twice the previous one. What is the sum of the perimeters of the eight nested triangles?

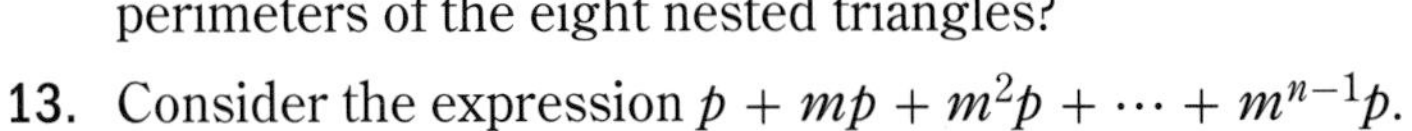

13. Consider the expression $p + mp + m^2p + \cdots + m^{n-1}p$.
    a. Rewrite the expression using sigma notation.
    b. Use a formula from this lesson to write the sum as a single fraction.

14. a. Give an example of a geometric sequence with common ratio 1.
    b. Explain why the formula for finding the sum of the first n terms of a geometric sequence cannot be applied to this sequence.
    c. Give an expression for the sum of the first n terms of your answer to Part a.

15. The sum of the first four terms of a geometric sequence with common ratio 3 is 200. Find an explicit formula for the nth term of this sequence.

16. Suppose a superball bounces to 75% of the height from which it falls. The superball is dropped straight down from 10 feet above the ground. What is the total vertical distance traversed by the ball when it hits the ground for the eighth time? (The bounces are shown spread out to help you visualize the problem. The actual bounces would be closer to vertical.)

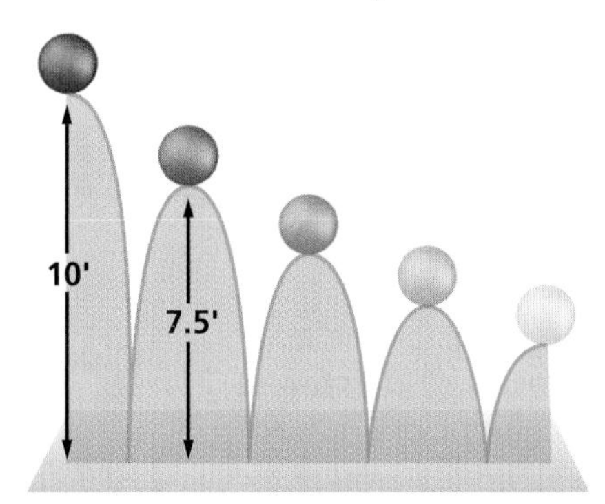

REVIEW

17. Suppose the 99th term of an arithmetic sequence is 200 and the 101st term is 38. Find the 100th term. **(Lesson 8-1)**

18. Evaluate. **(Lesson 8-4)**
    a. $\sum_{n=1}^{5} (2^n) - 3$ b. $\sum_{n=1}^{5} (2^n - 3)$

19. Let $t_n = \frac{3^n}{3^n + 37}$.
    a. Write the first five terms in the sequence generated by this formula.
    b. Does $\lim_{n \to \infty} t_n$ exist? If so, what is it? **(Lessons 8-3, 8-2)**

20. In a laboratory experiment on the growth of insects, there were 52 insects four days after the beginning of the experiment and 93 insects six days after the beginning of the experiment.
    a. Find an exponential model for the data.
    b. Find the initial number of insects.
    c. Predict the number of insects 10 days after the beginning of the experiment.
    d. Predict the number of insects after 100 days. Does the prediction seem realistic? Explain why or why not. **(Lesson 2-5)**

Born yesterday A fruit fly reproduces about 10 days after birth, or longer if the temperature is colder.

EXPLORATION

21. The chess story on pages 529 and 530 gives a volume for a grain of wheat.
    a. Explain how the volume 230 km^3 was obtained.
    b. Kansas is the U.S. state that harvests the most wheat, producing about 20% of the nation's wheat. If 230 cubic kilometers of wheat were spread evenly over Kansas, how many meters deep would the wheat be?

QY ANSWERS

1. -624
2. a. 25; 35 b. 40%

Lesson

8-6 How Much Does a Loan Cost?

Vocabulary

principal
interest rate
term of a loan

▶ **BIG IDEA** The total amounts received or spent from car loans, house mortgages, and other periodic payments are sometimes the sums of finite geometric series.

In a loan, the amount borrowed is the **principal**. To pay back the loan, you must pay back both the principal and some *interest*. The **interest rate** is the percentage of borrowed money that the borrower is charged by the lender for borrowing the money. It is typical for a borrower to have to pay something back to the lender each month. The **term** of the loan is the length of time the person has to pay back the entire loan.

For instance, suppose a person borrows \$6000 with a term of 36 months to buy a used car. If the interest rate is 9% per year, then each month the rate is $\frac{9}{12}$%, which equals $\frac{3}{4}$% or 0.0075. So, the first month, the borrower would owe

$$0.0075 \cdot \$6000 = \$45$$

in interest. But the borrower also has to pay back some of the principal. The lender could say: Pay back $\frac{1}{36}$ of the principal plus \$45, that is, \$166.67 + \$45, or \$211.67. Then the next month, the borrower would have \$5833.33 left in principal, and would pay something less than \$211.67 back. Each month the borrower would pay a little less back than the previous month until, in the last month, there would be only \$166.67 principal left, with an interest payment of about \$1.25.

If the procedure of the preceding paragraph were followed, a borrower would pay back different amounts each month. To simplify the payment process, most agreements between lender and borrower require that the loan be paid back in equal monthly amounts.

Mental Math

Identify the formula for compound interest.

A $A = r(P + 1)^t$

B $A = Pr^t$

C $A = P(r + 1)^t$

D $A = P(t + 1)^r$

How Is the Monthly Payment Calculated?

If the lender had not loaned the money, but invested it immediately at a 9% annual (0.0075 monthly) rate for 36 months, using the compound interest formula $A = P(1 + r)^t$, the principal would grow to

$$6000(1 + 0.0075)^{36} = \$7851.87.$$

So you might think you should just divide \$7851.87 by 36 to get the monthly payment. But this would ignore the fact that the lender can make money from each monthly payment. This makes the calculation a little more complicated.

Let M be the monthly payment. The first payment is paid after 1 month, so it grows to $M(1.0075)^{35}$ by 36 months. The second monthly payment is also M, but it grows to $M(1.0075)^{34}$ by the end of 36 months. This continues until 36 payments have been made. The second-to-last has only one month to grow, so it grows to $M(1.0075)$. The last payment is simply M. The sum of these payments plus the interest the lender can earn is consequently

$$M(1.0075)^{35} + M(1.0075)^{34} + \dots + M(1.0075)^{1} + M.$$

This is a finite geometric series with 36 terms, the first term being M and the constant ratio being 1.0075. Its sum is

$$\frac{M((1.0075)^{36} - 1)}{1.0075 - 1} = \frac{M((1.0075)^{36} - 1)}{0.0075}.$$

This sum must equal what the lender would earn with this money, had none of it been loaned, which we found to be $6000(1.0075)^{36} = \$7851.87$.

$$\frac{M((1.0075)^{36} - 1)}{0.0075} = 6000(1.0075)^{36}$$

The numbers may be complicated, but this is just an equation of the form $aM = b$. Solving for M,

$$M = \frac{6000(1.0075)^{36}(0.0075)}{1.0075^{36} - 1} = \$190.80.$$

QY

QY

Do the calculation for M on your calculator to verify that $M = \$190.80$.

The above analysis can be done with any loan amount P paid back in n equal installments at a rate of r per installment. The result is the following theorem.

Theorem (Periodic Installment Formula for a Loan)

The periodic payment required to pay back a loan of P dollars in n equal installments at a rate of r per installment is $\frac{P(1 + r)^n r}{(1 + r)^n - 1}$.

The spreadsheet below shows how the payment of \$190.80 each month for 36 months pays off a car loan of \$6000 at 9% interest.

◇	A	B	C	D
1	Month	Principal \$6000	Interest at Rate 0.0075	Monthly Payment
2	1	6000.00	45.00	\$190.80
3	2	5854.20	43.91	\$190.80

The key to the spreadsheet is how the principal in month 2 is calculated. The lender adds the interest of 0.75% of \$6000 = \$45.00 to the loan amount of \$6000 for month 1, then subtracts the monthly payment. The result is the principal for month 2.

$$\$6000.00 + \$45.00 - \$190.80 = \$5854.20$$

The borrower then has \$5,854.20 left to pay. The loan charge of 0.75% of \$5854.20 = \$43.91 is added to \$5854.20, and then the payment of \$190.80 is subtracted to obtain the principal for month 3.

$$\$5854.20 + \$43.91 - \$190.80 = \$5707.31$$

These calculations are highlighted in the spreadsheet below where

$$B3 + C3 - D3 = B4.$$

Continuing this process, the principal is lowered each month until the 36th month when the principal plus the interest is \$190.73 and paid off by the final monthly payment. Thus, the borrower makes 35 payments of \$190.80 and a last payment of \$190.73 for a total of \$6868.73 paid.

◇	A	B	C	D
1	Month	Principal \$6000	Interest at Rate 0.0075	Monthly Payment
2	1	\$6,000.00	45.00	\$190.80
3	2	\$5,854.20	43.91	\$190.80
4	3	\$5,707.31	42.80	\$190.80
5	4	\$5,559.31	41.69	\$190.80
6	5	\$5,410.21	40.58	\$190.80
7	6	\$5,259.98	39.45	\$190.80
8	7	\$5,108.63	38.31	\$190.80
9	8	\$4,956.15	37.17	\$190.80
10	9	\$4,802.52	36.02	\$190.80
11	10	\$4,647.74	34.86	\$190.80
12	11	\$4,491.80	33.69	\$190.80
13	12	\$4,334.68	32.51	\$190.80
14	13	\$4,176.39	31.32	\$190.80
15	14	\$4,016.92	30.13	\$190.80
16	15	\$3,856.24	28.92	\$190.80
17	16	\$3,694.37	27.71	\$190.80
18	17	\$3,531.27	26.48	\$190.80
19	18	\$3,366.96	25.25	\$190.80
20	19	\$3,201.41	24.01	\$190.80

◇	A	B	C	D
1	Month	Principal \$6000	Interest at Rate 0.0075	Monthly Payment
⋮	⋮	⋮	⋮	⋮
21	20	\$3,034.62	22.76	\$190.80
22	21	\$2,866.58	21.50	\$190.80
23	22	\$2,697.28	20.23	\$190.80
24	23	\$2,526.71	18.95	\$190.80
25	24	\$2,354.86	17.66	\$190.80
26	25	\$2,181.72	16.36	\$190.80
27	26	\$2,007.28	15.05	\$190.80
28	27	\$1,831.54	13.74	\$190.80
29	28	\$1,654.48	12.41	\$190.80
30	29	\$1,476.08	11.07	\$190.80
31	30	\$1,296.35	9.72	\$190.80
32	31	\$1,115.28	8.36	\$190.80
33	32	\$932.84	7.00	\$190.80
34	33	\$749.04	5.62	\$190.80
35	34	\$563.86	4.23	\$190.80
36	35	\$377.28	2.83	\$190.80
37	36	\$189.31	1.42	\$190.73
38	37	\$0.00	0.00	

Activity

MATERIALS spreadsheet

Step 1 Imagine you are buying a car and have to borrow \$6000 as in the spreadsheet above, but you want to spread payments over 5 years (60 months) rather than 3 years. You go online and find that the monthly payment would be \$124.55. Verify that \$124.55 is the correct monthly payment

a. using a spreadsheet.

b. using the formula.

Step 2 Suppose you can get a car loan to borrow \$15,000 over 3 years at an annual rate of 10%.

a. What is the corresponding monthly interest rate?

b. Show with a spreadsheet that if you paid \$800 a month you would pay off this loan in less than 3 years.

c. Show with a spreadsheet that if you paid \$400 a month you would not be able to pay off this loan in 3 years.

Step 3 Experiment with the spreadsheet from Step 2 to find the amount (to the nearest dollar) that you would need to pay each month in order to pay off the loan in 36 months.

Step 4 Use the installment formula for a loan to compute the amount (to the nearest dollar) that you would need to pay each month in order to pay off the loan described in Step 2 in 36 months.

Mortgages

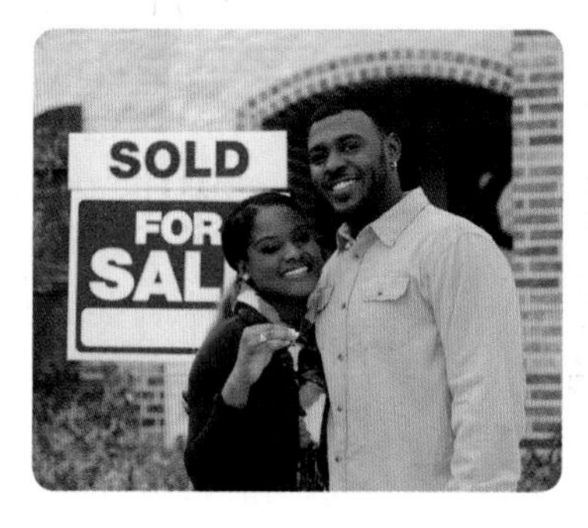

The largest loans most people ever take are *mortgages* for purchasing homes. The amounts to be paid back in a mortgage are calculated just like loans for autos or other purposes. Typical lengths of mortgages are 15, 20, or 30 years, and rates on shorter-term mortgages are typically lower than rates on longer-term mortgages.

GUIDED

Example

A couple takes a \$250,000 mortgage at an annual rate of 5.85% to be paid back in monthly installments over 20 years. What is the monthly payment?

Solution Use the formula. Since 5.85% is the annual interest rate, the monthly rate is __?__. So $r =$ __?__. Because there are monthly payments for 20 years, $n =$ __?__. The principal $P =$ __?__. From the formula, the monthly payment is

$$\frac{P(1+r)^n r}{(1+r)^n - 1} = \frac{\underline{?}\,(1+\underline{?})^{?}\,0.004875}{(1+\underline{?})^{?} - 1}.$$

The monthly payment should be \$1769.51.

Check Use the financial function on a CAS.

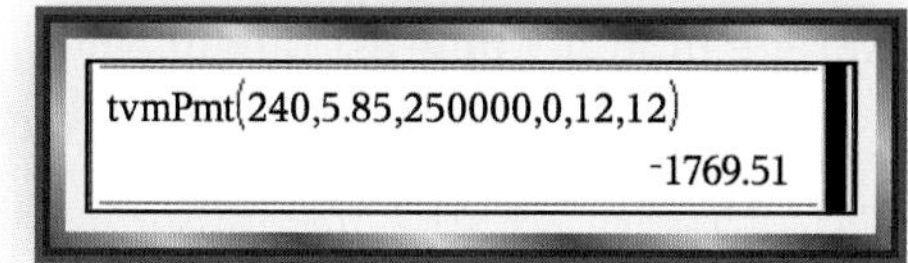

Many CAS, spreadsheets, and graphing calculators have financial functions such as the one shown above; there are also many web pages that will do these calculations. Each has its own requirements for data entry. In the check for the Example, the number 240 stands for the total number of payments, 5.85 for the annual percentage rate, 250000 for the initial loan amount, 0 for the final amount (the loan is exactly paid off), and 12 for both the number of payments per year and the number of compounding periods per year.

Questions

COVERING THE IDEAS

In 1–3, suppose a person borrows \$2500 at 8% interest to be paid back in 36 monthly installments.

1. Identify each.
   a. the term of the loan
   b. the principal of the loan
   c. the monthly interest rate
2. In the formula $M = \frac{P(1+r)^n r}{(1+r)^n - 1}$, what do M, P, r and n represent?
3. a. What amount would an investment of \$2500 grow to if it earned an annual rate of 8% for 3 years, compounded monthly?
   b. What amount would a borrower have to pay back each month in order to pay off this loan?
4. Consider the finite geometric series $M(1.0075)^{47} + M(1.0075)^{46} + \dots + M(1.0075)^1 + M$, similar to one that appears in this lesson.
   a. Explain what the term $M(1.0075)^{47}$ represents.
   b. What are the first term and the common ratio?
   c. Written in reverse order, this sum is a geometric series with what first term and what common ratio?
5. Calculate the monthly payment necessary to pay off a 30-year mortgage for \$175,000 at an annual rate of 6.6%.
6. a. Calculate the monthly payment necessary to pay off a \$200,000 15-year mortgage at 6%.
   b. Calculate the monthly payment necessary to pay off a \$100,000 15-year mortgage at 6%.
   c. Is your answer to Part b *greater than*, *equal to*, or *less than* half of your answer to Part a? Explain why this is so.

APPLYING THE MATHEMATICS

7. Between 1973 and 1990, the monthly national average of interest rates for conventional single-family mortgages in the United States was around 10.43%. The lowest rate since 1990 was in 2003, with an average monthly interest rate of 5.36%.
   a. Calculate the monthly payment necessary to pay off a \$200,000 15-year mortgage at 10.43%.
   b. Calculate the monthly payment necessary to pay off a \$200,000 15-year mortgage at 5.36%.
   c. Is your answer to Part b *greater than*, *equal to*, or *less than* half of your answer to Part a? Explain why this is so.
8. Use algebra to explain why doubling the amount of a loan, while keeping the same interest rate and term, doubles the amount of the required payment.

9. A school takes a \$25,000 loan to buy a grand piano. Suppose the school has the choice to pay this loan back over either 3 years or 4 years at an 8.5% interest rate.
   a. Which term will result in the lower monthly payment and by how much?
   b. Which term will result in the lower total amount spent by the school and by how much?

REVIEW

In 10 and 11, refer to the St. Ives poem on page 500. (Lessons 8-5, 8-2)

10. Suppose that the number 11 replaced the number 7 in the poem.
    a. Write the sequence of man, wives, sacks, cats, and kittens.
    b. Write the series and compute the sum.
11. Suppose a number other than seven appeared in the St. Ives riddle. What would be the largest number that would give a sum of man, wives, sacks, cats, and kittens under 1000?

In 12 and 13, evaluate the expression. (Lesson 8-5)

12. $\sum_{i=1}^{5} 16\left(\frac{1}{4}\right)^{i-1}$

13. $\sum_{i=1}^{9} 8\left(\frac{3}{4}\right)^{i-1}$

14. Consider the arithmetic series $87 + 70 + 53 + \cdots - 729$.
    a. Use $a_n = a_1 + (n - 1)d$ to determine how many terms are in the series.
    b. Write a formula for the value of the series.
    c. Evaluate the series. **(Lesson 8-4)**
15. **True or False** The harmonic sequence diverges to 0. **(Lesson 8-3)**

In 16 and 17, find an exact value without a calculator. (Lesson 4-4)

16. $\sin\left(\frac{5\pi}{6}\right) + \cos\left(\frac{5\pi}{6}\right)$

17. $\sin^2\left(-\frac{\pi}{4}\right) + \cos^2\left(-\frac{\pi}{4}\right)$

EXPLORATION

18. Look on the Internet or at another source for the latest rates on auto loans. Pick a vehicle that you would like to have and suppose that you would be able to make a down payment of 20%, but would have to borrow the remaining 80% of its cost. If you allowed yourself 5 years to pay back the loan, how much would you have to pay back monthly?

QY ANSWER

$\frac{6000\cdot(1.0075)^{36}\cdot 0.0075}{(1.0075)^{36}-1}$ 190.798

Lesson 8-7

Infinite Series

Vocabulary

infinite series
sum of an infinite series, S_∞
convergent series
divergent series

▸ **BIG IDEA** Some infinite series have finite sums; there exists a remarkably simple formula for the sum of an infinite geometric series.

In Lessons 8-4 and 8-5, you found the sum of the terms of a finite series. Now we consider adding an infinite number of terms.

Mental Math

Evaluate.

a. $1 - \frac{1}{2} - \frac{1}{4}$

b. $1 - \frac{1}{2} - \frac{1}{4} - \frac{1}{8}$

c. $1 - \frac{1}{2} - \frac{1}{4} - \frac{1}{8} - \frac{1}{16}$

d. $1 - \frac{1}{2} - \frac{1}{4} - \frac{1}{8} - \frac{1}{16} - \frac{1}{32}$

What is an Infinite Series?

In general, an **infinite series** is an indicated sum that can be expressed in the form

$$\sum_{i=1}^{\infty} a_i = a_1 + a_2 + a_3 + \cdots.$$

An important question about infinite series is which series have finite sums. One way to find out whether an infinite series has a finite sum is to examine the *sequence of partial sums*. The *n*th partial sum of an infinite series is given by $S_n = a_1 + a_2 + \cdots + a_n$. That is,

$$S_1 = a_1 = \sum_{i=1}^{1} a_i$$

$$S_2 = a_1 + a_2 = \sum_{i=1}^{2} a_i$$

$$S_3 = a_1 + a_2 + a_3 = \sum_{i=1}^{3} a_i$$

$$\vdots$$

$$S_n = a_1 + a_2 + \cdots + a_n = \sum_{i=1}^{n} a_i.$$

If the sequence $S_1, S_2, S_3, ..., S_n, ...$ converges, then its limit $\lim_{n\to\infty} S_n$ is called the **sum** of the infinite series. The symbol for this limit is $\mathbf{S_\infty}$. That is,

$$S_\infty = \sum_{i=1}^{\infty} a_i = \lim_{n\to\infty} S_n = \lim_{n\to\infty} \sum_{i=1}^{n} a_i.$$

The infinite series of the positive integers $1 + 2 + 3 + 4 + \cdots$ illustrated at the right is an arithmetic series because it is formed from an arithmetic sequence. Clearly the partial sums S_n have no limit. This is true for all infinite arithmetic series except when $a_i = 0$ for all i.

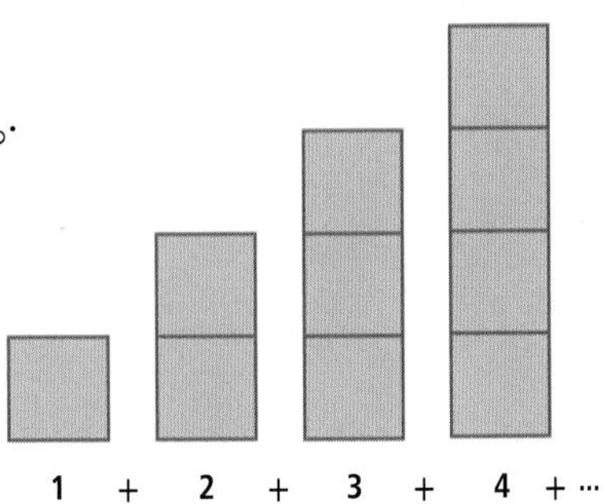

A Finite Sum of Infinitely Many Numbers

You may think that adding an infinite number of terms will always produce a sum that is infinitely large. Activity 1 shows otherwise.

Activity 1

Begin with a unit square as shown at the left below.

Step 1 Examine these three diagrams. The shaded regions represent the sums S_n of the first n terms of the infinite geometric series $\frac{1}{2} + \frac{1}{4} + \frac{1}{8} + \cdots$.

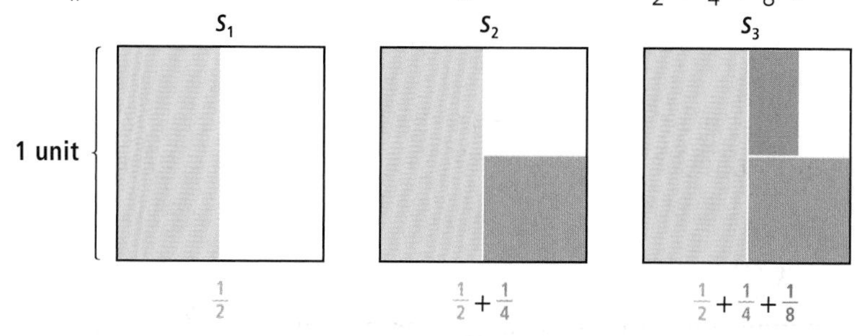

Step 2 Write S_2 as a single fraction.

Step 3 Write S_3 as a single fraction.

Step 4 Draw a figure for S_4 and write S_4 as a single fraction.

Step 5 Find a pattern for S_1 through S_4 and use the pattern to conjecture a formula for S_n.

Step 6 Use your formula to conjecture the value of S_{10}.

As we increase the number of stages, the shaded area is increasing, but you can see that successive shaded regions do not spill out beyond the outer square. The area of the square is larger than the area of the shaded region within it for any particular value of n, but equal to the limit of the areas of the shaded region. We say that

$$\sum_{i=1}^{\infty} \frac{1}{2^i} = \lim_{n \to \infty} \sum_{i=1}^{n} \frac{1}{2^i} = 1.$$

That is, the infinite sum is the limit of the finite sums, which is 1.

Here is a numerical example of an infinite geometric series. Recall that $\frac{7}{9}$ is equivalent to the repeating decimal 0.777777… . This decimal can be written as an infinite series.

$$\frac{7}{9} = 0.\overline{7} = \frac{7}{10} + \frac{7}{10^2} + \frac{7}{10^3} + \frac{7}{10^4} + \cdots$$

This is another illustration of the fact that the sum of an infinite series can be finite, in this case $\frac{7}{9}$.

As with sequences, if $\lim_{n \to \infty} S_n$ exists, the series is called **convergent**. If the limit does not exist, the series is **divergent**.

Example 1

Consider the infinite sequence $\frac{4}{3}, \frac{8}{3}, \frac{16}{3}, \ldots, \frac{2^{n+1}}{3}, \ldots$ and its associated series $\frac{4}{3} + \frac{8}{3} + \frac{16}{3} + \cdots + \frac{2^{n+1}}{3} + \cdots$.

a. What type of sequence is this? Justify your answer.

b. Find the first five partial sums of the series.

c. Does the infinite series seem to converge?

Solution

a. The sequence is geometric because each term is 2 times the previous term. There is a common ratio $r = 2$.

b. $S_1 = \frac{4}{3}$

$S_2 = \frac{4}{3} + \frac{8}{3} = \frac{12}{3}$

$S_3 = \frac{4}{3} + \frac{8}{3} + \frac{16}{3} = \frac{28}{3}$

$S_4 = \frac{4}{3} + \frac{8}{3} + \frac{16}{3} + \frac{32}{3} = \frac{60}{3}$

$S_5 = \frac{4}{3} + \frac{8}{3} + \frac{16}{3} + \frac{32}{3} + \frac{64}{3} = \frac{124}{3}$

c. The sequence of partial sums appears to increase without a limit. So the infinite series does not converge.

Activity 2

Consider the infinite geometric series $\sum_{i=1}^{\infty} 500\left(\frac{4}{5}\right)^{i-1} = 500 + 400 + 320 + \cdots$

Step 1 Find S_1, S_2, S_3, S_4 and S_5, the first five partial sums. Give any noninteger answers as fractions.

n	Partial Sums S_n
1	$S_1 = 500$
2	$S_2 = 500 + 400 = 900$
3	$S_3 = $?
4	$S_4 = $?
5	$S_5 = $?

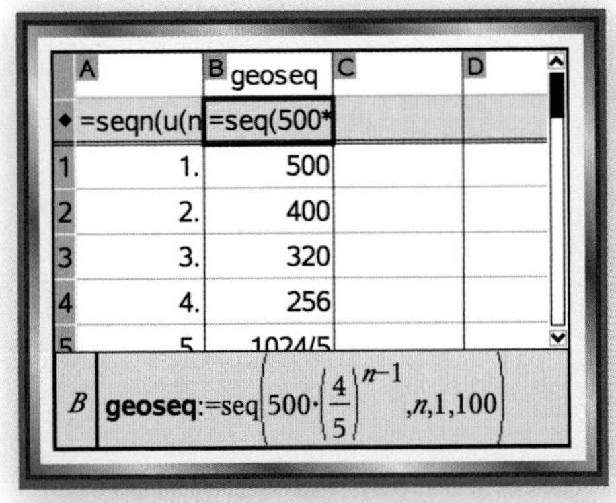

	A	B geoseq	C	D
◆	=seqn(u(n	=seq(500*		
1	1.	500		
2	2.	400		
3	3.	320		
4	4.	256		
5	5.	1024/5		

B **geoseq**:=seq$\left(500\cdot\left(\frac{4}{5}\right)^{n-1},n,1,100\right)$

Step 2 Do you think the sequence of partial sums has a limit? Explain your answer.

Step 3 On your calculator, generate the first 100 terms of the sequence of partial sums in a list. The cumulative sum `cumsum` command can be used to compute the partial sums of a list. In the screenshot shown at the right, column C was created using `cumsum(B)`. Column D contains the decimal approximations for the values in column C.

	A	B geoseq	C partsum	D
◆	=seqn(u(n	=seq(50(	=cumsum	=approx
1	1.	500	500	500.
2	2.	400	900	900.
3	3.	320	1220	1220.
4	4.	256	1476	1476.
5	5.	1024/5	8404/5	1680.8
6	6.	4096/25	46116/25	1844.64

C **partsum**:=cumsum(b)

Step 4 Look at the values of the partial sums. Do they appear to approach a limit? Graph the partial sums (n, S_n). Does the graph appear to have a horizontal asymptote? Do you think it is possible to find a value for n such that $S_n > 2500$?

A Formula for the Sum of a Convergent Infinite Geometric Series

You have now seen one infinite geometric series that does not converge (in Example 1) and three that do (the series in Activities 1 and 2 and the series for $0.\overline{7}$). How can you determine whether an infinite geometric series converges without adding terms? The key is the common ratio r. Consider the infinite geometric series

$$g_1 + g_1 r + g_1 r^2 + \cdots + g_1 r^{n-1} + \cdots.$$

If r is between -1 and 1, then each new term $g_1 r^{n-1}$ gets closer and closer to 0. However, if $r > 1$ or $r < -1$, then each successive term gets farther from 0. This suggests that the convergence or divergence of the geometric series depends on the value of the common ratio r.

Theorem (Sum of an Infinite Geometric Series)

Consider the infinite geometric series

$S_n = g_1 + g_1 r + g_1 r^2 + \cdots + g_1 r^{n-1} + \cdots$, with $g_1 \neq 0$.

a. If $-1 < r < 1$, the series converges and

$S_\infty = \lim_{n \to \infty} S_n = \frac{g_1}{1 - r}$.

b. If $r > 1$ or $r < -1$, the series diverges.

Proof From the formula for the sum of a finite geometric series,

$$S_n = \frac{g_1(1 - r^n)}{1 - r}.$$

$$S_\infty = \lim_{n \to \infty} S_n = \lim_{n \to \infty} \frac{g_1(1 - r^n)}{1 - r}$$

Since g_1 and $1 - r$ are independent of n, by Limit Property (4) in Lesson 8-3, $\lim_{n \to \infty} S_n = \frac{g_1}{1 - r} \cdot \lim_{n \to \infty}(1 - r^n)$. Now, when $-1 < r < 1$, $\lim_{n \to \infty} r^n = 0$. Then $\lim_{n \to \infty} (1 - r^n) = 1$. So, when $-1 < r < 1$,

$$\lim_{n \to \infty} S_n = \frac{g_1}{1 - r} \cdot 1 = \frac{g_1}{1 - r}.$$

However, when $r > 1$ or $r < -1$, the term $g_1 r^n$ is farther from zero for every successive value of n, and the infinite series is divergent.

GUIDED

Example 2

Which of the following could be convergent geometric series? For those that could be, give the limit.

a. $\frac{5}{3}, \frac{5}{9}, \frac{5}{27}, \frac{5}{81}, \ldots$

b. $\frac{5}{3} + \frac{5}{9} + \frac{5}{27} + \frac{5}{81} + \cdots$

c. $640 + 160 + 40 + 10 + \cdots$

d. $\frac{1}{2} + \frac{5}{2} + \frac{9}{2} + \frac{13}{2} + \cdots$

(continued on next page)

Solution

a. This is not a series because the terms are not added. It is a sequence.

b. This could be a geometric series with r = __?__. By the theorem above, the series converges, and the sum is __?__.

c. This could be a geometric series with r = __?__. The series __?__, and the limit is __?__.

d. This could be an arithmetic series with a constant difference of $\frac{4}{2} = 2$. Because it is arithmetic, the series diverges.

Finding a Simple Fraction Equal to a Repeating Decimal

With the formula for the sum of an infinite geometric series, you can find a simple fraction equal to any repeating decimal.

Example 3

Find a simple fraction equal to $39.4\overline{65} = 39.4656565...$, where the 65 repeats forever.

Solution Separate the repetend (the repeating part) from the rest of the decimal.

$$39.4\overline{65} = 39.4 + 0.0\overline{65}$$

Convert the finite decimal into a simple fraction.

$$39.4 = \frac{394}{10}$$

Write the repeating decimal as an infinite geometric series.

$$0.0\overline{65} = 0.065 + 0.00065 + 0.0000065 + \cdots$$

In the series, notice that the first term g_1 is 0.065 and the constant ratio r is $\frac{1}{100}$ or 0.01. Apply the formula for the sum of an infinite geometric series and convert the sum to a simple fraction.

$$0.0\overline{65} = \frac{0.065}{1 - 0.01} = \frac{0.065}{0.99} = \frac{65}{990}$$

Bring together the two parts of the original decimal.

$$39.4\overline{65} = \frac{394}{10} + \frac{65}{990} = \frac{39071}{990} = 39\,\frac{461}{990}$$

 QY

▶ QY

Consider the infinite repeating decimal $0.\overline{27}$.

a. Write this as an infinite series. Give the decimals for S_1, S_2, and S_3.

b. What simple fraction does $\lim_{n\to\infty} S_n$ equal?

Questions

COVERING THE IDEAS

1. What is the difference between an infinite sequence and an infinite series?

2. For the series $S = \frac{3}{5} + \frac{9}{10} + \frac{27}{20} + \frac{81}{40} + \cdots$,

   a. find S_1, S_2, S_3, and S_4.

   b. find S_∞, if it exists.

3. Consider the infinite geometric series with first term $\frac{8}{5}$ and common ratio $\frac{1}{2}$.
   a. Write the first three partial sums of the series.
   b. Find the sum of the infinite series.

4. The screenshot at the right shows term numbers of a sequence in column A, the terms of the sequence in column B, and their partial sums in column C.
   a. Find an explicit formula for the sequence.
   b. The value in the 6th row of column C is missing. Find this value and explain what it means.
   c. Does the infinite series $5 + 8 + 11 + 14 + \cdots$ converge? Explain your answer.

	A	B	C	D	E	F	G
◆							
1	1	5	5				
2	2	8	13				
3	3	11	24				
4	4	14	38				
5	5	17	55				
6	6	20					
C6							

5. Consider $\sum_{k=1}^{\infty} \frac{1}{k^4}$. Enter the sequence into a spreadsheet on your calculator and compute the first 100 partial sums. Does the sequence of partial sums appear to converge? If yes, give what appears to be the limit, rounded to 6 decimal places.

In 6–11, an infinite series is given. Find the constant ratio *r* if the series is geometric. State whether or not the series is convergent. If the series is convergent, give its sum.

6. $3 + \frac{9}{4} + \frac{27}{16} + \frac{81}{64} + \cdots$

7. $100 + 25 + 6.25 + \cdots$

8. $0.\overline{21} = 0.2121212121\ldots$

9. $4.00\overline{8}$

10. $\sum_{j=0}^{\infty} 7 \cdot \left(\frac{5}{9}\right)^j$

11. $2 - 4 + 8 - 16 + 32 - \cdots$

APPLYING THE MATHEMATICS

12. Consider the sequence $\begin{cases} t_1 = 40 \\ t_n = \frac{2}{5}t_{n-1}, \text{ for } n > 1 \end{cases}$.
    a. Is the sequence arithmetic, geometric, or neither?
    b. Does the sequence t converge? If it converges, give its limit.
    c. Does the infinite series $\sum_{n=1}^{\infty} t_n$ converge? If it does, give its limit.

13. At a construction site, a pile driver drives a 3-meter post into the ground. The first hit drives the post in 60 cm, the second hit drives the post in 40 cm further, and successive distances driven form a geometric sequence.
    a. How far will the post be driven into the ground if the pile driver is allowed to run forever?
    b. What percent of the total distance driven is reached after 20 hits?

14. When the movie *The Dark Knight* was released in 2008, a news report stated that 62% of the people who saw the movie once in the theater planned to see it again. Suppose that 62% of those people see it a third time, and the pattern continues infinitely. How many total tickets will the movie sell, assuming 10,000,000 first-time viewers? (Of course, in reality a person cannot see a movie an infinite number of times.)

15. In a circus, a jumping flea is at the center of a circular ring of radius 1 meter. Suppose the flea jumps along a radius toward the edge of the ring, but that each jump is half the length of the previous jump (as the flea gets tired). Suppose the flea's first jump is $\frac{1}{2}$ m long.

a. Write a sequence showing the length of the first five jumps.

b. What was the length of the flea's eighth jump?

c. How far was the flea from the center after the eighth jump?

d. What is the smallest number of jumps that will get the flea is within 0.001 m of the edge of the ring?

e. Will the flea ever reach the edge of the ring? Justify your answer.

16. a. Does the series $1 + 1 + 1 + 1 + 1 + \cdots$ converge? Explain your answer.

b. Does the series $1 - 1 + 1 - 1 + 1 - 1 + \cdots$ converge? Justify your answer.

17. Consider the harmonic series $\frac{1}{1} + \frac{1}{2} + \frac{1}{3} + \frac{1}{4} + \cdots$. A graph of the first 50 partial sums is shown at the right. Although the graph appears to be increasing slowly enough that it would approach a horizontal asymptote to the right of 50, the series diverges. Use a spreadsheet on your calculator to find

a. the smallest value of n so that $S_n > 2$.

b. the smallest value of n such that $S_n > 5$.

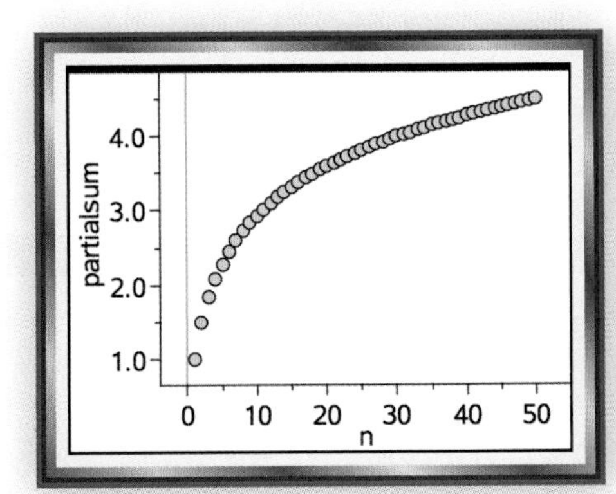

18. Find the sum of the series $S = p + 2p^2 + 3p^3 + 4p^4 + \cdots$ where $-1 < p < 1$. (Hint: Consider $S - Sp$.)

REVIEW

19. Calculate the monthly payment necessary to pay off a 20-year mortgage for \$600,000 at an annual rate of 7.11%. **(Lesson 8-6)**

20. What does doubling the interest rate do to the monthly payment on a fixed rate mortgage? Be specific. **(Lesson 8-6)**

21. Use a formula to find the sum of the series $\frac{4}{7} + \frac{4}{49} + \frac{4}{343} + \frac{4}{2401} + \frac{4}{16807}$. **(Lesson 8-5)**

22. Find the sum of the integers between 500 and 1000 that are divisible by 3. **(Lesson 8-4)**

23. **True or False** When a person has repeatedly flipped tails with a fair coin, the person is now more likely to flip heads. Justify your answer. **(Lesson 6-8)**

EXPLORATION

24.

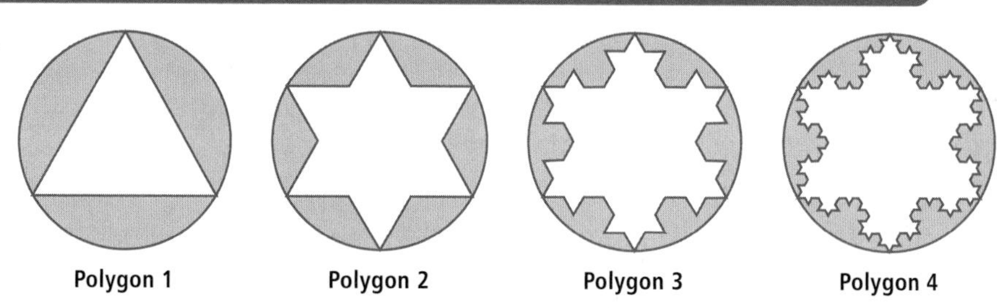

The pictures above show the first four of an infinite sequence of polygons, starting with an equilateral triangle inscribed in a circle. Each new polygon is formed by replacing the middle third of each side by two segments that have the same length as each third.

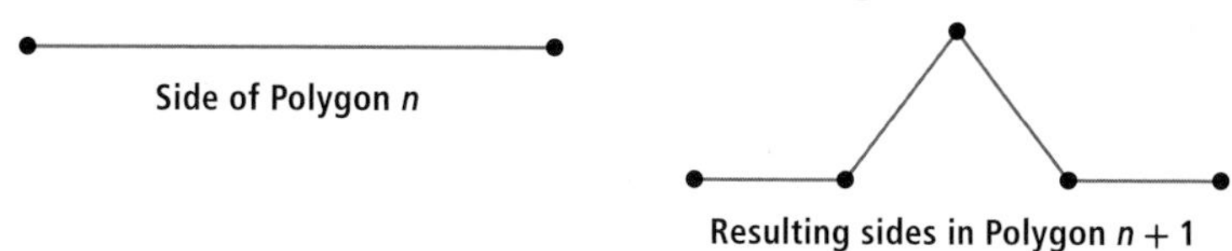

As n increases, the area of Polygon n increases but remains less than the area of the circle. The area of Polygon 1 is A.

a. Explain why
$\lim_{n \to \infty}$ (Area of Polygon n) $= A + A\left(\frac{1}{3} + \frac{4}{27} + \frac{16}{243} + \cdots\right)$.

b. Let S_n be the area of Polygon n. Find the limit of S_n.

QY ANSWERS

a. $0.27 + 0.0027 + 0.000027 + \cdots$;
$S_1 = 0.27$;
$S_2 = 0.2727$;
$S_3 = 0.272727$

b. $\frac{27}{99} = \frac{3}{11}$

Chapter 8 Projects

1 The Four Numbers Game

A mathematical game as taught by Paul Sally, a professor of mathematics at the University of Chicago, is played as follows.

Step 1 Draw a square and put an integer at each corner as shown at the right.

Step 2 Now compute the absolute differences between pairs of corners and make a new square from these numbers. For instance, with the numbers above, the new square has vertices 1, 7, 1, 7.

Step 3 Repeat Step 2 with the new square until you get a square with all zeros. At this point the game is over and you count the number of new squares made, not including the first one. The goal of the Four Numbers game is to choose original numbers that require the most new squares before the game is over. The number of new squares is your score. (In the example above, the score is 3.)

Play the game with a classmate a few times to get the hang of it.

It turns out that *tribonacci numbers* take a very long time to converge in the Four Numbers Game. *Tribonacci numbers* are a generalization of the Fibonacci numbers defined by the recurrence equation for all $n \geq 4$,
$T_n = T_{n-1} + T_{n-2} + T_{n-3}$ where $T_1 = 1$, $T_2 = 1$, and $T_3 = 2$.

a. Generate the first 15 tribonacci numbers.

b. Choose 4 consecutive tribonacci numbers and play the Four Numbers Game again with a classmate. What is your score?

2 A Test for Convergence

If an infinite series containing only positive terms is to converge, it is necessary that the terms approach 0 as n approaches infinity. However, this is not sufficient; the text points out that the harmonic series diverges even though the terms approach 0.

a. Explain why the following series diverges.

$$A = 1 + \frac{1}{2} + \underbrace{\frac{1}{4} + \frac{1}{4}}_{\text{2 terms}} + \underbrace{\frac{1}{8} + \frac{1}{8} + \frac{1}{8} + \frac{1}{8}}_{\text{4 terms}} + \underbrace{\frac{1}{16} + \frac{1}{16} + \frac{1}{16} + \frac{1}{16} + \frac{1}{16} + \frac{1}{16} + \frac{1}{16} + \frac{1}{16}}_{\text{8 terms}} + \cdots$$

b. Show that each term of the harmonic series $H = 1 + \frac{1}{2} + \frac{1}{3} + \frac{1}{4} + \frac{1}{5} + \frac{1}{6} + \cdots$ is greater than or equal to the corresponding term of series A. Explain how this result shows that the harmonic series diverges.

c. The following variation of the harmonic series converges:
$B = 1 - \frac{1}{2} + \frac{1}{3} - \frac{1}{4} + \cdots + (-1)^{n+1}\frac{1}{n} + \cdots$
Use a calculator or computer to approximate B to six decimal places.

d. If every term of the harmonic series that contains a 9 is deleted (such as $\frac{1}{9}, \frac{1}{95}, \frac{1}{409}$, etc.), the resulting series converges. Use a calculator or computer to approximate the sum of this series to four decimal places.

3 Geometric Means

Find at least five theorems about geometric figures that mention geometric means. Draw an accurate figure for each theorem.

4 How Do Credit Cards Work?

Many incoming college freshmen are bombarded with credit cards offering "0% APR for the first six months" with fine print similar to the following (taken from an actual credit card web site for college students):

*After the promotional period ends, your standard purchase APR will be applied to any unpaid balance transfer and promotional purchase balances, and your standard cash advance APR will be applied to any unpaid cash advance balances. As of March 21, 2008 the standard variable purchase APR is 12.24%, and the standard variable cash advance APR is 20.24%. However, if you are in default under any Card Agreement that you have with us, we may automatically increase the rate on all balances (including any promotional balances) to a variable default rate of 29.24%. The minimum finance charge: $0.50. The minimum monthly payment is $5.00 or 3.0% of the outstanding balance, whichever is higher. Subject to credit approval. Additional terms and conditions apply.

a. If you buy a $1000 laptop immediately after signing up for the card, what is the outstanding balance on the card after the first six months, assuming no payments have been made?

b. What is the minimum monthly payment at the end of the seventh month?

c. If you miss two payments and are deemed "in default," what is your new interest rate?

d. What is the minimum finance charge?

e. Use a spreadsheet to organize by month (column 1) your balance (column 2), charges by month (column 3), and minimum payments required (column 4). Assume only the minimum payment due is what is paid on the balance and charges each month.

f. How much of the balance is paid off after 100 months? Sum the payments column for the first 100 months. How much has your $1000 laptop cost so far?

g. Paying the minimum payment due, how long does it take to pay off the laptop and how much is paid in total?

h. Repeat Parts e–g, if in the 9th month you accidentally miss a payment ($0) and your interest rate is bumped up to the "default rate."

i. Add a column to your spreadsheet computing the cumulative payments made as the months go by. Graph remaining balance versus total paid each month. What does the graph suggest about the situation?

5 Recursively Defined Curves

Many interesting curves can be created by recursive definitions. For instance, begin with a square with a given side, say ————. At each stage, replace each side of the square with this shape:

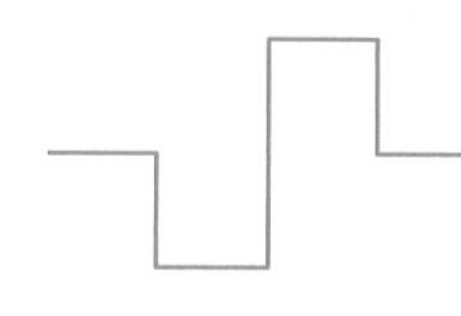

The original square and subsequent stages leading to a limit curve sometimes called the *dragon curve* are shown below.

a. Determine the perimeter and area of the nth curve in the sequence and the dimension of the dragon curve. To determine the dimension, you will need to learn something about *fractals*.

b. Find out about other curves that are defined recursively. Some famous ones are called snowflake curves and space-filling curves. Repeat Part a for one other such curve.

c. Design your own recursively-defined curve.

Chapter 8 Summary and Vocabulary

- A sequence is a function whose domain is the set of consecutive integers greater than or equal to a given number k, where k is most often 1. An **explicit formula** allows you to find any specified **term** of a sequence efficiently. A **recursive formula** describes each term in relation to previous ones and is often more efficient for generating the terms.

- Two important kinds of sequences are **arithmetic** and **geometric** sequences. In an arithmetic sequence, the difference between successive terms is constant. In a geometric sequence, the ratio of successive terms is constant. If an arithmetic sequence has first term a_1 and constant difference d, the nth term $a_n = a_1 + (n - 1)d$. If a geometric sequence has first term a_1 and constant ratio r, the nth term $g_n = g_1 r^{n-1}$.

- An arithmetic sequence is the best model for a real situation when a quantity is increasing or decreasing by a constant or nearly constant amount. A geometric sequence is the best model when a quantity is increasing or decreasing at a nearly constant rate.

  Some examples of geometric sequences and series are:

  (1) the successive powers of any number;

  (2) the repeating parts in an infinite repeating decimal: For $0.\overline{36}$, 0.36, 0.0036, 0.000036, ...;

  (3) the successive values $A(1 + r)^t$ of an investment A at a rate r compounded when the values of t are consecutive integers or vary in an arithmetic sequence.

- A **series** is the indicated sum of the terms of a sequence. Explicit formulas for sums of finite arithmetic and geometric series exist. The sum S_n of the first n terms of an arithmetic series is given by $S_n = \frac{n}{2}(2a_1 + (n - 1)d)$ or $S_n = \frac{n}{2}(a_1 + a_n)$. For the sum of the first n terms of a geometric series, $S_n = \frac{g_1(1 - r^n)}{1 - r}$. Using this last formula with the investment sequence (3) above gives the formula $M = \frac{P(1 + r)^n r}{(1 + r)^n - 1}$ for the amount M required to be paid back in each of n installments on a loan of P dollars at rate r in the installment time.

Vocabulary

Lesson 8-1
Fibonacci sequence
term of a sequence
*harmonic sequence
*explicit formula for a sequence
sequence
initial condition
recurrence relation
*recursive formula for a sequence
*arithmetic sequence, linear sequence

Lesson 8-2
geometric sequence, exponential sequence
arithmetic mean
geometric mean

Lesson 8-3
end behavior of a sequence
limit of a sequence c_n as n approaches infinity, $\lim_{n \to \infty} c_n$
diverge, divergent
converge, convergent
alternating harmonic sequence

Lesson 8-4
*series
value, sum of a series
finite series
*arithmetic series
infinite series

- Some infinite sequences **converge** to a limit, while others **diverge**. A computer or calculator is useful for deciding whether or not a sequence is convergent, but it is not an infallible tool. Limits of some sequences, such as those formed by adding, subtracting, multiplying, or dividing two convergent sequences, can be found by applying properties of limits.

- The **sum of an infinite series** is defined as the limit of a sequence $S_1, S_2, \ldots, S_n$ of **partial sums** as n approaches infinity. Some infinite series have a limit; some do not. For an infinite geometric series with first term g_1 and constant ratio r, the limit exists when $-1 < r < 1$. Then $S_\infty = \frac{g_1}{1-r}$. The series is said to **converge** in this case. It **diverges** when $r > 1$ or $r < -1$.

Vocabulary

Lesson 8-5
*geometric series
nth partial sum, S_n

Lesson 8-6
principal
interest rate
term of a loan

Lesson 8-7
infinite series
sum of an infinite series, S_∞
convergent series
divergent series

Theorems and Properties

Theorem (Formulas for Arithmetic Sequences) (p. 505)
Theorem (Formulas for Geometric Sequences) (p. 509)
Limit of Harmonic Sequence Property (p. 516)
Properties of Limits and Constants (p. 517)
Limit of a Sum Property (p. 518)
Limit of a Difference Property (p. 518)
Limit of a Product Property (p. 518)
Limit of a Quotient Property (p. 518)
Theorem (Sum of an Arithmetic Series) (p. 524)
Theorem (Sum of a Finite Geometric Series) (p. 530)
Theorem (Periodic Installment Formula for a Loan) (p. 537)
Theorem (Sum of an Infinite Geometric Series) (p. 545)

Chapter 8 Self-Test

Take this test as you would take a test in class. You will need a calculator. Then use the Selected Answers section in the back of the book to check your work.

In 1 and 2, consider this formula for a sequence:
$\begin{cases} v_1 = 0 \\ v_n = v_{n-1} + n^2, \text{ for all integers } n > 1 \end{cases}$.

1. Find the first four terms of the sequence.
2. Is the sequence arithmetic, geometric, or neither? Explain your reasoning.
3. The first term of an arithmetic sequence is -3 and the constant difference is d. Find the 12th term.

In 4 and 5, consider the sequence t which begins $t_1 = 8$, and $t_2 = 24$.

4. Give the next 3 terms
   a. if t is an arithmetic sequence.
   b. if t is a geometric sequence.
5. a. Write an explicit formula for t_n if t is an arithmetic sequence.
   b. Write a recursive formula for t if t is a geometric sequence.

In 6 and 7, tell whether the sequence has a limit, and if it does, determine the limit.

6. the sequence g with $g_n = \frac{27}{n^3}$
7. the arithmetic sequence 213, 206, 199, 192, …
8. When $S_n = \sum_{k=1}^{n} 5(2)^{k-1}$, find the value of S_4.
9. Find an explicit formula for the nth term of the sequence 2, $7r$, $12r^2$, $17r^3$,… .
10. Does the infinite geometric series with first term 5 and common ratio $\frac{1}{3}$ converge? If the series is convergent, give its sum.
11. Rewrite $5.4\overline{321}$ as the sum of a finite decimal and a geometric series. Explain how you know the series converges.
12. Lester began running every day for exercise. The first day he ran 0.5 mile. The 10th day he ran 5 miles. If each day he increased his distance by a constant amount, how many miles did he run in all 10 days combined?
13. Calculate the monthly payment necessary to pay off a $45,000 college loan when given an annual rate of 6% and 10 years to make payments.
14. Suppose that a particular computer loses 50% of its value each year, and has an original value of $2,600. In what year will the value of the computer first become $100 or less?
15. Below is a scatterplot of a sequence d defined by $d_n = \frac{2n + (-1)^n}{n}$.

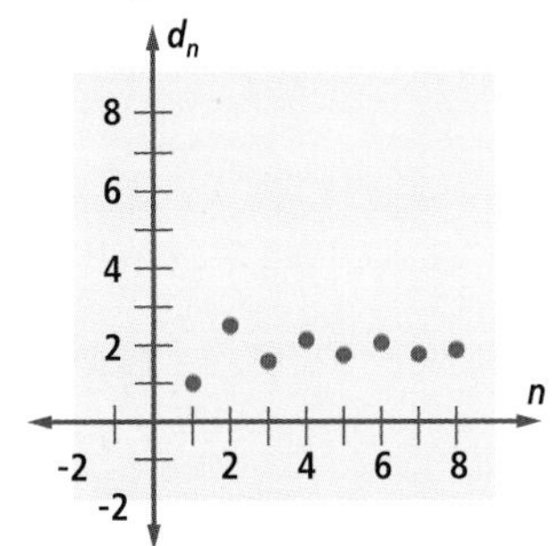

   a. Express the end behavior of the sequence using limit notation.
   b. Is d convergent? If so, find the limit.
   c. Explain your answer to Part b.
16. Consider the integers from 100 to n.
   a. Write the sum of the integers using sigma notation.
   b. Find the sum.

Chapter 8 Chapter Review

SKILLS
PROPERTIES
USES
REPRESENTATIONS

SKILLS Procedures used to get answers

OBJECTIVE A Find terms of sequences given explicit or recursive formulas for them. (Lessons 8-1, 8-2)

In 1 and 2, a sequence is defined. Find the first 5 terms.

1. $\begin{cases} B_1 = 6 \\ B_n = 5B_{n-1}, n > 1 \end{cases}$

2. $\begin{cases} A_1 = -4 \\ A_n = A_{n-1} + 2n, n \geq 2 \end{cases}$

In 3 and 4, a sequence is defined. Find the 5th term and the 10th term of the sequence.

3. $t_n = 2 \cdot 3^{n-1}$

4. $q_n = -9 + 2(n - 3)$

OBJECTIVE B Find explicit or recursive formulas for the nth term of an arithmetic or geometric sequence. (Lessons 8-1, 8-2)

5. The first four terms of an arithmetic sequence are 96, 77, 58, and 39. Find
   a. the nth term.
   b. the 50th term.

6. The first three terms of a geometric sequence are 36, -72, and 144. Find
   a. the nth term.
   b. the 15th term.

7. Give an explicit formula for the sequence defined by
$\begin{cases} r_1 = 11000 \\ r_n = 0.7r_{n-1}, \text{ for } n > 1 \end{cases}$.

8. Give a recursive formula for a sequence defined by $A_n = 64 - 6n$.

In 9 and 10, the 3rd term of a sequence b is 12 and the 5th term is 24. Find the 4th term and a formula for the nth term if the sequence is

9. arithmetic.

10. geometric.

OBJECTIVE C Evaluate arithmetic or geometric series. (Lessons 8-4, 8-5)

In 11–14, an arithmetic or geometric series is given.

a. Write the series using sigma notation.

b. Evaluate it.

11. $59 + 75 + 91 + 107 + \cdots + 171$

12. $(4a - b) + (3a + 2b) + (2a + 5b) + \cdots + (20b - 3a)$

13. $1 + 4 + 16 + 64 + \cdots + 4^{11}$

14. $1 + r + r^2 + \cdots + r^{12}$

15. Let $s_n = \sum_{k=1}^{n} (3k - 11)$. Evaluate s_{81}.

16. Let $s_n = \sum_{k=1}^{n} 10(0.7)^{k-1}$. Evaluate s_{14}.

PROPERTIES Principles behind the mathematics

OBJECTIVE D Determine whether a sequence is arithmetic or geometric. (Lessons 8-1, 8-2)

In 17–22, classify the sequence as possibly arithmetic, definitely arithmetic, possibly geometric, definitely geometric, or definitely neither.

17. 7, 20, 33, 46, ...

18. 33, 39, 46, 54, ...

19. $-3p, -p - q, -2q + p, -3q + 3p, \ldots$

20. $\begin{cases} a_1 = -3 \\ a_n = (a_{n-1})^3 - 1, \text{ for } n > 1 \end{cases}$

21. $t_n = -7(-0.8)^n$

22. a sequence in which each term after the first is 2 greater than the previous term

OBJECTIVE E Determine the end behavior of certain sequences. (Lesson 8-3)

In 23–26, decide whether the sequence has a limit, and if it does, determine the limit.

23. the geometric sequence 75, 62.5, 52.08, 43.40, ...

24. the arithmetic sequence with first term 75 and constant difference -1

25. the sequence g with $g_n = \frac{9}{n}$

26. the sequence r defined by
$$\begin{cases} r_1 = 1 \\ r_n = (-2)^n\, r_{n-1}, \text{ for } n > 1 \end{cases}$$

27. Describe the end behavior of $-\frac{3}{32}, -\frac{17}{125}, -\frac{11}{16}, \ldots, \frac{5n+7}{3n-131}, \ldots$ using limit notation.

28. Find $\lim\limits_{n \to \infty} \frac{n^2+9}{10n}$, if it exists.

OBJECTIVE F Tell whether an infinite series converges. If it does, give a limit. (Lesson 8-7)

In 29 and 30, does the series converge? If the series is convergent, give its sum.

29. the infinite geometric series with first term 3 and common ratio 2

30. the infinite arithmetic series with first term 3 and common ratio 2

31. How can you to tell whether an infinite geometric series converges?

32. Evaluate $\sum\limits_{k=1}^{\infty} 10(0.7)^{k-1}$.

33. Find the simple fraction in lowest terms equal to $0.2\overline{81}$.

USES Applications of mathematics in real-world situations

OBJECTIVE G Use arithmetic sequences and series in real-world situations. (Lessons 8-1, 8-4)

34. Suppose you have \$850 from a graduation party. Each week after that, your allowance adds an additional \$20 per week of savings. How much total savings will you have at the end of the 12th week?

35. A theater has 15 seats in the first row and 3 more seats in each subsequent row. If the last row has 27 seats, how many rows are in the theater?

In 36 and 37, suppose that a display of cans in a supermarket is built with one can on top, two cans in the next row, and one more can in each succeeding row.

36. If there are 15 rows of cans, how many cans are in the display?

37. If 360 cans are available to be displayed, how many rows are needed, and how many cans will be left over?

38. In training for a marathon, an athlete runs 7000 meters on the first day, 7600 meters on the next day, 8200 meters the third day, and so on, each day running 600 meters more than on the previous day. How far will the athlete have run in all at the end of thirty days?

OBJECTIVE H Use geometric sequences and series in real-world situations. (Lessons 8-2, 8-5, 8-7)

39. A certain car loses 21% of its value each year, with an original cost of \$24,800.

a. Find the value of the car after one year.

b. Write an explicit formula for the value of the car after n years.

40. Suppose that on each bounce, a superball bounces to 72% of the height from which it falls. What is the total vertical distance traveled by the superball, dropped from 16 feet above the ground, when it hits the ground for the 7th time?

41. Consider a pile driver driving a 3-meter pile into the ground. The first hit drives the pile in 50 cm, the second hit drives the pile in 40 cm further, and successive distances driven form a geometric sequence.

a. How far will the pile be driven into the ground if the pile driver is allowed to run forever?

b. What percent of the total distance driven into the ground is reached after 25 hits?

42. In a certain housing complex, rents are increased by 6% every year. Consider a family that has been living at this complex for 17 years, and assume that they paid \$550 per month in the first year.

a. Determine the rent they are paying this year.

b. At the end of this year, how much will the family have paid during their 17 years of residency? (Hint: Consider the total rent the family has paid in each complete year.)

OBJECTIVE I **Determine the total payment made and installments for car loans, mortgage payments, and other long term periodic payments. (Lesson 8-6)**

In 43 and 44, suppose someone in your class borrows \$8000 for a used car at 8.5% interest to be paid back in 30 monthly installments.

43. a. What is the term of the loan?

b. What is the principal of the loan?

c. What is the monthly interest rate?

44. a. How much would the bank have made if it had not lent the borrower the \$8000 and instead let the money compound at the same interest monthly?

b. What will the monthly payment be?

45. Calculate the monthly payment necessary to pay off a 20-year mortgage for \$400,000 at an annual rate of 5.5%.

46. Consider two 30-year mortgages of \$160,000 and \$320,000 at 6.7%. Explain the relationship between the two monthly payments and why this is so.

REPRESENTATIONS Pictures, graphs, or objects that illustrate concepts

OBJECTIVE J **Graph and interpret graphs of sequences. (Lessons 8-1, 8-3)**

47. Explain why the points on the graph of an arithmetic sequence lie on a line.

In 48–50, determine whether the scatterplot shows an arithmetic, a geometric, or a harmonic sequence. There may be more than one possible answer.

48.

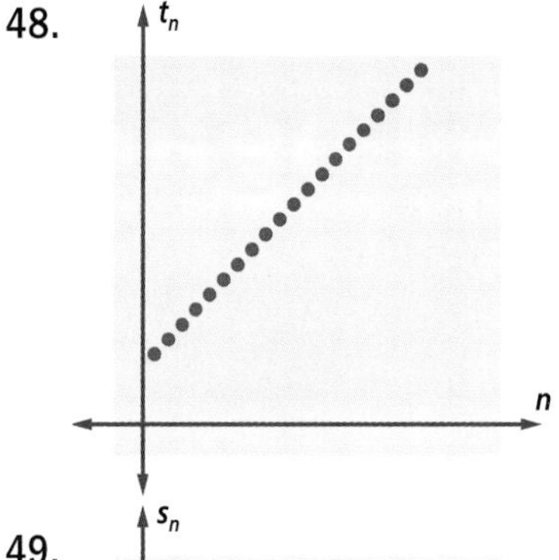

49.

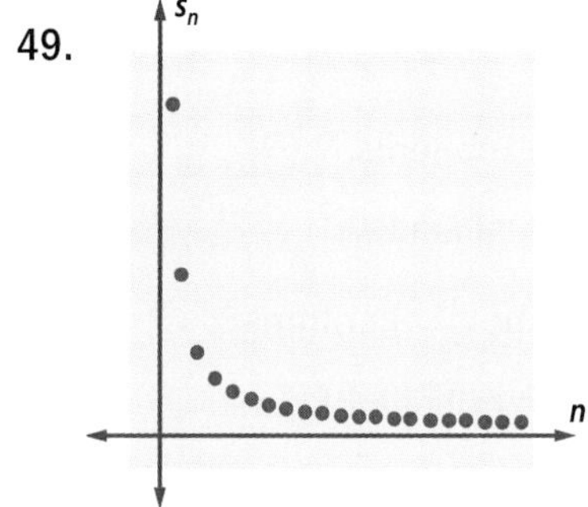

50.

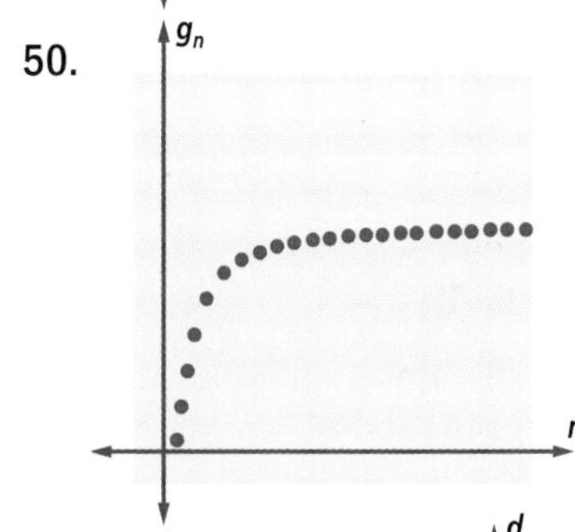

51. Multiple Choice Consider the scatterplot of the sequence d at the right. Which of the following best defines d? $(t > 0)$

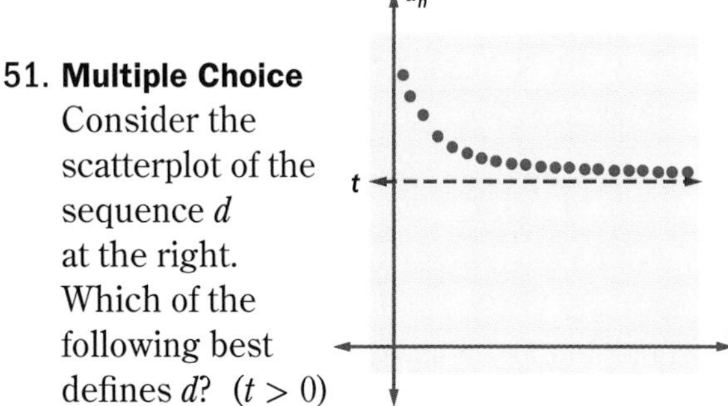

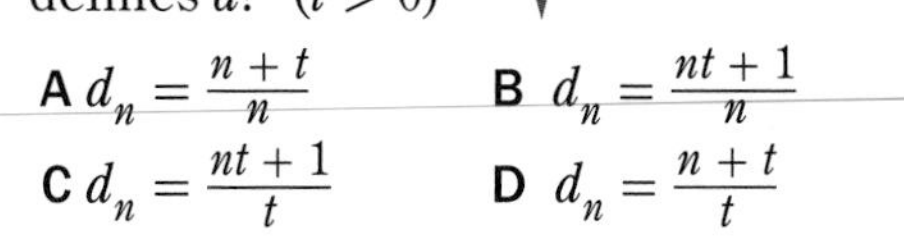

A $d_n = \frac{n + t}{n}$ B $d_n = \frac{nt + 1}{n}$

C $d_n = \frac{nt + 1}{t}$ D $d_n = \frac{n + t}{t}$

In 52 and 53, graph the first five terms of the sequence with the given formula.

52. $w_n = 4n - 1$ **53.** $f_n = 24\left(\frac{1}{3}\right)^{n-1}$

Chapter

9 Roots, Powers, and Logarithms

Contents

Object	Approximate Size
quarks	$< 10^{-18}$ meter
protons	$\approx 10^{-15}$ meter
carbon C_{12} atom	$\approx 10^{-10}$ meter
complete coil of DNA	$\approx 10^{-8.5}$ meter
medium-size virus	$\approx 10^{-6.8}$ meter
smallest bacterium	$\approx 10^{-6.6}$ meter
thickness of notebook paper	$\approx 10^{-4}$ meter
diameter of a penny	$\approx 10^{-1.7}$ meter
meter stick	10^{0} meter

Object	Approximate Size
diameter of the Moon	$\approx 10^{6.3}$ meters
distance between NY and LA	$\approx 10^{6.7}$ meters
diameter of Earth	$\approx 10^{7.6}$ meters
diameter of the Sun	$\approx 10^{9.2}$ meters
diameter of the Earth's orbit	$\approx 10^{11.5}$ meters
length of a light year	$\approx 10^{16}$ meters
diameter of our Milky Way galaxy	$\approx 10^{21}$ meters
diameter of universe	$> 10^{26}$ meters

Estimates of the sizes of various objects in the universe are shown on the previous page.

Each size is written in exponential form with base 10. By writing the numbers in this consistent way we have avoided the various multiples of meters, such as nanometers (1 nm = 10^{-9} meter), kilometers (1 km = 10^3 meters), or centimeters (1 cm = 10^{-2} meter). We can compare the sizes easily by division. For instance, the diameter of a penny is about $\frac{10^{-1.7}}{10^{-4}}$ or $10^{2.3}$ times the thickness of notebook paper. However, writing in this way requires that the reader understand the meaning of power such as $10^{-8.5}$ and $10^{6.3}$.

Rational powers such as these are values of x that solve equations of the form $x^n = a$ and $b^x = a$. Solutions to $x^n = a$ are called nth roots of a. The solution to $b^x = a$ is called the logarithm of a to base b.

In this chapter, you will work with logarithm functions and logarithmic models and scales, and you will use logarithms to estimate the sizes of numbers and to solve equations. You will see how these powers, roots, and logarithms are related.

Lesson

9-1 *n*th Roots

Vocabulary

square root
cube root
*n*th root, $\sqrt[n]{x}$, $x^{\frac{1}{n}}$
*n*th root functions
radical, $\sqrt{}$

BIG IDEA Taking the *n*th root $\left(x^{\frac{1}{n}} \text{ or } \sqrt[n]{x}\right)$ of nonnegative numbers is the inverse operation of taking the *n*th power.

Mental Math

Evaluate exactly over the complex numbers.

a. $\sqrt{36}$ **b.** $\sqrt{-25}$
c. $\sqrt[3]{27}$ **d.** $\sqrt[3]{-8}$

Powers and roots are intimately connected. If $x^2 = k$, then x is called a **square root** of k. If $x^3 = k$, then x is a **cube root** of k. Higher powers lead to 4th roots, 5th roots, and so on.

Definition of *n*th Root

Let n be an integer greater than 1. Then x is an ***n*th root** of k if and only if $x^n = k$.

In Chapter 7 you saw that when $k \neq 0$, the equation $x^n = k$ has n solutions in the set of complex numbers. Therefore, every nonzero real number has n nth roots. Some of these are real. This chapter focuses on the real roots rather than the nonreal. Here are some examples.

17 is a cube root of 4913 because $17^3 = 4913$.
$\frac{1}{2}$ is a fifth root of $\frac{1}{32}$ because $\left(\frac{1}{2}\right)^5 = \frac{1}{32}$.
8 and -8 are both fourth roots of 4096 because $8^4 = 4096$ and $(-8)^4 = 4096$.

0 is a fourth root of 0 because $0^4 = 0$. There is no other fourth root of 0.

Roots of negative numbers are discussed later in this lesson.

Expressing Roots as Powers

You know from algebra that subtraction can be expressed in terms of addition, its inverse operation. In a similar way, finding a root can be expressed in terms of taking a power.

$\frac{1}{n}$ Exponent Theorem

When $x \geq 0$ and n is an integer greater than 1, $x^{\frac{1}{n}}$ is an nth root of x.

Proof To prove that $x^{\frac{1}{n}}$ is an nth root of x, we must show that raising $x^{\frac{1}{n}}$ to the nth power results in x. By the Power of a Power Property, $\left(x^{\frac{1}{n}}\right)^n = x^{\left(\frac{1}{n}\right) \cdot n} = x^1 = x$. So $x^{\frac{1}{n}}$ is an nth root of x.

To ensure that $x^{\frac{1}{n}}$ has exactly one value, we restrict the base x to be a nonnegative real number, and let $x^{\frac{1}{n}}$ stand for the ***unique nonnegative*** *n*th root. For example, $25^{\frac{1}{2}}$ is 5, the positive square root of 25. $-25^{\frac{1}{2}}$ is -5, the negative square root of 25. Similarly, $8^{\frac{1}{3}}$ is the positive cube root of 8, so $8^{\frac{1}{3}} = 2$.

Example 1

Evaluate without a calculator.

a. $125^{\frac{1}{3}}$ b. $\left(\frac{1}{36}\right)^{\frac{1}{2}}$ c. $-16^{\frac{1}{4}}$

Solution

a. The positive cube root of 125 is 5, since $5^3 = 125$.

b. Since $\frac{1}{6} \cdot \frac{1}{6} = \frac{1}{36}$, the positive square root of $\frac{1}{36}$ is $\frac{1}{6}$.

c. $-16^{\frac{1}{4}} = -\left(16^{\frac{1}{4}}\right) = -2$

QY1

> **QY1**
>
> a. Estimate $6^{\frac{1}{50}}$ to the nearest millionth using a calculator.
>
> b. What is $-6^{\frac{1}{50}}$?

*n*th Root Functions

Taking the *n*th power and taking the *n*th root of a number are *inverse operations*. Each undoes the result of the other. For instance, if you start with the number 10, raise it to the 4th power (to get $10^4 = 10{,}000$), and then take the 4th root of the result (to get $10{,}000^{\frac{1}{4}}$), you end up with the original number, 10. Similarly, on the domain $\{x \mid x \geq 0\}$, the functions f and g with $f(x) = x^n$ and $g(x) = x^{\frac{1}{n}}$ are inverses of each other.

Activity 1

Step 1 Use a graphing utility to plot $f(x) = x^4$ $(x \geq 0)$ and $g(x) = x^{\frac{1}{4}}$ $(x \geq 0)$ on the window $-0.5 \leq x \leq 4$, $-0.5 \leq y \leq 4$.

Step 2 Add the dotted line $y = x$ to your graph. Zoom to square the grid in the window. Sketch the graph. What confirms that f and g are inverses?

Step 3 Trace the graph of f to approximate $f(1.2)$.

Step 4 Use the graph of g to approximate $g(x)$ when $x = f(1.2)$.

The functions with equations of the form $y = x^{\frac{1}{n}}$, where n is an integer and $n \geq 2$, are called ***n*th root functions.** Because $x^{\frac{1}{n}}$ is defined only when $x \geq 0$, the domain of all these functions is the set of nonnegative real numbers. The range is also the set of nonnegative reals. Some *n*th root functions are graphed at the right.

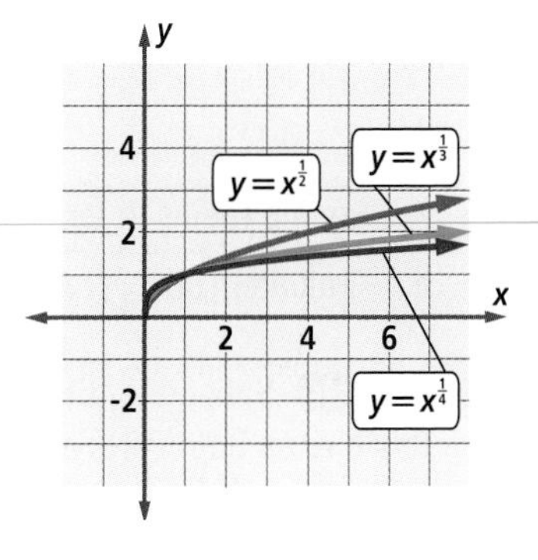

QY2

> **QY2**
> Name four points on the graph of $y = x^{\frac{1}{4}}$ for which both coordinates are integers.

Applications of *n*th Roots

Because of the inverse relationship between *n*th powers and *n*th roots, every formula with *n*th powers can lead to the calculation of *n*th roots. For instance, in any situation of exponential growth or decay, there is an equation of the form $y = ag^x$. In this equation, x is the length of time, a is the starting amount, g is the growth factor in a unit time period, and y is the ending amount.

Example 2

A bowling game that cost 50¢ in 1960 might cost $3.00 in 2010. What is the annual growth rate?

Solution In the formula $y = ag^x$, here $y = \$3.00$, $a = \$0.50$, $x = 50$ years, and we want g. Substitute.

$$3.00 = 0.50 \cdot g^{50}$$
$$6 = g^{50}$$
$$6^{\frac{1}{50}} = (g^{50})^{\frac{1}{50}}$$
$$g = 6^{\frac{1}{50}} \approx 1.036$$

The cost grew at a rate of about 3.6% annually.

In 2005, 53.5 million Americans participated in bowling at least once.

Cube roots originated historically from problems involving volume.

GUIDED

Example 3

The volume V of a cone with radius r and height equal to the radius is given by $V = 1/3\pi r^3$.

a. Give a formula for the radius in terms of V.

b. A cone with radius r and height equal to its radius has volume 60 mm^3. Use the formula in Part a to find its height.

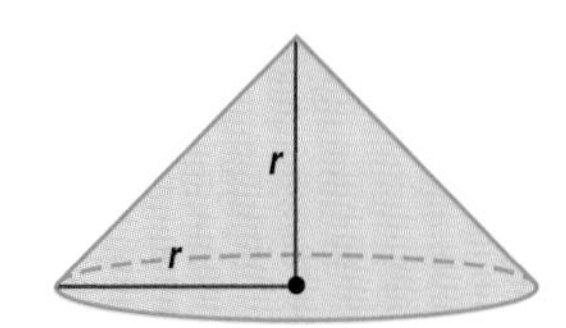

Solution

a. Solve for r, given $V = 1/3\pi r^3$.

$$r^3 = \underline{\ ?\ }$$
$$r = (\underline{\ ?\ })^{\frac{1}{3}}$$

So, the radius in terms of V is given by $r = f(V) = \underline{\ ?\ }$.

b. Evaluate $f(60)$. $f(60) = (\underline{\ ?\ })^{\frac{1}{3}} \approx 3.86$ mm.

Check Substitute 3.86 into the original formula. $V = \frac{1}{3}\pi(3.86)^3 \approx 60$. This checks both Parts a and b.

How Many Real *n*th Roots Does a Real Number Have?

The graphs of $y = x^n$ are of two different forms, one when n is odd and the other when n is even. Remember that the number of intersections of $y = k$ with $y = x^n$ is the *number of solutions to* $x^n = k$. Therefore the number of intersections is the number of real roots of k. The nth root situation is simpler when n is odd than it is when n is even.

$y = x^n$ when n is odd

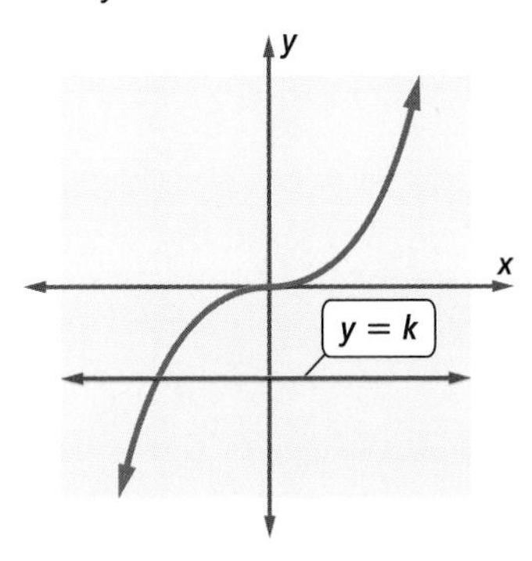

$x^n = k$ has exactly one real solution.

$y = x^n$ when n is even

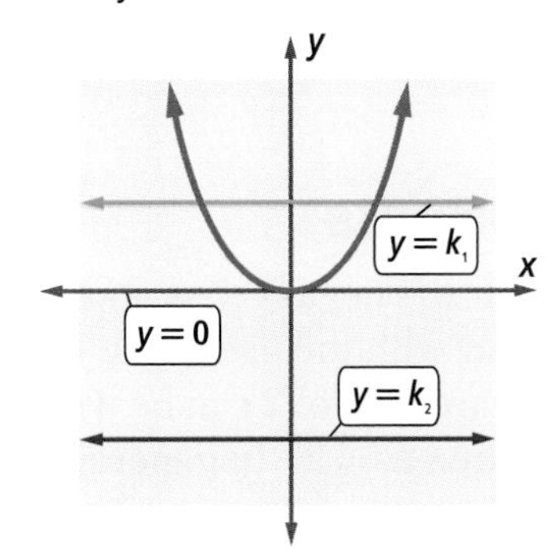

When $k > 0$, $x^n = k$ has two real solutions.
$x^n = 0$ has one solution.
When $k < 0$, $x^n = k$ has no real solutions.

When n is odd, every value of k has exactly one real nth root.

Example: -3 is the only real 5th root of -243.

When n is even:

1. If k is positive, k has two real nth roots.
   Example: Both 3 and -3 are the real 4th roots of 81.
2. If $k = 0$, there is exactly one nth root of k.
   Example: 0 is the only 4th root of 0.
3. If k is negative, k has no real nth roots.
   Example: There is no real 4th root of -81.

QY3

The theorem on the next page describes the relationship between the number of real nth roots that a real number has and whether the real number is positive, negative, or zero. It is easy to remember if you realize that when n is even, the number of real nth roots is the same as the number of real square roots, and when n is odd, the number of real nth roots is the same as the number of real cube roots.

▶ **QY3**

How many real 50th roots does 6 have?

Number of Real Roots Theorem

Every positive real number has:
- 2 real nth roots, when n is even.
- 1 real nth root, when n is odd.

Every negative real number has:
- 0 real nth roots, when n is even.
- 1 real nth root, when n is odd.

Zero has: 1 real nth root.

Radical Notation for *n*th Roots

The expression $x^{\frac{1}{n}}$ is not defined when x is negative, so another symbol for nth roots is useful. You are familiar with the use of the **radical** symbol $\sqrt{}$ to express roots. The square root of 2 can be written as $\sqrt{2}$. The cube root of 2 can be written as $\sqrt[3]{2}$. In general, when x is positive, the radical symbol $\sqrt[n]{x}$ stands for its unique positive nth root. The symbol $\sqrt[n]{x}$ is also used when x is negative, provided that n is odd. For instance, -4 is the unique 3rd root of -64, so we write $\sqrt[3]{-64} = -4$. But -64 has no square root in the real number system.

The use of the radical symbol is summarized below.

Definitions of Radical Symbol

Let n be an integer with $n \geq 2$.

When $x \geq 0$, the nonnegative nth root of x is $\sqrt[n]{x}$ (also written $x^{\frac{1}{n}}$).

When $x < 0$ and n is odd, the real nth root of x is $\sqrt[n]{x}$.

These definitions do not address an even root of a negative number, like $\sqrt{-64}$ or $\sqrt[4]{-81}$. If a CAS is set to real number mode, entering one of these expressions is likely to result in an error message, alerting you that the result is a nonreal number. However, setting the calculator to complex mode will allow these calculations. Complex values that are not real numbers will contain the symbol i.

Example 4

Evaluate $\sqrt{-64}$ on a calculator set for complex numbers and confirm that the result is a square root of –64.

Solution Set your calculator to complex mode. $\sqrt{-64}$ is found to be $8i$.

Since $(8i)^2 = 8i \cdot 8i = 64i^2 = 64(-1)$, $8i$ is a square root of -64.

Activity 2

Step 1 Examine the following five numbers. Which are real numbers?

a. $\sqrt[4]{33}$ b. $33^{\frac{1}{4}}$ c. $-33^{\frac{1}{4}}$ d. $(-33)^{\frac{1}{4}}$ e. $\sqrt[4]{-33}$

Step 2 Evaluate each number from Step 1 using a calculator set to complex mode.

Questions

COVERING THE IDEAS

1. Show that 11 is a fourth root of 14,641.
2. a. Show that $\frac{3}{4}$ is a fifth root of $\frac{243}{1024}$.
   b. Does $\frac{243}{1024}$ have another real fifth root?
3. According to the text, what values of x are allowed in the expression $x^{\frac{1}{3}}$?

In 4–7, evaluate without a calculator.

4. $27^{\frac{1}{3}}$ 5. $\left(\frac{1}{25}\right)^{\frac{1}{2}}$ 6. $\sqrt[5]{-32}$ 7. $\sqrt[4]{81}$

In 8 and 9, a number *x* and a value of *n* are given.
a. Give the number of real *n*th roots *x* has.
b. Find all the real *n*th roots of *x*.

8. $x = 4096$, $n = 12$ 9. $x = \frac{1}{729}$, $n = 3$
10. Sketch a graph of $y = x^6$ and the line $y = 3$. Use the graph to write a statement about the number of solutions to $x^6 = 3$.
11. **True or False** $-49^{\frac{1}{2}} = -7$.
12. a. Calculate $\left(19^{\frac{1}{8}}\right)^8$ without technology.
    b. Generalize Part a.
13. Find all real solutions of $n^{10} = 80$ to the nearest thousandth.
14. From 1987 to 2007, the national debt of the United States grew from 2.35 to 9.01 trillion dollars, a growth factor of approximately 383%, or 3.83.
    a. **Multiple Choice** To find the annual growth factor g of this debt, which of the following equations can you solve?
       A $2.35^g = 9.01$ B $g^{20} = 3.83$
       C $3.83^g = 20$ D $9.01^g = 2.35$
    b. Solve the equation you chose in Part a, using technology.
15. The volume of a sphere is given by the formula $V = \frac{4}{3}\pi r^3$.
    a. Express the radius r of the sphere in terms of its volume V.
    b. Use the formula from Part a to find the radius of a sphere with volume 120 cm^3.

16. Consider the functions f and g with $f(x) = x^{\frac{1}{3}}$ and $g(x) = x^3$, for $x \geq 0$.
   a. Show that f and g are inverse functions.
   b. For what values of x is $f(x) > g(x)$?
   c. Graph f and g on the same axes.

APPLYING THE MATHEMATICS

17. a. Estimate $4.6^{\frac{1}{10}} \cdot 4.6^{\frac{1}{10}}$ to the nearest hundredth.
   b. **Fill in the Blank** The answer to Part a is the __?__ root of __?__.

18. Place a $<$ or $>$ sign in the blank.
   a. When $0 < x < 1$, $x^{\frac{1}{5}}$ __?__ $x^{\frac{1}{8}}$.   b. If $x > 1$, $x^{\frac{1}{5}}$ __?__ $x^{\frac{1}{8}}$.

19. For what values of x and n are $x^{\frac{1}{n}}$ and $\sqrt[n]{x}$ equal?

20. It has been estimated that the speed s in knots of a ship is a function of the horsepower p developed by its engines, with $s = 6.5\,p^{\frac{1}{7}}$. How fast will a ship travel with engines producing 575 horsepower?

21. The ancient Greeks wanted to construct a cube that would have twice the volume of a cube at the altar at Delos. If the cube at Delos had an edge of length one unit, what would be the length of an edge of the constructed cube, to the nearest thousandth?

22. From 1900 to 1950, the population of Cleveland, Ohio, grew from about 382,000 to about 915,000. In 2000, Cleveland's population was about 478,000.
   a. What was the annual growth factor from 1900 to 1950?
   b. What was the annual growth factor from 1950 to 2000?
   c. What was the annual growth factor from 1900 to 2000?

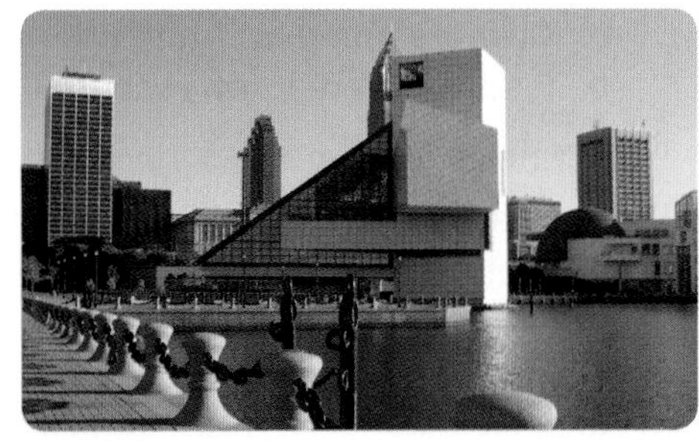

The Rock and Roll Hall of Fame and Museum, Cleveland, OH

In 23 and 24, use a graphing utility.

23. a. Plot $y = x^{\frac{1}{2}}$ and $y = -x^{\frac{1}{2}}$ on one set of axes.
   b. What single equation describes the union of these two graphs?

24. a. Plot $f(x) = \sqrt[5]{x}$ and $g(x) = \sqrt[5]{x-3} + 4$ on one set of axes.
   b. What transformation maps the graph of f onto the graph of g?
   c. What transformation maps the graph of g onto the graph of f?
   d. What is the domain of g?

REVIEW

25. Are the functions f and g with $f(x) = \frac{7}{x}$ and $g(x) = \frac{x}{7}$ inverse functions? Explain your answer. **(Lesson 3-8)**

26. Given $r(x) = 6x$ and $t(x) = x^4$, find each composite. **(Lesson 3-7)**
   a. $r \circ r(a)$ b. $t(t(b))$ c. $t \circ r(x)$ d. $t \circ r(x + 1)$

27. Women in Japan have among the greatest life expectancies at birth of any national group. The following table presents their life expectancy for various years from 1980 to 2002. Use linear regression to predict the life expectancy of a Japanese girl born in 2020. How accurate do you think this prediction is? **(Lesson 2-3)**

Year	1980	1990	1995	2000	2001	2002
Life Expectancy	78.8	81.9	82.9	84.6	84.9	85.2

Source: U. S. Department of Health and Human Services, Health, United States, 2006

28. In Question 27, is the prediction an example of interpolation or extrapolation? **(Lesson 2-2)**

29. ***Skill Sequence*** Solve. **(Previous Course)**
 a. $4^t = 256$
 b. $4^{(r-2)} = 256$
 c. $4^{x^2-3x} = 256$

30. ***Skill Sequence*** Simplify. **(Previous Course)**
 a. $x^6 \cdot x^{11}$
 b. $(2x^6)^{11}$
 c. $\frac{x^6}{x^{11}}$

EXPLORATION

31. Let x be a positive number.
 a. The cube root of the square root of x is what root of x?
 b. Generalize Part a and prove your generalization.

QY ANSWERS

1. a. 1.036485
 b. −1.036485

2. Answers vary. Sample: (1,1), (16, 2), (81, 3), (256, 4)

3. 2

Lesson 9-2 Rational Exponents

Vocabulary

rational power

BIG IDEA The exponent $\frac{m}{n}$, where m and n are integers and $n \neq 0$, stands for the mth power of an nth root, or, equivalently, for an nth root of the mth power.

On page 558, the diameter of the Moon was said to be about $10^{6.3}$ meters. Rewriting the decimal 6.3 as the fraction $\frac{63}{10}$, the diameter of the moon is about $10^{\frac{63}{10}}$ meters. In this lesson, you will see how to interpret powers with rational exponents.

Mental Math

Write in exponential form.

a. $x \cdot x \cdot x$

b. $\frac{1}{y} \cdot \frac{1}{y} \cdot \frac{1}{y}$

c. $x \cdot x \cdot x \cdot \frac{1}{y} \cdot \frac{1}{y}$

d. $\frac{x}{y} \cdot \frac{x}{y} \cdot \frac{x}{y} \cdot x$

The Meaning of Rational Exponents

An advantage of indicating roots with exponents instead of radical notation is that familiar properties of exponents can dicate how calculations can be performed.

In Lesson 9-1, you saw that $x^{\frac{1}{n}}$ indicates the nth root of a nonnegative number x. An exponent of the form $\frac{m}{n}$, where m and n are both integers and $n \neq 0$, is a rational exponent and $x^{\frac{m}{n}}$ is called a **rational power** of x.

For instance, consider the expression $x^{\frac{4}{3}}$ where $x \geq 0$. The exponent equals both $\frac{1}{3} \cdot 4$ and $4 \cdot \frac{1}{3}$. Remember that $x^{mn} = (x^m)^n$ for all exponents m and n and all positive bases x. So,

$$x^{\frac{4}{3}} = x^{\frac{1}{3} \cdot 4} = \left(x^{\frac{1}{3}}\right)^4, \text{ the 4th power of the 3rd root of } x.$$

$$x^{\frac{4}{3}} = x^{4 \cdot \frac{1}{3}} = (x^4)^{\frac{1}{3}}, \text{ the 3rd root of the 4th power of } x.$$

So $x^{\frac{4}{3}} = \left(x^{\frac{1}{3}}\right)^4 = (x^4)^{\frac{1}{3}}$. This result can be generalized.

Rational Exponent Theorem

For any nonnegative real number x and positive integers m and n, $x^{\frac{m}{n}} = \left(x^{\frac{1}{n}}\right)^m$, the mth power of the positive nth root of x, and $x^{\frac{m}{n}} = (x^m)^{\frac{1}{n}}$, the positive nth root of the mth power of x.

Written in radical notation instead of using exponents, the Rational Exponent Theorem states that $x^{\frac{m}{n}} = (\sqrt[n]{x})^m = \sqrt[n]{x^m}$. Thus, by writing $10^{6.3}$ as $10^{\frac{63}{10}}$, $10^{6.3}$ can be interpreted either as the 10th root of 10^{63} or as the 63rd power of the 10th root of 10. You would not compute $10^{6.3}$ by hand, but there are some rational powers that can be computed by hand.

 QY1

QY1

Use your calculator to estimate $(100^4)^{\frac{1}{3}}$ and $\left(100^{\frac{1}{3}}\right)^4$.

Example 1

Evaluate $64^{\frac{2}{3}}$ without a calculator.

Solution 1 First take the cube root, and then square the result.

$64^{\frac{2}{3}} = \left(\sqrt[3]{64}\right)^2 = 4^2 = 16$

Solution 2 First, square the base, and then take the cube root.

$64^{\frac{2}{3}} = \sqrt[3]{64^2} = \sqrt[3]{4096} = 16$

Solution 3 Rewrite 64 as an exponent with base 2. Substitute 2^6 for 64.

$64^{\frac{2}{3}} = (2^6)^{\frac{2}{3}} = 2^{6 \cdot \frac{2}{3}} = 2^4 = 16$

When evaluating rational powers, it is generally easier to take roots before powering, as in Solution 1 to Example 1.

An indirect proof can be used to show why we do not use noninteger rational exponents with negative bases. By the Rational Exponent Theorem, if $(-8)^{\frac{1}{3}}$ were defined, it must equal $\sqrt[3]{-8}$, which is -2. But $\frac{1}{3} = \frac{2}{6}$, so by substitution, $(-8)^{\frac{1}{3}} = (-8)^{\frac{2}{6}}$. We calculate $(-8)^{\frac{2}{6}}$ using the Power of a Power Property: $(-8)^{\frac{2}{6}} = ((-8)^2)^{\frac{1}{6}} = (64)^{\frac{1}{6}} = 2$. So $-2 = (-8)^{\frac{1}{3}} = (-8)^{\frac{2}{6}} = 2$. This is a contradiction. This kind of contradiction can occur each time a negative base is used with a rational exponent.

Twelfth Roots and the Musical Scale

Guitars, harps, violins, mandolins, cellos, banjos, double basses, and violas are some of the many string instruments.

A string instrument player makes notes by pressing on a string to shorten the vibrating section of the string. The Pythagoreans discovered a simple property relating different notes produced by a string instrument. If the length of the vibrating part of a string is cut in half, the tone is raised an octave. Most of our music today divides that octave into 12 notes called the *chromatic scale*. Let g be the factor that relates the length of a string for two adjacent notes, like C and C#. Then $g^{12} = \frac{1}{2}$. Raising both sides to the $\frac{1}{12}$ power gives $g = \left(\frac{1}{2}\right)^{\frac{1}{12}} \approx 0.944$. So to move from one note in a chromatic scale to the next, the string must be shortened to about 94.4% of its previous length.

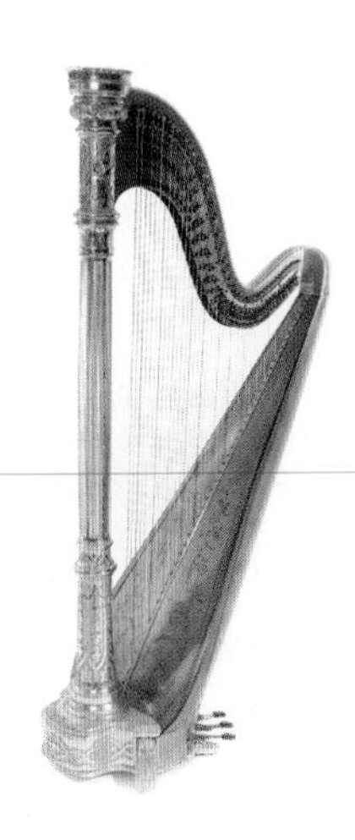

For example, suppose that a string that is 650 mm long plays a C. Then to play the E, four notes higher on the scale, the vibrating part should be $650\left(\frac{1}{2}\right)^{\frac{4}{12}} = 650\left(\frac{1}{2}\right)^{\frac{1}{3}} \approx 516$ mm long.

When this is done for 12 notes, $\left(\left(\frac{1}{2}\right)^{\frac{1}{12}}\right)^{12} = \left(\frac{1}{2}\right)^{1} = \frac{1}{2}$ and the length of the string for the tone 12 notes higher is half the length for the first note.

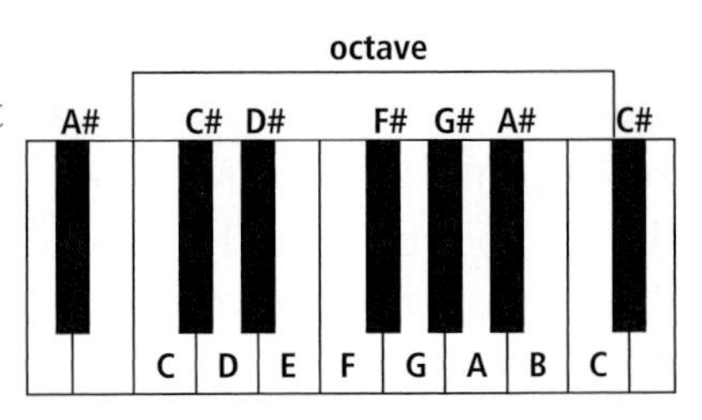

GUIDED

Example 2

The G that is 7 notes above C sounds very good with the C because the strings are close to vibrating together. This is because the ratio of the lengths of the strings is very close to a simple fraction with small whole numbers in its numerator and denominator. What is that ratio?

Solution Suppose the string for C has length L. Then, since G is 7 notes above C, the length of the string for G is $L \cdot \left(\frac{1}{2}\right)^{\frac{?}{}}$.

$$\frac{L \cdot \left(\frac{1}{2}\right)^{\frac{?}{}}}{L} \approx 0.6674199271\ldots .$$

This number is very close to the fraction __?__.

Activity

Step 1 Estimate $\left(\frac{1}{2}\right)^{\frac{n}{12}}$ for $n = 1, 2, \ldots, 11$.

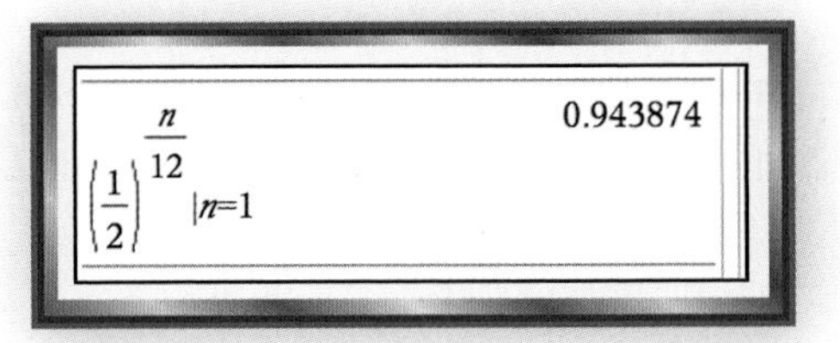

Step 2 In Example 2, one of the values from Step 1 was found to be within 0.001 of the value of a simple fraction with small whole numbers in its numerator and denominator. Find another value from Step 1 with this property.

Properties of Powers

All the properties of powers can be deduced from a small number of basic properties. You should know them from previous courses. We have already applied the first of these properties.

Properties of Powers

For any nonnegative bases x and y and nonzero exponents a and b, or any nonzero bases and integer exponents:

Power of a Power Property	$(x^a)^b = x^{ab}$
Product of Powers Property	$x^a \cdot x^b = x^{a+b}$
Quotient of Powers Property	$\frac{x^a}{x^b} = x^{a-b}$
Power of a Product Property	$(xy)^b = x^b y^b$
Power of a Quotient Property	$\left(\frac{x}{y}\right)^b = \frac{x^b}{y^b}$

 QY2

> **QY2**
>
> Compute $2^{\frac{4}{3}} \cdot 2^{\frac{2}{3}}$.

You can deduce the meaning of zero exponents and negative exponents from the properties on the previous page.

Zero Exponent Theorem

If x is any nonzero real number, $x^0 = 1$.

Proof Let x be any nonzero real number. Then $\frac{x}{x} = 1$.
But by the Quotient of Powers Property, $\frac{x}{x} = \frac{x^1}{x^1} = x^{1-1} = x^0$.
Thus, by substitution, $x^0 = 1$.

Notice that x cannot be zero in the proof of the Zero Exponent Theorem. 0^0 is not defined. (See Question 27.)

Negative Exponent Theorem

For all $x > 0$ and real numbers n, or for all $x \neq 0$ and integers n, $x^{-n} = \frac{1}{x^n}$.

Proof Consider x^{-n}. By the Product of Powers Property, whenever x^{-n} is defined,

$$x^n \cdot x^{-n} = x^{n+(-n)}$$
$$= x^0.$$

So $x^n \cdot x^{-n} = 1$ by the Zero Exponent Theorem.

Dividing both sides of this equation by x^n yields

$$x^{-n} = \frac{1}{x^n}.$$

Because of the Negative Exponent Theorem, x^n and x^{-n} are reciprocals. For example, $64^{\frac{1}{3}}$ and $64^{-\frac{1}{3}}$ are reciprocals: $64^{\frac{1}{3}} = 4$ and $64^{-\frac{1}{3}} = \frac{1}{4}$.

 QY3

> **QY3**
>
> What simple fraction equals $8^{-\frac{1}{3}}$?

Example 3

Find the simple fraction equal to $49^{-\frac{3}{2}}$.

Solution Use properties of powers to simplify.

$$49^{-\frac{3}{2}} = \left(49^{\frac{1}{2}}\right)^{-3} = 7^{-3} = \frac{1}{343}$$

Check Find the decimal equivalents of $49^{-\frac{3}{2}}$ and $\frac{1}{343}$ on a calculator.

$49^{\frac{-3}{2}}$	0.0029
$\frac{1}{343}$	0.0029

 QY4

> **QY4**
>
> If $x > 1$, put in increasing order: x^2, $\sqrt[3]{x^5}$, $(\sqrt{x})^7$, $x^{2.8}$.

Example 4

Write $\left(\frac{1}{2}\right)^{\frac{n}{12}}$ as a power of 2.

Solution $\frac{1}{2} = \frac{1}{2^1} = 2^{-1}$. So $\left(\frac{1}{2}\right)^{\frac{n}{12}} = (2^{-1})^{\frac{n}{12}}$.

By the Power of a Power Property, $(2^{-1})^{\frac{n}{12}} = 2^{-\frac{n}{12}}$. So $\left(\frac{1}{2}\right)^{\frac{n}{12}} = 2^{-\frac{n}{12}}$.

A Geometric Application of Rational Exponents

Two formulas with integer powers used together may result in a formula with a noninteger rational power, as Example 5 illustrates.

Example 5

a. Find the volume of a cube with surface area 105 square inches.

b. Express the volume V of a cube as a function of its surface area S.A.

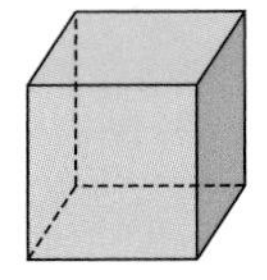

Solution

a. Let s be the length of one edge of the cube. You should be familiar with formulas for the volume and surface area.

$$V = s^3 \text{ and } S.A. = 6s^2$$

Substitute 105 for S.A. in the second formula, and solve for s.

$$105 = 6s^2, \text{ so } s = \sqrt{\frac{105}{6}} = \sqrt{\frac{35}{2}} = \sqrt{17.5}.$$

Now use the first formula.

$$V = s^3 = \left(\sqrt{17.5}\right)^3 = 17.5^{\frac{3}{2}} \approx 73.21 \text{ in}^3$$

b. Solve the second formula for s: $s = \sqrt{\frac{S.A.}{6}}$.

Then $V = s^3 = \left(\sqrt{\frac{S.A.}{6}}\right)^3 = \left(\frac{S.A.}{6}\right)^{\frac{3}{2}}$

Check Let S.A. = 105 in the formula found in Part b. Then $V = \left(\frac{105}{6}\right)^{\frac{3}{2}}$. A calculator gives the same value as found in Part a.

The exponent $\frac{3}{2}$ in the formula in the Solution for Example 5b signifies the change in dimension from surface area (a 2-dimensional measure) to volume (a 3-dimensional measure).

Questions

COVERING THE IDEAS

In 1–4, the variables represent positive numbers. Rewrite each expression using positive rational exponents in fraction form.

1. $3x^{2.6}$
2. $8r^{-0.7}$
3. $(\sqrt{w})^7$
4. $\sqrt[6]{t^9}$

In 5 and 6, rewrite the expression using a radical sign. Assume $x > 0$.

5. $x^{\frac{3}{4}}$
6. $x^{-\frac{4}{5}}$

In 7–10, simplify without using a calculator. Check with a calculator.

7. a. $27^{\frac{2}{3}}$ b. $27^{-\frac{1}{3}}$

8. a. $36^{\frac{1}{2}}$ b. $36^{\frac{3}{2}}$

9. a. $64^{\frac{3}{2}}$ b. $64^{-\frac{2}{3}}$ c. $64^{\frac{11}{6}}$

10. a. $81^{-\frac{5}{4}}$ b. $81^{-\frac{1}{4}}$ c. $81^{-\frac{3}{4}}$

11. Suppose the vibrating part of a string on a musical instrument is 18 inches long and the note produced is a C. How long a vibrating part will produce a D?

12. Write $\left(\frac{1}{3}\right)^{\frac{3}{4}}$ as a power of 3.

13. To the nearest cubic inch, find the volume of a cube whose surface area is 1000 square inches.

APPLYING THE MATHEMATICS

14. a. Graph the equation $y = x^{1.5}$.
    b. Find five points with integer coordinates on this graph.

15. a. To the nearest 0.01 square meter, find the surface area of a cube whose volume is 200 m^3.
    b. Express the surface area of a cube as a function of its volume.
    c. Interpret the exponent in the formula you find in Part b as a change in dimension.

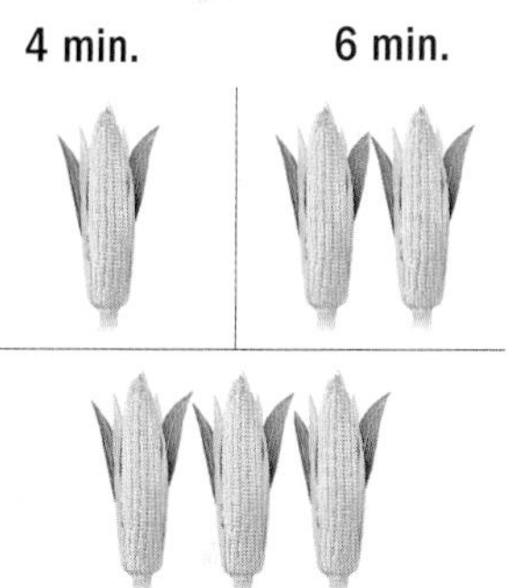

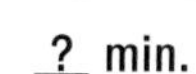

16. According to one microwave cookbook, whenever you double the amount being cooked, the cooking time should be multiplied by 1.5. For example, suppose that one portion of a certain food takes 4 minutes to cook. Then two portions cook in 6 minutes and four portions cook in 9 minutes. The approximate time $T(p)$ it takes to cook p portions is given by $T(p) = 4p^{0.585}$. What is the time needed to cook 3 portions?

17. According to Kepler's Laws, each planet orbits the sun in an ellipse with the Sun at one focus. The orbital period P (the time it takes to make one revolution) is proportional to the $\frac{3}{2}$ power of the length a of the major axis. In symbols, $P = k \cdot a^{\frac{3}{2}}$.
    a. Earth orbits the Sun with a major axis of 186,000,000 miles and a period of 365.24 days. Determine the value of k. Express your answer in scientific notation.
    b. Mars' major axis is about 283,000,000 miles long. Determine the period of Mars' orbit.
    c. Write a formula giving a in terms of P.

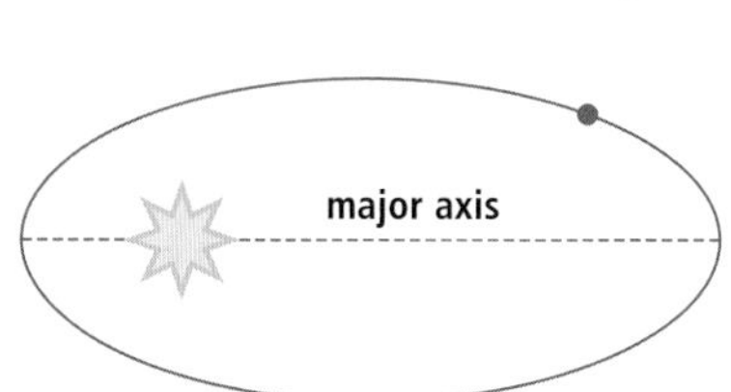

18. a. Without a calculator, decide which is larger, $(0.2)^{0.9}$ or $(0.2)^{-0.9}$.
    b. Explain how you made your decision.

19. Write each number as an integer power of 2.
    a. $8^{\frac{5}{3}}$ b. $8^{-\frac{5}{3}}$ c. $64^{\frac{5}{3}}$ d. $64^{-\frac{5}{3}}$

20. a. Plot $f(x) = x^{\frac{3}{4}}$ and $g(x) = x^{\frac{4}{3}}$ on the same axes for $x \geq 0$.
   b. How are the graphs related? How are the functions related?
   c. Find expressions for $f(g(x))$ and for $g(f(x))$.

21. Another musical relationship involving 12th roots relates the frequency of the sound waves for various notes of the scale. Each piano key (white *and* black) has a frequency $2^{\frac{1}{12}}$ times that of the previous key. The A above (to the right of) middle C is often tuned to a frequency of 440 Hz.
   a. Find the frequency of middle C.
   b. Find the frequency of the A below middle C, which is one octave (12 notes) lower than the A above middle C.

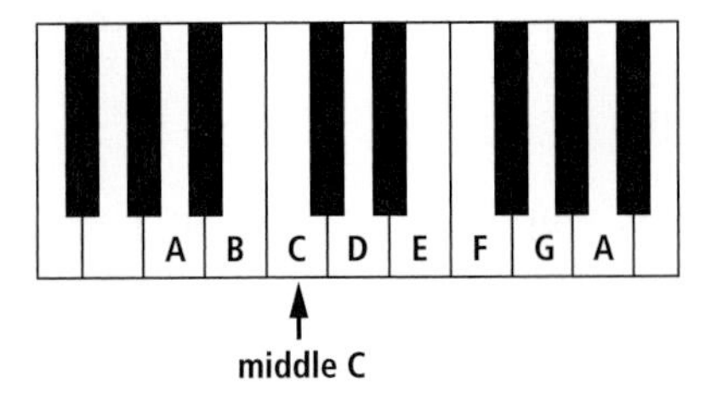

REVIEW

22. Bill has some money to invest. He wants to double the amount in 10 years. So his growth factor g will need to satisfy $g^{10} = 2$.
   a. Find g to the nearest thousandth.
   b. What yearly growth *rate* does Bill need? **(Lesson 9-1)**

23. a. Give an equation for the inverse of f when $f(x) = \sqrt[5]{x} - 3$.
   b. Sketch f and its inverse on the same set of axes.
   c. What is the largest domain on which both f and its inverse can be defined so each is a function? **(Lessons 9-1, 3-8)**

24. Order from least to greatest: $4^3, 4^{-3}, 3^{-4}, 3^{\frac{1}{4}}, \frac{3}{4}$.
   (Lesson 9-1, Previous Course)

25. Evaluate without a calculator. **(Lessons 5-4, 4-4)**
   a. $\sin^{-1}\left(\frac{1}{2}\right)$ b. $\sin^{-1}\left(-\frac{1}{2}\right)$ c. $\sin\left(-\frac{\pi}{6}\right)$

26. Malik scored 92 on his math exam. The class mean on the math exam was 83 and the standard deviation was 6.5. That same week he scored 79 on a chemistry exam while the class mean was 66.5 and the standard deviation was 7.2. On which exam did he score higher relative to his classmates? Justify your answer. **(Lesson 3-9)**

EXPLORATION

27. The expression 0^0 is sometimes called an *indeterminate form* because more than one value might be reasonable for 0^0.
   a. Compute the values of the sequence $0^2, 0^1, 0^{\frac{1}{2}}, 0^{\frac{1}{4}}, 0^{\frac{1}{8}}, 0^{\frac{1}{16}}, \ldots$. What does this imply as a reasonable value for 0^0?
   b. Compute the values of the sequence $2^0, 1^0, \left(\frac{1}{2}\right)^0, \left(\frac{1}{4}\right)^0, \left(\frac{1}{8}\right)^0, \ldots$. What does this imply as a reasonable value for 0^0?
   c. Compute the values of the sequence $2^2, 1^1, \left(\frac{1}{2}\right)^{\frac{1}{2}}, \left(\frac{1}{4}\right)^{\frac{1}{4}}, \left(\frac{1}{8}\right)^{\frac{1}{8}}, \ldots$. What does this sequence imply as a reasonable value for 0^0?
   d. Evaluate 0^0 using the powering key on different calculators. What do you get?

QY ANSWERS

1. ≈ 464.16; ≈ 464.16
2. 4
3. $\frac{1}{2}$
4. $\sqrt[3]{x^5}, x^2, x^{2.8}, (\sqrt{x})^7$

Lesson 9-3 Logarithm Functions

Vocabulary

logarithm of x to the base b, $\log_b x$
common logarithm
logarithm function with base b

BIG IDEA When a number is written in exponential form with base b, the logarithm of that number to the base b is the exponent.

What is a Logarithm?

According to the table on page 558, the diameter of the Sun is about $10^{9.2}$ meters. $10^{9.2} = 10^{\frac{92}{10}}$, so this number is the 92nd power of the positive 10th root of 10. A calculator shows that $10^{9.2} \approx 1{,}584{,}893{,}192$. The exponent 9.2 is called the *logarithm* of 1,584,893,192 to the base 10. In general, any positive number except 1 can be the base for logarithms.

Our Sun with black dot representing the size of Earth

Mental Math

Write each number as a power of 10.

a. 1,000
b. 0.0001
c. $10^{25} \cdot 10^{18}$
d. $(1{,}000)^{11}$

Definition of Logarithm

Let $b > 0$ and $b \neq 1$. Then y is the **logarithm of x to the base b**, written $y = \log_b x$, if and only if $b^y = x$.

Logarithm is often abbreviated as "log" in both spoken and written mathematics. You can read $\log_b x$ as "log of x to the base b" or "log base b of x." The definition essentially says that the log of x to the base b is the exponent of the power of b that equals x.

 QY1

QY1

Multiple Choice Which statement is read "p is the log of q to the base s"?

A $\log_q p = s$
B $\log_p q = s$
C $p = \log_q s$
D $p = \log_s q$

Example 1

Evaluate.

a. the log of 512 to the base 8

b. $\log_8 \frac{1}{64}$ c. $\log_8 \sqrt[7]{8}$

Solution

a. Think: What power of 8 equals 512? $512 = 8^3$, so $\log_8 512 = 3$.

b. Think: What power of 8 equals $\frac{1}{64}$? $\frac{1}{64} = \frac{1}{8^2} = 8^{-2}$, so $\log_8 \frac{1}{64} = -2$.

c. Think: What power of 8 equals $\sqrt[7]{8}$? $\sqrt[7]{8} = 8^{\frac{1}{7}}$ so $\log 8 \sqrt[7]{8} = \frac{1}{7}$.

Logarithms to the Base 10 (Common Logarithms)

Logarithms to the base 10 historically have been used more often than any other logarithm. For this reason, they are sometimes called **common logarithms**. Common logarithms are often written without indicating the base 10. In this book, when you see an expression like log 100, you can assume it means $\log_{10} 100$.

GUIDED

Example 2

Find the logarithm of each number to the base 10.

a. 100 b. 0.00001 c. 1 trillion d. $10\sqrt{10}$

Solution In each case, write the number as a power of 10. Then the exponent of the power is the logarithm.

a. $100 = 10^2$, so the log of 100 to the base 10 is $\log_{10} 100 = 2$.

b. $0.00001 = 10^{-5}$, so the log of 0.00001 to the base 10 is $\log_{10}$ __?__ $= -5$.

c. 1 trillion $= 10^{?}$, so $\log_{10}$ __?__ $=$ __?__.

d. Use the fact that $\sqrt{10} = 10^{\frac{1}{2}} = 10^{0.5}$.
Then $10\sqrt{10} = 10^1 \cdot 10^{0.5} = 10^{?}$, so $\log_{10} 10\sqrt{10} =$ __?__.

Example 2 shows that if you can write a number as a power of 10, then you can easily write its logarithm. Part d of Example 2 shows that multiplying a number by 10 adds 1 to its common logarithm. This property means that shifting the decimal point in the base 10 representaton of a number changes its common logarithm by an integer. For instance, since $\sqrt{10} = 10^{0.5} \approx 3.162$,

$$\begin{aligned} &\log_{10} 3.162 \approx \log_{10}(10^{0.5}) = 0.5, \\ &\log_{10} 31.62 \approx \log_{10}(10^{1.5}) = 1.5, \\ &\log_{10} 316.2 \approx \log_{10}(10^{2.5}) = 2.5, \\ \text{and } &\log_{10} 3162 \approx \log_{10}(10^{3.5}) = 3.5. \end{aligned}$$

This property of logarithms made them useful in computation. If you can write a set of digits as a power of 10, then you can write any number with that set of digits as a power of 10. Before electronic calculators, tables of logarithms thus needed only to give the logs of numbers between 1 and 10. This process was the origin of scientific notation. The first table of common logarithms was produced by Henry Briggs (1561–1630) of England. In 1624 he published *Arithmetica Logarithmica*, which contain common logarithms to fourteen decimal places.

Today you can find the common log of any number using a calculator.

QY2

QY2

Use a calculator to find the common log of each to the nearest thousandth. Check by evaluating a power of 10.

a. 45.8 **b.** 458
c. 0.00458

Common logarithms can be used to solve equations of the form $10^x = a$, where $a > 0$.

Example 3

Solve $10^x = 7$ to the nearest thousandth.

Solution Use the definition of a logarithm to rewrite the equation as $x = \log_{10} 7$. Find $\log_{10} 7$ with a calculator.

$$\log_{10} 7 \approx 0.8451$$

Check Use a calculator to verify that $10^{0.845} \approx 6.998 \approx 7$.

To find the number that has a given common log, use the definition of logarithm to rewrite the log equation as an exponential equation.

Example 4

Solve $\log x = 2.873$ to the nearest tenth.

Solution Because no base is written, the base is 10. Apply the definition.

$$\log_{10} x = 2.873 \text{ if and only if } x = 10^{2.873}.$$

Use a calculator to compute the power. $x \approx 746.4$

Check Does $\log 746.4 \approx 2.873$? A calculator verifies that it does.

Examples 1 and 2 evaluate $\log_b x$ where x is a power of b. In Example 5, this is not the case, but the base and the argument are both powers of another number.

Example 5

Evaluate $\log_4 8$.

Solution From the definition, $\log_4 8$ is the power of 4 that equals 8. So, write 8 as a power of 4. That is, if $y = \log_4 8$, then $4^y = 8$.

To solve this equation, express each side as a power of the same base. In this case, we use 2: $(2^2)^y = 2^3$ substitution

$2^{2y} = 2^3$ Power of a Power Property

The exponents must be the same if the two expressions are equal. So,

$$2y = 3 \text{ and so } y = \frac{3}{2}.$$

Thus, $\log_4 8 = \frac{3}{2}$.

Check Go back to the definition of $\log_b x$. Does $4^{\frac{3}{2}} = 8$?

$4^{\frac{3}{2}} = (\sqrt{4})^3 = 2^3 = 8$. Yes, it checks.

STOP QY3

QY3

Evaluate $\log_{125} 25$ without a calculator.

Graphs of Logarithm Functions

The inverse of the exponential function f with $f(x) = 2^x$ has equation $x = 2^y$. By the definition of logarithm, this exponential equation is equivalent to $y = \log_2 x$. That is, the inverse of the exponential function with base 2 is the logarithm function with base 2. In general, the function that maps x onto $\log_b x$ is called the **logarithm function with base *b***. It is the inverse of the exponential function with base b. That is,

if for all real numbers x, $f(x) = b^x$,

then for all positive x, $f^{-1}(x) = \log_b x$.

At the right are graphed three exponential functions and their inverses, the corresponding logarithm functions.

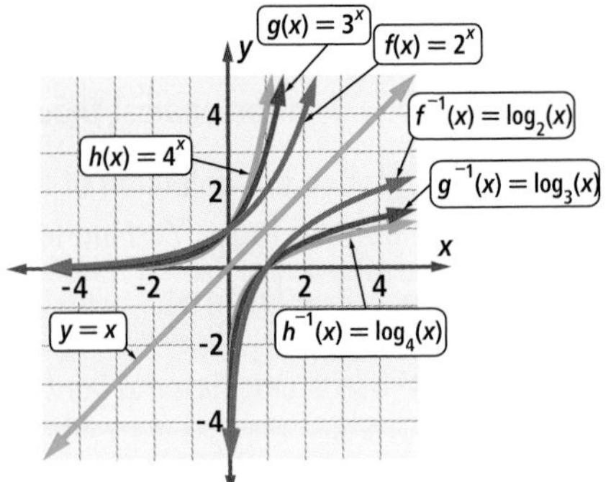

Activity

MATERIALS graphing utility with slider capability

Step 1 On a graphing utility, graph $y = b^x$. Create a slider for b whose domain is $1 \le b \le 20$.

Step 2 Manipulate the graph of $y = b^x$ for various values of b by moving the slider. As b changes, how does the graph change? What properties are invariant (remain the same)?

Step 3 Graph $y = \log_b x$. As b changes, how does the graph change? What properties are invariant?

Step 4 Match each changing or invariant property from Step 2 with a corresponding property from Step 3. For example, the *domain* of the exponential function corresponds to the *range* of the corresponding logarithm function.

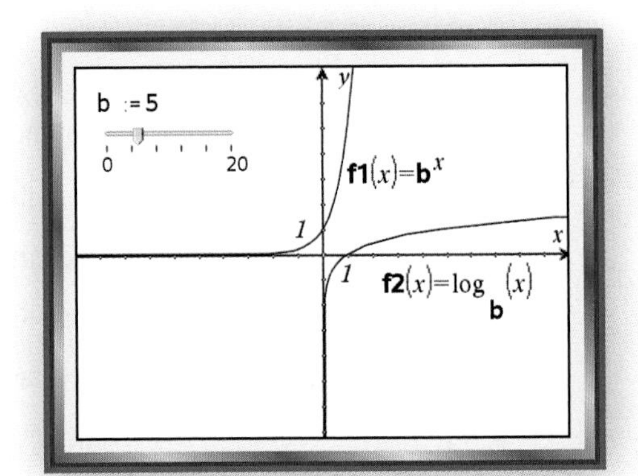

Like the family of exponential functions studied in Chapter 2, the graphs of logarithm functions also form a family of curves with a characteristic shape and similar properties. From the graphs of the various functions above, you can see that for any log function with base $b > 1$:

(1) The domain is the set of positive real numbers.

(2) The range is the set of all real numbers.

(3) The graph contains (1, 0).

(4) The function is strictly increasing.

(5) The end behavior, as x gets larger, is to increase without bound.

(6) As x gets smaller and approaches 0, the values of the function are negative with larger and larger absolute values.

(7) The y-axis is a vertical asymptote of the graph.

Log functions are strictly increasing, so if $x_2 > x_1$, then $\log_b x_2 > \log_b x_1$. This property allows you to solve equations with logarithms in them because if $\log_a x = \log_a y$, then it must be true that $x = y$.

To see another consequence of the "strictly increasing" property, note that $\log_{10} 1000 = 3$, $\log_{10} 10000 = 4$, and so on. Consequently, all 4-digit integers have common logarithms between 3 and 4. For example, since 8726 is between 10^3 and 10^4, $\log_{10} 8726$ is between 3 and 4. In general, the common logarithm of numbers between successive powers of 10 is a number between successive integers.

This property can be used in reverse. A number with a common logarithm between 2 and 3 must be between 100 and 1000, an instance of which you saw in Example 4. This relationship can be used to determine the number of digits in a number (see Question 21).

Questions

COVERING THE IDEAS

1. Write an equation equivalent to $3^6 = 729$ using logarithms.
2. Write an equation equivalent to $\log_7 343 = 3$ using exponents.
3. Evaluate without a calculator.
   a. $\log_6 36$ b. $\log_6 6$ c. $\log_6 1$ d. $\log_6 \frac{1}{6}$

In 4–7, use the definition of logarithm to evaluate without a calculator.

4. $\log_5 625$
5. $\log_6 5^0$
6. $\log_5 0.2$
7. $\log_{32} 256$

In 8–10, evaluate without a calculator.

8. $\log 10{,}000$
9. $\log(100 \text{ million})$
10. $\log 0.001$

In 11–14, solve.

11. $\log_x 625 = 4$
12. $10^x = 6$
13. $\log_{10} x = 5.432$
14. $\log_{10} x = -1.8$
15. **Fill in the Blank** The function with equation $y = \log_6 x$ is the inverse of the function with equation _?_.
16. Consider the function described by $f(x) = \log_{5.5} x$.
   a. State the domain of f.
   b. **True or False** The graph of f contains the point (1, 0).
   c. Which axis is an asymptote of the graph?
17. **Multiple Choice** If $\log_{10} p = 2.179$, then p must be between which of the following? (Do not use a calculator.)
   A 2 and 3 B 1 and 10 C 10 and 100 D 100 and 1000

APPLYING THE MATHEMATICS

18. **Multiple Choice** In which interval is $\log_8 73$? Justify your answer.
   A $0 \le x < 1$ B $1 \le x < 2$ C $2 \le x < 3$ D $9 \le x < 10$

In 19 and 20, use this information. The pH level of a solution of pure water with bicarbonate concentration *b* and carbonic acid concentration c is given by pH $= 7 + \log\left(\frac{b}{c}\right)$, where *b* and c are given in moles/liter. Find the pH level of a solution with the given concentrations.

19. $b = 32, c = 3$

20. $b = 20c$

21. a. Use a calculator to find the common logarithm of 3^{100}.
    b. How many digits does 3^{100} have?

22. a. Use a calculator to find the common logarithm of -3.
    b. Explain your calculator's answer in Part a.

23. Consider the numbers 1776, 177,600, and 17,760,000,000,000.
    a. Write each number in scientific notation.
    b. Find the common logarithm of each number.
    c. How are the answers to Parts a and b related?

24. Solve for x: $\log_8(4x - 3) = \log_8 33$.

25. The intensity of an earthquake is often described using the Richter scale. The values on the scale are common logarithms of measures of the amplitude of the ground's vibrations. An earthquake rated 5.5 is considered moderate, while one rated 6.5 is quite severe. How many times as large are the vibrations for an earthquake rated 6.5 than for one rated 5.5?

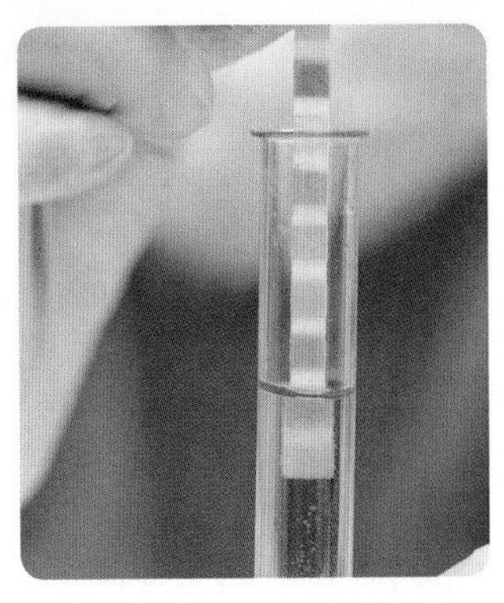

A pH test measures acidity on a scale from 1 to 14. A low number means the solution is acidic; a high number means it is more alkaline.

REVIEW

26. Suppose f is defined by $f(x) = x^{\frac{5}{7}}$. **(Lesson 9-2)**
    a. State the domain and range of f.
    b. Write an equation for f^{-1}.
    c. Graph f and f^{-1} on the window $-2.5 \le x \le 2.5$, $-2.5 \le y \le 2.5$.

27. Give a decimal equal to $343^{-\frac{2}{3}}$. **(Lesson 9-2)**

28. Rewrite $\sqrt[7]{xy^2}$ without a radical sign. **(Lesson 9-2)**

29. Graph the functions f and g with $f(x) = x^{\frac{2}{3}}$ and $g(x) = \sqrt[3]{x^2}$. Are they the same function? Why or why not? **(Lesson 9-2)**

30. The power P of a radio signal varies inversely as the square of the distance d from the transmitter. That is, $P = \frac{k}{d^2}$. **(Lessons 9-2, 9-1)**
    a. Solve for d, writing your solution with radicals.
    b. Write your solution without radical notation.

31. A five-year certificate of deposit is purchased for $5000. If the certificate paid 5.5% interest compounded quarterly, find the value of the certificate at maturity. **(Lesson 2-5, Previous Course)**

EXPLORATION

32. Find out what prompted John Napier (1550–1617) of Scotland to invent logarithms.

QY ANSWERS

1. D
2. a. 1.661; $10^{1.661} \approx 45.8$
   b. 2.661; $10^{2.661} \approx 458$
   c. -2.339; $10^{-2.339} \approx 0.00458$
3. $\frac{2}{3}$

Lesson 9-4 e and Natural Logarithms

Vocabulary

e
compounded continuously
natural logarithm
ln *x*
natural logarithm function

BIG IDEA The irrational number e is commonly used as a base for logarithms because of its special properties related to growth.

Mental Math

Solve each equation.

a. $2^x = 512$

b. $2^y = \frac{1}{1024}$

c. $2^z = 1$

Recall the Compound Interest Formula $A = P\left(1 + \frac{r}{n}\right)^{nt}$ where A is the value of an investment of P dollars earning an interest rate r, compounded n times per year for t years. Suppose you invest P dollars for 1 year at a huge 100% annual interest rate. Then $r = 1$. What is the balance after 1 year? The value depends entirely on n, the number of times the investment is compounded during the year. When $P = \$1$, the formula gives the value of the investment per dollar invested.

Activity 1

Step 1 **a.** In a spreadsheet, use the Compound Interest Formula to find the value of A, given $P = 1$ and $r = 1$, for $n = 1, 2, 4, 6, 12, 24, 36$, and 365 compounding periods per year. Assume $t = 1$.

b. Draw a conclusion about the value of A as n increases.

c. Try a value for n greater than 365. Is your conclusion confirmed?

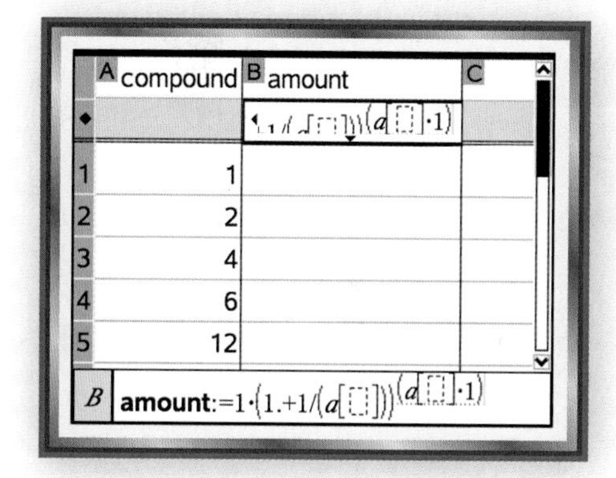

Step 2 As n becomes very large, A gets closer to the real number called e. Your calculator will display an approximate value for e. Record this value to the nearest thousandth.

Step 3 As n increases, because $\left(1 + \frac{1}{n}\right)^n$ becomes closer and closer to e, $y = \left(\left(1 + \frac{1}{n}\right)^n\right)^x$ comes closer and closer to $y = e^x$.

a. Graph $f_n(x) = \left(\left(1 + \frac{1}{n}\right)^n\right)^x$ on the window $-3 \leq x \leq 3, -1 \leq y \leq 10$, for $n = 1, 2, 4, 12$.

b. Graph $g(x) = e^x$ on the same axes.

c. When $x = 0$, what is true about all the graphs?

d. Try to find a value of n so that $y = e^x$ and $y = \left(\left(1 + \frac{1}{n}\right)^n\right)^x$ are indistinguishable on the window used in Part a.

e. Do you think that $\left(1 + \frac{1}{n}\right)^n$ is ever equal to e? Why or why not?

In Activity 1, you used the Compound Interest Formula $A = P\left(1 + \frac{r}{n}\right)^{nt}$ with $P = \$1$, $r = 100\%$, $t = 1$ year, for values of n ranging from 1 to 365. These values correspond to interest being compounded from annually to daily. Values for A as the number of compoundings increases from 1 per day to 1 per second appear in the table below.

Compound Schedule	n	$P\left(1 + \frac{r}{n}\right)^{nt}$	Balance in \$ after 1 year
every day	365	$1\left(1 + \frac{1}{365}\right)^{365}$	2.71457
every hour	8,760	$1\left(1 + \frac{1}{8760}\right)^{8760}$	2.71813
every minute	525,600	$1\left(1 + \frac{1}{525600}\right)^{525600}$	2.71828
every second	31,536,000	$1\left(1 + \frac{1}{31,536,000}\right)^{31,536,000}$	2.71828

Leonhard Euler (1707–1783)

The balance after one year increases with greater numbers of compounding periods. However, no number of compoundings yields an infinite sum of money. As the number of compoundings increases, the value of the investment approaches the number *e*.

The number **e** is the limit of the sequence $a_n = \left(1 + \frac{1}{n}\right)^n$. The symbol *e* was first used to represent this number by Leonhard Euler, who discovered its importance. The number *e* is irrational; its decimal expansion, neither repeats nor terminates, but has been computed to many places. A decimal approximation of *e* to 150 decimal places is

$$e \approx 2.71828182845904523536028747135266249775724709369995\\95749669676277240766303535475945713821785251664274274663\\91932003059921817413596629043572900334295260595630738\ldots$$

e and Continuous Change

Examine the banking advertisement at the right carefully. Notice that a distinction is made between the interest *rate* and the annual *yield*.

Here is how the yield is calculated from the rate. The ad states that interest compounds monthly, so on an investment of P dollars from \$9 to \$24,999, the bank uses a monthly interest rate of $\frac{0.0575}{12}$. After one month, the interest earned is $P \cdot \frac{0.0575}{12}$. So, after one month the total in the account is $P + P \cdot \frac{0.0575}{12} = P\left(1 + \frac{0.0575}{12}\right)$. Thus, the monthly scale factor is $\left(1 + \frac{0.0575}{12}\right)$. If no deposits or withdrawals are made, the balance after one year is

$$P\left(1 + \frac{0.0575}{12}\right)^{12} \approx 1.05904P.$$

Because the investment grows at 5.904% per year, we say that its annual yield is 5.904%.

TENTH SAVINGS BANK

introduces Money Market Plus

BALANCE	RATE	YIELD
\$9 to \$9,999	5.75%	5.904%
\$10,000 to \$24,999	5.75%	5.904%
\$25,000 to \$49,999	5.90%	6.062%
\$50,000 Plus	6.10%	6.273%

All you need to open a Money Market Plus is \$1,000.00. Tenth Savings Bank starts giving right away, with current rates* paid on your entire balance and with interest that compounds monthly.

Tenth Savings Bank knows how to give.

*Rates subject to change without notice.

In general, when the initial investment P is invested at an annual interest rate r compounded n times per year, the interest rate in each period is prorated as $\frac{r}{n}$. The balance at the end of one year is $P\left(1 + \frac{r}{n}\right)^n$ and the annual yield is $\left(1 + \frac{r}{n}\right)^n - 1$. At the end of t years, the amount A the investment is worth is given by $A = P\left(1 + \frac{r}{n}\right)^{nt}$.

Activity 2

Step 1 Calculate to the nearest millionth.

a. $\left(1 + \frac{0.05}{365}\right)^{365}$ and $e^{0.05}$

b. $\left(1 + \frac{0.06}{365}\right)^{365}$ and $e^{0.06}$

c. $\left(1 + \frac{0.085}{365}\right)^{365}$ and $e^{0.085}$

Step 2 What do you notice about the pairs of calculated values in Step 1?

Activity 2 shows that when n is large, $\left(1 + \frac{r}{n}\right)^n \approx e^r$. Since $\left(1 + \frac{r}{n}\right)^n - 1$ is the annual yield, the limit indicates that, as the number of compounding periods increases, the annual yield is $e^r - 1$. When this yield is used to calculate interest, the interest is said to be **compounded continuously**.

Example 1

If \$500 is put in a saving account at an 8% annual rate compounded continuously, calculate

a. the annual yield.

b. the value of the investment after one year.

Solution

a. The annual yield at an interest rate r is $e^r - 1$, and here $r = 0.08$.
So the yield is $e^{0.08} - 1 \approx 1.083287 - 1 \approx 8.329\%$.

b. After one year the account grows to
$500 + 500(e^{0.08} - 1) = 500(e^{0.08}) \approx \$541.64$.

After t years, the account in Example 1 would contain $500(e^{0.08})^t = 500e^{0.08t}$. This is an instance of the following general formula.

Continuous Change Model

If a quantity P grows or decays continuously at an annual rate r, the amount A after t years is given by $A = Pe^{rt}$.

GUIDED

Example 2

Suppose \$500 is invested in an account paying an 8% annual interest rate.

a. Give the balance after 4.5 years if interest is compounded continuously.

b. How does this balance compare with quarterly compounding?

(continued on next page)

Solution

a. Use the Continuous Change Model $A = Pe^{rt}$ with $P = 500$, $r = 0.08$, and $t = 4.5$.

$$A = 500e^{0.08 \cdot \underline{?}} \approx 500(\underline{\ ?\ }) \approx \underline{\ ?\ }$$

The balance after 4.5 years is $\underline{\ ?\ }$.

b. Use the formula $A = P\left(1 + \frac{r}{n}\right)^{nt}$, with $P = 500$, $r = 0.08$, $n = 4$, and $t = 4.5$.

$$A = 500(1 + \underline{\ ?\ })^{?} = 500(\underline{\ ?\ }) = \$714.12$$

Continuous compounding earns $2.54 more in this case.

The Graph of $y = e^x$

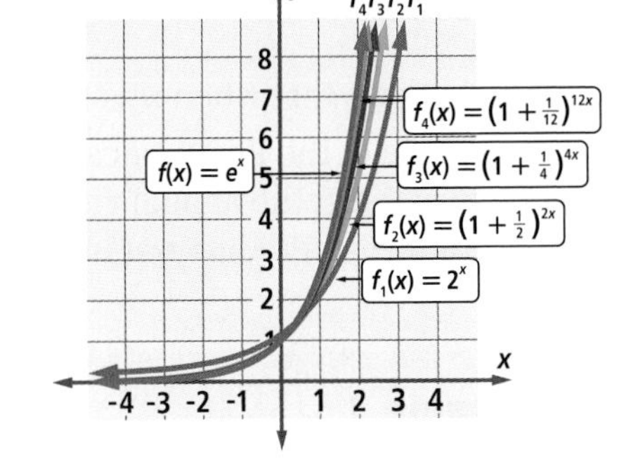

In Activity 1, you should have found that as n increases, the graphs of the exponential functions f_n where

$$f_n(x) = \left(1 + \frac{1}{n}\right)^{nx} = \left(\left(1 + \frac{1}{n}\right)^n\right)^x$$

approach the graph of the function with equation $f(x) = e^x$. This function is the exponential function with base e.

Like other exponential functions, the exponential function with base e has the set of real numbers as its domain and the set of positive real numbers as its range. The x-axis is a horizontal asymptote for negative x-values. Although its graph gets steep for large positive x-values, the graph of $y = e^x$ does not have any vertical asymptotes.

Also, like other exponential functions, this function has an inverse with equation $x = e^y$. This equation is equivalent to $y = \log_e x$. That is, the inverse of $f(x) = e^x$ has equation $f^{-1}(x) = \log_e x$.

Natural Logarithms

The inverse of the exponential function with base e is the logarithm function to the base e. The values of this function are so important that they have a special name and symbol.

Definition of Natural Logarithm

The **natural logarithm** of x is the logarithm of x to the base e, written **ln x**. That is, for all x, $\ln x = \log_e x$.

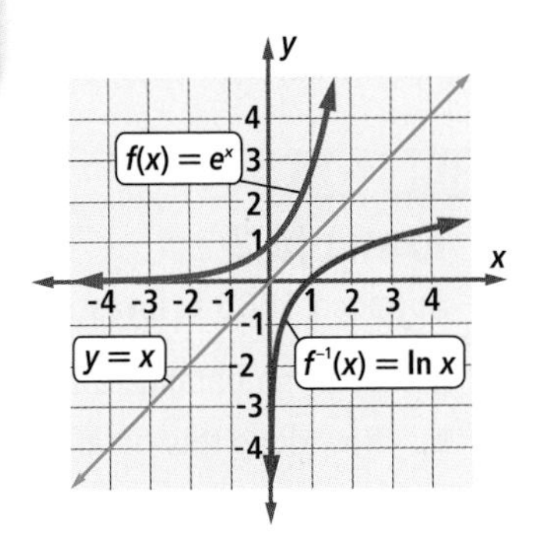

The function with equation $y = \ln x$ is the **natural logarithm function**. Note that $y = \ln x$ if and only if $e^y = x$. The graphs of the exponential function with base e and its inverse, the natural logarithm function, are shown at the right. As with all inverse functions, the graphs are reflection images of each other over the line with equation $y = x$.

Natural logarithms are used in many applications and in calculus. Example 3 illustrates one situation.

Example 3

Under certain conditions, the height h in feet above sea level can be approximated from the atmospheric pressure P in pounds per square inch (psi) using the equation $h = \frac{\ln P - \ln 14.7}{-0.000039}$. Human blood "boils" at 0.9 psi, and such a situation is fatal. At what height above sea level will blood "boil" in an unpressurized airplane cabin?

Solution Substitute 0.9 for P, and evaluate.

$$h = \frac{\ln 0.9 - \ln 14.7}{-0.000039} \approx \frac{-0.10536 - 2.6878}{-0.000039} \approx 71{,}621$$

Without a pressurized cabin, heights greater than 72,000 feet above sea level would be fatal.

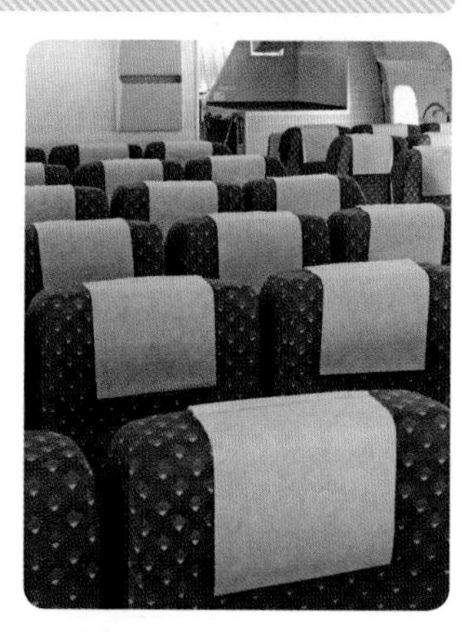

Under pressure Airplane cabins are pressurized during flights above 10,000 feet to maintain a sufficient level of oxygen in the cabin.

Questions

COVERING THE IDEAS

1. a. Estimate $\left(1 + \frac{1}{n}\right)^n$ to the nearest millionth when n equals
   i. 1000. ii. 1,000,000.
   b. As n gets larger, what number does $\left(1 + \frac{1}{n}\right)^n$ approach?
2. a. Evaluate $\left(1 + \frac{7}{1000}\right)^{1000}$.
   b. Evaluate $\left(1 + \frac{7}{1{,}000{,}000}\right)^{1{,}000{,}000}$.
   c. As n increases, what number does $\left(1 + \frac{7}{n}\right)^n$ approach?
   d. About how close is the value in Part b to the answer in Part c?

In 3 and 4, suppose \$3,000 is invested in an account paying 5.5% annual interest for 5 years.

3. If interest is compounded continuously, find each of the following.
   a. the annual yield
   b. the amount the investment is worth after one year
4. Suppose the account is left for 5 years. Give the balance if interest is compounded as follows.
   a. continuously b. monthly

In 5–7, evaluate using a calculator.

5. $e^{1.82}$ 6. $e^{-0.6}$ 7. $\ln 0.55$

In 8–13, evaluate without a calculator.

8. e^0 9. $\ln 1$ 10. $\ln e$
11. $\ln(e^7)$ 12. $\log 10$ 13. $\ln(\log 10)$

14. a. State the domain and range of $f: x \rightarrow e^x$.
    b. State the domain and range of $g: x \rightarrow \ln x$.
15. Refer to Example 3. A geologist measured atmospheric pressure to be about 10 psi. Estimate the altitude of the geologist.

APPLYING THE MATHEMATICS

16. a. To the nearest thousandth, compute log 15, log 150, log 1500, log 15000. When the argument is multiplied by 10, how does its common logarithm change?
    b. Repeat Part a using natural logarithms.
    c. Find ln 10. How is it related to your answers in Part b?

17. The population of Ireland in 2008 was 4,156,119 with an average annual growth rate of 1.3%. Use this growth rate and the Continuous Change Model to predict Ireland's population in 2020.

In 18 and 19, a particular satellite has a radioisotope power supply, with power output given by $P = 60e^{-\frac{t}{300}}$ where t is the time (in days) and P is the power output (in watts).

18. What is the power output at the end of its first year?

19. When the power output drops below 6 watts, there will be insufficient power to operate the satellite. Use a calculator to estimate how long the satellite will remain operable.

20. It can be shown that the velocity V in kilometers per second reached by a rocket when all its propellant is burned is given by $V = c \ln R$, where c is the exhaust velocity of the engine and R is the ratio $\frac{\text{takeoff weight}}{\text{burnout weight}}$ for the rocket. Space shuttle engines produce exhaust velocities of 4.6 kilometers per second. If the ratio R for a space shuttle is 3.5, what velocity can it reach from its own engines?

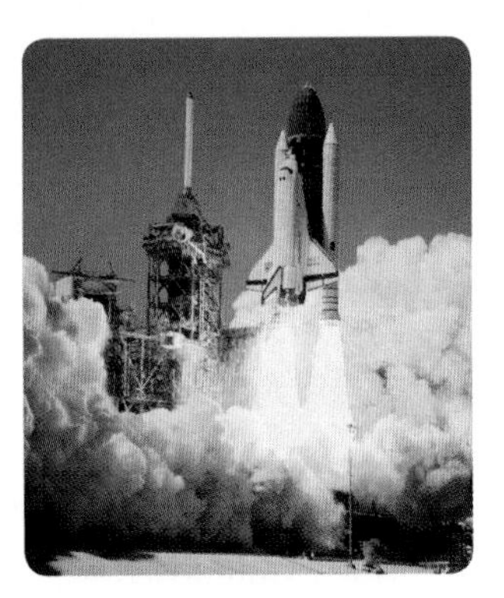

Launch of Space Shuttle Challenger from the Kennedy Space Center in Florida on April 4, 1983

REVIEW

In 21 and 22, give an equation for the inverse of the function f.

21. $f(x) = 2^x$

22. $f(x) = \log_8 x$ **(Lesson 9-3)**

23. Write a logarithmic equation equivalent to $\left(\frac{1}{8}\right)^{\frac{1}{3}} = \frac{1}{2}$. **(Lesson 9-3)**

24. Evaluate $\log_{10} \sqrt[3]{10}$ without a calculator. **(Lesson 9-3)**

25. **Fill in the Blank** Use =, >, or <: $900^{\frac{5}{7}}$ __?__ $\left(\sqrt[5]{900}\right)^7$. **(Lesson 9-1)**

26. Evaluate tan π without a calculator. **(Lesson 4-4)**

27. If x varies directly as y^2 and y varies directly as z^3, what can be said about how x varies with z? **(Previous Course)**

EXPLORATION

28. The value of e^x can be given by the following infinite sum.

$$e^x = 1 + \sum_{n=1}^{\infty} \frac{x^n}{n!} = 1 + \frac{x}{1!} + \frac{x^2}{2!} + \frac{x^3}{3!} + \frac{x^4}{4!} + \cdots$$

    a. Evaluate e^1 using the five terms shown in the sum. How close is your answer to the value given in this lesson?
    b. How many terms of the sum need to be used to obtain an estimate for e accurate to the thousandths place?

Lesson

9-5 Properties of Logarithms

BIG IDEA Logarithms turn products into sums, quotients into differences, and powers into products.

The word "logarithm" was first used by John Napier of Scotland, who is usually credited as the inventor of logarithms. In his 1614 brochure, *Mirifici logarithmorum canonis descriptio* (A Description of the Wonderful Law of Logarithms), Napier outlined the properties of logarithms that have made them so useful for four centuries. Because logarithm functions are the inverses of exponential functions, for every property of powers, there is a corresponding property of logarithms.

MIRIFICI
Logarithmorum
Canonis descriptio,
Ejusque usus, in utraque
Trigonometria; ut etiam in
omni Logistica Mathematica,
Authore ac Inventore,
IOANNE NEPERO,
Barone Merchistonii,
EDINBURGI,
Ex officinâ ANDREÆ HART

Mental Math

Simplify.

a. x^1

b. x^0

c. $x^a \cdot x^b$

d. $\frac{x^a}{x^b}$

e. $(x^a)^b$

Logarithms of Specific Numbers

Recall that any positive number except 1 can be the base b of a logarithm, and that $x = \log_b m$ if and only if $b^x = m$. From this definition, you can deduce properties of logarithms corresponding to each of the properties of exponents listed in Lesson 9-2.

For example, because $b^0 = 1$ for any nonzero b, applying the definition of logarithm gives $\log_b 1 = 0$. In words, the logarithm of 1 to *any* base is zero. For this reason, the graph of $f(x) = \log_b x$ contains (1, 0) for any base b. You can check this on the graphs in Lesson 9-3.

Theorem (Logarithm of 1)

For any base b, $\log_b 1 = 0$.

Consider next the property that $b^1 = b$ for any real number b. From the definition of logarithm, we obtain a corresponding property of logarithms: $\log_b b = 1$.

Theorem (Logarithm of the Base)

For any base b, $\log_b b = 1$.

For instance, let $b = 10$ in this theorem. Then $\log 10 = 1$. If you let $b = e$, then the theorem shows that $\log_e e = \ln e = 1$.

A third basic property that leads to specific values of logarithms of numbers is the most subtle. Consider the property that $b^n = b^n$. Think of the left side as a single number. Think of the right side as a number raised to a power. That is, think $(b^n) = (b)^{(n)}$. Now apply the definition of logarithm to get $\log_b(b^n) = n$.

Theorem (Logarithm of a Power of the Base)

For any base b and any real number n, $\log_b (b^n) = n$.

QY1

QY1

Evaluate $\log_2 32$.

Logarithms of Products and Quotients

Activity 1

MATERIALS graphing utility with slider capability

Step 1 Make a slider with domain from –4 to 4 and label its value r. Graph $f(x) = \log(10x)$ and $g(x) = r + \log x$. How does changing the value of r affect the graph of g?

Step 2 Adjust the value of r until the graphs of f and g appear to coincide. Record the value of r.

Step 3 Redefine $f(x)$ as $\log(100x)$ and repeat Step 2. Record the new value of r and make a conjecture: how is the value of r related to the coefficient of x? Test your conjecture with $f(x) = \log(1000x)$, $f(x) = \log(\sqrt{10}x)$, and $f(x) = \log\left(\frac{1}{10}x\right)$.

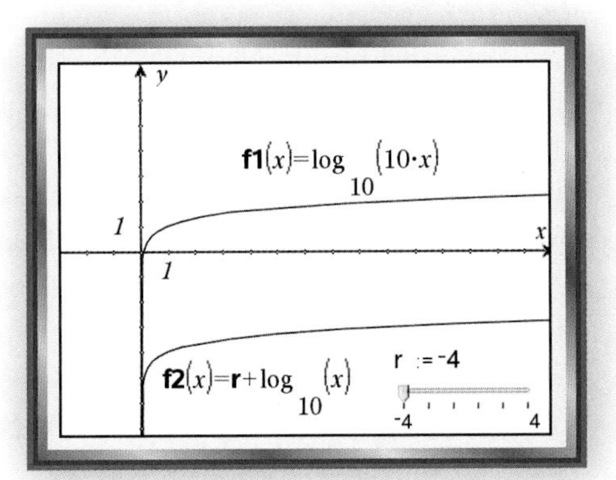

In Activity 1, you worked with the equations for functions until their graphs were identical. This should have led you to some conjectures involving the values of these functions. By applying properties of powers, we now prove what may have been your conjecture.

Theorem (Logarithm of a Product)

For any base b and for any positive real numbers x and y,
$$\log_b(xy) = \log_b x + \log_b y.$$

In words, this theorem says that the logarithm of a product is the sum of the logarithms of its factors.

Proof Suppose you know two positive numbers x and y and their logarithms to base b. Call these logarithms m and n. That is, let $\log_b x = m$ and $\log_b y = n$. Then by the definition of logarithm, $x = b^m$ and $y = b^n$.

To find the logarithm of the product xy, express xy as a power of the base b.

$xy = b^m \cdot b^n$ — Multiplication Property of Equality

$xy = b^{m+n}$ — Product of Powers Property

Therefore, by the definition of logarithm, $\log_b(xy) = m + n$.

Substituting back for m and n gives the theorem.

People in the early 1600s viewed logarithms as a miracle for computation because this theorem says that logarithms convert multiplication into addition. Until the late 1900s, people routinely used tables of logarithm values to multiply and divide numbers, and take powers and roots.

Example 1

Using $\log 2 \approx 0.301$ and $\log 3 \approx 0.477$, compute without a calculator.

a. $\log 6$ **b.** $\log 20$

Solution

a. Because $6 = 2 \cdot 3$, you can use the Logarithm of a Product Theorem.

$$\log 6 = \log(2 \cdot 3) = \log 2 + \log 3$$
$$= 0.301 + 0.477 = 0.778$$

b. $20 = 2 \cdot 10$. So, $\log 20 = \log(2 \cdot 10) = \log 2 + \log 10$
$= 0.301 + 1 = 1.301$.

QY2

> **QY2**
>
> Compute using the approximations from Example 1 and no technology.
>
> **a.** log 12 **b.** log 30

The Logarithm of a Product Theorem sets up a correspondence between two operations: multiplying two numbers is equivalent to adding their logarithms. This leads to other logarithm properties. For example, just as adding logs is equivalent to multiplying numbers, subtracting logs is equivalent to dividing numbers.

Theorem (Logarithm of a Quotient)

For any base b and for any positive real numbers x and y,

$$\log_b\left(\frac{x}{y}\right) = \log_b x - \log_b y.$$

This property can be proved using the corresponding exponential property $\frac{b^m}{b^n} = b^{m-n}$. You are asked to write a proof in Question 6.

Activity 2

Given that $\log_5 2 = r$ and $\log_5 3 = s$, copy the table below and fill in as many logarithms as possible, without using technology to compute unknown values.

n	$\log_5 n$	n	$\log_5 n$	n	$\log_5 n$	n	$\log_5 n$
1	?	6	?	11	?	25	?
2	r	7	?	15	?	50	?
3	s	8	?	18	?	$\frac{1}{5}$	?
4	$r + r$	9	?	20	?	$\frac{1}{2}$	?
5	?	10	?	24	?	$\frac{3}{5}$	?

Logarithms of Powers

Examine these computations.

$$\log_b 9 = \log_b(3^2) = \log_b(3 \cdot 3) = \log_b 3 + \log_b 3 = 2\log_b 3$$
$$\log_b 16 = \log_b(2^4) = \log_b(2 \cdot 2 \cdot 2 \cdot 2)$$
$$= \log_b 2 + \log_b 2 + \log_b 2 + \log_b 2 = 4\log_b 2$$

The general property is the Logarithm of a Power Theorem.

Theorem (Logarithm of a Power)

For any base b, any positive real number x, and any real number m, $\log_b x^m = m \log_b x$.

The proof of this theorem uses the Power of a Power Property: $(b^m)^n = b^{mn}$. You are asked to prove the theorem in Question 7.

Historically, one of the most important applications of logarithms was finding powers and roots of numbers.

Example 2

Using ln 17 ≈ 2.833, find the value without using technology.

a. $\ln 17^{10}$ b. $\ln \sqrt[4]{17}$

Solution

a. By the Logarithm of a Power Theorem,
$\ln 17^{10} = 10 \ln 17 \approx 10(2.833) = 28.33$.

b. Notice that $\sqrt[4]{17} = 17^{\frac{1}{4}}$. Therefore,
$\ln \sqrt[4]{17} = \ln 17^{\frac{1}{4}} = \frac{1}{4} \cdot \ln 17 = 0.708$.

It is often helpful to use combinations of these properties to simplify expressions containing logarithms.

GUIDED

Example 3

Rewrite as a logarithm of a single number.

a. $\log 40 + 2 \log 5$ b. $\log_2 \frac{3}{5} + \log_2 \frac{5}{2} - \log_2 3$

Solution

a. First by The Logarithm of a Power Theorem, $2 \log 5 = \log 5^2 = \log 25$.
Then by The Logarithm of a Product Theorem,
$\log 40 + \log 25 = \log$ __?__.

b. First by The Logarithm of a Product Theorem, $\log_2 \frac{3}{5} + \log_2 \frac{5}{2} = \log_2 \frac{3}{2}$.
Then by The Logarithm of a Quotient Theorem,
$\log_2 \frac{3}{2} - \log_2 3 = \log_2$ __?__.

You can also use these properties in expressions with variables.

Example 4

Rewrite $\frac{1}{3}\ln t + \ln 5$ as a single natural logarithm.

Solution

$\frac{1}{3}\ln t + \ln 5n = \ln t^{\frac{1}{3}} + \ln 5$ Logarithm of a Power Theorem

$= \ln(t^{\frac{1}{3}} \cdot 5)$ Logarithm of a Product Theorem

$= \ln(5t^{\frac{1}{3}})$ Commutative Property of Multiplication

 QY3

QY3

Substitute a value for t in $\frac{1}{3}\ln t + \ln 5$ and $\ln\left(5t^{\frac{1}{3}}\right)$ to see if they are equal.

The Number of Digits in a Number

One of the uses of the Logarithm of a Power Theorem is to estimate the size of numbers that are too large even for many calculators or computers to write out. For example, large prime numbers have always been of interest to mathematicians, computer scientists, and cryptographers. In 1999, the Electronic Frontier Foundation announced a $100,000 prize to the first person or team who found a prime number with more than 10 million digits.

Example 5

The number $2^{32,582,657} - 1$ was the largest known prime when it was discovered in 2006 by a GIMPS (Greatest Internet Mersenne Prime Search) team headed by Curtis Cooper and Steven Boone of Central Missouri State University. Did they win the $100,000 prize?

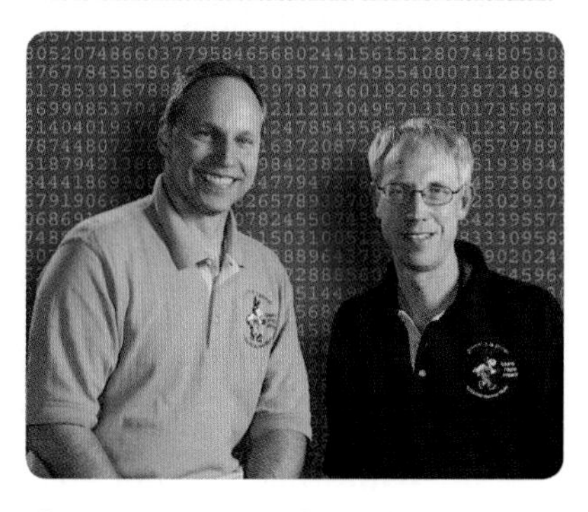

Steven Boone and Curtis Cooper

Solution Let $x = 2^{32,582,657}$. If you try to evaluate $2^{32,582,657}$ directly on a typical calculator, an "overflow" error will occur. However, the Logarithm of a Power Theorem can be used to approximate this large power. To estimate this number as a decimal (base 10), take the common log of each side.

$\log x = \log(2^{32,582,657})$

$\log x = 32,582,657 \log 2$ Logarithm of a Power Theorem

Using $\log 2 \approx 0.301029995664$,

$\log x \approx 32,582,657(0.301029995664) = 9,808,357.0954316$

$x \approx 10^{9,808,357.0954316}$ definition of $\log x$

$= 10^{9,808,357} \cdot 10^{0.0954316}$ Product of Powers Property

$= 10^{9,808,357} \cdot 1.24575$

So $2^{32,582,657}$ is about $1.24575 \cdot 10^{9,808,357}$, a 9,808,358-digit number beginning with 124575... . This number was about 192,000 digits short of winning the $100,000 prize.

Questions

COVERING THE IDEAS

In 1–5, assume that all variables are positive.

a. Rewrite as the logarithm of a single number.

b. Evaluate without using a calculator.

1. $\log 50 + \log 2$
2. $\log 100^7$
3. $\log_b xb^2 - \log_b x$
4. $\log_b \frac{b}{y} + \log_b \frac{y}{z} + \log_b \frac{z}{b}$
5. $\log_b \frac{b}{3} + \log_b 3b$
6. Prove the Logarithm of a Quotient Theorem by mimicking the proof of the Logarithm of a Product Theorem.
7. Copy and complete this proof of the Logarithm of a Power Theorem. We want to prove that $\log_b(x^n) = n \log_b x$. Suppose $\log_b x = m$. Then, because of __?__, $b^m = x$. Taking each side of $b^m = x$ to the nth power, $(b^m)^n = x^n$. By the __?__ Property, $(b^m)^n = b^{mn}$. So by substitution, $\log_b(x^n) = \log_b(b^{mn}) = mn = n \log_b x$.

In 8–10, use the facts that $\log_7 11 \approx 1.232$ and $\log_7 4 \approx 0.712$ to estimate without a calculator.

8. $\log_7 44$
9. $\log_7\left(\frac{11}{4}\right)$
10. $\log_7(11^4)$
11. If $\log a = 3 \log b - 2 \log c$, find an expression for a that does not involve logarithms.
12. In August 2008, a team headed by Edson Smith at UCLA discovered that $2^{43,112,609} - 1$ is prime.
    a. How many digits does this number have?
    b. With what digit does it start?
    c. Did this team win the $100,000 prize from the Electronic Frontier Foundation?
13. a. How many digits does 17^{100} have?
    b. With what digit does it start?

APPLYING THE MATHEMATICS

14. Using $\ln 8 \approx 2.08$, use properties (not a calculator) to estimate $\ln \frac{e^5}{8}$.
15. Given $F = \frac{GmM}{d^2}$, write $\ln F$ in terms of $\ln G$, $\ln m$, $\ln M$, and $\ln d$.

In 16 and 17, write each expression as a single logarithm.

16. $\log x + a \log q$
17. $\log_2 p - \log_2 q + 1$
18. In Activity 2, logarithms base 5 of eight of the integers from 1 to 10 were written in terms of $r = \log_5 2$ and $s = \log_5 3$. You can also write the logarithms base 5 of five integers from 26 to 40 in terms of r and s.
    a. Identify the five integers.
    b. Write the logarithm base 5 of each integer in terms of r and s.

19. Using the information in Questions 8–10, estimate $\log_7 \sqrt[3]{121}$.

20. Which is larger, 700^{800} or 800^{700}? Justify your answer.

21. Prove that for all bases b and positive n, $\log_b\left(\frac{1}{n}\right) = -\log_b n$.

22. A portion of a table of common logarithms is shown at the right. Use the values in the table to find each value to the nearest 0.001.

a. log 493 b. log 0.493 c. $\log 4.93^3$

x	log(x)
4.900	0.690
4.910	0.691
4.920	0.692
4.930	0.693
4.940	0.694
4.950	0.695

23. Many CAS use properties of logarithms to simplify expressions. In each case shown below, state which property of logarithms justifies the equivalence, and show the underlying algebra steps.

	Input	Output
a.	$\log_{10}(6)+\log_{10}(5)$	$\log_{10}(30)$
b.	$\log_{10}(36)$	$2\cdot\log_{10}(6)$
c.	$\log_5\left(\frac{1}{125}\right)$	-3
d.	$\log_4(\sqrt{2})$	$\frac{1}{4}$

24. Find the value of each logarithm below without using technology. Assume that all variables are positive.

a. $\log_{10} 0.0001$ b. $\log_b \sqrt[2x]{b^x}$

c. $\log_7 49^2$ d. $\log(1000(x+y)) - \log(x+y)$

REVIEW

In 25 and 26, evaluate without a calculator. (Lessons 9-4, 9-3)

25. $\log_2 64$ 26. $\ln e^{27}$

27. If t is time in years and r is the annual interest rate, match the interest formulas to their compounding periods. **(Lessons 9-4, 2-4)**

a. $A(t) = Pe^{rt}$ I. annual compounding

b. $A(t) = P\left(1 + \frac{r}{n}\right)^{nt}$ II. periodic compounding

c. $A(t) = P(1 + r)^t$ III. continuous compounding

28. ***Skill Sequence*** Solve for x. **(Lesson 9-3)**

a. $\log_4 x = 3$ b. $\log_4(x + 2) = 2$ c. $\log_4(2x) = \log_4(4x - 1)$

29. Let A = the set of even integers from 30 to 45 and B = the set of integers from 40 to 50. **(Previous Course)**

a. List the elements of $A \cap B$. b. List the elements of $A \cup B$.

EXPLORATION

30. Use the Internet to find out what a Mersenne prime is, how many Mersenne primes are known, and why they are so named.

QY ANSWERS

1. 5

2. a. log 12 = log(2 • 2 • 3) = log 2 + log 2 + log 3 = 1.079

b. log 30 = log(10 • 3) = log 10 + log 3 = 1 + 0.477 = 1.477

3. Answers vary. Sample: Let $t = 2$. $\frac{1}{3}\ln 2 + \ln 5 \approx 1.8405 \approx \ln(5 \cdot 2^{\frac{1}{3}})$

Lesson

9-6 Solving Exponential Equations

Vocabulary

exponential equation

▶ **BIG IDEA** By using the definition of logarithm, you can solve equations in which the variable is in an exponent.

You already know how to solve an equation of the form $x^a = b$. Take each side to the $\frac{1}{a}$th power. For example, $x^4 = 45$ can be solved by taking each side to the one-fourth power. However, different techniques are needed if the variable is in the exponent rather than the base, such as in the equation $4^x = 45$. An equation with a variable in an exponent is called an **exponential equation**.

Mental Math

Evaluate.

a. $\log 5 + \log 4 + \log 50$

b. $\log 300 - \log 75 + \log 25$

c. $\log_2 \sqrt[3]{64}$

Using Logarithms to Solve Exponential Equations

Any equation of the form $b^x = a$, where a and b are positive, can be solved by using logarithms. If you take the logarithm of both sides of an exponential equation, the exponent will become a factor in a linear equation.

Example 1

Solve $5^x = 46$ to four decimal places.

Solution

$5^x = 46$ Given

$\log(5^x) = \log 46$ Take the logarithm of each side.

$x \log 5 = \log 46$ Logarithm of a Power Theorem

$x = \frac{\log 46}{\log 5}$ Divide each side by log 5.

$x \approx 2.37887$

Check Use a calculator in approximate mode to estimate $5^{2.37887}$.

Although any base for logarithms can be used to solve an exponential equation, sometimes one base is preferred over others. For example, the Continuous Change Formula from Lesson 9-4 gives rise to equations of the form $A(t) = Pe^{rt}$. When you start from equations of this form, using natural logarithms makes the solution easier because $\ln e = \log_e e = 1$.

Example 2

A family has \$34,500 in a savings account that is paying interest at a rate of 0.71%. If the interest is compounded continuously, how long would it take to grow to \$40,000?

Solution 1 Assume the growth rate stays constant. Use the Continuous Change Formula $A = Pe^{rt}$, where t represents the number of years. Then $A = 40{,}000$, $P = 34{,}500$, and $r = 0.0071$. Substitute these values.

$$40{,}000 = 34{,}500e^{0.0071t}$$

$$\frac{40{,}000}{34{,}500} = e^{0.0071t}$$

Take the natural logarithm of each side.

$$\ln\left(\frac{40{,}000}{34{,}500}\right) = \ln(e^{0.0071t})$$

Apply the Logarithm of a Power Theorem.

$$\ln\left(\frac{40{,}000}{34{,}500}\right) = (0.0071t) \cdot \ln e$$

Use the fact that $\ln e = 1$ and solve for t.

$$t = \frac{\ln\left(\frac{40{,}000}{34{,}500}\right)}{0.0071} \approx 20.8$$

So, if the interest rate remains constant, the money would grow to \$40,000 in about 20.8 years! Clearly, 0.71% is a very low interest rate.

Solution 2 Follow Solution 1 until the equation $\frac{40{,}000}{34{,}500} = e^{0.0071t}$.

Use the definition of natural logarithm.

$$0.0071t = \ln\left(\frac{40{,}000}{34{,}500}\right)$$

$$t = \frac{\ln\left(\frac{40{,}000}{34{,}500}\right)}{0.0071} \approx 20.8$$

Finding Exponential Models

In an exponential situation, two data points are sufficient to determine an exponential function that models the situation. In many cases, one of the points is measured at the start of an experiment.

The assumption of continuous change is appropriate in physical situations such as the decay of radioactive isotopes of elements. Similar to $A(t) = Pe^{rt}$, radioactive decay models follow the formula $y = ab^t$ where a is the original concentration and y is the concentration of the isotope after t years. In a continuous change model, we use e as the base. However, because the decay factor b depends on the substance decaying, we have to calculate the decay factor for an isotope just as we did in the Litvinenko investigation in Lesson 2-5.

Suppose that initially, there is some amount a of a substance with a half-life t_{half}. This is the amount of time after which $\frac{a}{2}$ of the substance remains. Therefore, $\frac{a}{2} = ab^{t_{\text{half}}}$, and so $\frac{1}{2} = b^{t_{\text{half}}}$. Thus, $\left(\frac{1}{2}\right)^{\frac{1}{t_{\text{half}}}} = b$.

Therefore, when a substance has half-life t_{half}, a formula for the amount of that substance that left after time t is

$$y = ab^t = a\left(\left(\frac{1}{2}\right)^{\frac{1}{t_{\text{half}}}}\right)^t = a\left(\frac{1}{2}\right)^{\frac{t}{t_{\text{half}}}}.$$

Example 3

Carbon-14 has a half-life of 5730 years. From this assumption, find the approximate age of a skull found by an archaeologist if the skull has 67% of its original carbon-14 concentration.

Solution We start with the half-life formula $y = ab^t$, where $y = 0.67a$.

So, $0.67a = a\left(\frac{1}{2}\right)^{\frac{t}{t_{\text{half}}}}$, and

$$0.67 = \left(\frac{1}{2}\right)^{\frac{t}{t_{\text{half}}}}.$$

Take the logarithm of both sides.

$$\log 0.67 = \log\left((0.5)^{\frac{t}{5730}}\right)$$

$$\log 0.67 = \frac{t}{5730}\log 0.5$$

$$\frac{t}{5730} = \frac{\log 0.67}{\log 0.5}$$

$$t = 5730 \cdot \frac{\log 0.67}{\log 0.5} \approx 3310$$

So, the skull is about 3,310 years old.

Check More than half the carbon-14 was left, so the age should be less than the half-life. It is. It checks.

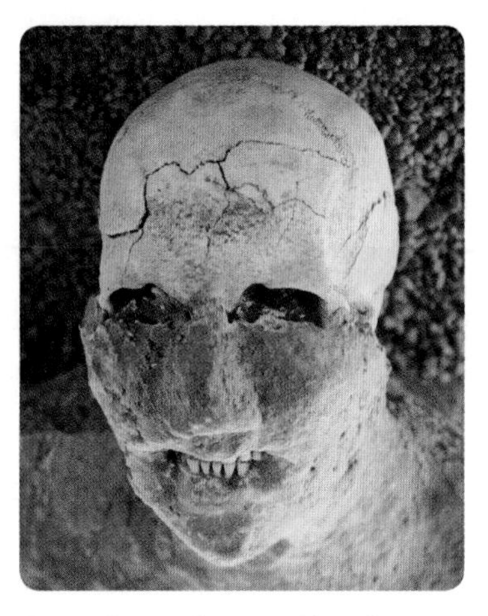

Cast of a skull, an artifact from the eruption of Mount Vesuvius, Italy, that destroyed Pompeii and Herculaneum, in 79 C.E.

The Change of Base Theorem

The equation $b^x = a$ can be solved for x using any base c. Take the logarithm of each side of the equation to that base.

$b^x = a$	Given
$\log_c(b^x) = \log_c a$	Take $\log_c$ of each side.
$x \log_c b = \log_c a$	Log of a Power Theorem
$x = \frac{\log_c a}{\log_c b}$	Divide both sides by $\log_c b$.

So the solution to $b^x = a$ is $x = \frac{\log_c a}{\log_c b}$. However, by the definition of log, $b^x = a$ is also equivalent to $x = \log_b a$. This proves a theorem about the quotient of logarithms with the same base.

Change of Base Theorem

For all values of a, b, and c for which the logarithms exist,

$$\log_b a = \frac{\log_c a}{\log_c b}.$$

The Change of Base Theorem allows you to evaluate a logarithm to any base on your calculator, as long as the calculator has either a common logarithm key or a natural logarithm key.

GUIDED

Example 4

Evaluate $\log_5 71$ to the nearest millionth.

Solution 1 Use the Change of Base Theorem. You may choose any base c. Choose $c = 10$ to use common logs, or $c = e$ to use natural logs. We first use common logs.

$$\log_5 71 = \frac{\log_{10} 71}{\underline{\ ?\ }} \approx 2.648552$$

Solution 2 Use natural logs.

$$\log_5 71 = \frac{\ln \underline{\ ?\ }}{\underline{\ ?\ }} \approx 2.648552$$

Check Use a calculator to show that $5^{2.648552} \approx 71.000009$.

Your CAS may have a built-in command to find logarithms to bases other than 10 or e, such as those shown below. Note that the base is the first argument on one machine, but the second argument in the other.

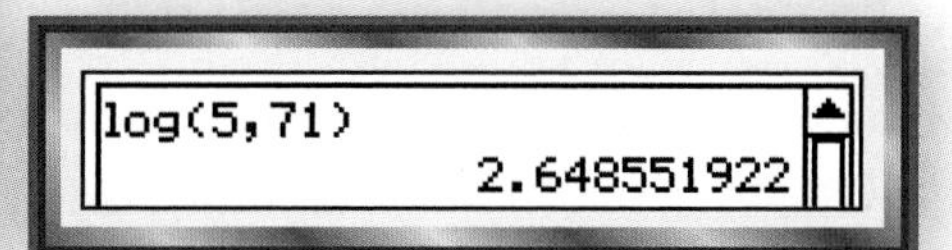

Questions

COVERING THE IDEAS

1. a. Solve the equation $5^x = 46$ using natural logs.
   b. Check your solution against that in Example 1.
2. Consider the equation $7^x = 690$.
   a. Estimate the solution graphically.
   b. Solve the equation using logs.

In 3 and 4, solve and check.

3. $7^x = 28$
4. $4^t = 0.4$
5. Suppose an art dealer in 2009 used carbon-14 dating to determine whether a painting was likely to have been painted by the great Italian artist and scientist Leonardo da Vinci (1452–1519). A specimen of paint was found to have 93% of the original amount of carbon-14. Is it plausible that the painting could be a da Vinci? Justify your answer.

6. In 2007, the population of Ecuador was estimated at 13.8 million, with an annual growth rate of 1.6%. If this rate continues, estimate when the population will reach 15 million.

A street scene in Quito, the capital of Ecuador

In 7 and 8, evaluate using the Change of Base Theorem.

7. $\log_5 16$

8. $\log_8 23$

APPLYING THE MATHEMATICS

9. The internet auction company eBay was started in 1995. In 1997, after two years in business, it had 341,000 registered users. After 11 years it had grown to 222 million registered users.
   a. Use exponential regression to find a function that models the number of users $n(t)$ after eBay had been in business t years.
   b. According to this model, how many registered users would eBay predict for their 15th year in business?

10. **Multiple Choice** Which expression(s) could be entered into a graphing utility to produce a graph of $y = \log_{13} x$?

    A $y = \frac{\log 13}{\log x}$ B $y = \frac{\log x}{\log 13}$ C $y = \frac{\ln x}{\ln 13}$ D $y = \ln\left(\frac{x}{13}\right)$

11. If money is invested at 6% interest compounded continuously, how long will it take for the money to double in value?

12. a. Find $\log_{11} 539$.
    b. Find $\log_{539} 11$.
    c. How are the answers to Parts a and b related?
    d. Use the Change of Base Theorem to explain the relationship in Part c.

13. State and prove the general theorem underlying Question 12.

14. The following formula is commonly used for estimating the age t in years of a specimen using the original number N_0 of radioactive atoms in the sample, the number N of radioactive atoms in the sample today, and the half-life $t_{0.5}$ of the substance: $t = \frac{t_{0.5}}{0.693} \ln\left(\frac{N_0}{N}\right)$.
    a. Check that this formula works for Example 3 by replacing $\frac{N_0}{N}$ with $\frac{1}{\frac{2}{3}}$.
    b. Prove that the formula is correct.

15. Solve the equation $A = Pe^{rt}$ for t, given that $P > 0$.

REVIEW

In 16 and 17, evaluate without using a calculator. (Lesson 9-5)

16. $\log 4 + \log 25$

17. $\log r - \log 100r$

In 18–20, evaluate without a calculator. (Lesson 9-3)

18. $6 \log_2 128$

19. $0.1 \log 0.1$

20. $\log 10^{0.05}$

In 21 and 22, use this situation from psychology. Learning and forgetting are often modeled with logarithm functions. In one experiment, subjects studied nonsense syllables (like "gpl") and were asked to recall them after *t* seconds. A model for remembering was found to be $P = 92 - 25 \ln t$, where *P* is the percent of students who remembered a syllable after *t* seconds. (Lesson 9-4)

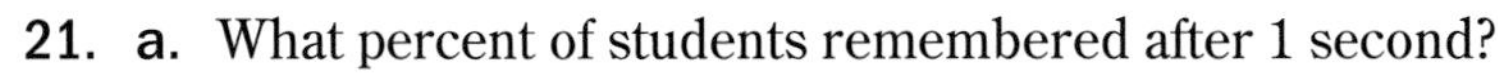

21. a. What percent of students remembered after 1 second?
    b. What percent remembered after 10 seconds?
22. Use a graphing utility to find the approximate time after which only half of the students remembered a syllable.
23. Consider the function g with $g(x) = \frac{x^3 - 12x^2 + 48x - 64}{x^2 - 8x + 16}$.
    a. **True or False** For all $x \neq 4$, g is a linear function.
    b. Justify your answer to Part a. **(Lessons 7-7, 7-3, 7-1)**
24. **True or False** When I have repeatedly flipped tails with a fair coin, I am now more likely to flip heads. Explain your answer. **(Lesson 6-8)**
25. Mean lengths of Illinois channel catfish of various ages below.

Age (years)	1	2	3	4	5	6	7	8	9	10
Length (inches)	6.4	9.6	12.6	14.3	16.7	18.5	21.0	22.6	25.6	26.6

Source: Illinois Department of Conservation

    a. Find the line of best fit for predicting length from age.
    b. Interpret the slope of the line.
    c. Use the line to predict the length of a twelve-year-old channel catfish in Illinois.
    d. Suggest a reason for being cautious about your prediction in Part c.
    e. Find the correlation coefficient r between the age of the fish and its mean length.
    f. Interpret the sign of the correlation. **(Lesson 2-3)**

No strings attached
Tying a string around your finger is an example of prospective memory, which uses cues or events to help a person remember to do something.

EXPLORATION

26. a. Generalize the answer to Question 11. That is, give a rule for the length of time t it takes to double an investment at interest rate r% assuming continuous compounding.
    b. A rule of thumb used by some people to determine the length of time t in years it would take to double their money at interest rate r is called the *Rule of 72*. The time t is estimated to be $t = \frac{72}{r}$. To see how this compares with your rule for Part a, graph the two rules with a graphing utility on the same axes.
    c. On investments at rates from 2% to 12%, what is the greatest difference between the actual doubling time and the time given by the *Rule of 72*?

Lesson 9-7

Linearizing Data to Find Models

BIG IDEA Some data can be modeled by logarithm functions.

You have seen linear, quadratic, and exponential functions used to model sets of data. Some of these models are theory-based because they are deduced from theoretical assumptions, such as population growth being exponential or area being quadratic. Others are based only on the shape of the scatterplot of the data set.

Mental Math

Given that $\log 2 \approx 0.301$, estimate each value.

a. log 20 **b.** log 4

c. log 8 **d.** log 0.2

Modeling with a Logarithm Function

An experiment was conducted to determine the effect of study time on the ability to recall the meanings of unfamiliar words. Subjects in the experiment were given t seconds to study each word and its meaning. The value of t ranged from 1 to 30 seconds. After a short time, each subject was asked for the meanings of the words. The mean percent P of correctly recalled word meanings was recorded. For instance, the point (5, 53) indicates that when subjects were given 5 seconds to concentrate on each word, on average they recalled 53% of the meanings correctly. The relationship between P and t is nonlinear.

t	P
1	30
2	43
5	53
10	65
15	74
20	76
30	85

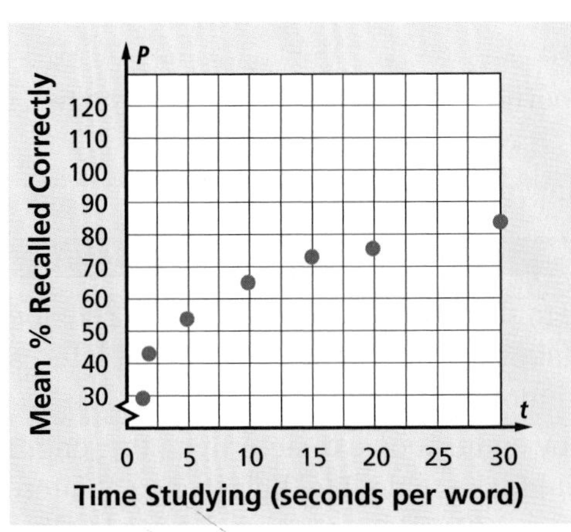

The shape of the graph suggests that an appropriate model may be logarithmic:

$$P = a \log_b t + k$$

for some positive a and b. You can choose any base b for the logarithm, so the problem is to determine a and k.

Example 1

a. **Find an equation of the form $P = a \log_b t + k$ to model the study time data on the previous page.**

b. **Use your model to predict the amount of study time needed for the mean percentage of word meanings recalled to exceed 95%.**

Solution

a. Let $x = \log_b t$. Then the formula $P = a \log_b t + k$ becomes $P = ax + k$, the equation for a line. Since you know how to find a line of best fit, you can use that line to find values of a and k that will provide a close-fitting logarithm function. Choose a base for the logarithm. We choose $b = 10$. Calculate $\log t$ for the values of t in the table on the previous page. The results of the calculations are below. Plot the points (x, P), which are also $(\log t, P)$.

t	log t	P
1	0	30
2	0.3	43
5	0.7	53
10	1	65
15	1.18	74
20	1.30	76
30	1.48	85

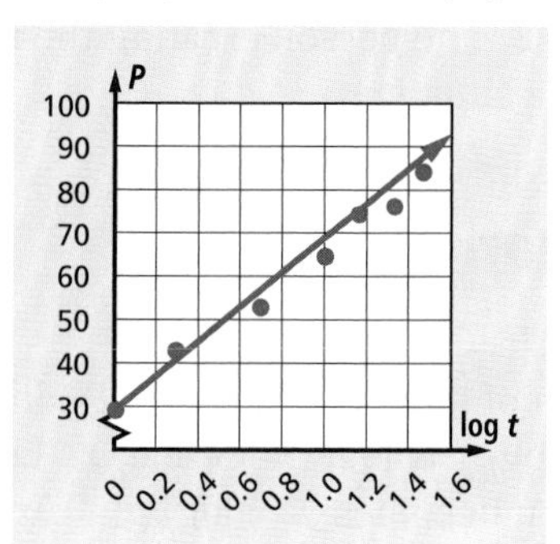

The graph shows a very strong linear relationship between P and $\log t$. This verifies that a logarithmic model is appropriate for these data. A statistics utility gives $y = 36.2x + 30.1$ as an equation for the line of best fit, with $r \approx 0.996$. Substituting P for y and $\log t$ for x, gives an equation that models the study time data:

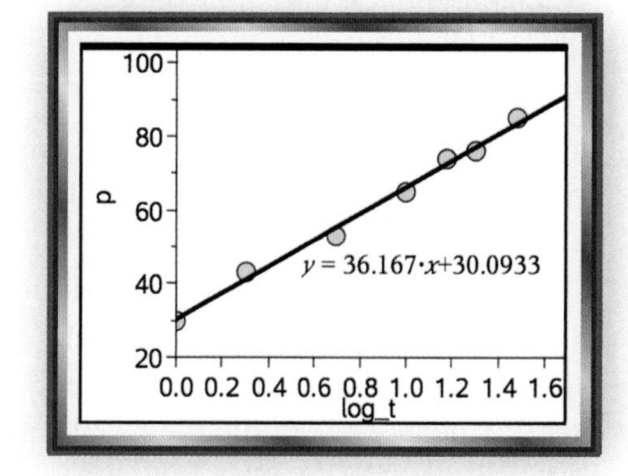

$$P \approx 36.2 \log t + 30.1 .$$

b. Substitute 95 for P and solve for t.

$$95 \approx 36.2 \log t + 30.1$$
$$64.9 \approx 36.2 \log t$$
$$\log t \approx 1.79282$$

So by the definition of $\log t$, $t \approx 10^{1.79282} \approx 62.1$.

The model predicts that on average people will exceed 95% accuracy after about 62 seconds of study time per word.

Using a Scatterplot to Choose a Model

The shape of a scatterplot offers clues to the appropriate function to model the data set. For instance, you can tell whether a power function $f(x) = x^n$ with $n > 1$, a root function $g(x) = x^{\frac{1}{n}}$ with $n > 1$, or a logarithm function $h(x) = \log_b x$ with $b > 1$ is the most appropriate model for the data by examining the shape of the graph. Examples are shown on the next page.

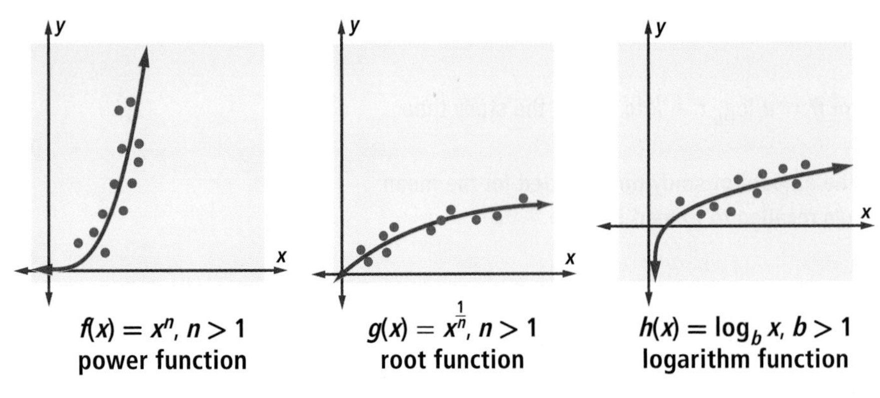

$f(x) = x^n,\ n > 1$
power function

$g(x) = x^{\frac{1}{n}},\ n > 1$
root function

$h(x) = \log_b x,\ b > 1$
logarithm function

However, the scatterplot cannot always tell which power function to use or whether an exponential function might be more appropriate. For example, the graphs of $y = x^2$, $y = x^3$, and $y = 2^x$ at the right each show about the same growth when $0 \le x \le 2$. The person creating the model must make a choice appropriate to the situation.

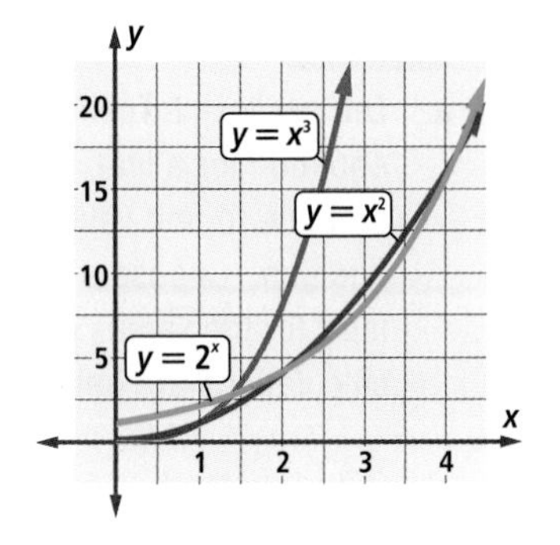

Linearizing using the Logarithm of Both Variables

Every power function has an equation of the form $y = ax^n$. Taking the logarithm base b of both sides gives $\log_b y = \log_b a + n \log_b x$. Letting $y' = \log_b y$ and $x' = \log_b x$, the equation becomes $y' = \log_b a + nx'$. This is a linear equation relating x' to y'. You can use linear regression to find a line of best fit from x' to y'. Then you can use properties of logarithms to transform the regression equation in x' and y' to an equation in x and y.

Consider this situation. One way to compare the world-record times between different events is to measure the mean time per 100 meters in each event. Consider the table and scatterplot at the right, which display r, the mean $\frac{\text{time}}{100\text{ m}}$ for thirteen world-record times for various distances d in men's track and field events. As the first event is the 100 m, the time r is the same as the record itself, namely 9.69 seconds. Also interesting is that the mean $\frac{\text{time}}{100\text{ m}}$ of the men's 200-meter race is faster than the 100-meter world record!

Mean Time per 100 m for Men's Track World Records

d	r
100	9.69
200	9.65
400	10.80
800	12.64
1000	13.20
1500	13.73
2000	14.24
3000	14.69
5000	15.15
10000	15.78
20000	16.93
25000	17.74
30000	17.86

Source: International Association of Athletics Federation 2008

The scatterplot shows that the relationship between the variables is not linear. However, it is not clear what type of function fits the data. It could be a root function or logarithm function.

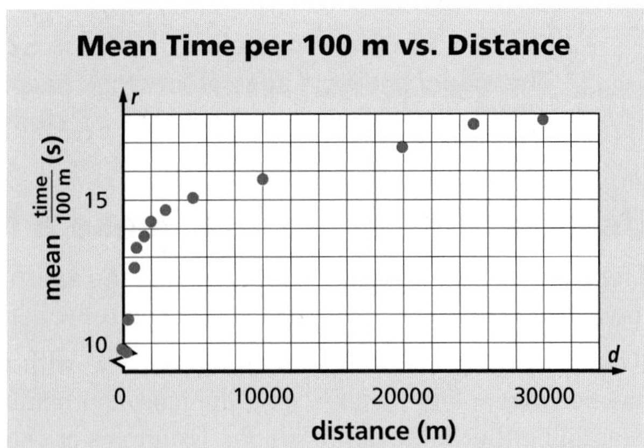

Linear regression can be used to test whether the data agree with a power function model $d = ar^b$. Take the natural logarithm of both sides: $\ln d = \ln a + b \ln r$. This is a linear equation $y = a + bx$ with y-intercept a and slope b. Instead of (x, y) pairs providing data for regression analysis, the $(\ln d, \ln r)$ pairs are used.

Example 2

Consider the distance d and the mean $\frac{\text{time}}{100 \text{ m}}$ r of the world record holders in various men's track and field events shown on the previous page.

a. Take the natural logarithms of d and r and find the line of best fit for predicting $\ln r$ from $\ln d$.

b. Use the line of best fit in Part a to find an equation relating d and r.

c. If a world record holder were to run a 15,000 meter race, estimate what his mean time per 100 m would be given the line of best fit from Part b.

Solution

a. Find the logarithm base e of the distance d and of the mean $\frac{\text{time}}{100 \text{ m}}$ r using a spreadsheet or the list capability of your technology. The result is shown in the spreadsheet below. Then make a scatterplot of these data as shown below at the right.

◇	A	B	C	D
1	d	r	ln d	ln r
2	100	9.69	4.61	2.27
3	200	9.65	5.30	2.27
4	400	10.80	5.99	2.38
5	800	12.64	6.68	2.54
6	1000	13.20	6.91	2.58
7	1500	13.73	7.31	2.62
8	2000	14.24	7.60	2.66
9	3000	14.69	8.01	2.69
10	5000	15.15	8.52	2.72
11	10000	15.78	9.21	2.76
12	20000	16.93	9.90	2.83
13	25000	17.74	10.13	2.88
14	30000	17.86	10.31	2.88

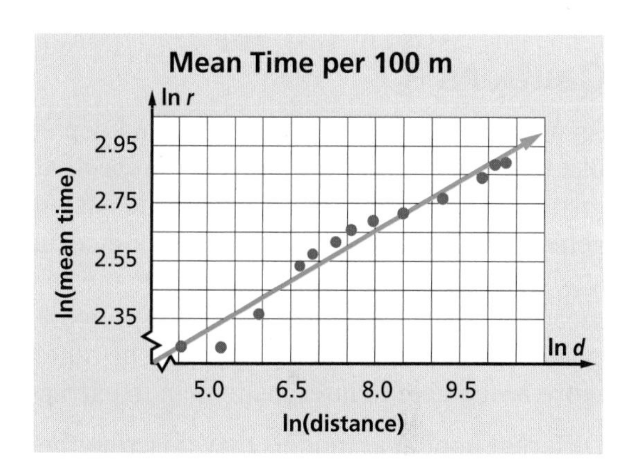

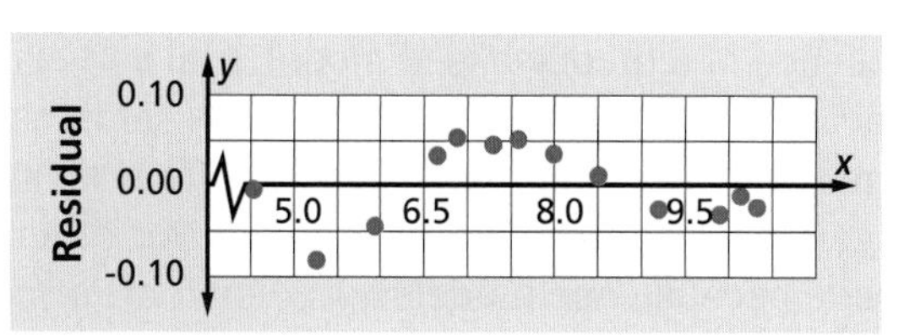

A statistics utility gives $\ln r = 0.1113 \ln d + 1.7596$ as the line of best fit for $\left(\ln(\text{distance}), \ln\left(\text{mean } \frac{\text{time}}{100 \text{ m}}\right)\right)$ with $r \approx 0.979488$. The linear fit is very good. The scatterplot of $(\ln d, \ln r)$ looks approximately linear, and the residual plot clusters around the x-axis, as shown above at the right.

(continued on next page)

b. The line of best fit has the equation
$\ln r = 0.1113 \ln d + 1.7596$.
So $\ln r = \ln(d^{0.1113}) + \ln 5.81011$ by the Logarithm of a Power Theorem and because $e^{1.7596} = 5.81011$.
Then $\ln r = \ln(5.81011d^{0.1113})$ by the Log of a Product Theorem.
Thus, $r = 5.81011d^{0.1113}$.

c. Substitute 15000 for d in the equation in Part b.
$r = 5.81011(15000^{0.1113})$
$r \approx 16.94 \frac{\text{seconds}}{\text{100 meters}}$

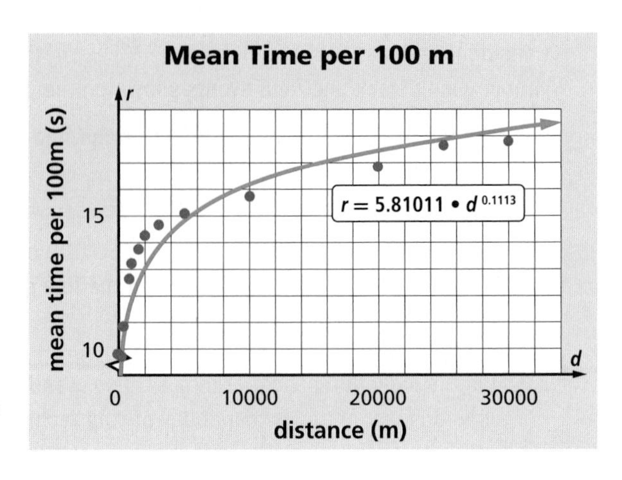

Check A scatterplot of the original data and with the graph of $r = 5.81011d^{0.1113}$ on it is at the right. The fit is good.

Notice that the data in Example 2 fits a 9th root function!

Cautions

As with all real data, it is important to keep equations and predictions like those in Examples 1 and 2 in perspective. In Example 1, for instance, it might not be reasonable to assume most people will ever achieve 95% recall of the words. After a certain number of seconds of study, it is possible that boredom becomes a factor and the percentage of meanings recalled actually declines. However, for study times within the range observed, the model seems to be quite accurate. As always, it is more acceptable to interpolate than to extrapolate.

Different nonlinear models may describe the same data set. In particular, a square root function models the data in Example 1. In absence of a theory-based model, trial and error, the goodness of the linear fit, and an analysis of residuals serve as guidelines for choosing one model over another. A decision to prefer an exponential or logarithm or power function model over other possible models is often made because one model is predicted by, or consistent with, other theories, hypotheses, or past research.

Questions

COVERING THE IDEAS

In 1 and 2, refer to the word-recall experiment and Example 1.

1. According to the model, what is the predicted percent recall if a person has 4 seconds of study time on each word?
2. About how much time should a subject have to study each word in order to achieve 80% recall?

In 3–5, refer to the world-record data and the model in Example 2.

3. Why was a linear equation in ln d and ln r used to model these data?
4. A men's world-record runner decides to run the Boston Marathon, which is approximately 41,843 meters long.
   a. Estimate his mean time per 100 meters using the model in Example 2.
   b. The best men's time for the Boston Marathon in 2008 averaged 18.32 seconds per 100 m for the race. How far off is the model?
5. Estimate the length of a race run by a men's world record holder who averages 24.32 seconds per 100 meters.

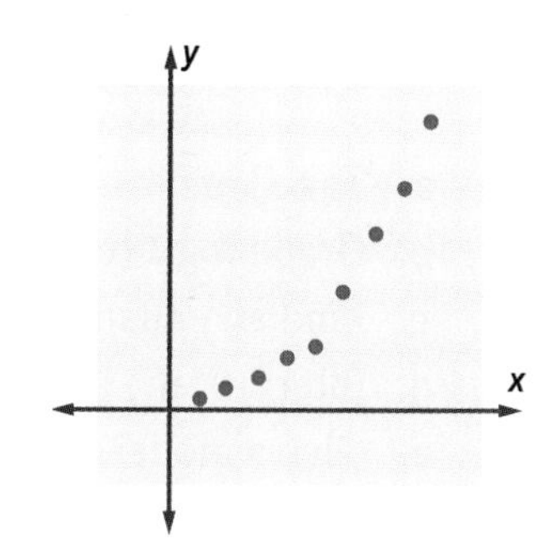

6. **Multiple Choice** Data from an experiment are plotted at the right. Which equation certainly does *not* model these data?
   A $y = ax^2$ B $y = a \cdot 2^x$ C $y = ax^3$ D $y = a \log_2 x$

APPLYING THE MATHEMATICS

7. The table at the right lists the estimates of expected life span in years for people in the United States at different ages.

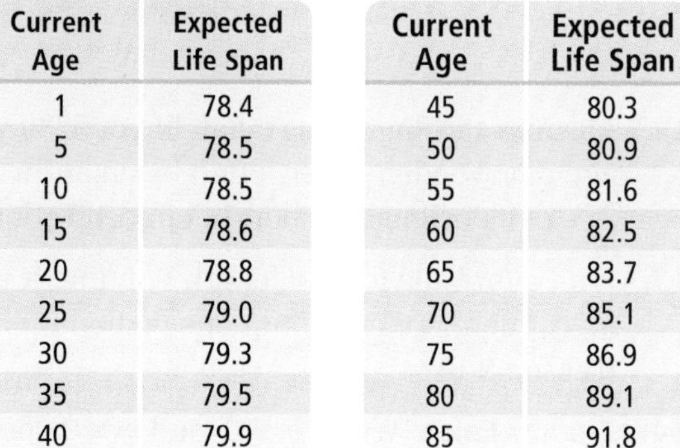

Current Age	Expected Life Span	Current Age	Expected Life Span
1	78.4	45	80.3
5	78.5	50	80.9
10	78.5	55	81.6
15	78.6	60	82.5
20	78.8	65	83.7
25	79.0	70	85.1
30	79.3	75	86.9
35	79.5	80	89.1
40	79.9	85	91.8

Source: World Almanac and Book of Facts, 2008

   a. Draw a scatterplot of current age x versus expected life span y.
   b. Which of the two models, $y = ab^x$ or $y = a \log x$, seems to fit these data better?
   c. Determine an equation for the better-fitting model in Part b.
   d. For the data above, a statistics utility calculated the equation $y = 0.135x + 76.044$ for the line of best fit. Does the linear model or the model from Part c predict the expected life span of a 40-year old with greater accuracy?
   e. Comment on the suitability of the models in Parts c and d for predicting the expected life span of a 90-year old.

8. The table below shows the trend in life expectancy for a child born in the given year in the United States.

Year t	1900	1915	1930	1945	1960	1990	2005
Life Expectancy L (years)	47.3	54.5	59.7	65.9	69.7	75.4	77.9

Source: Centers for Disease Control

   a. Find a logarithm equation of the form $L = a \ln t + b$ to fit these data. Use it to estimate the life expectancy in 1975.
   b. Find a square-root equation of the form $L = a\sqrt{t} + b$ to fit these data. Use it to estimate the life expectancy in 1975.
   c. Based on an analysis of correlation coefficients and residuals, which of the two models do you think is better? Why?
   d. Based on the model you chose in Part c, predict the life expectancy for a child born in the year of your birth.

Sarah Knauss (1880–1999) with her great-great-great-grandson. She lived 119 years and 97 days, making her the oldest American on record.

9. It has been suggested that the time t (taken in seconds) for a rowing shell to cover 2000 m is related to the number x of oarsmen in the boat, according to the power function $t = cx^n$. Here are the winning times for various rowing events at the 2004 Olympics.

x	1	2	3	4
t	409.3	389.00	356.85	342.48

Source: BBC News

a. Take logs of each side of the equation $t = cx^n$.

b. Graph $\log t$ (vertical axis) against $\log x$ (horizontal axis).

c. Find an equation for the line of best fit for the graph in Part b.

d. Interpret the slope of your line in Part c.

e. Give a model of the form $t = cx^n$ for the 2004 Olympics data.

10. Solve for y.

a. $\log y = 0.6x$ b. $\log y = 0.6x + 1$ c. $\log y = 0.6 \log x + 1$

REVIEW

11. Solve $2^n = 578$. **(Lesson 9-6)**

12. Assume carbon-14 has a half-life of 5730 years. Approximate the age of a wooden spear found by an anthropologist if the spear has 37% of its original carbon-14 concentration. **(Lesson 9-6)**

13. Find a formula for the length t of time it takes to triple an investment under continuous compounding at r% interest. **(Lesson 9-6)**

14. **True or False** When $b > 0$, and $c > 0$, $\log\left(\frac{b}{c}\right)^2 = 2 \log b - \log c$. If true, why? If false, correct the right side of the equation. **(Lesson 9-5)**

15. a. For what values of x is $x^{\frac{1}{3}}$ less than $x^{\frac{1}{5}}$?

b. For what values of x is $x^{\frac{1}{3}}$ greater than $x^{\frac{1}{5}}$?

c. Generalize your answer to Parts a and b. **(Lesson 9-2)**

16. Use a graphing utility to graph the functions $f\colon x \rightarrow e^{5x}$, $g\colon x \rightarrow \ln(5x)$, and $h\colon x \rightarrow \frac{1}{5}\ln x$. **(Lesson 9-4, 3-8)**

a. Which function, g or h, is the inverse of f? How do you know?

b. Use function composition to show that your answer to Part a is correct.

EXPLORATION

17. Collect data on the number of cases reported for some disease in recent years for your city or state. Which type of function (linear, exponential, power, or logarithm) best models the data over time?

Chapter 9 Projects

1 Semilog Graph Paper

The graph below shows the U.S. population from 1790 to 1860 on a special kind of paper called semilog graph paper. The scale on the y-axis is graduated in proportion with the logarithms of the numbers represented. The x-axis is scaled uniformly, which is why the paper is only "semi"-logarithmic. With such paper, you do not need a calculator or computer to do a logarithmic transformation. You merely graph the populations directly. (Watch the scale carefully; the halfway mark between vertical measures is the geometric mean of those values, not the arithmetic mean.) Similarly, you can read values directly from either scale. Extending the line in this case, the predicted 1920 population is about 200 million.

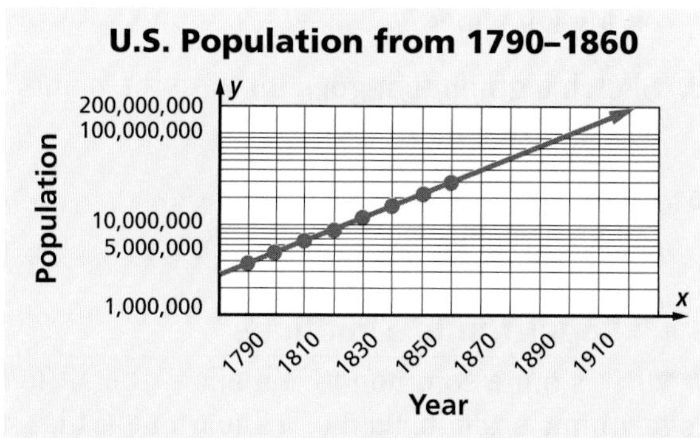

Source: US Census Bureau

a. Obtain some semilog graph paper. Carefully examine the scale on the vertical axis. Label the vertical axis from 1 to 100 and the horizontal axis from 1 to 10. Graph $f(x) = 1.6^x$. Describe the shape of the graph. Graph $g(x) = 3(1.6)^x$ on the same set of axes. What kind of transformation maps the graph of f onto the graph of g? Explain why.

b. Graph two other exponential functions on semilog paper, for example $k(x) = e^x$, $m(x) = e^{2x}$, or $n(x) = 1.8^x$. Describe the graph of any exponential function that is plotted on semilog graph paper.

c. Log-log paper has logarithmic scales on both the x- and y-axis. Find some log-log paper and graph $f(x) = x^3$, $g(x) = x^{1.5}$, and $h(x) = x^{-2.2}$ on the same set of axes. What do you notice? What is true about the graph of any function of the form $y = x^m$ when it is plotted on log-log paper?

2 Using Logs to Calculate: the Old Way

Logarithms and the slide rule were once used to simplify complex calculations. Do *one* of the following.

a. Find an advanced algebra or trigonometry book published before 1965. Find out how logarithms were used. Write a description of what you find.

b. Locate a slide rule and instruction book. Find out how to multiply, divide, and take powers of numbers. Demonstrate this skill to your class.

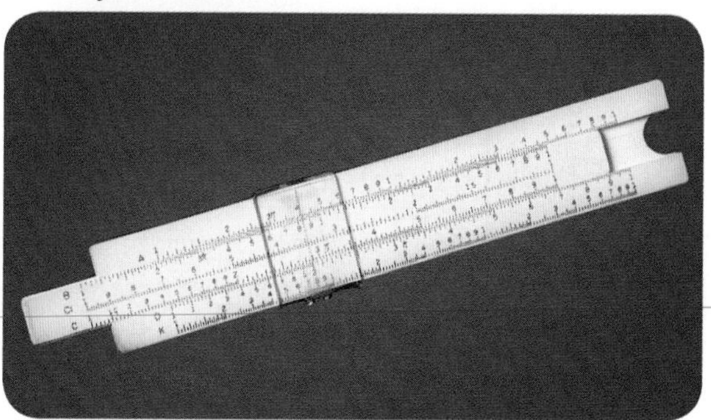

3 Kepler's Third Law and Log-log Graphs

Twelve years after discovering two major planetary laws, Johnnes Kepler made a third discovery in March 1618. The third law is that the square of the orbital periods of the planets varies directly with the cube of their mean distances from the Sun. In symbols, the third law can be stated as $p^2 = kd^3$, which can be simplified to $p = kd^{\frac{3}{2}}$, where p is the period of a planet's orbit and d is the mean distance of the planet from the Sun.

How did Kepler discover this law? Some of the credit is due to the discovery of logarithms by Napier. The table below gives the lengths d of the semimajor axes (the longer axis of the elliptical orbit) scaled so that the length is in astronomical units (AU), and the orbital periods p for all of the planets known in Kepler's time. 1 AU is about 93,000,000 miles, the mean distance of Earth from the Sun. The data in the table are shown in the graph below it.

Planet	Semimajor Axis (AU)	Orbital Period (years)
Mercury	0.3871	0.2408
Venus	0.7233	0.6152
Earth	1.000	1
Mars	1.5273	1.8809
Jupiter	5.2028	11.862
Saturn	9.5388	29.458

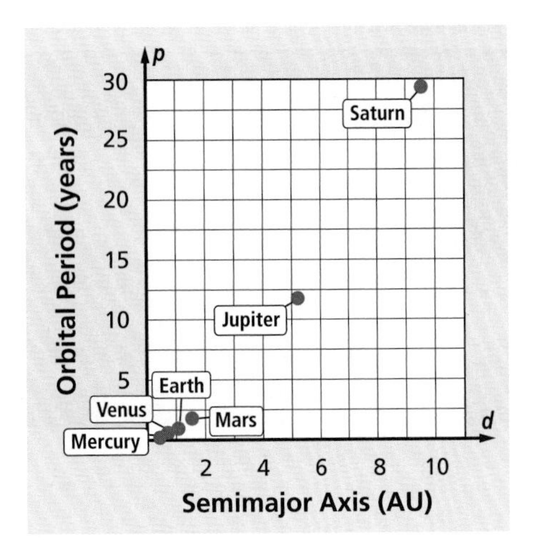

Graphing is a luxury we have today. It was not known in Kepler's time.

If you take logarithms (to any base) of the numbers in the green table, the result is startling.

a. Complete the table below as Kepler did in 1618. Use common logarithms.

Planet	x = log(semimajor axis)	y = log(period)
Mercury	-0.4122	-0.6183
Venus	?	?
Earth	?	?
Mars	?	?
Jupiter	?	?
Saturn	?	?

b. Explain how the table shows that $p = d^{\frac{3}{2}}$, where d is the semi-major axis and p is the orbital period. This, in essence, is what Kepler discovered.

c. Log-log graphs have both axes scaled logarithmically. Graph the data from Part a on log-log paper.

d. Sketch a graph that goes through the points and describe its shape.

e. Identify an advantage to using log-log graphs.

4 Logarithmic Timelines

It is quite common for timelines to use a logarithmic scale instead of a linear one. This most often occurs when there are large spans of time between events being charted. Research how to create a logarithmic scale and what it means to create a logarithmic timeline. Show at least two examples of logarithmic timelines used by other people.

Chapter 9 Summary and Vocabulary

- If n is a positive integer ≥ 2, an equation of the form $p(x) = x^n$ describes an nth power function and $r(x) = x^{\frac{1}{n}}$ describes an **nth root function**. Also, if m and n are integers and $n \neq 0$, $s(x) = x^{\frac{m}{n}}$ describes a *rational power function.* nth root and nth power functions are defined only for nonnegative values of x. The nth root functions are inverses of the corresponding nth power functions with the domain restricted to $x \geq 0$.

- An exponent that is the reciprocal of a positive integer n is equivalent to an **nth root**.

$$x^{\frac{1}{n}} = \sqrt[n]{x} \text{ for all } x \geq 0$$

- A rational exponent $\frac{m}{n}$ signifies the nth root of the mth power, or the mth power of the nth root.

$$x^{\frac{m}{n}} = \sqrt[n]{x^m} = (\sqrt[n]{x})^m \text{ for all } x \geq 0$$

- **Logarithm functions** are inverses of exponential functions. That is, the **logarithm base b** of a number is an exponent: $y = \log_b x$ if and only if $b^y = x$. The expression **$\log_b x$** requires that $x > 0$, $b > 0$, and $b \neq 1$. Although any positive base other than 1 is possible, **common logarithms** (base 10) are especially convenient because the decimal number system also has base 10, so $\log 100 = 2$, $\log 1000 = 3$, and so on.

- Before calculators were invented, logarithms were the main tool for computing powers, products, and roots. Log functions are still used today to solve equations of the form $b^x = a$, where b and a are positive and $b \neq 1$, and to transform relations involving multiplication, division, or powers into relations involving addition, subtraction, or products.

- **Natural logarithms** (base e) are often used in modeling because they represent continuous change nicely. The Continuous Change Model $A(t) = Pe^{rt}$, with initial value P and annual growth rate r, gives the amount $A(t)$ of a substance after t years when growth or decay is continuous. Several natural phenomena follow this model.

Vocabulary

Lesson 9-1
square root
cube root
nth root, $\sqrt[n]{x}$, $x^{\frac{1}{n}}$
nth root functions
radical, $\sqrt{}$

Lesson 9-2
rational power

Lesson 9-3
logarithm of x to the base b, $\log_b x$
common logarithm
logarithm function with base b

Lesson 9-4
e
compounding continuously
natural logarithm
$\ln x$
natural logarithm function

Lesson 9-6
exponential equation

- The basic properties of logarithms correspond to properties of powers. If you rewrite a power property letting $b^m = x$ and $b^n = y$, taking the log of both sides will result in a log property.

Power Property	Logarithm Property
$b^0 = 1$	$\log_b 1 = 0$
$b^m \cdot b^n = b^{m+n}$	$\log_b(xy) = \log_b x + \log_b y$
$\frac{b^m}{b^n} = b^{m-n}$	$\log_b\left(\frac{x}{y}\right) = \log_b x - \log_b y$
$(b^m)^a = b^{am}$	$\log_b(x^a) = a \log_b x$
$b^x = a$	$\log_b a = \frac{\log_t a}{\log_t b}$

- Properties of exponents can be used to prove theorems about logs of products, quotients, and powers, and these in turn can be used to solve exponential equations. The Change of Base Theorem allows you to convert logs from one base to another and to graph $y = \log_b x$ for values of b other than 10 and e.

- Logarithm functions model many types of real situations. When a relationship between two variables is modeled by a curve, a log transformation can re-express one or both variables. When the transformation results in an approximately linear relationship, a line of best fit can be calculated. Properties of logarithms transform the regression equation to model a power or exponential relationship in the original measurement system of the variables.

Properties

$\frac{1}{n}$ Exponent Theorem (p. 560)
Number of Real Roots Theorem (p. 564)
Rational Exponent Theorem (p. 568)
Properties of Powers (p. 570)
Zero Exponent Theorem (p. 571)
Negative Exponent Theorem (p. 571)
Continuous Change Model (p. 583)
Logarithm of 1 Theorem (p. 587)
Logarithm of the Base Theorem (p. 587)
Logarithm of a Power of the Base Theorem (p. 588)
Logarithm of a Product Theorem (p. 588)
Logarithm of a Quotient Theorem (p. 589)
Logarithm of a Power Theorem (p. 590)
Change of Base Theorem (p. 596)

Chapter 9 Self-Test

Take this test as you would take a test in class. You will need a calculator. Then use the Selected Answers section in the back of the book to check your work.

In 1–3, simplify and show your work.

1. $8^{-\frac{4}{3}}$ 2. $\sqrt[5]{x^{10}y^5}$, $x > 0, y > 0$ 3. $\ln e^{18}$

4. Solve $\log_{999} x = 0$.

5. Show how $\log 225 + \log 4 - \log 9$ can be evaluated without using a calculator.

6. If $\log 30 = 1.48$, find the value of
 a. $\log 300$. b. $\log 900$. c. $\log \frac{1}{30}$.

7. Use the facts that $\log_{11} 7 \approx 0.8115$ and $\log_{11} 3 \approx 0.4582$ to estimate $\log_{11}\left(\frac{7}{3}\right)$.

8. What restrictions can be made on the domain of the function $g : x \rightarrow x^9$ so that its inverse is an nth root function?

9. Find the solution to the nearest hundredth: $6^x = 1.48$.

10. **True or False** The natural logarithm function has base e.

11. a. Find the volume of a cube with surface area 130 square inches.
 b. Express the volume V of a cube as a function of its surface area.

12. Strontium-90 has a half-life of 25 years. How long will it take 10 grams of Strontium-90 to decay to 3 grams?

13. Consider $g(t) = \log_5 t$.
 a. Graph $y = g(t)$ and $y = g^{-1}(t)$ for $0 \le t \le 10$.
 b. Give equations of any asymptotes of g.
 c. Find an equation for g^{-1}.

14. Use properties to write an equivalent equation that does not contain logarithms.
$$\frac{1}{2}\log a + \log b = \frac{1}{3}\log c$$

15. **Multiple Choice** Which of the following is an equation for the inverse of $f: x \rightarrow 3^x$?
 A $y = 3 \log x$ B $y = 3 \ln x$
 C $y = \log_3 x$ D $y = \frac{1}{3^x}$

16. Rewrite $A = 3.75(0.054)^t$ as a linear equation with points $(t, \ln A)$.

17. At the beginning of an experiment, a cup of coffee at 100°C was placed in a 20°C room. At one-minute intervals, the temperature of the coffee was measured, and the difference y in temperature from room temperature recorded. The results are in the table below.

Minutes A	Difference y (°C)
1	69
3	54
5	40
7	30
8	26
10	21

 a. Transform the data by finding the natural log of each y.
 b. Draw a graph of the ordered pairs $(A, \ln y)$.
 c. Find an equation for the line of best fit for the points in Part b.
 d. Use the line of best fit to find an exponential equation to fit the original set of data.

18. a. How are the graphs of $y = x^{\frac{7}{4}}$ and $y = x^{\frac{4}{7}}$ related?
 b. Rewrite $\left(\frac{1}{x}\right)^{\frac{7}{4}}$ as a rational power of x.

Chapter 9 Chapter Review

SKILLS
PROPERTIES
USES
REPRESENTATIONS

SKILLS Procedures used to get answers

OBJECTIVE A Evaluate $b^{\frac{m}{n}}$ for $b > 0$. (Lessons 9-1, 9-2)

In 1–4, evaluate without a calculator.

1. a. $8^{\frac{1}{3}}$ b. $8^{\frac{5}{3}}$ c. $8^{-\frac{2}{3}}$
2. a. $256^{\frac{7}{8}}$ b. $256^{0.5}$ c. $256^{\frac{1}{4}}$
3. a. the fifth root of 243
   b. the fifth root of 32
   c. $\sqrt[5]{\frac{32}{243}}$
4. $\sqrt[3]{-8}$

In 5 and 6, assume each base is positive. Rewrite using a radical sign.

5. a. $p^{\frac{3}{5}}$ b. $2r^{\frac{3}{4}}$
6. a. $q^{-\frac{7}{3}}$ b. $\frac{1}{5}n^{\frac{1}{6}}$

In 7 and 8, assume that all variables are positive. Write without a radical sign.

7. $\sqrt[7]{ab^5}$ 8. $\frac{1}{\sqrt[15]{m^3n^5}}$
9. Insert the appropriate symbol (>, <, or =): $(100{,}000)^{\frac{4}{5}}$ _?_ $(10{,}000)^{\frac{5}{4}}$.

In 10 and 11, a number x and a value of n are given.

a. Give the number of real nth roots x has.
b. Find all real nth roots of the number.

10. $x = 6561$, $n = 8$
11. $x = 512$, $n = 9$

OBJECTIVE B Use the definition of logarithm to evaluate some logarithms without a calculator. (Lessons 9-3, 9-4, 9-5)

In 12–18, evaluate. Assume all variables are positive.

12. $\log_7 7^5$ 13. $\log_{10}(0.001)$ 14. $\log_b \frac{1}{b^4}$
15. $\log_{64} 512$ 16. $\ln(e^n)$ 17. $e^{\ln \pi^2}$
18. $\log_m\left(\sqrt[6]{m^3}\right)$

OBJECTIVE C Use logarithms to solve exponential equations. (Lesson 9-6)

In 19–22, solve to the nearest thousandth.

19. $e^n = 56$ 20. $2^x = \frac{1}{40}$
21. $9(4.5)^t = 10$ 22. $4.6 \cdot 10^y = 7$

PROPERTIES Principles behind the mathematics

OBJECTIVE D Describe properties of rational power and nth root functions. (Lessons 9-1, 9-2)

In 23 and 24, order the numbers from least to greatest.

23. $\log 4$, $\ln 4$, $\log_7 4$
24. $\log_2 13$, $\log_2 1$, $\log_2 \frac{1}{13}$
25. Give the domain and range of the function f, where $f(x) = x^{\frac{8}{5}}$.
26. a. What restrictions can be made on the domain of the function $g: x \rightarrow x^7$ so that its inverse is an nth root function?
    b. Describe g^{-1}.

In 27 and 28, tell whether the statement is *true* or *false* for all real x and nonzero integers n. If it is false, provide a counterexample.

27. $\sqrt[n]{x} = x^{\frac{1}{n}}$ 28. $x^{-n} = \frac{1}{x^n}$

OBJECTIVE E Describe properties of logarithm functions. (Lessons 9-3, 9-4)

In 29 and 30, h is the logarithm function with base 3.

29. State the domain and range of h.
30. Find an equation for the inverse of h.
31. **Multiple Choice** Which of the following is an equation for the inverse of $f: x \rightarrow \log_4 x$?
    A $y = 4^x$ B $y = \sqrt[4]{x}$
    C $y = 4 \log x$ D $y = x^4$

32. Define e.

33. What point(s) is (are) on the graph of the function f with equation $f(x) = \log_b x$, regardless of the value of b?

OBJECTIVE F **Use properties of logarithms to find values of logarithm functions. (Lesson 9-5)**

In 34–37, if $\log 15 = v$, give the logarithm of each number in terms of v.

34. $\log 150$

35. $\log \frac{1}{15}$

36. $\log 225$

37. $\log(15^x)$

In 38–40, use the fact that $\log_4 11 \approx 1.73$ and $\log_4 7 \approx 1.40$ to evaluate the expression.

38. $\log_4 77$

39. $\log_4\left(\frac{121}{7}\right)$

40. $\log_4 \sqrt[14]{7}$

In 41–43, write as a single logarithm.

41. $\ln(500) + \ln(200)$

42. $\log(500) + 2\log(20)$

43. $\ln(100) - \ln(25)$

In 44–47, simplify.

44. $\ln y^2 + \ln y^3$

45. $\frac{\log 64}{\log 2}$

46. $\log_\pi 1$

47. $\log_{123456789} 1$

48. Show how $\log 75 + \log 4 - \log 3$ can be evaluated without using a calculator.

49. 18^{150} is an n-digit number. Find n.

50. Use the definition of logarithm to show that $\log_{\frac{1}{2}} 16 = \log_2 \frac{1}{16}$.

OBJECTIVE G **Rewrite equations including exponents as equivalent equations including logarithms, and vice-versa. (Lessons 9-3, 9-6)**

51. If $A = \frac{C^3}{D}$, express $\log A$ in terms of $\log C$ and $\log D$.

52. The height of a cone is given by the formula $h = \frac{3V}{\pi r^2}$. Write $\ln h$ in terms of $\ln V$ and $\ln r$.

53. If $y = 6(4)^x$, then write $\ln y$ as a linear function of x.

In 54 and 55, rewrite as an equation that does not involve logarithms.

54. $\log z = 2\log w + \log 3$

55. $\ln v = \frac{4}{7}\ln b - \ln c$

56. A line of best fit for $\ln b$ in terms of c is given by $\ln b = 0.24c - 1.3$. Rewrite the equation to give b as a function of c.

57. Explain why $\log_{10} x = (\log_{10} e)(\ln x)$ for all $x > 0$.

58. Find the value of $\log_a\left(\frac{x}{y}\right) + \log_a\left(\frac{y}{z}\right) + \log_a\left(\frac{z}{x}\right)$ when a, x, y, and z are positive.

USES Applications of mathematics in real-world situations

OBJECTIVE H **Use rational exponents to model situations. (Lesson 9-2)**

59. Express the radius r of a spherical ball bearing as a function of the volume V, given $V = \frac{4}{3}\pi r^3$.

60. The volume V of a cone with radius r and height r is given by $V = \frac{1}{3}\pi r^3$.
    a. Give a formula for r in terms of V.
    b. A cone with equal radius and height has volume 60 mm^3. Use the formula in Part a to find its height.

61. Earth's atmospheric pressure decreases as you ascend from the surface. It can be shown that, at an altitude of h kilometers ($0 < h < 80$), the pressure P in grams per square centimeter is approximately given by the formula $P = 1035 \cdot 2^{\frac{-10h}{58}}$. Give the approximate pressure at each altitude.
    a. sea level
    b. 30 km above Earth
    c. 50 km above Earth

62. a. Find the volume of a cube with surface area 97 square inches.
    b. Express the surface area S.A. of a cube as a function of its volume V.

OBJECTIVE I Solve problems arising from exponential or logarithmic models. (Lessons 9-4, 9-6)

63. The half-life of a certain substance is 24 hours. About how many hours will it take for 6 grams to decay to 1.3 grams?

64. A family has \$16,000 saved for college tuition. They will need \$64,000 for college expenses five years from now. If they place their money in an account with continuous compounding, what annual rate must they find to assure them of the needed \$64,000?

65. The intensity I of sunlight at points below sea level is thought to decrease exponentially with the depth d of the water. One model for this situation is $I = 100(0.325)^d$. If special equipment is needed for divers to see when the intensity drops below 0.2, at what depth is this equipment needed?

66. The population of Iceland in 2008 was about 304,000, with an annual growth rate of 0.78%. Assume the growth rate is constant.

a. Give an equation for the population n years after 2008.

b. Predict the population in 2020.

c. Predict when the population will reach 400,000.

67. A certain radioactive substance has a half-life of 5 hours. Let A be the original amount of the substance and L be the amount left after h hours.

a. Give a formula for L in terms of A and h.

b. What percent of the original amount of the substance will remain after 11 hours?

OBJECTIVE J Use logarithmic functions to model data. (Lesson 9-7)

68. The formula $\log w = -2.866 + 2.722 \log h$ estimates the mean weight w in pounds of a girl h inches tall. Estimate the mean weight of a girl who is 51" tall.

69. The length a of the semimajor axis of a planet's elliptical orbit is directly related to the time T it takes for the planet to complete a revolution around the Sun. When T is measured in days and a in millions of kilometers, an equation modeling the situation is $\ln T = \frac{3}{2} \ln a - 1.72$. Solve for a as a function of T.

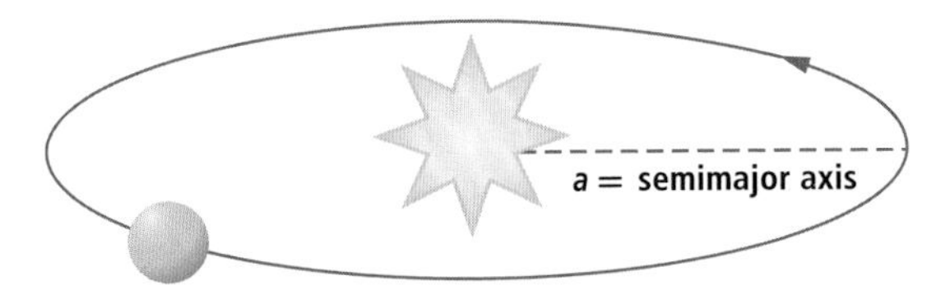

70. When Julie began to practice the darts game Cricket, she recorded her score s after w weeks. The results are shown below.

week (w)	1	3	5	7	15	20	25	30
average score (s)	105	140	156	160	178	179	212	213

a. Draw a scatterplot of the ordered pairs (w, s).

b. Which of the following might be used to linearize the data?

i $(\log w, s)$

ii $(w, \log s)$

iii $(\log w, \log s)$

c. Draw a scatterplot of $(\ln w, s)$.

d. Find a line of best fit for the scatterplot in Part c, and give the correlation coefficient.

e. Use the relationship in Part d to predict Julie's average darts score after a year of practice, if it continues to improve in a similar way to the first 30 weeks.

REPRESENTATIONS Pictures, graphs, or objects that illustrate concepts

OBJECTIVE K Graph and interpret *n*th root and rational power functions. (Lessons 9-1, 9-2)

71. a. Plot $g(x) = x^{\frac{1}{2}}$ and $h(x) = -x^{\frac{1}{2}}$ on one set of axes.

b. What single equation describes the union of these two graphs?

72. a. Plot $r(x) = \sqrt[5]{x}$ and $t(x) = \sqrt[5]{x+3} + 2$ on one set of axes.

b. What transformation maps *r* to *t*?

c. What transformation maps *t* to *r*?

73. Suppose $f: x \rightarrow x^{\frac{5}{3}}$.

a. State the range of *f*.

b. Find an equation for the inverse of *f*.

c. Graph *f* and its inverse.

d. Is the inverse a function? Explain why or why not.

74. a. Sketch the graphs of $f(x) = x^{\frac{3}{2}}$ and $g(x) = \left(\frac{1}{x}\right)^{\frac{3}{2}}$.

b. Are *f* and *g* inverses of each other? How do you know?

c. Rewrite $g(x)$ as a rational power of *x*.

OBJECTIVE L Graph logarithm functions. (Lessons 9-3, 9-4)

75. a. Graph $f(x) = \log x$ for $0 \leq x \leq 10$.

b. Use the graph in Part a to estimate log 2 and log 7 to the nearest tenth.

76. How is the graph of $f(t) = \log_4 t$ related to the graph of $g(t) = 4^t$?

77. a. Sketch the graph of $f(x) = \log_3 x$ for $0 < x \leq 81$.

b. Sketch a graph of f^{-1} on the same axes, and give an equation for f^{-1}.

78. a. Identify 3 points on the graph of $f(x) = \ln e^x$ without using technology.

b. Check using a graphing utility.

79. What line is an asymptote of the graph of every equation of the form $y = \log_b x$, $b > 0$, $b \neq 1$?

Chapter

10 Binomial Distributions

Contents

On a downtown street in a large city, there can be a stoplight at almost every intersection. Suppose on one street there are six such lights in a row. As you drive down the street, you are likely to have to stop at a few of the lights. How many times you stop will depend on the probability that each light is red when you reach the intersection.

If the probability is $\frac{1}{2}$ that each light will be red, and the lights are independent events, then this situation can be modeled by a Galton board. A Galton board is an upright board with a triangular array of pins and slots at the bottom. The front of the board is covered with glass or transparent plastic. If the board is carefully constructed, when a ball is dropped at the top of the array, at each horizontal row the ball will go left or right with probability $\frac{1}{2}$.

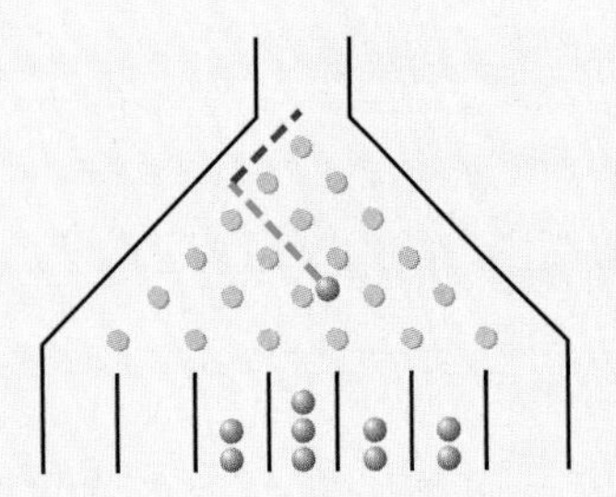

Think of a ball falling to the left of a pin as meeting a red light. In the situation pictured on the previous page, the ball coming down the board has hit two red lights and then three green lights and has one light yet to meet. The balls already in the slots mean that no ball has hit five or six red lights, 2 balls have hit four red lights, 3 balls have hit three red lights, 2 balls have hit two red lights, 2 balls have hit one red light, and no balls have hit all green lights.

Over the long run, how will the balls be distributed in the slots? The answer to this question can be represented by a type of probability distribution known as a *binomial distribution*. The numbers in binomial distributions are related to the number of subsets of a set, to the numbers in the array known as Pascal's Triangle, and to the powers of a binomial of the form $(a + b)$, from which the distribution gets its name.

Binomial distributions have an extraordinary number and variety of applications. They can apply whenever a situation with two possible outcomes has probability of success p and occurs repeatedly, say n times, and you want to know how often a particular outcome will occur r of those times. How often will a team win against a particular rival? How often will it rain at the time of July 4th fireworks? How often will a child have the same color eyes as his or her father? How often will a person have to stop for four of six traffic lights on their way to school?

In this chapter you will study ideas like these. The chapter begins where many discussions of probability begin, with a particular type of counting problem.

Lesson 10-1 Combinations

Vocabulary

combination

number of combinations of n things taken r at a time, ${}_nC_r$, $\binom{n}{r}$

▶ **BIG IDEA** With a set of n elements, it is often useful to be able to compute the number of subsets of size r.

In earlier lessons, you found answers to counting problems involving permutations. Recall that a *permutation* of r of n items is an ordering of r objects taken from the n objects. Numbers of permutations can be found using the Multiplication Counting Principle from Lesson 6-3.

In contrast, a **combination** is a collection of objects in which order does not matter. For instance, in making trail mix, if ingredients are added in different orders, the end result is the same. The Activity exhibits a process that can be used to determine numbers of combinations.

Mental Math

In the 26-letter English alphabet, how many different 2-letter initials are possible

a. if the same letter can be repeated?

b. if the same letter cannot be repeated?

Activity

Sierra is making trail mix for a hike with friends. She has five available ingredients: granola, peanuts, raisins, sunflower seeds, and dried blueberries. Knowing that her friends have different food preferences, she decides to make small bags of mix, each with a different selection of three ingredients, rather than a big bag with all five ingredients. How many bags can she make so that each bag has a different mix?

Step 1 Using the Multiplication Counting Principle, Sierra reasons that there are 5 choices for the first ingredient in the mix, 4 choices for the second ingredient, and 3 choices for the third. How many mixes does she think she can make?

Step 2 Worried that she is going to lose track of what was put in each bag, Sierra makes labels. Her label GPB means granola-peanuts-blueberries. Make a list of the possibilities. Make sure that you avoid duplicate mixes. BGP is the same mix as GPB. Check your list with another student's list to make sure that you have listed all the possibilities. How many did you find?

Step 3 How should Sierra adjust her answer from Step 1 to produce the number of mixes you listed in Step 2? (Hint: If you have chosen three ingredients, say granola-peanuts-blueberries, how many different ways could you write the three-letter label?)

Step 4 Repeat Steps 1–3 as if Sierra had six ingredients instead of 5. Check your answer by listing all the possibilities.

On the trail Trail mix is a nutritious, high energy, and lightweight snack eaten while hiking

A Formula for Counting Combinations

Step 1 of the Activity asks for the number of permutations of 3 of 5 things. Step 2, by contrast, asks for the number of combinations of 5 things taken 3 at a time. Notice the difference between permutations and combinations. In permutations, order matters and so different orders are counted as unique. In combinations, BGP and GPB are considered the same. They are different permutations of 3 letters but they are the same combination of 3 letters.

 QY1

The Activity shows how to compute a number of combinations by dividing two numbers of permutations. In Step 1, the number of ways of choosing three ingredients in order is ${}_5P_3 = \frac{5!}{2!}$ or $5 \cdot 4 \cdot 3$. In Step 3, the number of ways of reordering any set of three ingredients is ${}_3P_3 = 3!$. The number 3! is the number of times each trail mix was counted in Step 1. Thus, to find the number of different unordered mixes, you can divide the number of ordered mixes by the number of times each unordered mix appears.

$$\frac{{}_5P_3}{{}_3P_3} = \frac{\left(\frac{5!}{2!}\right)}{3!} = \frac{5!}{3!2!} = \frac{5 \cdot 4}{2} = 10$$

The number of *combinations* of 5 ingredients taken 3 at a time is written ${}_5C_3$. In general, the **number of combinations of n things taken r at a time** is written ${}_nC_r$ or $\binom{n}{r}$. Some people read this as "n choose r." A formula for ${}_nC_r$ can be derived from the formula for ${}_nP_r$ by generalizing the argument above.

Theorem (Formula for ${}_nC_r$)

For all integers n and r, with $0 \le r \le n$, ${}_nC_r = \frac{{}_nP_r}{r!} = \frac{n!}{(n-r)!r!}$.

Proof Any combination of r objects can be arranged in $r!$ ways. So $r! \cdot {}_nC_r = {}_nP_r$.

Solve for ${}_nC_r$. ${}_nC_r = {}_nP_r \cdot \frac{1}{r!}$

Substitute from the Formula for ${}_nP_r$ Theorem.

Thus, ${}_nC_r = \frac{n!}{(n-r)} \cdot \frac{1}{r!} = \frac{n!}{(n-r)!r!}$.

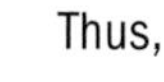 QY2

Because order does not matter in counting combinations, you can consider ${}_nC_r$ as the number of subsets of size r of a set of n objects.

GUIDED

Example 1

A restaurant menu has ten appetizers and twelve main courses. A group of people decides they want to order six appetizers and four main courses. In how many ways can they do this?

(continued on the next page)

QY1

Which situation involves combinations? Which situation involves permutations?

a. Three scholarships worth \$1500, \$1000, and \$500 are distributed to students from a class of 800.

b. Three \$1000 scholarships are distributed to the students of a class of 800.

QY2

Write the formula for the number of ways of making bags of 4-ingredient trail mix if 9 ingredients are available. Do not compute an answer.

Solution First consider the appetizers. There are 10 appetizers, from which 6 are to be chosen. This is like choosing a subset of 6 objects from a set of 10 objects, so use the formula for ${}_{10}C_6$.

$$ {}_{10}C_6 = \frac{10!}{4!6!} = \frac{10 \cdot 9 \cdot 8 \cdot 7 \cdot 6 \cdot 5 \cdot 4 \cdot 3 \cdot 2 \cdot 1}{4 \cdot 3 \cdot 2 \cdot 1 \cdot 6 \cdot 5 \cdot 4 \cdot 3 \cdot 2 \cdot 1} $$

$$ = \frac{10 \cdot 9 \cdot 8 \cdot 7}{4 \cdot 3 \cdot 2 \cdot 1} = 10 \cdot 3 \cdot 7 = 210 $$

For the main courses, there are 12 dishes, from which ___?___ are to be chosen. Since all the courses come out at once, use the formula for $\underline{?}\,C\,\underline{?}$ to get $\frac{12!}{\underline{?}!\,\underline{?}!}$ = ___?___ total main course combinations.

To compute the total number of meals, use the Multiplication Counting Principle:

210 appetizer combinations • ___?___ main course combinations = ___?___ different meals.

Check Use a calculator.

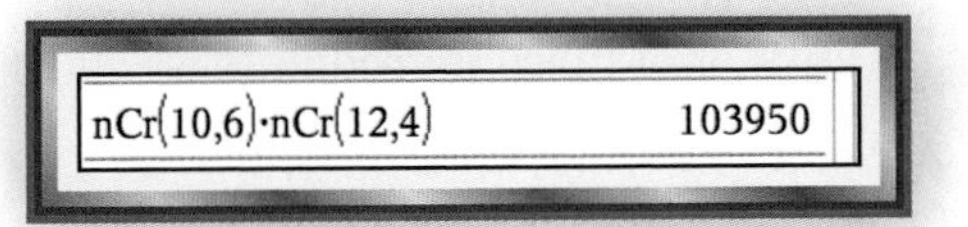

 QY3

> **QY3**
> How many different ways are there of ordering 7 main courses from a list of 10?

Using Combinations to Count Arrangements

As you will see in this chapter, combinations are very important in the calculation of certain probabilities. For instance, suppose five pennies are tossed. If you want to know the probability of getting three tails, then you need to know how many ways three tails can occur in five tosses. To answer this question, think of the pennies as occupying five positions along a line, numbered 1 through 5.

Each way of getting three tails is like picking three of the five numbers. For instance, if coins 3, 4, and 5 are tails, as pictured above, then you have picked the numbers 3, 4, and 5. Do you see that this is just like Sierra's problem from the Activity? The number of ways of getting three tails in five tosses is the number of combinations of five things taken three at a time, ${}_5C_3$.

 QY4

> **QY4**
> How many ways are there to get 3 heads on a toss of 7 coins?

Each time you get exactly three tails in five tosses, the other two coins are heads. So the number of ways of getting three tails in five tosses is the same as the number of ways of getting two heads in five tosses. That is the idea behind the two solutions to Example 2.

Example 2

A vocabulary test has 30 items. Each correct answer is worth 1 point and each wrong answer is worth 0 points. How many ways are there of scoring 25 points?

Solution 1 Think: How many ways are there of selecting 25 questions out of 30 to be correct?

$$_{30}C_{25} = \frac{30!}{5!25!} = 142{,}506$$

Solution 2 Think: How many ways are there to get exactly 5 out of 30 questions wrong?

$$_{30}C_5 = \frac{30!}{25!5!} = 142{,}506$$

"Thinking Factorially": Computing Combinations by Hand

Sometimes numbers resulting from computing a combination are too large for a calculator's memory or display. Even when a calculator can compute and display the exact number of combinations, it can be easier and more illuminating to verify the calculator's results by doing some computations by hand. When working with quotients of factorials, common pairs of factors appear and cancel each other. The result is a more manageable fraction. We describe this method of simplifying calculations as "thinking factorially."

Example 3

A school requires that all participants in sports submit to random drug testing. During one round of tests, 25 students are randomly selected from among the 250 students participating in sports. What is the probability that the quarterback of the football team and the goalies of both the boys' and girls' soccer teams are three of the students selected?

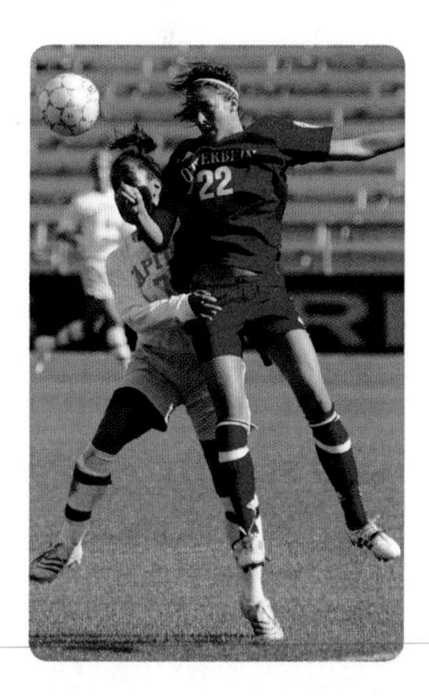

Solution There are $_{250}C_{25}$ combinations of 25 players that could be picked. This number is huge, about $1.65 \cdot 10^{34}$. This number is the denominator of the probability to be calculated. The numerator is the number of these $1.65 \cdot 10^{34}$ combinations of 25 players that include the three given players. Notice that each such combination includes 22 players from the remaining 247 athletes. There is only one way to select the three given players ($_3C_3 = 1$) and there are $_{247}C_{22}$ ways to select the other 22 players. Thus, the probability that these three players are selected, assuming randomness, is

$$\frac{_3C_3 \cdot {}_{247}C_{22}}{_{250}C_{25}} = \frac{1 \cdot \left(\frac{247!}{225!22!}\right)}{\left(\frac{250!}{225!25!}\right)} = \frac{247!25!}{250!22!} = \frac{247! \cdot 25 \cdot 24 \cdot 23 \cdot 22!}{250 \cdot 249 \cdot 248 \cdot 247! \cdot 22!}$$

$$= \frac{\overset{1}{\cancel{25}} \cdot \overset{3}{\cancel{24}} \cdot 23}{\underset{10}{\cancel{250}} \cdot 249 \cdot \underset{31}{\cancel{248}}} \approx 0.000894.$$

Questions

COVERING THE IDEAS

In 1–4, tell whether the situation uses permutations or combinations. You do not have to compute an answer.

1. number of ways to assign 25 students to 30 desks
2. number of committees of 5 students out of a class of 30 students
3. number of ways of picking 3 books out of a reading list of 10 books
4. number of different orders in which a family could have five girls out of seven children

5. Match each item on the left with one on the right.
   a. combination — i. an arrangement of objects in which different orderings are distinguishable
   b. permutation — ii. an arrangement of objects in which different orderings are equivalent

6. a. How many combinations of the letters UNESCO, taken five at a time, are possible?
   b. Explain why the result of Part a is the same as the number of combinations taken one at a time.
   c. Which would be easier to list, the combinations in Part a or those in Part b?
   d. Explain what would be different if you were asked to count all 5-letter strings from UNESCO.

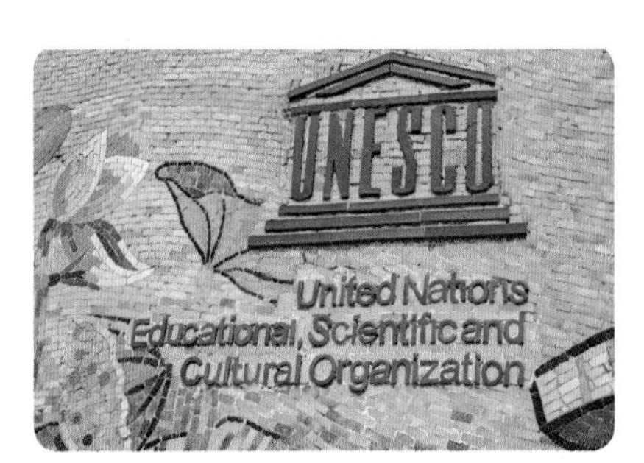

Wall of a United Nations Educational, Scientific and Cultural Organization (UNESCO) building in Hanoi, Vietnam

In 7 and 8, evaluate.

7. $\binom{28}{4}$

8. ${}_{34}C_1$

9. a. How many ways can you flip 10 coins and get exactly 3 heads?
   b. Using the Multiplication Counting Principle, how many total outcomes are possible?
   c. Using your answers from Parts a and b, what is the probability of getting three heads in ten flips?

10. Suppose all 4 members of the school golf team are among the 30 athletes chosen for drug testing at a school with 150 athletes.
    a. Write an expression for the number of combinations of players that could be chosen.
    b. Write an expression for the number of combinations that include the four golf team members.
    c. What is the probability that all 4 golf team members are chosen?

APPLYING THE MATHEMATICS

In 11–13, evaluate the expression. Then explain your answer in terms of choosing objects from a collection.

11. $\binom{n}{1}$

12. ${}_nC_{n-1}$

13. ${}_nC_n$

14. Pete's A Pizza restaurant menu lists fifteen different toppings.
 a. How many different pizzas could be ordered with six toppings?
 b. How many different pizzas could be ordered with nine toppings?

15. Give an example of a situation that would lead to the number ${}_{100}C_{10}$.

In 16–18, solve for *x*.

16. ${}_{x}C_5 = \frac{{}_{13}P_5}{5!}$ 17. $7! \cdot {}_{x}C_7 = {}_{31}P_7$ 18. $\binom{x}{x-2} = 325$

19. A lottery picks five balls from a bin of 55 white balls labeled 1 to 55 and one ball from a bin of 42 green balls labeled 101 to 142. To win the jackpot, a participant must correctly guess all six numbers, although the order of the white numbers is irrelevant. Tickets cost \$1. If someone were to buy all possible tickets, how much would it cost?

20. Six points are in a plane with no three of them collinear, as shown.
 a. How many triangles can be formed having these points as vertices?
 b. Generalize your result in Part a to the case of *n* points.

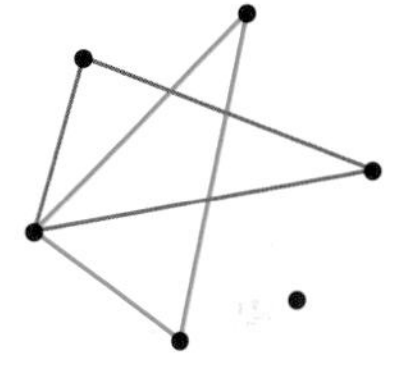

REVIEW

21. A theater has 17 seats in the first row and there are eight more seats in each subsequent row. If the last row has 217 seats, how many rows are there in the theater? **(Lesson 8-1)**

22. Solve for n: $23! = n \cdot 21!$ **(Lesson 6-4)**

23. When one coin is tossed there are two possible outcomes—heads or tails. How many outcomes are possible when the following numbers of coins are tossed? **(Lesson 6-3)**
 a. 2 b. 5 c. 8 d. n

In 24 and 25, two bags each contain five slips of paper. An angle is drawn on each slip. The measures of the angles in each bag are 10°, 30°, 45°, 60°, and 90°. Let *a* and *b* be the measures of the angles pulled from bags 1 and 2, respectively. (Lessons 6-1, 4-2)

24. Find $P(a + b \geq 90°)$. 25. Find $P(\sin(a + b) \geq \sin 90°)$.

The Stanley Cup, pictured here, is given to the champion of the National Hockey League after a series like that in Question 26.

EXPLORATION

26. In professional baseball, basketball, and ice hockey, the champion is determined by two teams playing a first-to-four wins out of seven games series. Call the two teams *X* and *Y*.
 a. In how many different ways can the series occur if team *X* wins? For instance, two different 6-game series are *XXYXYX* and *XYYXXX*. (Hint: Determine the number of 4-game series, 5-game series, 6-game series, and 7-game series.)
 b. There is an ${}_{n}C_r$ with n and r both less than 10 which equals the answer for Part a. Find n and r.
 c. Explain why the combination of Part b answers Part a.

QY ANSWERS

1. a. permutations
 b. combinations
2. ${}_{9}C_4 = \frac{{}_{9}P_4}{4!} = \frac{9!}{5!4!}$
3. 120
4. ${}_{7}C_3 = 35$

Lesson 10-2 Pascal's Triangle

Vocabulary

Pascal's Triangle

$(r + 1)$st term in row n

▶ **BIG IDEA** Pascal's Triangle displays numbers of combinations in an easy-to-use way.

Suppose you toss a coin five times. As you saw in Lesson 10-1, you can determine the number of possible arrangements of heads and tails using combinations. A systematic consideration of the problem leads to *Pascal's Triangle*, an array of numbers that has remarkable properties and applications in different areas of mathematics.

Mental Math

Expand.

a. $(a - b)^2$

b. $(a - 2b)^2$

c. $(3a - 2b)^2$

d. $(3a^2 - 2b)^2$

Activity

Consider five tosses of a coin. Follow the steps below to complete the table with a count of the number of possible arrangements of heads.

Heads	0	1	2	3	4	5	Total
Arrangements	?	?	?	?	?	?	32
$_nC_r$	$_5C_0$	$_5C_1$	?	?	?	?	

Step 1 First, suppose there are 0 heads. Then there are 5 tails. There is only one possible arrangement, which is shown below.

Write this number in the table above. Notice that $1 = {}_5C_0$.

Step 2 If there is 1 head, then there are 4 tails. The 1 head can be in any one of the five places, so there are 5 possible arrangements. Record this number in the table. Notice that $5 = {}_5C_1$. This makes sense, as we are choosing one position from the five as the location of the 1 head.

Step 3 If there are 2 heads, then there are 3 tails. How many arrangements are there with 2 heads? Write your answer in the table and explain why this can be calculated as $_5C_2$.

Step 4 Continue with 3 heads, 4 heads, and 5 heads. Fill in the remaining entries in the table.

Step 5 To check your work, add the numbers in the second row of the table. There should be 32 arrangements in all. Each coin can be either heads or tails, so by the Multiplication Counting Principle, there are $2 \cdot 2 \cdot 2 \cdot 2 \cdot 2 = 2^5 = 32$ possible arrangements.

What Is Pascal's Triangle?

To count all the possible arrangements of n coins, you can calculate ${}_nC_0$, ${}_nC_1$, ${}_nC_2$, ..., ${}_nC_n$, a total of $n + 1$ calculations, and find the sum of those values. For instance, for $n = 4$, the 5 values are

${}_4C_0 = 1$ ${}_4C_1 = 4$ ${}_4C_2 = 6$ ${}_4C_3 = 4$ ${}_4C_4 = 1$

and the total number of arrangements is $1 + 4 + 6 + 4 + 1 = 16$.

When the values of ${}_nC_r$ are displayed systematically, a beautiful pattern emerges. At the left below, the values of ${}_nC_r$ with values of n and r from 0 to 6 are arranged in an array in the form of a right triangle. At the right below is the more common isosceles triangle arrangement.

Pascal's Triangle

(right triangle format)

	${}_nC_r$	0	1	2	3	4	5	6
	0	1						
	1	1	1					
	2	1	2	1				
n	3	1	3	3	1			
	4	1	4	6	4	1		
	5	1	5	10	10	5	1	
	6	1	6	15	20	15	6	1

(isosceles triangle format)

row	
0	1
1	1 1
2	1 2 1
3	1 3 3 1
4	1 4 6 4 1
5	1 5 10 10 5 1
6	1 6 15 20 15 6 1

Blaise Pascal (1623-1662)

In the Western world, the array shown in two ways above is called *Pascal's Triangle*, after Blaise Pascal, a French mathematician and philosopher. Notice that the row numbering starts with row 0 at the top.

Because entries in Pascal's Triangle can be identified by the row number n and column number r, the array can be considered as a two-dimensional sequence. The following definition provides an explicit formula for the terms in the nth row of Pascal's Triangle, where n can be any whole number. It is an algebraic definition for the triangle.

Definition of Pascal's Triangle

Let n and r be integers with $0 \le r \le n$. **Pascal's Triangle** is the two-dimensional sequence in which the $(r + 1)$st term in row n is ${}_nC_r$ or $\binom{n}{r}$.

Example 1

Find the first four terms in row 9 of Pascal's Triangle.

Solution By the definition of Pascal's Triangle, the first term of row 9 is ${}_9C_0$, the second term is ${}_9C_1$, the third term is ${}_9C_2$, and the fourth term is ${}_9C_3$. From the formula for ${}_nC_r$, these are $\frac{9!}{9!0!}$, $\frac{9!}{8!1!}$, $\frac{9!}{7!2!}$, and $\frac{9!}{6!3!}$ or 1, 9, 36, and 84.

 QY1

Properties of Pascal's Triangle

Looking closely at Pascal's Triangle, you can find many patterns and sequences in the rows, columns and diagonals. Here are some properties that can be proved true for every row of Pascal's Triangle. The properties are described both in words and in symbols.

1. The first and last terms in each row are ones. That is, for each whole number n, ${}_nC_0 = {}_nC_n = 1$.
2. The second and next-to-last terms in the nth row equal n. For each whole number n, ${}_nC_1 = {}_nC_{n-1} = n$.
3. Each row is symmetric. For any whole number n, ${}_nC_r = {}_nC_{n-r}$.
4. The sum of the terms in row n is 2^n. For any whole number n, $\sum_{r=0}^{n} {}_nC_r = 2^n$.

 QY2

Because the entries in Pascal's Triangle can be interpreted in terms of choosing or arranging objects, you have already seen some justifications for these properties based on counting principles. It is also possible to give formal algebraic proofs using the definition ${}_nC_r = \frac{n!}{(n-r)!r!}$. You are asked to verify Property 1 in Question 6. Properties 2 and 3 are proved in and just after Example 2, respectively. In the next lesson you are asked to prove Property 4.

If you do not understand the statement you are asked to prove, it may help to use a specific example. For instance, in Property 2, let $n = 5$. Then the statement to be proved is that the 2nd and 5th terms in row 5 equal 5, that is, that ${}_5C_1 = {}_5C_{5-1} = 5$. From the formula for ${}_nC_r$,

$$ {}_5C_1 = \frac{5!}{4!1!} = 5 \text{ and } {}_5C_{5-1} = {}_5C_4 = \frac{5!}{1!4!} = 5. $$

Example 2 generalizes the specific example computed above.

> **QY1**
>
> In which row of Pascal's Triangle is ${}_{10}C_6$? Which term is ${}_{10}C_6$ in that row?

> **QY2**
>
> **a.** Give the first two and last two terms of row 95 of Pascal's Triangle.
>
> **b.** Find the sum of all the entries in row 9 of Pascal's Triangle.

Example 2

Prove that the second and next-to-last terms in the nth row of Pascal's Triangle equal n.

Solution In any row n, the second entry is ${}_nC_1$. By the formula for ${}_nC_r$,

$$ {}_nC_1 = \frac{n!}{(n-1)!1!} = \frac{n!}{(n-1)!} = n. $$

The next-to-last entry in the nth row is ${}_nC_{n-1}$. Using the ${}_nC_r$ formula again,

$$ {}_nC_{n-1} = \frac{n!}{(n-(n-1))!(n-1)!} = \frac{n!}{(n-n+1)!(n-1)!} = \frac{n!}{1!(n-1)!} = n. $$

So, ${}_nC_1 = {}_nC_{n-1} = n$.

The proof of Example 2 can be broadened to prove Property 3, which is so important that we label it as a theorem.

Theorem (Pascal Triangle Symmetry)

If n and r are any whole numbers with $n \geq r$, ${}_nC_r = {}_nC_{n-r}$.

Proof Use the Formula for ${}_nC_r$ Theorem.

$${}_nC_r = \frac{n!}{(n-r)!r!}$$

$${}_nC_{n-r} = \frac{n!}{(n-(n-r))!(n-r)!}$$

$$= \frac{n!}{(n-n+r)!(n-r)!}$$

$$= \frac{n!}{r!(n-r)!}$$

So for whole numbers n and r, where $r \leq n$, ${}_nC_r = {}_rC_{n-r}$.

 **QY3**

A fifth property of Pascal's Triangle gives a recursion relation for the two-dimensional array. It shows how each row is related to the preceding row.

5. Each element in Pascal's Triangle is the sum of the two elements nearest it in the preceding row. Specifically, for whole numbers n and r with $1 \leq r \leq n$, ${}_nC_{r-1} + {}_nC_r = {}_{n+1}C_r$.

For example, the first 4 and 6 in row 4 add to 10, which is the entry just below these numbers in row 5, as shown below.

row n	
0	1
1	1 1
2	1 2 1
3	1 3 3 1
4	1 4—6 4 1
5	1 5 10 10 5 1
6	1 6 15 20 15 6 1

In Question 14, you are asked to give an algebraic proof of this property. Here is an argument using combinations that shows why the property works for a specific case. Suppose you wished to know ${}_7C_3$, the number of combinations of 7 objects taken 3 at a time. Call the objects A, B, C, D, E, F, and G. Each of the ${}_7C_3$ combinations either contains G or it doesn't. The combinations that contain G include 2 objects from A–F, so ${}_6C_2$ combinations contain G. The number of combinations that do not contain G is the number of ways to pick 3 objects from A–F, so ${}_6C_3$ do not contain G. In this way, ${}_7C_3 = {}_6C_2 + {}_6C_3$.

QY3

What does the Pascal Triangle Symmetry Theorem say when $n = 17$ and $r = 11$?

GUIDED

Example 3

Use Properties 1, 3, and 5 to construct row 7 in Pascal's Triangle.

Solution Start by copying row 6.

row 6 1 6 ___?___ ___?___ ___?___ ___?___ ___?___

For row 7, Property 1 states that its first and last entries are 1. To find the second, third, and fourth entries in row 7, add consecutive pairs of terms in row 6 as described in Property 5; the results are $1 + 6 = 7$, $6 + 15 =$ ___?___, and $15 + 20 =$ ___?___. Use the Pascal Triangle Symmetry Theorem (or continue adding pairs of terms of row 6) to find the rest of the terms. Copy the pattern below and write in the result.

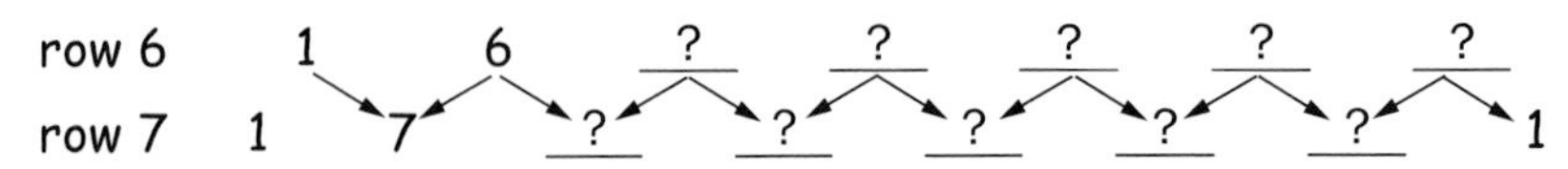

Check 1 Do the entries equal ${}_7C_0, {}_7C_1, {}_7C_2, ...$? Check with a calculator.

Check 2 Use Property 4. Does the sum of the elements equal $2^7 = 128$? Yes, it checks.

A Brief History of Pascal's Triangle

The array we call "Pascal's Triangle" was actually discovered many centuries before Pascal. The Persian mathematician and poet Omar Khayyam (c. 1048–1122) used the triangle of numbers around the year 1100. The Chinese mathematician Chu Shih-Chieh wrote about the array in books published in 1299 and 1303. Perhaps the earliest description of the triangle comes from the Jain poet and mathematician Pingala, who lived in India about 200 B.C.E. and wrote in Sanskrit, the language of sacred Hindu and Jain texts. Pingala discovered methods for calculating the entries in Pascal's Triangle, and referred to the array as the *meruprastāra*, or "Tower of Jewels."

The first Western discoverer of the triangle was Pascal himself, who wrote extensively about this triangular array of numbers and its properties in a 1653 publication, *Treatise on the Arithmetic Triangle*. It is for this reason that Westerners refer to the triangle by his name.

Questions

COVERING THE IDEAS

1. a. How many terms are in row 6 of Pascal's Triangle?
   b. The fifth term is ${}_6C_r$. What is the value of r? Of ${}_6C_r$?
   c. The middle term is ${}_6C_s$. What is the value of s? Of ${}_6C_s$?
   d. What is the sum of the numbers in this row? Express your answer as a power of 2.

2. In how many ways can six coins be arranged with two heads and four tails?

In 3–5, match the English description of the pattern in Pascal's Triangle to the description using combination notation.

A For each n, $\binom{n}{1} = \binom{n}{n-1} = n$.
B For each n and each r ($r \le n$), $\binom{n}{r} = \binom{n}{n-r}$.
C For each n, $\binom{n}{0} = \binom{n}{n} = 1$.

3. Each row in the isosceles triangle is symmetric with respect to a vertical line.
4. The second and next-to-last entries in each row are equal to the row number.
5. The first and last entries in a row are 1.
6. Prove that for all integers $n \ge 1$, ${}_nC_0 = {}_nC_n = 1$.
7. Let $n = 11$, $r = 7$.
   a. Compute ${}_nC_{r-1} + {}_nC_r$ and ${}_{n+1}C_r$.
   b. Which property stated in this lesson is illustrated by Part a?
8. a. Give the entries in row 8 of Pascal's Triangle.
   b. Check your answer by showing that the entries add to 2^8.
9. The first nine entries of row 16 of Pascal's Triangle are 1, 16, 120, 560, 1820, 4368, 8008, 11440, 12870.
   a. How many other entries are there in row 16?
   b. Without doing any calculations, list the remaining entries.
   c. What is the sum of the entries in row 16?
   d. List the first four entries in row 17.
10. What is the 10th term in row 20 of Pascal's Triangle?

APPLYING THE MATHEMATICS

11. What are the last two terms in the row of Pascal's Triangle whose terms add to 2^{34}?
12. a. Expand the following.
   i. $(x + y)^1$ ii. $(x + y)^2$ iii. $(x + y)^3$
   b. Relate the coefficients in Part a to Pascal's Triangle.
   c. Verify your conjecture by expanding $(x + y)^6$ on a CAS.
13. The arrows at the right indicate the *diagonals* of Pascal's Triangle.
   a. What is the nth term in the first diagonal?
   b. What is the nth term in the second diagonal?
   c. What is the nth term in the third diagonal?

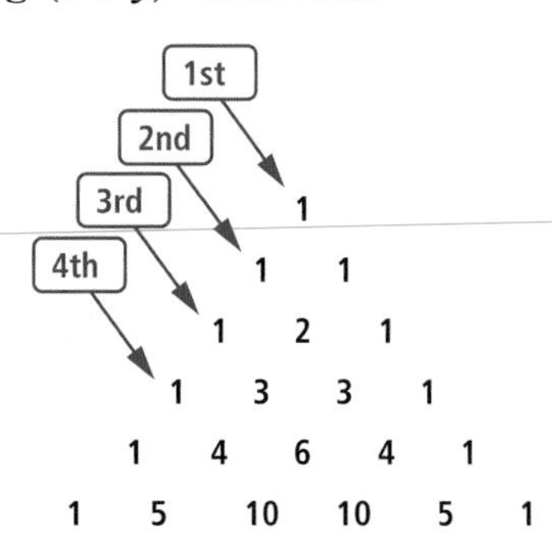

14. Complete Parts a–c to prove Property 5 in the lesson, which states that any entry in Pascal's Triangle (presented in an isosceles triangle array) is the sum of the two entries closest to it in the row above it. That is, show that for all positive integers r and n with $1 \le r \le n$, ${}_{n+1}C_r = {}_nC_{r-1} + {}_nC_r$.
   a. Write an expression for ${}_nC_{r-1} + {}_nC_r$.
   b. Use a CAS to rewrite the expression from Part a with a single denominator.
   c. Explain why the expression in Part b is equivalent to ${}_{n+1}C_r$.

15. Suppose that five fair coins are tossed. As in the Activity, the number of arrangements of 0 heads is 1, the number of arrangements of 1 head is 5, and so on.
   a. Suppose that each arrangement of heads and tails is equally likely. Explain why the probability of each arrangement is $\frac{1}{32}$.
   b. Explain why the probability of getting 1 head in 5 tosses is $\frac{5}{32}$.
   c. Find the probability of getting 2 heads in 5 tosses.

REVIEW

16. An 18-team softball league split into two divisions of 9 teams each. In how many ways can this be done? **(Lesson 10-1)**

17. A 20-person law firm, managed by 4 senior partners, holds a random drawing in which 4 employees each get a laptop computer. What is the probability that the four employees chosen are the 4 senior partners? **(Lesson 10-1)**

In 18 and 19, solve for x. (Lesson 9-6)

18. $0.3^x = 0.07$

19. $4(0.2)^x = 0.9$

20. ***Skill Sequence*** Find all real solutions. **(Lesson 7-8)**
   a. $x^4 + x^3 = 0$
   b. $x^4 + x^3 - x - 1 = 0$

EXPLORATION

21. Although Pascal is most often associated with the triangle in this lesson, he had many other interests in science, philosophy, and religion. Research those interests and describe one of his ideas.

QY ANSWERS

1. 7th term in row 10
2. a. 1, 95, 95, and 1
   b. $2^9 = 512$
3. ${}_{17}C_{11} = {}_{17}C_6$

Lesson 10-3 The Binomial Theorem

Vocabulary

expansion of $(x+y)^n$

binomial coefficients

▶ **BIG IDEA** The Binomial Theorem shows that the coefficients of terms in the expansion $(a + b)^n$ are combinations that can be read from Pascal's Triangle.

Recall that to *expand* a product or power of polynomials means to write the expression as a sum.

Mental Math

Multiply.

a. $(4 + 8)(4 + 8)$

b. $(x + y)(x + y)$

c. $(40 + 1)(40 + 1)$

Activity

Step 1 Use the expand command on a CAS to expand $(x + y)^n$ for $n = 1, 2, 3, 4$, and 5.

Step 2 Write the coefficients as a table, with the coefficients of $(x + y)^1$ in the first row, of $(x + y)^2$ in the second row, and so on, with the coefficients written in order of the decreasing powers of x. What patterns do you notice? Describe them as clearly and specifically as you can.

Degree n	Coefficients (decreasing powers of x)						
1	1	1					
2	1	2	?				
3	?	?	?	?			
4	?	?	?	?	?		
5	?	?	?	?	?	?	
6	?	?	?	?	?	?	?

Step 3 Based on your answer to Step 2, predict what coefficients will result for $(x + y)^6$. Use a CAS to check your prediction.

Step 4 Does $(x + y)^0$ fit into this pattern? Explain.

The Activity illustrates that the coefficients of the terms in the expansion of $(x + y)^n$ are the numbers in the nth row of Pascal's Triangle. Consider $(x + y)^3 = (x + y)(x + y)(x + y)$. You can find this product by multiplying each term of the first factor by each term of each of the other factors and then adding those partial products. In all, eight different partial products are computed. The terms that are multiplied to get each partial product are highlighted in red at the right. Adding like terms gives $(x + y)^3 = 1x^3 + 3x^2y + 3xy^2 + 1y^3$. The coefficients are the entries row 3 of Pascal's Triangle.

Choice of Factors (in red)	Partial Product
$(x + y)(x + y)(x + y)$	$xxx = x^3$
$(x + y)(x + y)(x + y)$	$xxy = x^2y$
$(x + y)(x + y)(x + y)$	$xyx = x^2y$
$(x + y)(x + y)(x + y)$	$xyy = xy^2$
$(x + y)(x + y)(x + y)$	$yxx = x^2y$
$(x + y)(x + y)(x + y)$	$yxy = xy^2$
$(x + y)(x + y)(x + y)$	$yyx = xy^2$
$(x + y)(x + y)(x + y)$	$yyy = y^3$

To see why each coefficient is a combination, examine the eight partial products. Each partial product is a sequence of x's and y's resulting from choosing x's from some factors and y's from other factors. The result x^3 occurs when x is used as a factor three times and y is used as a factor zero times. There are ${}_3C_0$ ways to choose 0 y's from the three y terms, so x^3 occurs in ${}_3C_0 = 1$ way. That is, x^3 occurs once, so its coefficient in the expansion is 1.

The product x^2y occurs when x is chosen from two of the three factors and y from one. There are three y terms from which to choose, so this can be done in ${}_3C_1 = 3$ ways. So x^2y occurs three times and its coefficient in the expansion is 3. Similarly, xy^2 occurs when x is chosen from one factor and y from two, so the coefficient of xy^2 is ${}_3C_2 = 3$.

Finally, y^3 occurs when all three y terms are chosen, and this occurs in ${}_3C_3 = 1$ way. Thus,

$$(x + y)^3 = x^3 + 3x^2y + 3xy^2 + y^3.$$

Using the language of combinations, we can write

$$(x + y)^3 = {}_3C_0x^3 + {}_3C_1x^2y + {}_3C_2xy^2 + {}_3C_3y^3.$$

Notice that each coefficient ${}_nC_r$ can be viewed as the answer to a problem in counting strings. For instance, ${}_3C_1$ is the number of ways 2 x's and 1 y can be arranged.

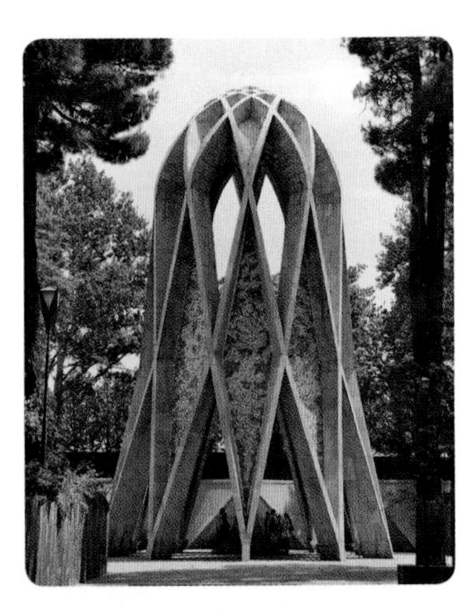

Tomb of Omar Khayyam (1048-1123) in the mausoleum of Imamzadeh Mahruq in Neishabur, Iran

What Is the Binomial Theorem?

In general, the **expansion of $(x + y)^n$** has ${}_nC_0x^n = x^n$ as its first term and ${}_nC_ny^n = y^n$ as its last. The second term is ${}_nC_1x^{n-1}y = nx^{n-1}y$, and the second from the last is ${}_nC_{n-1}xy^{n-1} = nxy^{n-1}$. The sum of the exponents in each term is n, and the coefficient of $x^{n-k}y^k$ is ${}_nC_k$. As the preceding discussion suggests, an easy way to remember the coefficient is that ${}_nC_k$ is the number of strings of x's and y's in which y occurs exactly k times and x occurs exactly $(n - k)$ times. These results are found in the following important theorem, first proved by Omar Khayyam.

Binomial Theorem

For any nonnegative integer n,

$$(x + y)^n = {}_nC_0x^ny^0 + {}_nC_1x^{n-1}y^1 + {}_nC_2x^{n-2}y^2 + \cdots + {}_nC_kx^{n-k}y^k + \cdots + {}_nC_nx^0y^n$$

$$= \sum_{k=0}^{n} {}_nC_kx^{n-k}y^k.$$

For example, $(x + y)^4 = {}_4C_0x^4 + {}_4C_1x^3y + {}_4C_2x^2y^2 + {}_4C_3xy^3 + {}_4C_4y^4$

$$= x^4 + 4x^3y + 6x^2y^2 + 4xy^3 + y^4.$$

Because of their application in this theorem, the combinations ${}_nC_k$ are sometimes called **binomial coefficients.**

Some Uses of the Binomial Theorem

The Binomial Theorem can be used to expand the power of any binomial. Even if you have a CAS, it can sometimes be faster to use the Binomial Theorem to find the coefficient of a specific term rather than to work out the entire expansion.

GUIDED

Example 1

Find the power of y and the coefficient of the x^3 term in $(x + y)^8$.

Solution From the eight factors of $(x + y)^8$, x is to be chosen three times, so y will be chosen __?__ times, and the power of y is __?__. Three x's can be chosen in ${}_8C_{?} =$ __?__ ways, so the coefficient of the x^3 term is __?__.

Check Use a CAS to expand $(x + y)^8$.

 QY1

> **QY1**
>
> Find the coefficient of x^4y^2 in $(x + y)^6$.

Recall from Lesson 7-8 that chunking is the process of grouping small bits of information into a single piece of information. That method is useful when applying the Binomial Theorem.

Example 2

Expand $(3w - 2)^4$ without a CAS.

Solution Use the Binomial Theorem or Pascal's Triangle to expand $(x + y)^4$.

$$(x + y)^4 = x^4 + 4x^3y + 6x^2y^2 + 4xy^3 + y^4$$

Let $x = 3w$ and $y = -2$ and substitute.

$$\begin{aligned}(3w - 2)^4 &= (3w)^4 + 4(3w)^3(-2) + 6(3w)^2(-2)^2 \\ &\quad + 4(3w)(-2)^3 + (-2)^4 \\ &= 81w^4 + 4(27)(-2)w^3 + 6(9)(4)w^2 \\ &\quad + 4(3)(-8)w + 16 \\ &= 81w^4 - 216w^3 + 216w^2 - 96w + 16\end{aligned}$$

Check Let $w = 1$. Then $(3w - 2)^4 = 1^4 = 1$. The right side of the equation is $81 - 216 + 216 - 96 + 16 = 1$. This checks the coefficients.

 QY2

> **QY2**
>
> Find the coefficient of x^3y^3 in $(x - 2y)^6$.

As you have seen, the Binomial Theorem links counting problems to algebra.

Example 3

a. A coin is flipped five times. How many of the possible arrangements of heads and tails have at least two heads?

b. Find all terms in $(H + T)^5$ in which the power of H is at least two.

The U.S. Mint estimates that a coin stays in circulation for at least 30 years.

Solution

a. "At least two" means there could be 2, 3, 4, or 5 heads. Count according to the number of heads. Make a table to organize your work.

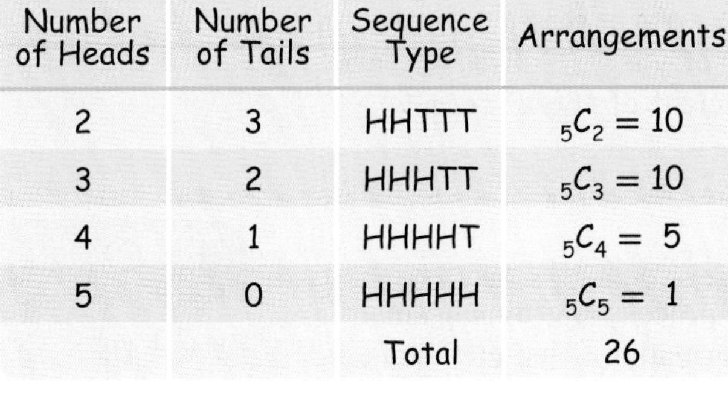

Number of Heads	Number of Tails	Sequence Type	Arrangements
2	3	HHTTT	${}_5C_2 = 10$
3	2	HHHTT	${}_5C_3 = 10$
4	1	HHHHT	${}_5C_4 = 5$
5	0	HHHHH	${}_5C_5 = 1$
		Total	26

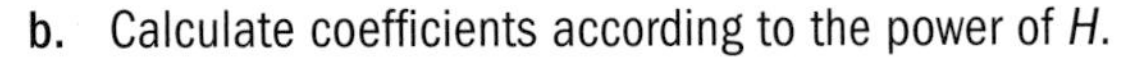

b. Calculate coefficients according to the power of H.

Power of H	Power of T	Product Type	Coefficients
2	3	H^2T^3	${}_5C_2 = 10$
3	2	H^3T^2	${}_5C_3 = 10$
4	1	H^4T	${}_5C_4 = 5$
5	0	H^5	${}_5C_5 = 1$

The terms in which the power of H is at least two are $10H^2T^3$, $10H^3T^2$, $5H^4T$, and H^5.

Check Of the total $2^5 = 32$ possible arrangements, 1 has no heads and 5 have one head, so 6 have fewer than 2 heads. That leaves $32 - 6 = 26$ outcomes with at least 2 heads. It checks.

Example 3 shows that the coefficient of H^rT^{n-r} in the expansion of $(H + T)^n$ is the number of ways of obtaining exactly r heads in n flips of a coin.

Questions

COVERING THE IDEAS

1. a. In the product $(x + y)(x + y)(x + y)(x + y)$, what is the coefficient of x^3y?
   b. How can this coefficient be derived from a combination problem?
2. Use the Binomial Theorem to find the coefficient of x^4 in $(x - 3)^7$.
3. How many ways are there to get at least 4 heads in 7 flips of a coin?
4. a. Write an expression for the coefficient of x^{997} in $(x + 1)^{1008}$.
   b. Find its value.
5. Find the term containing p^3 in $(6p + 2q)^4$.

In 6–8, expand with a CAS. Then check by letting $x = 2$ and evaluating the given binomial power and the expansion.

6. $(x - 2)^6$
7. $(1 - 3x)^4$
8. $(2x + 1)^5$
9. How many of the possible arrangements of heads and tails in six tosses of a coin have the following?
   a. exactly 3 tails
   b. at least 3 tails
10. **True or False** Suppose kx^py^q is a term in the expansion of $(x + y)^n$.
    a. $p + q = n$
    b. $k = {}_pC_q$

APPLYING THE MATHEMATICS

11. Calculate 1001^4 by expanding $(10^3 + 1)^4$ using the Binomial Theorem.
12. Calculate 0.9^4 by expanding $(1 - 0.1)^4$ using the Binomial Theorem.
13. A coin is flipped six times.
    a. In how many ways is it possible to get exactly four heads?
    b. One such outcome is *HHHHTT*. If heads and tails are equally likely on each flip, what is the probability of this outcome?
    c. If the probability of heads is $\frac{3}{4}$ on each flip, what is the probability of the outcome *HHHHTT*?
14. Mr. and Mrs. Ippy hope to have four children.
    a. In how many orders can they have two boys and two girls?
    b. Assuming boys and girls are equally likely, what is the probability that they will have two boys and two girls? $\frac{3}{8}$

15. Use the Binomial Theorem to prove Property 4 from Lesson 10-2, that $\sum_{k=0}^{n} {}_nC_k = 2^n$. Hint: Let $2^n = (1 + 1)^n$.

16. a. Use a CAS to find the coefficient of $x^3y^2z^2$ in $(x + y + z)^7$.
    b. What counting problem involving strings does this coefficient answer?

REVIEW

In 17 and 18, use the numbers below, which are the first six terms of row 11 of Pascal's Triangle. (Lesson 10-2)

1 11 55 165 330 462

17. Which term represents ${}_{11}C_5$?

18. a. Write the rest of row 11.
    b. Write row 12 in full.
    c. Write row 10 in full.

In 19 and 20, suppose a charity sells 1000 raffle tickets for $100 each.
   a. Tell whether the question asks for a combination or a permutation.
   b. Answer the question. (Lessons 10-1, 6-4)

19. The first place winner receives $25,000, the second place winner receives $15,000, the third place winner receives $10,000, and the fourth place winner receives $5,000. In how many different ways could the prizes be distributed?

20. All four winners share $55,000 evenly. How many ways could the prizes be distributed?

21. Three different integers are chosen at random from the integers 1 to 36. What is the probability that they are consecutive integers? **(Lessons 10-1, 6-8)**

EXPLORATION

22. a. Expand $(1 + 0.001)^5$ to obtain the decimal for 1.001^5.
    b. How many terms of the expansion are needed to get an estimate to 1.001^5 accurate to the nearest thousandth?
    c. Use your results from Parts a and b to estimate 1.002^8 to the nearest millionth.
    d. Give the complete decimal expansion of 1.002^8.
    e. How close is your calculator value of 1.002^8 to the value of your answer in Part c?

QY ANSWERS

1. 15

2. –160

Lesson 10-4 Probability Distributions

Vocabulary

probability distribution
random variable
mean of a random variable
mode of a distribution
variance of a random variable
standard deviation of a random variable

BIG IDEA Statistics can be calculated for a probability distribution by considering the values of the distribution as a data set.

From the first chapter of this book, you have seen relative frequency distributions. A *probability distribution* is very similar. A **probability distribution** is a function that maps each outcome or event onto its probability rather than its relative frequency. The probability distribution is the *theoretical* estimation of chance for the *experimental* relative frequency distribution. Every probability distribution has the following characteristics:

(1) Its domain is the set of mutually exclusive outcomes or non-overlapping events whose union is the sample space of the experiment.

(2) It maps each outcome or event onto its probability, so its range is the set of probabilities corresponding to its domain.

(3) The sum of the probabilities is 1.

Mental Math

In an advanced placement class of 20 students, 5 students scored 2 on the AP test, 4 students scored 3 on the AP test, 5 students scored 4, and 6 students scored 5. What was the mean score for the class?

A Familiar Example of a Probability Distribution

In Lesson 6-1, the 36 possible outcomes for the tossing of two dice are listed. If the dice are fair, then the probability of obtaining each possible sum is given in the table below.

x (sum of dice)	2	3	4	5	6	7	8	9	10	11	12
P(x) (probability)	$\frac{1}{36}$	$\frac{2}{36}$	$\frac{3}{36}$	$\frac{4}{36}$	$\frac{5}{36}$	$\frac{6}{36}$	$\frac{5}{36}$	$\frac{4}{36}$	$\frac{3}{36}$	$\frac{2}{36}$	$\frac{1}{36}$

 QY

QY

Verify that no possible outcome has been excluded by checking that the sum of the probabilities is 1.

In this situation, the domain variable x, the sum of the top faces of two dice, is a *random variable*. In general, a **random variable** is a variable whose values are numbers determined by the outcome of an experiment. You can think of a probability distribution as a function that maps each value of a random variable onto its probability.

The probability distribution P for the sum of two dice is graphed at the right. Both the table and the graph, show that the domain of P is the set of integers between 2 and 12, inclusive; the range is the set $\left\{\frac{1}{36}, \frac{2}{36}, \frac{3}{36}, \frac{4}{36}, \frac{5}{36}, \frac{6}{36}\right\}$.

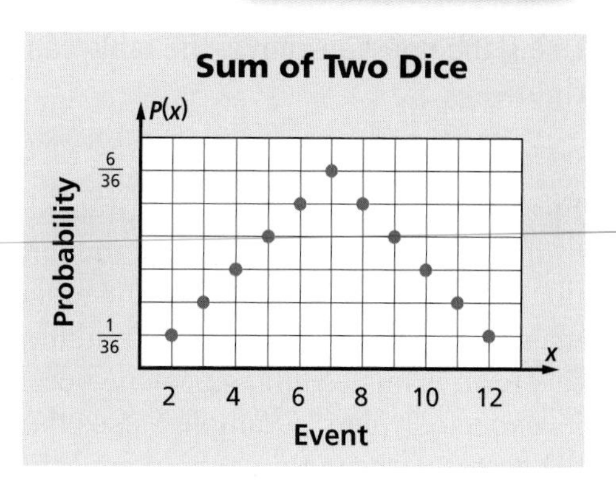

Mean of a Random Variable

In Chapter 1, you studied the standard mean (μ or $\bar{x}$) of a data set $\{x_1, x_2, x_3, \ldots, x_n\}$. Why, then, do we need a *mean of a random variable*? In the calculation of a standard mean, all values are given equal weight. In contrast, the **mean of a random variable x** is the *weighted* mean of a variable x whose *weights are the probabilities* associated with each possible value of the variable. The critical difference between the standard mean of a data set and the mean of a random variable whose values are the elements of that data set is that all elements of the data set are not necessarily equally likely and so not necessarily given equal weight. The common example of a lottery makes the difference clear.

A lottery in which a random four-digit number is picked pays out $8,000 to the winner. There are 10,000 four-digit numbers, so the probability of winning is 0.0001. If we let the random variable x be the amount won, then the probability distribution can be organized as shown at the right.

Winnings x	$0	$8000
Probability	0.9999	0.0001

The unweighted mean of the set of possible outcomes would be $\frac{\$0 + \$8000}{2} = \$4{,}000$. However, intuitively, this doesn't make sense because you know that a player is much less likely to win $8,000 than to win no money at all. In the long run, after say 10,000 tickets, you would expect to win $8,000 once and win nothing the other 9,999 times! Therefore, your long run average winnings from a ticket would be

$$\$8000(0.0001) + \$0(0.9999) = \$0.80,$$

a significantly different result from the unweighted mean of $4,000. $0.80 is the mean of the random variable x in this case. The mean of a random variable is denoted by μ, which is also used for the mean of a population as was seen in Chapter 1. The following generalizes the situation above to any probability distribution.

Definition of the Mean of a Random Variable

Suppose x is a random variable whose probability distribution is $\{(x_1, P(x_1)), (x_2, P(x_2)), \ldots, (x_n, P(x_n))\}$. The **mean** of x is

$$\mu = \sum_{i=1}^{n} (x_i \cdot P(x_i)).$$

Using the notation above, the table can be relabeled as shown at the right.

Winnings x_i	$0 x_1	$8000 x_2
Probability $P(x_i)$	0.9999 $P(x_1)$	0.0001 $P(x_2)$

So $\mu = x_1 \cdot P(x_1) + x_2 \cdot P(x_2) = \$0 \cdot 0.9999 + \$8000 \cdot 0.0001 = \$\,0.80$.

Just as probabilities are the theoretical estimates of long-run relative frequencies, so is the mean of a random variable the expected theoretical estimate of long-run average outcomes. For this reason, many people call the mean of a random variable its *expected value*, but this term is misleading, since one occurrence of x may not be close to its expected value. Example 1 uses the mean of a random variable and illustrates a way to think about expected value.

Example 1

In the card game of bridge, a standard 52-card deck is used. That deck has 13 cards (Ace, King, Queen, Jack, 10, 9, 8, 7, 6, 5, 4, 3, 2) in each of 4 suits (spades, hearts, diamonds, clubs). In bridge, the deck is distributed randomly among 4 players. Because certain cards beat other cards during the game, it is common for a bridge player to evaluate the 13 cards in a hand according to the following scheme: Ace, 4 points; King, 3 points; Queen, 2 points; Jack, 1 point; all other cards from 10 down to 2, 0 points. For example, a hand with A, Q, Q, J, 10, 9, 6, 6, 6, 5, 5, 3, and 2 would be worth 9 points.

a. Let x be a random variable standing for the number of points a card is worth. Make a table of the probability distribution for x.

b. Calculate the mean of the random variable x.

c. What is the expected value of x?

Solution

a. There are 4 cards with each point value from 1 to 4 and the other 36 cards have no point value. Each possible point value has a probability that is estimated using relative frequencies. For example, since 4 out of the 52 cards are worth 3 points, the probability that $x = 3$ is $\frac{4}{52}$.

x (point value)	0	1	2	3	4
P(x) (probability of point value)	$\frac{36}{52}$	$\frac{4}{52}$	$\frac{4}{52}$	$\frac{4}{52}$	$\frac{4}{52}$

b. Use the definition of the mean of a random variable and the table of values of x_i and $P(x_i)$.

$$\mu = \sum_{i=1}^{5} (x_i \cdot P(x_i))$$
$$= 0 \cdot \frac{36}{52} + 1 \cdot \frac{4}{52} + 2 \cdot \frac{4}{52} + 3 \cdot \frac{4}{52} + 4 \cdot \frac{4}{52}$$
$$= 0 + \frac{1}{13} + \frac{2}{13} + \frac{3}{13} + \frac{4}{13}$$
$$= \frac{10}{13} \approx 0.769$$

c. The mean of the random variable is the point value you would expect for any chosen card in the long run, so you would expect 0.769 points even though you would never be able to choose a card with that value!

Frequently, for discrete outcomes, a probability distribution is graphed as a histogram. It is common to center each bar over the individual value of the random variable. The height of each bar is the corresponding probability, as shown at the right for Example 1.

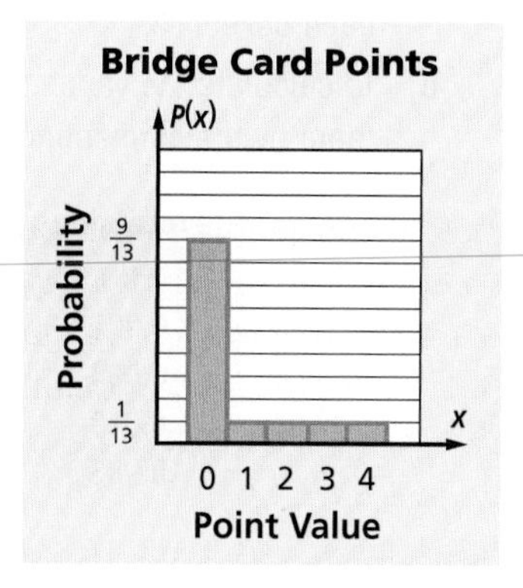

In the case of the bridge card points, the distribution is asymmetric. Immediately striking is the tallest bar for the 0-point value outcome. In a probability distribution, the value of the random variable whose probability is the highest is the **mode of the distribution**. In this case, the mode is the 0-point outcome.

Variance and Standard Deviation of a Random Variable

Recall that the variance σ^2 of a population is the average squared deviation of the members of the data set from the mean. The variance of a random variable x is denoted by σ^2 or $\sigma_x{}^2$ and uses values of the form $(x - \mu)^2$ weighted by the probabilities $P(x)$.

Definition of Variance of a Random Variable

The **variance** of a random variable x is $\sigma_x{}^2 = \sum_{i=1}^{n} (x_i - \mu)^2 \cdot P(x_i)$,

where μ is the mean of the random variable.

The **standard deviation of a random variable** is the square root of its variance, so it equals σ or σ_x.

Example 2

A charity holds a raffle. Tickets cost $10 and all 1000 tickets are expected to be sold. First prize is a $2000 TV set, there are four $500 second prizes, and ten third prizes are worth $100 each.

a. If you buy a ticket, what is the expected value of your winnings?

b. What is the variance and standard deviation of the random variable in this situation?

a. First calculate the probability of winning each prize. $P(\text{winning } \$2000) = \frac{1}{1000} = 0.001$. The other probabilities are easily calculated. Remember to calculate the probability of winning nothing! In the table below, the probabilities are recorded in the second column from the left. To obtain the expected value, multiply the prizes by their probabilities and add:

$$\mu = 2000 \cdot 0.001 + 500 \cdot 0.004 + 100 \cdot 0.010 + 0 \cdot 0.985$$
$$= 5.$$

This means that the average prize is $5. But remember that each person has spent $10 for a ticket. So, on average, the charity gains $5 from each ticket.

b. To calculate the variance, calculate the squared deviations and multiply them by the corresponding probabilities.

Prize x	Probability $P(x)$	Squared Deviation $(x - \mu)^2$	Deviation • Probability $(x - \mu)^2 \cdot P(x)$
2000	0.001	$(2000 - 5)^2 = 1995^2$	$1995^2 \cdot 0.001$
500	0.004	$(500 - 5)^2 = 495^2$	$495^2 \cdot 0.004$
100	0.010	$(100 - 5)^2 = 95^2$	$95^2 \cdot 0.010$
0	0.985	$(0 - 5)^2 = 5^2$	$5^2 \cdot 0.985$

Then variance σ^2

$= 1995^2 \cdot 0.001 + 495^2 \cdot 0.004 + 95^2 \cdot 0.010 + 5^2 \cdot 0.985$

$= 3980.03 + 980.1 + 90.25 + 24.625$

$= 5075.01$

and standard deviation $\sigma = \sqrt{5075.01} \approx \$71.24$.

Check Use a calculator with lists or spreadsheet capabilities. Place the prize values (x) in the first column and the probabilities in the second column. Compute one-variable statistics on the first column with frequencies in the second column. Your calculator may use $\bar{x}$ for the mean rather than μ. The standard deviation is likely to appear as σ_x.

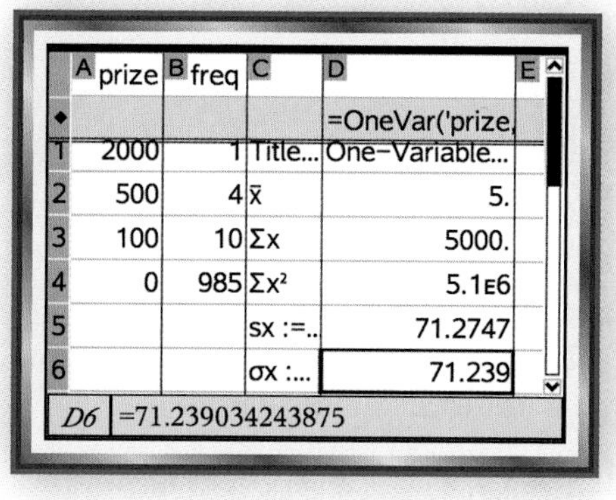

	A prize	B freq	C	D	E
◆				=OneVar('prize,	
1	2000	1	Title...	One-Variable...	
2	500	4	$\bar{x}$	5.	
3	100	10	Σx	5000.	
4	0	985	Σx²	5.1E6	
5			sx :=..	71.2747	
6			σx :...	71.239	

D6 =71.239034243875

Algebra can be used to transform the variance formula into a form that does not require calculating the deviations.

Alternate Formula for Variance of a Random Variable

The variance of a random variable x is $\sigma_x{}^2 = \sum_{i=1}^{n} x_i{}^2 \cdot P(x_i) - \mu^2$, where μ is the mean of the random variable.

Proof In this proof, to shorten the work, we have written $\sum$ for $\sum_{i=1}^{n}$ and not written indices for the x_i.

$\sigma^2 = \sum (x - \mu)^2 \cdot P(x)$	definition of variance of a random variable
$= \sum (x^2 - 2x\mu + \mu^2) \cdot P(x)$	Expand the binomial.
$= \sum x^2 \cdot P(x) - \sum 2x\mu \cdot P(x) + \sum \mu^2 \cdot P(x)$	Distribute $P(x)$ and then distribute the summation over its terms.
$= \sum x^2 \cdot P(x) - 2\mu \sum x \cdot P(x) + \mu^2 \cdot \sum P(x)$	Factor constants from each sum.
$= \sum x^2 \cdot P(x) - 2\mu^2 + \mu^2 \cdot 1$	definition of μ and property that $\sum P(x) = 1$
$= \sum x^2 \cdot P(x) - \mu^2$	Distributive Property (collect terms)

Questions

COVERING THE IDEAS

1. What is the total height of the bars in a histogram for a probability distribution?

2. **True or False** The mean of a random variable is always a value of the random variable of the experiment. Explain your answer.

In 3–6, tell whether *P* is a probability distribution. If it is, find the mean of the random variable *x*.

3.

x	1	2	3	4
P(x)	75%	25%	25%	25%

4.

x	1	2	3	4
P(x)	0	0	1	0

5.

x	1	2	3	4
P(x)	$\frac{1}{5}$	$\frac{7}{24}$	$\frac{1}{4}$	$\frac{1}{3}$

6.

x	1	2	3	4
P(x)	0.18	0.27	0.45	0.10

7. A researcher collects the following data about the incubation time for a certain disease.

x = Number of Days	1	2	3	4	5	6	7
P(x)	$\frac{1}{14}$	$\frac{3}{28}$	$\frac{5}{21}$	$\frac{1}{7}$	$\frac{1}{3}$	$\frac{1}{14}$	$\frac{1}{28}$

a. What is the random variable?

b. Find the mean incubation time.

c. Find the mode of the distribution represented by the table.

8. Consider the experiment of tossing a fair 6-sided die twice.

a. Construct a probability distribution table in which the value of the random variable is calculated by subtracting the value showing on the second toss from the value showing on the first toss.

b. Graph the distribution in Part a as a scatterplot.

c. Find the mean of the random variable.

9. Consider the probability distribution of bridge card points in Example 1.

a. What is the variance of the point values?

b. Find the standard deviation of the point values.

APPLYING THE MATHEMATICS

In 10 and 11, an experiment involves rolling two dice and recording the absolute value of the difference between the numbers showing. For example, if you roll a 4 and a 6, the outcome is $|4 - 6| = 2$.

10. The following table of relative frequencies was formed after 360 trials.

Difference	0	1	2	3	4	5
Relative Frequency	$\frac{62}{360}$	$\frac{98}{360}$	$\frac{77}{360}$	$\frac{60}{360}$	$\frac{38}{360}$	$\frac{25}{360}$

a. Make a scatterplot of these data.

b. Find the mean difference.

11. Let x = the absolute value of the difference between the numbers showing on two dice.
    a. Make a table of values of the probability distribution for x.
    b. Find the mean of x. Why is this value different than your answer to Question 10, Part b?
    c. Calculate the variance of x.
    d. Find the standard deviation of x.

12. In a lottery, the value of a ticket is a random variable, defined to be the amount of money you win less the cost of playing. Suppose that in a lottery with 125 tickets, each ticket costs \$1. First prize is \$50, second prize is \$30, and third prize is \$20. Then the possible values of the random variable are \$49, \$29, \$19, and -\$1.
    a. Why is one of the values negative?
    b. Find the probability of winning nothing.
    c. What is the mode of the distribution in this case?
    d. Find the expected value of a ticket.

13. An ecologist collected the data shown in the table below on the life span of a species of deer.

Age at Death (years)	1	2	3	4	5	6	7	8
Number	30	86	132	173	77	40	10	2

    a. Based on this sample, what is the expected life span of this species?
    b. Find the variance for the life span of this species of deer.
    c. Find the standard deviation for the life span of this species of deer.

14. **Multiple Choice** Which two of the following could be the graph of a probability distribution?

A
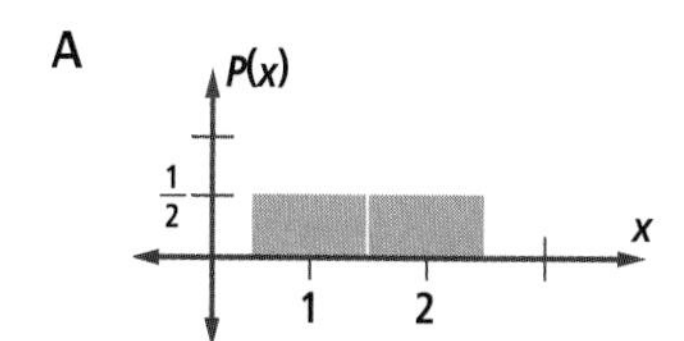

B
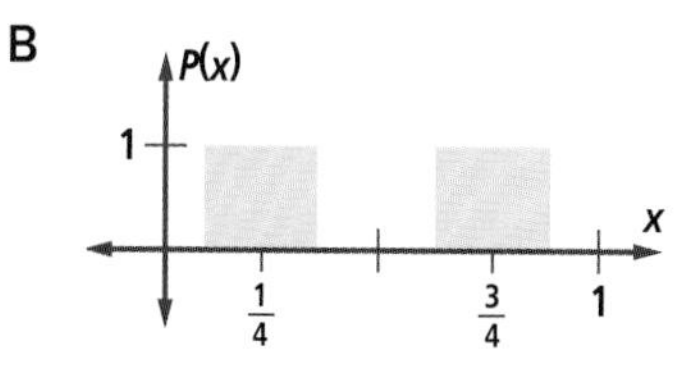

C
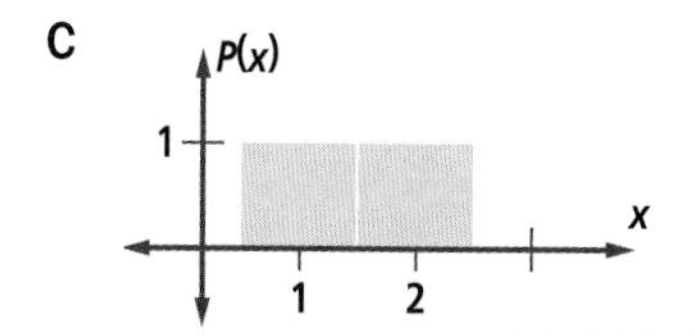

D
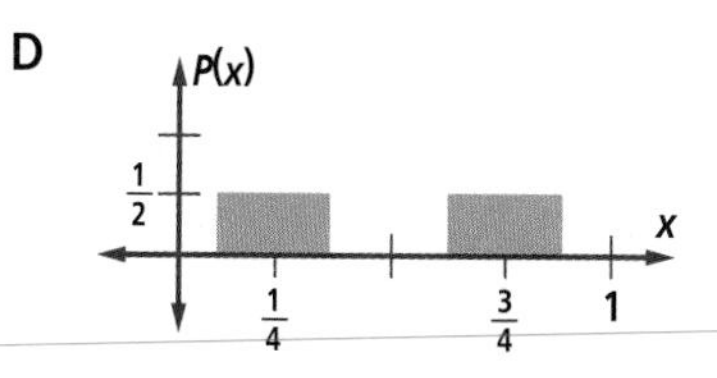

REVIEW

15. Notice that except for the first and last entries, every entry in row 3 of Pascal's Triangle is divisible by 3; except for the first and last entries, every entry in row 5 of Pascal's Triangle is divisible by 5. **(Lesson 10-2)**

a. Verify that the pattern continues for row 7.

b. Verify that the pattern does not hold for row 9.

c. Explain why if p is prime, $r \neq 0$, and $r \neq p$, ${}_pC_r$ is divisible by p.

16. What are the 17th and 21st terms in row 43 of Pascal's Triangle? **(Lesson 10-2)**

In 17 and 18, compute. (Lessons 10-1, 6-4)

17. the number of ways to choose 3 colleges out of 6 considered by a student

18. the number of batting orders for a 10-person softball team

19. a. Find the chi-square statistic for an experiment in which a fair coin is tossed 10 times and 9 heads occur.

b. Using technology, find the probability of such a chi-square value and test this against a 0.01 significance level.

c. What conclusion can you draw from your answer to Part b? **(Lesson 6-9)**

20. Solve Peppermint Patty's math problem using permutations. **(Lesson 6-4)**

PEANUTS: © United Feature Syndicate, Inc.

EXPLORATION

21. a. Find the life expectancy for a person of your age and sex.

b. Find out how life expectancies are determined.

QY ANSWER

$\frac{1}{36} + \frac{2}{36} + \frac{3}{36} + \frac{4}{36} + \frac{5}{36} + \frac{6}{36} + \frac{5}{36} + \frac{4}{36} + \frac{3}{36} + \frac{2}{36} + \frac{1}{36} = \frac{36}{36} = 1$

Lesson 10-5 Binomial Probabilities

Vocabulary

binomial experiment

BIG IDEA In a binomial experiment, if an event E has probability p, then the probability that E occurs r times in n trials is a term in the expansion of $(p + q)^n$, where $q = 1 - p$.

In Lesson 10-3, you used the expansion of $(H + T)^5$ to find the probability of getting 2 heads in 5 tosses of a fair coin. This situation is an example of a *binomial experiment*. A **binomial experiment** has the following features:

(1) There are repeated situations, called *trials*.
(2) There are a fixed number of trials.
(3) For each trial, there are only two possible outcomes, often called *success* (S) and *failure* (F).
(4) The probability of success is the same in each trial.
(5) The trials are independent events.

Listed below are all possible outcomes for a binomial experiment with four trials.

Outcomes for Binomial Experiment with Four Trials

	0 Successes	1 Success	2 Successes	3 Successes	4 Successes
	FFFF	FFFS FFSF FSFF SFFF	FFSS FSFS FSSF SFFS SFSF SSFF	FSSS SFSS SSFS SSSF	SSSS
Number of Outcomes	$1 = {}_4C_0$	$4 = {}_4C_1$	$6 = {}_4C_2$	$4 = {}_4C_3$	$1 = {}_4C_4$

Notice that there are ${}_4C_k$ ways to obtain exactly k successes among the 4 trials. That is, the number of outcomes in the columns above (1, 4, 6, 4, and 1) are precisely the numbers from row 4 of Pascal's Triangle.

Typical Binomial Experiment Situations

Some binomial experiments are theoretical. In Example 3 of Lesson 10-3, a coin was tossed 5 times. This situation is a binomial experiment because (1) each toss is a trial; (2) there were 5 trials, a fixed number; (3) the two possible outcomes were heads (success) and tails (failure); (4) the probability of heads was the same $\left(\frac{1}{2}\right)$ for each toss; and (5) the results of one toss did not affect the results of any other toss.

Mental Math

A bag contains r red marbles, b blue marbles, and g green marbles. If selections are made randomly, what is the probability of selecting

a. a red marble?

b. a blue or a green marble?

c. a red marble, returning it to the bag, and then selecting a second red marble?

Other binomial experiments are done with real data. Suppose past experience has shown that the probability that an implanted heart pacemaker operates in a patient for the first year without flaw is 0.85. This is not yet a binomial experiment situation because there are no repeated trials. However, if 4 patients in a hospital are given these pacemakers in a particular week, there is a binomial experiment situation. (1) Giving a patient a pacemaker is a trial. (2) There are 4 trials. (3) Success means that the pacemaker works without a flaw for a year; failure means that the pacemaker has a flaw within the year. (4) The probability of success in each trial is 0.85. (5) The trials are independent because how one pacemaker works does not influence any other pacemaker. Example 1 shows how binomial coefficients help calculate probabilities in this kind of situation.

Example 1

Suppose the probability of a heart pacemaker working flawlessly for one year is 0.85. What is the probability that two patients out of four receiving pacemakers this week will have some kind of problem with their pacemakers over the next year?

Solution For each patient,
P(pacemaker works flawlessly for one year) = $p = 0.85$.
So P(pacemaker has problem) = $1 - 0.85 = 0.15 = q$.

Since there are 4 pacemakers, this is a binomial experiment with 4 trials. The middle column of the table on the previous page shows that there are six ways in which two of the four pacemakers can fail. The probability of each of these outcomes is $p^2q^2 = (0.85)^2(0.15)^2$. So
P(exactly 2 successes in 4 trials) $= 6(0.85)^2(0.15)^2 \approx 0.0975$.
The chance of exactly 2 of 4 pacemakers not working flawlessly for a year is about 0.0975, or about 10%.

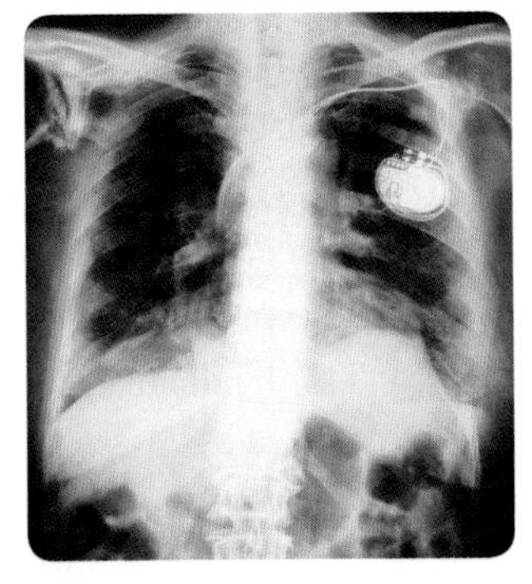

X-ray of a human chest with a pacemaker

Activity

Compute the probability that two *or more* pacemakers have problems during the year.

Step 1 Follow Example 1 to find the probability that 3 pacemakers have problems (which is equivalent to 1 pacemaker working flawlessly).

Step 2 Find the probability that all 4 pacemakers have problems over the year.

Step 3 Add the probabilities in Steps 1 and 2 to the result of Example 1 to find the probability that 2, 3, or 4 of the pacemakers will have problems during the year. Write a sentence summarizing this result, rounding the probability to the nearest percent.

Step 4 Compare your answer in Step 3 to that of Example 1 and explain your findings.

The reason you can add the probabilities in Step 3 of the Activity is that the three events are mutually exclusive. So P(at least two failures) = P(2 failures) + P(3 failures) + P(4 failures).

The General Binomial Experiment

The computation of probabilities in Example 1 and the Activity can be generalized. In a binomial experiment, if the probability of success is p, the probability of failure is $1 - p = q$. Then the probability of exactly k successes in 4 trials is ${}_4C_k p^k q^{4-k}$.

These ideas can be generalized still further.

Binomial Probability Theorem

Suppose that in a binomial experiment with n trials, the probability of success is p in each trial and the probability of failure is q, where $q = 1 - p$. Then

$$P(\text{exactly } k \text{ successes}) = {}_nC_k \cdot p^k q^{n-k} = \binom{n}{k} p^k q^{n-k}.$$

Remember that, in the Binomial Theorem from Lesson 10-3, ${}_nC_k \cdot p^k q^{n-k}$ is the $(k + 1)$st term of the expansion of $(q + p)^n$. This is why the above theorem is called the Binomial Probability Theorem.

Most statistics utilities are programmed to be able to compute binomial probabilities given the values of p, n, and k.

 QY

> **QY**
>
> Estimate $\binom{n}{k} p^k q^{n-k}$ to the nearest thousandth when $p = \frac{1}{3}$, $n = 20$, and $k = 5$.

If you know the probability of getting each question right on a test, you can calculate the probability of getting *all* the questions right or none of the questions right. With the Binomial Probability Theorem, you can calculate the probability of getting ***any particular number*** of questions right.

Example 2

Suppose a quiz has 10 multiple-choice questions, each with 5 choices. If a student guesses randomly on every question, what is the probability the student will get seven or more correct?

Solution This is a binomial experiment. Answering a question is one trial. The 10 questions form an experiment with 10 trials ($n = 10$). Success in any one trial is getting a question correct.

$P(S) = \frac{1}{5} = 0.2$, so $P(F) = 1 - P(S) = \frac{4}{5} = 0.8$.

Because the student is guessing randomly, answers are independent.

(continued on next page)

Getting 7 or more correct is equivalent to the union of the mutually exclusive events of getting exactly 7 or exactly 8 or exactly 9 or exactly 10 correct. So

P(7 or better)

$= P(\text{exactly } 7) + P(\text{exactly } 8) + P(\text{exactly } 9) + P(\text{exactly } 10)$

$= {}_{10}C_7(0.2)^7(0.8)^3 + {}_{10}C_8(0.2)^8(0.8)^2 + {}_{10}C_9(0.2)^9(0.8) + {}_{10}C_{10}(0.2)^{10}$

$\approx 120(0.0000065536) + 45(0.0000016384) + 10(0.0000004096) + 1(0.0000001024)$

≈ 0.000864

Check 1 Use a calculator or a spreadsheet and the Binomial Theorem to add the probabilities. The sum from 7 to 10 is shown at the right in two ways.

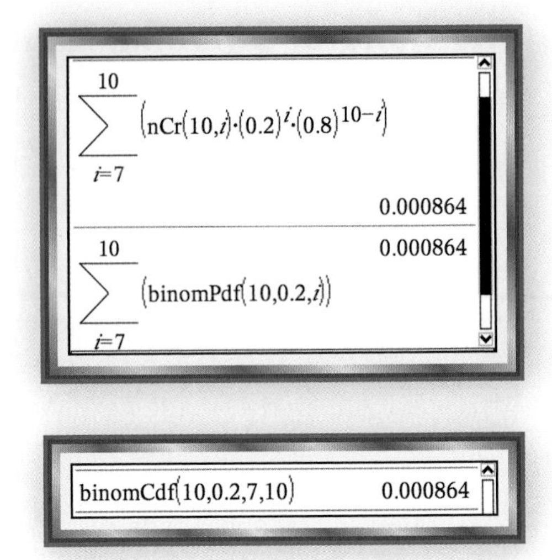

Check 2 You may be able to use a binomial probability distribution command on a calculator as shown here. The cumulative binomial distribution command `binomCdf` uses the arguments (number of trials, probability of success, lower bound of successes, upper bound of successes).

So, random guessing on a 10-item multiple-choice quiz yields a probability of less than 1 chance in 1000 of getting 7 or more items correct. (Chances improve substantially with study!)

Questions

COVERING THE IDEAS

1. State the five characteristics of a binomial experiment.

In 2–5, determine if the situation is a binomial experiment. If not, identify the missing property or properties.

2. A student records that he successfully shot a basket during a game.
3. A student takes the SAT three times, the first time with no preparation and each other time with more preparation than the previous time. A success is defined as a total of 1800 or more.
4. A fair spinner has four colors (red, blue, green, yellow). It is spun 5 times, and the color is recorded.
5. Two fair 6-sided dice are rolled 5 times and "doubles" (the same number showing on both dice) is considered a success.

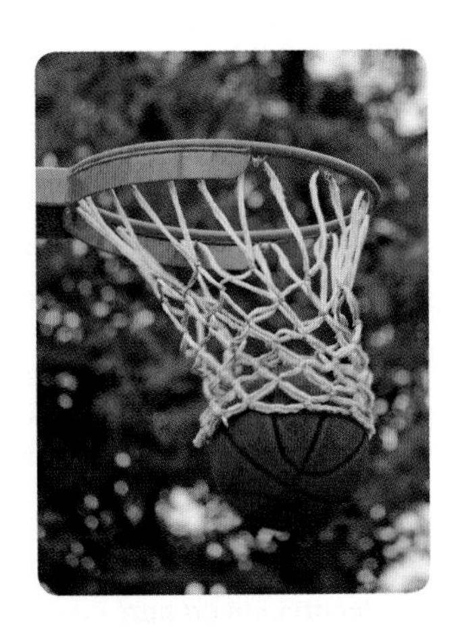

6. a. List the possible outcomes if a coin is tossed 3 times.
   b. How is your answer to Part a related to the Binomial Theorem?
7. Consider tossing a coin 10 times.
   a. Find the probability of getting exactly 6 heads if the coin is fair.
   b. Find the probability of getting exactly 6 heads if the probability of heads each time is 0.7.

In 8 and 9, refer to Example 1 and the Activity. Suppose that the probability of a pacemaker functioning flawlessly for a year has increased to 0.92 and that the pacemaker is given to six patients.

8. What is the probability that five of the six of the pacemakers will function flawlessly for a year?

9. What is the probability that at least three of the six pacemakers will function flawlessly for a year?

10. A quiz has 10 always-sometimes-never questions. A student guesses randomly on every question.
    a. What is a trial in this situation?
    b. How do we know the trials are independent?

11. Suppose a true-false quiz has twelve questions. If a student answers all questions by randomly guessing, what is the probability that at least six questions are answered correctly?

APPLYING THE MATHEMATICS

12. Refer to the spinner at the right with six congruent sectors. On any one spin, the spinner is equally likely to land in any of the six sectors.

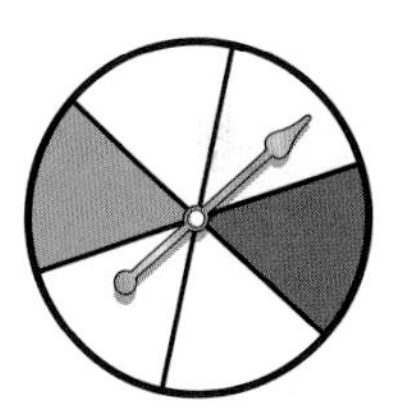

    a. What is the probability of the spinner landing in a white sector?
    b. Suppose a trial is one spin and success means the spinner lands in a white sector. Complete the following probability distribution for the number of successes in four spins.

w = exact number of successes	0	1	2	3	4
$P(w)$	?	?	?	?	?

    c. Find P(at least 3 successes).
    d. Find P(at least 1 success).

13. A baseball player has a batting average of 0.250. This can be interpreted to mean that the probability of a hit is $0.250 = \frac{1}{4}$. Some people think this means that in 4 times at bat, the batter is *sure* to get a hit. Find the probability that this batter gets the following number of hits in 4 at-bats.
    a. exactly 1 hit
    b. at least 1 hit

14. Suppose that in a binomial experiment P(success on one trial) = a.
    a. Express P(failure on one trial) in terms of a.
    b. Find P(success on exactly 8 of 16 trials).

15. Suppose that, in the past, 80% of clients of an interior decorating company have requested white or pastel colors for the walls in the kitchen. Suppose 12 people at random come to choose colors for their kitchen walls.
    a. Find the probability that exactly 10 of them want pastel or white.
    b. Find the probability that no more than 3 of them request colors other than white or pastel.

REVIEW

16. The table below gives the estimated probabilities of family size in the United States based on data from the U.S. Census Bureau. **(Lessons 10-4, 1-2)**

S = Family Size	2	3	4	5	6	7	8 or more
$P(S)$	0.04	0.11	0.18	0.38	0.15	0.10	0.04

a. Graph the probability distribution.

b. What is the mean family size? Use 8 for "8 or more."

c. What is the standard deviation of the random variable representing family size?

In 17–19, expand. (Lesson 10-3)

17. $(x + y)^2$

18. $(3a - 2b)^5$

19. $(\sin \theta + 2 \cos \theta)^3$

20. In Florida's Lotto game, you must pick 6 numbers correctly (in any order) out of 53 without replacement to win. What is your probability of winning? **(Lesson 10-1)**

In 21 and 22, find an exact solution. (Lesson 9-6)

21. $e^y = 117$

22. $\log r^2 = 4$

23. Written in the Latin alphabet, the Hawaiian language has only the following letters: A, E, H, I, K, L, M, N, O, P, U, W. Every Hawaiian word ends with a vowel, and some words have no consonants. Two consonants never occur without a vowel between them. How many four-letter words are possible? **(Lessons 6-4, 6-3)**

The hula dance originated as a part of Hawaiian religious ceremonies.

EXPLORATION

24. Binomial trials are sometimes called *Bernoulli trials.*

a. Who was Bernoulli?

b. What work of Bernoulli is related to binomial probabilities?

QY ANSWER

0.146

Lesson

10-6 Binomial Probability Distributions

Vocabulary

binomial probability distribution

binomial random variable

▶ **BIG IDEA** The graph of a binomial probability distribution provides information about the likelihood of an event occurring r times in n trials.

In Lesson 10-5, you saw that the probability of k successes in n trials in a binomial experiment is ${}_nC_k \cdot p^k(1-p)^{n-k}$. In this lesson, we use the variable x instead of the constant k in this formula because we are interested in how this probability changes as x varies from 0 to n, its only possible values.

Mental Math

Suppose z varies jointly as x and y. What happens to the value of z if

a. x is multiplied by 5 and y is multiplied by 10?

b. x is multiplied by 5 and y is divided by 10?

c. x is kept the same and y is multiplied by c?

Examples of Binomial Probability Distributions

The expression ${}_nC_x \cdot p^x(1-p)^{n-x}$ has three variables: n, p, and x. For a given value of n, x can take on any integer value from 0 to n.

Airplane delays are caused by air traffic congestion, weather, and other factors that vary, essentially randomly, from day to day. Suppose a certain flight is scheduled daily. Then it is reasonable to consider each day's trip as an independent trial of a binomial experiment in which arriving on time is considered a success, and arriving late is considered a failure. If the experiment is carried out over n trips, the number x of on-time arrivals can range from 0 to n. The function B that maps x onto the probability ${}_nC_x \cdot p^x(1-p)^{n-x}$ of x successful arrivals is called a **binomial probability distribution** B where $B(x) = {}_nC_x \cdot p^x(1-p)^{n-x}$. In a binomial probability distribution B, the numbers n and p are *parameters* of the binomial distribution, while x is a **binomial random variable**.

GUIDED

Example 1

Assume that the probability of an on-time arrival of any flight is 75% and that different trips are independent of one another. Determine and graph the probability distribution for the number of on-time arrivals in 8 trips.

Solution 1 Let x be the number of on-time arrivals in 8 trips. Then x can be any whole number from 0 to 8. The probability p of success for each trip is 0.75. By the Binomial Probability Theorem, $B(x) = {}_8C_x \cdot (0.75)^x(0.25)^{8-x}$. Evaluate this expression for each possible value of x to generate the numbers in the table on the next page. Then graph the data. *(continued on next page)*

The calculation of $B(5)$ is shown below and the bar of the graph for $B(5)$ has been drawn at the right. Notice that the bar is centered over the 5. Use the values in the table to draw a complete graph.

x	B(x)
0	$\underline{?} \cdot (0.75)^{\underline{?}} \cdot (0.25)^{\underline{?}} \approx \underline{?}$
1	$\underline{?} \cdot (0.75)^{\underline{?}} \cdot (0.25)^{\underline{?}} \approx \underline{?}$
2	$\underline{?} \cdot (0.75)^{\underline{?}} \cdot (0.25)^{\underline{?}} \approx \underline{?}$
3	$\underline{?} \cdot (0.75)^{\underline{?}} \cdot (0.25)^{\underline{?}} \approx \underline{?}$
4	$\underline{?} \cdot (0.75)^{\underline{?}} \cdot (0.25)^{\underline{?}} \approx \underline{?}$
5	$56 \cdot (0.75)^5 \cdot (0.25)^3 \approx 0.208$
6	$\underline{?} \cdot (0.75)^{\underline{?}} \cdot (0.25)^{\underline{?}} \approx \underline{?}$
7	$\underline{?} \cdot (0.75)^{\underline{?}} \cdot (0.25)^{\underline{?}} \approx \underline{?}$
8	$\underline{?} \cdot (0.75)^{\underline{?}} \cdot (0.25)^{\underline{?}} \approx \underline{?}$

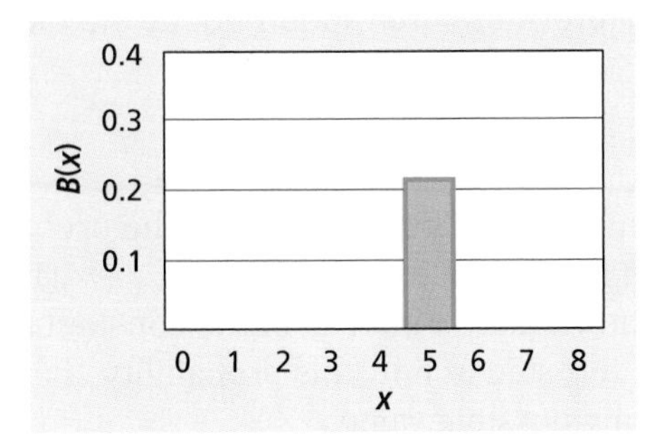

Solution 2 Use a calculator to compute individual binomial probabilities as shown below.

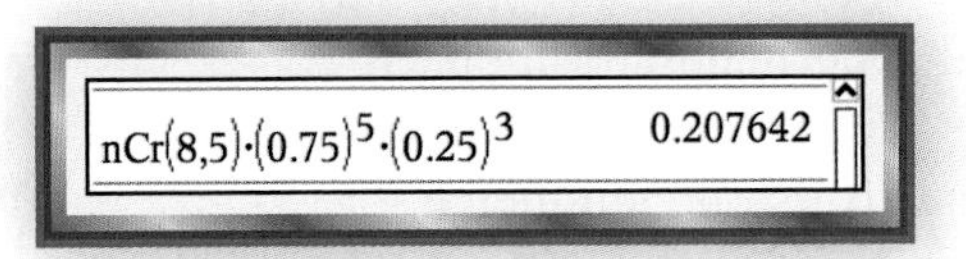

Check The nine probabilities in the table in Solution 1 should add to 1. The screen at right shows that they do.

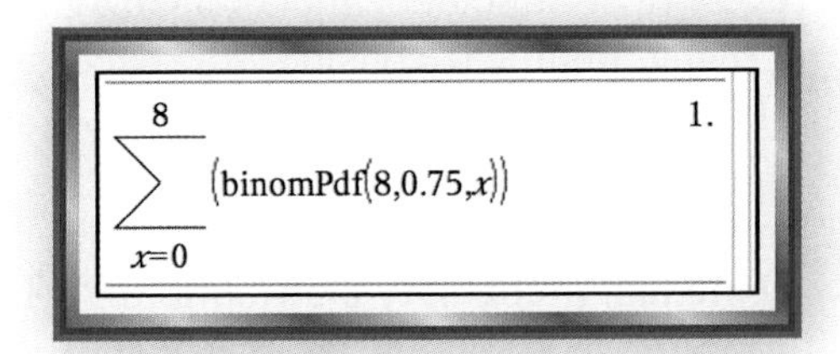

QY1

What If the Probability of Success Changes?

How do changes in the probability p of success affect the distribution? Intuitively, you should expect that when p increases, the expected count of successes would increase. In fact, we know from Chapter 6 that for an outcome of probability p in an experiment of n trials, the expected count of successes is np. So the expected count varies jointly as n and p. As we will see in Lesson 10-7, this expected count is in fact the *mean of a binomial random variable.* In Example 1, we would expect 6 of the 8 trials to be on time. But would we expect 6 if p were equal to 0.01? Probably not. Activity 1 explores how skewness, the mean, and the mode change as p changes.

▸ QY1

Suppose the situation in Example 1 were modified so that $B(x) = {}_5C_x \cdot (0.20)^x(0.80)^{5-x}$. What then, is the meaning of $B(0)$?

Activity 1

MATERIALS binomial probability distribution application provided by your teacher

Using the applet or file provided by your teacher, set the sliders so that the number of trials is 8 and the probability of success is 0.5.

Step 1 According to the graph shown, for what value of x is $B(x)$ greatest? What does this value represent?

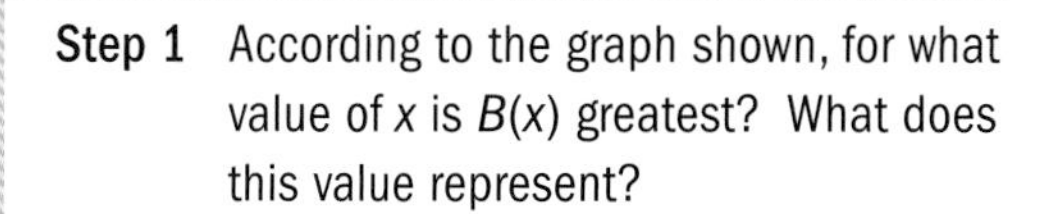

Step 2 Make a sketch predicting what the distribution would look like for $p = 0.8$, and another sketch for $p = 0.2$.

Step 3 Change p using the slider to the following values (or values near them): 0.01, 0.10, 0.25, 0.40, 0.5, 0.8, 0.99. How accurate were your sketches in Step 2? What values of p give distributions that closely resemble the sketches you drew?

Step 4 Describe how the graph changes as you increase p from 0.01 to 0.99. How does the mode change? How does the mean change? When is the graph symmetric?

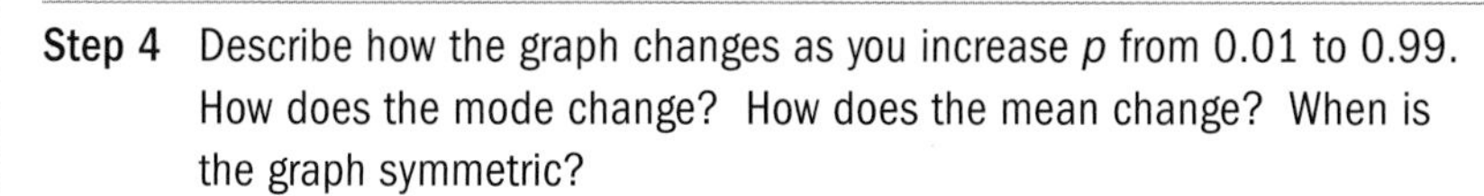

Step 5 Does the total area of the bars change as p changes? Why or why not?

STOP QY2

> **QY2**
>
> Suppose in Activity 1 that p represents the probability of success for a particular kind of surgery. What, then, does each bar in the distribution represent?

For each of Graphs I–VI below, a different value of p has been chosen, ranging from $p = 0.05$ to $p = 0.95$. For instance, Graph V, with $p = 0.75$, represents the situation in Example 1, showing the probability of x on-time arrivals in 8 trips on a flight with an on-time arrival rate of 75%. The other graphs represent similar situations with other success rates.

I.
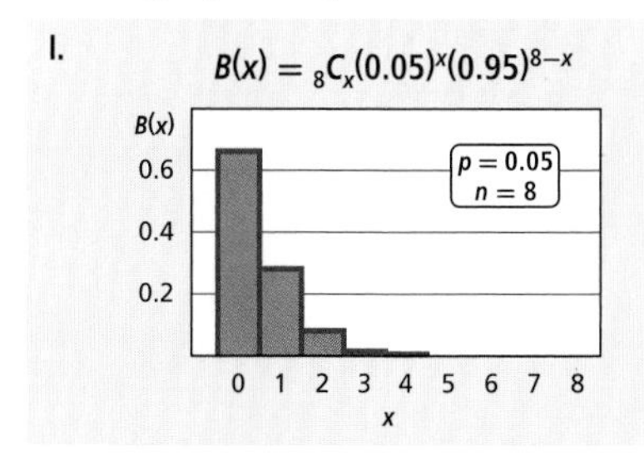

II.
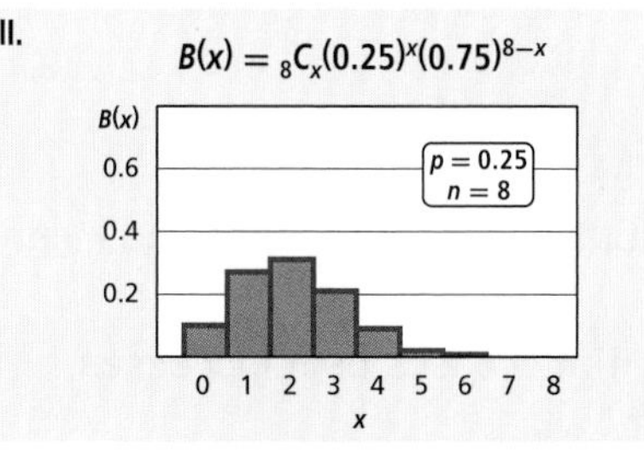

III.
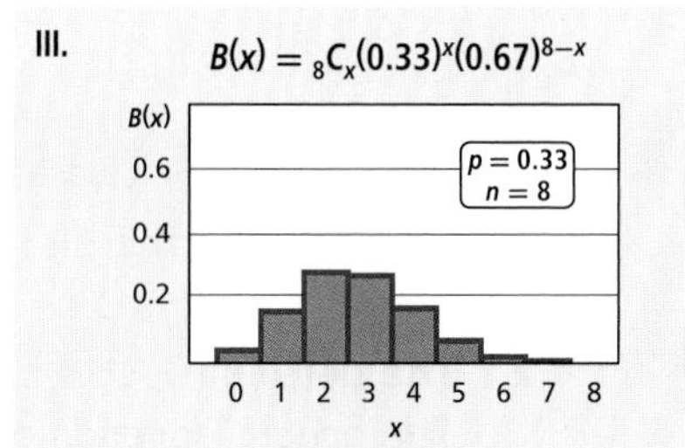

IV.
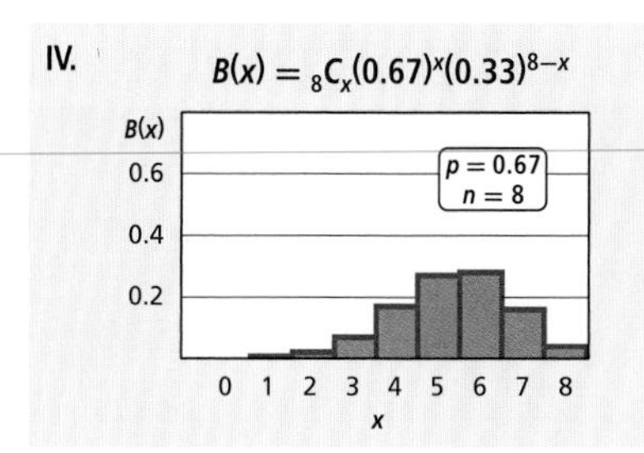

V.
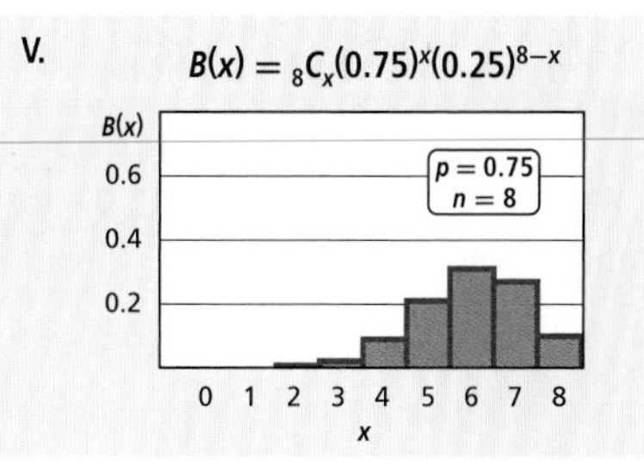

VI.
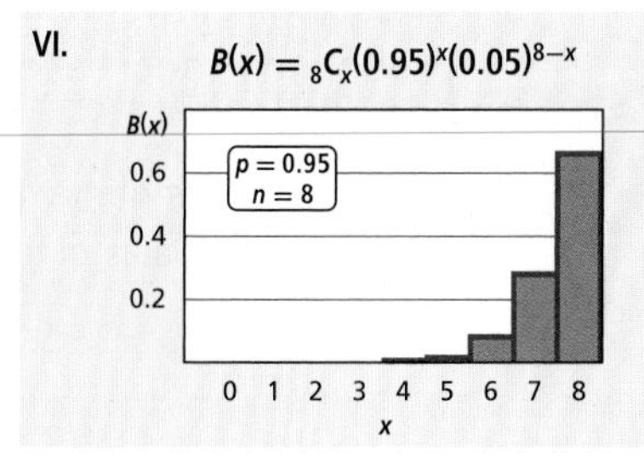

You can see that the value of the parameter p in a binomial probability distribution affects whether the distribution is skewed or symmetric. Also, as the value of p increases from very small values to very large values, you can see the distribution's mode shift from left to right. For the on-time arrival flights, increasing p from 0 to 1 causes the expected number of on-time arrivals out of eight trials to increase from 0 to 8.

What if the Number of Trials Changes?

Now we reverse the situation for the two parameters n and p. We fix the value of p and look at the different distributions that result from different values of n, the number of trials.

Example 2

a. Let $p = 0.5$. Draw graphs of the binomial probability distributions for $n = 2, 6$, and 10.

b. Describe the effect of increasing n on the probability distribution B when $p = 0.5$.

Solution

a. Generate graphs of binomial distributions by hand or with technology.

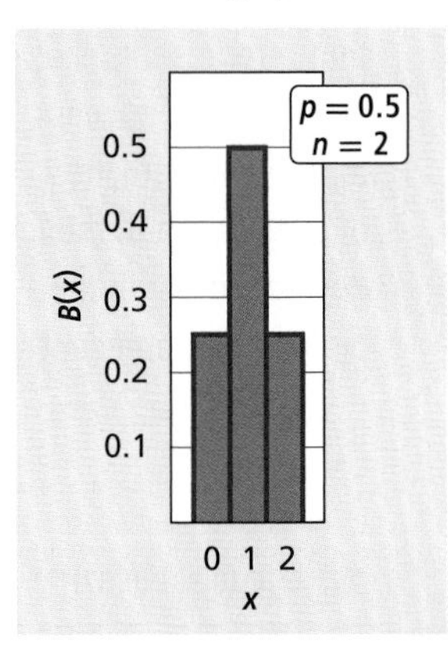

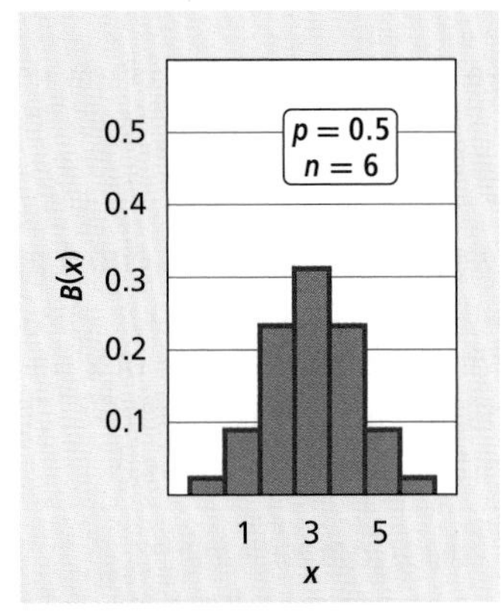

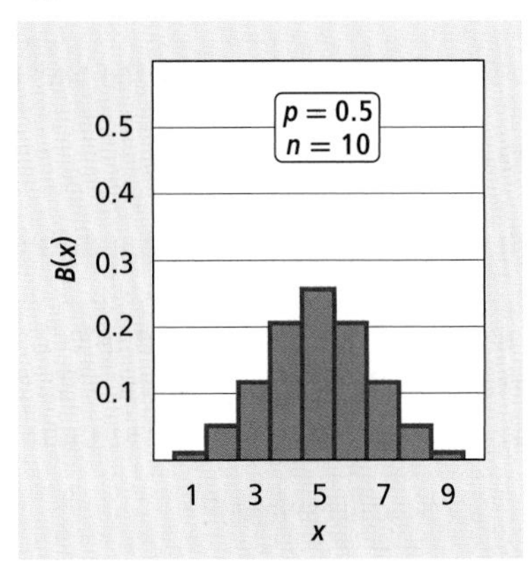

b. When $p = 0.5$, as n increases, the following patterns occur.

1. The mean of the distribution is np or $\frac{1}{2}$n, so the mean varies directly as n.
2. The domain {0, 1, 2, ..., n} increases as n increases. The graph "spreads out."
3. The maximum value in the range, which is the maximum probability, decreases. That is, the distribution "flattens."
4. The graph is symmetric with respect to the bar at $\frac{n}{2}$. This is the expected value of the random variable x.

As Example 2 and Activity 1 show, changes to either n or p affect the shape, center, and spread of a binomial distribution. The mode and mean depend on the values of n and p.

When $p = 0.5$, the distribution is symmetric. When $p \neq 0.5$, the distribution is not symmetric. Activity 2 investigates the effect of changing n when $p \neq 0.5$.

Activity 2

MATERIALS: binomial probability distribution application provided by your teacher

Using the applet or file provided by your teacher, set the sliders so that $n = 10$ trials and $p = 0.4$.

Step 1 Use the graph to estimate $B(6)$.

Step 2 Keeping p constant at 0.4, change n using the slider to the following values: 15, 20, 30, 40, 100. Watch how the distribution changes with changes in n. What happens to the graph as you increase n from 15 to 100, while keeping p constant?

Step 3 Repeat Step 2 using $p = 0.9$.

Step 4 Repeat Step 2 using $p = 0.1$. Compare the results of Steps 3 and 4.

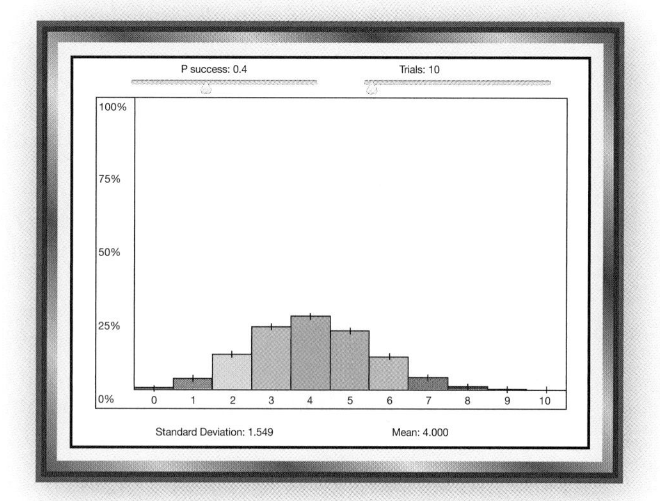

Questions

COVERING THE IDEAS

1. **Fill in the Blanks** The expression ${}_{25}C_{18}(0.77)^{18}(0.23)^{25-18}$ gives the probability of getting exactly __?__ successes in __?__ trials of a binomial experiment in which the probability of success on each trial is __?__.

2. Suppose that as part of an advertising campaign, the Breakfast Treat Cereal Company randomly puts prizes in 30% of its boxes of cereal. During the campaign, the Foster family buys 4 boxes.
   a. Write a formula for the probability that the Fosters got x prizes in their cereal boxes.
   b. Make a table for the probability distribution.
   c. Make a graph of the probability distribution.

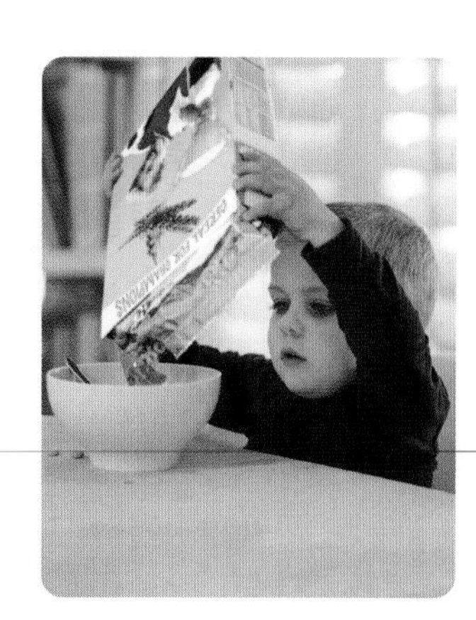

3. Dave plays on a softball team and usually bats 5 times during a game. Using his batting average as the probability p of a hit, the table at the right shows the binomial probability distribution for x, the number of hits that Dave gets when he bats 5 times.

x	$B(x)$
0	0.0916
1	0.2808
2	0.3441
3	0.2109
4	0.0646
5	0.0079

   a. What is the meaning of $B(2)$?

   b. Find the probability that Dave has at least 4 hits in a game.

4. Use the graph at the right, which shows a binomial probability distribution when $n = 6$ and $p = 0.80$.

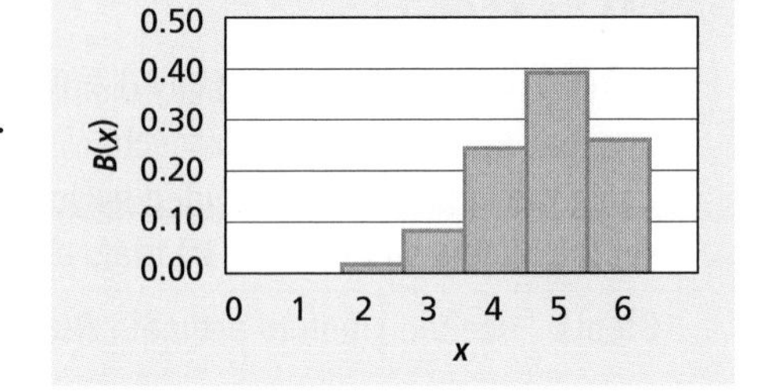

   a. Give the mode of this binomial probability distribution.

   b. Use the graph to estimate the individual values of $B(x)$ and $\sum_{x=0}^{6} B(x)$.

   c. Describe a situation that would lead to the distribution graphed here.

   d. Sketch the graph of the binomial probability distribution with $n = 6$ and $p = 0.20$ and describe a situation that would lead to that distribution.

5. In a binomial experiment, suppose that the probability p of success is 0.4.

   a. Graph the probability distribution for the number of successes in 6 trials.

   b. Describe the effect on the graph of the distribution as p increases and the number of trials remains constant.

6. **Matching** Match each equation with its graph.

   a. $B(x) = {}_{10}C_x\,(0.25)^x(0.75)^{10-x}$

   b. $B(x) = {}_{10}C_x(0.50)^x(0.50)^{10-x}$

   c. $B(x) = {}_{10}C_x(0.85)^x(0.15)^{10-x}$

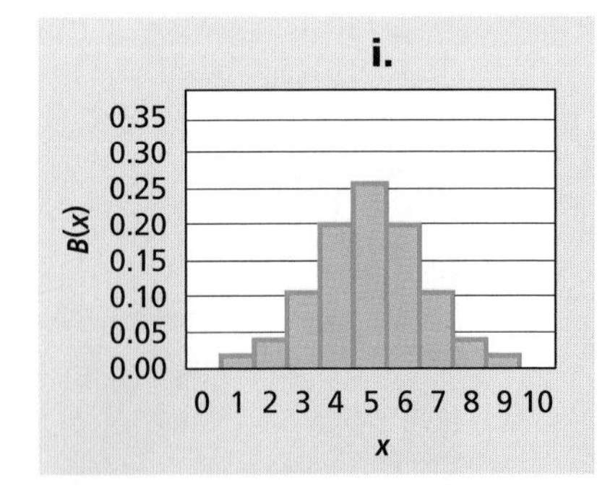

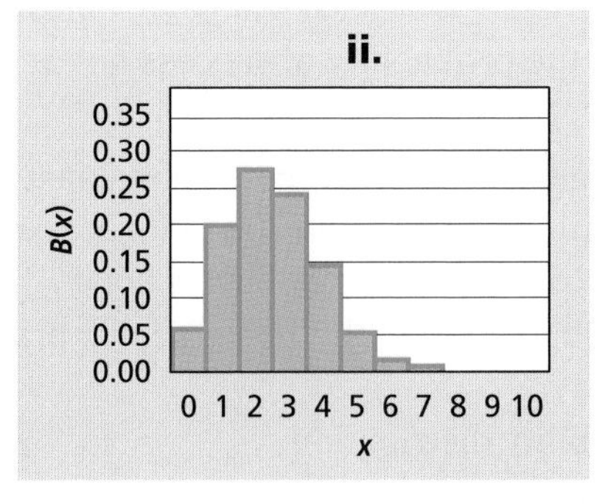

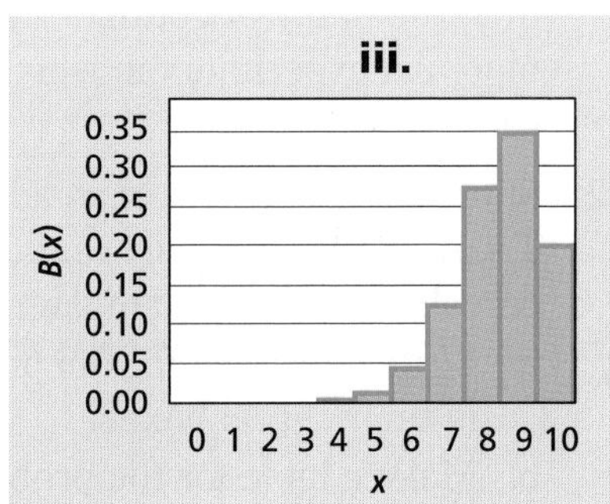

7. a. For the binomial probability distribution with equation $B(x) = {}_{30}C_x(p)^x(1-p)^{30-x}$, what value of p produces a symmetric distribution?

   b. Where exactly is the line of symmetry? (You should be able to answer this without graphing.)

8. a. Graph the binomial probability distributions when $p = 0.3$ and $n = 5, 10, 25,$ and 100.

   b. Describe the effect of fixing p and increasing n.

APPLYING THE MATHEMATICS

9. On a given day, the probability that a particular flight arrives on time is 80%. Is it more likely that it will arrive on time 3 times out of 4 days, or 6 times out of 8 days? Justify your answer.

10. How would a situation leading to Expression 1 differ from a situation leading to Expression 2?

    a. Expression 1: ${}_{15}C_{10}(0.82)^{10}(0.18)^{15-10}$,
    Expression 2: ${}_{20}C_{10}(0.82)^{10}(0.18)^{20-10}$.

    b. Expression 1: ${}_{15}C_{10}(0.82)^{10}(0.18)^{15-10}$,
    Expression 2: ${}_{15}C_{2}(0.82)^{2}(0.18)^{15-2}$.

11. A gardener buys a package of flower seeds that contains 16 seeds. When planted, each seed has an 80% chance of germinating. Determine the following probabilities.

    a. *P*(exactly 12 seeds germinate)

    b. *P*(more than 12 seeds germinate)

    c. *P*(no more than 12 seeds germinate)

12. The wheel used on a certain TV game show has 19 sections. Nine sections have big prizes, nine sections have small prizes, and one section throws the player out of the game. The scatterplot at the right shows the probability of x big prizes being awarded in 50 spins of this wheel.

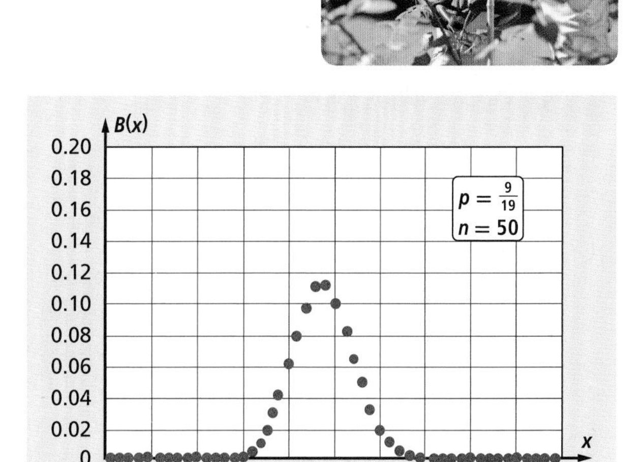

    a. From the graph, estimate the probability of winning 30 or more big prizes in 50 spins.

    b. The graph is nearly reflection-symmetric. Why?

    c. If the wheel were spun more than 50 times, how would you expect the graph to change?

REVIEW

13. What is the probability that a student will get a perfect score by randomly guessing on an exam with 15 multiple-choice questions, each of which has four possible answers? **(Lesson 10-5)**

14. Explain why the given table does not define a probability distribution. **(Lesson 10-4)**

x	Red	Blue	Green
f(*x*)	0.4	0.3	0.2

15. Expand $(x^3 + y^2)^6$. **(Lesson 10-3)**

16. Find the tenth term in row 15 of Pascal's Triangle. **(Lesson 10-2)**

17. Use the explicit formula for ${}_nC_r$ to write an expression equivalent to $\binom{n-1}{x-1}$ that involves factorials where n and x are positive integers and $x \le n$. **(Lesson 10-1)**

18. a. Calculate the mean and standard deviation for the set {6, 7, 8, 8, 9, 10, 10, 10}.

 b. Make up another data set of eight numbers, using each of 6, 7, 8, 9, and 10 at least once, that has the same mean but a smaller standard deviation. **(Lesson 1-6)**

EXPLORATION

19. A *cumulative binomial probability table* gives, for fixed n, the probabilities of k or fewer successes. The number in row k and column p of the table below represents the probability of k or fewer successes in 7 binomial trials, each of which has probability p. For instance, the probability of no more than 2 successes in 7 binomial trials, each of which has probability of success 0.8, is about 0.0047.

k	0.01	0.05	0.1	0.2	0.3	0.4	0.5	0.6	0.7	0.8	0.9	0.95	0.99
0	.9321	.6983	.4783	.2097	.0824	.0280	.0078	.0016	.0002	.0000	.0000	.0000	.0000
1	.9980	.9556	.8503	.5767	.3294	.1586	.0625	.0188	.0038	.0004	.0000	.0000	.0000
2	1.0000	.9962	.9743	.8520	.6471	.4199	.2266	.0963	.0288	.0047	.0002	.0000	.0000
3	1.0000	.9998	.9973	.9667	.8740	.7102	.5000	.2898	.1260	.0333	.0027	.0002	.0000
4	1.0000	1.0000	.9998	.9953	.9712	.9037	.7734	.5801	.3529	.1480	.0257	.0038	.0000
5	1.0000	1.0000	1.0000	.9996	.9962	.9812	.9375	.8414	.6706	.4233	.1497	.0444	.0020
6	1.0000	1.0000	1.0000	1.000	.9998	.9984	.9922	.9720	.9176	.7903	.5217	.3017	.0679

 a. Use the table to determine the probability of five or fewer successes in a binomial experiment with 7 trials if in each trial, the probability of success is 0.8. Compare this answer to the one you get by applying the method of Lesson 10-5 directly.

 b. Cumulative binomial probability tables can also be used to calculate probabilities for single events. Figure out how to use a cumulative binomial table to determine the probability of exactly two successes in 7 trials, each with 0.4 probability of success. Verify your answer by direct calculation using the Binomial Probability Theorem.

 c. Some calculators have built-in features for generating cumulative binomial probabilities. Find out whether your calculator has such a feature. If it does, describe what you need to do to answer Part a of this question using the calculator.

QY ANSWERS

1. $B(0)$ is the probability that none of 5 arrivals will be on time given that the probability of an on-time arrival is 0.2.

2. Each bar represents the probability that a certain number of surgeries out of 8 surgeries will be successful.

Lesson 10-7 Mean and Standard Deviation of a Binomial Random Variable

▸ **BIG IDEA** The mean and standard deviation of the number of times an event will occur in n trials of a binomial experiment can be easily calculated from the probability p of the event.

In the previous lesson, you studied the shapes of the graphs of binomial probability distributions produced under two conditions: (1) fixing the number n of trials and allowing the probability p of success on an individual trial to vary; or (2) fixing the value of p and allowing n to vary. In the former case, intuitively, we posited that increases in the probability of success p would increase the number of successes we expect on average. The notion of expected count (np) in Chapter 6 was, therefore, linked to the *mean of the binomial random variable.* This was true because the experiments mentioned in Chapter 6 were binomial experiments with two outcomes repeated over some number of trials n. Kerrich's famous experiments with coin flips were therefore binomial experiments.

You also saw in the last lesson that if p is fixed, then as n increases, the domain of the distribution becomes larger. Because of this increase in the domain, the distribution spreads out; that is, the standard deviation of the associated binomial random variable increases. Thus, both the mean and standard deviation of a binomial random variable increase as n increases.

The mean and standard deviation of a binomial random variable become useful when answering polling questions, which we cover in Chapter 11. But before we can answer such questions, we need formulas for the mean and standard deviation. We know that these quantities increase as n increases. But how much do they increase? In this lesson, we will specify the relationships between these numbers.

Mental Math

A spinner is divided into 4 unequal sectors worth different numbers of points summarized in the following table. What is the mean number of points gained from a random spin?

% of total area	Points
10	50
20	20
30	10
40	5

The Mean of a Binomial Random Variable

We have asserted that the mean of a binomial random variable with n trials and probability p of success is np. Now we state this as a theorem.

Theorem (Mean of a Binomial Random Variable)

The mean μ of a binomial random variable with n trials and probability p of success on each trial is given by $\mu = np$.

Proof In a binomial distribution with n trials, the possible values of x are the numbers of successes 0, 1, ..., n. So the sum ranges from 0 to n.
Let $q = 1 - p$.

$$\mu = \sum_{x=0}^{n} x \cdot P(x) = \sum_{x=0}^{n} x \cdot \binom{n}{x} p^x q^{n-x}$$ P is a binomial distribution.

When $x = 0$, the term in the sum is 0, so the summation can start with $x = 1$.

$$= \sum_{x=1}^{n} x \cdot \frac{n!}{x!(n-x)!} p^x q^{n-x}$$ formula for $\binom{n}{x}$

The goal now is to rewrite the expression on the right with n and p outside the summation.

$$= \sum_{x=1}^{n} \frac{n(n-1)!}{(x-1)!(n-x)!} p^x q^{n-x}$$ rewriting factorials and powers

$$= np \sum_{x=1}^{n} \frac{(n-1)!}{(x-1)!(n-x)!} p^{x-1} q^{n-x}$$ Since n and p do not depend on x, they can be factored out of the summation.

$$= np \sum_{x=1}^{n} \binom{n-1}{x-1} p^{x-1} q^{n-x}$$ formula for $\binom{n-1}{x-1}$

$$= np(p+q)^{n-1}$$ Binomial Theorem for $n - 1$ trials

$$= np \cdot 1^{n-1}$$ $p + q = 1$

$$= np$$

As was the case with expected count in Lesson 6-8, the mean of a binomial random variable does not have to be an integer because it measures what happens on average and does not have to be a possible outcome of the experiment.

QY1

Example 1 verifies this theorem by calculating the mean of a binomial random variable using two methods. The results are the same.

> **QY1**
>
> If the probability of success on one trial of a binomial experiment is 0.4, what is the mean of the random variable representing the number of successes in 4 trials?

GUIDED

Example 1

Each day for four days, a radio station is holding a raffle to give away a pair of tickets to an upcoming concert. In the past, the station has found that 80% of the raffle entries are from the local area. If this is the case, find the mean of the random variable x, where x is the number of winners from the local areas.

Solution 1 Use the Mean of a Binomial Random Variable Theorem.
In this situation, there are 4 trials and there is an 80% chance of success. So, $n = 4$, $p =$ __?__, and the mean of the random variable is $\mu = 4 \cdot$ __?__ $=$ __?__.

Solution 2 Use the definition of the mean of a random variable from Lesson 10-4: $\mu = \sum_{i=1}^{n} (x_i \cdot P(x_i))$.

First calculate the probability for each possible value of x: 0, 1, 2, 3, and 4.

$P(x \text{ local winners}) = \binom{n}{x} p^x q^{n-x}$ where $p = 0.80$ and $q =$ __?__

Copy and complete the table.

x local winners	0	1	2	3	4
P(x)	__?__	$\binom{4}{1}(.80)^1(.20)^3$	__?__	__?__	$\binom{4}{4}(.80)^4(.20)^0$

Compute the probabilities exactly. (Use exact arithmetic on a CAS.)

x local winners	0	1	2	3	4
P(x)	__?__	0.0256	__?__	__?__	0.4096

Now substitute these values into the formula for the mean.

$\mu = 0 \cdot$ __?__ $+ 1 \cdot 0.0256 + 2 \cdot$ __?__ $+ 3 \cdot$ __?__ $+ 4 \cdot$ __?__ $=$ __?__

The Variance and Standard Deviation of a Binomial Random Variable

Lesson 10-4 gave two ways for computing variance of a random variable. The first was the squared deviation form:

$$\sigma^2 = \sum (x - \mu)^2 \cdot P(x).$$

The alternate formula did not require computation of deviations:

$$\sigma^2 = \sum (x^2 \cdot P(x)) - \mu^2.$$

The standard deviation of a random variable is found by taking the square root of the variance, regardless of which variance formula is used. The second formula is used in Example 2.

Example 2

Suppose a darts player has probability p of hitting the bull's-eye with a single dart, and all attempts are independent. Find the variance and standard deviation for the random variable x, where x is the number of bull's-eyes the player will hit in 3 attempts.

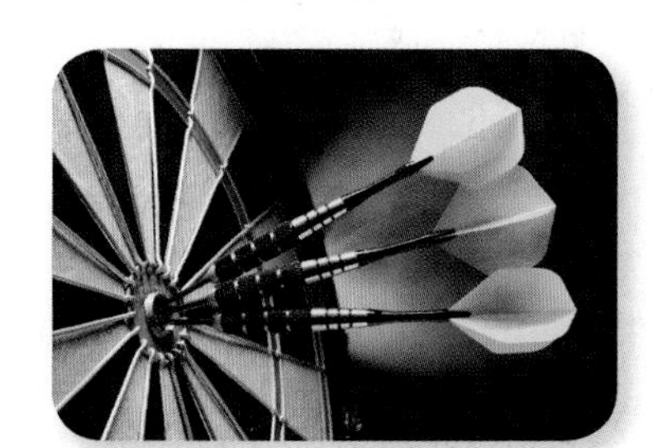

Solution This is a binomial experiment with $n = 3$ trials, probability p of success, and probability $q = 1 - p$ of failure. Make a probability distribution table for the number of bull's-eyes. There are 4 possible numbers of bull's-eyes.

number of bull's-eyes = x	3	2	1	0
probability = P(x)	p^3	$3p^2q$	$3pq^2$	q^3

By the Mean of a Binomial Random Variable Theorem, $\mu = 3p$. So by the second variance formula above,

$$\sigma^2 = \sum_{x=0}^{3} (x^2 P(x)) - \mu^2$$

$$= (3^2p^3 + 2^2(3p^2q) + 1^2(3pq^2) + 0^2q^3) - (3p)^2.$$

(continued on next page)

Evaluate the powers.

$$\sigma^2 = 9p^3 + 12p^2q + 3pq^2 - 9p^2$$

Rewrite $12p^2q$ as $9p^2q + 3p^2q$, and factor using grouping of terms.

$$\begin{aligned}\sigma^2 &= 9p^3 + 9p^2q + 3p^2q + 3pq^2 - 9p^2 \\ &= 9p^2(p + q) + 3pq(p + q) - 9p^2 \\ &= 9p^2 + 3pq - 9p^2 \text{ since } (p + q = 1)\end{aligned}$$

Thus the variance is $\sigma^2 = 3pq$.

The standard deviation is the square root of the variance.

$$\sigma = \sqrt{3pq}$$

Check Use a CAS to find the variance.

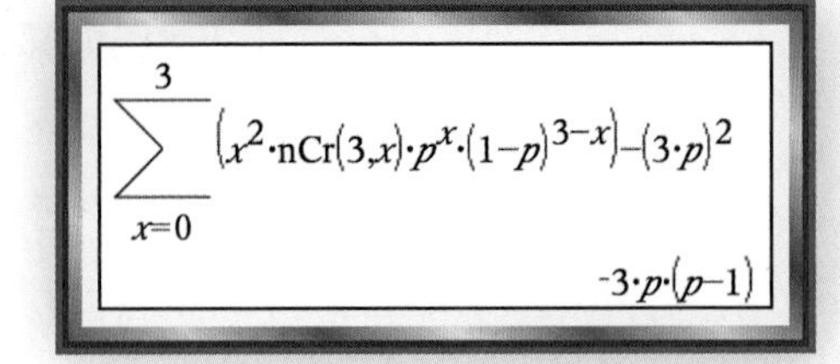

 **QY2**

The algebra of Example 2 can be applied to any binomial random variable. The general result follows.

▶ QY2

Why is the CAS answer for variance equivalent to the variance in Example 2?

Theorem (Variance and Standard Deviation of a Binomial Random Variable)

In a binomial distribution with n trials, probability p of success on each trial, and probability q of failure on each trial, the variance is $\sigma^2 = npq$, and the standard deviation is $\sigma = \sqrt{npq}$.

Example 3

In the United States, about 49% of newborn children are female. Suppose a large hospital has 2500 live births in a given year. Let $P(x)$ be the probability that x female babies are born in this hospital.

a. What is the mean of the random variable x?

b. What is the standard deviation of the random variable x?

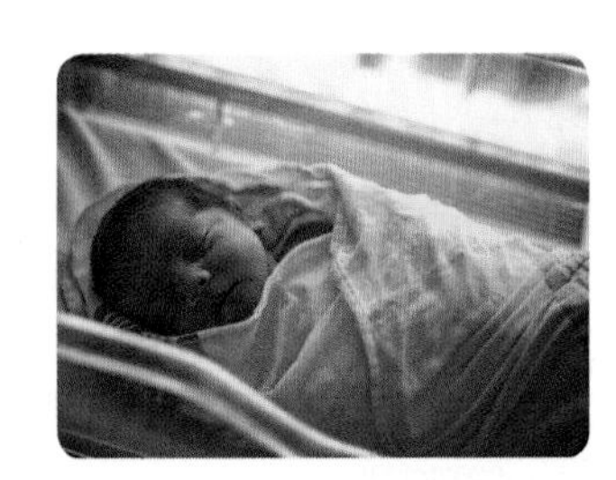

Solution Let x be the number of female children in 2500 live births. If births are independent, then the probability distribution P is binomial with n = 2500 and p = 0.49. By the previous theorems,

a. $\mu = 2500(0.49) = 1225$

b. $\sigma = \sqrt{2500(0.49)(0.51)} \approx 25$

In the next chapter, you will see how the standard deviation of a binomial random variable can be used to determine the likelihood that the mean of a population will fall within a specified interval around the mean obtained from a sample of the population. For example, you can use the standard deviation to determine how likely it is that the results of a poll of a random sample of people reflect the preferences of the entire population.

To increase their accuracy, polls usually use large sample sizes. These large sample sizes make calculating probabilities using binomial distributions a tedious and cumbersome process. Fortunately, as the sample size increases, the binomial distribution representing the poll results becomes closer and closer to another kind of distribution, called a *normal distribution*. This distribution has many useful properties and is the topic of the next chapter.

Questions

COVERING THE IDEAS

In 1 and 2, suppose a particular dog catches a thrown stick 65% of the time. Suppose a stick is thrown four times to this dog.

1. Let μ be the expected number of times the dog catches the stick.
   a. Find μ by calculating a sum using the definition of the mean of a random variable.
   b. Find μ by using the formula $\mu = np$.
2. a. Find the variance for the random variable of caught sticks.
   b. Find the standard deviation for the same random variable.
3. Suppose a darts player has probability $p = \frac{3}{4}$ of hitting a bull's-eye on any toss of a single dart.
   a. Suppose the player throws 60 darts. Find the mean number of bull's-eyes.
   b. Let $P(x)$ be the probability of x bull's-eyes. Find the standard deviation for the random variable x.

In 4 and 5, characteristics of a binomial experiment are given.

a. Find the mean of the random variable representing the number of successes.

b. Find the variance of the random variable.

c. Find the standard deviation of the random variable.

4. $n = 25, p = 0.13$
5. $n = 500, p = 0.95$
6. Consider the hospital and year referred to in Example 3. Let $Q(x)$ be the probability that x males are born. What is
   a. the mean of x?
   b. the standard deviation of x?
7. Consider the Galton board on page 616. How many red lights would you expect to hit in a string of six lights if $P(\text{red light}) = \frac{1}{3}$?

APPLYING THE MATHEMATICS

8. A baseball player has a batting average a, where $0 \le a \le 1$. Let x be the number of hits made in the next 25 times at bat. For the random variable x, find
   a. the mean.
   b. the variance.

9. In the United States, about 45% of people have blood type O. Suppose 5 people are giving blood. Let $P(x)$ be the probability that x of the donors have type O blood.

a. Construct a table for the probability distribution P.

b. Draw a graph of the distribution in Part a.

c. Find the mean number of donors with type O blood.

d. Find the standard deviation of the random variable x.

10. If, for some binomial random variable, the mean is 40 and the standard deviation is 6, find the number n of trials and the probability p of success on each trial.

11. Determine whether each statement is true or false.

a. The mean of a binomial random variable is directly proportional to n, the number of trials.

b. The standard deviation of a binomial random variable varies directly as n, the number of trials.

12. According to a study by a federal agency, the probability is about 0.2 that a polygraph (lie detector) test given to a truthful person suggests that the person is deceptive. (That is, in about 20% of cases a truthful person will be wrongly labeled as deceptive.) Suppose that a security-guard firm asks 20 job applicants about their experience on previous jobs, and uses a polygraph to judge their truthfulness. Suppose also that all 20 answer truthfully.

a. What is the probability that the polygraph tests show that at least one person is deceptive?

b. What is the mean number among 20 truthful persons who will be classified as deceptive? What is the standard deviation of the number of applicants classified as deceptive?

c. What is the probability that the number classified as deceptive is less than the mean?

REVIEW

13. Fill in the Blanks The expression ${}_nC_k p^k(1 - p)^{n-k}$ gives the probability of getting exactly __?__ successes in __?__ trials of a __?__ experiment in which the probability of success on each trial is __?__. **(Lesson 10-6)**

14. Assume a basketball player makes 45% of the shots taken, and that the shots are independent. What is the probability that the player will make at least 5 of the next 10 shots? **(Lesson 10-6)**

In 15 and 16, consider a probability distribution in which $p = 0.5$. Determine whether each statement is true or false. Justify your answer. (Lesson 10-6)

15. The graph of the distribution is symmetric with respect to the line $x = \frac{n}{p}$.

16. As n increases, the probability of getting exactly the mean number of successes increases.

17. Suppose that a certain disease has a 65% recovery rate without any treatment. **(Lesson 10-5)**

a. What is the probability that in a study of five people with the disease, all five will recover?

b. In one experiment, a drug is tested on five people with the disease and all five recover. Is this unlikely to happen without the drug, given your answer in Part a? What does this suggest about the effectiveness of the drug?

c. In a second experiment, 25 people get the drug and 25 people recover. What is the probability that all 25 recover, given no treatment?

d. Consider your answers to Parts b and c. Does the drug appear to be effective? Which experiment might be considered more conclusive and why?

e. Find the probability that 18 or more people out of 25 recover even if the drug has no effect on the illness.

18. Multiple Choice When $0 \le r \le 1$, the expression $\sum_{i=1}^{n} a_1 r^{i-1}$ is equivalent to which expression? **(Lesson 8-5)**

A $\frac{a_1(1-r)^n}{1-r}$ B $\frac{a_1(1-r^n)}{1-r}$ C $\frac{a_1(1-r)^n}{r-1}$ D $\frac{a_1(1-r^n)}{r-1}$

In 19 and 20, find the exact value without a calculator. (Lesson 4-4)

19. $\sin\left(-\frac{11\pi}{4}\right)$

20. $\tan \frac{\pi}{6}$

EXPLORATION

21. Use a binomial probability distribution application or web applet provided by your teacher to investigate probability distributions by: (1) fixing the number n of trials and allowing the probability p of success on an individual trial to vary; or (2) fixing the value of p and allowing n to vary.

a. Copy and complete the tables below.

n	p	μ	σ
100	0.2	?	?
100	0.4	?	?
100	0.6	?	?
100	0.8	?	?

n	p	μ	σ
10	0.6	?	?
20	0.6	?	?
50	0.6	?	?
100	0.6	?	?

b. Describe how the mean and standard deviation of the distribution change as p increases for a given n, and as n increases for a given p.

QY ANSWERS

1. 1.6

2. $-3p \cdot (p - 1) = -3p \cdot (-1)(1 - p) = 3p \cdot (1 - p) = 3pq$

Lesson

10-8 Is That Coin Fair?

Vocabulary

null hypothesis, H_0

▶ **BIG IDEA** You can accumulate evidence to determine whether a hypothetical probability is consistent with your data by comparing the data with probabilities in a binomial distribution.

Can You Tell Whether a Coin Is Fair or Biased?

When discussing probability, it is common to assume that a coin is fair. However, you can never tell for certain whether a coin is fair or biased. To see why this is so, consider the five binomial probability distributions graphed below. They show the probabilities of getting x heads in 100 tosses of a coin with the probability of heads on a single toss being 0.3, 0.4, 0.5, 0.6, and 0.7, respectivley. $p = 0.5$ is the situation for a fair coin; the others are biased.

Mental Math

Write each number as a fraction in lowest terms.

a. 0.01 **b.** 0.05

c 0.04 **d.** 0.001

e. 10^{-5} **f.** $-2 \cdot 10^{-1}$

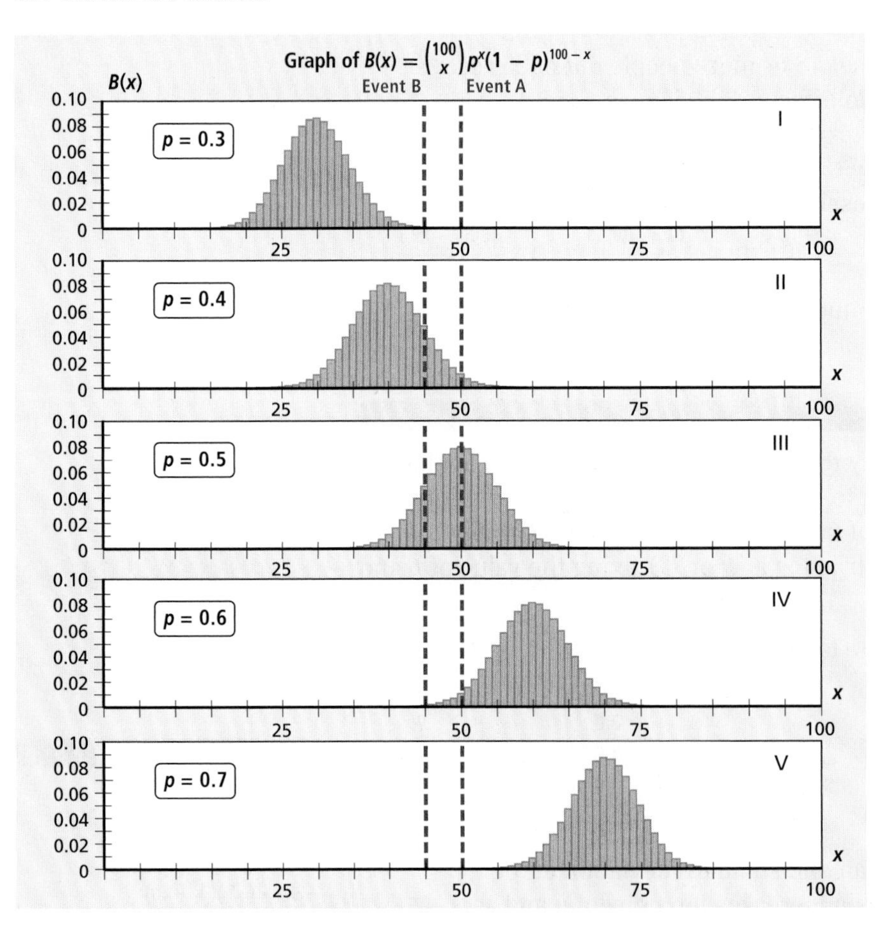

Suppose that you toss a coin 100 times in an attempt to determine whether it is fair. Two possible events are listed in the table at the right and denoted in red on the distributions on the previous page. If Event A occurs, you might immediately think that the coin is fair, but from examination of the binomial probability distributions, you can see that many different p values make this event possible. As seen on the graph, a result of 50 heads and 50 tails happens most often when the coin is fair.

Event	Heads	Tails
A	50	50
B	45	55

If the coin were so weighted that the probability of heads was 0.7 or 0.3, there is still a possibility, though remote, that there could be 50 heads in 100 tosses. *Because of the variability associated with randomness, even a biased coin might turn up heads exactly half of the time.*

Suppose Event B occurs. Is the coin biased? Not necessarily. When $p = 0.5$, Event B is not an unusual situation. As seen on the graph and calculated using binomial probabilities, 45 heads and 55 tails occur with a fair coin with probability ${}_{100}C_{45}(0.5)^{45}(0.5)^{55} \approx 0.0485$, or about once in every 21 experiments. *Because of the variability associated with randomness, even a fair coin might turn up heads less than half of the time.*

The Best We Can Do

Since we can never be sure whether a coin is fair or biased, the best we can do is to determine whether a particular result is very unlikely given our assumptions of the original probability. Unless you have some reason to think otherwise, you would typically believe a coin to be fair. But what do you do when you get results from tossing a coin that call into question the fairness of that coin?

Example 1

A coin is tossed 10 times and 9 heads occurred. What is the probability that this event or any event like it would occur if the coin were fair?

Solution In this case, we are interested in the probability that 9 or more heads occurred or 9 or more tails occurred. We include these events in our calculations because they are the same kind of event as the one that actually occurred. That is, all these events suggest that the coin might be biased, either towards heads or towards tails. We ask: What is the probability that when a fair coin is tossed 10 times, 9 or more heads occur or 9 or more tails occur?

(continued on next page)

P(exactly 9 heads in 10 tosses of a fair coin) $= {}_{10}C_9 \left(\frac{1}{2}\right)^9\left(\frac{1}{2}\right)^1 = \frac{10}{1024}$

P(10 heads in 10 tosses of a fair coin) $= {}_{10}C_{10} \left(\frac{1}{2}\right)^{10}\left(\frac{1}{2}\right)^0 = \frac{1}{1024}$

P(exactly 9 tails in 10 tosses of a fair coin) $= {}_{10}C_1 \left(\frac{1}{2}\right)^1\left(\frac{1}{2}\right)^9 = \frac{10}{1024}$

P(10 tails in 10 tosses of a fair coin) $= {}_{10}C_{10} \left(\frac{1}{2}\right)^0\left(\frac{1}{2}\right)^{10} = \frac{1}{1024}$

Since these outcomes are mutually exclusive, the total probability of 9 or more heads or 9 or more tails is $\frac{22}{1024} = 0.0215$. This means that an event as unusual as this outcome would occur about 2.15% of the time if the coin were fair. Therefore, the probability that this outcome or one like it occurred is 2.15%.

More Exact than the Chi-Square Statistic

Recall from Lesson 6-9 that the chi-square statistic indicates the extent to which observed frequencies differ from expected frequencies. When we tested whether a coin was fair with the chi-square statistic, we only *estimated* the extent to which the result of a binomial experiment might differ than what we expected. Now, using binomial probabilities, we are able to calculate the exact probability of a particular outcome and outcomes like it in a binomial experiment. With this probability, we can make more confident inferences about our assumptions.

An assumption about a situation is called a *hypothesis*, just as in Chapter 6. In Example 1, our hypothesis is that the coin is fair, meaning that the probability p of success is 0.5. Instead of asking, "Is that coin fair?" we ask "What is the likelihood that a binomial experiment of 10 coin flips returns an outcome of 9 heads (or any similar outcome) given that the probability of success is $p = 0.5$ (the coin is fair)?" Before we can make a decision about the hypothesis, we need to choose a *significance level* just as in chi-square. If we can show that the probability of the given event or one like it is less than the chosen significance level, then we can *reject* the hypothesis. If the probability is greater than the significance level, we do not accept the hypothesis. Instead, we conclude there is *insufficient evidence against* the hypothesis.

Example 2

Should you reject the hypothesis that the coin in Example 1 is fair

a. at the 0.05 significance level? **b. at the 0.01 significance level?**

Solution

a. The result of Example 1 shows that the probability of 9 heads or an event like it is 2.15%. Thus, at the 5% significance level, we reject the hypothesis that the coin is fair because $0.0215 < 0.05$. At the 5% significance level, we have reason to believe that the coin is not fair.

b. At the 1% significance level, the probability 2.15% is not rare enough to reject the notion that the coin is fair since $0.0215 > 0.01$. Therefore, we do not have sufficient evidence against the hypothesis.

In Example 1, the probability of the outcome "exactly 9 heads in 10 tosses" is $\frac{10}{1024} \approx 0.009766$. This is below the 1% significance level given in Example 2 Part b. While that particular outcome is unusual, we are testing whether or not the coin is fair, not whether one outcome is unusual. So we must consider all possible similarly unusual outcomes, as we did in order to get our probability of 2.15%. It is not always easy to determine which outcomes are similar to a given outcome. In the coin experiment in Example 1, it makes sense to include the additional outcomes of 10 heads, 9 tails, and 10 tails as they are all at least as unlikely as the original result of 9 heads in 10 tosses.

Activity

MATERIALS technology with spreadsheet capabilities

Step 1 Suppose you toss a coin 100 times. Write an equation for a binomial probability distribution that models the probabilities of the possible outcomes of your experiment, assuming that the coin is fair.

Step 2 In a spreadsheet, enter the integers from 0 to 100 (the possible frequencies of heads) in column A. One way to do this is to enter 0 in cell A1, enter the formula =A1+1 in cell A2, and then use the fill down function.

Step 3 Enter your equation from Step 1 as the formula for column B to calculate the probability of each possible outcome. Your display should look similar to the one at the right.

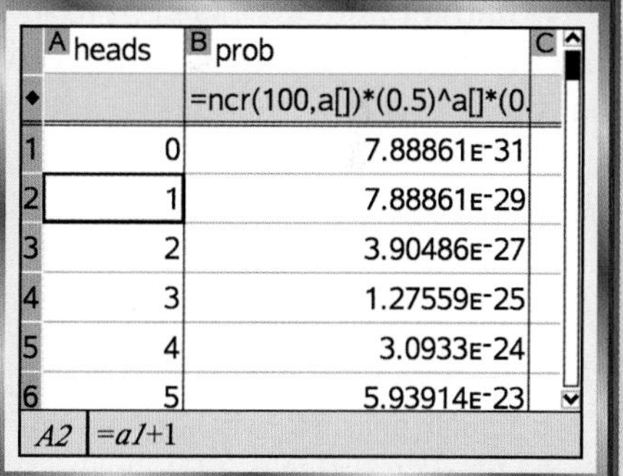

	A heads	B prob	C
◆		=ncr(100,a[])*(0.5)^a[]*(0.	
1	0	7.88861E-31	
2	1	7.88861E-29	
3	2	3.90486E-27	
4	3	1.27559E-25	
5	4	3.0933E-24	
6	5	5.93914E-23	
A2	=a1+1		

Step 4 Suppose that when you tossed your coin 100 times, you got 60 heads. Use the sum function of your spreadsheet to calculate the probability that a fair coin, when tossed 100 times, will give you at least 60 heads or at least 60 tails.

Step 5 Assuming a significance level of 0.05, do you have reason to suspect that your coin is not fair? Explain your answer.

Significance Tests and Hypothesis Testing

The choice of a reasonable significance level depends on the situation. If the hypothesis is that a coin is fair, then you might choose 0.05 or 0.01 as the level below which you would reject it. If the hypothesis is that a person is innocent of a serious crime, you would choose a much lower level, say 0.000001, or $\frac{1}{1{,}000{,}000}$, as the level that would cause a belief that the person is guilty.

Caution: Significance levels are like goal lines in a game. The significance level must be chosen before analyzing data. Changing a significance level after computing the probability of your result is like moving a goal line to win a game. We used two different significance levels in Example 2 only to illustrate an instance of rejecting a hypothesis and an instance of having insufficient evidence against a hypothesis. In a real-world situation, one significance level would be chosen before calculations were made, and the level would not be altered.

Examples 1 and 2 together show an instance of statistical *hypothesis testing*. We hypothesized that the coin was fair or, said differently, that the probability p of success was 0.5. Such an assumption is called a *null hypothesis*, abbreviated $\mathbf{H_0}$ and pronounced "H-naught." The word "null" suggests that the coin is not out of the ordinary. A **null hypothesis** is a statement of "no effect" or "no difference." In Examples 1 and 2, the implied null hypothesis was H_0: The probability that the coin lands heads is 0.5 (meaning the coin is fair).

The goal of hypothesis testing is to measure the strength of the evidence against H_0. In testing a hypothesis:

(1) You suppose a null hypothesis H_0 and a significance level.

(2) Using a probability distribution, you reason from H_0.

(3) If you arrive at a situation whose probability is less than your significance level, you reject H_0. If you cannot reject H_0, you conclude that there is *insufficient evidence* against H_0.

Note that in Example 2 Part b, we could not explicitly say that H_0 is actually true or that we accepted the hypothesis. All we could say was that there was insufficient evidence to reject the hypothesis.

Example 3

A manufacturer delivers 20,000 switches to a contractor. The manufacturer guarantees that no more than 3% of the switches are defective. The contractor randomly pulls 70 switches from the delivery and tests those switches. He finds that 6 fail. Should the contractor accept the switches? The contractor conducts this test at the 0.05 significance level.

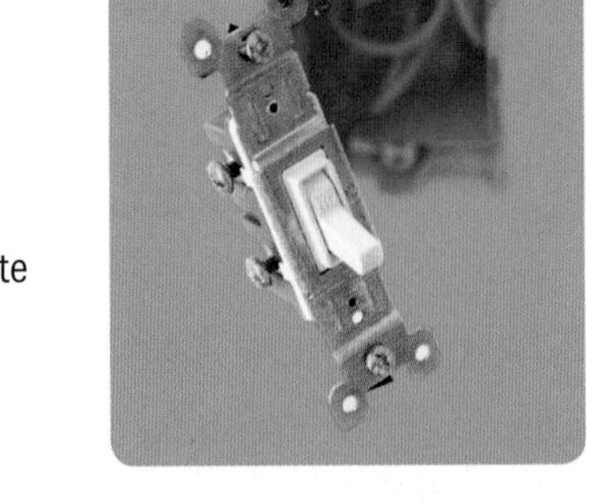

Solution State the hypotheses and the significance level.

Let H_0 be that the proportion of defective switches is no more than 0.03. Use a significance level of 0.05. The problem is to compute the probability that 6 or more of the 70 tested switches would fail under the assumption that $P(\text{failure}) = 0.03$.

The situation is modeled by a binomial probability distribution with $p = 0.03$ and $n = 70$. We need to find

$\sum_{x=6}^{70} ({}_{70}C_x(0.03)^x(0.97)^{70-x})$. The screenshot at the right shows that this sum equals about 0.019.

binomCdf(70,0.03,6,70)	0.018637

So the probability of 6 or more of the 70 tested switches failing under hypothesis H_0 is about 0.019. This is smaller than the significance level of 0.05. Reject the null hypothesis H_0. The contractor should reject the shipment of switches.

The use of probability to test hypotheses has applications well beyond coins and games of chance. In medicine, it is very unusual for all patients to react the same way to a new treatment. For some the treatment may be helpful, while for others it is not. We can never be certain that a new treatment is better than an old one, so we are forced into hypothesis testing. In elections, seldom do all people favor the same candidate. A sample poll can never tell for certain whether one candidate will win. Instead, hypothesis testing is used. A typical null hypothesis is that the candidates are equally popular. Then, when the poll's results are found, pollsters ask: If the candidates were equally popular, how unusual would it be to get poll results like these?

Questions

COVERING THE IDEAS

1. What does it mean to say a result is significant at the 0.05 level?
2. Explain why, in an experiment, a biased coin might turn up heads exactly half of the time.

In 3–6, a situation is described.

a. What is an appropriate null hypothesis for the situation?

b. What outcomes should be considered along with the given outcome?

c. What is an appropriate significance level?

3. A coin is tossed 15 times and 13 tails occur.
4. A coin is tossed 15 times and 2 heads occur.
5. A manufacturer's guarantee claims that no more than 1.1% of potato chip bags have less than 12 ounces of chips. A sample of 1100 bags gives 77 having less than 12 ounces.
6. In a lineup of 10 people, 17 out of 20 witnesses identify a defendant as the perpetrator of a crime.
7. In Example 3, tell what the conclusion would be if 5 out of the 70 sampled switches failed.
8. A coin is tossed 9 times and 8 heads occur. Determine whether the coin is fair at the 0.01 significance level by answering the following questions.
   a. What is the null hypothesis?
   b. Assuming the null hypothesis, what is the probability of the given event (or of one like it)?
   c. Is there sufficient evidence to reject the null hypothesis?

9. In 2008, a news magazine reported that a survey showed 18% of all drivers text message while driving. In the fall, a high school principal surveyed 39 seniors and 19 admitted to text messaging while driving. The principal wishes to test a claim that seniors in his high school who admit to text messaging while driving is greater than that of the population of all drivers. His null hypothesis is that 18% of the seniors in his school text message while driving.
   a. What is the probability of 19 or more of the 39 seniors responding affirmatively to texting while driving, assuming the principal's null hypothesis?
   b. If the school district instructed the principal to use a significance level of 0.01 to test the null hypothesis, write the principal's conclusion.

APPLYING THE MATHEMATICS

10. A pollster takes an exit poll of 8 voters in a particular precinct after an election between two candidates. All of them voted for candidate A. At the 0.05 significance level, test the claim that candidate A is more popular than candidate B.

And the winner is... Exit polls are used to predict the election outcome by asking some voters who they voted for as they leave the polling place.

11. Under the current treatment, a certain disease has a recovery rate of 25%. A virologist has developed a new treatment for the disease and tests it on 211 patients. He finds that 75 of the 211 patients recover. The virologist wants to test the claim that his new treatment is more effective than the old treatment.
    a. Considering the meaning of "null," name an appropriate null hypothesis.
    b. Test the claim at the 0.01 significance level.

12. Suppose a coin is tossed 10 times and 9 heads result.
    a. At the 1% significance level, test the hypothesis that the coin is biased with the probability of heads on each toss being $\frac{1}{3}$.
    b. In Examples 1 and 2b, you used the same experimental results and significance level to test the hypothesis that the coin was fair. Why were the outcomes you considered in Part a different than the outcomes you considered in Examples 1 and 2?

13. a. A coin is tossed 12 times and 6 heads occur. At the 0.05 significance level, test the null hypothesis that the probability of heads is $\frac{1}{3}$.
    b. In another experiment, the same coin is tossed 12 times and 9 heads occur. Test the same null hypothesis at the 0.05 significance level.
    c. Why do the conclusions in Parts a and b make sense?

14. A field-goal kicker in football is thought to be able to make 85% of field-goal attempts from within 20 yards. On the first 5 attempts of the season from within 20 yards, the kicker makes only 2 field goals. Is the coach justified in saying that the kicker seems no longer able to make 85% of field-goal attempts from within 20 yards?

REVIEW

15. Determine the mean and standard deviation of the random variable representing the number of heads in Question 13, assuming the null hypothesis is true. **(Lesson 10-7)**

16. Consider a multiple-choice test of 25 questions, each of which has 5 choices.
    a. Find the mean number of questions a student guessing on all answers is expected to get right.
    b. Find the standard deviation of the corresponding random variable. **(Lesson 10-7)**

17. A circular archery target is 122 cm in diameter and consists of a bull's-eye of diameter 12.2 cm, surrounded by nine evenly spaced concentric circles. Assume that an arrow that hits a target from 90 meters away is equally likely to land anywhere on the target. Find the probability that the arrow hits within the given region.

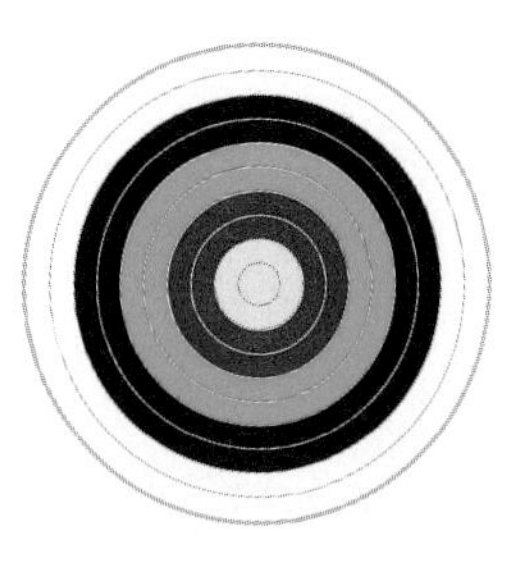

    a. the bull's-eye
    b. somewhere in the outermost 4 rings
    c. within the fifth ring counting from the outer edge **(Previous Course)**

EXPLORATION

18. In hypothesis testing, there are two types of errors that can be made. They are labeled Type I error and Type II error. Find out what these errors are and how they are related.

Chapter 10 Projects

Sir Francis Galton, 1822-1911

1 The Quincunx

The board pictured on the first page of this chapter was designed in the early 1870s by the English physician, explorer, and scientist, Sir Francis Galton, to illustrate binomial experiments. Sometimes the Galton board is called a *quincunx*. The original quincunx had a glass face and a funnel at the top. Small balls were poured through the funnel and cascaded through an array of pins. Each ball struck a pin at each level; theoretically, the ball had an equal probability of falling to the right or to the left. The balls collected in compartments at the bottom. Turning the quincunx upside down sent all the balls back to their original position. Galton's original quincunx still survives in England, large replicas exist in many science museums, and smaller ones can be purchased from science suppliers. A diagram of a quincunx is on page 616.

a. Obtain an example of a quincunx with at least ten rows of pins, or build your own. Release the balls and observe the distribution. Repeat this experiment 20–30 times and describe any trends.

b. What binomial probability distribution does your quincunx represent? Discuss the relation between your observations in Part a and this probability distribution.

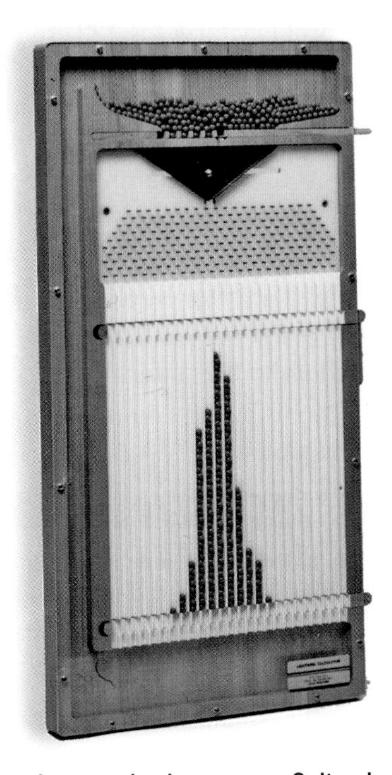

The quincunx, also known as a Galton board

2 Leibniz Harmonic Triangle

There are triangles other than Pascal's Triangle that have interesting properties. One such triangle is the *Leibniz harmonic triangle.* Do research to determine the first 7 rows of the harmonic triangle, and how binomial coefficients are related to it. Also, research what the *pronic numbers* are, and how they are related to the Leibniz harmonic triangle.

3 Probabilities and the Lottery

a. Obtain the rules, entry sheets, and an information sheet for a lottery. Use this information to find the number of different entries possible for each game, the probability that you will win first prize with a single entry, and the probability of winning *any* prize with a single entry. Would you advise someone to buy lottery tickets? Why or why not?

b. Repeat the steps in Part a for another lottery. Which lottery does the person have a better chance of winning? Suggest some reasons why a particular game might be played.

(*Note*: Do not attempt to buy a ticket in your local lottery game without first checking that it is legal for you to do so. All states restrict sales to persons over a certain age.)

4 Multinomial Probability Distributions

A *multinomial experiment* consists of n repeated trials where each trial has more than 2 possible outcomes. As in a binomial experiment, the probabilities of each outcome are constant and the trials are independent. A *multinomial distribution* is a probability distribution of the outcomes of this kind of experiment.

a. Find out how the probabilities in a multinomial experiment are calculated.

b. Consider the following situation. Suppose a card is drawn randomly from a standard 52-card deck and then put back in the deck. This exercise is repeated five times. What is the probability of drawing at least one of each suit?

5 The Binomial Theorem for Rational Exponents

Isaac Newton generalized the Binomial Theorem to all rational exponents. That is, he derived series expansions for expressions such as $(x + y)^{-3}$, $(x + y)^{\frac{2}{3}}$, and $(x + y)^{-\frac{5}{6}}$.

a. What did Newton find?

b. What are the first four terms of the series expansions of the binomials above?

c. How can this extended Binomial Theorem be used to aid in calculations?

6 Sierpiński Sieve

a. Draw the first 10 rows of Pascal's Triangle.

b. Color all the odd numbers blue and all the even numbers orange. What appears is part of an infinite array called the *Sierpiński sieve*.

c. Research the Sierpinski sieve and describe its history and some of its properties.

Sierpiński Quilt

Chapter 10 Summary and Vocabulary

- The number of ways to select a set of r items from a set of n elements is ${}_nC_r = \frac{{}_nP_r}{r!} = \frac{n!}{(n-r)!r!}$. Each selection is called a **combination**. A famous configuration of combinations is **Pascal's Triangle**. Each row of Pascal's Triangle gives coefficients of terms in the **expansion of $(x + y)^n$**. For all positive integers n,

$$(x + y)^n = {}_nC_0 x^n + {}_nC_1 x^{n-1}y + {}_nC_2 x^{n-2}y^2 + \cdots + {}_nC_n y^n.$$

- A **probability distribution** is formed from the related probabilities of a **random variable** representing each outcome of an experiment. A special case is a **binomial probability distribution**, which gives the probability of getting exactly k successes in n binomial trials, each of which has probability p of success.

- A **binomial experiment** has a fixed number of trials, each with only two possible outcomes (success and failure), the probabilities of which are fixed and sum to 1. In a binomial experiment with n trials and probability of success p, the probability of exactly k successes is ${}_nC_k\, p^k q^{n-k}$, where $q = 1 - p$. By the Binomial Theorem, this is the $(k + 1)$st term in the expansion of $(q + p)^n$. Binomial probabilities are used to analyze events with repeated independent trials.

- The formulas for **mean** and **variance of a random variable** apply to binomial random variables. The mean μ and standard deviation σ of a binomial random variable with n trials each having a probability p of success are given by $\mu = np$ and $\sigma = \sqrt{np(1 - p)}$, respectively. Below is a graph of a binomial distribution with $n = 50$ and $p = 0.5$. As n increases, the distribution flattens and gets wider. As p decreases, the graph moves to the left and is no longer symmetric.

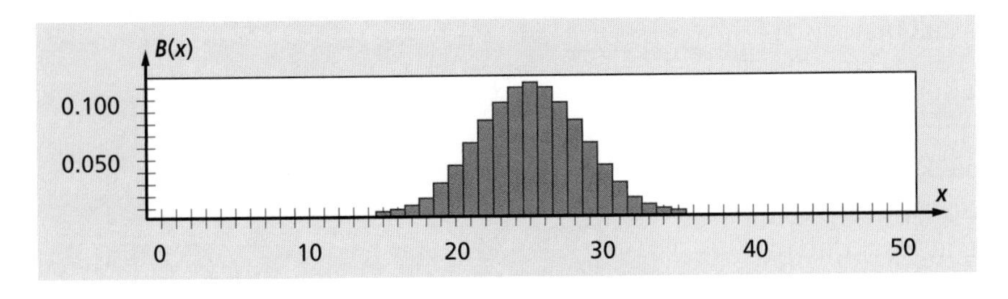

Vocabulary

10-1
combination
number of combination of n things taken r at a time, ${}_nC_r$, $\binom{n}{r}$

10-2
Pascal's Triangle
$(r + 1)$st term in row n

10-3
expansion of $(x + y)^n$
binomial coefficient

10-4
probability distribution
random variable
mean of a random variable
mode of a distribution
variance of a random variable
standard deviation of a random variable

10-5
binomial experiment

10-6
binomial probability distribution
binomial random variable

10-8
null hypothesis, H_0

- Assuming a binomial distribution, one can determine the probability of an observed event. Based on predefined significance levels, you can decide whether or not the calculated probability is sufficient evidence to reject the **null hypothesis** that the event *is* a reasonable result given the assumed distribution. Such statistical inferences are not definitive claims about what has or has not happened; rather, they are descriptions of the probability of what could have happened.

Theorems

Formula for ${}_nC_r$ Theorem (p. 619)
Pascal Triangle Symmetry (p. 627)
Binomial Theorem (p. 632)
Alternate Formula for Variance of a Random Variable (p. 641)
Binomial Probability Theorem (p. 647)
Mean of a Binomial Random Variable Theorem (p. 659)
Variance and Standard Deviation of a Binomial Random Variable(p. 662)

Chapter 10 Self-Test

Take this test as you would take a test in class. You will need a calculator. Then use the Selected Answers section in the back of the book to check your work.

1. What is the 16th term in row 53 of Pascal's Triangle?

2. The fifth terms in rows 25 and 26 of Pascal's Triangle are 12650 and 14950, respectively. What is the fourth term in row 25?

3. a. Use the Binomial Theorem to expand $(x - 2y)^4$, but do not simplify.
   b. Simplify your expression in Part a.

4. Find the mean and standard deviation of a binomial random variable with 300 trials and a chance of success of 0.35.

5. a. Prove that for all integers n greater than or equal to 1, ${}_nC_1 = n$.
   b. What property of Pascal's Triangle does this illustrate?

6. A new committee of 4 U.S. Senators is chosen at random. What is the probability that the committee will include the two senators from your state? (There are a total of 100 U.S. Senators.)

7. Suppose the probability that a student will pass a driving test on the first attempt is 68%. Find the probability that exactly 14 students in a class of 20 pass on their first attempt.

8. Find the mean and variance of the random variable x, where $P(x)$ is the probability that a student in a particular class spends x hours per week on math homework.

x	3	6	8	9	11	13
P(x)	0.24	0.41	0.15	0.04	0.04	0.12

9. A coin is tossed 20 times and 16 heads occur. At the 0.05 level, test the hypothesis that the coin is fair.

10. A restaurant dessert menu has three custards and seven pastries. A group of patrons decide to order two custards and five pastries. In how many ways can they do this?

11. The 3rd number in row 7 of Pascal's Triangle is equal to the coefficient of the $x^c y^d$ term of the expansion $(x + y)^7$. What are c and d?

12. a. Create a table of values for the function B where $B(x) = {}_{10}C_x(0.20)^x(0.80)^{10-x}$.
    b. Graph the function B.

13. What happens to the shape of the graph of $B(x) = {}_nC_x\,p^x(1 - p)^{n-x}$ as p increases from 0 to 1?

Chapter 10 Chapter Review

SKILLS
PROPERTIES
USES
REPRESENTATIONS

SKILLS Procedures used to solve problems

OBJECTIVE A Expand binomials using the Binomial Theorem. (Lesson 10-3)

In 1–4, expand the binomial.

1. $(a + b)^3$
2. $(x - 3)^6$
3. $(3x - 4)^4$
4. $\left(\frac{r}{3} + 3t\right)^5$
5. Find the second term in the binomial expansion of $(w - v)^{18}$.
6. Find the middle term in the binomial expansion of $(p + 2q)^{14}$.

OBJECTIVE B Calculate the mean and standard deviation of a binomial random variable. (Lesson 10-7)

In 7 and 8, find the mean and standard deviation of the binomial random variable with the given probability p of success and number n of trials.

7. $p = 0.3$, $n = 100$
8. $p = 0.6$, $n = 80$
9. If $\mu = 72$ and $\sigma = 6$ for a binomial random variable, determine n and p.

PROPERTIES Principles behind the mathematics

OBJECTIVE C Prove and apply properties involving combinations. (Lesson 10-2)

In 10 and 11, prove the given identity.

10. ${}_nC_r = {}_nC_{n-r}$ for all positive integers n and r, $r \le n$.
11. ${}_nC_{n-1} = n$
12. **True or False** For all positive integers n and r such that $r \le n$, ${}_nP_r > {}_nC_r$.
13. Solve for x using a CAS: ${}_9C_3 + {}_9C_4 = {}_xC_4$.

OBJECTIVE D: Interpret and describe properties of binomial coefficients combinatorially and algebraically. (Lessons 10-2, 10-3)

14. Explain how the number of ways to choose 3 snacks from 10 possible snacks is the same as the number of ways to choose 7 snacks from 10 possible snacks.
15. Why are combinations called binomial coefficients?
16. The 5th number in row 8 of Pascal's Triangle is equal to the coefficient of the x^cy^d term of the expansion of $(x + y)^8$. What are c and d?
17. In the expansion of $(x + y)^7$, the coefficient of x^3y^4 equals the coefficient of x^ay^b. If $a \ne 3$, what are a and b?

OBJECTIVE E Identify and apply characteristics of binomial probability distributions. (Lesson 10-6)

In 18–22, let B be a binomial distribution with n trials, each having a probability p of success.

18. Give a formula for $B(x)$, the probability of exactly x successes.
19. If n remains constant, what happens to the graph of B as p increases from 0 to 1?
20. If n remains constant, what happens to the mean and mode as p increases from 0 to 1?
21. If p remains constant, what happens to the graph of B as n increases from 2 to 2000?
22. When will the graph of B be symmetric?

In 23 and 24, consider a binomial distribution B for $n = 10$ and $p = 0.15$.

23. Find $\sum_{x=0}^{10} B(x)$.
24. Describe a situation that could lead to this distribution.

25. Express the definition of mean of a binomial random variable x using sigma notation.

26. Fill in the Blank As the number n of trials increases, the maximum value of a probability distribution __?__.

27. True or False The sum of the values of a probability distribution is always less than 1.

USES Applications of mathematics in real-world situations

OBJECTIVE F Use combinations to compute the number of ways of selecting objects. (Lesson 10-1)

28. A waffle house offers 13 different kinds of waffles. How many combinations of waffles can a family of four order if no two family members choose the same kind of waffle?

29. There are 18 members in a pocket billiards club. For a tournament, each member must play each of the others twice. How many games must be played?

30. A sheepshead hand consists of six cards from a deck of 32 different cards. How many different sheepshead hands are there?

31. How many sets of fifteen hits can a DJ select from the top 50 to play in the next hour?

32. To pack for a trip you can choose from the following items: 10 pairs of socks, 8 shirts, 6 pants, 5 pairs of shoes, and 3 belts. How many different combinations are there of 4 pairs of socks, 4 shirts, 3 pants, 2 pair of shoes, and a belt?

OBJECTIVE G Compute probabilities involving combinations in non-binomial problems. (Lesson 10-1)

33. Refer to Question 30. Half the cards in a sheepshead deck are red and the other half are black. What is the probability of being dealt a sheepshead hand with only black cards?

In 34 and 35, a box contains 8 red, 3 green, and 9 white balls. Three balls are drawn at random without replacement.

34. Determine the probability that all three balls are red.

35. Determine the probability that one ball of each color is drawn.

In 36 and 37, 5 cards are drawn from a well-shuffled deck of 52 cards. Each deck contains four each of aces, Kings, Queens, Jacks, 10s, 9s, 8s, 7s, 6s, 5s, 4s, 3s, and 2s.

36. Find the probability that 4 of the 5 cards are aces.

37. Find the probability that 3 cards are Jacks and 2 are Queens.

OBJECTIVE H Determine probabilities in situations involving binomial experiments. (Lesson 10-5)

38. Determine the probability that there are 7 sixes in 10 tosses of a fair die.

39. Using the Binomial Probability Theorem, determine the probability of getting at least 6 tails when a fair coin is flipped 10 times.

40. A coin is tossed 7 times. If $P(\text{heads}) = 0.25$, what is the probability of getting exactly 5 tails?

41. Refer to Example 3 of Lesson 10-3. What does the coefficient of H^rT^{n-r} divided by 2^n mean in the context of the problem?

In 42 and 43, imagine rolling a fair six-sided die 7 times. Let x be the number of times a 5 occurs. Calculate each probability.

42. $P(x = 2 \text{ or } x = 3)$ **43.** $P(x \geq 6)$

In 44 and 45, consider a multiple-choice quiz with 10 questions that have four options each.

44. What is the probability that a student will get a perfect score if he or she guesses randomly?

45. What is the probability a student gets at least 5 right by guessing randomly?

46. The green sea turtle lays a clutch of 100 eggs but 98% of the hatchlings do not survive to maturity. What is the probability that at least 4 of the hatchlings from the clutch will survive to maturity?

47. Suppose that the probability that a randomly selected heart transplant patient will survive more than a year is 0.88. Find the probability that at least 4 of 5 randomly selected heart transplant patients will survive for longer than a year.

OBJECTIVE I Compute the mean and standard deviation of a random variable in a situation modeled by a probability distribution. (Lesson 10-4)

In 48–51, find the mean and standard deviation of the random variable *x*, *y*, *z* or *w*.

48.

x	2	4	20	40
P(x)	90%	9.6%	0.3%	0.1%

49.

y	2	5	7	8	10	11
P(y)	0.25	0.5	0.125	0.0625	0.03125	0.03125

50. $P(z)$ = the probability that a fair six-sided die lands with z showing.

z	1	2	3	4	5	6
P(z)	$\frac{1}{6}$	$\frac{1}{6}$	$\frac{1}{6}$	$\frac{1}{6}$	$\frac{1}{6}$	$\frac{1}{6}$

51. $P(w)$ = the probability that the residual for a model on observed data equals w.

w	9	6	3	0	-3
P(w)	$\frac{1}{10}$	$\frac{1}{10}$	$\frac{1}{10}$	$\frac{2}{3}$	$\frac{1}{30}$

In 52 and 53, consider a test with 40 items.

52. If all items are true-false and students guess randomly, what is the mean and standard deviation of the random variable representing the number of questions answered correctly?

53. If all questions are multiple-choice with five options and students guess randomly, what is the mean and standard deviation of the random variable giving the number of questions answered correctly?

54. How many questions with 4 options should be on a multiple-choice test to have an expected mean score of 25 correct answers when students guess randomly?

OBJECTIVE J Use binomial distributions to test hypotheses. (Lesson 10-8)

55. Suppose a coin is tossed 50 times, and 9 tails occur.

a. Calculate the probability that this event or an event like it would occur if the coin was fair.

b. List the event you included in your calculations in Part a. Describe what makes these event similar to the given event of 9 out of 50 tails.

c. If a 0.01 significance level is used to test the hypothesis that the coin is fair, should the hypothesis be rejected?

56. A certain disease has a recovery rate of 75% with current treatments. A researcher has developed a new treatment for the disease and tests it on 140 patients. She finds that 121 of the 140 patients recovered from the disease. The researcher claims that the new treatment is more effective against the disease than current treatments.

a. State a null hypothesis that could be used to test the researcher's claim.

b. State an appropriate significance level for this situation.

c. Test the researcher's claim at the significance level you chose. Should the null hypothesis be rejected?

d. Interpret your answer to Part c in the context of the problem.

57. A deck of 50 cards contains 25 red cards and 25 blue cards. In an experiment on extrasensory perception (ESP), one person shuffles the deck and turns over the cards one at a time. In another room, the subject, who does not know how many red and blue cards are in the deck, states the color of each card. The subject identifies 33 of the 50 cards correctly. At the 0.01 level of significance, does the subject appear to have ESP?

REPRESENTATIONS Pictures, graphs, or objects that illustrate concepts

OBJECTIVE K Represent combinations and binomial coefficients by Pascal's Triangle. (Lessons 10-2, 10-3)

58. If the third and fourth term of a row of Pascal's Triangle are 91 and 364, what does the fourth term of the next row equal?

59. The second, fourth, fifth, middle, eighth, eleventh, and thirteenth terms in a row of Pascal's Triangle are 12, 220, 495, 924, 792, 66, and 1, respectively. Fill in the other terms.

60. Find the first four terms in row 15 of Pascal's Triangle.

61. a. What is the 22nd term in row 47 of Pascal's Triangle?

b. What term of what power of $(a + b)$ is this coefficient?

62. The sum of the elements in one of the rows in Pascal's Triangle is 8192. Which row is this?

63. What are the last two terms of the row in Pascal's Triangle whose terms add to 2^{29}?

In 64-66, copy the display below. Show three places where the given property is represented.

1
1 1
1 2 1
1 3 3 1
1 4 6 4 1
1 5 10 10 5 1
1 6 15 20 15 6 1
1 7 21 35 35 21 7 1
1 8 28 56 70 56 28 8 1
1 9 36 84 126 126 84 36 9 1
1 10 45 120 210 252 210 120 45 10 1

64. ${}_nC_r = {}_nC_{n-r}$

65. ${}_nC_r = {}_{n-1}C_r + {}_{n-1}C_{r-1}$

66. $\sum_{r=0}^{n} {}_nC_r = 2^n$

OBJECTIVE L Graph and interpret binomial probability distributions. (Lesson 10-6)

67. Graph the binomial probability distribution where $p = 0.5$ and $n = 4$.

68. Tell why this table does not describe a probability distribution.

x	–1	0	1	2	4
P(x)	$\frac{1}{6}$	$\frac{1}{12}$	$\frac{7}{12}$	$\frac{1}{12}$	$\frac{1}{3}$

69. Let x be the number of girls in a family of 5 children. Let $P(x)$ = the probability that a family of 5 has x girls in it. Assume that the births of boys and girls are equally likely.

x = number of girls	0	1	2	3	4	5
P(x)	?	?	?	?	?	?

a. Fill in the table of probabilities.

b. Graph the distribution.

70. Let x be the number of items that a person gets right by guessing on a 3-item multiple-choice test. If each question has 4 possible choices, then the probability $P(x)$ of getting x items correct is $P(x) = {}_3C_x \left(\frac{1}{4}\right)^x \left(\frac{3}{4}\right)^{3-x}$.

a. Construct a table of probabilities for $x = 0$, 1, 2, and 3. (Round values to the nearest thousandth.)

b. Graph the probability distribution as a scatterplot.

71. The graph of a binomial probability distribution is given below.

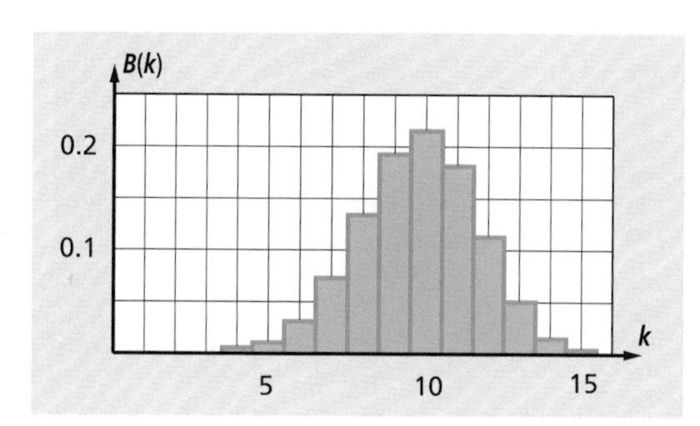

a. What is the mode of B?

b. What happens to the mode as n increases?

c. What happens to the spread of the distribution as n decreases?

72. Construct a binomial probability distribution for flipping a fair coin 10 times.

73. Multiple Choice Graphs of three binomial probability distributions with equations of the form $y = \binom{n}{x} p^x (1 - p)^{n-x}$, with the same number n of trials, are shown below. In which distribution is p smallest, and why?

A

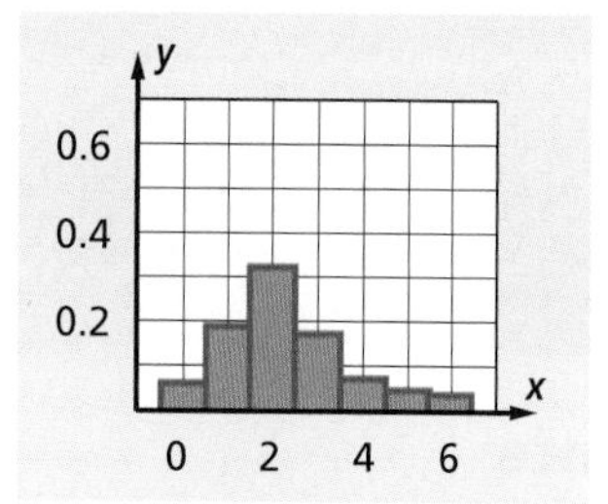

B

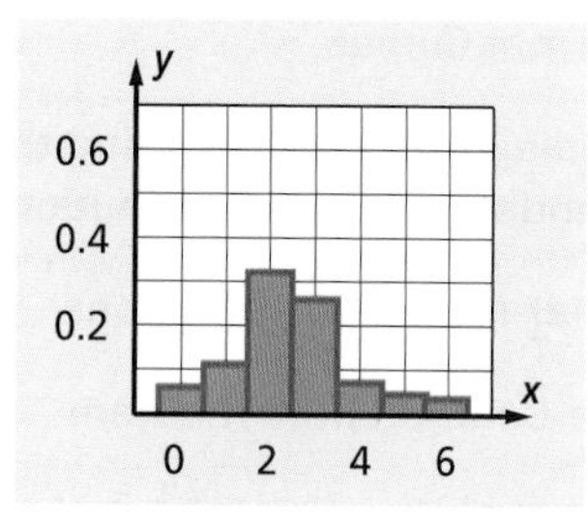

C

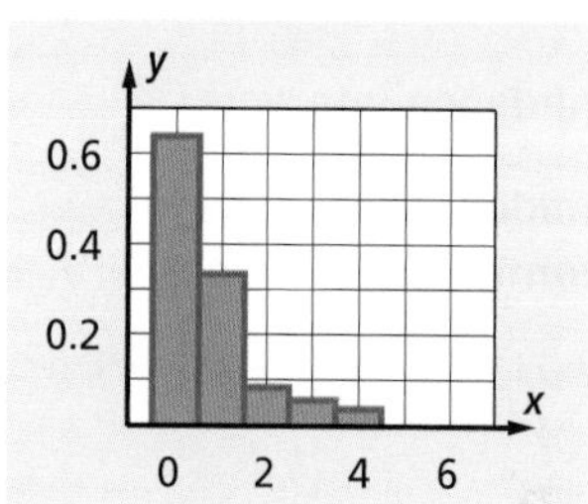

Chapter 11

Normal Distributions

Contents

Liberty Bell, Philadelphia, PA

In earlier chapters, you have seen three kinds of distributions:

- frequency distributions—mapping an event *x* onto the number of times *x* has occurred
- relative frequency distributions—mapping an event *x* onto the proportion of times *x* has occurred
- probability distributions—mapping an event *x* onto its probability given some assumptions about the situation

This chapter discusses the most important distribution in statistics, *the normal probability distribution*. Its graph, shown on the next page, is called a bell curve because it resembles a bell, such as the Liberty Bell.

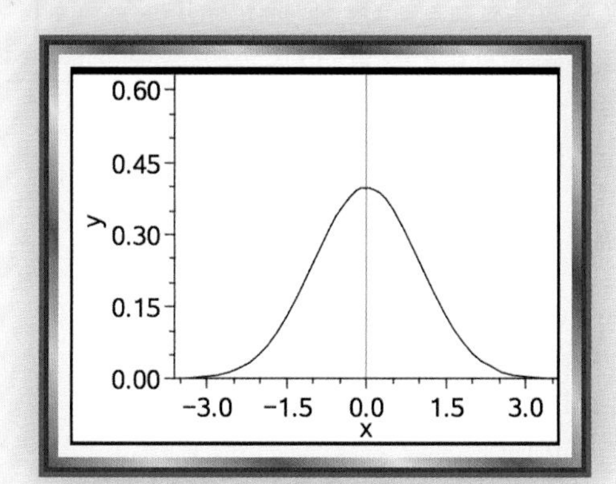

There are many reasons to study normal probability distributions. A normal distribution is the limit of a binomial probability distribution as the number of trials gets large without bound. Measures of many natural phenomena, such as the heights of people of the same gender from the same population, are approximately normally distributed.

The binomial distributions presented in Chapter 10 and the normal distributions presented here are very important functions in statistics. They are often employed to make or to support decisions in situations involving uncertainty. Even when phenomena are not normally distributed, in many circumstances, the distribution of the means of samples of size *n* from a population may be very close to a normal distribution.

In this chapter, you will see distributions of sample means of data reported as frequencies and as percents. You will also see how these distributions are related to the testing of hypotheses, to standardized tests, and to the degree of confidence a person can have in the results of a poll involving random sampling.

Lesson

11-1 Normal Curves

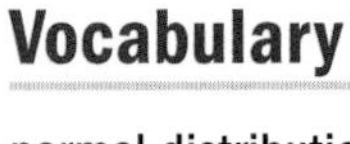

Vocabulary

normal distribution
normal curve
concave down
concave up
inflection point
standard normal curve
standard normal distribution

BIG IDEA All normal distributions are offspring of the function $f: x \rightarrow e^{-x^2}$, which can be transformed to model data distributions and to construct the standard normal distribution.

Consider a binomial experiment. For a specific probability p of success, as the number n of trials increases, the graph of a binomial probability distribution approaches a curve that has the shape of a bell.

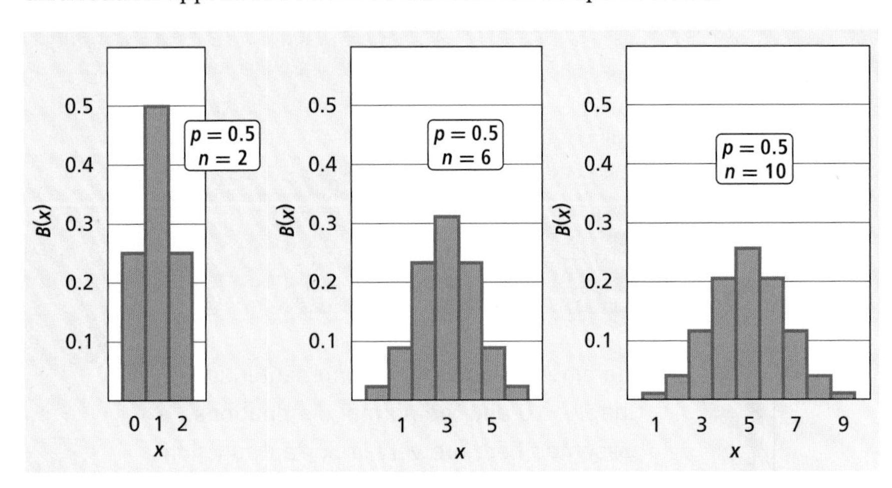

Mental Math

Consider the rectangle below. What percent of the total area is in each interval?

a. $0 \leq x \leq 5$

b. $0 \leq x \leq 7$

c. $7 \leq x \leq 10$

d. $3 \leq x \leq 6$

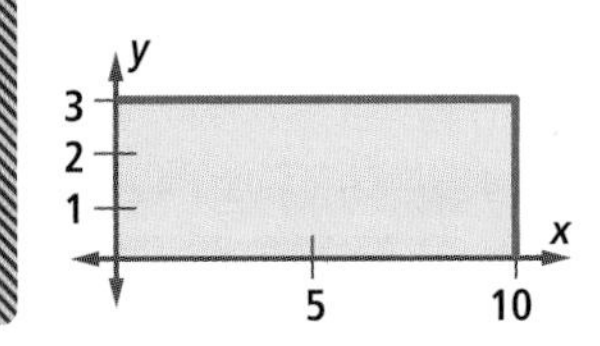

The values of the binomial distribution approach those of a continuous function called a **normal distribution**. Normal curves were first discovered by Abraham De Moivre (1664–1727) in conjunction with binomial distributions. A century later, Gauss noted that errors made in astronomical measurements were distributed like normal distributions. Many everyday data sets of natural phenomena have approximately normal distributions, such as the heights of people from the same population, the amounts of annual rainfall in a city over a long period of time, the weights of individual fruits from a particular orchard in a year, and standardized test scores.

The Parent Normal Curve

The graph of a normal distribution is called a **normal curve**. The parent normal curve is the graph of $f(x) = e^{-x^2}$, shown at the right. Some properties of the parent normal curve are listed on the next page.

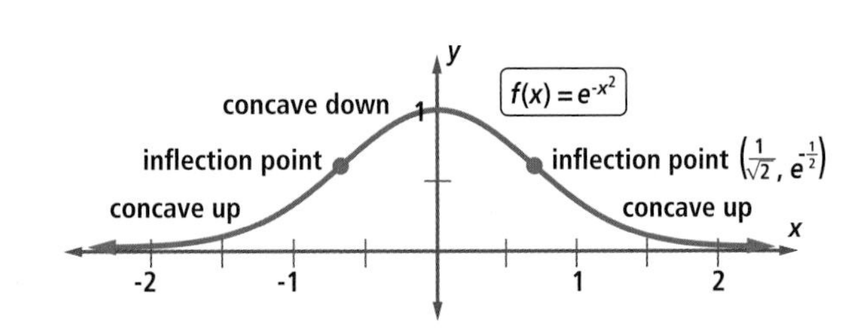

1. The domain is the set of real numbers; the range is $\{y \mid 0 < y \leq 1\}$.
2. The maximum value of the function is 1, which occurs when $x = 0$.
3. f is an even function, so the y-axis is a line of symmetry of the graph.
4. $\lim_{x \to \infty} f(x) = 0$ and $\lim_{x \to -\infty} f(x) = 0$, so the x-axis is an asymptote of the graph.
5. Near the y-intercept, the graph is curved downward, called **concave down**. Further away from the y-axis, the graph is curved upward, called **concave up**. When $x = \pm\frac{1}{\sqrt{2}} = \pm\frac{\sqrt{2}}{2}$, the graph changes concavity. These points are called **inflection points**.

QY

> **QY**
>
> What are the coordinates of the inflection point to the left of the y-axis?

Even though the graph of $f(x) = e^{-x^2}$ never touches the x-axis, it can be shown that the area under the parent of the normal curve is finite. Finding the exact area requires calculus, but it can be approximated using the Monte Carlo methods from Lesson 6-7. Imagine a rectangle containing the part of the graph of $f(x) = e^{-x^2}$ between $x = -3$ and $x = 3$, and between $y = 0$ and $y = 1$, as shown below.

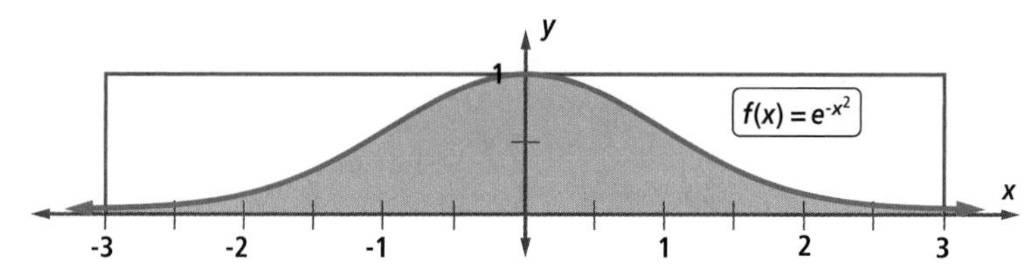

Activity 1

MATERIALS random number generator, spreadsheet software

Step 1 Use a random number generator to randomly pick a point (a, b) in the rectangle $-3 \leq x \leq 3$, $0 \leq y \leq 1$. You will need to randomly pick a and then randomly pick b. Pick each to six decimal places.

=6 * RAND() – 3
=RAND()
=EXP(-(A3A3))
=IF(B4 > C4, 0, 1)

◇	A	B	C	D
1	a	b	e^-(a^2)	
2	0.804755	0.283301	0.523284	1
3	2.735951	0.642282	0.000561	0
4	-0.761265	0.056688	0.560165	1
5	0.914332	0.381742	0.43344	1
6	2.675742	0.10391	0.000777	0
⋮	⋮	⋮	⋮	⋮
30000	2.512731	0.103207	0.001811	0

Step 2 Determine whether (a, b) is above or below the curve $y = e^{-x^2}$ by comparing b with e^{-a^2}.

(continued on next page)

Step 3 Repeat Steps 1 and 2 n times, where $n \geq 1000$. Use a spreadsheet to organize your data. (On some calculator spreadsheets, you may choose to enter a command in the column entry line so that you do not have to fill down cells manually.) Count the number m of times the point is below the curve. To do this on some calculators, find the sum of column D. A sample spreadsheet is shown on the previous page.

Step 4 The area under the curve is approximately $\frac{m}{n}$ times the area of the rectangle. What approximation does your Monte Carlo simulation give for this area?

We did Activity 1 with $n = 30{,}000$ and found $m = 8{,}887$. This gave us the estimate $\frac{8887}{30000} \cdot 6$, or 1.7774 for the area under the curve. Calculus shows that the area between the complete graph of $f(x) = e^{-x^2}$ and the x-axis is exactly $\sqrt{\pi} \approx 1.7725$.

The Standard Normal Curve

The graph of an important offspring of the function $y = e^{-x^2}$ is the bell-shaped curve shown below. It has inflection points at $x = 1$ and $x = -1$ and an area of 1 between the curve and the x-axis. This curve is known as the **standard normal curve**. Because the area under the curve is 1, the corresponding function can be viewed as a probability distribution.

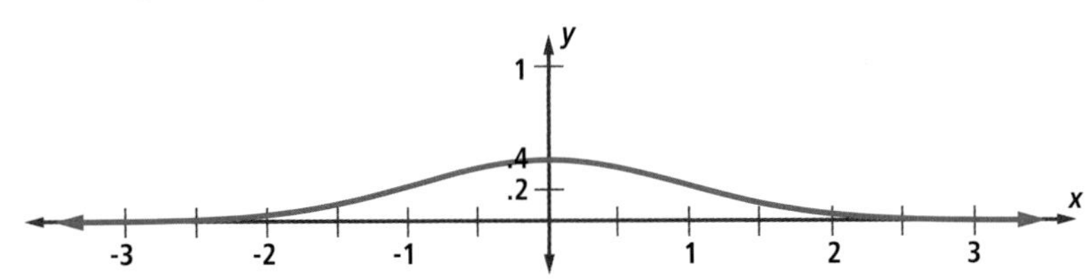

Activity 2

MATERIALS normal curve application provided by your teacher

The application shows the graph of a normal curve $y = b \cdot e^{-(ax)^2}$ in red. When $a = b = 1$, this is the parent normal curve.

Step 1 Set the value of a to 1 and vary the value of b. Describe how the graph changes. Changing the value of b is equivalent to applying what type of transformation?

Step 2 Set the value of b to 1 and vary the value of a. Describe how the graph changes. Changing the value of a is equivalent to applying what type of transformation?

Step 3 Use both sliders to match the parent normal curve to the standard normal curve, which is shown in black. What values do you get for a and b?

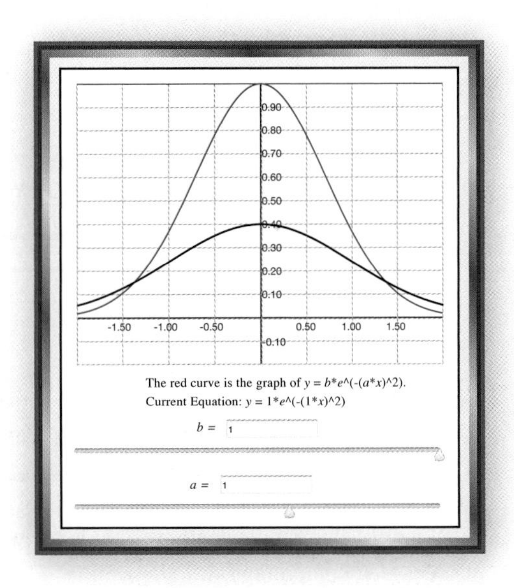

The values of a and b you found in Activity 2 are approximate. To find the exact transformation that gives the graph of the standard normal curve, note that the inflection points of the standard normal curve are at $x = \pm 1$ and the inflection points of the parent curve are at $x = \pm \frac{1}{\sqrt{2}}$. So the scale change S_1: $(x, y) \to (\sqrt{2}x, y)$ will modify the graph of $y = e^{-x^2}$ so that the inflection points of the image are at $x = \pm 1$. An equation for the image is $y = e^{-\left(\frac{x}{\sqrt{2}}\right)^2}$, which simplifies to $y = e^{-\frac{x^2}{2}}$. Your value for a in Activity 2 should have been close to $0.71 \approx \frac{1}{\sqrt{2}}$.

The parent curve (blue) and image (red) under S_1 are graphed below. However, the image is not yet the standard normal curve. The fact that the area between the curve and the x-axis is 1 determines the exact value of the vertical scale change needed.

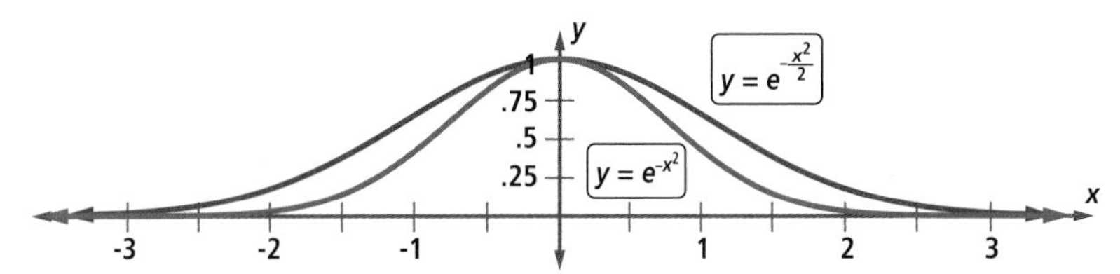

We have noted that the area between the graph of $y = e^{-x^2}$ and the x-axis is $\sqrt{\pi}$. The scale change S_1 multiplies area by $\sqrt{2}$, so the area between the graph of $y = e^{-\frac{x^2}{2}}$ and the x-axis is $\sqrt{\pi} \cdot \sqrt{2} = \sqrt{2\pi}$. This is why your value for b in Activity 2 should have been close to $0.4 \approx \frac{1}{\sqrt{2\pi}}$. When a second scale change S_2: $(x, y) \to \left(x, \frac{y}{\sqrt{2\pi}}\right)$ is applied to the graph of $y = e^{-\frac{x^2}{2}}$, the points of inflection are unchanged and the area under the image is 1. An equation of the image of $y = e^{-\frac{x^2}{2}}$ under S_2 is $\sqrt{2\pi}\, y = e^{-\frac{x^2}{2}}$.

Solving for y gives an equation of this curve: $y = \frac{1}{\sqrt{2\pi}} e^{-\frac{x^2}{2}}$.

This curve represents a probability distribution called the **standard normal distribution**. It is the most important of all probability distributions. Shown below is a graph of the standard normal distribution and the parent curve.

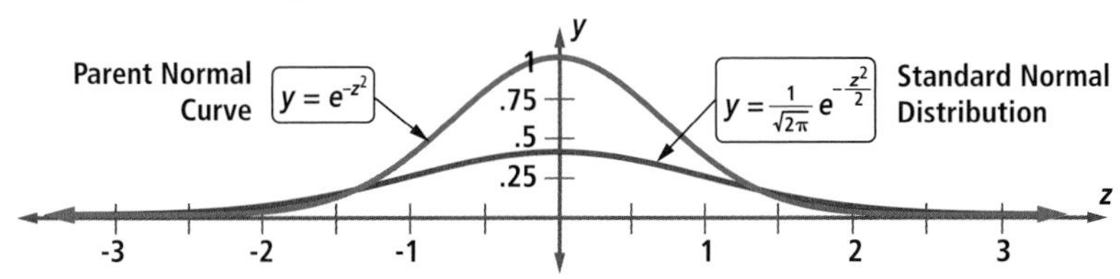

Note that we have labeled the independent variable z and the horizontal axis as the z-axis. The letter z is used instead of x because the standard normal distribution is associated with the z-scores you computed from data in Lesson 3-9. For this reason it is customary to use z to name the horizontal axis when discussing the standard normal curve. If you are working with a different normal distribution, which is common when you graph histograms of raw data, use the letter x.

Properties of the Standard Normal Distribution

Properties of the standard normal distribution f where $f(z) = \frac{1}{\sqrt{2\pi}} e^{-\frac{z^2}{2}}$ follow from properties of the parent function with equation $y = e^{-x^2}$ and scale transformations.

1. Its domain is the set of real numbers; its range is $\left\{y \mid 0 < y \leq \frac{1}{\sqrt{2\pi}}\right\}$.
2. Its maximum value is $f(0) = \frac{1}{\sqrt{2\pi}} \approx 0.3989$.
3. It is an even function, so the y-axis is an axis of symmetry for its graph.
4. $\lim_{z \to \infty} f(z) = 0$ and $\lim_{z \to -\infty} f(z) = 0$, so the z-axis is an asymptote to the graph of the function.
5. Its graph is concave down where $-1 < z < 1$ and concave up where $|z| > 1$; its inflection points occur when $z = 1$ and $z = -1$.
6. The area between the curve and the z-axis is 1.

Questions

COVERING THE IDEAS

In 1–4, consider the functions g with equation $g(x) = e^{-x^2}$ and f with equation $f(x) = \frac{1}{\sqrt{2\pi}} e^{-\frac{x^2}{2}}$.

1. Are the functions even, odd, or neither?
2. Identify any lines of symmetry for their graphs.
3. Identify any asymptotes of their graphs.
4. What is the area between the graphs of these functions and the x-axis?
5. What is meant by a point of inflection?
6. Give the coordinates for the indicated points of the parent normal curve.
   a. points of inflection
   b. maximum point
7. Answer Question 6 for the standard normal curve.
8. **Fill in the Blanks** The scale change $S: (x, y) \to (\underline{\ ?\ }, \underline{\ ?\ })$ maps the parent normal curve onto the standard normal curve.
9. Give an instance of a variable in the real world with an approximately normal distribution.

APPLYING THE MATHEMATICS

10. About 34% of the area between a normal distribution curve and the x-axis lies between $x = 0$ and $x = 1$. Given this fact, about what percent lies in each of these intervals?
    a. $-1 \leq x \leq 1$
    b. $x \geq -1$
    c. $x \leq -1$

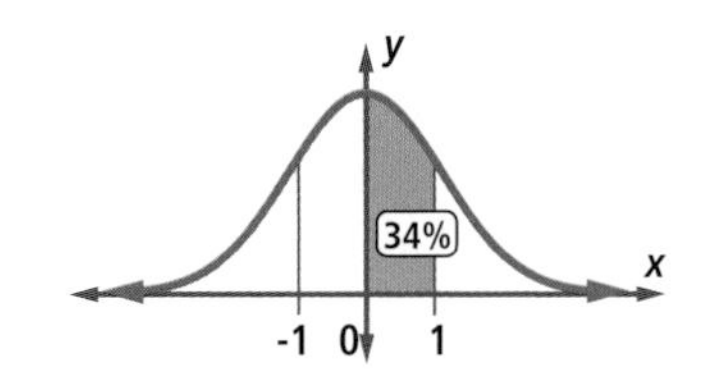

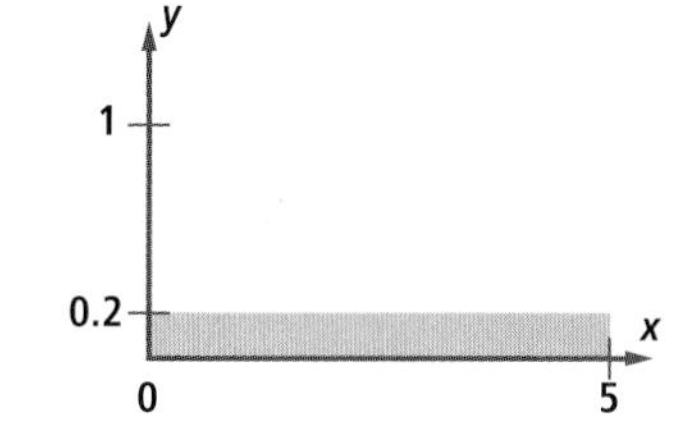

11. a. Suppose a point (x, y) is selected randomly from the shaded rectangular region at the right. What is the probability that $1 \le x \le 4$?

    b. If four points are selected randomly from the rectangular region, what is the probability that all four will have x-coordinates between 1 and 4?

In 12–14, what transformation maps $f(x) = \frac{1}{\sqrt{2\pi}} e^{-\frac{1}{2}x^2}$ onto the graph of the given equation?

12. $f(x) = \frac{1}{\sqrt{2\pi}} e^{-\frac{1}{2}(x-\mu)^2}$

13. $f(x) = \frac{1}{\sqrt{2\pi}\sigma} e^{-\frac{1}{2}\left(\frac{x}{\sigma}\right)^2}$

14. $f(x) = \frac{1}{\sqrt{2\pi}\sigma} e^{-\frac{1}{2}\left(\frac{x-\mu}{\sigma}\right)^2}$

REVIEW

15. Suppose $g: x \rightarrow x^{\frac{7}{5}}$ for all nonnegative real numbers x.

    a. Find an equation for the inverse of g.

    b. Is the inverse of g a function? Why or why not? **(Lessons 9-2, 3-8)**

16. Tell whether each number is a real number. **(Lesson 9-1)**

    a. $\sqrt[6]{43}$ b. $43^{\frac{1}{6}}$ c. $-43^{\frac{1}{6}}$ d. $(-43)^{\frac{1}{6}}$

17. A fair die is tossed three times. Consider these events:

    A: at most one 3 occurs

    B: a 3 and a 4 occur at least once.

    Are A and B independent? Justify your answer using the definition of independence. **(Lessons 6-3, 6-1)**

18. A function f is defined by $f(x) = \begin{cases} 3x + 2, \text{ for } 0 \le x \le 9 \\ 0, \text{ otherwise} \end{cases}$.

    a. Graph the function.

    b. Suppose a dart lands in a random location in the region between the graph of f and the x-axis. What is the probability that the x-coordinate of the location is between 6 and 7? **(Lessons 6-1, 2-1)**

19. A school compares its students' performance on a college-entrance examination to the performance of students in the nation.

    a. What is the population?

    b. What is the sample? **(Lesson 1-1)**

20. A pie fits snugly in a box with a 10-inch square base. If a speck of dust randomly floats into the box, what is the probability that it lands on the pie? **(Previous Course)**

EXPLORATION

21. Find out what astronomical event led Gauss to realize that measurement errors might be normally distributed.

QY ANSWER

$\left(-\frac{1}{\sqrt{2}}, e^{-\frac{1}{2}}\right)$

Lesson

11-2 Finding Probabilities Using the Standard Normal Distribution

Vocabulary

Standard Normal Distribution Table

▶ **BIG IDEA** Areas under the curve for the standard normal distribution provide probabilities for inequality statements about *z*-values.

Mental Math

Two mutually exclusive events have probabilities *p* and *q*, respectively. What is the probability of neither event happening?

Below is a graph of the standard normal curve.

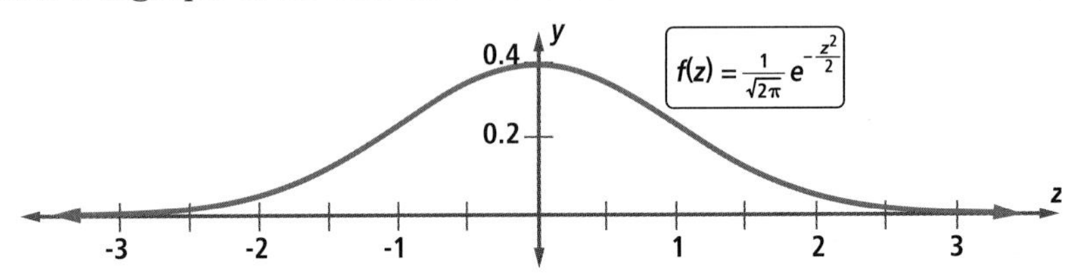

In this lesson you will use areas under the standard normal curve to compute probabilities. In the next lesson, you will transform distributions of raw data into *z*-scores so that you can compute probabilities for events.

The standard normal curve can represent a probability distribution because the area between it and the *z*-axis is 1. The associated function is called the ***standard*** normal distribution because the mean and standard deviation of its domain variable are 0 and 1, respectively. Recall from Chapter 3 that these are exactly the mean and standard deviation of a set of data that has been transformed into *z*-scores. The standard normal probability distribution is a distribution of *z*-scores.

From Area to Probability

The calculation of probabilities using the standard normal distribution is different in a major way from the calculation of binomial probabilities in Chapter 10. A binomial distribution is discrete, so you can calculate the probability that *x* takes on a particular value. For instance, you can calculate the probability that there are 12 heads in 20 tosses of a fair coin. The probability is the area of the histogram bar above 12 in the graph of the binomial distribution with mean 10 and standard deviation $\sqrt{5}$, as shown at the right.

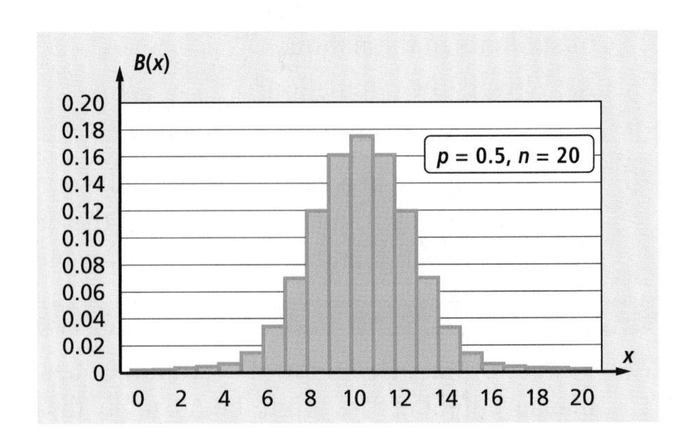

However, a normal distribution is continuous. The area directly above 12 in the graph below is 0.

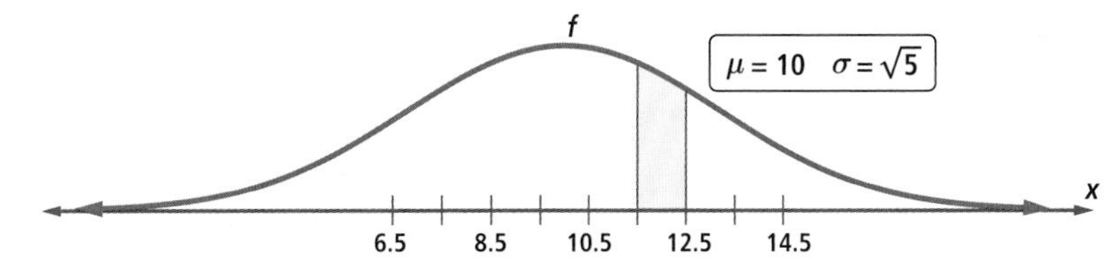

When we approximate the previous binomial distribution by the normal distribution with mean 10 and standard deviation $\sqrt{5}$ we find the probability of 12 heads by calculating the probability that there are between 11.5 and 12.5 heads, even though exactly 12 heads is the only possibility. The probability is the area under the normal curve between $x = 11.5$ and $x = 12.5$.

Calculating Specific Probabilities

When calculating areas under normal curves, we always refer back to the standard normal distribution. This is done by converting the raw data to z-scores, as you will see in the next lesson. For now, we examine the connection between probability and the area under the standard normal curve.

Consider first the probability that a value in a normally distributed data set is less than the mean. Then the corresponding z-score is less than 0. This is represented by the shaded area in the graph below. Because of the symmetry of the standard normal curve and the fact that the area between the curve and the z-axis is 1, $P(z < 0) = 0.5$. This is what you would expect. Also $P(z > 0)$ is $1 - 0.5$, which is also 0.5.

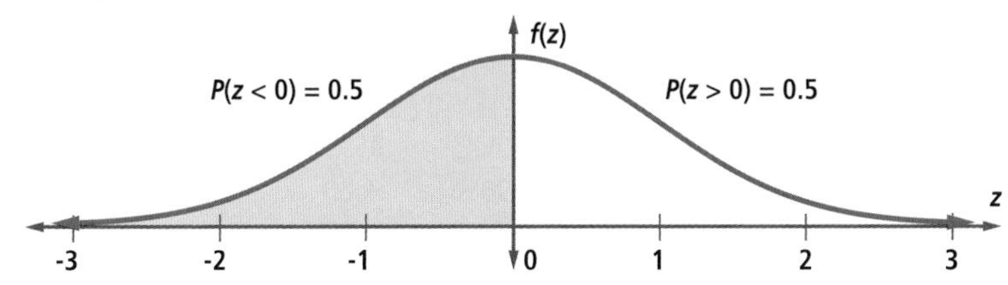

The standard normal distribution is so important that probabilities derived from it have been recorded in tables and preprogrammed into many calculators. On the next page is a **Standard Normal Distribution Table.** It gives the area under the standard normal curve to the left of a given positive number a. The number a is how many standard deviations a z-score is above the mean. The table is read in the following way. For $a = 1.83$, first start with the row $a = 1.8$, then look under the column entry 3 for the value 0.9664.

Standard Normal Distribution Table

This table gives the area under the standard normal curve to the left of a given positive number a.

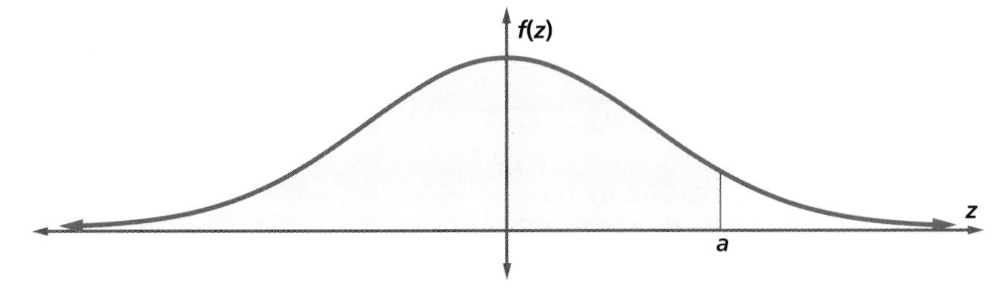

$P(z < a)$ for $a \geq 0$

a	0	1	2	3	4	5	6	7	8	9
0.0	.5000	.5040	.5080	.5120	.5160	.5199	.5239	.5279	.5319	.5359
0.1	.5398	.5438	.5478	.5517	.5557	.5596	.5636	.5675	.5714	.5753
0.2	.5793	.5832	.5871	.5910	.5948	.5987	.6026	.6064	.6103	.6141
0.3	.6179	.6217	.6255	.6293	.6331	.6368	.6406	.6443	.6480	.6517
0.4	.6554	.6591	.6628	.6664	.6700	.6736	.6772	.6808	.6844	.6879
0.5	.6915	.6950	.6985	.7019	.7054	.7088	.7123	.7157	.7190	.7224
0.6	.7257	.7291	.7324	.7357	.7389	.7422	.7454	.7486	.7517	.7549
0.7	.7580	.7611	.7642	.7673	.7704	.7734	.7764	.7794	.7823	.7852
0.8	.7881	.7910	.7939	.7967	.7995	.8023	.8051	.8078	.8106	.8133
0.9	.8159	.8186	.8212	.8238	.8264	.8289	.8315	.8340	.8365	.8389
1.0	.8413	.8438	.8461	.8485	.8508	.8531	.8554	.8577	.8599	.8621
1.1	.8643	.8665	.8686	.8708	.8729	.8749	.8770	.8790	.8810	.8830
1.2	.8849	.8869	.8888	.8907	.8925	.8944	.8962	.8980	.8997	.9015
1.3	.9032	.9049	.9066	.9082	.9099	.9115	.9131	.9147	.9162	.9177
1.4	.9192	.9207	.9222	.9236	.9251	.9265	.9279	.9292	.9306	.9319
1.5	.9332	.9345	.9357	.9370	.9382	.9394	.9406	.9418	.9429	.9441
1.6	.9452	.9463	.9474	.9484	.9495	.9505	.9515	.9525	.9535	.9545
1.7	.9554	.9564	.9573	.9582	.9591	.9599	.9608	.9616	.9625	.9633
1.8	.9641	.9649	.9656	.9664	.9671	.9678	.9686	.9693	.9699	.9706
1.9	.9713	.9719	.9726	.9732	.9738	.9744	.9750	.9756	.9761	.9767
2.0	.9772	.9778	.9783	.9788	.9793	.9798	.9803	.9808	.9812	.9817
2.1	.9821	.9826	.9830	.9834	.9838	.9842	.9846	.9850	.9854	.9857
2.2	.9861	.9864	.9868	.9871	.9875	.9878	.9881	.9884	.9887	.9890
2.3	.9893	.9896	.9898	.9901	.9904	.9906	.9909	.9911	.9913	.9916
2.4	.9918	.9920	.9922	.9925	.9927	.9929	.9931	.9932	.9934	.9936
2.5	.9938	.9940	.9941	.9943	.9945	.9946	.9948	.9949	.9951	.9952
2.6	.9953	.9955	.9956	.9957	.9959	.9960	.9961	.9962	.9963	.9964
2.7	.9965	.9966	.9967	.9968	.9969	.9970	.9971	.9971	.9973	.9974
2.8	.9974	.9975	.9976	.9977	.9977	.9978	.9979	.9979	.9980	.9981
2.9	.9981	.9982	.9982	.9983	.9984	.9984	.9985	.9985	.9986	.9986
3.0	.9987	.9987	.9987	.9988	.9988	.9989	.9989	.9989	.9990	.9990

When calculating probabilities using the Standard Normal Distribution Table, it helps to sketch the standard normal curve to get an idea of the probability desired.

Example 1

a. **Consider a data set with a standard normal distribution. Find the probability that a randomly chosen observation is less than 0.85, that is, less than 0.85 standard deviation above the mean.**

b. **Write the meaning of the result in Part a in words, estimating the probability to the nearest percent.**

Solution 1

a. Imagine or sketch a standard normal curve. The desired probability is the area under the curve to the left of the line $z = 0.85$. Use the table directly. Read down column a until you get to 0.8. Go across this row until you get to column 5. The entry there is 0.8023. So $P(z < 0.85) \approx 0.8023$.

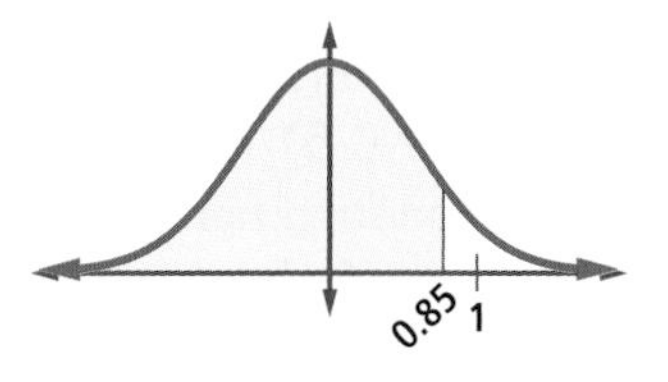

b. 0.8023 is about 80%. About 80% of the values in a standard normal distribution are less than 0.85.

Solution 2

Use a calculator with a standard normal distribution calculation function. The calculator shown here uses the command `normCdf`, with parameters lower bound, upper bound, mean, and standard deviation. Since the left tail of the standard normal distribution is unbounded, you must use the symbol $-\infty$ (negative infinity) for the lower bound.

By adding and subtracting areas and using symmetry properties of the standard normal curve, the Standard Normal Distribution Table can be used to estimate nearly any needed probability.

Activity

Step 1 From the result of Example 1, find $P(z > 0.85)$ and sketch a picture of the area under a standard normal curve represented by this probability.

Step 2 Find $P(0 < z < 0.85)$ and sketch a picture of the area under a standard normal curve represented by this probability.

Step 3 Find $P(-0.85 < z < 0.85)$ and sketch a picture as in Steps 1 and 2.

Step 4 Find another probability from the information of Example 1 and sketch a picture of it.

Because of so many applications to industry, business, education, and testing, it is good to know what percent of normally-distributed data are within one, two, or three standard deviations of the mean. We look at the region one standard deviation around the mean in Example 2, and you are asked to examine others in Questions 9 and 10.

Example 2

About what percent of the data in a standard normal distribution are within one standard deviation of the mean?

Solution Being within one standard deviation of the mean indicates that $-1 < z < 1$. From the symmetry of the standard normal curve,

$$\begin{aligned} P(-1 < z < 1) &= 2 \cdot P(0 < z < 1) \\ &= 2(P(z < 1) - P(z < 0)) \\ &\approx 2(0.8413 - 0.5) \approx 0.6826. \end{aligned}$$

Thus, about 68% of data in a standard normal distribution are within one standard deviation of the mean.

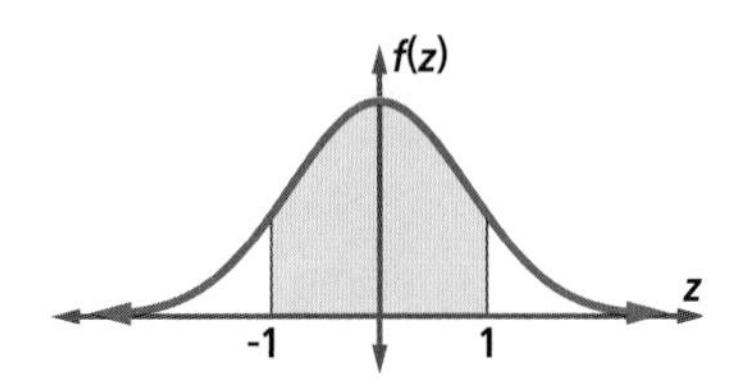

Check Use a calculator to find `normCdf(-1,1,0,1)`. This calculator gives 0.682689.

Recall from earlier courses that the double inequality $-1 < z < 1$ can be rewritten using absolute value as the single inequality $|z| < 1$. In general, for any positive number x, we can write that x is between $-a$ and a as a double inequality: $-a < x < a$; or as a single inequality: $|x| < a$.

QY

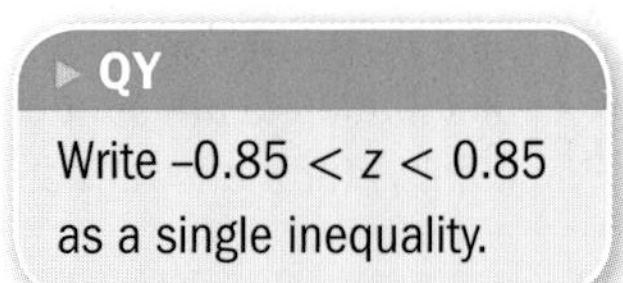

Questions

COVERING THE IDEAS

1. State the mean and standard deviation of the domain variable of the standard normal distribution.

In 2 and 3, use the table on page 694.

2. The graph at the right shows a standard normal curve.
   a. What is the area of the shaded region?
   b. What is the area of the unshaded region below the curve?

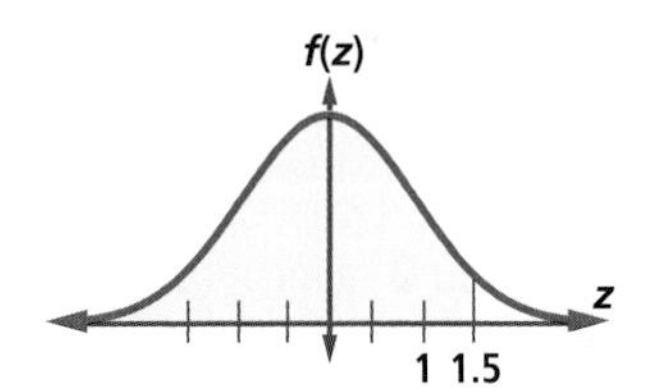

3. Calculate the following probabilities.
   a. $P(0 < z < 1.5)$ b. $P(|z| < 1.5)$ c. $P(z > -1.5)$

In 4–7, evaluate using the Standard Normal Distribution Table on page 694 or a statistics utility. Draw a sketch to support your answer.

4. $P(z < 1.45)$
5. $P(z > 0.06)$
6. $P(-1.4 < z < 0.6)$
7. $P(z > -3)$

8. Find the area between the standard normal curve and the z-axis for $-0.6 \le z \le 0.6$.

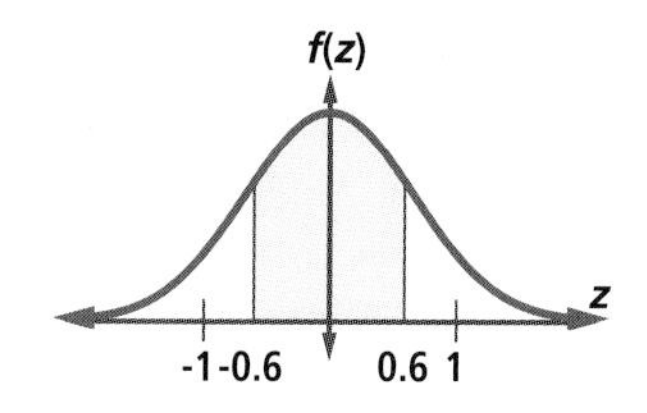

9. To the nearest tenth of a percent, what percent of normally distributed data lies within two standard deviations of the mean?

10. Repeat Question 9 for three standard deviations.

APPLYING THE MATHEMATICS

11. Find the answer to Example 2 by calculating $P(z < 1) - P(z < -1)$.

In 12 and 13, find the value of c satisfying the equation.

12. $P(z < c) = 0.1736$

13. $P(z > c) = 0.2514$

14. Which value of the random variable in a standard normal distribution is exceeded by about 25% of the distribution?

15. **Fill in the Blank** 90% of the observations of a standard normal distribution fall within __?__ standard deviations of the mean.

16. SAT scores were once normally distributed with mean 500 and standard deviation 100. About what percent of scores were above 700?

REVIEW

17. How do the areas under the parent normal curve and the standard normal curve compare? **(Lesson 11-1)**

18. **True or False** All normal curves have a finite area under them equal to 1. **(Lesson 11-1)**

19. If, for a particular binomial probability distribution, $\mu = 40$ and $\sigma = 4.6$, find n and p. **(Lesson 10-7)**

20. Find the z-score for the number 68 in a data set whose mean is 75 and whose standard deviation is 5. **(Lesson 3-9)**

21. The number of calories in one ounce of each of five different kinds of cereal is: 155, 90, 110, 120, 135. **(Lesson 1-6)**
    a. Find the mean and standard deviation of these data.
    b. Let x_i = the number of calories in the ith cereal. Let $z_i = \frac{x_i - m}{s}$, where m is the mean and s the standard deviation in Part a. Find the mean and standard deviation of the z_i-value.

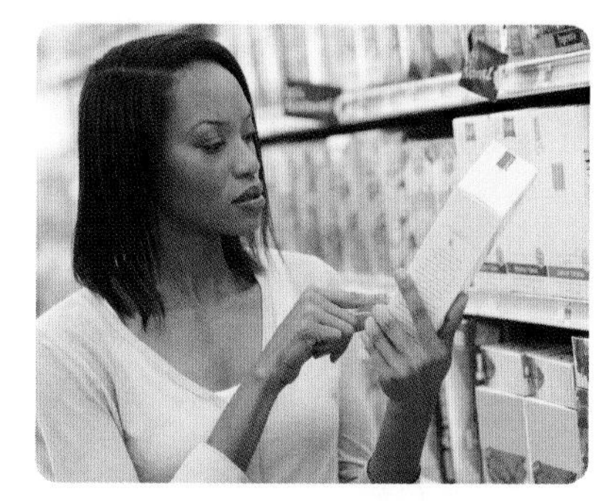

The U.S. Food and Drug Administration has required easy-to-read nutrition labels on food since 1994.

EXPLORATION

22. a. Examine the differences between successive values in the first row of the Normal Distribution Table. Estimate the following.
    i. $P(z < 0.035)$
    ii. $P(z < 0.068)$
    b. Why are the differences referred to in Part a almost constant?
    c. Why are differences between successive values in other rows of the table not constant?

QY ANSWER

$|z| < 0.85$

Lesson 11-3 Other Normal Distributions

Vocabulary

standardizing a variable

▶ **BIG IDEA** A *z*-score transformation of raw data will transform any normal distribution into the standard normal distribution.

Transforming Any Normal Curve into the Standard Normal Curve

Recall that if each number in a data set is translated by a constant, the mean of the data set is also translated by that constant, but the standard deviation remains unchanged. For instance, consider the fact that the heights of adult men in the Netherlands are approximately normally distributed with mean $\mu = 183$ cm and standard deviation $\sigma = 6.7$ cm. If $\mu = 183$ cm is subtracted from each height, the mean of the resulting data set is 183 cm − 183 cm = 0 cm. There is no change in the standard deviation.

 QY

If each translated height is divided by 6.7 cm, the standard deviation of the resulting data set is 1. Thus, the transformation $x \to \frac{x - 183}{6.7}$ maps the data set of heights with $\mu = 183$ cm and $\sigma = 6.7$ cm, whose distribution is at the right in the graph below, onto a data set with mean $\mu = 0$ cm and standard deviation $\sigma = 1$ cm, whose distribution is at the left in the graph below.

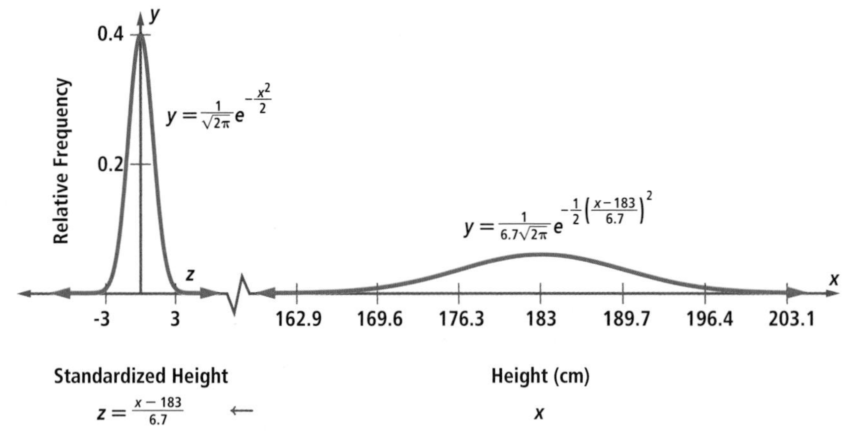

But notice that this transformation maps *x* onto its *z*-score! The Graph-Standardization Theorem shows that the distribution of *z*-scores, shown above, is the standard normal curve. The same argument could be used with any mean and any standard deviation, leading to the following theorem.

Mental Math

Calculate the *z*-score of a score of 83 on an exam on which the mean is 80 and the standard deviation of scores is 4.

▶ **QY**

What heights are within 1 standard deviation of the mean height of men in the Netherlands?

Standardization Theorem

If a variable x has a normal distribution with mean μ and standard deviation σ, then the variable

$$z = \frac{x - \mu}{\sigma}$$

has the standard normal distribution.

The process of getting z-values from an original data set by applying the transformation $x \rightarrow \frac{x - \mu}{\sigma}$ is often referred to as **standardizing** the variable. By standardizing the domain of normal distributions, you can determine probabilities by using the procedures in Lesson 11-2.

Caution: the Standardization Theorem applies only to sets of raw data that are normally distributed. The z-score transformation of data preserves the shape of the raw-score distribution. If the original distribution is *not* normal, then the z-score transformation will *not* provide the standard normal distribution, even though the z-score mean is 0 and standard deviation is 1.

Standardizing Variables to Find Probabilities

Many measurements of humans provide normal distributions in appropriate populations. When adults in a country are separated by gender and ethnicity, men's heights and women's heights each form distributions that are modeled by a normal distribution.

Example 1

Use the information given on page 698. An adult Dutch male is selected randomly. What is the probability that the man is less than 188 cm tall?

Solution Because the set of heights x of adult Dutch males has an approximately normal distribution with a mean of 183 cm and standard deviation of 6.7 cm, the variable $z = \frac{x - 183}{6.7}$ has a standard normal distribution. The z-score for $x = 188$ is $z = \frac{188 - 183}{6.7} \approx 0.75$. This indicates that a height of 188 cm is about 0.75 standard deviation above the mean of 183 cm.

Note that the two curves on the next page are centered around their mean values, 183 and 0, respectively. The two shaded areas in the graphs are equal. Because each area represents a probability, the two probabilities are equal.

(continued on next page)

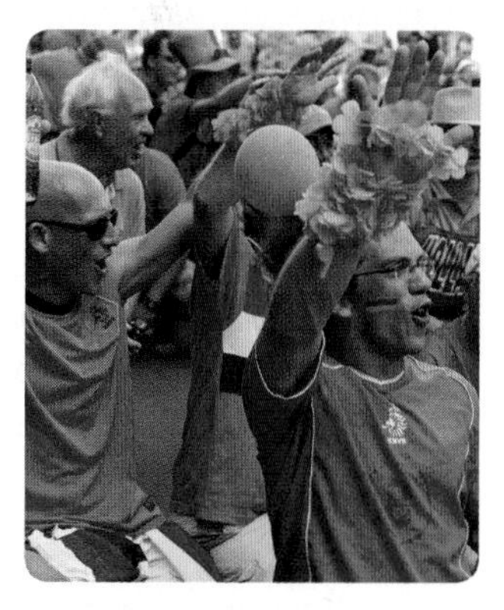

Dutch men, seen here supporting their football team, tend to be taller than men in most other countries.

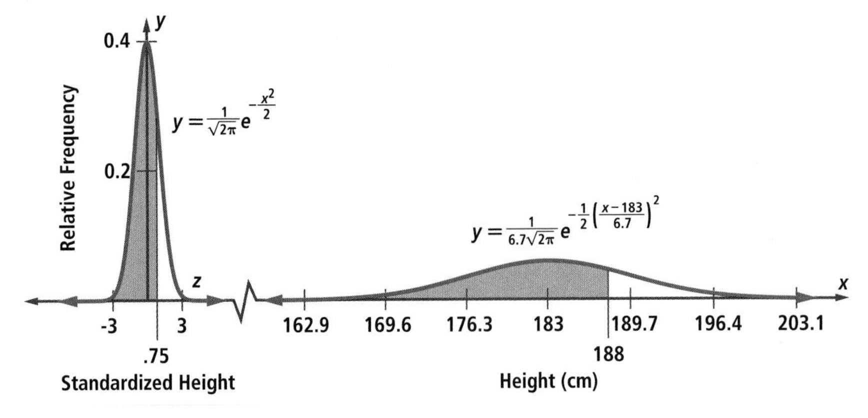

From the table of values for the standard normal distribution,

$$P(x < 188) \approx P(z < 0.75)$$
$$\approx 0.7734.$$

The probability of a randomly selected adult Dutch male being less than 188 cm tall is about 0.77. In other words, just over $\frac{3}{4}$ of all adult males in the Netherlands are less than 188 cm tall.

Check Use a calculator to compute standard normal distribution probabilities. The first line at the right shows the answer for transformed data. The second line shows the command applied to raw data.

$\text{normCdf}\left(-\infty, \frac{188-183}{6.7}, 0, 1\right)$	0.772248
$\text{normCdf}(-\infty, 188, 183, 6.7)$	0.772248

In Example 2, we want to find the probability that the amount of popcorn in filled popcorn packets lies in a particular interval.

GUIDED

Example 2

A machine fills bags with popcorn. If the bags are underfilled, there will be complaints from consumers. If the bags are overfilled, they cannot be closed and sealed by the next machine in the production line. The bag states that there are 7 ounces in a bag. The company has chosen to fill each bag with 8 ounces of popcorn on average. Assume the quantity of popcorn in each bag is normally distributed with a standard deviation $\sigma = 0.6$ ounces.

a. **What percent of the bags are likely to be under the stated weight of 7 ounces?**

b. **If a manager purchases empty bags that can hold 10 ounces fully filled, then what proportion would be filled over capacity?**

c. **What proportion of bags would contain between 7 and 10 ounces of popcorn?**

d. **If a manager decides that at most 1% of bags should be overfilled, then what should be the capacity of the bag used in the packaging process?**

One of the oldest American foods
Archeologists have found 4,000-year-old ears of popcorn in New Mexico.

Solution 1

a. Sketch a diagram like the one below. Mark the value for 7 ounces on the right graph and shade to the left.

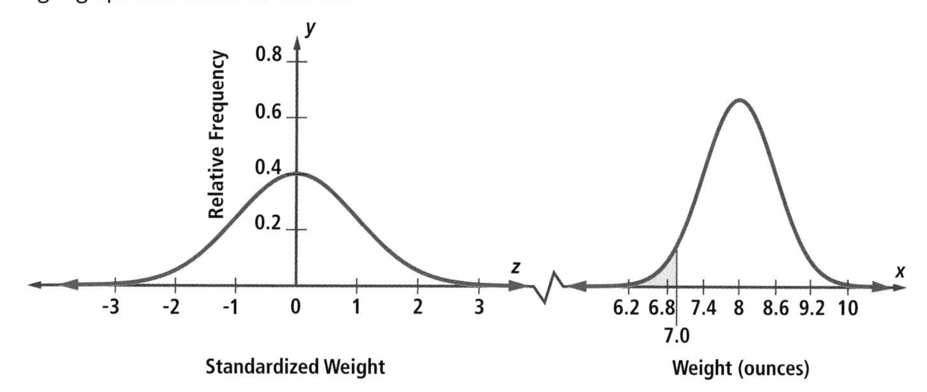

To find the z-score, calculate $\frac{7-?}{?}$. So $z \approx$ _?_. Mark this value on the left graph and shade to the left. Consult the Standard Normal Distribution Table to find the probability that a bag is underweight.

$P(x < 7) = P(z <$ _?_$) \approx$ _?_

b. $P(x > 10) = P(z >$ _?_$) \approx 1 -$ _?_ $\approx$ _?_

c. In Part a you found the lower tail and in Part b you found the upper tail. So in this part, $P(7 < x < 10) = 1 - P(x <$ _?_$) - P(x >$ _?_$) \approx$ $1 -$ _?_ $-$ _?_ $\approx$ _?_.

d. If there is 1% in the upper tail, then there is _?_% below the tail. Look for this percent in the Standard Normal Distribution Table to find that the corresponding z-score is _?_.

Since $z = \frac{x-8}{0.6}$; _?_ $= \frac{x-8}{0.6}$, from which $x =$ _?_. So the manager can use _?_ ounce bags and have at most 1% be overfilled.

Solution 2 Use a statistics utility to find the various probabilities. Note that the solution to Part d uses a calculator command whose arguments are p, μ, and σ, respectively.

normCdf(-∞,7,8,0.6)	0.04779
normCdf(10,∞,8,0.6)	0.000429
normCdf(7,10,8,0.6)	0.951781
invNorm(0.99,8,0.6)	9.39581

Approximating a Binomial Distribution with a Normal Distribution

In the binomial probability distribution B with $n = 100$ and $p = 0.2$, the random variable has mean $\mu = 100(0.2) = 20$ and standard deviation

$$\sigma = \sqrt{npq} = \sqrt{100(0.2)(0.8)} = \sqrt{16} = 4.$$

At the right is the graph for this binomial distribution. The curve for the normal distribution with $\mu = 20$ and $\sigma = 4$ is superimposed on the graph. For these values of n and p, the normal distribution appears to approximate the binomial distribution quite well.

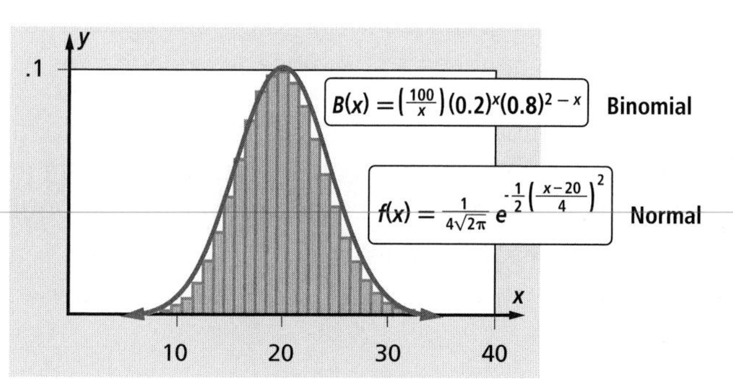

In general, if n is quite large, or when p is close to 0.5, a normal curve approximates a binomial distribution quite well. But, if n is small and p is near 0 or 1, then the binomial distribution is not well approximated by a normal distribution. A rule-of-thumb is that a binomial distribution can be approximated by a normal distribution with mean np and standard deviation $\sqrt{npq}$ provided np and nq are each at least 10.

The graphs of binomial distributions in Chapter 10 provide evidence for this approximation. We return to the situation of testing whether or not a coin is fair. Notice how the use of the normal approximation to the binomial distribution allows us to deal with the large numbers of trials that are more realistic in real tests.

Example 3

A coin is tossed 100,000 times and 50,482 tails result. Using the 0.01 significance level, test the hypothesis that the coin is fair.

Solution Write the null hypothesis first. H_0: The coin is a fair coin.

The number of tails is binomially distributed. If the coin is fair, the probability of a tail is $p = 0.5$. In this experiment, $n = 100{,}000$. The experiment has both np and nq greater than 10, so a normal distribution can approximate the binomial. The binomial random variable has mean $np = 50{,}000$ and standard deviation $\sqrt{npq} = \sqrt{100{,}000 \cdot 0.5 \cdot 0.5} = \sqrt{25000} \approx 158.11$.

Change to z-scores to model the binomial distribution with a standard normal distribution. The z-score equivalent of 50,482 is $\frac{50{,}482 - 50{,}000}{158.11}$, which is about 3.049. That is, 50,482 is about 3.049 standard deviations above the mean. Find $P(z \geq 3.049)$. To use the Standard Normal Distribution Table, round the z-score to the nearest hundredth.

The table indicates $P(z < 3.05) \approx 0.9989$.

So $\quad P(z \geq 3.05) \approx 1 - 0.9989 = 0.0011$.

If the coin is fair, then the probability of tossing 50,482 or more tails in 100,000 tosses is 0.0011. Additionally, the probability of 50,482 or more heads is also 0.0011 and is included as in Lesson 10-8. The total probability is 0.0022, which is less than the significance level 0.01, so reject the null hypothesis H_0. The coin appears to be biased towards tails.

Notice how much simpler the calculation is for the normal approximation than is the calculation for the exact answer using the binomial distribution. The exact answer would require calculating a sum of hundreds of terms!

Questions

COVERING THE IDEAS

1. **Multiple Choice** Each of the data sets graphed below can be converted to z-scores. Which has a z-score distribution that can be approximated by the standard normal distribution?

A
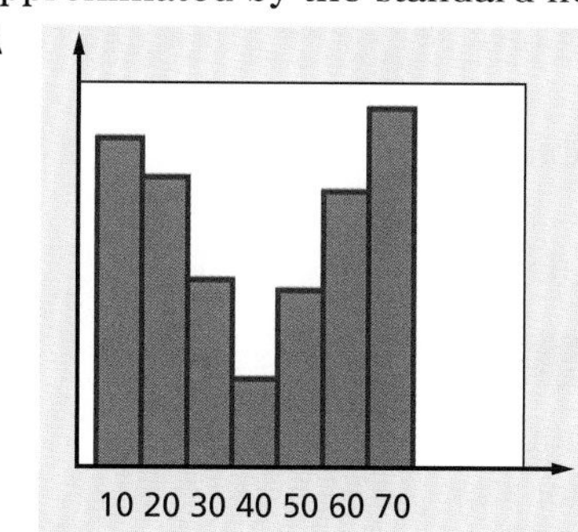

B
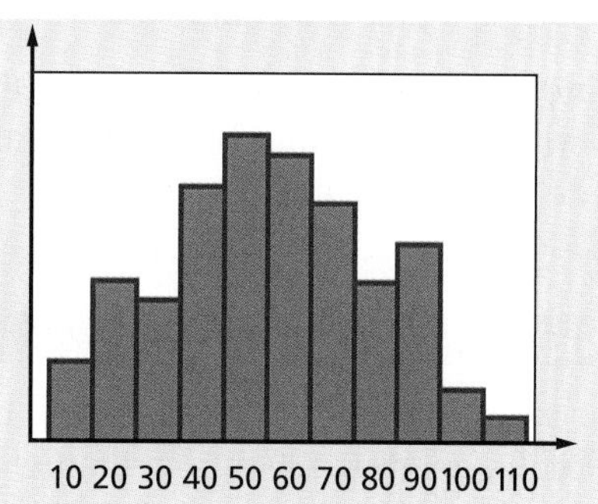

C
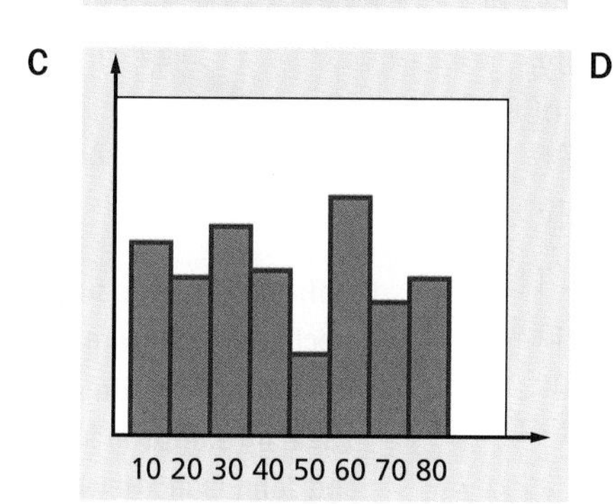

D
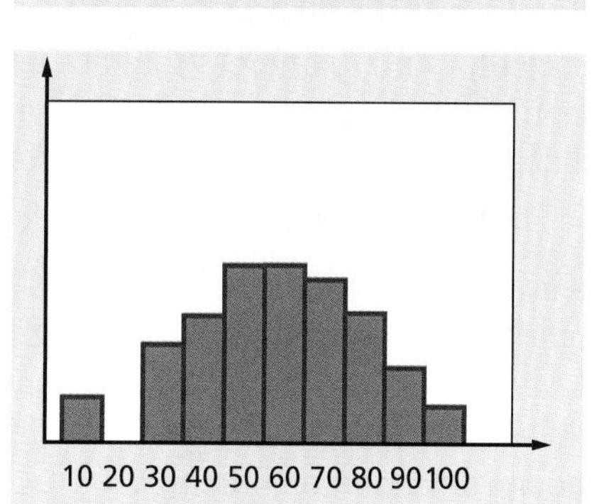

In 2–6, use this information. In Serbia/Montenegro, adult male heights have $\bar{x} = 186$ cm and $s = 6.0$ cm. In Vietnam, $\bar{x} = 163$ cm and $s = 6.7$ cm. Assume heights in each country are normally distributed.

2. If the transformation $x \to \frac{x - 186}{6.0}$ is applied to each data point for Serbia/Montenegro, what type of distribution results?
3. What proportion of men are less than 180 cm tall in
   a. Serbia/Montenegro? b. Vietnam?
4. Out of 200 male employees of a company in Serbia/Montenegro, how many can the coach of the company basketball team expect to be taller than 198 cm?
5. One airline's flight attendants must be 152.4 to 190.5 cm tall. What percent of Vietnamese men are likely to be an appropriate height?
6. Give an equation for the normal distribution of adult male heights in each country.

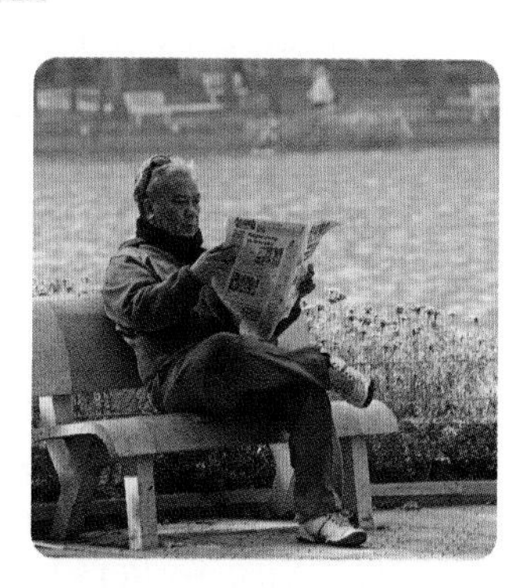

Reading the news in Vietnam

7. The advertisement on a package of peanuts says: "Money back guaranteed if you have an underweight packet." Suppose the amount of peanuts is normally distributed with a mean of 8 ounces and standard deviation of 0.2 ounces. What amount should be printed on the package so that only 1% of packets are under the advertised weight?

8. Suppose that the length ℓ of a speech is normally distributed with mean 3 minutes and standard deviation of $\frac{1}{2}$ minute.
   a. Identify a transformation that can map each ℓ to a z-score.
   b. What are the mean and standard deviation of the z-scores?

9. A coin is tossed 100,000 times and 50,297 heads result. At the 0.05 level, test the hypothesis that the coin is fair.

10. Use the $np \geq 10$ and $nq \geq 10$ criteria to decide whether the binomial distribution with the given number of trials and probability of success can be modeled by a normal distribution.
    a. $n = 100, p = 0.89$
    b. Example 3 in Lesson 10-8

APPLYING THE MATHEMATICS

11. Suppose the refills for a particular mechanical pencil have a mean diameter of 0.5 mm. Refills below 0.485 mm in diameter do not stay in the pencil, while those above 0.520 mm do not fit in the pencil at all. If a firm makes refills with diameters that are normally distributed with a mean of 0.5 mm and a standard deviation of 0.007 mm, what is the probability that a randomly chosen refill will
    a. be too large?
    b. be too small?
    c. fit correctly?

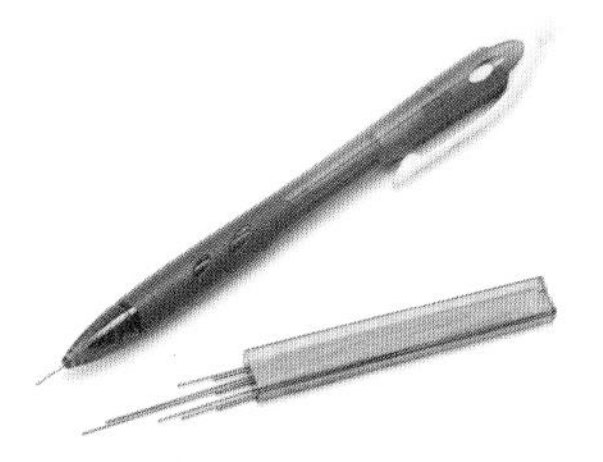

Pencil It In The first mechanical pencil mechanism was patented in Britain in 1822 by Sampson Mordan and John Isaac Hawkins.

12. A theme park's rides are rated as mild, moderate, and max. They have restrictions requiring that passengers have heights of at least 42 inches, 48 inches, and 54 inches, respectively. Suppose the population of children attending the park has a mean height of 53 inches with a standard deviation of 4 inches.
    a. What percent of children can go on all rides?
    b. What percent of children can participate in only mild and moderate rides?
    c. If a child is chosen randomly, what is the probability that that child can only go on mild rides?
    d. What percent of children are excluded from all rides?

13. At the 0.01 level, test the hypothesis that 85% of New Yorkers believe traffic congestion is a problem ($p = 0.85$). Use a poll of 1162 people conducted by Quinnipiac University, which found that 1057 New Yorkers believed that traffic congestion is a problem. State any conditions necessary to compute your conclusion.

In 14 and 15, assume that the time for a seed to germinate is normally distributed with a mean of 14 days and a standard deviation of 2 days.

14. What percent of seeds germinate within 17 days?

15. What is the probability that a randomly chosen seed will not have germinated within 21 days?

16. A coin is tossed 500 times. Suppose the coin is fair.
    a. Find the probability that there are between 243 and 257 heads.
    b. Find the probability that the number of heads differs from 250 by more than 25.

17. The makers of the SAT reported that among the 1,465,744 SAT test takers in 2006, 454,705 actually scored from 500 to 590 on the mathematics section. Suppose that the makers of the SAT did not publicize this data but we knew the data to be normally distributed. Test takers of the SAT know that scores are rounded to the nearest ten points (meaning a score such as 518 is not possible).
    a. In 2006, the scaled scores on the SAT mathematics section were reported to have mean 518 and standard deviation 115. From this information, estimate the probability that a randomly selected student in 2006 had an SAT mathematics score from 495 to 594. (We use these numbers because we are making an estimate using a continuous model rather than the discrete values actually used by the SAT).
    b. Calculate the residual between your estimate and the actual relative frequency of students who scored between 500 and 590.

REVIEW

In 18 and 19, suppose the variable z has a standard normal distribution.

18. Find the area of the shaded region at the right.

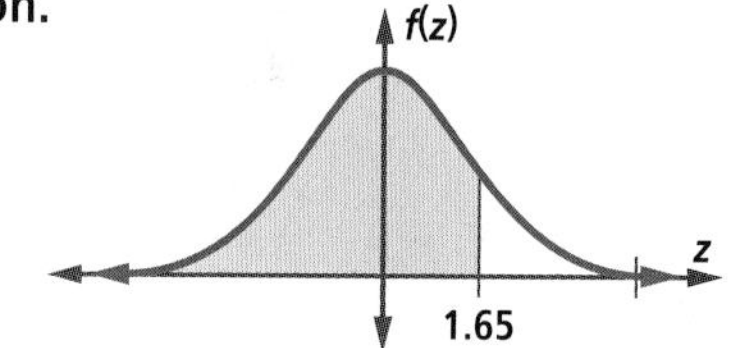

19. Determine $P(z \geq 2.67)$. **(Lesson 11-2)**

20. **Fill in the Blank** In a standard normal distribution, 95% of the observations are within __?__ standard deviations of the mean. **(Lesson 11-2)**

21. How many combinations of four faculty and two students can be formed from a group of 20 faculty and 300 students? **(Lesson 10-1)**

In 22 and 23, suppose that a system on a spacecraft is composed of three independent subsystems, X, Y, and Z. The probability that these will fail during a mission is 0.002, 0.006, and 0.003, respectively. (Lessons 6-3, 6-2, 6-1)

22. If the three are connected in series, as shown at the right, a failure in any one of the three will lead to a failure in the whole system. What is the probability that the system will be reliable—that is, it will *not* fail?

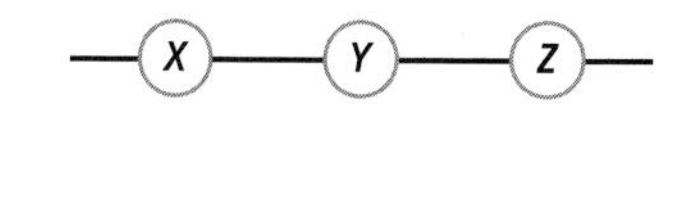

23. Suppose the subsystems are connected as shown at the right. In this case, both X and either Y or Z must be reliable for the whole system to be reliable. What is the probability that this whole system will *not* fail?

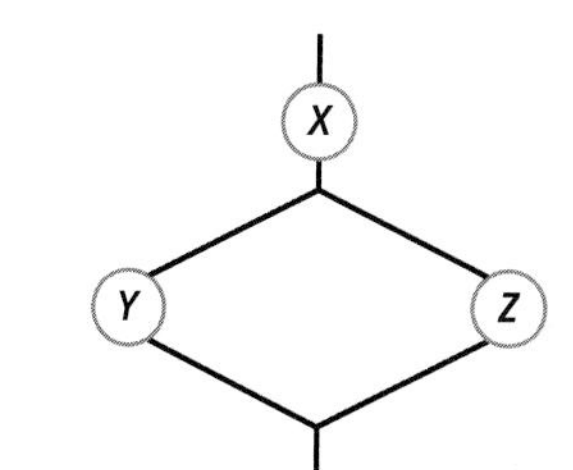

24. Consider the set {0, 1, 2, 3, 4, 5, 6, 7, 8, 9}. **(Lessons 1-6, 1-2)**
    a. Find the mean of the numbers in the set.
    b. Find the standard deviation of the numbers in the set.

EXPLORATION

25. Some books give the following rule of thumb for finding the standard deviation of a normally distributed variable: $s \approx \frac{\text{range}}{6}$. Explain why this rule of thumb works.

QY ANSWER

176.3 cm to 189.7 cm

Lesson

11-4 The Central Limit Theorem

Vocabulary

simple random sample, SRS
sampling distribution of the mean

▶ BIG IDEA For a specific sample size, means of many samples from a population are normally distributed even though the distribution of the population may not be normal.

It is often too costly, too difficult, or impossible to study an entire population. For these reasons, usually a sample is taken and inferences about the population are made from the sample data. However, the characteristics of the sample are not likely to be identical to the corresponding characteristics of the population, even if the sample has been randomly selected or carefully selected to be representative of the population. For instance, the means of various random samples from the same population are likely to differ somewhat from each other and from the population mean. To understand how measures of a sample are related to the corresponding measures of the population from which it was taken, it is helpful to study how different samples drawn randomly from the same population vary, that is, to study the distribution of sample means.

Mental Math

To the nearest percent, what percent of values in a normal distribution are within

a. 1 standard deviation of the mean?

b. 2 standard deviations of the mean?

c. 3 standard deviations of the mean?

An Example of a Distribution of Sample Means

In Chapter 1, you saw population pyramids for various countries. Below is a graph of the percents of the U.S. population age 15 or over in 2007, categorized into various age groups, to the nearest percent. It is the kind of distribution that a person wanting to sell products or ideas might use in designing advertisements.

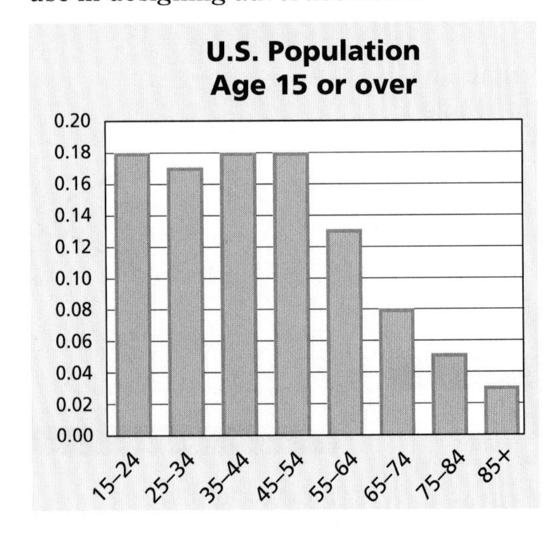

Age Range	% of Total
15–24	0.18
25–34	0.17
35–44	0.18
45–54	0.18
55–64	0.13
65–74	0.08
75–84	0.05
85+	0.03

If we take 20 as the mean age for individuals aged 15–24, 30 as the mean age for ages 25–34, and so on, with 90 as the mean age for ages 85+, then we can calculate the mean and standard deviation of the ages for this population. (Note that the U.S. population age 15 or over is the statistical *population* for this situation.)

$$\mu = 0.18 \cdot 20 + 0.17 \cdot 30 + 0.18 \cdot 40 + 0.18 \cdot 50 + 0.13 \cdot 60 + 0.08 \cdot 70 + 0.05 \cdot 80 + 0.03 \cdot 90 = 45.00$$

$$\sigma = \sqrt{\sum_{i=1}^{n} x_i^2 \cdot P(x_i) - \mu^2}$$

$$= \sqrt{400 \cdot 0.18 + 900 \cdot 0.17 + \cdots + 8100 \cdot 0.03 - 45.0^2} \approx 19.$$

So the mean age of a person in this 15-and-over population is 45.0 and the standard deviation is 19. But how do these compare to means and standard deviations of samples taken from this population?

A **simple random sample** or **SRS** is a sample of size n chosen from a population in such a way that every sample of size n has an equal chance of being the sample selected. If we take many SRS of size n and find the mean of each sample, we will have a set of sample means. Activity 1 explores how the size n of the samples affects the relationship between the mean and standard deviation of the population and the mean and standard deviation of a set of sample means.

Activity 1

MATERIALS sampling distribution application provided by your teacher

Step 1 Set the sample size to 5 at the top left corner of the screen. Use the application to sample 5 people randomly from the 15-and-over population and record the mean of the 5 ages. This is equivalent to one trial. You can choose the number of trials at the bottom of the screen.

Step 2 a. Reset the trials, and run a simulation of 250 trials where the sample size is 5.

b. A graph of your 250 sample means is created. An example is shown at the right. Does the distribution of sample means appear normal? Make a rough sketch of the graph.

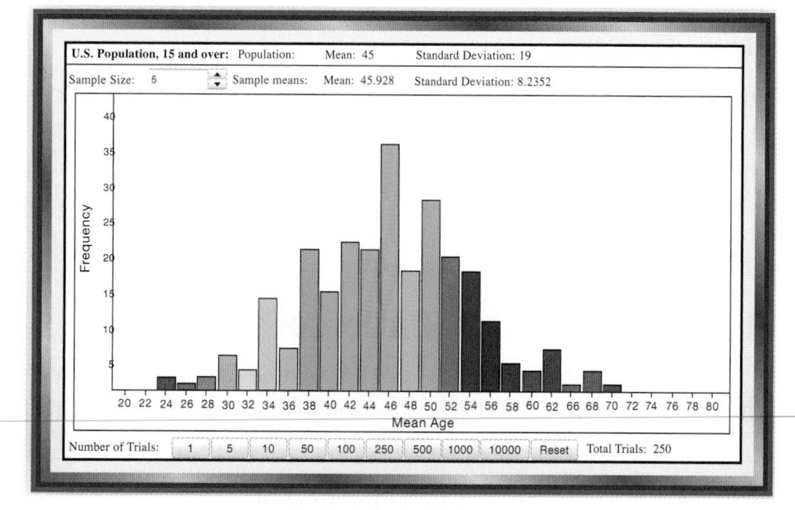

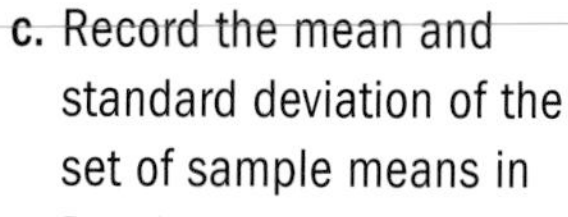

c. Record the mean and standard deviation of the set of sample means in Part b (shown at the top of the screen). Compare these statistics with the population mean and standard deviation.

(continued next page)

Step 3 **a.** Reset the trials and choose a sample size of 15. Again run 250 trials.

b. Does the histogram appear normal? Compare this histogram (of sample size 15) with the rough sketch of the graph from Step 2b. What do you notice about the two graphs? Compare your results to others in your class. How did the distribution of sample means change when the sample size increased from 5 to 15?

c. Record the mean and standard deviation of the set of 250 sample means of size 15. Compare these statistics to the population mean and standard deviation. How does this contrast with your answer from Step 2c?

d. Make a conjecture about the mean and standard deviation of the set of sample means in comparison to the mean and standard deviation of the population based on your answer to Part b. How does the sample size affect this relationship?

Step 4 Use the sampling distribution application to simulate a set of 10,000 sample means when 30 people are randomly selected. Record the mean and standard deviation of the set of sample means. Compare these statistics with the population statistics. Do these results support the conjecture you made in Step 3d?

In Activity 1, you saw three different means, notated as follows:

- the mean μ of the original population or random variable (calculated on page 707);
- the means $\bar{x}$ of individual samples (e.g., the mean in Step 1); and
- the means $\mu_{\bar{x}}$ of the sets of sample means (Steps 2c, 3c, and 4).

The notation for standard deviations follows the same pattern. The original population has standard deviation σ, each separate sample has a standard deviation denoted s, and $\sigma_{\bar{x}}$ indicates the standard deviation of the set of sample means.

Notice that although the distribution of the original population (the ages) used in Activity 1 is not normal, the distribution of sample means appears to be. Furthermore, the mean $\mu_{\bar{x}}$ of the set of sample means is almost equal to the mean μ of the population and the standard deviation of the set of sample means decreases as the sample size increases.

The Sampling Distribution of the Mean

In Activity 1, you examined distributions of sets of sample means, but you examined only a subset of the possible samples. For example, in Step 2, you took only 250 samples of size 5, but there are about 239 million people in this population for a total of $\binom{239{,}000{,}000}{5}$ or about $6.50 \cdot 10^{39}$ possible samples! When *all* the SRS of a specific size are taken, the distribution of means from all possible samples is called the **sampling distribution of the mean.**

Because of the large number of possible samples, it is difficult to study the complete sampling distribution of the mean. However, technology can be used to generate a large number of samples, allowing you to see the important characteristics of the sampling distribution.

Activity 2

MATERIALS spreadsheet software or statistics utility

Draw samples from a uniform distribution of digits from 1 to 6, as in a fair die. Note that the probability of rolling any number on a fair die is $\frac{1}{6}$.

Step 1 Define the population to be the digits 1–6. Generate 1000 random samples of size 15 and compute the mean of each sample. List the 1000 means in the column of a spreadsheet. Label this column "smallsamps," as shown at the right.

◇	A	B	C
1	Smallsamps		
2	3.8		
3	2.8		
4	4.1333		
5	3.7333		
6	3.5333		
7	3.0667		
8	3.2		
⋮	⋮		
1000	3.4		

The spreadsheet evaluates the mean of a random sample of size 15.

Step 2 Compute the mean and standard deviation of the "smallsamps" data.

Step 3 Make another column called "largesamps"; compute the means of 1000 random samples each of size 100.

Step 4 Compute the mean and standard deviation of the "largesamps" data.

Step 5 Draw histograms of both distributions of means on comparable scales (1 to 6) using bins of width 0.2. How do the means, the standard deviations, and the shapes of the histograms compare?

Your results from Step 5 of Activity 2 may be similar to the graphs below. Notice that as the size n of the sample increases, the mean $\mu_{\bar{x}}$ of the sample means does not vary much (3.501 to 3.502), but the standard deviation $\sigma_{\bar{x}}$ decreases (0.424 to 0.162). This supports the relationships discovered in Activity 1.

We have superimposed the normal curve with the corresponding mean and standard deviation on each histogram. Notice that both of the distributions are nearly bell-shaped, but the histogram with larger samples fits its normal curve more closely.

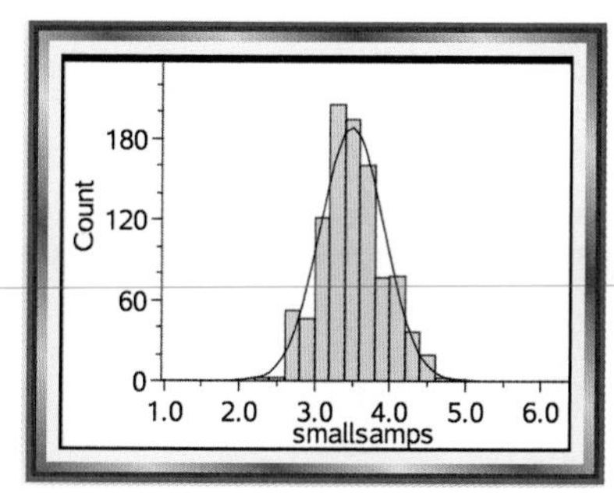

$n = 15$
$\mu_{\bar{x}} = 3.501$, $\sigma_{\bar{x}} = 0.424$

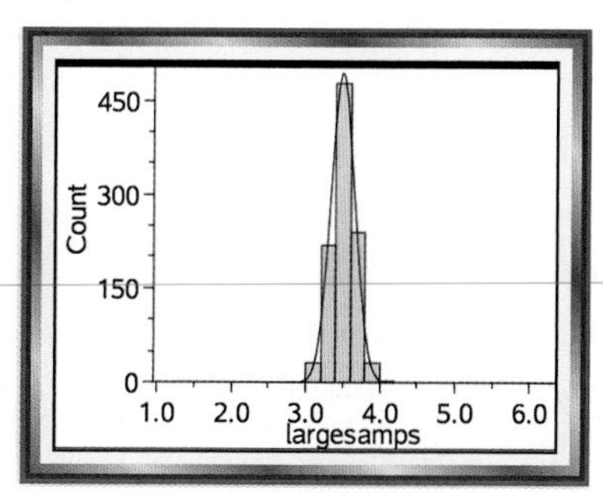

$n = 100$
$\mu_{\bar{x}} = 3.502$, $\sigma_{\bar{x}} = 0.162$

These relationships among the sample size n, the means μ of the population and $\mu_{\bar{x}}$ of the set of sample means, and the standard deviations σ of the population and $\sigma_{\bar{x}}$ of the set of sample means are quite simple. They were first proved by the French mathematician Pierre Simon Laplace (1749–1827), and are summarized in a result called the *Central Limit Theorem.* The proof of this theorem is beyond the scope of this course.

Central Limit Theorem (CLT)

Suppose *all* random samples of size n are selected from a population with mean μ and standard deviation σ. Then, as n increases, the distribution of sample means approaches a normal distribution with mean μ and standard deviation $\frac{\sigma}{\sqrt{n}}$.

We know from Activities 1 and 2 that the standard deviation of the set of sample means is less than the standard deviation of the population and decreases as n increases. The Central Limit Theorem tells us that, in fact, as n increases, $\sigma_{\bar{x}}$ approaches the value $\frac{\sigma}{\sqrt{n}}$. Activity 3 verifies this part of the Central Limit Theorem with the statistics generated in Activity 1.

Activity 3

Refer to your data from Activity 1.

Step 1 For the sample sizes 5, 15, and 30, give the values for each $\sigma_{\bar{x}}$.

Step 2 Calculate the value of $\frac{\sigma}{\sqrt{n}}$ for $n = 5$, 15, and 30. How do these calculated values relate to each $\sigma_{\bar{x}}$ calculated in Activity 1?

The Central Limit Theorem is *central* because it deals with the mean of a population, a measure of *center.* It is a *limit* theorem because it deals with what can be expected for samples of larger and larger size.

The importance of the Central Limit Theorem is its wide applicability. Even if the underlying population distribution is not normal, as in Activity 1, the Central Limit Theorem states that the means of samples of a particular sufficient size will be normally distributed. However, the Central Limit Theorem does not specify exactly what that sufficient size is. Fortunately, there is a rule of thumb for determining how large the samples should be so that the distribution of sample means is close enough to normal. Its proof is also beyond the scope of this course.

Central Limit Theorem Corollary

Consider a population from which random samples of size n are taken. Then the distribution of sample means is approximately normal whenever one of the following occurs:

a. the population itself is normally distributed, or

b. the sample size $n \geq 30$.

Example

A manufacturer claims that its fluorescent light bulbs have a mean life of 2000 hours with standard deviation of 300 hours. A laboratory purchases 64 bulbs to test the claim. What is the shape, mean, and standard deviation for the sampling distribution of the mean for samples of size $n = 64$?

Fluorescent lights were introduced to the U.S. public in 1939 at the New York World's Fair and the San Francisco Golden Gate Exposition.

Solution From the given information, we know that $\mu = 2000$ hours and $\sigma = 300$ hours.

The sample size $n = 64$ is larger than 30, so the sampling distribution of the mean is approximately normal.

The mean of the set of sample means is $\mu_{\overline{x}} = \mu = 2000$ hours.

The standard deviation of the set of sample means is $\sigma_{\overline{x}} = \frac{300}{\sqrt{64}} = 37.5$ hours.

Questions

COVERING THE IDEAS

1. Explain the difference between a sample and a sampling distribution.

In 2 and 3, consider the distribution of the U.S. population aged 15 or over. Suppose two simulations *A* and *B* are conducted where *A* has 500 random samples each of size 3, and *B* has 500 random samples each of size 25.

2. **Fill in the Blank** The means of the sample means from the two simulations will be approximately __?__.
3. Which simulation is likely to yield the set of sample means with the larger standard deviation?

In 4–6, True or False. Consider a population from which a very large number of simple random samples of size 40 is taken. Explain your choice.

4. The mean of the sample means approximates the population mean.
5. The standard deviation of the sample means approximates the standard deviation of the population divided by $\sqrt{40}$.
6. The shape of the distribution of the sample means approximates the shape of the original population distribution.
7. The distribution of heights of young adult women in the United States can be modeled by a normal distribution with $\mu = 64.1$ inches and $\sigma = 2.6$ inches. A medical study is conducted in which women are randomly selected from this population in sample sizes of 200, and the mean height of the women in each sample is recorded. Give the shape, mean, and standard deviation for the sampling distribution of the mean for samples of size 200.

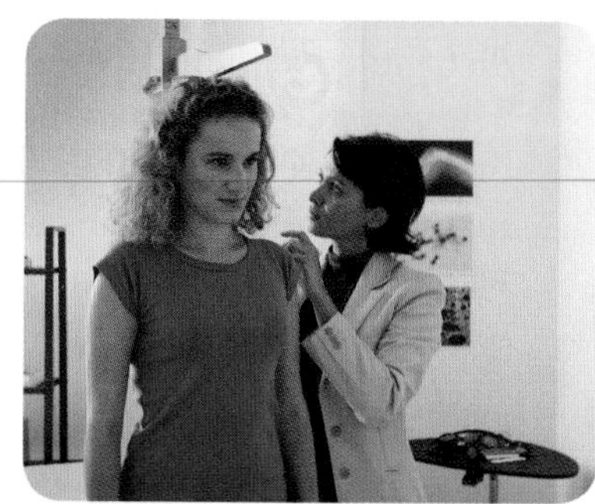

8. The dates on pennies in a sack are recorded. The distribution is graphed at the right. The dates run from a minimum of 1948 to a maximum of 2001. The mean date for all pennies in the bag is 1992.4 and the standard deviation is 9.99. Samples of pennies are drawn from the bag and the mean date of each sample is recorded.

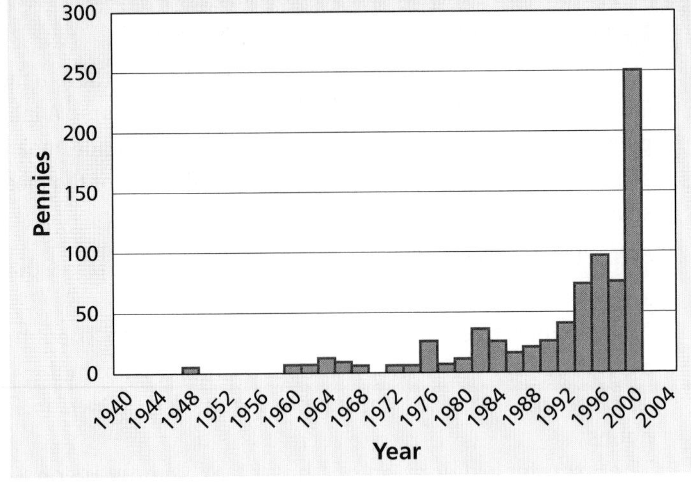

   a. Why might the distribution of sample means for samples of size 10 not be normal?
   b. What is the smallest sample size that assures a normal distribution approximates the sampling distribution of the mean?
   c. What would be the mean and standard deviation for the set of sample means for samples of size 45?

APPLYING THE MATHEMATICS

9. Consider a bowler in a league who keeps track of individual games, as well as average scores for three games played each time the league meets. The league meets 15 weeks during a season.
   a. How does the mean of the 45 games bowled compare to the mean of the 15 weekly averages?
   b. Which will have the larger standard deviation, the scores of the 45 games or the 15 weekly averages?

10. To truly know if the mean of the set of sample means will eventually equal the population mean, you need to consider all possible samples of size n from a given population.
    a. How many samples of size 4 would need to be taken from a population of size 1000?
    b. How does knowing the mean of the set of sample means help avoid having to take all possible samples of size n?

REVIEW

11. An appliance manufacturer estimates the time before a new refrigerator needs new parts to be approximately normally distributed with mean 6 years and standard deviation 11 months. The manufacturer guarantees to provide spare parts to the original purchaser within eight years of the purchase date. If the manufacturer's estimates are correct, what percent of refrigerators will need parts before the guarantee expires? **(Lesson 11-3)**

12. In Lesson 6-9, the fairness of the coin used in Dr. Kerrich's coin flipping experiment, when 5067 heads appeared in 10,000 flips, was tested using the chi-square statistic. Now, test the hypothesis that the coin is fair at the 0.01 significance level using
    a. binomial probabilities.
    b. a normal approximation for the binomial. **(Lessons 11-3, 10-8)**

13. Consider the standard normal curve for variable z. Determine the probability that z is between
    a. -1 and 1.
    b. -1.65 and 1.65.
    c. -1.96 and 1.96.
    d. -2.58 and 2.58. **(Lesson 11-2)**

14. **True or False** The area between the x-axis and the parent normal curve is $\sqrt{\pi}$ whereas the area between the x-axis and the standard normal curve is 1. **(Lesson 11-1)**

15. A researcher is testing babies for color preference. There are twenty balls of identical size in a box; 10 are red and 10 are blue. A baby is observed picking up four balls from the box. She chooses three red balls and one blue ball.
    a. State a null hypothesis H_0 about the baby's color preference.
    b. What outcomes should be considered in deciding whether the null hypothesis should be rejected?
    c. Determine whether the null hypothesis should be rejected at the 0.05 significance level. **(Lessons 10-8, 10-1, 6-4)**

16. Factor $10a^2 - 8a + 5ab - 4b$, if possible. **(Lesson 7-8)**

17. The following are the average monthly high temperatures in °F for Sarasota, FL.

Jan.	Feb.	Mar.	Apr.	May	June	July	Aug.	Sept.	Oct.	Nov.	Dec.
72	74	77	82	87	90	91	91	90	85	80	74

Source: The Weather Channel

    a. Draw a scatterplot of these data, using x = month of the year (January = 1, February = 2, and so on.)
    b. Use a quadratic regression to model the data.
    c. Graph the function you found in Part b on the same axes as the scatterplot.
    d. Explain why a quadratic function is probably not the best choice for modeling data on weather. **(Lessons 2-6, 1-1)**

18. Solve for x: $x + 5 < 40 < x + 10$. **(Previous Course)**

EXPLORATION

19. While Pierre-Simon Laplace laid the foundation for the Central Limit Theorem, a number of other mathematicians made significant contributions to further develop the Central Limit Theorem. Research the history of the development of the Central Limit Theorem from Laplace to the present and write a brief summary of your findings.

Lesson

11-5 Making Inferences about Means

▶ **BIG IDEA** The Central Limit Theorem provides a way to use *z*-scores and the standard normal distribution to test hypotheses about means.

In every situation discussed in Lesson 11-4, you were given or could compute the mean μ and standard deviation σ of the *population* from which you were sampling. You were given the distribution of ages in Activity 1. Activity 2 involved the distribution of outcomes from tossing a fair die, from which the mean and standard deviation could be calculated. And in the Example, you were told what the mean and standard deviation of light bulb lifetimes were.

You were given this information because the purpose of the Activities in that lesson was to show how the mean $\mu_{\bar{x}}$ and standard deviation of the *set of the means of random samples of size n* compares to the mean μ and standard deviation σ of the *population*. The Central Limit Theorem and its rule of thumb state that, no matter what the population distribution, when $n \geq 30$, the distribution of the means from samples of size n approximates a normal distribution.

Mental Math

Suppose *y* varies inversely as the square root of *x*. By what is *y* multiplied if *x* is multiplied by

a. 100? **b.** 9?

c. 2? **d.** $\frac{1}{4}$?

Using the Central Limit Theorem to Determine Probabilities

Because the distribution of the means of samples of size n is normal, if you know μ and σ, you can determine how likely it is that a mean of a single sample of size n will be a certain distance from the population mean.

Example 1

In a certain country, the mean family income (when converted to U.S. dollars) is known to be $13,500 with a standard deviation of $2,000. Suppose a random sample of 100 families is to be chosen. Let $\bar{x}$ be the mean family income of the sample. Determine each of the following.

a. the mean $\mu_{\bar{x}}$ and standard deviation $\sigma_{\bar{x}}$ of the set of sample means

b. the probability that the sample mean is less than $13,000

c. the probability that the sample mean is between $13,300 and $13,700

Solution

a. Use the Central Limit Theorem. In the sampling distribution of the mean for samples of size n from this population of family incomes,
$\mu_{\bar{x}} = \mu = \$13{,}500$ and $\sigma_{\bar{x}} = \frac{\sigma}{\sqrt{n}} = \frac{\$2000}{\sqrt{100}} = \frac{\$2000}{10} = \$200.$

b. Because the sample size $n = 100$, n is large enough to assume that the distribution of the sample means is approximately normal. So, the sample means are normally distributed with $\mu_{\bar{x}} = \$13{,}500$ and $\sigma_{\bar{x}} = \$200$. Converting the sample means to z-scores where $z = \frac{x - 13500}{200}$ makes it possible to calculate probabilities using the Standard Normal Distribution Table. The probability that $\bar{x}$ is less than $\$13{,}000$ is

$$P(\bar{x} < 13000) = P\left(\frac{\bar{x} - 13500}{200} < \frac{13000 - 13500}{200}\right) \quad \text{standardizing } \bar{x}$$
$$= P(z < -2.5) \quad \text{substitution}$$
$$\approx 0.0062. \quad \text{value from table}$$

The probability is about 0.62%. Less than 1% of samples of size 100 would have average incomes less than $13,000.

c. From Part b, $\sigma_{\bar{x}} = \$200$, so $13,300 and $13,700 are one standard deviation below and above the mean, respectively. $P(-1.00 < z < 1.00) = 0.6826$. About 68% of means of samples of size 100 will be between $13,300 and $13,700.

 QY

In Example 1, notice that in the distribution of the population of incomes, $13,300 has a z-score of $\frac{13300 - 13500}{2000} = -0.1$, so it is $\frac{1}{10}$ of a standard deviation lower than the mean. But in the distribution of sample means, $13,300 is a full standard deviation below the mean. The CLT says that sample means are more tightly bunched than the original data.

> **QY**
> Suppose that, in Example 1, random samples of 400 families were chosen instead. Compute the mean and standard deviation of the set of sample means.

Testing Hypotheses Using the Central Limit Theorem

In the Example of Lesson 11-4, a manufacturer of fluorescent bulbs had knowledge about the distribution of lifetimes of those bulbs. In the next example, a manufacturer *makes a claim* about the amount of time it takes to perform maintenance on a copy machine. The Central Limit Theorem enables that claim to be tested.

GUIDED

Example 2

A manufacturer of copy machines claims that the amount x of time needed by a technician to perform routine maintenance on a copy machine is modeled by a distribution with $\mu = 1$ hour and $\sigma = 1$ hour. From examining a sample of 60 machines, a local supplier finds that the average maintenance time is 72 minutes, and believes that the manufacturer is underestimating the maintenance time needed. Test the manufacturer's claim at the 0.01 level.
(continued on next page)

Solution

State the claim as a null hypothesis to be tested. H_0: The mean maintenance time for this brand of copy machine is 1 hour.

Using the manufacturer's model, compute the probability that the average maintenance time for a sample of 60 machines is greater than 72 minutes. The Central Limit Theorem implies that the distribution for samples of size 60 machines is approximately normally distributed, with mean _?_ and standard deviation _?_. The probability that the average maintenance time is greater than 72 minutes can be computed using a normal distribution command. One calculator uses the command below with the mean and standard deviation you computed. (Note that 72 minutes is 1.2 hours.)

normCdf(1.2, ∞, _?_, _?_) ≈ .061

Interpret the answer. Assuming the manufacturer's model, in _?_% of all samples of 60 machines, the mean maintenance time would be greater than 72 minutes. This value is not less than 0.01, so the null hypothesis H_0 is not rejected at the 1% confidence level. On the basis of his sample, the supplier is not justified in claiming that the manufacturer is underestimating the average maintenance time.

In Examples 1 and 2, we used a specified population mean μ to determine the distribution of means for samples of size n. In each case you had to know the population mean. In Example 2, you tested a hypothesis that a particular sample was unlikely given the population statistics. Lesson 11-6 considers the converse problem, that of estimating an unknown population mean from the mean of a single sample.

Questions

COVERING THE IDEAS

In 1 and 2, consider the samples taken in Example 1.

1. What is the probability that a sample of 100 families would have a mean income of under $13,400?
2. Suppose you wanted the distribution of the sample means to have standard deviation $50. Approximately how many families would you have to survey?

In 3 and 4, refer to Example 2. Assume the manufacturer's claim is correct.

3. Consider the distribution of the average amount of time needed by a technician to perform maintenance on a random sample of 50 copy machines.
   a. Give the mean of the set of sample means.
   b. Give the standard deviation.
   c. Describe the distribution.

4. Suppose another supplier maintains a random sample of 100 such machines. Find the probability that the average maintenance time for these machines exceeds 45 minutes.

APPLYING THE MATHEMATICS

In 5 and 6, consider that in a recent year, the distribution of scores of students on the ACT college entrance exam was modeled by a normal distribution with $\mu = 20.9$ and $\sigma = 4.7$.

5. What is the probability that a randomly chosen student from the population who took the exam has a score of 23 or higher?
6. The mean score of the 84 students at Acme High who took the ACT that year was 23.
   a. What is the probability that the mean score for 84 students selected randomly from all who took the test nationally is 23 or higher?
   b. The people at Acme believe they scored significantly higher than the nation as a whole. Test this hypothesis at the 0.01 level.

In 7 and 8, suppose it is reported that the mean $\mu = 12.5$ and standard deviation $\sigma = 0.67$ for a set of data.

7. Suppose the data are *not* normally distributed. Find the probability that a random sample of size 5 from this data set will have a mean $\bar{x} > 13$, or explain why you do not have the tools to do so.
8. Asume the data *are* normally distributed. At the 0.01 significance level, test the hypothesis that the mean μ is in fact 12.5 as reported, given a sample of size 45 has a mean of 10.3.

REVIEW

9. To what kinds of samples does the Central Limit Theorem not apply? **(Lesson 11-4)**
10. Suppose that the average credit card balance for couples between the ages of 25 and 30 is \$6350 with a standard deviation of \$3910. Random samples of 100 couples are taken and the mean credit card balance for each sample is recorded. Give the shape, mean, and standard deviation for the sampling distribution of the sample means. **(Lesson 11-4)**
11. The strength of paper coming from a machine at a paper plant is normally distributed with a mean of 25 pounds per square inch and a standard deviation of 0.43. **(Lesson 11-3)**
    a. Paper with strength of either less than 24 pounds or greater than 26 pounds is considered waste. What percent of the machine's output is waste paper?
    b. If the machine is cleaned and overhauled, the standard deviation will be reduced. What standard deviation is needed to reduce the machine's waste to $\frac{1}{2}$%?

In 12 and 13, find the value of c satisfying the equation. (Lesson 11-2)

12. $P(z < c) = 0.1376$

13. $P(z > c) = 0.3417$

14. The distribution of weights of chicken eggs is nearly normal. Assume the mean weight is 2 ounces, and 5% of eggs weigh more than 2.5 ounces. Estimate the probability that a randomly chosen egg will have weight w in ounces as follows. **(Lesson 11-1)**
   a. $w > 1.5$
   b. $w < 2.5$
   c. $1.5 < w < 2.5$

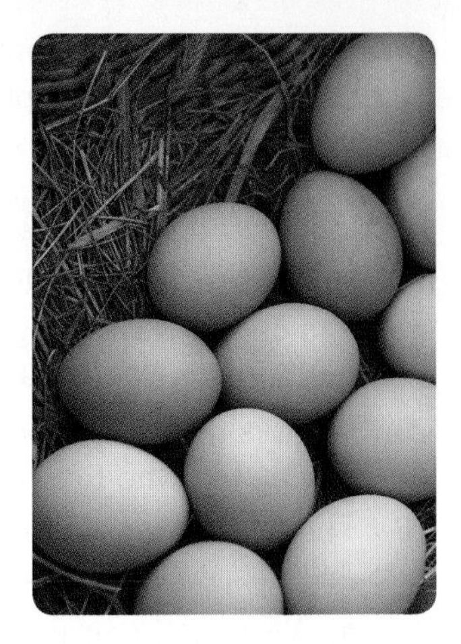

EXPLORATION

15. Randomly choose a page of text in a novel, and then choose 15 consecutive sentences on that page. Count the number of words in each sentence. Then calculate the mean of these counts. Repeat this process with different pages ten times.
   a. Determine the mean $\mu_{\bar{x}}$ and standard deviation $\sigma_{\bar{x}}$ of the ten means.
   b. Based on the Central Limit Theorem, what can you predict about the mean and standard deviation of the number of words in all the sentences in the novel?
   c. State one assumption which allows the use of the Central Limit Theorem in making the prediction in Part b.
   d. Discuss how different (if at all) the results would be for Parts a and b if you used each of the following.
      i. a physics textbook
      ii. a book for readers under the age of 10

QY ANSWER

$\mu_{\bar{x}} \approx \$13{,}500$ and
$\sigma_{\bar{x}} \approx \$100$

Lesson

11-6 Confidence Intervals

Vocabulary

confidence level
confidence interval
margin of error

BIG IDEA The Central Limit Theorem provides a way of using an appropriate sample to estimate a possible range of values for the population mean.

People very often sample when they want to estimate a population mean. Since sample means of size $n \geq 30$ are approximately normally distributed, we can tell when a sample mean is likely to be within a certain number of standard deviations of the population mean.

Mental Math

Estimate each fraction to the nearest percent.

a. $\frac{2}{3}$ **b.** $\frac{19}{30}$

c. $\frac{1}{75}$ **d.** $\frac{6}{1000}$

Benchmarks in Normal Distributions

If the mean of a normally distributed data set is μ and the standard deviation is σ, then you have seen in Lesson 11-2 that

$$\text{About} \begin{Bmatrix} \text{68\% of the data falls within } 1\sigma \\ \text{95.5\% of the data falls within } 2\sigma \\ \text{99.7\% of the data falls within } 3\sigma \end{Bmatrix} \text{of the mean } \mu.$$

For testing hypotheses, it is useful to know how far 90%, 95%, and 99% of the data fall from the mean of a normally distributed data set.

$$\text{About} \begin{Bmatrix} \text{90\% of the data falls within } 1.645\sigma \\ \text{95\% of the data falls within } 1.96\sigma \\ \text{99\% of the data falls within } 2.58\sigma \end{Bmatrix} \text{of the mean } \mu.$$

These statements can be described using the language of probability. If a random variable x is normally distributed with mean μ and standard deviation σ, then one of the above statements is

$P(\mu - 1.645\sigma < x < \mu + 1.645\sigma) \approx 0.90$ using a double inequality, or

$P(|x - \mu| < 1.645\sigma) \approx 0.90$ using absolute value.

For instance, if the time you spend on homework for a class is normally distributed with $\mu = 45$ minutes and $\sigma = 20$ minutes, then you should expect that 90% of the time you will spend between $\mu - 1.645\sigma$ and $\mu + 1.645\sigma$ minutes on homework, that is, between $45 - 1.645 \cdot 20$ and $45 + 1.645 \cdot 20$ minutes, or between 12.1 and 77.9 minutes. The other 10% of the time you should expect to spend less than 12.1 or more than 77.9 minutes on homework.

QY1

QY1

Write the statement that 99% of data falls within 2.58σ of the mean μ

a. using a double inequality.

b. using absolute value.

Contrasting Intervals for μ and $\mu_{\bar{x}}$

The benchmarks on the previous page are true for any normal distribution, so they are true both for normally-distributed populations and for the distribution of the means of the samples of size $n \geq 30$ from any population. But the standard deviations of these distributions are different. Example 1 is about that difference.

Example 1

A candy company produces 60,000 bags of candy each day. From the data on its production, the company has learned that the weights of the bags are normally distributed with mean weight $\mu = 7.5$ ounces and standard deviation $\sigma = 0.04$ ounces.

a. **Between what weights symmetric to the mean do 95% of the bags fall?**

b. **Each day 25 packages are randomly selected for quality control of weight, taste, and appearance. The mean weight of these packages is recorded. Over time, 95% of the mean weights will fall between what limits?**

Solution

a. Because the distribution is normal, 95% of the bags will fall within 1.96 standard deviations of the mean. Calculating the lower bound gives $\mu - 1.96\sigma = 7.5 - 1.96 \cdot 0.04 = 7.4216$. Similarly, the upper bound is $\mu + 1.96\sigma = 7.5784$. 95% of the bags will have weights between about 7.42 ounces and 7.58 ounces.

b. Because the original population of weights is normally distributed, the sampling distribution of the mean for sample size 25 will be normal, with mean $\mu_{\bar{x}} = 7.5$ ounces and standard deviation $\sigma_{\bar{x}} = \frac{0.04}{\sqrt{25}} = 0.008$.
The lower bound is $\mu_{\bar{x}} - 1.96\,\sigma_{\bar{x}} = 7.5 - 1.96 \cdot 0.008 = 7.48432$.
The upper bound is $\mu_{\bar{x}} + 1.96\,\sigma_{\bar{x}} = 7.5 + 1.96 \cdot 0.008 = 7.51568$.
95% of the sample mean weights $\bar{x}$ will be between about 7.48 ounces and 7.52 ounces.

Note the difference between Parts a and b. Example 1a asks about the distribution of weights of individual bags. Example 1b asks about the distribution of weights of means of samples of bags.

Estimating the Population Mean from a Sample Mean

The statement about the means $\bar{x}$ of samples in Example 1 can be written symbolically as $P(7.48 < \bar{x} < 7.52) = 0.95$. In general,

$$P(\mu - 1.96\tfrac{\sigma}{\sqrt{n}} < \bar{x} < \mu + 1.96\tfrac{\sigma}{\sqrt{n}}) = 0.95.$$

We now solve this double inequality for μ.

$$\mu - 1.96\tfrac{\sigma}{\sqrt{n}} < \bar{x} < \mu + 1.96\tfrac{\sigma}{\sqrt{n}} \qquad \text{Add } -\mu \text{ to all three parts.}$$

$$-1.96\tfrac{\sigma}{\sqrt{n}} < -\mu + \bar{x} < 1.96\tfrac{\sigma}{\sqrt{n}} \qquad \text{Add } -\bar{x} \text{ to all parts.}$$

$$-\bar{x} - 1.96\tfrac{\sigma}{\sqrt{n}} < -\mu < -\bar{x} + 1.96\tfrac{\sigma}{\sqrt{n}}$$

Multiplying all three parts by -1 reverses the sense of the inequalities.

$$\bar{x} - 1.96\frac{\sigma}{\sqrt{n}} < \mu < \bar{x} + 1.96\frac{\sigma}{\sqrt{n}}$$

So $P(\bar{x} - 1.96\frac{\sigma}{\sqrt{n}} < \mu < \bar{x} + 1.96\frac{\sigma}{\sqrt{n}}) = 0.95$. Thus, *if you know the standard deviation* σ of the population, the size of a sample n, and the mean of one sample $\bar{x}$, you can compute an interval in which the population mean μ has a 95% chance of falling.

Example 2

A superintendent of a large school district is worried that juniors will not do well on the required state test in reading. Past years of students provided widely varying means for the district, but the standard deviation has been close to 9.2 points. The superintendent uses a released form of the exam to test 72 randomly-selected students in different high schools two months before the state test. The students score an average $\bar{x} = 135$. Estimate the mean score μ of all juniors in the district.

Solution Because the sample size is larger than 30, the sampling distribution of the mean will be normal with mean μ and standard deviation $\frac{\sigma}{\sqrt{n}} = \frac{9.2}{\sqrt{72}}$.

$$\bar{x} + 1.96\frac{\sigma}{\sqrt{n}} = 135 + 1.96\left(\frac{9.2}{\sqrt{72}}\right) \approx 135 + 2.13 = 137.13$$

$$\bar{x} - 1.96\frac{\sigma}{\sqrt{n}} = 135 - 1.96\left(\frac{9.2}{\sqrt{72}}\right) \approx 132.87$$

So, $P(132.87 < \mu < 137.13) = 0.95$. There is a 95% probability that the mean μ for all juniors will be between 132.87 and 137.13. In the language of absolute value, $P(|\mu - 135| < 2.13) = 0.95$.

What Is a Confidence Interval?

In Example 2, the interval $132.87 < \mu < 137.13$ is called the 95% *confidence interval* for the population mean μ. That is, we are 95% *confident* that the interval 132.87 to 137.13 captures the unknown population mean. The distance from the mean, 2.13 in this case, is called the *margin of error* for the confidence interval. 95% is called the **confidence level**. In 95% of random samples of size 72, $\bar{x}$ will fall within 2.13 of the population mean μ. This is why we can be 95% confident that μ falls between 132.87 and 137.13 in Example 2.

The statements at the beginning of this lesson can be interpreted for confidence intervals. For a population with mean μ, where the standard deviation σ and a sample mean $\bar{x}$ from a sample of size n are known, the **confidence interval** for μ is $\bar{x} \pm z\frac{\sigma}{\sqrt{n}}$, where z is the z-score for the confidence level. The **margin of error** is $z\frac{\sigma}{\sqrt{n}}$. For example, the 95% confidence interval is $\bar{x} \pm 1.96\frac{\sigma}{\sqrt{n}}$ and the margin of error is $1.96\frac{\sigma}{\sqrt{n}}$.

QY2

QY2

Write the expression for the 90% confidence interval.

Confidence intervals were invented in 1937 by Jerzy Neyman (1884–1981), a Polish mathematician who moved to the United States and was professor of mathematics for many years at the University of California at Berkeley. Statisticians have invented techniques for finding confidence intervals for many different parameters based on various assumptions. But all confidence intervals share two properties: there is an interval of possible values constructed from the sample data, and there is a confidence level that gives the probability that the method produces an interval containing the true value of the parameter.

Choices that Affect the Confidence Interval

There are two things that affect the margin of error, and therefore the confidence interval. First, an increase in sample size will decrease the margin of error. However, since margin of error is inversely proportional to the square root of the sample size, doubling the sample size will not halve the margin of error. To halve the margin of error, you must increase the sample size by a factor of four. Increasing the sample size to reduce the margin of error can be expensive in practice.

Second, as the level of confidence increases, so does the width of the confidence interval. This makes intuitive sense; if you want to place a higher probability on your estimate, you need a wider interval to "cover your bases." The graph below compares the sizes of 90% and 95% confidence intervals for the district state test data in Example 2.

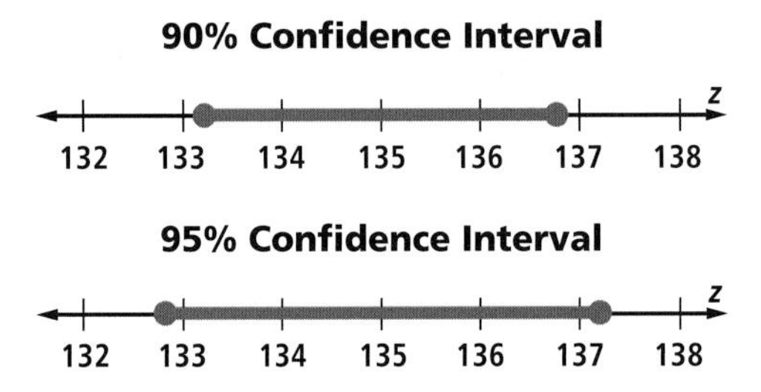

What does this mean if you would like a "finer tuning" or closer estimate of the mean? You have two choices: choose a lower confidence level or use a larger sample. Generally, people who use statistics do not want to use a confidence level below 90%, so in order to get within a desired margin of error, they must increase the sample size.

Cautions about Statistical Inferences

Just as medications carry warnings about their potential harmful effects, the techniques described in this lesson to make inferences about populations must be applied with caution. First, the data collected must be from a random sample of the population. The methods used here to determine a confidence interval for μ do not apply to other types of samples such as volunteers or similar "convenient" samples.

Second, you need to have large samples or be confident that your variable is distributed normally in the population of interest. As a practical matter, if the sample size is relatively small, outliers can have a large effect on the confidence interval. You should always look for outliers in a sample, and try to correct them or justify their removal before computing a sample mean.

Third, the margin of error in a confidence interval covers only random sampling errors. That is, the margin of error describes typical error expected because of chance variation in random selection. It does not describe errors arising from sloppy data collection or data entry. Fancy formulas can never compensate for sloppy data.

Last, the techniques developed here require that you know the standard deviation σ of the population. To know σ, you usually need to know the mean that you are trying to estimate. This is circular reasoning. However, in a binomial experiment, there is a way to deal with not knowing the population standard deviation, as you will see in the next lesson.

Questions

COVERING THE IDEAS

In 1 and 2, the population standard deviation of scores on the SAT mathematics section test is 98. One high school principal samples 400 students, and they score a mean of 603. A confidence level is given.

a. Find the confidence interval for the population mean.

b. Find the margin of error.

1. 95.5% level

2. 99.7% level

3. Find a 90% confidence interval for the mean SAT-M scores of the seniors in the state of Minnesota if a random sample of 400 seniors has a mean score of 536 and the standard deviation for the population is 123.

4. **True or False** In most statistical applications, a confidence level of at least 90% is used.

5. **True or False** As the confidence level increases, the width of the confidence interval increases.

6. At the right are shown 95% and 99% confidence intervals for a population with mean μ.

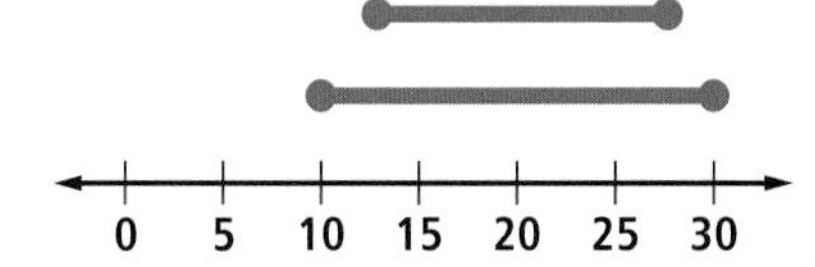

   a. Estimate μ.

   b. Which is the 95% confidence interval? How can you tell?

7. Name two ways to decrease the range of a confidence interval.

In 8 and 9, a sample of size n is taken from a population whose standard deviation is known to be σ. The sample mean is $\bar{x}$.

8. **Fill in the Blanks** We are __?__ confident that the 99% confidence interval between __?__ and __?__ captures the __?__ (sample/population) mean.
9. **Fill in the Blanks** In about 99% of random samples of size __?__ taken from the population, the sample mean $\bar{x}$ will be within __?__ of the population mean μ.
10. Rewrite $\bar{x} - 1.96\frac{\sigma}{\sqrt{n}} < \mu < \bar{x} + 1.96\frac{\sigma}{\sqrt{n}}$ using absolute value.

APPLYING THE MATHEMATICS

11. Refer to the SAT data in Questions 1 and 2. Suppose you wanted to estimate μ with 95% confidence to within 10 points. How large a sample would be needed?
12. When opinion polls are conducted, the results are often reported as a percent in favor of something $\pm$ a margin of error as a percent. In most polls it is standard practice to report the margin of error for a 95% confidence interval unless stated otherwise.
    a. Suppose you read that 37% $\pm$ 3% of Americans surveyed recently felt that homelessness was the most serious problem in the United States. Describe the 95% confidence interval with a double inequality.
    b. Can you be certain that the true population percent falls within the interval in Part a?
13. Medical tests for cholesterol level are not perfectly precise. Moreover, a person's total cholesterol level varies from day to day. Suppose that repeated measurements for an individual on different days vary normally with $\sigma = 5$.
    a. On one test, Alisa's total cholesterol level is reported to be 180. Find a 95% confidence interval for her mean cholesterol level.
    b. A cholesterol level under 200 is considered desirable. Should Alisa be concerned about the results of her cholesterol test?
    c. A sample of 15 test results from Alisa's medical history is randomly selected and returns a mean of 174.2. Find a 95% confidence interval for the population mean of all her tests.

REVIEW

14. Suppose that in the past year the monthly cell phone bills from a company under a particular plan have a mean of \$33.25 and a standard deviation of \$8.65. **(Lessons 11-5, 11-3)**
    a. What is the probability that a single cell phone bill is greater than \$36.00?
    b. Suppose that in the past few months, a random sample of 100 bills has shown a mean charge of \$36.00. Test at the 0.01 level the hypothesis that the average monthly cost has not increased.

15. Consider a factory that makes earbud headphones with a target diameter of $\mu = 0.5$ inch. The machines that produce the earbuds have some variability, so the standard deviation of the diameters is $\sigma = 0.095$ inch. A quality control program inspects samples of 100 earbuds every 3 hours and records the sample mean diameter. **(Lesson 11-5)**

a. What will be the mean and standard deviation of the quality control records?

b. What is the probability that a sample of 100 earbuds has a mean diameter greater than 0.515 inch?

c. What is the probability that a sample of 100 earbuds has a mean diameter between 0.395 and 0.495 inch?

16. The height of a certain species of plant is known to be normally distributed with mean 61 cm and standard deviation 5 cm. If a nursery grows 3000 plants of this species, how many should they expect to be less than 48 cm tall (and consequently, too short to sell as top-quality)? **(Lesson 11-3, 10-7, 6-8)**

Multiple Choice In 17-20, which equation corresponds to the graph? **(Lessons 11-3, 11-2, 11-1)**

A $y = e^{-x}$ B $y = e^{-x^2}$ C $y = \frac{1}{\sqrt{2\pi}} e^{\frac{-x^2}{2}}$ D $y = \frac{1}{\sqrt{2\pi}} e^{-\frac{1}{2}(x-1)^2}$

17.

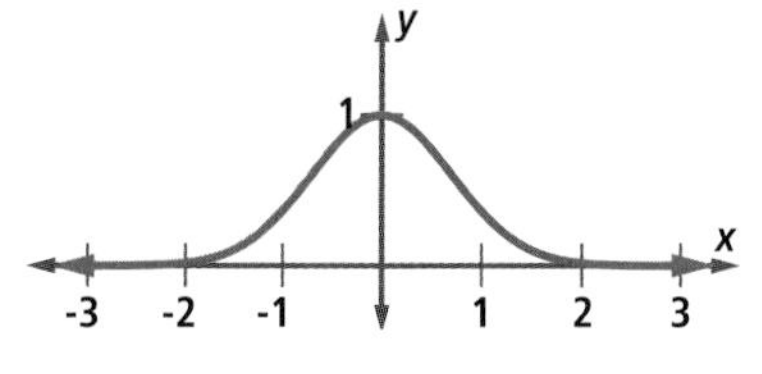

18.

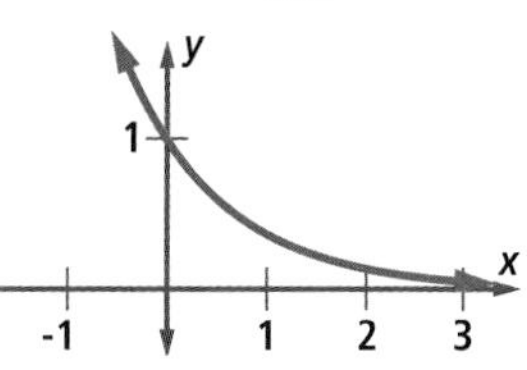

19.

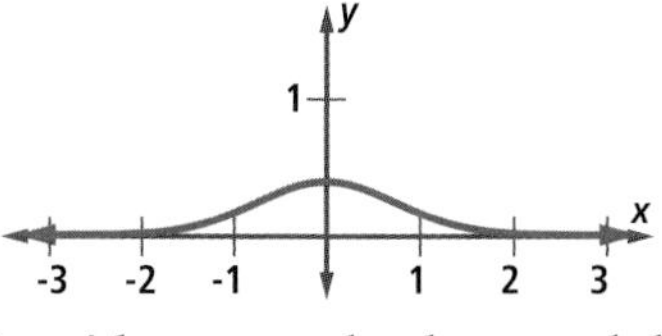

20.

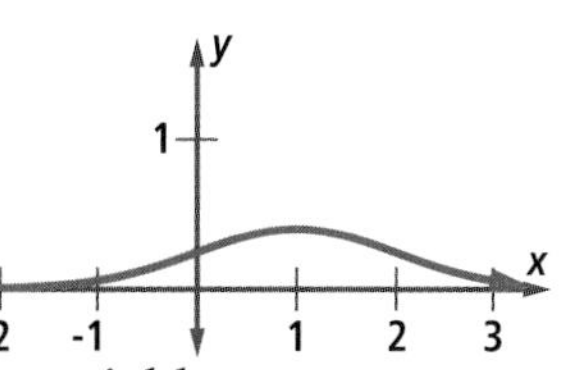

21. Consider a standard normal distribution of a variable z. Suppose $P(z < a) = t$. Express each of the following in terms of t. **(Lesson 11-2)**

a. $P(z > a)$ b. $P(z < -a)$ c. $P(-a < z < a)$

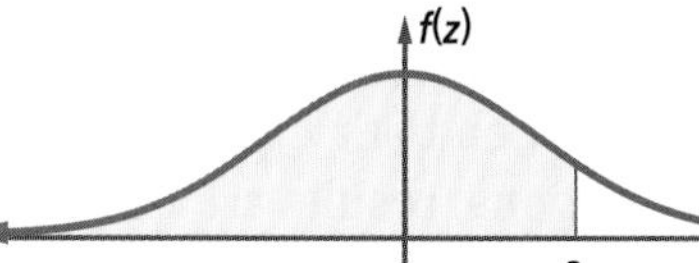

22. **Multiple Choice** Which question does a hypothesis test answer? **(Lesson 10-8)**

A Is the sample random? B Is the experiment properly designed?

C Is the result due to chance? D Is the result important?

23. Suppose $y = x(1 - x)$ and x is a real number. What value of x gives the largest possible value of y? **(Previous Course)**

EXPLORATION

24. What size sample is used in ratings of TV programs? What is the accuracy of these ratings? What is the confidence level?

QY ANSWERS

1. a. $P(\mu - 2.58\sigma < x < \mu + 2.58\sigma) \approx 0.99$.
   b. $\left(|x - \mu| < 2.58\sigma\right) \approx 0.99$

2. $\bar{x} \pm 1.645\frac{\sigma}{\sqrt{n}}$

Lesson

11-7 Confidence Intervals in Binomial Experiments

▶ BIG IDEA Reports from opinion polls are based on confidence intervals, and the formula for margin of error determines necessary sample sizes for an accurate survey.

Mental Math

Compute the value of $p(1 - p)$ when p is

a. $\frac{4}{5}$. **b.** $\frac{1}{2}$.

c. $\frac{2}{3}$. **d.** $\frac{1}{6}$.

e. 0.

Opinion polls have become a fact of life worldwide. The following two polls were taken during the first week in January 2008.

- A Quinnipiac University poll of 1162 voters found that 91% of New Yorkers believe that traffic congestion is either a very serious or somewhat serious problem, with a margin of error of $\pm$ 2.9%.
- In Germany, an Infratest poll of 1000 people showed that support for the Christian Democratic Union Party fell to 35%, with a margin of error of $\pm$ 3.1%.

The polls above have only two possible responses, so we can think of them as binomial experiments. Let asking one person the question be a trial, and think of a particular response as success. Then if n people are polled and x respond in a particular way (success), the poll is a binomial experiment with n trials. The probability p of success can be approximated by $\frac{x}{n}$, the relative frequency of success. For instance, since 35% of Germans supported the Christian Democratic Union out of the 1000 polled in the Infratest poll, $p = \frac{x}{n} = 35\%$.

STOP **QY1**

▶ QY1

What is $\frac{x}{n}$ in the Quinnipiac study?

In Example 3 of Lesson 11-3, you saw that a frequency distribution arising from a binomial experiment can be approximated by a normal distribution when both np and $n(1 - p)$ are greater than 10. But in polling situations, results are usually reported as *relative* frequencies, as illustrated above. To move from a distribution of frequencies to a distribution of relative frequencies in an experiment with n trials, we divide each frequency by n.

The random variable in a binomial distribution with n trials and probability p of success has mean np and standard deviation $\sqrt{np(1 - p)}$. The scale change that transforms the data from frequencies x to relative frequencies $\frac{x}{n}$ divides both the mean and standard deviation by n. Thus, for the distribution of relative frequencies, the mean $\mu_{\bar{x}} = \frac{np}{n} = p$, and the standard deviation $\sigma_{\bar{x}} = \frac{\sqrt{np(1-p)}}{n} = \sqrt{\frac{p(1-p)}{n}}$. These are the formulas to use when looking at opinion polls.

 **QY2**

▶ QY2

What are the mean and standard deviation for the distribution of relative frequencies in the Infratest poll?

Confidence Intervals for Polls

The key difference between the confidence intervals in Lesson 11-6 and confidence intervals for polls is that in the former, the Central Limit Theorem could be invoked. We were given the population standard deviation, so we could calculate the confidence interval using $\sigma_{\bar{x}}$ in the margin of error calculation. With polls, we need not know the population standard deviation. Instead, we use the fact that the sampling distribution for an opinion poll is from a binomial experiment, and use the mean p and standard deviation $\sqrt{\frac{p(1-p)}{n}}$ to calculate the confidence interval. Thus, the 95% confidence interval is $\frac{x}{n} \pm 1.96\sqrt{\frac{p(1-p)}{n}}$.

 QY3

Consequently, the margin of error of a poll of sample size n for a 95% confidence level is $1.96\sqrt{\frac{p(1-p)}{n}}$. In general, the margin of error is the corresponding z-score for the confidence level multiplied by the standard deviation of the sampling distribution being used. In polls, the margin of error is $z\sqrt{\frac{p(1-p)}{n}}$.

Example 1

Compute the 95% confidence interval in the Quinnipiac study for the percent of voters who thought traffic congestion was a very serious or somewhat serious problem.

Solution The relative frequency of success is $\frac{x}{n} \approx 0.91 = p$ as found in QY1. Because $np \approx 1057$ and $n(1-p) \approx 105$ are both greater than 10, we can approximate the distribution with a normal curve and use the z-score of 1.96 for the 95% confidence level.

$$\text{Margin of error} = z\sqrt{\frac{p(1-p)}{n}}$$

$$\approx 1.96\sqrt{\frac{0.91 \cdot 0.09}{1162}} \approx 0.016455 \approx 1.6\%.$$

The 95% confidence interval is 0.91 ± 0.016, or $91\% \pm 1.6\%$. There is a 95% probability that the percent of all New Yorkers who think traffic congestion is a problem is between 89.4% and 92.6%.

In Example 1, you might have noticed that the margin of error differed from the one quoted on page 726 of 2.9%. When pollsters are designing polls, they may not have any notion of what the true value of p is for the population. Using $p = 0.5$ makes the confidence interval the widest possible, because the expression $p(1-p)$ is largest when $p = 0.5$. Thus, using $p = 0.5$ allows the pollsters to make the most conservative estimate of the margin of error. Calculating the margin of error using $p = 0.5$, we obtain $1.96\sqrt{\frac{0.5 \cdot 0.5}{1162}} \approx 0.02875 \approx 2.9\%$ as the pollsters reported.

 QY4

QY3

Write an expression for the 99% confidence interval for $\frac{x}{n}$.

QY4

How do you know from the graph of $y = p(1-p)$ that y is largest when $p = 0.5$?

GUIDED

Example 2

a. **Calculate the margin of error for the 95% confidence level in the Infratest poll.**

b. **Compute the 95% confidence interval in the Infratest poll.**

c. **Why does the value in Part a differ from the margin of error reported by the pollster?**

Solution Before calculating a margin of error or confidence interval for a result in a poll, you must first determine whether the poll satisfies the conditions to be able to use the *z*-score for the confidence interval desired. Because $np = 350$ and $n(1 - p) = 650$ are both greater than 10, you can use *z*-scores.

a. We again use z = __?__ for the 95% confidence interval. In this poll, n = __?__, and p = __?__. Therefore, the margin of error $= z\sqrt{\frac{p(1-p)}{n}} = \underline{?}\sqrt{\frac{? \cdot ?}{?}} \approx \underline{?}$.

b. The 95% confidence interval is __?__ ± __?__, or __?__% ± __?__%.

c. The margin of error found in Part a was __?__ %. The margin of error reported by the Infratest poll was 3.1%. The pollsters before the poll was taken, decided to report the most conservative or wide confidence interval to cover themselves for any sampling errors. That is, they used p = 0.5 in their calculation of the margin of error, $z\sqrt{\frac{p(1-p)}{n}}$.

However, once the results of a sample or poll are known, the consumer of the poll can use the p found in that poll, in this case p = 0.35.

In Examples 1 and 2, the pollsters used a 95% confidence interval to get their reported margins of error. This is a widely accepted and commonly used confidence level in polls today. However, at times higher confidence is desired and the polls are designed accordingly. Instead of taking the perspective of the consumer when examining polling data and statistics, as we did in Examples 1 and 2, we now take the perspective of the pollster to determine how many responses a poll needs in order to have a margin of error with a particular confidence.

Population Size and Margin of Error

When Quinnipiac polled 1162 people from over 8 million New Yorkers there was a 2.9% margin of error. It seems intuitive that the Infratest poll should have a much larger margin of error since only 1000 people were polled out of 82 million Germans. Why was the margin of error on the Infratest poll only 3.1% when the population of Germany is 10 times that of New York City?

Look at the expression for the margin of error for a poll, $z\sqrt{\frac{p(1-p)}{n}}$. This expression is largest when $p = 0.5$ as discussed on page 727. Then $z\sqrt{\frac{0.5 \cdot 0.5}{n}} = \frac{z}{2}\left(\frac{1}{\sqrt{n}}\right)$. Since z is determined by the level of confidence required, $\frac{1}{\sqrt{n}}$ determines the margin of error. This means that the size of the population from which the sample is taken has *no effect* on the margin of error. Pollsters realize this and decide on the sample size based on the desired confidence level and margin of error. The margin of error for the Quinnipiac poll is smaller than that of the Infratest poll not because the population was smaller, but because the sample was larger.

Example 3 shows how the sample size would change if a pollster desired a different margin of error.

GUIDED

Example 3

China has a population of over 1.3 billion.

a. What sample size is needed for a poll if a 3% margin of error is desired with 99% confidence?

b. What sample size is required to reduce the margin of error to 1%?

Solution

a. For a 99% confidence level, $z \approx$ __?__ , and for safety, a pollster will use $p = 0.5$. The margin of error is 0.03. Solve for n.

So,

$$\frac{z}{2}\left(\frac{1}{\sqrt{n}}\right) = \underline{\ ?\ }$$

$$\frac{1}{\sqrt{n}} = 2\left(\frac{?}{?}\right) \approx \underline{\ ?\ }$$

$$\sqrt{n} = \underline{\ ?\ }$$

$$n = \underline{\ ?\ }$$

b. The margin of error is inversely proportional to the square root of the sample size. To divide the margin of error by 3, the sample size needs to be multiplied by __?__. So the required sample size is __?__ • n = __?__ .

Note that in the formula, n is the sample size, not the population size. The population size has no bearing on the margin of error calculation.

Effect of Confidence Level on Sample Size

While a 95% confidence level is conventional, there are occasions when a higher or lower level of confidence is desired. Higher confidence levels require larger samples while lower confidence levels require smaller samples. Networks will rarely call an election based on the results of an exit poll. The decision desk analysts will require a very high degree of confidence (usually at least 99%) before they would consider calling a winner. Setting the confidence level this high may be due to network concerns about misleading poll results.

Example 4

A TV network conducted an exit poll in 2004 using a 95% confidence level and a 1% margin of error. In 2008 they conducted a similar poll using a 99% confidence level and a 1% margin of error. About how many more voters did they poll in 2008?

Solution For 2004, use the margin of error formula with $z = 1.96$, and solve for n.

$$\frac{1.96}{2}\left(\frac{1}{\sqrt{n}}\right) = 0.01$$

$$n = \left(\frac{0.98}{0.01}\right)^2 = 9604$$

For 2008, use the margin of error formula with $z = 2.58$.

Then $\frac{2.58}{2}\left(\frac{1}{\sqrt{n}}\right) = 0.01$, so $n = \left(\frac{1.29}{0.01}\right)^2 = 16{,}641$.

So about 7,000 more voters were polled in 2008 than in 2004.

Questions

COVERING THE IDEAS

1. Explain why the size of the population for a particular poll has no bearing on the margin of error.
2. a. Explain why the assumption $p = 0.5$ is the safest assumption that can be made when p is not known.
   b. Verify that the margin of error reported on the Quinnipiac study was the most conservative calculation.
3. **True or False** Increasing the confidence level of a poll reduces the margin of error. Explain your reasoning.

In 4 and 5, the Gallup organization polled 3057 people between October and December, 2007. They found that 73% of U.S. adults planned to reduce their level of debt in 2008.

4. a. Compute the 95% confidence interval for the Gallup poll.
   b. What is the margin of error, given the Gallup poll results above?
   c. What margin of error was reported given that the pollsters had no reasonable guess as to the percent of U. S. adults planning to reduce their debt in 2008?
5. a. If Gallup had used a 90% confidence level with the same reported margin of error, about how many people would need to be sampled?
   b. How many people would need to be polled if the reported margin of error was halved at the 90% confidence level?
6. By what number do you need to multiply the sample size in order to multiply the margin of error by $\frac{3}{4}$?

7. **True or False** At the same level of confidence, a poll of 100 people in a city of 100,000 will have the same margin of error as a poll of 1000 people in a city of 1,000,000. Justify your answer.

APPLYING THE MATHEMATICS

8. A poll on women's issues was conducted with 1032 women and 476 men. The poll reported the men's and women's opinions separately. The margin of error for the results from the men was ± 4%. The margin of error for results from the women was ± 3%.
   a. Why was the margin of error for men larger?
   b. Estimate the confidence level used for the men's poll.
   c. Estimate the confidence level used for the women's poll.

9. Consider again Dr. Kerrich's coin flipping experiment in which 5067 heads appeared in 10,000 flips.
   a. Test the null hypothesis that Kerrich's coin had probability 0.5 of heads at the 0.01 significance level.
   b. Find a 99% confidence interval for the probability of heads for Kerrich's coin.

10. A MSN-Zogby poll of 4103 people in December, 2007, found that 83% of adults in the United States associate unsafe toys with China. Using the polling data, a student calculates a margin of error of about 1.6%. What confidence level did the student use?

11. What is the largest confidence level for a result in a poll of 1000 people with a reported 2% margin of error?

Sojourner Truth (1797-1883), a former slave, fought for abolition, women's rights, and religious tolerance.

REVIEW

In 12–14, a factory produces 72,000 bags of microwave popcorn each day. Based on historical data, the company knows the weight of the filled bags is normally distributed with a mean weight of 3.0 oz and a standard deviation of 0.08 oz. (Lessons 11-6, 11-5, 11-4)

12. Between what weights do 95% of the popcorn bags fall?

13. Each day 30 bags are randomly selected for quality control tests of weight, taste, and appearance. The average weight of these packages is checked. Over time, 95% of the mean weights will fall between what limits?

14. a. If the weights of filled popcorn bags were not normally distributed, what condition would have to be met in order to answer Question 13?
   b. Does the situation meet this condition?

15. In 1995, the mean total income for physician assistants in the U.S. was \$61,750 with a standard deviation of \$18,546. Suppose a random sample of 100 physician assistants is to be chosen. Let $\bar{x}$ be the mean income of physician assistants in the sample. Determine each of the following. **(Lesson 11-5)**
   a. the mean $\mu_{\bar{x}}$ of the set of all possible sample means
   b. the standard deviation $\sigma_{\bar{x}}$ of the set of all sample means
   c. the probability that the sample mean is less than \$57,750
   d. the probability that the sample mean is between \$55,000 and \$68,500

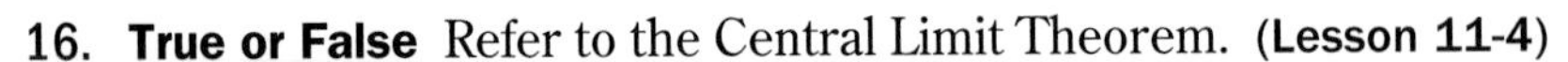

16. **True or False** Refer to the Central Limit Theorem. **(Lesson 11-4)**
   a. As sample size increases, the mean of the set of sample means approaches the population mean.
   b. As sample size increases, the standard deviation of the sample means approaches the standard deviation of the population.

In 17 and 18, consider the set of two-digit numbers from 10 to 99. Use a graphing utility to simulate and answer the following questions. (Lessons 11-4, 10-4)

17. Suppose one number x, is chosen at random.
   a. Name the type of probability distribution that best models this situation.
   b. Find the mean and standard deviation of the random variable x.

18. a. Simulate 100 random samples of size 5.
   b. Find the mean and standard deviation of the set of sample means.
   c. What type of distribution models the distribution of sample means?
   d. Suppose 100 random samples of size 60 are selected, and their means calculated. How would you expect the mean and standard deviation of the sample means of this sample size to compare to the mean and standard deviation found in Part b?

In 19 and 20, evaluate the probability involving z-scores and draw a sketch to support your answer (Lesson 11-3)

19. $P(z < 1.85)$

20. $P(-0.85 < z < 1.35)$

EXPLORATION

21. Opinion polling has become a major business over the past several decades. Research and write a few paragraphs about the history of opinion polling.

QY ANSWERS

1. $p = \frac{x}{n} = \frac{1057}{1162} \approx 0.91$

2. $p = 0.35$; $\sqrt{\frac{p(1-p)}{n}} \approx 0.015$

3. $\frac{x}{n} \pm 2.58 \sqrt{\frac{p(1-p)}{n}}$

4. The vertex of the parabola with equation $y = p(1 - p)$ has p-coordinate $\frac{1}{2}$.

Chapter 11 Projects

1 Jar of Pennies

A jar of 610 pennies has the following distribution. Age is the number of years since the coin has been minted.

Age	Count	Age	Count
1	73	21	7
2	102	22	16
3	84	23	5
4	58	24	5
5	33	25	2
6	27	26	4
7	30	27	1
8	20	28	3
9	23	29	2
10	19	30	2
11	9	31	4
12	14	32	1
13	12	33	0
14	8	34	2
15	10	35	0
16	3	36	0
17	9	37	1
18	8	38	0
19	8	39	0
20	4	40	1

a. Calculate the mean μ and standard deviation σ of the ages of the pennies in the jar.

b. Design a simulation to randomly select 30 pennies from the jar.

c. Run the simulation 40 times and record the mean age in each trial.

d. Calculate the mean $\mu_{\bar{x}}$ and standard deviation $\sigma_{\bar{x}}$ of the set of sample means from Part c.

e. Compare μ and $\mu_{\bar{x}}$. Also compare $\frac{\sigma}{\sqrt{n}}$ and $\sigma_{\bar{x}}$, where n is the sample size. Do your comparisons support the Central Limit Theorem and its Corollary?

2 Is Your Class Typical?

Refer to Project 5 of Chapter 1 (page 68) and the related projects in Chapters 2 and 3.

a. Determine a way to choose a random sample of at least 30 students in your school. Conduct a survey of that sample with the same variables.

b. Use displays and descriptive statistics to describe a typical student in the sample.

c. Make inferences to describe a typical student in the school.

d. Compare your results from Part b above to the results you obtained in earlier chapters. How typical is your class compared to your school as a whole?

3 How Common Is a Particular Letter?

a. Describe a way to pick letters at random from a book, newspaper, or other source with a lot of prose. Choose a particular letter you will study. Randomly pick any 300 letters from your source and calculate the relative frequency of occurrence of the particular letter you have chosen.

b. Repeat Part a at least 9 times and use your collected data to determine a 95% confidence interval for the percentage of your letter in the English language.

c. Repeat Parts a and b with another source of prose.

d. Compare your answers to Parts b and c. Why might these differences exist?

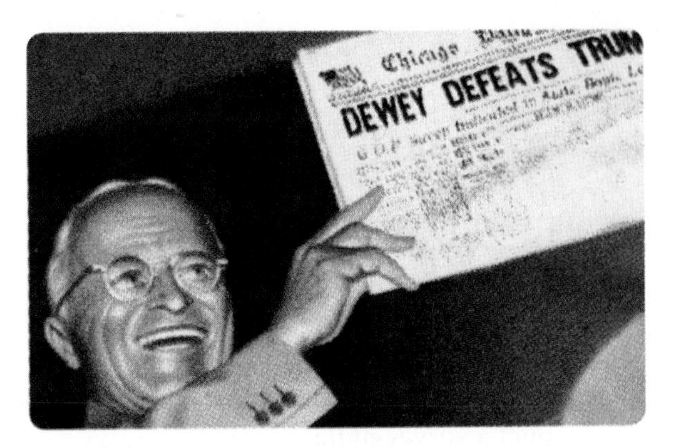

Harry Truman, the 33rd U.S. President, holds up the mistaken headline.

4 Dewey Defeats Truman!!

This famously incorrect headline from the front page of the Chicago Tribune on November 3, 1948, proclaimed victory for Thomas Dewey the day after Harry Truman in reality won the election. Probably the most famous gaffe in statistical polling history, this headline laid the foundation for modern polling techniques. Research the history behind the mistake and how major pollsters were affected. Also, research what different sources say may have led to the mistake.

5 The Bell Curve

Historically, social scientists refer to the normal curve as the *bell curve* because of its shape. Research the significance of the bell curve in three different contexts. Describe why many people refer to the bell curve as a fundamental law of natural science.

6 Sums of Random Digits

a. Use a computer, calculator, or table of random numbers to choose two random integers from 0 to 9. Then calculate their sum. Repeat this until you have 500 such random sums. For instance, your first ten sums might be 14, 10, 5, 3, 15, 3, 6, 7, 7, 12. Let $s =$ the sum of the two random digits. Then s can be any integer from 0 to 18. Find $P(s = n)$ for all $n \in \{0, 1, 2, ..., 18\}$. Express your answer as a table of values and as a histogram with bin width 1. What do you think the shape of the histogram would be if you had 500,000 random sums?

b. Modify Part a to choose two random numbers from 0 to 9. Express your answer as a table of values and as a histogram with bin width $\frac{1}{2}$. Compare the histogram in Parts a and b.

c. Transitioning from Part a to Part b can be thought of as the transition from a discrete random variable to a continuous random variable. Using the random numbers you found in Part b, reduce the bin width, yet again, to $\frac{1}{4}$. How does bin width affect the shape of the histogram?

d. As the bin width approaches 0, what shape does the histogram approach?

Chapter 11 Summary and Vocabulary

- Many naturally occurring phenomena can be approximated by normal distributions. The parent of all these curves has equation $f(x) = e^{-x^2}$. The **standard normal distribution** is an offspring with mean 0 and standard deviation 1; its equation is $f(z) = \frac{1}{\sqrt{2\pi}}e^{\frac{-z^2}{2}}$.

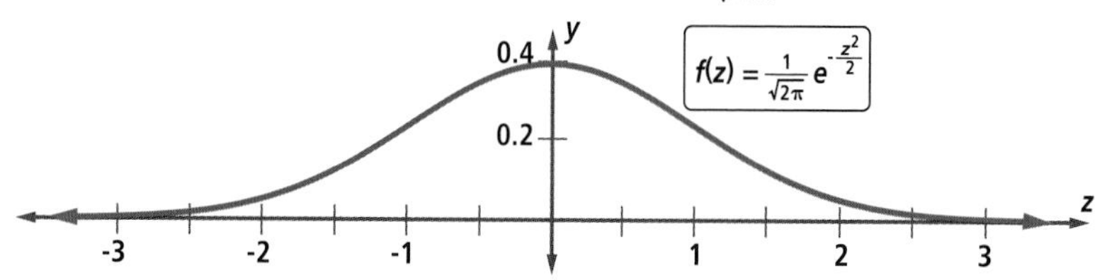

The area between the graph of the standard normal distribution (shown above) and the z-axis is 1. Areas under the **standard normal curve** between various values of z represent probabilities. A random variable that has a normal distribution with mean μ and standard deviation σ has equation $y = \frac{1}{\sqrt{2\pi}\sigma}e^{-\frac{(x-\mu)^2}{2\pi^2}}$. Probabilities in this distribution can be estimated by first **standardizing the variable** with z-scores, where $z = \frac{x-\mu}{\sigma}$, and then using tabulated values for the standard normal distribution from a table or calculator.

- Normal distributions are often used to make judgments or inferences about issues. Assuming a normal distribution, one can determine how probable a certain observed event is. Then, based on predefined significance levels, one can decide whether the calculated probability refutes the null hypothesis that the event *is* a reasonable result given the assumed distribution. If the null hypothesis is rejected, the hypothesized value is not a good model.

- The Central Limit Theorem states that if all random samples of size n are selected from any population with mean μ and standard deviation σ, then as n increases, the distribution of sample means approaches a normal distribution with mean μ and standard deviation $\frac{\sigma}{\sqrt{n}}$. The theorem can be applied if the original population is itself normally distributed, or if the sample size $n \geq 30$.

Vocabulary

Lesson 11-1
normal distribution
normal curve
concave down
concave up
inflection point
*standard normal curve
*standard normal distribution

Lesson 11-2
Standard Normal Distribution Table

Lesson 11-3
standardizing a variable

Lesson 11-4
simple random sample, SRS
sampling distribution of the mean

Lesson 11-6
confidence level
confidence interval
margin of error

- Based on the Central Limit Theorem, **confidence interval** estimates can be determined for the mean of a population using sample results. The intervals may be narrow or wide, depending on the desired **confidence level** and the sample size. Using the *z*-score, one can determine how large a sample size is needed to ensure a certain **margin of error** at a desired confidence level. As with all experimental results, caution must be taken when making inferences about populations.
- When considering opinion polls in which there are two options, binomial experiment statistics can be used to calculate the margin of error and the confidence interval. When given a sample relative frequency, use $p = \frac{x}{n}$ in the margin-of-error calculation. When there are no sample data to work from, use $p = 0.5$ to have the safest estimate.

Properties and Theorems

Standardization Theorem (p. 699)
Central Limit Theorem (p. 710)
Central Limit Theorem Corollary (p. 710)

Chapter 11 Self-Test

Take this test as you would take a test in class. You will need a calculator. Then use the Selected Answers section in the back of the book to check your work.

1. The graph below models the test scores of students. What transformation is necessary to change the graph to the standard normal curve given a mean score of 76.9 and standard deviation of 13.6?

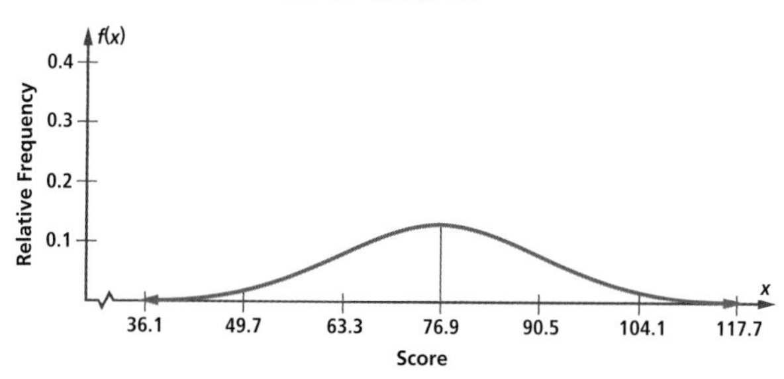

2. The area between the standard normal curve and the z-axis equals what number?
3. The area shaded below, which lies between the standard normal curve and the z-axis, is about 0.8212 . Determine $P(|z| > 0.92)$.

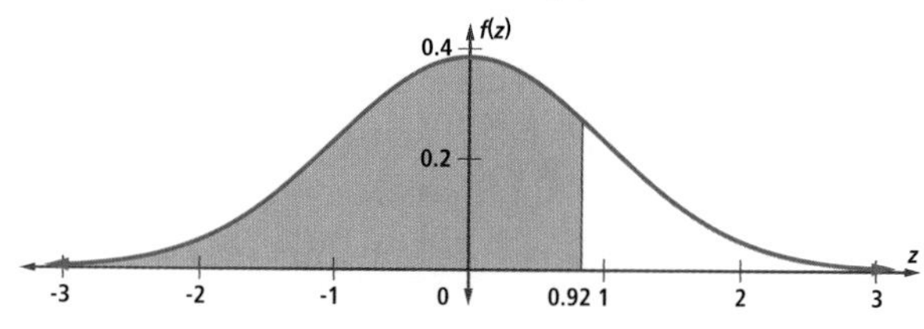

4. Find the probability that a randomly chosen observation z from a standard normal distribution is in the given interval.
   a. $z < 1.7$
   b. $-1.1 < z < 1.1$
5. A sample of 521 students yields a mean SAT Verbal score of 490. Given a population standard deviation of 93, find the 95% confidence interval for the population mean.
6. A November 2007 Harris Poll found that 88% of 2,455 pet owners considered their pets as members of their family.
   a. At the 95% confidence level, what is the margin of error?
   b. If the pollsters had no data on pet owners before the poll, what margin of error would have been quoted?
7. A census of the students at Peaks High School shows that they travel an average of 3.12 miles to school every day with a standard deviation of 0.89 miles. Assuming the distances to be normally distributed, determine the probability that a randomly chosen student at Peaks High travels over 4 miles to school.
8. It is commonly thought that average "normal" body temperature is 98.6°F. Assume we know the population standard deviation of body temperature to be 1.2°F. A doctor's office randomly samples 130 of its patients to gauge this "normal" body temperature estimate. The mean body temperature is found to be 98.2°F. State a null hypothesis H_0 and test whether this sample goes against what is thought to be average "normal" body temperature at the 0.01 level.
9. Suppose the maximum power outputs of a certain brand of speakers are normally distributed with a mean of 130.2 watts and a standard deviation of 7.3 watts. Predict the shape, mean, and standard deviation for the sampling distribution of the mean for samples of size $n = 42$.
10. An Associated Press-Ipsos poll for January 15–17, 2008, found that among 471 Democrats polled, 40% would vote for Hillary Clinton if she ran for the presidency. Compute the 95% confidence interval for the percent of voters who would vote for Hillary Clinton.
11. a. What increase in sample size is necessary to divide the margin of error by 9?
   b. **True or False** If false, change the statement so it is true. For a given sample size, larger populations will have larger margins of error.

Chapter 11 Chapter Review

SKILLS
PROPERTIES
USES
REPRESENTATIONS

SKILLS Procedures used to get answers

OBJECTIVE A Find probabilities in the standard normal distribution. (Lesson 11-2)

In 1–6, use the standard normal distribution command of a CAS or the Standard Normal Distribution Table on page 694.

1. a. What is the probability that a random variable z with a standard normal distribution takes on a value between -1.35 and 1.35?
   b. Determine the area under the standard normal curve from $z = -1.35$ to $z = 1.35$.

In 2–5, evaluate the given probability concerning a z-score.

2. $P(z < 2)$
3. $P(1 < z < 2)$
4. $P(z < -2.15)$
5. $P(z > 2.73)$
6. About 99.7% of the area under the standard normal curve is between $z = -k$ and $z = k$. What is k?

PROPERTIES Principles behind the mathematics

OBJECTIVE B Use properties of normal distributions. (Lessons 11-1, 11-2, 11-3)

7. a. Show that $f(x) = e^{-x^2}$ describes an even function.
   b. What kind of symmetry does this imply for the graph of any normal distribution?

In 8 and 9, fill in the blank.

8. The graph of the parent normal curve has inflection points when $x =$ _?_.
9. The total area between the standard normal curve and the x-axis equals _?_.
10. Give an equation for the normal distribution with mean 120 and standard deviation 8.
11. What transformation maps a normally-distributed data set with mean 120 and standard deviation 8 onto a data set modeled by the standard normal distribution?
12. Match the percent of observations under a normal distribution in the left column with its distance in standard deviations from the mean in the right column.

A	68%	i.	3
B	90%	ii.	1.645
C	95%	iii.	2
D	99.7%	iv.	1.96
E	95.45%	v.	2.58
F	99%	vi.	1

OBJECTIVE C Identify the relationships between sample size, margin of error, and confidence level. (Lessons 11-6, 11-7)

In 13 and 14, tell whether the statement is *true* or *false*. If the statement is false, change it so that it is true.

13. The size of a population does not affect the margin of error calculation in a poll.
14. If the sample size is multiplied by 8 while the confidence level is kept the same, then the margin of error is divided by 8.

In 15 and 16, fill in the blank.

15. The _?_ confidence level is typically used in opinion polls.
16. As the confidence level decreases, the confidence interval _?_.
17. A network was not satisfied with the confidence level in a particular poll with a margin of error of 4%. A pollster was asked to design a new poll, changing the confidence level from 95% to 99% at the same margin of error. If the pollster used the same value for p in the new poll, how many more people were polled?

18. A public opinion organization polled 928 people. If the organization wanted to divide the margin of error by 2.5, how many people would need to be polled?

USES Applications of mathematics in real-world situations

OBJECTIVE D Solve probability problems using normal distributions. (Lesson 11-3)

19. Some movie theaters give patrons a free beverage if they wait more than 10 minutes at the concession stand. If the concession wait times are normally distributed with mean 6.8 minutes and standard deviation 1.9 minutes, what percentage of movie-goers will receive a free beverage?

20. An allergy test requires that small samples of chemicals be injected under the skin of patients. Individuals sensitive to any one of the chemicals have a reaction time that is approximately normally distributed with a mean of 7 hours and a standard deviation of 2 hours. How often can you expect a reaction time of more than 12 hours with such individuals?

21. Based on factory and field tests, a car company determines that the trouble-free mileage for their new model is normally distributed with a mean of 55,700 miles and a standard deviation of 19,600 miles. The car is marketed with a 60,000-mile guarantee. Determine the percentage of cars of this make the company should expect to have trouble before 60,000 miles.

22. The elevator weight limit in a particular building is marked at 3000 pounds. The total weight of a random sample of 15 elevator users is modeled by a normal distribution with mean 2315 pounds and standard deviation 211 pounds. What is the probability that 15 elevator users exceed the weight limit?

OBJECTIVE E Use normal distributions to test hypotheses. (Lesson 11-5)

23. A manufacturing process produces cables that have a breaking strength with a mean of 1700 pounds and a standard deviation of 120 pounds. It is claimed that a new process will increase the breaking strength. A sample of 80 cables was used to test this claim. The sample mean breaking strength was 1750 pounds. Test the hypothesis that the breaking strength has not changed with the new process at the 0.01 level.

24. An airline states that the average weight of checked luggage per person on a domestic flight is 30 kg with standard deviation of 12.3 kg. For flight safety purposes, they randomly sample 100 of their customers and study the weight of their checked luggage. The mean weight from this sample is found to be 27.9 kg. Can the airline officers still claim that the average weight of checked luggage is 30 kilograms? Use a 0.01 significance level.

25. A highway patrol group decides to check the results they have read in a report, which claimed that the speed of cars in their region of the highway has a mean of 75 miles per hour and standard deviation of 11 miles per hour. They track 50 cars every day for a month and record the average speed observed at the end of each day. Suppose that on a particular day they calculate an average speed of 69.4 mph. Can the highway patrol claim that the mean speed of cars on the highway given by the report has not changed using a 0.05 significance level?

OBJECTIVE F Apply the Central Limit Theorem. (Lessons 11-4, 11-5)

In 26 and 27, fill in the blank.

26. If a distribution of population data is known not to be normal, one needs to use large samples of $n \geq$ __?__ to apply the Central Limit Theorem.

27. Given a population with mean μ and standard deviation σ, the standard deviation of the sampling distribution of the mean for samples of size $n \geq 30$ from this population is approximately __?__.

28. Suppose that the weights of full-grown oranges in a citrus grove are normally distributed with mean 8.4 ounces and standard deviation 2.2 ounces.
    a. Determine the mean and standard deviation of the set of means of all possible samples of 75 oranges randomly picked from this grove.
    b. Find the probability that a sample of 75 oranges has a mean weight of 7.9 ounces or less.

29. A vitamin manufacturer reports that the average weight of their vitamin C tablets is 500 mg with a standard deviation of 2 mg. To check the veracity of this claim, a consumer protection group randomly samples 100 tablets from each lot produced and records the mean weight of each sample. Predict the shape, mean, and standard deviation for the sampling distribution of the mean of the sample data collected by this group.

OBJECTIVE G Apply confidence intervals to real-world problems. (Lessons 11-6, 11-7)

In 30 and 31, suppose a sample of 500 SAT-M scores for freshmen entering Fictitious University had $\overline{x} = 590$. The College Board reported a standard deviation of all tests takers of $\sigma = 50$.

30. Find the confidence interval for the population mean μ for a 95% level of confidence.

31. Find the margin of error for μ for a 90% level of confidence.

In 32–35, fill in the blank.

32. When the population standard deviation is known, we can use a sample mean to estimate a confidence interval for the __?__.

33. The 99% confidence interval for the population mean when σ is known is __?__.

34. In polls, the confidence interval $\frac{x}{n} \pm z\sqrt{\frac{p(1-p)}{n}}$ has margin of error __?__.

35. When pollsters design polls and don't have any prior sample or estimate to work from, they use $p =$ __?__.

In 36–38, a Gallup poll in November of 2008 asked 1009 American adults aged 18 or older, "Within the last 12 months, have you or a member of your family put off any sort of medical treatment because of the cost you would have to pay?"

36. **True or False** If the poll concluded that 29% responded yes, then you can use $p = 0.29$ in the confidence interval and margin of error calculations.

37. 29% of those polled answered yes. Therefore, there is a 95% chance that if all Americans were polled, the percent of those who responded yes would fall between what limits?

38. If the margin of error was *reported* as $\pm$ 3.1%, what confidence level does this poll have?

In 39 and 40, the Associated Press (AP) conducted a poll of 1029 adults in February 2008. 48% supported Democratic presidential candidate Barack Obama over other candidates.

39. Use the sample data to compute the 95% confidence interval for the percent of voters who supported Obama.

40. The AP reported a margin of error for the 520 Democrats polled of 4.3 percent. The margin of error for the 357 Republicans was 5.2 percent. Verify that the AP used the same confidence level for both the Republicans and Democrats.

REPRESENTATIONS Pictures, graphs, or objects that illustrate concepts

OBJECTIVE H Interpret graphs of normal distributions. (Lessons 11-1, 11-2)

41. Below are the graphs of three normal distributions. Identify each.
    a. the standard normal distribution
    b. the distribution with $\sigma = 2$
    c. the distribution with $\sigma = \frac{1}{2}$

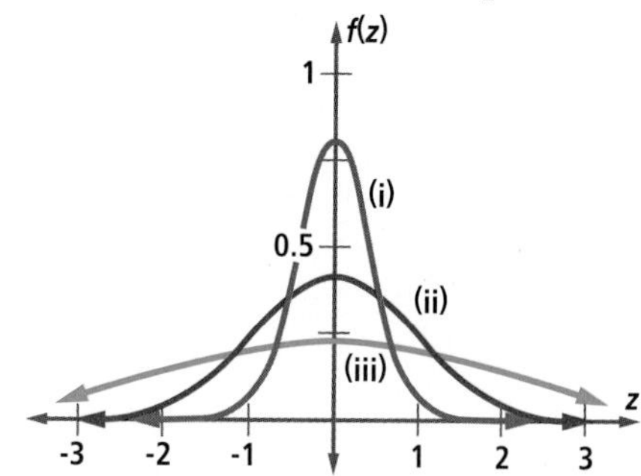

In 42–44, the region shaded below between the standard normal curve and the *z*-axis has an area of about 0.9545. Determine the following probabilities.

42. $P(0 < z < 2)$
43. $P(z < -2)$
44. $P(z < 2)$

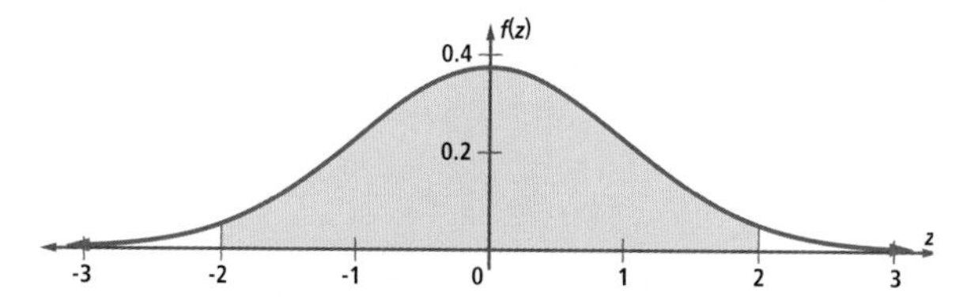

In 45–47, suppose that the variable *z* is normally distributed with mean 0 and standard deviation 1, and that the probability that *z* is between –*a* and *a* is *r*. Express each of the following in terms of *r*.

45. $P(0 < z < a)$
46. $P(z < -a)$
47. $P(z < a)$

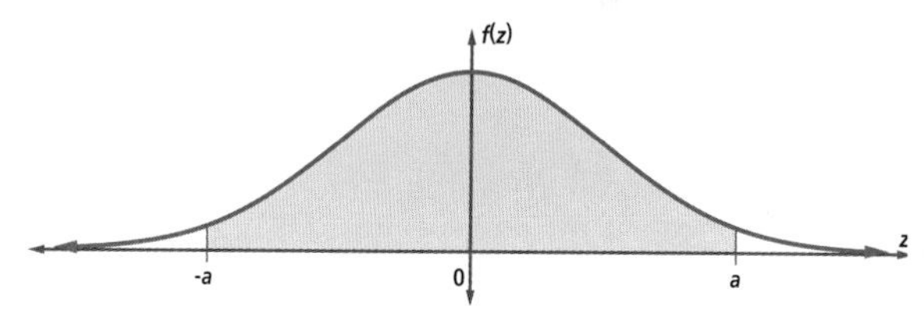

Chapter

12

Matrices and Trigonometry

Contents

The rectangular arrays of numbers called matrices have numerous applications. In earlier courses, you may have seen matrices used to solve systems, to generate and decipher codes, and to represent geometric transformations.

Have you ever wondered how computer games produce animated images? A standard programming technique uses matrices and trigonometry.

For instance, to generate an animation with monkeys, a single monkey might be described by a set of points. These data are stored in a matrix in which each column represents a key point on the monkey. Then the figure is changed by transformations such as scale changes, translations, or rotations.

These transformations can also be stored as matrices. Multiplying a matrix representing a transformation by the matrix representing a figure gives a matrix for the image of the original figure.

For example, the matrix below represents a transformation that translates the figure 82 pixels to the right and 37 pixels up and rotates it 22.5° clockwise about the center of the monkey, as was done repeatedly with the monkey shown below.

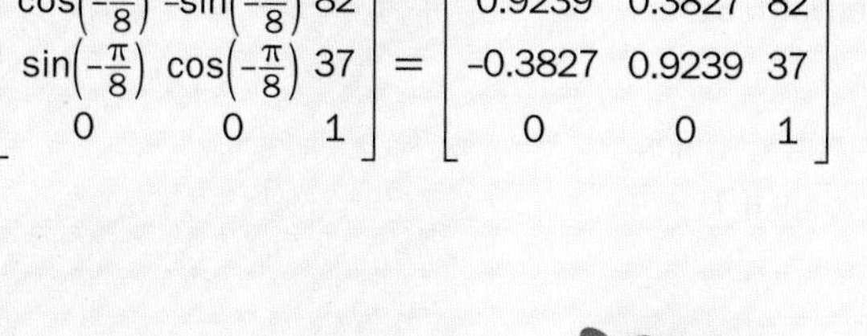

$$\begin{bmatrix} \cos\left(-\frac{\pi}{8}\right) & -\sin\left(-\frac{\pi}{8}\right) & 82 \\ \sin\left(-\frac{\pi}{8}\right) & \cos\left(-\frac{\pi}{8}\right) & 37 \\ 0 & 0 & 1 \end{bmatrix} = \begin{bmatrix} 0.9239 & 0.3827 & 82 \\ -0.3827 & 0.9239 & 37 \\ 0 & 0 & 1 \end{bmatrix}$$

In this chapter, you will study matrices for several transformations, including some described using trigonometric functions. These matrices are applied to transform figures and to derive some important trigonometric identities.

Lesson 12-1 Matrix Multiplication

Vocabulary

matrix, matrices
element of a matrix
dimensions of a matrix
row matrix
column matrix
product matrix

BIG IDEA Matrices organize data in useful ways to make computation methodical.

Mental Math

If A's are worth 4 points, B's 3 points, and C's 2 points, what is the grade point average of a person with 3 A's, 2 B's, and 1 C, to the nearest tenth?

A **matrix** (plural **matrices**) is a rectangular arrangement of objects. We identify a matrix by enclosing its rectangular array in square brackets. Each object in the array is called an **element**. Matrices are often used to store data. Here is an example.

Peter Piper's Pizza Place peddles personal pizzas. The matrix Q below contains the quantities of each type of pizza Peter Piper's Pizza Place sold last Saturday at lunch and dinner.

$$\begin{array}{r} \\ \\ \text{lunch} \\ \text{dinner} \end{array} \begin{array}{cccc} \text{four} & & \text{sausage and} & \\ \text{cheese} & \text{veggie} & \text{mushroom} & \text{supreme} \end{array}$$
$$\begin{matrix} \text{lunch} \\ \text{dinner} \end{matrix} \begin{bmatrix} 75 & 60 & 90 & 80 \\ 90 & 100 & 80 & 140 \end{bmatrix} = Q$$

The matrix Q has 2 rows, labeled lunch and dinner, and 4 columns, one for each type of pizza. The labels are not part of the matrix. Each element may be identified by its row and column. For instance, the element in row 2, column 1 is 90. Q is said to have *dimensions* 2 by 4, written 2×4. In general, a matrix with m rows and n columns has **dimensions** $m \times n$. It is an $m \times n$ matrix.

Example 1

Peter Piper's pizza prices are \$3.00, \$3.50, \$4.00, and \$5.00 for the four cheese, veggie, sausage and mushroom, and supreme, respectively. Store these values in a matrix.

Solution There are two possible matrices. One is a 4×1 matrix. It has 4 rows and 1 column, and is matrix U below at the left. The other is a 1×4 matrix, and is matrix V below at the right.

$$U = \begin{bmatrix} 3.00 \\ 3.50 \\ 4.00 \\ 5.00 \end{bmatrix} \quad \text{or} \quad V = [3.00 \quad 3.50 \quad 4.00 \quad 5.00]$$

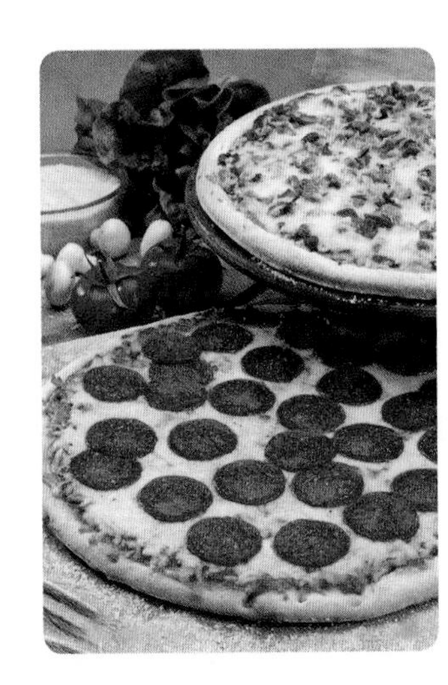

It is important to note that in Example 1, changing the order of prices would not be consistent with the original matrix above. The prices should be in the same order as the pizza types in matrix Q. It is important to always be consistent in setting up matrices.

Multiplying a Row and a Column

To calculate the total amount of money (the *revenue*) Peter Piper's received during lunch, you can multiply the quantity sold of each item by its cost, and then add the results. For lunch, the revenue is $1195 because 75(3.00) + 60(3.50) + 90(4.00) + 80(5.00) = 1195.00. This can be considered as the product of a matrix of quantities and a matrix of prices. The quantity matrix is called a **row matrix**, because it consists of a single row. The price matrix is a **column matrix**, because it consists of a single column. The product of a row matrix and a column matrix is a matrix with a single element whose dimensions are 1 × 1.

$$\underbrace{[75 \quad 60 \quad 90 \quad 80]}_{\text{row matrix}} \cdot \underbrace{\begin{bmatrix} 3.00 \\ 3.50 \\ 4.00 \\ 5.00 \end{bmatrix}}_{\text{column matrix}} = [1195.00]$$

 QY1

▶ QY1

a. Represent the dinner revenue as a product of a row matrix and a column matrix.

b. Show that the revenue received during dinner is $1640.00.

Multiplying Entire Matrices

The separate lunch and dinner matrix revenue calculations can be done together. The result is the product of the original 2 × 4 matrix Q for quantities and the 4 × 1 cost matrix U from Example 1. The product $Q \cdot U$ represents the revenue for Saturday.

$$\begin{array}{c} \\ \text{lunch} \\ \text{dinner} \end{array} \overset{\begin{array}{cccc} \text{four cheese} & \text{veggie} & \text{sausage and mushroom} & \text{supreme} \end{array}}{\begin{bmatrix} 75 & 60 & 90 & 80 \\ 90 & 100 & 80 & 140 \end{bmatrix}} \cdot \begin{bmatrix} 3.00 \\ 3.50 \\ 4.00 \\ 5.00 \end{bmatrix} \begin{array}{l} \text{four cheese} \\ \text{veggie} \\ \text{sausage and mushroom} \\ \text{supreme} \end{array}$$

$$Q \cdot U$$

In general, *matrix multiplication* is done using rows from the left matrix and columns from the right matrix. Multiply the first element in the row by the first element in the column, the second element in the row by the second element in the column, and so on. Finally, add the resulting products. The shading below shows the first row times the column, which gives a total of $1195.00. Similarly, the second row times the column gives $1640.00. Both results are represented in the 2 × 1 matrix R.

$$\underbrace{\begin{bmatrix} 75 & 60 & 90 & 80 \\ 90 & 100 & 80 & 140 \end{bmatrix}}_{Q} \cdot \underbrace{\begin{bmatrix} 3.00 \\ 3.50 \\ 4.00 \\ 5.00 \end{bmatrix}}_{U} = \underbrace{\begin{bmatrix} 1195.00 \\ 1640.00 \end{bmatrix}}_{R} \begin{array}{l} \text{revenue from lunch} \\ \text{revenue from dinner} \end{array}$$

Notice how the dimensions of the original matrices Q and U are related. The number of elements in each row of Q equals the number of elements in each column of U. Each element of Q's rows is the quantity of a kind of pizza, and each element of U's column is the corresponding price.

In general, the product $A \cdot B$ of two matrices A and B exists if and only if the number of columns of A equals the number of rows of B.

The dimensions of the product matrix R can be determined from the dimensions of the original two matrices. Each row of Q corresponds to a row in the product, because each row is a different meal. Similarly, the columns in the product correspond to products in the second factor, matrix U.

We can write:

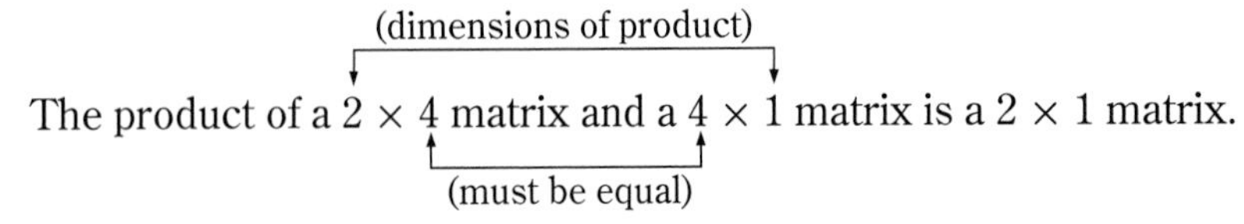

Definition of Product Matrix

Suppose A is an $m \times n$ matrix and B is an $n \times p$ matrix. Then the **product matrix** $A \cdot B$ is an $m \times p$ matrix. The element in row i and column j of the product is the sum of the products of elements in row i of A and corresponding elements in column j of B.

As is the case with real or complex number multiplication, often we write AB for the product $A \cdot B$.

To apply the definition and find the product of two matrices by hand:

First, determine whether the matrices can be multiplied.

Next, find the dimensions of the product matrix.

Then, make space for the elements of the product matrix.

Last, find the value of each element of the product matrix using the appropriate row-by-column multiplication.

This process is described in Example 2.

GUIDED

Example 2

Consider $A = \begin{bmatrix} 2 & 0 \\ 1 & 3 \end{bmatrix}$ and $B = \begin{bmatrix} 5 & -3 & 2 \\ 1 & 0 & 4 \end{bmatrix}$. Find the matrix if it exists.

a. AB

b. BA

Solution

a. By the definition of matrix multiplication, the product of a 2×2 and a 2×3 matrix is a __?__ $\times$ __?__ matrix. This means you should expect to find 6 elements in the product matrix.

The product of row 1 of A and column 1 of B is $2 \cdot 5 + 0 \cdot 1 = 10$.

Write this in the 1st row and 1st column of the product matrix.

$$\begin{bmatrix} 2 & 0 \\ 1 & 3 \end{bmatrix} \cdot \begin{bmatrix} 5 & -3 & 2 \\ 1 & 0 & 4 \end{bmatrix} = \begin{bmatrix} 10 & & \\ & & \end{bmatrix}$$

The product of row 1 of A and column 2 of B is $2 \cdot$ __?__ $+ 0 \cdot$ __?__ $=$ __?__. This is the element in row 1 and column 2 of the product.

$$\begin{bmatrix} 2 & 0 \\ 1 & 3 \end{bmatrix} \cdot \begin{bmatrix} 5 & -3 & 2 \\ 1 & 0 & 4 \end{bmatrix} = \begin{bmatrix} 10 & ? & \\ & & \end{bmatrix}$$

The other four elements of the product matrix are found by using the same row by column pattern. For instance, the element in the 2nd row, 3rd column of AB is found by multiplying the 2nd row of A by the 3rd column of B.

$$\begin{bmatrix} 2 & 0 \\ 1 & 3 \end{bmatrix} \cdot \begin{bmatrix} 5 & -3 & 2 \\ 1 & 0 & 4 \end{bmatrix} = \begin{bmatrix} 10 & -6 & ? \\ ? & ? & ? \end{bmatrix}$$

Find the remaining elements of the product AB.

b. To calculate the element in the 1st row, 1st column of BA you would have to multiply the shaded row and column.

$$\begin{bmatrix} 5 & -3 & 2 \\ 1 & 0 & 4 \end{bmatrix} \cdot \begin{bmatrix} 2 & 0 \\ 1 & 3 \end{bmatrix}$$

The product BA does not exist. A 2×3 matrix cannot be multiplied by a 2×2 matrix in this order because the number of columns in the left matrix does not equal the number of rows in the second matrix.

Check Check the products using a calculator.

a.

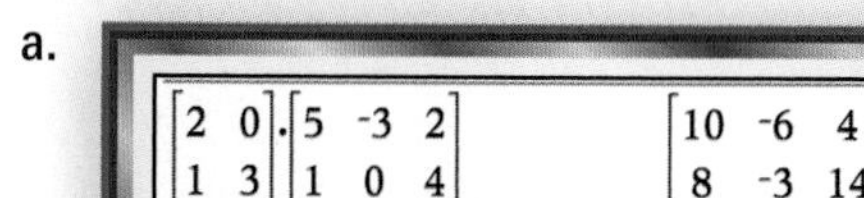

b.

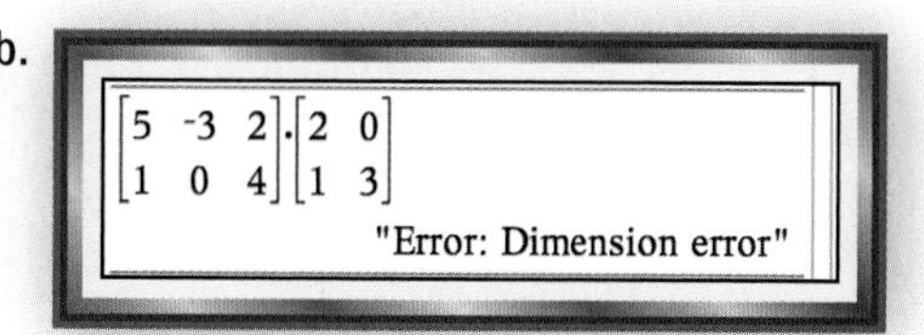

Properties of Matrix Multiplication

As Example 2 illustrates, matrix multiplication is not commutative. However, matrix multiplication is associative. That is, for matrices A, B, and C where the applicable products exist, $(AB)C = A(BC)$. In Example 3, one product of three matrices is calculated. In Question 14, you will verify associativity for this case.

Example 3

Suppose that Peter Piper's profit margin on Saturday is 32% during lunch hours and 39% during dinner hours. Let $M = [0.32 \quad 0.39]$ represent the profit margins. Calculate $M(QU)$, and describe what this product represents.

Solution Earlier we calculated

$$QU = \begin{bmatrix} 75 & 60 & 90 & 80 \\ 90 & 100 & 80 & 140 \end{bmatrix} \cdot \begin{bmatrix} 3.00 \\ 3.50 \\ 4.00 \\ 5.00 \end{bmatrix} = \begin{bmatrix} 1195 \\ 1640 \end{bmatrix}.$$

Thus, $M(QU) = [0.32 \quad 0.39] \cdot \begin{bmatrix} 1195 \\ 1640 \end{bmatrix} = [1022].$

The product matrix [1022] represents the total profit of $1022.00 earned by Peter Piper's during lunch and dinner on Saturday.

QY2

> **QY2**
> Why is the matrix M in Example 3 $[0.32 \quad 0.39]$ rather than $\begin{bmatrix} 0.32 \\ 0.39 \end{bmatrix}$?

Questions

COVERING THE IDEAS

1. What is a *matrix*?

In 2–5, use the matrix at the right. It displays the number of members of the U.S. Congress of each gender at the conclusion of that Congress. For the 110th Congress, figures are current as of 2008.

	male	female
107th	460	75
108th	458	77
109th	450	85
110th	445	90

2. What are the dimensions of this matrix?
3. What is the element in row 3, column 2?
4. If a_{ij} represents the element in row i and column j, what is a_{21}?
5. What does the sum of the elements in each row represent?
6. If A and B are matrices, under what circumstances does the product AB exist?

In 7 and 8, give the dimensions of AB if it exists.

7. A is 3×7, B is 7×4
8. A is 1×6, B is 8×1

In 9–12, multiply.

9. $[2 \quad 8] \cdot \begin{bmatrix} 6 \\ 2 \end{bmatrix}$
10. $\begin{bmatrix} 5 & 3 & 3 \\ 0 & 7 & 2 \end{bmatrix} \cdot \begin{bmatrix} -1 & 0 & 100 \\ 1 & 0 & 10 \\ -2 & 1 & 1 \end{bmatrix}$
11. $\begin{bmatrix} a & b \\ c & d \end{bmatrix} \cdot \begin{bmatrix} x \\ y \end{bmatrix}$
12. $\begin{bmatrix} 1 & 0 \\ 0 & 1 \end{bmatrix} \cdot \begin{bmatrix} -2 & 3 \\ 1 & 4 \end{bmatrix}$
13. Suppose M is a 4×5 matrix. Give the dimensions of a matrix N for which both MN and NM exist.
14. In Example 3, the elements of the matrix $M(QU)$ were found. Calculate $(MQ)U$. That is, find the product matrix MQ, then multiply this result by U. Does $(MQ)U = M(QU)$?

House Financial Services Chairman Barney Frank and Speaker Nancy Pelosi in 2008

APPLYING THE MATHEMATICS

In 15 and 16, the matrix N gives the number of tickets sold for a play. The matrix C gives the unit cost in dollars for each ticket.

$$N = \begin{matrix} & \text{Adults} & \text{Children} \\ \text{Weekday Matinee} & 274 & 335 \\ \text{Weekend Matinee} & 273 & 315 \\ \text{Weekend Evening} & 165 & 395 \end{matrix} \qquad C = \begin{bmatrix} 40.00 \\ 24.00 \end{bmatrix} \begin{matrix} \text{Adults} \\ \text{Children} \end{matrix}$$

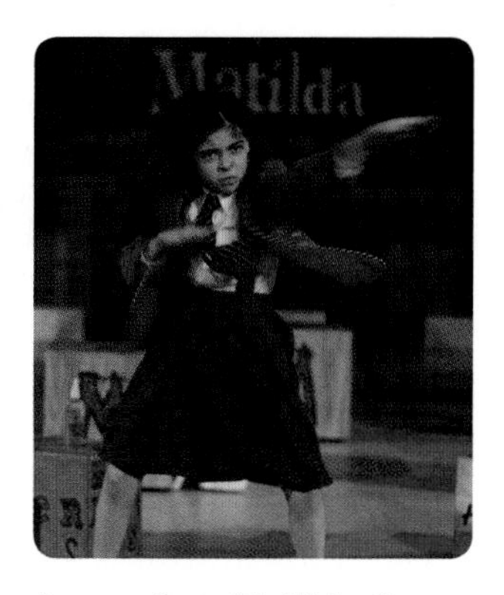

A scene from ***Matilda the Musical***

15. a. Find NC.
    b. What do the elements of NC represent?
    c. What was the theater's total revenue for the weekday performance?

16. A portion of the receipts for each performance goes to a children's health charity: 40% for the weekday performance and 30% for each of the weekend performances.
    a. Define an appropriate matrix for these portions.
    b. Find the total contribution to charity for all three performances.

In 17 and 18, solve for the variables.

17. $\begin{bmatrix} 2 & 3 \\ 1 & 2 \end{bmatrix} \cdot \begin{bmatrix} x \\ 1 \end{bmatrix} = \begin{bmatrix} -1 \\ 0 \end{bmatrix}$

18. $\begin{bmatrix} 3 & a \\ 2 & b \end{bmatrix} \cdot \begin{bmatrix} 5 \\ 6 \end{bmatrix} = \begin{bmatrix} -7 \\ 8 \end{bmatrix}$

In 19 and 20, a *square matrix* is one with the same number of rows as columns. Square matrices can be multiplied by themselves. If M is a square matrix, then $M^2 = M \cdot M$, $M^3 = M \cdot M^2$, and so on.

19. Let $M = \begin{bmatrix} 0 & -3 \\ -3 & 0 \end{bmatrix}$.
    a. Calculate M^2.
    b. Calculate M^3.
    c. Find the formula for the elements of M^n, where n is a positive integer.

20. Calculate A^2 and A^4 when $A = \begin{bmatrix} 1 & b \\ b & 1 \end{bmatrix}$.

21. The diagram at the right shows the major highways connecting four cities.

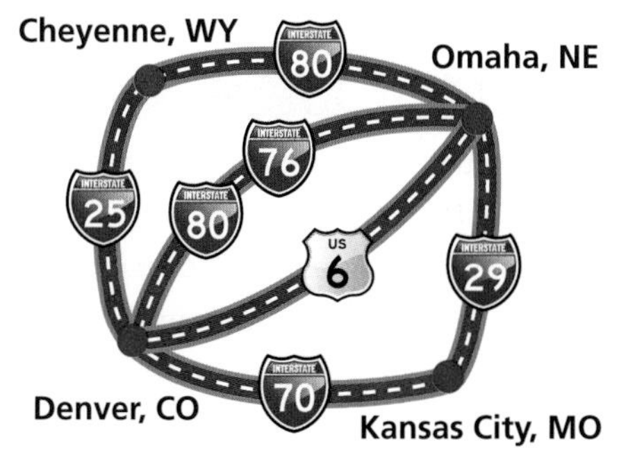

    a. Write the number of direct routes (not through any other city on the diagram) between each pair of cities in a matrix as begun below. (I-80 and I-76 are considered to be the same route.)

$$\begin{matrix} & \text{Denver} & \text{Cheyenne} & \text{Omaha} & \text{Kansas City} \\ \text{Denver} & 0 & 1 & 2 & 1 \\ \text{Cheyenne} & ? & ? & ? & ? \\ \text{Omaha} & ? & ? & ? & ? \\ \text{Kansas City} & ? & ? & ? & ? \end{matrix}$$

    b. Multiply the matrix from Part a by itself and interpret what the product signifies.

REVIEW

22. a. Use a CAS to expand $\left(\frac{p}{2} + 2q\right)^5$.
    b. Check the middle terms using the Binomial Theorem. **(Lesson 10-3)**

23. Match each function with a characteristic. **(Lesson 7-1)**

a. $A: x \rightarrow 5x - x^2 - 2$	i no relative extrema
b. $C: x \rightarrow -8x - 3$	ii one relative maximum
c. $D: x \rightarrow (x - 9)^2 - 4$	iii one relative minimum
d. $E: x \rightarrow x^3 - 7x$	iv both a relative maximum and a relative minimum

24. **True or False** Suppose that $(x - 6)$ is a factor of the polynomial $f(x)$.
    a. 6 must be a solution to the equation $f(x) = 0$.
    b. -6 must be a solution to the equation $f(x) = 0$.
    c. The graph of $f(x)$ intersects the x-axis at (6, 0).
    d. $f(6) = 0$. **(Lesson 7-4)**

25. **Fill in the Blank** If A and B are mutually exclusive events in the same sample space, then $P(A \cup B) =$ __?__. **(Lesson 6-2)**

26. Match the equation for each transformation with the *best* description. **(Lessons 3-5, 3-4, 3-3)**

a. $M(x, y) = (x + 5, y - 1)$	i reflection over the x-axis
b. $N(x, y) = (x, -y)$	ii reflection over the line $y = x$
c. $P(x, y) = (y, x)$	iii scale change
d. $Q(x, y) = \left(\frac{x}{7}, \frac{y}{7}\right)$	iv size change
e. $V(x, y) = (0.2x, 20y)$	v translation

27. Captain Jack's map showed treasure buried 130 paces north of Penguin Rock, then 50 paces east. To the nearest tenth, how many paces is the treasure from Penguin Rock? **(Previous Course)**

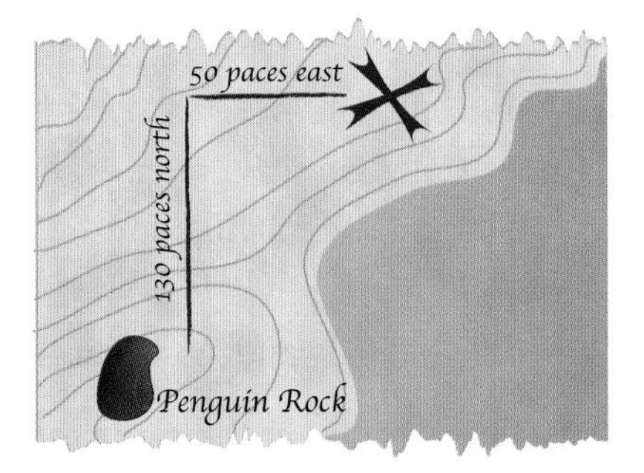

EXPLORATION

28. Matrices frequently appear in newspapers, though they are not usually identified with brackets.
    a. Find an example of a matrix in a newspaper.
    b. State its dimensions.
    c. Describe what each row and column represents.

QY ANSWERS

1. a. $[90 \ \ 100 \ \ 80 \ \ 140] \cdot \begin{bmatrix} 3.00 \\ 3.50 \\ 4.00 \\ 5.00 \end{bmatrix}$

b. $90(\$3.00) + 100(\$3.50) + 80(\$4.00) + 140(\$5.00) = \$1640.00$

2. Since QU is a 2×1 matrix, M must be a 1×2 matrix for their product to exist.

Lesson 12-2 Matrices for Transformations

Vocabulary

point matrix
matrix representing a transformation
identity transformation
2 x 2 identity matrix for multiplication

BIG IDEA Matrix operations can translate to geometric transformations on the coordinate plane.

Representing a Geometric Figure by a Matrix

Matrices can be used to represent geometric figures as well as numerical data. To do so, let the point (a, b) be written as the matrix $\begin{bmatrix} a \\ b \end{bmatrix}$. Such a matrix is sometimes called a **point matrix**. Then a polygon with n sides can be represented by a $2 \times n$ matrix in which the columns are the coordinates of consecutive vertices. For instance, $\triangle QRS$ at the right can be represented by the matrix $\begin{bmatrix} -1 & 2 & 4 \\ 0 & -2 & 3 \end{bmatrix}$ (columns Q, R, S). It is important that the order of the columns corresponds to the order of the vertices. If the triangle were named $\triangle SRQ$, it would be represented by $\begin{bmatrix} 4 & 2 & -1 \\ 3 & -2 & 0 \end{bmatrix}$.

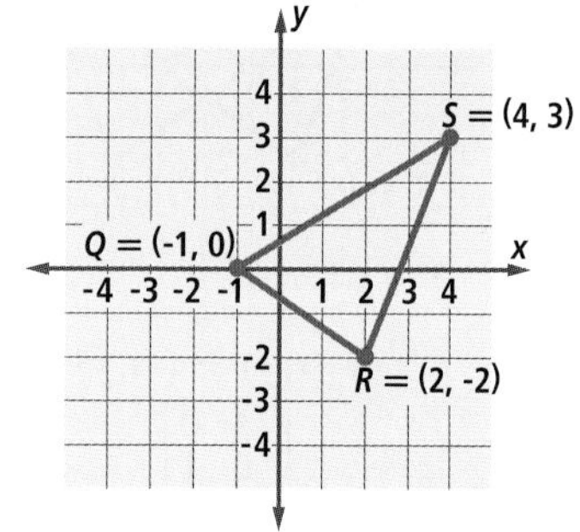

STOP QY1

Mental Math

Give the image of the point (x, y) under each transformation.

a. $r_{x\text{-axis}}$

b. $r_{y\text{-axis}}$

c. $r_{y=x}$

QY1

In the diagram above, name the triangle represented by $\begin{bmatrix} 2 & -1 & 4 \\ -2 & 0 & 3 \end{bmatrix}$.

Representing a Transformation by a Matrix

Matrices may also represent transformations.

Example 1

a. Multiply the matrix for $\triangle QRS$ above by the matrix $\begin{bmatrix} 1 & 0 \\ 0 & -1 \end{bmatrix}$ and graph the resulting image, $\triangle Q'R'S'$.

b. Describe the transformation represented by the matrix.

Solution

a. $\begin{bmatrix} 1 & 0 \\ 0 & -1 \end{bmatrix} \cdot \begin{bmatrix} -1 & 2 & 4 \\ 0 & -2 & 3 \end{bmatrix} = \begin{bmatrix} -1 & 2 & 4 \\ 0 & 2 & -3 \end{bmatrix}$

(columns Q, R, S map to Q', R', S')

b. $\triangle Q'R'S'$ is the reflection image of $\triangle QRS$ over the x-axis.

The triangle and its image are graphed at the right.

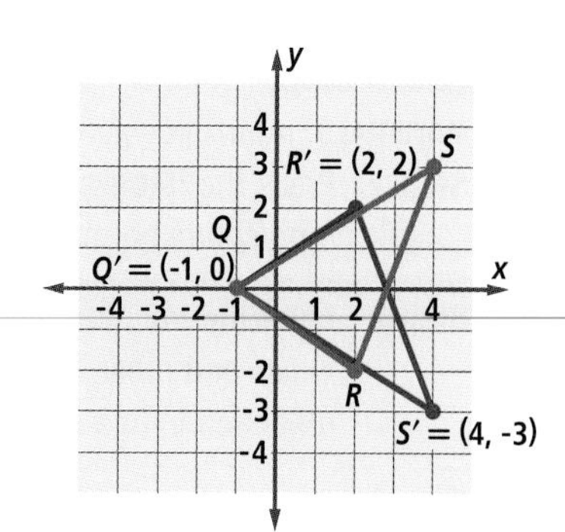

In general, multiplying $\begin{bmatrix} 1 & 0 \\ 0 & -1 \end{bmatrix}$ by any point matrix $\begin{bmatrix} x \\ y \end{bmatrix}$ gives $\begin{bmatrix} 1 & 0 \\ 0 & -1 \end{bmatrix} \cdot \begin{bmatrix} x \\ y \end{bmatrix} = \begin{bmatrix} x \\ -y \end{bmatrix}$. This means that multiplication by $\begin{bmatrix} 1 & 0 \\ 0 & -1 \end{bmatrix}$ maps (x, y) to $(x, -y)$. This is the transformation r_x, reflection over the x-axis. (Notice that multiplying in the opposite order is impossible.)

The above result can be generalized. Let T be a transformation and F be a matrix for a geometric figure. If, for all figures F, the product MF is a matrix for $T(F)$, then M is called the **matrix representing the transformation** T, or the matrix for T.

Matrices for Reflections

Below are the matrices for three important reflections.

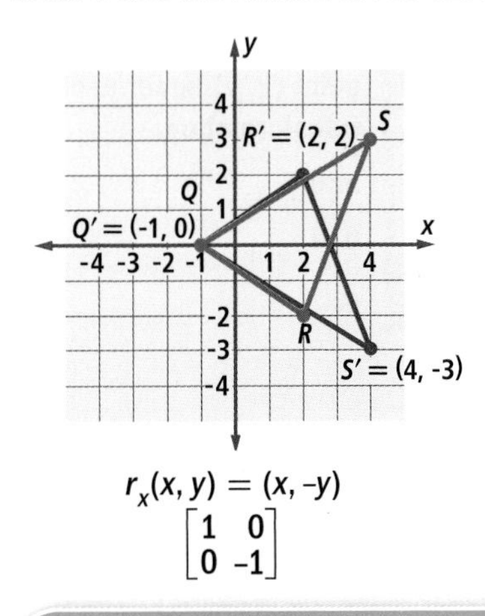

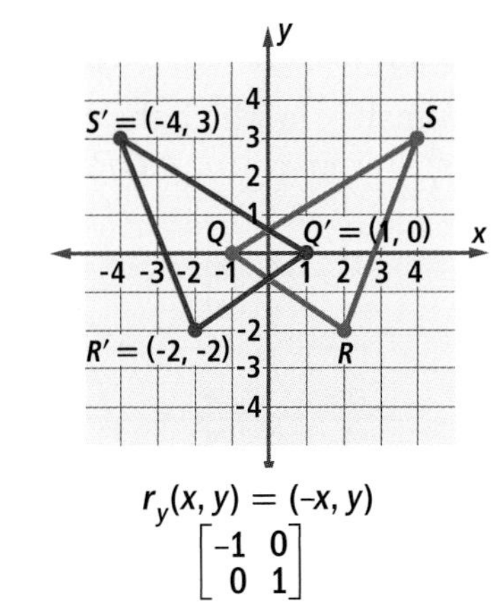

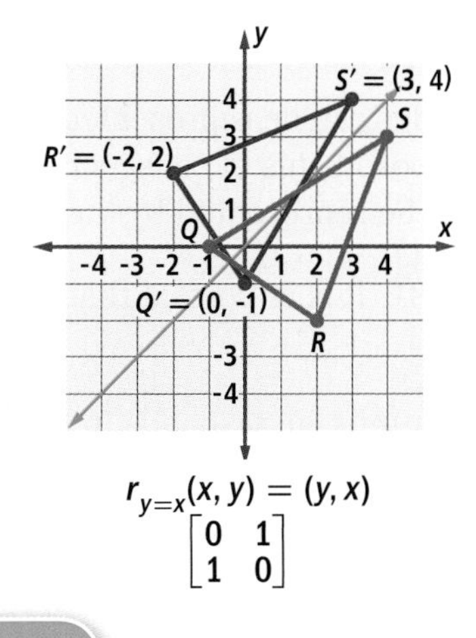

Theorem (Matrices for Reflections)

is the matrix for r_x, reflection over the x-axis.

is the matrix for r_y, reflection over the y-axis.

is the matrix for $r_{y=x}$, reflection over the line $y = x$.

Activity 1

MATERIALS matrix polygon application provided by your teacher

Step 1 Press the "Monkey!" button to plot the stored 2×44 matrix representing a monkey.

Step 2 Enter the transformation matrix for a reflection over the x-axis and press the "Transform" button to graph the reflection image of the monkey.

Step 3 Clear the grid. Repeat Steps 1 and 2 using the matrix for a reflection over the line $y = x$.

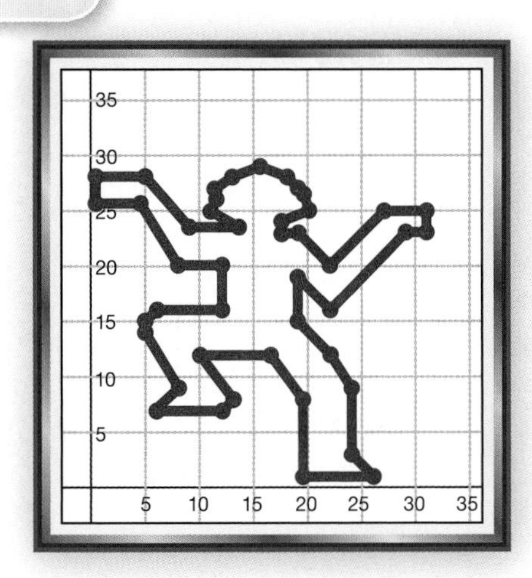

Matrices for Size and Scale Changes

Matrices can represent transformations other than reflections.

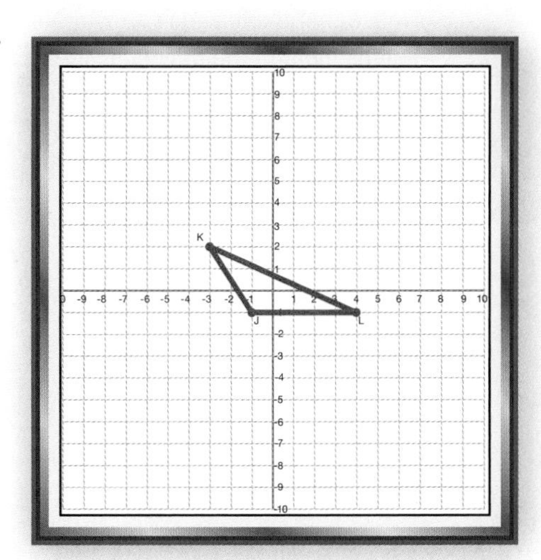

Activity 2

MATERIALS matrix polygon application or graph paper and pencil

Step 1 Graph the preimage $\triangle JKL$ with $J = (-1, -1)$, $K = (-3, 2)$, and $L = (4, -1)$.

Step 2 Graph the product of $\begin{bmatrix} 2 & 0 \\ 0 & 2 \end{bmatrix}$ and the matrix for $\triangle JKL$.

Step 3 Repeat Step 2 for the products of

a. $\begin{bmatrix} 3 & 0 \\ 0 & 3 \end{bmatrix}$ and the matrix for $\triangle JKL$.

b. $\begin{bmatrix} 0.5 & 0 \\ 0 & 0.5 \end{bmatrix}$ and the matrix for $\triangle JKL$.

Step 4 What transformation does $\begin{bmatrix} k & 0 \\ 0 & k \end{bmatrix}$ represent?

The Activity shows that a size change of magnitude k for $k \neq 0$ is represented by the matrix $\begin{bmatrix} k & 0 \\ 0 & k \end{bmatrix}$. The matrix $\begin{bmatrix} 1 & 0 \\ 0 & 1 \end{bmatrix}$ is special. Because $\begin{bmatrix} 1 & 0 \\ 0 & 1 \end{bmatrix} \cdot \begin{bmatrix} x \\ y \end{bmatrix} = \begin{bmatrix} x \\ y \end{bmatrix}$, the transformation represented by this matrix maps any point (x, y) to itself, and is the **identity transformation**. Thus, $\begin{bmatrix} 1 & 0 \\ 0 & 1 \end{bmatrix}$ is called the **2 × 2 identity matrix for multiplication**. A matrix of the form $\begin{bmatrix} a & 0 \\ 0 & b \end{bmatrix}$ with $a \neq 0$, $b \neq 0$ represents the scale change $S(x, y) = (ax, by)$. Recall that this transformation stretches a preimage horizontally by a and vertically by b and is denoted $S_{a,b}$.

QY2

QY2

Calculate $\begin{bmatrix} 1 & 0 \\ 0 & 1 \end{bmatrix} \cdot \begin{bmatrix} a & b \\ c & d \end{bmatrix}$.

GUIDED

Example 2

a. Apply the transformation represented by $\begin{bmatrix} 3 & 0 \\ 0 & 2 \end{bmatrix}$ to the figure below.

b. Identify the transformation.

Solution

a. The large rectangle can be represented by $\begin{bmatrix} 6 & 6 & -5 & ? \\ 6 & -3 & -3 & ? \end{bmatrix}$.

The image of the rectangle is represented by this product:

$\begin{bmatrix} 3 & 0 \\ 0 & 2 \end{bmatrix} \cdot \begin{bmatrix} 6 & 6 & -5 & ? \\ 6 & -3 & -3 & ? \end{bmatrix} = \begin{bmatrix} 18 & ? & ? & ? \\ 12 & ? & ? & ? \end{bmatrix}$.

The images of some of the interior figures are given by:

right eye $\begin{bmatrix} 3 & 0 \\ 0 & 2 \end{bmatrix} \cdot \begin{bmatrix} -1 & -3 & -3 & -1 \\ 4 & 4 & 2 & 2 \end{bmatrix} = \begin{bmatrix} -3 & -9 & -9 & -3 \\ 8 & 8 & 4 & 4 \end{bmatrix}$

nose $\begin{bmatrix} 3 & 0 \\ 0 & 2 \end{bmatrix} \cdot \begin{bmatrix} 0 & 1 & 2 \\ 1 & 2 & 1 \end{bmatrix} = \begin{bmatrix} ? & ? & ? \\ ? & ? & ? \end{bmatrix}$

mouth $\begin{bmatrix} 3 & 0 \\ 0 & 2 \end{bmatrix} \cdot \begin{bmatrix} -2 & 0 & 5 \\ 0 & -2 & 0 \end{bmatrix} = \underline{\quad ? \quad}$

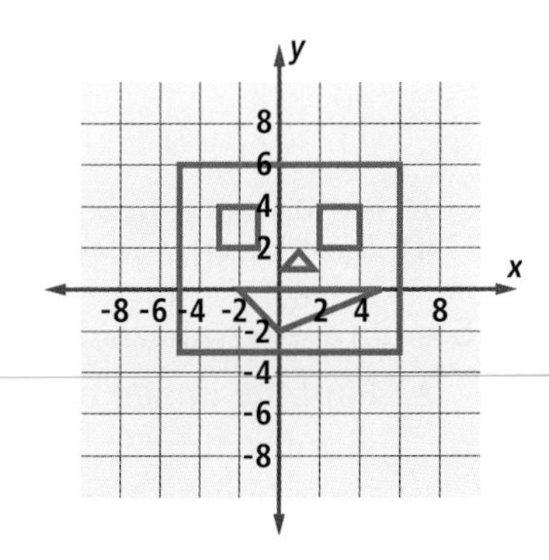

Question 4 asks for the image of the left eye.

(continued on next page)

b. Graph the image.

Notice that the image is a distortion, not a simple enlargement of the original figure. Every part of the original figure is stretched by a factor of 3 horizontally and by a factor of 2 vertically. The transformation is the scale change $S_{3,2}$.

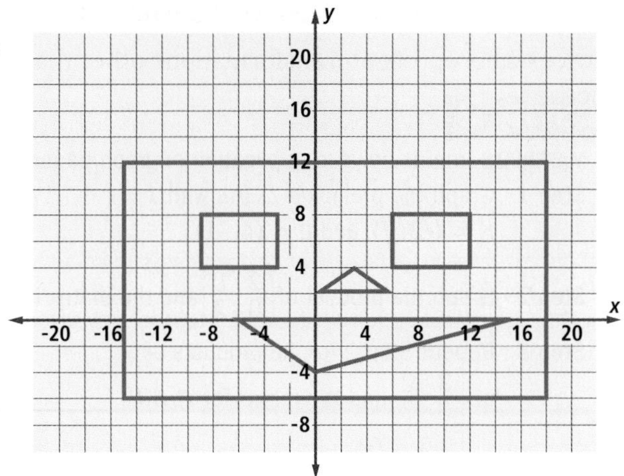

Questions

COVERING THE IDEAS

1. **Multiple Choice** Which matrix represents the point (12, –4)?

A $\begin{bmatrix} 12 & -4 \end{bmatrix}$ B $\begin{bmatrix} -4 & 12 \end{bmatrix}$ C $\begin{bmatrix} 12 \\ -4 \end{bmatrix}$ D $\begin{bmatrix} -4 \\ 12 \end{bmatrix}$

2. Refer to the figure at the right.

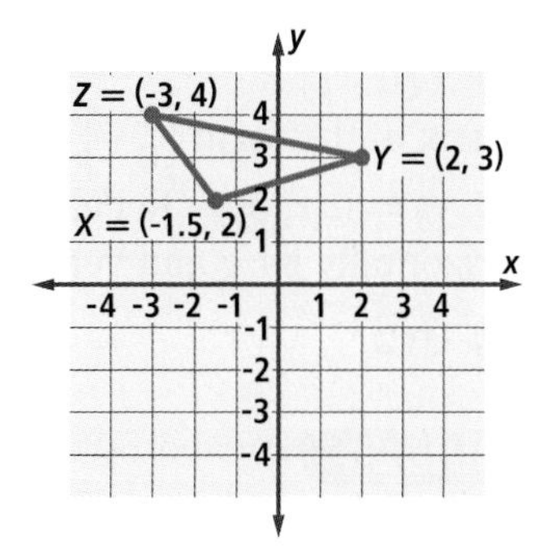

a. Write $\triangle XYZ$ as a matrix.

b. Multiply the matrix for $\triangle XYZ$ by $\begin{bmatrix} 0 & 1 \\ 1 & 0 \end{bmatrix}$.

c. Graph the image matrix from Part b.

d. The matrix in Part b represents what transformation?

3. Match the matrix with its transformation.

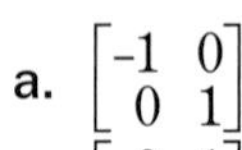

a. $\begin{bmatrix} -1 & 0 \\ 0 & 1 \end{bmatrix}$ i identity

b. $\begin{bmatrix} 0 & -1 \\ -1 & 0 \end{bmatrix}$ ii reflection over x-axis

c. $\begin{bmatrix} 1 & 0 \\ 0 & -1 \end{bmatrix}$ iii reflection over y-axis

d. $\begin{bmatrix} 1 & 0 \\ 0 & 1 \end{bmatrix}$ iv reflection over $y = x$

e. $\begin{bmatrix} 0 & 1 \\ 1 & 0 \end{bmatrix}$ v none of the above

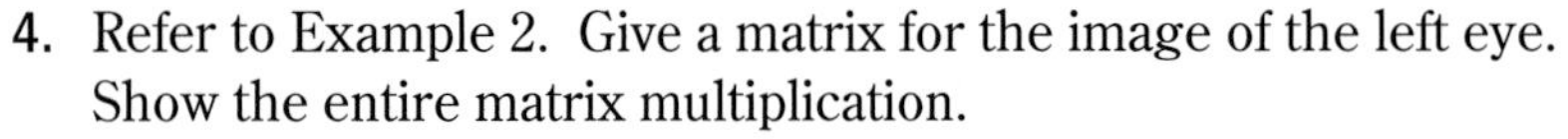
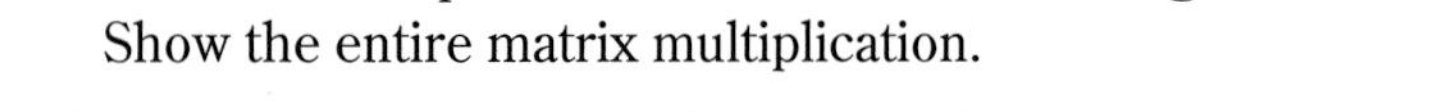

4. Refer to Example 2. Give a matrix for the image of the left eye. Show the entire matrix multiplication.

In 5 and 6, use trapezoid *TRZD* at the right.

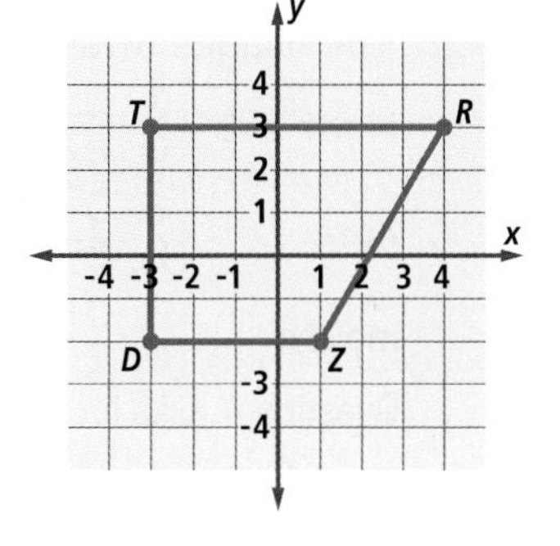

a. Find the image of *TRZD* under the transformation with the given matrix.

b. Describe the transformation.

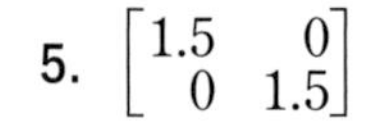

5. $\begin{bmatrix} 1.5 & 0 \\ 0 & 1.5 \end{bmatrix}$

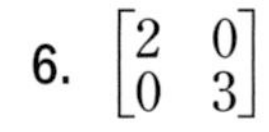

6. $\begin{bmatrix} 2 & 0 \\ 0 & 3 \end{bmatrix}$

In 7 and 8, state a matrix for the transformation.

7. size change of magnitude a 8. scale change $S_{a,b}$

APPLYING THE MATHEMATICS

9. Prove that $\begin{bmatrix} -1 & 0 \\ 0 & 1 \end{bmatrix}$ is the matrix for r_y, reflection over the y-axis.

10. a. Write the matrix for $r_{y=x}$.
    b. Square your answer from Part a.
    c. What transformation does the matrix in Part b represent?

11. Refer to Activity 1. Experiment to find what 2×2 matrix will create a reflection of the monkey over the line with equation $y = -x$.

12. a. Find $T(ABC)$ when $\triangle ABC$ is represented by $\begin{bmatrix} -1 & 2 & 4 \\ 2 & 3 & 0 \end{bmatrix}$ and T has matrix $\begin{bmatrix} -1 & 0 \\ 0 & -1 \end{bmatrix}$.
    b. Describe T.

In 13 and 14, consider the drawing of Pat the Bunny at the right.

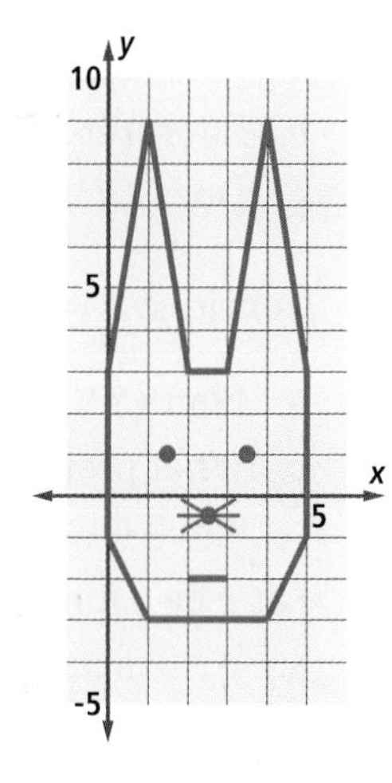

13. Write a matrix F to represent the polygon outlining Pat's face.

14. Find a matrix for a transformation that, when applied to Pat, produces the desired image.
    a. A face similar to Pat's with four times the area.
    b. A face that is shorter and wider than Pat's.
    c. A face that is longer and thinner than Pat's.

15. Here are matrices for three collinear points A, B, and C, and a matrix T for a transformation:

$$A = \begin{bmatrix} 1 \\ 4 \end{bmatrix},\ B = \begin{bmatrix} 0 \\ 2 \end{bmatrix},\ C = \begin{bmatrix} -2 \\ -2 \end{bmatrix};\ T = \begin{bmatrix} 2 & 7 \\ -1 & -4 \end{bmatrix}.$$

    a. Find the images A', B', and C' of the points under the transformation.
    b. Are the image points collinear?
    c. Find the distances AB and BC. Compare the results to the corresponding distances $A'B'$ and $B'C'$. Does the transformation preserve distance?

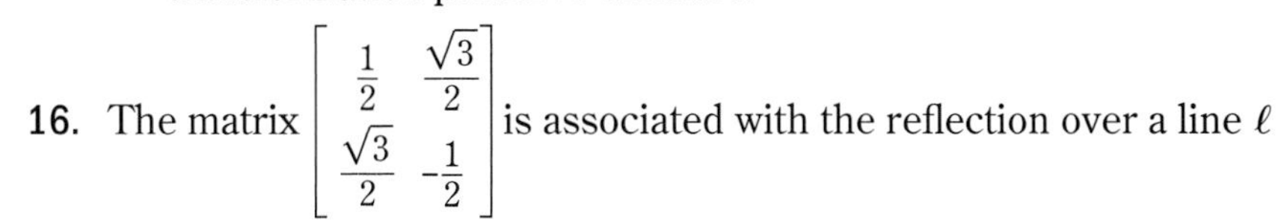

16. The matrix $\begin{bmatrix} \frac{1}{2} & \frac{\sqrt{3}}{2} \\ \frac{\sqrt{3}}{2} & -\frac{1}{2} \end{bmatrix}$ is associated with the reflection over a line ℓ

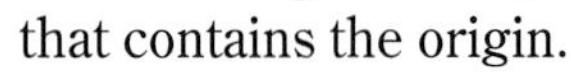

that contains the origin.

    a. Find the image of (1, 0) under this transformation.
    b. Find the image of (0, 1) under this transformation.
    c. What is the measure of the acute angle formed by ℓ and the x-axis?

REVIEW

17. Matrix X has dimensions 7×4 and matrix Y has dimensions 4×6.

a. Which product, XY or YX, exists, and what are its dimensions?

b. How many elements are in the product? **(Lesson 12-1)**

In 18 and 19, multiply, if the product exists. (Lesson 12-1)

18. $\begin{bmatrix} -2 & 3 \\ 0 & -4 \end{bmatrix} \cdot \begin{bmatrix} x \\ y \end{bmatrix}$

19. $\begin{bmatrix} 4 & 0 & 3 \\ 3 & -1 & -2 \end{bmatrix} \cdot \begin{bmatrix} 5 & 0 \\ 1 & 2 \end{bmatrix}$

20. A clothing manufacturer has factories in Oakland, California, and Charleston, South Carolina. The quantities (in thousands) of each of three products manufactured are given in the production matrix P below. The costs in dollars for producing each item during three years are given in the cost matrix C below.

$$\begin{array}{r} \\ \text{Oakland} \\ \text{Charleston} \end{array} \begin{array}{c} \text{Coats} \quad \text{Pants} \quad \text{Shirts} \\ \begin{bmatrix} 38 & 21 & 19 \\ 42 & 18 & 0 \end{bmatrix} \end{array} = P \qquad \begin{array}{r} \\ \text{Coats} \\ \text{Pants} \\ \text{Shirts} \end{array} \begin{array}{c} 2005 \quad 2006 \quad 2007 \\ \begin{bmatrix} 39 & 44 & 43 \\ 8 & 11 & 10 \\ 6 & 7 & 7.5 \end{bmatrix} \end{array} = C$$

a. Calculate PC.

b. Interpret PC by telling what each element represents.

c. Does CP exist? Why or why not? **(Lesson 12-1)**

Apparel manufacturing employs over 200,000 workers in the U.S.

21. Consider $X = \begin{bmatrix} -1 & 3 \\ 2 & 1 \end{bmatrix}$, $Y = \begin{bmatrix} 4 & -3 \\ 0 & -1 \end{bmatrix}$, and $Z = \begin{bmatrix} 0 & 2 \\ 1 & -1 \end{bmatrix}$.

a. Find $(XY)Z$.

b. Find $X(YZ)$.

c. What property of matrix multiplication do the results of Parts a and b illustrate? **(Lesson 12-1)**

22. Let g be a function with $g(x) = x^2 + 1$.

a. Write a polynomial for $(g \circ g \circ g)(x)$.

b. Evaluate $g(g(g(2)))$. **(Lesson 3-7)**

EXPLORATION

23. A unit cube in a three-dimensional coordinate system can be represented by the matrix C below.

$$C = \begin{bmatrix} 0 & 0 & 0 & 1 & 1 & 1 & 1 & 0 \\ 0 & 1 & 1 & 1 & 1 & 0 & 0 & 0 \\ 0 & 0 & 1 & 1 & 0 & 0 & 1 & 1 \end{bmatrix}$$

a. Graph the cube.

b. Let T be the transformation represented by the matrix $M = \begin{bmatrix} 2 & 0 & 0 \\ 0 & 3 & 0 \\ 0 & 0 & 4 \end{bmatrix}$. Calculate and graph MC.

c. Describe the transformation T.

QY ANSWERS

1. $\triangle RQS$

2. $\begin{bmatrix} a & b \\ c & d \end{bmatrix}$

Lesson 12-3 Matrices for Composites of Transformations

▶ **BIG IDEA** To find the image of a figure under a composite of transformations, you need only multiply in order by the matrices for the individual transformations.

Mental Math

Give the coordinates of

a. $R_{90}(1, 0)$.
b. $R_{-90}(1, 0)$.
c. $R_{180}(-1, 0)$.
d. $R_{-180}(0, 1)$.

As you know from Chapter 3, when two transformations are composed, the result is another transformation. Composites of transformations can be represented by matrices.

Activity

MATERIALS pencil and graph paper or matrix polygon application

Step 1 Plot $\triangle ABC$, represented by $\begin{bmatrix} 4 & 6 & 6 \\ 1 & 1 & 5 \end{bmatrix}$.

Step 2 Let $\triangle A'B'C' = r_x(\triangle ABC)$. Find the coordinates of $\triangle A'B'C'$, and plot the triangle.

Step 3 Let $\triangle A''B''C''$ be the reflection image of $\triangle A'B'C'$ over the line $y = x$. Find the coordinates of $\triangle A''B''C''$, and plot the triangle.

In the Activity, $\triangle A''B''C'' = r_{y=x}(r_x(\triangle ABC))$. That is, $\triangle A''B''C''$ is the image of $\triangle ABC$ under the *composite* of the reflections $r_{y=x}$ and r_x. To find a single matrix for the composite $r_{y=x} \circ r_x$, notice that the matrix for $\triangle A''B''C''$ comes from the product $\begin{bmatrix} 0 & 1 \\ 1 & 0 \end{bmatrix} \cdot \left(\begin{bmatrix} 1 & 0 \\ 0 & -1 \end{bmatrix} \cdot \begin{bmatrix} 4 & 6 & 6 \\ 1 & 1 & 5 \end{bmatrix} \right)$.
Because matrix multiplication is associative, the preceding expression may be rewritten as $\left(\begin{bmatrix} 0 & 1 \\ 1 & 0 \end{bmatrix} \cdot \begin{bmatrix} 1 & 0 \\ 0 & -1 \end{bmatrix} \right) \cdot \begin{bmatrix} 4 & 6 & 6 \\ 1 & 1 & 5 \end{bmatrix}$. Thus, multiplying the matrices for $r_{y=x}$ and r_x gives the matrix for the composite $r_{y=x} \circ r_x$:
$\begin{bmatrix} 0 & 1 \\ 1 & 0 \end{bmatrix} \cdot \begin{bmatrix} 1 & 0 \\ 0 & -1 \end{bmatrix} = \begin{bmatrix} 0 & -1 \\ 1 & 0 \end{bmatrix}$.

QY1

▶ QY1

Verify that the product $\begin{bmatrix} 0 & -1 \\ 1 & 0 \end{bmatrix} \cdot \begin{bmatrix} 4 & 6 & 6 \\ 1 & 1 & 5 \end{bmatrix}$ gives the matrix for $\triangle A''B''C''$ found in the Activity.

Compare the positions of the preimage and the final image in the Activity. The transformation that maps $\triangle ABC$ to $\triangle A''B''C''$ is R_{90}, the rotation of 90 degrees counterclockwise with the origin as its center. (We usually omit the degree symbol in the subscript for rotations.) So $R_{90} = r_{y=x} \circ r_x$.

This result is generalized in the theorem on the next page.

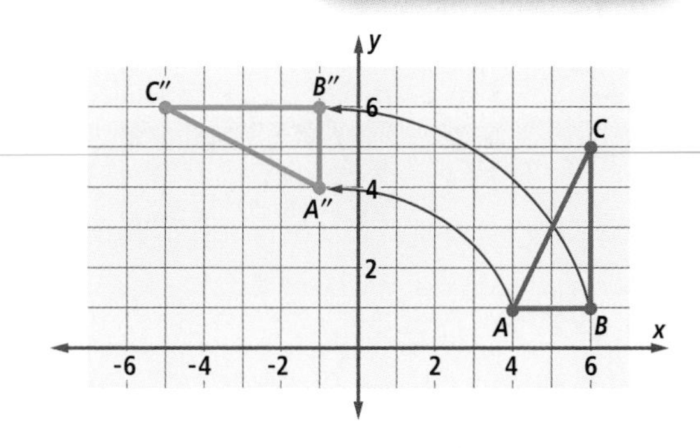

Matrices and Composites Theorem

If M is the matrix associated with a transformation t, and N is the matrix associated with a transformation u, then MN is the matrix associated with the transformation $t \circ u$.

How to Find the Matrix for a Composite of Transformations

The above theorem shows how to find the matrix for a composite transformation $t \circ u$ if you know the matrices for t and u. Just multiply!

GUIDED

Example 1

Find a matrix for R_{180}.

Solution Use the fact that $R_{180} = R_{90} \circ R_{90}$. The product of the matrix for R_{90} with itself equals the matrix for R_{180}.

$$R_{90} \circ R_{90} = R_{180}$$

$$\begin{bmatrix} 0 & -1 \\ 1 & 0 \end{bmatrix} \cdot \begin{bmatrix} 0 & -1 \\ 1 & 0 \end{bmatrix} = \begin{bmatrix} ? & ? \\ ? & ? \end{bmatrix}$$

So $\begin{bmatrix} ? & ? \\ ? & ? \end{bmatrix}$ is a matrix for R_{180}.

Check Apply the matrix for R_{180} to $\triangle ABC$ of the Activity.

$$\begin{bmatrix} -1 & 0 \\ 0 & -1 \end{bmatrix} \cdot \begin{bmatrix} 4 & 6 & 6 \\ 1 & 1 & 5 \end{bmatrix} = \begin{bmatrix} -4 & -6 & -6 \\ -1 & -1 & -5 \end{bmatrix}$$

The product should represent the image of $\triangle ABC$ under a rotation of 180° about the origin. The graph below illustrates that it does.

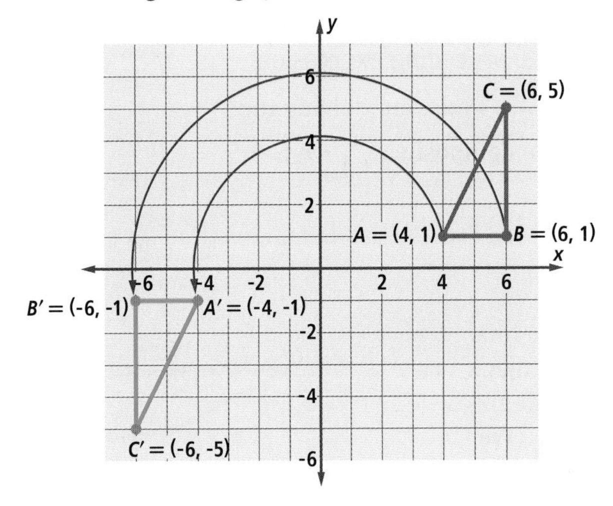

Turkish carpets often display designs with 180° rotation symmetry.

QY2

QY2

Use the fact that $R_{270} = R_{180} \circ R_{90}$ to find a matrix for R_{270}.

The theorem on the next page summarizes these results.

Theorem (Matrices for Rotations)

$\begin{bmatrix} 0 & -1 \\ 1 & 0 \end{bmatrix}$ is the matrix for R_{90}, the rotation of 90° about the origin.

$\begin{bmatrix} -1 & 0 \\ 0 & -1 \end{bmatrix}$ is the matrix for R_{180}, the rotation of 180° about the origin.

$\begin{bmatrix} 0 & 1 \\ -1 & 0 \end{bmatrix}$ is the matrix for R_{270}, the rotation of 270° about the origin.

From the Matrices for Rotations Theorem, you can immediately determine the image of any point under these rotations. For instance, since $\begin{bmatrix} 0 & -1 \\ 1 & 0 \end{bmatrix}\begin{bmatrix} x \\ y \end{bmatrix} = \begin{bmatrix} 0 \cdot x + -1 \cdot y \\ 1 \cdot x + 0 \cdot y \end{bmatrix} = \begin{bmatrix} -y \\ x \end{bmatrix}$, $R_{90}(x, y) = (-y, x)$. Similar arguments prove the other two parts of this corollary:

For any point (x, y), $R_{90}(x, y) = (-y, x)$, $R_{180}(x, y) = (-x, -y)$, and $R_{270}(x, y) = (y, -x)$.

The Matrix Basis Theorem

You may wonder how you will ever remember matrices for each of the transformations you have studied. The key is a simple and beautiful result: if a transformation T can be represented by a 2×2 matrix, then the first column of the matrix is $T(1, 0)$, and the second column of the matrix is $T(0, 1)$.

For instance, suppose you forget the matrix for R_{90}. If you visualize rotating (1, 0) and (0, 1) 90°, you can see that $R_{90}(1, 0) = (0, 1)$ and $R_{90}(0, 1) = (-1, 0)$. So the matrix for R_{90} is

$$\begin{matrix} R_{90}(1, 0) & R_{90}(0, 1) \\ \downarrow & \downarrow \end{matrix}$$
$$\begin{bmatrix} 0 & -1 \\ 1 & 0 \end{bmatrix}$$

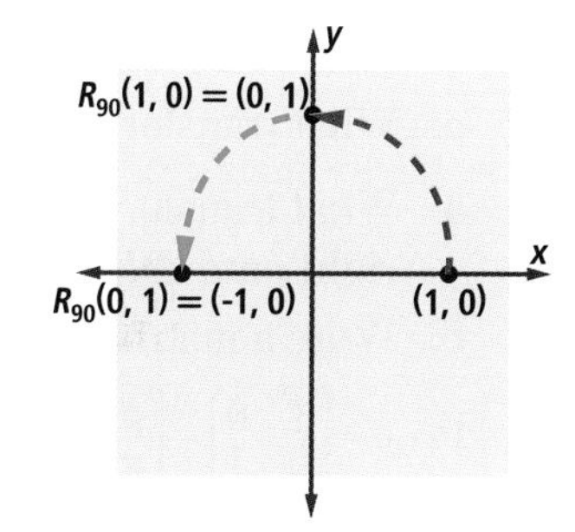

This property of transformations and matrices is called the Matrix Basis Theorem because the matrix is *based* on the images of the points (1, 0) and (0, 1).

Matrix Basis Theorem

Suppose T is a transformation represented by a 2×2 matrix. If $T(1, 0) = (x_1, y_1)$ and $T(0, 1) = (x_2, y_2)$, then T has the matrix $\begin{bmatrix} x_1 & x_2 \\ y_1 & y_2 \end{bmatrix}$.

Proof Let M be the 2×2 matrix for T. Because $T(1, 0) = (x_1, y_1)$ and $T(0, 1) = (x_2, y_2)$, $M \cdot \begin{bmatrix} 1 & 0 \\ 0 & 1 \end{bmatrix} = \begin{bmatrix} x_1 & x_2 \\ y_1 & y_2 \end{bmatrix}$.

(↑ 1st point, ↑ 2nd point, ↑ image of 1st point, ↑ image of 2nd point)

But $\begin{bmatrix} 1 & 0 \\ 0 & 1 \end{bmatrix}$ is the identity matrix for multiplication, so $M = \begin{bmatrix} x_1 & x_2 \\ y_1 & y_2 \end{bmatrix}$.

Example 2

Find the 2 × 2 matrix M for the transformation $S_{4,3}$ using the Matrix Basis Theorem.

Solution $S_{4,3}$ is a scale change. $S_{4,3}(1, 0) = (4, 0)$ and $S_{4,3}(0, 1) = (0, 3)$. So by the Matrix Basis Theorem, $M = \begin{bmatrix} 4 & 0 \\ 0 & 3 \end{bmatrix}$.

Questions

COVERING THE IDEAS

1. In the composite $r_{y=x} \circ r_x$, which reflection is done first?
2. What is the matrix for $r_{y=x} \circ r_x$?
3. Find the image of $UVWX$ under R_{270} when $U = (7, 0)$, $V = (8, 1)$, $W = (11, -6)$, and $X = (9, -2)$.
4. Refer to $\triangle ABC$ from the Activity.
   a. Reflect $\triangle ABC$ over the x-axis. Then reflect its image $\triangle A'B'C'$ over the y-axis to get a second image $\triangle A''B''C''$.
   b. Give the matrix for each of $\triangle A'B'C'$ and $\triangle A''B''C''$.
   c. What transformation maps $\triangle ABC$ directly onto $\triangle A''B''C''$?
   d. The composite $r_y \circ r_x$ is represented by $\begin{bmatrix} -1 & 0 \\ 0 & 1 \end{bmatrix} \cdot \begin{bmatrix} 1 & 0 \\ 0 & -1 \end{bmatrix}$. Find this product.
   e. What transformation is represented by the product in Part d?
5. a. What transformation that you have studied maps $\triangle RQS$ at the right onto $\triangle R'Q'S'$?
   b. Write a matrix for that transformation.

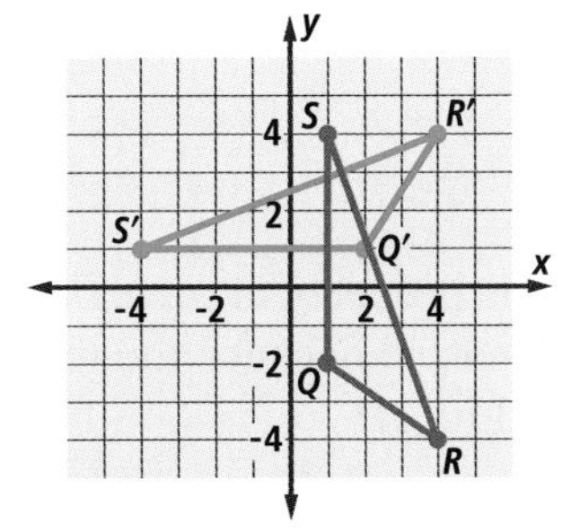

6. Prove $\begin{bmatrix} 1 & 0 \\ 0 & 1 \end{bmatrix} \cdot \begin{bmatrix} p & q \\ n & m \end{bmatrix} = \begin{bmatrix} p & q \\ n & m \end{bmatrix} \cdot \begin{bmatrix} 1 & 0 \\ 0 & 1 \end{bmatrix}$.
7. Solve this matrix equation for a, b, c, and d:
   $\begin{bmatrix} a & b \\ c & d \end{bmatrix} \cdot \begin{bmatrix} 1 & 0 \\ 0 & 1 \end{bmatrix} = \begin{bmatrix} w & x \\ y & m \end{bmatrix}$.

In 8 and 9, a transformation is given.

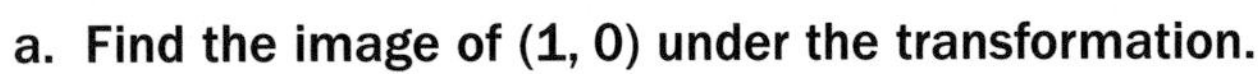

a. Find the image of (1, 0) under the transformation.
b. Find the image of (0, 1) under the transformation.
c. Find the 2 × 2 matrix for the transformation using the Matrix Basis Theorem.

8. R_{-90}
9. $r_{y=-x}$

10. Let $\triangle DEF$ be represented by the matrix $\begin{bmatrix} 0 & 8 & 8 \\ 0 & 0 & 4 \end{bmatrix}$.
    a. Find the reflection image of $\triangle DEF$ over the line $y = x$.
    b. Transform the image in Part a by the scale change $S_{2,0.6}$.
    c. Give a matrix for the composite $S_{2,0.6} \circ r_{y=x}$.

APPLYING THE MATHEMATICS

In 11 and 12, transformation t has matrix $\begin{bmatrix} 2 & 5 \\ 1 & 3 \end{bmatrix}$ and transformation u has matrix $\begin{bmatrix} 4 & 2 \\ -1 & 3 \end{bmatrix}$. Let $\triangle ABC$ be represented by $\begin{bmatrix} 8 & 3 & 2 \\ -2 & 0 & -5 \end{bmatrix}$. Calculate and graph $\triangle ABC$ and the indicated image.

11. $t \circ u(\triangle ABC)$

12. $u \circ t(\triangle ABC)$

13. Using the fact that a matrix for R_{30} is $\begin{bmatrix} \frac{\sqrt{3}}{2} & -\frac{1}{2} \\ \frac{1}{2} & \frac{\sqrt{3}}{2} \end{bmatrix}$, find the matrix for the following transformations.
   a. R_{60}
   b. R_{150}

14. Refer to the matrix monkey in Lesson 12-2 and R_{30} in Question 13. Find the image of the matrix monkey under $R_{30} \circ R_{270}$.

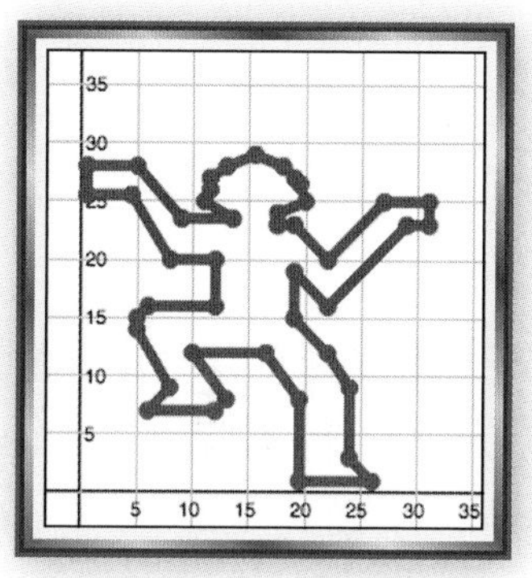

15. a. Calculate a matrix for $r_x \circ r_{y=x}$.
    b. Describe the composite transformation.

16. a. Find the image of (7, 5) under each of R_{90}, R_{180}, and R_{270}.
    b. Graph (7, 5) and the three images.
    c. The four points are vertices of what kind of polygon?
    d. Find the center and radius of a circle passing through these points.

17. Find the 2×2 matrix for T if $T(x, y) = (2x - 3y, x + y)$.

REVIEW

18. Write the 2×2 matrix for the single scale change that includes a horizontal scale change of magnitude $\frac{1}{300}$ and a vertical scale change of magnitude 50. **(Lesson 12-2)**

19. **Fill in the Blank** About __?__ of the observations from a normal distribution are within two standard deviations of the mean. Explain your answer. **(Lesson 11-2)**

20. Vera Vittles knows that each of her four children likes roughly 85% of the meals she serves. Assume that all the children's likes and dislikes are independent. What is the probability that
    a. all four children will like a given meal?
    b. at least one child will dislike a given meal? **(Lesson 10-5)**

21. Decide whether the sequence is convergent or divergent. If it is convergent, give its limit. **(Lesson 8-7)**
    a. $\frac{8}{3}, \frac{11}{15}, \frac{16}{35}, \ldots, \frac{n^2+7}{4n^2-1}, \ldots$
    b. $\frac{8}{3}, \frac{11}{7}, \frac{16}{11}, \ldots, \frac{n^2+7}{4n-1}, \ldots$

22. The menu at a Thai restaurant contains 41 main dishes. How many different meals with six different dishes are possible? **(Lesson 10-1)**

23. Simplify $\frac{\sin\theta\cos^2\theta}{(1-\sin^2\theta)\tan\theta}$. **(Lesson 4-3)**

EXPLORATION

24. Write a Matrix Basis Theorem for 3×3 matrices and their transformations. Verify your theorem with an example.

QY ANSWERS

1. $\begin{bmatrix} -1 & -1 & -5 \\ 4 & 6 & 6 \end{bmatrix}$

2. $R_{270} = \begin{bmatrix} 0 & 1 \\ -1 & 0 \end{bmatrix}$

Lesson

12-4 The General Rotation Matrix

▶ **BIG IDEA** All rotations about (0,0) can be represented by a 2×2 transformation matrix.

Computer animators use matrices to produce rotation images of objects for any magnitude θ. It may surprise you that all rotations about (0, 0) can be represented by 2×2 matrices. Developing a rotation matrix requires only the Matrix Basis Theorem and a little trigonometry.

Mental Math

Give the coordinates of (cos 45°, sin 45°) under a rotation of 180° about the origin.

Finding the General Rotation Matrix

Let θ be the magnitude of a rotation around the origin. The matrix for R_θ is quite simple.

Rotation Matrix Theorem

The 2×2 matrix for R_θ, the rotation of magnitude θ about the origin, is $\begin{bmatrix} \cos\theta & -\sin\theta \\ \sin\theta & \cos\theta \end{bmatrix}$.

Proof Let A' and B' be the images of the points $A = (1, 0)$ and $B = (0, 1)$ under R_θ, as shown at the right. By definition of the cosine and sine, $A' = R_\theta(A) = R_\theta(1, 0) = (\cos\theta, \sin\theta)$. Now we need to find the coordinates of B'. Because $B = (0, 1)$, $B = R_{90}(A)$. $B' = R_\theta(B)$, so by substitution, $B' = R_\theta(R_{90}(A))$. But $R_\theta \circ R_{90} = R_{90+\theta} = R_{90} \circ R_\theta$. So $R_\theta \circ R_{90}(A) = R_{90} \circ R_\theta(A) = R_{90}(A')$. Thus, $B' = R_{90}(A')$.

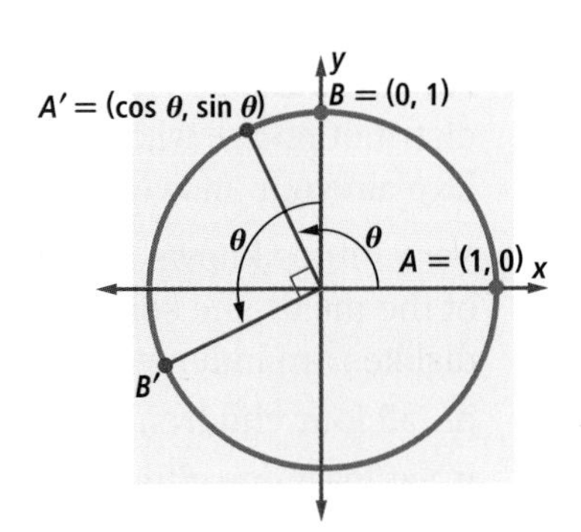

Now by the Matrices for Rotations Theorem,

$$\begin{matrix} R_{90} & (A') & = & B' \\ \begin{bmatrix} 0 & -1 \\ 1 & 0 \end{bmatrix} & \cdot \begin{bmatrix} \cos\theta \\ \sin\theta \end{bmatrix} & = & \begin{bmatrix} -\sin\theta \\ \cos\theta \end{bmatrix}. \end{matrix}$$

So $B' = R_\theta(B) = R_\theta(0, 1) = (-\sin\theta, \cos\theta)$.

With the images of (1, 0) and (0, 1), the Matrix Basis Theorem tells us what the matrix for R_θ is. The first column is $R_\theta(1, 0)$ or $(\cos\theta, \sin\theta)$. The second column is $R_\theta(0, 1)$ or $(-\sin\theta, \cos\theta)$.

The next Activity is a simplification of what happens in computer graphics to transform a complicated figure. You are asked to transform 44 points on a figure to create its image. A computer might work with thousands of pixels.

Activity

MATERIALS matrix polygon application provided by your teacher

Step 1 Press the "Monkey!" button to plot the stored 2×44 matrix of the monkey used in Lesson 12-2.

Step 2 Enter the matrix for a 67.5° rotation about (0, 0).

Step 3 Press the "Transform" button to multiply the monkey matrix by the rotation matrix and create the rotation image of the monkey.

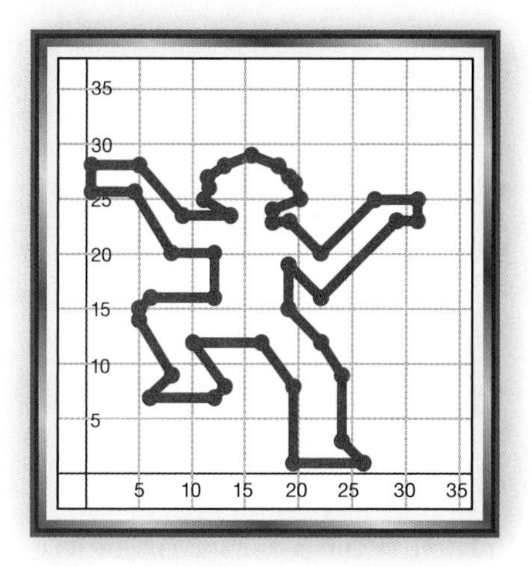

The matrices you have seen for R_{90}, R_{180}, and R_{270} are specific instances of the Rotation Matrix Theorem.

GUIDED

Example 1

Find the 2×2 matrix for R_{60}.

Solution Use the Rotation Matrix Theorem.

The matrix for R_{60} is $\begin{bmatrix} \cos 60° & ? \\ \sin 60° & ? \end{bmatrix}$.

Since $\cos 60° = \frac{1}{2}$ and $\sin 60° = \underline{\ \ ?\ \ }$, the matrix is $\begin{bmatrix} \frac{1}{2} & ? \\ ? & ? \end{bmatrix}$.

Check Compute the third power of your matrix to see if three rotations of 60° gives the same matrix as R_{180}. You are asked to do this in Question 7.

Finding Images of Figures

The Rotation Matrix Theorem enables you to rotate any figure about the origin.

GUIDED

Example 2

In a computer game, the wheel at the right (pictured on a coordinate grid) spins counterclockwise around the origin at the rate of 5.32 radians per second. The T in the word TO has endpoints at (1.7, 4.8), (1.8, 8), and (4, 7). Approximate the coordinates of the endpoints of the T if a player hits a key to stop the wheel in 5.78 seconds.

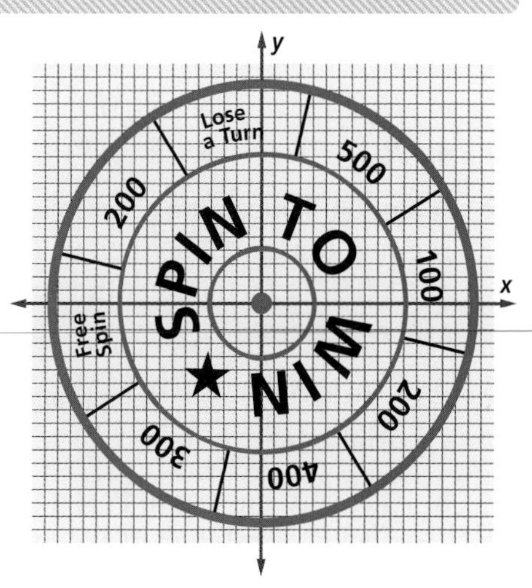

Solution Here we consider everything in radians.

In 5.78 seconds, the wheel turns about 30.75 radians.

The matrix for $R_{30.75}$ is $\begin{bmatrix} \cos(30.75) & -\sin ? \\ \sin ? & \cos ? \end{bmatrix} \approx \begin{bmatrix} 0.786 & ? \\ ? & ? \end{bmatrix}$.

(continued on next page)

Multiply this matrix by the matrix for the preimage T.

$$\begin{bmatrix} 0.786 & 0.618 \\ ? & ? \end{bmatrix} \cdot \begin{bmatrix} 1.7 & 1.8 & 4 \\ 4.8 & 8 & 7 \end{bmatrix} \approx \begin{bmatrix} 4.3 & ? & ? \\ ? & 5.2 & ? \end{bmatrix}$$

Thus, the coordinates of the endpoints of the image of the T are about (4.3, __?__), (__?__, 5.2), and (__?__, __?__).

Questions

COVERING THE IDEAS

1. What is the 2 × 2 matrix for R_t, the rotation of magnitude t about the origin?
2. Give the 2 × 2 matrix for R_{135}.
3. Given $\sin 33° \approx 0.545$ and $\cos 33° \approx 0.839$, what is an approximate matrix for R_{33}?
4. Given $\sin(-17°) \approx -0.292$ and $\cos(-17°) \approx 0.956$, what is an approximate matrix for R_{-17}?

In 5 and 6, verify the matrix for the rotation given in the Matrices for Rotations Theorem is a special instance of the Rotation Matrix Theorem.

5. R_{180}
6. R_{270}
7. Complete the Check for Example 1.
8. Refer to Example 2. The endpoints of the I in "SPIN" are given by the matrix $\begin{bmatrix} -4.2 & -6.8 \\ 3.2 & 5 \end{bmatrix}$. Approximate the coordinates of the image of the I after a rotation of 7.3π radians counterclockwise.

APPLYING THE MATHEMATICS

9. a. Find a matrix for the 30°-60°-90° triangle at the right.
   b. Find a matrix for its image under R_{200}.
   c. Verify that this image is a 30°-60°-90° triangle.

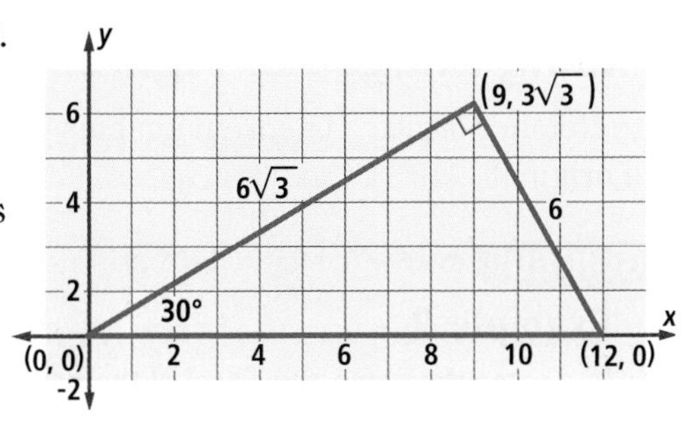

10. Find the exact 2 × 2 matrix for the rotation of $\frac{\pi}{4}$ radians about the origin.

11. Let $P = (-100, 500)$. Compute to the nearest tenth.
    a. $R_6(P)$
    b. $R_{96}(P)$
    c. $R_{186}(P)$
    d. $R_{276}(P)$
    e. $R_{366}(P)$

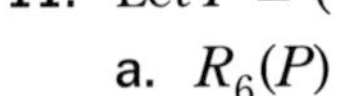

12. A student is designing a club logo on a computer. The outline is a regular nonagon inscribed in a circle with radius 5. One vertex is at (0, 5). To the nearest thousandth, find the coordinates of the other vertices.

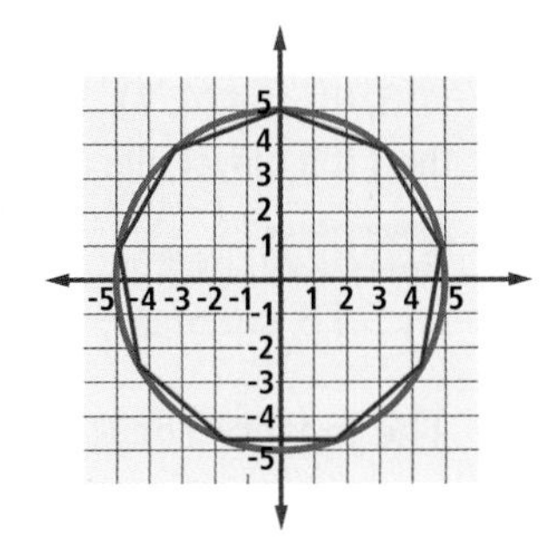

13. a. Prove that a matrix for $R_{-\theta}$, a *clockwise* rotation of θ about the origin, is $\begin{bmatrix} \cos\theta & \sin\theta \\ -\sin\theta & \cos\theta \end{bmatrix}$.

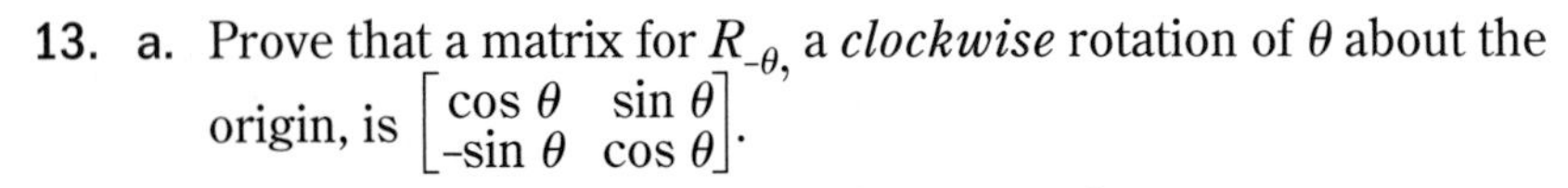

    b. Use your answer to Part a to determine the matrix for a clockwise rotation of 136° about the origin.

REVIEW

14. Let $\begin{bmatrix} 8 & -5 \\ -4 & 1 \end{bmatrix}$ represent the transformation t, $\begin{bmatrix} 7 & 8 \\ 2 & 1 \end{bmatrix}$ represent the transformation u, and $\begin{bmatrix} 4 & 0 & 2 \\ 2 & 1 & 0 \end{bmatrix}$ represent $\triangle CAP$. Calculate the matrix for $t \circ u(\triangle CAP)$. **(Lesson 12-3)**

15. Is the given transformation a description of the identity transformation; that is, does it map a figure to itself? **(Lesson 12-2)**
    a. rotation of 360°
    b. reflection over the y-axis
    c. size change of magnitude 1
    d. size change of magnitude -1

16. Find x and y so that the following is true. **(Lesson 12-1)**

$$\begin{bmatrix} 1 & -1 \\ x & 3 \end{bmatrix} \cdot \begin{bmatrix} 2 & y \\ 0 & 1 \end{bmatrix} = \begin{bmatrix} 2 & 3 \\ 6 & 15 \end{bmatrix}$$

17. **True or False** Refer to the Central Limit Theorem. Explain your reasoning. **(Lesson 11-4)**
    a. The mean of the set of all sample means of size n from a population is equal to the population mean.
    b. The standard deviation of the set of all sample means of size n from a population is equal to the standard deviation of the population.

18. The heights of adult men in the United States are approximately normally distributed with mean 70" and standard deviation 3".
    a. If the transformation $x \rightarrow \frac{x - 70}{3}$ is applied to each height x, what type of distribution results?
    b. What is the median height of adult men?
    c. About what percent of men have heights less than or equal to five feet, nine inches tall? **(Lessons 11-3, 11-2)**

In 19 and 20, rewrite as an expression that does not involve logarithms. (Lesson 9-5)

19. $\log y = 2 \log x + \log 13$

20. $\ln t = \frac{4}{3} \ln a - \ln p$

In 21 and 22, evaluate. (Lessons 8-5, 8-4)

21. the arithmetic series $(5u - v) + (4u + v) + (3u + 3v) + \cdots + (23v - 7u)$

22. $\sum_{n=1}^{20} 10(0.6)^{n-1}$

EXPLORATION

23. a. Find the rotation matrices in three dimensions for rotating a figure θ radians about the x-axis, y-axis, and z-axis.
    b. How do the terms roll, pitch, and yaw relate to rotations in three dimensions?

Lesson 12-5 Formulas for $\cos(\alpha + \beta)$ and $\sin(\alpha + \beta)$

BIG IDEA Exact values of $\sin \theta$ and $\cos \theta$ can be found for many values of θ using formulas for $\cos(\alpha \pm \beta)$ and $\sin(\alpha \pm \beta)$.

Mental Math

Evaluate.

a. cos 45°

b. sin 120°

c. cos (–45°)

d. sin (–120°)

You have seen exact decimal or radical expressions for $\sin \theta$ and $\cos \theta$ when $\theta = 30°, 45°, 60°, 90°$, as well as multiples of these arguments. You may wonder if there are similar expressions for other values of θ. Recall that if you know $\log x$ and $\log y$ you can find $\log(xy)$, since $\log(xy) = \log x + \log y$. Suppose you wanted to know the exact value of sine or cosine for a θ-value that is not a multiple of one you already know. How can you find it?

The Point $(\cos(\alpha + \beta), \sin(\alpha + \beta))$

It turns out that both $\sin(\alpha + \beta)$ and $\cos(\alpha + \beta)$ can be calculated, but for each you need to know the values of $\sin \alpha$, $\cos \alpha$, $\sin \beta$, and $\cos \beta$. To find formulas for $\sin(\alpha + \beta)$ and $\cos(\alpha + \beta)$, rotate the point $P = (\cos \beta, \sin \beta)$ by a magnitude α about the origin. The image of P will then be Q, where $Q = R_{\alpha+\beta}(1, 0) = (\cos(\alpha + \beta), \sin(\alpha + \beta))$, as illustrated here.

You can use rotation matrices to compute $R_{\alpha+\beta}(1, 0) = (R_\alpha \circ R_\beta)(1, 0)$, as we do in the proof of the theorem below.

Theorem (Addition Formulas for the Cosine and Sine)

For all real numbers α and β,

$$\cos(\alpha + \beta) = \cos \alpha \cos \beta - \sin \alpha \sin \beta, \text{ and}$$
$$\sin(\alpha + \beta) = \sin \alpha \cos \beta + \cos \alpha \sin \beta.$$

Proof By definition of the sine and cosine functions,

$$(\cos(\alpha + \beta), \sin(\alpha + \beta)) = R_{\alpha+\beta}(1, 0).$$

But it is also true that $R_{\alpha+\beta} = R_\alpha \circ R_\beta$.

Thus,
$$\begin{aligned}(\cos(\alpha + \beta), \sin(\alpha + \beta)) &= R_\alpha \circ R_\beta(1, 0)\\ &= R_\alpha(R_\beta(1, 0))\\ &= R_\alpha(\cos \beta, \sin \beta).\end{aligned}$$

Now translate this last sentence into matrices.

$$\begin{bmatrix} \cos(\alpha + \beta) \\ \sin(\alpha + \beta) \end{bmatrix} = \begin{bmatrix} \cos \alpha & -\sin \alpha \\ \sin \alpha & \cos \alpha \end{bmatrix} \cdot \begin{bmatrix} \cos \beta \\ \sin \beta \end{bmatrix}$$

$$= \begin{bmatrix} \cos \alpha \cos \beta - \sin \alpha \sin \beta \\ \sin \alpha \cos \beta + \cos \alpha \sin \beta \end{bmatrix}$$

Two matrices are equal if and only if their corresponding elements are equal,

so $\cos(\alpha + \beta) = \cos \alpha \cos \beta - \sin \alpha \sin \beta$

and $\sin(\alpha + \beta) = \sin \alpha \cos \beta + \cos \alpha \sin \beta$.

Using the exact values of the cosine and sine of 30°, 45°, 60°, and 90°, these formulas lead to the exact values for the sine and cosine of many other angles.

Example 1

Find an exact value for sin 105° by hand.

Solution Find two numbers that add to 105° whose cosines and sines you know.

$$\sin 105° = \sin(45° + 60°)$$

Use the Addition Formula for the Sine with $\alpha = 45°$ and $\beta = 60°$.

$$\sin(45° + 60°) = \sin 45° \cos 60° + \cos 45° \sin 60°$$
$$= \frac{\sqrt{2}}{2} \cdot \frac{1}{2} + \frac{\sqrt{2}}{2} \cdot \frac{\sqrt{3}}{2}$$
$$= \frac{\sqrt{2} + \sqrt{6}}{4}$$

Check Evaluate sin 105° on a calculator. The expression $\frac{(\sqrt{3} + 1)\sqrt{2}}{4}$ is equivalent to $\frac{\sqrt{2} + \sqrt{6}}{4}$. It checks.

QY

QY

Find the exact value for cos 105° by hand.

Formulas for $\sin(\alpha - \beta)$ and $\cos(\alpha - \beta)$ can be derived from the Addition Formulas.

Theorem (Subtraction Formulas for the Cosine and Sine)

For all real numbers α and β,

$\cos(\alpha - \beta) = \cos \alpha \cos \beta + \sin \alpha \sin \beta$, and

$\sin(\alpha - \beta) = \sin \alpha \cos \beta - \cos \alpha \sin \beta$.

Proof To derive the formula for $\sin(\alpha - \beta)$, rewrite $\alpha - \beta$ as $\alpha + (-\beta)$. Then for all α and β,

$\sin(\alpha - \beta) = \sin(\alpha + (-\beta))$	algebraic definition of subtraction
$= \sin \alpha \cos(-\beta) + \cos \alpha \sin(-\beta)$	Addition Formula for the Sine
$= \sin \alpha \cos \beta - \cos \alpha \sin \beta$.	Opposites Theorem

You are asked to derive the corresponding formula for $\cos(\alpha - \beta)$ in Question 6.

The formulas for $\cos(\alpha \pm \beta)$ and $\sin(\alpha \pm \beta)$ provide another way to prove some theorems about trigonometric functions that you saw in Chapter 4.

GUIDED

Example 2

Prove the Supplements Theorem for the sine function: $\sin(\pi - \theta) = \sin\theta$ for all real numbers θ.

Solution From the Subtraction Formula for the Sine, for all real numbers θ,

$$\begin{aligned}\sin(\pi - \theta) &= \underline{\ ?\ } \cos\theta - \underline{\ ?\ } \sin\theta \\ &= \underline{\ ?\ } \bullet \cos\theta - (\underline{\ ?\ })\sin\theta \\ &= \underline{\ ?\ }.\end{aligned}$$

You can use the formulas to find cosines and sines of sums and differences of 0°, 30°, 45°, ..., and also of other arguments related to known arguments.

GUIDED

Example 3

Find an exact value for sin 15° by hand.

Solution Think of two multiples of 30° and 45° whose difference is 15°.

$$\begin{aligned}\sin 15° &= \sin(45° - \underline{\ ?\ }) \\ &= \sin 45° \cos \underline{\ ?\ } - \underline{\ ?\ } \\ &= \frac{\sqrt{2}}{2} \bullet \underline{\ ?\ } - \frac{\sqrt{2}}{2} \bullet \underline{\ ?\ } \\ &= \underline{\ ?\ }\end{aligned}$$

Example 4

If $0 < \theta < \frac{\pi}{2}$ and $\cos\theta = \frac{3}{5}$, compute $\cos\left(\theta - \frac{\pi}{6}\right)$.

Solution $\cos\left(\theta - \frac{\pi}{6}\right) = \cos\theta\cos\frac{\pi}{6} + \sin\theta\sin\frac{\pi}{6}$.

So we need to find $\sin\theta$. By the Pythagorean Identity, $\cos^2\theta + \sin^2\theta = 1$.

So $\quad \left(\frac{3}{5}\right)^2 + \sin^2\theta = 1,$

and $\quad \sin^2\theta = \frac{16}{25}.$

Because $0 < \theta < \frac{\pi}{2}$, $\sin\theta > 0$. So $\sin\theta = \sqrt{\frac{16}{25}} = \frac{4}{5}$.

Thus,
$$\begin{aligned}\cos\left(\theta - \frac{\pi}{6}\right) &= \cos\theta\cos\frac{\pi}{6} + \sin\theta\sin\frac{\pi}{6} \\ &= \frac{3}{5} \bullet \frac{\sqrt{3}}{2} + \frac{4}{5} \bullet \frac{1}{2} \\ &= \frac{3\sqrt{3} + 4}{10}.\end{aligned}$$

Questions

COVERING THE IDEAS

1. In the proof of the Addition Formulas, it is stated that $R_\alpha \circ R_\beta(1, 0) = R_\alpha(\cos \beta, \sin \beta)$. Explain why this is true.

In 2 and 3, simplify by using the appropriate addition or subtraction formula.

2. $\sin 68^\circ \cos 22^\circ + \cos 68^\circ \sin 22^\circ$
3. $\cos\left(\frac{21\pi}{12}\right)\cos\left(\frac{3\pi}{12}\right) + \sin\left(\frac{21\pi}{12}\right)\sin\left(\frac{3\pi}{12}\right)$

In 4 and 5, find an exact value by using an appropriate addition or subtraction formula.

4. $\cos 15^\circ$
5. $\sin 195^\circ$
6. Prove the Subtraction Formula for the Cosine: For all real numbers α and β, $\cos(\alpha - \beta) = \cos \alpha \cos \beta + \sin \alpha \sin \beta$.

In 7 and 8, complete the statement and use the formula for $\cos(\alpha + \beta)$ to prove the statement.

7. Supplements Theorem for the cosine function: $\cos(\pi - x) = \ldots$.
8. Complements Theorem for the cosine function: $\cos\left(\frac{\pi}{2} - x\right) = \ldots$.
9. If $\frac{\pi}{2} < \theta < \pi$ and $\sin \theta = \frac{7}{25}$, compute $\sin\left(\theta + \frac{\pi}{4}\right)$.

APPLYING THE MATHEMATICS

In 10–12, give exact values by using an appropriate formula.

10. $\sin 285^\circ$
11. $\sin\left(-\frac{13\pi}{12}\right)$
12. $\tan 75^\circ$
13. A student thought: *For all α and β, $\cos(\alpha + \beta) = \cos \alpha + \cos \beta$.*
    a. Give a counterexample to show the statement is not true.
    b. Give one pair of values for which the statement is true.
14. **True or False** Exact values of $\sin \theta$ and $\cos \theta$ can be found for all integer multiples of $\frac{\pi}{12}$ between 0 and 2π. Justify your answer.
15. a. Graph the following on the same set of axes and identify which function is not equivalent to the others.
    $f(x) = 2\sin\left(x + \frac{\pi}{6}\right)$; $g(x) = 2\sin x + 2\sin\frac{\pi}{6}$; $h(x) = \sqrt{3}\sin x + \cos x$
    b. Explain the results of Part a using a theorem in this lesson.
16. Simplify without a calculator: $\cos 22^\circ \cos 38^\circ - \sin 22^\circ \sin 38^\circ$.
17. Use the formula for $\sin(\alpha - \beta)$ to show that $\sin(x - 2\pi n) = \sin x$ for all integers n.
18. Give the elements of the matrix for $R_{\alpha+\beta}$ in terms of sines and cosines of α and β.
19. $\triangle PQR$ has acute angles P and Q with $\sin P = \frac{1}{3}$ and $\sin Q = \frac{1}{2}$.
    a. Find $\sin(P + Q)$.
    b. Find $\sin R$.
    c. Draw a possible $\triangle PQR$.

REVIEW

20. Determine the coordinates of the image of point (5, 2) under a rotation of -24° about the origin. **(Lesson 12-4)**

In 21 and 22, consider the polygon P represented by the matrix $P = \begin{bmatrix} 1 & 7 & 9 & 3 \\ 2 & 2 & 5 & 5 \end{bmatrix}$. **(Lessons 12-4, 12-3, 12-2)**

21. a. Calculate $\begin{bmatrix} \cos 30° & -\sin 30° \\ \sin 30° & \cos 30° \end{bmatrix} \cdot P$, the matrix for polygon P^*.
    b. Describe the transformation mapping P onto P^*.

22. a. Calculate $\begin{bmatrix} -2 & 0 \\ 0 & 1 \end{bmatrix} \cdot P$, the matrix for polygon P'.
    b. Describe the transformation mapping P onto P'.

23. Carey, Larry, and Sari work part-time after school. The matrix R below gives the hourly salary for each. The matrix H shows the number of hours each worked on the days of a certain week. **(Lesson 12-1)**

$$R = \begin{bmatrix} 7.25 \\ 6.70 \\ 7.10 \end{bmatrix}$$

$H =$

	Carey	Larry	Sari
Monday	3	5	4
Tuesday	5	4	4
Wednesday	3	3	4
Thursday	4	3	4
Friday	2	0	4

    a. How many hours did Larry work in this week?
    b. On which day was the total hours the three worked the least?
    c. Which product matrix, HR or RH, gives the total money earned by Carey, Larry, and Sari on each day?
    d. How much did the three of them earn on Thursday?

In 2002, almost half of the teenage workforce in the state of Washington worked in restaurants.

24. The Gallup organization polled 1015 American adults aged 18 and older in early May of 2009. They found that 6% of U.S. adults thought that it was *essential* that the next Supreme Court Justice be a woman. **(Lesson 11-7)**
    a. Compute the 95% confidence interval for the Gallup poll.
    b. What is the margin of error, given Part a?
    c. What margin of error was reported, given that the pollsters had no reasonable guess as to the results of the poll?

25. Explain why a polynomial function of degree seven with real coefficients must have at least one real zero. **(Lesson 7-6)**

26. **Fill in the Blank** An angle whose radian measure is $\frac{\pi}{2}$ is about  ? times as large as an angle whose measure is $\frac{\pi}{2}°$. **(Lesson 4-1)**

EXPLORATION

27. a. Use the fact that $\tan(\alpha + \beta) = \dfrac{\sin(\alpha + \beta)}{\cos(\alpha + \beta)}$ to derive a formula for $\tan(\alpha + \beta)$ in terms of tangents of α and β. (Hint: divide both numerator and denominator of the fraction by $\cos \alpha \cos \beta$.)
    b. Check your formula using some values of tangents you know.

QY ANSWER

$\cos 105° = \cos(45° + 60°) = \cos 45° \cos 60° - \sin 45° \sin 60° = \frac{\sqrt{2}}{2} \cdot \frac{1}{2} - \frac{\sqrt{2}}{2} \cdot \frac{\sqrt{3}}{2} = \frac{\sqrt{2} - \sqrt{6}}{4}$

Lesson 12-6 Formulas for cos 2θ and sin 2θ

▶ **BIG IDEA** Formulas for sin 2θ and cos 2θ are easily derived from the Addition Formulas for the Sine and Cosine.

Ignoring air resistance and wind, the path of an object thrown or kicked into the air from ground level will be part of a parabola. For instance, if a golf ball is driven off the ground with velocity v at an initial angle of θ degrees to the ground, the horizontal distance d that the ball travels is given by $d = \frac{v^2 \sin 2\theta}{g}$, where g is the acceleration due to gravity. That is, the distance varies jointly as the square of the velocity and the sine of twice the initial angle of flight.

Hitting from a tee, a good golfer can typically hit a ball with a clubhead speed of 110 $\frac{\text{miles}}{\text{hr}}$ or about 160 $\frac{\text{ft}}{\text{sec}}$. If the ball leaves the ground at an initial angle of 50°, it travels about

$$\frac{\left(160 \frac{\text{ft}}{\text{sec}}\right)^2 \cdot \sin(2 \cdot 50°)}{32 \frac{\text{ft}}{\text{sec}^2}} \approx 788 \text{ ft} \approx 263 \text{ yd.}$$

Mental Math

Complete the sentence to make the equation an identity.

a. $\log(2x) = \log x$ ...

b. $e^{2x} = e^x$ ...

c. $(2x)^n = x^n$ ...

d. $|2x| = |x|$ ...

Activity

Step 1 Suppose a golfer wants to know the angle to hit the ball that will give the most distance. Let $f(\theta) = \frac{\left(160 \frac{\text{ft}}{\text{sec}}\right)^2 \cdot \sin 2\theta}{32 \frac{\text{ft}}{\text{sec}^2}}$, and graph f for $0° \le \theta \le 90°$.

Step 2 For what value of θ is $f(\theta)$ largest? How large is $f(\theta)$ for this value of θ?

Step 3 What advice would you give the golfer?

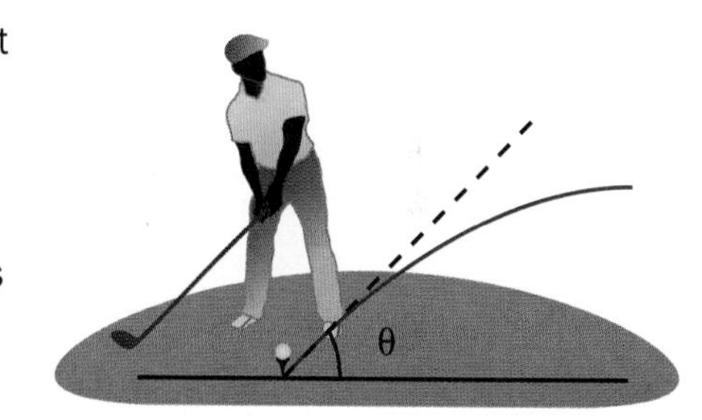

Proving the Double Angle Formulas

Expressions involving sin 2θ and cos 2θ occur often in mathematics and science. Formulas expressing sin 2θ and cos 2θ as functions of θ are often called *Double Angle Formulas*. They can be derived directly using the Addition Formulas for the Sine and Cosine.

Theorem (Double Angle Formulas)

For all real numbers θ,

$$\sin 2\theta = 2 \sin\theta \cos\theta, \text{ and}$$
$$\cos 2\theta = \cos^2\theta - \sin^2\theta$$
$$= 2\cos^2\theta - 1 = 1 - 2\sin^2\theta.$$

Proof Set $\alpha = \theta$ and $\beta = \theta$ in the formulas for $\sin(\alpha + \beta)$ and $\cos(\alpha + \beta)$.

Then
$$\begin{aligned} \sin 2\theta &= \sin(\theta + \theta) \\ &= \sin\theta\cos\theta + \cos\theta\sin\theta \\ &= (\sin\theta)(\cos\theta + \cos\theta) \\ &= 2\sin\theta\cos\theta. \end{aligned}$$

Similarly,
$$\begin{aligned} \cos 2\theta &= \cos(\theta + \theta) \\ &= \cos\theta\cos\theta - \sin\theta\sin\theta \\ &= \cos^2\theta - \sin^2\theta. \end{aligned}$$

Substituting $\sin^2\theta = 1 - \cos^2\theta$ from the Pythagorean Identity yields the second form of the Double Angle Formula for the cosine function.

$$\begin{aligned} \cos 2\theta &= \cos^2\theta - (1 - \cos^2\theta) \\ &= \cos^2\theta - 1 + \cos^2\theta \\ &= 2\cos^2\theta - 1 \end{aligned}$$

You are asked to derive the third form of the Double Angle Formula for the cosine function in Question 3.

Example 1

If $\cos\theta = \frac{5}{8}$, compute $\sin 2\theta$ and $\cos 2\theta$.

Solution From the Double Angle Formulas, $\sin 2\theta = 2\sin\theta\cos\theta$. So we need to know $\sin\theta$. Since $\cos^2\theta + \sin^2\theta = 1$, $\left(\frac{5}{8}\right)^2 + \sin^2\theta = 1$, from which $\sin\theta = \pm\frac{\sqrt{39}}{8}$.

So $\sin 2\theta = 2 \cdot \pm\frac{\sqrt{39}}{8} \cdot \frac{5}{8} = \pm\frac{10\sqrt{39}}{64} = \pm\frac{5\sqrt{39}}{32}$.

To find $\cos 2\theta$, use the Double Angle Formula that involves $\cos\theta$.

$\cos 2\theta = 2\cos^2\theta - 1 = 2 \cdot \left(\frac{5}{8}\right)^2 - 1 = \frac{50}{64} - 1 = -\frac{14}{64} = -\frac{7}{32}$.

Check Does $\left(-\frac{7}{32}\right)^2 + \left(\frac{5\sqrt{39}}{32}\right)^2 = 1$? Verify with a calculator.

QY

> **QY**
>
> Why are there two answers for $\sin 2\theta$ in Example 1?

Using the Double Angle Formulas to Find $\sin\frac{1}{2}\theta$ or $\cos\frac{1}{2}\theta$

You can use the Double Angle Formulas to find exact values of the trigonometric functions for $\frac{1}{2}\theta$ if you know the values for θ.

Example 2

Use a formula for $\cos 2\theta$ to find the exact value of $\sin\frac{\pi}{12}$.

Solution Notice that $\frac{\pi}{12}$ is half of $\frac{\pi}{6}$, an argument with known trigonometric values. Let $\theta = \frac{\pi}{12}$. Use $\cos 2\theta = 1 - 2\sin^2\theta$.

$$\cos 2\left(\frac{\pi}{12}\right) = 1 - 2\sin^2\left(\frac{\pi}{12}\right) \quad \text{Substitute for } \theta.$$

$$\cos\left(\frac{\pi}{6}\right) = 1 - 2\sin^2\left(\frac{\pi}{12}\right) \quad \text{arithmetic}$$

$$\frac{\sqrt{3}}{2} = 1 - 2\sin^2\left(\frac{\pi}{12}\right) \quad \text{Use the known value of } \cos\frac{\pi}{6}.$$

Now we solve the equation for $\sin\left(\frac{\pi}{12}\right)$.

$$\frac{\sqrt{3}}{2} - 1 = 1 - 1 - 2\sin^2\left(\frac{\pi}{12}\right)$$

$$\frac{\sqrt{3} - 2}{2} = -2\sin^2\left(\frac{\pi}{12}\right)$$

$$\frac{2 - \sqrt{3}}{4} = \sin^2\left(\frac{\pi}{12}\right)$$

$$\sqrt{\frac{2 - \sqrt{3}}{4}} = \sin\left(\frac{\pi}{12}\right)$$

Check Evaluate $\sqrt{\frac{2 - \sqrt{3}}{4}}$ and $\sin\frac{\pi}{12}$ on your calculator.

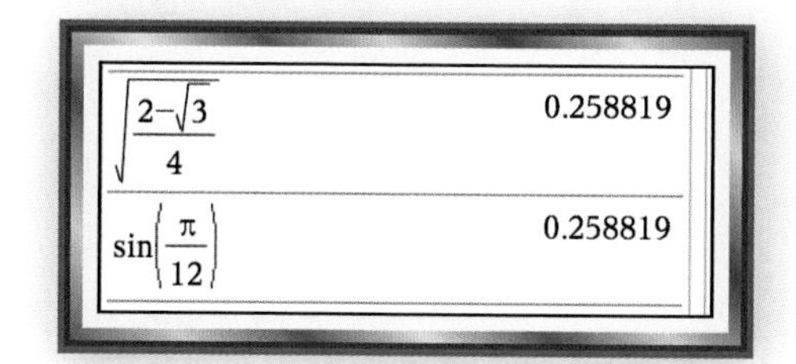

Extending the Double Angle Formulas

From the identities for $\cos 2\theta$ and $\sin 2\theta$, identities for the cosine and sine of other multiples of θ can be found.

Example 3

Express $\cos 4\theta$ as a function of $\cos\theta$ only.

Solution First, use the Double Angle Formula to write $\cos 4\theta$ in terms of $\cos 2\theta$.

$$\begin{aligned}\cos 4\theta &= \cos 2(2\theta)\\ &= 2\cos^2(2\theta) - 1\end{aligned}$$

Now use the Double Angle Formula again to write $\cos 2\theta$ in terms of $\cos\theta$.

$$\begin{aligned}\cos 4\theta &= 2(2\cos^2\theta - 1)^2 - 1\\ &= 2(4\cos^4\theta - 4\cos^2\theta + 1) - 1\\ &= 8\cos^4\theta - 8\cos^2\theta + 2 - 1\\ &= 8\cos^4\theta - 8\cos^2\theta + 1\end{aligned}$$

Check Substitute a value for θ for which the values of $\cos\theta$ and $\cos 4\theta$ are known. We use $\theta = 45°$. Then $\cos 4\theta = \cos 180° = -1$ and

$$\begin{aligned}8\cos^4\theta - 8\cos^2\theta + 1 &= 8\left(\frac{\sqrt{2}}{2}\right)^4 - 8\left(\frac{\sqrt{2}}{2}\right)^2 + 1\\ &= 8\left(\frac{4}{16}\right) - 8\left(\frac{2}{4}\right) + 1\\ &= -1.\end{aligned}$$

It checks.

Questions

COVERING THE IDEAS

In 1 and 2, use the formula $d = \frac{v^2 \sin 2\theta}{g}$, with $g = \frac{9.8 \text{ m}}{\text{sec}^2}$.

1. About how far will a soccer ball travel with respect to the ground if a player kicks it at 30 $\frac{\text{m}}{\text{sec}}$ at an initial angle of 40° with the ground?

The modern soccer ball is based on a design by architect Buckminster Fuller.

2. How far will a tennis ball travel horizontally if it is thrown from the ground at 20 $\frac{\text{m}}{\text{sec}}$ at an angle of 80°?

3. Prove this Double Angle Formula for the cosine: For all real numbers θ, $\cos 2\theta = 1 - 2\sin^2\theta$.

4. Let $f(x) = \cos^2 x - \sin^2 x$, $g(x) = 2\cos^2 x - 1$, and $h(x) = 1 - 2\sin^2 x$.
   a. Give the amplitude and period for the graphs of the functions f, g, and h.
   b. What is the simplest formula for a function whose graph coincides with those of f, g, and h?

5. Let θ be the smaller acute angle in a 3-4-5 right triangle. Find $\sin 2\theta$ and $\cos 2\theta$.

In 6–9, simplify each expression using a Double Angle Formula.

6. $2\cos^2 35° - 1$
7. $2 \sin 25° \cos 25°$
8. $1 - 2\sin^2 \frac{5\pi}{8}$
9. $\cos^2 \frac{4\pi}{9} - \sin^2 \frac{4\pi}{9}$

10. Use $\theta = 135°$ and a double argument formula for $\cos 2\theta$ to find an exact value for sin 67.5°.

11. Express $\sin 4\theta$ as a function of $\cos\theta$ and $\sin\theta$.

APPLYING THE MATHEMATICS

12. A study of a softball player's swing speed indicated that she hits the ball at a velocity of about 125 $\frac{\text{ft}}{\text{sec}}$. When the player hits the ball at the angle that will cause it to go the maximum horizontal distance, how far will the ball travel?

13. Suppose $\sin A = \frac{5}{13}$. Find the exact value of $\sin 2A$ when
   a. $\angle A$ is acute.
   b. $\angle A$ is obtuse.

14. Suppose $\cos B = \frac{2}{7}$. Find $\cos 2B$, if B is in the given interval.
   a. $0 < B < \frac{\pi}{2}$
   b. $\frac{3\pi}{2} < B < \frac{7\pi}{4}$

15. Use the Law of Cosines and the triangle at the right to show $\cos 2\theta = 1 - 2\sin^2\theta$.

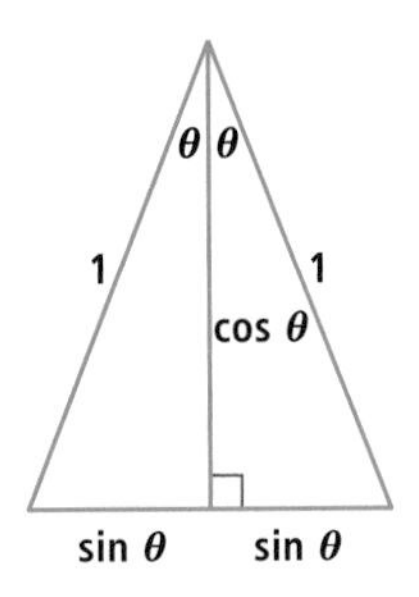

In 16–18, consider the line ℓ which passes through the origin at an angle θ with the positive x-axis. Then the matrix $\begin{bmatrix} \cos 2\theta & \sin 2\theta \\ \sin 2\theta & -\cos 2\theta \end{bmatrix}$ represents r_ℓ, the reflection over ℓ.

16. Suppose ℓ is the x-axis. Verify that the matrix for r_ℓ equals the matrix you know for r_x.

17. a. Suppose ℓ is the line with equation $y = x$. What is θ in this case?
    b. Verify that the matrix for r_ℓ equals the matrix you know for $r_{y=x}$.

18. a. If ℓ is the line with equation $y = \frac{1}{\sqrt{3}}x$, find a matrix of exact values for r_ℓ.
    b. Use your matrix from Part a to find the image of (1, 0) when reflected over $y = \frac{1}{\sqrt{3}}x$. Check your answer with a drawing.

In 19 and 20, the Double Angle Formulas are employed to derive half-angle formulas.

19. Consider $\cos 2\theta = \cos^2\theta - \sin^2\theta$ and $1 = \cos^2\theta + \sin^2\theta$.
    a. Prove: For all θ, $\cos \theta = \pm\sqrt{\frac{1 + \cos 2\theta}{2}}$. (The sign used is determined by the quadrant in which θ lies.) Hint: As a first step, add the two given equations.
    b. Use the formula in Part a to show that $\cos \frac{\pi}{8} = \frac{\sqrt{2 + \sqrt{2}}}{2}$.

20. a. Prove: For all θ, $\sin \theta = \pm\frac{\sqrt{1 - \cos 2\theta}}{2}$.
    b. Find an exact value for $\sin \frac{\pi}{8}$.

21. Write $\cos 3\theta$ as a function of $\sin \theta$ and $\cos \theta$.

REVIEW

22. A and B are acute angles with $\sin A = 0.4$ and $\cos B = 0.7$. Find $\cos(A - B)$. **(Lesson 12-5)**

23. **Multiple Choice** For all real numbers α and β, the expression $\cos \alpha \cos \beta - \sin \alpha \sin \beta$ equals what? **(Lesson 12-5)**
    A $\cos(\alpha + \beta)$ B $\sin(\alpha + \beta)$ C $\cos(\alpha - \beta)$ D $\sin(\alpha - \beta)$

24. A transformation that can be represented by a matrix maps a square with vertices at (0, 0), (0, 1), (1, 1), and (1, 0) to a parallelogram with corresponding vertices at (0, 0), (-3, 2), (-2, 6), and (1, 4). What is the matrix? **(Lesson 12-4)**

25. If $\begin{bmatrix} 8 & 4 \\ -1 & 1 \end{bmatrix}\begin{bmatrix} a \\ 6b \end{bmatrix} = \begin{bmatrix} 31 \\ 38 \end{bmatrix}$, find a and b. **(Lesson 12-1)**

26. Suppose you toss a fair coin ten times. Find the probability of
    a. exactly seven heads.
    b. at least seven heads. **(Lesson 10-5)**

27. The figure at the right represents the way a golfer played a hole to avoid two water hazards. To the nearest yard, how far is it from the tee to the green directly over the water? **(Lesson 5-3)**

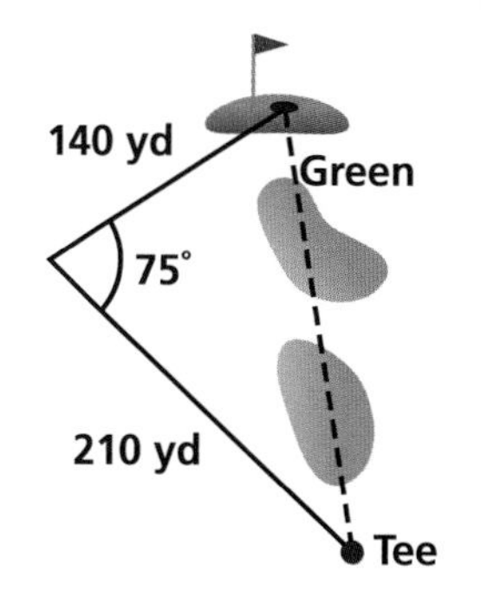

EXPLORATION

28. In this lesson there are proofs of the identities below.
    $\cos(2\theta) = 2\cos^2\theta - 1$ $\cos(4\theta) = 8\cos^4\theta - 8\cos^2\theta + 1$
    Can $\cos(n\theta)$, where n is a positive integer, always be expressed as a polynomial in $\cos \theta$? If it can be, what can you say about the degree and coefficients of the polynomial? Justify your answers.

QY ANSWERS

When $\cos \theta = \frac{5}{8}$, θ can be in either Quadrant I or Quadrant IV. Then 2θ is in either Quadrant II or Quadrant III. In II, the sine is positive. In III, the sine is negative.

Chapter 12 Projects

1 Shear Mapping

A *shear* is a particular kind of transformation that has a slanting effect on objects in the coordinate plane. It leaves all points on one axis fixed while other points are shifted parallel to the axis by a distance proportional to their perpendicular distance from the axis. A matrix for a shear of magnitude k is $\begin{bmatrix} 1 & k \\ 0 & 1 \end{bmatrix}$. Find the image of the rectangle $HOPE$, represented by $\begin{bmatrix} 0 & 0 & 5 & 5 \\ 0 & 4 & 4 & 0 \end{bmatrix}$, under a shear of

a. magnitude 3;

b. magnitude -2.

c. Graph the preimage and the images in Parts a and b on one set of axes.

d. *Shear stress* is a force commonly studied in physics and mechanical engineering. Research shear stress and find three examples of matrices that are used to organize properties of stresses in physics and/or engineering.

2 Triple-Angle Formulas and Beyond

Using the Addition and Double Angle Formulas in the chapter, use transformation matrices to find formulas for the following in terms of sin θ and cos θ.

a. $\sin 4\theta$ b. $\cos 4\theta$ c. $\sin 5\theta$

d. $\cos 5\theta$ e. $\sin 6\theta$ f. $\cos 6\theta$

3 Bug Walk

Consider the grid below, with some key points labeled. Imagine a bug randomly starts at the indicated point on the grid. Create a series of transformation matrices the bug could use to traverse the three key points labeled x on the graph. Using these matrices, draw and label a path on top of the grid, creating a story about the bug's journey. Write out the short story and share it with your teacher and classmates.

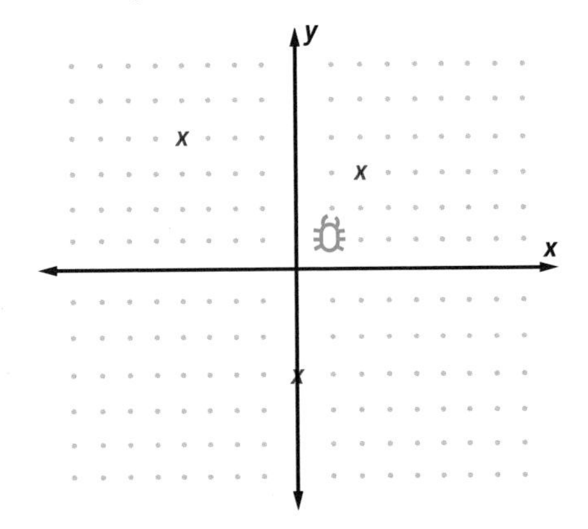

4 Translation using Matrix Multiplication

In this chapter, you saw matrices for reflections and rotations that map (0, 0) onto (0, 0). Translating points a certain distance in the coordinate plane can be done using 3 × 3 matrices. An example was shown on page 743 and is repeated below.

$$\begin{bmatrix} \cos\left(-\frac{\pi}{8}\right) & -\sin\left(-\frac{\pi}{8}\right) & 82 \\ \sin\left(-\frac{\pi}{8}\right) & \cos\left(-\frac{\pi}{8}\right) & 37 \\ 0 & 0 & 1 \end{bmatrix}$$

a. What familiar 2 × 2 matrix appears in the 3 × 3 matrix above?

b. Show that the matrix $\begin{bmatrix} a & b & 0 \\ c & d & 0 \\ 0 & 0 & 1 \end{bmatrix}$ represents the same transformation as the two-dimensional matrix $\begin{bmatrix} a & b \\ c & d \end{bmatrix}$. (Hint: If the point $\begin{bmatrix} x \\ y \\ 1 \end{bmatrix}$ is transformed by this 3 × 3 matrix, what matrix results?)

c. Rotate the point (3, 4) 45° by assigning values to a, b, c, d, x, and y and using the matrices in Part b. What are the coordinates of the image?

d. Show that the matrix $\begin{bmatrix} 1 & 0 & h \\ 0 & 1 & k \\ 0 & 0 & 1 \end{bmatrix}$ can represent the translation T: $(x, y) \rightarrow (x + h, y + k)$.

e. Translate the point (3, 4) six units to the right and 2 units down using the 3 × 3 matrix in Part d.

f. Multiply the matrices $\begin{bmatrix} 1 & 0 & h \\ 0 & 1 & k \\ 0 & 0 & 1 \end{bmatrix}$ and $\begin{bmatrix} a & b & 0 \\ c & d & 0 \\ 0 & 0 & 1 \end{bmatrix}$. What transformation does the product matrix describe?

g. In the photo at the right, a base jumper is rotating as he falls. If you wanted to create an animation of his fall, what matrix could you use to map the first image of the jumper onto the last image of him?

5 Trigonometric Identities

Use the identities from this and previous chapters to prove, for all x and y, six of the following eight identities.

a. $\sin^2 x = \frac{1 - \cos 2x}{2}$

b. $\cos^2 x = \frac{1 + \cos 2x}{2}$

c. $\sin x + \sin y = 2 \sin\left(\frac{x + y}{2}\right) \cos\left(\frac{x - y}{2}\right)$

d. $\cos x - \cos y = -2 \sin\left(\frac{x + y}{2}\right) \sin\left(\frac{x - y}{2}\right)$

e. $\sin x \sin y = \frac{1}{2}(\cos(x - y) - \cos(x + y))$

f. $\cos x \sin y = \frac{1}{2}(\sin(x + y) - \sin(x - y))$

g. $\cos 3x = 4 \cos^3 x - 3 \cos x$

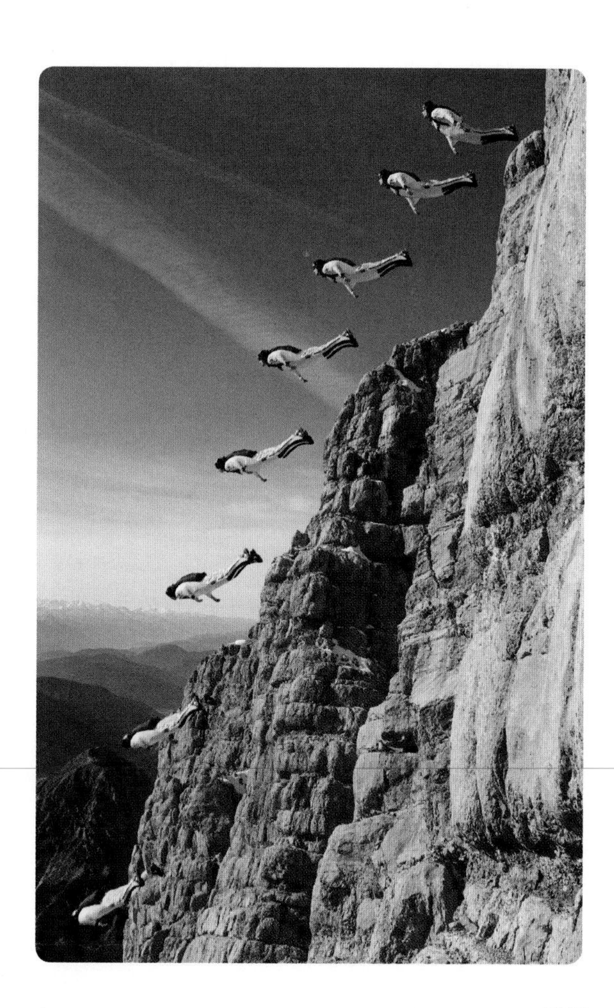

Chapter 12 Summary and Vocabulary

- An $m \times n$ **matrix** is a rectangular array of elements arranged in m rows and n columns.
- If A and B are matrices, the product $C = AB$ exists if the number of columns of A equals the number of rows of B. The element in the ith row and jth column of C is the sum of the term-by-term product of the ith row of A and the jth row of B. Matrix multiplication has many applications.
- Among the many transformations that can be represented by 2×2 matrices are: reflections over the x-axis, the y-axis, and the lines $y = \pm x$; all size changes with center (0, 0); all scale changes of the form $(x, y) \rightarrow (ax, by)$; and all rotations with center at the origin.

Reflections:

over x-axis

$$\begin{bmatrix} 1 & 0 \\ 0 & -1 \end{bmatrix}$$

$r_x: (x, y) \rightarrow (x, -y)$

over y-axis

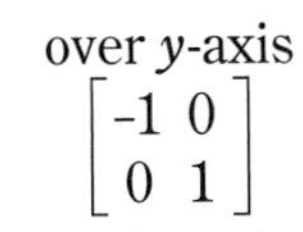

$$\begin{bmatrix} -1 & 0 \\ 0 & 1 \end{bmatrix}$$

$r_y: (x, y) \rightarrow (-x, y)$

over the line $y = x$

$$\begin{bmatrix} 0 & 1 \\ 1 & 0 \end{bmatrix}$$

$r_{y=x}: (x, y) \rightarrow (y, x)$

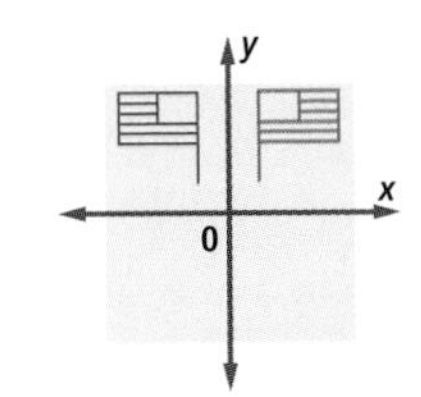

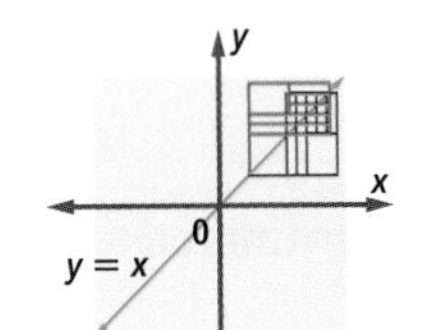

Other Transformations:

Size changes with center (0, 0), magnitude k:

$$\begin{bmatrix} k & 0 \\ 0 & k \end{bmatrix}$$

$S_k: (x, y) \rightarrow (kx, ky)$

Scale changes with horizontal magnitude a and vertical magnitude b:

$$\begin{bmatrix} a & 0 \\ 0 & b \end{bmatrix}$$

$S_{a,b}: (x, y) \rightarrow (ax, by)$

Rotations of magnitude θ about the origin:

$$\begin{bmatrix} \cos\theta & -\sin\theta \\ \sin\theta & \cos\theta \end{bmatrix}$$

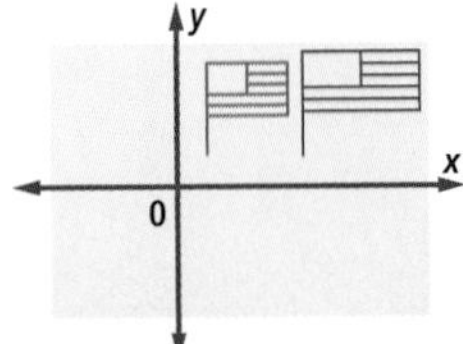

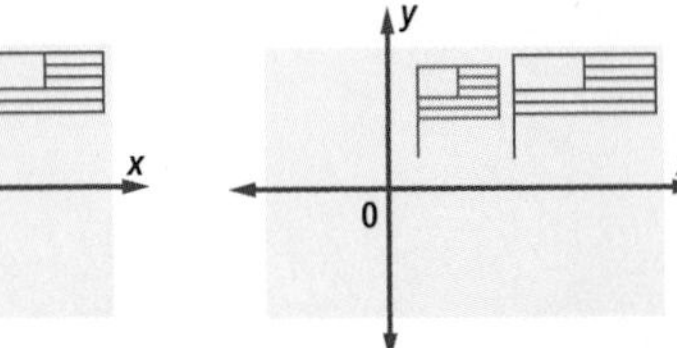

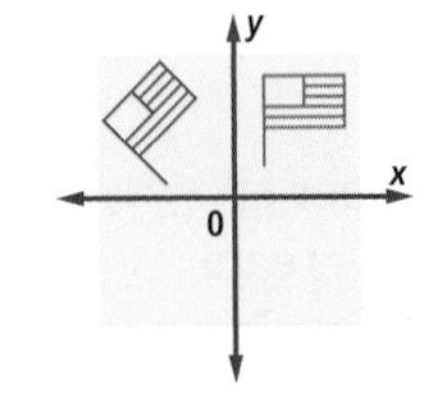

Vocabulary

12-1
*matrix, matrices
element of the matrix
*dimensions of a matrix
row matrix
column matrix
product matrix

12-2
point matrix
matrix representing a transformation
identity transformation
2 × 2 identity matrix for multiplication

- If a transformation has a 2×2 matrix, then the Matrix Basis Theorem provides an easy way to recall the matrix; you need only know the images of (1, 0) and (0, 1) under that transformation.
- Any n-gon can be represented by a $2 \times n$ matrix whose columns are its vertices. If M represents a transformation T and P represents a polygon, then the product MP represents $T(P)$. This enables figures to be transformed using matrices. If M_1 and M_2 are matrices for transformations T_1 and T_2, then M_1M_2 represents $T_1 \circ T_2$.
- From the matrix for a rotation, formulas for $\cos(\alpha + \beta)$ and $\sin(\alpha + \beta)$ can be rather quickly derived. From these formulas, formulas for $\cos(\alpha - \beta)$, $\sin(\alpha - \beta)$, $\cos 2\theta$, $\sin 2\theta$, $\cos \frac{\theta}{2}$, and $\sin \frac{\theta}{2}$ follow. These formulas enable exact values for certain cosines and sines to be determined and many other identities to be proved.

Properties and Theorems

Algorithm for Matrix Multiplication (p. 746)
Theorem (Matrices for Reflections) (p. 752)
Matrices and Composites Theorem (p. 758)
Theorem (Matrices for Rotations) (p. 759)
Matrix Basis Theorem (p. 759)
Rotation Matrix Theorem (p. 762)
Theorem (Addition Formulas for the Cosine and Sine) (p. 766)
Theorem (Subtraction Formulas for the Cosine and Sine) (p. 767)
Theorem (Double Angle Formulas) (p. 771)

Chapter 12 Self-Test

Take this test as you would take a test in class. You will need a calculator. Then use the Selected Answers section in the back of the book to check your work.

In 1 and 2, use the following matrices. P gives the number of registered voters (in millions) in each part of a state. V represents the percent of registered voters in each area that would vote for each party in an election.

$$\begin{matrix}\text{Urban}\\ \text{Rural}\\ \text{Suburban}\end{matrix}\begin{bmatrix}4.8\\0.8\\3.2\end{bmatrix} = P$$

$$\begin{matrix} & \text{Urban} & \text{Rural} & \text{Suburban} \\ \text{Democrat} & 0.57 & 0.24 & 0.29 \\ \text{Republican} & 0.21 & 0.59 & 0.48 \\ \text{Independent} & 0.22 & 0.17 & 0.23 \end{matrix} = V$$

1. Find VP.
2. What does the matrix VP represent?
3. Find the product $\begin{bmatrix}6 & -1\\3 & 0\end{bmatrix}\begin{bmatrix}3 & 7\\1 & -1\end{bmatrix}$.
4. Suppose M is a 3×5 matrix, N is a $5 \times t$ matrix, and Z is a 3×6 matrix. If $MN = Z$, find the value of t.

In 5–7, use the figure below.

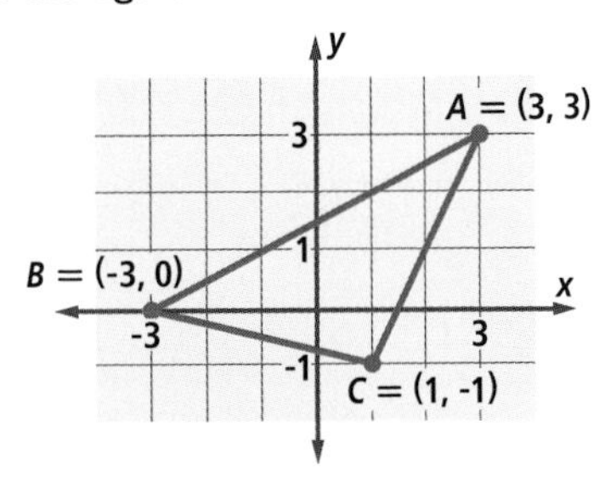

5. Write a matrix M for $\triangle ABC$.
6. a. Find XM, where $X = \begin{bmatrix}2 & 0\\0 & -1\end{bmatrix}$.
   b. Graph the image represented by XM.
7. Give a matrix representing the image of $\triangle ABC$ after a rotation of 120°.
8. a. What single matrix represents the composite $R_{90} \circ r_x$?
   b. What single transformation is the composite?
9. Give a matrix representing the reflection over the line with equation $y = x$.
10. **Multiple Choice** Which matrix represents R_{60}?

    A $\begin{bmatrix}\frac{60}{2} & \frac{\sqrt{3}}{2}\\ \frac{\sqrt{3}}{2} & \frac{1}{2}\end{bmatrix}$ B $\begin{bmatrix}\frac{1}{2} & -\frac{\sqrt{3}}{2}\\ \frac{\sqrt{3}}{2} & \frac{1}{2}\end{bmatrix}$ C $\begin{bmatrix}-\frac{1}{2} & -\frac{\sqrt{3}}{2}\\ \frac{\sqrt{3}}{2} & -\frac{1}{2}\end{bmatrix}$

    D $\begin{bmatrix}\frac{\sqrt{3}}{2} & -\frac{1}{2}\\ \frac{1}{2} & \frac{\sqrt{3}}{2}\end{bmatrix}$ E $\begin{bmatrix}\frac{\sqrt{3}}{2} & \frac{1}{2}\\ -\frac{1}{2} & \frac{\sqrt{3}}{2}\end{bmatrix}$

11. Let $P = (1, 0)$, $Q = (4, 7)$ and $R = (0, -3)$. What is the image of $\triangle PQR$ under a rotation of magnitude -56°?

In 12 and 13, write the expression as the sine or cosine of a single argument.

12. $\sin 79° \cos 37° - \cos 79° \sin 37°$
13. $2\cos^2 18° - 1$
14. Using an addition formula, simplify $\sin(\pi + \theta)$.
15. If $\cos A = 0.30$ and A is acute, find $\sin 2A$.
16. Let X be the matrix representing $r_{x\text{-axis}}$.
    a. Compute X^3.
    b. What transformation does X^3 represent?
17. A transformation that can be represented by a 2×2 matrix maps (0, 0) to (0, 0), (1, 0) to (-1, 0), and (0, 1) to (0, 7).
    a. Give the matrix for the transformation.
    b. What is the image of (1, 1)?
18. Tell in words what composite of transformations is represented by

$$\begin{bmatrix}\cos\frac{\pi}{3} & -\sin\frac{\pi}{3}\\ \sin\frac{\pi}{2} & \cos\frac{\pi}{3}\end{bmatrix} \cdot \begin{bmatrix}0 & -1\\1 & 0\end{bmatrix} \cdot \begin{bmatrix}0 & 1\\1 & 0\end{bmatrix}.$$

Chapter 12 Chapter Review

SKILLS
PROPERTIES
USES
REPRESENTATIONS

SKILLS Procedures used to get answers

OBJECTIVE A Multiply matrices, when possible. (Lesson 12-1)

In 1–4, use the following matrices.

$A = \begin{bmatrix} \frac{1}{3} & 2 \\ -1 & 5 \end{bmatrix}$ $B = \begin{bmatrix} -1 & 7 & 1 \\ 1 & 0 & -1 \end{bmatrix}$ $C = \begin{bmatrix} 5 \\ x \end{bmatrix}$ $D = [-3, 1]$

1. **Multiple Choice** Which of the following products does not exist?
   A AC B CD C BA D A^2C
2. Calculate AB.
3. Calculate DC.
4. Calculate A^2.

PROPERTIES Principles behind the mathematics

OBJECTIVE B Apply properties of matrices and matrix multiplication. (Lessons 12-1, 12-3)

In 5 and 6, an equation relating the matrices M, N, and P is given. The matrix M has dimensions 5×3, and the matrix P has dimensions 5×6. Give the dimensions of N.

5. $MN = P$
6. $PN = M$
7. **Multiple Choice** Which of the following statements about any 2×2 matrices R, S, and T is false?
   A RS is also a 2×2 matrix.
   B $(RS)T = R(ST)$
   C There is a matrix I such that $RI = IR = R$.
   D $RS = SR$
8. Name three types of transformations that can be represented by 2×2 matrices.

OBJECTIVE C Apply the Addition, Subtraction, and Double Angle Formulas. (Lessons 12-5, 12-6)

In 9 and 10, compute the exact value using an Addition or Subtraction Formula.

9. $\cos 105^\circ$
10. $\sin \frac{5\pi}{12}$

In 11 and 12, A and B are acute angles with $\sin A = 0.7$ and $\cos B = 0.4$. Find the exact value.

11. $\cos(A - B)$
12. $\sin 2A$

In 13 and 14, use an Addition or Subtraction Formula to simplify.

13. $\sin(\pi - \theta)$
14. $\cos\left(\theta - \frac{\pi}{2}\right)$
15. Give a formula for $\cos 4\theta$ in terms of $\sin \theta$ only.

In 16–18, write as the sine or cosine of a single argument.

16. $2 \sin x \cos x$
17. $\cos \frac{3\pi}{5} \cos \frac{\pi}{3} + \sin \frac{3\pi}{5} \sin \frac{\pi}{3}$
18. $2 \cos^2 35^\circ - 1$
19. Express $\sin 2\theta$ as a function of $\sin \theta$ for $0 \le \theta \le \frac{\pi}{2}$.
20. Express $\sin 3\theta$ as a function of $\sin \theta$ and $\cos \theta$.

USES Applications of mathematics in real-world situations

OBJECTIVE D Use a matrix to organize information. (Lesson 12-1)

In 21 and 22, use matrix F below of Foreign Exchange rates on the first banking day in January. The amounts in each row are equal in value to one U.S. dollar.

	United Kingdom (pounds)	Canada (dollar)	Japan (yen)	United States (dollar)
1990	1.61	0.86	0.0068	1.00
1995	1.56	0.71	0.0099	1.00
2000	1.63	0.69	0.0098	1.00
2005	1.91	0.83	0.0097	1.00

$= F$

21. **a.** Describe the information given by the first row of F.

b. How many U.S. dollars were 1000 Japanese yen worth in 2005?

22. **a.** If a tourist entered the U.S. in 2005 with 212 pounds U.K., 150 dollars Canadian, and 50 dollars U.S., what is the U.S. equivalent of that cash?

b. **True or False** To compare the total cash value of the tourist in Part a at five-year intervals between 1990 and 2005, you could look at the product matrix FC, where $C = \begin{bmatrix} 212 \\ 150 \\ 0 \\ 50 \end{bmatrix}$.

In 23 and 24, use the production matrix P and the cost matrix S shown below.

	Melons	Lettuce (heads)
Farm A	1000	10,000
Farm B	1400	12,000
Farm C	500	18,000
Farm D	500	5,000

$= P$

	Cost to produce	Cost to consumer
Melons	0.55	1.10
Lettuce	0.70	1.25

$= S$

23. **a.** Calculate PS. What does PS represent?

b. Find the total cost of producing melons and lettuce on Farm C.

c. What element of PS contains this cost?

24. Find the total cost to the consumer of the melons and lettuce produced on Farm A.

REPRESENTATIONS Pictures, graphs, or objects that illustrate concepts

OBJECTIVE E Find the image of a figure under a transformation using a matrix. (Lesson 12-2)

25. Write a matrix for $\triangle FRY$ when $F = (2, 4)$, $R = (6, 1)$, and $Y = (-1, 0)$.

26. Write a matrix for $\triangle ABC$ shown in Question 27.

27. Refer to $\triangle ABC$ below.

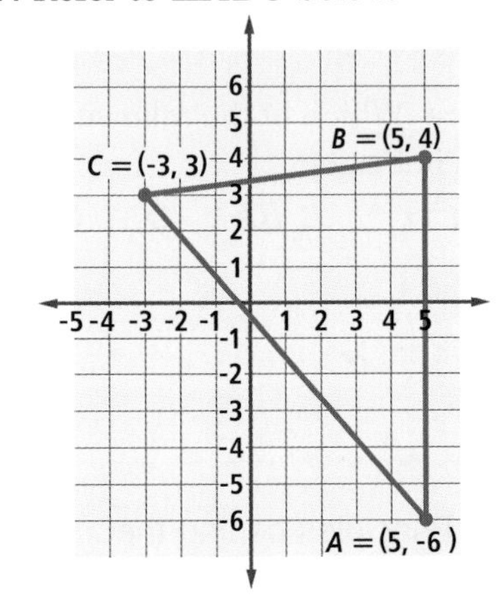

a. Find a matrix for the image $\triangle A'B'C'$ under the transformation represented by $\begin{bmatrix} 1 & 1 \\ 0 & -1 \end{bmatrix}$.

b. Draw the image.

28. **a.** Find a matrix for the vertices of the image of the square having opposite vertices (3, 3) and (5, 5) under the transformation with matrix $\begin{bmatrix} -3 & 1 \\ 2 & -1 \end{bmatrix}$.

b. Draw the square and its image on the same axes.

OBJECTIVE F Represent reflections, rotations, scale changes, and size changes by matrices. (Lessons 12-2, 12-3, 12-4)

29. Write the matrix representing r_x, reflection over the x-axis.

30. Write the matrix for the scale change $T: (x, y) \rightarrow \left(\frac{1}{5}x, y\right)$.

31. Write the matrix that represents a 270° rotation about the origin.

In 32–34, describe the transformation represented by the matrix.

32. $\begin{bmatrix} 19 & 0 \\ 0 & 19 \end{bmatrix}$ 33. $\begin{bmatrix} 0 & 1 \\ 1 & 0 \end{bmatrix}$ 34. $\begin{bmatrix} -1 & 0 \\ 0 & -1 \end{bmatrix}$.

35. Give the matrix for R_{27}.

36. Give exact values of the matrix for R_{240}.

37. If $T(1, 0) = (q, t)$, $T(0, 1) = (e, z)$, and T can be represented by a 2×2 matrix, what is this matrix?

38. A transformation that can be represented by a matrix takes (0, 0) to (0, 0), (1, 0) to (2, 6), and (0, 1) to (0, 3).
   a. What is the 2×2 matrix for the transformation?
   b. What is the image of (-1, 2)?

39. A two-dimensional figure has been rotated 75° about the origin. Write the matrix for the transformation that returns the figure to its original position.

40. Let $A = (-4, 7)$, $B = (5, 1)$, and $C = (0, -8)$.
   a. Find the exact coordinates of the image of $\triangle ABC$ under a rotation of 110° about the origin.
   b. Round the coordinates you found in Part a to the nearest hundredth.

OBJECTIVE G Represent composites of transformations by matrix products. (Lessons 12-3, 12-4)

41. A figure is rotated 90° clockwise, then reflected over the y-axis.
   a. Write a matrix product for the composite transformation.
   b. Compute the matrix product from Part a.

42. Let $A = (9, 0)$, $B = (0, 3)$, and $C = (-1, -1)$.
   a. Describe $(r_x \circ R_{90})(\triangle ABC)$ as the product of three matrices.
   b. Find a single matrix for the image.

43. a. Tell, in words, what composite of transformations is represented by $\begin{bmatrix} \cos 27^\circ & -\sin 27^\circ \\ \sin 27^\circ & \cos 27^\circ \end{bmatrix} \cdot \begin{bmatrix} 0 & 1 \\ 1 & 0 \end{bmatrix} \cdot \begin{bmatrix} 0 & -1 \\ 1 & 0 \end{bmatrix}$.
   b. What single matrix represents the composite?

44. The matrix representing R_{135} is $A = \begin{bmatrix} \frac{\sqrt{2}}{2} & -\frac{\sqrt{2}}{2} \\ \frac{\sqrt{2}}{2} & -\frac{\sqrt{2}}{2} \end{bmatrix}$
   a. Compute A^2.
   b. What transformation does A^2 represent?

45. Let Y be the matrix representing r_y.
   a. Compute Y^3.
   b. What transformation does Y^3 represent?

Chapter 13

Further Work with Trigonometry

Contents

The Clematis, Trillium and Dogwood flowers resemble polar graphs, sometimes called petal curves or rose curves.

This chapter covers three topics in trigonometry. First, we derive and prove some important relationships involving values of trigonometric functions.

Second is an introduction to a coordinate system different from rectangular coordinates, the *polar coordinate system*. When plotted with polar coordinates, the graphs of trigonometric functions often are quite beautiful. The patterns shown on the previous page are examples.

The third topic is complex numbers. Complex numbers can be represented in a variety of ways and graphed in both the rectangular and polar coordinate systems.

Consider the identities for cos 2α and cos 4α found in Chapter 12. You may have noticed that the double-angle identity involved squaring, and the quadruple-angle identity involved raising to the fourth power. This pattern suggests a deep underlying connection between powers and trigonometry. Expressing complex numbers in a polar coordinate system reveals this connection, and creates an elegant representation for addition, multiplication, and powers of complex numbers. These operations culminate in a beautiful theorem discovered by De Moivre that you can use to calculate *n*th powers and *n*th roots of any complex number.

Lesson

13-1 Proving Trigonometric Identities

Vocabulary

domain of an identity

▸ **BIG IDEA** Using trigonometric identities and algebra, equivalent trigonometric statements can be verified.

Recall that an *identity* is an equation that is true for all values of the variables for which the expression is defined. In this lesson, you will see various ways to derive and prove trigonometric identities. Although the proof techniques are illustrated using only trigonometric functions, they are applicable to any function.

Mental Math

State whether the equation is true. If not, correct the equation to make it true.

a. $\begin{bmatrix} 1 & 0 \\ 0 & 1 \end{bmatrix} \cdot \begin{bmatrix} 3 & -7 \\ 2 & 6 \end{bmatrix} = \begin{bmatrix} 3 & -7 \\ 2 & 6 \end{bmatrix}$

b. $3(0!) = 10$

c. $3 \log 100 = 6$

d. $x^{\frac{n}{7}} = \sqrt[n]{x^7}$

Testing Whether an Equation Might Be an Identity

You have seen many instances in which graphs are used to determine whether a particular statement is true. In particular, if you want to see whether an equation in one variable is an identity, you can consider each side of the equation as a separate function of x, and graph the two functions. If the graphs coincide, the original equation is *likely* to be an identity.

As an example, consider the Pythagorean Identity,

$$\cos^2 x + \sin^2 x = 1.$$

Treat each side of the equation as a separate function of x, and graph both.

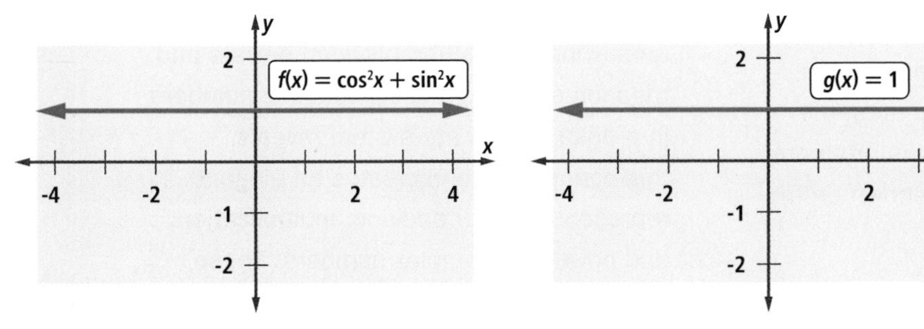

As expected, the graphs *appear* to be identical. If they had been graphed on the same axes, the graphs would have coincided.

In contrast, the equation $(\cos x - \sin x)^2 = \cos^2 x - \sin^2 x$ is *not* an identity. The graphs of $f(x) = (\cos x - \sin x)^2$ and $g(x) = \cos^2 x - \sin^2 x$ are shown on the next page. It is apparent that f and g are not the same function because the graphs do not coincide.

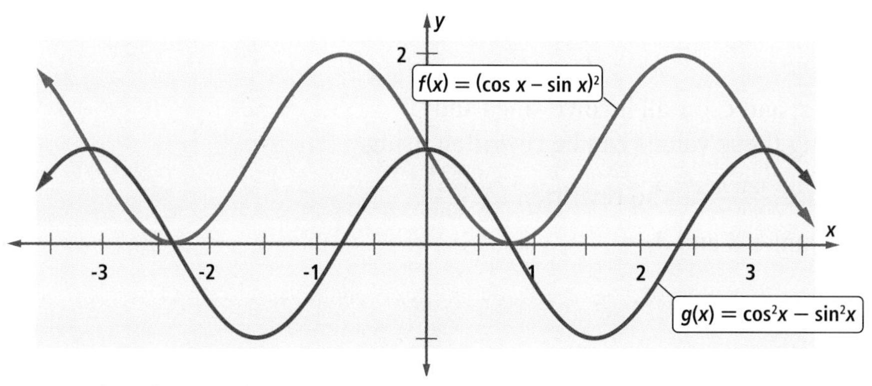

Notice that the graph of g appears to be a transformation image of the graph of the cosine function with amplitude 1 and period π. This suggests that an expression identically equal to $\cos^2 x - \sin^2 x$ is $\cos 2x$. This agrees with the Double Angle Formula in Lesson 12-6. For all x, $\cos 2x = \cos^2 x - \sin^2 x$.

Deducing New Identities by Manipulating Expressions

Here are some identities that you have seen in earlier chapters. These identities are true for all x and y.

Pythagorean Identity	$\cos^2 x + \sin^2 x = 1$
Addition and Subtraction Formulas	$\cos(x \pm y) = \cos x \cos y \mp \sin x \sin y$
	$\sin(x \pm y) = \sin x \cos y \pm \cos x \sin y$
Double Angle Formulas	$\sin 2x = 2 \sin x \cos x$
	$\cos 2x = \cos^2 x - \sin^2 x$

Using these identities, you can derive new identities.

Example 1

Find an identity with $(\cos x - \sin x)^2$ on one side.

Solution Expand the expression, then make use of the identities above.

$(\cos x - \sin x)^2$

$= \cos^2 x - 2 \cos x \sin x + \sin^2 x$ Expand the binomial.

$= \cos^2 x + \sin^2 x - 2 \cos x \sin x$ Commutative Property of Addition

$= 1 - 2 \cos x \sin x$ Pythagorean Identity

$= 1 - \sin 2x$ Double Angle Formula

So an identity is $(\cos x - \sin x)^2 = 1 - \sin 2x$.

 QY

In Example 1, one expression was manipulated to equal a second expression. This technique can often be used to prove that a given equation is an identity. Start with one side of a proposed identity, and rewrite it using definitions, known identities, or algebraic properties until it equals the other side.

QY

a. Check Example 1 by graphing $f(x) = (\cos x - \sin x)^2$ and $g(x) = 1 - \sin 2x$ to see if the graphs appear to coincide.

b. Check the result by evaluating both sides of the identity when $x = \frac{\pi}{2}$.

Proving Identities by Writing All Expressions as Sines or Cosines

The definitions of $\tan x$, $\cot x$, $\sec x$, and $\csc x$ all involve sines and cosines. So an expression involving these values can be rewritten using only sines and cosines. For instance, $\frac{\tan x}{\sin x}$ can be rewritten as $\frac{\frac{\sin x}{\cos x}}{\sin x}$. This technique is used in the Examples 2 and 3.

Example 2

Prove the identity $\frac{\sec x}{\sin x} - \frac{\sin x}{\cos x} = \cot x$.

Solution

$$\frac{\sec x}{\sin x} - \frac{\sin x}{\cos x} = \frac{\frac{1}{\cos x}}{\sin x} - \frac{\sin x}{\cos x} \quad \text{definition of } \sec x$$

$$= \frac{\frac{1}{\cos x}}{\sin x} \bullet \frac{\cos x}{\cos x} - \frac{\sin x}{\cos x} \bullet \frac{\sin x}{\sin x} \quad \text{Get a common denominator.}$$

$$= \frac{1}{\sin x \cos x} - \frac{\sin^2 x}{\cos x \sin x} \quad \text{Multiply.}$$

$$= \frac{1 - \sin^2 x}{\sin x \cos x} \quad \text{Subtract the fractions.}$$

$$= \frac{\cos^2 x}{\sin x \cos x} \quad \text{Use the Pythagorean Identity.}$$

$$= \frac{\cos x}{\sin x} \quad \text{Divide numerator and denominator by } \cos x.$$

$$= \cot x \quad \text{definition of } \cot x$$

GUIDED

Example 3

Prove that for all x for which $\tan x$ and $\sec x$ are defined, $1 + \tan^2 x = \sec^2 x$.

Solution Substitute into the left expression and rewrite until you get the expression on the right.

$$1 + \tan^2 x = 1 + \left(\frac{\sin x}{\cos x}\right)^2 \quad \text{definition of tangent}$$

$$= 1 + \frac{?}{?} \quad \text{Square the fraction.}$$

$$= \frac{?}{\cos^2 x} + \frac{?}{\cos^2 x} \quad \text{Form a common denominator.}$$

$$= \frac{? + ?}{\cos^2 x} \quad \text{Add fractions with a common denominator.}$$

$$= \frac{1}{\cos^2 x} \quad \text{Pythagorean Identity}$$

$$= \sec^2 x \quad \text{definition of secant}$$

So, $1 + \tan^2 x = \sec^2 x$ for all x in the domains of $\tan x$ and $\sec x$.

Check 1 Graph $f(x) = 1 + \tan^2 x$ and $g(x) = \sec^2 x$ on the same set of axes to see if the graphs appear to coincide.

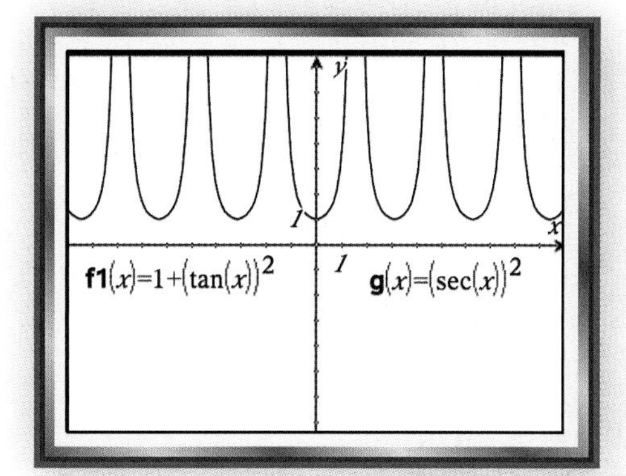

Check 2 Using a CAS, set the expressions equal to each other. The CAS validates that the statement is true.

The **domain of an identity** is the set of values for which the identity is true. In Example 3, since $\tan x = \frac{\sin x}{\cos x}$ and $\sec x = \frac{1}{\cos x}$, the identity is not defined when $\cos x = 0$, or when $x = \frac{\pi}{2} + k\pi$, where k is any integer.

Activity

Use graphs to classify each statement as either *sometimes true, always true,* or *never true.* If the statement is *sometimes true*, describe all values of x that make it true. If the statement is *always true*, prove it.

a. $\sin x = 2 \sin x \cos x$

b. $(\sin x + \cos x)^2 + (\sin x - \cos x)^2 = 2$

c. $\sin x - \cos^2 x = \sin x(1 + \sin x)$

d. $2 \sin^2 x + \cos 2x = 1$

Questions

COVERING THE IDEAS

1. a. Find a value of x for which $(\cos x - \sin x)^2 = \cos^2 x - \sin^2 x$.
   b. Is $(\cos x - \sin x)^2 = \cos^2 x - \sin^2 x$ an identity?
2. a. Explain why $(\cos x + \sin x)^2 = \cos^2 x + \sin^2 x$ is not an identity.
   b. Write $(\cos x + \sin x)^2$ as an expression involving a single trigonometric function.
3. Show that $\frac{\pi}{4}$ satisfies $1 + \tan^2 x = \sec^2 x$, the identity in Example 3.

In 4–7, prove the identity.

4. $1 + \cot^2 x = \csc^2 x$ (Hint: Use the idea of Example 3.)
5. $\frac{1 - \cos 2\theta}{2} = \sin^2\theta$ (Hint: Use the idea of Example 1.)
6. $\cos(A + B) + \cos(A - B) = 2 \cos A \cos B$ (Hint: Use identities from Chapter 12.)
7. $\frac{\sec x}{\sin x} - \frac{\sin x}{\cos x} = \cot x$ (Hint: Use the idea of Example 2.)

In 8 and 9, the graph of a product of trigonometric functions is drawn.

a. **What identity is suggested by the graph? (Hint: The answer involves one of the six parent trigonometric functions.)**

b. **Prove your answer to Part a.**

8. $y = \tan x \cos x$

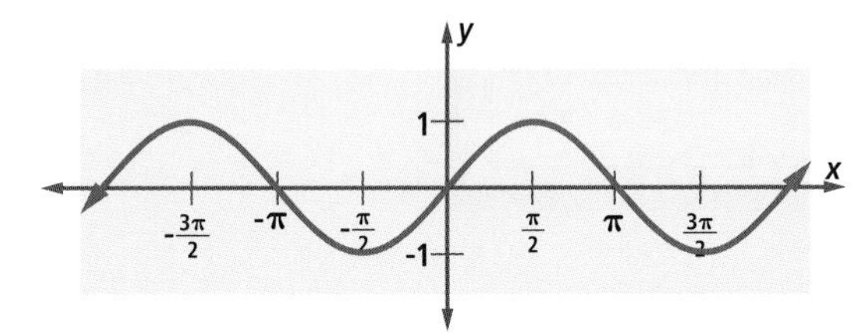

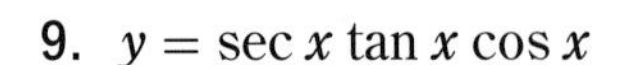

9. $y = \sec x \tan x \cos x$

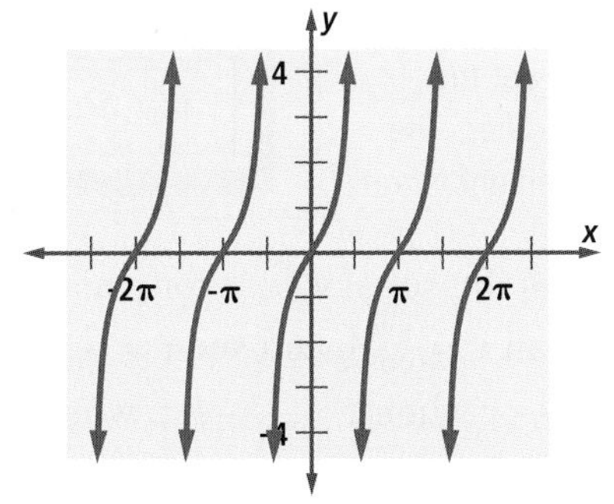

APPLYING THE MATHEMATICS

10. a. Simplify $\cos(x + \pi)$ to a single trigonometric function. What identity results?

    b. Use $x = \frac{\pi}{6}$ to check the identity.

In 11–16, a trigonometric equation is given.

a. **Use a graph to test whether the equation may be an identity.**

b. **Prove the identity or give a counterexample.**

11. $\cos x = \sin\left(\frac{\pi}{2} - x\right)$
12. $\sin 3\theta = 3 \sin\theta \cos\theta$
13. $\frac{\cot y}{\cos y} = \csc y$
14. $\cot^2 x \sin^2 x = 1 - \sin^2 x$
15. $\cot x - \tan x = \frac{\cos 2x}{\sin x \cos x}$
16. $\frac{1 + \tan^2 x}{\tan^2 x} = \csc^2 x$

In 17 and 18, find an identity involving the given expression.

17. $\tan x(\sin x + \cot x \cos x)$
18. $\frac{\sin(x + y)}{\cos x \cos y}$

In 19 and 20, use the techniques of this lesson to test whether the statement is or is not an identity. Justify your answer.

19. $x^2 + x^5 = x^7$
20. $\log(x^2) + \log(x^5) = \log(x^7)$

21. Suppose $90° < \theta < 180°$, and $\tan \theta = -\frac{4}{7}$. Use the Pythagorean Identity or one of its corollaries to find the exact value.

a. $\cot \theta$ b. $\sec \theta$ c. $\sin \theta$

REVIEW

In 22 and 23, let $\triangle ABC$ be a right triangle with right angle C. Express each function in terms of the sides a, b, or c. (Lesson 5-9)

22. $\csc A$ 23. $\cot B$

24. Give the number of solutions to the equation for $0 \le x < 2\pi$. **(Lesson 5-7)**

a. $7 \sin x = 4$ b. $4 \sin x = 7$ c. $4 \tan x = 7$

In 25 and 26, an equation is given.

a. Find all solutions such that $0 \le x \le 2\pi$.

b. Describe the general solution. (Lesson 5-7)

25. $\tan^2 x - 1 = 0$ 26. $2 \cos^2 x - \cos x = 1$

In 27 and 28, evaluate in radians. (Lessons 5-6, 5-4)

27. $\sin^{-1}(1)$ 28. $\tan^{-1}(-\sqrt{3})$

29. The height h in meters of a tide in a harbor is given by the equation $h = \cos\left(\frac{\pi t}{6}\right) + 5$, where t is the time in hours after high tide. **(Lessons 4-8, 4-7)**

a. Sketch a graph of this function for $0 \le t \le 24$.

b. What is the maximum height of the tide during this 24-hour period?

c. At what time during the 24-hour period does the highest tide occur?

EXPLORATION

30. Refer to Questions 8 and 9.

a. Find another pair of parent functions whose product seems to be a parent function, and prove the identity.

b. Find another set of three parent functions whose product seems to be another parent function, and prove the identity.

QY ANSWERS

a.

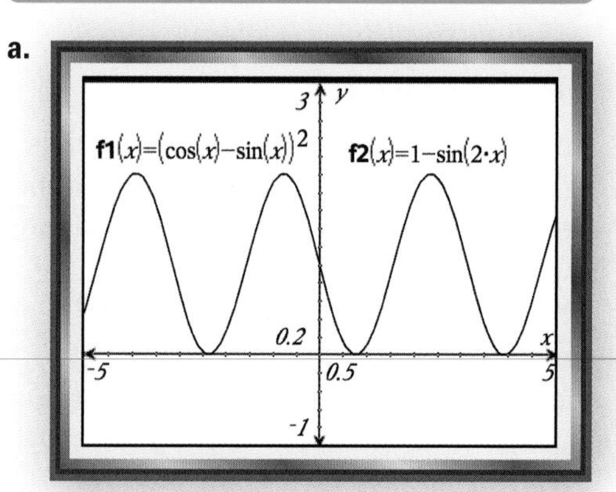

b. $\left(\cos \frac{\pi}{2} - \sin \frac{\pi}{2}\right)^2 = (0 - 1)^2 = 1$,
$1 - \left(\sin 2 \cdot \frac{\pi}{2}\right) = 1 - \sin \pi = 1 - 0 = 1$

Lesson 13-2 Restrictions on Trigonometric Identities

Vocabulary

singularity
discontinuity
removable singularity
removable discontinuity

BIG IDEA Some trigonometric identities have restricted domains in order to avoid zero denominators and values for which a trigonometric function is undefined.

In Example 3 of Lesson 13-1, we noted that $1 + \tan^2 x = \sec^2 x$ is an identity for "all x in the domain of $\tan x$ and $\sec x$."

Only two of the parent trigonometric functions, sine and cosine, are defined for all real numbers. The other four have the following *restrictions* on their domains, where n is any integer.

$f(x)$	Restriction on domain
$\tan x = \frac{\sin x}{\cos x}$	$x \neq \frac{\pi}{2} + n\pi$
$\cot x = \frac{\cos x}{\sin x}$	$x \neq n\pi$
$\sec x = \frac{1}{\cos x}$	$x \neq \frac{\pi}{2} + n\pi$
$\csc x = \frac{1}{\sin x}$	$x \neq n\pi$

Mental Math

For what values of x is the expression not defined?

a. $\frac{x-2}{x-3}$ **b.** $\sqrt{x^2}$

c. $\tan x$ **d.** $\frac{\cos x}{\sin x}$

In general, the domain of an identity cannot be larger than the largest domain on which all the relevant functions are defined. The domain can also not include any value of a variable for which a denominator would be zero. Thus, the domain of the identity $\sin^2 x + \cos^2 x = 1$ is the set of all real numbers, but the domain of the identity $\frac{\sin^2 x + \cos^2 x}{\sin^2 x} = \frac{1}{\sin^2 x}$ does not include any value of x for which $\sin x$ equals 0.

Singularities

An isolated value for which a function is undefined is called a **singularity** or **discontinuity**. The singularities of the parent tangent, cotangent, secant, and cosecant functions are signaled graphically by vertical asymptotes. For instance, for any integer n, $n\pi$ is a singularity of f where $f(x) = \cot x$, and the lines with equations $x = n\pi$ are vertical asymptotes of the graph of $f(x) = \cot x$. Singularities of other functions may not be represented by asymptotes, and may not be obvious on a graph.

QY1

QY1

Suppose f is defined by $f(x) = 1 + \frac{\cos^2 x}{\sin^2 x}$. Identify all singularities.

GUIDED

Example 1

a. **Prove the identity $\frac{\cos 2\theta}{\cos\theta - \sin\theta} = \cos\theta + \sin\theta$.**

b. **What is the domain of the identity?**

Solution

a. Begin with the left expression and apply a Double Angle Formula for $\cos 2\theta$.

$\frac{\cos 2\theta}{\cos\theta - \sin\theta} = \frac{\cos^2\theta - \sin^2\theta}{\cos\theta - \sin\theta}$ Double Angle Formula

$= \frac{(\underline{\ ?\ })(\underline{\ ?\ })}{\cos\theta - \sin\theta}$ Factor the numerator.

$= \underline{\ ?\ } \cdot \frac{\underline{\ ?\ }}{\cos\theta - \sin\theta}$

$= \underline{\ ?\ }$

Therefore, $\frac{\cos 2\theta}{\cos\theta - \sin\theta} = \cos\theta + \sin\theta$ for all θ for which the expressions are defined.

b. The right side, $\cos\theta + \sin\theta$, is defined for all θ. However, when $\cos\theta - \sin\theta = 0$, the left side $\frac{\cos 2\theta}{\cos\theta - \sin\theta}$ is not defined. $\cos\theta - \sin\theta = 0$ when $\cos\theta = \sin\theta$. To determine when $\cos\theta = \sin\theta$, consider the unit circle. The two points with equal first and second coordinates are images of (1, 0) under rotations of $\underline{\ ?\ }$ and $\underline{\ ?\ }$. So the domain of the identity is the set of all $\theta \neq \underline{\ ?\ } + 2n\pi$ or $\theta \neq \underline{\ ?\ } + 2n\pi$.

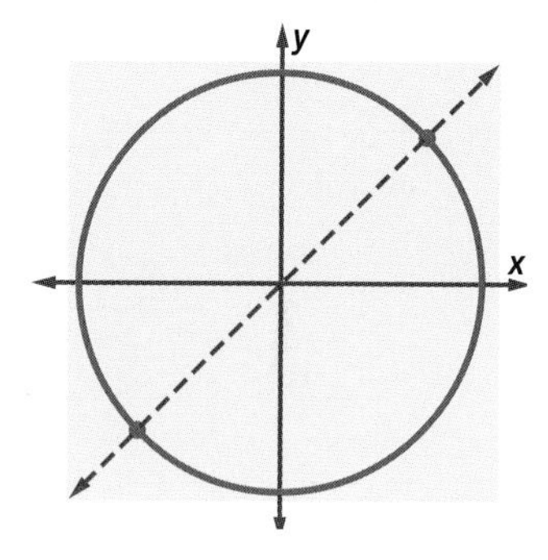

Often you cannot discern a singularity from a graph. Then you have to examine the algebra in the expressions.

Example 2

Consider the equation $\cos x \tan x = \sin x$. Use a graphing utility to test whether or not the equation seems to be an identity. If it seems to be, prove the identity over the largest possible domain.

Solution The graphs of $f(x) = \cos x \tan x$ and $g(x) = \sin x$, as they appear on one graphing utility, are shown at the right. It appears that the graphs coincide, so an algebraic proof of the identity is worth pursuing. Rewrite the left side.

$$\cos x \tan x = \cos x \cdot \frac{\sin x}{\cos x} = \sin x$$

Since the tangent function is not defined when $\cos x = 0$, $\cos x \tan x = \sin x$ has singularities when $x = \frac{\pi}{2} + n\pi$, for all integers n.

So $\cos x \tan x = \sin x$ when $x \neq \frac{\pi}{2} + n\pi$, for all integers n.

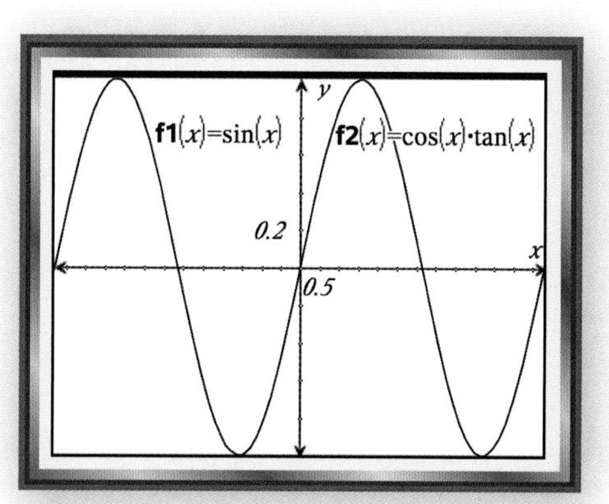

Identities are equalities between expressions. Sometimes, a CAS will change one expression into another and may even alert you about problems with the domain. In the screenshot at the right, the warning indicates that the domain of the identity $\cos x \cdot \tan x = \sin x$ has singularities.

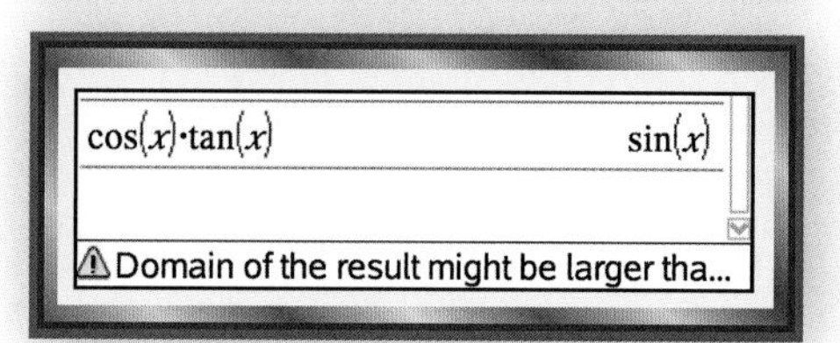

QY2

Removable Singularities

Geometrically, you can think of the singularities of the function f with $f(x) = \cos x \tan x$ as "holes" in the graph of $g(x) = \sin x$ at the points where x is an odd multiple of $\frac{\pi}{2}$. A singularity of this type, where the break in the graph can be "removed" by adding a single point, is called a **removable singularity** or **removable discontinuity**. Below is the graph of f, with open dots to indicate the holes. On some graphing utilities, holes do not appear. On others, if you plot $f(x) = \cos x \tan x$ and repeatedly zoom in around a singularity, you eventually see a small hole in the graph.

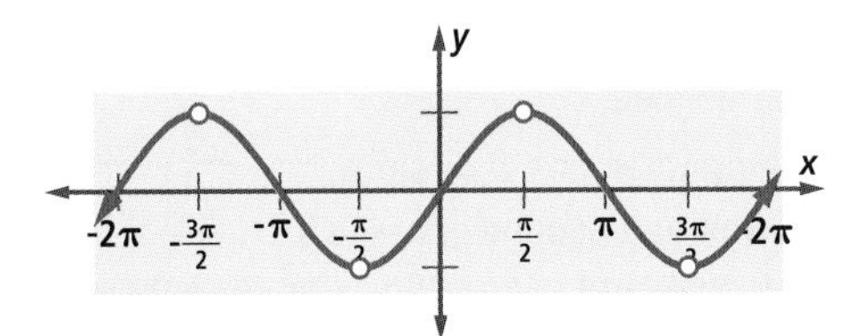

> **QY2**
>
> Suppose f is defined by $f(x) = \frac{x^2 - 9}{x^2 - 4}$. For what value(s) of x is $f(x)$ not defined?

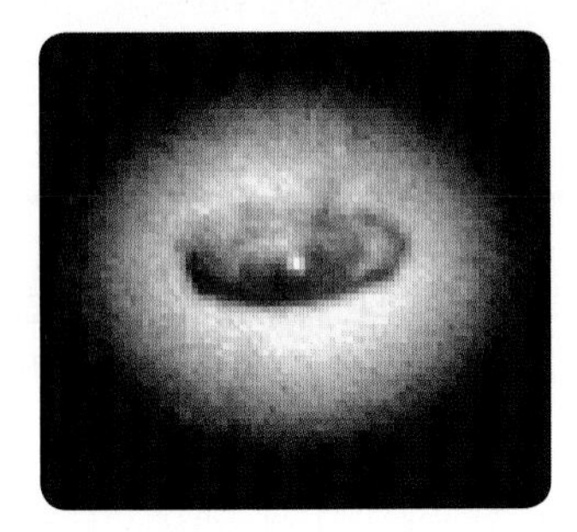

The object at the center is a black hole, located 100 million light-years away. Black holes are like singularities in that they exhibit different behavior than the world around them.

Example 3

Consider the equation $\frac{2 \cot \theta}{1 + \cot^2\theta} = \sin 2\theta$.

a. Determine any restrictions on the domain.

b. Prove that the equation is an identity.

Solution

a. $\cot \theta$ is undefined when $\theta = n\pi$, for all integers n. Since $\cot^2\theta$ is never negative, $1 + \cot^2\theta > 0$ for all θ. So there are no restrictions due to the denominator of the fraction because $1 + \cot^2\theta$ will never equal 0. Thus, the domain of the identity is $\{\theta \mid \theta \neq n\pi\}$.

b. $$\frac{2 \cot \theta}{1 + \cot^2\theta} = \frac{2 \cot \theta}{\csc^2\theta}$$
$$= \frac{\frac{2 \cos \theta}{\sin \theta}}{\frac{1}{\sin^2\theta}}$$
$$= \frac{2 \cos \theta}{\sin \theta} \cdot \frac{\sin^2\theta}{1}$$
$$= 2 \sin \theta \cos \theta$$
$$= \sin 2\theta$$

Questions

COVERING THE IDEAS

1. What is a *singularity* of a function?

In 2 and 3, identify any singularities of the function described.

2. $f(x) = \csc x$
3. $g(x) = \sin x \tan x$

4. **True or False** All singularities of functions are signaled by asymptotes.

In 5–8, a trigonometric equation is given.

a. **Use a graphing utility to test whether the equation may be an identity.**

b. **If the equation is an identity, prove it and give its domain.**

5. $\sin^2 x(\csc^2 x - 1) = \cos^2 x$
6. $\cot x \sec x = \csc x$
7. $\csc x = \frac{\cot x}{\sec x}$
8. $\sin 2x = \frac{2 \cos x}{1 + \cos^2 x}$

In 9 and 10, a trigonometric equation is given.

a. **State any restrictions on the domain of the proposed identity.**

b. **If the equation is an identity, prove it over the domain you stated in Part a.**

9. $\tan x + \cot x = \sec x \csc x$
10. $\sec^2 x \cos x = \frac{\csc^2 x - \cot^2 x}{\cos x}$

APPLYING THE MATHEMATICS

In 11 and 12, an equation for a function is given.

a. **Graph the function.**

b. **Identify its domain.**

c. **Propose an identity based on your graph.**

d. **Prove the identity.**

11. $f(x) = \frac{1 - \tan^2 x}{1 + \tan^2 x}$
12. $g(x) = \frac{1 + \cos 2x}{\sin 2x}$

13. a. Graph $f(x) = \frac{x^2 - x - 12}{x - 4}$ and $g(x) = x + 3$ on the same set of axes.

    b. Find the singularity in the graph of $y = f(x)$.

    c. What is the largest possible domain for the identity $\frac{x^2 - x - 12}{x - 4} = x + 3$?

    d. Prove the identity.

REVIEW

14. Suppose α is in the interval $\frac{\pi}{2} < \alpha < \pi$, and $\cot \alpha = -5$. Determine each of the following exactly. **(Lessons 13-1, 5-9)**

    a. $\csc \alpha$
    b. $\sin \alpha$
    c. $\sin 2\alpha$

15. Prove: For all t for which the functions are defined, $\tan \frac{t}{2} = \frac{\sin t}{1 + \cos t}$. **(Lesson 13-1)**

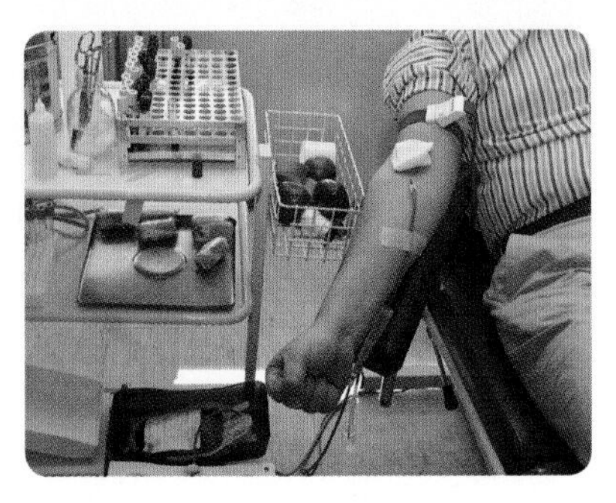

Eleven tests for infectious diseases are performed on every unit of donated blood.

In 16 and 17, consider the ELISA test, which was introduced in the mid-1980s to screen donated blood for the presence of antibodies to the AIDS virus. (Lessons 10-7, 10-5, 6-8, 6-6, 6-3)

16. According to one source, when presented with AIDS-contaminated blood, ELISA gave a positive response in about 99.7% of all cases. Suppose that among the blood that passed through a blood bank in a year, there were 25 units containing AIDS antibodies.
    a. What is the probability that ELISA detected all 25 of these units?
    b. What is the probability that more than 2 of the 25 contaminated units escaped detection?
    c. What is the mean number of contaminated units among the 25 that ELISA detected?
    d. What is the standard deviation of the number detected?
17. ELISA claims that AIDS antibodies are present in uncontaminated blood about 1.5% of the time.
    a. What are these kinds of claims known as?
    b. Suppose a blood bank contains 20,000 units of blood. What is the expected number of such claims among this group?
    c. What is the standard deviation of the number of such claims?
18. Consider the row of Pascal's Triangle that begins with 1, 12, ...
    a. What is the next entry in the row?
    b. What is the sum of all the numbers in the row? **(Lesson 10-2)**
19. Find all solutions to $5 \sin x + 1 = 0, 0 \le x < 2\pi$. **(Lesson 5-7)**
20. Solve the equation $\tan \theta = 1.7$ on the given domain. **(Lesson 5-7)**
    a. $\left\{\theta : -\frac{\pi}{2} \le \theta < \frac{\pi}{2}\right\}$
    b. $\{\theta : 0 \le \theta < 2\pi\}$
    c. the set of all real numbers

EXPLORATION

21. The sentence $\sqrt{x^2 - 1} = x - \frac{1}{2x}$ is not an identity for $x \ge 1$. In fact, the expressions on the two sides never have the same value.
    a. Evaluate the two expressions in the sentence at $x = 1, x = 5$, $x = 9$, and $x = 10$ to support the preceding claims.
    b. Notice that the two expressions $\sqrt{x^2 - 1}$ and $x - \frac{1}{2x}$ have closer and closer values as x gets bigger. Use a graphing utility to determine the values of x for which $\sqrt{x^2 - 1}$ is within 0.01 of $x - \frac{1}{2x}$.

QY ANSWERS

1. $n\pi$, for any integer n
2. $x = \pm 2$

Lesson 13-3 Polar Coordinates

Vocabulary

pole

polar axis

polar coordinates, $[r, \theta]$

BIG IDEA Polar coordinates identify a point by its distance and direction from a fixed point called the pole.

The rectangular coordinate system that you have been using for many years dates back to the early 1600s, when René Descartes and Pierre de Fermat worked to develop analytic geometry. Later in that same century, other mathematicians, notably Isaac Newton and Jakob Bernoulli, introduced other coordinate systems. One of these is the *polar coordinate* system.

Mental Math

Recall that the matrix for R_θ is $\begin{bmatrix} \cos\theta & -\sin\theta \\ \sin\theta & \cos\theta \end{bmatrix}$. Find a 2×2 rotation matrix for $R_{30°}$.

The Polar Coordinate System

Recall that on a rectangular coordinate grid, every point in the plane is identified by a unique ordered pair of numbers (x, y) representing the point's horizontal and vertical distance and direction from a fixed point, the origin. In a polar coordinate grid, a pair of numbers $[r, \theta]$ again represents a unique point. Here r or $-r$ is a distance, but θ is a magnitude of rotation measured in degrees or radians. Square brackets are used to distinguish polar coordinates from rectangular coordinates.

To construct a polar coordinate grid, first select a fixed point O as the **pole** of the system. Then select any line through O as the **polar axis**. Usually the polar axis is drawn horizontally as shown below. Coordinatize this line so O has coordinate 0. Any point P has **polar coordinates $[r, \theta]$** if and only if P is the image of the point on the polar axis with coordinate r under a rotation θ about the pole O. Below, $P \approx [2, 60°]$.

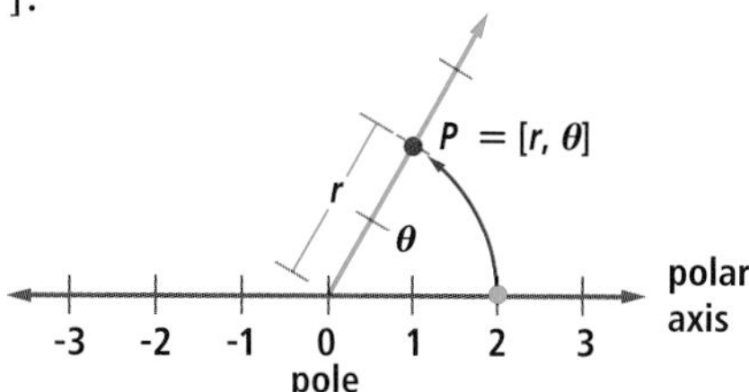

The points on the compass rose originally represented the division of the horizon into 32 points of direction.

The Polar Grid

Polar coordinate graph paper is very helpful for plotting points and sketching curves in the polar plane. A *polar grid* is pictured at the right. Each of the concentric circles in the grid represents a value of r, and each ray from the pole represents a value of θ.

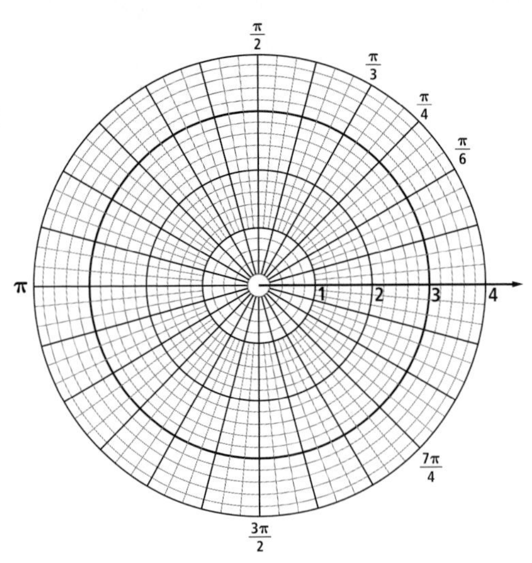

When plotting using polar coordinates, you should identify the positive polar axis with an arrow and put a scale on it to indicate values of r.

Example 1

Plot $\left[2.5, \frac{5\pi}{6}\right]$ and $\left[-4, \frac{3\pi}{2}\right]$ on a polar coordinate grid.

Solution To plot $\left[2.5\ \frac{5\pi}{6}\right]$, locate 2.5 on the polar axis. Rotate this point $\frac{5\pi}{6}$ radians. To plot $\left[-4, \frac{3\pi}{2}\right]$, locate –4 on the axis and rotate it $\frac{3\pi}{2}$.

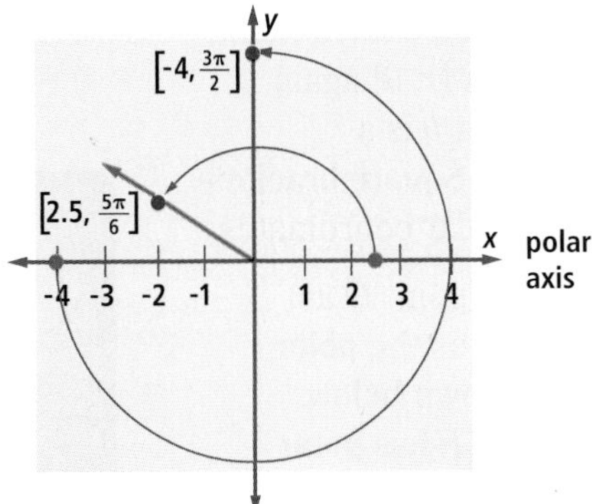

QY1

Every Point Has Many Polar Coordinates

A particular ordered pair $[r, \theta]$ identifies a unique point in the polar plane. However, there are an infinite number of polar coordinates for any given point. Example 2 shows how some of those coordinates can be found.

QY1

Give polar coordinates for point P graphed below.

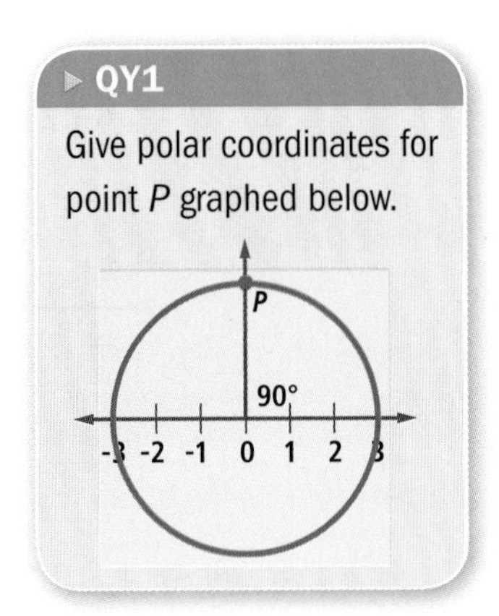

GUIDED

Example 2

Give four different polar coordinate pairs for the point *A* graphed at the right.

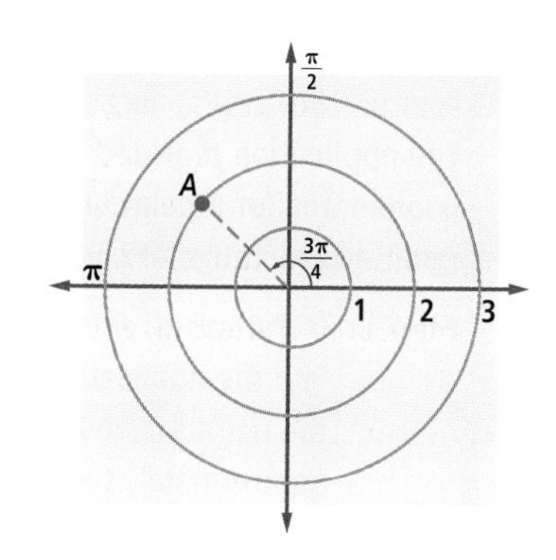

Solution There are infinitely many possible answers. One of them is $\left[2, \frac{3\pi}{4}\right]$ because *A* is the image of the point on the polar axis with coordinate 2 under a rotation of $\frac{3\pi}{4}$ radians. Rotating *A* another 2π yields [? , $\frac{?}{?}$]. Point *A* can also be considered as the image of the point -2 on the polar axis under a rotation of either $-\frac{\pi}{4}$ or ? , so [? , $\frac{?}{?}$] and [? , $\frac{?}{?}$] are also polar coordinates for *A*.

The results of Example 2 are generalized in the following theorem.

Equivalent Polar Coordinates Theorem

For any particular values of *r* and θ, the following polar coordinate pairs name the same point.

- $[r, \theta]$
- $[r, \theta + 2\pi n]$, for all integers *n*
- $[-r, \theta + (2n + 1)\pi]$, for all integers *n*

Notice that values of θ in the second part of the above theorem can be generated using even multiples of π, whereas those in the third part involve odd multiples of π.

The simplest polar equations have simple graphs.

Example 3

a. **Sketch all solutions $[r, \theta]$ to the equation $r = 2$.**

b. **Sketch all solutions $[r, \theta]$ to the equation $\theta = -\frac{\pi}{3}$.**

Solution

a.

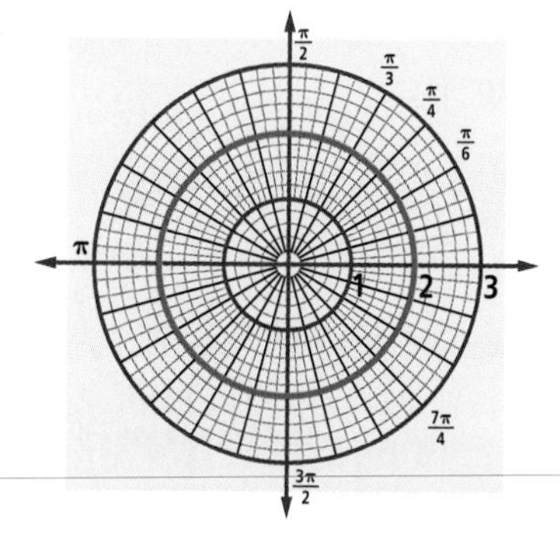

The equation $r = 2$ describes all points 2 units from the pole. The graph is a circle of radius 2 centered at the pole.

b.

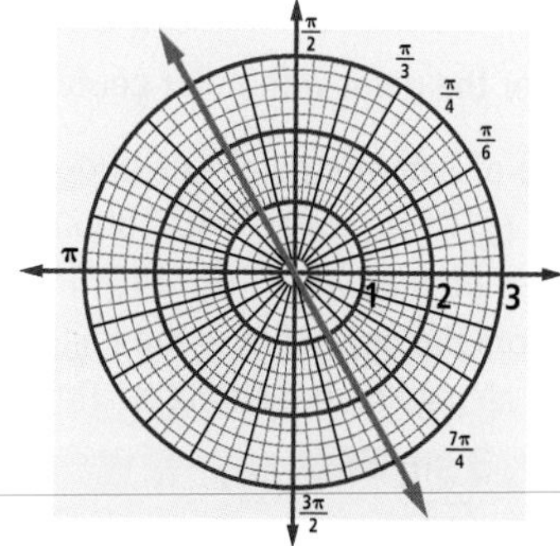

The graph of $\theta = -\frac{\pi}{3}$ is the line obtained by rotating the polar axis $-\frac{\pi}{3}$ about the pole.

Activity

MATERIALS polar coordinate and rectangular coordinate application provided by your teacher or graphing utility

The application provided by your teacher allows you to click or enter coordinates for a point on a polar or rectangular graph and display both its polar and rectangular coordinates.

Step 1 In Parts a–d, enter the rectangular coordinates into the application to plot on the polar grid the point whose rectangular coordinates are approximately those given. Write the polar coordinates of each point. Round r to the nearest tenth and θ to the nearest degree.

a. (–4, 2) b. (–6, –1)
c. (0, –2) d. (5, –3)

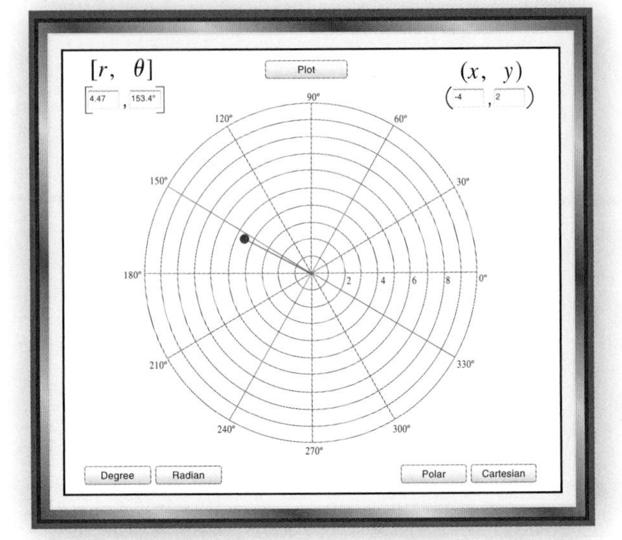

Step 2 In Parts e–h, enter the polar coordinates to plot on the rectangular grid the point whose polar coordinates are given. Assume θ is in radians. Write the rectangular coordinates of this point, rounding to the nearest tenth.

e. [3, 1] f. [4, 3.142]
g. $\left[3, \frac{\pi}{3}\right]$ h. $\left[1, \frac{5\pi}{4}\right]$

Converting from Polar to Rectangular Coordinates

Often polar and rectangular coordinate systems are superimposed on the same plane. Then, the polar axis coincides with the x-axis and the pole is at the origin. When this is done, you can use trigonometry to find the unique rectangular coordinate representation for any point for which polar coordinates are known.

Example 4

Find the rectangular coordinates for the point with polar coordinates [3, 240°].

Solution [3, 240°] is the image of 3 on the polar axis under a rotation of 240° about the pole. Think: When (1, 0) is rotated 240° about the origin, the rectangular coordinates of its image are (cos 240°, sin 240°). The point [3, 240°] is 3 times as far from the pole, so it is the image of (cos 240°, sin 240°) under a size change of magnitude 3. Thus,

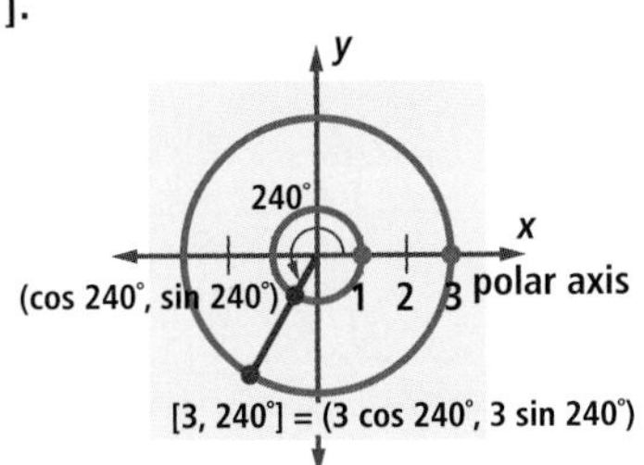

$$[3, 240°] = (3 \cos 240°, 3 \sin 240°)$$
$$= \left(3 \cdot -\frac{1}{2}, 3 \cdot -\frac{\sqrt{3}}{2}\right)$$
$$= \left(-\frac{3}{2}, -\frac{3}{2}\sqrt{3}\right).$$

The process in Example 4 can be generalized. When the point on the polar axis with coordinate 1 is rotated θ about the origin, the rectangular coordinates of its image are $(\cos\theta, \sin\theta)$. A size change of magnitude r maps this image onto $(r\cos\theta, r\sin\theta)$. This image has polar coordinates $[r, \theta]$.

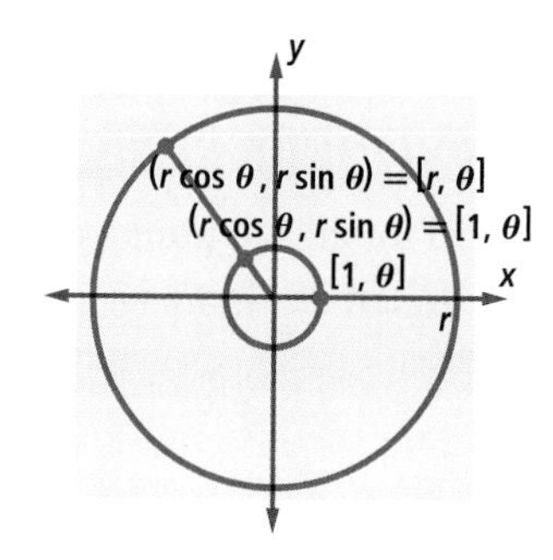

Polar-Rectangular Conversion Formula

If P has polar coordinates $[r, \theta]$, then the rectangular coordinates (x, y) of P are given by $x = r\cos\theta$ and $y = r\sin\theta$.

This theorem is true even when r is negative.

QY2

QY2

Estimate the rectangular coordinates for the point with polar coordinates [-26, 43°] to the nearest tenth.

Converting from Rectangular to Polar Coordinates

When the rectangular coordinates of a point are known, polar coordinates can be found.

Example 5

Find polar coordinates for the point with rectangular coordinates (−5, −12).

Solution Plot the point and draw a right triangle. Think of the length of the hypotenuse as the radius r of the circle containing (−5, −12). Find r.

$r^2 = 5^2 + 12^2$, so $r = \sqrt{169} = \pm 13$.

Also $\tan\theta = \frac{-12}{-5} = 2.4$. One value of θ satisfying $\tan\theta = 2.4$ is $\tan^{-1}(2.4) \approx 67°$. The choice of particular values of θ and r depends on matching the angle to the quadrant. Because (−5, −12) is in the third quadrant, a value of θ that can be used with $r = 13$ is $67° + 180° = 247°$.

So a set of polar coordinates for (−5, −12) is about [13, 247°].

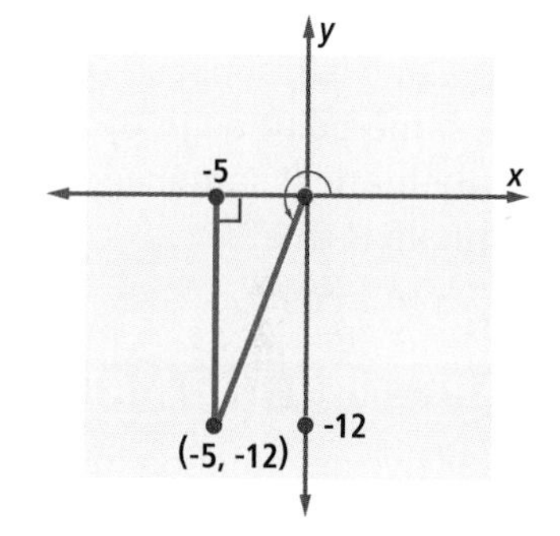

To generalize Example 5, suppose a point has polar coordinates $[r, \theta]$ and rectangular coordinates (x, y). Using the Polar-Rectangular Conversion Formula,

$$\begin{aligned} x^2 + y^2 &= r^2\cos^2\theta + r^2\sin^2\theta \\ &= r^2(\cos^2\theta + \sin^2\theta) \\ &= r^2. \end{aligned}$$

So $$|r| = \sqrt{x^2 + y^2},$$

and $$r = \pm\sqrt{x^2 + y^2}.$$

To find θ, divide the second equation from the Polar-Rectangular Conversion Formula by the first:

$$\frac{y}{x} = \frac{r\sin\theta}{r\cos\theta} = \tan\theta.$$

So if $(x, y) = [r, \theta]$, θ must be an angle whose tangent is $\frac{y}{x}$.

Questions

COVERING THE IDEAS

1. Plot all the points on the same polar grid.
   a. $A = \left[1, \frac{\pi}{3}\right]$
   b. $B = \left[-1, \frac{7\pi}{6}\right]$
   c. $C = \left[3.8, -\frac{\pi}{6}\right]$
   d. $D = [-5, -75°]$
2. Suppose $P = \left[7, \frac{5\pi}{6}\right]$.
   a. Give a set of polar coordinates $[r, \theta]$ for P with $r = 7$, but $\theta \neq \frac{5\pi}{6}$.
   b. Give a set of polar coordinates $[r, \theta]$ for P with $r = -7$.

In 3–5, find the rectangular coordinates for the point whose polar coordinates are given.

3. $\left[5, \frac{5\pi}{2}\right]$
4. $[6, -45°]$
5. $\left[-4, \frac{\pi}{6}\right]$

In 6 and 7, give one pair of polar coordinates for the (*x*, *y*) pair.

6. $(15, 15\sqrt{3})$
7. $(6, -8)$
8. **Multiple Choice** Which polar coordinates do *not* identify the same point as $[9, -37°]$?
   A $[9, 323°]$ B $[9, -397°]$ C $[-9, 37°]$ D $[-9, 143°]$

In 9 and 10, sketch all solutions to the equation on a polar grid.

9. $r = 5$
10. $\theta = \frac{5\pi}{3}$

In 11 and 12, suppose (*x*, *y*) in a rectangular coordinate system names the same point as [*r*, *θ*] in a polar coordinate system. Determine if the mathematical sentence is true or false. If false, correct the statement to make it true.

11. $x = r\cos\theta$
12. $|r| = x^2 + y^2$

APPLYING THE MATHEMATICS

13. Give an equation in polar coordinates for the figure containing all points of the form $\left[r, \frac{11\pi}{4}\right]$ and describe the figure.
14. Give an equation in polar coordinates for the figure containing all points of the form $[12, \theta]$ and describe the figure.

In 15 and 16, consider the point $P = \left[8, \frac{2\pi}{3}\right]$. State one pair of polar coordinates for the image of *P* under the given transformation.

15. reflection over the polar axis
16. reflection over the line $\theta = \frac{\pi}{2}$
17. Airport runways are often numbered in a way that is related to polar coordinates. If you land from the north, you see a runway numbered 0. If you land from the west, you see a runway numbered 9. Each 1-unit increase in runway number corresponds to 10° counterclockwise, the highest number being 35. From what direction do you land on a runway with the given number?
    a. 15 b. 1 c. 11 d. 5

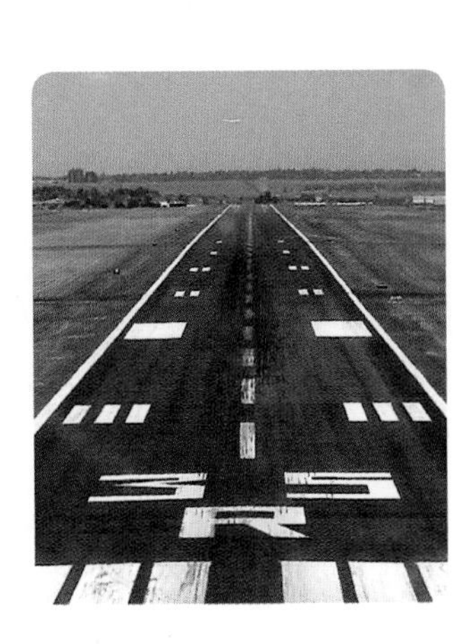

18. a. Plot the points in this table on polar graph paper.

r	0	$\frac{1}{2}(\sqrt{6} - \sqrt{2})$	1	$\sqrt{2}$	$\sqrt{3}$	$\frac{1}{2}(\sqrt{2} + \sqrt{6})$	2
θ	0°	15°	30°	45°	60°	75°	90°

b. The points above all satisfy the equation $r = 2 \sin \theta$. Let $\theta = 105°, 120°, \ldots$, and find six more points satisfying this equation. Plot these points.

c. Make a conjecture about the graph of all points $[r, \theta]$ satisfying $r = 2 \sin \theta$.

REVIEW

In 19 and 20, if the equation is an identity, give a proof and state the domain on which it is true. If the equation is not an identity, provide a counterexample. (Lesson 13-2)

19. $\sin\left(\frac{\pi}{2} + x\right) = \cos x$

20. $\sec x + \cot x \csc x = \sec x \csc x$

In 21 and 22, simplify the expression using identities you have studied. (Lessons 13-1, 4-3)

21. a. $\sin^2\theta + \cos^2\theta$

b. $36 \sin^2\theta + 36 \cos^2\theta$

22. a. $1 + \tan^2\theta$

b. $r^2 + r^2\tan^2\theta$

23. **True or False** The equation $4x^4 + 16x^3 + 5x^2 + 25 = 0$ has exactly 4 complex solutions. **(Lesson 7-6)**

24. Prove that the quadrilateral with consecutive vertices $(0, 0)$, (m, n), (p, q), and $(p - m, q - n)$ is a parallelogram. **(Previous Course)**

EXPLORATION

25. How might points in three-dimensional space be described with the help of polar coordinates?

QY ANSWERS

1. $[3, 90°]$
2. $(-26 \cos 43°, -26 \sin 43°) \approx (-19.0, -17.7)$

Lesson 13-4 Polar Graphs

▸ **BIG IDEA** The graphs of equations in polar coordinates have a variety of interesting shapes.

Any equation which involves only the variables r and θ can be graphed using polar coordinates. For instance, in the previous lesson the equations $r = 2$ and $\theta = -\frac{\pi}{3}$ were graphed. Some graphing utilities allow you to graph polar relations. The relation is usually entered in the form $r = f(\theta)$ or $r = f(t)$. Also, it is common that you must specify both the domain and step value of the graph.

Mental Math

Convert to rectangular coordinates.

a. $\left[1, \frac{\pi}{2}\right]$

b. $[2, \pi]$

c. $[1, 60°]$

d. $[1, 225°]$

Graphs of Simple Relations

Activity 1 explores how changing a parameter in a polar equation affects the graph of that equation.

Activity 1

MATERIALS polar graph paper and a graphing utility with sliders

Step 1 Graph $r = 2\theta$ by using a calculator to approximate the values in the table below and plotting the points $[r, \theta]$ on polar graph paper. Connect those points.

θ	0	$\frac{\pi}{4} \approx 0.7854$	$\frac{\pi}{2} \approx 1.571$	$\frac{3\pi}{4} \approx 2.356$	$\pi \approx 3.142$	$\frac{7\pi}{4} \approx 5.498$
r (approximate)	?	1.57	?	?	?	?

Step 2 Describe the graph of $r = 2\theta$ in words. Add 5 points to your graph corresponding to θ-values of your own choosing. At least one point should have $\theta < 0$.

Step 3 Put your graphing utility into polar graphing mode and radian mode. Create a slider for m, and graph $r = m\theta$ for $0 \le \theta \le 2\pi$. If a θ-step value is required, set it to 0.15. Describe how the graph of $r = m\theta$ changes when m changes.

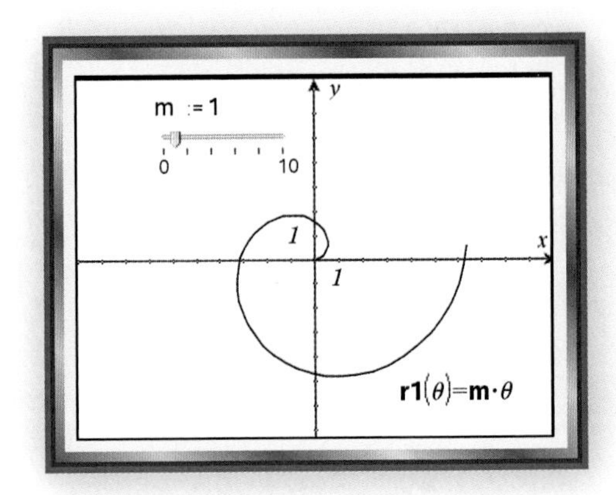

Step 4 On a new screen, add a slider for a and graph $r = \theta^2 - a$. Sketch the graph for $a = 4$ on polar graph paper.

Activity 1 demonstrates some relationships of properties of functions described by polar equations to properties of their graphs. They are summarized in the table on the next page.

Property of equation $r = f(\theta)$	Property of graph of $r = f(\theta)$
r increases as θ increases	Graph moves counterclockwise and outward
$\theta = 0$	Graph crosses line $\theta = 0$
$r = 0$	Graph crosses pole
relative extrema	Graph is closest to or furthest from origin

The Graph of the Sine Function in Polar Coordinates

You might think that polar graphs of equations involving $\sin\theta$ would be more complicated than polar graphs of equations involving θ, but this is not the case.

Example 1

Sketch a graph of the polar equation $r = \sin\theta$.

Solution Make a table of values for $[r, \theta]$ when $0 \le \theta \le 2\pi$. When making the table, consider θ as the independent variable. Some values are given in the table below. The table is identical to a table for $y = \sin x$, but with θ and r in place of x and y.

θ	0	$\frac{\pi}{6}$	$\frac{\pi}{4}$	$\frac{\pi}{3}$	$\frac{\pi}{2}$	$\frac{2\pi}{3}$	$\frac{3\pi}{4}$	$\frac{5\pi}{6}$	π
r	0	0.5	0.707	0.866	1	0.866	0.707	0.5	0

Plot these points. Notice that as θ increases from 0 to $\frac{\pi}{2}$, r increases from 0 to 1, and points to the right of the line $\theta = \frac{\pi}{2}$ are generated. As θ increases from $\frac{\pi}{2}$ to π, r decreases from 1 to 0, which produces points left of the line $\theta = \frac{\pi}{2}$.

When θ is between π and $\frac{3\pi}{2}$, $\sin\theta$, and thus each r-value, is negative. Therefore, all such points are drawn in the first quadrant instead of the third quadrant, and coincide with the points generated when $0 < \theta < \frac{\pi}{2}$. For instance, when $\theta = \frac{5\pi}{4}$, $r = \sin\frac{5\pi}{4} \approx -0.707$. The point $\left[-0.707, \frac{5\pi}{4}\right]$ coincides with the point $\left[0.707, \frac{\pi}{4}\right]$, which has already been plotted.

Similarly, when $\frac{3\pi}{2} \le \theta \le 2\pi$, the points generated coincide with those generated by $\frac{\pi}{2} \le \theta \le \pi$. The complete graph is drawn at the right. The graph appears to be a circle.

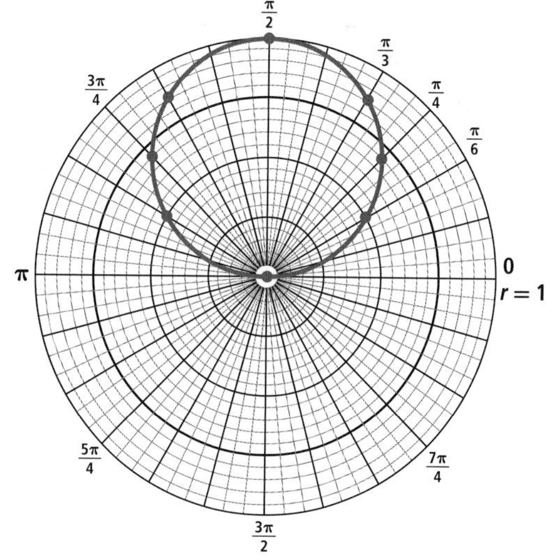

By converting from polar to rectangular coordinates, the graph of $r = \sin\theta$ can be proved to be a circle.

Example 2

Prove that the graph of $r = \sin\theta$ is a circle.

Solution The idea is to derive a rectangular equation for $r = \sin\theta$, and show that it is of the form $(x-h)^2 + (y-k)^2 = r^2$.

In general, $y = r\sin\theta$, or $\sin\theta = \frac{y}{r}$, so substitute $\frac{y}{r}$ for $\sin\theta$ in the equation $r = \sin\theta$ and rewrite.

If $\quad r = \sin\theta,$

then $\quad r = \frac{y}{r}.$

So $\quad r^2 = y.$

But $r^2 = x^2 + y^2$, and substituting for r^2 in the preceding equation gives

$$x^2 + y^2 = y$$

or, equivalently, $\quad x^2 + y^2 - y = 0.$

Now complete the square in y.

$$x^2 + \left(y^2 - y + \frac{1}{4}\right) = \frac{1}{4}$$

$$x^2 + \left(y - \frac{1}{2}\right)^2 = \frac{1}{4}$$

The equation $x^2 + \left(y - \frac{1}{2}\right)^2 = \frac{1}{4}$ is an equation in the rectangular coordinate system for the circle with center at $\left(0, \frac{1}{2}\right)$ and radius $\frac{1}{2}$.

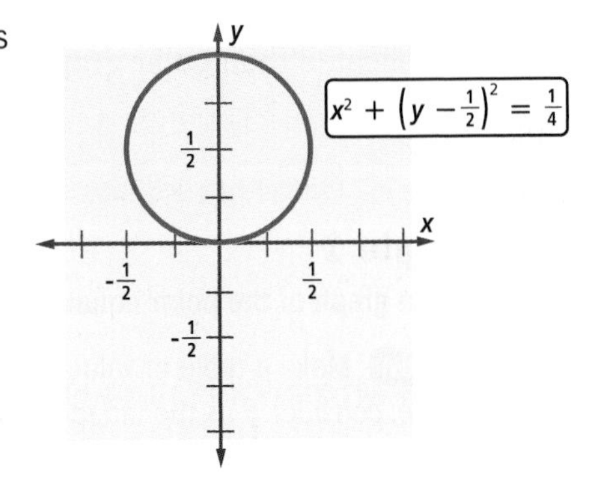

 QY

> **QY**
>
> What do you think the graph of $r = 2\sin\theta$ looks like?

Rose Curves

Recall that in the rectangular coordinate system, graphs of equations of the form $y = \cos b\theta$ or $y = \sin b\theta$, where b is a positive integer, are sine waves with amplitude 1 and period $\frac{2\pi}{b}$. The graphs of polar equations in the form $r = \cos b\theta$, where b is a positive integer, are quite different and beautiful. Below are graphs for $b = 2$, 3, and 4.

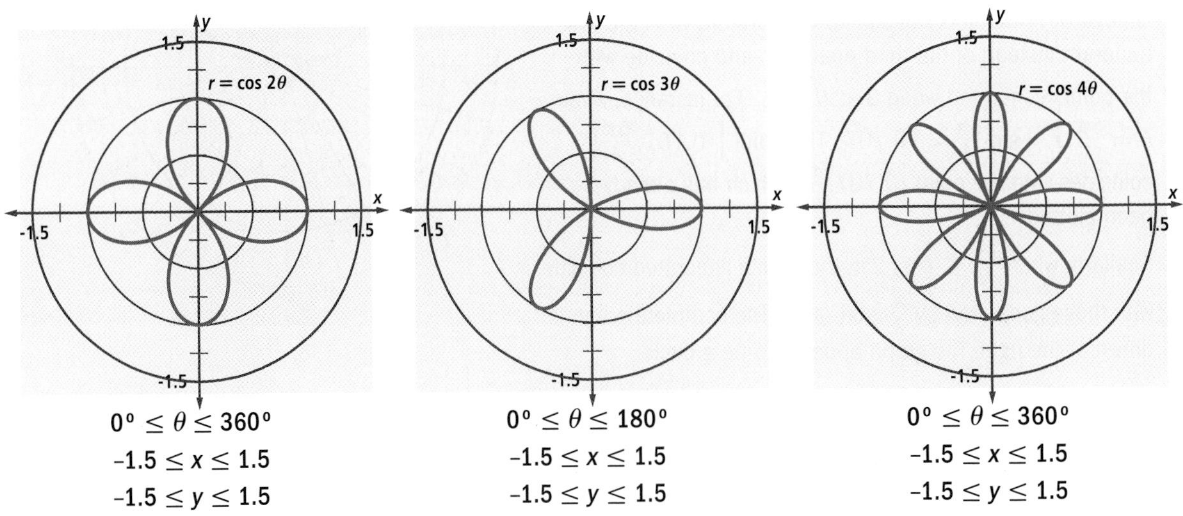

$0° \le \theta \le 360°$	$0° \le \theta \le 180°$	$0° \le \theta \le 360°$
$-1.5 \le x \le 1.5$	$-1.5 \le x \le 1.5$	$-1.5 \le x \le 1.5$
$-1.5 \le y \le 1.5$	$-1.5 \le y \le 1.5$	$-1.5 \le y \le 1.5$

These graphs are part of a family of polar graphs called *rose curves* or *petal curves*. Activity 2 shows how to draw a rose curve from its corresponding rectangular graph.

Clematis flower

Activity 2

MATERIALS polar tracer application provided by your teacher

Step 1 Using the polar tracer application, graph the equation $y = 3 \sin 2x$ for $0 \leq x \leq 2\pi$ in rectangular coordinates. What is the amplitude and period of the rectangular graph?

Step 2 Check the `Trace` and `Partial Draw` boxes at the top of the screen. Click on the rectangular coordinate graph so that a red point appears at the point (0, 0). What are the coordinates of the corresponding point on the polar graph?

Step 3 Drag the red point along the graph to the right until it lands on the next maximum or minimum point of the function. What are the corresponding values of r and t on the polar graph?

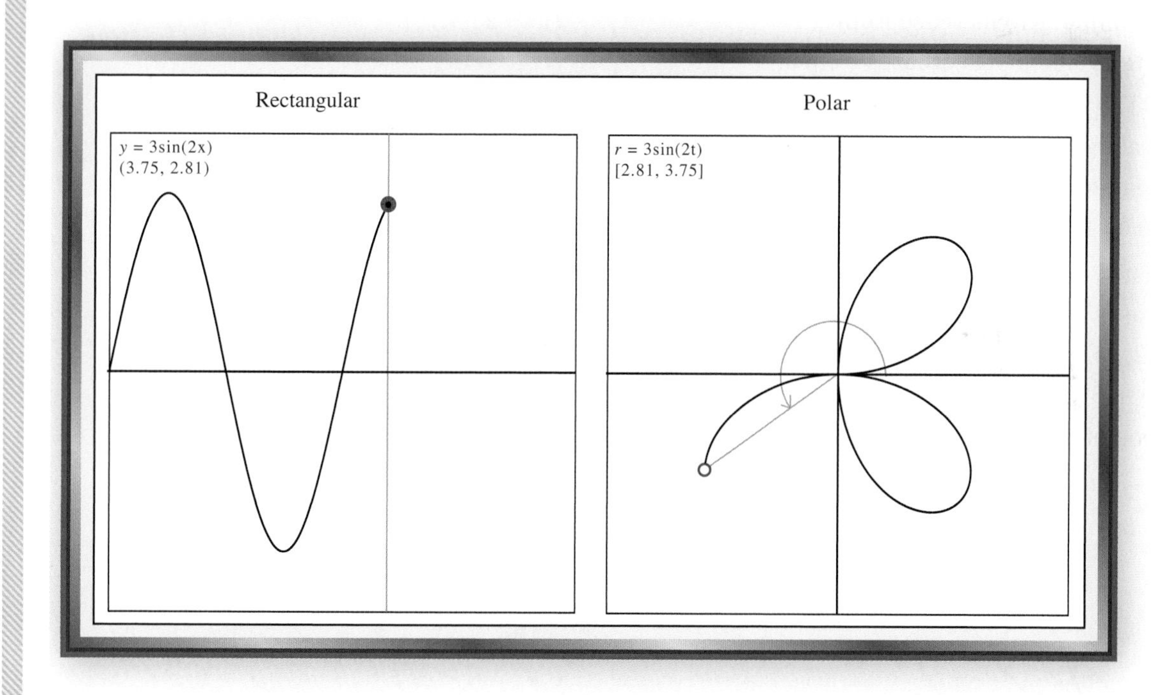

Step 4 Continue to drag the red point until it lands on the next x-intercept of the rectangular coordinate graph. What are the corresponding values of r and t on the polar graph?

Step 5 Repeat Steps 3 and 4 until you have traced the entire rectangular graph.

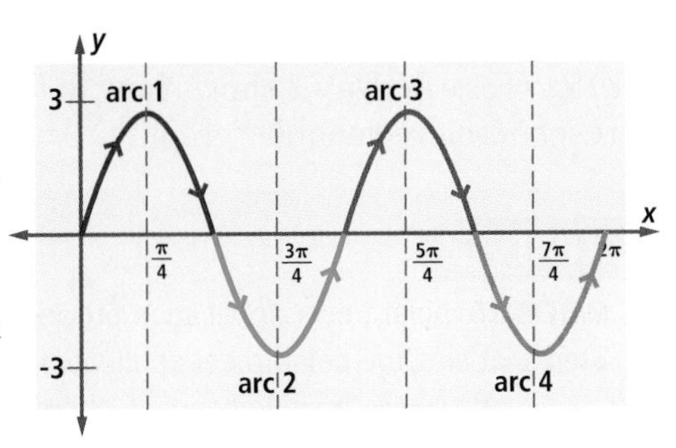

As Activity 2 shows, the possible y-values of the function with equation $y = 3 \sin 2x$ on the rectangular coordinate graph are the possible values of r in the polar coordinate graph. Thus, $-3 \leq r \leq 3$. The x-intercepts of the rectangular graph indicate when $r = 0$, that is, when the polar graph passes through the pole. Notice also that when $0 \leq x \leq 2\pi$, the rectangular coordinate graph has 4 congruent arcs, each symmetric to a vertical line where x is an odd multiple of $\frac{\pi}{4}$. In the polar graph, this reflection symmetry gives rise to symmetry in the "petals."

To sketch the polar graph, begin with the point $[0, \theta]$. As θ increases from 0 to $\frac{\pi}{4}$, r increases from 0 to 3. This part of the graph of $r = 3 \sin 2\theta$ is pictured at the right and corresponds to the first half of arc 1 on the rectangular coordinate graph above.

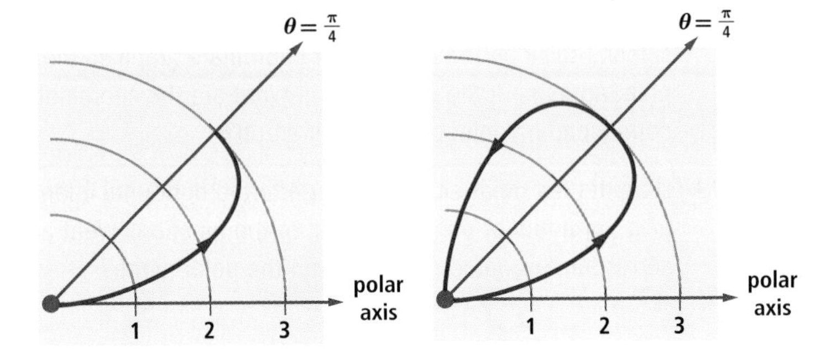

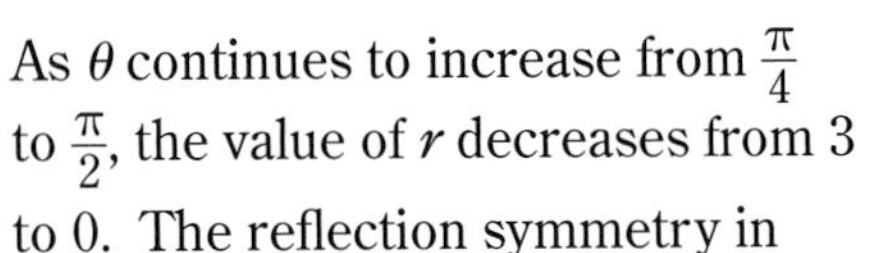
As θ continues to increase from $\frac{\pi}{4}$ to $\frac{\pi}{2}$, the value of r decreases from 3 to 0. The reflection symmetry in arc 1 above results in symmetry over the line $\theta = \frac{\pi}{4}$ for the corresponding arc in the polar plane. Thus, the loop shown at the far right above has been completed.

Similarly, as θ increases from $\frac{\pi}{2}$ to $\frac{3\pi}{4}$, r decreases from 0 to -3; and as θ increases from $\frac{3\pi}{4}$ to π, r increases from -3 to 0. These points are on loop 2, as noted on the graph at the left below. A similar analysis of arcs 3 and 4 indicates that there are two more loops in the polar graph.

A complete graph of $r = 3 \sin 2\theta$ is below. It is a 4-petaled rose.

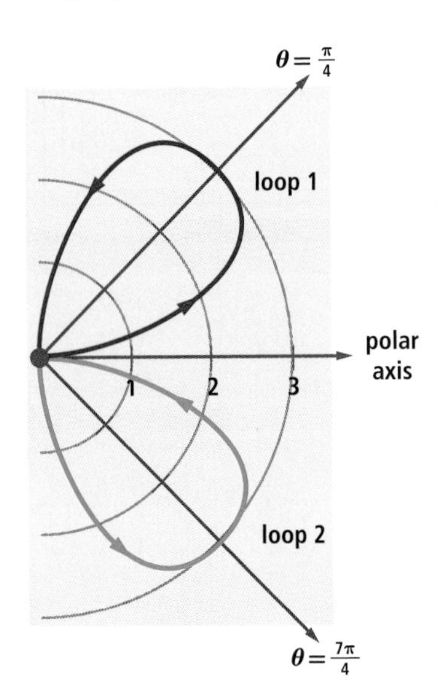

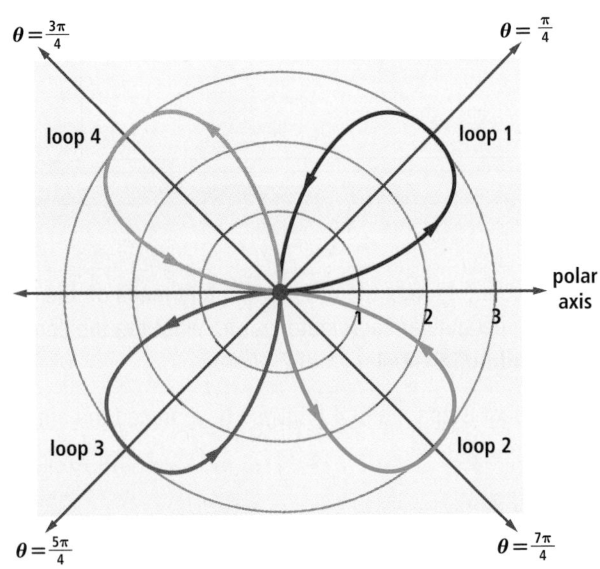

In general, polar graphs of trigonometric functions with equations of the form $r = c + a \sin b\theta$ or $r = c + a \cos b\theta$, where b is a positive integer and $a \neq 0$, are beautiful curves. You are asked to draw some of these in the Questions.

Questions

COVERING THE IDEAS

1. Consider the polar equation $r = 3$.
   a. Sketch a graph of the equation.
   b. Give an equation in rectangular coordinates for this graph.
2. Describe the graph of the polar equation $r = a$, where a is a constant nonzero real number.
3. Consider the polar equation $r = 2 \sin \theta$.
   a. Find and plot at least six points on its graph in polar coordinates.
   b. Prove that the graph is a circle.
4. Consider $r = \cos \theta$.
   a. Sketch a graph of this equation on a polar coordinate grid.
   b. Derive an equation for this curve in rectangular coordinates.

In 5 and 6, refer to the rose curves on page 806.

5. Is the point $[4, \pi]$ on the graph of $r = \cos 4\theta$? Why or why not?
6. Is the point $[0, 0]$ on the graph of $r = \cos 2\theta$? Why or why not?
7. Consider the equation $r = 2 \cos 3\theta$.
   a. Sketch a graph of this equation in the rectangular coordinate plane, with θ on the horizontal axis and r on the vertical axis.
   b. What are the maximum and minimum values of r?
   c. Find r when $\theta = 0, \frac{\pi}{6}$, and $\frac{\pi}{2}$, and plot these points in the polar plane.
   d. Use the technique described in this lesson to sketch a complete graph of this equation in polar coordinates.

APPLYING THE MATHEMATICS

8. At the right is a polar graph of $r = \sin 2\theta$. Give the coordinates of four points on the graph.

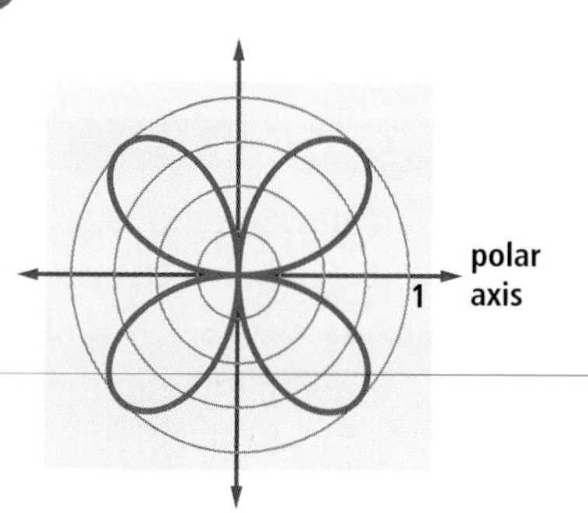

9. Sketch a graph of $r = a \sin 2\theta$, where a is a positive real number, in polar coordinates.
10. Refer to the graphs of $r = \cos 2\theta$, $r = \cos 3\theta$, and $r = \cos 4\theta$ on page 806.
    a. Make a conjecture regarding the number of petals on the polar curve $r = \cos n\theta$, where n is a positive integer.
    b. Test your conjecture by drawing graphs of $r = \cos 5\theta$ and $r = \cos 6\theta$ with a graphing utility.

11. The graph of $r = a(1 + \cos \theta)$ in the polar plane is called a *cardioid*.
    a. Graph this curve when $a = 1$.
    b. Why is *cardioid* an appropriate name for this curve?

12. The graph of $r = \theta$, for $\theta > 0$, in the polar plane is called an *Archimedean spiral*, named after Archimedes. Graph this curve.

13. The graph of $r = a^\theta$ in the polar plane, where $a > 0$, is called a *logarithmic spiral*.
    a. For $r = 2^\theta$, find the coordinates (to the nearest hundredth) of the points in the interval $0 \le \theta < 2\pi$ where θ is a multiple of $\frac{\pi}{4}$.
    b. Graph the curve.

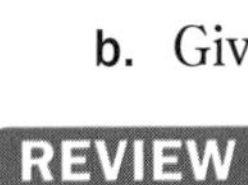

14. Consider the equation $r = \sec \theta$.
    a. Graph the equation in polar coordinates.
    b. Give an equation for the graph in rectangular coordinates.

REVIEW

15. **Multiple Choice** Which of these polar pairs does not describe the same point as the others? **(Lesson 13-3)**

A $\left[0.5, \frac{5\pi}{6}\right]$ B $\left[-\frac{1}{2}, -\frac{\pi}{6}\right]$ C $\left[\frac{1}{2}, -\frac{7\pi}{6}\right]$ D $\left[-0.5, -\frac{5\pi}{6}\right]$

In 16 and 17, a trigonometric equation is given.
a. Use a graphing utility to test whether the equation seems to be an identity.
b. Prove your conclusion from Part a. (Lessons 13-2, 13-1)

16. $\sin^2 x = \frac{1}{2}(1 - \cos 2x)$ 17. $\tan 2x = 2 \tan x$

18. A March 2009 Zogby poll of 3365 people nationwide revealed that 58% of likely voters supported a push for action on climate change outlined in the President's address to Congress. **(Lesson 11-7)**
    a. Compute the margin of error at a 95% confidence level.
    b. Compute the 95% confidence interval.

19. Evaluate the given probabilities concerning a *z*-score. **(Lesson 11-2)**
    a. $P(z < 3)$ b. $P(-3 < z < 1)$
    c. $P(z < -1.96)$ d. $P(z > -2.76)$

EXPLORATION

20. Consider equations of the form $r = a + b \sin \theta$, where $a > 0$ and $b > 0$. Experiment with a graphing utility in polar mode using various values of a and b.
    a. Find an equation whose graph looks like the one at the right.
    b. In general, what is true about a and b if the graph of $r = a + b \sin \theta$ has a loop as shown at the right?

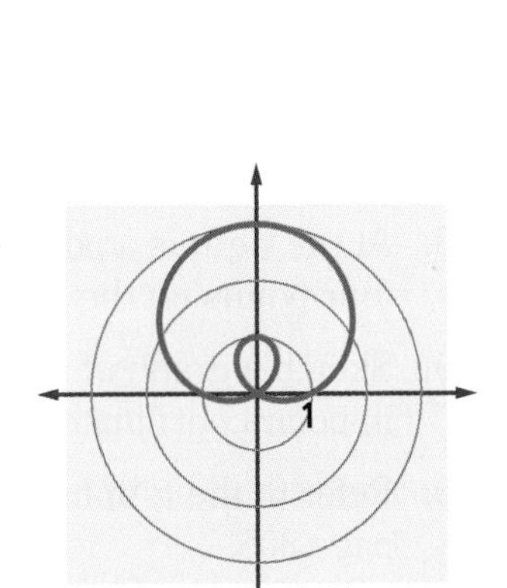

QY ANSWER

It is a circle of radius 1, centered at (0, 1).

Lesson 13-5 The Geometry of Complex Numbers

Vocabulary

complex plane
real axis
imaginary axis
absolute value of a complex number, modulus of a complex number
argument of a complex number
polar form of a complex number

▶ **BIG IDEA** Complex numbers can be represented and transformed geometrically.

Mental Math

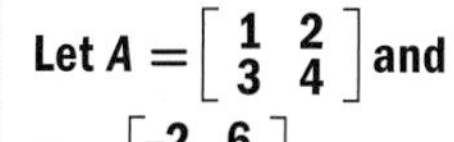

Let $A = \begin{bmatrix} 1 & 2 \\ 3 & 4 \end{bmatrix}$ and $B = \begin{bmatrix} -2 & 6 \\ 3 & -5 \end{bmatrix}$.

a. Find $A + B$.

b. Find $A - B$.

Graphing Complex Numbers in the Rectangular Coordinate Plane

Around the year 1800, Caspar Wessel, a Norwegian surveyor, and Jean Robert Argand, a Swiss mathematician, independently invented a geometric representation of complex numbers. Their diagrams, sometimes called *Argand diagrams*, represent each complex number as a point in a **complex plane**. These diagrams can be illuminating and beautiful. The diagrams changed the way people viewed complex numbers.

In a complex plane, the horizontal axis is called the **real axis** and the vertical axis is called the **imaginary axis**. To graph the complex number $a + bi$ in the complex plane, first write it as the ordered pair (a, b). Then plot (a, b) as you normally would in a rectangular coordinate system. Notice that b is a real number.

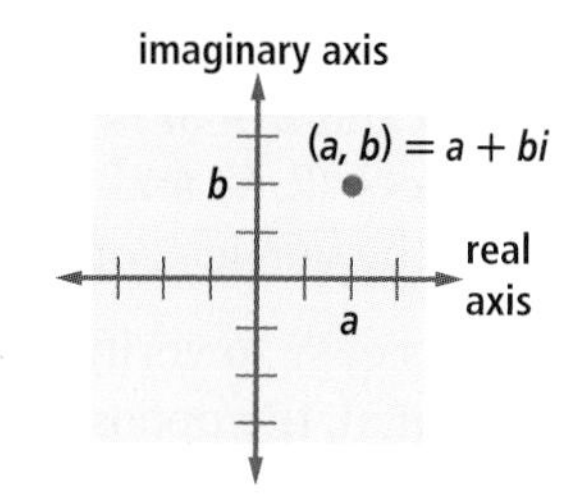

Each real number is of the form $a + 0i$, so it can be written as the ordered pair $(a, 0)$ and is plotted on the real axis in the complex plane. Similarly, every pure imaginary number is of the form $0 + bi$ and $(0, b)$ is plotted on the imaginary axis. The complex number $0 = 0 + 0i$ can be written as $(0, 0)$ and is graphed at the origin.

Example 1

Graph in the complex plane.

a. $3 + 4i$

b. $-5 - i$

c. $-5i$

Solution

a. $3 + 4i$ corresponds to $(3, 4)$.

b. $-5 - i = -5 + -1 \cdot i$ corresponds to $(-5, -1)$.

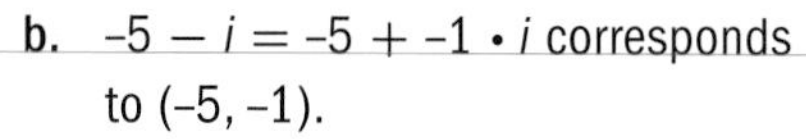

c. $-5i$ can be rewritten as $0 - 5i$, which corresponds to $(0, -5)$.

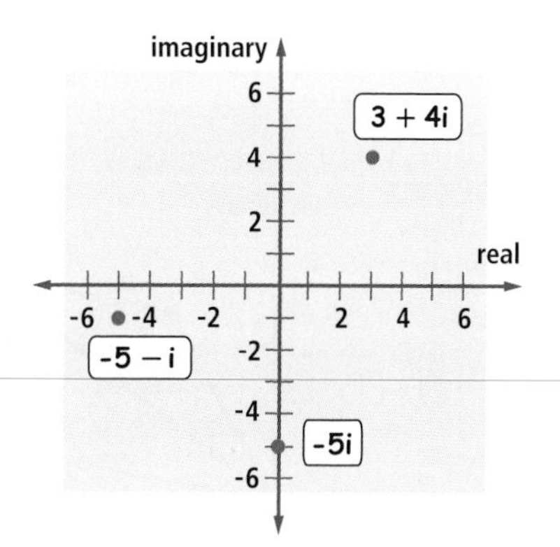

STOP QY

▶ **QY**

Which complex number corresponds to each point plotted below?

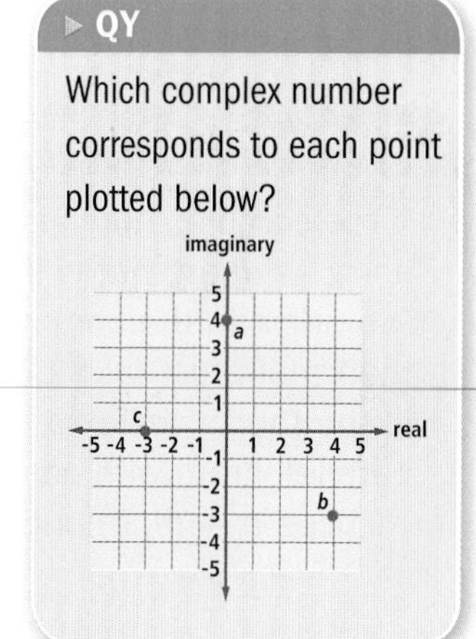

Picturing Addition of Complex Numbers

Activity

MATERIALS graph paper or a dynamic graphing utility

Step 1 Plot two points, not collinear with the origin, and label them z and w. Find their x- and y-coordinates and label them. Recall the point (a, b) represents the complex number $a + bi$.

Step 2 Compute $z + w =$ _?_ $+$ _?_ i. Then plot the ordered pair representing $z + w$. Draw the quadrilateral whose vertices are the origin, z, $z + w$, and w.

Step 3 Repeat Steps 1 and 2 using different points. What kind of quadrilateral did you draw?

Step 4 Suppose $z = 4 + 2i$. Plot several values of w satisfying each condition below, and describe the set of all such w.

a. $z + w$ lies on the real axis.

b. $z + w$ lies on the imaginary axis.

c. $z + w$ lies on the line through 0 and z.

The Activity shows that addition of complex numbers has an elegant geometric interpretation in the complex plane. For instance, the sum of $2 + 5i$ and $2 - 3i$ is $4 + 2i$. As shown in the diagram at the right, the numbers (2, 5) and (2, -3), their sum (4, 2), and the origin (0, 0) appear to be the vertices of a parallelogram.

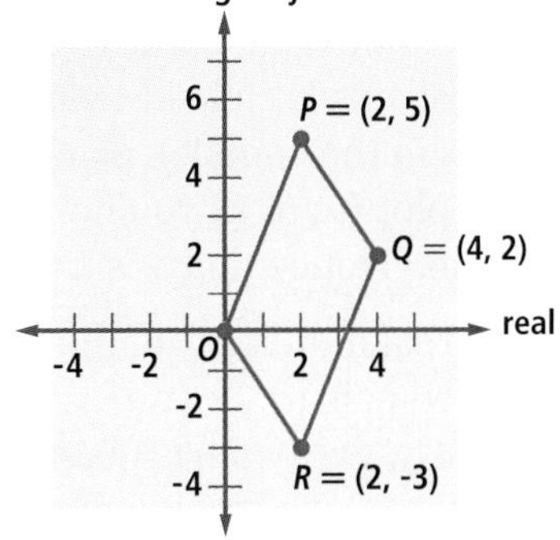

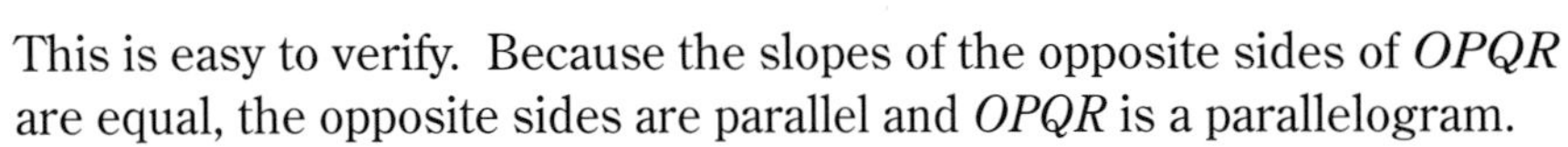

This is easy to verify. Because the slopes of the opposite sides of $OPQR$ are equal, the opposite sides are parallel and $OPQR$ is a parallelogram.

$$\text{slope of } \overline{OP} = \frac{5-0}{2-0} = \frac{5}{2}$$

$$\text{slope of } \overline{PQ} = \frac{2-5}{4-2} = -\frac{3}{2}$$

$$\text{slope of } \overline{QR} = \frac{-3-2}{2-4} = \frac{-5}{-2} = \frac{5}{2}$$

$$\text{slope of } \overline{OR} = \frac{-3-0}{2-0} = -\frac{3}{2}$$

This proves one instance of the following theorem. You are asked to prove the general case in Question 6.

Geometric Addition Theorem

If the complex numbers $a + bi$ and $c + di$ are not collinear with the origin in the complex plane, then their sum $(a + c) + (b + d)i$ is the fourth vertex of a parallelogram with consecutive vertices $a + bi$, 0, and $c + di$.

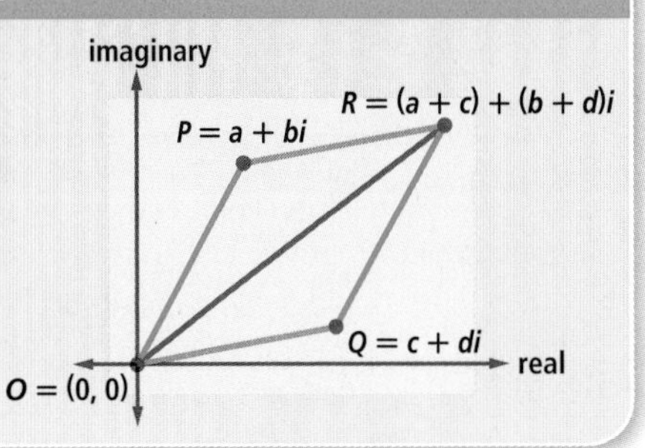

When you add the same complex number to each element in a set of complex numbers, the result is a translation.

GUIDED

Example 2

Let $z_1 = 4 + i$, $z_2 = 3 - i$, and $z_3 = 2 + 2i$, and let $f(z) = z + (-4 + i)$.

a. Draw $\triangle z_1z_2z_3$ and its image under f.

b. Describe the transformation in symbols and in words.

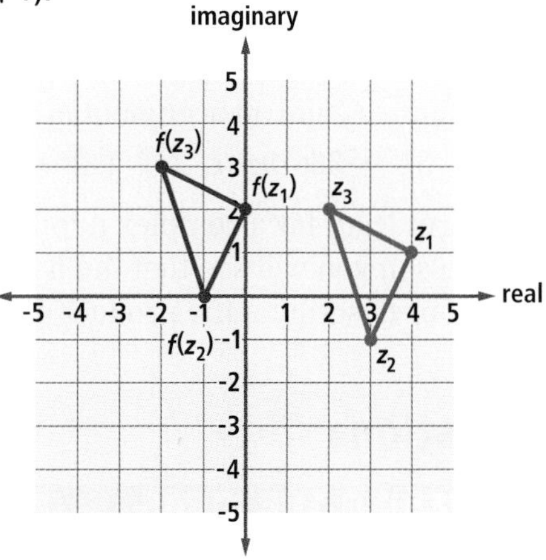

Solution

a. Plot the points $z_1 = (4, 1)$, $z_2 = (3, -1)$, and $z_3 = (2, 2)$.

$f(z_1) = (4 + i) + (-4 + i) = 0 + 2i$

$f(z_2) = (3 + -i) + (-4 + i) =$ ___?___

$f(z_3) = (2 + 2i) + (-4 + i) =$ ___?___

Plot $f(z_1)$, $f(z_2)$, and $f(z_3)$ and connect them to complete the graph.

b. $f:(x, y) \rightarrow (x - 4,$ ___?___$)$. f translates points left 4 units and up ___?___.

Graphing Complex Numbers in the Polar Coordinate Plane

Complex numbers can also be represented with polar coordinates. Let $[r, \theta]$ with $r \geq 0$ be polar coordinates for (a, b). From Lesson 13-3, the Polar-Rectangular Conversion Formula indicates that $r = \sqrt{a^2 + b^2}$ (because here r is not negative) and $\tan \theta = \frac{b}{a}$. There are usually two values of θ between 0 and 2π (or 0° and 360°) that satisfy $\tan \theta = \frac{b}{a}$. The correct value can be determined by examining the quadrant in which (a, b) is located.

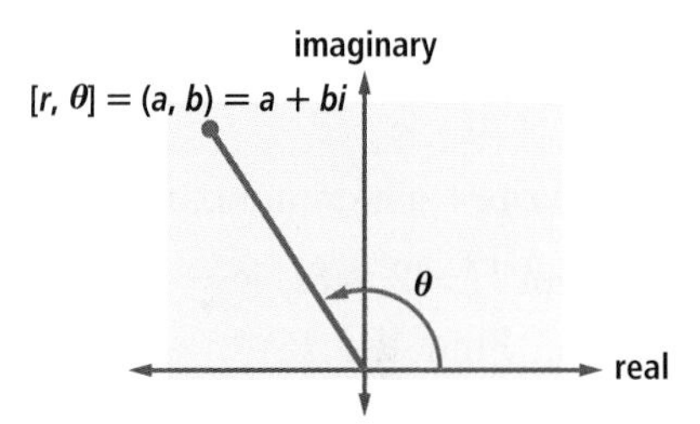

Example 3

Find polar coordinates for the complex number $5 - 3i$.

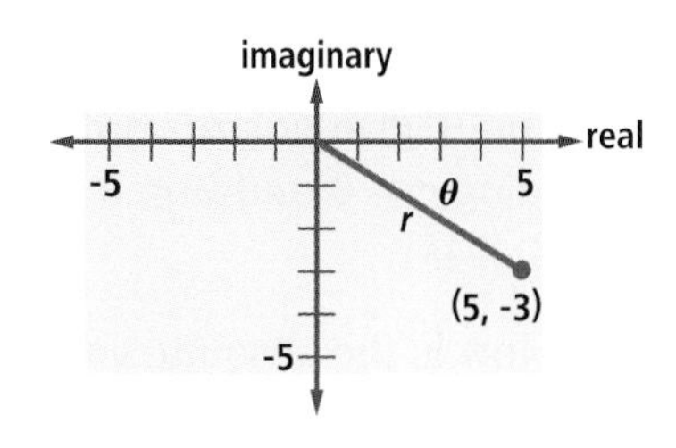

Solution The rectangular coordinates for $5 - 3i$ are $(5, -3)$. Let $[r, \theta]$ be polar coordinates for this point.

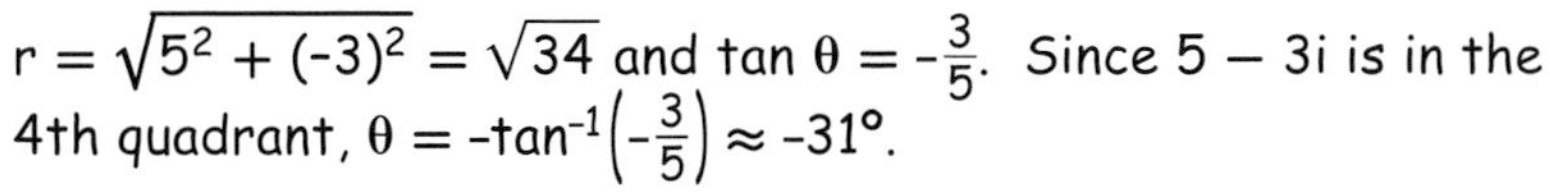

$r = \sqrt{5^2 + (-3)^2} = \sqrt{34}$ and $\tan \theta = -\frac{3}{5}$. Since $5 - 3i$ is in the 4th quadrant, $\theta = -\tan^{-1}\left(-\frac{3}{5}\right) \approx -31°$.

Thus one pair of polar coordinates for $(5, -3)$ is about $[\sqrt{34}, -31°]$, or about $[5.83, -31°]$.

Check Use a CAS. Set the complex format to polar. The screen at the right shows how one calculator displays the result in exact mode (at the top) and approximate mode (on the bottom).

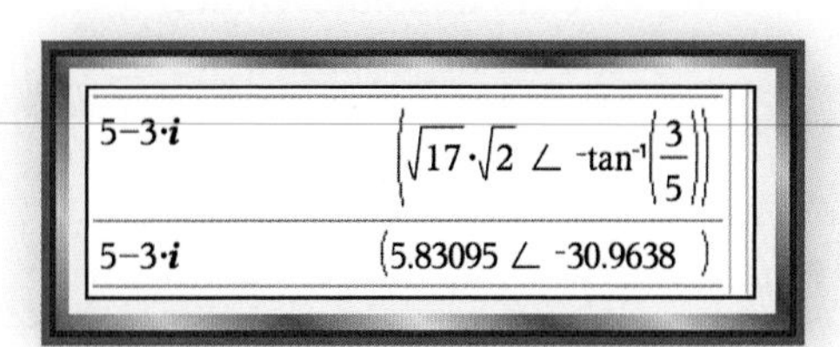

For any complex number $z = [r, \theta]$ with $r \geq 0$, r is the **absolute value**, or **modulus**, of the complex number and is written $|z|$. As with real numbers, $|z|$ is the distance from z to the origin. If $z = a + bi$, then $|z| = |a + bi| = \sqrt{a^2 + b^2}$. The direction θ is called an **argument** of the complex number and can be found using trigonometry: $\tan \theta = \frac{a}{b}$. An argument may be measured in degrees or radians. Because of periodicity, more than one argument exists for each complex number. In Example 3 the modulus of the complex number $5 - 3i$ is $\sqrt{34}$, and an argument is about -31°. Other arguments are about $-31° + 360°n$, or $-\frac{31\pi}{180} + 2\pi n$, where n is an integer.

The form $[r, \theta]$ for a complex number is called its **polar form**. In the next lesson you will see that the polar form of complex numbers is very useful for describing the product or quotient of two complex numbers.

Questions

COVERING THE IDEAS

1. **Fill in the Blanks** In a complex plane the horizontal axis is called the __?__ axis, and the vertical axis is called the __?__ axis.

In 2 and 3, a complex number is given.

a. Rewrite each number as an ordered pair (*a*, *b*).

b. Graph the number in the complex plane.

2. $6 - 4i$
3. $-4 + 5i$

4. Write each complex number pictured at the right in $a + bi$ form.

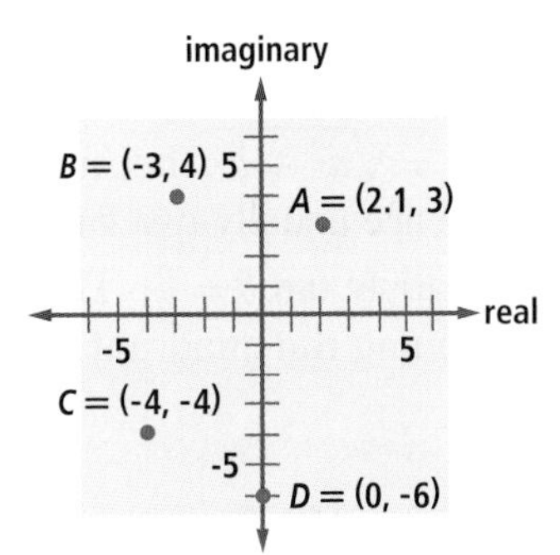

5. Let $U = 2 - 3i$, $V = -6 + 5i$, and $C = 0 + 0i$.
   a. Find $U + V$.
   b. Graph U, V, and $U + V$ in the same coordinate plane.
   c. Verify that U, C, V, and $U + V$ are the vertices of a parallelogram.
6. Prove the Geometric Addition Theorem.

In 7 and 8, give polar coordinates $[r, \theta]$ for each complex number, assuming $r \geq 0$ and $0 \leq \theta < 2\pi$.

7. $3 - 4i$
8. $\frac{1}{2} + \frac{\sqrt{3}}{2}i$

9. How is the absolute value of the complex number $a + bi$ calculated?

In 10 and 11, a complex number is given.

a. Find its absolute value.

b. Find its argument θ, if $r \geq 0$ and $0° \leq \theta < 360°$.

10. $13i$
11. $-6 + 3i$

APPLYING THE MATHEMATICS

12. Name and graph five complex numbers with modulus 1.

13. Let $z = 6 + 7i$ and $w = 4 - 2i$.

    a. Graph z, w, and the following numbers on the same axes.

    i. $z + w$ ii. $z - w$ iii. $w - z$ iv. $-(z + w)$

    b. Prove that the quadrilateral whose vertices are given in Part a is a parallelogram.

14. a. Draw the quadrilateral with vertices $P = 4 + i$, $Q = 4 - i$, $P + Q$, and $(0, 0)$.

    b. What special type of parallelogram is this?

    c. Determine the length of the longer diagonal of the quadrilateral.

15. On the real number line, the distance between points with coordinates u and v is $|u - v|$. Is $|u - v|$ the distance from $u = a + bi$ to $v = c + di$ in the complex plane? Justify your answer.

16. Consider $u = -4 + 3i$ and $v = -11i$.

    a. Find $|u| + |v|$. b. Find $|u + v|$.

    c. Use the Triangle Inequality from geometry to explain why $|u + v| \leq |u| + |v|$ in general.

REVIEW

In 17–20, graph each equation in the polar plane. (Lesson 13-4)

17. $\theta = \frac{3\pi}{2}$

18. $r = 3\theta, 0 \leq \theta \leq 4\pi$

19. $r = 2 \cos \theta$

20. $r = \sin 3\theta$

21. a. If a relation graphed in the rectangular coordinate plane fails the vertical-line test, what can you conclude?

    b. Can the graph of a function with equation $r = f(\theta)$ on a polar grid fail the vertical-line test? **(Lessons 13-4, 2-1)**

22. **Multiple Choice** Which pairs describe the same point? **(Lesson 13-3)**

    a. $\left[3, \frac{\pi}{4}\right]$ b. $\left[-3, \frac{5\pi}{4}\right]$ c. $\left[-3, -\frac{3\pi}{4}\right]$ d. $\left[3, -\frac{5\pi}{4}\right]$

In 23–25, state whether or not the geometric series is convergent. If the series is convergent, give its sum. (Lesson 8-7)

23. $3 - \frac{9}{2} + \frac{27}{4} - \frac{81}{8} + \cdots$

24. $5 + 4 + 3.2 + 2.65 + \cdots$

25. $16 - 12 + 9 - 6.75 + \cdots$

26. The Keeling curve is a graph that gained popularity in 2006 when former American Vice President Al Gore included it in his documentary on global warming, *An Inconvenient Truth*. The curve shows the variation in concentration of atmospheric carbon dioxide since 1958. It shows that some factors (including human activities) are increasing the greenhouse effect with implications for global warming.

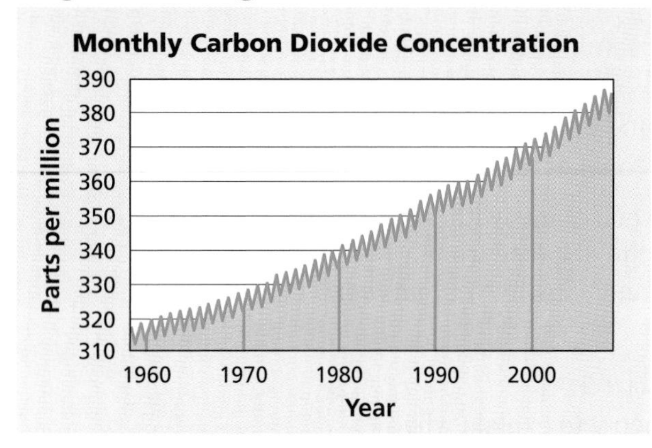

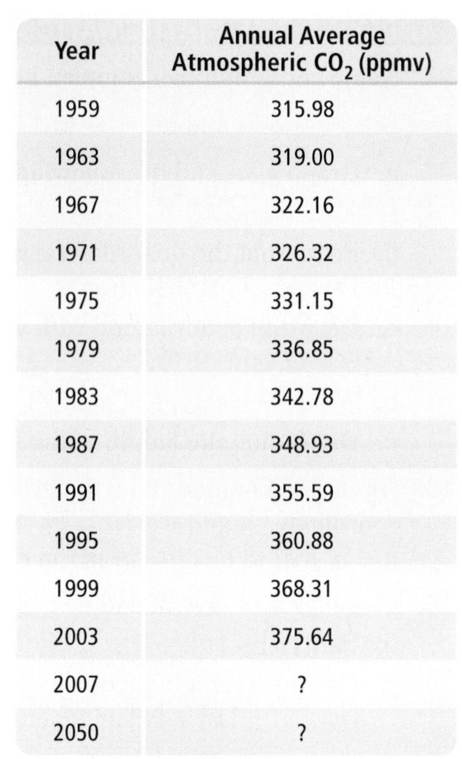

Year	Annual Average Atmospheric CO_2 (ppmv)
1959	315.98
1963	319.00
1967	322.16
1971	326.32
1975	331.15
1979	336.85
1983	342.78
1987	348.93
1991	355.59
1995	360.88
1999	368.31
2003	375.64
2007	?
2050	?

Source: Scripps institute of Oceanography

Refer to the table of average annual carbon dioxide concentration at the right. **(Lesson 2-8)**

a. Estimate the annual average for 2007 and 2050 using linear and exponential regressions.

b. Make a scatterplot of the data.

c. Write a sentence indicating which of these models, exponential or linear, you would choose and why.

EXPLORATION

27. Let $w = 1 + i$.

a. Write w^0, w^2, w^3, and w^4 in $a + bi$ form.

b. Graph w^0, w, w^2, w^3, and w^4 in the complex plane.

c. Describe the pattern that emerges in the graph of w^n for positive integers n.

Thermagraphic photo of the energy from a dog.

QY ANSWER

$a = 0 + 4i = 4i$;

$b = 4 - 3i$;

$c = -3 + 0i = -3$

Lesson

13-6 Trigonometric Form of Complex Numbers

Vocabulary

trigonometric form of a complex number

▶ **BIG IDEA** Complex numbers in $a + bi$ form can be represented in trigonometric form using r and θ.

Mental Math

Expand.

a. $3(2x + 7)$

b. $4x(2x + 7)$

c. $(4x + 3)(2x + 7)$

d. $(4 + 3i)(2 + 7i)$

You have now seen many ways of representing complex numbers. These forms are listed below.

z	single letter
$a + bi$	$a + bi$ form
(a, b)	rectangular coordinate form
$[r, \theta]$	polar coordinate form

You can convert from $a + bi$ form or rectangular coordinate form to polar coordinate form using the relationships $r = \sqrt{a^2 + b^2}$ and $\tan \theta = \frac{b}{a}$. You can convert back to rectangular or $a + bi$ form using the relationships $a = r \cos \theta$ and $b = r \sin \theta$.

These conversions are useful because each form has advantages. The single letter form z is the shortest and helps shorten formulas such as the distance $|w - z|$ between two complex numbers w and z. In $a + bi$ form, operations with complex numbers can be performed as if the complex numbers are polynomials in i. In rectangular form, addition of complex numbers can be seen graphically as in finding the fourth vertex of a parallelogram or as a translation. In this lesson, we introduce a fifth form for complex numbers, one that is quite useful for picturing multiplication, powers, and roots of complex numbers.

Arithmetic Operations as Transformations

In the previous lesson, you saw that adding a complex number $a + bi$ to another complex number has the effect of translating that number a units to the right and b units up. The Activity below investigates the geometry of multiplication of complex numbers.

Activity

MATERIALS dynamic geometry system or a complex transformations application provided by your teacher

Let $A = 2i$, $B = 3 - 2i$, and $C = 4$.

Step 1 Graph $\triangle ABC$.

(*continued on next page*)

Step 2 Create or locate the sliders for a and b. Set $b = 0$. Move the slider for a to compute the values of aA, aB, and aC for $a = 2, 3$, and –1 and graph the corresponding triangles. What transformation results from multiplying by a real number?

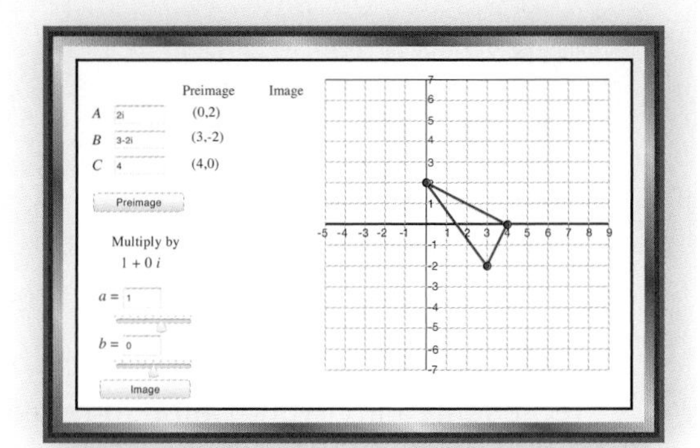

Step 3 Set $a = 0$. Move the slider for b to compute the value of biA, biB, and biC for $b = 1, 2, 3$, and –1 and graph the corresponding triangles. What transformation results from multiplying by a pure imaginary number?

Step 4 Find $(a + bi)A$, $(a + bi)B$, and $(a + bi)C$ for $b = 1$ and $a = 1, 2, 3$, and –1, and then graph the corresponding triangles. Repeat this step with $b = -2$ and $b = -3$. What transformation results from multiplying by a complex number of the form $(a + bi)$?

The Activity shows that multiplication by a complex number involves a combination of a size change and a rotation about the origin. Thus, connecting complex multiplication to geometry requires knowing a point's angle and distance from the origin, as shown in polar form. However, complex multiplication was defined in terms of rectangular coordinate form: $(a + bi)(c + di) = (ac - bd) + (ad + bc)i$. We introduce the *trigonometric form* of a complex number to bridge polar and rectangular forms.

What Is the Trigonometric Form of a Complex Number?

Consider the complex number $z = a + bi$ with polar coordinates $[r, \theta]$ and with $r \geq 0$, as graphed at the right.

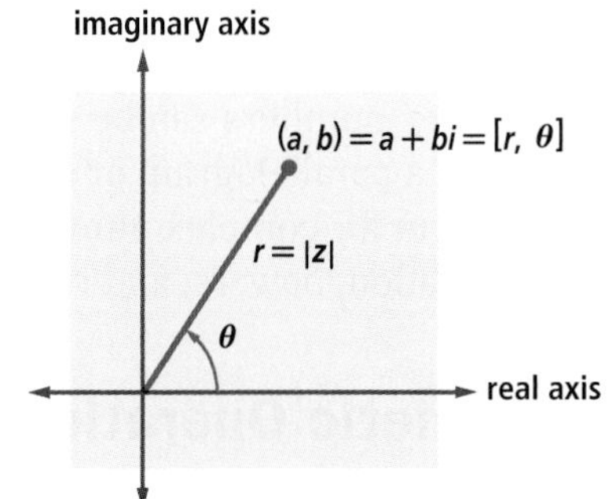

By the Polar-Rectangular Conversion Formula, $a = r\cos\theta$ and $b = r\sin\theta$. So

$$\begin{aligned} z &= a + bi \\ &= r\cos\theta + (r\sin\theta)i \\ &= r(\cos\theta + i\sin\theta). \end{aligned}$$

The expression $r(\cos\theta + i\sin\theta)$ is called the **trigonometric form of a complex number** $a + bi$ because it uses the cosine and sine of the argument θ. Like polar form, the trigonometric form denotes a complex number in terms of r and θ, its absolute value and argument. But unlike polar form, a complex number in trigonometric form is still in the rectangular coordinate form $a + bi$. Because of this link between polar and rectangular coordinates, the trigonometric form of complex numbers is quite useful.

Converting from $a + bi$ Form to Trigonometric Form

Example 1

Write the complex number $2 - 2\sqrt{3}i$ in trigonometric form, with θ in the interval $0 \le \theta < 2\pi$.

Solution The process is quite similar to that used in finding polar coordinates for a point. First sketch a graph.

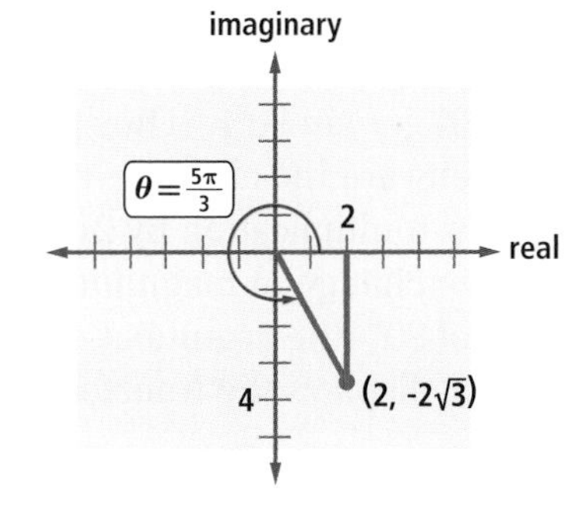

Let $2 - 2\sqrt{3}i = a + bi$. Then $a = 2$ and $b = -2\sqrt{3}$.

$$r = \sqrt{a^2 + b^2} = \sqrt{(2)^2 + (-2\sqrt{3})^2}$$
$$= \sqrt{4 + 12} = 4$$

$$\tan\theta = \frac{b}{a} = \frac{-2\sqrt{3}}{2} = -\sqrt{3}$$

Since $2 - 2\sqrt{3}i$ is in the fourth quadrant, $\theta = 2\pi - \frac{\pi}{3} = \frac{5\pi}{3}$.

So in polar coordinate form, $2 - 2\sqrt{3}i = \left[4, \frac{5\pi}{3}\right]$.

Therefore, in trigonometric form,

$2 - 2\sqrt{3}i = 4\left(\cos\frac{5\pi}{3} + i\sin\frac{5\pi}{3}\right)$.

Look again at the graph in Example 1. Notice how the trigonometric form relates to the polar form $\left[4, \frac{5\pi}{3}\right]$ of the number. Other pairs of polar coordinates of this number are $\left[4, \frac{5\pi}{3} + 2n\pi\right]$, where n is any integer. Accordingly, $2 - 2\sqrt{3}i = 4\left(\cos\left(\frac{5\pi}{3} + 2n\pi\right) + i\sin\left(\frac{5\pi}{3} + 2n\pi\right)\right)$, where n is any integer. In general, every complex number has infinitely many trigonometric forms.

Converting From Trigonometric Form to $a + bi$ Form

Example 2

Write the complex number $3\left(\cos\frac{4\pi}{3} + i\sin\frac{4\pi}{3}\right)$ in $a + bi$ form.

Solution Distribute the 3 and evaluate the sine and cosine.

$$3\left(\cos\frac{4\pi}{3} + i\sin\frac{4\pi}{3}\right) = 3\cos\frac{4\pi}{3} + i\left(3\sin\frac{4\pi}{3}\right)$$
$$= -\frac{3}{2} - \frac{3\sqrt{3}}{2}i$$

Check Use a CAS in radian and complex rectangular mode.

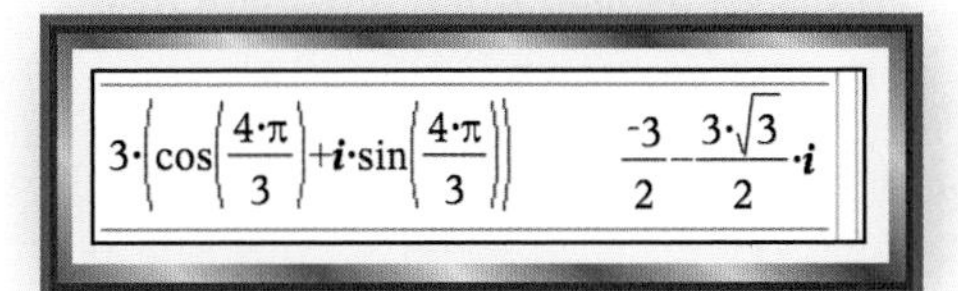

Picturing Multiplication of Complex Numbers

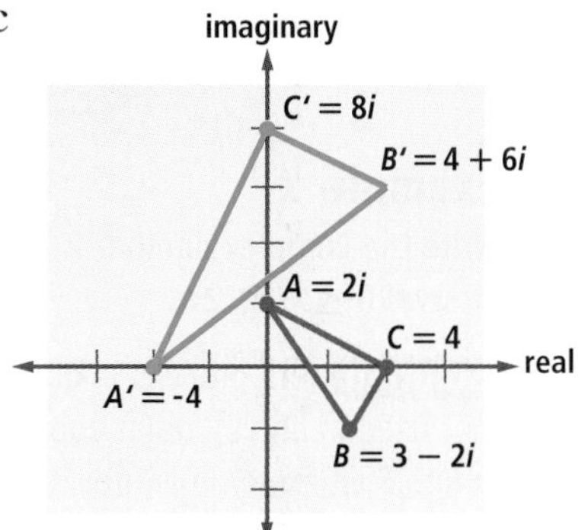

The trigonometric form of complex numbers explains a nice geometric property of complex number multiplication. Some instances were given in the Activity in which three complex numbers, A, B, and C, were each multiplied by a real number, a pure imaginary number, and a more general complex number. Consider the multiplication by $2i$.

The graphs of $\triangle ABC$ and $\triangle A'B'C'$ are shown at the right. $\triangle A'B'C'$ is the image of $\triangle ABC$ under the composite of a size change of magnitude 2 and a rotation of 90°.

In trigonometric form, $2i = 2(\cos 90° + i \sin 90°)$. Thus the magnitudes of the two transformations are the absolute value and argument of $2i$, respectively. That is, multiplication by $2i$ can be considered as the composite of a size change of magnitude 2 (the absolute value of $2i$) and a rotation of 90° (the argument of $2i$) about the origin. The polar form, $2i = [2, 90°]$, displays the transformation even more clearly.

In general, multiplying a complex number z_1 by the complex number $z_2 = [r_2, \theta_2] = r_2(\cos \theta_2 + i \sin \theta_2)$ applies to the graph of z_1 the composite of a size change of magnitude $|z_2|$ and a rotation of magnitude θ_2 about the origin.

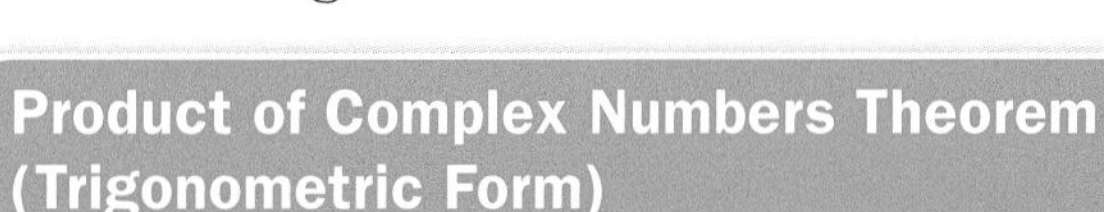

Product of Complex Numbers Theorem (Trigonometric Form)

If $z_1 = r_1(\cos \theta_1 + i \sin \theta_1)$ and
$z_2 = r_2(\cos \theta_2 + i \sin \theta_2)$,
then $z_1 z_2 = r_1 r_2(\cos(\theta_1 + \theta_2) + i \sin(\theta_1 + \theta_2))$.

Proof First, multiply z_1 and z_2 like any other polynomials.

$$z_1 z_2 = (r_1(\cos \theta_1 + i \sin \theta_1))(r_2(\cos \theta_2 + i \sin \theta_2))$$
$$= r_1 r_2(\cos \theta_1 + i \sin \theta_1)(\cos \theta_2 + i \sin \theta_2)$$
$$= r_1 r_2(\cos \theta_1 \cos \theta_2 + i \cos \theta_1 \sin \theta_2 + i \sin \theta_1 \cos \theta_2 + i^2 \sin \theta_1 \sin \theta_2)$$

Now use the commutative and associative properties of addition and the distributive property of multiplication over addition.

$$= r_1 r_2((\cos \theta_1 \cos \theta_2 + i^2 \sin \theta_1 \sin \theta_2) + i(\cos \theta_1 \sin \theta_2 + \sin \theta_1 \cos \theta_2))$$

Now use the fact that $i^2 = -1$ and the sum formulas for sine and cosine.

$$= r_1 r_2((\cos \theta_1 \cos \theta_2 - \sin \theta_1 \sin \theta_2) + i(\sin \theta_1 \cos \theta_2 + \cos \theta_1 \sin \theta_2))$$
$$= r_1 r_2(\cos(\theta_1 + \theta_2) + i \sin(\theta_1 + \theta_2))$$

The product is the complex number with absolute value $r_1 r_2$ and argument $\theta_1 + \theta_2$, written in trigonometric form.

This also proves the polar form of the theorem, which is the shortest version.

Product of Complex Numbers Theorem (Polar Form)

If $z_1 = [r_1, \theta_1]$ and $z_2 = [r_2, \theta_2]$, then $z_1z_2 = [r_1r_2, \theta_1 + \theta_2]$.

GUIDED

Example 3

If $z_1 = 6i$ and $z_2 = 3(\cos 135° + i \sin 135°)$, find the product z_1z_2 in trigonometric form.

Solution In trigonometric form, $z_1 = 6(\cos \underline{\ ?\ }° + i \sin \underline{\ ?\ }°)$.

Now use the Product of Complex Numbers Theorem.

$$z_1z_2 = \underline{\ ?\ } \cdot 3(\cos(\underline{\ ?\ } + 135°) + i \sin(\underline{\ ?\ } + 135°))$$
$$= \underline{\ ?\ } (\cos 225° + i \sin 225°)$$

Division of Complex Numbers

Recall from Lesson 7-5 that to divide a number z by a complex number in $a + bi$ form, it was useful to multiply the numerator and denominator of $\frac{z}{a + bi}$ by the complex conjugate $a - bi$ of $a + bi$. In trigonometric form, the conjugate of $r(\cos \theta + i \sin \theta)$ is $r(\cos \theta - i \sin \theta)$. This fact is needed to prove the following theorem, which you are asked to do in Question 17.

Quotient of Complex Numbers Theorem (Trigonometric Form)

if $z_1 = r_1(\cos \theta_1 + i \sin \theta_1)$ and
$z_2 = r_2(\cos \theta_2 + i \sin \theta_2)$,

then $\frac{z_1}{z_2} = \frac{r_1}{r_2}(\cos(\theta_1 - \theta_2) + i \sin(\theta_1 - \theta_2))$.

Geometrically, division of the complex number $z_1 = [r_1, \theta_1]$ by the complex number $z_2 = [r_2, \theta_2]$ applies to z_1 the composite of a size change of magnitude $\frac{1}{|z_2|} = \frac{1}{r_2}$ and a rotation of $-\theta_2$ about the origin.

Again the polar form description is shortest.

Quotient of Complex Numbers Theorem (Polar Form)

If $z_1 = [r_1, \theta_1]$ and $z_2 = [r_2, \theta_2]$, then $\frac{z_1}{z_2} = \left[\frac{r_1}{r_2}, \theta_1 - \theta_2\right]$.

Example 4

If $z_1 = 12\left(\cos \frac{\pi}{2} + i \sin \frac{\pi}{2}\right)$ and $z_2 = 4\left(\cos \frac{\pi}{6} + i \sin \frac{\pi}{6}\right)$, write $\frac{z_1}{z_2}$

a. in trigonometric form.

b. in polar form.

Solution

a. From the Quotient of Complex Numbers Theorem (Trigonometric Form),

$$\frac{z_1}{z_2} = \frac{12}{4}\left(\cos\left(\frac{\pi}{2} - \frac{\pi}{6}\right) + i \sin\left(\frac{\pi}{2} - \frac{\pi}{6}\right)\right)$$
$$= 3\left(\cos \frac{\pi}{3} + i \sin \frac{\pi}{3}\right).$$

b. For z_1, $r_1 = 12$ and $\theta_1 = \frac{\pi}{2}$, so $z_1 = \left[12, \frac{\pi}{2}\right]$ in polar form. Similarly, $z_2 = \left[4, \frac{\pi}{6}\right]$. From the Quotient of Complex Numbers Theorem (Polar Form), $\frac{z_1}{z_2} = \left[\frac{12}{4}, \frac{\pi}{2} - \frac{\pi}{6}\right] = \left[3, \frac{\pi}{3}\right]$.

Check Evaluate the trigonometric functions in z_1, z_2, and $\frac{z_1}{z_2}$.

$$\frac{z_1}{z_2} = \frac{12(0 + i(1))}{4\left(\frac{\sqrt{3}}{2} + i\frac{1}{2}\right)}$$
$$= \frac{12i}{2\sqrt{3} + 2i}$$
$$= \frac{12i(2\sqrt{3} - 2i)}{(2\sqrt{3} + 2i)(2\sqrt{3} - 2i)}$$
$$= \frac{24\sqrt{3}i + 24}{16}$$
$$= \frac{3 + 3\sqrt{3}i}{2}$$
$$= 3\left(\frac{1}{2} + \frac{\sqrt{3}}{2}i\right)$$
$$= 3\left(\cos \frac{\pi}{3} + i \sin \frac{\pi}{3}\right) \quad \text{It checks.}$$

In Example 4, notice how using the trigonometric form greatly simplifies division of complex numbers. The check was far more involved than the solution!

Questions

COVERING THE IDEAS

In 1–4, a complex number is given.

a. Graph each number on the complex plane.

b. Convert it to trigonometric form with $0° \le \theta < 360°$.

1. $5 + 5i$
2. $7\sqrt{3} - 7i$
3. -3
4. $-5 - 2i$

In 5 and 6, a complex number in trigonometric form is given.

a. **Graph each number on the complex plane.**

b. **Convert to $a + bi$ form.**

5. $13\left(\cos \frac{3\pi}{2} + i \sin \frac{3\pi}{2}\right)$

6. $2\left(\cos\left(-\frac{4\pi}{3}\right) + i \sin\left(-\frac{4\pi}{3}\right)\right)$

7. a. Let $z_1 = 3(\cos 55° + i \sin 55°)$ and $z_2 = 2(\cos 25° + i \sin 25°)$. Find z_1z_2 and express the result in trigonometric form.

   b. The composite of which two transformations maps z_1 to z_1z_2?

   c. Illustrate the multiplication with a diagram showing the appropriate size transformation and rotation.

   d. Do Part a by changing z_1, z_2, and z_1z_2 to polar form.

In 8–10, two complex numbers z_1 and z_2 are given.

a. **Find z_1z_2.**

b. **Give the absolute value of the product.**

c. **Give an argument of the product. Use exact values if possible.**

8. $z_1 = 3(\cos 330° + i \sin 330°),\ z_2 = 2(\cos 60° + i \sin 60°)$

9. $z_1 = \left[8, \frac{19\pi}{4}\right], z_2 = \left[4, \frac{\pi}{2}\right]$

10. $z_1 = 2 + 3i,\ z_2 = -4 + i$

In 11 and 12, write the conjugate of the complex number in the form given.

11. $7\left(\cos \frac{5\pi}{6} + i \sin \frac{5\pi}{6}\right)$

12. $[3, 165°]$

13. Let $z_1 = 9(\cos 200° + i \sin 200°)$ and $z_2 = 5(\cos 100° + i \sin 100°)$.

    a. Divide z_1 by z_2 and express the result in trigonometric form.

    b. Repeat Part a by expressing z_1, z_2, and $\frac{z_1}{z_2}$ in polar form.

    c. The composite of which two transformations maps z_1 to $\frac{z_1}{z_2}$?

In 14 and 15, two complex numbers z_1 and z_2 are given.

a. **Find $\frac{z_1}{z_2}$ and express the result in trigonometric form.**

b. **Check your result by converting to $a + bi$ form.**

14. $z_1 = 15(\cos \pi + i \sin \pi), z_2 = 3\left(\cos \frac{\pi}{3} + i \sin \frac{\pi}{3}\right)$

15. $z_1 = [20, 240°], z_2 = [4, 60°]$

APPLYING THE MATHEMATICS

16. A complex number z has absolute value 5 and argument $\frac{4\pi}{3}$.

    a. Express z in polar coordinate form.

    b. Express z in $a + bi$ form.

17. Prove the Quotient of Complex Numbers Theorem.

18. The complex number $z = 6(\cos 35° + i \sin 35°)$ undergoes a transformation that multiplies its absolute value by 5 and rotates it 75° about the origin.
    a. What is the image of z under the transformation?
    b. Identify the arithmetic operation and the complex number that will accomplish the transformation.

19. Let $z = 7(\cos 20° + i \sin 20°)$.
    a. Calculate z^2, z^3, z^4, and z^5. (Hint: Use the facts that $z^2 = z \cdot z$, $z^3 = z^2 \cdot z$, and so on.)
    b. Look for a pattern in your answers to Part a. Use the pattern to predict what z^{12} should be.

REVIEW

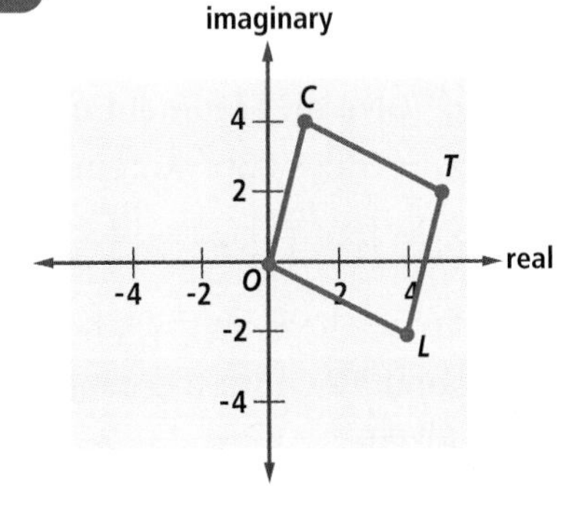

20. Refer to the graph at the right. What is the length of the diagonal $\overline{OT}$ of the parallelogram $COLT$ in the complex plane formed by $O = 0$, $C = 1 + 4i$, $L = 4 - 2i$, and T? **(Lesson 13-5)**

21. Sketch a graph of $r = -5 \cos \theta$ in polar coordinates. **(Lesson 13-4)**

22. A man owns L long-sleeved shirts and S short-sleeved shirts. He also owns P pairs of pants. How many outfits does he have
    a. for warm weather (short-sleeved shirts)?
    b. for cold weather (long-sleeved shirts)?
    c. altogether? **(Lesson 6-3)**

In 23 and 24, an equation is given.
   a. Solve for θ when $0 \le \theta < 2\pi$.
   b. Solve for θ over the real numbers. (Lesson 5-7)

23. $\sin 2\theta = \frac{1}{2}$

24. $\cos 3\theta = \frac{-\sqrt{3}}{2}$

25. If you hear a tone from a tuning fork at a frequency of 329.6 Hz, what is the period with which the air pressure is causing the hair cells in your inner ear to move back and forth? **(Lesson 4-7)**

EXPLORATION

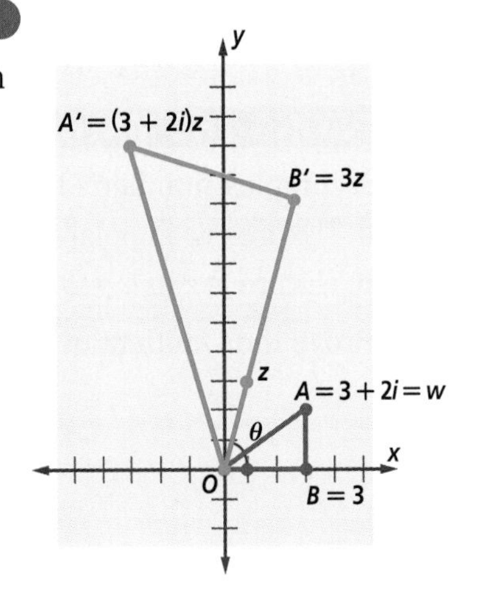

26. The diagram at the right suggests another proof that multiplication by a complex number is a composite of a rotation and a scale change. In the diagram, $z = 1 + 3i$, and $w = 3 + 2i$.
    a. Use geometry to explain why $\triangle OAB \sim \triangle OA'B'$.
    b. Use those similar triangles to explain why $m\angle A'OB = m\angle AOB + \theta$.
    c. Use those similar triangles to explain why $|zw| = |z| \cdot |w|$.

Lesson 13-7 De Moivre's Theorem

▶ BIG IDEA De Moivre's theorem enables powers and roots of complex numbers to be easily found.

Mental Math

Expand.

a. $(3i)^2$ **b.** $(2 + 3i)^2$

c. $(3i)^4$ **d.** $(3i)^{10}$

Abraham de Moivre, (1667-1754)

Gauss (1777–1855) was the first person to call the numbers you have been studying in the last two lessons "complex." He applied complex numbers to the study of electricity. To honor his work, the gauss, a unit of electromagnetism, is named after him. But during the century before Gauss, several mathematicians explored complex numbers and discovered many remarkable properties of them. This lesson presents a theorem about powers of complex numbers which is named after Abraham de Moivre (1667–1754), a mathematician who was born in France and lived most of his life in England.

Finding a Power of a Complex Number (Quickly)

Consider expanding $(2\sqrt{3} + 2i)^4$. One way to do it is to use the Binomial Theorem.

For all x and y, $(x + y)^4 = x^4 + 4x^3y + 6x^2y^2 + 4xy^3 + y^4$.

So $$\begin{aligned}(2\sqrt{3} + 2i)^4 &= (2\sqrt{3})^4 + 4 \cdot (2\sqrt{3})^3(2i)^1 + 6 \cdot (2\sqrt{3})^2(2i)^2 + 4 \cdot (2\sqrt{3})^1(2i)^3 + (2i)^4 \\ &= 144 + 192\sqrt{3}i + 72(-4) + 8\sqrt{3}(-8i) + 16 \\ &= -128 + 128\sqrt{3}i.\end{aligned}$$

This is very tedious.

Another approach is to rewrite the number $2\sqrt{3} + 2i$ in trigonometric form and use the theorems of the previous lesson. You may use the formulas or geometric inspection to find r and θ. On the next page we show angle measures in degrees.

Let $\quad z = 2\sqrt{3} + 2i.$

Then $\quad r = \sqrt{(2\sqrt{3})^2 + 2^2} = 4 \text{ and } \theta = \tan^{-1}\left(\frac{2}{2\sqrt{3}}\right) = 30°.$

So $\quad z = 4(\cos 30° + i \sin 30°).$

Then
$$z^2 = 4(\cos 30° + i \sin 30°) \cdot 4(\cos 30° + i \sin 30°)$$
$$= 4^2(\cos 60° + i \sin 60°),$$

because when complex numbers are multiplied, their arguments are added. Similarly,

$$z^3 = 4^2(\cos 60° + i \sin 60°) \cdot 4(\cos 30° + i \sin 30°)$$
$$= 4^3(\cos 90° + i \sin 90°),$$

and
$$z^4 = 4^3(\cos 90° + i \sin 90°) \cdot 4(\cos 30° + i \sin 30°)$$
$$= 4^4(\cos 120° + i \sin 120°)$$
$$= 256\left(-\frac{1}{2} + i\frac{\sqrt{3}}{2}\right)$$
$$= -128 + 128\sqrt{3}i.$$

This second approach may seem more tedious than the first, but it reveals a useful pattern: for each value of n, $z^n = 4^n(\cos(n \cdot 30°) + i \sin(n \cdot 30°)$. This result is an example of *De Moivre's Theorem.*

De Moivre's Theorem

If $\quad z = r(\cos \theta + i \sin \theta)$ and n is an integer,
then $z^n = r^n(\cos n\theta + i \sin n\theta)$.

In polar form, De Moivre's Theorem states that if $z = [r, \theta]$, then $z^n = [r^n, n\theta]$ for all integers n. According to some historians, De Moivre only proved the theorem when $r = 1$, but it is true for any r. The proof is beyond the scope of this course.

QY

QY

If $z = 4(\cos 60° + i \sin 60°)$, write z^5 in trigonometric form.

Graphing Powers of a Complex Number

Activity

MATERIALS graph paper, paper and pencil

Work with a partner if possible. Let $z = 2\left(\cos \frac{\pi}{3} + i \sin \frac{\pi}{3}\right)$ and let $w = \frac{3}{4}\left(\cos\left(-\frac{\pi}{3}\right) + i \sin\left(-\frac{\pi}{3}\right)\right)$.

Step 1 Find z^n for $n = 1, 2, 3, 4, 5, 6$, and 7 to the nearest hundredth. Plot the powers in the complex plane.

Step 2 Repeat Step 1 using w instead of z.

Step 3 Compare the graphs made in Steps 1 and 2. Describe their similarities and differences. Try to explain why the differences occur.

The Activity shows that, when graphed, powers of a complex number form a spiral either outward from or inward toward the origin unless $|z| = 1$. The smooth curve connecting them, shown below at the right, is a *logarithmic spiral*.

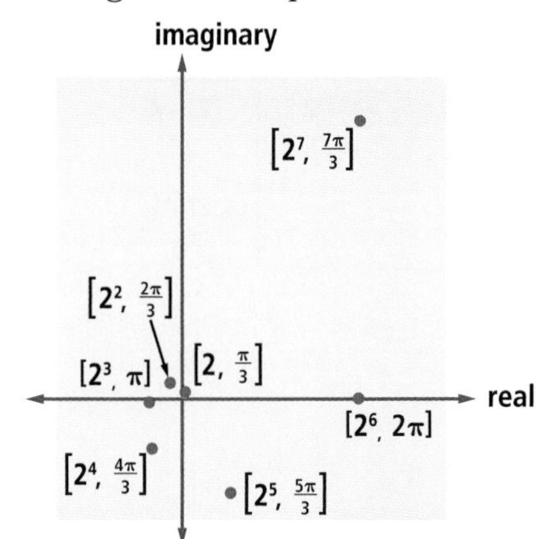

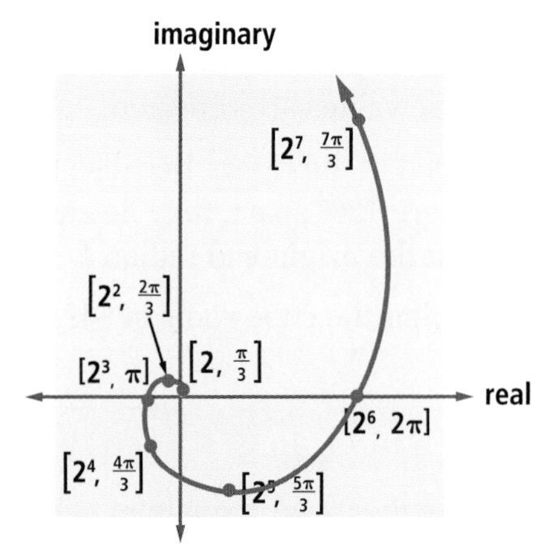

The most common type of galaxy in the universe is the "spiral galaxy," closely resembling a logarithmic spiral.

Finding the *n*th Roots of a Complex Number

In Chapter 9, you learned that for any complex number a and any positive integer n, the equation $z^n = a$ has exactly n complex solutions, the complex nth roots of a. Sometimes you can find those solutions by factoring. For example, to find the cube roots of 8, you can factor the polynomial $z^3 - 8 = (z - 2)(z^2 + 2z + 4)$. Then $z^3 - 8 = 0$ when $z = 2$ or $z^2 + 2z + 4 = 0$. Using the quadratic formula, the solutions to the second equation are $z = -1 \pm \sqrt{3}i$.

De Moivre's Theorem allows you to find the complex nth roots of a complex number even when you do not know how to factor the corresponding polynomial. For instance, consider trying to find a cube root z of the complex number $64i$. By definition of cube root, $z^3 = 64i$. To use De Moivre's Theorem, rewrite $64i$ in trigonometric form. By examining the graph of $64i$ at the right, you can see that the argument of $64i$ is 90° and its absolute value is 64. So in trigonometric form, $64i = 64(\cos 90° + i \sin 90°)$. Substituting this and $z = r(\cos\theta + i\sin\theta)$ into $z^3 = 64i$ gives

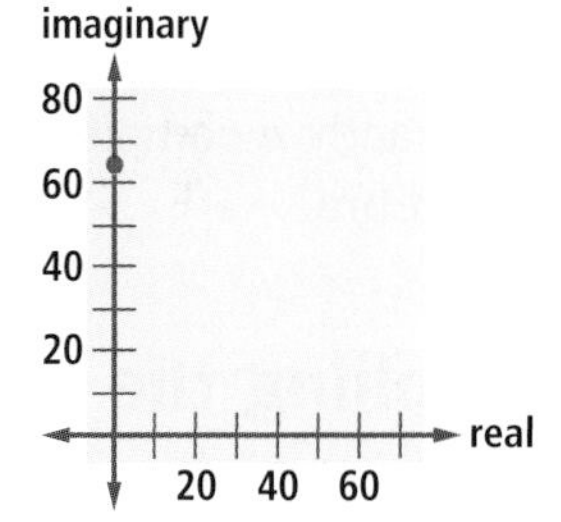

$$(r(\cos\theta + i\sin\theta))^3 = 64(\cos 90° + i \sin 90°).$$

Applying De Moivre's Theorem, we get

$$r^3(\cos 3\theta + i \sin 3\theta) = 64(\cos 90° + i \sin 90°).$$

Two complex numbers in polar or trigonometric form are equal if and only if their absolute values are equal and their arguments differ by an integer multiple of 360°. Therefore,

$$r^3 = 64 \text{ and } 3\theta = 90° + 360°n.$$

Thus $\quad r = 4 \quad$ and $\quad \theta = 30° + 120°n$, where n is an integer.

Therefore, the cube roots of $64i$ are of the form

$$z = 4(\cos(30° + 120°n) + i \sin(30° + 120°n)).$$

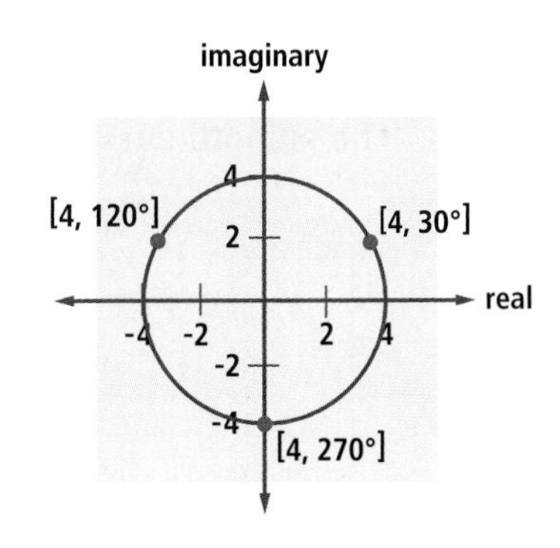

This solution may seem complicated, but actually there are only three distinct roots. For $n = 0$, 1, and 2, the cube roots are

$$4(\cos 30° + i \sin 30°) = 2\sqrt{3} + 2i,$$
$$4(\cos 150° + i \sin 150°) = -2\sqrt{3} + 2i,$$
and
$$4(\cos 270° + i \sin 270°) = -4i.$$

For any $n > 2$, you will find that these values are repeated.

The three cube roots of $64i$ are plotted at the right. Because they all have the same absolute value 4 and are 120° apart, they lie equally spaced around a circle with center at the origin and radius 4.

Generalizing the technique used to find the cube roots of $64i$ leads to a proof of the following theorem.

Roots of a Complex Number Theorem

Let z be any nonzero complex number, r be a positive number, and n be any positive integer. Then there are n distinct roots of $z^n = r(\cos \theta + i \sin \theta)$. They are

$$r^{\frac{1}{n}}\left(\cos\left(\frac{\theta}{n} + k\frac{360°}{n}\right) + i\sin\left(\frac{\theta}{n} + k\frac{360°}{n}\right)\right)$$

where $k \in \{0, 1, 2, ..., n - 1\}$.

In polar form, when $r > 0$, the nth roots of $[r, \theta]$ are $\left[r^{\frac{1}{n}}, \frac{\theta}{n} + \frac{2\pi k}{n}\right]$, where $k \in \{0, 1, 2, ..., n - 1\}$. Since n and θ are fixed, as k increases, the arguments $\frac{\theta + 2\pi k}{n}$ form an arithmetic sequence with common difference $\frac{2\pi}{n}$. When $k = n$, $\frac{\theta}{n} + \frac{2\pi n}{n} = \frac{\theta}{n} + 2\pi$, which is equivalent to the argument when $k = 0$. Thus, when $k \geq n$, the arguments are equivalent to arguments with $k \in \{0, 1, 2, ..., n - 1\}$. Therefore, there are exactly n complex nth roots, as stated in the Fundamental Theorem of Algebra.

Graphing the *n*th Roots of a Complex Number

In general, for $n > 2$, the graphs of the nth roots of any nonzero complex number are vertices of a regular n-gon centered at the origin!

Example

Find the 5th roots of $16 + 16\sqrt{3}i$ and graph them in the complex plane.

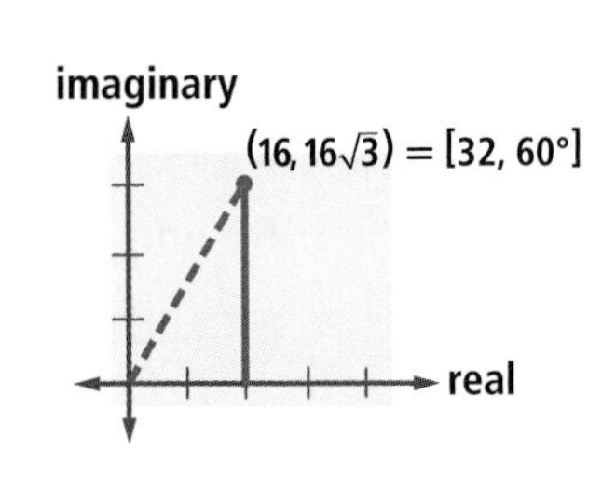

Solution Let $[r, \theta] = 16 + 16\sqrt{3}i$. Calculate the absolute value and an argument of $16 + 16\sqrt{3}i$ or examine its graph to determine that $r = 32$ and $\theta = 60°$.

In trigonometric form, $16 + 16\sqrt{3}i = 32(\cos 60° + i \sin 60°)$.

Use the Roots of a Complex Number Theorem. The fifth roots are

$\sqrt[5]{32}\left(\cos\left(\frac{60°}{5} + k\frac{360°}{5}\right) + i \sin\left(\frac{60°}{5} + k\frac{360°}{5}\right)\right) = 2(\cos(12° + 72°k) + i \sin(12° + 72°k))$, where $k = 0, 1, 2, 3$, and 4.

Thus the roots are
2(cos 12° + i sin 12°),
2(cos 84° + i sin 84°),
2(cos 156° + i sin 156°),
2(cos 228° + i sin 228°), and
2(cos 300° + i sin 300°).

They are graphed at the right. Note that the five fifth roots of $16 + 16\sqrt{3}i$ are equally spaced on a circle centered at the origin with radius $\sqrt[5]{32} = 2$.

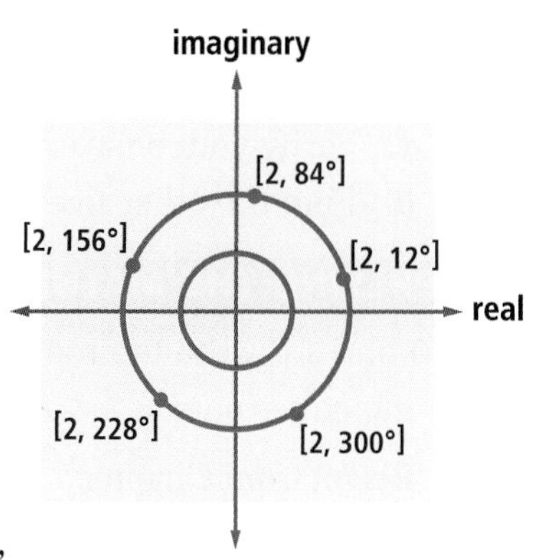

The graphs of the nth roots of a complex number show that polynomials, powers, trigonometry, complex numbers, equation solving, and regular polygons are all connected to each other. Earlier in this book, you saw how normal probability distributions involve π, square roots, and powers of e, the base of natural logarithms. One of the projects for this chapter connects powers of e with sines and cosines. The many connections among mathematical topics are a source of wonder—why should such diverse ideas be connected?

Functions, statistics, and trigonometry may seem like three diverse topics, but we hope that in reading this book and doing its activities and problems you have seen that they are connected aspects of the giant universe of mathematics. The authors hope you have enjoyed this book and learned many things from it.

Questions

COVERING THE IDEAS

In 1 and 2, use De Moivre's Theorem to write each power in trigonometric form.

1. $\left(3\left(\cos \frac{\pi}{7} + i \sin \frac{\pi}{7}\right)\right)^4$
2. $\left(2\left(\cos \frac{4\pi}{5} + i \sin \frac{4\pi}{5}\right)\right)^3$

In 3 and 4, use De Moivre's Theorem to find each power in $a + bi$ form.

3. $(4i)^4$
4. $(-\sqrt{3} + i)^6$
5. Consider $z = 2\left(\cos \frac{5\pi}{6} + i \sin \frac{5\pi}{6}\right)$.
   a. Use De Moivre's Theorem to calculate z^n for $n = 2$ to 6.
   b. Plot z^1, z^2, z^3, z^4, z^5, and z^6 in the complex plane.
   c. Describe the pattern in the graph of the powers of z.
6. a. Solve the equation $z^3 = 8(\cos 150° + i \sin 150°)$.
   b. Plot the solutions in the complex plane.
   c. Describe the graph in Part b.
7. One cube root of $125(\cos 60° + i \sin 60°)$ is $5(\cos 20° + i \sin 20°)$. What are the other two cube roots of this number?
8. Describe the 5 fifth roots of $[r, \theta]$.

9. a. The 6th roots of $64i$, when graphed, are vertices of what figure?
   b. Write $64i$ in polar form to find one 6th root of $64i$.
   c. Verify your answer to Part b by multiplication.
   d. Find the other five 6th roots of $64i$.

APPLYING THE MATHEMATICS

In 10 and 11, write the roots in polar form and graph them.

10. the cube roots of $8 + 8\sqrt{3}i$

11. the fourth roots of -16

12. Recall from Chapter 7 that $z^3 - 216$ can be factored into polynomials with integer coefficients, because $216 = 6^3$.
    a. Factor the polynomial.
    b. Compute and graph its zeros.
    c. The zeros are the m nth roots of q. What are m, n, and q?

13. Consider the polynomial equation $z^5 = 32$.
    a. According to the Fundamental Theorem of Algebra, how many solutions exist in the complex numbers?
    b. Use De Moivre's Theorem to find those solutions in polar form.

14. A ninth root of z is $3(\cos 60^\circ + i \sin 60^\circ)$. Find z in polar form.

REVIEW

15. Write $100(\cos 35^\circ + i \sin 35^\circ)$ in $a + bi$ form. **(Lesson 13-6)**

16. The complex number $z = 5(\cos 55^\circ + i \sin 55^\circ)$ undergoes a transformation that multiplies its absolute value by 4 and rotates it 35° about the origin. **(Lesson 13-6)**
    a. What is the image of z under the rotation?
    b. Identify the arithmetic operation and the complex number that will accomplish this operation.

17. Prove or disprove: For all α, $\sin^2\alpha = \frac{1}{2}(1 - \cos^2\alpha)$. **(Lesson 13-1)**

18. a. Find the product. $\begin{bmatrix} -1 & 0 \\ 0 & 1 \end{bmatrix} \cdot \begin{bmatrix} 1 & 2 \\ 2 & 3 \end{bmatrix}$ **(Lesson 12-2)**
    b. What transformation is represented by $\begin{bmatrix} -1 & 0 \\ 0 & 1 \end{bmatrix}$?

19. Explain why $\log_{10} 3 = \frac{1}{\log_3 10}$ without computing either logarithm. **(Lesson 9-5)**

20. Sylvester invested x dollars in a CD that paid 4% for a year, then paid 3.5% the next year, and paid 4.75% in the third year. At the end of the three years, he had $2085.93. What is x? **(Previous Course)**

EXPLORATION

21. By De Moivre's Theorem, $(\cos \theta + i \sin \theta)^2 = (\cos 2\theta + i \sin 2\theta)$.
    a. Expand the left side of this equation and derive identities for $\cos 2\theta$ and $\sin 2\theta$.
    b. Find identities for $\cos 3\theta$ and $\sin 3\theta$.

QY ANSWER

$z^5 = 1024(\cos 300^\circ + i \sin 300^\circ)$

Chapter 13 Projects

1 Euler's Theorem

Using calculus, it can be shown that if x is in radians, $\cos x = 1 - \frac{x^2}{2!} + \frac{x^4}{4!} - \frac{x^6}{6!} + \cdots$ and $\sin x = x - \frac{x^3}{3!} + \frac{x^5}{5!} - \frac{x^7}{7!} + \cdots$. These series are sometimes the ones used in calculators to approximate values of sine and cosine.

a. Approximate cos 0.2 using the first three terms of the appropriate series. Show that the difference between your approximation and the calculator value of cos 0.2 is less than the absolute value of the fourth term.

b. Repeat Part a for sin 0.2.

c. Give an explicit formula for the nth term of each series.

d. Use the series expansions for $\sin x$ and $\cos x$ to find series expansions for $\sin 2x$ and $\cos 2x$. Check your answers with a calculator.

e. Using calculus, it can be shown also that, for all complex numbers x,
$$e^x = 1 + x + \frac{x^2}{2!} + \frac{x^3}{3!} + \frac{x^4}{4!} + \cdots$$
Find a series expansion for e^{ix}.

f. Use the result of Part e to prove that $e^{ix} = \cos x + i \sin x$.

g. Use the result of Part f to show that $i = e^{i \cdot \frac{\pi}{2}}$.

h. Find a complex number in the form $a + bi$ for $e^{i\pi}$. (The answer to this is known as *Euler's Theorem* and is one of the most extraordinary results in mathematics.)

i. Find i^i. (It may surprise you that i^i is a real number.)

2 Famous Polar Equations

Explore at least five of the following polar equations with a graphing utility. If the graphing utility does not have an option for entering equations in the form $r^2 = f(\theta)$, graph $r = \sqrt{f(\theta)}$ and $r = -\sqrt{f(\theta)}$ simultaneously. In all the equations, a may be any nonzero real number. Graph each equation a few times, with different values of a. Describe the patterns that you find.

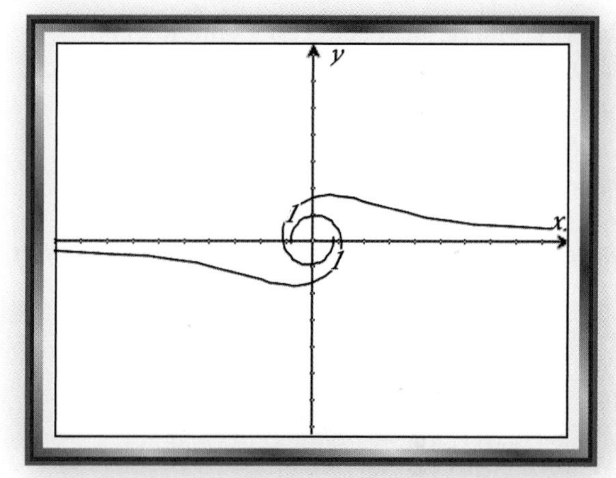

a. Cardioids: $r = a(\cos \theta - 1)$

b. Cissoid of Diocles: $r = a \sin \theta \tan \theta$

c. Cochleoid: $r = \frac{a \sin \theta}{\theta}$

d. Folium of Descartes: $r = \frac{3a \sin \theta \cos \theta}{\sin^3\theta + \cos^3\theta}$

e. Strophoid: $r = a \cos 2\theta \sec \theta$

f. Lemniscate of Bernoulli: $r^2 = a^2 \cos 2\theta$; $r^2 = a^2 \sin 2\theta$

g. Lituus: $r^2 = \frac{a^2}{\theta}$

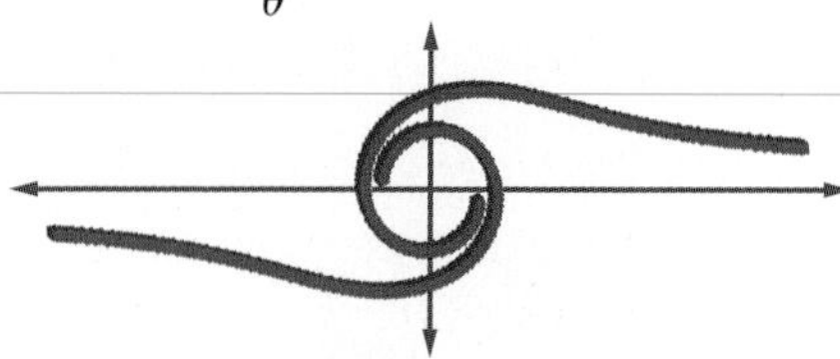

3 The Mandelbrot Set

In recent years, a new field of mathematics has arisen, called dynamical systems, in which complex numbers play an important role. Some of the products of this field are beautiful computer-generated drawings that have won awards in art competitions. Among these are drawings of the *Mandelbrot set* named after Benoit Mandelbrot, a mathematician at IBM, who discovered it around 1980. Three such drawings are shown below.

a. Research and describe how complex numbers are used to generate the Mandelbrot set.

b. The Mandelbrot set is an example of a *fractal*. Use the Internet, an encyclopedia, or another mathematics text to define *fractal*, and give some other examples.

4 Graphics Using Polar Equations

Create and color a graphic design using polar equations. Include a listing of your equations below your graphic.

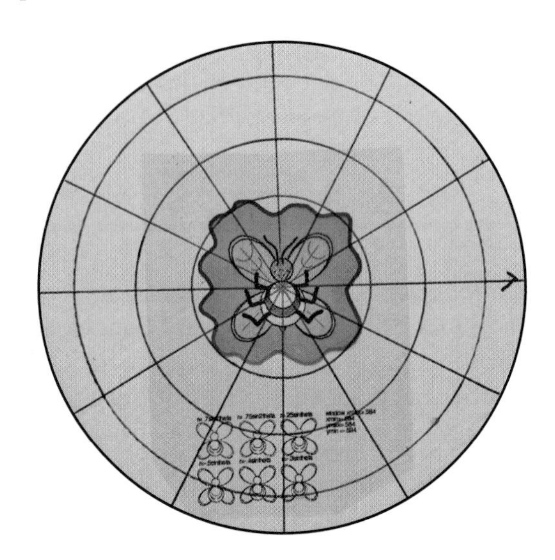

Benoit Mandelbrot (1924–2010) coined the phrase "fractal."

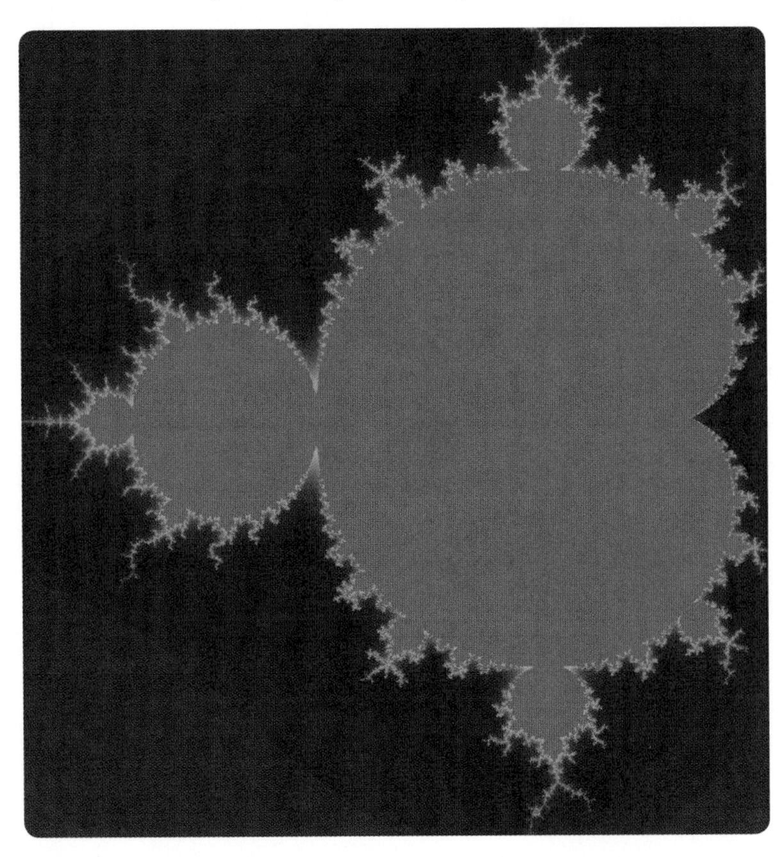

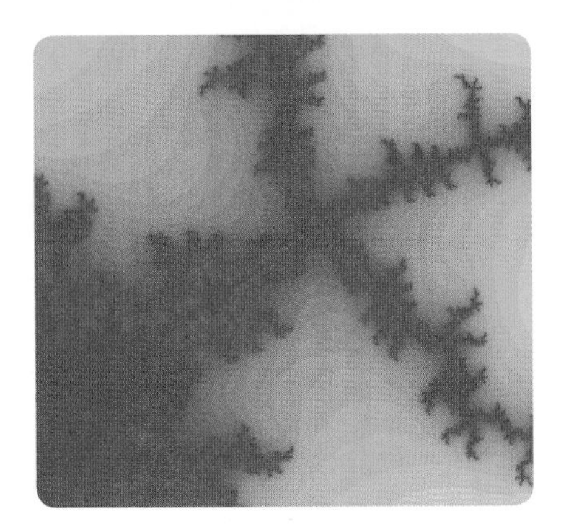

Chapter 13 Summary and Vocabulary

- Identities are statements that are true for all values of the variables in them. To test whether an equation of the form $f(x) = g(x)$ is an identity, check if the graphs of f and g coincide. While the graphs cannot prove an identity, they can help in finding a counterexample. To prove an identity, you can start with one side and rewrite it until it equals the other side, rewrite each side independently until equal expressions are obtained on both sides, or begin with a known identity and derive equivalent statements until the proposed identity appears.

- Many identities involving trigonometric functions hold only on restricted domains, excluding points where one or more of the functions in the identity are not defined. A point where a function is undefined is called a **singularity**. Singularities of functions are identified by vertical asymptotes or missing single points on the graphs.

- In a polar coordinate system, a point is identified by **$[r, \theta]$**, where r is the distance of the point from the **pole** on a ray which is the image of a horizontal ray, called the **polar axis**, under a rotation of θ. Every point has infinitely many **polar coordinate** representations. Use the relationships in the table below to convert from one system to the other.

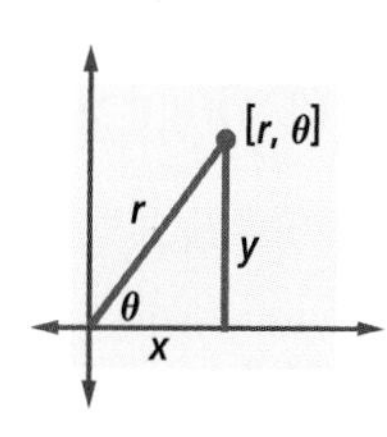

From Polar [, θ] to Rectangular (,):	$x = r\cos\theta$, $y = r\sin\theta$
From Rectangular (,) to Polar [, θ]:	$r = \sqrt{x^2 + y^2}$, $\tan\theta = \frac{y}{x}$

- Graphs of sets of points which satisfy equations involving r and θ include familiar figures, such as lines and circles, and beautiful spirals, rose curves, and other curves that do not have simple descriptions in terms of rectangular coordinates.

Vocabulary

Lesson 13-1
domain of an identity

Lesson 13-2
singularity
discontinuity
removable singularity
removable discontinuity

Lesson 13-3
pole
polar axis
polar coordinates, $[r, \theta]$

Lesson 13-5
complex plane
real axis
imaginary axis
*absolute value, modulus of a complex number
*argument of a complex number
polar form of a complex number

Lesson 13-6
*trigonometric form of a complex number

- The complex number $a + bi$ is represented in rectangular coordinates as the point (a, b). It can also be represented in polar coordinates by the point $[r, \theta]$. A third representation, the **trigonometric form** of this number, is $r(\cos \theta + i \sin \theta)$.

- If $z = a + bi$, the **absolute value** or **modulus** of z is $|z| = \sqrt{a^2 + b^2}$. The sum $(a + c) + (b + d)i$ of the complex numbers $P = a + bi$ and $Q = c + di$ is the fourth vertex of a parallelogram with consecutive vertices at point P, the origin, and point Q. $z_1 + z_2$ is the image of z_1 under a translation. When $z_1 = r_1(\cos \theta_1 + i \sin \theta_1)$ and $z_2 = r_2(\cos \theta_2 + i \sin \theta_2)$, $z_1 z_2 = r_1 r_2 \cdot \left(\cos(\theta_1 + \theta_2) + i \sin(\theta_1 + \theta_2)\right)$ and $\frac{z_1}{z_2} = \frac{r_1}{r_2} \cdot \cos(\theta_1 - \theta_2) + i \sin(\theta_1 - \theta_2)$. In polar form, $[r_1, \theta_1] \cdot [r_2, \theta_2] = [r_1 r_2, \theta_1 + \theta_2]$ and $\frac{[r_1, \theta_1]}{[r_2, \theta_2]} = \left[\frac{r_1}{r_2}, \theta_1 - \theta_2\right]$. The product $z_1 z_2$ is the image of z_1 under the composite of a size change with magnitude $|z_2|$ and a rotation with magnitude θ_2.

- Repeated multiplications of the complex number $z = r(\cos \theta + i \sin \theta) = [r, \theta]$ leads to De Moivre's Theorem: For all positive integers n,

  $$z^n = r^n(\cos n\theta + i \sin n\theta) = [r^n, n\theta].$$

  Working backwards leads to a theorem for finding nth roots:

  If $z^n = r^n(\cos n\theta + i \sin n\theta)$ with $r > 0$,

  then $z = r^{\frac{1}{n}} \cos\left(\left(\frac{\theta}{n} + k\frac{360°}{n}\right) + i \sin\left(\frac{\theta}{n} + k\frac{360°}{n}\right)\right) = \left[r^{\frac{1}{n}}, \frac{\theta}{n} + \frac{2\pi k}{n}\right]$.

Theorems and Properties

Equivalent Polar Coordinates Theorem (p. 799)
Polar-Rectangular Conversion Formula (p. 801)
Geometric Addition Theorem (p. 812)
Product of Complex Numbers Theorem (p. 820, 821)
Quotient of Complex Numbers Theorem (p. 821)
De Moivre's Theorem (p. 826)
Roots of a Complex Number Theorem (p. 828)

Chapter 13 Self-Test

Take this test as you would take a test in class. You will need a calculator. Then use the Selected Answers section in the back of the book to check your work.

1. A point is located at $(-5\sqrt{2}, 5\sqrt{2})$ in a rectangular coordinate system. Find three pairs of polar coordinates $[r, \theta]$ that name this same point. Assume θ is in radians.

In 2 and 3, convert the complex number to each of the other forms ($a + bi$, polar, and trigonometric).

2. $60 - 40i$

3. $5 \cos \frac{\pi}{6} + 5i \sin \frac{\pi}{6}$

In 4–6, plot on an appropriate coordinate system.

4. $[r, \theta] = [4, -185°]$

5. $-4 + 3i$

6. $2\left(\cos \frac{\pi}{2} + i \sin \frac{\pi}{2}\right)$

In 7 and 8, represent $z_1 z_2$ and $\frac{z_1}{z_2}$ in the same form as the original numbers.

7. $z_1 = \left[4, \frac{\pi}{2}\right]; z_2 = \left[\frac{1}{2}, \frac{\pi}{3}\right]$

8. $z_1 = 20 \cos \frac{5\pi}{4} + 20i \sin \frac{5\pi}{4}$;
   $z_2 = \frac{1}{5} \cos \frac{\pi}{6} + \frac{1}{5} i \sin \frac{\pi}{6}$

9. a. Give the coordinates of two points on the graph of $r = 2 \cos 3\theta$.
   b. Trace the graph below and label on your copy the points you gave in Part a.

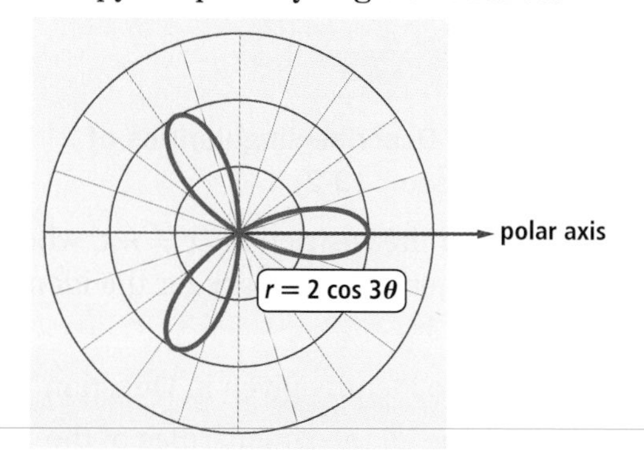

10. Name the singularities of the cotangent function graphed below.

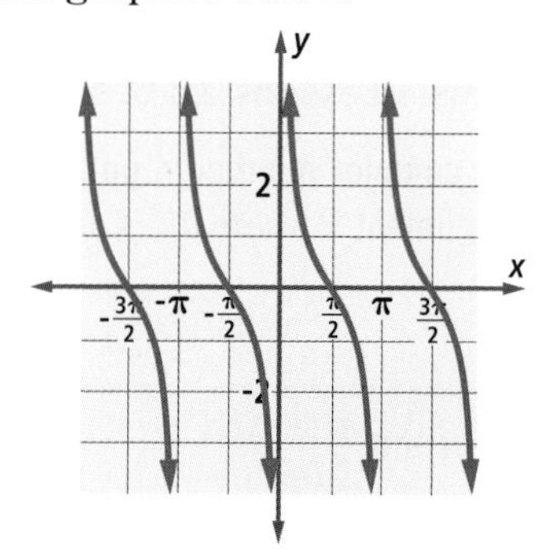

11. Graph $r = 3 \sin \theta$ in the polar coordinate system.

In 12 and 13, consider the equation $\sin^2 x\,(2 \cot^2 x - \csc^2 x) = \cos 2x$.

12. Explain why the graphs of $f(x) = \sin^2 x\,(2 \cot^2 x - \csc^2 x)$ and $g(x) = \cos 2x$ look the same but are not.

13. Prove that the equation is an identity for all values of x for which the trigonometric functions are defined.

14. Convert $[-8, 60°]$ to rectangular coordinates.

15. Write $(3(\cos 20° + i \sin 20°))^4$ in $a + bi$ form.

16. Find the 4 fourth roots of $16(\cos 250° + i \sin 250°)$.

17. Prove $(\sin x + \cos x)^2 + (\sin x - \cos x)^2 = 2$.

Chapter 13 Chapter Review

SKILLS
PROPERTIES
USES
REPRESENTATIONS

SKILLS Procedures used to get answers

OBJECTIVE A Represent complex numbers in different forms. (Lessons 13-5, 13-6)

In 1–3, write the complex number in polar coordinate form. Use an argument θ where $0° \le \theta < 360°$.

1. $5 + 3i$
2. $-2\sqrt{3} + i$
3. $\frac{1}{6}\left(\cos \frac{\pi}{6} + i \sin \frac{\pi}{6}\right)$

In 4 and 5, write the complex number in $a + bi$ form.

4. $5(\cos(-240°) + i \sin(-240°))$
5. $\left[8, \frac{3\pi}{5}\right]$
6. Give two different representations of each complex number in trigonometric form.
   a. $[-3, 135°]$ b. $\left[7, \frac{\pi}{6}\right]$ c. $2 - i$

In 7 and 8, a complex number is given. Give its modulus and its argument.

7. $-9 + 11i$
8. $12 \cos \frac{2\pi}{3} + i \sin \frac{2\pi}{3}$

OBJECTIVE B Perform operations with complex numbers in polar or trigonometric form. (Lesson 13-6)

In 9 and 10, determine z_1z_2. Leave your answer in the form of the original numbers.

9. $z_1 = 4(\cos 23° + i \sin 23°)$;
   $z_2 = 3(\cos 175° + i \sin 175°)$
10. $z_1 = [6, 0]$; $z_2 = \left[1.4, \frac{\pi}{3}\right]$

In 11 and 12, given z_1 and z_2, determine $\frac{z_1}{z_2}$. Leave your answer in the form of the original numbers.

11. $z_1 = \left[4, \frac{\pi}{8}\right]$, $z_2 = \left[5, \frac{7\pi}{8}\right]$
12. $z_1 = 3\left(\cos \frac{2\pi}{3} + i \sin \frac{2\pi}{3}\right)$,
    $z_2 = 6\left(\cos \frac{5\pi}{6} + i \sin \frac{5\pi}{6}\right)$

OBJECTIVE C Find powers and roots of complex numbers. (Lesson 13-7)

In 13 and 14, find z^n for the given z and n.

13. $z = \sqrt{2} + \sqrt{2}i$, $n = 6$
14. $z = 2(\cos 240° + i \sin 240°)$, $n = 5$

In 15 and 16, find and graph the indicated roots.

15. the 4 fourth roots of $81(\cos 16° + i \sin 16°)$
16. the 6 sixth roots of -3

PROPERTIES Principles behind the mathematics

OBJECTIVE D Prove trigonometric identities. (Lessons 13-1, 13-2)

17. Prove that $(1 - \cos^2 x)(1 + \cot^2 x) = 1$ is an identity by starting with the left side and rewriting it until it equals the right side.
18. Prove that $\cos\left(\theta - \frac{\pi}{3}\right) = \sin\left(\theta + \frac{\pi}{6}\right)$ for all θ by expanding each side using identities from this chapter.
19. Prove that $\csc^2 x - \cot^2 x = 1$ for all $x \ne n\pi$, where n is an integer.
20. Prove: For all θ for which the equation is defined, $\frac{\cos \theta}{\sin \theta \cdot \cot \theta} = 1$.

OBJECTIVE E Describe singularities of functions. (Lesson 13-2)

21. Explain why the restriction "$x \ne n\pi$, where n is an integer" is necessary for the identity in Question 19.
22. Consider the identity given in Question 20.
    a. Determine all the singularities of the functions mentioned in the equation.
    b. Give the biggest domain on which the identity holds.

In 23 and 24, determine the singularities (if any) of the function with the given equation.

23. $f(x) = \frac{x^3 - 64}{x - 4}$ 24. $g(x) = \frac{x^3 - 64}{x^2 + 4x + 16}$

25. **True or False** The proposed identity is true for all real numbers x. Explain your answer.

a. $\frac{x^3 - 64}{x - 4} = x^2 + 4x + 16$

b. $\frac{x^3 - 64}{x^2 + 4x + 16} = x - 4$

USES Applications of mathematics in real world situations

There are no uses objectives for this chapter.

REPRESENTATIONS Pictures, graphs, or objects that illustrate concepts

OBJECTIVE F Plot points in a polar coordinate system. (Lesson 13-3)

In 26-28, plot $[r, \theta]$, where θ is in radians.

26. $\left[5, \frac{\pi}{5}\right]$ 27. $[-9, 0]$ 28. $\left[2, \frac{7}{6}\right]$

29. **Multiple Choice** Which polar coordinates do not name the same point as $[3, 240°]$?

A $[3, -120°]$ B $[-3, 120°]$

C $[-3, 60°]$ D $[3, 960°]$

OBJECTIVE G Given polar coordinates of a point, determine its rectangular coordinates, and vice-versa. (Lesson 13-3)

In 30-32, convert from polar coordinates to rectangular coordinates.

30. $\left[6, \frac{3\pi}{4}\right]$ 31. $[-7, 35°]$ 32. $[-1, 26°]$

In 33-35, give two pairs of polar coordinates for each (x, y) pair.

33. $(2, 5)$ 34. $(-3, -2)$ 35. $(-4, 9)$

36. A point is located at $\left(-3\sqrt{3}, 3\right)$ in a rectangular coordinate system. Find three pairs of polar coordinates $[r, \theta]$ that name this same point. Assume θ is in radians.

37. In one set of polar coordinates for P, $\theta = \frac{\pi}{5}$. When the coordinates of P are written in rectangular form, $x = 6$. Find the polar and rectangular coordinates for P.

OBJECTIVE H Graph and interpret graphs of polar equations. (Lesson 13-4)

38. Verify that $\left[2, \frac{2\pi}{3}\right]$ is on the graph of $r = 2 \cos 6\theta$.

39. Give the coordinates of two points on the graph of $r = \csc \theta$.

In 40 and 41, graph the equation in the polar coordinate system.

40. $r = \sin(2\theta)$ 41. $r = 4 \cos 3\theta$

In 42 and 43, a polar equation is given.

a. Use a graphing utility to graph the equation.

b. Verify the shape of the graph in Part a by finding a rectangular coordinate equation for the relation.

42. $r = 3 \sin \theta$ 43. $r = \frac{1}{4} \cos \theta$

OBJECTIVE I Graph complex numbers. (Lessons 13-5, 13-6, 13-7)

In 44 and 45, graph in the complex plane.

44. $5 - 3i$ 45. $5(\cos 140° + i \sin 140°)$

46. a. Graph the origin, $A = 4 + 2i$, $B = 2 - 4i$, and $A + B$ on one coordinate system.

b. Prove that the figure with vertices A, $(0, 0)$, B, and $A + B$ is a parallelogram.

47. Consider $z = 2(\cos 72° + i \sin 72°)$.

a. Graph z, z^2, z^3, z^4, and z^5 on one complex coordinate system.

b. What shape do these points form?

c. Verify that z is a solution to $z^5 = 32$.

48. a. One fourth root of z is $3 + i$. Graph this and the other three fourth roots of z.

b. These four points are the vertices of what figure?

49. Let $z_1 = [2, 40]$ and $z_2 = [3, 75]$.

a. graph z_1 and z_1z_2.

b. The composite of which two transformations maps z_1 onto z_1z_2?

Properties

Algebra Properties from Earlier Courses

Selected Properties of Real Numbers

For any real numbers a, b, and c:

Postulates of Addition and Multiplication (Field Properties)

	Addition	*Multiplication*
Closure property	$a + b$ is a real number.	ab is a real number.
Commutative property	$a + b = b + a$	$ab = ba$
Associative property	$(a + b) + c = a + (b + c)$	$(ab)c = a(bc)$
Identity property	There is a real number 0 with $0 + a = a + 0 = a$.	There is a real number 1 with $1 \cdot a = a \cdot 1 = a$.
Inverse property	There is a real number $-a$ with $a + -a = -a + a = 0$.	If $a \neq 0$, there is a real number $\frac{1}{a}$ with $a \cdot \frac{1}{a} = \frac{1}{a} \cdot a = 1$.
Distributive property	$a(b + c) = ab + ac$	

Postulates of Equality

Reflexive property	$a = a$
Symmetric property	If $a = b$, then $b = a$.
Transitive property	If $a = b$ and $b = c$, then $a = c$.
Substitution property	If $a = b$, then a may be substituted for b in any arithmetic or algebraic expression.
Addition property	If $a = b$, then $a + c = b + c$.
Multiplication property	If $a = b$, then $ac = bc$.

Postulates of Inequality

Trichotomy property	Either $a < b$, $a = b$, or $a > b$.
Transitive property	If $a < b$ and $b < c$, then $a < c$.
Addition property	If $a < b$, then $a + c < b + c$.
Multiplication property	If $a < b$ and $c > 0$, then $ac < bc$. If $a < b$ and $c < 0$, then $ac > bc$.

Postulates of Powers

For any nonzero bases a and b and integer exponents m and n:

Product of Powers property	$b^m \cdot b^n = b^{m+n}$
Power of a Power property	$(b^m)^n = b^{mn}$
Power of a Product property	$(ab)^m = a^m b^m$
Quotient of Powers property	$\frac{b^m}{b^n} = b^{m-n}$
Power of a Quotient property	$\left(\frac{a}{b}\right)^m = \frac{a^m}{b^m}$

Selected Theorems of Graphing

The set of points (x, y) satisfying $Ax + By = C$, where A and B are not both 0, is a line.

The line with equation $y = mx + b$ has slope m and y-intercept b.

Two nonvertical lines are parallel if and only if they have the same slope.

Two nonvertical lines are perpendicular if and only if the product of their slopes is -1.

The set of points (x, y) satisfying $y = ax^2 + bx + c$ is a parabola.

If $A = (x_1, y_1)$ and $B = (x_2, y_2)$, then $AB = \sqrt{|x_2 - x_1|^2 + |y_2 - y_1|^2}$.

Selected Theorems of Algebra

For any real numbers a, b, c, and d (with denominators of fractions not equal to 0):

Multiplication Property of Zero	$0 \cdot a = 0$
Multiplication Property of -1	$-1 \cdot a = -a$
Opposite of an Opposite Property	$-(-a) = a$
Opposite of a Sum	$-(b + c) = -b + -c$
Distributive Property of Multiplication over Subtraction	$a(b - c) = ab - ac$
Addition of Like Terms	$ac + bc = (a + b)c$
Zero Product	$ab = 0$ if and only if $a = 0$ or $b = 0$.
Addition of Fractions	$\frac{a}{c} + \frac{b}{c} = \frac{a+b}{c}$
Multiplication of Fractions	$\frac{a}{b} \cdot \frac{c}{d} = \frac{ac}{bd}$
Equal Fractions	$\frac{ac}{bc} = \frac{a}{b}$
Means-Extremes	If $\frac{a}{b} = \frac{c}{d}$, then $ad = bc$.
Binomial Square	$(a + b)^2 = a^2 + 2ab + b^2$
Difference of Squares Factoring	$a^2 - b^2 = (a + b)(a - b)$
Extended Distributive Property	To multiply two polynomials, multiply each term in the first polynomial by each term in the second, and then add the products.
Zero Exponent	If $b \neq 0$, then $b^0 = 1$.
Negative Exponent	If $b \neq 0$, then $b^{-n} = \frac{1}{b^n}$.
Absolute Value-Square Root	$\sqrt{a^2} = \|a\|$
Product of Square Roots	If $a \geq 0$ and $b \geq 0$, then $\sqrt{ab} = \sqrt{a} \cdot \sqrt{b}$.
Quadratic Formula	If $ax^2 + bx + c = 0$ and $a \neq 0$, then $x = \frac{-b \pm \sqrt{b^2 - 4ac}}{2a}$.
Square Root of a Negative Number	If $k < 0$, $\sqrt{k} = i\sqrt{-k}$
Discriminant Theorem	Let $D = b^2 - 4ac$. Then $ax^2 + bx + c = 0$ has two real solutions if $D > 0$, one real solution if $D = 0$, and no real solutions if $D < 0$.
Factorial Product	For all integers $n \geq 1$, $n! = n(n - 1)!$

Geometry Properties from Earlier Courses

In this book, the following symbols are used:

Symbol	Meaning
a, b, c	sides
A	area
B	area of base
b_1, b_2	bases
C	circumference
d	diameter
d_1, d_2	diagonals
h	height
ℓ	length
ℓ	slant height (in conics)
L.A.	lateral area
n	number of sides
p	perimeter
r	radius
s	side
S.A.	surface area
V	volume
w	width

Two-Dimensional Figures

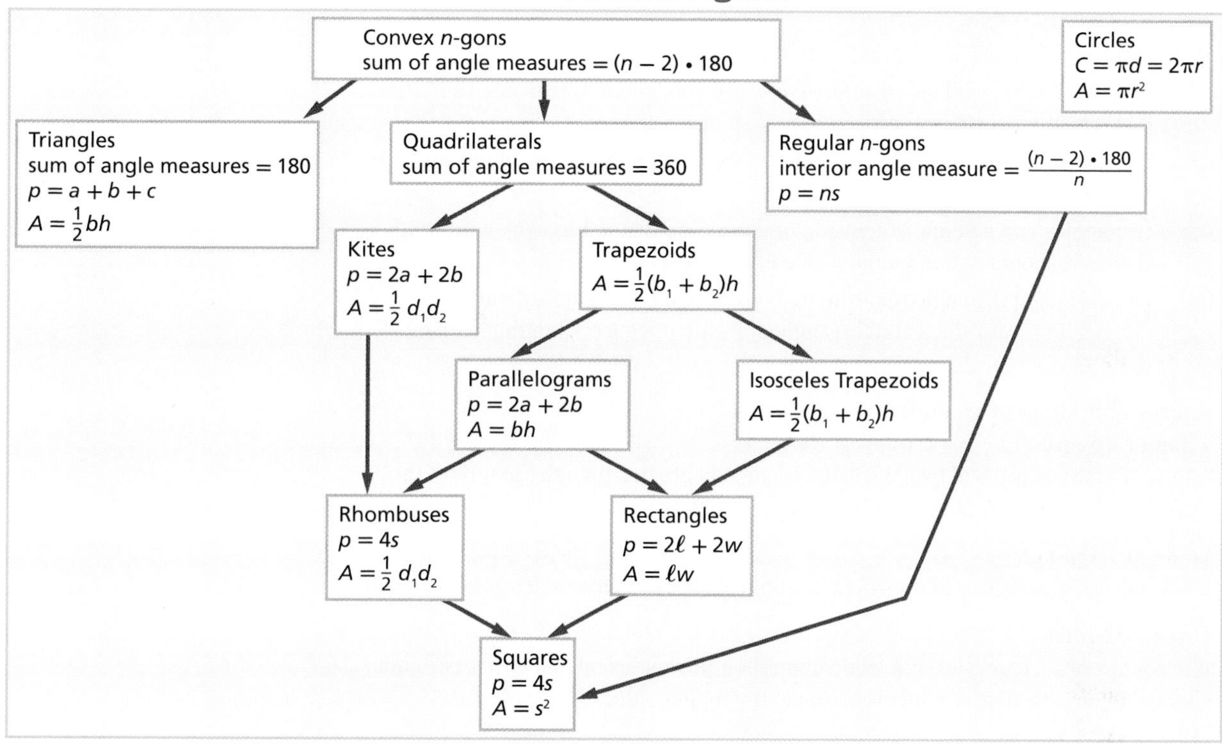

Three-Dimensional Figures

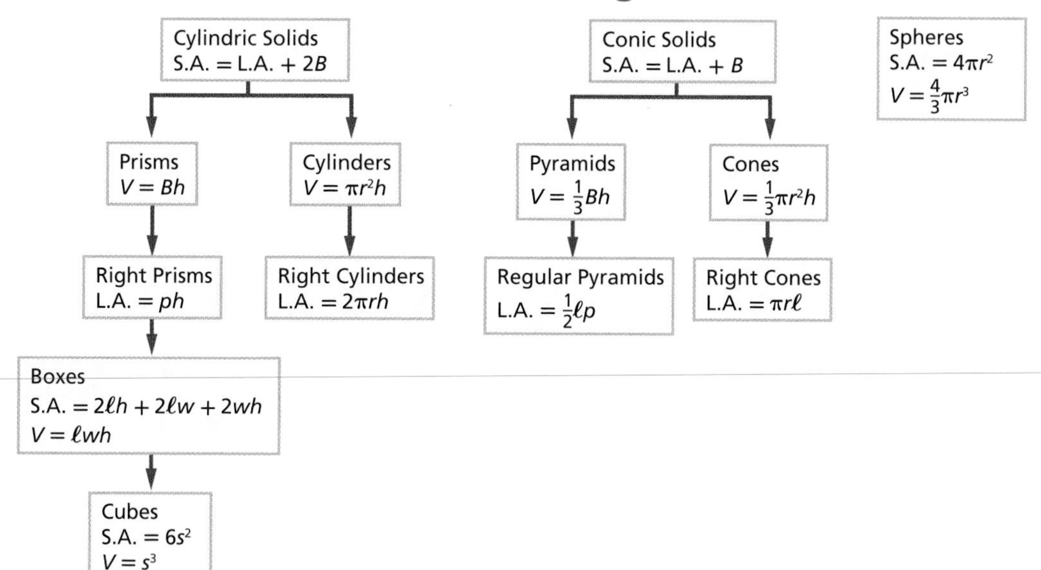

Selected Theorems of Geometry

Parallel Lines

Two lines are parallel if and only if:

1. corresponding angles have the same measure.
2. alternate interior angles are congruent.
3. alternate exterior angles are congruent.
4. they are perpendicular to the same line.

Triangle Congruence

Two triangles are congruent if:

SSS three sides of one are congruent to three sides of the other.

SAS two sides and the included angle of one are congruent to two sides and the included angle of the other.

ASA two angles and the included side of one are congruent to two angles and the included side of the other.

AAS two angles and a nonincluded side of one are congruent to two angles and the corresponding nonincluded side of the other.

SsA two sides and the angle opposite the longer of the two sides of one are congruent to two sides and the angle opposite the corresponding side of the other.

Angles and Sides of Triangles

Triangle Inequality

The sum of the lengths of two sides of a triangle is greater than the length of the third side.

Isosceles Triangle

If two sides of a triangle are congruent, the angles opposite those sides are congruent.

Unequal Sides

If two sides of a triangle are not congruent, then the angles opposite them are not congruent, and the larger angle is opposite the longer side.

Unequal Angles

If two angles of a triangle are not congruent, then the sides opposite them are not congruent, and the longer side is opposite the larger angle.

Pythagorean Theorem

In any right triangle with legs a and b and hypotenuse c, $a^2 + b^2 = c^2$.

30-60-90 Triangle

In a 30-60-90 triangle, the sides are in the extended ratio $x : x\sqrt{3} : 2x$.

45-45-90 Triangle

In a 45-45-90 triangle, the sides are in the extended ratio $x : x : x\sqrt{2}$.

Parallelograms

A quadrilateral is a parallelogram if and only if:

1. one pair of sides is both parallel and congruent.
2. both pairs of opposite sides are congruent.
3. both pairs of opposite angles are congruent.
4. its diagonals bisect each other.

Quadrilateral Hierarchy
If a figure is of any type in the hierarchy pictured at the right, it is also of all types above it to which it is connected.

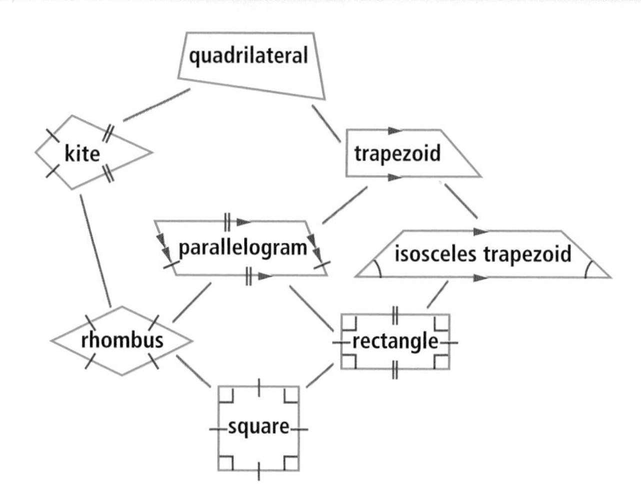

Properties of Transformations

A-B-C-D
Every isometry preserves angle measure, betweenness, collinearity, and distance.

Two-Reflection for Translations
If $m \parallel n$, the translation $r_n \circ r_m$ has magnitude two times the distance between m and n in the direction from m perpendicular to n.

Two-Reflection for Rotations
If m intersects ℓ, the rotation $r_m \circ r_\ell$ has a center at the point of intersection of m and ℓ, and has a magnitude twice the measure of an angle formed by these lines, in the direction from ℓ to m.

Isometry
Every isometry is a transformation that is a reflection or a composite of reflections.

Size-Change
Every size change with magnitude k preserves angle measure, betweenness, and collinearity; a line is parallel to its image; distance is multiplied by k.

Fundamental Theorem of Similarity
If two figures are similar with ratio of similitude k, then:

1. corresponding angle measures are equal.
2. corresponding lengths and perimeters are in the ratio k.
3. corresponding areas and surface areas are in the ratio k^2.
4. corresponding volumes are in the ratio k^3.

Triangle Similarity
Two triangles are similar if:

1. three sides of one are proportional to three sides of the other (SSS).
2. the ratios of two pairs of corresponding sides are equal and the included angles are congruent (SAS).
3. two angles of one are congruent to two angles of the other (AA).

Coordinate Plane Formulas
For all $A = (x_1, y_1)$ and $B = (x_2, y_2)$:

Distance formula: $AB = \sqrt{(x_2 - x_1)^2 + (y_2 - y_1)^2}$

Midpoint formula: The midpoint of $\overline{AB}$ is $\left(\frac{x_1 + x_2}{2}, \frac{y_1 + y_2}{2}\right)$.

For all points (x, y):

reflection over the x-axis	$(x, y) \rightarrow (x, -y)$
reflection over the y-axis	$(x, y) \rightarrow (-x, y)$
reflection over $y = x$	$(x, y) \rightarrow (y, x)$
size change of magnitude k, center (0, 0)	$(x, y) \rightarrow (kx, ky)$
translation h units horizontally, k units verically	$(x, y) \rightarrow (x + h, y + k)$

Symbols

Symbol	Meaning
$\mathbb{N}$	set of natural numbers
$\mathbb{Z}$	set of integers
$\mathbb{Q}$	set of rational numbers
$\mathbb{R}$	set of real numbers
$\{...\}$	set
$\in$	is an element of
$\{\ \}, \varnothing$	null set, empty set
$\{x \mid x...\}$, $\{x : x...\}$	the set of all x such that ...
$\cup$	set union
$\cap$	set intersection
$\approx$	is approximately equal to
$\overleftrightarrow{AB}$	line containing A and B
$\overrightarrow{AB}$	ray with endpoint A and containing B
$\overline{AB}$	line segment with endpoints A and B
AB	distance between points A and B
$\parallel$	is parallel to
$\perp$	is perpendicular to
$m\angle ABC$	measure of angle ABC
$m\widehat{AB}$	measure of arc AB
A'	image of A
$T_{h,k}$	translation h units horizontally and k units vertically
r_x	reflection over the x-axis
r_y	reflection over the y-axis
r_m	reflection over the line m
R_θ	rotation of magnitude θ counterclockwise with center (0,0)
S_k	size change of magnitude k
$S_{a,b}$	scale change with horizontal magnitude a and vertical magnitude b
$\lvert x \rvert$	absolute value of x
a^b, a^b	bth power of a
$\sqrt{}$	radical sign; square root
$\sqrt[n]{x}$	the largest real nth root of x
$\log_b a$	logarithm of a to the base b
e	2.71828...
$\ln x$	natural logarithm of x
(x, y)	rectangular coordinates
$[r, \theta]$	polar coordinates
$f: x \rightarrow y$	the function that maps x onto y
$f(x)$	the value of the function f at x
f^{-1}	inverse function of a function f
$\lim_{x\to\infty} f(x)$	the limit of $f(x)$ as x gets larger and larger
$g \circ f$	composite of function f followed by g
$g(f(x))$	value at x of the composite of function f followed by g
i	$\sqrt{-1}$
$a + bi$	a complex number, where a and b are real numbers
a_n	"a sub n"; the nth term of a sequence a
$\sum_{i=m}^{n} x_i$	the sum of $x_m + x_{m+1} + \ldots + x_n$
$\sin\theta$	sine of θ
$\cos\theta$	cosine of θ
$\tan\theta$	tangent of θ
$\cot\theta$	cotangent of θ
$\sec\theta$	secant of θ
$\csc\theta$	cosecant of θ
$x!$	x factorial
$N(A)$	number of elements in set A
${}_nP_r$	the number of permutations of r objects from n different objects
$\binom{n}{r}$, ${}_nC_r$	the number of sets of r objects from n different objects
$P(E)$	probability of event E
$P(B \mid A)$	conditional probability of an event B given A has occurred
μ	population mean
$\bar{x}$	sample mean
σ^2	population variance
σ	population standard deviation
S^2	sample variance
S	sample standard deviation
Q_n	nth quartile (n = 1, 2, or 3)
r	correlation coefficient
χ^2	chi-square statistic

CAS Commands

The Computer Algebra System (CAS) commands used in this course and examples of their use are given below. Each command must be followed by a number, variable, expression, or equation, usually enclosed in parentheses.

Command	Description	Example
Define	A rule for a function is stored under the name indicated. Values of that function can then be calculated by entering the function's name followed by the value of the independent variable in parentheses.	Define $f(n)=2000\cdot n-1400$ *Done* $f(4)$ 6600
| (such that)	Variable values that appear after the symbol are substituted into an expression, inequality, or equation that appears before the symbol.	$r=\frac{\sqrt{v}}{\sqrt{h\cdot\pi}}\|v=500000$ $r=\frac{500\cdot\sqrt{2}}{\sqrt{h\cdot\pi}}$
solve	An equation, inequality, or system is solved for an indicated variable or variables. All real solutions are given.	$\text{solve}\left(y=\frac{1}{2}\cdot x-5 \text{ and } y=2\cdot x-1,\{x,y\}\right)$ $x=\frac{^-8}{3} \text{ and } y=\frac{^-19}{3}$
expand	The Distributive Property is applied to products and powers of mathematical expressions.	$\text{expand}((72+w)\cdot(24+2\cdot w))$ $2\cdot w^2+168\cdot w+1728$
DelVar	Any stored values for the indicated variable are deleted from memory.	DelVar a *Done*
cSolve	An equation or inequality is solved for an indicated variable. All complex solutions are given.	$\text{cSolve}(z^4=81,z)$ $z=3\cdot i \text{ or } z=^-3\cdot i \text{ or } z=^-3 \text{ or } z=3$
factor	A polynomial is factored over the rational numbers. On some CAS, if ",x" is added to the end of the polynomial, it is factored over the real numbers.	$\text{factor}(x^4-14\cdot x^2+45)$ $(x-3)\cdot(x+3)\cdot(x^2-5)$ $\text{factor}(x^4-14\cdot x^2+45,x)$ $(x-3)\cdot(x+3)\cdot(x+\sqrt{5})\cdot(x-\sqrt{5})$
cFactor	A polynomial is factored over the complex numbers.	$\text{cFactor}(x^2+36,x)$ $(x+^-6\cdot i)\cdot(x+6\cdot i)$
propFrac	Division is applied to a rational expression to return the sum of the resulting quotient and a ratio of the remainder over the divisor.	$\text{propFrac}\left(\frac{x^3+4\cdot x^2-10}{x+2}\right)$ $\frac{^-2}{x+2}+x^2+2\cdot x-4$

A Table of Random Numbers

row	col. 1	2	3	4	5	6	7	8	9	10	11	12	13	14
1	10480	15011	01536	02011	81647	91646	69719	14194	62590	36207	20969	99570	91291	90700
2	22368	46573	25595	85393	30995	89198	27982	53402	93965	34095	52666	19174	39615	99505
3	24130	48360	22527	97265	76393	64809	15179	24830	49340	32081	30680	19655	63348	58629
4	42167	93093	06423	61680	17856	16376	39440	53537	71341	57004	00849	74917	97758	16379
5	37570	39975	81837	16656	06121	91782	60468	81305	49684	60672	14110	06927	01263	54613
6	77921	06907	11008	42751	27756	53498	18602	70659	90655	15053	21916	81825	44394	42880
7	99562	72905	56420	69994	98872	31016	71194	18738	44013	48840	63213	21069	10634	12952
8	96301	91977	05463	07972	18876	20922	94595	56869	69014	60045	18425	84903	42508	32307
9	89579	14342	63661	10281	17453	18103	57740	84378	25331	12566	58678	44947	05585	56941
10	85475	36857	43342	53988	53060	59533	38867	62300	08158	17983	16439	11458	18593	64952
11	28918	69578	88231	33276	70997	79936	56865	05859	90106	31595	01547	85590	91610	78188
12	63553	40961	48235	03427	49626	69445	18663	72695	52180	20847	12234	90511	33703	90322
13	09429	93969	52636	92737	88974	33488	36320	17617	30015	08272	84115	27156	30613	74952
14	10365	61129	87529	85689	48237	52267	67689	93394	01511	26358	85104	20285	29975	89868
15	07119	97336	71048	08178	77233	13916	47564	81056	97735	85977	29372	74461	28551	90707
16	51085	12765	51821	51259	77452	16308	60756	92144	49442	53900	70960	63990	75601	40719
17	02368	21382	52404	60268	89368	19885	55322	44819	01188	65255	64835	44919	05944	55157
18	01011	54092	33362	94904	31272	04146	18594	29852	71585	85030	51132	01915	92747	64951
19	52162	53916	46369	58586	23216	14513	83149	98736	23495	64350	94738	17752	35156	35749
20	07056	97628	33787	09998	42698	06691	76988	13602	51851	46104	88916	19509	25625	58104
21	48663	91245	85828	14346	09172	30168	90229	04734	59193	22178	30421	61666	99904	32812
22	54164	58492	22421	74103	47070	25306	76468	26384	58151	06646	21524	15227	96909	44592
23	32639	32363	05597	24200	13363	38005	94342	28728	35806	06912	17012	64161	18296	22851
24	29334	27001	87637	87308	58731	00256	45834	15398	46557	41135	10367	07684	36188	18510
25	02488	33062	28834	07351	19731	92420	60952	61280	50001	67658	32586	86679	50720	94953
26	81525	72295	04839	96423	24878	82651	66566	14778	76797	14780	13300	87074	79666	95725
27	29676	20591	68086	26432	46901	20849	89768	81536	86645	12659	92259	57102	80428	25280
28	00742	57392	39064	66432	84673	40027	32832	61362	98947	96067	64760	64584	96096	98253
29	05366	04213	25669	26422	44407	44048	37937	63904	45766	66134	75470	66520	34693	90449
30	91921	26418	64117	94305	26766	25940	39972	22209	71500	64568	91402	42416	07844	69618
31	00582	04711	87917	77341	42206	35126	74087	99547	81817	42607	43808	76655	62028	76630
32	00725	69884	62797	56170	86324	88072	76222	36086	84637	93161	76038	65855	77919	88006
33	69011	65797	95876	55293	18988	27354	26575	08625	40801	59920	29841	80150	12777	48501
34	25976	57948	29888	88604	67917	48708	18912	82271	65424	69774	33611	54262	85963	03547
35	09763	83473	73577	12908	30883	18317	28290	35797	05998	41688	34952	37888	38917	88050
36	91567	42595	27958	30134	04024	86385	29880	99730	55536	84855	29080	09250	79656	73211
37	17955	56349	90999	49127	20044	59931	06115	20542	18059	02008	73708	83517	36103	42791
38	46503	18584	18845	49618	02304	51038	20655	58727	28168	15475	56942	53389	20562	87338
39	92157	89634	94824	78171	84610	82834	09922	25417	44137	48413	25555	21246	35509	20468
40	14577	62765	35605	81263	39667	47358	56873	56307	61607	49518	89656	20103	77490	18062

Selected Answers

Chapter 1

Lesson 1-1 (pp. 6-13)

Guided Example 1 **b.** amount of protective gear worn. **c.** 40.7 **d.** 188; 709; 26.5 **e.** 242; 754; 754 **f.** 242; 140; 382

Questions

1. Answers vary. Sample: eye color, arm span, birth month **3.** population: the batch of cookies; yes; number of raisins per cookie **5.** in decreasing order: falls, motor-vehicle, and suffocation by ingestion or inhalation **7.** A representative sample is a sample that accurately reflects the important features of the population. **9. a.** about 21.5% **b.** Guided Example 1d asks for the percentage of beginner skaters in the sample who wore no gear, whereas Part a asked for the percent of no gear skaters who were beginners. **11. a.** categorical **b.** categorical **c.** numerical **d.** numerical **13.** 67.6% **15.** Answers vary. Sample: The median income of households headed by someone who has at least a bachelor's degree is more than twice as high as the median income of households headed by someone with only a high school diploma. **17. a.** the age distribution of trout in a certain lake **b.** sample; The population is all of the trout in the lake. The sample is representative of the population because the trout were captured from different parts of the lake. **c.** about 14.6% **19. a.** $x = 20$ **b.** $x = 2000$ **c.** $x = 0.05$ **d.** $x = 0.0005$ **21. a.** Answers vary. Sample: (0, -13), (1, 16) **b.** $y = -3x - 13$

Lesson 1-2 (pp. 14-21)

Guided Example 1 **a.** 65, \$65,000; 15, \$1,170,000 **b.** 11

Questions

1. mean and median **3.** A weighted average is an average calculated when some elements in the set are assigned a larger or smaller weight. **5.** There are two employees with salaries of \$90,000 and four people with salaries of \$65,000. **7. a.** $\frac{3}{16}$ **b.** \$75,000 **c.** \$50,000 **9.** $x_4 + x_5 + x_6 + x_7 + x_8$ **11. a.** 28 **b.** 4 **c.** $\frac{0(4) + 1(9) + 2(7) + 3(4) + 4(2) + 5(1) + 7(1)}{28} \approx 1.96$ **13.** about 9 **15. a.** 80.45 **b.** 85 **17.** $\sum_{i=1}^{100}(x_i + 2)$, by 198 **19.** $n = 4$ **21.** 3,392,750 **23.** 23.9%

Lesson 1-3 (pp. 22-30)

Guided Example 2 **b.** 60–64 **c.** 29; 5; 13; 2020, 1980

Questions

1. Answers vary. Sample: The graph is sharply concentrated around the 2–3 minute time interval. The graph also seems skewed with a tail on the right.

3. a.

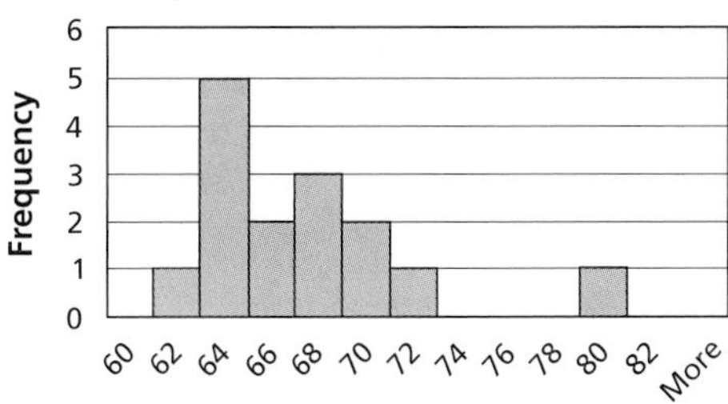

b. Answers vary. Sample: A bin width of 2 was used because it shows trends in the data accurately. **5.** 50 **7.** about 18% **9.** The median height of African American males born 1920–1929 is in the 175 ≤ height < 180 cm interval and the median height of African American males born 1980–1989 is in the 180 ≤ height < 185 cm interval. The second group has the larger median height.

11. a.

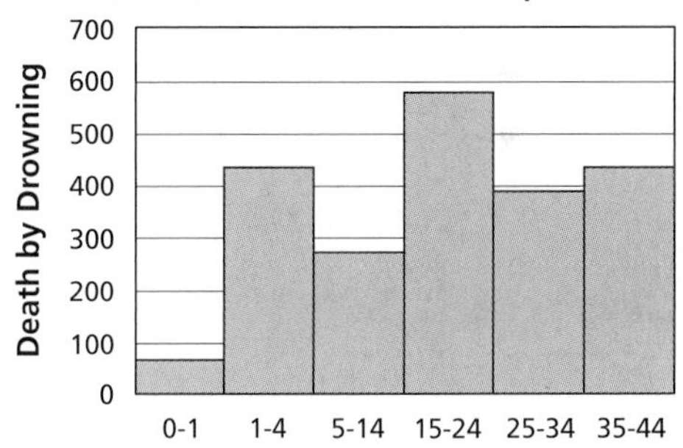

b. between ages 1–4 because the number of deaths per age is the greatest **c.** The data are organized into bins that are not all the same size. **13.** Answers vary. Sample: This graph might be skewed with a right tail because most workers earn at the low end of a payroll scale. There will likely be a long, high-paid tail. **15. a.** The distribution is skewed with a right tail. A roll of 1 is clearly favored. **b.** about 48% **c.** No; if the die is fair, then each number should have appeared about 17% of the time. **17.** about \$2,528.57 **19.** -205 **21. a.** v **b.** ii **c.** iv **d.** i

Lesson 1-4 (pp. 31-37)

Guided Example 1 **a.** 10; 29; 50; 37; 46 **b.** 37; 46; 29; 50

Guided Example 2 38; 157; 5; 157; 5

Questions

1. A **3.**

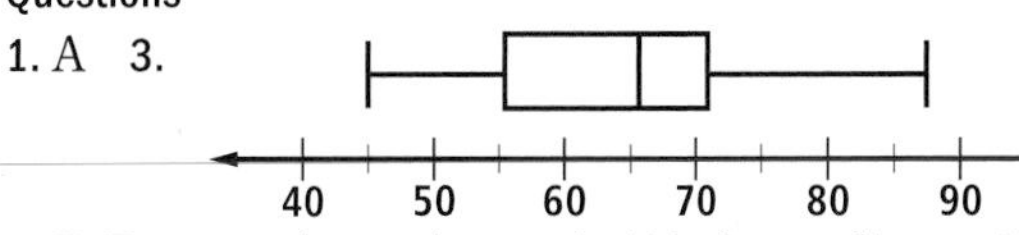

5. B; Because the total count is 100, the median resides in the 5–10 bin. The only plot whose median is in this range is plot B. **7. a.** Answers vary. Sample: the frequency of scores in each interval **b.** Answers vary. Sample:

the values of Q_1, the median, and Q_3 **9.** Answers vary. Sample: {0, 0, 0, 0, 2, 2, 4, 4, 4, 8} **11. a.** about 53 **b.** about $\frac{100}{500}$ or $\frac{1}{5}$ c. about $\frac{345}{500}$ or 69% **d.** Answers vary. Sample: The median wait time is between 5 and 6 minutes. The range of wait times is about 19 minutes. The data are skewed with a tail on the right. **13. a.** about 3.82 **b.** about 3.53 **c.** by about 0.28

Lesson 1-5 (pp. 38-44)

Guided Example 2 a. 14; 6; 8; 48; 34; 16

b.

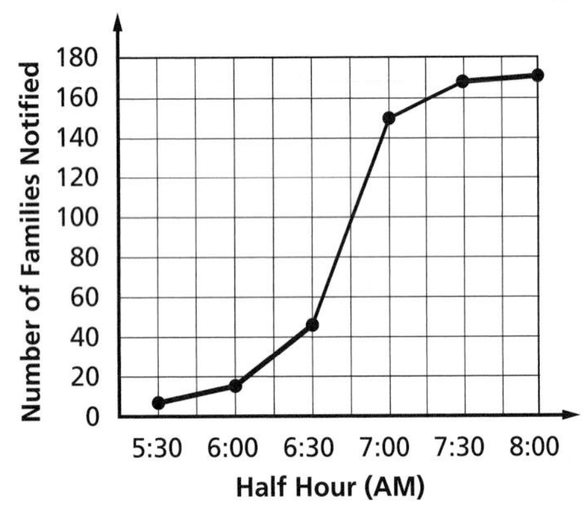

c. 85, around 6:40 **d.** 128, around 6:55

Questions

1.

Year	Number of cases	Total Cases	Relative Frequency	Cumulative Relative Frequency
2003	4	4	0.011	0.011
2004	46	50	0.131	0.142
2005	98	148	0.2792	0.4217
2006	115	263	0.3276	0.7493
2007	88	351	0.2507	1

3. a.

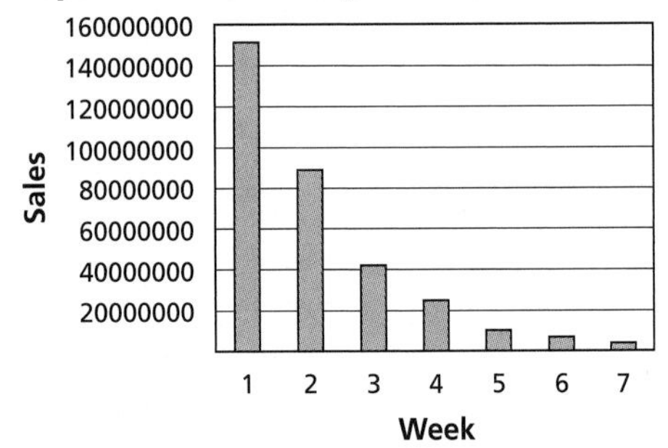

3. b.

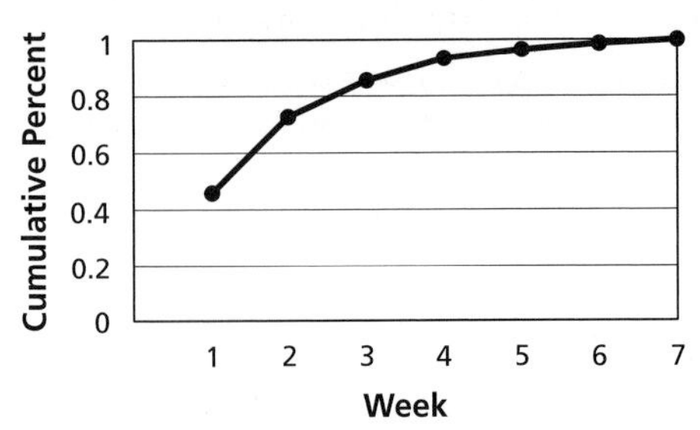

5. a. false **b.** false **c.** true **7. a.** Roger Maris hits more and more homeruns until he peaks in his 5th year, at which point he starts hitting fewer homeruns. **b.** The cumulative graph increases the quickest in the middle of Roger Maris' career, in agreement with the trend from Part a.

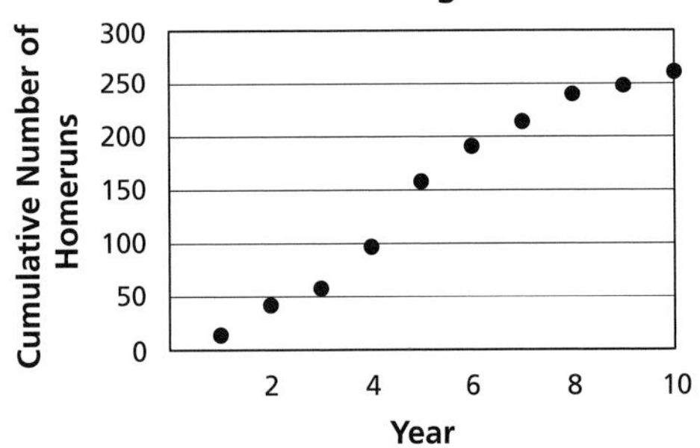

9. Answers vary. Sample: A retailer's cumulative sales receipts rise during the month of December, but fall after Christmas as gifts are returned. **11.** Answers vary. Sample: {0, 0, 2, 2, 2, 2, 2, 2, 2, 2, 4, 4} **13.** \$65.48 per share **15.** 12.155 **17. a.** 13 **b.** 13 **c.** 169 **d.** 169 **e.** -2197 **f.** 2197

Lesson 1-6 (pp. 45-53)

Guided Example 2 10.14; 4.58; -4.86; 12; 18; 26.42; 90; $\sum_{i=1}^{7}(x_i - \bar{x})$; 244.86; 6; $\frac{244.86}{6}$; 40.81; 40.81; 6.39

Questions

1. a. center **b.** spread **c.** spread **d.** spread **e.** spread **f.** center **3. a.** variance = 5.20; standard deviation ≈ 2.28 **b.** 0 **c.** 0 **d.** 3.58 – 2.28 = 1.3 **e.** Answers vary. Sample: The heights of the Dolphins and Sweet Peppers have the same mean and range, but the Dolphins have a greater standard deviation. Therefore the Dolphins' heights are spread out more than the Sweet Peppers' heights from the shared center, though still within the shared range. **5.** 20.25 cm^2 **7.** $\bar{x} = 118$; $s \approx 5.30$ **9.** B and D **11.** The variance is always positive. **13. a.** 60.84 kg^2 **b.** about 4.36 kg **15.** $16 \le x \le 26$ **17.** Answers vary. Sample: The standard deviation of {70, 80, 80, 150} is about 36.97. **19. a.** U.S. Total Imports in Goods and Services (\$ billions):

Cumulative Total for Jan., 187; Feb., 373; Mar., 565; Apr., 756; May, 949; Jun., 1144; Jul., 1341; Aug., 1538; Sept., 1736; Oct., 1936; Nov., 2141; Dec., 2344

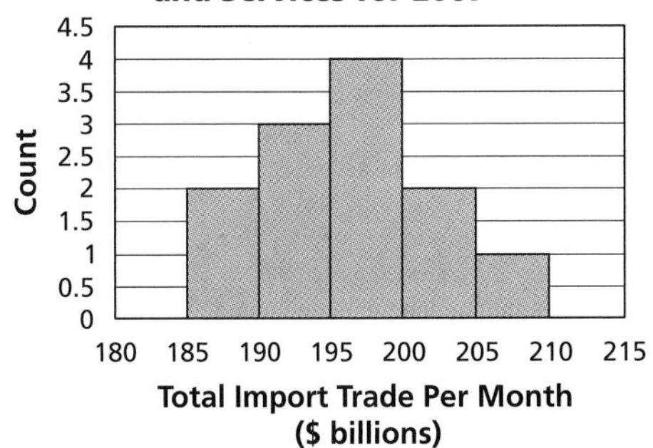

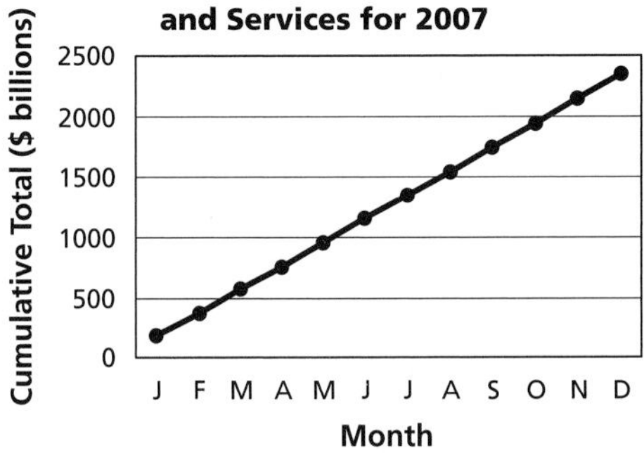

19. b. 2344 **21.** Answers vary. Sample: The maximum influx of permanent legal residents into one state was between 260 and 270 thousand, and the minimum influx of residents was between 0 and 10 thousand; 30 states had the minimum level of influx. The mean value is between 25 and 35 thousand and the median is between 0 and 10 thousand. Most of the states are clustered in the 0 to 10 thousand interval, and the graph is skewed with a tail on the right. **23. a.** about 8504 women/year **b.** about 10,548 men/year

Lesson 1-7 (pp. 54-60)

Questions

1. The variability of the Eastern states is greater than that of the Western states; the interquartile range of 299.4 is greater than 55.4. Whereas both the distributions of the Western and Eastern states have tails to the right, that of the Eastern states is longer, corresponding to its larger standard deviation of 298.3 compared to 54.4. The Eastern states also have a greater mean and median than the Western states, 301.4 and 172.3 compared to 52.4 and 38.6, respectively. **3.** shape, center, and spread **5.** Answers vary. Sample: The mean of the class data is 19.35 states, which is less than the mean of 24.7 for the sample. The standard deviation of the class data is 12.8893 states, which is larger than that of the sample, at 10.32 states. Again, this shows that the variability of the states visited is greater for the class. **7.** Answers vary. Sample: The distribution of the sample of U.S. bridge lengths appears to be bimodal. The distribution of the sample of the outside U.S. bridge lengths appears to have an outlier on the right. Both the mean and median of the bridges outside the U.S. (30,864 m and 29,250 m, respectively) are higher than those inside the U.S. (19,889 m and 15,289 m). The IQR of bridges in the U.S. is 18,403 m, which is much bigger than the IQR of bridges outside the U.S., at 14,110 m. The standard deviation of bridges inside the U.S. is also higher with 11,479 instead of 10,255 outside the U.S.
9. a. 18.49 in^2 **b.** 3.098 in.
11. a.

◇	A	B	C
1	Score Interval	Frequency	Relative Frequency
2	50-59	1	0.033
3	60-69	6	0.200
4	70-79	10	0.333
5	80-89	6	0.200
6	90-99	7	0.233

b.

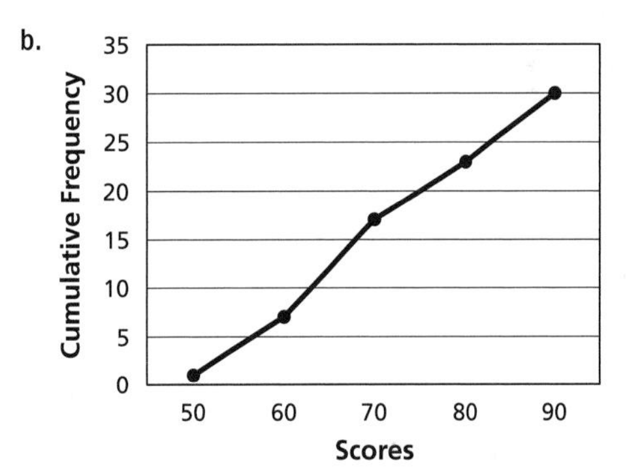

13. -18,778 **15.** $y = -\frac{\pi}{5}x + \pi$

Lesson 1-8 (pp. 61-66)

Questions

1. The Federalist papers were written between 1787 and 1788 in order to persuade the citizens of the State of New York to ratify the Constitution. **3.** 48 papers by Hamilton and 50 papers by Madison **5.** about 32% **7.** Inferential reasoning gives evidence for what is likely, but does not prove findings with certainty. **9.** This disputed paper has rates of 8.07, 3.80, and 46.04 per 1000 words for "by," "from," and "to," respectively. This pattern more closely matches Hamilton's word usage than Madison's.
11. a. Marco likely has the flu since two data points, especially his fever, match the data for the flu better than for a cold. The data about coughing is inconclusive.
b. Annette likely has a cold since both her temperature and the magnitude of her muscle pain match the median values for a cold better than the flu. The data about coughing is inconclusive. **13.** Answers vary. Sample: This is likely symmetric, centered around some average winning score. **15. a.** 3 **b.** 13 **c.** 64 **d.** the average of shoe

ownership per person **17. a.** Answers vary. Sample: The domain of both functions is all real numbers x and the range of both functions is $y \leq 0$. Both functions have an abolute maximum at $x = 0$. **b.** Answers vary. Sample : The slope of $y = -x^2$ changes continuously, while the slope of $y = -x$ is 1 for $x < 0$, -1 for $x > 0$ and undefined at $x = 0$.

Self-Test (pp. 70-71)

1. variance: $\frac{\sum_{i=1}^{14}(x_i - \bar{x})^2}{14} \approx \frac{\sum_{i=1}^{14}(x_i - 21.6)^2}{14} \approx 195.245$; standard deviation: $\sqrt{195.245} \approx 13.973$ **2.** $\frac{\sum_{i=1}^{14} x_i}{14}$ **3.** The median is the middle value when the data are ordered, halfway between 19 and 20, so is 19.5. The box goes from the first quartile to the third quartile. The first quartile Q_1 is the median of numbers less than the median, or 14. The third quartile Q_3 is the median of numbers greater than the median, or 31. The line in the box is at the median. The whiskers go from the least to greatest value, from 0 to 52. The result is the box plot below.

High School Football Team Scores

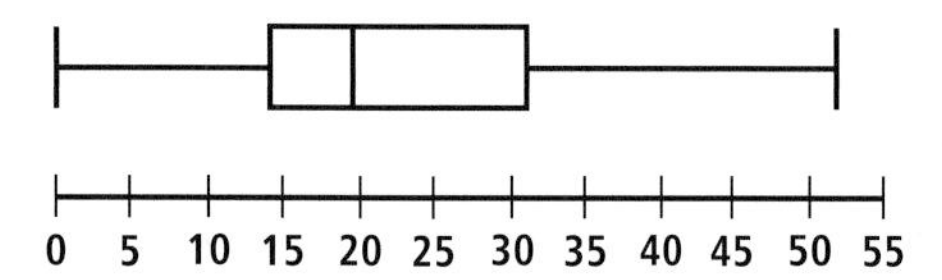

4. IQR $= Q_3 - Q_1 = 17$; $Q_1 - 1.5 \cdot 17 = -11.5$, $Q_3 + 1.5 \cdot 17 = 56.5$. No values in the data set are less than -11.5 or greater than 56.5, so there are no outliers. **5.** The first team would have more variability since its data have a larger standard deviation.

6.

Position	Frequency	Relative Frequency
Principals	13,340	0.025
Teachers	304,311	0.563
Aides	69,201	0.128
Office/Clerical	36,116	0.067
Counselors	6640	0.012
Librarians	1218	0.002
Other	109,381	0.202

7. True, because even if there are a few extreme data points in a set the median still is the central value.
8. a. The minimum 0 is indicated by the bar to the left. Q_1 = the location of the left side of the rectangle = 2; median = location of segment splitting the rectangle $\approx$ 4; Q_3 = location of the right side of the rectangle $\approx$ 9; maximum = location of rightmost dot $\approx$ 57. **b.** between Q_1 and Q_3, that is, between 2 and 9 inclusive **9. a.** The range is the maximum minus the minimum and therefore is not calculated using the mean. **b.** The variance is the mean of squared deviations from the mean; therefore, it is calculated using the mean. **c.** The IQR is not calculated using the mean. **d.** The standard deviation is calculated with the mean because it is the square root of the variance. **10. a.** The sum of the counts is 14. **b.** The distribution is skewed with a tail on the right. **11. a.** Subtract the actual 2006 receipts from the estimated 2007 receipts. Personal income taxes have the largest difference: 1,168.8 – 1,043.9 = 124.9 billion dollars. **b.** For each category, subtract the estimated 2007 receipts from the estimated 2008 receipts and divide by the amount of the 2007 receipts. The greatest percent increase is for excise taxes, $\frac{68.1 - 57.1}{57.1} = \frac{11}{57.1} \approx 19\%$.

12. a.

Month	Cumulative Snowfall (inches)
January	5.5
February	5.5 + 11.4 = 16.9
March	16.9 + 23.3 = 40.2
April	40.2 + 2.9 = 43.1
May	43.1 + 0.1 = 43.2
June	43.2
July	43.2
August	43.2
September	43.2
October	43.2 + 15.2 = 58.4
November	58.4 + 12.0 = 70.4
December	70.4 + 45.5 = 115.9

b.

Cumulative Average Snowfall in Boulder, Colorado

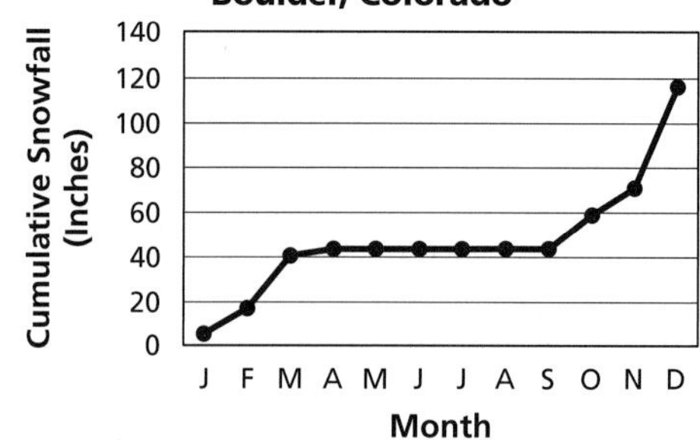

c. $\frac{\text{Sum of January to March}}{\text{total snowfall during year}} = \frac{40.2}{115.9} \approx 0.347$ or 34.7%

13. $\frac{(2 \cdot 3.3) + (4 \cdot 4) + (6 \cdot 2.3) + (4 \cdot 3)}{2 + 4 + 6 + 4} = 3.025$

14. Adding up the last two intervals, 15% of students scored at least 80 on the psychology test.

15. The median score for Psychology is higher than for Biology, indicating that Psychology scores are generally higher than Biology. Both distributions seem to be slightly skewed with a tail on the right, especially if the outlier in Psychology is thrown out. Lastly, the spread of the Biology scores is greater, as indicated by a greater IQR.

Self-Test Correlation Chart

Question	**1**	**2**	**3**	**4**	**5**	**6**	**7**	**8**	**9**	**10**
Objective(s)	A	C	I	E	E	B	D	I	D	H
Lesson(s)	1-6	1-2	1-4	1-4	1-6	1-2	1-2	1-4	1-6	1-3
Question	**11**	**12**	**13**	**14**	**15**					
Objective(s)	F	J	B	H	G					
Lesson(s)	1-1	1-5	1-2	1-3	1-7, 1-8					

Chapter Review (pp. 72-77)

1. a. 16 **b.** 15 **c.** 6 **3. a.** 26.5 years **b.** about 28.3 years **c.** 9 and 42 years **d.** 59 years **e.** about 19.1 years **5.** minimum = 1,045, Q_1 = 49,372, median = 56,194, Q_3 = 150,755.5, maximum = 571,951 (all answers in square miles) **7.** 35 cm per month **9.** 5.4 **11.** 89 **13. a.** $\frac{\sum_{i=1}^{8} g_i}{8}$ **b.** 18.125 **c.** 75 **15. a.** $\frac{\sum_{i=1}^{8} n_i p_i}{\sum_{i=1}^{8} n_i}$ **b.** about $5.63 **17. a.** false **b.** Answers vary. Sample: Consider the data set {1, 1, 2, 3, 4, 5, 6, 6}. The mean is 3.5 and modes 1 and 6. **19.** C **21.** C **23. a.** minimum: 48, Q_1: 55.5, median: 61.5, Q_3: 80, maximum: 94 **b.** 1.5 • IQR = 36.75; No values fall below 18.75 or above 116.75; therefore, there are no outliers. **25. a.** minimum: 97.9, Q_1: 127.1, median: 191.8, Q_3: 310.6, maximum: 630 (all answers in 1000's of dollars) **b.** Answers vary. Sample: There is a large range of home prices in U.S. metropolitan areas. **27. a.** mean: $188.4, median: $151.5, standard deviation: $138.16 **b.** 440, 445, 500 **c.** mean: $140.18, median: $140, standard deviation: $77.83 **29.** year, voting age population (millions), percent of voting age population reporting they voted, age, gender, education **31.** 67.25 million **33.** 65 years old and over **35.** Answers vary. Sample: The mean, median, IQR and standard deviation differ very little between the males and females. These statistics, as well as the box plots and histograms, indicate little to no difference in the running times of males and females.

37.

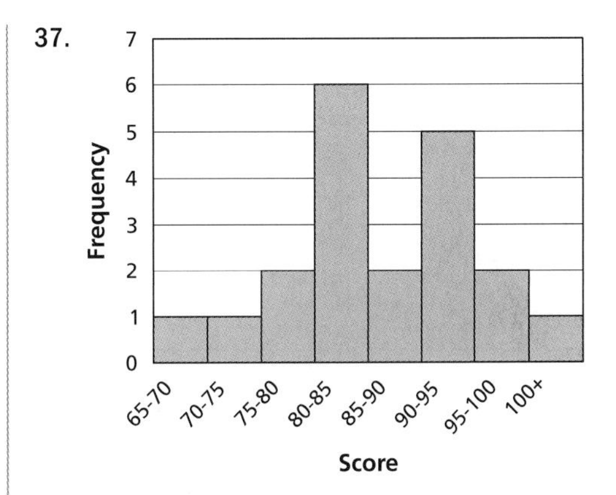

39. a. 42 **b.** false **c.** about 35.7%

41.

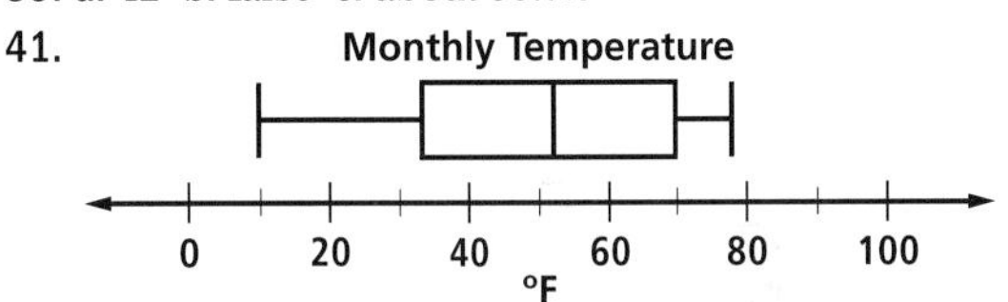

43. a. See table. **b.** month 4

Month	Cumulative Wages
1	$3250
2	$7375
3	$10,125
4	$15,500
5	$20,000
6	$23,800

c.

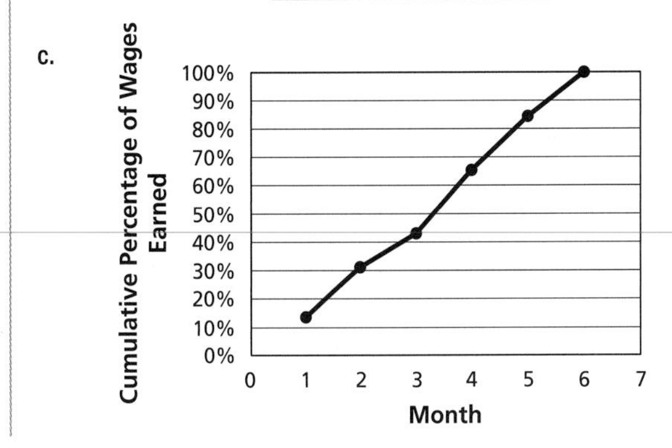

Chapter 2

Lesson 2-1 (pp. 80-86)

Guided Example 2 Answers vary. Sample:

n	Sample set (ordered)	Count *C* of numbers larger than the median
2	{2, 5}	1
3	{24, 46, 68}	1
4	{13, 24, 35, 47}	2
5	{1, 2, 3, 4, 5}	2
6	{1, 2, 4, 6, 7, 9}	3
7	{2, 5, 6, 8, 9, 10, 15}	3
8	{1, 3, 4, 5, 7, 9, 10, 12}	4
9	{3, 4, 5, 7, 8, 10, 11, 12, 20}	4
10	{1, 2, 5, 6, 8, 11, 13, 17, 21, 30}	5

2; 2; 3; 3; 4; 4; 5

Questions

1. independent variable: number of chores completed; dependent variable: allowance earned **3.** independent variable: sunlight and water; dependent variable: tree growth **5.** The range is the set of all possible values for the dependent variable. **7. a.** f **b.** y **c.** $f(x)$ **d.** x **9. a.** $\sqrt{q+2}$ **b.** 7 **11.** function **13. a.** domain: $\{x \mid -2 \leq x \leq 5.9\}$ range: $\{y \mid -2.5 \leq y \leq 2.6\}$ **b.** -0.4 **c.** $x = 0$ or $x = 4$

15.

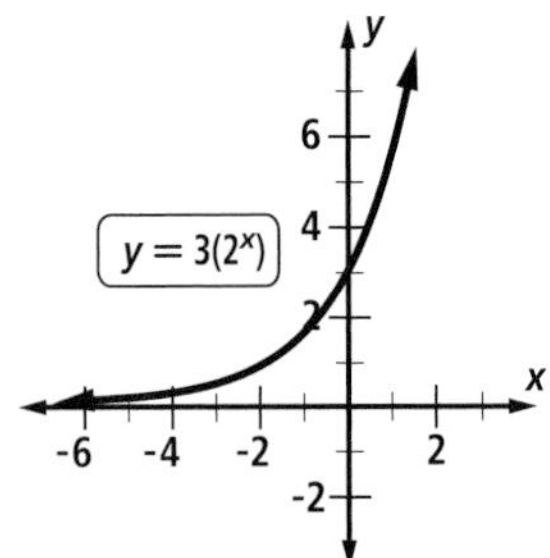

Yes, this relation passes the vertical line test and is therefore a function. domain: $\{x \mid x \in \mathbb{R}\}$; range: $\{y \mid y > 0\}$

17.

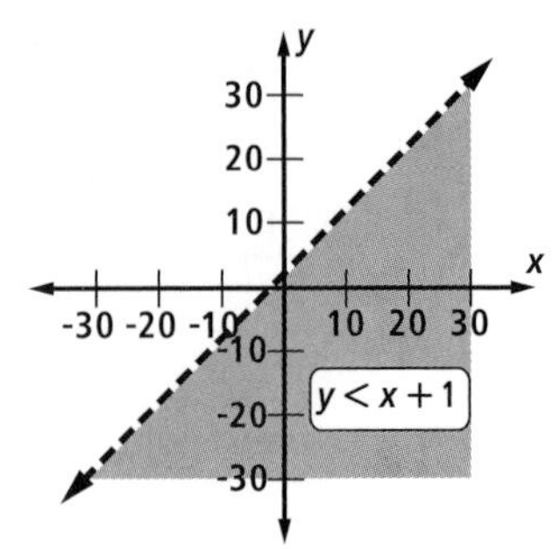

No, this relation does not pass the vertical line test. For example, there are infinitely many y–values that satisfy the expression for $x = 0$. **19. a.** about \$14,758.33 **b.** mean **21. a.** 36 **b.** $10x$ **c.** $x + 11$ **d.** $x + a$ **23.** $y - 4 = -\frac{3}{2}(x + 6)$

Lesson 2-2 (pp. 87-93)

Guided Example 1 a. $\frac{1400 - 600}{0.32 - 0.18}$; 5714.29; 600; 0.18; $5714.29x - 428.57$ **b.** 5714.29 **c.** \$1285.72 \$1285.78 **d.** -\$142.86

Questions

1. A *linear function* is a set of ordered pairs (x, y) which can be described by an equation of the form $y = mx + b$, where m and b are constants. **3.** observed; predicted **5. a.** \$1,200 **b.** \$90 **7.** "first square each residual and then sum the squares" **9. a.** 0.55; The number of states visited increases by 0.55 for every one year increase in age.

b.

◇	A	B	C	D
1	Age	States Visited	Predicted	Residual^2
2	16	15	12.3	7.29
3	16	18	12.3	32.49
4	40	27	25.5	2.25
5	70	42	42.0	0
6	10	9	9.0	0
7	45	19	28.25	85.56

c. 127.59 **d.** interpolation **e.** 20 states

11a and b. This line appears to model the data fairly well.

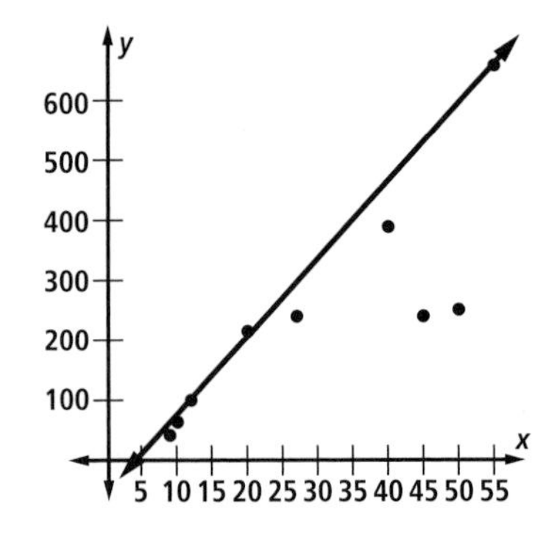

c. $y = \frac{560}{43}x - \frac{2420}{43}$

d. The average gestation period increases by about 13.0 days for every increase of one year in an animal's expected life span. e. –55.35 f. about 986 days; extrapolation g. about 211,453 13. a. about 10.59 b. Because there are only five data points and the range is 29, the spread is large.

Lesson 2-3 (pp. 94-101)

Questions

1. least squares line, line of best fit 3. A 5. C 7. This does not affect r because swapping the data does not affect its degree of correlation. 9. $(\bar{x}, \bar{y})$ 11. $r \approx \pm 0.6325$ 13. Answers vary. Sample: $\{(1, 4), (3, 4), (4, 4), (5, 4), (5.5, 4), (6, 4), (9, 4)\}$. The correlation is undefined. This is because in the equation for computing the correlation coefficient, s_y appears in the denominator, and for the chosen data set $s_y = 0$ and division by 0 is undefined. 15. zero; There is no correlation between a person's height and how far he/she lives from school. 17. $r = 1$ 19. No; both likely correlate with Florida's increasing population, but an increase in the rate of sales tax would not cause an increase in shark attacks. 21. $y = x$ 23. $t = 3 \pm \sqrt{15}$ 25. a. $5t^2$ b. $4y$ c. $25r^2$

Lesson 2-4 (pp. 102-109)

Guided Example 2 a. 1.053 b. 2500; 1.053; $2500(1.053)^t$ c. 1.053; \$3588.71 d. 1.053; 2254.67; \$2254.67

Guided Example 3 a. all real numbers; positive real numbers b. 8 c. $y = 0$ d. increasing

Questions

1. a. 4,203,000 b. $P(n) = 4{,}156{,}119(1.0113)^n$ c. 4,756,000 3. a. $A = 4000(1.08)^t$ b. about \$118,223.89 c. about \$2,940.12 5. a. false b. decay 7. a. g b. f

c.

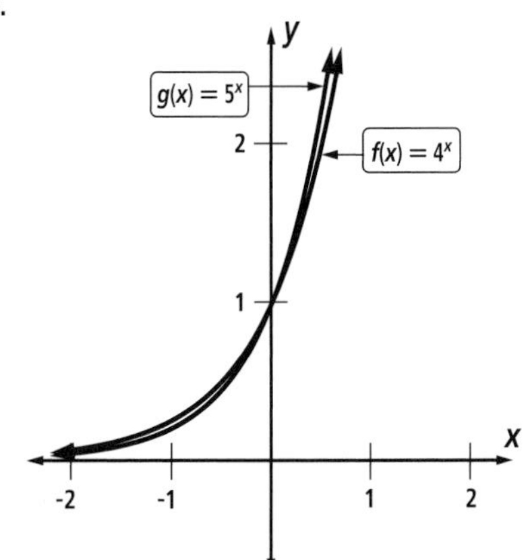

9. a. B b. $y = 12$ when $x = 0$ is the initial value. The growth factor is $\frac{1}{2}$ because as x increases by 1, y is divided by 2. 11. a. 18, 16.2, and 14.53 cubic feet b. $20(0.9)^n$ c. false 13. False, correlation does not mean causation. For instance, both a and b may be causally related to a third variable c, and thus correlate to each other without being directly related. 15. a. –1 b. 2 c. False; $f(1) + f(2) = -1 + 0 = -1$, but $f(1 + 2) = f(3) = \frac{3}{5}$ d. False; the domain is $\{x| x \neq -2\}$, while the range is $\{y| y \neq 3\}$ e. $\frac{3p - 15}{p - 1}$ 17. $r = \pm\frac{20}{27}, s = \pm 3$

Lesson 2-5 (pp. 110-116)

Guided Example 2 a. $296.177(1.079)^x$

b.

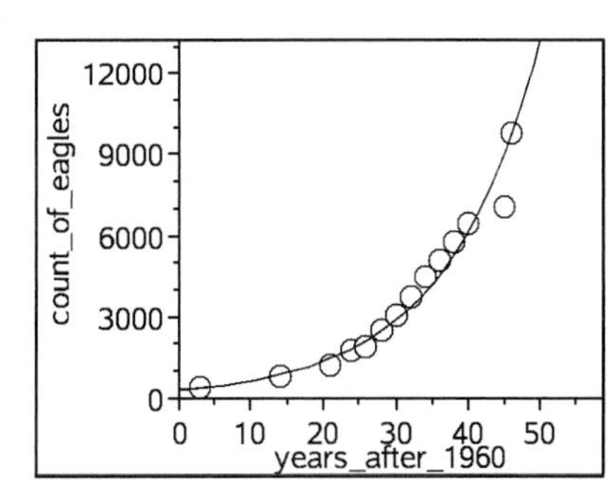

c. 296; 1960; 1.079; 7.9 d. 6200; 6471; 6200; 271; 9068; 7066; 9068; -2002

Questions

1. a. $\begin{cases} 20 = ab^3 \\ 156 = ab^{10} \end{cases}$ b. $f(t) = 8.293(1.341)^t$ c. $8.293 \cdot 1.341^3 \approx 20$, $8.293 \cdot 1.341^{10} \approx 156$ 3. a. $y = 47.979(1.372)^x$ b. about 48 c. 233 5. a. $y = 10(0.766)^x$ b. about 0.000001 units 7. a. $a \approx 736$, $b \approx 1.398$; $y = 736(1.398)^t$

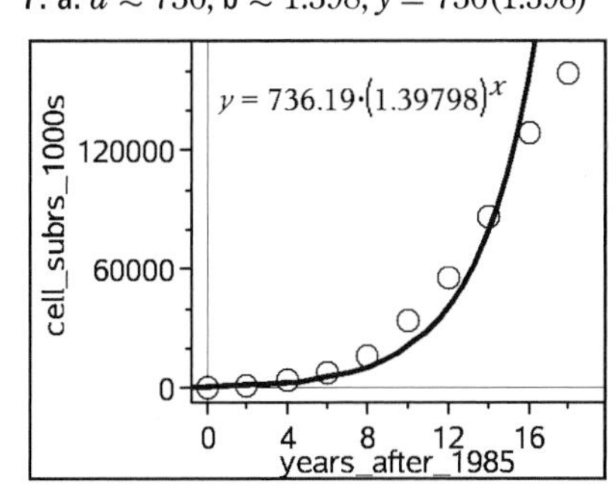

b. The model increases much more quickly than the data in the years after 1997, but otherwise fits well. c. 1985

9. a.

Number of Half-Lives	t	$f(t)$
0	0	3
1	1620	1.5
2	3240	0.75
3	4860	0.375

b. $y = 3(0.9995722)^x$ c. about 542 mg 11. a. g b. f 13. x-axis 15. Answers vary. Sample: $y = -3^x$ domain: the set of all real numbers 17. a. about 30,400 thousand barrels per day **b.** about 197.3 thousand barrels per day **c.** about 68.2% **d.** The U.S. petroleum consumption is an extreme data point.

Lesson 2-6 (pp. 117-124)

Guided Example 1 **a.** -2; 3; -3; 2; -2; 2; 3; 25; 4 **b.** positive; $\frac{3}{4}$; $\frac{3}{4}$; $\frac{3}{4}$; $\frac{3}{4}$; 3; 4
Guided Example 2 **a.** 20; 15; 9.8; 20 **b.** 30.9; 30.9 **c.** 0; –0.65; 4.73; 4.73

Questions

1. $f(x) = ax^2 + bx + c$ **3.** $\{y| \, y \geq -3.125\}$
5. a. minimum **b.** minimum **c.** maximum **d.** maximum
7. $3(16) - 20(4) + 8 = -24$ **9. a.** 21.45% **b.** interpolation
11. a. $f(n) = 0.5n^2 + 0.5n + 1$ **b.** $f(5) = 16$

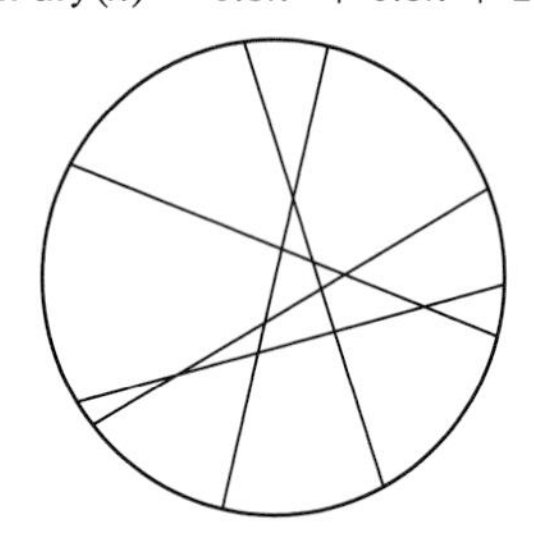

13. a.

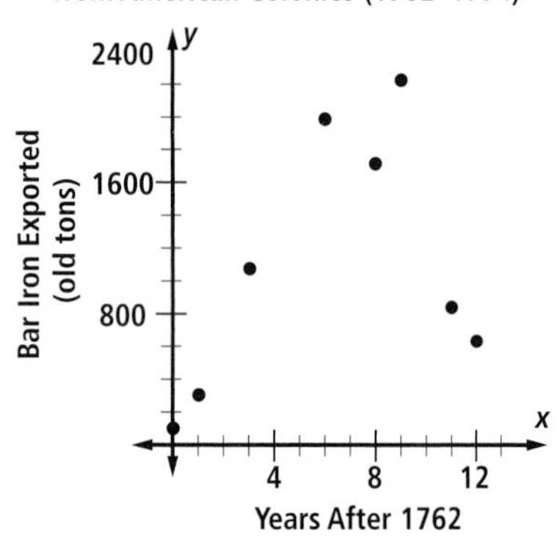

b. $f(x) = -46.393x^2 + 620.48x - 120.6$; x is the number of years after 1762 **c.** $f(14) = -526.9$ **d.** The negative value indicates that the colonies were importing bar iron from England in 1776. However, in 1776 the colonies declared war on England and active trade between the colonies and England stopped. **15. a.** $f(t) = 100(0.952)^t$ **b.** $f(13.7) \approx 50.87$ g **17.** about 23,953,000 **19. a.** $x = 2$ or $x = 3$ **b.** $x = -1$ or $x = \frac{1}{3}$ **c.** $x = \pm\frac{\sqrt{3}}{3}$

Lesson 2-7 (pp. 125-131)

Guided Example 1 **a.** 60; 84.6; 60; $\frac{5076}{V}$ **c.** decreasing **d.** 126.9

Questions

1. C **3. a.** $k = 450$ **b.** 225 **5.** an inverse-square variation **7. a.** $k = 28{,}704$ kPa • mL **b.** $P = \frac{28{,}704}{V}$

c.

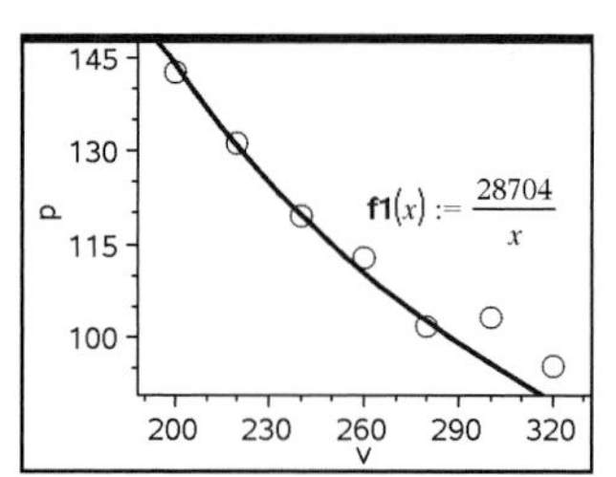

d. about 97.52 **9. a.** about $3.987 \cdot 10^8 \frac{\text{m}^3}{\text{s}^2}$ **b.** about $1.49 \frac{\text{m}}{\text{s}^2}$ **c.** about $0.00270 \frac{\text{m}}{\text{s}^2}$ **11. a.** $y = 0.021x^2 - 1.2x + 511$ **b.** sum of squared residuals ≈ 78.2969

13. a. interpolation **b.** The 2016 prediction would probably have a larger error since it is a result of extrapolation.
15. a. median **b.** median

Lesson 2-8 (pp. 132-139)

Questions

1. The residuals are clustered around the x-axis.

3. a.

Days (d)	Predicted Day Length (p)
0	792.63
10	763.88
20	736.17
30	709.47
40	683.73
50	658.93
60	635.03
70	611.99
80	589.79
90	568.40
100	547.78

b.

Days (d)	Predicted Day Length (p)	Computed Day Length (L)	Residual
0	792.63	793	0.37
10	763.88	766	2.12
20	736.17	739	2.83
30	709.47	711	1.53
40	683.73	684	0.27
50	658.93	657	-1.93
60	635.03	631	-4.03
70	611.99	607	-4.99
80	589.79	586	-3.79
90	568.40	568	-0.40
100	547.78	556	8.22

c. The exponential model is not a good fit for this data because there is a distinct pattern in the residual plot.

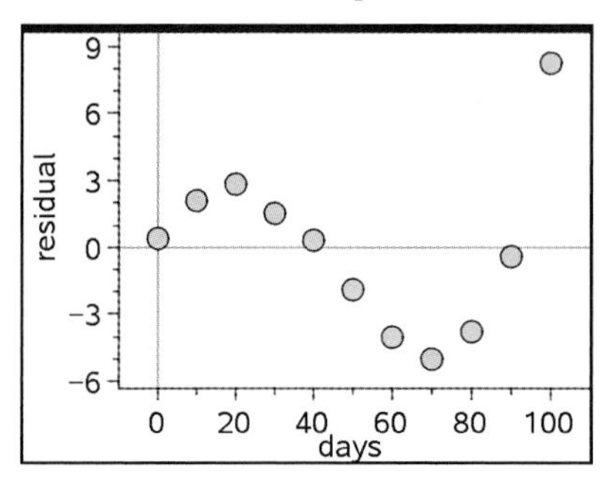

5 a. $y = (3.99866)1.0292^x$

b.

◇	A	B	C	D	E
1	Year (Y)	Year After 1790	Population (m)	Predicted Populaton (p)	Residual (R = m-p)
2	1790	0	4	4.00	0.00
3	1800	10	5	5.33	-0.33
4	1810	20	7	7.11	-0.11
5	1820	30	10	9.48	0.52
6	1830	40	13	12.64	0.36
7	1840	50	17	16.86	0.14
8	1850	60	23	22.49	0.51
9	1860	70	31	29.98	1.02
10	1870	80	40	39.98	0.02
11	1880	90	50	53.32	-3.32

c.

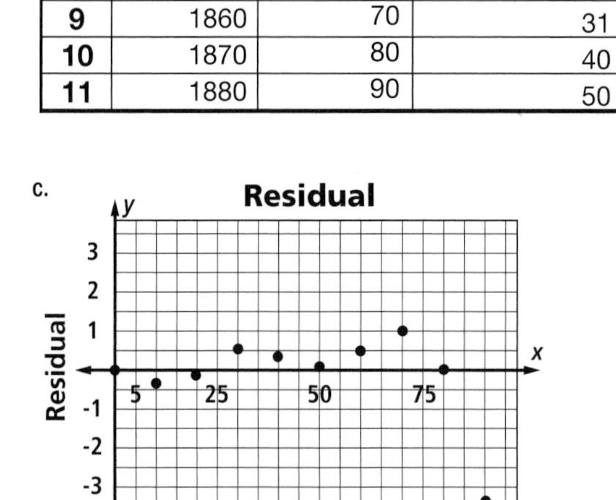

d. It seems to be appropriate for the time period because the residuals are scattered closely about the x-axis. However, the data for 1880 diverge unexpectedly, so this model probably should not be used to extrapolate into the future. 7. Answers vary. Sample: The prices of stocks fluctuate based on many factors that cannot be accurately forecasted or predicted. 9. a. Answers vary. Sample: The amount of driving and motor vehicle ownership generally depends on the population, so theoretically an exponential model might be a good fit. However, because there seems to be no immediate relationship behind time and miles driven in motor vehicles, this would most likely be considered impressionistic and therefore have no theoretical basis. b. Answers vary. Sample: Eyeballing the scatterplot, it seems that any of the three standard models (linear, exponential, or quadratic) might fit the data.

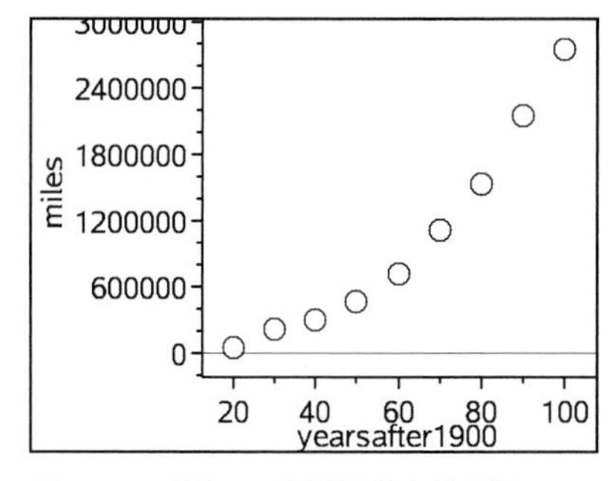

c. Exponential: $y = 38720(1.04667)^x$

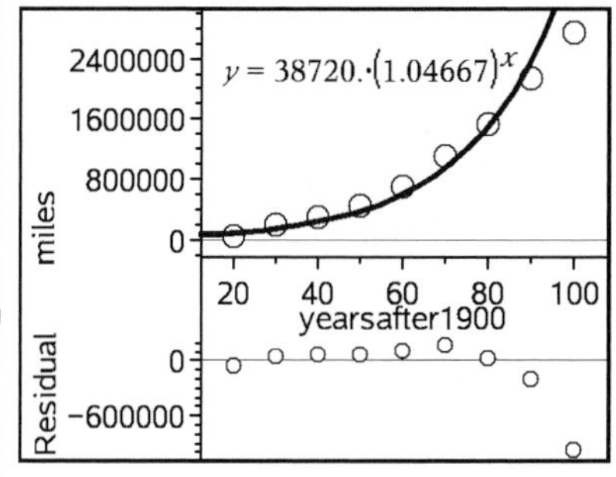

Linear: $y = 32855.2x - 942265$

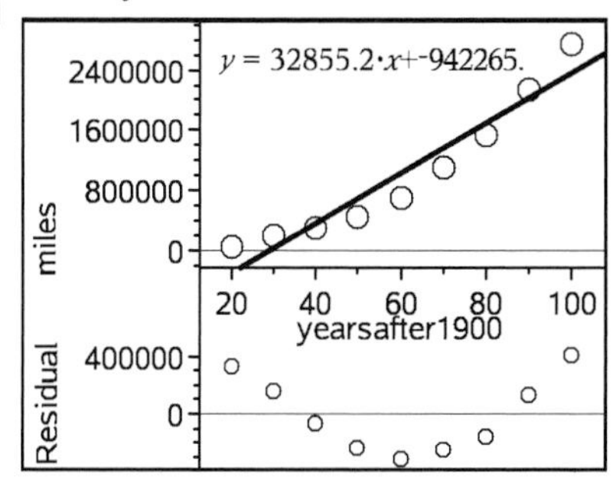

d. Quadratic: $y = 422.455x^2 - 17839.5x + 296938$

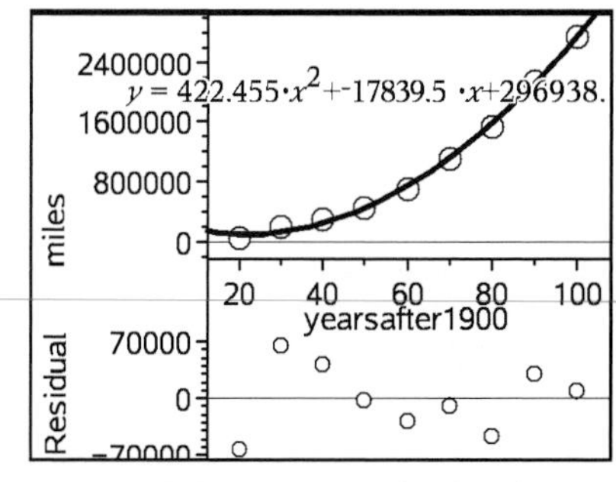

The quadratic regression model seems to fit the data best, and the residuals all lie in a relatively small range.

11. 320 N

13. a.

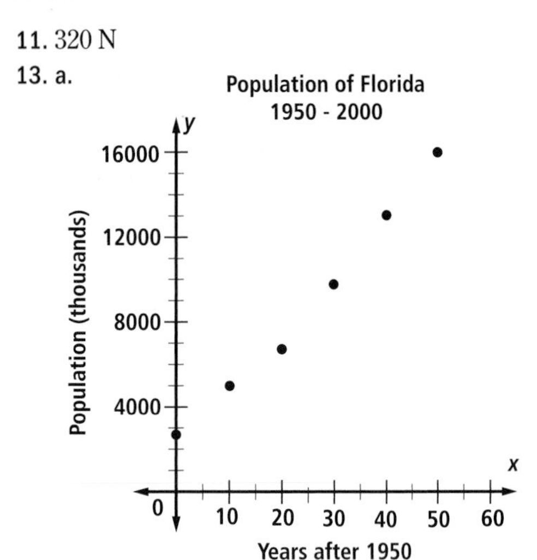

b. $y = 1.736x^2 + 178.813x + 2801.61$ **c.** about 21,760 thousand people

Self-Test pages 144-145

1. a. $f(2) = -9.8(2)^2 - 5.2(2) + 12.7 = -39.2 - 10.4 + 12.7 = -36.9$ **b.** $g(2n) = 5(2n)^2 - 3 = 20n^2 - 3$ **2.** It seems that any number can be an argument for this function, so domain: $\{x|x \in \mathbb{R}\}$. The value of the function is never less than or equal to 1, so range: $\{y|y > 1\}$. **3.** x can be any real number except 0 so domain: $\{x|x \neq 0\}$; $-\frac{2}{x}$ can be any real number except 0 so range: $\{y|y \neq 0\}$. **4. a.** D and E; the points with greater x-values have or tend to have smaller y-values. **b.** C and F; C has no pattern to the data points and F is symmetric about a vertical line. **c.** A and D; all of the points are essentially on a line. **5.** residual = observed – predicted, so observed $= 17{,}000 + 3{,}424{,}000(1.013)^{15} \approx$ 4,172,999 people. **6.** The larger the sum of the squared residuals, the farther the data points are from the line. Because Stephen's line has a smaller sum of squared residuals it is a better fit for the data. **7. a.** Substituting $h_0 = 200$, $v_0 = 44$, and $g = 32$, $h = -16t^2 + 44t + 200$. **b.** Use the Quadratic Formula on the equation $-16t^2 + 44t + 200 = 0$. $t = \frac{-44 \pm \sqrt{44^2 - 4(-16)(200)}}{2(-16)} \approx 1.375 \pm 3.793 \approx$ 5.168 seconds **8. a.** The form is $y = \frac{k}{x}$ and $60 = \frac{k}{10}$, so $k = 600$. **b.** Substitute $x = 3$. $y = \frac{600}{3} = 200$ **9. a.** The form is $y = \frac{k}{x^2}$ and $60 = \frac{k}{100}$, so $k = 6000$ **b.** Substitute $x = 3$. $y = \frac{6000}{3^2} = \frac{2000}{3}$ **10.** The growth factor is greater than 1 (and a in $a(b)^x$ is positive), so the function models exponential growth. **11.** Use 115 for x and evaluate. $y = 2.51(115) + 471.1 = \$759.75$ million; The residual is observed – predicted $= 524 - 759.75 = -\$235.75$ million. **12.** Find the growth rate by using the half-life. $\frac{a}{2} = ab^{30}$ so $b \approx 0.977$. The initial value a is 10 so the equation is $y = 10(0.977)^x$. When $x = 20$, $y \approx 6.3$ g, so 6.3 g will be left after 20 years. **13.** By the graphs of the residuals, the linear model seems to be a poor fit because the residuals are far from the x-axis. The inverse-square model seems to fit the data very well because the residuals are clustered near the x-axis and don't seem to make a pattern. Therefore, the inverse-square model is more appropriate. **14. a.** Quadratic regression on a calculator with t as the independent variable and h as the dependent variable shows $h = -15.786t^2 + 52.5t + 261$. **b.** This is a theory-based model because it is based on the theory of gravity. **c.** Substitute 4 for t and evaluate. $h = -15.786(4)^2 + 52.5(4) + 261 \approx 218.4$ feet **15. a.** To find the line of best fit, use linear regression on a calculator. The slope, -0.265 means that the percent of 18-24 year-olds who vote has decreased by about 0.265% per year. **b.** The correlation coefficient, -0.905, indicates that there is a strong negative correlation between the year and the percent of young people who vote.

16. a. Answers vary. Sample: A quadratic model looks appropriate for the data.

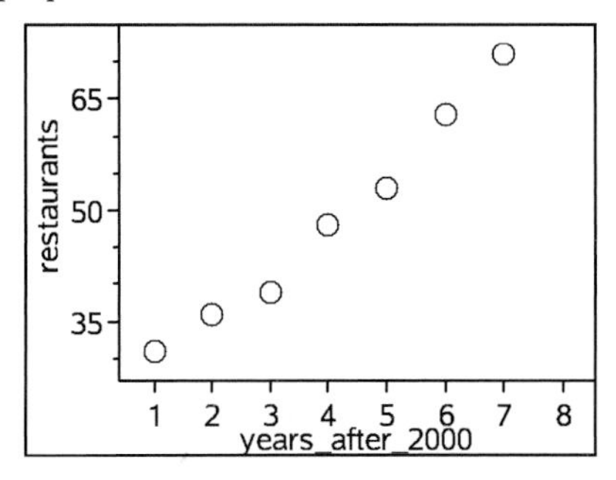

b. Linear: $y = 6.71429x + 21.8571$

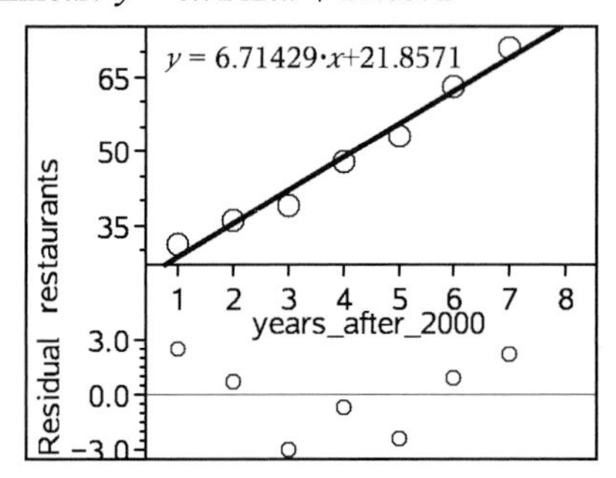

with correlation coefficient of about 0.99

Quadratic: $y = 0.5x^2 + 2.71429x + 27.8571$

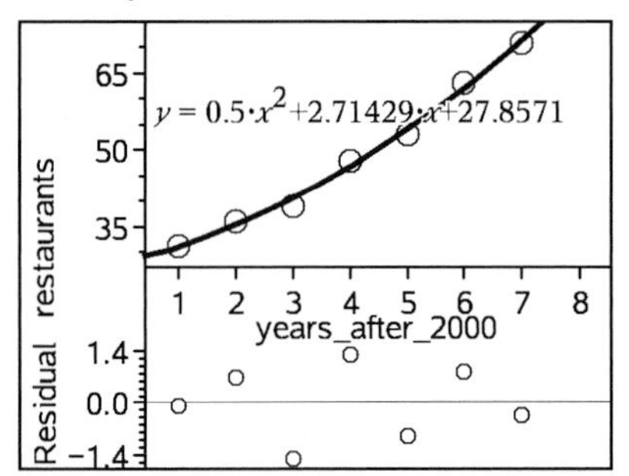

Exponential: $y = 26.7905(1.14995)^x$

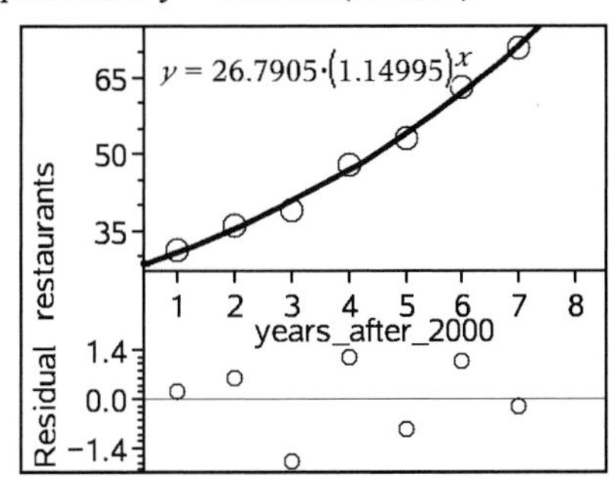

c. Answers vary. Sample: Neither theory nor an eyeball of the residual plot lead to any particular model. The linear residual plot has an almost parabolic pattern with larger residuals than the quadratic and exponential regressions. Both the quadratic and exponential regressions have very similar residual plots, so either model could be used.

Self-Test Correlation Chart

Question	**1**	**2**	**3**	**4**	**5**	**6**	**7**	**8**
Objective(s)	A	H	C	I	B	B	F	E
Lesson(s)	2-1	2-1	2-1	2-3	2-2	2-2	2-6	2-7
Question	**9**	**10**	**11**	**12**	**13**	**14**	**15**	**16**
Objective(s)	E	E	B	F	I	F	D	G
Lesson(s)	2-7	2-4	2-2	2-5	2-8	2-6	2-3	2-8

Chapter Review (pp. 146–149)

Questions

1. a. 4 **b.** $\frac{1}{4}$ **c.** 64 **3. a.** $x = 1$ **b.** no, $(4^3 + 2) - (2^3 + 2) = 56 \neq 2^3 + 2 = 10$

5. a.

n	$F(n)$	model	residuals
1	1	-4.82	5.82
2	1	-1.17	2.17
3	2	2.48	-0.48
4	3	6.13	-3.13
5	5	9.78	-4.78
6	8	13.43	-5.43
7	13	17.08	-4.08
8	21	20.73	0.27
9	34	24.38	9.62

b. 937.07 **c.** 210.206

7. a.

n	$F(n)$	model	residuals
1	1	2.521	-1.521
2	1	0.664	0.336
3	2	0.381	1.619
4	3	1.672	1.328
5	5	4.537	0.463
6	8	8.976	-0.976
7	13	14.989	-1.989
8	21	22.576	-1.576
9	34	31.737	2.263

b. 610 **9. a.** $\{t \mid t \in \mathbb{R}\}$ **b.** $\{y \mid y > 0\}$ **11. a.** $\{d \mid d \neq 0\}$ **b.** $\{I \mid I > 0\}$ **13.** t; y **15.** true **17.** true **19.** $r = 1$ **21.** Most of the observed data fall close to a linear model with positive slope; this indicates that the two variables have a strong positive relationship. **23.** a models exponential growth, b models exponential decay **25. a.** I, III **b.** II, IV **27.** $f(x)$ decreases **29.** 400 **31. a.** $y = 0.335(6.108)^x$ **b.** -2.34 **33. a.** $s = 0.5h - 24.5$ **b.** 2 **35.** 225.4 mg **37.** exponential; Population growth is often described by exponential growth equations. **39.** linear; Since peoples' body parts are proportional to each other, their measurements can be described by a linear model. **41. a.** Answers vary. Sample: Cell division splits or doubles the amount of cells after each division. Exponential growth models describe this type of process. **b.** The scatterplot appears to be exponentially growing. The rate of change of bacteria per hour is increasing.

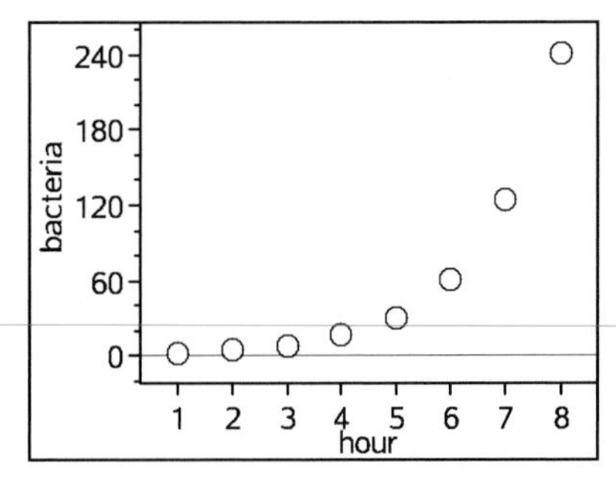

c. Exponential: $y = 1.01688(1.98128)^x$; The residual plot supports the choices made in Parts a and b.

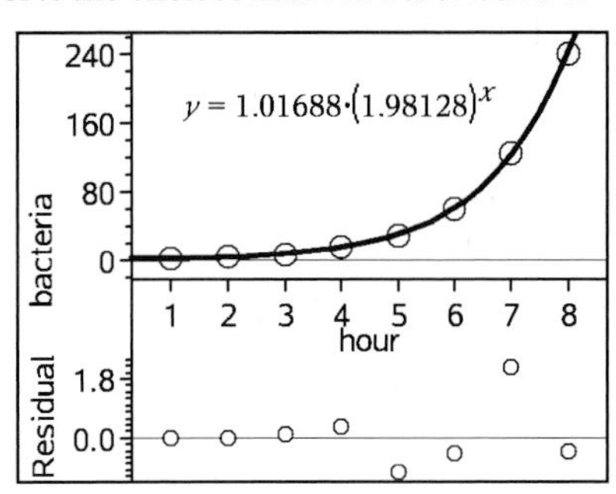

43. function; $\{x \mid x \in \mathbb{R}\}$; $\{y \mid y \leq 0.5\}$ **45.** function; $\{x \mid x \in \mathbb{R}\}$; $\{y \mid y \leq 3\}$ **47. a.** 83% **b.** 82% **c.** 1% **49.** about 8 or 9 hours; Answers vary. Sample: This might make sense because a student who studies excessively for a test may negatively affect his performance through stress or lack of sleep. **51.** positive **53.** negative

Chapter 3

Lesson 3-1 (pp. 152-158)

Questions

1. linear, cubic, inverse variation **3.** quadratic, inverse square, absolute value **5.** domain: $\{x | x \in \mathbb{R}\}$; range: $\{y | y \geq 12\}$ **7. a.**

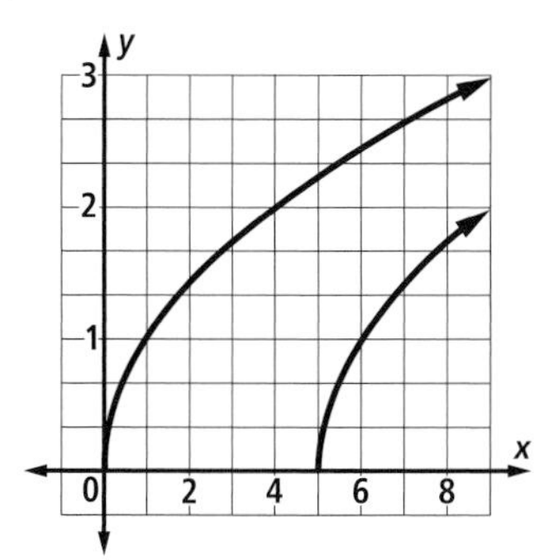

b. The graph of $y = \sqrt{x - 5}$ is the image of the graph of $y = \sqrt{x}$ under the translation 5 units to the right.
9. Answers vary. Sample: $-10 \leq x \leq 10$, $-10 \leq y \leq 10$
11.a.

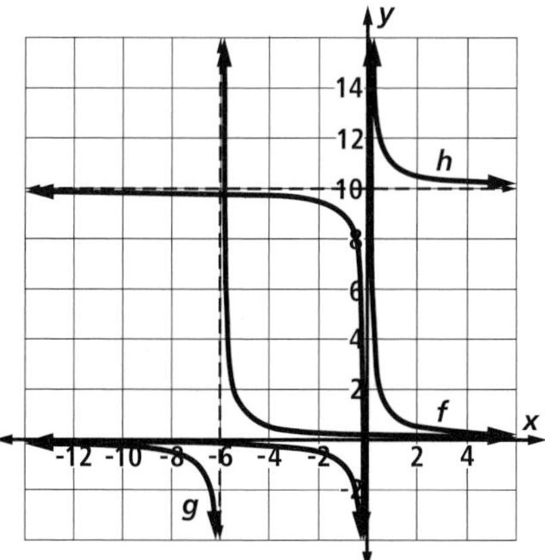

b. $x = 0$, -6, and 0 for f, g, and h, respectively **c.** $x = 0$, $x = -6$, and $x = 0$ for f, g, and h, respectively **d.** g is a translation of f 6 units to the left and h is a translation of f 10 units up. **13. a.** domain: $\{x | x \leq -20 \text{ or } x \geq -10\}$
b.

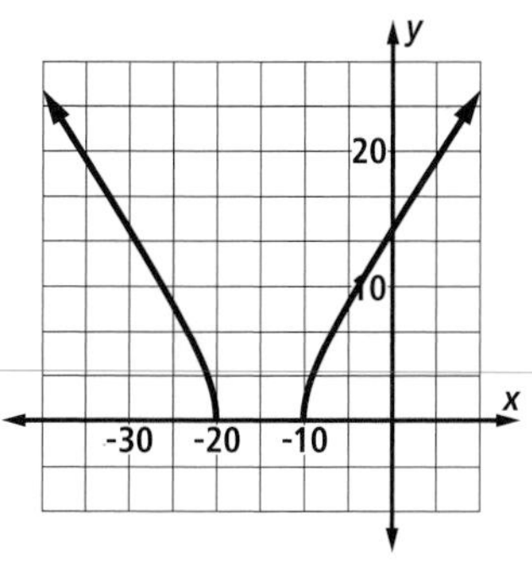

15. a.

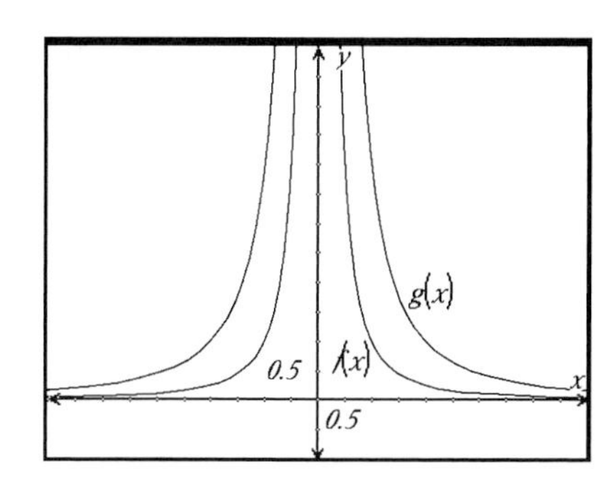

b. true **c.** The y values of g are greater than those of f in the graph above. Since $x^2 > 0$, when $x \neq 0$ and $1 < 4$, then $\frac{1}{x^2} < \frac{4}{x^2}$. So, $f(x) < g(x)$, when $x \neq 0$. **17.** not a function **19.** function
21.

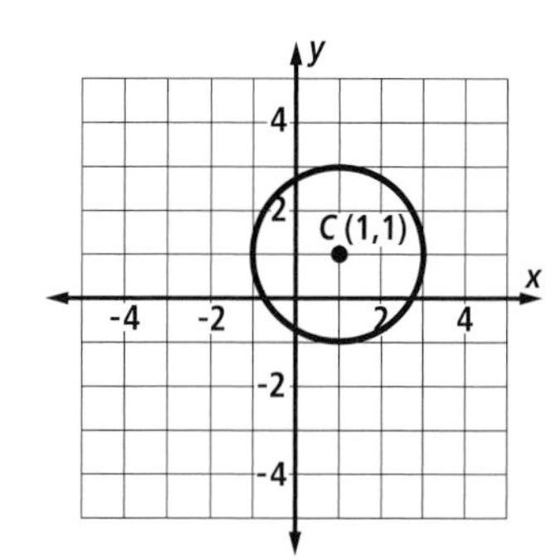

23. a.

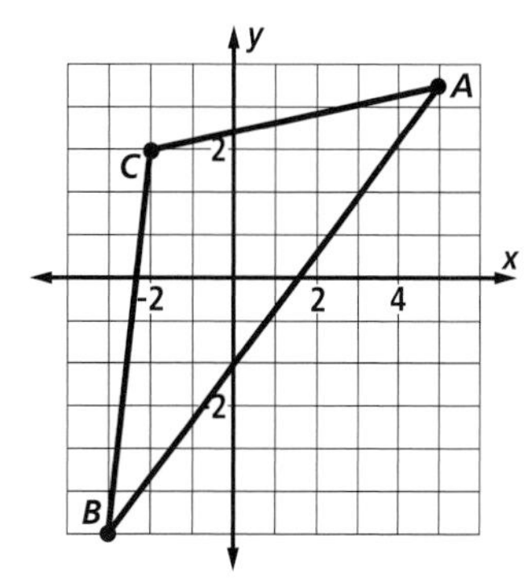

b.

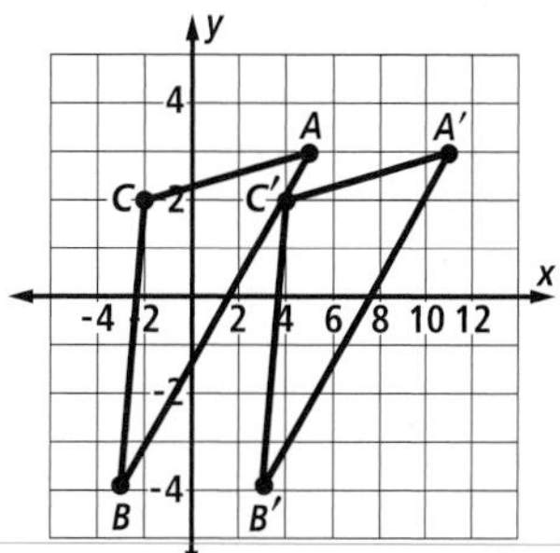

c. $\triangle A'B'C'$ is the image of $\triangle ABC$ under the translation of 6 units to the right.

Lesson 3-2 (pp. 159–164)

Guided Example 2 $y + 4; x - 5; y + 4 = \sqrt{25 - (x - 5)^2}; \sqrt{-x^2 + 10x - 4}; 2; 0; 0 + 4 = \sqrt{-(2^2) + 10 \cdot 2}; 4 = \sqrt{16}; 4 = 4$

Questions

1. (-1, 1) **3.** $(p + a, q + b)$ **5.** $(r - 3, s + 4)$ **7. a.** (4, 5), (2, 3), (3.5, 6) **b.** $5 - 4 = 1 = \frac{1}{4-3}$; $3 - 4 = -1 = \frac{1}{2-3}$; $6 - 4 = 2 = \frac{1}{3.5-3}$ **9.** $y = |x + 4| + 6$ **11.a.** $T(x, y) = (x - 2, y + 1)$ **b.**

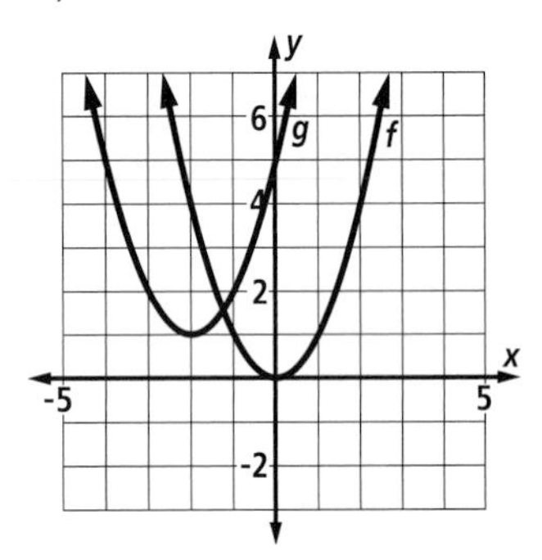

13. $h = 5, k = -2$ **15. a.** about $8,288,000,000,000 **b.** $N = 6.36 \cdot 10^{-72}(1.09)^t$ **17. a.** $y = 1.621x + 6.413$ where y = length and x = age. **b.** For every year that a largemouth bass ages, it grows 1.621 inches. **c.** about 25.87 inches **d.** Answers vary. Sample: Extrapolation may not be appropriate in this context, because largemouth bass may stop growing in length when they are 10 years old. **e.** 0.97 **f.** Age and length are positively correlated for the largemouth bass.

Lesson 3-3 (pp. 165–171)

Guided Example

3 oz; 3 oz; 3 lb 5 oz; 3 oz; 3 oz; 3 lb 3.4 oz; 8 oz; 1 lb

Questions

1. translation **3. a. i.** $70 **ii.** $100 **iii.** $100 **iv.** $90 **v.** $700 **vi.** ≈$26.46 **b.i.** $70 **ii.** $101.50 **iii.** $101.50 **iv.** $91.50 **v.** $700 **vi.** ≈$26.46 **5. a.** $\bar{x} + b$ **b.** s

7.

	Original Data	Transformed Data
cases	10	10
mean	63.5	53
sd	8.03	8.03
median	70	59.5
range	23	23
IQR	12	12

9. a.

score1	score2
84	79
84	81
66	61
98	98
70	67
86	83

A score1

b.

	$\bar{x}$	s
score 1	80.33	12.34
score 2	77.93	16.10

c.

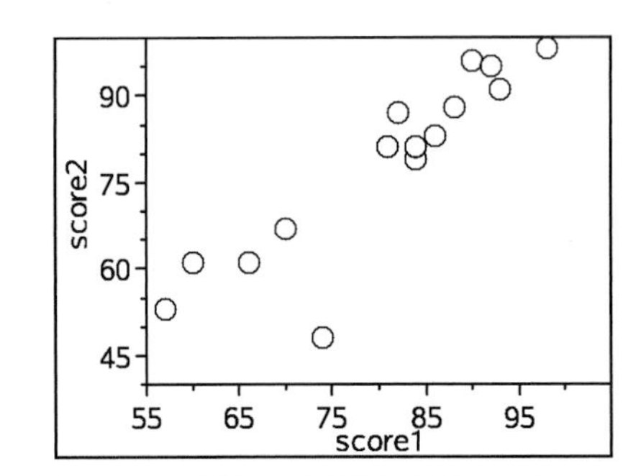

d. 0.899 **e.** $y = 1.173x - 16.333$ **f.** The points are shifted on the x-axis 5 units to the right.

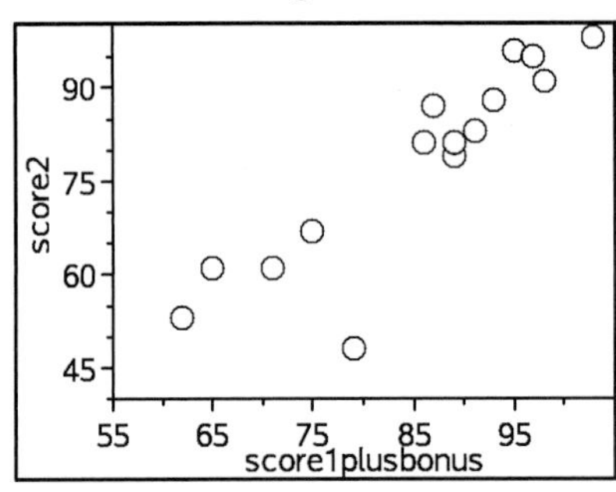

g. $r = 0.899$, $y = 1.173x - 22.2$; The correlation coefficient and slope of the regression line are invariant; the y-intercept changed. **h.** For any point (x, y) in Part c, the point is translated to $(x + 5, y - 3)$. For the point (x, y) in Part f, the point is translated to $(x, y - 3)$.

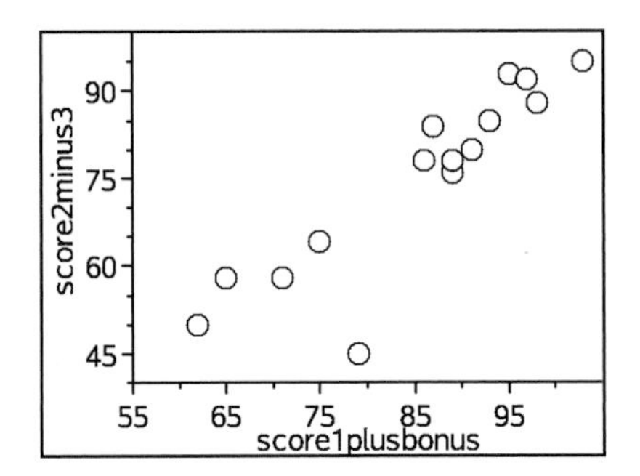

i. $r = 0.899$, $y = 1.173x - 25.2$ j. The correlation coefficient and slope are invariant, the y-intercept changed. 11. 6′3″

13. a. and b.

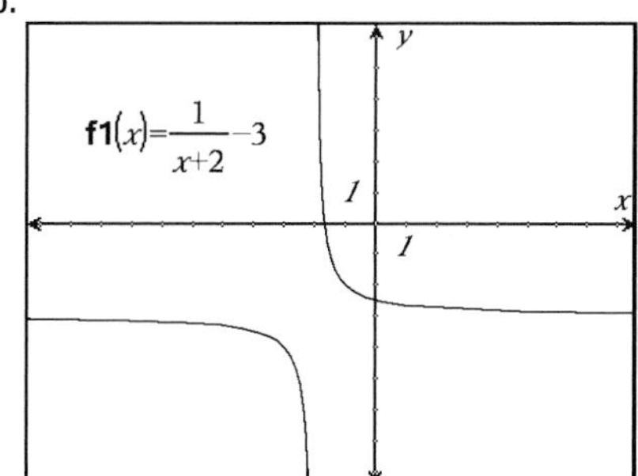

c. $T(x, y) = (x - 2, y - 3)$ 15. $\{x \mid x \geq -99\}$

17. a.

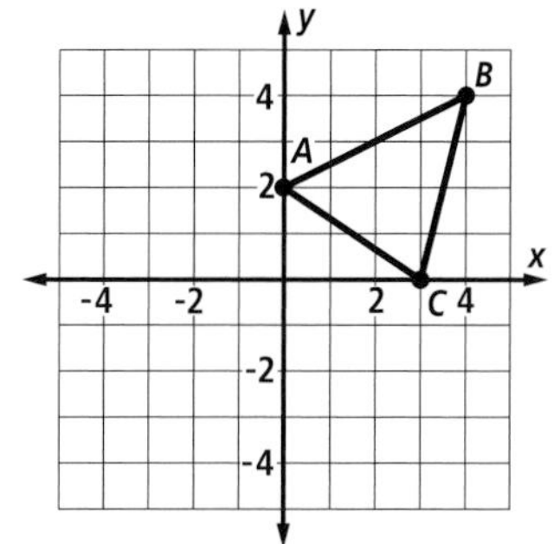

b.

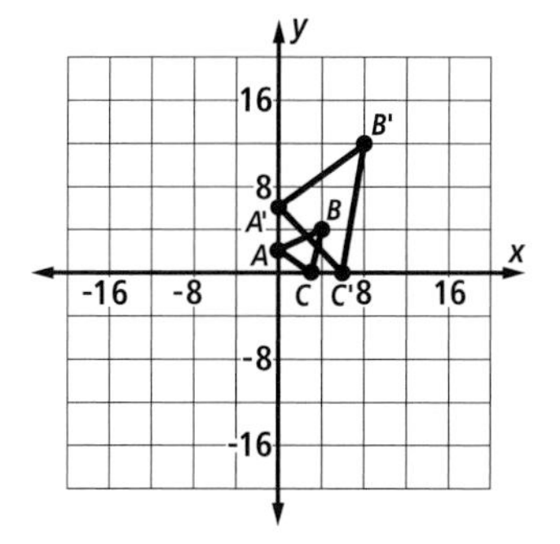

c. the scale change $S_{2,3}$

Lesson 3-4 (pp. 172–178)

Questions

1. a.

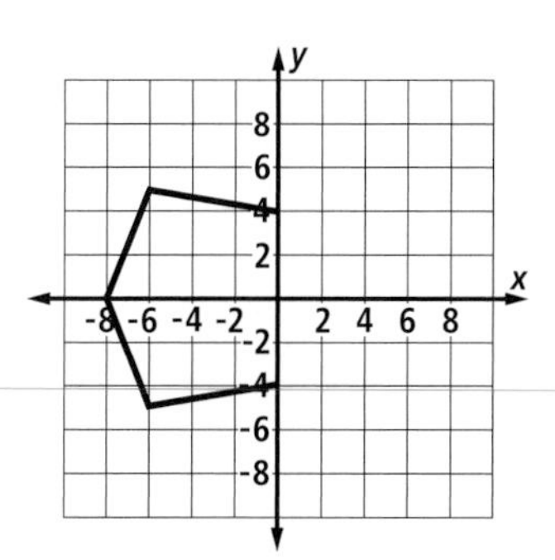

b. (-6, -5) 3. (-4, -2) 5. a. $-9 \neq 3|3|$; the graph is not symmetric with respect to the y-axis. b. $9 \neq -3|-3|$; the graph is not symmetric with respect to the x-axis. c. $9 = 3|3|$; the graph could be symmetric with respect to the origin. 7. $f(-x) = \frac{8}{2 + x^2} = f(x)$, $-f(x) = -\frac{8}{2 + x^2} \neq f(x)$, and $-f(-x) = -\frac{8}{2 + x^2} \neq f(x)$, so f is only symmetric with respect to the y-axis.

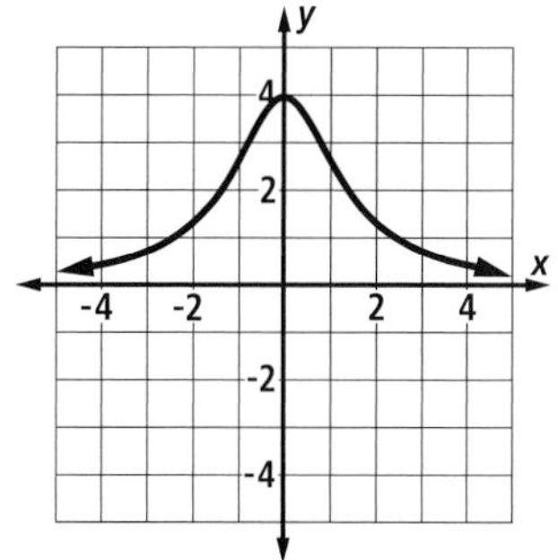

9. The function is odd.

11. a.

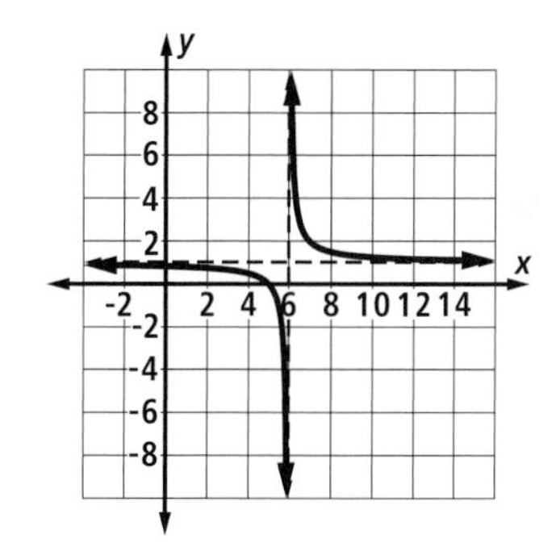

b. $x = 6$, $y = 1$ c. They are translated 6 units to the right and 1 unit up. 13. Answers vary. Sample: The point (2, 6) is on the graph but the point (-2, 6) is not, so f cannot be even. 15.

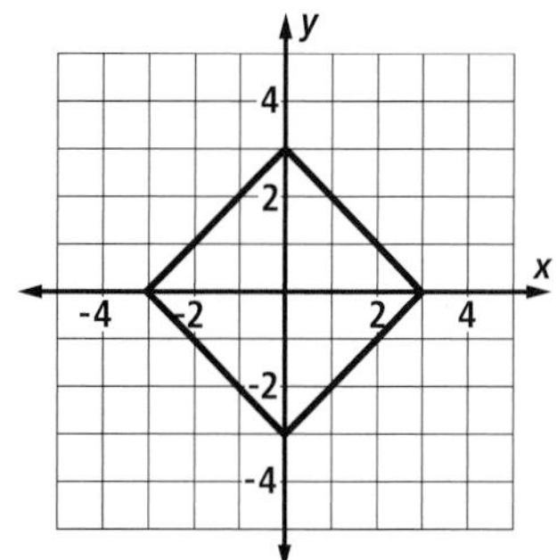

17. Suppose $f(x)$ and $g(x)$ are odd functions, then $f(-x) = -f(x)$ and $g(-x) = -g(x)$. Let $h(x) = f(x) + g(x)$. Then $h(-x) = f(-x) + g(-x) = -f(x) - g(x) = -(f(x) + g(x)) = -h(x)$, therefore, $h(x)$ is odd. Thus, the sum of two odd functions is odd. 19. a. $\bar{x} = 71$ b. $s = 13$ 21. a. $y = (x - 2)^5 - 5$ b. yes; Translations preserve length and angle measure. 23. $\sqrt{a^2 + b^2}$ 25. $-b$

Lesson 3-5 (pp. 179–184)

Guided Example 1

a.

	Preimage		Image	
Point	x	y	$2x$	y
A	5	0	10	0
B	4	3	8	3
C	3	4	6	4
D	0	5	0	5
E	–3	4	–6	4
F	4	–3	8	–3

b. 

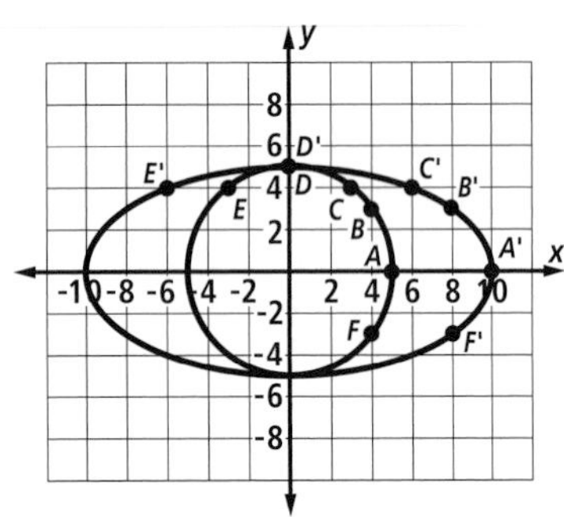

c. $\frac{x}{2}$; $\left(\frac{x}{2}\right)^2 + y^2 = 25$

Questions

1. false **3.** If $a = 0$, every x-coordinate is mapped to 0, which maps the graph to a subset of the y-axis; if $b = 0$, every y-coordinate is mapped to 0, which maps the graph to a subset of the x-axis. **5. a.** $f1\left(\frac{x}{3}\right) = \frac{1}{27}x^3 + \frac{1}{3}x^2 - \frac{4}{3}x$ **b.** $f1\left(\frac{x}{3}\right)$ is the image of $f(x)$ stretched horizontally by a factor of 3. **7. a.** $S(x, y) = \left(x, \frac{1}{2}y\right)$ **b.** $x = -\frac{1}{2}, 3$ **c.** The y-intercept of g is $\frac{1}{2}$ the value of the y-intercept of f. **d.** (1.25, 3.0625)

9.a. 

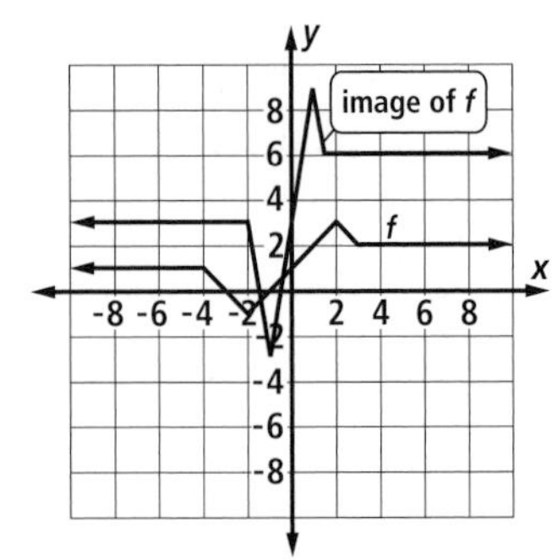

b. $x = -1.5, -0.5$; $y = 3$ **c.** (1, 9) **11.** $S(x, y) = (12x, y)$, or a horizontal scale change of magnitude 12. **13. a.** $y = x + \frac{4}{x}$ **b.** $y = -3x - \frac{1}{3x}$ **15.** $S(x, y) = (2x, -y)$ **17.** neither **19.** They are reflections of each other over the x-axis. **21. a.** The table confirms this statement: the number of injuries on children's rides decreased from 277 in 2003 to 219 to 192. **b.** The table refutes this statement: There were 613 injuries on roller coasters in 2004, more than the 504 injuries in 2003. **c.** The table partially confirms this statement: roller coasters had more injuries than children's rides in every year. On the other hand, the table cannot say whether the chance of being injured is higher on a roller coaster or on a children's ride.

Lesson 3-6 (pp. 185–192)

Questions

1. A scale change of a set of data is a transformation that maps each x_i to ax_i, where a is a nonzero constant. **3.** Between 1975 and 1980, the CPI increased by 85.36%. **5. a.** 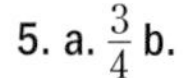 $\frac{3}{4}$ **b.**

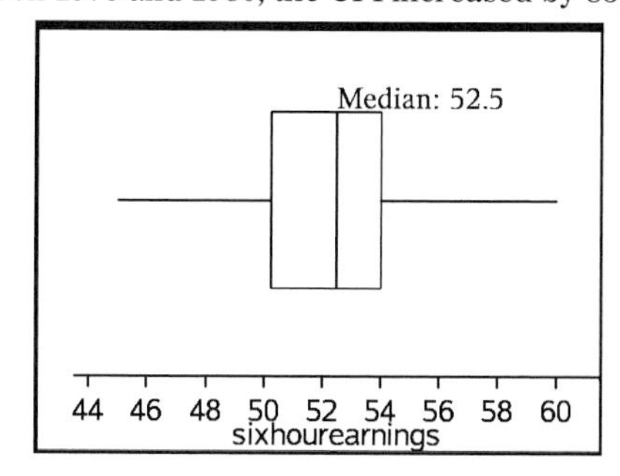

7. C **9.** For n odd, the median is the middle number when the set of numbers is put in increasing or decreasing order. If a set is $\{x_1, x_2, x_3, ..., x_n\}$, then under a scale change with scale factor a, the set becomes $\{ax_1, ax_2, ax_3, ..., ax_n\}$. Therefore, if the median of a set is x_m, the median of the set subject to the scale change is ax_m. For n even, using the same data set, the median is the mean of the middle two numbers, $\frac{x_j + x_k}{2}$. For the scaled set the median is $\frac{ax_j + ax_k}{2} = a\left(\frac{x_j + x_k}{2}\right)$, which satisfies the Centers of Scale Changes of Data Theorem. **11. a.** $\bar{x}r + b$ **b.** $(rs)^2$ **c.** rs
13. a.

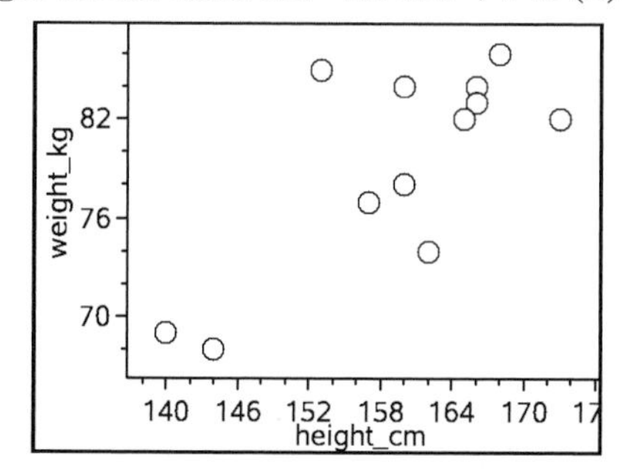

b. $w = 3.628 + 0.4746h$ **c.** $r = 0.7499$ **d.** The scatterplot appears the same but the scale of the axes is different.

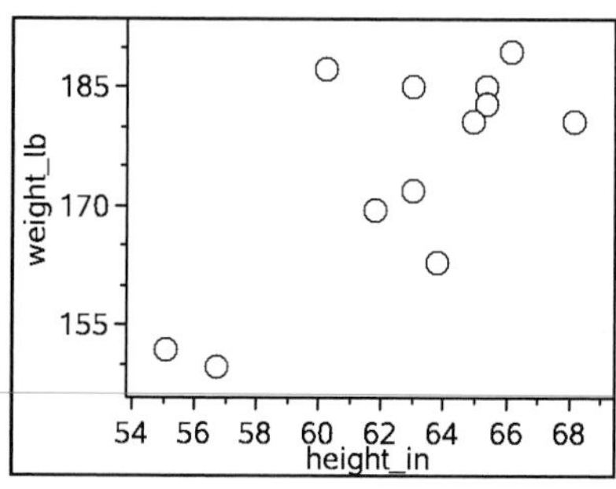

e. $w = 7.991 + 2.655h$; $r = 0.7499$ **f. i.** invariant **ii.** not invariant **iii.** not invariant **15.** B **17.** A **19.** $g(x) = |3x|$, $h(x) = \frac{1}{3}|x|$ **21.** $6x^2(x - 3)$ **23.** $(3r - 5)(r + 1)$

Lesson 3-7 (pp. 193–198)

Guided Example 2

a. $x - 7$; $x - 7$; x^2; $14x$; $2x^2 - 25x + 77$

Guided Example 4

a. 2; translation 4 units to the right and 3 units down; **c.** $x + 4$; $y - 3$; $x + 4$; $y - 3$; $2x + 8$; $y - 3$

Questions

1. maternal grandmother **3.** $f(g(0)) = f(-7) = 77$; $g(f(0)) = g(0) = -7$ **5. a.** $(M \circ N)(t) = \frac{6}{t+1} - 1$ **b.** all real numbers except -1 **7.** 36; 14 **9. a.** -55 **b.** -1 **11. a.** $(x + 3, y + 1)$; $\left(x, \frac{y}{4}\right)$ **b.** $(T \circ S)(x, y) = \left(x + 3, \frac{y}{4} + 1\right)$; $(S \circ T)(x, y) = \left(x + 3, \frac{y + 1}{4}\right)$ **13.** $(S \circ S)(x, y) = (-x, -y)$ **15.** Discounts are usually based on some percentage of the original price and rebates are usually a fixed dollar amount reduction in price. **17.** $(D \circ R)(x) = 0.9x - 90$; $(R \circ D)(x) = 0.9x - 100$; $0.9x - 90 \neq 0.9x - 100$ **19. a.** $m(x) = \frac{x}{5280}$ **b.** $k(x) = 1.609344x$ **c.** $(k \circ m)(x) = 0.0003048x$ **d.** 16,404.2 feet
21.

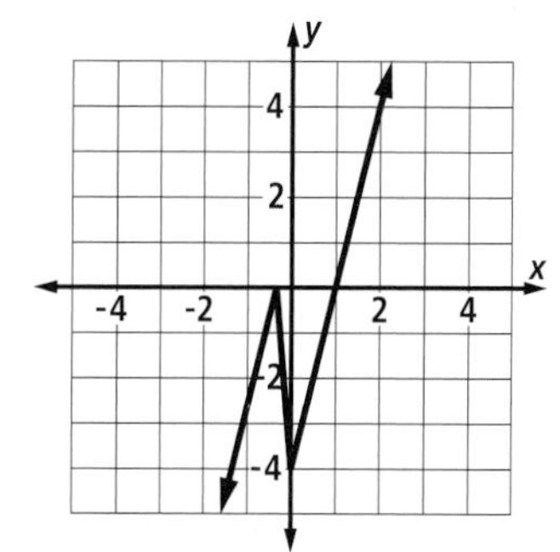

23. $y = 16x^2$

Lesson 3-8 (pp. 199–205)

Questions

1. The relation found by switching the x- and y-coordinates of all points. **3. a.**

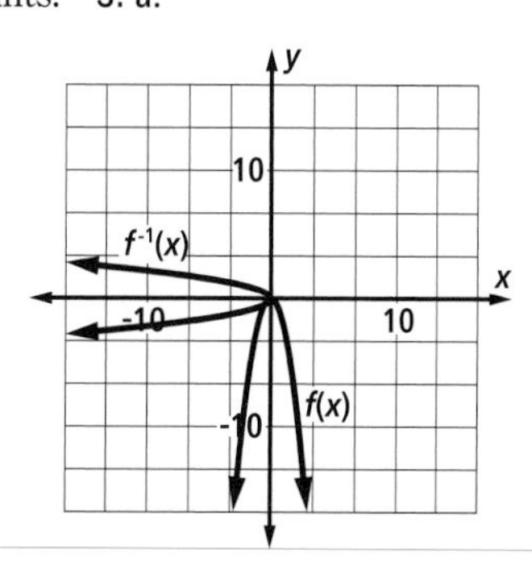

b. $f^{-1}(x)$ is equivalent to the relation $y = \pm\sqrt{\frac{-x}{2}}$; therefore some x-values are assigned two y-values, which is not allowed in a function. **5.** $x = 2y - 4$; $y = \frac{x}{2} + 2$; yes **7.** $x = \sqrt{y}$; $y = x^2$ for $x \geq 0$; yes

9. $g(f(x)) = x$ for all x, $f(g(x)) = x$ for all x, therefore the functions are inverses.

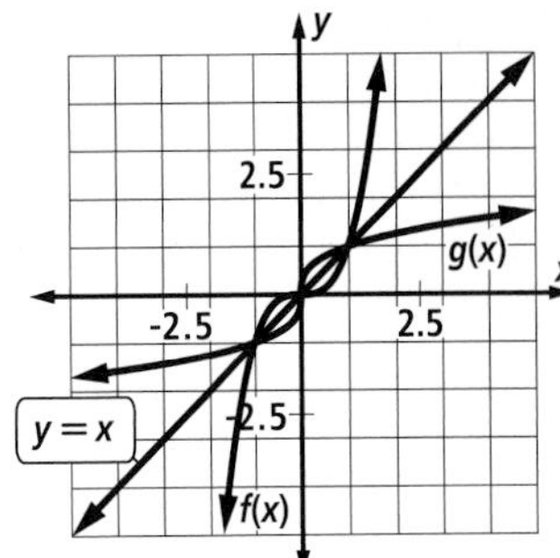

11. a. $M(x) = 10.033x$, $U(x) = \frac{x}{10.033}$ b. \$1,993.42 c. yes, $M(U(x)) = \frac{10.033x}{10.033} = x$ for all real x, $U(M(x)) = \frac{10.033x}{10.033} = x$ for all real x 13. a.

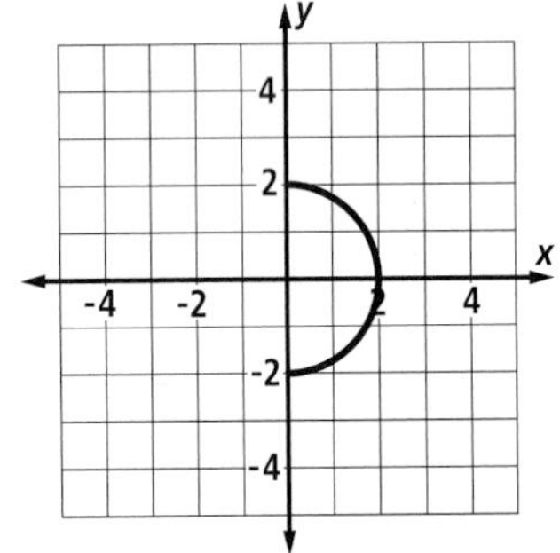

b. no 15. To find the inverse $h^{-1}(p)$, switch p and $h(p)$: $p = \frac{1}{h(p)}$, $h^{-1}(p) = \frac{1}{p}$; $h^{-1}(h(p)) = \frac{1}{\frac{1}{p}} = p$ for all $p \neq 0$, therefore the function is its own inverse.
17. $n(w) = \frac{w - 11}{8}$ 19. a. $v(r(t)) = 39(t + 17) = 39t + 663$ b. $v(v(t))\, 39(39t) = 1521t$ 21. a. $p(x) = -|x| + 7$, so $p(-x) = -|-x| + 7 = -x + 7 = p(x)$. Therefore p is an even function. b. The graph is symmetric over the y-axis.
23. a.

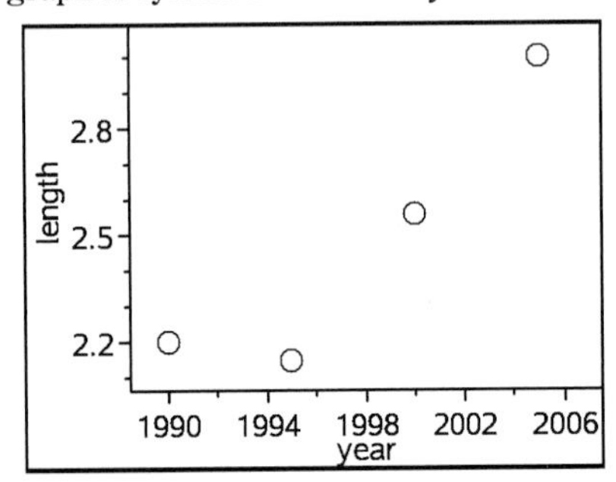

b. $y = 0.0049x^2 - 19.5193x + 19441.1$ c. 3.79 min
25. $\sum_{i=1}^{n} p_i q_i$

Lesson 3-9 (pp. 206-210)

Guided Example 2
78, 0.7, 62, 6, 1, chemistry

Questions
1. L_2 mean $= 0.0029$, L_2 standard deviation $= 10.7$, L_3 mean $= 0.014$, L_3 standard deviation $= 1.019$ 3. \$17.57 5. Part 4b's test score of 84 is better compared to others who took the test. 7. for all data points x, $\frac{x - 19}{4.3}$
9. English: 2.39, Math: 0.5, Reading: -1.56, Science: -0.22
11. 6 minutes 16 seconds
13.

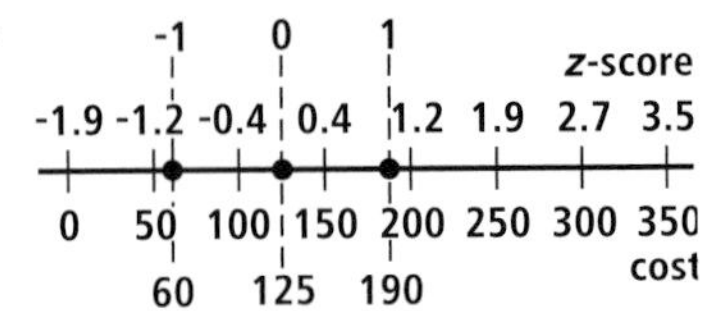

15. a. $\frac{x}{s} - \bar{x}$ b. mean: $\approx$ -20.76, standard deviation: ≈ 1 c. The mean is different because $T \circ S \neq S \circ T$. The standard deviations are the same because it remains invariant under translations. 17. Physics 19. ≈ 42
21. a. true b. false 23. false; Counterexample: $f(x) = x^3 - x$ is odd, but its inverse doesn't pass the vertical line test.
25. B

Self-Test (p. 215)

Questions
1. The graph of h is the image of the graph of $f: x \to \frac{1}{x}$ under the translation $T(x, y) = (x + 7, y - 9)$. The asymptotes of the graph of f are the x-axis and the y-axis. By the Graph-Translation Theorem, the asymptotes of the graph of h are 7 units to the right and 9 units below, that is, $x = 7$ and $y = -9$. 2. Under T, the graph of $y = x^2$ is translated 4 units to the right and 2 units down. So, by the Graph-Translation Theorem, an equation for the image is $y + 2 = (x - 4)^2$ 3. Since the vertex of the graph of $y = x^2$ is (0, 0), the vertex of the image is 4 units to the right and 2 units down, or (4, -2). 4. Under S, the image of (4, 0) is (8, 0). The image of (-1, 2) is (-2, -2), and the image of (-4, -4) is (-8, 4). Connecting these points in order as in the preimage yields the graph here.

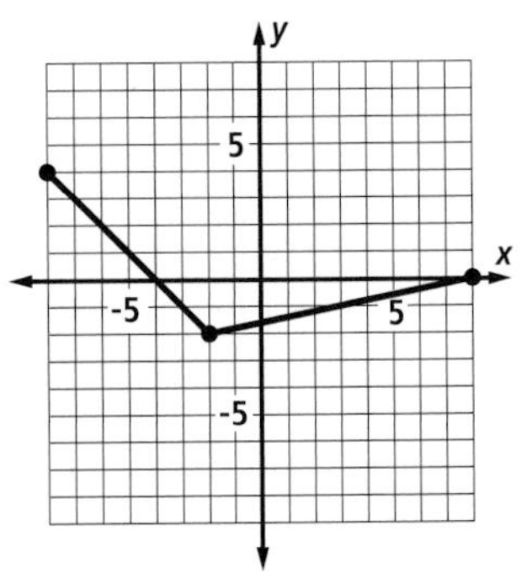

5. a. The median is 150 larger: 3.8 + 150 = 153.8 g b. The range is the same: 13.6 – 0.4 = 13.2 g c. The IQR is the same: 4.6 – 3.6 = 1 g 6. Convert grams to ounces. median: $(153.8 \text{ g})\left(0.035 \frac{\text{oz}}{\text{g}}\right) \approx 5.383$ oz; range: $(13.2 \text{ g})\left(0.035 \frac{\text{oz}}{\text{g}}\right) \approx 0.462$ oz; IQR: 1 g $\approx$ 0.035 oz. 7. Let $y = f(x)$ and switch x and y: $x = 4\sqrt[3]{y + 2}$. 8. Solve the equation in Question 7 for y: $\left(\frac{x}{4}\right)^3 = y + 2$, from which $y = \left(\frac{x}{4}\right)^3 - 2$. This is an equation for a function because

each value of x yields exactly one value of y. **9.** f and f^{-1} are reflection images of each other over the line $y = x$. **10.** $n(m(x)) = n(16 - 5x)$ $= 16 - 5x + \sqrt{16 - 5x}$ **11.** The domain of $n \circ m$ is the set of all values x in the domain of m for which $m(x)$ is in the domain of n. The domain of m is the set of all real numbers, and so is the range of m. Thus $m(x)$ can be any real number. But if $m(x)$ is negative, then $\sqrt{m(x)}$ is not real. So the domain of $n \circ m$ is the set of all x with $m(x) \geq 0$, that is, with $16 - 5x \geq 0$, the set $\left\{x \middle| x \leq \frac{16}{5}\right\}$. **12.** Use the Graph Scale-Change Theorem with $a = 3$ and $b = 0.5$. An equation for the image is $\frac{y}{b} = \frac{1}{\left(\frac{x}{a}\right)^2}$. Substituting for a and b, $2y = \frac{1}{\left(\frac{x}{3}\right)^2}$ or $y = \frac{4.5}{x^2}$.

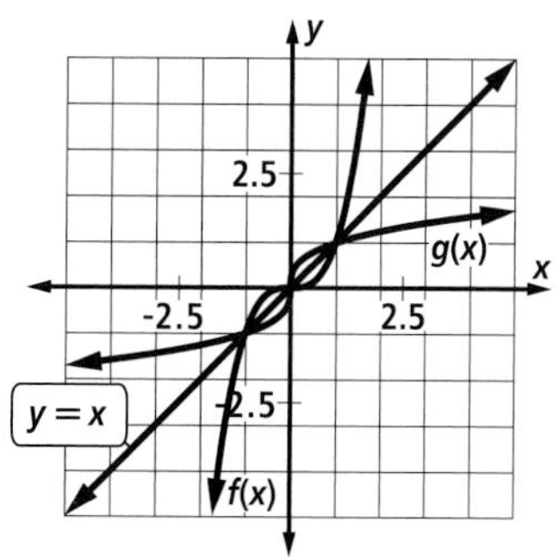

13. Find the z-score using $\bar{x} = 83$ and $s = 7$. The transformation is $x \rightarrow \frac{x - 83}{7}$. **14.** Since $j(-x) = (-x)^2 + 5 = x^2 + 5 = j(x)$, j is an even function. **15. a.** The parent function is the function whose graph was transformed to get the graph of g. That is the absolute value function A with $A(x) = y = |x|$. **b.** The graph of g can be viewed as the image of the graph of A under the scale change $S_{2, 0.5}$. So, by the Graph Scale-Change Theorem, an equation for g is $\frac{y}{0.5} = \left|\frac{x}{2}\right|$. **c.** The graph is symmetric with respect to the y-axis. **d.** $\{y | y \geq 0\}$ **16.** Let T be the transformation. When a point (x, y) is reflected over the line $y = x$, its image is (y, x). So a rule is $T(x, y) = (y, x)$. **17.** Standardize these scores to compare them. For the mouse, $z = \frac{37 - 30}{3.4} \approx 2.06$; for the moose, $z = \frac{1260 - 910}{185} \approx 1.89$. Since the mouse has a higher z-score, it is heavier relative to its population than the moose.

Self-Test Correlation Chart

Question	1	2	3	4	5	6	7	8	9	10
Objective(s)	E	C	D	J	H	H	B	B	K	A
Lesson(s)	3-1	3-2	3-2	3-5	3-3	3-6	3-8	3-8	3-8	3-7
Question	**11**	**12**	**13**	**14**	**15**	**16**	**17**			
Objective(s)	F	C	G	E	I	F	H			
Lesson(s)	3-7	3-5	3-9	3-4	3-4	3-8	3-9			

Chapter Review (pp. 216–219)

Questions

1. 27 **3. a.** 33 **b.** 552 **5. a.** $-\frac{1}{4}$ **b.** -24 **7. a.** $x = |y|$ **b.** The inverse is not a function. **9. a.** $\{(4, 3), (12, 4), (9, 5), (0, 6), (11, 7)\}$ **b.** The inverse is a function. **11.** B **13.** $y = 12x^2$ **15.** $y = |x + 1|$ **17.** $T: (x, y) \rightarrow (x, y + 4)$ **19.** The graph of the image is translated 15.3 units down. **21.** The graph of the image is stretched by a factor of 8 vertically. **23.** true **25.** neither **27.** even **29.** $x = \frac{1}{2}$, $y = 0$ **31.** $\{x | x \geq 2\}$ **33.** false **35.** true **37.** false **39.** 0 **41.** The data point is 1.3 standard deviations less than the mean. **43. a.** 5 **b.** 20 **c.** Spread of Scaled Data Theorem **45. a.** 46.2 mm **b.** 68.8 mm **c.** 136.52 mm^2 **47.** The porpoise is heavier relative to its group. **49.** No, some outputs have more than one input, for example, $y = 0$. **51.** B **53.** A **55.** F **57.** odd **59. a.**

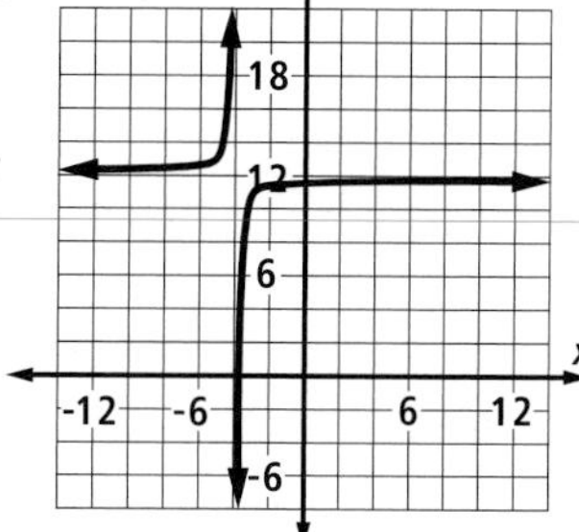

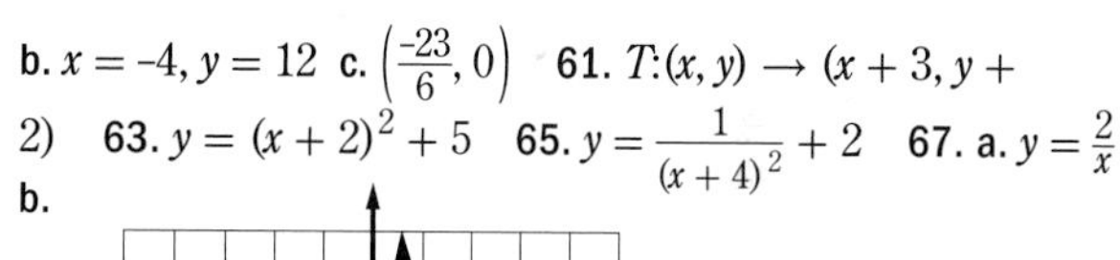

b. $x = -4$, $y = 12$ **c.** $\left(\frac{-23}{6}, 0\right)$ **61.** $T: (x, y) \rightarrow (x + 3, y + 2)$ **63.** $y = (x + 2)^2 + 5$ **65.** $y = \frac{1}{(x + 4)^2} + 2$ **67. a.** $y = \frac{2}{x}$ **b.**

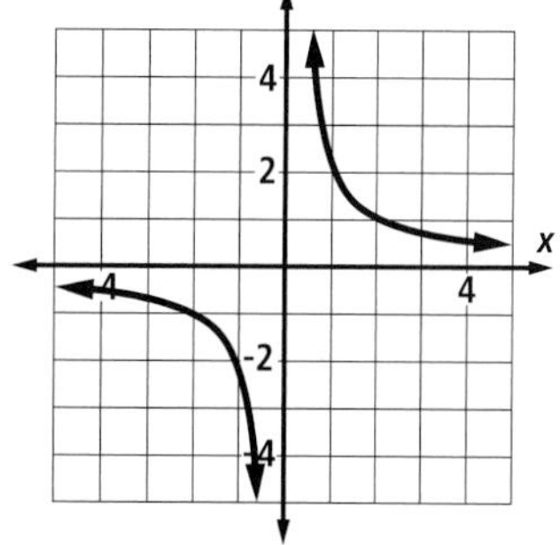

c. The inverse is a function. **69.** They are inverses of each other. We can tell because they are symmetric over the line $y = x$, which implies that $f(g(x)) = g(f(x)) = x$.

Chapter 4

Lesson 4-1 (pp. 222-228)

Guided Example 3 12π; 12π; $\frac{5\pi}{3}$; 5.24

Questions

1. 324° **3. a.** π **b.** π **5. a.**

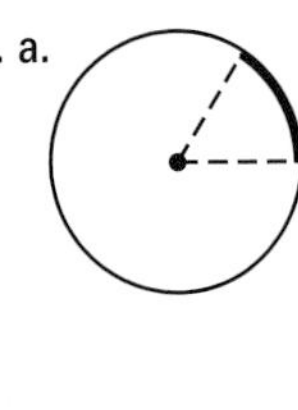

b. about 57.3°

7.

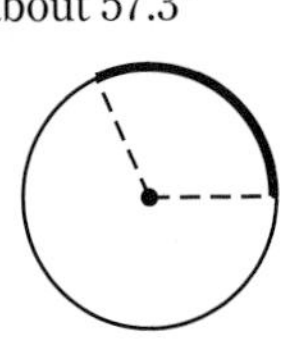

9. a. 1.5 revolutions **b.** 3π **11.** $-\frac{4\pi}{9}$ **13.** $\frac{55\pi}{2}$ **15.** about 4.398 cm **17.** about 7.3 ft **19. a.** about 4.19 in. **b.** about 40,212.4 in. **21. a.** 22π in. or about 69.1 in. **b.** 3300π in. or about 10,367.3 in. **c.** $\frac{25\pi}{8}$ mph or about 9.8 mph **23.** $y = (x - 3)^2 - 2$ **25. a.** Pros mean: 3.64; Pros median: -2.5; Pros standard deviation: 21.1; Darts mean: 3.97; Darts median: 3.6; Darts standard deviation: 6.33 b. The Darts minimum, Q_1, and median are greater than the Pros minimum, Q_1, and median. Nearly all of the Darts data is contained between the Pros median and Q_3. The Pros Q_3 is nearly the same as the Darts maximum. The Pros range is much greater than the Darts range.

Lesson 4-2 (pp. 229-234)

Guided Example 2 a. -1; 0; 0 **b.** 0; 1; undefined

Questions

1. a. x-coordinate **b.** y-coordinate **3.** A **5.** $\tan\theta = \frac{\sin\theta}{\cos\theta}$ **7. a.** 0 **b.** -1 **c.** 0 **9. a.** Answers vary. Sample: 90°, 270° **b.** Answers vary. Sample: $\frac{\pi}{2}$, $\frac{3\pi}{2}$ **11.** (0.540, 0.841) **13. a.** positive **b.** negative **c.** negative **15.** 212 ft **17.** about 1.1 ft **19.** Answers vary. Sample: 180°, –180°, 540° **21.** decreases **23.** -300° **25.** $\frac{\sqrt{2}}{2}$ **27. a.** $(-x, y)$ **b.** $(x, -y)$ **c.** $(-x, -y)$

Lesson 4-3 (pp. 235-241)

Questions

1. true **3.** -3 **5.** the x-coordinates of P' and P'': c and e. **7.** false **9.** $\frac{1}{3}$ **11.** $\sin 33.5° \approx 0.552 \approx \sin 146.5°$; $\cos 33.5° \approx 0.834 \approx -\cos 146.5°$; $\tan 33.5° \approx 0.662 \approx -\tan 146.5°$ **13.** $\frac{5}{13}$ **15.** $-k$ **17.** Half-Turn Theorem and Supplements Theorem **19.** Half-Turn Theorem **21.** Half-Turn Theorem **23.** $\frac{\sqrt{5}-1}{4}$ **25.** $-\frac{\sqrt{5}-1}{4}$ **27.** 1 **29.** 0 **31. a.** 36° **b.** 36°

Lesson 4-4 (pp. 242-246)

Guided Example 1 $\overline{PF}$; $\frac{1}{2}$; b

Guided Example 2 1; $\frac{1}{2}$; $\left(\frac{1}{2}\right)^2$; $\frac{3}{4}$; $\frac{\sqrt{3}}{2}$; $\left(\frac{\sqrt{3}}{2}, \frac{1}{2}\right)$; $\frac{\sqrt{3}}{2}$; $\frac{1}{2}$

Guided Example 4 $-\cos 60°$; $\sin 60°$

Guided Example 5 b. 150; $-\frac{\sqrt{3}}{2}$ **c.** $\frac{0}{-1}$; 0

Questions

1. $\overline{OF}$ **3.** $\overline{RH}$ **5. a.** $-\frac{\sqrt{3}}{2}$ **b.** $-\frac{1}{2}$ **c.** $\sqrt{3}$ **7. a.** $-\frac{1}{2}$ **b.** $\frac{\sqrt{3}}{2}$ **c.** $-\frac{\sqrt{3}}{3}$

9. $\frac{1}{2}$

11.

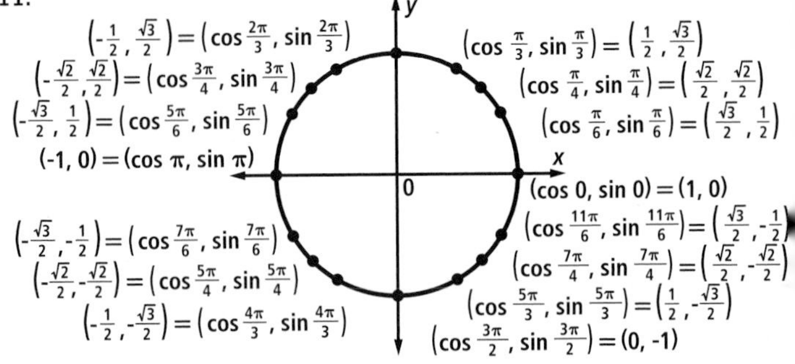

13. a.

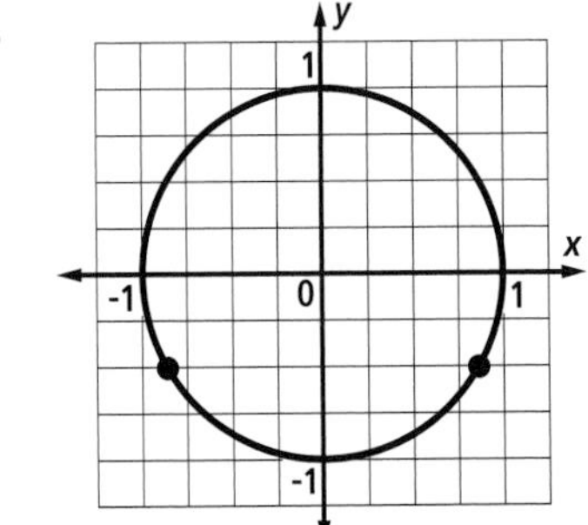

b. 210° and 330° **c.** $\frac{7\pi}{6}$ and $\frac{11\pi}{6}$ **15.** true; When $\theta = 45n$ for some odd n, the triangle made from an extension to the unit circle and a vertical line to the x-axis is isosceles. This means that the leg representing $\sin\theta$ is the same length as the leg representing $\cos\theta$. Because $\tan\theta = \frac{\sin\theta}{\cos\theta}$, this makes $|\tan\theta| = 1$, or $\tan\theta = \pm 1$. **17. a.** -0.788 **b.** 0.788 **c.** -0.788 **d.** 0.788 **19.** -1 **21. a.** $\frac{3\pi}{4}$ **b.** $\frac{13\pi}{6}$ **c.** $-\frac{43\pi}{36}$ **d.** $-\frac{3\pi}{2}$ **23.** $(f \circ g)(t) = 3(t^2 + 1) - 1$

Lesson 4-5 (pp. 247-251)

Questions

1. a. domain: $\{x | x \in \mathbb{R}\}$; range: $\{y | -1 \le y \le 1\}$ **b.** Answers vary. Sample: $-\pi$, 0, π, 2π, 3π

3. a.

x	$\frac{2\pi}{3}$	$\frac{3\pi}{4}$	$\frac{7\pi}{6}$	$\frac{5\pi}{4}$	$\frac{4\pi}{3}$	$\frac{3\pi}{2}$	$\frac{5\pi}{3}$	$\frac{7\pi}{4}$	$\frac{11\pi}{6}$	2π
cos x (exact)	$-\frac{1}{2}$	$-\frac{\sqrt{2}}{2}$	$-\frac{\sqrt{3}}{2}$	$-\frac{\sqrt{2}}{2}$	$-\frac{1}{2}$	0	$\frac{1}{2}$	$\frac{\sqrt{2}}{2}$	$\frac{\sqrt{3}}{2}$	1
cos x (approx)	-0.5	-0.707	-0.866	-0.707	-0.5	0	0.5	0.707	0.866	1

b.

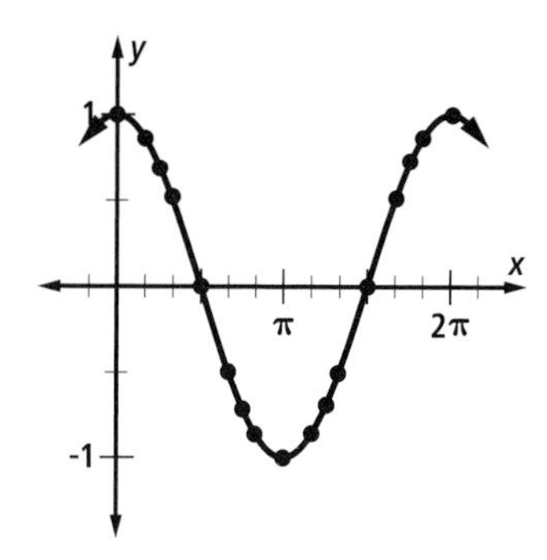

5. Answers vary. Sample: The graphs of both sin θ and cos θ have a cycle of length 2π (or 360°), a maximum value of 1, and a minimum value of -1. The graph of cos θ crosses the y-axis at (0, 1) and the graph of sin θ crosses the y-axis at (0, 0). The graph of cos θ is symmetric about the y-axis and the graph of sin θ is not. **7. a.** $\sin\left(\theta + \frac{\pi}{2}\right) = \sin\left(\frac{\pi}{2} - \theta\right)$ **b.** Answers vary. Sample: $x = \frac{3\pi}{2}$ and $x = -\frac{3\pi}{2}$ **9. a.** -1 **b.** $\frac{21\pi}{2}$ **c.** f is the sine function, because $\sin(10\pi) = 0$ whereas $\cos(10\pi) = 1$. **11.** $\left(-\frac{1}{2}, \frac{\sqrt{3}}{2}\right)$ **13.** 3π **15.** $\left(\frac{k180}{\pi}\right)$

Lesson 4-6 (pp. 252-256)

Guided Example 3 a. about 128 mm; about 87 mm **b.** about 41 mm **c.** about 0.167 seconds

Questions

1. a. 90°, 270° **b.** undefined **c.** The graph of the tangent function has a vertical asymptote at values where $\cos\theta = 0$. **3.** $\sin 495° = \sin(135° + 1 \cdot 360°) = \sin 135° = \frac{\sqrt{2}}{2}$ **5.** $\tan 3750° = \tan(330° + 9 \cdot 360°) = \tan 330° = -\frac{\sqrt{3}}{3}$ **7. a.** 2π **b.** 2π **c.** π **9.** 450° **11. a.** $x = \frac{\pi}{2}, x = -\frac{\pi}{2}$ **b.** $x = 90°, x = -90°$

13. a.

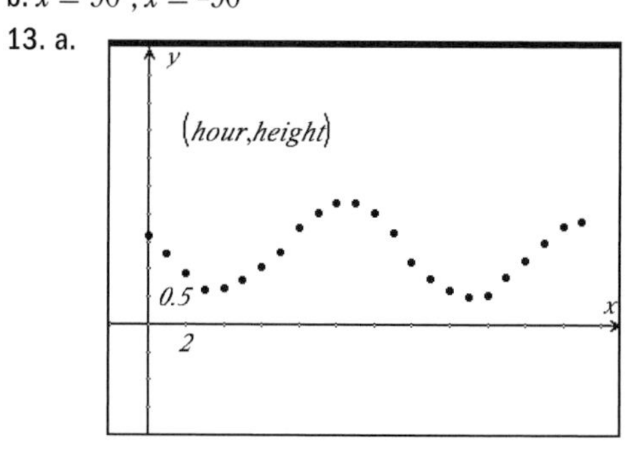

b. $\{h \mid 0.5 \le h \le 2.25\}$ where h is the height after t hours **c.** about 14 hours **15. a.** predator: $\{t \mid t > 0\}$; prey: $\{t \mid t > 0\}$ where t is months after the first measurement **b.** predator: maximum ≈ 710, minimum ≈ 500; prey: maximum ≈ 1,025, minimum ≈ 450 **c.** predator: 20 months; prey: 20 months **17.** even; $\cos(-x) = \cos x$ **19. a.** $T(x, y) \rightarrow (x + 6, y + 5)$ **b.** (15, 14)

Lesson 4-7 (pp. 257-262)

Questions

1. a. true **b.** 2π **c.** $\frac{1}{5}$ **d.**

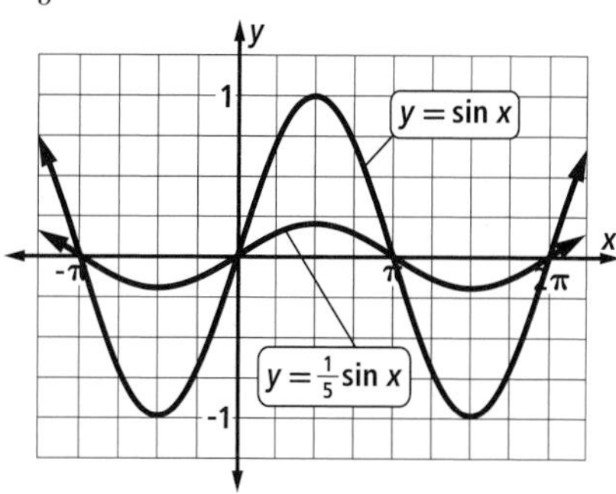

e. The graph of $y = \frac{1}{5}\sin x$ is the image of the graph of $y = \sin x$ under the scale change that maps (x, y) to $\left(x, \frac{y}{5}\right)$, a vertical shrink by a factor of $\frac{1}{5}$. **3. a.** 4 **b.** 6π **5. a.** $y = \sin\left(\frac{x}{5}\right)$ **b.** 1; 10π **7.** $y = 6\sin\left(\frac{x}{4}\right)$ **9. a.** the tone with frequency of 660 cycles per second **b.** It is one octave higher. **11.** Answers vary. Sample:

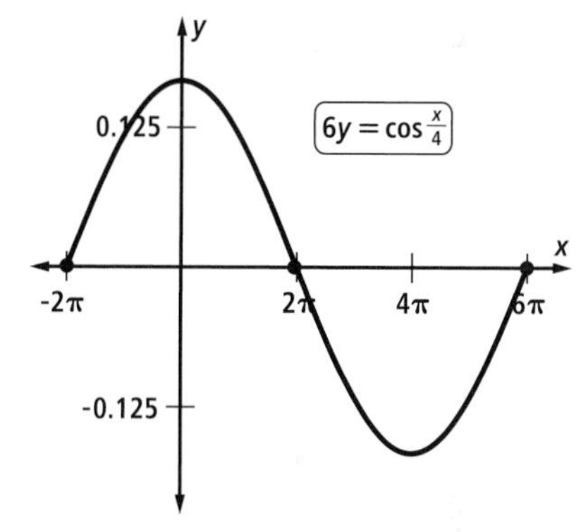

13. $y = 32\sin(880\pi x)$ **15.** f and h **17.** A **19.** Answers vary. Sample: -2.467; 3.817; 6.958 **21. a.** -4π **b.** $\frac{5\pi}{4}$ **c.** $\frac{7\pi}{4}$

23.a.

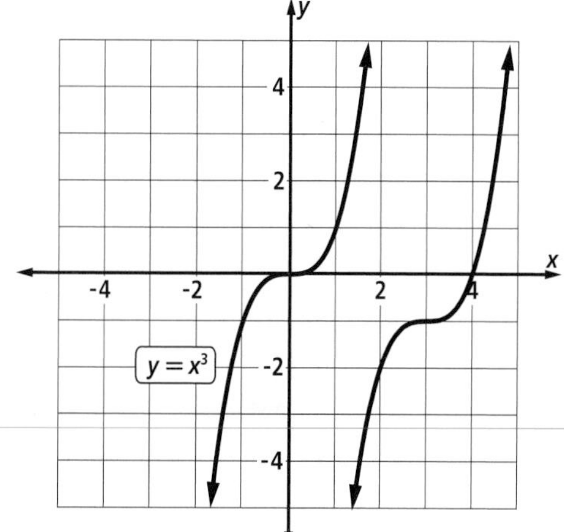

b. $y + 1 = (x - 3)^3$

Lesson 4-8 (pp. 263-268)

Guided Example 1 –60°; 60° left; –60°

Questions

1. true **3. a.** $-\frac{3\pi}{4}$ **b.** 5 **5.** Answers vary. Sample: $y = \sin x - 4$ **7.** $y = \sin\left(x + \frac{3\pi}{4}\right) - 1$ **9. a.** Answers vary. Sample: $T(x, y) = \left(x - \frac{\pi}{2}, y\right)$ **b.** Answers vary. Sample: $\cos x = \sin\left(x + \frac{\pi}{2}\right)$

11. a. π **b.** phase shift to the right by $\frac{\pi}{4}$
c. $\left(\frac{\pi}{2}, 1\right)$ **d.** $x = \frac{3\pi}{4}, x = \frac{7\pi}{4}$ **e.**

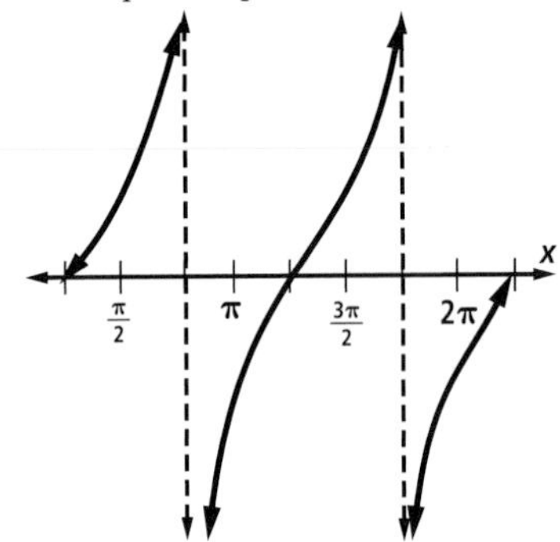

13. a. 4 **b.** 2π

15.

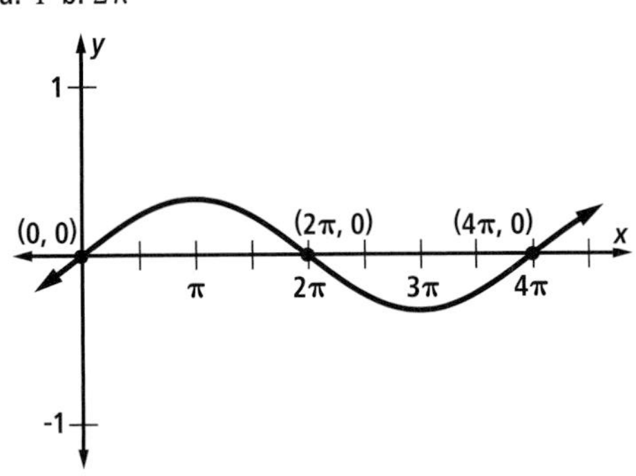

17. 225° **19. a. and b.**

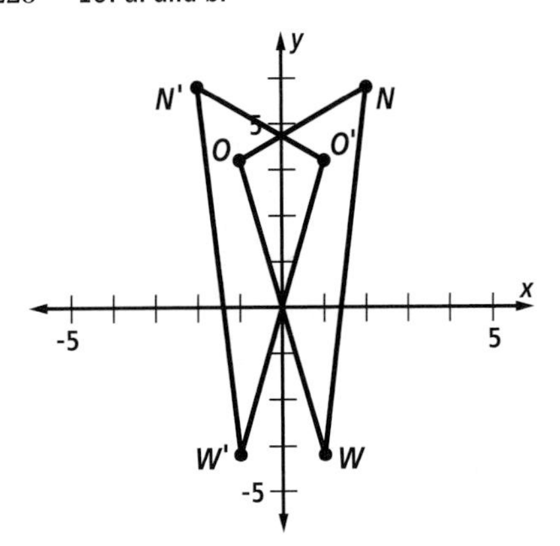

c. $(-x, y)$

Lesson 4-9 (pp. 269-275)

Questions

1. a. (1, –2) **b.** (21, 298) **c.** (–19, 298) **3. a.** amplitude: 1; period: 6π; phase shift: π units to the right **b.** A scale change of $(x, y) \rightarrow (3x, y)$, followed by a translation of $(x, y) \rightarrow (x + \pi, y)$. **5. a.** $\frac{y+5}{7} = \sin\left(\frac{x}{\frac{1}{\pi}}\right)$ **b.** amplitude: 7; period: 2; phase shift: none; vertical shift: 5 units down

7. minimum: 9 ft; maximum: 271 ft **9.** $\frac{y - 140}{131} = \sin\left(\frac{x + \frac{9}{4}}{\frac{9}{2\pi}}\right)$

11. The graphs of functions with equations $\frac{y-k}{b} = \tan\left(\frac{x-h}{a}\right)$, with $a \neq 0$ and $b \neq 0$, have vertical scale change $= b$, period $= \pi|a|$, phase shift $= h$, and vertical shift $= k$.

13. $\frac{y+1}{5} = \sin\left(\frac{x - \frac{\pi}{4}}{3}\right)$ **15.** Answers vary. Sample: $y - 5 = 10\sin\left(x - \frac{\pi}{2}\right)$ **17.** $y - 2 = \left|x - \frac{\pi}{3}\right|$ **19.** $c = 0, \pi, 2\pi, 3\pi, 4\pi, 5\pi, 6\pi$ **21. a.** 10^5 **b.** $3 \cdot 10^8$ **c.** $2 \cdot 10^{-5}$ **d.** $\left(\frac{a}{b}\right) \cdot 10^{m-n}$

Lesson 4-10 (pp. 276-281)

Guided Example 2 6; $\frac{10}{9}$ **a.** 6; 425°F; 431°F; 419°F **b.** $\frac{20\pi}{9} \approx 7$ minutes

Questions

1. the 174th day, or June 23rd; 14.086 hours **3.** $t = 425 + 6\cos 0.45m$ **5. a. i.** half of the range of the pendulum's distance from the motion detector; 0.75 m **ii.** the time the pendulum takes to complete a full swing; 3.8 sec.
iii. the distance of the midpoint of the pendulum's swing from the motion detector; 1.3 m **b.** $\frac{d - 1.3}{0.75} = \sin\left(\frac{\pi t}{1.9}\right)$
7. period: about 0.033 sec, amplitude: 1.6 **9. a.** 28 ft **b.** 2 ft **c.** 30 ft **d.** 2 minutes **e.** $\frac{y - 30}{28} = \sin\left(\frac{x}{\frac{1}{\pi}}\right)$
11. a. and b.

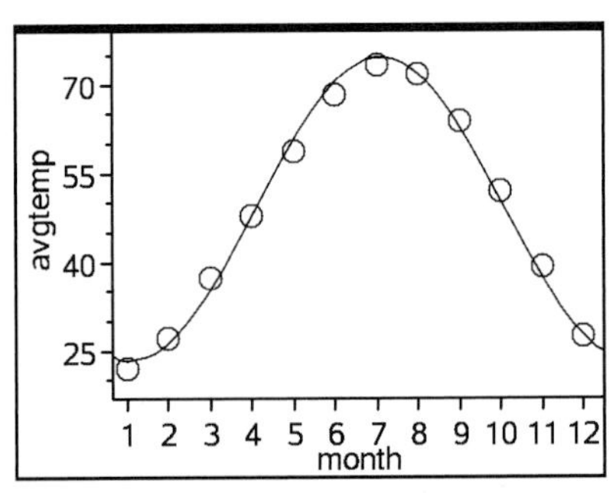

Answers vary. Sample: about 25.63 **c.** 12 **d.** B **e.** Answers vary. Sample: $\frac{y - 49}{-25.63} = \sin\left(\frac{\pi x - 3.5 + 3\pi}{6}\right)$

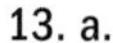

13. a.

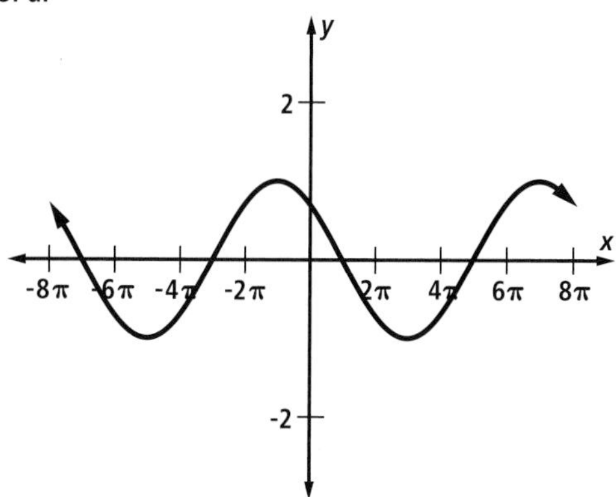

b. amplitude: 1, period: 8π, phase shift: π units to the left **c.** a scale change of $(x, y) \rightarrow (4x, y)$ followed by a translation of $(x, y) \rightarrow (x - \pi, y)$ **15. a.** $c = \frac{\sqrt{2}}{2}$ **b.** $\frac{\pi}{4}$

17. a. $-\frac{49}{340}$ **b.** $-\frac{49}{25}$

Self-Test (pp. 286-287)

1. $\frac{7}{9}$ revolution $\cdot \frac{2\pi \text{ radians}}{1 \text{ revolution}} = \frac{14\pi}{9}$ radians **2.** $-\frac{5\pi}{2}$ radians $\cdot \frac{180^\circ}{\pi \text{ radians}} = -450^\circ$ **3.** $120^\circ \cdot \frac{\pi \text{ radians}}{180^\circ} = \frac{2\pi}{3}$ radians
4. $\sin\left(\frac{3\pi}{4}\right) = \sin\left(\pi - \frac{3\pi}{4}\right)$ by the Supplements Theorem; Therefore, $\sin\left(\frac{\pi}{4}\right) = \frac{\sqrt{2}}{2}$. **5.** $\tan 60^\circ = \frac{\sin 60^\circ}{\cos 60^\circ} = \frac{\frac{\sqrt{3}}{2}}{\frac{1}{2}} = \sqrt{3}$
6. Use a calculator; -0.3090 **7. a.** x can be any real number so domain $= \{x | x \in \mathbb{R}\}$ **b.** The parent sine function has range $\{y | -1 \leq y \leq 1\}$ and this sine functon is translated up 3 from the parent so its range is $\{y | 2 \leq y \leq 4\}$. **c.-f.** From the Characteristics of Sine Waves Theorem, when $\frac{y-k}{b} = \sin\left(\frac{x-h}{a}\right)$, then **c.** amplitude $= b = |1| = 1$. **d.** period $= 2\pi|a| = 2\pi|1| = 2\pi$. **e.** phase shift $= h = \pi$. **f.** vertical shift $= k = 3$. **8. a.** $\cos\theta$ is the x-coordinate of the rotation image of the point (1, 0) by θ. This will be positive in the first and fourth quadrants. $\tan\theta = \frac{\sin\theta}{\cos\theta}$ where $\sin\theta$ is the y-coordinate of the rotation of the point (1, 0) by θ. This will be positive in the first and third quadrants. Both are positive on the interval $\{\theta | 0 < \theta < \frac{\pi}{2}\}$. **b.** From Part a, $\cos\theta$ is negative in the second and third quadrants and $\tan\theta$ will be negative in the second and fourth quadrants. Both are negative on the interval $\{\theta | \frac{\pi}{2} < \theta < \pi\}$.
9. From the diagram, $\theta = \pi + x$ where $\sin x = 0.988$. By the Half-Turn Theorem, $\sin(\pi + x) = -\sin x$. Therefore, $\sin\theta = -0.988$. **10.** $R_{\theta+\pi}(1, 0) = (0.156, 0.988)$. Therefore $\cos(\theta + \pi) = 0.156$. **11.** False. By the Periodicity Theorem, $\cos(\theta + 3\pi) = \cos(\theta + \pi)$ and by the Half-Turn Theorem, $\cos(\theta + \pi) = -\cos\theta \neq \cos\theta$ for all θ.
12. $f(x) = \cos x$ because by the Periodicity Theorem, $\cos(x + 2\pi k) = \cos x$ and so $\cos(-4\pi) = \cos(0 - 4\pi) = \cos 0 = 1$. **13.** b is the amplitude which is half of the difference between the maximum and minimum values, so $b = 2$. k is the vertical shift so $k = -3$. h is the phase shift so $h = \frac{\pi}{2}$. $2\pi|a|$ is the period, and the period is 2π so $|a| = 1$. By checking the equation, $a = 1$.
14. By the Pythagorean Identity, $\cos^2\theta + \sin^2\theta = \cos^2\theta + \left(\frac{5}{13}\right)^2 = 1$. So $\cos^2\theta = \frac{144}{169}$. Thus $\cos\theta = \pm\frac{12}{13}$. Since $\frac{\pi}{2} < \theta < \pi$, θ is in the 2nd quadrant. In that quadrant, $\cos\theta$ is negative, so $\cos\theta = -\frac{12}{13}$. Since for all θ, $\cos(-\theta) = \cos\theta$, $\cos(-\theta) = -\frac{12}{13}$. **15.** We first apply the scale change $(x, y) \rightarrow \left(\frac{x}{3}, y\right)$ resulting in an image with equation $y = \tan 3x$. Next, we apply the translation $(x, y) \rightarrow (x + \pi, y)$ resulting in the final image $y = \tan(3(x - \pi))$. **16.** B; A vertical scale-change affects the height of a graph, which affects the amplitude. **17. a.** The maximum is 70 and the minimum is 22 so the amplitude is $\frac{70 - 22}{2} = 24^\circ$. **b.** The period is from minimum (January) to minimum (January), or 12 months. **c.** Answers vary. Sample: The vertical shift is about $22 + 24 = 46$ and the phase shift from the sine function is about 3 months. The equation, using the Characteristics of Sine Waves Theorem, is $\frac{T - 46}{24} = \sin\left(\frac{n - 3}{\frac{12}{2\pi}}\right)$ or $T = 24\sin\frac{\pi}{6}(n - 3) + 46$. **d.** Substitute 1 into the equation. $T \approx 25.2^\circ$ **18. a.** The maximum of $\cos\theta$ is 1 so the maximum of $12\cos(14\pi t)$ is 12 volts. **b.** The maximum of $\cos\theta$ is achieved when $\theta = 2n\pi$ for any integer n, so $\cos(14\pi t)$ is at a maximum when $14\pi t = 2\pi n$ for some integer n, or when $t = \frac{n}{7}$. Answers vary. Sample: $t = \frac{1}{7}, \frac{2}{7}, \frac{3}{7}$ **19.** D; The graph shows no phase shift from a cosine function with its maximum when $x = 0$. The amplitude is 5 and period is 4π, so the equation must be D. **20.** Answers vary. Sample: The frequency is the reciprocal of the period, so the period is $\frac{1}{60}$. Because no starting point

Self-Test Correlation Chart

Question	1	2	3	4	5	6	7	8	9	10
Objective(s)	A	A	A	B	B	B	E	C	H	H
Lesson(s)	4-1	4-1	4-1	4-4	4-4	4-5	4-8	4-2	4-2	4-3
Question	**11**	**12**	**13**	**14**	**15**	**16**	**17**	**18**	**19**	**20**
Objective(s)	D	I	J	D	J	E	F	G	J	F
Lesson(s)	4-3	4-6	4-9	4-3	4-8	4-7	4-7	4-10	4-7	4-7

is specified, either a sine or a cosine wave will work. Using the Characteristics of Sine Waves Theorem, an equation is $y = 0.1\sin(120\pi x)$ or $y = 0.1\cos(120\pi x)$.

Chapter Review (pp. 288-291)

1. a. 144° **b.** $\frac{4\pi}{5}$ **3.** -150° **5.** $\frac{5\pi}{4}$ **7.** $\frac{1}{3}$ of a revolution **9.** $\frac{\sqrt{2}}{2}$ **11.** $\frac{\sqrt{3}}{3}$ **13.** 0.996 **15.** -0.99 **17.** 2.90 **19.** $-\frac{\sqrt{3}}{2}$ **21.** 0 **23.** (-1, 0) **25.** $x = \frac{2\pi}{3}, \frac{4\pi}{3}$ **27.** $\frac{\pi}{2} + \pi n$, for $n \in \mathbb{Z}$ **29.** true; the range of the sine function is $\{y \mid -1 \leq y \leq 1\}$ as determined by the unit circle **31.** D **33.** By the Complements Theorem, $-\sin\left(\frac{\pi}{2} - \theta\right) = -\cos\theta$. By the Supplements Theorem, $-\cos\theta = \cos(\pi - \theta)$. **35.** $-\sqrt{1 - k^2}$ **37.** $-\sqrt{1 - k^2}$ **39.** true; by the Periodicity Theorem, $\sin(\theta + 4\pi) = \sin\theta$ **41. a.** $\frac{\sqrt{15}}{4}$ and $-\frac{\sqrt{15}}{4}$ **b.** $\sqrt{15}$ and $-\sqrt{15}$ **43. a.** $\frac{2}{3}$ **b.** 2 **c.** 0 **45. a.** $\frac{\pi}{3}$ **b.** does not exist **c.** 0 **47.** maximum: 15; minimum: 5 **49. a.** $y = -3\cos(4(x - 7))$ **b.** $(x, y) \rightarrow \left(\frac{x}{4} + 7, -3y\right)$ **51. a.** $\frac{1}{150}$ seconds ≈ 0.00667 seconds **b.** Answers vary. Sample: $t = \frac{1}{25}, \frac{2}{25}, \frac{3}{25}$ **53. a.** $\frac{1}{2}$ radians **b.** $\frac{3}{2\pi} \approx 0.477$ **c.** $\frac{10\pi}{3} \approx$ 10.472 seconds **55. a.** $y = 2.233\sin(0.017x - 1.317) + 12.133$ **b.** about 372.9 **c.** about 9.90 hours **57. a.** $y = 3\sin(5\pi t)$ **b.** 0 cm **c.** $h = 3\cos(5\pi t) + 3\sin(5\pi t)$ **59. a.** g **b.** -0.940 **61. a.** f **b.** 0.342 **63.** $-b$ **65.** $-b$ **67.** $f(x)$ must be $-\cos(x)$ because $-\sin(20\pi) = 0$ **69. a.**

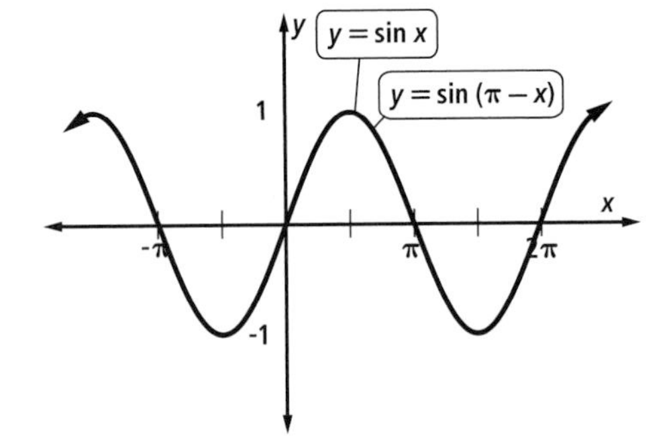

b. the graphs are identical **c.** Supplements Theorem **71.** C **73.** A **75.**

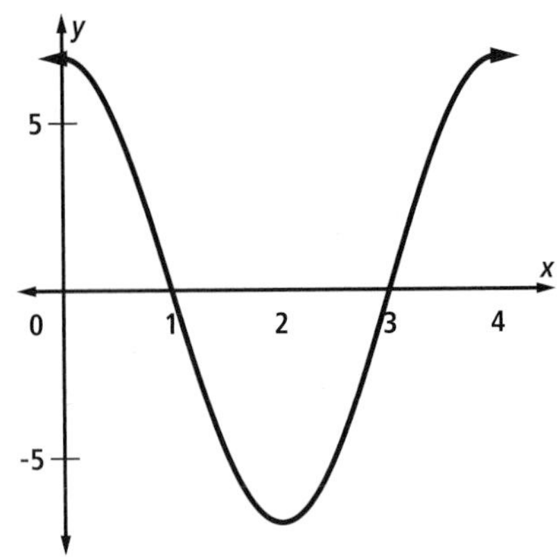

77. $y = \cos\left(\frac{x + \frac{\pi}{3}}{2}\right)$

Chapter 5

Lesson 5-1 (pp. 294-298)

Guided Example 1

hypotenuse; $\frac{20}{29}$; leg adjacent to A; $\frac{21}{29}$; $\frac{\text{leg opposite } A}{\text{leg adjacent to } A}$; $\frac{20}{21}$

Guided Example 3

EP; 60; 46.9; 52.1

Questions

1. $\tan\theta = \frac{\sin\theta}{\cos\theta} = \frac{\frac{\text{leg opposite } \theta}{\text{hypotenuse}}}{\frac{\text{leg adjacent to } \theta}{\text{hypotenuse}}} = \frac{\text{leg opposite } \theta}{\text{leg adjacent to } \theta}$ **3.** about 10.2 **5.** about 290 ft **7. a.** $\frac{3}{5}, \frac{3}{5}$ **b.** The two values are equal because the leg opposite D is the same as the leg adjacent to B. **9.** $24\frac{1}{8}$ in. **11.** (11.3, 22.3) **13.** $\cos\theta$ **15. a.** about 0.28 **b.** -0.292 **c.** 0.96

Lesson 5-2 (pp. 299-303)

Questions

1. Answers vary. Sample: The function does not take every value in the range of cosine. **3.** Answers vary. Sample: The function is not continuous on the restricted domain. **5.** $\frac{\pi}{4}$; 45° **7.** $\frac{5\pi}{6}$; 150° **9.** about 19.8° **11.** 50° **13.** Answers vary. Sample: -338°, -22°, 22°, 338°, 382° **15.** about 22.9 miles north, about 32.8 miles east **17. a.** $\left(400, -\frac{1}{2}\right)$ **b.** $\left(0, \frac{1}{8}\right)$ **c.** $\left(60, \frac{1}{16}\right)$

Lesson 5-3 (pp. 304-308)

Questions

1. a. $t = \sqrt{a^2 + b^2 - 2ab\cos T}$ **b.** $t = \sqrt{65 + 28\sqrt{2}}$ **3.** A **5. a.** slope of $\overline{CD} = \tan C$; slope of $\overline{CA} = \tan C$ **b.** The lines through both $\overline{CA}$ and $\overline{CD}$ have the same slope ($\tan C$) and y-intercept (0, 0) so the points are collinear. **7.** $BC = 15\sqrt{2} + \sqrt{3}$ **9.** about 111.4 ft **11.** about 18.459 m **13.** Let A be the largest angle. By the Law of Cosines, $7^2 = 2^2 + 3^2 - 2 \cdot 3 \cdot 2 \cdot \cos A$; $36 = -12\cos A$; $\cos A = -3 < -1$. But $\cos\theta$ must lie between -1 and 1. So, the triangle cannot exist. **15.** about 105 feet **17. a.** $d = \sqrt{289 - 240\cos\theta}$ **b.** $\theta = \cos^{-1}\left(\frac{d^2 - 289}{-240}\right)$ **19. a.** 0° or 0 radians **b.** 150° or $\frac{5\pi}{6}$ **21. a.**

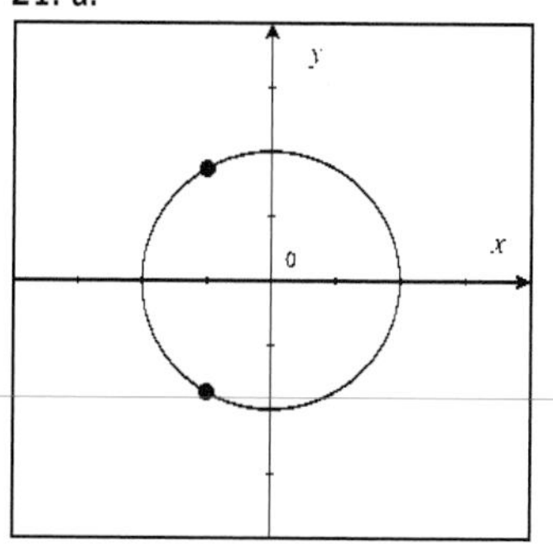

b. Two known values that satisfy the equation are 120° and -120°. By the Periodicity Theorem, adding 360° to each will result in the same cosine. Doing this twice gives us solutions 600°, 840°, which are both in the range. In radians these are $\frac{10\pi}{3}$, and $\frac{14\pi}{3}$. **c.** -120° or $-\frac{2\pi}{3}$

Lesson 5-4 (pp. 309-313)

Guided Example 2

46.17°; 133.83°; 46.17°

Questions

1. Answers vary. Sample: A value in the domain of the inverse sine function could have more than one value in the range, so it cannot be a function. **3.a.**

Point on $y = \sin x$	$\left(-\frac{\pi}{2}, -1\right)$	$\left(-\frac{\pi}{3}, -\frac{\sqrt{3}}{2}\right)$	$\left(-\frac{\pi}{4}, -\frac{\sqrt{2}}{2}\right)$	$\left(-\frac{\pi}{6}, -\frac{1}{2}\right)$	(0, 0)	$\left(\frac{\pi}{6}, \frac{1}{2}\right)$	$\left(\frac{\pi}{4}, \frac{\sqrt{2}}{2}\right)$	$\left(\frac{\pi}{3}, \frac{\sqrt{3}}{2}\right)$	$\left(\frac{\pi}{2}, 1\right)$
Corresponding point on $y = \sin^{-1} x$	$\left(-1, -\frac{\pi}{2}\right)$	$\left(-\frac{\sqrt{3}}{2}, -\frac{\pi}{3}\right)$	$\left(-\frac{\sqrt{2}}{2}, -\frac{\pi}{4}\right)$	$\left(-\frac{1}{2}, -\frac{\pi}{6}\right)$	(0, 0)	$\left(\frac{1}{2}, \frac{\pi}{6}\right)$	$\left(\frac{\sqrt{2}}{2}, \frac{\pi}{4}\right)$	$\left(\frac{\sqrt{3}}{2}, \frac{\pi}{3}\right)$	$\left(1, \frac{\pi}{2}\right)$

b.

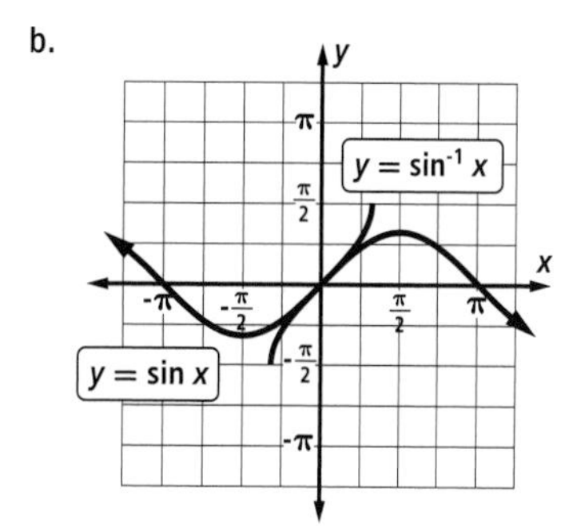

c. domain: $\{x \mid -1 \le x \le 1\}$ range: $\left\{y \mid -\frac{\pi}{2} \le y \le \frac{\pi}{2}\right\}$ **d.** reflection over the line $y = x$; $T(x, y) = (y, x)$ **5.** $\frac{\pi}{2}$; 90°

7. a. $\theta = \sin^{-1}\left(\frac{6}{m}\right)$ **b.**

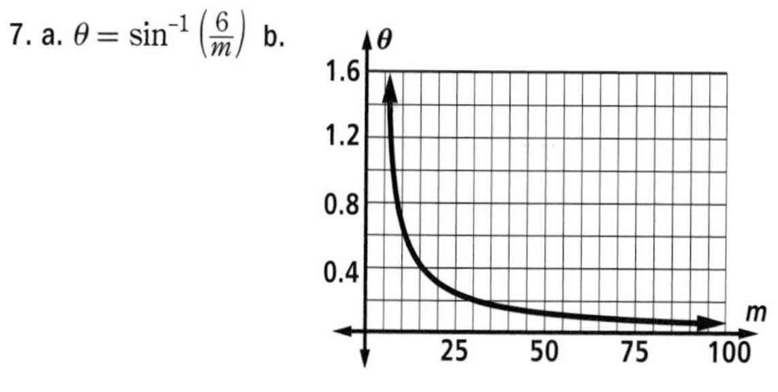

c. 0.2; This is the angle in radians at which the plane is descending. **9.** B is false because, since the domain is restricted for the inverse sine function, there are values of θ that are not in the domain of the inverse sine function. **11.** For $-1 \le x \le 1$, let $y = \sin^{-1} x$. By the definition of $\sin^{-1} x$, $\sin y = x$. Therefore, $\sin(\sin^{-1} x) = x$ for all x such that $-1 \le x \le 1$. **13. a.** $t = \frac{1}{60\pi}\sin^{-1}\left(\frac{E}{4}\right)$ **b.** The graph of t is the image of the graph of the inverse sine function under $S: (x, y) \rightarrow \left(4x, \frac{y}{60\pi}\right)$. **15.** $\frac{7\pi}{6}$ **17.** about 365 yards **19. a.** Answers vary. Sample: 504° **b.** Answers vary. Sample: -216° **c.** 36° **21. a.** y-intercept: -15; x-intercepts: -3 and 2.5 **b.** It has a minimum at $\left(-\frac{1}{4}, -\frac{121}{8}\right)$.

Lesson 5-5 (pp. 314-319)

Guided Example 3

sin 35°; 8; 8 sin 35°; 0.7648; 0.7648; 49.89°; 49.89°; 130.11°; 49.89°, 130.11

Questions

1. about 405.4 **3.** its opposite side **5.** about 9.2 **7.** $a \approx 7.07$, $b \approx 13.66$ **9. a.** about 74.6° or 105.4°

b.

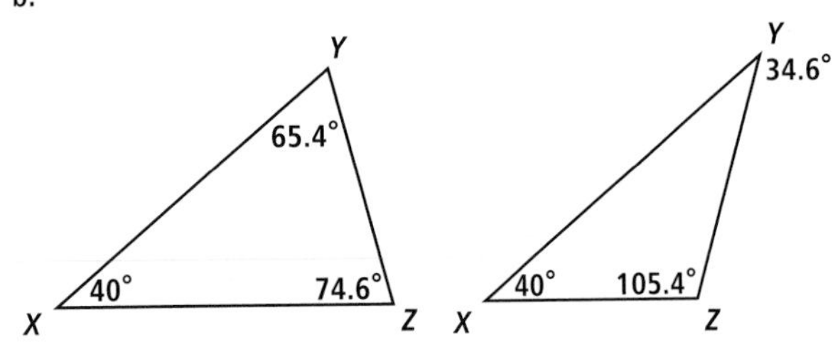

11. false **13.** about 0.55 cm **15. a.** $50\sqrt{2} < XY < 100$ **b.** $100 \leq XY$ or $XY = 50\sqrt{2}$ **c.** $0 \leq XY < 50\sqrt{2}$ **17.** 30°, $\frac{\pi}{6}$ **19.** $\frac{\sqrt{3}}{2}$ **21.** $-t$ **23.** $y + 4 = |x + 3|$ **25. a.** 3920 **b.** $490k$ **c.** $\frac{490}{k}$ **d.** 434

Lesson 5-6 (pp. 320-324)

Questions

1. a. 4.0° **b.** If a plane had an altitude of 5.5 miles and was about 78.6 miles from the airport, it would need to descend at an angle of about 4°.

3. a.

Point on $y = \tan x$	(-1.5, -14.1)	$\left(-\frac{\pi}{3}, -\sqrt{3}\right)$	$\left(-\frac{\pi}{4}, -1\right)$
Corresponding point on $y = \tan^{-1} x$	(-14.1, -1.5)	$\left(-\sqrt{3}, -\frac{\pi}{3}\right)$	$\left(-1, -\frac{\pi}{4}\right)$

$\left(-\frac{\pi}{6}, -\frac{\sqrt{3}}{3}\right)$	(0, 0)	$\left(\frac{\pi}{6}, \frac{\sqrt{3}}{3}\right)$	$\left(\frac{\pi}{4}, 1\right)$	$\left(\frac{\pi}{3}, \sqrt{3}\right)$	(1.5, 14.1)
$\left(-\frac{\sqrt{3}}{3}, -\frac{\pi}{6}\right)$	(0, 0)	$\left(\frac{\sqrt{3}}{3}, \frac{\pi}{6}\right)$	$\left(1, \frac{\pi}{4}\right)$	$\left(\sqrt{3}, \frac{\pi}{3}\right)$	(14.1 , 1.5)

b.

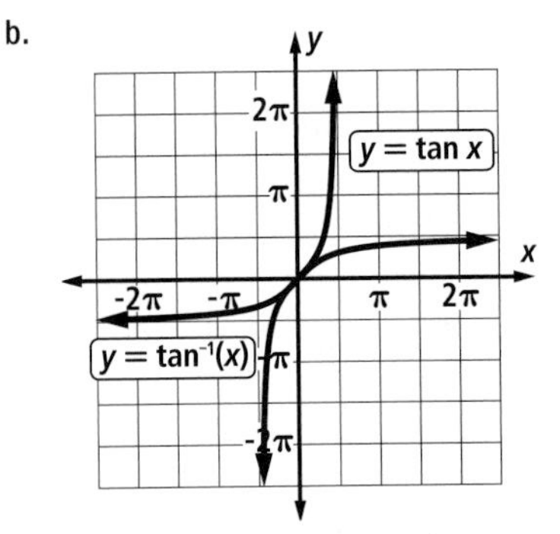

c. domain: $\{x \mid x \in \mathbb{R}\}$ range: $\left\{y \mid -\frac{\pi}{2} < y < \frac{\pi}{2}\right\}$ **d.** $T{:}(x, y) \rightarrow (y, x)$ **5.** $\frac{\pi}{4}$ or 45°

7. a.

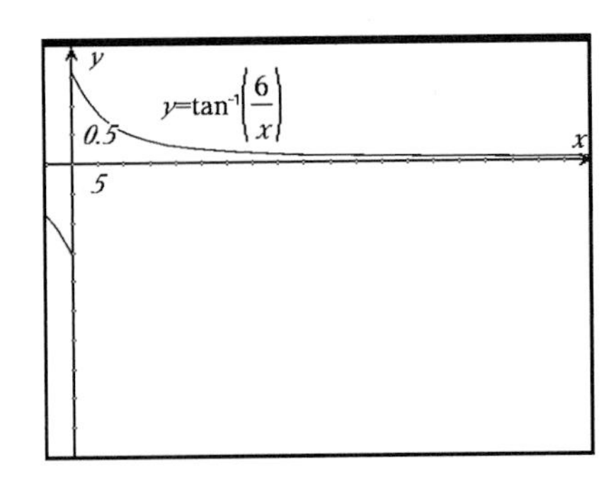

b. about 0.086

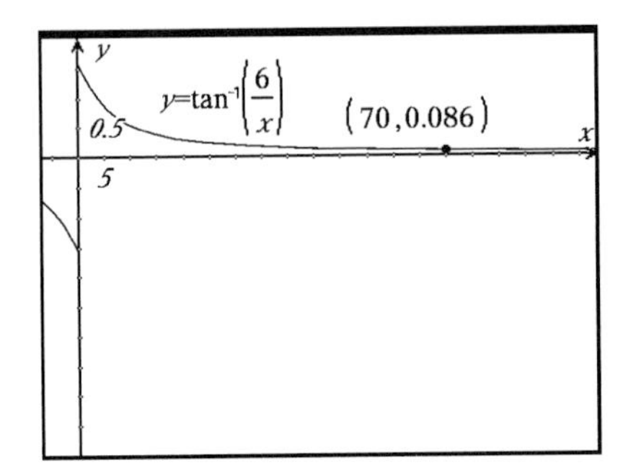

c. When $x = 70$ miles, $\tan^{-1}\left(\frac{6}{70}\right) \approx 0.086 \approx 4.9°$. The angle of descent is almost 5°.

9. a.

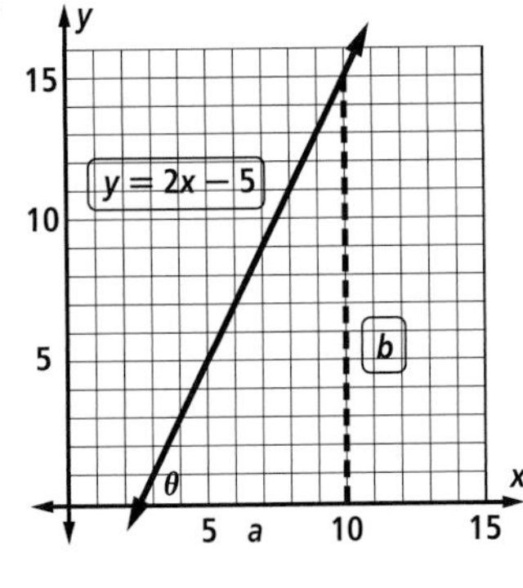

The slope of the line is 2, so $\frac{b}{a} = 2$ and $\tan\theta = \frac{b}{a} = 2$. Thus $\tan^{-1}(2) = \theta$. **b.** If $y = mx + b$, then $\tan\theta = m$ and $\theta = \tan^{-1} m$. **11. a.** $y = \tan^{-1}\left(\frac{80}{x}\right)$ **b.** $y = \tan^{-1}\left(\frac{110}{x}\right) - \tan^{-1}\left(\frac{30}{x}\right)$ **13.** $-\frac{\pi}{3}$ **15.** about 883.7 feet **17.** B **19.** C **21.** B **23. a.** $x = 1, x = 3$ **b.** $x = \frac{1}{2}$, $x = -\frac{5}{3}$ **c.** no real solutions

Lesson 5-7 (pp. 325-330)

Guided Example 2

9; $\frac{3}{4}$; 2.2935; 2.2935; 1.1350; 2.1350; 0.3650; 1.3650; 2.3650; 9.00; 9.00

Questions

1. The restricted domains of the trigonometric functions used in obtaining their inverse functions; an interval equal in size to the period of the function under study; and the set of all real numbers on which the function is defined.

3. a. $\theta \approx 0.767$, or $\theta \approx 5.516$

b.

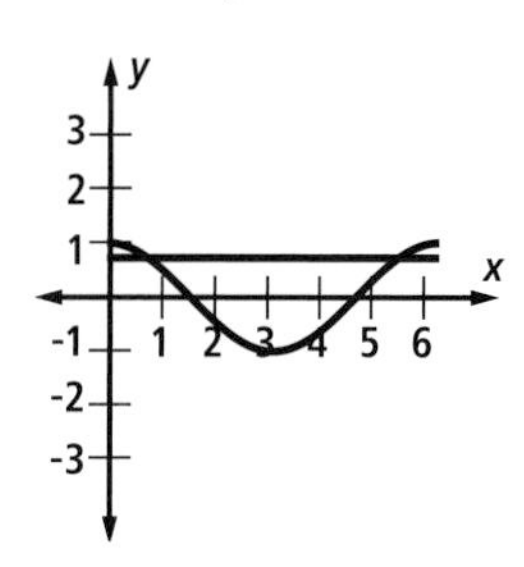

c. $\theta \approx 0.767 + 2\pi n$, or $\theta \approx 5.516 + 2\pi n$, where n is any integer **5. a.** 10

b.

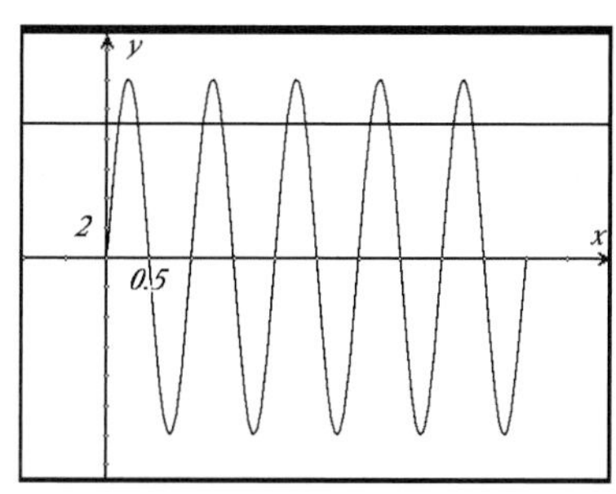

c. 3.1350, 4.1350, 3.3650, and 4.3650 seconds **7.** ≈ 3.343, or ≈ 6.082 **9. a.** 82° **b.** 82°, 278° **c.** $82° + 360n°$ or $278° + 360n°$ for all integers n **11. a.** 194 days or 353 days **b.** October 1 or March 9 **13.** $\theta \approx 1.25$ or $\theta \approx 4.39$ **15.** $\approx 35.8°$ **17. a.** $y = 17.5 \cos\left(\frac{\pi x}{60}\right) + 12.5$ **b.** 30, 29.40, 27.66, 24.87, 21.25, 17.03, 12.50, 7.97, 3.75, and 0.13 meters **19.** neither **21.** odd **23.** odd

Lesson 5-8 (pp. 331-335)

Guided Example 1

a.

t	$x = 2\cos t$	$y = 5\sin t$
0°	2.00	0
30°	1.73	2.50
60°	1.00	4.33
90°	0	5.00
180°	-2.00	0
270°	0	-5.00

b. $\left(\frac{x}{2}\right)^2$; $\left(\frac{y}{5}\right)^2$; $\frac{y}{5}$; $\frac{x^2}{4}$; $\frac{y^2}{25}$

Guided Example 2

$2\cos t$; $2\sin t$; -4; 5; -4; $2\sin t$

Questions

1. $\begin{cases} x = 25\cos t + 3 \\ y = 25\sin t - 2 \end{cases}$ **3.** $\left(\frac{x}{4}\right)^2 + y^2 = 1$ **5.** $(x - 3)^2 + (y + 2)^2 = 25$ **7.** $\begin{cases} x = 3\cos t + 5 \\ y = 3\sin t + 5 \end{cases}$, $\begin{cases} x = 2\cos t + 8 \\ y = 2\sin t + 5 \end{cases}$

9. The point moves counterclockwise on an ellipse centered at (0, 0). **11.** A size change and translation of $(x, y) \to \left(\frac{x}{2} + 8, \frac{y}{2} - 3\right)$

13. a.

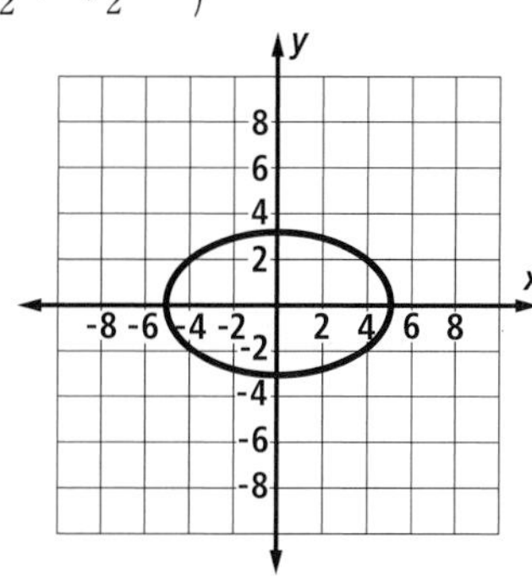

b. The graph is an ellipse that has a semi-major axis of 5 and a semi-minor axis of 3. **c.** Horizontal scale change of $S(x, y) \to (5x, y)$; vertical scale change of $S(x, y) \to (x, 3y)$ **d.** The y-intercepts of the graph are closer to the origin than the graph of the circle with radius 5. **e.** $\left(\frac{x}{5}\right)^2 + \left(\frac{y}{5}\right)^2 = 1$ **15. a.** Traces a circle by moving counterclockwise around the origin, from an original position of (1, 0). **b.** Traces a circle by moving clockwise around the origin, from an original position of (0, 1). **c.** Traces a circle by moving counterclockwise around the origin, from an original position of (1, 0). Traces twice as fast as the other two graphs and goes around twice. **17.** 24 **19.** Amplitude: 3, period: 12π, vertical shift: 7 units up, phase shift: π units to the right **21. a.** $h(x) = |1 - x^2|$ **b.** domain: all real numbers, range: $\{y \mid y \geq 0\}$

Lesson 5-9 (pp. 336-340)

Questions

1. cot **3. a. and b.**

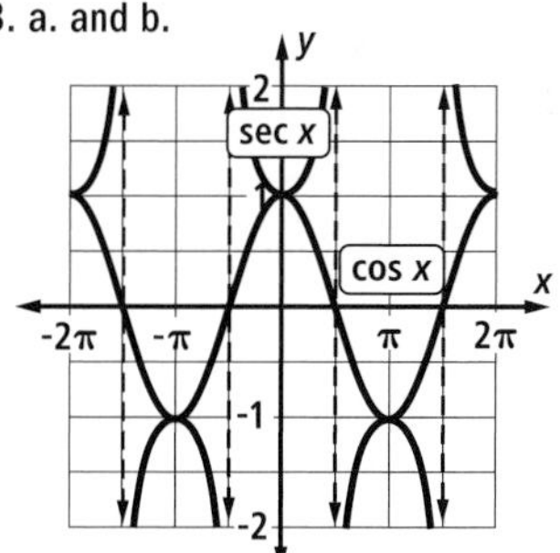

5. $-\frac{2\sqrt{3}}{3}$ **7.** $-\frac{2\sqrt{3}}{3}$ **9.** $\sqrt{2}$ **11.** domain: all real numbers x such that $x \neq n\pi$, where n is an integer; range: all real numbers y such that $y \geq 1$ or $y \leq -1$; period: 2π; asymptotes: any vertical line with equation $x = n\pi$ where n is an integer. **13. a.** odd **b.** not odd **c.** odd **d.** odd **15. a.** $\frac{10}{7}$ **b.** $-\frac{10}{7}$ **c.** $\frac{10}{7}$ **d.** $\frac{\sqrt{51}}{10}$ **e.** $\frac{10\sqrt{51}}{51}$ **f.** $\frac{\sqrt{51}}{7}$ **17.** No. The square of secant or cosecant each must be greater than 1 because the absolute value of secant and cosecant are each greater than 1. Two numbers greater than one will sum to a number greater than one. **19.** By the Supplements Theorem, $\tan(\pi - x) = -\tan x$ for all x.

Therefore, $\cot(\pi - x) = \frac{1}{\tan(\pi - x)} = -\frac{1}{\tan x} = -\cot x$ for all x in the domain of the cotangent function. **21.** 1, where $x \neq \frac{\pi n}{2}$ for any integer n **23.** $\begin{cases} x = \sqrt{17}\cos t \\ y = -6 + \sqrt{17}\sin t \end{cases}$ **25.** $\left(3x, \frac{y}{5}\right)$

Lesson 5-10 (pp.341-348)

Questions

1. always **3.** Greenwich meridian or prime meridian **5.** equator **7.** about 517 mi **9.** about 2 hours and 21 minutes **11. a.** about 873.2 mi **b.** about 871.7 mi **c.** about 1.5 mi **13.** about 6152 mi **15. a.** $S: (x, y) \rightarrow (3x, y)$; $T: (x, y) \rightarrow (x - \pi, y)$; $T \circ S: (x, y) \rightarrow (3x - \pi, y)$

b.

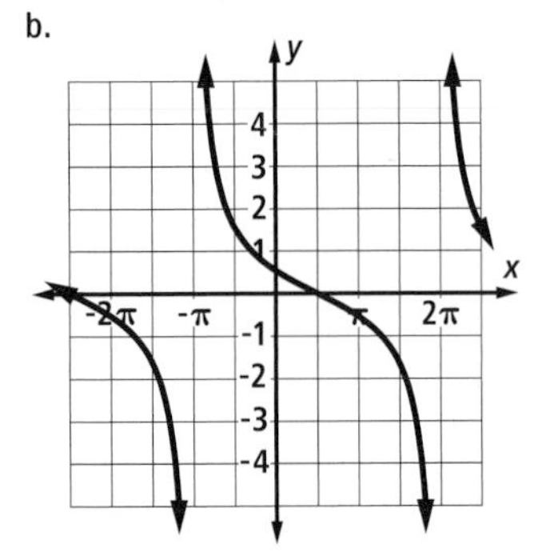

c. period: 3π; phase shift: $-\pi$ **17.** The point moves from (0, 2) clockwise along the ellipse with equation $\frac{x^2}{25} + \frac{y^2}{4} = 1$. **19.** 20.8° **21.** car **23. a.** $x = 0$ and $x = 6$ **b.** $y = -23$ and $y = 7$ **c.** $z = -3$, $z = 7$, and $z = -15$

Self-Test (p. 353)

Questions

1. $\sin B = \frac{\text{opposite}}{\text{hypotenuse}} = \frac{13}{\sqrt{205}}$ **2.** $\tan C = \frac{\text{opposite}}{\text{adjacent}} = \frac{6}{13}$ **3.** $\cos B = \frac{6}{\sqrt{205}}$, so $m\angle B = \cos^{-1}\left(\frac{6}{\sqrt{205}}\right) \approx 65.2°$ **4.** This is the number y whose sine is $\frac{1}{2}$. In the restricted range $-\frac{\pi}{2} \leq y \leq \frac{\pi}{2}$, or $-90° \leq y \leq 90°$, that happens at $\frac{\pi}{6}$, or 30°. **5.** This is the number y whose tangent is 1, which happens when the sine and cosine are equal. In the restricted range $-\frac{\pi}{2} \leq y \leq \frac{\pi}{2}$, or $-90° \leq y \leq 90°$, this happens when $y = \frac{\pi}{4}$, or 45°. **6.** This is the reciprocal of the tangent of $\frac{3\pi}{4}$. $\frac{1}{\tan\left(\frac{3\pi}{4}\right)} = \frac{1}{-1} = -1$. **7.** $\sin\left(-\frac{2\pi}{3}\right) = -\frac{\sqrt{3}}{2}$. But $\sin^{-1}\left(-\frac{\sqrt{3}}{2}\right) = -\frac{\pi}{3} \neq -\frac{2\pi}{3}$. This is because $-\frac{2\pi}{3}$ is not in the range of the inverse sine function, it can never be a value of $\sin^{-1}(x)$.

8. a.

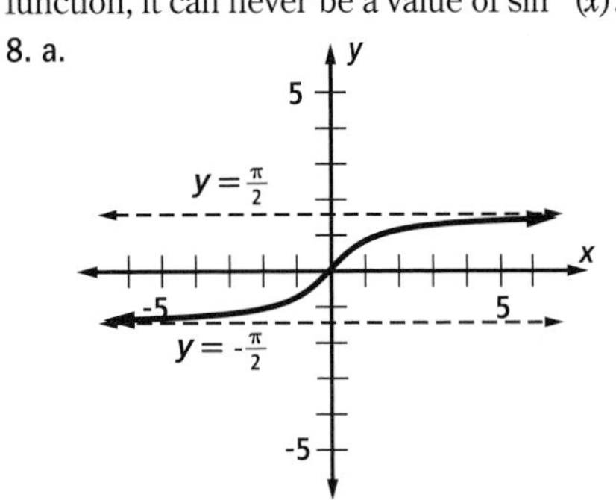

b. The graph, the domain is $\{x| x \in \mathbb{R}\}$, and the range is $\{y| -\frac{\pi}{2} < y < \frac{\pi}{2}\}$. These are the reverse of the domain and range of the tangent function. **9. a.** $\theta = \cos^{-1}(-0.125) \approx 1.696$ **b.** One solution is about 1.696, from Part a. By the Opposites Theorem, -1.696 is a solution to $\cos\theta = -0.125$, but it is not in the set of allowable values of θ. Using the Periodicity Theorem, $\theta \approx -1.696 + 2\pi \approx 4.587$ is a second solution. **c.** Each of the solutions from Part b, plus $2\pi n$ for some integer n, is a solution, so $\theta \approx 1.696 + 2\pi n$ or $\theta \approx 4.587 + 2\pi n$ for all integers n. **10.** Solving for $\cos\theta$, $\cos\theta = -\frac{\sqrt{3}}{2}$, so we are looking for the number whose cosine is $-\frac{\sqrt{3}}{2}$. One solution is $\frac{5\pi}{6}$. By the Opposites Theorem, another solution is $-\frac{5\pi}{6}$. Each of these solutions plus $2\pi n$ for some integer n is a solution, so $\theta = \frac{5\pi}{6} + 2\pi n$ or $\theta = -\frac{5\pi}{6} + 2\pi n$ for all integers n. **11.** By the Law of Sines, $\frac{6}{\sin 135°} = \frac{x}{\sin 20°}$, so $x = \frac{6\sin 20°}{\sin 135°} \approx 2.90$. **12.** Answers vary. Sample: The Law of Cosines finds either the square of the length of a side given SAS, or the cosine of an angle given SSS. Since these are sufficient conditions for congruence, the length and angle are uniquely determined. **13.** First use the Quadratic Formula to find $\cos\theta$. $\cos\theta = \frac{-3 \pm \sqrt{3^2 - 4(1)(-3)}}{2} = \frac{-3 \pm \sqrt{21}}{2}$. $\cos^{-1}\left(\frac{-3 - \sqrt{21}}{2}\right)$ is not defined because $\frac{-3 - \sqrt{21}}{2} \approx -3.79$ is not in the domain of the inverse cosine function. $\theta = \cos^{-1}\left(\frac{-3 + \sqrt{21}}{2}\right) \approx 0.66$. By the Opposites Theorem, $\theta \approx \pm 0.66$. **14.** The height of the ladder is the length of the side opposite θ, and the length of the ladder is the hypotenuse, so they are related by the sine function. $\sin\theta = \frac{h}{12}$, so $h = 12\sin\theta$. **15.** Substituting into the given equation, $13.6 = 11.65 + 2.35\sin\left(\frac{d}{\frac{365}{2\pi}}\right)$. Then solve for d. $\frac{13.6 - 11.65}{2.35} = \sin\left(\frac{d}{\frac{365}{2\pi}}\right)$, so $\left(\frac{d}{\frac{365}{2\pi}}\right) = \sin^{-1}\left(\frac{13.6 - 11.65}{2.35}\right)$, and $d = \frac{365}{2\pi}\sin^{-1}\left(\frac{13.6 - 11.65}{2.35}\right) \approx 57$ days after March 21. Using the Supplements Theorem, another solution is $\frac{365}{2\pi}\left(\pi - \sin^{-1}\left(\frac{13.6 - 11.65}{2.35}\right)\right) \approx 126$. The next two times will be one period later. A period is 365 days, so they will be $365 + 57 = 422$, and $365 + 126 = 491$ days after March 21. **16.** Use the Law of Cosines. Let the High Street side $= H$. $H^2 = 79^2 + 63^2 - 2(79)(63)\cos 86°$, so $H \approx 97.55$. The fencing needed will be $79 + 63 + 97.55 = 239.55$ ft. **17.** This is the image of the unit circle under a scale change of 3 in the x- and y-directions, and then a translation of 2 in the x-direction and -3 in the y-direction. $\begin{cases} x = 3\cos\theta + 2 \\ y = 3\sin\theta - 3 \end{cases}$, $0 \leq \theta \leq 2\pi$. To turn this into rectangular form, start with the Pythagorean Identity, $\cos^2\theta + \sin^2\theta = 1$. Solve the equations to find $\cos\theta = \frac{x-2}{3}$ and $\sin\theta = \frac{y+3}{3}$ and substitute. $(x-2)^2 + (y+3)^2 = 9$. **18.** Answers vary. Sample: $\sec x$ is undefined when $\cos x = 0$ because secant is the reciprocal of cosine. $\cos x = 0$ when $x = \frac{\pi}{2} + \pi k$ for any integer k. Two possible values are $\frac{\pi}{2}$ and $\frac{3\pi}{2}$. **19.** Extend meridians from the north pole N through each city to the equator as shown here. Draw the great circle that goes between the cities W and L. $m\widehat{WN} = 90° - 41.35° = 48.65°$. $m\widehat{NL} = 90° - 51.32° = 38.68°$. $m\angle N$ is the same as the angle made at the center of Earth between the two cities, and is also the same as the arc made at the equator between the two cities. This measure is $72.11° - 0.5° = 71.61°$. Using the Spherical Law of Cosines we find that $\cos n = \cos(48.65)\cos(38.68) + \sin(48.65)\sin(38.68)\cos(71.61) \approx 0.6638$. Using the inverse cosine function, $n \approx 48.41°$. This arc is a portion of a great circle of radius 3960, so the length of the arc is $\frac{48.41°}{360°}(2\pi(3960)) \approx 3346$ mi.

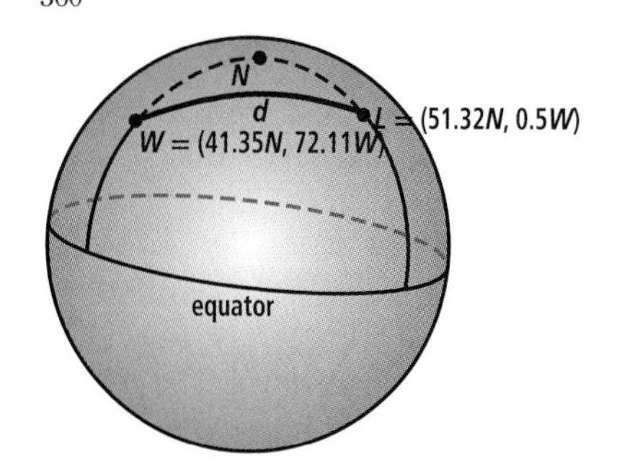

Self-Test Correlation Chart

Question	1	2	3	4	5	6	7	8	9	10
Objective(s)	A	A	B	B	B	E	G	L	D	D
Lesson(s)	5-1	5-1	5-2	5-4	5-6	5-9	5-4	5-6	5-2	5-7
Question	**11**	**12**	**13**	**14**	**15**	**16**	**17**	**18**	**19**	
Objective(s)	C	F	D	H	J	I	M	E	K	
Lesson(s)	5-5	5-3	5-7	5-1	5-7	5-3	5-8	5-9	5-10	

Chapter Review (pp. 354-357)

Questions

1. $\sin A = \frac{4}{\sqrt{41}}$; $\cos A = \frac{5}{\sqrt{41}}$; $\tan A = 0.8$ **3.** $\frac{\sqrt{3}}{2}$ **5.** $-\frac{\sqrt{3}}{2}$ **7.** 0.402 **9.** $-\frac{\pi}{4}$, -45° **11.** $\frac{\pi}{3}$, 60° **13.** $\frac{\pi}{3}$ **15.** ≈ 11.46 **17.** ≈ 5.06 **19.** $\approx 38.9°$, $\approx 141.1°$ **21.** $A \approx 124.6°$; $B \approx 19.0°$; $C \approx 36.4°$ **23.** ≈ 1.13 **25.** $\frac{\pi}{3}$ **27.** ≈ 2.51 and ≈ 3.77 **29.** ≈ 0.61 and ≈ 2.53 **31.** ≈ 7.39 and ≈ 10.53 **33.** $y = -\frac{\pi}{3} + 2\pi k$ and $y = -\frac{2\pi}{3} + 2\pi k$ for any integer k **35.** $w \approx 0.42 + \frac{\pi}{3}k$ for any integer k **37.** $\theta = \frac{\pi}{6} + \frac{2\pi}{3}k$ and $\theta = -\frac{\pi}{18} + \frac{2\pi}{3}k$ and $\theta = -\frac{5\pi}{18} + \frac{2\pi}{3}k$ **39.** $\sqrt{\frac{10}{19}}$ **41.** -1 **43.** $\sec x$ **45.** The Pythagorean Theorem is the Law of Cosines for a right triangle. Since $\cos\frac{\pi}{2} = 0$, $-2ab\cos C = 0$, leaving $c^2 = a^2 + b^2$. **47.** In such a triangle, $\sin\angle ACB \approx 1.88$, but since this is outside the range of sine, this triangle is not possible. **49.** D **51.** domain: $\{x \mid -1 \le x \le 1\}$; range: $\{y \mid 0 \le y \le \pi\}$ **53.** Answers vary. Sample: $\cos\left(-\frac{\pi}{4}\right) = \frac{\sqrt{2}}{2}$, but $\cos^{-1}\left(\frac{\sqrt{2}}{2}\right) = \frac{\pi}{4} \neq -\frac{\pi}{4}$ **55.** Answers vary. Sample: -4.323, -1.961, 1.961, 4.323 **57.** about 27.17 m **59. a.** about 18.01 miles **b.** about 9.005 mph **c.** about 4:51 P.M. **61.** about 670 ft **63. a.** about 17.64 mi **b.** about 9.83 mi

65. a. and b. When the graph is below zero the paddle is in the water.

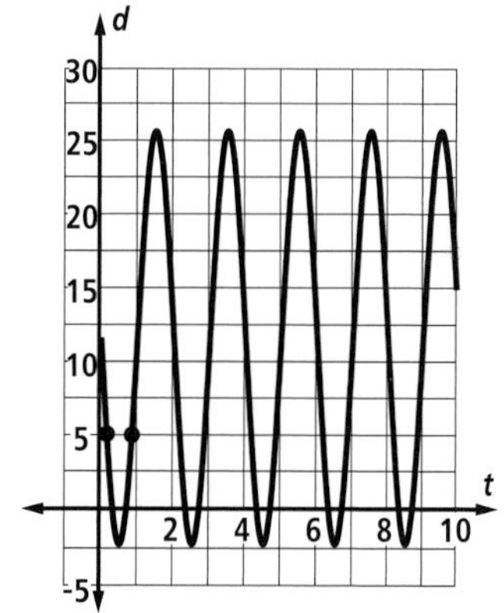

c. $t = \frac{\sin^{-1}\left(\frac{5-11}{15}\right)}{\pi} + 3$; about 0.87 and about 0.13 minutes

67. a. $t = \frac{\sin^{-1}\left(\frac{d}{6}\right)}{\pi}$ **b.** about 0.27 sec

69. about 430 mi

71. about 9244 mi

73. D

75.

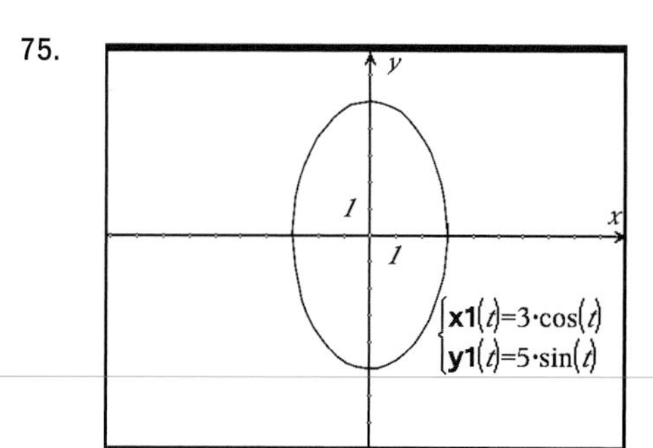

77. a. $(x - 7)^2 + (y + 5)^2 = 1$ **b.** original: $\begin{cases} x = \cos t \\ y = \sin t \end{cases}$, $0 \le t \le 2\pi$; image: $\begin{cases} x = \cos t + 7 \\ y = \sin t - 5 \end{cases}$, $0 \le t \le 2\pi$

Chapter 6

Lesson 6-1 (pp. 360-366)

Guided Example 2

HT; TH; TT; HH; HT; TH; TT; 1; 2; 1; are not

Guided Example 3 **a.** GBB; BGG; GBG; GGB; GGG b. 8; 3; $\frac{3}{8}$

Questions

1. **a.** {Always, Sometimes, Never} **b.** {Sometimes, Never} **c.** 2 **d.** $\frac{2}{3}$ **3.** **a.** {(1, 3), (2, 4), (3, 5), (4, 6), (3, 1), (4, 2), (5, 3), (6, 4)} **b.** $\frac{2}{9}$ **5.** **a.** Ø **b.** 0 **7.** **a.** {HT, TH, TT} **b.** $\frac{3}{4}$ **9.** **a.** {BBBB, BBBG, BBGB, BBGG, BGBB, BGBG, BGGB, BGGG, GBBB, GBBG, GBGB, GBGG, GGBB, GGBG, GGGB, GGGG} **b.** $\frac{1}{4}$

11. **a.**

n	$P(n)$
1	$\frac{1}{2}$
2	$\frac{1}{4}$
3	$\frac{1}{8}$
4	$\frac{1}{16}$

b. $\frac{1}{32}$ **13.** 414

15. **a.**

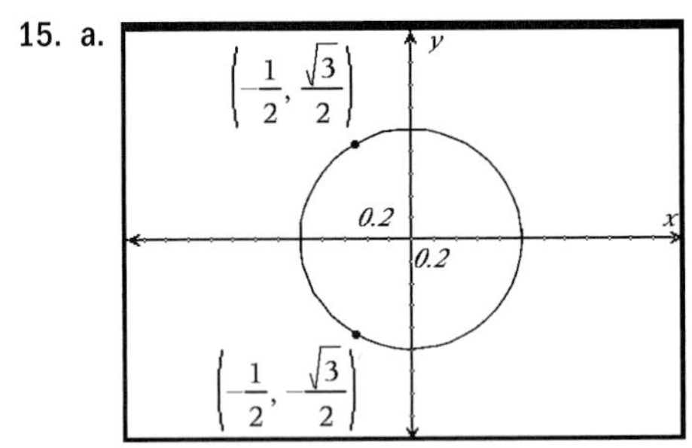

b. 600°, 840° **c.** Answers vary. Sample: -120°

17. **a.** ≈ 79.3 **b.** about 59.2 **19.** **a.** union **b.** union **c.** intersection

Lesson 6-2 (pp. 367-373)

Guided Example 3 10; 6; 0; P(sum > 9); P(has 3 and sum > 9); 10; 0; 16

Questions

1. **a.** {(1, 2), (2, 1), (1, 5), (2, 4), (3, 3), (4, 2), (5, 1), (3, 6), (4, 5), (5, 4), (6, 3), (6, 6)} **b.** $\frac{1}{3}$

3. **a.** {(1, 2), (2, 1), (3, 6), (4, 5), (5, 4), (6, 3)}
b. $\frac{1}{6}$ **5.** when $P(A \cap B) = 0$ **7.** true; Complementary events never intersect. Therefore, they are always mutually exclusive.

9. about 90% of males in the U.S. are not left-handed **11.** $\frac{N(A \cup B)}{N(S)} = \frac{N(A)}{N(S)} + \frac{N(B)}{N(S)} - \frac{N(A \cap B)}{N(S)}$. So, $P(A \cup B) = P(A) + P(B) - P(A \cap B)$.

13. $\frac{21}{36}$ or $\frac{7}{12}$ **15.** $\frac{14}{36}$ or $\frac{7}{18}$ **17.** No, because $P(A \cap B) = 0.05 \neq 0$. **19.** **a.** {H1, H2, H3, H4, H5, H6, T1, T2, T3, T4, T5, T6} **b.** 8; $\frac{8}{12} = \frac{2}{3}$ **c.** Answers vary. Sample: heads and a number greater than 3 **21.** **a.** $\frac{48}{55}$ **b.** $-\frac{73}{55}$ **c.** $-\frac{73}{48}$ **d.** $\frac{55}{48}$ **e.** 3.86 radians **23.** \$3.385

Lesson 6-3 (pp. 374-380)

Guided Example 3 26; 26^2; 10; 10^4; 676; 10,000

Guided Example 5 10; $\frac{10}{36} = \frac{5}{18}$; 6; 6; $\frac{6}{36} = \frac{1}{6}$

Questions

1. 90 vacations **3.** 260,000 possible 5-character ID numbers **5.** Independent: $P(A) \cdot P(B) = P(A \cap B) = \frac{1}{30}$. **7.** $2^6 \cdot 4^{12}$ = 1,073,741,824 ways of answering **9.** **a.** 3,910 sundaes **b.** 6,670 sundaes **c.** 4,080 sundaes **11.** $P(A) = \frac{1}{2}$, $P(B) = \frac{6}{8} = \frac{3}{4}$, so $P(A) \cdot P(B) = \frac{3}{8}$, and $P(A \cap B) = \frac{3}{8}$. So A and B are independent. **13.** **a.** $\frac{1}{7}$ **b.** $\frac{17}{30}$ **15.** $\frac{7}{36}$ **17.** $x(2x - 3)(4x + 1)$ **19.** 3,628,800 **21.** 220

Lesson 6-4 (pp. 381–386)

Guided Example 2 10; 6; 10; 6; 6; 7 • 6 • 5; 151,200; 6; 6

Questions

1. BAT; BTA; ATB; ABT; TAB; TBA **3.** **a.** 117 • 116 • 115 • 114 **b.** $\frac{117!}{113!}$ **5.** $\frac{317!}{315!}$ = 100,172 **7.** **a.** ${}_8P_3 = 336$ **b.** ${}_5P_3 = 60$ **9.** n; There are n choices when you choose 1 element from a set with n elements. **11.** By one formula, ${}_nP_n = n!$. By the alternate formula, ${}_nP_n = \frac{n!}{(n-n)!} = \frac{n!}{0!}$. These formulas are only equal if 0! = 1, so it is defined to equal 1. **13.** **a.** ${}_5P_4 = 120$ **b.** $\frac{1}{120}$ **15.** ${}_3P_3 \cdot {}_2P_2 = 12$ different seating arrangements **17.** **a.** ${}_nP_0 = \frac{n!}{(n-0)!} = \frac{n!}{n!} = 1$ **b.** ${}_nP_0$ is the number of permutations of 0 objects selected from n objects. There can only be one because there is only one empty set possible. **19.** 67,600,000 **21.** **a.** $\frac{\sqrt{3}}{3}$ **b.** $\frac{1}{2}$ **23.** **a.** $\frac{1}{3}$ **b.** $\pm\frac{4\sqrt{3}}{3}$

Lesson 6-5 (pp. 387-394)

Guided Example 2 **a.** 583; 61; 513 **b.** have no tattoo **c.** do not have Hepatitis C **d.** $\frac{8}{61} \approx 13.1$; more

Questions

1. **a.** about 36.1% **b.** about 81.3% **c.**

	Freshman	Sophomore	Junior	Senior	Total
Spanish	8	11	5	2	26
French	2	5	0	1	8
No Language	5	8	10	13	36
Total	15	24	15	16	70

3. **a.** developed polio: about 0.02%; did not develop polio: about 99.98% **b.** vaccinated: about 22.3%; unvaccinated: about 77.7% **c.** about 22.3% **5.** **a.** 6,231.6; the number (in thousands) of passenger cars involved in accidents **b.** 47.9;

the number (in thousands) of fatalities in all motor vehicle accidents **c.** about 0.5% **d.** about 16.3% **7. a.** about 8.3% **b.** about 44.4% **c.** about 25.0% **9. a.** The population for this question is children in first class; 100% **b.** The population for this question is all survivors; about 0.8% **c.** The population for this question is all people on the Titanic; about 0.3% **11. a.** $98 \cdot 97 \cdot 96 = 912{,}576$ **b.** $\frac{98!}{95!} = 912{,}576$ **13.** $\frac{212!}{209!} = 9{,}393{,}720$ **15. a.** {(1,1), (1,3), (2,2), (3,1), (6,2), (5,3), (4,4), (3,5), (2,6), (6,4), (5,5), (4,6)} **b.** $\frac{1}{3}$

Lesson 6-6 (pp. 395–401)

Guided Example 2 0.0049; 0.05, 0.995; 0.04975; 0.02, 0.0001; 0.995, 0.94525; 99.5; *D*; not *A*; 0.0001

Questions

1. about 62.5% **3. a.** let D = has the disease and A = tests positive

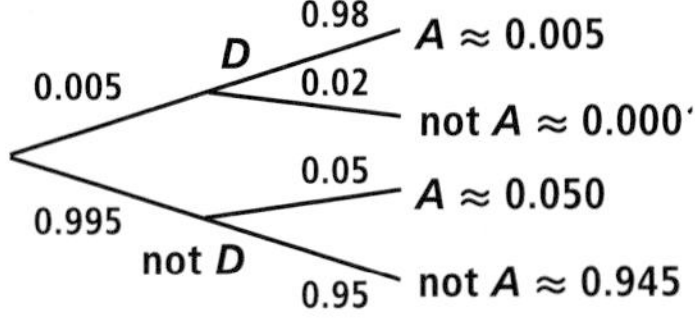

b. $P(\text{not } D|A) = \frac{P(\text{not } D \cap A)}{P(A)} = \frac{0.050}{0.50 + 0.0050} \approx 0.91$

c. Answers vary. Sample: 91% of positives are false positives. The reason for this surprising result is that very few people in the population have the disease, so a low percentage of false positives is enough to overtake a high percentage of true positives. **5. a.** A true statement is detected as a lie. **b.** A lie is undetected. **7. a.** $P(\text{false positive}|\text{positive test}) = \frac{0.03}{0.36 + 0.03} = 0.077 \approx 7.7\%$

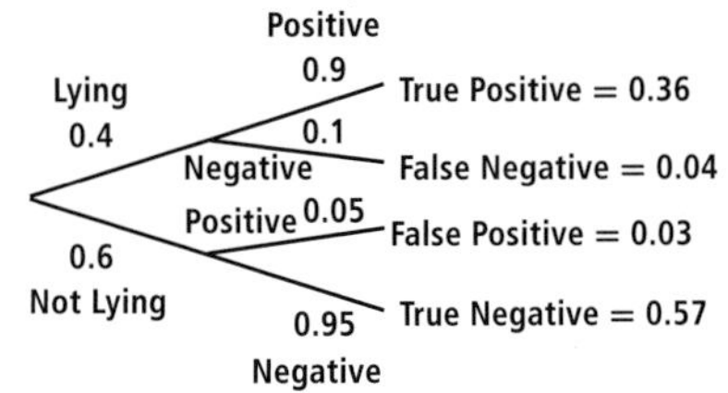

b. $P(\text{false positive}|\text{positive test}) = \frac{0.2125}{0.2125 + 0.1125} = 0.654 = 65.4\%$

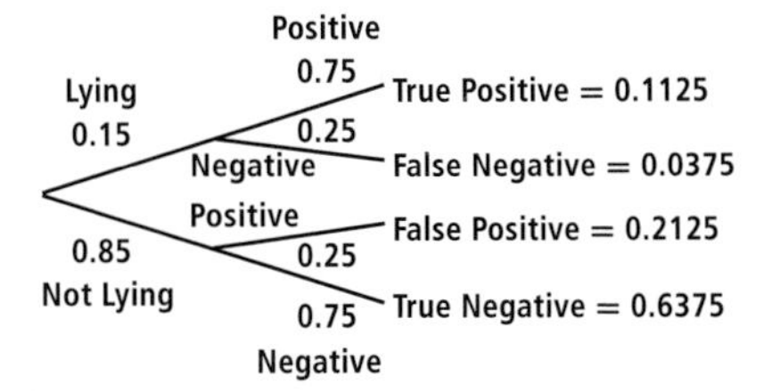

c. Answers vary. Sample: About 65.4% of people given a polygraph test whose result indicated a lie were actually telling the truth. **9.** about 98% **11. a.** $P(M)$ **b.** $P(C|M)$ **c.** 3 **d.** $\frac{1}{3}$ **e.** Answers vary. Sample: The prosecutor computed the chance of a random person matching the description, not the probability that someone who matched the description actually committed the crime. Since more than one person in California matches the description, the probability that the accused is guilty is $\frac{P(M \cap C)}{P(M)} = \frac{1}{3}$. **13. a.** 14.2% **b.** 60.9% **15. a.** ${}_6P_4 = 360$ **b.** ${}_2P_1 \cdot {}_5P_3 = 120$

Lesson 6-7 (pp. 402–409)

Questions

1. A simulation is an experimental model that attempts to capture all aspects of the original situation that affect the outcomes. **3. a.** Assign 1, 2, and 3 to the coupon colors and pick them at random to represent visits to the store. **b.** Answers vary. Sample:

Trial #	Simulated Coupons	# needed for ice cream
1	213	3
2	3312	4
3	222113	6
4	1213	4
5	3231	4

c. Answers vary. Sample: Based on the simulation, 4 visits.

5. a.

Trial #	Success?
1	No
2	No
3	No
4	No
5	Yes
6	No

Answers vary. Sample: A success equals a 10 digit number in which all the digits are less than 7 as in trial #5. **b.** Answers vary. Sample: Based off of the simulation, if 10 adults are chosen at random, about $\frac{1}{6}$ of the time all 10 would be in favor of raising the driving age.

7. a. Answers vary. Sample: Generate numbers from 1 to 100, once for men and once for women in each trial. For men, numbers 1–10 are left-handed, and all others are right-handed. For women, 1–8 are left-handed, and all others are right-handed.The relative frequency for 10 trials was 20% in this simulation, so the probability should be about 20%.

Trial #	Man	Woman	Success?
1	82	39	No
2	94	37	No
3	50	24	No
4	10	19	Yes
5	41	24	No
6	65	79	No
7	36	46	No
8	82	34	No
9	73	72	No
10	3	95	Yes

b. 17.2%

9. **a.** Answers vary. Sample: Let a random number from 1-100 represent one official at-bat; let 1-30 represent a hit and 31-100 represent no hits. One trial is made up of five such random numbers. **b.** Answers vary. Sample:

Trial #	Simulated Hits	Had Hits?
1	73, 14, 38, 57, 59	Yes
2	34, 23, 28, 90, 20	Yes
3	53, 33, 71, 97, 93	No
4	65, 70, 2, 11, 80	Yes
5	37, 51, 49, 92, 40	No
6	18, 44, 99, 24, 49	Yes
7	8, 98, 32, 51, 41	Yes
8	24, 11, 97, 23, 16	Yes
9	95, 39, 39, 63, 23	Yes
10	56, 27, 74, 36, 73	Yes

c. about 20% **11. a.** In one trial, generate two numbers from 1-10. The first number indicates the accuracy of the test. A 1 is a false result, and 9-10 are accurate results. The second number represents the health of the patient. A 1 means the patient is diabetic, and 9-10 mean the patient is not diabetic. **b.** Answers vary. Sample: In 50 trials, the relative frequency of a positive test was $\frac{13}{50}$. **c.** Answers vary. Sample: In the same experiment, 6 non-diabetics tested positive for diabetes. **13.** ${}_{700}P_3 = 341{,}531{,}400$ **15.** $n!$ **17. a.**

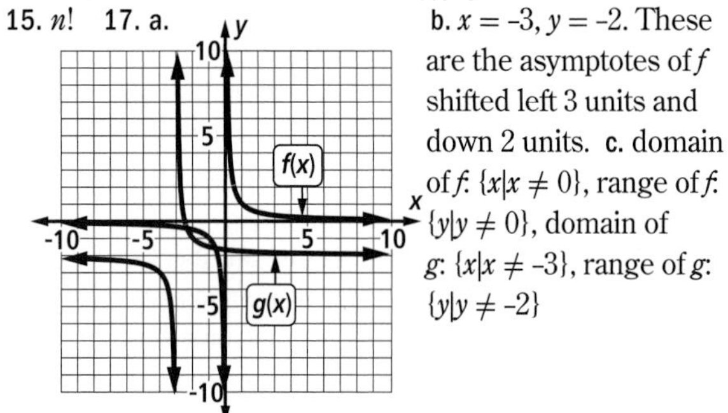

b. $x = -3, y = -2$. These are the asymptotes of f shifted left 3 units and down 2 units. **c.** domain of f: $\{x | x \neq 0\}$, range of f: $\{y | y \neq 0\}$, domain of g: $\{x | x \neq -3\}$, range of g: $\{y | y \neq -2\}$

Lesson 6-8 (pp. 410-416)

Guided Example 2 a. 0.5; 0.5; 5000 **b.** 5000; 67 **c.** 10,000; 0.5067 **d.** 0.5067; 0.5

Questions

1. 3.6 **3. a.** $\frac{23}{3} \approx 7.67$ **b.** $\frac{23}{1650} \approx 0.014$ **c.** Yes, the difference between the probability and the relative frequency is not large. **5.** Answers vary. Sample: She is basing this prediction on the invalid "law of averages." The probability of rolling a 6 on each turn is still $\frac{1}{6}$. **7.** They should attempt a touchdown, because the expected outcome of a field goal attempt is 1.05 points and the expected outcome of a touchdown attempt is 1.2 points. **9.** 32.5 minutes **11. a.** The expected count of points for 1 question is 0. Therefore, 10 questions also have an expected count of 0. **b.** The outcome from guessing on n random questions is 0 points. **c.** No.

13. a.

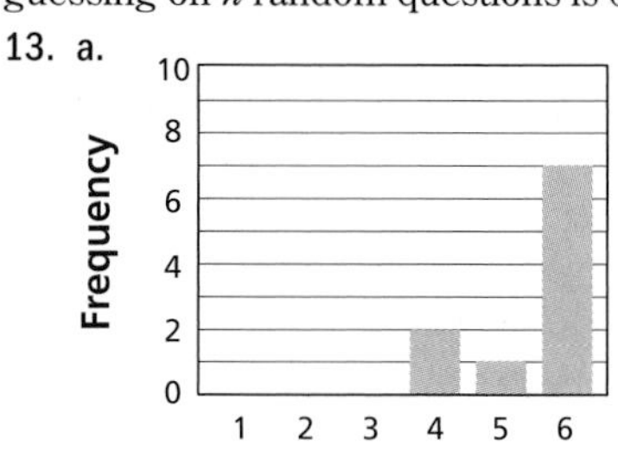

b. No, she has only performed 10 trials, this is not enough to make any long-run conclusions. **c.** No, in fact the new data refute her assertion because the relative frequencies are now closer to $\frac{1}{6}$. **d.** The Law of Large Numbers **15. a.** Answers vary. Sample: Generate numbers from 1 to 100, once for each woman. For women, numbers 1-12 have attached ear lobes and 13-100 do not. A random simulation of 10 trials returned a relative frequency of $\frac{2}{10}$ where at least one of the two has attached ear lobes. **b.** $1 - (0.88)^2 \approx 0.2256$ **17. a.** 210 **b.** 60 **19.** $y = \left(x - \frac{1}{3}\right)^2 - 2$. **21.** Die 3 is green, die 1 is red, and die 2 is blue.

Lesson 6-9 (pp. 417-425)

Guided Example 2 30; 20; 25; 25; 1; 20; 25; 25; 1; 1; 1

Questions

1. a. the observed number of survivors in a given class **b.** the expected number of survivors in a given class **3.** $\frac{(178 - 228)^2}{228} \approx 11.0$ **5. a.** 5.12 **b.** about 0.024 **c.** Because $0.024 < 0.05$, we rejected the hypothesis that the coin is fair. **7.** There is no reason to believe the die is biased because the probability of χ^2 as large as 3.1 is about 0.685.

9. a.

Year	2001	2002	2003	2004	2005	2006	2007	Total
Observed Number of Cyclones	15	12	16	15	28	10	15	111.0
Expected Number of Cyclones	15.86	15.86	15.86	15.86	15.86	15.86	15.86	111.0

b. about 12.54 **c.** about 0.0509 **d.** The results do not provide sufficient evidence to reject the hypothesis at the 0.05-significance level because $0.051 > 0.05$. **11.** The probability of a chi-square value as large as 0.38 is about 0.941 so because $0.941 > 0.05$, then at the 0.05-significance level there is insufficient evidence to reject the claim that point scoring by quarter is the same. **13.** 12

15. Answers vary. Sample: Run a simulation with 100 trials. In each trial you generate a random number from 1 to 100, where 1-45 means the patient has recovered fully, and 46-100 means the patient has not. Add the number of successes to find the relative frequency of recovery.

17. a. $-\frac{7}{9} \approx -0.78$ **b.** $\frac{2}{9} \approx 0.22$ **c.** 1

Self-Test (pp. 430-431)

Questions

1. a. {HHH, HHT, HTH, HTT, THH, THT, TTH, TTT} **b.** HTT, THT, TTT, TTH **2. a.** ${}_{15}P_6 = \frac{15!}{(15-6)!} = \frac{15!}{9!} = 15 \cdot 14 \cdot 13 \cdot 12 \cdot 11 \cdot 10 = 3{,}603{,}600$. **b.** ${}_{15}P_6$ is the number of all permutations of 6 objects from 15 objects. **3.** ${}_9P_4 = \frac{9!}{(9-4)!} = \frac{9!}{5!} = 9 \cdot 8 \cdot 7 \cdot 6 = 3024$ **4.** Tossing the coin has 2 possible outcomes and rolling the die has 6 possible outcomes, so there are 12 possible outcomes in the sample space, by the Multiplication Counting Priciple. In five of these outcomes, the sum of the outcomes of the two events is less than 5: {1 + 3, 1 + 2, 1 + 1, 2 + 2, 2 + 1}. Therefore the probability is $\frac{5}{12}$. **5. a.** The lengths of flosses in the packages are independent, so the probability is $(0.15)^4 \approx 0.0005$. **b.** The probability that a package has at least the advertised amount is $1 - 0.15 = 0.85$ because the two events are complements. The probability that all four packages have at least the stated amount is $(0.85)^4 \approx 0.5220$. **6.** Use the Multiplication Counting Principle. $4 \cdot 5 \cdot 2 = 40$ different outfits. **7. a.** By the Addition Counting Principle, $P(A \cup B) = P(A) + P(B) - P(A \cap B) = 0.7 + 0.8 - 0.6 = 0.9$ **b.** $P(B|A) = \frac{P(A \cap B)}{P(A)} = \frac{0.6}{0.7} \approx 0.86$
8. a. Answers vary. Sample:

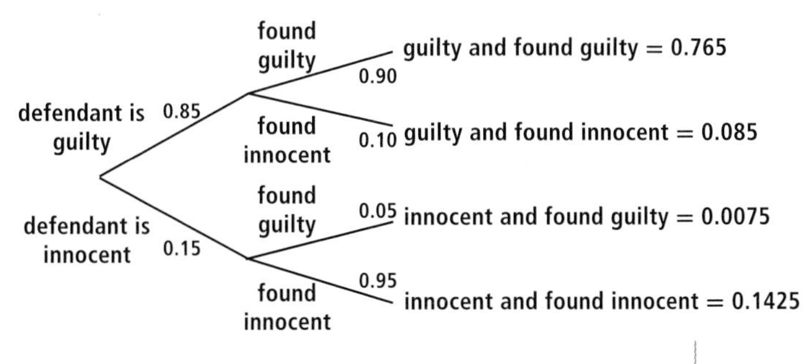

b. $\frac{P(\text{innocent and found guilty})}{P(\text{found guilty})} = \frac{0.0075}{0.0075 + 0.765} \approx 0.97\%$
c. $\frac{P(\text{guilty and found innocent})}{P(\text{found innocent})} = \frac{0.085}{0.085 + 0.1425} \approx 37.36\%$

9. true; If A and B are complementary, then they are mutually exclusive and thus make up the entire sample space. Since $P(S) = 1$, $P(A) + P(B) = 1$, and therefore $P(B) = 1 - P(A) = 1 - k$. **10.** false; For a fair die, the probability of rolling a 2 is $\frac{1}{6}$. The expected number of "2"s in 600 rolls is $\frac{1}{6} \cdot 600 = 100$. **11.** No. A and B are independent events if and only if $P(A \cap B) = P(A) \cdot P(B)$. Here, $P(A) \cdot P(B) = 0.21 \cdot 0.17 = 0.0357 \neq P(A \cap B) = 0.02$. **12.** Calculate the expected distribution of majors and use this with the observed survey results to do the chi-square test: $\chi^2 = \frac{(125-156)^2}{156} + \frac{(312-351)^2}{351} + \frac{(138-117)^2}{117} + \frac{(205-156)^2}{156} \approx 29.65$. The probability of a χ^2 value of 29.65 or larger is less than 0.001, so the new survey yields sufficient evidence to reject the hypothesis that the proportions by major are still what the guidance office states. **13.** By totaling the rows and columns we find that the total dorm population is 1020 and the total car-owning population is 290. $\frac{290}{1020} \approx 28.4\%$. **14.** The total car-owning population is 290, and there are 70 juniors who own cars. $\frac{70}{290} \approx 24.1\%$. **15.** Since $.325 = \frac{13}{40}$, simulate this situation by generating random numbers between 1 and 40, where 1-13 represents a hit and 14-40 represents not a hit. Generating eight random numbers equals one trial. A trial is successful if 4 or more of the numbers generated are a hit (1-13). Run this simulation for forty trials and use the relative frequency of success to estimate the probability. **16. a.** If the die is fair, the probability of rolling a 2 is $\frac{1}{6}$, so the expected count is $\frac{1}{6} \cdot 12{,}000 = 2{,}000$. **b.** Answers vary. Sample: You might conclude that the die is unfair because 1 occurs very frequently, and 4 occurs rarely. You could do a chi-square test to test your hypothesis.
c. No, the sample is too small. By the Law of Large Numbers, the relative frequency of each of the outcomes should approach their probability as the number of trials increases. A chi-square test might be inaccurate on a sample this small.

Self-Test Correlation Chart

Question	1	2	3	4	5	6	7	8	9	10
Objective(s)	G	C	B	A	I	J	D	I	D	A
Lesson(s)	6-1	6-4	6-4	6-1	6-3	6-3	6-2, 6-6	6-6	6-2	6-8
Question	**11**	**12**	**13**	**14**	**15**	**16**				
Objective(s)	E	L	M	H	K	F				
Lesson(s)	6-3	6-9	6-5	6-5	6-7	6-8				

Chapter Review (pp. 432-435)

Questions

1. $\frac{5}{12}$ **3.** $\frac{1}{6}$ **5.** $\frac{1}{8}$ **7.** $\frac{7}{8}$ **9.** $38.\overline{3}$ **11.** $8! = 40{,}320$ **13.** ${}_7P_5 = 2520$ **15.** $2^8 = 256$ **17.** 21 **19.** $54 \cdot 53 \cdot 52 \cdot 51$ **21.** 1320 **23.** n **25.** 0 **27.** 0.45 **29.** independent **31.** independent **33.** The announcer is incorrectly applying the "law of averages", assuming that the observed count should approach the expected count. **35.** expected winnings: $(0)(0.999) + (700)(0.001) = \$0.70$; Someone who plays everyday has expected winnings of $(\$0.70)(365) = \$255.5$ in one year but spent $365 during the year. In the long run, the winnings will approach $0.70 per lottery ticket and therefore yield a profit of $0.30 per lottery ticket for the state. **37.** integers from 0 to 359 **39.** 366 days **41.** about 26.7%; about 28.0% **43.** $\frac{15}{1024} \approx 0.015$ **45.** 0.96% **47. a.** about 45.2% **b.** about 24.0% **49.** 36 **51.** 1440 **53.** Answers vary. Sample: The experiment in 52 can be modified by changing the value of one of the penny columns to 5 and running another 100 trials. **55. a.** In the same order: 375, 1125, 1125, 3375 **b.** The degree of freedom is 3.

c.

$$\frac{(315-375)^2}{375}+\frac{(1202-1125)^2}{1125}+\frac{(1146-11}{1125}$$

15.6901

$\chi^2\text{Cdf}(15.69,\infty,3)$ 0.001313

The probability of a chi-square value as high as 15.69 or greater is 0.0013. Because $0.0013 < 0.01$ we can reject the geneticist's claim.

57. The chi-square simulation web applet gives a chi-square value of 4.296 and a probability of 11.8%. Because $0.118 > 0.05$, we have insufficient evidence to reject the claim of the hypothesized ratios of morning rush hour traffic.

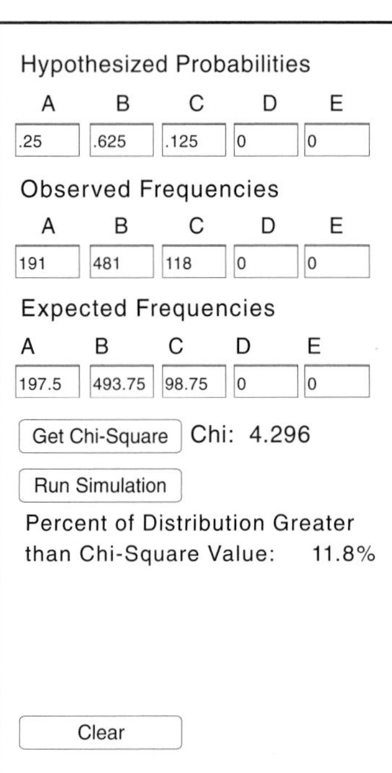

59. a. about 12.5%, about 13.1%
b. about 27.6%, about 28.9%
c. No, whether a student studies more or less than three hours, the relative frequency of a score increase of less than 25 points appears to be the same. The same is true for students whose scores increased more than 25 points.
d.

	< 25	> 25	Total
Less Study	107	376	483
More Study	63	348	411
Total	170	724	894

Score increase of < 25 points with less study: 22.2%; score increase of < 25 points with more study: 15.3% **e.** These data exemplify Simpson's Paradox, since there appears to be a correlation between amount of study and performance when you combine the school districts, but not when you examine them separately.

Chapter 7

Lesson 7-1 (pp. 438–445)

Guided Example 3 **a.** $x < a; x > g$ **b.** $b < x < d; x > f$

Questions

1. The expression has a variable with a negative integer exponent, $8x^{-3}$. **3.** A box with height 30 cm would have length $60 - 2 \cdot 30 = 0$ cm, and thus zero volume.

5. a. $a < x < c$ and $x > e$ **b.** $x < a$ and $c < x < e$ **c.** $x < b$ and $x > d$ **d.** $b < x < d$

7. a.

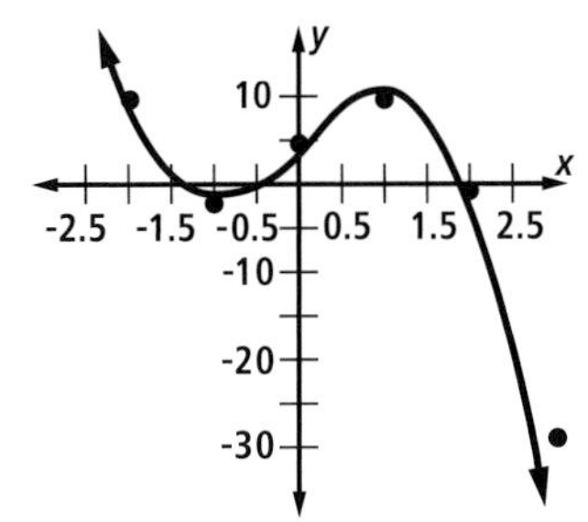

b. -2 and -1; -1 and 0; 1 and 2

9. a. II

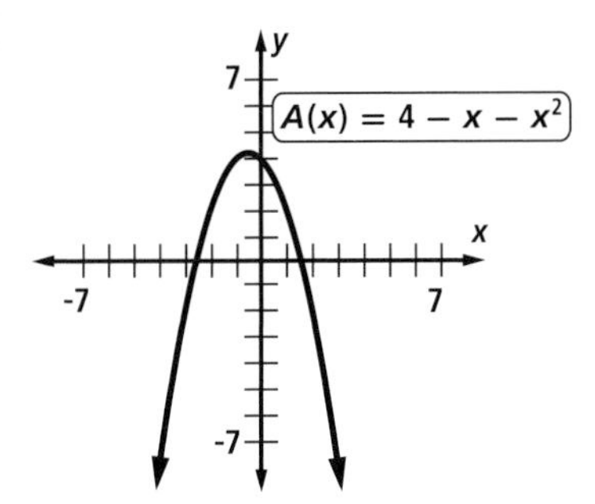

b. I

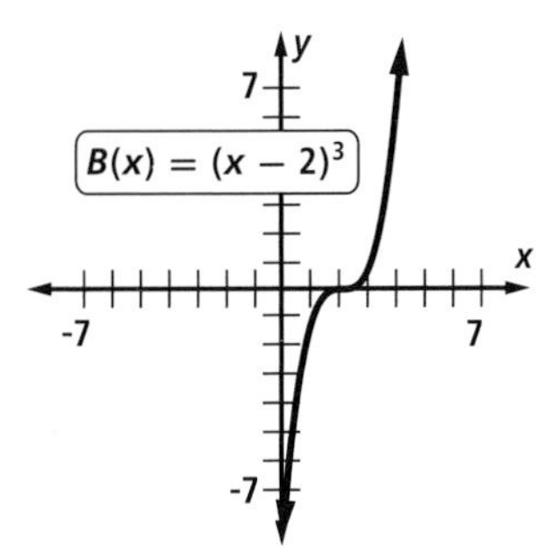

c. I

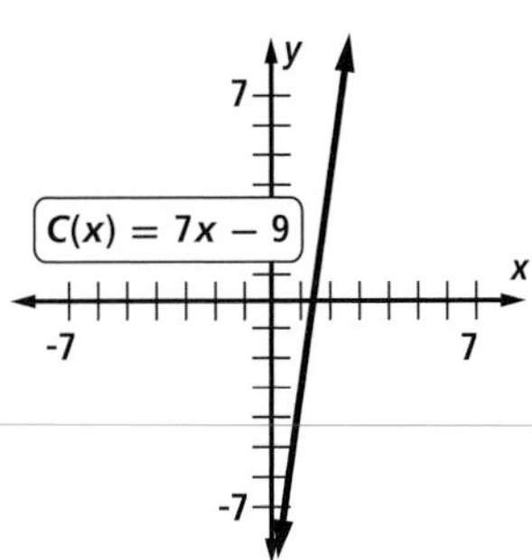

d. III

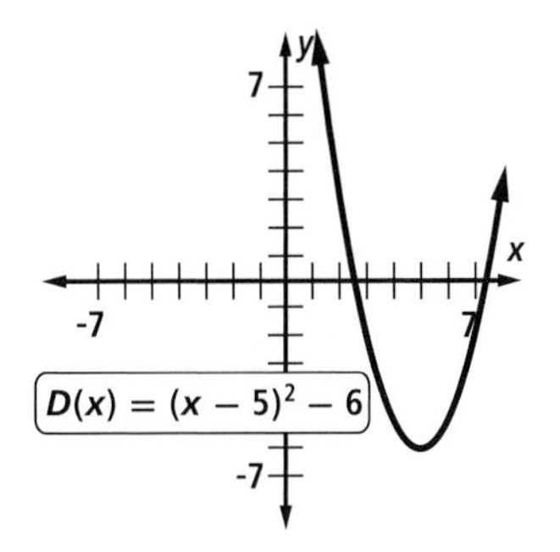

e. IV

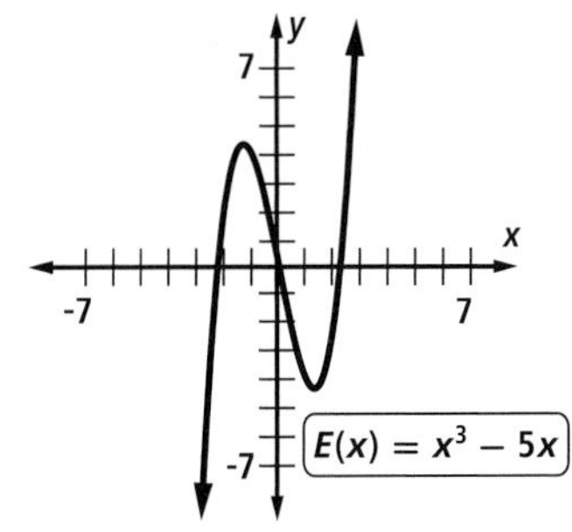

11. One minimum value at (2, 3)

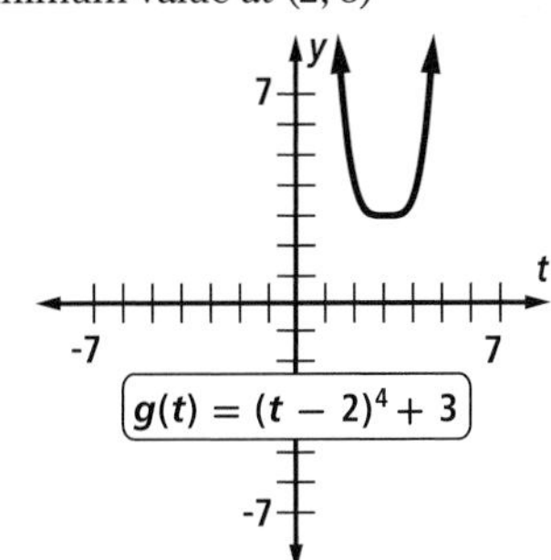

13. One minimum value at about (5.3, -254.2)

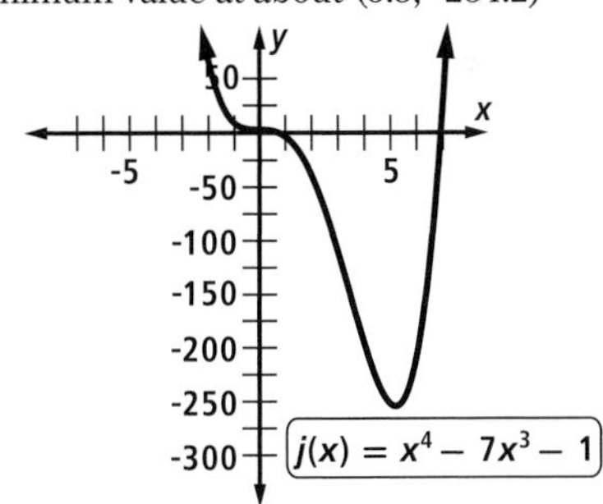

15. zeros: $x = -3, -\sqrt{2}, \sqrt{2}$; relative minimum at $\left(\frac{\sqrt{15}-3}{3}, \frac{-10\sqrt{15}}{9} - 2\right)$; relative maximum at $\left(\frac{-\sqrt{15}-3}{3}, \frac{10\sqrt{15}}{9} - 2\right)$

Define $g(x)=x^3+3\cdot x^2-2\cdot x-6$	*Done*
$\text{zeros}(g(x),x)$	$\{-3,-\sqrt{2},\sqrt{2}\}$
$\text{fMax}(g(x),x,-3,-\sqrt{2})$	$x=\frac{-(\sqrt{15}+3)}{3}$
$\text{fMin}(g(x),x,-\sqrt{2},\sqrt{2})$	$x=\frac{\sqrt{15}-3}{3}$

17. about 6400 miles 19. $T(x, y) = (x - 3, y - 7)$
21. $(x - 9)^2$

Lesson 7-2 (pp. 446-452)

Guided Example 1 6; 10; 15; 21; 4; 5; 6; 1; 1

Questions

1. a. $T(7) = T(6) + t(7) = 56 + 28 = 84$ b. $T(7) = \frac{1}{6}(7)^3 + \frac{1}{2}(7)^2 + \frac{1}{3}(7) = 84$ 3. The x-values must form a sequence of consecutive terms of an arithmetic sequence. 5. no 7. a. 3 b. $f(x) = 2x^3 - x^2 + 8x$ c. 693 9. 161.2; 376.4; 718.8; 1217.2; 116.8; 215.2; 342.4; 498.4; 98.4; 127.2; 156.0; $a = 194.6$; $b = 571$; $c = 1289.8$; $d = 2507$; $n = 4$ 11. a. Let $n = 1, 2, 3$. $f(n) = \frac{3}{2}n^2 - \frac{3}{2}n + 1$; $f(1) = 1 = T(1)$; $f(2) = 4 = T(2)$; $f(3) = 10 = T(3)$ b. $f(4) = 19 \neq T(4)$ 13. $f(x)$ has no relative extrema. 15. a. $P(n) = 141{,}400{,}000(0.995)^n$ b. 132.5 million c. 2032

Lesson 7-3 (pp. 453-458)

Guided Example 2 3, 7, 6; 3; 3; -7; -7; -6; -6; 3, 7, 6; 3, 7, 6;

Questions

1. $2x + 3$ 3. a. $x^2 + 7x - 1$ b. $x - 4$ 5. Answers vary. Sample: $q(x) = x$, $r(x) = 13$ 7. a. -13

b.

$$
\begin{array}{r}
y^2 - 9 \\
y + 3\overline{)y^3 + 3y^2 - 9y - 40} \\
\underline{y^3 + 3y^2} \qquad\qquad \\
-9y - 40 \\
\underline{-9y - 27} \\
-13
\end{array}
$$

9. 0 11. a. $q(a) = 3a + 5$; $r(a) = 12a + 10$ b. degree of remainder = 1; degree of divisor = 2 c. $(3a + 5)(a^2 - 4) + (12a + 10) = 3a^3 + 5a^2 - 10$ 13. $x^2 + 3xy + y^2$

15. a.

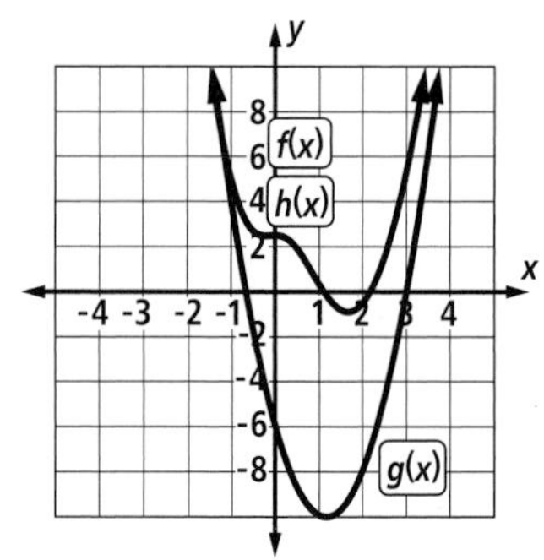

b. Answers vary. Sample: The graphs of $f(x)$ and $h(x)$ are identical. The graph of $g(x)$ differs from them both, particularly about the origin, suggesting that the remainder is important to describing the behavior of the function.

17. a.

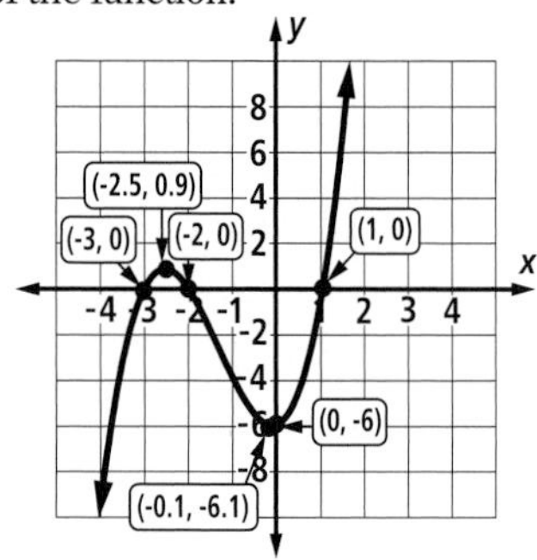

b. $-3 < s < -2$ and $s > 1$
c. $s < -3$ and $-2 < s < 1$
d. $s < -2.5$ and $s > -0.1$
e. $-2.5 < s < -0.1$
19. $7! = 5040$
21. a. $x = 0$ or $x = -8$
b. $y = 4$ or $y = 7$
c. $z = 6$, $z = -2$, or $z = -11$

Lesson 7-4 (pp. 459-464)

Questions

1. For a polynomial $p(x)$, $(x - c)$ is a factor of $p(x)$ if and only if $p(c) = 0$. 3. $f(x) = (x - 7)(x - 3)(x + 2)(x + 5)$ 5. Answers vary. Sample: $p(x) = x(x - 6)(x + 4) = x^3 - 2x^2 - 24x$ 7. Answers vary. Any two of: $(x + 2)$ is a factor of $g(x)$; $g(-2) = 0$; -2 is an x-intercept of the graph of $y = g(x)$; the remainder when $g(x)$ is divided by $(x + 2)$ is 0. 9. Answers vary. Any two of: $(t - 2)$ is a factor of $f(t)$; $f(2) = 0$; 2 is a zero of $f(t)$; 2 is an x-intercept of the graph of $y = f(t)$. 11. a.

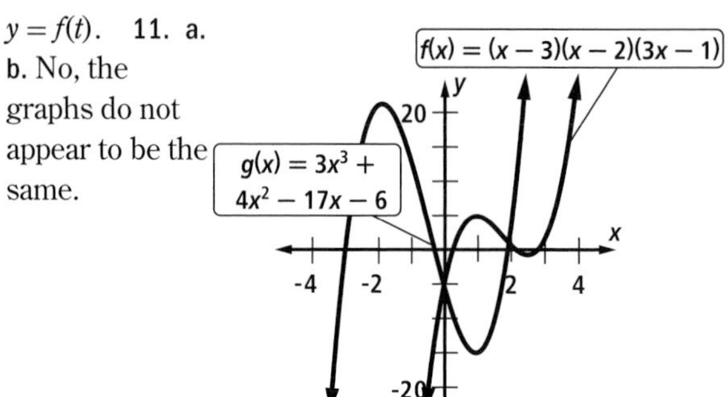

b. No, the graphs do not appear to be the same.

13. i. a.

expand((x–a)·(x–b)) $x^2 - a \cdot x - b \cdot x + a \cdot b$

b. The opposites of the zeros give coefficients for the linear terms, and the constant term is the product of the zeros.

ii. a. Expansion: $x^3 - (a + b + c)x^2 + (ab + ac + bc)x - abc$

expand((x–a)·(x–b)·(x–c))
$x^3 - a \cdot x^2 - b \cdot x^2 - c \cdot x^2 + a \cdot b \cdot x + a \cdot c \cdot x + b \cdot c \cdot x$▸

b. The opposites of the zeros give the coefficients for the quadratic terms; the three possible products of two zeros give coefficients for the linear terms; and the opposite of the product of all the zeros gives the constant term.

iii. a. Expansion: $x^4 - (a + b + c + d)x^3 + (ab + ac + ad + bc + bd + cd)x^2 - (abc + abd + acd + bcd)x + abcd$

expand((x–a)·(x–b)·(x–c)·(x–d))
$x^4 - a \cdot x^3 - b \cdot x^3 - c \cdot x^3 - d \cdot x^3 + a \cdot b \cdot x^2 + a \cdot c \cdot x$▸

b. The opposites of the zeros give the coefficients for the cubic terms; the possible products of two zeros give coefficients for the quadratic terms; the possible products of the opposites of three zeros give coefficients for linear terms; and the product of all the zeros gives the constant term. 15. Answers vary. Sample: $g(x) = (x + 8)(x)(x - 5.1)(x - 10) = x^4 - 7.1x^3 - 69.8x^2 + 408x$
17. $3z^2 - \frac{3}{2}z - 3$ 19. a. 1st differences: 4, 7, 10; 2nd differences: 3, 3; The 2nd differences are constant, so p has degree 2. b. $p(n) = \frac{3}{2}n^2 - \frac{1}{2}n$

c.

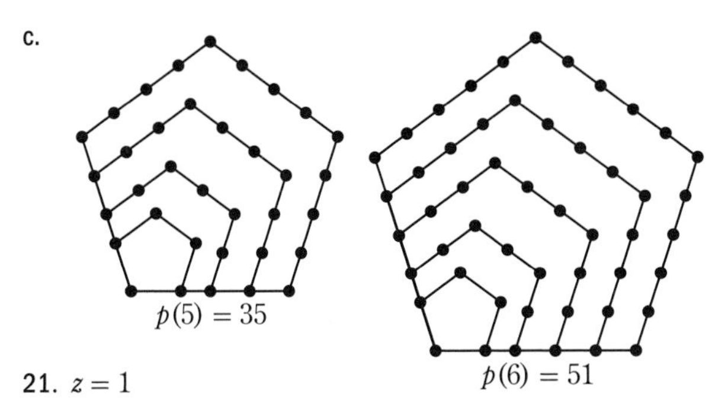

21. $z = 1$

Lesson 7-5 (pp. 465-472)

Guided Example 3 a. $2 - 3i$; 2; $3i$ b. $2 - 3i$; 2; $3i$; $5i + 3i$; $8i$ c. $10i$; $15i^2$; $2i$; 15 d. 2; $\frac{5}{2}$

Questions

1. $(i\sqrt{13})^2 = i^2 \cdot 13 = -1 \cdot 13 = -13$; $(-i\sqrt{13})^2 = (-i)^2 \cdot 13 = -1 \cdot 13 = -13$ 3. $3i\sqrt{2}$ 5. a. $3 + 2i$ b. 13 7. a. $10i$ b. -10 9. a. -91 b. $x = \frac{3}{2} + \frac{\sqrt{91}}{2}i$; $x = \frac{3}{2} - \frac{\sqrt{91}}{2}i$ 11. They are complex conjugates. 13. a. $7(z - i\sqrt{2})(z + i\sqrt{2})$ b. $7(z - i\sqrt{2})(z + i\sqrt{2}) = 7(z^2 + i\sqrt{2}z - i\sqrt{2}z - 2i^2) = 7(z^2 + 2) = 7z^2 + 14$ 15. a. False; Only complex numbers of the form $a + 0i$ are real. b. True; Real numbers are complex numbers of the form $a + 0i$. 17. $\frac{12 - i}{12 - i}$ 19. $\frac{8}{5} - \frac{4}{5}i$ 21. $-115 - 236i$ 23. $i^2 = -1$; $i^3 = -i$; $i^4 = 1$; $i^5 = i$ 25. a. $f(3 + i) = 0$; $f(3 - i) = 0$ b. $f(x) = (x - (3 + i))(x - (3 - i)) = (x - 3 - i)(x - 3 + i)$ 27. a. -1 b. -1 c. -1 d. -1 29. Answers vary. Sample: $f(x) = (x + 1)(x - 3)(4x + 3)$; $g(x) = 2(x + 1)(x - 3)(4x + 3)$; $h(x) = (x + 1)^2(x - 3)(4x + 3)$ 31. 2 33. a. 3; 0; 4; 2; 5; 5; 6; 9 b. $d(n) = \frac{1}{2}n^2 - \frac{3}{2}n$ c. exact d. 1175

Lesson 7-6 (pp. 473-480)

Guided Example 1 $x^2 - 24x + 144$; 3; 1; 2

Questions

1. $x^2 - 4wz$ 3. If $p(x)$ is any polynomial of degree $n \geq 1$ with complex coefficients, then $p(x)$ has at least one complex zero. 5. true 7. true 9. true 11. a. $x = 13$ has multiplicity 2 and $x = -6$ has multiplicity 3.

b.

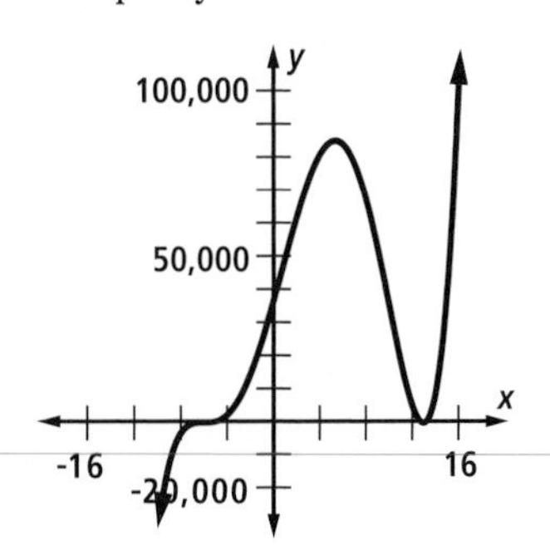

13. a. 15 b. 1 15. 4 17. 1, -1, i, $-i$ 19. $p(x) = x^3 - 4x^2 + 14x - 20$ 21. a. Yes. Answers vary. Sample: The graph shown could be part of, for example, the graph of $p(x) = (x - 1)(x + 1)(x - 4)(x + 4)$. b. Yes. Answers vary. Sample: Any horizontal line $y = d$ crosses the graph no more than 4 times. c. No. Answers vary. Sample: The graphed polynomial has 6 x-intercepts and so is at least of degree 6. 23. a. $6 + 5i$ b. $12 + 0i$ c. $0 + 10i$ d. $61 + 0i$ e. $\frac{11}{61} + \frac{60}{61}i$ 25. about 127.2° 27. $\frac{7\pi}{4}$

Lesson 7-7 (pp. 481-485)

Guided Example 3 10; 3; 100; 30; 9; 13; 79

Questions

1. $(-1 - i\sqrt{3})^3 = (-1 - i\sqrt{3})(1 + i\sqrt{3} + i\sqrt{3} - 3) = (-1 - i\sqrt{3})(-2 + 2i\sqrt{3}) = (2 + 2i\sqrt{3} - 2i\sqrt{3} - 6i^2) = 2 + 6 = 8$ 3. a. $m(x) = (x - 5)(x^2 + 5x + 25)$; zero at $x = 5$

3. b.

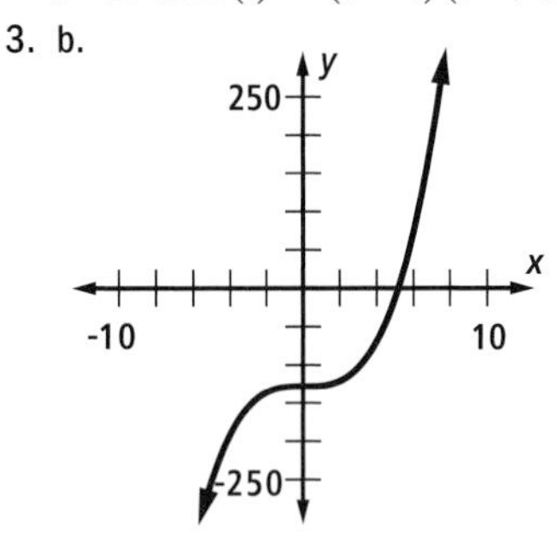

5. a. $(t - 2)(t^6 + 2t^5 + 4t^4 + 8t^3 + 16t^2 + 32t + 64)$ b. $(t - 2)(t^6 + 2t^5 + 4t^4 + 8t^3 + 16t^2 + 32t + 64) = t^7 + 2t^6 + 4t^5 + 8t^4 + 16t^3 + 32t^2 + 64t - 2t^6 - 4t^5 - 8t^4 - 16t^3 - 32t^2 - 64t - 128 = t^7 - 128$

7. a. Difference of squares b. $(a + 8b)(a - 8b)$ 9. a. Difference of cubes b. $(50 - 4d)(2500 + 200d + 16d^2)$ 11. $1001 = (10^3 + 1^3) = (10 + 1)(10^2 - 10 + 1) = 11(91) = 11(7)(13)$; Its prime factorization is $7 \cdot 11 \cdot 13$. 13. False. $x + y$ is not a factor because $x^4 + y^4$ is the sum of two even-degree powers. 15. $6x^2(2y - x)(2y + x)$ 17. a. $(x^2 - 4)(x^4 + 4x^2 + 16)$ b. $(x - 2)(x + 2)(x^2 + 2x + 4)(x^2 - 2x + 4)$ 19. False. The Fundamental Theorem of Algebra guarantees the existence of at least one complex solution. 21. $\frac{101}{116} + \frac{16}{29}i$ 23. Answers vary. Sample: $f(x) = (x + 2)(6x - 7)(x - 9)$

Lesson 7-8 (pp. 486-490)

Guided Example 4 a. $x^2 + x + 1$; $x + 1$; $x^2 + x + 1$; $x + 1$ b. $\frac{-1 \pm \sqrt{1^2 - 4(1)(1)}}{2(1)}$; $-\frac{1}{2} \pm \frac{\sqrt{3}}{2}i$

Questions

1. a. $t^3 - 3t^2 - 4t + 12 = t^3 - 4t + (-3t^2 + 12) = t(t^2 - 4) - 3(t^2 - 4) = (t - 3)(t^2 - 4) = (t - 3)(t - 2)(t + 2)$ b. $(t - 2)(t + 2)(t - 3) = (t^2 - 4)(t - 3) = t^3 - 4t - 3t^2 + 12$

3. a. $2(x - 1)(x - 2)(x + 1)$; zeros at 1, 2, and -1 b. zeros at 1, 2, and -1

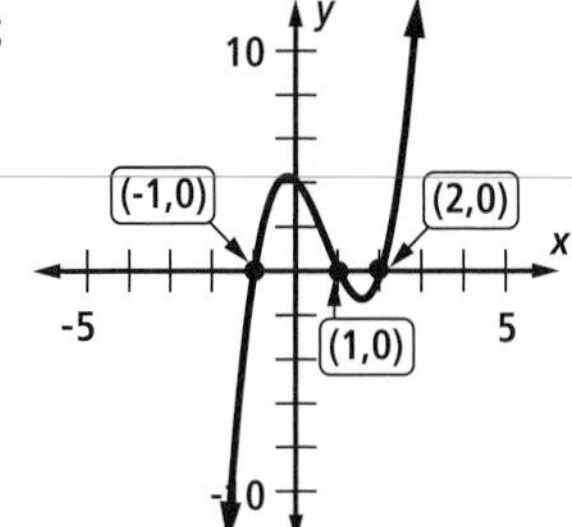

5. a.

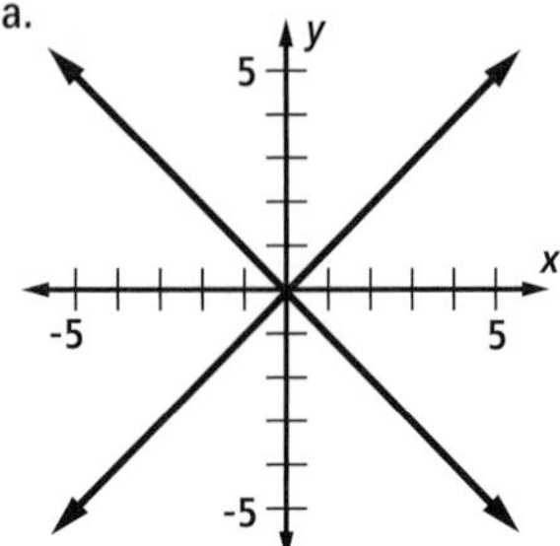

b. Answers vary. Sample: The graph is the union of the lines $y = -x$ and $y^2 = x^2$, and thus, the union of the lines $y = x$ and $y = -x$.
7. $(a - b)(x + y + z)$

9.

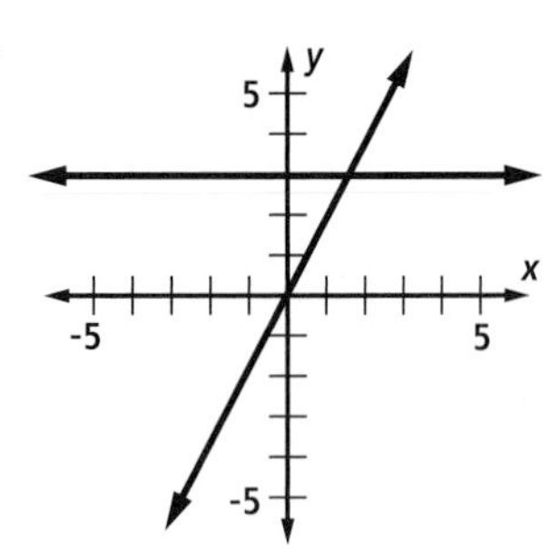

11. $x + b + 5$ when $x \neq b$. 13. a. $x = \frac{1}{2}, 1$ b. For any integer n, $\theta = 2\pi n, \frac{\pi}{3} + 2\pi n, -\frac{\pi}{3} + 2\pi n$ 15. $(t + u)(t^2 - tu + u^2)$
17. $-(4x^2 - y^2z)(16x^4 + 4x^2y^2z + y^4z^2)$
19. Answers vary. Sample: Let the digits 0, 1, and 2 denote rain on one day, and let the digits 3, 4, 5, 6, 7, 8, and 9 denote no rain. Choosing one random digit between 0 and 9 twice represents a simulation of the weather over the weekend.
21. interpolation; extrapolation

Self-Test (pp. 494–495)

1.

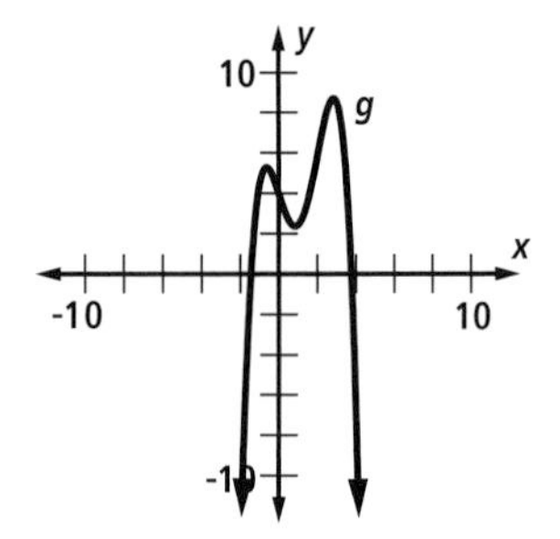

2. By tracing the graph, a relative maximum point is about (–0.64, 5.31). The maximum point of the graph is about (2.81, 8.77). A relative minimum point is about (0.83, 2.42).

3. By tracing, using the intersect command on a graphing utility, or solving $g(x) = 0$, the zeros are about -1.45 and 3.72. 4. Since the largest degree of any term is 4, the polynomial is of degree 4. 5. On the graph, two real zeros can be seen where the graph crosses the axes. Therefore, there are $4 - 2 = 2$ nonreal zeros.

6.

$$
\begin{array}{r}
3x^3 - 13x^2 + 29x - 59 \\
x + 2 \overline{)\, 3x^4 - 7x^3 + 3x^2 - x - 2} \\
\underline{3x^4 + 6x^3} \\
-13x^3 + 3x^2 \\
\underline{-13x^3 - 26x^2} \\
29x^2 - x \\
\underline{29x^2 + 58x} \\
-59x - 2 \\
\underline{-59x - 118} \\
116
\end{array}
$$

The quotient is $3x^3 - 13x^2 + 29x - 59$ and the remainder is 116. 7. A; $f(1) = 0$, so by the Factor Theorem, $(x - 1)$ is a factor of f. 8. Answers vary. Sample: Because $2 - i$ is a zero, its conjugate, $2 + i$, must also be a zero. By the Factor Theorem, the polynomial must have factors $(x - 5)$, $(x - 2 + i)$ and $(x - 2 - i)$. Multiply these to get one possible polynomial, $x^3 - 9x^2 + 25x - 25$. 9. a. $3 - 2i$, the conjugate of $3 + 2i$. b. $\frac{2 - 3i}{3 + 2i} = \frac{(2 - 3i)(3 - 2i)}{(3 + 2i)(3 - 2i)} = \frac{6 - 13i + 6i^2}{9 - 4i^2} = 0 - i = -i$ 10. a. $2(2 + 4i) + (3 + 7i) = 7 + 15i$ b. $(2 + 4i)(3 + 7i) = 6 + 14i + 12i + 28i^2 = -22 + 26i$ 11. a. The first differences are 4, 7, 10, 13, 16. The second differences are 3, 3, 3, 3. Because the 2nd differences are constant, the polynomial is of degree 2. Using quadratic regression, $P_n = \frac{3}{2}n^2 - \frac{1}{2}n$.
b. $P_{20} = \frac{3}{2} \cdot 20^2 - \frac{1}{2} \cdot 20 = 590$ 12. a. $\ell + 2w = 500$, so $w = 250 - \frac{1}{2}\ell$. b. $A = \ell w = \ell(250 - \frac{1}{2}\ell) = 250\ell - \frac{1}{2}\ell^2$
c. Because the length and width must be positive, $\ell > 0$ and $w = 250 - \frac{1}{2}\ell > 0$. Then $\ell > 0$ and $\ell < 500$, so the domain is $0 < \ell < 500$. 13. $9ab + 18b + 5a + 10 = (9ab + 18b) + (5a + 10) = 9b(a + 2) + 5(a + 2) = (9b + 5)(a + 2)$ 14. Sample: $64m^6 + 27k^3 = (4m^2)^3 + (3k)^3 = (4m^2 + 3k)((4m^2)^2 - (4m^2)(3k) + (3k)^2) = (4m^2 + 3k)(16m^4 - 12m^2k + 9k^2)$ 15. Answers vary. Sample: Let $u = x^5$. Then $x^{10} - 1 = u^2 - 1 = (u - 1)(u + 1) = (x^5 - 1)(x^5 + 1) = (x - 1)(x + 1)(x^4 + x^3 + x^2 + x + 1)(x^4 - x^3 + x^2 - x + 1)$.
16. By the Polynomial Difference Theorem, the polynomial has degree 5. 17. Count the number of x-intercepts of each graph: R has 3 real zeros; S has 1 real zero; T has 1 real zero. 18. (1, 0); $R(1) = S(1) = T(1) = 0$, so by the Factor Theorem (1, 0) is a point on each graph. 19. C; S has degree 3, and because it has no nonreal zeros, by the Number of Zeros of a Polynomial Theorem, it must have 3 real zeros. Because 1 is the only x-intercept, and every real zero is an x-intercept, 1 must be a zero of multiplicity 3. 20. Because function T has degree 3 and has only 1 real zero, by the Number of Zeros of a Polynomial Theorem, it has 2 nonreal zeros.
21. There is a horizontal line that the graph of $p(x)$ appears to cross 5 times, so by the Polynomial Graph Wiggliness Theorem, $p(x)$ has degree at least 5.

Self-Test Correlation Chart

Question	1	2	3	4	5	6	7	8	9	10
Objective(s)	J	A	A	F	H	C	G	H	E	E
Lesson(s)	7-1	7-1	7-1	7-1	7-6	7-3	7-4	7-6	7-5	7-5
Question	**11**	**12**	**13**	**14**	**15**	**16**	**17**	**18**	**19**	**20**
Objective(s)	I	I	D	D	D	B	J	G	F	J
Lesson(s)	7-2	7-1	7-8	7-7	7-8	7-2	7-1	7-4	7-6	7-6
Question	**21**									
Objective(s)	J									
Lesson(s)	7-6									

Chapter Review (pp. 496-499)

1. about -0.414 **3.** Because $t = 4$ is a solution to $g(t) = 0$, 4 must be a t-intercept of the graph of g. **5.** about -6.2 **7.** 5 **9. a.** y is a polynomial of degree 2. **b.** $y = 2x^2 - 3x + 5$ **11. a.** $h_n = \frac{5}{2}n^2 - \frac{3}{2}n$ **b.** 66178 **13.** Quotient: $x^3 + 2x^2 + 4x + 14$; Remainder: 0

$$\begin{array}{r} x^3 + 2x^2 + 4x + 14 \\ x - 3\overline{)x^4 - x^3 - 2x^2 + 2x - 42} \\ \underline{x^4 - 3x^3} \qquad\qquad\qquad\quad \\ 2x^3 - 2x^2 \qquad\qquad\quad \\ \underline{2x^3 - 6x^2} \qquad\qquad\quad \\ 4x^2 + 2x \qquad\quad \\ \underline{4x^2 - 12x} \qquad\quad \\ 14x - 42 \\ \underline{14x - 42} \\ 0 \end{array}$$

15. Quotient: $4x^3 - 4x^2 + 12x - 36$; Remainder: 225

$$\begin{array}{r} 4x^3 - 4x^2 + 12x - 36 \\ 2x + 6\overline{)8x^4 + 16x^3 + 0x^2 + 0x + 9} \\ \underline{8x^4 + 24x^3} \qquad\qquad\qquad\quad \\ -8x^3 + 0x^2 \qquad\qquad\quad \\ \underline{-8x^3 - 24x^2} \qquad\qquad\quad \\ 24x^2 + 0x \qquad\quad \\ \underline{24x^2 + 72x} \qquad\quad \\ -72x + 9 \\ \underline{-72x - 216} \\ 225 \end{array}$$

17. $x = y$, $x \approx (-0.81 \pm 0.59i)y$, $(0.31 \pm 0.95i)y$ **19.** $-\sqrt{5}$ **21.** Answers vary. Sample: $x^3 - 4x^2 + 6x - 4$ **23. a.** $(x - \sqrt{6})(x + \sqrt{6})(x + 1)(x^2 - x + 1)$ **b.** $(x - \sqrt{6})(x + \sqrt{6})(x + 1)\left(x - \frac{1 + i\sqrt{3}}{2}\right)\left(x - \frac{1 - i\sqrt{3}}{2}\right)$ **25. a.** $(x + 2)(x - 1)(x + 1)(x^2 + 1)$ **b.** $(x + 2)(x - 1)(x + 1)(x - i)(x + i)$

27. $(x - 1 - i\sqrt{2})(x - 1 + i\sqrt{2})$ **29.** $-4 + 11i$ **31.** $\frac{101}{3} - \frac{23}{3}i$ **33.** $216 + 0i$ **35.** $-\frac{20}{13} - \frac{9}{13}i$ **37.** Answers vary. Sample: $5x^3 + 2x^2 + x$ **39.** False; The roots of a polynomial function p refer to its x-intercepts. **41. a.** 4 **b.** -6 **c.** 8 **43.** Answers vary. Sample: $p(x) = x^4 - 13x^3 + 53x^2 - 83x + 42$ **45.** Answers vary. Sample: $k(5) = 0$ and $k(x)$ has 2 complex zeros. **47.** 324 **49.** Every real number a can be written in complex form as $a + 0i$. Its complex conjugate is $a - 0i = a$. **51.** All four zeros must be nonreal. **53.** True. If $2 + i$ is a zero, then $2 - i$ must also be a zero because nonreal zeros occur as conjugate pairs. If $1 + i$ were a zero, then $1 - i$ would also have to be a zero. Since the polynomial has degree 3, it only has 3 zeros, meaning that $1 + i$ cannot be a zero.
55. $V(x) = 2x^3 - 600x^2 + 40000x$
57. a. $A(x) = 350x^{13} + 350x^{12} + \dots + 350x + 350$ **b.** $\approx$ \$7102.40 **59.** $A(\ell) = \ell(\ell - 25)$
61. a.

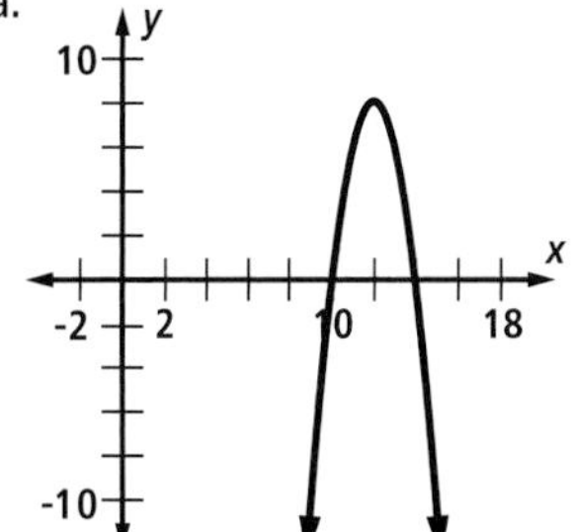

b. Relative maximum at (12, 8) **c.** $x = 10, 14$
63. a. 3 **b.** f only has 3 real zeros, which means it must be 2 nonreal zeros. **65.** true

Chapter 8

Lesson 8-1 (pp. 502-508)

Guided Example **a.** 4; 5; -2.5; -2.5 **b.** 16.5; 16.5, -2.5

Questions

1. a. t_{n-1} **b.** t_{n+1} **3.** false; The domain of a sequence is the set of consecutive integers greater than or equal to some integer k. **5. a.** 2, -3, -8 **b.** yes **7. a.** 1, 9, 36 **b.** no **9. a.** $a_n = n$ **b.** $\begin{cases} a_1 = 1 \\ a_n = a_{n-1} + 1 \text{ for } n > 1 \end{cases}$ **c.** yes **11. a.** 9.61 years **b.** 10 **13.** $-\frac{1}{6}$ **15.** 22 **17.** yes; the sequence has a constant difference $2d$ **19.** no; the difference between two consecutive terms h_n and h_{n+1} is $\frac{-d}{a_n(a_n + d)}$. Because a_n varies, there is no constant difference. **21.** $\frac{624}{74088} \approx 0.008$ **23.** false **25.** $f(a) + f(b) = a^2 + b^2 - 4$; $f(a + b) = a^2 + 2ab + b^2 - 2$ **27.** $f(a) + f(b) = \sqrt{a+3} + \sqrt{b+3}$; $f(a + b) = \sqrt{a + b + 3}$

Lesson 8-2 (pp. 509-514)

Guided Example 1 **a.** 10,000; 0.0483; t; 10,000; t **b.** 10,000; 1.0483

Guided Example 2 **a.** 6; 48; $18 + 6(n - 1)$ **b.** $\frac{3}{4}; \frac{3}{4}; \frac{3}{4}; \frac{3}{4}$; 5.0625; 3.796875; $16 \cdot \left(\frac{3}{4}\right)^{n-1}$

Questions

1. a. 14, 12, $\frac{72}{7}$, $\frac{432}{49}$ **b.** $\frac{6}{7}$ **c.** $n = 23$

	A	B
◆	=seqn(6*u(n−1)/7,{14},255)	
19	0.873144	
20	0.748409	
21	0.641493	
22	0.549852	
23	0.471301	
A23	$=\frac{6 \cdot a22}{7}$	

3. $\begin{cases} g_1 = 16 \\ g_n = \frac{g_{n-1}}{32} \text{ for } n > 1 \end{cases}$; $g_n = 16\left(\frac{1}{32}\right)^{n-1}$

5. $\begin{cases} g_1 = 1000 \\ g_n = \frac{g_{n-1}}{2} \text{ for } n > 1 \end{cases}$; $g_n = 1000\left(\frac{1}{2}\right)^{n-1}$

7. $v_n = 1200(1.063)^{n-1}$ **9. a.** $a_n = 15 + 18(n - 1)$ **b.** $g_n = 15\left(\frac{11}{5}\right)^{n-1}$ **11.** about 4.83 miles **13 a.** $11,250.65 **b.** $v_n = 13{,}555(0.83)^{n-1}$ **c.** ≈15 years

	A	B	C	D
◆	=seqn(0.83*u			
11	2103.2			
12	1745.66			
13	1448.89			
14	1202.58			
15	998.143			
16	828.459			
A	=*seqn*(0.83·*u*(*n*−1),{13555},255)			

15. $8\sqrt{5}$ **17. a.** $R_3 = 9$; $R_4 = 13$ **b.** $\begin{cases} R_1 = 1 \\ R_n = R_{n-1} + 4 \text{ for } n > 1 \end{cases}$; $R_n = 1 + 4(n-1)$ **19.** 360 **21.** true

Lesson 8-3 (pp. 515-521)

Questions

1. a description of what happens to the values of a sequence as n gets very large **3.** false

5. Yes, it converges to 1.

	A n	B tn	C	D
◆		=a[]/(a[]+1)		
3	3	3/4		
4	4	4/5		
5	5	5/6		
6	50	50/51		
7	1000	1000/1001		
B7	$=\frac{1000}{1001}$			

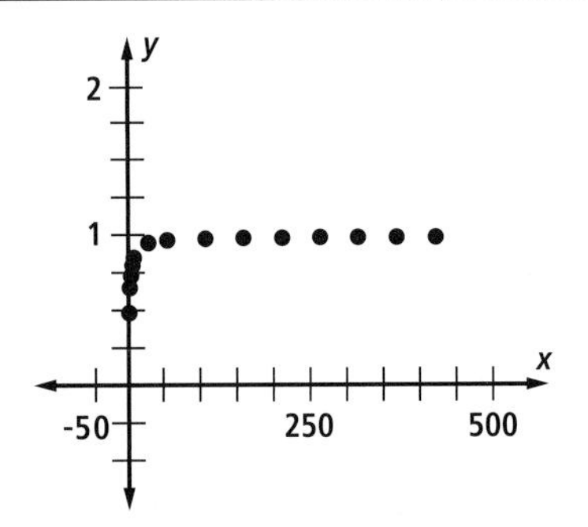

7. does not converge **9. a.** -4, 4, -4, 4, -4 **b.** No; as n grows larger, the terms do not stop alternating from 4 to -4. **11. a.** limit exists; 5 **b.** limit does not exist **c.** limit does not exist **13. a.** 6, 2, $\frac{2}{3}, \frac{2}{9}, \frac{2}{27}$ **b.** h converges; $\lim_{n\to\infty} h_n = 0$ **15.** Sequence c does not converge; the cosine function is cyclical and therefore, the limit does not exist. **17.** $a_n + b_n$ and $\frac{b_n}{a_n}$ converge; $\lim_{n\to\infty} a_n + b_n = 15$; $\lim_{n\to\infty} \frac{b_n}{a_n} = -1$. **19. a.** $g_n = 100\left(\frac{7}{10}\right)^{n-1}$ **b.** $\begin{cases} g_1 = 100 \\ g_n = \frac{7}{10} g_{n-1} \text{ for } n > 1 \end{cases}$ **c.** $n = 16$

21. a.

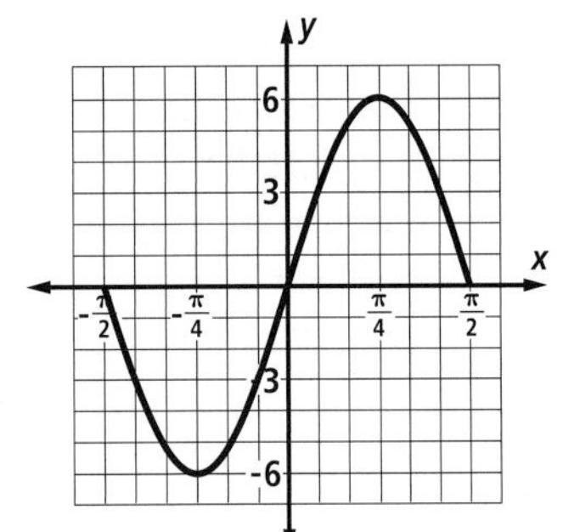

b. $y = 6 \sin 2x$ **c.** Answers vary. Sample: $-\frac{\pi}{4}, \frac{3\pi}{4}, \frac{15\pi}{4}$

Lesson 8-4 (pp. 522-528)

Guided Example 2 7500; 500; (2(7500) + (29)500); 442,500; 442,500

Questions

1. D **3.** A series is an indicated sum of consecutive terms of a sequence. **5.** 156

7.

a_1	d	n	a_n	Sum	Summation Notation
15	3	5	27	15 + 18 + 21 + 24 + 27	$\sum_{n=1}^{5}(12+3n)$

9.

a_1	d	n	a_n	Sum	Summation Notation
43	5	8	78	43 + 48 + 53 +···+ 78	$\sum_{n=1}^{8}(38+5n)$

11. 500,500 **13.** the natural numbers from 1 to 1000

15.

a_1	d	n	a_n	Sum (written out)	Summation Notation	Sum
2.5	1.5	13	20.5	2.5 + 4 + 5.5 +···+ 20.5	$\sum_{n=1}^{13}(1+1.5n)$	149.5

17. Substitute $a_n = a_1 + (n-1)d$ into the formula for S_n: $S_n = \frac{n}{2}(a_1 + (a_1 + (n-1)d)) = \frac{n}{2}(2a_1 + (n-1)d)$ **19.** 78 **21. a.** $w + 2(k-1)$ **b.** $\frac{k}{2}(2w + 2(k-1))$ **23.** divergent **25. a.** $Q_n = n(n+2)$ **b.** 9999

Lesson 8-5 (pp. 529-535)

Guided Example 2 24; 6; $\frac{1}{3}$; $\frac{2912}{81}$; $35\frac{77}{81}$; $\frac{2912}{81}$

Questions

1. a. $S_{25} = \sum_{i=1}^{25} 2^{i-1} = 1 + 2 + 4 + \cdots + 2^{24}$ **b.** 33,554,431 **3. a.** -2, -8, -32, -128, -512, -2048 **b.** -2730 **c.** -2730 **5. a.** $\sum_{i=1}^{9} 10(-2)^{i-1}$ **b.** 1710 **7.** ≈ 29,537.84 **9. a.** 32 **b.** 62 **11.** $6 + \frac{6}{5} + \frac{6}{25} + \frac{6}{125}$ **13. a.** $\sum_{i=1}^{n} pm^{i-1}$ **b.** $\frac{p(1-m^n)}{1-m}$ **15.** $g_n = 5(3)^{n-1}$ **17.** 119 **19. a.** $\frac{3}{40}, \frac{9}{46}, \frac{27}{64}, \frac{81}{118}, \frac{243}{280}$ **b.** yes; 1

Lesson 8-6 (pp. 536-541)

Guided Example 0.4875%; 0.004875; 240; $250,000; 250,000; 0.004875; 240; 0.004875; 240

Questions

1. a. 36 months (3 years) **b.** $2500 **c.** about 0.667% **3. a.** about $3,175.59 **b.** about $78.34 **5.** about $1,117.65 **7. a.** about $2,202.13 **b.** about $1,619.35 **c.** greater than half of the answer to Part a; The monthly payment is not directly proportional to the interest rate. **9. a.** 4 years; $172.98 **b.** 3 years; $1,167.24 **11.** 5 **13.** $\frac{242461}{8192} \approx 29.6$ **15.** false **17.** 1

Lesson 8-7 (pp. 542-549)

Guided Example 2 b. $\frac{1}{3}$; $\frac{5}{2}$ **c.** $\frac{1}{4}$; converges; $\frac{2560}{3}$

Questions

1. An infinite sequence is a list of an infinite number of terms. An infinite series is the sum of an infinite number of terms. **3. a.** $S_1 = \frac{8}{5}$; $S_2 = \frac{8}{5} + \frac{4}{5} = \frac{12}{5}$; $S_3 = \frac{8}{5} + \frac{4}{5} + \frac{2}{5} = \frac{14}{5}$ **b.** $\frac{16}{5}$ **5.** yes; 1.082323

	A	B	C
◆	=seq(	=seq(1/k^4,k,1.,1(	=cumsum(b[])
1	1	1.	1.
2	2	0.0625	1.0625
3	3	0.012346	1.07485
4	4	0.003906	1.07875
5	5	0.0016	1.08035

B =seq$\left(\frac{1}{k^4}, k, 1., 100\right)$

7. $r = \frac{1}{4}$; convergent; $\frac{400}{3}$ **9.** not geometric; convergent; $\frac{902}{225}$ **11.** $r = -2$; divergent **13. a.** 180 cm **b.** 99.97% **15. a.** $\frac{1}{2}, \frac{1}{4}, \frac{1}{8}, \frac{1}{16}, \frac{1}{32}$ **b.** $\frac{1}{256}$ m **c.** $\frac{255}{256}$ m **d.** 10 **e.** No; The flea would have to jump an infinite number of times to reach the outside of the ring. The flea can only jump a finite number of times so it will not make it. **17. a.** 4 **b.** 83 **19.** $4,691.49

21. $S_5 = \frac{\frac{4}{7}\left(1 - \left(\frac{1}{7}\right)^5\right)}{1 - \frac{1}{7}} = \frac{11,204}{16,807} \approx 0.667$

23. false; Answers vary. Sample: The Law of Large Numbers says that the relative frequency of heads will approach $\frac{1}{2}$, but this does not mean that the probability of any flip is affected by the outcome of the flip before it.

Self-Test (p. 554)

1. $v_1 = 0$, $v_2 = v_1 + 2^2 = 0 + 2^2 = 4$, $v_3 = v_2 + 3^2 = 4 + 3^2 = 13$, $v_4 = v_3 + 4^2 = 13 + 4^2 = 29$ **2.** Neither; there is no constant difference or constant ratio between consecutive terms of this sequence. **3.** There are 11 terms between the first term and the 12th term, so in an arithmetic sequence this is equivalent to 11*d*. Therefore the 12th term is -3 + 11*d*. **4. a.** If *t* is an arithmetic sequence, the sequence has a constant difference of 16, so the next three terms are 24 + 16 = 40, 40 + 16 = 56, and 56 + 16 = 72. **b.** If *t* is a geometric sequence, the sequence has a

constant ratio of 3, so the next three terms are $24 \cdot 3 = 72$, $72 \cdot 3 = 216$, and $216 \cdot 3 = 648$. **5. a.** t is an arithmetic sequence with first term 8 and constant difference 16. So, by a Formula for an Arithmetic Sequence, $t_n = 8 + 16(n - 1)$. **b.** t is a geometric sequence with first term 8 and constant ratio 3. So, by a Formula for a Geometric Sequence, $\begin{cases} t_1 = 8 \\ t_n = 3t_{n-1}, \text{ for } n > 1 \end{cases}$.
6. yes; This sequence is 27 times the cube of the harmonic sequence. By the Limit of a Product Property the limit of a product of two sequences is the product of their limits. Since the limit of the harmonic sequence is 0, the limit of its cube is 0, and the limit of 27 times its cube is 0.
7. no; the terms of the sequence decrease without bound, and therefore the sequence is divergent.
8. $S_4 = 5 + 10 + 20 + 40 = 75$ **9.** The coefficients of these terms of the sequence can be described by the arithmetic sequence with first term 2 and constant difference 5, and the powers of the sequence are described by the geometric sequence with first term 1 and constant ratio r. Therefore, the nth term can be written as $(5n - 3)r^{n-1}$ **10.** yes, the terms of the series converge to 0; $\lim_{n\to\infty} S_n = \frac{g_1}{1-r} = \frac{5}{1-\frac{1}{3}} = \frac{15}{2}$

11. $5.43 + 0.00\overline{21} = 5.43 + 0.0021\sum_{i=1}^{\infty}(.01)^{n-1}$; The geometric series converges because $r = .01$ is between 1 and -1. **12.** 27.5 mi; In nine days, Lester increased his distance by 4.5 miles. Therefore the constant difference in the arithmetic sequence representing his runs is 0.5 miles. The first term is 0.5 miles. The distance is $\sum_{n=1}^{10} 0.5n = 27.5$.
13. \$499.59; The principal $P = \$45{,}000$, the term $n = 120$ months, and the monthly rate $r = \frac{.06}{12} = 0.005$, so each monthly payment is: $\frac{P(1+r)^n r}{(1+r)^n - 1} = \frac{45000(1.005)^{120}0.005}{(1.005)^{120} - 1} \approx$ \$499.59 **14.** the 5th year; We can write this problem as a geometric sequence with first term \$2600 and constant ratio 0.5; the value in the nth year is $2600(0.5)^n$. Solve $100 = 2600(0.5)^n$ to find $n = 5$. **15. a.** The end behavior is $\lim_{n\to\infty} a_n = \lim_{n\to\infty} \frac{2n + (-1)^n}{n}$. **b.** yes; $\lim_{n\to\infty} \frac{2n + (-1)^n}{n} = \lim_{n\to\infty} 2 + \frac{(-1)^n}{n}$. By the Limit of a Constant property and the Limit of a Sum property, $\lim_{n\to\infty} 2 + \frac{(-1)^n}{n} = 2 + 0 = 2$. **c.** As n increases, the terms of the sequence oscillate about and get ever closer to 2.
16. a. $\sum_{k=100}^{n} k$ **b.** $S_n = \frac{(n - 100 + 1)}{2}(100 + n) = \frac{(n - 99)(n + 100)}{2}$

Self-Test Correlation Chart

Question	1	2	3	4	5	6	7	8	9	10
Objective(s)	A	D	B	A	B	E	E	C	B	F
Lesson(s)	8-1	8-2	8-1	8-1	8-2	8-3	8-3	8-5	8-2	8-7
Question	**11**	**12**	**13**	**14**	**15**	**16**				
Objective(s)	F	G	I	H	J	C				
Lesson(s)	8-7	8-4	8-6	8-2	8-3	8-4				

Chapter Review (pp. 555-557)

Questions

1. 6, 30, 150, 750, 3750 **3.** 162; 39,366 **5. a.** $96 - 19(n-1)$ **b.** -835 **7.** $r_n = 11{,}000(0.7)^{n-1}$ **9.** 18; $6n - 6$ **11. a.** $\sum_{n=1}^{8} 43 + 16n$ **b.** 920 **13. a.** $\sum_{k=1}^{12} 4^{k-1}$ **b.** 5,592,405 **15.** 9072 **17.** possibly arithmetic **19.** possibly arithmetic **21.** definitely geometric **23.** yes; 0 **25.** yes; 0 **27.** $\lim_{n\to\infty} \frac{5n+7}{3n-131} = \frac{5}{3}$ **29.** diverges **31.** If $|r| < 1$, the series converges. **33.** $\frac{31}{110}$ **35.** 5 rows **37.** 26 rows with 9 cans left over **39. a.** \$19,592 **b.** $v_n = 24{,}800(0.79)^n$ **41. a.** 250 cm **b.** about 99.6% **43. a.** 30 months **b.** \$8,000 **c.** about 0.00708 **45.** about \$2,751.46

47. The difference between consecutive terms of an arithmetic sequence is a constant. **49.** harmonic or geometric **51.** B

53.

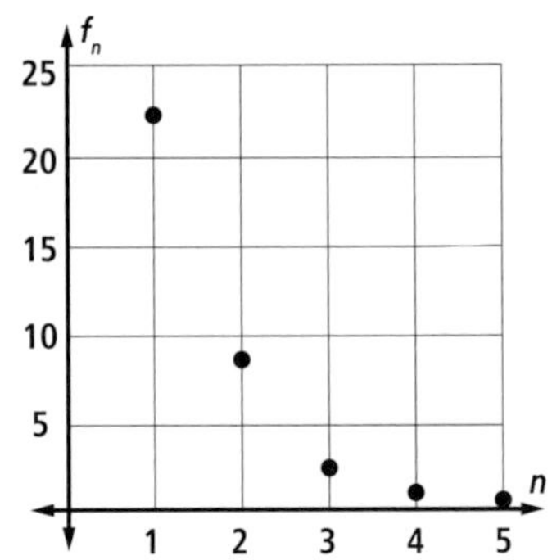

Chapter 9

Lesson 9-1 (pp. 560-567)

Guided Example 3 **a.** $\frac{3V}{\pi}$; $\frac{3V}{\pi}$; $\left(\frac{3V}{\pi}\right)^{\frac{1}{3}}$ **b.** $\frac{3 \cdot 60}{\pi}$

Questions

1. 11 is a fourth root of 14,641 because $11^4 = 14{,}641$.
3. $\{x \mid x \geq 0\}$ **5.** $\frac{1}{5}$ **7.** 3 **9. a.** 1 **b.** $\frac{1}{9}$ **11.** true
13. ± 1.550 **15. a.** $r = \sqrt[3]{\frac{3V}{4\pi}}$ **b.** about 3.06 cm **17. a.** 1.36
b. fifth; 4.6 **19.** $x \geq 0$ and n an integer with $n \geq 2$
21. about 1.260 units
23. a.

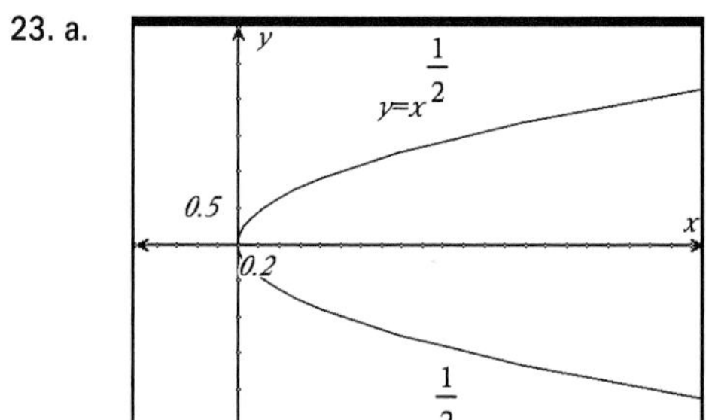

b. $x = y^2$ **25.** $f(g(x)) = \frac{7}{\frac{x}{7}} = \frac{49}{x}$. Because $f(g(x)) \neq x$, f and g are not inverses. **27.** about 90.34 years. The accuracy is questionable because the estimate is based on an extrapolation well beyond the range of the data.
29. a. $t = 4$ **b.** $r = 6$ **c.** $x = 4$ or -1

Lesson 9-2 (pp. 568-574)

Guided Example 2 $\frac{7}{12}$; $\frac{7}{12}$; $\frac{2}{3}$

Questions

1. $3x^{\frac{13}{5}}$ **3.** $w^{\frac{7}{2}}$ **5.** $\sqrt[4]{x^3}$ **7. a.** 9 **b.** $\frac{1}{3}$ **9. a.** 512 **b.** $\frac{1}{16}$
c. 2048 **11.** about 16.04 in. **13.** 2152 cubic inches
15. a. 205.20 square meters **b.** S.A. $= 6V^{\frac{2}{3}}$ **c.** The exponent $\frac{2}{3}$ implies that you must switch from 3 dimensions to 2 dimensions. **17. a.** $k \approx 1.44 \cdot 10^{-10}$
b. about 685 days **c.** $a = \left(\frac{P}{1.44 \cdot 10^{-10}}\right)^{\frac{2}{3}}$ **19. a.** 2^5 **b.** 2^{-5}
c. 2^{10} **d.** 2^{-10} **21. a.** 262 Hz **b.** 220 Hz **23. a.** $y = (x + 3)^5$
b.

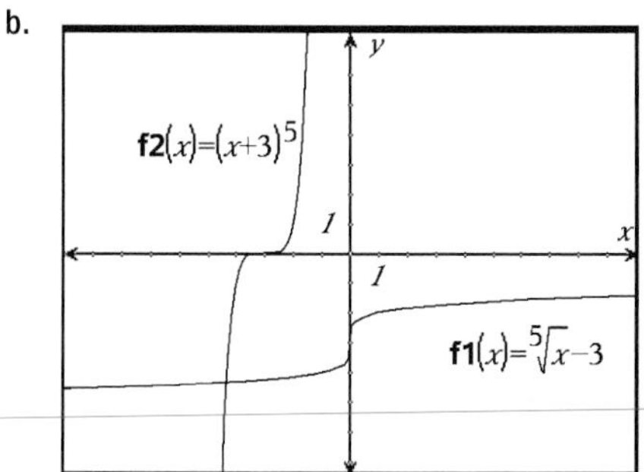

c. all real numbers **25. a.** $\frac{\pi}{6}$ **b.** $-\frac{\pi}{6}$ **c.** $-\frac{1}{2}$

Lesson 9-3 (pp. 575-580)

Guided Example 2 **b.** 0.00001 **c.** 12, 10^{12}, 12 **d.** 1.5, 1.5

Questions

1. $\log_3 729 = 6$ **3. a.** 2 **b.** 1 **c.** 0 **d.** -1 **5.** 0 **7.** $\frac{8}{5}$
9. 8 **11.** $x = 5$ **13.** $x \approx 270{,}396$ **15.** $y = 6^x$ **17.** D
19. about 8.03 **21. a.** about 47.71 **b.** 48
23. a. 1.776×10^3; 1.776×10^5; 1.776×10^{13} **b.** about 3.25, about 5.25, about 13.25 **c.** The integer part of the logarithms and the exponent of the powers of 10 are the same. The decimal part of the logarithm is log 1.776.
25. 10 times as large **27.** about 0.0204
29. no; The graphs are not the same because they have different domains.

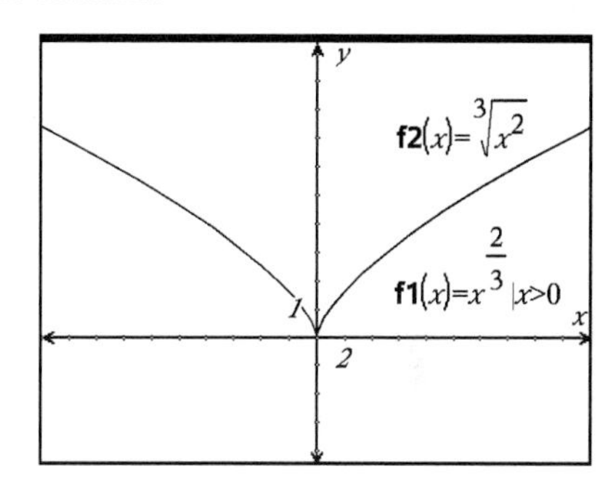

31. about \$6,570.33

Lesson 9-4 (pp. 581-586)

Guided Example 2 **a.** 4.5, $e^{0.36}$, 716.6647; 716.66
b. $\frac{0.08}{4}$, 4(4.5); 1.428246

Questions

1. a. i. 2.716924 **ii.** 2.718280 **b.** e **3. a.** about 5.654%
b. \$3169.62 **5.** ≈ 6.17 **7.** ≈ -0.60 **9.** 0 **11.** 7 **13.** 0
15. about 9879 feet above sea level **17.** about 4,857,781
19. about 691 days **21.** $f^{-1}(x) = \log_2 x$
23. $\log_{\frac{1}{8}} \frac{1}{2} = \frac{1}{3}$ **25.** < **27.** x varies directly as z^6

Lesson 9-5 (pp. 587-593)

Guided Example 3 **a.** 1000 **b.** $\frac{1}{2}$

Questions

1. a. log 100 **b.** 2 **3.a.** $\log_b \frac{xb^2}{x}$ **b.** 2 **5. a.** $\log_b \frac{3b^2}{3}$ **b.** 2
7. the definition of logarithm; Power of a Power Property
9. 0.520 **11.** $a = \frac{b^3}{c^2}$ **13. a.** 124 **b.** 1 **15.** $\ln F = \ln G + \ln m + \ln M - 2\ln d$ **17.** $\log_2 \frac{2p}{q}$ **19.** 0.8213
21. $-\log_b n = -1 \cdot \log_b n$; by the Logarithm of a Power Theorem, $-\log_b n = \log_b n^{-1}$. By the Negative Exponent Theorem, $n^{-1} = \frac{1}{n}$, so $-\log_b n = \log_b \frac{1}{n}$. **23.** Logarithm of a Product Theorem, $\log_{10} 6 + \log_{10} 5 = \log_{10} (6 \cdot 5)$; Logarithm of a Power Theorem, $\log_{10} 6^2 = 2 \log_{10} 6$; Logarithm of a Power Theorem, $\log_5 (5^{-3}) = -3 \log_5 5 = -3$; Logarithm of a Power of the Base Theorem, $\log_4 4^{\frac{1}{4}} = \frac{1}{4} \log_4 4 = \frac{1}{4}$ **25.** 6 **27. a.** III **b.** II **c.** I

29. a. $\{40, 42, 44\}$ **b.** $\{30, 32, 34, 36, 38, 40, 41, 42, 43, 44, 45, 46, 47, 48, 49, 50\}$

Lesson 9-6 (pp. 594-599)

Guided Example 4 $\log_{10} 5$; 71; $\ln 5$

Questions

1. a. $x = \frac{\ln 46}{\ln 5} \approx 2.3789$ **b.** They are the same. **3.** $x = 1.712$; $7^{1.712} \approx 28$ **5.** No, carbon dating shows that the painting is about 600 years old, so it was painted sometime around 1409, but da Vinci was born in 1452. **7.** $\frac{\ln 16}{\ln 5} \approx 1.723$ **9. a.** $n(t) \approx 80{,}818.4e^{0.7198t}$ **b.** about 3,950,000,000 **11.** about 11.6 years **13.** For positive real numbers x and b, $\log_b x = \frac{1}{\log_x b}$. $\log_b x = \frac{\log x}{\log b}$ and $\log_x b = \frac{\log b}{\log x}$, so $\log_b x = \frac{1}{\log_x b}$. **15.** $t = \frac{\ln A - \ln P}{r}$ **17.** -2 **19.** -0.1 **21. a.** 92% **b.** about 34.44% **23. a.** true **b.** For $x \neq 4$, $g(x)$ simplifies to $x - 4$, which is linear. **25. a.** $y = 2.2x + 5.2$ **b.** channel catfish grow about 2.2 inches per year **c.** about 31.6 inches **d.** Extrapolations from data are less certain than interpolations. **e.** about 0.996 **f.** Catfish get longer over time.

Lesson 9-7 (pp. 600-606)

Questions

1. about 51.9% **3.** To provide data for linear regression analysis to see if a power function model is appropriate. **5.** about 382,533 meters

7. a.

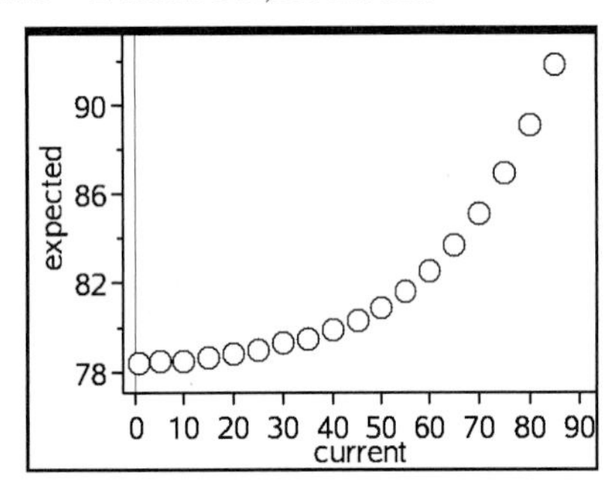

b. $y = ab^x$ **c.** $y \approx 76.2582(1.00162)^x$ **d.** the model from Part c **e.** Both models predict an estimated life span of 88 years for a 90 year-old! Neither model is suitable for this age. **9. a.** $\log t = \log(cx^n) = \log c + n \log x$

b.

logt
2.60
2.57
2.54
0.00 0.15 0.30 0.45 0.60
logx

c. Answers vary. Sample: $\log t \approx -0.13 \log x + 2.62$ **d.** The time to cover 2000 m decreases with more oarsmen. **e.** $t \approx 416.9x^{-0.13}$ **11.** ≈ 9.1749 **13.** $t \approx \frac{\ln 3}{r}$ years **15. a.** $0 < x < 1$ **b.** $x > 1$ **c.** For $0 < x < 1$, if $m > n$, then $x^{\frac{1}{n}} < x^{\frac{1}{m}}$. For $x > 1$, if $m > n$, then $x^{\frac{1}{n}} > x^{\frac{1}{m}}$.

Self-Test (p. 611)

1. $8^{-\frac{4}{3}} = (2^3)^{-\frac{4}{3}} = 2^{-4} = \frac{1}{2^4} = \frac{1}{16}$ **2.** $\sqrt[5]{x^{10}y^5} = x^{\frac{10}{5}}y^{\frac{5}{5}} = x^2y$ **3.** $\ln e^{18} = 18 \ln e = 18 \cdot 1 = 18$ **4.** For all $n > 0$, $\log_n 1 = 0$, so $x = 1$. **5.** $\log 225 + \log 4 - \log 9 = \log\left(\frac{225 \cdot 4}{9}\right) = \log \frac{900}{9} = \log 100 = 2$ **6. a.** $\log 300 = \log(30 \cdot 10) = \log 30 + \log 10 \approx 1.48 + 1 = 2.48$ **b.** $\log 900 = \log(30^2) = 2 \log 30 \approx 2 \cdot 1.48 = 2.96$ **c.** $\log \frac{1}{30} = \log 30^{-1} = -\log 30 \approx -1.48$ **7.** $\log_{11}\left(\frac{7}{3}\right) = \log_{11} 7 - \log_{11} 3 \approx 0.8115 - 0.4582 = 0.3533$ **8.** Restrict the domain to $x > 0$. Then the inverse, $g^{-1}: x \to x^{\frac{1}{9}}$, is a function. **9.** $\log 6^x = \log 1.48$; $x \log 6 = \log 1.48$; $x = \frac{\log 1.48}{\log 6} \approx 0.22$ **10.** true, $\ln x = \log_e x$ by definition of natural log. **11. a.** From the given, S.A. $= 130 = 6x^2$, so $x^2 = \frac{65}{3}$ and $x^3 = (x^2)^{\frac{3}{2}} = \left(\frac{65}{3}\right)^{\frac{3}{2}} = \frac{65\sqrt{195}}{9}$ cubic inches **b.** $V = x^3$ and $S.A. = 6x^2$. So $x = \sqrt{\frac{S.A.}{6}}$ and $V = \left(\sqrt{\frac{S.A.}{6}}\right)^3 = \left(\frac{S.A.}{6}\right)^{\frac{3}{2}}$ **12.** $3 = 10(0.5)^{\frac{t}{25}}$; $0.3 = 0.5^{\frac{t}{25}}$; $\log 0.3 = \log 0.5^{\frac{t}{25}}$; $\log 0.3 = \frac{t}{25} \log 0.5$; $\frac{25 \log 0.3}{\log 0.5} \approx 43.4$ years

13. a.

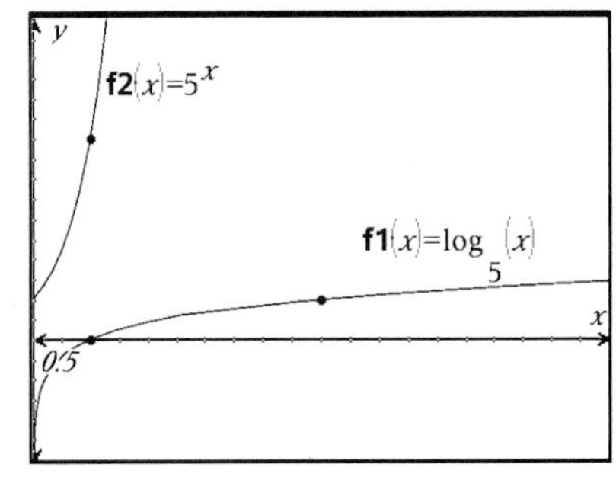

b. As t approaches 0, $\log_5 t$ gets smaller and smaller. Therefore $t = 0$ is an asymptote of g. **c.** Let $g(t) = y = \log_5 t$. Then $5^y = t$. So g^{-1} has equation $5^t = y$. Thus $g^{-1}(t) = 5^t$. **14.** $\frac{1}{2} \log a + \log b = \frac{1}{3} \log c$ given; $\log a^{\frac{1}{2}} + \log b = \log c^{\frac{1}{3}}$ Logarithm of a Power Theorem; $\log\left(a^{\frac{1}{2}}b\right) = \log c^{\frac{1}{3}}$ Logarithm of a Product Theorem; $a^{\frac{1}{2}}b = c^{\frac{1}{3}}$ Apply $\log^{-1}$ to both sides of the equation. **15.** C; Switching x and y, an equation for the inverse is $x = 3^y$; this equation can be rewritten as $y = \log_3 x$. **16.** Take the natural log of both sides: $\ln A = \ln\left((3.75)(0.054)^t\right)$, then simplify using the Logarithm of a Power and Logarithm of a Product Theorems: $\ln A = \ln 3.75 + t \ln 0.054$.

17.a.

Minutes (A)	ln y (y = Difference °C)
1	ln 69 ≈ 4.234
3	ln 54 ≈ 3.989
5	ln 40 ≈ 3.689
7	ln 30 ≈ 3.401
8	ln 26 ≈ 3.258
10	ln 21 ≈ 3.045

b.

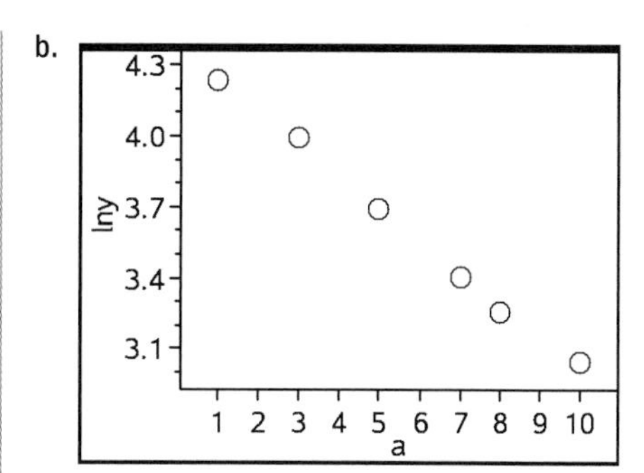

c. Use linear regression on a statistics utility; $\ln y \approx -0.136A + 4.373$ **d.** Raise both sides as a power of e: $y = e^{\ln y} = e^{-0.136A+4.373}$; $y = 79.281e^{-0.136A}$ **18. a.** The functions are inverses of each other, so their graphs are reflection images of each other over the line $y = x$. **b.** $\left(\frac{1}{x}\right)^{\frac{7}{4}} = (x^{-1})^{\frac{7}{4}} = x^{-\frac{7}{4}}$

Self-Test Correlation Chart

Question	**1**	**2**	**3**	**4**	**5**	**6**	**7**	**8**	**9**	**10**
Objective(s)	A	A	B	B	F	F	F	D	C	E
Lesson(s)	9-2	9-2	9-5	9-3	9-5	9-5	9-5	9-1	9-6	9-4
Question	**11**	**12**	**13**	**14**	**15**	**16**	**17**	**18**		
Objective(s)	H	I	L	G	E	G	J	K		
Lesson(s)	9-2	9-6	9-3	9-6	9-3	9-6	9-7	9-2		

Chapter Review (pp. 612–615)

1. a. 2 **b.** 32 **c.** $\frac{1}{4}$ **3. a.** 3 **b.** 2 **c.** $\frac{2}{3}$ **5. a.** $\sqrt[5]{p^3}$ **b.** $2\sqrt[4]{r^3}$ **7.** $a^{\frac{1}{7}}b^{\frac{5}{7}}$ **9.** $<$ **11. a.** 1 **b.** 2 **13.** -3 **15.** $\frac{3}{2}$ **17.** π^2 **19.** ≈ 4.025 **21.** ≈ 0.070 **23.** log 4, $\log_7 4$, ln 4 **25.** domain: $\{x \mid x \geq 0\}$; range: $\{y \mid y \geq 0\}$ **27.** false; let $n = 1$, then $\sqrt[1]{x}$ is not defined and $x^{\frac{1}{1}} = x$ **29. a.** domain: $\{x \mid x > 0\}$ **b.** range: the set of real numbers **31.** A **33.** (1, 0) **35.** $-v$ **37.** xv **39.** 2.06 **41.** ln(100,000) **43.** ln(4) **45.** 6 **47.** 0 **49.** 63 **51.** $\log A = 3\log C - \log D$ **53.** $\ln y = \ln 6 + x\ln 4$ **55.** $v = \frac{b^{\frac{4}{7}}}{c}$ **57.** By the Change of Base Theorem, $\log_{10} e = \frac{\log_{10} e}{\log_{10} 10} = \log e$ and $\ln x = \frac{\log_{10} x}{\log_{10} e}$, so $\log_{10} e \ln x = \log e \cdot \frac{\log_{10} x}{\log_{10} e} = \log_{10} x$ **59.** $r = \left(\frac{3V}{4\pi}\right)^{\frac{1}{3}}$ **61. a.** $1035 \frac{\text{g}}{\text{cm}^2}$ **b.** about $28.70 \frac{\text{g}}{\text{cm}^2}$ **c.** about $2.63 \frac{\text{g}}{\text{cm}^2}$ **63.** about 52.95 hours **65.** deeper than about 5.53 units **67. a.** $L = A\left(\frac{1}{2}\right)^{\frac{h}{5}}$ **b.** about 21.76% **69.** $a = \left(Te^{1.72}\right)^{\frac{2}{3}}$

71. a.

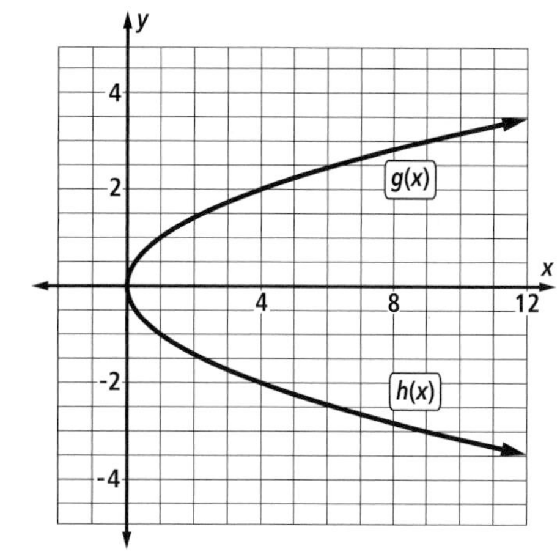

b. $x = y^2$

73. a. range: $\{y \mid y \geq 0\}$ b. $f^{-1}: x \rightarrow x^{\frac{3}{5}}$

c.

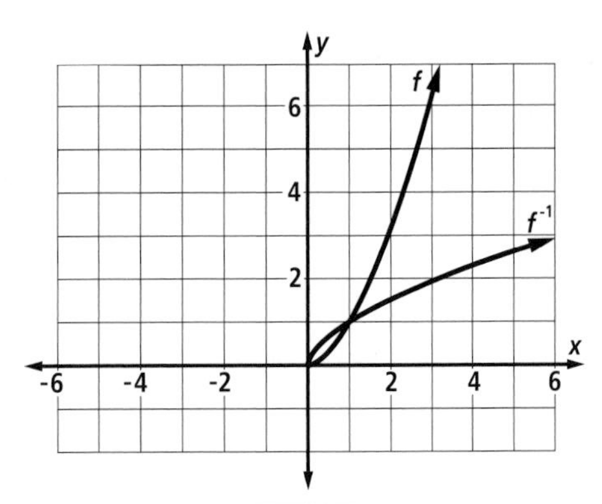

d. The inverse is a function because in the domain (the range of the original function) every x is mapped to exactly one y.

75. a.

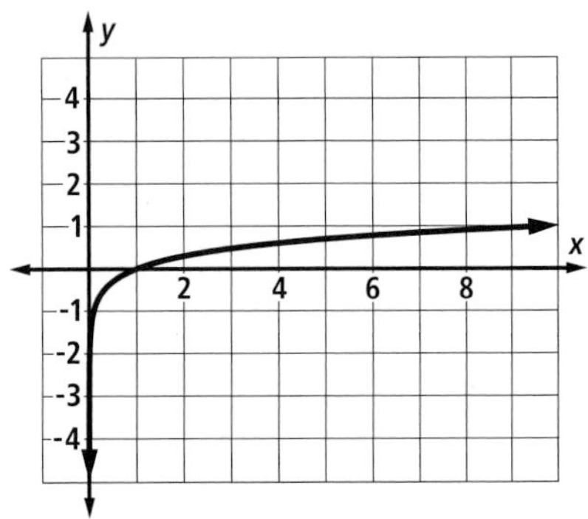

b. 0.3; 0.8

77. a and b.

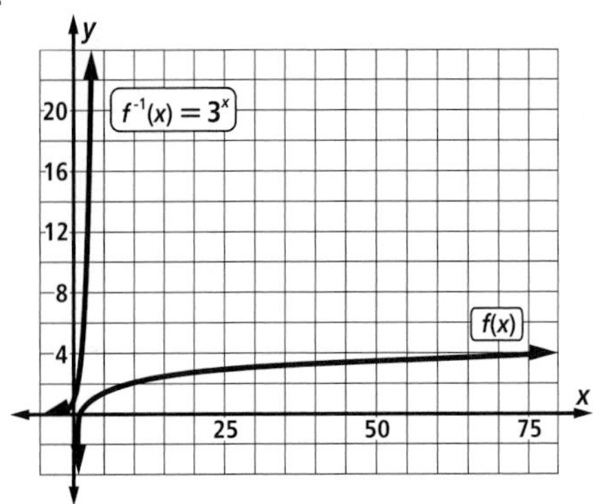

79. $x = 0$

Chapter 10

Lesson 10-1 (pp. 618-623)

Guided Example 1 4; 12, 4; 8, 4, 495; 495; 103,950

Questions

1. permutations **3.** combinations **5.a.** (ii) **b.** (i) **7.** 20,475 **9. a.** 120 **b.** 1024 **c.** $\frac{15}{128} \approx 0.117$ **11.** n; each group has one object and each group must be different so there are n groups. **13.** 1; There is only one way to choose all n of the objects. **15.** Answers vary. Sample: Out of a class of 100 students, a company randomly chooses 10 to complete a survey. How many different groups of 10 people are possible? **17.** 31 **19.** $146,107,962 **21.** 26 **23. a.** 4 **b.** 32 **c.** 256 **d.** 2^n **25.** 0.12

Lesson 10-2 (pp. 624-630)

Guided Example 3 15; 20; 15; 6; 1; 21; 35; 15; 20; 15; 6; 1; 21; 35; 35; 21; 7

Questions

1. a. 7 **b.** $r = 4$; 15 **c.** $s = 3$; 20 **d.** 2^6 **3.** B **5.** C **7. a.** 792; 792 **b.** Property 5: Each element in Pascal's Triangle is the sum of the two elements nearest it in the preceding row. Specifically, for whole numbers n and r with $1 \le r \le n$, ${}_nC_{r-1} + {}_nC_r = {}_{n+1}C_r$. **9. a.** 8 **b.** 11,440, 8008, 4368, 1820, 560, 120, 16, 1 **c.** $2^{16} = 65{,}536$ **d.** 1, 17, 136, 680 **11.** 34, 1 **13. a.** 1 **b.** n **c.** $\sum_{i=1}^{n} i$ **15. a.** The sum of row 5 in Pascal's triangle is 32; therefore there are 32 total outcomes possible. So the probability of each outcome is $\frac{1}{32}$. **b.** ${}_5C_1 = 5$, and there are 32 possible outcomes. So 5 out of 32 is $\frac{5}{32}$. **c.** $\frac{{}_5C_2}{32} = \frac{10}{32} = \frac{5}{16}$ **17.** $\frac{1}{4845}$ **19.** about 0.927

Lesson 10-3 (pp. 631-636)

Guided Example 1 5; 5; 3; 56; 56

Questions

1. a. 4 **b.** We can use the combination ${}_4C_1 = 4$ to find the coefficient because the coefficients of the terms in the binomial expansion are the numbers in the nth row of Pascal's Triangle. **3.** 64 **5.** $1728p^3q$

7.

expand$((1-3\cdot x)^4)$

$81\cdot x^4 - 108\cdot x^3 + 54\cdot x^2 - 12\cdot x + 1$

$(1 - 3 \cdot 2)^4 = (-5)^4 = 625$; $1296 - 864 + 216 - 24 + 1 = 625$ **9. a.** 20 **b.** 42 **11.** $(10^3)^4 + 4(10^3)^3 + 6(10^3)^2 + 4(10^3) + 1 = 1{,}004{,}006{,}004{,}001$ **13. a.** 15 **b.** $\frac{1}{64}$ **c.** $\frac{81}{4096}$ **15.** Let $2^n = (1 + 1)^n$. By the Binomial Theorem, for any nonnegative integer n, $(1 + 1)^n = {}_nC_0 + \cdots + {}_nC_{n-1} + {}_nC_n = \sum_{k=0}^{n} {}_nC_k$ **17.** 462 **19. a.** permutation **b.** 994,010,994,000 **21.** $\frac{34}{{}_{36}C_3} = \frac{1}{210}$

Lesson 10-4 (pp. 637-644)

Questions

1. 1 **3.** Not a probability distribution. **5.** Not a probability distribution. **7. a.** the number of days of incubation time **b.** about 3.92 days **c.** $x = 5$ **9. a.** about 1.72 **b.** about 1.31

11. a.

x (Difference)	0	1	2	3	4	5
$P(x)$	$\frac{1}{6}$	$\frac{5}{18}$	$\frac{2}{9}$	$\frac{1}{6}$	$\frac{1}{9}$	$\frac{1}{18}$

b. about 1.94; because 1.97 was the mean of the relative frequencies **c.** about 2.05 **d.** about 1.43 **13. a.** about 3.64 years **b.** about 1.88 **c.** about 1.37 years **15. a.** 1, 7, 21, 35, 35, 21, 7, 1; all are divisible by 7 except the first and last **b.** 1, 9, 36, 84, 126, 126, 84, 36, 9, 1; 84 is not divisible by 9 **c.** ${}_nC_r = \frac{n!}{r!(n-r)!}$; We know that n is prime and r and $(n - r)$ are smaller than n, so any part of $r!$ multiplied by any part of $(n - r)!$ will never equal (and thus never cancel) n from the numerator. **17.** ${}_6C_3 = 20$

19. a. and b.

χ²GOF {9,1},{5,5},1: *stat.results*

"Title"	"χ² GOF"
"χ²"	6.4
"PVal"	0.011412
"df"	1.
"CompList"	"{...}"

c. There is insufficient evidence to reject the hypothesis that the coin is fair because 0.011 > 0.01.

Lesson 10-5 (pp. 645-650)

Questions

1. 1. There are repeated situations, called trials. 2. There are a fixed number of trials. 3. For each trial, there are only two possible outcomes, often called success (S) and failure (F). 4. The probability of success is the same in each trial. 5. The trials are independent events. **3.** no; the probability of success is greater with preparation than without **5.** yes **7. a.** about 0.2051 **b.** about 0.2001 **9.** about 0.9995 **11.** about 0.6128 **13. a.** about 0.4219 **b.** about 0.6836 **15. a.** about 0.2835 **b.** about 0.7946 **17.** $x^2 + 2xy + y^2$ **19.** $\sin^3\theta + 6\sin^2\theta\cos\theta + 12\sin\theta\cos^2\theta + 8\cos^3\theta$ **21.** $y = \ln 117$ **23.** 4,475

Lesson 10-6 (pp. 651-658)

Guided Example 1 1; 0; 8; 0.000015; 8; 1; 7; 0.000366; 28; 2; 6; 0.003845; 56; 3; 5; 0.023071; 70; 4; 4; 0.086517; 28; 6; 2; 0.311462; 8; 7; 1; 0.266968; 1; 8; 0; 0.100113

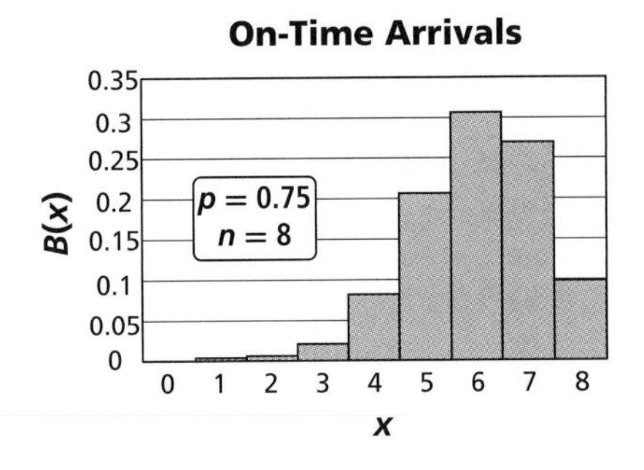

Questions

1. 18; 25; 0.77 **3. a.** the probability of 2 hits in 5 at-bats **b.** 0.0725 **5. a.** $B(x) = {}_6C_x(0.4)^x(0.6)^{6-x}$

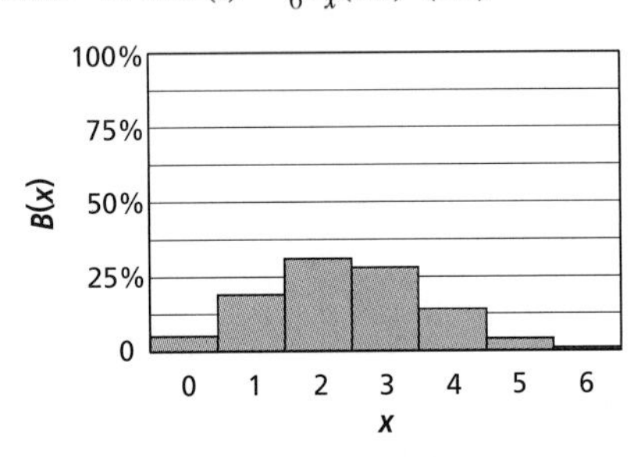

b. As f increases, the mode and mean increase (shifting to the right). This moves the bars further right with a tail on the left. **7. a.** 0.5 **b.** $x = 15$ **9.** 3 times out of 4 days, because ${}_4C_3(0.8)^3(0.2)^1 > {}_8C_6(0.8)^6(0.2)^2$. **11. a.** ≈ 0.200 **b.** ≈ 0.598 **c.** ≈ 0.402 **13.** about 9.313×10^{-10} **15.** $x^{18} + 6x^{15}y^2 + 15x^{12}y^4 + 20x^9y^6 + 15x^6y^8 + 6x^3y^{10} + y^{12}$ **17.** $\frac{(n-1)!}{(n-x)!(x-1)!}$

Lesson 10-7 (pp. 659-665)

Guided Example 1 Solution 1: 0.8; 0.8; 3.2; **Solution 2:** 0.20; $\binom{4}{0}(0.8)^0(0.2)^4$; $\binom{4}{2}(0.8)^2(0.2)^2$; $\binom{4}{3}(0.8)^3(0.2)^1$; 0.0016; 0.1536; 0.4096; 0.0016; 0.1536; 0.4096; 0.4096; 3.2

Questions

1. a. $\mu = 0 \cdot {}_4C_0(0.65)^0(0.35)^4 + 1 \cdot {}_4C_1(0.65)^1(0.35)^3 + 2 \cdot {}_4C_2(0.65)^2(0.35)^2 + 3 \cdot {}_4C_3(0.65)^3(0.35)^1 + 4 \cdot {}_4C_4(0.65)^4(0.35)^0 = 2.6$ **b.** $\mu = 4 \cdot 0.65 = 2.6$ **3. a.** 45 **b.** about 3.35 **5. a.** 475 **b.** 23.75 **c.** about 4.87 **7.** 2 **9. a.**

x "type O"	0	1	2	3	4	5
P(x)	≈ 0.05	≈ 0.21	≈ 0.34	≈ 0.28	≈ 0.11	≈ 0.02

b.

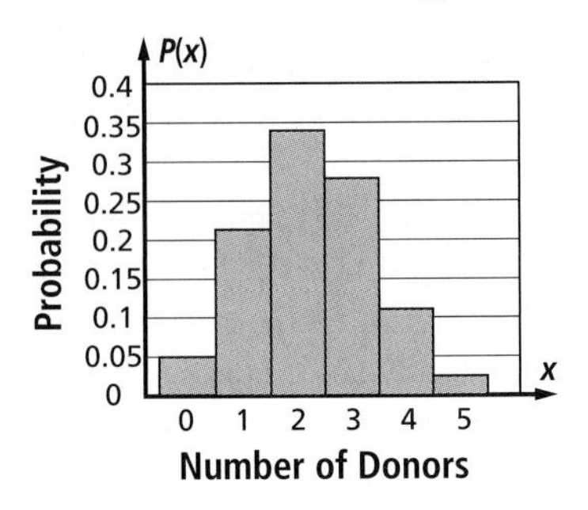

c. 2.25 **d.** about 1.11 **11. a.** true **b.** false **13.** k; n; binomial; p **15.** False. The distribution is symmetric to line $x = np$ when $p = 0.5$. **17. a.** ≈ 0.116 **b.** Given that 11.6% of the time, 5 of 5 patients will recover without any treatment, it seems difficult to make the case that the drug causes the recovery of the 5 patients. **c.** ≈ 0.000021 **d.** The probability of all patients recovering in the sample of 5 is high enough that it is not unlikely to see this outcome regardless of treatment. However, in the second experiment the probability of all 25 patients recovering given no treatment is so small that it leads us to believe that the drug had something to do with their recovery. **e.** ≈ 0.306 **19.** $-\frac{\sqrt{2}}{2}$

Lesson 10-8 (pp. 666-673)

Questions

1. A result is significant at the 0.05 level if the chance of that event occurring with the null hypothesis is less than 0.05. **3. a.** H_0: The coin is fair with $p = 0.5$. **b.** 13 or more tails as well as 13 or more heads in 15 trials **c.** Answers vary. Sample: 0.05 **5. a.** H_0: 1.1% of potato chip bags have less than 12 ounces. **b.** 77 or more bags out of a sample of 1100 bags **c.** Answers vary. Sample: 0.05 **7.** The shipment would not be rejected because the chance of this happening randomly is about 0.0592 which is greater than our significance level of 0.05. **9. a.** This situation is modeled by a binomial probability distribution with $p = 0.18$ and $n = 39$. We need to find $\sum_{x=19}^{39} \binom{39}{x}(0.18)^x(0.82)^{39-x}$. Technology shows that this sum equals about 0.000012, so the probability of 19 or more of the 39 tested students admitting to text messaging while driving is about 0.000012. **b.** This is smaller than the significance level of 0.01. Therefore, we reject the null hypothesis H_0 since $0.000012 < 0.01$. The principal should conclude that the percentage of seniors in his school who admit to text messaging while driving is greater than the population of all drivers.
11. a. H_0: The new treatment has no effect on the recovery rate of 25%. **b.** The probability of at least 75 out of 211

patients recovering under the null hypothesis is about 0.0004. This is smaller than our specified significance level so the virologist should reject the null hypothesis. The new treatment seems to be more effective than the old treatment. **13. a.** P(6 or more heads in 12 trials) $\approx$ 0.178. We therefore have insufficient evidence to reject the null hypothesis at the 0.05 level. **b.** P(9 or more heads in 12 trials) $\approx$ 0.0039. We can therefore reject the null hypothesis that the coin is biased against heads with probability $\frac{1}{3}$ since $0.0039 < 0.05$. **c.** Answers vary. Sample: For a coin that has probability $\frac{1}{3}$ of heads, the expected number of heads in 12 tosses is 4. Although a result of 6 heads, as in Part a, is higher than the expected value, it is not far off so it makes sense that the hypothesis is not rejected. However, an outcome of 9 heads, as in Part b, is farther from the expected value, so it makes sense that the hypothesis is rejected. **15.** $\mu = 4$; $\sigma = \frac{2\sqrt{6}}{3}$ **17. a.** 0.01 **b.** $\frac{16}{25}$ **c.** 0.11

Self-Test (p. 678)

1. The $(r + 1)$st term in row n is ${}_nC_r$. Therefore, the 16th term in row 53 of Pascal's Triangle is ${}_{53}C_{15} =$ 6,250,347,750,920. **2.** Given is ${}_{25}C_4 = 12650$ and ${}_{26}C_4 = 14950$ and we want ${}_{25}C_3$. Since ${}_nC_r = {}_{n-1}C_r + {}_{n-1}C_{r-1}$, ${}_{26}C_4 = {}_{25}C_4 + {}_{25}C_3$, so $14950 = 12650 + {}_{25}C_3$. Thus ${}_{25}C_3 = 2300$. **3. a.** ${}_4C_0x^4(-2y)^0 + {}_4C_1x^3(-2y)^1 + {}_4C_2x^2(-2y)^2 + {}_4C_1x^1(-2y)^3 + {}_4C_0x^0(-2y)^4$ **b.** $x^4 - 8x^3y + 24x^2y^2 - 32xy^3 + 16y^4$ **4.** $\mu = np = 300 \cdot 0.35 = 105$; $\sigma = \sqrt{npq} = \sqrt{300 \cdot 0.35 \cdot 0.65} \approx 8.26$ **5. a.** ${}_nC_1 = \frac{n!}{(n-1)!1!} = \frac{n!}{(n-1)!} = \frac{n(n-1)!}{(n-1)!} = n$. **b.** In any row n, the second entry is ${}_nC_1$. **6.** N(4-member committees from a pool of 100) $= {}_{100}C_4 = 3{,}921{,}225$; N(4-member committees from a pool of 100 of which 2 members are from my state) $= {}_2C_2 \cdot {}_{98}C_2 = 4753$; P(4-member committees from a pool of 100 of which 2 members are from my state) $= \frac{4753}{3{,}921{,}225} = \frac{1}{825}$.

7. ${}_{20}C_{14}(0.68)^{14}(0.32)^6 \approx 0.188$ **8.** mean $\mu = (3 \cdot 0.24) + (6 \cdot 0.41) + (8 \cdot 0.15) + (9 \cdot 0.04) + (11 \cdot 0.04) + (13 \cdot 0.12) = 6.74$; variance $\sigma^2 = 0.24(3 - 6.74)^2 + 0.41(6 - 6.74)^2 + 0.15(8 - 6.74)^2 + 0.04(9 - 6.74)^2 + 0.04(11 - 6.74)^2 + 0.12(13 - 6.74)^2 \approx 9.45$ **9.** H_0 = The coin is fair with probability of success of $p = 0.5$.

P(16 or more heads or 16 or more tails) $= \dfrac{\sum_{k=16}^{20}\binom{20}{k} + \sum_{k=0}^{4}\binom{20}{k}}{2^{20}} \approx 0.01$. Because $0.01 < 0.05$, there is enough evidence to reject the claim that the coin is fair. **10.** ${}_3C_2 \cdot {}_7C_5 = 63$ ways. **11.** $(x + y)^7 = x^7 + 7x^6y + 21x^5y^2 + \cdots$. Therefore, $c = 5$ and $d = 2$.

12. a.

x	0	1	2	3	4	5
B(x)	0.107	0.268	0.302	0.201	0.088	0.026
x	6	7	8	9	10	
B(x)	0.006	0.001	0.000	0.000	0.000	

b.

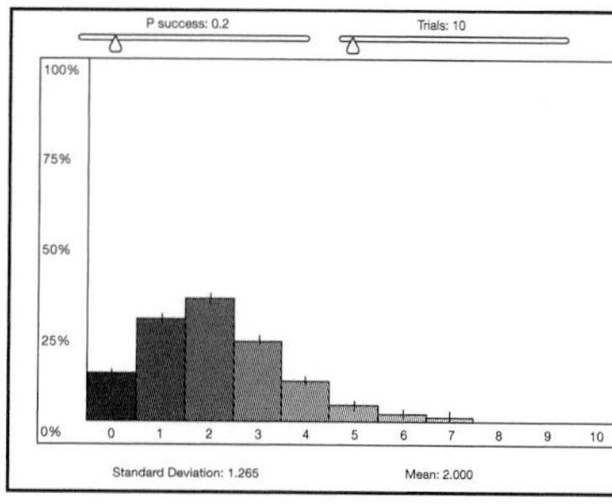

P success: 0.2
Trials: 10
100%
75%
50%
25%
0%
0 1 2 3 4 5 6 7 8 9 10
Standard Deviation: 1.265
Mean: 2.000

13. Answers vary. Sample: The graph starts out with a tail to the right. As p increases, the mean and mode increase and the graph shifts to the right. When $p = \frac{1}{2}$, the graph is symmetric to $x = \frac{n}{2}$. As p continues to increase, the graph develops a tail to the left.

Self-Test Correlation Chart

Question	1	2	3	4	5	6	7	8	9	10
Objective(s)	K	K	A	B	C	G	H	I	J	F
Lesson(s)	10-2	10-2	10-3	10-7	10-2	10-1	10-5	10-4	10-8	10-1
Question	**11**	**12**	**13**							
Objective(s)	D	L	E							
Lesson(s)	10-3	10-6	10-6							

Chapter Review (pp. 679-683)

1. $a^3 + 3a^2b + 3ab^2 + b^3$ **3.** $81x^4 - 432x^3 + 864x^2 - 768x + 256$ **5.** $-18w^{17}v$ **7.** $\mu = 30, \sigma = \sqrt{21} \approx 4.58$ **9.** $n = 144, p = 0.5$ **11.** ${}_nC_{n-1} = \frac{n!}{(n-(n-1))!(n-1)!} = \frac{n!}{1!(n-1)!} = n.$ **13.** $x = 10$

```
Solve (nCr(9,3) + nCr(9,4) = nCr(x,4), x)
                                x = 10 or x = -7
```

15. They are the coefficients of the terms of a binomial expansion. **17.** $a = 4, b = 3$ **19.** The peak of the graph moves from left to right. **21.** The distribution flattens and the domain increases, so the graph spreads out. **23.** 1 **25.** $\sum_{x=0}^{n} x \cdot {}_nC_x \cdot p^x \cdot (1-p)^{n-x}$ **27.** false **29.** 306 **31.** 2,250,829,575,120 **33.** ≈ 0.0088 **35.** ≈ 0.19 **37.** ≈ 0.00000923 **39.** ≈ 0.377 **41.** The probability of getting r heads and $n - r$ tails in n flips.
43. 0.000129 **45.** 0.078
47. 0.888 **49.** $\mu = 5.03125; \sigma \approx 2.28$ **51.** $\mu = 1.7;$ $\sigma \approx 3.16$ **53.** $\mu = 8, \sigma \approx 2.53$ **55. a.** 0.000006 **b.** 9 or fewer tails and 9 or fewer heads; All these outcomes are at least as unlikely as the given outcome. **c.** yes
57. H_0: The subject does not have ESP. The probability of guessing 33 or more cards correctly without ESP is 0.01642. Because $0.01642 > 0.01$, there is insufficient evidence at the 0.01 significance level to reject the null hypothesis. The subject does not appear to have ESP.
59. 1, 66, 792, 495, 220, 12 **61. a.** 12,551,759,587,422 **b.** This is the coefficient of the $a^{26}b^{21}$ term of the 47 power of $(a + b)$. **63.** 29, 1
65. Answers vary. Sample:

1
1 1
1 2 1
1 3 3 1
1 4 6 4 1
1 5 10 10 5 1
1 6 15 20 15 6 1
1 7 21 35 35 21 7 1
1 8 28 56 70 56 28 8 1
1 9 36 84 126 126 84 36 9 1

67.

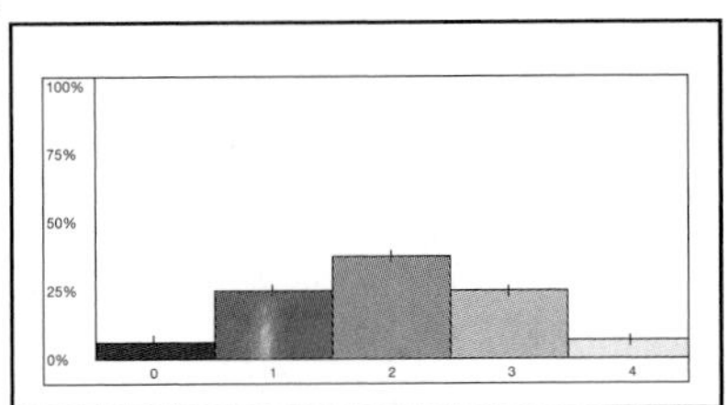

69. a.

x = number of girls	0	1	2	3	4	5
P(x)	$\frac{1}{32}$	$\frac{5}{32}$	$\frac{10}{16}$	$\frac{10}{16}$	$\frac{5}{32}$	$\frac{1}{32}$

b.

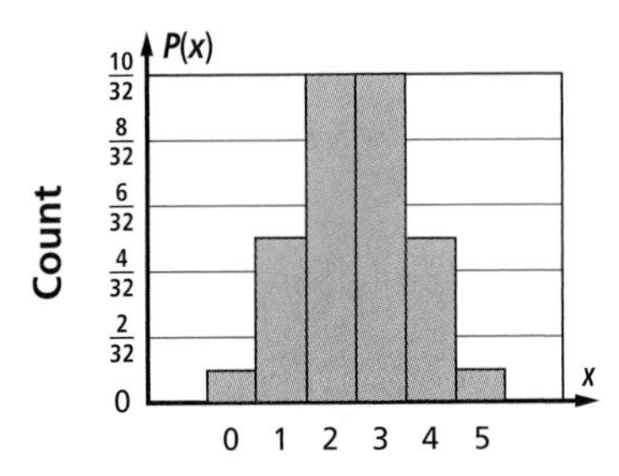

71. a. 10 **b.** The mode increases. **c.** The spread decreases. **73.** C, because the distribution is shifted the farthest towards the left.

Chapter 11

Lesson 11-1 (pp. 686-691)

Questions

1. both even **3.** both $y = 0$ **5.** the point where the graph goes from being concave up to concave down or vice versa **7. a.** $\left(1, \frac{e^{-\frac{1}{2}}}{\sqrt{2\pi}}\right), \left(-1, \frac{e^{-\frac{1}{2}}}{\sqrt{2\pi}}\right)$ **b.** $\left(0, \frac{1}{\sqrt{2\pi}}\right)$ **9.** Answers vary. Sample: the heights of adult men in the U. S. **11. a.** 0.6 **b.** $0.6^4 = 0.1296$ **13.** $(x, y) \to \left(\sigma x, \frac{y}{\sqrt{\sigma}}\right)$ **15. a.** $g^{-1}(x) = x^{\frac{5}{7}}$ **b.** Yes, $g(x)$ is one-to-one so the inverse is a function. **17.** $P(A) = \frac{200}{216}$, $P(B) = \frac{30}{216}$, and $P(A \cap B) = \frac{27}{216}$; $\frac{200}{216} \cdot \frac{30}{216} \neq \frac{27}{216}$, so the events are not independent. **19. a.** students in the nation taking the college-entrance examination **b.** students at one school taking the college-entrance examination

Lesson 11-2 (pp. 692-697)

Questions

1. mean 0; standard deviaton: 1 **3. a.** ≈ 0.4332 **b.** ≈ 0.8664 **c.** ≈ 0.9332 **5.** ≈ 0.4761

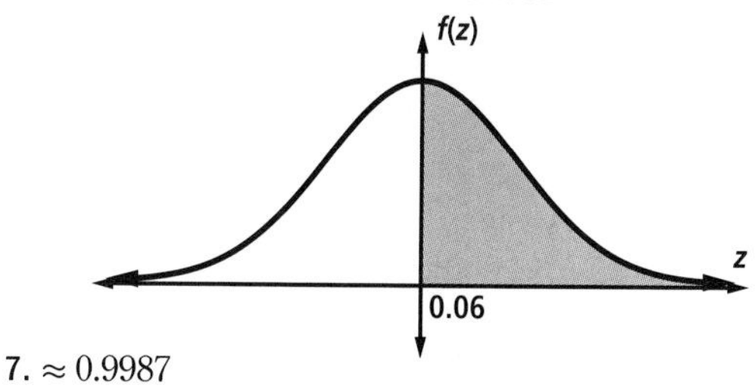

7. ≈ 0.9987

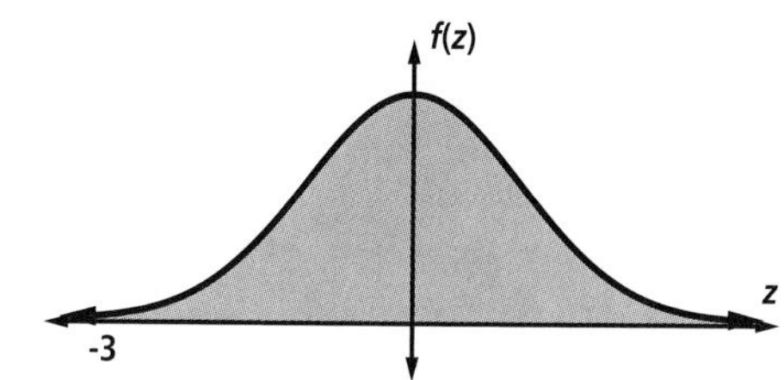

9. Answers vary. 95.5% **11.** $P(z < 1) - P(z < -1) = 0.8413 - 0.1587 = 0.6826$ **13.** $c \approx 0.67$ **15.** about 1.645 **17.** The area under the standard normal curve is 1. The area under the parent function $y = e^{-x^2}$ is $\sqrt{\pi}$. **19.** $n \approx 85$, $p \approx 0.471$ **21. a.** $\bar{x} = 122$, $s \approx 24.65$ **b.** mean is 0; standard deviation is 1

Lesson 11-3 (pp. 698-705)

Guided Example 2 a. 8; 0.6; -1.67; -1.67; 0.048 **b.** 3.33; Extrapolations from table vary. Sample: 0.9996 (should be near, but less than, 1); 0.0004 **c.** 7; 10; 0.048; 0.0004; 0.9516 **d.** 99; 2.33; 2.33; 9.398; 9.4

Questions

1. D **3. a.** about 15.9% **b.** about 99.5% **5.** about 94.3% **7.** about 7.53 oz **9.** Both np and nq are greater than 10. $P(z \geq 1.88) \approx 0.0301$. Additionally, the probability of 50,297 tails or more is counted. Therefore, we do not have sufficient evidence to reject the null hypothesis because $2(0.0301) \approx 0.060 > 0.05$. **11. a.** about 0.00214 **b.** about 0.01606 **c.** about 0.9818 **13.** $np = 987.7 > 10$; $nq = 174.3 > 10$; $z = 5.69$. We reject the hypothesis because $6.35(10^{-9}) < 0.01$. **15.** about 0.023% **17 a.** about 0.3249 **b.** about -0.0147 **19.** about 0.0038 **21.** 217,298,250 **23.** ≈ 0.998

Lesson 11-4 (pp. 706-713)

Questions

1. A sample is one selection of some number of elements of a population. A sampling distribution is the set of all possible samples of the same size n. **3.** A **5.** true; $\sigma_{\bar{x}} = \frac{\sigma}{\sqrt{n}}$ **7.** The shape of the distribution is approximately normal; $\mu_{\bar{x}} = 64.1$ in.; $\sigma_{\bar{x}} \approx 0.184$ in. **9. a.** They will be the same. **b.** the scores of the 45 games **11.** about 98.54% **13. a.** about 0.6827 **b.** about 0.9011 **c.** about 0.9500 **d.** about 0.9901 **15. a.** H_0: the baby has an equal preference for either color, so $p_{red} = 0.5$ **b.** There are $2^4 = 16$ possible outcomes. We first count all outcomes just as likely as three red balls which also includes the four red, no blue situation. Additionally, as this is analogous to a coin-flipping experiment, we include the outcomes three blue, one red and four blue, no red because they are just as likely. **c.** P(at least three blue) + P(at least three red) = $\frac{10}{16} = 0.625$. The null hypothesis cannot be rejected at the 0.05 level since $0.625 > 0.05$. This makes sense because when only four balls are chosen by a baby, variability is more likely and doesn't necessarily indicate bias.

17. a.

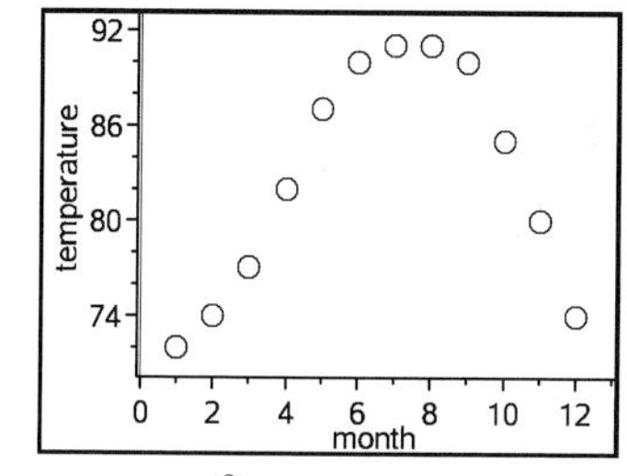

b. about $y = -0.59x^2 + 8.37x + 60.5$

c.

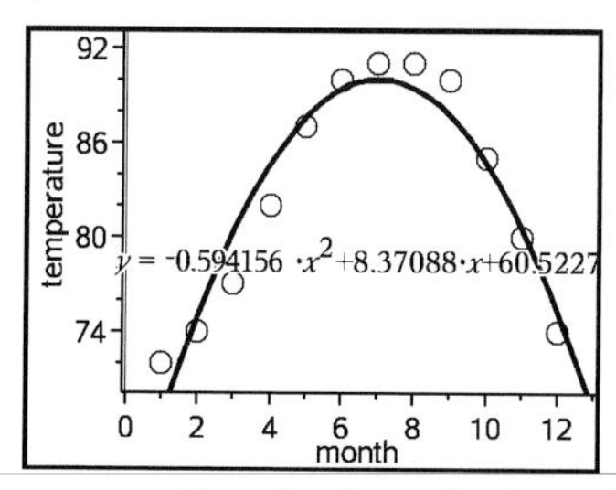

d. Answers vary. Sample: A quadratic model is not a good choice to model weather data because, while it may be accurate over the course of one year, the weather is cyclic and returns to similar temperatures the same time each year, while a quadratic model is not cyclic and thus would not show this.

Lesson 11-5 (pp. 714-718)

Guided Example 2 1 hour; $\frac{1}{\sqrt{60}}$ hour; 1; $\frac{1}{\sqrt{60}}$; 6.1

Questions

1. about 0.309 or 30.9% **3. a.** 1 hour **b.** about 0.1414 hours **c.** The distribution is approximately normal with data centered around 1 hour and a standard deviation of about 8.5 minutes. **5.** about 32.75% **7.** Because the data is not normally distributed, we rely on the sample size to determine whether the Central Limit Theorem can be used. In this case, the sample size is not large enough to use the Central Limit Theorem so we do not have the required tools to solve the problem. **9.** Answers vary. Sample: Samples of which the data was not originally normally distributed and the sample size is less than 30. **11. a.** about 2.00% **b.** about 0.356 **13.** about 0.4078

Lesson 11-6 (pp. 719-725)

Questions

1. a. between 593.2 and 612.8 **b.** 9.8 **3.** between 525.9 and 546.1 **5.** true **7.** increase the sample size or decrease the confidence level **9.** n; $\frac{2.58\sigma}{\sqrt{n}}$ **11.** 369 scores or more **13. a.** between 170.2 and 189.8 **b.** No. The likelihood of her actual mean cholesterol level being greater than or equal to 200 is less than 2.5 percent. **c.** between about 171.67 and 176.73 **15. a.** $\mu_{\bar{x}} = 0.5$ inch; $\sigma_{\bar{x}} = \frac{0.095}{\sqrt{100}} = 0.0095$ inch **b.** about 0.057 **c.** about 0.299 **17.** B **19.** C **21. a.** $1 - t$ **b.** $1 - t$ **c.** $2t - 1$ **23.** when $x = 0.5, y = 0.25$

Lesson 11-7 (pp. 726-732)

Guided Example 2 **a.** 1.96; 1000; 0.35; 1.96; $\frac{0.35 \cdot 0.65}{1000}$; 0.0296 **b.** 0.35; 0.0296; 35; 2.96 **c.** 2.96

Guided Example 3 **a.** 2.58; 0.03; $\frac{0.03}{2.58}$; 0.02; 43; 1849 **b.** 9; 9; 16641

Questions

1. The margin of error is $z\sqrt{\frac{p(1-p)}{n}}$ which means it is dependent only on the confidence level desired , the probability of success, and the size of the sample n. **3.** false. Increasing the value of z in the margin of error $z\sqrt{\frac{p(1-p)}{n}}$ results in a direct increase in the margin of error. **5. a.** about 2154 people **b.** $\approx$ 8614 or about 4 times more people **7.** false. Because population size has no bearing on the margin of error, the poll of 1000 people will have a smaller margin of error than the poll of 100 people. **9. a.** H_0: The coin is fair ($p = 0.5$). Because both np and $n(1 - p)$ are greater than 10, we can use mean of $p = 0.5$ and standard deviation of $\sqrt{\frac{p(1-p)}{n}} = 0.005$ to find the z-score for Kerrich's particular experiment result. That is, $z = \frac{\frac{5067}{10{,}000} - 0.5}{0.005} \approx 1.34$. So, $P(z > 1.34) \approx 0.09$ for the probability of at least 5067 heads. We double this probability to account for the probability of similar events which includes at least 5067 tails. Therefore, we have insufficient evidence to reject the null hypothesis since $0.18 > 0.01$. Note that in this case $p = 0.5$. **b.** Given Kerrich's experiment, we use $p = \frac{5067}{10{,}000} = 0.5067$. Because both np and $n(1 - p)$ are greater than 10, we can use $z = 2.58$ for the 99% confidence interval. Therefore, the margin of error is $z\sqrt{\frac{p(1-p)}{n}} = 2.58\sqrt{\frac{0.5067(1-0.5067)}{10{,}000}} \approx 0.012899$. So the confidence interval is 0.5067 ± 0.012899 or between 49.4% and 52.0%. That is, this is the range of values of p we can hypothesize in Part a and still not be rejected at the 0.01 significance level. Notice that $p = 0.5$ in Part a fell in this range and therefore the hypothesis was not rejected. **11.** about 79.4% **13.** $\sigma_{\bar{x}} \approx 0.0146$ so 95% of sample mean weights fall between 2.9714 oz. and 3.0286 oz. **15. a.** \$61,750 **b.** \$1,854.60 **c.** $z = \frac{57{,}750 - 61{,}750}{1{,}854.60} \approx -2.16$, so $P(z < -2.16) \approx 0.015$ **d.** $z = \frac{55{,}000 - 61{,}750}{1{,}854.60} \approx -3.64$ and $z = \frac{68{,}500 - 61{,}750}{1{,}854.60} \approx 3.64$, so $P(|z| < 3.64) \approx 0.9997 \approx 1$ **17. a.** uniform distribution **b.** $\mu = 54.5$; $\sigma \approx 25.98$ **19.** ≈ 0.9678

1.85

Self-Test (p. 737)

Questions

1. $x \rightarrow \frac{x-76.9}{13.6}$ **2.** 1; The standard normal curve pictures a probability distribution. **3.** The entire area under the standard normal curve and the x-axis is 1. Therefore, by symmetry, the area of the region $|z| > 0.92$ is equal in area to twice the unshaded region. That is, $P(|z| > 0.92) = 2(1 - 0.8212) = 0.3576$. **4. a.** Using the Standard Normal Distribution Table or technology $P(z < 1.7) \approx 0.9554$. **b.** Using the Standard Normal Distribution Table or technology $P(-1.1 < z < 1.1) = P(z < 1.1) - P(z < -1.1) = P(z < 1.1) - P(z > 1.1) = P(z < 1.1) - (1 - P(z < 1.1)) = 2P(z < 1.1) - 1 \approx 2(0.8643) - 1 \approx 0.7286$. **5.** The 95% confidence interval for μ given σ is $\bar{x} \pm 1.96\frac{\sigma}{\sqrt{n}}$. Therefore, the interval is $490 \pm 1.96\frac{93}{\sqrt{521}} \approx 490 \pm 7.99$, that is, between 482.01 and 497.99. **6. a.** $p = 0.88$; Therefore, the margin of error for a 95% confidence interval is $1.96\sqrt{\frac{0.88(0.12)}{2455}} \approx 0.013$ or 1.3%. **b.** $p = 0.5$; Therefore, the margin of error is $1.96\sqrt{\frac{0.5(0.5)}{2455}} \approx 0.020$ or 2.0%. **7.** A census gives population statistics. Therefore, $\mu = 3.12$ miles and $\sigma = 0.89$ miles. Because the distribution is normally distributed with $z = \frac{x-3.12}{0.89}$. When $x = 4$, $z \approx 0.99$. $P(z > 0.99) \approx 0.161$, so 16.1% of students at Peaks High would be expected to travel over 4 miles to school.

8. Null hypothesis H_0: the doctors' patients are a random sample of people with normal body temperature. The sampling distribution of sample means of samples of size 130 has $\mu_{\bar{x}} = 98.6°F$ and $\sigma_{\bar{x}} = \frac{1.2}{\sqrt{130}} \approx 0.105°F$. $z = \frac{x - 98.6}{0.105}$. When $x = 98.2$, $z \approx -3.81$. $P(|z| \geq 3.81) < 0.001$. Since $0.001 < 0.01$, we reject the null hypothesis. 9. Applying the Central Limit Theorem, the sampling distribution of sample means of size 42 will be normally distributed with mean 130.2 watts and standard deviation $\frac{7.3}{\sqrt{42}} \approx 1.13$ watts.

10. The 95% confidence interval for a poll is $\bar{x} \pm 1.96\sqrt{\frac{p(1-p)}{n}}$. Here $\bar{x} = 40\%$, $n = 471$, and $p = \frac{1}{2}$, so the interval is between about 35.5% and 44.5%. 11. a. The sample size needs to be multiplied by 81. b. false; For a given population, larger samples have smaller margins of error because the margin of error is inversely proportional to the square root of the sample size.

Self-Test Correlation Chart

Question	1	2	3	4	5	6	7	8	9	10
Objective(s)	B	B	H	A	G	G	D	E	F	G
Lesson(s)	11-3	11-1	11-2	11-2	11-6	11-7	11-3	11-5	11-4	11-7
Question	**11**									
Objective(s)	C									
Lesson(s)	11-7									

Chapter Review (pp. 738-741)

Questions

1. a. about 82.3% b. about 0.823 3. about 0.136 5. about 0.003 7. a. $f(-x) = e^{-(-x)^2} = e^{-x^2} = f(x)$ b. reflection symmetry over the line $x = \mu$ 9. 1 11. $x \rightarrow \frac{x-120}{8}$ 13. true 15. 95% 17. about 440 more people 19. about 4.6% 21. about 58.7% 23. Because the sample was greater than 30, we can invoke the Central Limit Theorem. We reject the hypothesis at the 0.01 level because P(of a sample of 80 cables giving a mean of at least 1750 given that the breaking strength has not changed) ≈ 0.000097, which is much less than 0.01.

25. Because the sample was greater than 30, we can invoke the Central Limit Theorem. We reject the hypothesis at the 0.05 level because P(of a sample of 50 cars have an average of 69.4 mph or less) ≈ 0.000159, which is much less than 0.05. 27. $\frac{\sigma}{\sqrt{n}}$ 29. Because the sample was greater than 30, we can invoke the Central Limit Theorem. The shape of the sampling distribution of sample means of size 100 is normally distributed with mean 500 mg and standard deviation 0.2 mg. 31. about 3.68 33. $\bar{x} \pm 2.58\frac{\sigma}{\sqrt{n}}$ 35. 0.5 37. between about 26.2% and 31.8% 39. between about 44.9% and 51.1% 41. a. (ii) b. (iii) c. (i) 43. ≈ 0.0228 45. $\frac{r}{2}$ 47. $0.5 + \frac{r}{2}$

Chapter 12

Lesson 12-1 (pp. 744-750)

Guided Example 2 2; 3; –3; 0; –6; –6; 4; 8; –3; 14

Questions

1. A matrix is an arrangement of elements in a rectangular array. 3. 85 5. total number of members in each Congress 7. 3 × 4 9. [28] 11. $\begin{bmatrix} ax+by \\ cx+dy \end{bmatrix}$ 13. 5 × 4 15. a. $\begin{bmatrix} 19{,}000 \\ 18{,}480 \\ 16{,}080 \end{bmatrix}$ b. The total revenue of each show. c. $19,000 17. $x = -2$ 19. a. $\begin{bmatrix} 9 & 0 \\ 0 & 9 \end{bmatrix}$ b. $\begin{bmatrix} 0 & -27 \\ -27 & 0 \end{bmatrix}$

c. $\begin{bmatrix} \frac{1+(-1)^n}{2}(-3)^n & \frac{1+(-1)^{n+1}}{2}(-3)^n \\ \frac{1+(-1)^{n+1}}{2}(-3)^n & \frac{1+(-1)^n}{2}(-3)^n \end{bmatrix}$

21. a. $\begin{bmatrix} 0 & 1 & 2 & 1 \\ 1 & 0 & 1 & 0 \\ 2 & 1 & 0 & 1 \\ 1 & 0 & 1 & 0 \end{bmatrix}$ b. $\begin{bmatrix} 6 & 2 & 2 & 2 \\ 2 & 2 & 2 & 2 \\ 2 & 2 & 6 & 2 \\ 2 & 2 & 2 & 2 \end{bmatrix}$ The matrix represents the number of exactly-two-step paths from one city to another. 23. a. ii. b. i. c. iii. d. iv. 25. $P(A) + P(B)$ 27. 139.3 paces

Lesson 12-2 (pp. 751-756)

Guided Example 2

–5; 6; $\begin{bmatrix} \cdots & 18 & -15 & -15 \\ & -6 & -6 & 12 \end{bmatrix}$; $\begin{bmatrix} 0 & 3 & 6 \\ 2 & 4 & 2 \end{bmatrix}$; $\begin{bmatrix} -6 & 0 & 15 \\ 0 & -4 & 0 \end{bmatrix}$

Questions

1. C 3. a. (iii) b. (v) c. (ii) d. (i) e. (iv) 5. a. $\begin{bmatrix} -4.5 & 6 & 1.5 & -4.5 \\ 4.5 & 4.5 & -3 & -3 \end{bmatrix}$ b. size change of magnitude 1.5 7. $\begin{bmatrix} a & 0 \\ 0 & a \end{bmatrix}$ 9. $\begin{bmatrix} -1 & 0 \\ 0 & 1 \end{bmatrix} \cdot \begin{bmatrix} x \\ y \end{bmatrix} = \begin{bmatrix} -x \\ y \end{bmatrix}$. So the matrix $\begin{bmatrix} -1 & 0 \\ 0 & 1 \end{bmatrix}$ maps each point (x, y) to the point $(-x, y)$, which is its reflection image over the y-axis. 11. $\begin{bmatrix} 0 & -1 \\ -1 & 0 \end{bmatrix}$

13. $\begin{bmatrix} 0 & 1 & 2 & 3 & 4 & 5 & 5 & 4 & 1 & 0 \\ 3 & 9 & 3 & 3 & 9 & 3 & -1 & -3 & -3 & -1 \end{bmatrix}$ 15. a. $A' = \begin{bmatrix} 30 \\ -17 \end{bmatrix}$; $B' = \begin{bmatrix} 14 \\ -8 \end{bmatrix}$; $C' = \begin{bmatrix} -18 \\ 10 \end{bmatrix}$ b. yes c. $AB \approx 2.236$; $BC \approx 4.472$; $A'B' \approx 18.358$; $B'C' \approx 36.715$; no 17. a. XY; 7 × 6 b. 42 19. Product does not exist. 21. a. $\begin{bmatrix} 0 & -8 \\ -7 & 23 \end{bmatrix}$

b. $\begin{bmatrix} 0 & -8 \\ -7 & 23 \end{bmatrix}$ c. The Associative Property

Lesson 12-3 (pp. 757-761)

Guided Example 1

–1; 0; 0; –1; –1; 0; 0; –1

Questions

1. r_x 3. $\begin{bmatrix} 0 & 1 & -6 & -2 \\ -7 & -8 & -11 & -9 \end{bmatrix}$ 5. a. R_{90} b. $\begin{bmatrix} 0 & -1 \\ 1 & 0 \end{bmatrix}$

7. $a = w$; $b = x$; $c = y$; $d = m$ 9. a. (0, –1) b. (–1, 0) c. $\begin{bmatrix} 0 & -1 \\ -1 & 0 \end{bmatrix}$ 11. $\begin{bmatrix} -14 & 9 & -89 \\ -14 & 3 & -53 \end{bmatrix}$

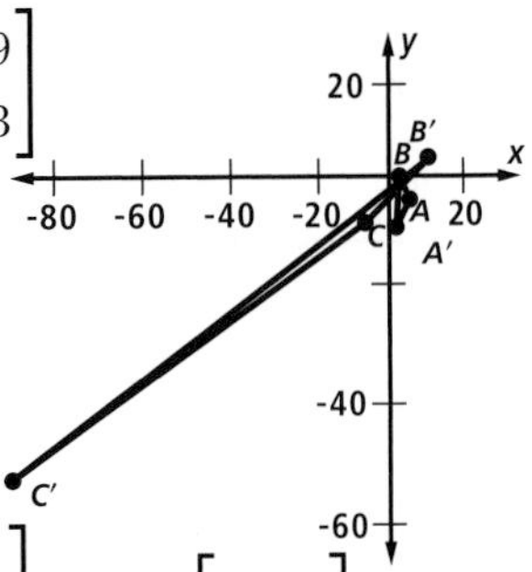

13. a. $\begin{bmatrix} \frac{1}{2} & -\frac{\sqrt{3}}{2} \\ \frac{\sqrt{3}}{2} & \frac{1}{2} \end{bmatrix}$ b. $\begin{bmatrix} -\frac{\sqrt{3}}{2} & -\frac{1}{2} \\ \frac{1}{2} & -\frac{\sqrt{3}}{2} \end{bmatrix}$ 15. a. $\begin{bmatrix} 0 & 1 \\ -1 & 0 \end{bmatrix}$ b. R_{270} 17. $\begin{bmatrix} 2 & -3 \\ 1 & 1 \end{bmatrix}$ 19. 95.45% 21. a. converges to $\frac{1}{4}$ b. diverges 23. $\cos\theta$

Lesson 12-4 (pp. 762-765)

Guided Example 1

–sin 60°; cos 60°; $\frac{\sqrt{3}}{2}$; $-\frac{\sqrt{3}}{2}$; $\frac{\sqrt{3}}{2}$; $\frac{1}{2}$

Guided Example 2

30.75; 30.75; 30.75; 0.618; –0.618; 0.786; –0.618; 0.786; 6.4; 7.5; 2.7; 3.0; 2.7; 6.4; 7.5; 3.0

Questions

1. $\begin{bmatrix} \cos t & -\sin t \\ \sin t & \cos t \end{bmatrix}$ 3. $\begin{bmatrix} 0.839 & -0.545 \\ 0.545 & 0.839 \end{bmatrix}$

5. $\begin{bmatrix} \cos 180° & -\sin 180° \\ \sin 180° & \cos 180° \end{bmatrix} = \begin{bmatrix} -1 & 0 \\ 0 & -1 \end{bmatrix}$

7. $(R_{60})^3 = \begin{bmatrix} \frac{1}{2} & -\frac{\sqrt{3}}{2} \\ \frac{\sqrt{3}}{2} & \frac{1}{2} \end{bmatrix} \cdot \begin{bmatrix} \frac{1}{2} & -\frac{\sqrt{3}}{2} \\ \frac{\sqrt{3}}{2} & \frac{1}{2} \end{bmatrix} \cdot \begin{bmatrix} \frac{1}{2} & -\frac{\sqrt{3}}{2} \\ \frac{\sqrt{3}}{2} & \frac{1}{2} \end{bmatrix} = \begin{bmatrix} -\frac{1}{2} & -\frac{\sqrt{3}}{2} \\ \frac{\sqrt{3}}{2} & -\frac{1}{2} \end{bmatrix} \cdot \begin{bmatrix} \frac{1}{2} & -\frac{\sqrt{3}}{2} \\ \frac{\sqrt{3}}{2} & \frac{1}{2} \end{bmatrix} = \begin{bmatrix} -1 & 0 \\ 0 & -1 \end{bmatrix} = R_{180}$

9. a. $\begin{bmatrix} 0 & 12 & 9 \\ 0 & 0 & 3\sqrt{3} \end{bmatrix}$ b. $\begin{bmatrix} 0 & -11.28 & -6.68 \\ 0 & -4.10 & -7.96 \end{bmatrix}$ c. The side lengths of the triangle are 6, $6\sqrt{3}$, and 12, which correspond to the sides of a 30°-60°-90° triangle. 11. a. (–151.7, 486.8)

b. (-486.8, -151.7) **c.** (151.7, -486.8) **d.** (486.8, 151.7) **e.** (-151.7, 486.8) **13. a.** A clockwise rotation of θ is equivalent to a counterclockwise rotation of $-\theta$, for which the matrix is

$$R_{-\theta} = \begin{bmatrix} \cos(-\theta) & -\sin(-\theta) \\ \sin(-\theta) & \cos(-\theta) \end{bmatrix} = \begin{bmatrix} \cos\theta & \sin\theta \\ -\sin\theta & \cos\theta \end{bmatrix}.$$

b. $\begin{bmatrix} -0.719 & 0.695 \\ -0.695 & -0.719 \end{bmatrix}$ **15. a.** yes **b.** no **c.** yes **d.** no

17. a. True; by the Central Limit Theorem, the mean of the sampling distribution of sample means is the population mean. **b.** False; by the Central Limit Theorem, the standard deviation of the sampling distribution of sample means is $\frac{\sigma}{\sqrt{n}}$, not σ. **19.** $y = 13x^2$
21. $143v - 13u$

Lesson 12-5 (pp. 766-770)

Guided Example 2

sin π; cos π; 0; –1; sin θ

Guided Example 3

30°; 30°; cos 45° sin 30°; $\frac{\sqrt{3}}{2}$; $\frac{1}{2}$; $\frac{\sqrt{6}-\sqrt{2}}{4}$

Questions

1. Using the Rotation Matrix Theorem, we know that $R_\beta(1, 0) = (\cos\beta, \sin\beta)$, so $R_\alpha \circ R_\beta(1, 0) = R_\alpha(R_\beta(1, 0))$ $= R_\alpha(\cos\beta, \sin\beta)$. **3.** $\cos\left(\frac{21\pi}{12} - \frac{3\pi}{12}\right) = \cos\left(\frac{3\pi}{2}\right) = 0$
5. $\frac{\sqrt{2}-\sqrt{6}}{4}$ **7.** $\cos(\pi - x) = \cos\pi\cos x + \sin\pi\sin x =$ $-\cos x$ **9.** $-\frac{17\sqrt{2}}{50}$ **11.** $\frac{\sqrt{6}-\sqrt{2}}{4}$ **13. a.** Answers vary. Sample: $\cos\pi = -1 \neq 0 = \cos\frac{\pi}{2} + \cos\frac{\pi}{2}$ **b.** Answers vary. Sample: $\alpha = \frac{\pi}{3}, \beta = -\frac{\pi}{3}$ **15. a.** $g(x)$,

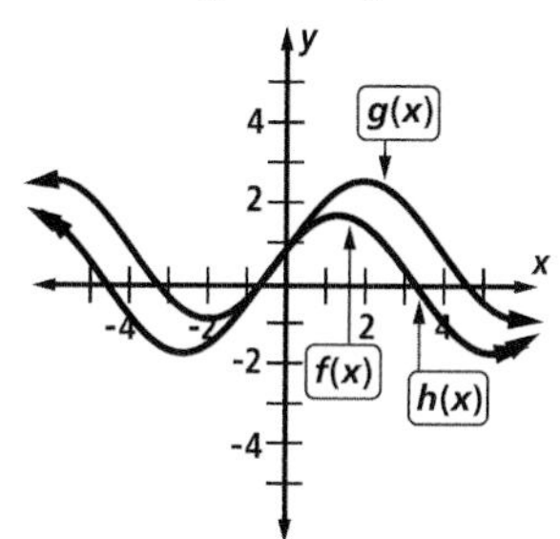

b. $f(x) = 2\sin\left(x + \frac{\pi}{6}\right) = 2\left(\sin x\cos\frac{\pi}{6} + \cos x\sin\frac{\pi}{6}\right) = 2\left(\frac{\sqrt{3}}{2}\sin x + \frac{1}{2}\cos x\right) = \sqrt{3}\sin x + \cos x = h(x)$
17. $\sin(x - 2\pi n) = \sin x\cos(2\pi n) - \cos x\sin(2\pi n) = 1(\sin x) - 0(\cos x) = \sin x$. **19. a.** $\frac{\sqrt{8}+\sqrt{3}}{6}$ **b.** $\frac{\sqrt{8}+\sqrt{3}}{6}$
c. Answers vary. Sample:

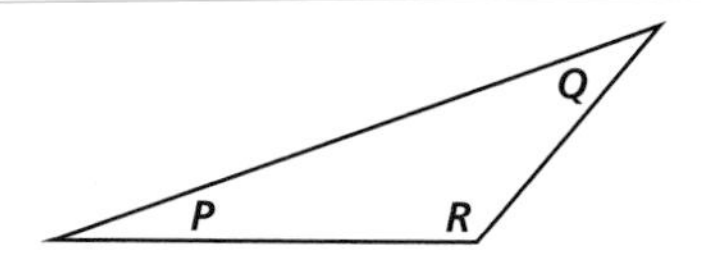

21. a. $\begin{bmatrix} \frac{\sqrt{3}-2}{2} & \frac{7\sqrt{3}-2}{2} & \frac{9\sqrt{3}-5}{2} & \frac{3\sqrt{3}-5}{2} \\ \frac{2\sqrt{3}+1}{2} & \frac{2\sqrt{3}+7}{2} & \frac{5\sqrt{3}+9}{2} & \frac{5\sqrt{3}+3}{2} \end{bmatrix}$

b. Counterclockwise rotation of 30° about the origin **23. a.** 15 **b.** Friday **c.** *HR* **d.** $77.50
25. Including multiplicities, a degree seven polynomial has seven solutions. If it has real coefficients, its complex solutions must come in conjugate pairs, so the number of complex solutions must be even. Because seven is odd, it must have at least one real zero.

Lesson 12-6 (pp. 771-775)

Questions

1. about 90.4 m **3.** By the Double Angle Formula, $\cos 2\theta = \cos^2\theta - \sin^2\theta$. By the Pythagorean Identity, $\cos^2\theta = 1 - \sin^2\theta$. Substitute to get $\cos 2\theta = (1 - \sin^2\theta) - \sin^2\theta$. **5.** $\frac{24}{25}$; $\frac{7}{25}$ **7.** sin 50° **9.** $\cos\frac{8\pi}{9}$
11. $4\sin\theta\cos^3\theta - 4\sin^3\theta\cos\theta$ **13. a.** $\frac{120}{169}$ **b.** $-\frac{120}{169}$
15. $(2\sin\theta)^2 = 1^2 + 1^2 - 2 \cdot 1 \cdot 1 \cdot \cos 2\theta$; $4\sin^2\theta = 2 - 2\cos 2\theta$; $\cos 2\theta = 1 - 2\sin^2\theta$ **17. a.** 45°

b. $\begin{bmatrix} \cos 90° & \sin 90° \\ \sin 90° & -\cos 90° \end{bmatrix} = \begin{bmatrix} 0 & 1 \\ 1 & 0 \end{bmatrix}$ **19. a.** $1 + \cos 2\theta = 2\cos^2\theta$; $\cos^2\theta = \frac{1+\cos 2\theta}{2}$; $\cos\theta = \pm\sqrt{\frac{1+\cos 2\theta}{2}}$; **b.** $\cos\frac{\pi}{8} = \pm\sqrt{\frac{1+\cos\frac{\pi}{4}}{2}} = \pm\sqrt{\frac{2+\sqrt{2}}{4}} = \pm\frac{\sqrt{2+\sqrt{2}}}{2}$. Since $\frac{\pi}{8}$ is in the first quadrant, $\cos\frac{\pi}{8} = \frac{\sqrt{2+\sqrt{2}}}{2}$. **21.** $\cos^3\theta - 3\sin^2\theta\cos\theta$
23. A **25.** $a = -\frac{121}{12}, b = \frac{335}{72}$ **27.** 220.19 yards

Self-Test (p. 780)

1. $\begin{bmatrix} 0.57 & 0.24 & 0.29 \\ 0.21 & 0.59 & 0.48 \\ 0.22 & 0.17 & 0.23 \end{bmatrix} \cdot \begin{bmatrix} 4.8 \\ 0.8 \\ 3.2 \end{bmatrix} =$

$$\begin{bmatrix} 4.8 \cdot 0.57 + 0.8 \cdot 0.24 + 3.2 \cdot 0.29 \\ 4.8 \cdot 0.21 + 0.8 \cdot 0.59 + 3.2 \cdot 0.48 \\ 4.8 \cdot 0.22 + 0.8 \cdot 0.17 + 3.2 \cdot 0.23 \end{bmatrix} = \begin{bmatrix} 3.856 \\ 3.016 \\ 1.928 \end{bmatrix}$$

2. the expected distribution of votes among the parties in an election, in millions of votes **3.** $\begin{bmatrix} 6 & -1 \\ 3 & 0 \end{bmatrix} \cdot \begin{bmatrix} 3 & 7 \\ 1 & -1 \end{bmatrix} =$

$\begin{bmatrix} 6 \cdot 3 + -1 \cdot 1 & 6 \cdot 7 + -1 \cdot -1 \\ 3 \cdot 3 + 0 \cdot 1 & 3 \cdot 7 + 0 \cdot -1 \end{bmatrix} \cdot \begin{bmatrix} 17 & 43 \\ 9 & 21 \end{bmatrix}$ **4.** Since M is a 3×5 matrix and N is a 5×t matrix, MN is a 3×t matrix (the number of rows of M by the number of columns of N). Since Z is a 3×6 matrix and $MN = Z$, N must have 6 columns, so $t = 6$. **5.** $M = \begin{bmatrix} 3 & -3 & 1 \\ 3 & 0 & -1 \end{bmatrix}$

6. a. $\begin{bmatrix} 2 & 0 \\ 0 & -1 \end{bmatrix} \cdot \begin{bmatrix} 3 & -3 & 1 \\ 3 & 0 & -1 \end{bmatrix} \cdot$ $\begin{bmatrix} 2 \cdot 3 + 0 \cdot 3 & 2 \cdot -3 + 0 \cdot 0 & 2 \cdot 1 + 0 \cdot -1 \\ 0 \cdot 3 + -1 \cdot 3 & 0 \cdot -3 + -1 \cdot 0 & 0 \cdot 1 + -1 \cdot -1 \end{bmatrix} =$ $\begin{bmatrix} 6 & -6 & 2 \\ -3 & 0 & 1 \end{bmatrix}$

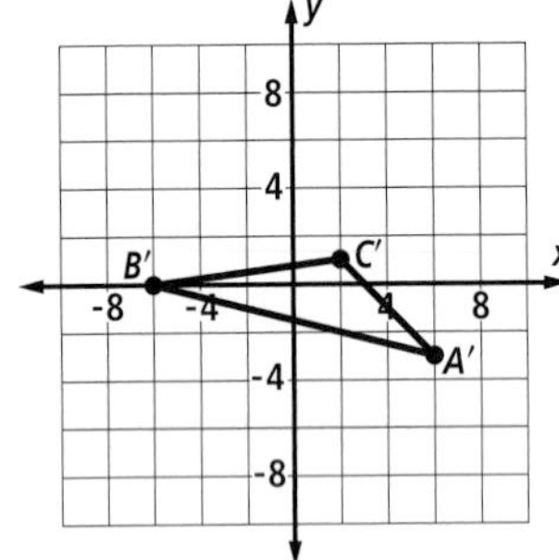

b.

7. The matrix representing R_{120} is $\begin{bmatrix} \cos 120° & -\sin 120° \\ \sin 120° & \cos 120° \end{bmatrix} =$ $\begin{bmatrix} -\frac{1}{2} & -\frac{\sqrt{3}}{2} \\ \frac{\sqrt{3}}{2} & -\frac{1}{2} \end{bmatrix}$. So $\begin{bmatrix} -\frac{1}{2} & -\frac{\sqrt{3}}{2} \\ \frac{\sqrt{3}}{2} & -\frac{1}{2} \end{bmatrix} \cdot \begin{bmatrix} 3 & -3 & 1 \\ 3 & 0 & -1 \end{bmatrix} =$ $\begin{bmatrix} \left(-\frac{1}{2} \cdot 3 + -\frac{\sqrt{3}}{2} \cdot 3\right) & \left(-\frac{1}{2} \cdot -3 + -\frac{\sqrt{3}}{2} \cdot 0\right) & \left(-\frac{1}{2} \cdot 1 + -\frac{\sqrt{3}}{2} \cdot -1\right) \\ \left(\frac{\sqrt{3}}{2} \cdot 3 + -\frac{1}{2} \cdot 3\right) & \left(\frac{\sqrt{3}}{2} \cdot -3 + -\frac{1}{2} \cdot 0\right) & \left(\frac{\sqrt{3}}{2} \cdot 1 + -\frac{1}{2} \cdot -1\right) \end{bmatrix}$ $= \begin{bmatrix} \frac{-3\sqrt{3}-3}{2} & \frac{3}{2} & \frac{-1+\sqrt{3}}{2} \\ \frac{3\sqrt{3}-3}{2} & \frac{-3\sqrt{3}}{2} & \frac{1+\sqrt{3}}{2} \end{bmatrix}$. **8. a.** The matrix representing r_x is $\begin{bmatrix} 1 & 0 \\ 0 & -1 \end{bmatrix}$, and the matrix representing R_{90} is $\begin{bmatrix} 0 & -1 \\ 1 & 0 \end{bmatrix}$. So the matrix representing $R_{90} \circ r_x$ is $\begin{bmatrix} 0 & -1 \\ 1 & 0 \end{bmatrix} \cdot \begin{bmatrix} 1 & 0 \\ 0 & -1 \end{bmatrix} = \begin{bmatrix} 0 & 1 \\ 1 & 0 \end{bmatrix}$. **b.** The matrix maps (x, y) onto (y, x). It represents the reflection over the line $y = x$. **9.** The reflection $r_{y=x}: (x, y) \rightarrow (y, x)$ maps (1, 0) to (0, 1) and (0, 1) to (1, 0). Thus the matrix representing $r_{y=x}$ is $\begin{bmatrix} 0 & 1 \\ 1 & 0 \end{bmatrix}$. **10.** B; The matrix representing R_{60} is $\begin{bmatrix} \cos 60° & -\sin 60° \\ \sin 60° & \cos 60° \end{bmatrix}$ which can be simplified to $\begin{bmatrix} \frac{1}{2} & -\frac{\sqrt{3}}{2} \\ \frac{\sqrt{3}}{2} & \frac{1}{2} \end{bmatrix}$.

11. $\triangle PQR$ is represented by the matrix $\begin{bmatrix} 1 & 4 & 0 \\ 0 & 7 & -3 \end{bmatrix}$ and R_{-56} is represented by the matrix $\begin{bmatrix} \cos(-56°) & -\sin(-56°) \\ \sin(-56°) & \cos(-56°) \end{bmatrix}$. So the image of $\triangle PQR$ is represented by $\begin{bmatrix} \cos(-56°) & -\sin(-56°) \\ \sin(-56°) & \cos(-56°) \end{bmatrix} \cdot \begin{bmatrix} 1 & 4 & 0 \\ 0 & 7 & -3 \end{bmatrix} \approx$ $\begin{bmatrix} 0.56 & 8.04 & -2.49 \\ -0.83 & 0.60 & -1.68 \end{bmatrix}$, thus $P' \approx (0.56, -0.83)$, $Q' \approx (8.04, 0.60)$, and $R' \approx (-2.49, -1.68)$. **12.** By the Subtraction Formula for the Sine, $\sin 79° \cos 37° - \cos 79° \sin 37° = \sin(79° - 37°) = \sin 42°$. **13.** By the Double Angle Formula, $2\cos^2 18° - 1 = \cos(2(18°)) = \cos 36°$. **14.** By the Addition Formula for the Sine, $\sin(\pi + \theta) = (\sin \pi)(\cos \theta) + (\cos \pi)(\sin \theta) = 0 \cdot \cos \theta + -1 \cdot \sin \theta = -\sin \theta$. **15.** Since $\cos A = 0.30$ and A is acute, $\sin A = \sqrt{1 - \cos^2 A} = \sqrt{1 - 0.30^2} = \sqrt{0.91}$. Then, by the Double Angle Formula, $\sin 2A = 2\sin A \cos A = 2\sqrt{0.91}(0.3) \approx 0.57$. **16. a.** $x = \begin{bmatrix} 1 & 0 \\ 0 & -1 \end{bmatrix}$, so $x^3 = \begin{bmatrix} 1 & 0 \\ 0 & -1 \end{bmatrix} \cdot$ $\begin{bmatrix} 1 & 0 \\ 0 & -1 \end{bmatrix} \cdot \begin{bmatrix} 1 & 0 \\ 0 & -1 \end{bmatrix} = \begin{bmatrix} 1 & 0 \\ 0 & 1 \end{bmatrix} \cdot \begin{bmatrix} 1 & 0 \\ 0 & -1 \end{bmatrix} = \begin{bmatrix} 1 & 0 \\ 0 & -1 \end{bmatrix}$. **b.** The matrix $\begin{bmatrix} 1 & 0 \\ 0 & -1 \end{bmatrix}$ maps (x, y) onto $(x, -y)$ and represents the reflection over the x-axis, r_x. **17. a.** Since (1, 0) is mapped to (-1, 0) and (0, 1) is mapped to (0, 7), the 2×2 matrix is $\begin{bmatrix} -1 & 0 \\ 0 & 7 \end{bmatrix}$. **b.** The point matrix for (1, 1) is $\begin{bmatrix} 1 \\ 1 \end{bmatrix}$. $\begin{bmatrix} -1 & 0 \\ 0 & 7 \end{bmatrix} \cdot \begin{bmatrix} 1 \\ 1 \end{bmatrix} = \begin{bmatrix} -1 \\ 7 \end{bmatrix}$ which is the point (-1, 7). **18.** This is a reflection over the line $y = x$ followed by a rotation of 90° followed by a rotation of 60°.

Self-Test Correlation Chart

Question	1	2	3	4	5	6	7	8	9
Objective(s)	A	D	A	B	D	E	F	G	F
Lesson(s)	12-1	12-1	12-1	12-1	12-2	12-2	12-4	12-3	12-2
Question	**10**	**11**	**12**	**13**	**14**	**15**	**16**	**17**	**18**
Objective(s)	F	F	C	C	C	C	G	F	G
Lesson(s)	12-4	12-4	12-5	12-6	12-5	12-6	12-3	12-3	12-3

Chapter Review (pp. 781-783)

1. C **3.** $[x - 15]$ **5.** 3×6 **7.** D **9.** $\frac{\sqrt{2}-\sqrt{6}}{4}$ **11.** $\frac{7\sqrt{21}+2\sqrt{51}}{50}$ **13.** $\sin\theta$ **15.** $8\sin^4\theta - 8\sin^2\theta + 1$ **17.** $\cos\frac{4\pi}{15}$ **19.** $2\sin\theta\sqrt{1-\sin^2\theta}$ **21. a.** The first row of F states that in 1990 one British pound was worth \$1.61, 1 Candian dollar was worth \$0.86, 1 Japanese yen was worth \$0.0068, and 1 dollar could by definition purchase exactly one dollar. **b.** \$9.70

23. a. $\begin{bmatrix} 7{,}550 & 13{,}600 \\ 9{,}170 & 16{,}540 \\ 12{,}875 & 23{,}050 \\ 3{,}775 & 6{,}800 \end{bmatrix}$; It is the total cost of production (first column) and total cost to the consumer (second column) resulting from the produce from each farm (the rows). **b.** \$12,875 **c.** the element in row 3, column 1

25. $\begin{bmatrix} 2 & 6 & -1 \\ 4 & 1 & 0 \end{bmatrix}$ **27. a.** $\begin{bmatrix} -1 & 9 & 0 \\ 6 & -4 & -3 \end{bmatrix}$

b.

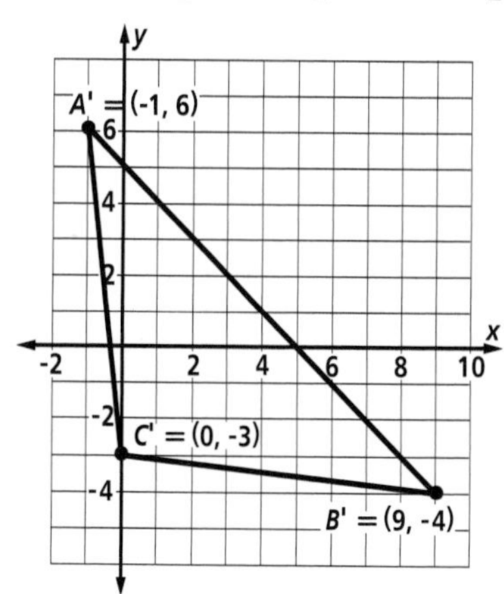

29. $\begin{bmatrix} 1 & 0 \\ 0 & -1 \end{bmatrix}$ **31.** $\begin{bmatrix} 0 & 1 \\ -1 & 0 \end{bmatrix}$ **33.** a reflection over the line $y = x$ **35.** $\begin{bmatrix} \cos 27^\circ & -\sin 27^\circ \\ \sin 27^\circ & \cos 27^\circ \end{bmatrix} \approx \begin{bmatrix} 0.89 & -0.45 \\ 0.45 & 0.89 \end{bmatrix}$

37. $\begin{bmatrix} q & e \\ t & z \end{bmatrix}$ **39.** $\begin{bmatrix} \cos 75^\circ & \sin 75^\circ \\ -\sin 75^\circ & \cos 75^\circ \end{bmatrix}$

41. a. $\begin{bmatrix} -1 & 0 \\ 0 & 1 \end{bmatrix} \cdot \begin{bmatrix} 0 & 1 \\ -1 & 0 \end{bmatrix}$ **b.** $\begin{bmatrix} 0 & -1 \\ -1 & 0 \end{bmatrix}$

43. a. a counterclockwise rotation of 90°, followed by a reflection over the line $y = x$, followed by a counterclockwise rotation of 27°
b. $\begin{bmatrix} \cos 27^\circ & \sin 27^\circ \\ \sin 27^\circ & -\cos 27^\circ \end{bmatrix}$ **45. a.** $\begin{bmatrix} -1 & 0 \\ 0 & 1 \end{bmatrix}$ **b.** r_y

Chapter 13

Lesson 13-1 (pp. 786-791)

Guided Example 3 $\sin^2 x$; $\cos^2 x$; $\cos^2 x$; $\sin^2 x$; $\cos^2 x$; $\sin^2 x$

Questions

1. a. Answers vary. Sample: $x = 0$ **b.** no **3.** $1 + \tan^2 \frac{\pi}{4} = 1 + 1^2 = 2 = \sec^2 \frac{\pi}{4}$ **5.** $\frac{1-\cos 2\theta}{2} = \frac{1-(1-2\sin^2\theta)}{2} = \frac{1-1+2\sin^2\theta}{2} = \sin^2\theta$ **7.** $\frac{\sec x}{\sin x} - \frac{\sin x}{\cos x} = \frac{1-\sin^2 x}{\sin x \cos x} = \frac{\cos^2 x}{\sin x \cos x} = \frac{\cos x}{\sin x} = \cot x$ **9. a.** $\sec x \tan x \cos x = \tan x$
b. $\sec x \tan x \cos x = \frac{1}{\cos x}\tan x \cos x = \tan x$

11. a. identity

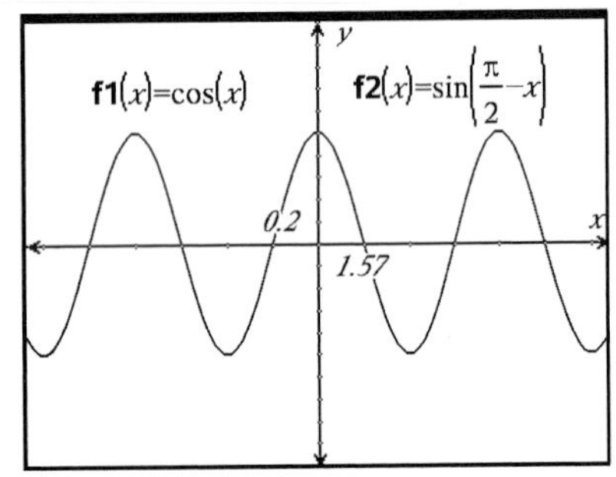

b. $\sin\left(\frac{\pi}{2} - x\right) = \sin\left(\frac{\pi}{2}\right)\cos x - \cos\left(\frac{\pi}{2}\right)\sin x = \cos x$

13. a. identity

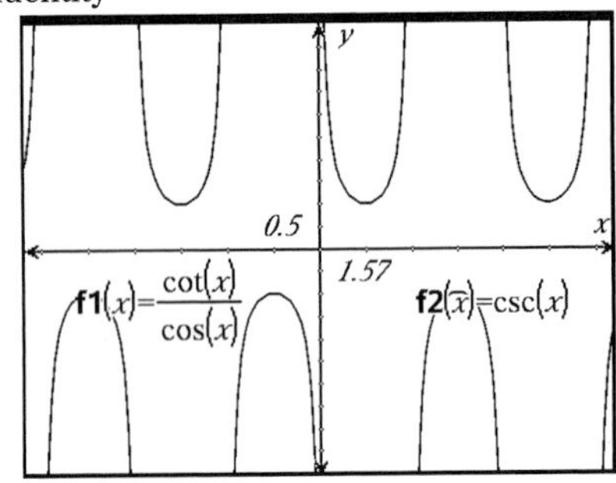

b. $\frac{\cot y}{\cos y} = \frac{\cos y}{\sin y \cos y} = \frac{1}{\sin y} = \csc y$

15. a. identity

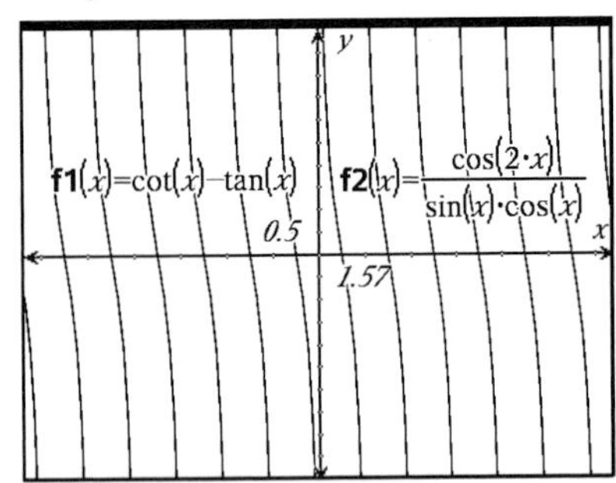

b. $\cot x - \tan x = \frac{\cos x}{\sin x} - \frac{\sin x}{\cos x} = \frac{\cos^2 x - \sin^2 x}{\sin x \cos x} = \frac{\cos(2x)}{\sin x \cos x}$

17. $\tan x(\sin x + \cot x \cos x) = \sec x$

19. not an identity

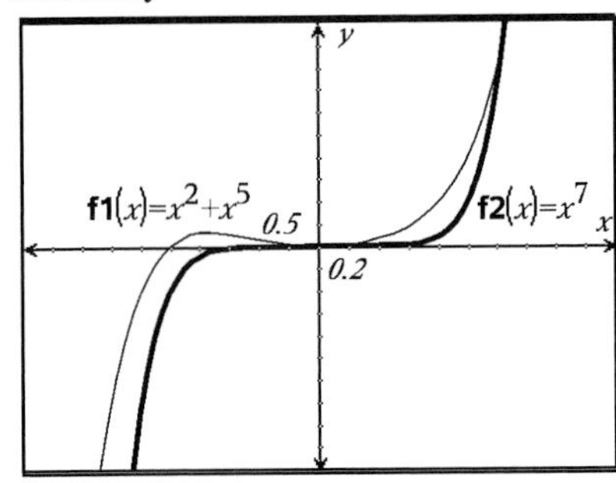

21. a. $-\frac{7}{4}$ **b.** $-\frac{\sqrt{65}}{7}$ **c.** $\frac{4\sqrt{65}}{65}$ **23.** $\frac{a}{b}$ **25. a.** $x = \frac{\pi}{4}, \frac{3\pi}{4}, \frac{5\pi}{4}, \frac{7\pi}{4}$ **b.** $x = \frac{(2n-1)\pi}{4}$, where n is an integer **27.** $\frac{\pi}{2}$

29. a.

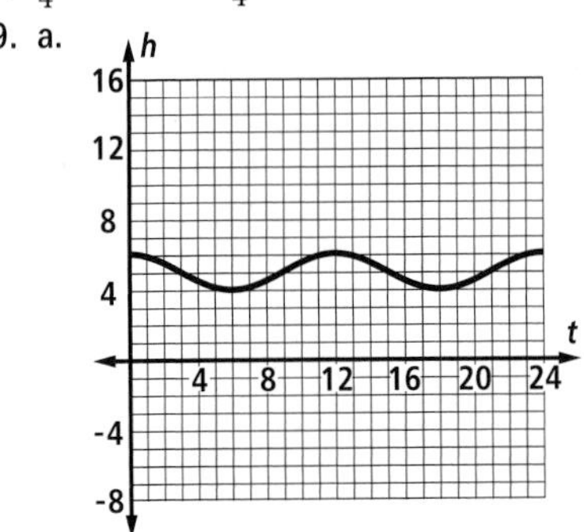

b. 6 m **c.** $t = 0$, $t = 12$, $t = 24$

Lesson 13-2 (pp. 792-796)

Guided Example 1 a. $(\cos\theta + \sin\theta)$; $(\cos\theta - \sin\theta)$; $(\cos\theta + \sin\theta)$; $(\cos\theta - \sin\theta)$; $(\cos\theta + \sin\theta)$ **b.** $\frac{\pi}{4}$; $\frac{5\pi}{4}$; $\frac{\pi}{4}$; $\frac{5\pi}{4}$

Questions

1. A singularity is an isolated value for which a function is undefined. **3.** $x = \frac{\pi}{2} + n\pi$, where n is an integer

5. a. identity

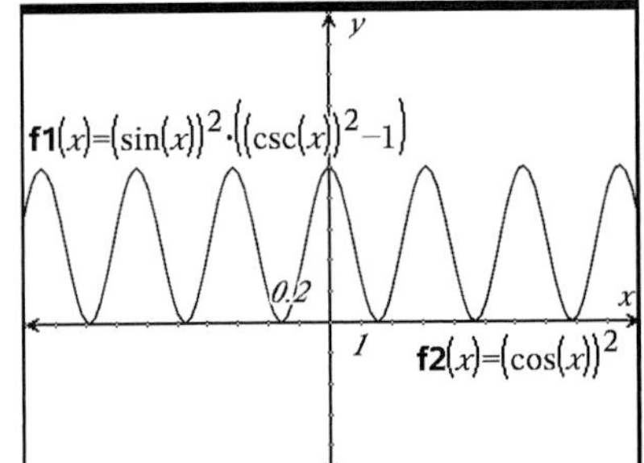

b. $\sin^2 x(\csc^2 x - 1) = \sin^2 x\left(\frac{1}{\sin^2 x} - 1\right) = 1 - \sin^2 x = \cos^2\theta$; domain: $x \neq n\pi$, where n is an integer.

7. a. not an identity

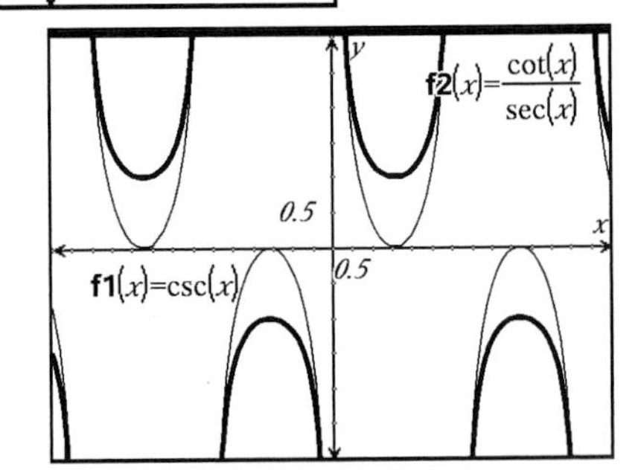

9. a. $x \neq \frac{\pi}{2} + n\pi$ and $x \neq n\pi$ where n is an integer.
b. $\tan x + \cot x = \frac{\sin x}{\cos x} + \frac{\cos x}{\sin x} = \frac{\sin^2 x+\cos^2 x}{\sin x \cos x} = \frac{1}{\sin x \cos x}$ $= \sec x \csc x$

11. a.

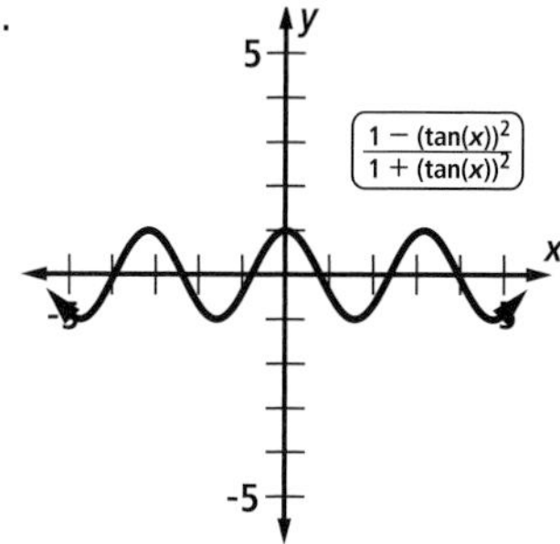

b. $x \neq \frac{\pi}{2} + n\pi$ where n is an integer. c. $\frac{1-\tan^2 x}{1+\tan^2 x} = \cos 2x$

d. $\frac{1-\tan^2 x}{1+\tan^2 x} = \frac{1-\frac{\sin^2 x}{\cos^2 x}}{\sec^2 x} = \frac{1-\frac{\sin^2 x}{\cos^2 x}}{\frac{1}{\cos^2 x}} = \cos^2 x\left(1 - \frac{\sin^2 x}{\cos^2 x}\right) =$ $\cos^2 x - \sin^2 x = \cos 2x$

13. a.

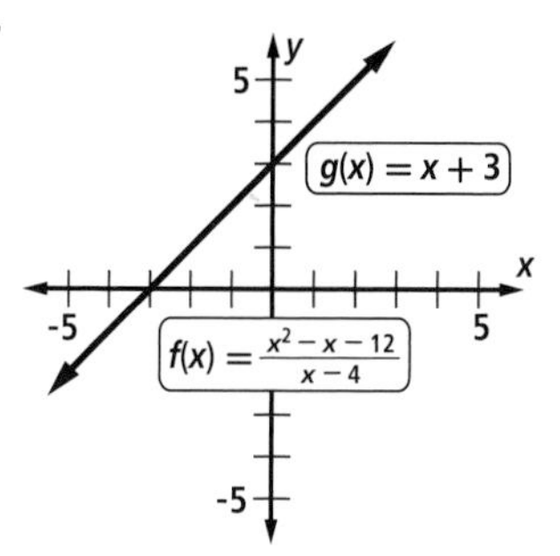

b. $x = 4$ c. $\{x \mid x \neq 4\}$ d. $\frac{x^2-x-12}{x-4} = \frac{(x-4)(x+3)}{x-4} = x - 3$

15. $\frac{\sin t}{1+\cos t} \cdot \frac{1-\cos t}{1-\cos t} = \frac{\sin t(1-\cos t)}{1-\cos^2 t} = \frac{\sin t - \sin t \cos t}{\sin^2 t} =$ $\frac{\sin t}{\sin^2 t} - \frac{\sin t \cos t}{\sin^2 t} = \frac{1}{\sin t} - \frac{\cos t}{\sin t} = \csc t - \cot t = \tan\left(\frac{t}{2}\right)$

17. a. false positives b. 300 c. ≈ 17.19 19. $x \approx 3.343$ or $x \approx 6.082$

Lesson 13-3 (pp. 797-803)

Guided Example 2 $2, \frac{11\pi}{4}; \frac{7\pi}{4}; -2, -\frac{\pi}{4}; -2, \frac{7\pi}{4}$

Questions

1. a. - d.

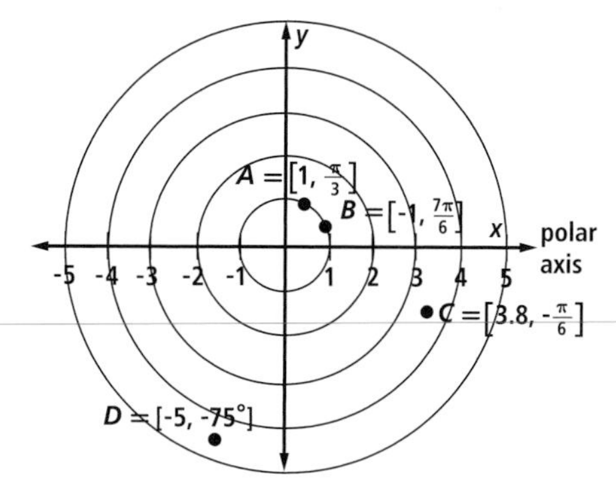

3. (0, 5) 5. (-3.46, -2) 7. [10, -53.1°]

9.

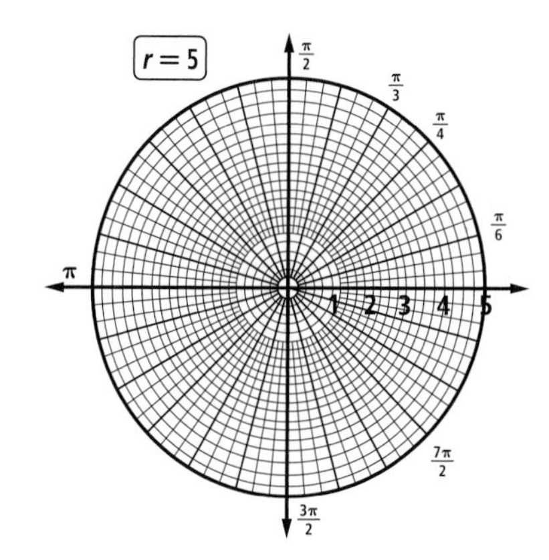

11. true 13. $\theta = \frac{11\pi}{4}$; It is a line extending from the origin through Quadrants II and IV. 15. $\left[-8, \frac{2\pi}{3}\right]$ 17. a. 30° west of south b. 10° west of north c. 20° south of west d. 50° west of north 19. The equation is an identity. $\sin\left(\frac{\pi}{2} + x\right) = \sin\left(\frac{\pi}{2}\right)\cos(x) + \cos\left(\frac{\pi}{2}\right)\sin(x) = 1 \cdot \cos x + 0 \cdot \sin x = \cos x$; domain: all real numbers. 21. a. 1 b. 36 23. true

Lesson 13-4 (pp. 804-810)

Questions

1. a.

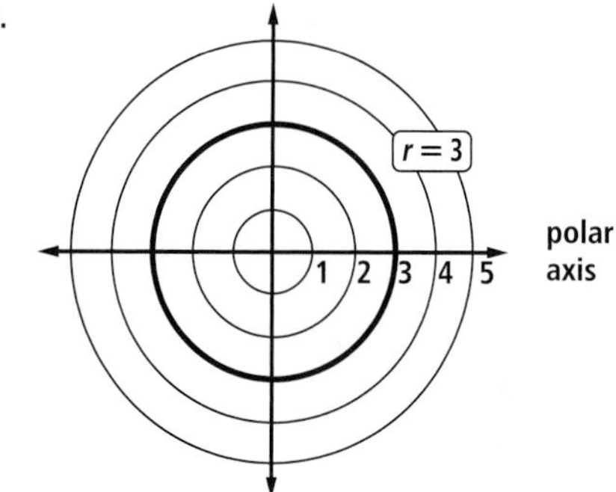

b. $x^2 + y^2 = 9$

3. a.

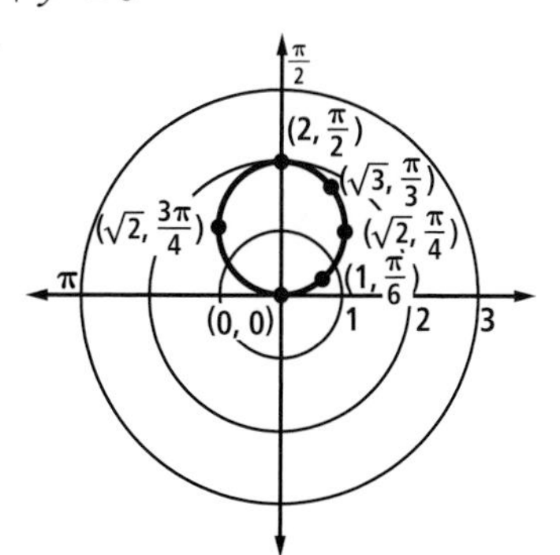

b. $r = 2\frac{y}{r} = 2\sin\theta; x^2 + y^2 = 2y; x^2 + (y-1)^2 = 1$

5. no; $\cos(4\pi) = 1 \neq 4$

7. a.

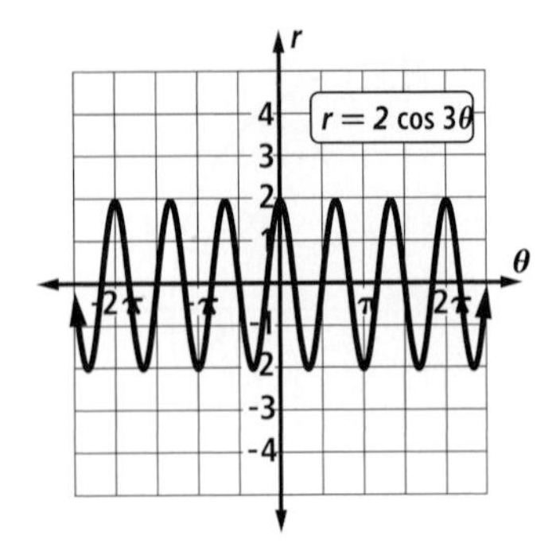

b. max: 2; min: -2 c. $r = 2, 0, 0$;

d.

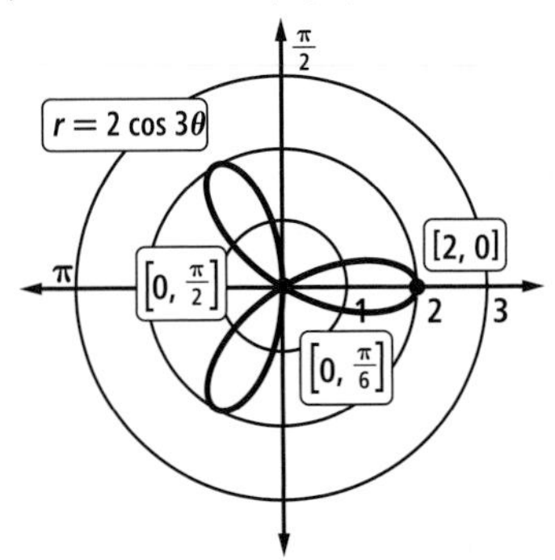

9. Answers vary. Sample: $a = 10$

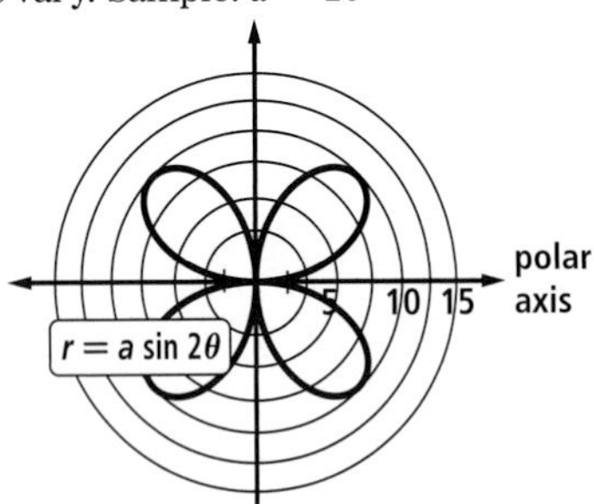

11. a.

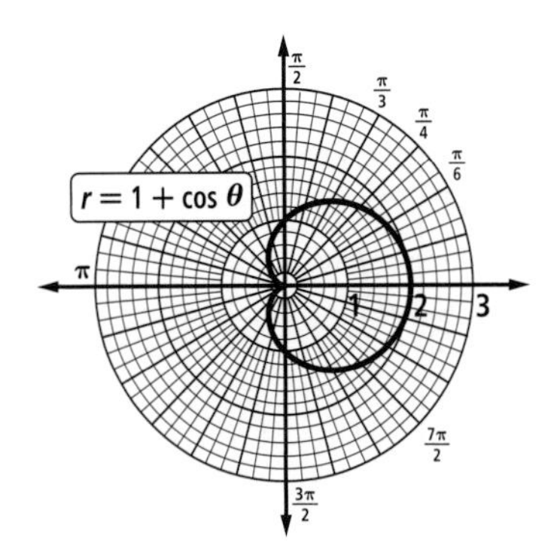

b. the curve is heart-shaped 13. a. $[1, 0]$, $\left[1.724, \frac{\pi}{4}\right]$, $\left[2.971, \frac{\pi}{2}\right]$, $\left[5.12, \frac{3\pi}{4}\right]$, $[8.83, \pi]$, $\left[15.21, \frac{5\pi}{4}\right]$, $\left[26.22, \frac{6\pi}{4}\right]$, $\left[45.19, \frac{7\pi}{4}\right]$, $\left[77.88, \frac{8\pi}{4}\right]$

b.

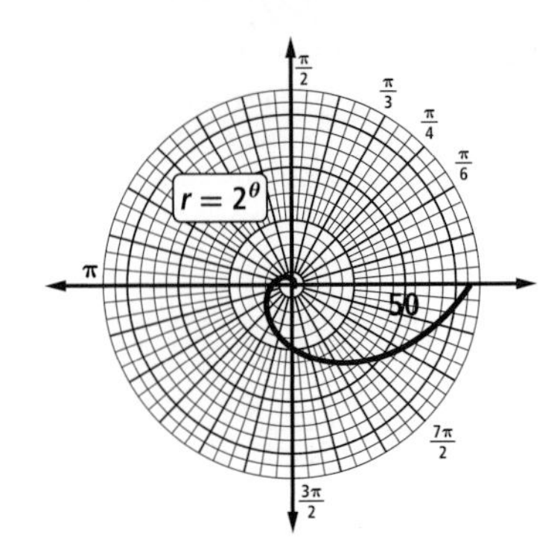

15. D 17. a. not an identity;

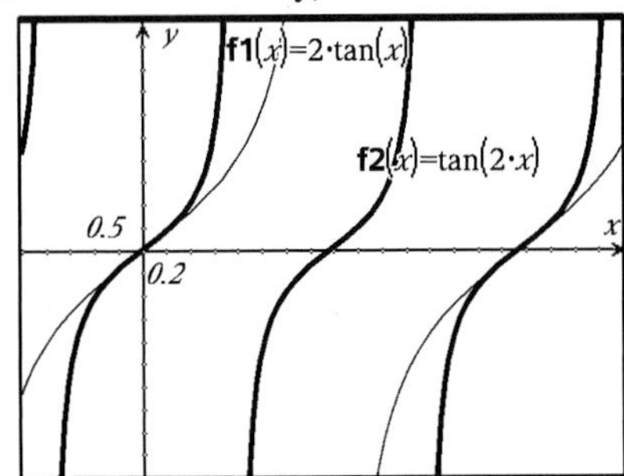

b. Answers vary. Sample: $\tan\left(\frac{2\pi}{3}\right) = -\sqrt{3}$, but $2\tan\left(\frac{\pi}{3}\right) = 2\sqrt{3}$ 19. a. 0.9987 b. 0.84 c. 0.025 d. 0.9971

Lesson 13-5 (pp. 811-816)

Guided Example 2 a. $-1 + 0i$; $-2 + 3i$ b. $y + 1$; 1 unit

Questions

1. real; imaginary 3. a. (-4, 5)

b.

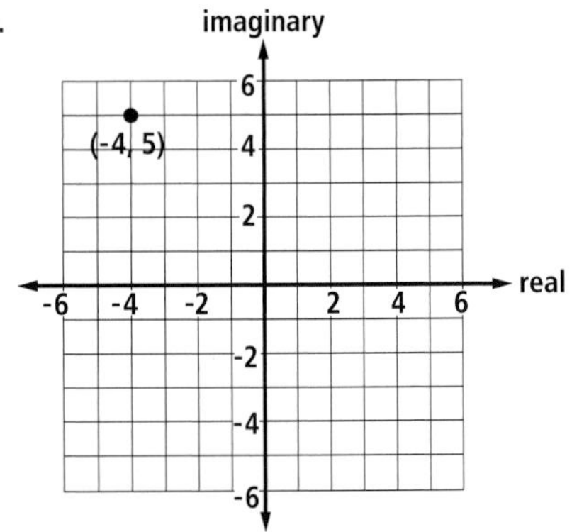

5. a. $U + V = -4 + 2i$ b and c.

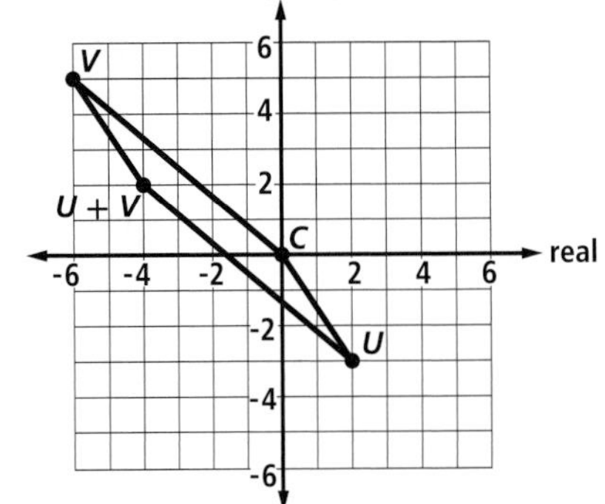

7. [5, -5.3559] 9. $|a + bi| = \sqrt{a^2 + b^2}$ 11. a. $3\sqrt{5}$ b. 153.4°
13. a. i. – iv.

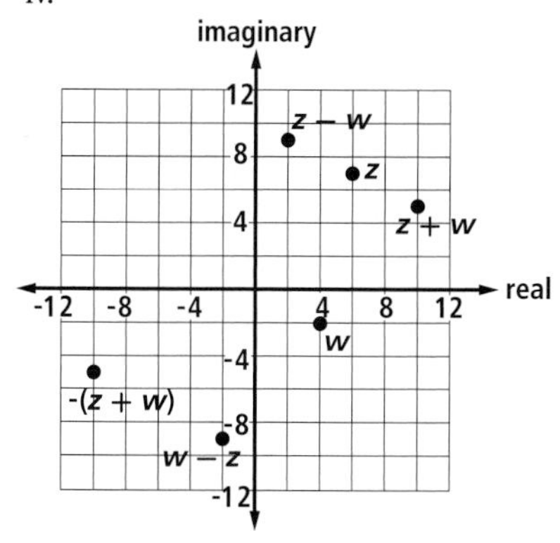

b. Let $A = z - w = (2, 9)$, $B = z + w = (10, 5)$, $C = w - z = (-2, -9)$, and $D = -(z + w) = (-10, -5)$. Slope of $\overline{AB} = -\frac{1}{2}$ and slope of $\overline{DC} = -\frac{1}{2}$, so $\overline{AB}$ and $\overline{DC}$ are parallel. Slope of $\overline{AD} = \frac{7}{6}$ and slope of $\overline{BC} = \frac{7}{6}$, so $\overline{AD}$ and $\overline{BC}$ are parallel. Thus the quadrilateral $ABCD$ is a parallelogram. 15. Yes; the distance from u to v is $\sqrt{(a-c)^2 + (b-d)^2}$ and $|u - v| = |(a - c) + i(b - d)| = \sqrt{(a-c)^2 + (b-d)^2}$.

17.

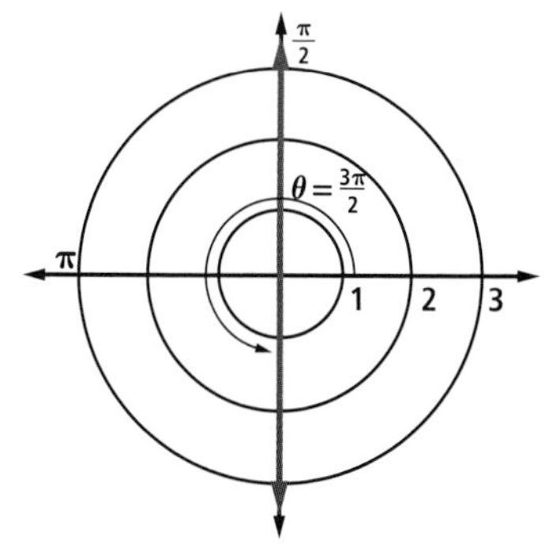

19.

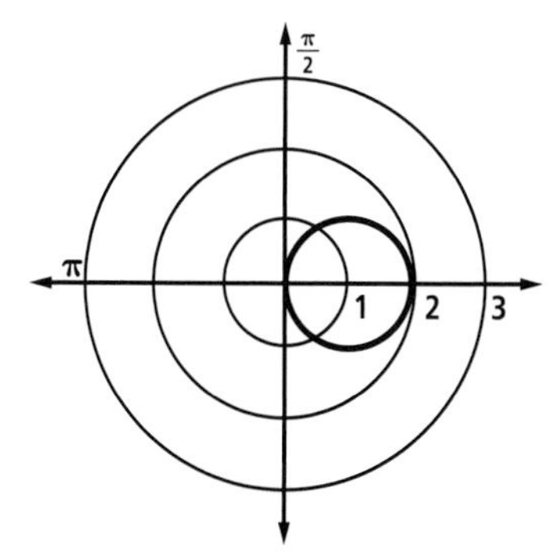

21. a. The relation is not a function. b. yes 23. divergent
25. convergent, $\frac{64}{7}$

Lesson 13-6 (pp. 817-824)

Guided Example 3 90; 90; 6; 90°; 90°; 18

Questions

1. a., 3. a..

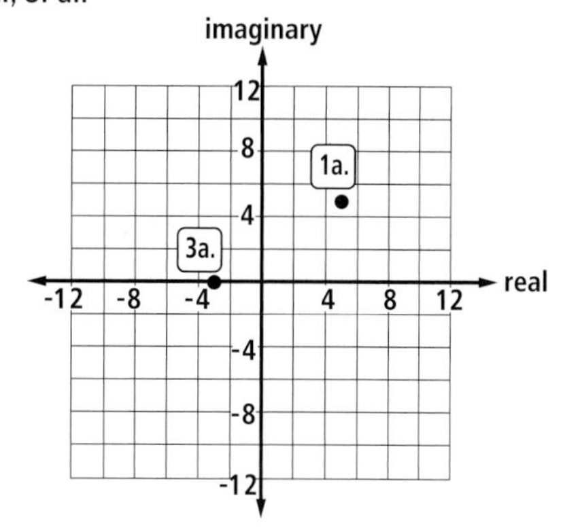

1 b. $5\sqrt{2}(\cos 45° + i \sin 45°)$ 3. b. $-3(\cos 0° + i \sin 0°)$
5. a.

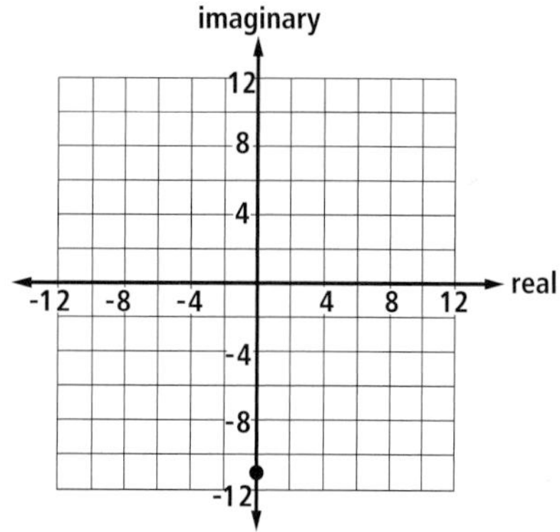

b. $0 - 13i$ 7. a. $6(\cos 80° + i \sin 80°)$ b. a size change of magnitude 2 and a rotation of 25° about the origin
c.

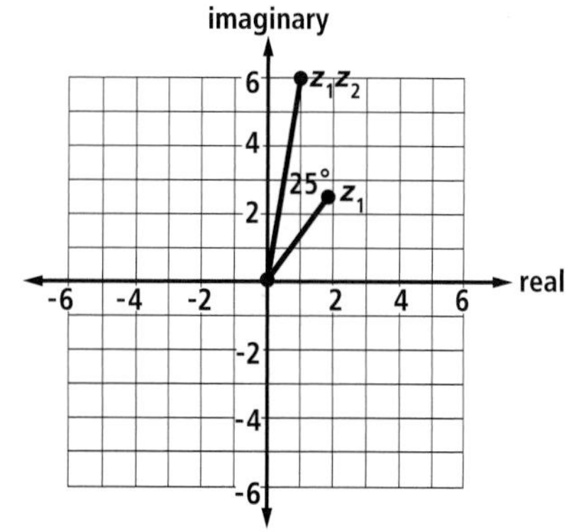

d. $z_1 = [3, 55°]$, $z_2 = [2, 25°]$ $z_1z_2 = [6, 80°]$
9. a. $\left[32, \frac{21\pi}{4}\right]$ b. 32 c. $\frac{21\pi}{4}$ 11. $7\left(\cos \frac{5\pi}{6} - i \sin \frac{5\pi}{6}\right)$
13. a. $\frac{9}{5}(\cos 100° + i \sin 100°)$ b. $z_1 = [9, 200°]$, $z_2 = [5, 100°]$, $\frac{z_1}{z_2} = \left[\frac{9}{5}, 100°\right]$ c. a size change of magnitude $\frac{1}{5}$ and a rotation of -100° about the origin
15. a. $5(\cos 180° + i \sin 180°)$ b. -5

17. Let $z_1 = r_1(\cos\theta_1 + i\sin\theta_1)$ and $z_2 = r_2(\cos\theta_2 + i\sin\theta_2)$. Then $\frac{z_1}{z_2} = \frac{r_1(\cos\theta_1 + i\sin\theta_1)}{r_2(\cos\theta_2 + i\sin\theta_2)}$ $= \frac{r_1}{r_2}\frac{(\cos\theta_1 + i\sin\theta_1)(\cos\theta_2 - i\sin\theta_2)}{(\cos\theta_2 + i\sin\theta_2)(\cos\theta_2 - i\sin\theta_2)} = \frac{r_1}{r_2}(\cos\theta_1\cos\theta_2 + \sin\theta_1\sin\theta_2 + i(\sin\theta_1\cos\theta_2 - \sin\theta_2\cos\theta_1)) = \frac{r_1}{r_2}(\cos(\theta_1 - \theta_2) + i\sin(\theta_1 - \theta_2))$, by the Addition Formulas for Cosine and Sine. **19. a.** $z^2 = 49(\cos 40° + i\sin 40°)$; $z^3 = 343(\cos 60° + i\sin 60°)$; $z^4 = 2401(\cos 80° + i\sin 80°)$; $z^5 = 16807(\cos 100° + i\sin 100°)$ **b.** $z^n = 7^n(\cos(n20°) + i\sin(n20°))$; $z^{12} = 7^{12}(\cos 240° + i\sin 240°)$

21.

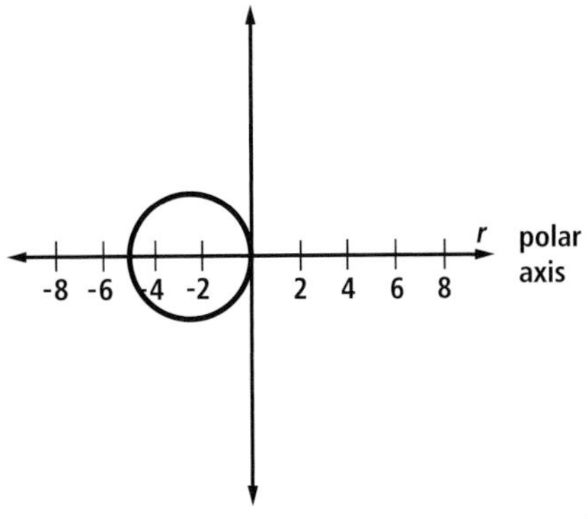

23. a. $\theta = \frac{\pi}{12}$ and $\theta = \frac{5\pi}{12}$ **b.** $\theta = \frac{\pi}{12} + \pi n$ and $\theta = \frac{5\pi}{12} + \pi n$
25. 0.00303 seconds

Lesson 13-7 (pp. 825-830)

Questions

1. $81\left(\cos\frac{4\pi}{7} + i\sin\frac{4\pi}{7}\right)$ **3.** $256 + 0i$ **5. a.** $2 - 2\sqrt{3}i$; $8i$; $-8 - 8\sqrt{3}i$; $16\sqrt{3} + 16i$; -64

b.

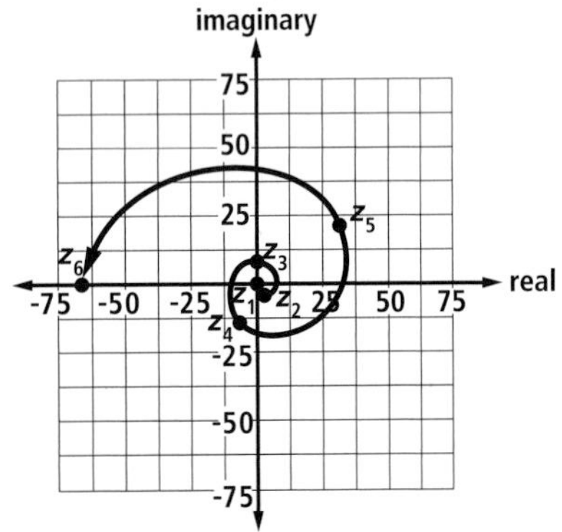

c. The points form a logarithmic spiral outward from the origin. **7.** $5(\cos 140° + i\sin 140°)$, $5(\cos 260° + i\sin 260°)$ **9. a.** a regular hexagon **b.** [64, 90°]; [2, 15°] **c.** $2^6 = 64$, $6 \cdot 15° = 90°$, $64(\cos 90° + i\sin 90°) = 64i$ **d.** [2, 75°], [2, 135°], [2, 195°], [2, 255°], [2, 315°]

11. $\left[2, \frac{\pi}{4}\right], \left[2, \frac{3\pi}{4}\right], \left[2, \frac{5\pi}{4}\right], \left[2, \frac{7\pi}{4}\right]$

13. a. 5 **b.** [2, 0°], [2, 72°], [2, 144°], [2, 216°], [2, 288°] **15.** $81.915 + 57.358i$ **17.** not an identity; By the Pythagorean Identity, $1 - \cos^2\alpha = \sin^2\alpha$. So, $\sin^2\alpha = \frac{1}{2}\sin^2\alpha$, which is only true for $\alpha = n\pi$, where n is an integer. **19.** If $\log_{10} 3 = \frac{1}{\log_3 10}$, then $(\log_{10} 3)(\log_3 10) = 1$. By the Logarithm of a Power Theorem this can be written as $\log_3\left(10^{\log_{10} 3}\right) = 1$. Because 10^x and $\log_{10} x$ are inverses we know that $10^{\log_{10} 3} = 3$. Thus we have $\log_3 3 = 1$, which is true by the Logarithm of the Base Theorem.

Self-Test (p. 835)

1. $r = \sqrt{(-5\sqrt{2})^2 + (5\sqrt{2})^2} = 10$, $\theta = \tan^{-1}\left(\frac{5\sqrt{2}}{-5\sqrt{2}}\right) = \tan^{-1}(-1) = \frac{3\pi}{4} + 2\pi n$, where n is an integer. So three pairs of polar coordinates are $\left[10, \frac{3\pi}{4}\right], \left[10, \frac{11\pi}{4}\right]$, and $\left[-10, \frac{7\pi}{4}\right]$. **2.** Polar form: $r = \sqrt{60^2 + (-40)^2} = 20\sqrt{13}$, $\theta = \tan^{-1}\left(-\frac{2}{3}\right) \approx 326.31°$. So $[r, \theta] = [20\sqrt{13}, 326.31°]$. Trigonometric form: $r(\cos\theta + i\sin\theta) = 20\sqrt{13}(\cos 326.31° + i\sin 326.31°)$. **3.** The number is in trigonometric form with $r = 5$ and $\theta = \frac{\pi}{6}$. Polar form: $[r, \theta] = \left[5, \frac{\pi}{6}\right]$. Since $5\sin\left(\frac{\pi}{6}\right) = \frac{5}{2}$ and $5\cos\left(\frac{\pi}{6}\right) = \frac{5\sqrt{3}}{2}$, in $a + bi$ form: $\frac{5\sqrt{3}}{2} + \frac{5}{2}i$.

4.

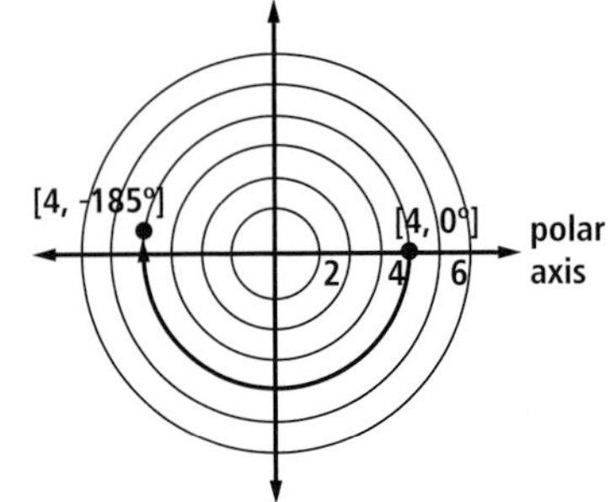

5. $-4 + 3i = (-4, 3)$

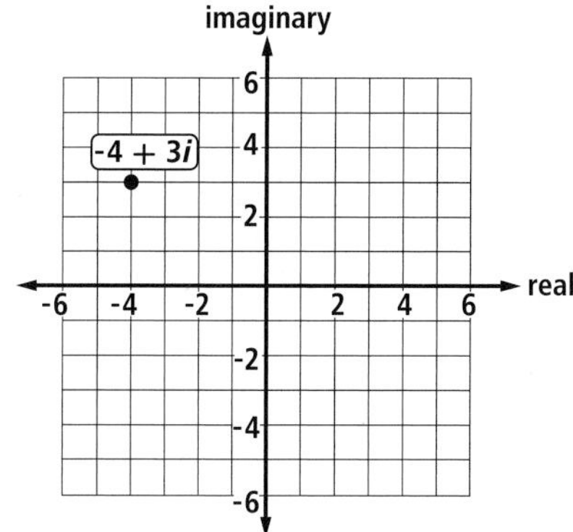

6. $2\left(\cos\frac{\pi}{2} + i\sin\frac{\pi}{2}\right) = \left[2, \frac{\pi}{2}\right]$

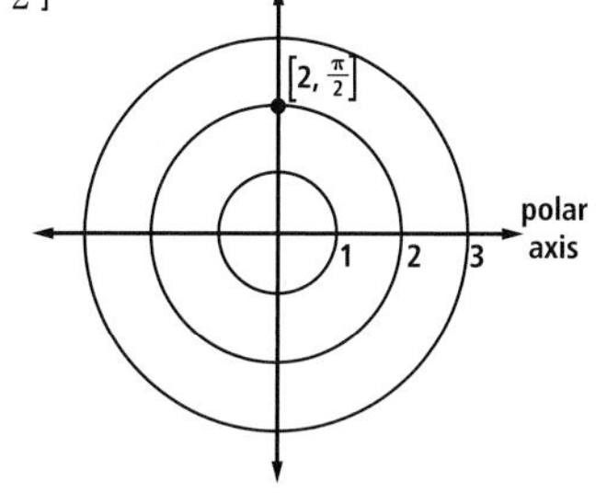

7. $z_1z_2 = \left[\left(4 \cdot \frac{1}{2}\right), \left(\frac{\pi}{2} + \frac{\pi}{3}\right)\right] = \left[2, \frac{5\pi}{6}\right]; \frac{z_1}{z_2} = \left[\left(4 \div \frac{1}{2}\right), \left(\frac{\pi}{2} - \frac{\pi}{3}\right)\right] = \left[8, \frac{\pi}{6}\right]$ **8.** $z_1z_2 = \left[20 \cdot \frac{1}{5}\left(\cos\frac{5\pi}{4} + \frac{\pi}{6}\right) + i\sin\left(\frac{5\pi}{4} + \frac{\pi}{6}\right) + i\sin\left(\frac{5\pi}{4} + \frac{\pi}{6}\right)\right] = 4\left(\cos\frac{17\pi}{12} + i\sin\frac{17\pi}{12}\right); \frac{z_1}{z_2} = \left[20 \div \frac{1}{5}\left(\cos\left(\frac{5\pi}{4} - \frac{\pi}{6}\right) + i\sin\left(\frac{5\pi}{4} - \frac{\pi}{6}\right)\right)\right] = 100\left(\cos\frac{13\pi}{12} + i\sin\frac{13\pi}{12}\right)$

9. a. Answers vary. Sample: Substitute values of θ to get the corresponding values for r: $[2, 0]$, $\left[0, \frac{3\pi}{2}\right]$.

b.

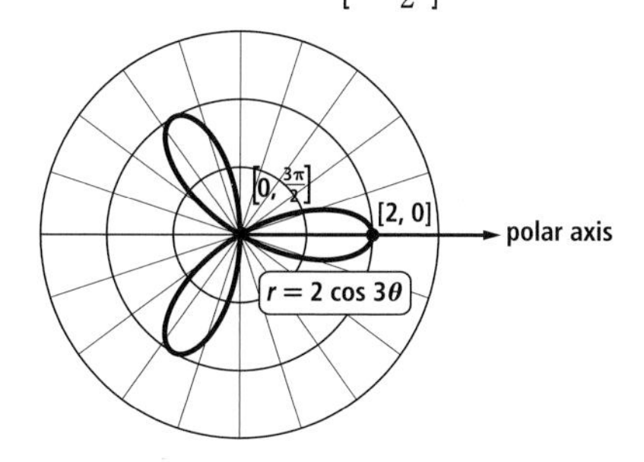

10. The singularities are the points at which the function is undefined. $\cot\theta = \frac{\cos\theta}{\sin\theta}$ and is undefined when $\sin\theta = 0$. The singularities are $\theta = n\pi$, where n is an integer, indicated by asymptotes on the graph at $x = n\pi$.
11. The graph is the size change image of the graph of $r = \sin\theta$, known to be a circle with center at $\left[\frac{1}{2}, \frac{\pi}{2}\right]$. This circle has center $\left[\frac{3}{2}, \frac{\pi}{2}\right]$.

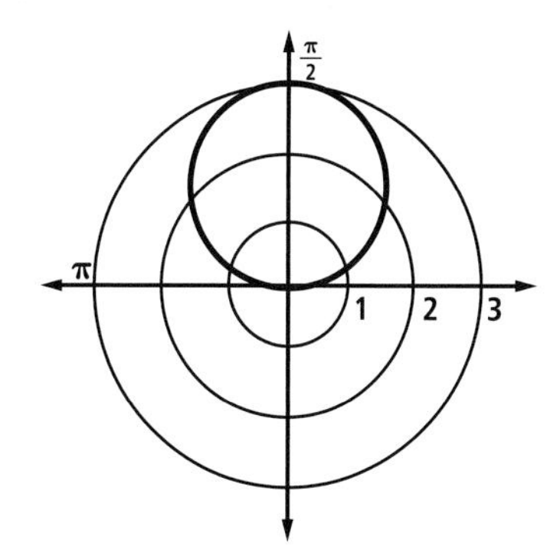

12. $\sin^2 x \cdot (2\cot^2 x - \csc^2 x)$ is not defined when $\sin^2 x = 0$, so there are holes in the graph of f, whereas the graph of g has no holes. **13.** $\sin^2 x(2\cot^2 x - \csc^2 x) = \sin^2 x\left(\frac{2\cos^2 x}{\sin^2 x} - \frac{1}{\sin^2 x}\right) = 2\cos^2 x - 1 = \cos 2x$ for $x \neq n\pi$, where n is an integer. **14.** $[-8, 60°] = -8\cos 60° - 8i \sin 60° = -4 - 4i\sqrt{3} = (-4, -4\sqrt{3})$ **15.** By DeMoivre's Theorem $3(\cos 20° + i\sin 20°))^4 = 3^4(\cos 80° + i\sin 80°) = 81(\cos 80° + i\sin 80°) \approx 14.066 + 79.769i$. **16.** The 4 fourth roots are $16^{\frac{1}{4}}\left(\cos\left(\frac{250°}{4} + k90°\right) + i\sin\left(\frac{250°}{4} + k90°\right)\right)$, where $k = 0, 1, 2$, and 3. Thus they are: $2(\cos 62.5° + i\sin 62.5°)$, $2(\cos 152.5° + i\sin 152.5°)$, $2(\cos 242.5° + i\sin 242.5°)$, and $2(\cos 332.5° + i\sin 332.5°)$.
17. $(\sin x + \cos x)^2 + (\sin x - \cos x)^2 = (\sin^2 x + 2\sin x\cos x + \cos^2 x) + (\sin^2 x - 2\sin x\cos x + \cos^2 x) = 2(\sin^2 x + \cos^2 x) = 2$.

Self-Test Correlation Chart

Question	1	2	3	4	5	6	7	8	9	10
Objective(s)	G	A	A	F	I	I	B	B	H	E
Lesson(s)	13-3	13-5	13-6	13-3	13-5	13-6	13-6	13-6	13-4	13-2
Question	**11**	**12**	**13**	**14**	**15**	**16**	**17**			
Objective(s)	H	E	D	G	C	C	D			
Lesson(s)	13-4	13-2	13-2	13-3	13-7	13-7	13-1			

Chapter Review (pp. 836-837)

1. $[\sqrt{34}, 30.96°]$ 3. $[\frac{1}{6}, 30°]$ 5. $-2.47 + 7.61i$ 7. $\sqrt{202}$, 129.29° 9. $12(\cos 198° + i \sin 198°)$ 11. $[\frac{4}{5}, -\frac{6\pi}{8}]$ 13. $-64i$ 15. $3(\cos 4° + i\sin 4°)$, $3(\cos 94° + i \sin 94°)$, $3(\cos 184° + i \sin 184°)$, $3(\cos 274° + i\sin 274°)$

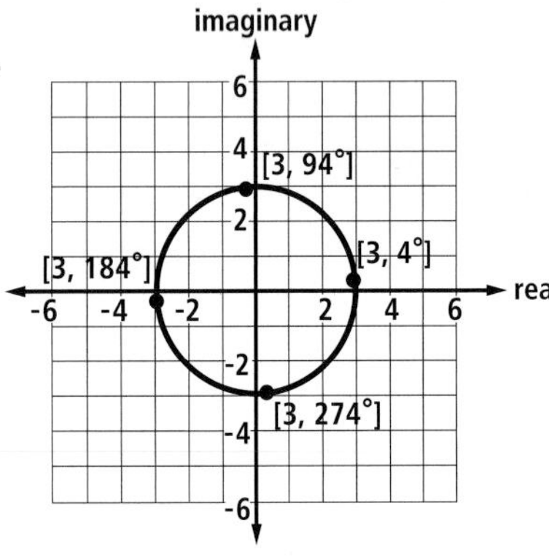

17. $(1 - \cos^2 x)(1 + \cot^2 x) = \sin^2 x(1 + \cot^2 x) = \sin^2 x + \sin^2 x \cot^2 x = \sin^2 x + \sin^2 x\left(\frac{\cos^2 x}{\sin^2 x}\right) = \sin^2 x + \cos^2 x = 1$ 19. $\csc^2 x - \cot^2 x = \frac{1}{\sin^2 x} - \frac{\cos^2 x}{\sin^2 x} = \frac{1-\cos^2 x}{\sin^2 x} = 1$ 21. $\sin(x) = 0$ for $x = n\pi$ where n is an integer, $\sin(x)$ occurs in the denominator of the expression in Question 19, and division by 0 is not defined. 23. $x = 4$ 25. a. False; it is true for all x, $x \neq 4$. b. True; $x^3 - 64 = (x - 4)(x^2 + 4x + 16)$, and the original expressions are both defined for all real numbers x.

27.

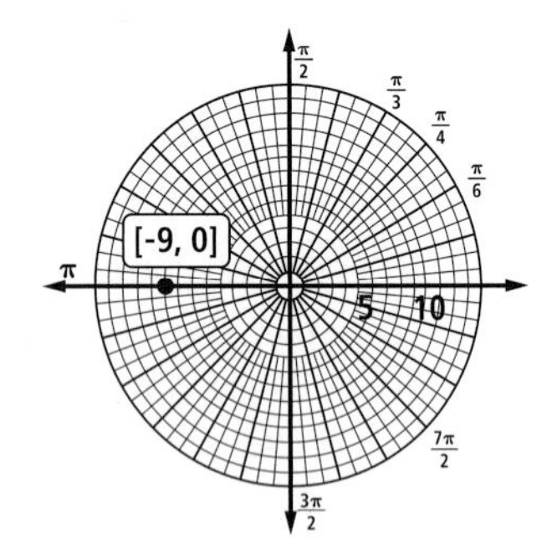

29. B 31. (-5.73, -4.02) 33. $[\sqrt{29}, 68.2°]$; $[-\sqrt{29}, 248.2°]$ 35. $[\sqrt{97}, 114°]$; $[\sqrt{97}, -246°]$ 37. (6, 4.36), $[7.42, \frac{\pi}{5}]$ 39. Answers vary. Sample: $[\sqrt{2}, \frac{\pi}{4}]$, $[1, \frac{\pi}{2}]$

41.

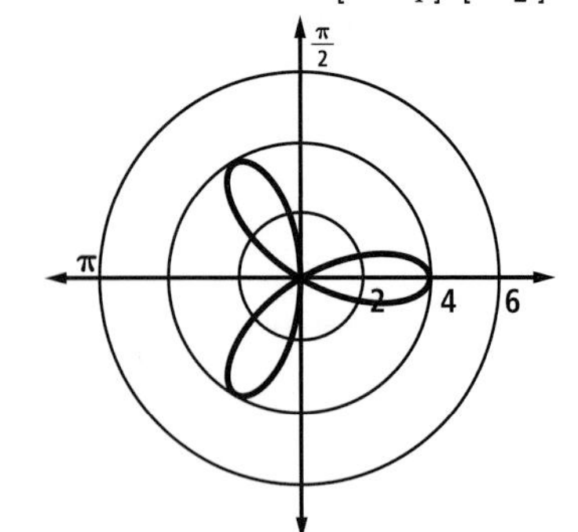

43. a.

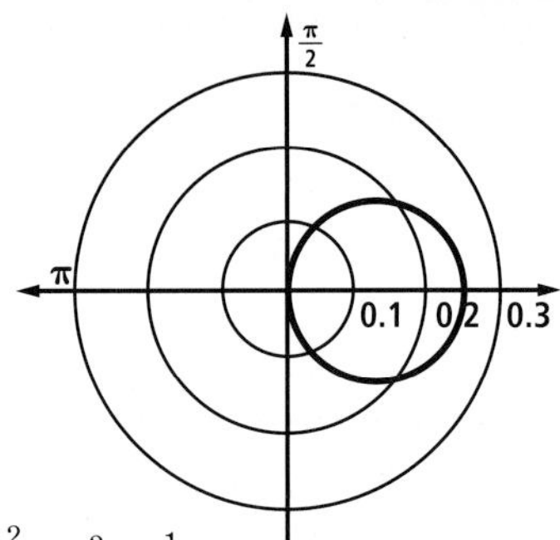

b. $\left(x - \frac{1}{8}\right)^2 + y^2 = \frac{1}{64}$

45.

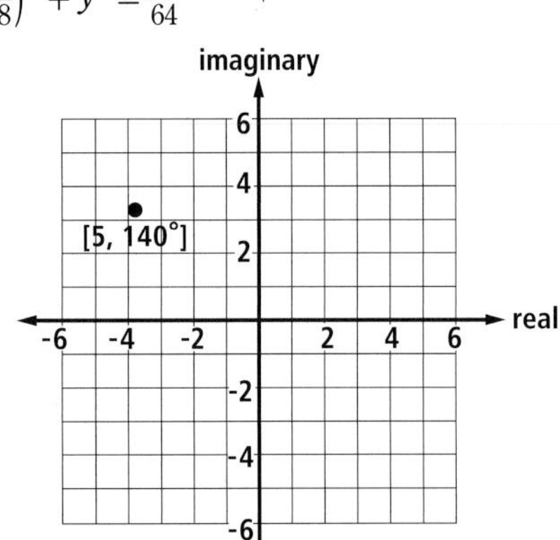

47. a.

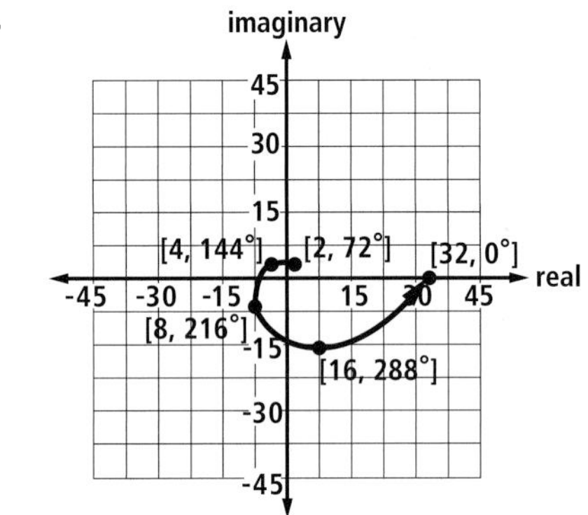

b. a logarithmic spiral c. $z^5 = (2(\cos 72° + i \sin 72°))^5 = 2^5(\cos(5 \cdot 72°) + i \sin(5 \cdot 72°)) = 32(1 + 0i) = 32$

49. a.

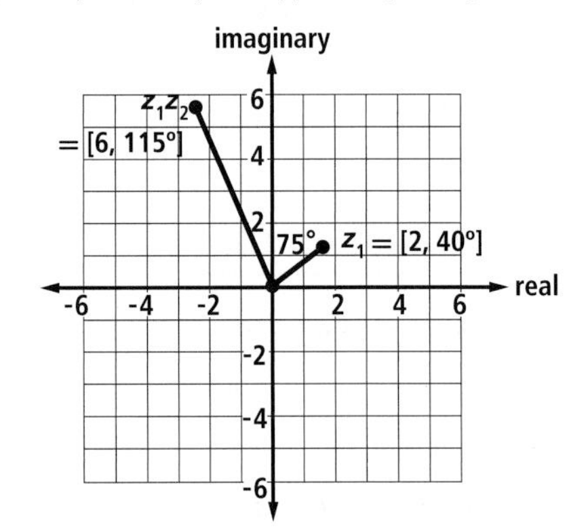

b. a size-change of magnitude 3 and a rotation of magnitude 75°.

Glossary

A

absolute value of a complex number, |z| The distance of the graph of a complex number from the origin or pole. Also called *modulus.* **(814)**

acceleration due to gravity The acceleration of a free-falling object toward another object caused by gravitational forces; on the surface of Earth equal to approximately 32 feet per second squared or 9.8 meters per second squared. **(119)**

Addition Counting Principle (General Form) For any finite sets A and B, $N(A \cup B) = N(A) + N(B) - N(A \cap B)$. **(369)**

Addition Counting Principle (Mutually Exclusive Form) If two finite sets A and B are mutually exclusive, then $N(A \cup B) = N(A) + N(B)$. **(367)**

Addition Formulas for the Cosine and Sine For all real numbers α and β, $\cos(\alpha + \beta) = \cos \alpha \cos \beta - \sin \alpha \sin \beta$, and $\sin(\alpha + \beta) = \sin \alpha \cos \beta + \cos \alpha \sin \beta$. **(766)**

Alternate Formula for $_nP_r$ Theorem The number of permutations of n objects taken r at a time is $_nP_r = \frac{n!}{(n - r)!}$. **(382)**

alternating harmonic sequence The sequence $-1, \frac{1}{2}, -\frac{1}{3}, \frac{1}{4}, \ldots$ in which the nth term is $c_n = \frac{(-1)^n}{n}$. **(518)**

amplitude One-half the difference between the maximum and minimum values of a sine wave. **(257)**

angle The union of two rays (its sides) with the same endpoint (its vertex). **(Previous Course)**

angle of depression An angle measured downward from a horizontal line. **(296)**

angle of elevation An angle measured upward from a horizontal line. **(296)**

Arccos function See *inverse cosine function.*

Archimedean spiral The graph of $r = k\theta$, for $\theta > 0$, in the polar plane. **(810)**

Arcsin function See *inverse sine function.*

Arctan function See *inverse tangent function.*

area under a curve The area between the curve and the x-axis. **(688)**

Argand diagram The graphical representation of the complex number $a + bi$ as (a, b) in the coordinate plane. **(811)**

argument of a complex number For the complex number $[r, \theta]$, θ. **(814)**

arithmetic sequence A sequence in which the difference between consecutive terms is constant. Also called *linear sequence.* **(505)**

arithmetic series An indicated sum of the terms of an arithmetic sequence. **(523)**

arithmetic mean See *mean.*

asymptote A line that the graph of a function $y = f(x)$ approaches as the variable x approaches a fixed value or increases or decreases without bound. **(105)**

at random See *randomly.*

average See *mean.*

axis of symmetry In a plane, a reflecting line ℓ over which a figure can be mapped onto itself. Also called line of symmetry. **(172)**

B

bar graph A two-dimensional display of data in which one axis labels categories or variables and the other is a numerical scale typically with counts or percentages. **(10)**

base (of an exponential function) The number b in the exponential function $f: x \rightarrow ab^x$. **(103)**

base (of a logarithmic function) The number b in the logarithmic function $f: x \rightarrow \log_b x$. **(575)**

Basic Properties of Probability Theorem Let S be the sample space associated with an experiment, and let $P(E)$ be the probability of E. Then $0 \leq P(E) \leq 1$; if $E = S$, then $P(E) = 1$; and if $E = \emptyset$ then $P(E) = 0$. **(363)**

bearing An angle measured counterclockwise from due north. **(298)**

bell-shaped curve See *normal curve.*

bins Non-overlapping intervals of equal width in the range of a numerical variable. **(22)**

binomial coefficients The coefficients in the expansion of $(x + y)^n$; the combinations $_nC_k$. **(632)**

binomial distribution function The function B where $B(x) = {}_nC_x p^x(1-p)^{n-x}$, giving the probability of getting exactly x successes in n binomial trials, each of which has a probability p of success. **(651)**

binomial experiment An experiment with a fixed number of independent trials, each with only two possible outcomes, often called success and failure, and each with the same probability of success. **(645)**

binomial probability distribution See *binomial distribution function.*

Binomial Probability Theorem Suppose that in a binomial experiment with n trials, the probability of success is p in each trial, and the probability of failure is q, where $q = 1 - p$. Then p(exactly x successes) $= {}_nC_x \cdot p^x q^{n-x}$. **(647)**

Binomial Theorem For any nonnegative integer, n, $(x+y)^n = {}_nC_0 x^n + {}_nC_1 x^{n-1}y + {}_nC_x x^{n-2}y^2 + \cdots + {}_nC_k x^{n-k}y^k + \cdots + {}_nC_n y^n = \sum_{k=0}^{n} {}_nC_k x^{n-k}y^k$. **(632)**

box plot A visual representation of the five-number summary of a data set in which a box represents the part of the data from the first to the third quartile, with another line segment crossing the box at the set's median, and two segments protruding from the box (called *whiskers*) to represent the rest of the data. **(31)**

box-and-whiskers plot See *box plot.*

C

cardioid The graph of $r = a(\cos\theta - 1)$ or $r = a(\sin\theta - 1)$ in the polar plane. **(810)**

categorical variable A variable whose values represent characteristics rather than measures. **(8)**

census A survey of an entire population; in the U.S., a survey, conducted every ten years, of the entire population of the United States. **(6)**

center of mass of a data set The point whose coordinates are the means of the corresponding coordinates of the points in the data set. **(94)**

center of symmetry for a figure The center of a rotation under which the figure is mapped onto itself. **(172)**

center of rotation A fixed point about which each point in the plane turns a fixed magnitude under a rotation. **(222)**

central angle of a circle An angle whose vertex is the center of the circle. **(222)**

Central Limit Theorem Suppose random samples of size n are selected from a population with mean μ and standard deviation σ. Then, as n increases the following occur: The mean $\mu_{\bar{x}}$ of the distribution of sample means approaches μ; the standard deviation $\sigma_{\bar{x}}$ of the distribution of sample means approaches $\frac{\sigma}{\sqrt{n}}$; and the distribution of sample means approaches a normal curve with mean μ and standard deviation $\frac{\sigma}{\sqrt{n}}$. **(710)**

Change of Base Theorem For all values of a, b, and c for which the logarithms exist, $\log_b a = \frac{\log_c a}{\log_c b}$. **(596)**

Circle Arc Length Formula If s is the length of the arc of a central angle of θ radians in a circle of radius r, then $s = r\theta$. **(226)**

circular functions The trigonometric functions, when defined in terms of the unit circle. **(229)**

circular motion Movement of a point around a circle. **(273)**

cluster In a distribution of a set of data, a place where a relatively large number of data are near each other. **(24)**

coefficient The numbers $a_n, a_{n-1}, \ldots, a_0$ of the polynomial $a_n x^n + a_{n-1}x^{n-1} + \cdots + a_1 x + a_0$; more generally, a constant factor of a variable term. **(438)**

column matrix A matrix consisting of a single column. **(745)**

combination A collection of objects in which the order of objects does not matter. **(618)**

combination of *n* things taken *r* at a time, $_nC_r$, $\binom{n}{r}$ A subset of r objects from a set of n objects. **(619)**

common logarithm A logarithm with base 10. **(576)**

complementary events Two events that are mutually exclusive and whose union is the entire sample space. **(370)**

Complements Theorem For all θ in radians, $\sin\left(\frac{\pi}{2} - \theta\right) = \cos\theta$ and $\cos\left(\frac{\pi}{2} - \theta\right) = \sin\theta$. **(239)**

complex conjugates A pair of complex numbers that can be written as $a + bi$ and $a - bi$. **(467)**

complex number A number that can be written in the form $a + bi$, where a and b are real numbers and $i = \sqrt{-1}$. **(466)**

complex plane A coordinate plane for representing complex numbers. **(811)**

composite The function $g \circ f$ defined by $(g \circ f)(x) = g(f(x))$, whose domain is the set of values of x in the domain of f for which $f(x)$ is in the domain of g. **(194)**

composition of functions The binary operation that maps two functions g and f onto their composite $g \circ f$. **(194)**

compounded continuously The limit of compounding interest as the number of compoundings grows larger and larger in a given time period. **(583)**

concave down Said of an interval that is curving downwards, that is, its slope is continuously decreasing. **(687)**

concave up Said of an interval that is curving upwards, that is, its slope is continuously increasing. **(687)**

confidence interval An interval within which a certain specified percentage (usually 90%, 95%, or 99%) of outcomes from an experiment is expected to occur. **(721)**

Conjugate Zeros Theorem Let $p(x) = a_n x^n + a_{n-1}x^{n-1} + \cdots + a_1 x + a_0$, where $a_n, a_{n-1}, \ldots, a_1$, and a_0 are all real numbers, and $a_n \neq 0$. If $z = a + bi$ is a zero of $p(x)$, then the complex conjugate of z, $a - bi$, is also a zero of $p(x)$. **(477)**

constant A variable whose values do not change in the course of a problem. **(Previous Course)**

constant function Any function f with equation $f(x) = k$, where k is a fixed value. **(Previous Course)**

constant of proportionality See *constant of variation.*

constant of variation The parameter k in a direct, inverse, or joint variation. **(125)**

Continuous Change Model If a quantity p grows or decays continuously at an annual rate r, the amount $A(t)$ after t years is given by $A(t) = Pe^{rt}$. **(583)**

convergent sequence A sequence which has a finite limiting value L. If this value exists, the sequence is said to be convergent to L. **(515)**

convergent series A series in which the limit of the sequence of partial sums exists. **(543)**

correlation coefficient, r A measure of the strength of the linear relation between two variables. **(96)**

$\cos^{-1}$ See *inverse cosine function.*

cosecant (csc) of a real number x For a real number x, $\csc x = \frac{1}{\sin x}$, when $\sin x \neq 0$. **(336)**

cosine (cos) of an acute angle θ in a right triangle $\frac{\text{leg adjacent to } \theta}{\text{hypotenuse}}$. **(294)**

cosine (cos) of a real number θ The first coordinate of the image of the point (1, 0) under a rotation of magnitude θ about the origin. **(229)**

cosine function The function that maps x onto $\cos x$ for all x in its domain. **(249)**

cotangent (cot) of a real number x For any real number x, $\cot x = \frac{\cos x}{\sin x}$, when $\sin x \neq 0$. **(336)**

cube root For a number k, a number x that satisfies $x^3 = k$. **(560)**

cumulative binomial probability table A table that gives, for fixed n (number of trials) and p (probability of success), the probabilities for k or fewer successes. **(658)**

cycle One period of a periodic function, such as a sine wave. **(254)**

D

data The plural of *datum*, the Latin word for fact; a piece of information. **(6)**

decreasing function A function is decreasing on an interval if the segment connecting any two points on the graph of the function over that interval has negative slope. **(441)**

deductive reasoning Reasoning adhering to strict principles of logic. **(64)**

degree A unit for measuring angles, arcs, or rotations. **(Previous Course)**

degree of a polynomial In a polynomial with a single variable, the number n in the polynomial $a_n x^n + a_{n-1}x^{n-1} + \cdots + a_1 x + a_0$; in a polynomial in more than one variable, the largest sum of the exponents of the variables in any term. **(438)**

De Moivre's Theorem If $z = r(\cos \theta + i \sin \theta)$ and n is an integer, then $z^n = r^n(\cos n\theta + i \sin n\theta)$. **(826)**

dependent events Events A and B for which $P(A \cap B) \neq P(A) \cdot P(B)$. **(378)**

dependent variable The second variable in a relation. **(80)**

deviation The difference of a value in a set from the set's mean. (46)

dimensions of a matrix The numbers of rows and columns of the matrix. (744)

discriminant of a quadratic equation For the equation $ax^2 + bx + c = 0$, the number $b^2 - 4ac$. (467)

disjoint sets Sets with no elements in common. Also called *mutually exclusive sets.* (367)

distribution A function whose values are the frequencies, relative frequencies, or probabilities of mutually exclusive events. (22)

divergent sequence A sequence with no finite limit. (515)

divergent series A series in which the sequence of partial sums does not have a finite limit. (543)

dividend The number or expression a when a is divided by b. (453)

divisor The number or expression b when a is divided by b. (453)

domain The set of first elements of ordered pairs of a function; more generally, the set of replacement values for a variable. (80)

domain of an identity The set of values for which the identity is true. (789)

Double Angle Formulas For all real numbers θ, $\sin 2\theta = 2 \sin \theta \cos \theta$ and $\cos 2\theta = \cos^2\theta - \sin^2\theta = 2\cos^2\theta - 1 = 1 - 2\sin^2\theta$. (771)

E

e $\lim_{x \to \infty}\left(1 + \frac{1}{n}\right)^n \approx 2.718$; the base of natural logarithms. (582)

element An object in an array. A member of a set. (81, 744)

end behavior In a sequence t, the behavior of the values of t_n as $n \to \infty$. (515)

error (in a prediction) The difference between the observed and the expected values of a variable. Also called *residual.* (89)

even function A function f such that for all x in its domain, $f(-x) = f(x)$. (174)

event Any subset of the sample space of an experiment. (361)

expansion of $(x + y)^n$ See *Binomial Theorem.*

expected count If an outcome in an experiment has probability p, then in n trials of the experiment, the expected count of the outcomes is np. (410)

expected value of a probability distribution See *mean of a random variable.*

experiment A situation that has several possible outcomes. (360)

explicit formula for a sequence A formula which gives the nth term of the sequence in terms of n. (503)

exponent The number n in the expression r^n. (Previous Course)

exponential decay A situation that can be modeled by $f(x) = ab^x$ with base $0 < b < 1$. (104)

exponential decay curve The graph of an exponential decay function. (104)

exponential decay function An exponential function $f(x) = ab^x$ in which $a > 0$ and $0 < b < 1$. (104).

exponential equation An equation to be solved for a variable in an exponent. (594)

exponential function with base *b* The exponential function with base b is a function with a formula of the form $f(x) = ab^x$, where $a \neq 0$, $b > 0$, and $b \neq 1$. (103)

exponential growth curve The graph of an exponential growth function. (103)

exponential growth function An exponential function $f = ab^x$ in which $a > 0$ and $b > 1$. (103)

exponential model A mathematical model of a situation in the form of an exponential function. (110)

exponential regression The method of fitting an exponential function to a set of data. (111)

exponential sequence See *geometric sequence.*

extrapolation Estimating a value beyond known values of data. (88)

extrema of a function The extreme values for the function, classified as either maxima or minima. (440)

F

f^{-1} see *inverse function.*

Factor Theorem For any polynomial $f(x)$, a number c is a solution to $f(x) = 0$ if and only if $(x - c)$ is a factor of $f(x)$. **(460)**

Factor-Solution-Intercept Equivalence Theorem For any polynomial $f(x)$, the following are logically equivalent statements: $(x - c)$ is a factor of $f(x)$; $f(c) = 0$; c is an x-intercept of the graph of $y = f(x)$; c is a zero of f; the remainder when $f(x)$ is divided by $(x - c)$ is 0. **(461)**

failure (binomial) See *binomial experiment.*

fair experiment An experiment in which all outcomes of the sample space are equally likely. Also called *unbiased experiment.* **(362)**

***Federalist* papers, The** Essays written between 1787 and 1788 by James Madison, Alexander Hamilton, and John Jay under the collective pen name "Publius" to persuade the citizens of the state of New York to ratify the U.S. Constitution. **(61)**

Fibonacci sequence The sequence beginning 1, 1, 2, 3, 5, ..., in which $t_n = t_{n-2} + t_{n-1}$, for all $n \geq 3$. **(502)**

finite arithmetic series An arithmetic series where the number of terms added is finite. **(523)**

finite series A series where the number of terms added is finite. **(523)**

first (lower) quartile In a data set, the median of the numbers less than the set's median. **(31)**

five-number summary The three quartiles, the maximum, and the minimum of a data set. **(31)**

Formula for $_nC_r$ Theorem For whole numbers n and r, with $r \leq n$, $_nC_r = \frac{_nP_r}{r!} = \frac{n!}{(n-r)!r!}$. **(619)**

Formula for $_nP_n$ Corollary There are $n!$ permutations of n different elements. **(384)**

Formula for $_nP_r$ Theorem The number of permutations of n objects taken r at a time is $_nP_r$, $n(n-1)(n-2) \cdot \ldots \cdot (n-r+1)$. **(382)**

frequency The number of times an event occurs. **(Previous Course)** The number of cycles in a period of a periodic function. **(259)**

frequency distribution A function that maps events onto their frequencies. **(22)**

frequency histogram A histogram that displays the number of values that fall into each interval of the histogram. **(22)**

function A set of ordered pairs (x, y) in which each value of x is paired with exactly one value of y. A correspondence between two sets A and B in which each element of A corresponds to exactly one element of B. **(81)**

function notation The notation $f(x)$ for the value of a function f when the value of the independent variable is x. **(83)**

Fundamental Theorem of Algebra If $p(x)$ is any polynomial of degree $n \geq 1$ with complex coefficients, then $p(x)$ has at least one complex zero. **(475)**

full turn A rotation of 360°, 2π, or one revolution. **(222)**

G

Galton board A device, invented by Sir Francis Galton, used to illustrate binomial experiments, consisting of a box in which balls striking an array of pins form, at the bottom of the box, a normal distribution. **(616)**

general form of a polynomial in one variable See *polynomial in x.*

general solution to a trigonometric equation The solution to a trigonometric equation for all real numbers for which the variable is defined. **(325)**

Geometric Addition Theorem If the complex numbers $a + bi$ and $c + di$ are not collinear with the origin in the complex plane, then their sum $(a + c) + (b + d)i$ is the fourth vertex of a parallelogram with consecutive vertices $a + bi$, 0, and $c + di$. **(812)**

geometric mean The geometric mean of x and y is $\sqrt{xy}$. **(511)**

geometric sequence A sequence in which the ratio of consecutive terms is constant. **(509)**

geometric series An indicated sum of the terms of a geometric sequence. **(530)**

Graph Scale-Change Theorem In a relation described by a sentence in x and y, replacing x by $\frac{x}{a}$ and y by $\frac{y}{b}$ in the sentence yields the same graph as applying the scale change $(x, y) \rightarrow (ax, by)$ to the graph of the original relation. **(181)**

Graph-Standardization Theorem In a relation described by a sentence in x and y, replacing x by $\frac{x-h}{a}$ and y by $\frac{y-k}{b}$ in the sentence yields the same graph as applying the scale change $(x, y) \rightarrow (ax, by)$, where $a \neq 0$ and $b \neq 0$, followed by applying the translation $(x, y) \rightarrow (x + h, y + k)$ to the graph of the original relation. **(270)**

Graph Translation Theorem In a relation described by a sentence in x and y, replacing x by $x - h$ and y by $y - k$ in the sentence yields the same graph as applying the translation $(x, y) \to (x + h, y + k)$ to the graph of the original relation. (161)

great circle A circle on a sphere that has the same center as the sphere. (342)

Greenwich meridian See *prime meridian.*

grouping A technique used to factor polynomials that contain groups of terms with common factors. (486)

growth factor In an exponential growth situation, the base of the exponential function. (103)

growth rate The factor by which a quantity changes during a given time period. (102)

H

Half-Angle Formulas For all θ, $\cos \theta = \pm \sqrt{\frac{1 + \cos 2\theta}{2}}$ and $\sin \theta = \pm \sqrt{\frac{1 - \cos 2\theta}{2}}$. The sign is determined by the quadrant in which θ lies. (775)

half-life The time it takes a quantity to decay to half its original amount, usually applied to exponential decay situations. (113)

half turn A rotation of 180°, π, or $\frac{1}{2}$ revolution. (222)

Half-Turn Theorem For all θ, $\cos(\pi + \theta) = -\cos \theta$, $\sin(\pi + \theta) = -\sin \theta$, and $\tan(\pi + \theta) = \tan \theta$. (237)

harmonic sequence The sequence h in which $h_n = \frac{1}{n}$ for all $n \geq 1$. (502)

histogram A bar graph in which the range of values of a numerical variable are broken into non-overlapping intervals of equal width, and side-by-side bars display the number of values that fall into each interval. (22)

homogeneous population A population in which members are very similar on some measure. (52)

horizontal scale change A transformation that maps (x, y) to (ax, y) for all (x, y), where $a \neq 0$. (180)

horizontal scale factor The number a in the transformation that maps (x, y) to (ax, by). (180)

hypothesis In statistics, a statement to be tested. (417)

hypothesis testing The process of using statistics to find if a given hypothesis fits a situation within a given significance level. (417)

I

identity An equation that is true for all values of the variable(s) for which the expressions are defined. (235)

identity function, *I* A function that maps each element in its domain onto itself, that is, $I(x) = x$ for all x. (202)

image The result of a transformation. (159)

imaginary axis The vertical axis in a complex plane. (811)

imaginary numbers The number $i = \sqrt{-1}$ and its nonzero real-number multiples. (465)

imaginary part of a complex number The real number b in the complex number $a + bi$. (466)

impressionistic model A model where no theory exists that explains why the model fits the data. Also called *non-theory-based model.* (121)

in-phase circuit An alternating current circuit in which the voltage and current flow coincide. (263)

increasing function A function is increasing on an interval if the segment connecting any two points on the graph of the function over that interval has positive slope. (441)

independent events Events A and B such that $P(A \cap B) = P(A) \cdot P(B)$. (378)

independent variable The first variable in a relation. (80)

index, *i* A variable indicating the position of a number in an ordered list or sequence. (15)

inductance A property of an alternating current circuit created when the current flow lags behind the voltage. (263)

inferential reasoning Reasoning based on principles of probability. (64)

infinite series An indicated sum of the terms of an infinite sequence. (523, 524)

infinity, ∞ The limit of a sequence whose terms after a given point become larger than any fixed number one might choose; greater than any given number. (516)

inflection point A point on a graph where the graph changes concavity. (687)

initial condition The description of the initial term or terms in a recursive formula. (504)

initial side (of an angle) The side of an angle from which the angle is considered to have been generated, and from which the angle is measured. **(Previous Course)**

initial value In a function *f* modeling a situation, the value *f*(0). **(104)**

interest rate In an investment, the percent by which the principal is multiplied to obtain the interest paid to the investor. **(536)**

International Date Line The meridian which is 180° W (and 180° E) of the prime meridian. **(343)**

interpolation Estimating a value between known values of data. **(88)**

interquartile range (IQR) The difference between the third quartile and the first quartile. **(32)**

intersection, $A \cap B$ The set of elements that are in both set A and set B. **(368)**

invariant Unchanged by a particular transformation. **(168)**

inverse cosine function, $\cos^{-1}$, Arccos The function described by $y = \cos^{-1} x = \text{Arccos}\, x$, if and only if $x = \cos y$ and $0 \leq y \leq \pi$. **(300)**

inverse function, f^{-1} The function which is the inverse of the function *f*. **(201)**

inverse of a function The relation formed by switching the coordinates of the ordered pairs of a given function. **(199)**

inverse sine function, $\sin^{-1}$, Arcsin The function described by $y = \sin^{-1} x = \text{Arcsin}\, x$, if and only if $x = \sin y$ and $-\frac{\pi}{2} \leq y \leq \frac{\pi}{2}$. **(310)**

inverse tangent function, $\tan^{-1}$, Arctan The function described by $y = \tan^{-1} x = \text{Arctan}\, x$, if and only if $x = \tan y$ and $\frac{-\pi}{2} < y < \frac{\pi}{2}$. **(320)**

Inverses of Functions Theorem Two functions *f* and *g* are inverse functions if and only if $f(g(x)) = x$ for all x in the domain of g, and $g(f(x)) = x$ for all x in the domain of *f*. **(202)**

inverse-square relationship A function with an equation the form $y = \frac{k}{x^2} = kx^{-2}$. **(127)**

inversely proportional y is inversely proportional to x if $y = \frac{k}{x}$ for all x, where k is a constant. y is inversely proportional to the square of x if $y = \frac{k}{x^2}$ for all x. **(125, 127)**

L

latitude A measure of the extent to which a point is north or south of the equator determined by the angle subtended at the center of Earth by an arc on a line of longitude. **(343)**

Law of Cosines In any triangle *ABC*, $c^2 = a^2 + b^2 - 2ab \cos C$. **(304)**

Law of Sines In any triangle *ABC*, $\frac{\sin A}{a} = \frac{\sin B}{b} = \frac{\sin C}{c}$. **(315)**

leading coefficient (of a polynomial) The coefficient of the term of the highest degree of the polynomial. **(438)**

least squares line See *line of best fit.*

limit of a sequence, $\lim\limits_{x \to \infty} s_n$ A number which the terms of a sequence approach as n increases without bound. **(515)**

line of best fit The line that fits a set of data points with the smallest value for the sum of the squares of the deviations (vertical distances) from the data points to the line. Also called *regression line* or *least-squares line.* **(94)**

line of longitude See *meridian.*

line of symmetry See *axis of symmetry.*

linear function A function with an equation of the form $y = mx + b$, where m and b are constants. **(87)**

linear model The model of one variable as a linear function of another variable. **(87)**

linear regression The method of finding a line of best fit for a given set of points. **(94)**

linear sequence See *arithmetic sequence.*

linearity A relation's degree of correlation to a linear model. **(98)**

ln See *natural logarithm.*

logarithm (log) Let $b > 0$ and $b \neq 1$. Then y is the logarithm of x to the base b, written $y = \log_b x$, if and only if $b^y = x$. **(575)**

logarithm function (with base *b*) The function that maps x onto $\log_b x$. **(578)**

Logarithm of 1 Theorem For any base b, $\log_b 1 = 0$. **(587)**

Logarithm of a Power of the Base For any base b and any real number n, $\log_b (b^n) = n$. **(588)**

Logarithm of a Power Theorem For any base b, and for any positive real number x and any real number p, $\log_b x^p = p \log_b x$. (590)

Logarithm of a Product Theorem For any base b, and for any positive real numbers x and y, $\log_b(xy) = \log_b x + \log_b y$. (588)

Logarithm of a Quotient Theorem For any base b and for any positive real numbers x and y, $\log_b \frac{x}{y} = \log_b x - \log_b y$. (589)

logarithmic spiral The polar graph of $r = ka^\theta$, where $a > 0$. (810)

longitude The number of degrees that a meridian is east or west of the prime meridian, used as a coordinate of a location on Earth. (343)

lower quartile See *first quartile.*

M

magnitude (of rotation) The amount by which every point in the plane is turned around the center of the rotation. (222)

margin of error Half the length of a confidence interval. (721)

mathematical model A mathematical description of a real situation, often involving some simplifications and assumptions about that situation. (78)

Matrices for Reflections Theorem The matrix for r_x, reflection over the x-axis, is $\begin{bmatrix} 1 & 0 \\ 0 & -1 \end{bmatrix}$. The matrix for r_y, reflection over the y-axis, is $\begin{bmatrix} -1 & 0 \\ 0 & 1 \end{bmatrix}$. The matrix for $r_{y=x}$, reflection over the line $y = x$, is $\begin{bmatrix} 0 & 1 \\ 1 & 0 \end{bmatrix}$. (752)

Matrices for Rotations Theorem The matrix for $R_{90°}$, the rotation of 90° around the origin, is $\begin{bmatrix} 0 & -1 \\ 1 & 0 \end{bmatrix}$. The matrix for $R_{180°}$, the rotation of 180° around the origin, is $\begin{bmatrix} -1 & 0 \\ 0 & -1 \end{bmatrix}$. The matrix for $R_{270°}$, the rotation of 270° around the origin, is $\begin{bmatrix} 0 & 1 \\ -1 & 0 \end{bmatrix}$. (759)

matrix A rectangular arrangement of objects into rows and columns. (744)

Matrix Basis Theorem Suppose T is a transformation represented by a 2×2 matrix. If $T(1, 0) = (x_1, y_1)$ and $T(0, 1) = (x_2, y_2)$, then T has the matrix $\begin{bmatrix} x_1 & x_2 \\ y_1 & y_2 \end{bmatrix}$. (759)

matrix multiplication An operation on an $m \times n$ matrix A and an $n \times p$ matrix B resulting in the product matrix $A \cdot B$, an $m \times p$ matrix whose element in row i and column j is the sum of the products of elements in row i of A and corresponding elements in column j of B. (745)

matrix representing a transformation A matrix M such that whenever F, a matrix for a geometric figure, is multiplied by M, the product is a matrix for the image of F under the transformation T. (752)

maximum The largest value in a set. (31)

maximum value of a function The largest value in the function's range. (440)

mean The sum of the elements of a numerical data set divided by the number of items in the data set. Also called *average* and *arithmetic mean.* (14)

Mean of a Binomial Random Variable Theorem The mean μ of a binomial random variable with n trials and probability p of success on each trial is given by $\mu = np$. (659)

mean of a random variable For the probability distribution $\{(x_1, P(x_1)), (x_2, P(x_2)), \ldots, (x_n, P(x_n))\}$, the number $\mu = \sum_{i=1}^{n}(x_i \cdot P(x_i))$. Also called *expected value of a probability distribution.* (638)

measure of an angle A number that represents the size and direction of rotation used to generate an angle. (222)

measure of center A statistic describing a typical value of a numerical data set. Measures of center include the mean and median, and sometimes the mode. Also called *measure of central tendency.* (14)

measure of central tendency See *measure of center.*

measure of spread A statistic that describes how far data are from a center of a distribution. (45)

median The middle value of a set of data placed in ascending or descending order. The median of a set with an even number of elements is the mean of the two middle values. (14)

member An element of a set. (81)

meridian A semicircle of a great circle on the surface of Earth from the north pole to the south pole. Also called *line of longitude.* (343)

method of least squares The process of finding the line of best fit. (94)

middle quartile See *second quartile.*

minimum The smallest value in a set. **(31)**

minimum value of a function The smallest value in the function's range. **(440)**

minute (of a degree) A unit for measuring angles. 60 minutes = 1 degree. **(228)**

mode The item(s) with the greatest frequency in a data set. **(15)**

mode of a distribution In a probability distribution, the value of the random variable whose probability is the highest. **(639)**

modulus See *absolute value of a complex number.*

Monte Carlo method The method of using random numbers and related probabilities to simulate events for the purpose of solving a problem. **(403)**

Multiplication Counting Principle Let A and B be any finite sets. The number of ways to choose one element from A and then one element from B is $N(A) \cdot N(B)$. **(374)**

multiplicity of a zero For a zero r of a polynomial, the highest power of $(x - r)$ that appears as a factor of that polynomial. **(474)**

mutually exclusive sets See *disjoint sets.*

N

***n* factorial, *n*!** For a positive integer n, the product of the positive integers from 1 to n. In symbols, $n! = n \cdot (n - 1) \cdot (n - 2) \cdot (n - 3) \cdot \cdots \cdot 3 \cdot 2 \cdot 1$. $0! = 1$. **(380)**

natural logarithm, ln The logarithm of x to the base e, written $\ln x$. **(584)**

natural logarithm function The function that maps x to $\ln x$. **(584)**

$_nC_r$ See *Formula for $_nC_r$ Theorem.*

Negative Exponent Theorem For all $x \neq 0$ and n for which x^n is defined, $x^{-n} = \frac{1}{x^n}$. **(571)**

negative association A relation between two variables where the larger values of one variable are associated with smaller values of the other. **(96)**

non-theory-based model See *impressionistic model.*

normal curve The graph of the function $f(x) = e^{-x^2}$. **(686)**

normal distribution The probability distribution represented by the *normal curve.* **(686)**

not *A* The complement of an event A, that is, the set of outcomes of the sample space that are not in A. **(370)**

***n*th partial sum, S_n** The sum of the first n terms of a sequence. **(530)**

***n*th root, $\sqrt[n]{k}$, $k^{\frac{1}{n}}$** x is an nth root of k if and only if $x^n = k$, where n is an integer ≥ 2. The positive nth root (if there are two) or the only nth root of a real number k is denoted by $\sqrt[n]{k}$. **(560)**

***n*th root function** A function with an equation of the form $y = x^{\frac{1}{n}}$, where n is an integer with $n \geq 2$. **(561)**

***n*th roots of unity** The zeros of $x^n - 1$. **(479)**

***n*th term of a sequence** The term in the nth domain value of a sequence. **(502)**

null hypothesis, H_0 The main hypothesis used in hypothesis testing of a situation. **(670)**

Number of Zeros of a Polynomial Theorem A polynomial of degree $n \geq 1$ with complex coefficients has exactly n complex zeros, if multiplicities are counted. **(476)**

numerical variable A variable that represents a numerical value. **(8)**

O

observed values Data collected from sources such as experiments or surveys. **(89)**

odd function A function f such that for all x in its domain, $f(-x) = -f(x)$. **(174)**

1.5 × IQR criterion A criterion under which those elements of a data set greater than 1.5 × IQR plus the third quartile or less than the first quartile minus 1.5 × IQR are considered to be outliers. **(33)**

Opposites Theorem For all θ, $\cos(-\theta) = \cos \theta$, $\sin(-\theta) = -\sin \theta$, and $\tan(-\theta) = -\tan \theta$. **(237)**

oscilloscope An instrument for representing the oscillations of varying voltage or current on the fluorescent screen of a cathode-ray tube. **(220)**

out-of-phase circuit An alternating current circuit in which the current flow lags behind the voltage. **(263)**

outcome A possible result of an experiment. **(360)**

outlier An element of a set of numbers which is very different from most or all of the other elements. (33)

overlapping sets Two sets that have one or more elements in common. (368)

P

parabola The graph of a quadratic function. (117)

parameter A variable that is constant in a particular situation but whose value can vary from situation to situation. (331)

parametric equation Two or more equations in which different variables are written in terms of a parameter. (331)

parent function A simple form or the simplest form of a class of functions, from which other members of the class can be derived by transformations. (152)

partial sum See *nth partial sum.*

Pascal's Triangle The values of ${}_nC_r$, arranged in an array in the form of a triangle; the $(r + 1)$st term in row n of Pascal's Triangle is ${}_nC_r$. (625)

pentagonal numbers The sequence of numbers 1, 5, 12, 22, ... , in which the nth term is $\frac{n(3n - 1)}{2}$. (464)

percentile The pth percentile of a set of numbers is the value in the set such that p percent of the numbers are less than or equal to that value, or alternatively, less than that value. (40)

perfect correlation A correlation coefficient of 1 or -1; a situation in which all data points lie on the same line. (96)

periodic function A function f in which there is a positive real number p such that $f(x + p) = f(x)$ for all x. The smallest positive value of p is the period of the function. (254)

Periodicity Theorem For all θ, and for every integer n, $\sin(\theta + 2\pi n) = \sin\theta$, $\cos(\theta + 2\pi n) = \cos\theta$, and $\tan(\theta + \pi n) = \tan\theta$. (253)

permutation An arrangement of a set of objects without replacement. (382)

permutation of n objects taken r at a time, ${}_nP_r$ An arrangement of r objects from a set of n objects. (382)

petal curve See *rose curve.*

phase shift The least positive or the greatest negative horizontal translation that maps the graph of a circular function or its scale change image onto a given sine wave. (263)

piecewise definition A definition of a function whose domain is broken into subsets with a rule for each subset. (82)

point matrix The matrix $\begin{bmatrix} a \\ b \end{bmatrix}$ when it represents the point (a, b). (751)

point symmetry 180° rotation symmetry. (172)

polar axis A ray, usually horizontal and drawn to the right, through the pole of a polar coordinate system, from which the magnitudes of rotations are measured. (797)

polar coordinate system A coordinate system in which a point is identified by polar coordinates $[r, \theta]$, where $|r|$ is the distance of the point from a fixed point (the pole), and θ is a magnitude of rotation from the polar axis. (797)

polar form of a complex number The description of a complex number using polar coordinates. (814)

polar grid A grid of rays and concentric circles radiating from a central point, used for plotting points and sketching curves in the polar plane. (798)

pole See *polar coordinate system.*

polynomial A sum of multiples of nonnegative integer powers of a variable or variables. (438)

polynomial function A function whose rule can be written as a polynomial. (439)

polynomial in x An expression of the form $a_nx^n + a_{n-1}x^{n-1} + a_{n-2}x^{n-2} + \dots + a_1x + a_0$ where n is a nonnegative integer and $a_n \neq 0$. Also called *general form of a polynomial in one variable.* (438)

Polynomial Difference Theorem The function $y = f(x)$ is a polynomial function of degree n if and only if, for any set of x-values that form an arithmetic sequence, the nth differences of corresponding y-values are equal and nonzero. (448)

population The set of all individuals or objects to be studied. (6)

population pyramid A double histogram with bin intervals along a central vertical axis and frequencies along the horizontal axis. (26)

Polynomial Graph Wiggliness Theorem Let $p(x)$ be a polynomial of degree $n \geq 1$. The graph of $p(x)$ can cross any horizontal line $y = d$ at most n times. (476)

population standard deviation, σ See *standard deviation.*

population variance, σ^2 See *variance.*

positive association A relation between two variables where larger values of one variable are associated with larger values of the other. **(96)**

power function A function f with an equation of the form $f(x) = ax^n$, where n is an integer greater than 1. **(128)**

predicted value A value predicted by a mathematical model. **(89)**

preimage The domain or set of domain values of a transformation. **(159)**

prime meridian The meridian through Greenwich, England from which all other meridians are measured. **(343)**

principal Of a loan, the amount of money borrowed. Of an investment, the amount of money that is earning interest. **(536)**

probability A number which indicates the measure of certainty of an event. **(361)**

probability distribution A function that maps each value of a random variable onto its probability. **(637)**

probability of an event, $P(E)$ If E is an event in a finite sample space S, and each outcome in S is equally likely, then the probability that E occurs, $P(E)$, is given by $P(E) = \frac{\text{number of outcomes in the event}}{\text{number of outcomes in the sample space}}$. **(362)**

probability theory The branch of mathematics that studies chance. **(360)**

Probability of the Union of Mutually Exclusive Events Theorem If A and B are mutually exclusive events in the same finite sample space, then $P(A \cup B) = P(A) + P(B)$. **(368)**

Probability of a Union of Events Theorem (General Form) If A and B are any events in the same finite sample space, then $P(A \text{ or } B) = P(A \cup B) = P(A) + P(B) - P(A \cap B)$. **(369)**

Probability of Complements Theorem If A is any event, then $P(\text{not } A) = 1 - P(A)$. **(371)**

product matrix See *matrix multiplication.*

Product of Complex Numbers Theorem (Trigonometric Form) If $z_1 = r_1(\cos \theta_1 + i \sin \theta_1)$ and $z_2 = r_2(\cos \theta_2 + i \sin \theta_2)$, then $z_1 z_2 = r_1 r_2(\cos(\theta_1 + \theta_2) + i \sin(\theta_1 + \theta_2))$. **(820)**

Pythagorean identity For every θ, $\cos^2 \theta + \sin^2 \theta = 1$. **(235)**

Q

quadratic function A function f with a rule of the form $f(x) = ax^2 + bx + c$ where $a \neq 0$. **(Previous Course)**

quadratic model A quadratic function used to estimate data in a set. **(117)**

quadratic regression A method of finding an equation for the best-fitting quadratic function through a data set. **(120)**

quartiles The three values which divide an ordered set into four subsets of approximately the same size. See *first (lower) quartile, second (middle) quartile,* and *third (upper) quartile.* **(31)**

quotient The answer to a division problem. For polynomials, the polynomial $q(x)$ when $f(x)$ is divided by $d(x)$, where $f(x) = q(x)d(x) + r(x)$, and either $r(x) = 0$ or the degree of $r(x)$ is less than the degree of $d(x)$. **(453)**

Quotient of Complex Numbers Theorem If $z_1 = r_1(\cos \theta_1 + i \sin \theta_1)$ and $z_2 = r_2(\cos \theta_2 + i \sin \theta_2)$, then $\frac{z_1}{z_2} = \frac{r_1}{r_2}(\cos(\theta_1 - \theta_2) + i \sin(\theta_1 - \theta_2))$. **(821)**

R

radian A unit for measuring an angle, arc, or the magnitude of a rotation. π radians $= 180°$. **(223)**

radical The symbol $\sqrt{\ }$ used to denote square roots or nth roots. **(564)**

random numbers A set of numbers such that each number has the same probability of occurring, each pair of numbers has the same probability of occurring, each trio of numbers has the same probability of occurring, and so on. **(402)**

random sample See *simple random sample.*

random variable A variable whose values are numbers determined by the outcome of an experiment. **(637)**

randomly A property of sampling a population so that every member of the population has an equal chance of being chosen. Also referred to as *at random.* **(362)**

range The difference between the highest and lowest values in a set. The set of possible values of the dependent variable in a function. **(80)**

Rational Exponent Theorem For all positive integers m and n, and all real numbers $x \geq 0$, $x^{\frac{m}{n}} = \left(x^{\frac{1}{n}}\right)^m = (\sqrt[n]{x})^m$ and $x^{\frac{m}{n}} = (x^m)^{\frac{1}{n}} = \sqrt[n]{x^m}$. **(568)**

rational power function A function f with an equation of the form $f(x) = ax^{\frac{m}{n}}$, where m and n are nonzero integers and $a \neq 0$. (568)

raw data Data that has not been transformed or statistically manipulated. (207)

real axis The horizontal axis (axis of first coordinates) in a complex plane. (811)

real function A function whose domain and range is a set of real numbers. (81)

real part of a complex number The real number a in the complex number $a + bi$. (466)

reciprocal trigonometric functions The secant, cosecant, and cotangent functions. (336)

recurrence relation A formula for a term of a sequence in terms of preceding terms. (504)

recursive formula A formula for a sequence in which the first term or the first few terms are given, and then the nth term is expressed using the preceding term(s). (504)

reflection-symmetric figure A figure that can be mapped onto itself by a reflection over some line ℓ. (172)

regression line See *line of best fit.*

relation A set of ordered pairs. (80)

relative extrema Relative maxima or relative minima. (440)

relative frequency The ratio of the number of times an event occurred to the number of times it could have occurred. (18)

relative frequency distribution A function mapping events onto their relative frequencies. (22)

relative frequency histogram A histogram that displays the proportion of values that fall into each interval of the histogram. (22)

relative maximum A point or value at which a function has a maximum on a specified interval. (440)

relative minimum A point or value at which a function has a minimum on a specified interval. (440)

remainder (in polynomial division) The polynomial $r(x)$ when $f(x)$ is divided by $d(x)$ and $f(x) = q(x)d(x) + r(x)$ when either $r(x) = 0$ or the degree of $r(x)$ is less than the degree of $d(x)$. (454)

Remainder Theorem If a polynomial $f(x)$ is divided by $x - c$, then the remainder is $f(c)$. (455)

removable singularity A point of discontinuity of a function that can be "removed" by adding a single point to the function. (794)

representative sample A sample that has the same characteristics as the population. (7)

rescaling See *scaling.*

residual The difference between the observed value and a value predicted by a model. (89)

residual plot A plot of each x-value in a data set and its residual. (133)

revolution A unit for measuring rotations. 1 revolution (counterclockwise) = 360°. Also called *full turn.* (222)

root function A function that maps x onto some root of x, such as its square root. (561)

root of a polynomial function For the polynomial function $f(x) = a_n x^n + a_{n-1}x^{n-1} + \cdots + a_1 x + a_0$, any value of x such that $f(x) = 0$. Also called *zero of the function.* (442)

Roots of a Complex Number Theorem For any positive integer n, the n distinct roots of $z^n = r(\cos\theta + i\sin\theta)$, $r > 0$, are $z = \sqrt[n]{r}\left(\cos\left(\frac{\theta}{n} + k\frac{360°}{n}\right) + i\sin\left(\frac{\theta}{n} + k\frac{360°}{n}\right)\right)$, where $k = 0, 1, 2, \ldots, n - 1$. (828)

rose curve The graph of the polar equation $r = a\sin b\theta$ or $r = a\cos b\theta$, where b is a positive integer and $a \neq 0$. Also called *petal curve.* (807)

rotation A transformation under which each point in the plane turns a fixed magnitude around a fixed point called its center. (222)

Rotation Matrix Theorem The matrix for R_θ, the rotation of magnitude θ about the origin, is $\begin{bmatrix} \cos\theta & -\sin\theta \\ \sin\theta & \cos\theta \end{bmatrix}$. (762)

row matrix A matrix consisting of a single row. (745)

S

sample The subset of a population that is studied in an experiment. (6)

sample space The set of all possible outcomes of an experiment. (360)

sample standard deviation, s See *standard deviation.*

sample variance, s^2 See *variance.*

sampling distribution A distribution of the means of all samples of a fixed size from the same population. (708)

SAS Area Formula for a Triangle The area of any triangle is one-half the product of the lengths of any two sides and the sine of their included angle. (314)

scale change (of data) A transformation that maps each data value x_i in a set of data $\{x_1, x_2, \ldots, x_n\}$ to ax_i, where a is a nonzero constant. (186)

scale change (in the plane) The transformation that maps (x, y) to (ax, by), where $a \neq 0$ and $b \neq 0$ are constants. (180)

scale factor The nonzero constant by which each data value is multiplied in a scale change. (186)

scale image The result of a scale change, or the point it represents. (186)

scaling Applying a scale change to a data set. Also called *rescaling.* (185)

scatterplot A graph of a finite set of ordered pairs in the coordinate plane. (Previous Course)

secant (sec) of a real number For any real number x, $\sec x = \frac{1}{\cos x}$, provided $\cos x \neq 0$. (336)

second (middle) quartile The median of a set of data. (31)

second (of a degree) A unit for measuring angles. 60 seconds = 1 minute, and 3600 seconds = 1 degree. (228)

sequence A function whose domain is a set of consecutive integers greater than or equal to a fixed integer k. (503)

sequence of partial sums The sequence whose nth term is the sum of the first n terms of a given sequence. (542)

series An indicated sum of the terms of a sequence. (523)

sides (of an angle) See *angle.*

Σ (sigma) A symbol for sum. (15)

Σ notation (sigma-notation) See *summation notation.* (15)

significance level The level of probability (often 0.05 or 0.01) that is chosen to test a hypothesis. (420, 669)

simple random sample A sample of size n chosen from a population such that every sample of size n has an equal chance of being selected. (707)

simulation An experimental model of a situation that attempts to capture all aspects of the situation that affect the outcomes. (402)

$\sin^{-1}$ See *inverse sine function.*

sine function The function that maps x onto $\sin x$ for all x in its domain. (248)

sine (sin) of an acute angle θ in a right triangle $\frac{\text{leg opposite } \theta}{\text{hypotenuse}}$. (294)

sine (sin) of a real number θ The second coordinate of the image of the point (1, 0) under a rotation of magnitude θ about the origin. (229)

sine wave The image of the graph of the sine or cosine function under a composite of translations and scale changes. (257)

singularity An isolated value for which a function is undefined. (792)

sinusoidal Varying in the manner of a sine wave. (221)

size change A scale change in which the scale factors are equal; a transformation that maps (x, y) to (kx, ky), where k is a nonzero constant. (180)

slope For the segment joining (x_1, y_1) and (x_2, y_2), the number $\frac{y_2 - y_1}{x_2 - x_1}$. (Previous Course)

spherical triangle A triangle on a sphere whose sides are arcs of great circles of that sphere. (345)

Spherical Law of Cosines If ABC is a spherical triangle with sides a, b, and c, then $\cos c = \cos a \cos b + \sin a \sin b \cos C$. (345)

spread An indication of how the value of data in a set vary. (45)

square root x is a square root of k if and only if $x^2 = k$. (560)

standard deviation, s, σ The square root of the sample variance (sample standard deviation s) or the population variance (population standard deviation σ). (46, 48)

standard normal curve The graph of the function $f(x) = \frac{1}{\sqrt{2\pi}} e^{-\frac{x^2}{2}}$. (688)

standard normal distribution The probability distribution represented by the *standard normal curve.* (689)

standard score See *z-score.*

Standard Normal Distribution Table A table that gives the area under the standard normal curve to the left of a given positive number a. (694)

standardized data Data that has been transformed into z-scores. (207)

standardizing a variable The process of getting z-values from an original data set. (699)

statistical inference Judgments using probabilities derived from statistical tests. (64)

statistic A number used to describe a set of numbers. (6)

statistics The branch of mathematics dealing with the collection, organization, analysis, and interpretation of information, usually numerical information. (6)

Strings With Replacement Theorem Let S be a set with n elements. Then there are n^k possible arrangements of k elements from S with replacement. (376)

strong correlation A relation for which the data in a data set falls close to a line (or another specified curve). (97)

Subtraction Formulas for the Cosine and Sine For all real numbers α and β, $\cos(\alpha - \beta) = \cos\alpha\cos\beta + \sin\alpha\sin\beta$, and $\sin(\alpha - \beta) = \sin\alpha\cos\beta - \cos\alpha\sin\beta$. (767)

success (binomial) See *binomial experiment.*

sum of a series The sum of the terms of a sequence. Also called *value of a series.* (523)

sum of an infinite series, S_∞, $\sum_{i=1}^{\infty} a_i$ The limit of the sequence of partial sums $S_n = \sum_{i=1}^{n} a_i$ of the series, provided the limit exists and is finite. (542)

sum of squared residuals The sum of the squares of the differences between observed values and predicted values. (90)

summation notation The use of the symbol Σ to represent a summation. Also called *sigma-notation* or *Σ-notation.* (15)

Sums and Differences of Cubes Theorem For all x and y, $x^3 + y^3 = (x + y)(x^2 - xy + y^2)$ and $x^3 - y^3 = (x - y)(x^2 + xy + y^2)$. (482)

Sums and Difference of Odd Powers Theorem For all x and y and for all odd positive integers n, $x^n + y^n = (x + y)(x^{n-1} - x^{n-2}y + x^{n-3}y^2 - \cdots - xy^{n-2} + y^{n-1})$ and $x^n - y^n = (x - y)(x^{n-1} + x^{n-2}y + x^{n-3}y^2 + \cdots + xy^{n-2} + y^{n-1})$. (483)

Supplements Theorem For all θ, $\sin(\pi - \theta) = \sin\theta$, $\cos(\pi - \theta) = -\cos\theta$, and $\tan(\pi - \theta) = -\tan\theta$. (238)

survey A gathering of facts or opinions through an interview or questionnaire; to gather such facts. (6)

symmetric distribution A distribution that is, or nearly is, reflection-symmetric. (24)

symmetric to the origin A property of a relation such that if (x, y) is on its graph, then so is $(-x, -y)$. (173)

symmetric with respect to the *x*-axis A property of a relation such that if (x, y) is on its graph, then so is $(x, -y)$. (173)

symmetric with respect to the *y*-axis A property of a relation such that if (x, y) is on its graph, then so is $(-x, y)$. (173)

symmetry to a point See *point symmetry.*

T

table of random numbers A listing of (pseudo-) random numbers used to simulate random situations. (404)

$\tan^{-1}$ See *inverse tangent function.*

tangent function The function that maps x onto $\tan x$ for all x in its domain. (252)

tangent (tan) of an acute angle in a right triangle $\frac{\text{leg opposite the angle}}{\text{leg adjacent to the angle}}$. (294)

tangent (tan) of a real number θ For all real numbers θ, $\frac{\sin\theta}{\cos\theta}$, if $\cos\theta \neq 0$. (230)

term (of a loan) The length of time a person has to pay back the entire amount of a loan. (536)

term (of a sequence) An element in the range of a sequence. (502)

tetrahedral array A three-dimensional array made up of layers of points arranged in triangular arrays. (446)

tetrahedral numbers The values $T(n)$ of the sequence defined by $\begin{cases} T(1) = 1 \\ T(n) = T(n-1) + t(n) \end{cases}$, for all integers $n > 1$, where $t(n)$ is the nth triangular number. (446)

theory-based model A model based on a concrete theory that explains why the model should fit the data. (121)

third (upper) quartile In a data set, the median of the numbers greater than the set's median. (31)

transformation A one-to-one correspondence between sets of points. (150, 159)

translation (of data) A transformation that maps each x_i of a data set to $x_i + h$, where h is some constant. (165)

translation (in the plane) A transformation that maps each point (x, y) to $(x + h, y + k)$, where h and k are constants. (159)

translation image The result of a translation. (159)

tree diagram A method of graphically presenting the possible outcomes of an experiment by using a network of branches, resembling a tree. (374)

trial One of the instances of an experiment. (402)

triangular numbers The values of the sequence $t(n) = \frac{n(n+1)}{2}$. (446)

trigonometric equation An equation in which the variable to be found is in an argument of a trigonometric function. (325)

trigonometric form of a complex number The form $r(\cos \theta + i \sin \theta)$ of the complex number $a + bi$. (818)

trigonometric functions The sine, cosine, tangent, cotangent, secant, and cosecant functions and their offspring. Also called *circular functions.* (229)

trigonometry The branch of mathematics that deals with the study of the circular functions, and the relations between sides and angles of triangles using these functions. (229)

U

unbiased experiment See *fair experiment.*

union, $A \cup B$ The set of all elements that are either in set A or set B, or in both (367)

unit circle The circle with center at the origin and radius 1. (229)

upper quartile See *third quartile.*

V

value One of the elements that a variable can represent. (83)

value (of a series) See *sum of a series.*

variable (in statistics) A characteristic of a person or thing which can be classified, counted, ordered, or measured. (6)

variance, s^2, σ^2 In a data set, the sum of the squared deviations of the data from the mean divided by one less than the number of elements in the set (sample variance s^2) or by the number of elements in the set (population variance σ^2). (46, 48)

variance of a random variable For a probability distribution $\{(x_i, P(x_i))\}$ with n outcomes and mean μ, $\sigma_x^2 = \sum_{i=1}^{n} (x_i - \mu)^2 \cdot P(x_i)$ or $\sigma_x^2 = \sum_{i=1}^{n} x_i^2 \cdot P(x_i) - \mu^2$. (640, 641)

Variance and Standard Deviation of a Binomial Random Variable Theorem In a binomial distribution with n trials, probability p of success and probability q of failure on each trial, the variance is $\sigma^2 = npq$, and the standard deviation is $\sigma = \sqrt{npq}$. (662)

varies inversely See *inversely proportional.*

Venn diagram A method of displaying unions and intersections of sets, using circles or ellipses. (368)

vertex (of an angle) See *angle.*

vertical line test A test to determine whether a set of ordered pairs in the rectangular coordinate plane is a function; if there exists a vertical line that intersects the set in more that one point, then the set is not a function. (83)

vertical scale change A transformation that maps (x, y) to (x, by), where $b \neq 0$ is a constant. (180)

vertical scale factor The number b in the transformation that maps (x, y) to (ax, by). (180)

viewing window The subset of the coordinate plane that appears on the screen of a graphing utility. (153)

W

weak correlation A relation for which, although a linear trend can be seen, many points are not very close to the line (or another specified curve). (97)

weighted average If x_i is a value in a data set and w_i is the weight of the value, the weighted average is $\frac{\sum_{i=1}^{n} w_i x_i}{\sum_{i=1}^{n} w_i} = \frac{w_1 x_1 + w_2 x_2 + \ldots + w_n x_n}{w_1 + w_2 + \cdots + w_n}$. **(17)**

window See *viewing window*.

whiskers Segments emanating from the box in a box-and-whiskers plot. **(32)**

Y

yield The actual percentage added to the principal in an investment annually, given by $\left(1 + \frac{r}{n}\right)^n$, where r is the annual interest rate and n is the number of compoundings per year. **(582)**

Z

z-score The value $z = \frac{x - \bar{x}}{s}$ for a member x of a data set with mean $\bar{x}$ and standard deviation s. Also called *standard score*. **(206)**

Zero Exponent Theorem If b is an nonzero real number, $b^0 = 1$. **(571)**

zero of a function For a function f, a value of x such that $f(x) = 0$. **(442)**

Index

B

C

D

G

H

N

Q

R

T

U

V

W

X

Y

Z

Photo Credits

Photo Credits

Cover, back: ©Digital Vision/Getty Images; **vi** (l) ©Peter Ginter/Science Faction/Getty Images, (r) ©Michael G. Smith/Shutterstock; **vii** (l) ©Frank Leung/iStock, (r) ©PHOTO 999/Shutterstock; **viii** (l) ©StephanHoerold/iStock, (r) ©markchentx/iStock; **ix** ©Ivan Cholakov/Shutterstock; **x** (l) ©NASA (r) ©Tony Tremblay/iStock; **xi** ©BestPhotoStudio/Shutterstock; **xii** ©Vlad G/Shutterstock.com; **4–5** (t) ©Peter Ginter/Science Faction/Getty Images; **5** (b) ©National Safety Council; **6** ©Pgiam/iStock; **7** © Tomasz Trojanowski/Shutterstock; **10** ©fstop123/iStock; **11** ©vanillaechoes/Shutterstock; **17** ©Charles Siegel; **22** ©Mykhaylo Palinchak/Shutterstock.com; **32** ©MidoSemsem/Shutterstock; **39** ©Yuliya Evstratenko/Shutterstock; **44** ©Jess Yu/Shutterstock; **47** ©Aspen Photo/Shutterstock.com; **49** ©diewis33/iStock; **50** ©photoquest7/iStock; **53** ©Wavebreak/iStock; **55** (t) ©Mayskyphoto/Shutterstock.com, (b) ©Alexey Stiop/Shutterstock.com; **58** ©sx70/iStock; **59** Public Domain / Wikimedia Commons; **61** Public Domain / Wikimedia Commons / The White House Collection; **65** ©Subbotina Anna/Shutterstock; **67** ©peterspiro/iStock; **68** ©Susan Chiang/iStock; **78–79** ©Michael G. Smith/Shutterstock; **81** ©sf_foodphoto/iStock; **85** ©Charles Siegel; **88** ©rgbdigital/iStock; **93** ©Menno Schaefer/Shutterstock; **94** ©bermau/iStock; **98** ©mediaphotos/iStock; **102** ©IS_ImageSource/iStock; **107** ©Artur Bogacki/Shutterstock; **108** ©cjp/iStock; **109** RIA Novosti archive, image #556242 / Yuriy Somov / CC-BY-SA 3.0; **110** ©Joe Karlovsky; **111** ©Frank Leung/iStock; **113** ©JacobH/iStock; **116** ©ndoeljindoel/Shutterstock; **119** ©ziche77/iStock; **121** ©Dmitry Kalinovsky/Shutterstock; **123** ©Aaliya Landholt/Shutterstock; **124** ©Alan_Lagadu/iStock; **126** ©bikeriderlondon/Shutterstock; **129** ©gordana jovanovic/iStock; **130** ©Gordon Galbraith/Shutterstock; **131** ©thelefty/Shutterstock.com; **133** ©Charles Siegel; **135** ©2002 Bill Amend. Reprinted with permission of UNIVERSAL UCLICK. All rights reserved; **136** ©EdStock/iStock; **138** ©Maudib/iStock; **139** ©mgahura/iStock; **140** ©T photography/Shutterstock.com; **141** ©nikkytok/iStock; **150–151** ©Frank Leung/iStock; **154** ©IPGGutenbergUKLtd/iStock; **157** ©PamelaJoeMcFarlane/iStock; **164** ©JLFCapture/iStock; **165** ©agaliza/iStock; **168** ©arekmalang/iStock; **169** ©RuslanDashinsky/iStock; **170** ©AVAVA/iStock; **178** ©Michael Vigliotti/Shutterstock; **184** ©tomalu/iStock; **185** ©Iakov Filimonov/Shutterstock; **186** ©mediaphotos/iStock; **188** ©kcline/iStock; **189** ©Kevin M. McCarthy/Shutterstock.com; **190** ©Mike Flippo/Shutterstock; **193** (t) ©Everett Historical/Shutterstock, (b) Library of Congress, LC-USZC4-5801; **198** ©Dejan Stanisavljevic/Shutterstock; **199** ©Stephen Mcsweeny/Shutterstock; **203** ©MikeyLPT/iStock; **204** (t) ©darrenpape/iStock, (b) ©ALXR/Shutterstock, Inc.; **205** ©Kuzma/Shutterstock; **206** ©dlewis33/iStock; **208** ©monkeybusinessimages/iStock; **210** ©7io/iStock; **211** (bl) ©Anthony Ross/Shutterstock, (tr) ©william87/iStock; **212** ©Maica/iStock; **220** (t) ©EHStock/iStock, (b) ©Tom Biegalski/Shutterstock; **220–221** ©PHOTO 999/Shutterstock; **223** ©Vernon Wiley/iStock; **226** ©archives/iStock; **227** ©pepbaix/iStock; **231** ©chictype/iStock; **232** ©Gunnar Rathbun/Shutterstock; **234** Courtesy Special Collections Research Center, University of Chicago; **241** ©Nastco/iStock; **246** ©sturti/iStock; **251** ©TimAbramowitz/iStock; **252** ©Rafal Cichawa/Shutterstock; **254** ©Sergey Nivens/Shutterstock; **257** ©polder/Shutterstock; **260** ©malerapaso/iStock; **261** ©Yuri/iStock; **263** ©sturti/iStock; **267** ©Veni/iStock; **268** ©pixinoo/Shutterstock; **273** Courtesy Special Collections Research Center, University of Chicago; **276** ©Davel5957/iStock; **278** ©Blend Images/Shutterstock; **280** ©baona/iStock; **282** ©webphotographeer/iStock; **283** ©revetina01/iStock; **292-293** ©StephanHoerold/iStock; **303** ©hamurishi/Shutterstock; **304** ©Hulton Archive/Getty Images; **316** ©Galyna Andrushko/Shutterstock; **322** ©RioPatuca/Shutterstock; **327** ©Vince Clements/Shutterstock; **329** (t) ©pagadesign/iStock, (b) ©Deejpilot/iStock; **341** NASA Earth Observatory; **342** (tr) NASA Earth Observatory, (c) ©Robert F. Balazik/Shutterstock; **345** ©steba2/iStock; **346** ©Galina Mikhalishina/Shutterstock; **347** (t) ©jqf/iStock, (b) ©Lori Martin/Shutterstock; **349** ©bouzou/Shutterstock; **350** ©Timothy Michael Morgan/Shutterstock; **358–359** ©markchentx/iStock; **362** ©James Steidl/Shutterstock; **363** ©Aldo Murillo/iStock; **364** ©Phil McDonald/iStock; **366** ©Lauri Patterson/iStock; **369** ©MoreISO/iStock; **372** ©noblige/iStock; **374** ©Sjo/iStock; **375** ©Cindy Hughes/Shutterstock; **376** ©Mike Flippo/Shutterstock; **379** ©sandocir/iStock; **381** ©aceshot1/Shutterstock; **382** ©Charles Siegel; **384** ©GlobalStock/iStock; **385** ©jane/iStock; **386** (t) ©AK2/iStock, (b) ©Paha_L/iStock; **387** National Archives and Records Administration; **391** ©Chao Kusollerschariya/Shutterstock; **394** ©Anthony Correia/Shutterstock.com; **396** ©asiseeit/iStock; **399** ©svengine/iStock; **400** ©Rich Legg/iStock; **402** (l) ©Debby Wong/Shutterstock.com, (lc) ©Photo Works/Shutterstock.com, (rc) ©Madookaboo/Dreamstime, (r) ©Aspen Photo/Shutterstock.com; **404** ©Elena Elisseeva/Shutterstock; **407** ©Lisa F. Young/Shutterstock; **408** ©John Barry de Nicola/Shutterstock.com; **409** ©1 design/iStock; **412** Courtesy Ronald Reagan Library; **413** ©Primeop76/iStock; **414** ©Grafissimo/iStock; **423** NASA Earth Observatory; **425** ©Djura Topalov/iStock; **426** ©nameinfame/iStock; **427** ©Daniel M. Silva/Shutterstock.com; **436–437** ©Ivan Cholakov/Shutterstock; **445** NASA; **446** ©dgstudiodg/iStock; **452** ©KlausRademaker/Shutterstock; **458** ©Radu Razvan/Shutterstock; **464** The Pentagon © David B. Gleason / Wikimedia Commons / CC-BY-SA-2.0; **465** ©Handy/Shutterstock; **471** Public Domain / Wikimedia Commons; **478** Derivative Work / Underlying Work: Public Domain / Wikimedia Commons; **480** ©Monkey Business Images/Shutterstock; **486** Family Circus ©Bil Keane, Inc., King Features Syndicate; **490** ©ra2studio/Shutterstock; **491** ©Kurt Tutschek/Shutterstock; **504** ©Palto/Shutterstock; **505** ©SergeyZavalnyuk/iStock; **507** ©Tito Wong/Shutterstock; **508** ©Gary Paul Lewis/Shutterstock; **511** ©twilightproductions/iStock; **513** ©American Spirit/Shutterstock.com; **514** ©monkeybusinessimages/iStock; **521** ©Everett Historical/Shutterstock; **522** Courtesy of the National Museum of Health and Medicine, Armed Forces Institute of Pathology (NCP 1603); **523** ©traveler1116/iStock; **525** ©Songquan Deng/Shutterstock.

com; **526** ©Graphic Design/Shutterstock; **528** ©Royalty Free/Getty Images; **529** ©Sadik Gulec/Shutterstock.com; **532** ©PeopleImages/iStock; **535** ©Studiotouch/Shutterstock; **539** ©Pamela Moore/iStock; **547** ©Shvaygert Ekaterina/shutterstock; **551** ©creisinger/iStock; **558–559** NASA / ESA / S. Beckwith (STScI) / The Hubble Heritage Team (STScI/AURA); **562** ©fonats/Shutterstock; **565** ©Harper 3D/Shutterstock; **566** ©Amy Nichole Harris/Shutterstock; **567** ©kevinruss/iStock; **568** ©Yongyut Kumsri/Shutterstock; **569** ©101cats/iStock; **573** ©a_Taiga/iStock; **575** NASA; **580** ©Bork/Shutterstock; **582** ©Georgios Kollidas/Shutterstock; **585** ©Mark Atkins/Shutterstock; **586** NASA Earth Observatory; **587** Public Domain / Wikimedia Commons; **591** ©University of Central Missouri; **596** ©Ary6/iStock; **598** ©Ildi Papp/Shutterstock.com; **599** ©Timothy Boomer/Shutterstock; **605** ©Theo Westenberger/Getty Images; **606** ©ben smith/Shutterstock; **607** ©TennesseePhotographer/iStock; **616–617** ©Tony Tremblay/iStock; **618** ©salez/iStock; **620** (t) ©Louis-Paul St-Onge/iStock, (b) ©3 song photography/Shutterstock.com; **621** ©9507848116/Shutterstock; **622** ©Claudine Van Massenhove/Shutterstock.com; **623** © J.delanoy / Wikimedia Commons / CC-BY-3.0; **624** ©mattesimages/iStock; **625** ©traveler1116/iStock; **630** ©alexsl/iStock; **632** Omar Khayyam Mausoleum © dynamosquito / Flickr; **634** ©dinycho/iStock; **635** ©sanneberg/Shutterstock; **640** ©grzym/Shutterstock; **642** ©CDC/James Gathany; **644** ©1974 Peanuts Worldwide LLC. Dist. By UNIVERSAL UCLICK. Reprinted with permission. All rights reserved; **646** ©MachineHeadz/iStock; **648** ©Spirins/iStock; **649** ©Lilac Mountain/Shutterstock; **650** ©Jose Gil/Shutterstock; **651** ©Norman Chan/Shutterstock; **655** ©PeopleImages/iStock; **657** ©Jacob Ammentorp Lund/iStock; **661** ©wragg/iStock; **662** ©shaunl/iStock; **663** ©Jeff Dalton/Shutterstock; **664** ©Dennis Cox/Shutterstock; **670** ©shaunl/iStock; **672** ©James Steidl/Shutterstock; **674** (l) ©SSPL/Getty Images, (r) ©Lightning Calculator, www.qualitytng.com; **675** ©Charles Siegel; **684** ©pmphoto/Shutterstock; **684–685** ©BestPhotoStudio/Shutterstock; **691** ©Marcie Fowler-Shining Hope Image/Shutterstock; **697** ©Monkey Business Images/Shutterstock; **699** ©maximult/iStock; **700** ©Esben_H/iStock; **703** ©VietnamPhotos/Shutterstock.com; **704** ©design 56/Shutterstock; **711** (t) ©JoLin/Shutterstock, (b) ©Image Point Fr/Shutterstock; **713** ©Elena Schweitzer/Shutterstock; **715** ©MSPhotographic/Shutterstock; **718** ©MagMos/iStock; **720** ©Catalin Petolea/Shutterstock; **725** ©Karramba Production/Shutterstock; **731** Public Domain / Wikimedia Commons / National Gallery of Art, Washington; **732** ©Hurst Photo/Shutterstock; **734** ©Ken Brown/istock; **744** ©bradwieland/iStock; **748** ©EdStock/iStock; **749** ©Anton Oparin/Shutterstock.com; **756** ©michaeljung/Shutterstock; **758** ©Vadim Besedin/Shutterstock; **761** ©video1/iStock; **765** ©gemenacom/iStock; **770** ©craftvision/iStock; **774** ©Le Do/Shutterstock; **776** ©Andrea Zabiello/Shutterstock; **777** ©Hermann Erber/Look-foto/Getty Images; **784** (l) ©Ron Blanton/Shutterstock, (c) ©Hamiza Bakirci/Shutterstock, (r) ©Andriy Solovyov/Shutterstock; **784–785** ©Vlad G/Shutterstock.com; **791** ©CristinaMuraca/Shutterstock.com; **794** L. Ferrarese (Johns Hopkins University) / NASA; **796** ©Picsfive/Shutterstock; **797** ©Jim Barber/Shutterstock; **802** ©Craig Mills/Shutterstock; **807** ©Andriy Solovyov/Shutterstock; **810** ©Nina Malyna/Shutterstock; **816** ©PandaWild/iStock; **825** Public Domain; **827** ©Wolfgang Kloehr/Shutterstock; **832** ©Charles Siegel